in
Private Practice

June 1985

Phase Equilibria in Chemical Engineering

Stanley M. Walas
Department of Chemical and Petroleum Engineering
University of Kansas
and
The C.W. Nofsinger Company

BUTTERWORTH PUBLISHERS
Boston • London
Sydney • Wellington • Durban • Toronto

To the memory of my parents,
Stanislaus and Apolonia,
and to my wife, Suzy Belle

Library of Congress Cataloging in Publication Data

Walas, Stanley M.
Phase equilibria in chemical engineering.

Bibliography: p.
Includes index.
1. Phase rule and equilibrium. I. Title.
TP156.E65W34 1984 660.2′963 83-15360
ISBN 0-409-95162-5

Butterworth Publishers
80 Montvale Avenue
Stoneham, MA 02180

10 9 8 7 6 5 4 3 2 1

Printed in the United States of America

Contents

LIST OF FIGURES

List of Tables

List of Examples

Preface

Phase equilibria exist or are being strived for all around us. Most industrial processes are designed for and operate near equilibrium conditions; even when this is not so, it is important to know what would happen at equilibrium. Chemical engineering processes of primary importance are those of mixing, conversion, and separation involving gases, liquids, and solids. This book is devoted to the thermodynamic basis and practical aspects of the calculation of equilibrium conditions of multiple phases that are pertinent to such processes.

At the present time, the theory and practice of process design based on phase equilibria are changing and growing significantly. Rigorous or unique solutions of phase equilibrium problems are not often attainable, and rarely is any particular method overwhelmingly superior. Accordingly, efforts have been made throughout the book to provide guidance to adequate theory and practice—although not always unambiguously because of the present state of the art. A brief summary guide immediately follows this preface.

Since much of the author's interest and experience has been in process design work, where the next step leads to construction of equipment, the book is slanted toward users rather than developers of phase equilibrium theory and correlations. To a certain extent the case method is used, whereby key theoretical results or algorithms are illustrated with numerical examples or problems for solution by the reader that are integrated with the text. These problems are within the capability of a programmable calculator or portable computer, but from an instructor's point of view many of them can be extended so as to make demands on 32K or more powerful computers. It is beyond the scope of this book to apply the material to the design of multistage equilibrium processes, a topic that is well covered elsewhere.

Although many results from classical and statistical thermodynamics are needed, particularly concerning mixtures, no detailed development of those subjects is attempted. Instead, a users' manual has been assembled with enough explanation to make the material plausible or to recall the background if it has been experienced by the reader. Ample references are made to standard works on the subject where the background material can be studied in any detail required. Care has been taken, however, to point out the local significance and limitations of the quoted results.

From the table of contents the pattern of the book should be clear. Some historical material has been included, both as a guide to the significance of what is now known or may be learned in the near future, and because it is interesting in itself. The equation of state is intimately bound up with the development of thermodynamics, so the book begins with a long chapter on this topic. Following material on basic thermodynamics and nonidealities in terms of fugacities and activities, individual chapters are devoted to equilibria primarily between pairs of phases. A few topics that do not fit into these categories and for which the state of the art is not yet developed quantitatively have been relegated to a separate chapter. The chapter on chemical equilibria is pertinent since many processes involve simultaneous chemical and phase equilibria.

A chapter on the evaluation of enthalpy and entropy changes of nonideal substances and mixtures, which is a part of any complete phase equilibrium design package, and a brief chapter on experimental methods finish off the body of the book. The latter topic has been included since many people wish to understand how the data they use have been obtained. Although experimental methods are diverse and may be very sophisticated, perhaps enough material has been presented to give the reader at least a rough idea of how such work can be done.

The book is intended as a reference and self-study as well as a textbook either for full courses in phase equilibria or as a supplement to related courses in the chemical engineering curriculum. Practicing engineers concerned with separation technology and process design also may find the book useful. In view of the importance of the subject in many areas of chemical engineering, it is likely that more full courses in phase equilibria will be taught as more textbooks of a pragmatic as well as a theoretical nature become available.

The author's interest in phase equilibria was stimulated by many industrial contacts where the need became apparent for the kind of material assembled here. The example of the experimental work being done at the Low Temperature Laboratory of the University of Kansas by Professor G.W. Swift also was an inspiration.

Many of the figures were drawn by Harold Bushnell of The C.W. Nofsinger Company. The computer graphs were prepared by graduate students Chyi-Gang Huang, Carlos Rocha, Lanny Schoeling, and particularly by Shahin Negahban, who contributed also particularly to the development of Chapter 7.

An early version of the manuscript was examined by Professor J.P. Kohn of the University of Notre Dame; his comments contributed significantly to its further development. Later, helpful reviews of the first six chapters of the manuscript were made by Professor Keith Johnston of the University of Texas, Professor Allan Myers of the University of Pennsylvania and Professor John P. O'Connell of the University of Florida. The encouragement of other academic and industrial colleagues and the assistance of editors Greg Franklin and Kathleen A. Benn are deeply appreciated.

A Guide for the Perplexed (with Apologies to Maimonides)

The abundance of viewpoints and procedures for evaluating various aspects of phase equilibrium relationships is evident to the informed, and perhaps to readers of this book. Rarely is any one method overwhelmingly superior; although this is frustrating when an answer must be obtained, it is the state of the art and must be tolerated. Nevertheless, a brief list and two sets of further guidelines are provided here of methods that usually afford at least modest accuracy, sufficient for a start.

1. For *PVT* relations of liquids and gases, the Soave or Peng-Robinson equations. When the parameters and the computer time are available, the BWRS.
2. For fugacities, the Soave or Peng-Robinson. For vapors only, alternately the *B*-truncated virial equation with the Tsonopoulos correlations.
3. For vaporization equilibrium ratios of hydrocarbons and associated inorganic gases, Soave or Peng-Robinson; sometimes the modified Chao-Seader. For spot checking, the NGPSA or API chart methods.
4. For liquid densities over wide ranges of T and P, the method of Thomson, Brobst, & Hankinson; Peng-Robinson is adequate when consistency with vapor properties is desired, as in distillation computations.
5. For correlation of activity coefficients of binary and multicomponent mixtures, the Wilson equations.
6. For activity coefficients from group contributions, the UNIFAC method is more highly developed but is about equivalent to the ASOG when data are available for both. The solubility parameter method has only limited applicability.
7. For liquid-liquid equilibria, the NRTL equation with $\alpha_{12} = 0.2$ generally does well, even with only binary parameters, but somewhat better when some multicomponent data are available for evaluation of the parameters.
8. For liquid-solid melt equilibria, the Schröder equation modified with experimental or structurally derived activity coefficients when available.
9. For chemical equilibria, the relaxation method is adequate when the number of stoichiometric equations is less than a half dozen or so. More complex problems require computer solution by minimization of the Gibbs energy.
10. Departures of enthalpy and entropy from ideality are handled best by the Lee-Kesler method, but the simpler Soave or Peng-Robinson usually are acceptable.

GUIDELINES FOR VAPOR-LIQUID EQUILIBRIA

Chao-Seader-Grayson-Streed Procedure

This method is suitable for the representation of hydrogen-rich systems such as hydrotreaters and reformers. It is satisfactory for the design of refinery topping and heavy end processing units and has been found useful for synthetic fuel systems. The temperature range is −20 to 450 C, pressure range below 200 atm.

Soave and Peng-Robinson Equations

For highest accuracy, binary interaction parameters are desirable. With this information the equations are satisfactory for the design of sweet and sour gas units (up to 25 percent H_2S). Phase behavior in the critical region can be predicted although the calculations are somewhat unstable at the critical point itself. Results for mixtures of hydrogen and saturated hydrocarbons are good; those for aromatics are less so but they can be improved with appropriate binary interaction parameters. Liquid compressibility predictions are accurate enough for fugacity calculations but not for liquid densities; the Soave equation can be 10-20 percent low, the P-R slightly more accurate. These methods are useful for cryogenic conditions, refinery mixtures and conditions up to about 350 atm.

BWR and BWRS

For the comparatively few substances for which the parameters have been determined, these are the standard of comparison. Because of their greater complexity, they are not as useful as the Soave or P-R when the design procedure involves many iterations.

Lee-Kesler and Lee-Kesler-Plöcker

These methods possess the advantage over the preceding of having universal parameters, but they also are cumbersome for making often repeated calculations.

GUIDELINES FOR LIQUID PHASE ACTIVITY COEFFICIENTS

The models are the Wilson, T-K-Wilson, NRTL, UNIQUAC, Margules, and van Laar equations. The last two cannot represent multicomponent mixtures with only binary parameters, and the Wilson cannot represent liquid-liquid equilibria. All models represent binary hydrocarbon equilibria more or less equally well. Systems with low-molecular-weight alcohols are represented best by Wilson, but with carbon atoms above three the superiority is less marked. For aqueous systems the NRTL is most often best, although the Wilson and Margules often are adequate. Mixtures of substances that interact by hydrogen bonding are represented best by Margules and least well by Wilson. Margules and van Laar are uncomplicated by multiple roots.

Even with those equations that incorporate temperature, the effect of temperature is best regressed separately. A form of correlation that has been used is

$$\ln \gamma_i = A_i + B_i/T + C_i/T^4.$$

Binary liquid-liquid equilibria can be modeled with all the equations except the Wilson. The recommendation of the DECHEMA Liquid-Liquid Equilibria project is the UNIQUAC and the NRTL with $\alpha_{12} = 0.2$, since they are applicable to multicomponent systems with only binary data, at least to a first approximation. The UNIQUAC may be preferable for mixtures of widely different molecular sizes.

1 Equations of State

The mechanical state of a substance, as distinguished from a thermal or thermodynamic one, is known when the pressure, temperature, and volume are fixed. Since these three properties are related by a so-called equation of state, $f(P, V, T) = 0$, however, only two of them are independent. Thermodynamic *fundamental equations* likewise are relations between certain groups of three properties—for instance, $f(U, S, V) = 0$—from which all other thermodynamic properties can be found, including the *PVT* relation. Conversely, it is shown in Chapter 2 that the *PVT* information, in combination with a knowledge of ideal gas heat capacity as a function of temperature, also constitutes a fundamental equation.

By itself, a suitable *PVT* equation of state can be used to evaluate many important properties of pure substances and mixtures, including the following:

1. Densities of liquid and vapor phases.
2. Vapor pressure.
3. Critical properties of mixtures.
4. Vapor-liquid equilibrium relations.
5. Deviation of enthalpy from ideality.
6. Deviation of entropy from ideality.

At present no one equation of state exists that is equally suitable for all these properties of any large variety of substances, but many useful results of limited scope have been achieved. Such applications will be covered in later chapters. In this one, a number of equations of state will be described, some of historical interest, but mostly those that are of current theoretical and practical importance.

1.1. HISTORICAL NOTES

A few key dates in the history of the quantitative relations between pressure, volume and temperature are listed in Table 1.1. This history begins with Boyle's experiments with air (1662), from which he deduced that at a given temperature the volume of a gas is inversely proportional to its pressure, or that $PV =$ Constant. The effect of temperature was quantified by Charles and by Gay-Lussac (1802), who found the relation to be linear, $V = V_0(1 + kT)$. These two results were combined by Clapeyron (1834) into the first statement of the *ideal gas law* as $PV = R(T + 267)$. Later work showed that the number should be 273.2 when the temperature is in °C. A plot in relief of the ideal gas equation of state is shown in Figure 1.1. For comparison, the same type of plot is shown of an improved equation of state, that of van der Waals, and of a real substance, water. These depict the existence of a liquid phase at appropriate temperatures and pressures, which the ideal equation cannot.

Investigations on mixed gases led to Dalton's law of partial pressures (1801): In a mixture, each gas behaves as though it alone occupied the entire volume. Amagat's law (1880) states that the volume of a mixture is the sum of the volumes of the components, each at the temperature and pressure of the mixture. Pure or mixed gases for which the relation $PV = RT$ and additivities of partial pressures and volumes hold are called *ideal* or *perfect* gases. Other properties such as pressure independence of the internal energy and enthalpy follow from these basic characteristics.

The transition between vapor and liquid phases received systematic attention from Faraday (1823), although it had been realized about two hundred years earlier (van Helmont) that some gases could be condensed by lowering of the temperature. At a certain pressure and temperature characteristic of each substance, the properties of the liquid and gas become indistinguishable, and some of the properties change very markedly as that condition is approached. This condition, called the *critical state,* was discovered by Cagniard de la Tour (1822). Comprehensive studies of the critical phenomena of pure substance and mixtures were made by Andrews (1863), and work continues to this day. Several properties of carbon dioxide that behave anomalously in the vicinity of the critical are shown in Figure 1.28.

From the beginning it was realized that the ideal gas law often is only a rough approximation of true behavior. Deviations were ascribed to the finite volumes occupied by the molecules themselves and to forces of repulsion and attraction between the molecules. Both these factors were taken into account quantitatively by van der Waals (1873) in an equation that is the basis for many currently accepted *PVT* relations. In a qualitative sense this equation predicts the coexistence of liquid and vapor phases and the critical state. A great achievement of this work is the *principle of corresponding states* (see Section 1.3).

Hundreds of equations representing the *PVT* behavior of gases have been proposed—a few before van der Waals but

Table 1.1. Early Dates in the History of the Gas Laws

1662	Boyle's Law: $PV =$ a constant at a fixed temperature and mass.
1787	Charles's Law: ΔV is proportional to ΔT at constant pressure.
1801	Dalton's Law of partial pressures. In a mixture, each gas behaves as though it alone occupied the entire volume of the vessel.
1802	Gay-Lussac: Verification of Charles's Law.
1822	Cagniard de la Tour: Discovery of the critical state.
1834	Clapeyron: Combined Boyle's and Charles's laws into $PV = R(t + 267)$.
1863	Andrews: Extensive investigation of the critical state.
1873	van der Waals: Dissertation on the continuity of the gas and liquid states. The equation of state and the concept of corresponding states.
1880	Amagat's Law: The volume of a mixture of gases equals the sum of the volumes of the component gases each at the temperature and pressure of the mixture.
1901	Onnes: Development of the virial equation as an empirical relation.
1901	G. N. Lewis: The concept of fugacity.
1927	Ursell: Statistical-mechanical development of the virial equation.
1937	Mayer: Further theoretical development of the virial equation.
1940	Benedict, Webb, & Rubin: An equation of state.
1949	Redlich & Kwong: An improved two-parameter equation of state.
1955	Pitzer: The acentric factor as a corresponding states parameter.

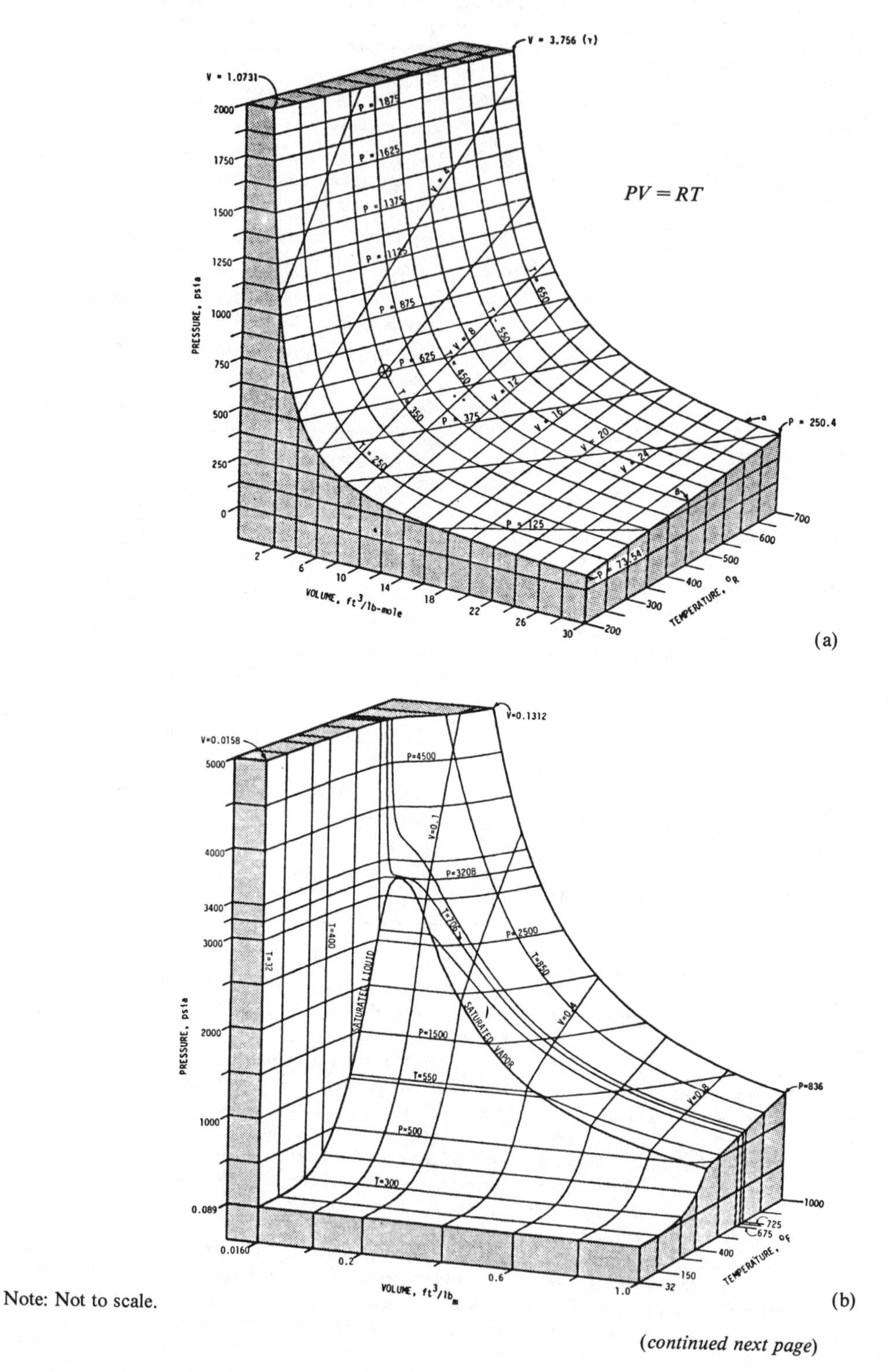

(*continued next page*)

Figure 1.1 Pressure-volume-temperature behavior of fluids: (a) The ideal gas (Jolls 1983; to appear in *J. Chem. Education*); (b) phase diagram of water and steam (Jolls et al. 1976); (c) phase diagram of carbon dioxide and its projections on the *P-T, P-V* and *V-T* planes. (Gerasimov, 1974); (d) behavior of ethylene according to the van der Waals equation of state (Jolls 1983; to appear in *J. Chem. Education*).

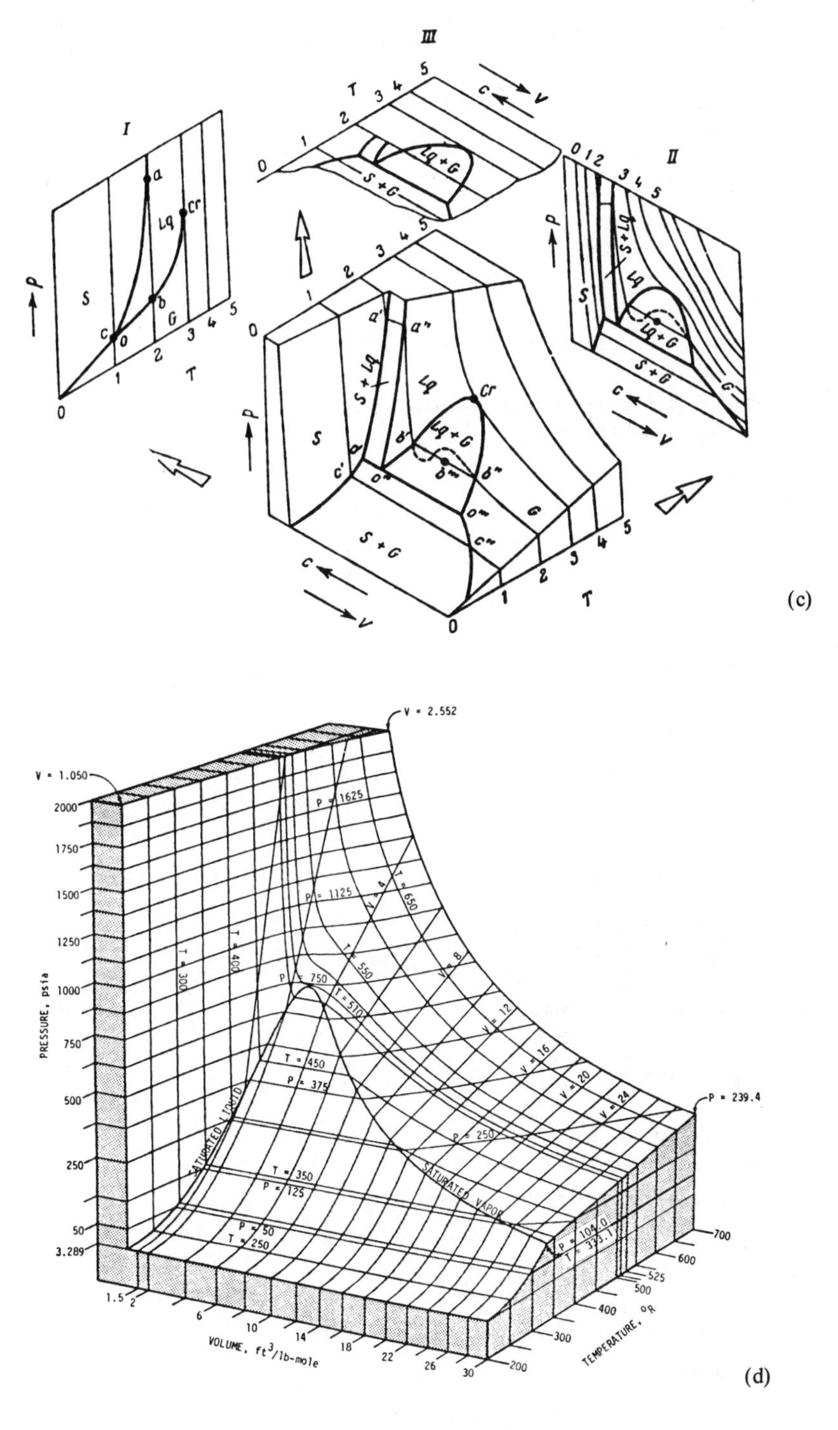

Figure 1.1 *(continued)*

mostly later, some associated with very eminent names (Rankine, Clausius, Boltzmann, Planck, and so forth). A collection of historical interest, dating back more than fifty years, is reproduced in Table 1.2.

Continued studies of forces between molecules led to a statistical mechanical derivation by Ursell (1927) of the *virial equation of state,* which actually had been proposed much earlier on a purely empirical basis by Thiesen (1885) and developed intensively by Onnes (1901). A particular merit of this derivation is a rational basis for relating the behavior of mixtures to the composition and properties of the pure components. An equation somewhat related to the virial equation was developed by Benedict, Webb, & Rubin (1940) following earlier work by Beattie & Bridgeman (1927). In its

Table 1.2. Some Early Equations of State (Partington & Shilling 1924)

1. $pv = RT$ (Boyle and Gay Lussac)[1]
   where R is a constant.
2. $pv = RT - a/Tv$ (Rankine)[2]
   where a is a constant.
3. $pv = RT + ap/T^2$ (Joule and Thomson)[3]
   where a is a constant.
4. $p = \dfrac{AT}{v+c} + \dfrac{f(T)}{v^2}$ (Dupré)[4]
   where c is the "co-volume" and A is a constant.
5. $(p + a)(v - b) = RT$ (Hirn)[5]
   where a = "sum of internal actions" ("internal pressure")
   and b = sum of volumes of molecules.
6. $pv = RT\left(1 - \dfrac{b}{v}\right)$ (Recknagel)[6]
   where $b = f(T)$.
7. $\left(p + \dfrac{a}{v^2}\right)(v - b) = RT$ (van der Waals)[7]
   where a, b, and R are constants.
8. $\left(p + \dfrac{a}{T(v+c)^2}\right)(v - b) = RT$ (Clausius)[8]
   where a, b, c, and R are constants.
9. $p = \dfrac{RT}{(v-b)}\left[1 - \dfrac{(AT^{-n} - B')(v-b)}{(v+c)^2}\right]$ (Clausius)[9]
   where A, b, B', c, R, and n are constants.
10. $\left(p + \dfrac{a}{v^2}\right)v = RT\left(1 + \dfrac{b}{v}\right)$ (Lorentz)[10]
    where a, b, and R are constants.
11. $(p + f(T))ve^{-b/v} = RT$ (Walter)[11]
    where e is the base of natural logarithms, and $b = f'(T)$.
12. $p(v - \alpha) = f(v)\left[T - \dfrac{M(v-a)}{v^m + av^{m-1} + \cdots + k}\right]$ (Amagat)[12]
    where α = atomic volume; m, k, a, etc., are constants.
13. $p = \dfrac{RT}{(v-\alpha)} - \dfrac{K\varepsilon^{-r}}{(v+\beta)^2}$ (Sarrau)[13]
    where R, α, β, K, and ε are constants; *cf.* equation (8) above.
14. $pv = RT\left(1 + \dfrac{T_1}{v} + \dfrac{T_2}{v^2} + \cdots\right)$ (Thiesen)[14]
    where T_1, T_2, etc., are functions of T.
15. $pv = RT(1 + a_1 + a_2 + a_3 + \cdots + b_1 + b_2 + b_3 + \cdots)$ (Natanson)[15]
    where $a_1, a_2 \ldots$ are functions of T depending on the nature of the gas, and $b_1, b_2 \ldots$ are integrals.

Table 1.2 (*continued*)

16. $$\left[p+\frac{1}{v(v+k)}\right]v=RT\left[1+\frac{2k}{v+k}\right]$$ (Sutherland)[16]

This is identical with equation (8) above with $f(T)=1-cT$ and $F(T)=b/\sqrt{T}$. The equation was deduced for CO_2.

17. $$p=\frac{aT}{b}\cdot\frac{(r/b)^2}{1-(r/b)^3}-A$$ (Lagrange)[17]

where a is a constant independent of the gas, $r=$ radius of the gas atoms, and $2b=$ mean distance between their centers; $A=$ const. depending on the molecular attraction.

18. $$\left[\frac{p+0.000004568M\sqrt[3]{np}-\dfrac{1}{13596p}}{2\{(v-b)(1+\alpha t)\}^2}\right](v-b)=RT$$ (Violi)[18]

where $n=$ number of atoms in molecule.

19. $pv=f(v)[A+T-B\sqrt{p}]$ (Antoine)[19]

where A and B are constants.

20. $$p(v-b)=RT-\frac{A}{v-c}+\frac{E}{v-d}$$ (Tait)[20]

where A, b, c, d, and E are constants.

21. Very complex; deduced for dissociating gases (Swart)[21]

22. $$\left(p+\frac{\lambda+\mu p}{T(v+\beta)^2}\right)(v-\gamma T)=RT$$ (Schiller)[22]

where λ, μ, β, and γ are constants depending on the gas.

23. $$\left(p+\frac{a}{v^2}\right)\left(v-\frac{b}{4}\right)^4=RTv^3$$ (Jäger)[23]

where a and b are constants.

24. $$p=\frac{RT}{v-\alpha}-\frac{mI^{-\mu}-nT^{v}}{(v+\beta)^2}$$ (Battelli)[24]

25. $$p=\frac{A}{v-\alpha}+\frac{B}{v-\beta}+\frac{C}{v^2-\gamma}$$ (Brillouin)[25]

26. Very complex (Weinstein)[26]

27. $$pv=RT\left(1+\frac{c}{v+k-gv^{-2}}\right)-\frac{av}{v+k}$$ (Rose Innes)[27]

where R, c, k, g, and a are constants.

28. $$\left(p+\frac{a}{v^2}\right)\left(v-b+\frac{17}{32}\cdot\frac{b^2}{v}\right)=RT$$ (Boltzmann)[28]

or

$$\left(p+\frac{a}{v^2}\right)\left(v-\frac{b}{3}\right)=RT\left(1+\frac{2}{3}\frac{b}{v}+\frac{7}{24}\frac{b^2}{v^2}\right)$$

where a, b, and R are constants.

29. $$\left(p+\frac{a}{v^2}\right)\left(v-b+\frac{c}{v^2+d}\right)=RT$$ (Boltzmann and Mache)[29]

Table 1.2 *(continued)*

30. $$\left[p+\frac{v-T[a+m(v-b)+c(v-b)^{-1}]}{kv^{2\cdot 85}-a+n\sqrt{(-\beta)^2+d^2}}\right]v=RT$$ (Amagat)[30]

Deduced for CO_2; a, b, c, m, n, k, d, α, and β are constants.

31. $$\left(p+\frac{\theta}{v^2}\right)(v-be^{C/T})=RT$$ (Reinganum)[31]

where $\theta=f(T)$, and b, C, R are constants.

32. $$\left(p+\frac{a}{v^{5/3}}\right)(v-b)=RT$$ (Dieterici)[32]

where a, b, and R are constants.

33. $p(v-b)=RTe^{-c/RTv}$ (Dieterici)[33]

where b, c, and R are constants.

34. $$\left[p+\frac{A}{Tv^{3/2}(v^{1/2}+\gamma)}\right](v-b)=RT$$ (Starkweather)[34]

35. $$\left(p+\frac{a}{Tv^2}\right)(v-b)=RT$$ (Berthelot)[35]

36. $$\left[p+\frac{k}{v^2}\left(1-\sqrt{\frac{b^2}{4\sqrt{2v^2}}}\right)\right](v-b)=RT$$ (Tumlirz)[36]

37. Same as (7), but

$$b=b_\infty\left(1-k_1\frac{b_\infty}{v}+k_2\frac{b_\infty^2}{v^2}\right)$$ (van der Waals)[37]

38. $p=CRTe^{-Q/RT}$ (Mie)[38]

39. $$\left(p+\frac{a}{(v-\alpha)^2}\right)(v-b)=RT$$ (Goebel)[39]

40. $$\left(p+\frac{a}{(v-\alpha)^2}\right)(v-b_0+b_1p)=RT$$ (Goebel; for CO_2)[40]

41. $$\left(p+a\rho^2+\frac{\rho}{A+B\rho+C\rho^2+D_T^\rho}\right)(v-b)=RT$$ (Smoluchowski)[40]

where $\rho=$ density.

42. $$pv=RT-\frac{\left(A+\frac{B}{v}\right)\left(\frac{v}{k}-1\right)}{v^2-s}$$ (Batschinski; for isopentane)[41]

where A, B, k, $s=f(T)$.

43. $$pv=a(T-\beta)-\frac{sT^{-1/2}(v-k)}{v(v-\lambda)}$$ (Batschinski; for ether)[41]

44. $$p=-\frac{kT}{\beta}\log\left(1-\frac{\beta}{v}\right)-\frac{\alpha}{v^2}$$ (Planck)[42]

where α, k, and β are constants.

45. $pv^2=f_1(T)-Tf_2(T)$ (Leduc)[43]

Table 1.2 *(continued)*

46. $pv = A + \frac{B}{v} + \frac{C}{v^2} + \frac{D}{v^4} + \frac{E}{v^6} + \frac{F}{v^8}$ (Onnes)[44]

or, reduced,

$$\pi\phi = k\theta\left[1 + \beta\frac{k}{\phi} + \gamma\left(\frac{k}{\phi}\right)^2 + \delta\left(\frac{k}{\phi}\right)^4 + \varepsilon\left(\frac{k}{\phi}\right)^6 + \eta\left(\frac{k}{\phi}\right)^8\right]$$

where $k = RT_c/p_cv_c$ and $\beta = b_1 + b_2/\theta + b_3/\theta^3 + b_4/\theta^4 + b_5/\theta^5$

47. $pv = Ae^{ap+bp^2+cp^3+\cdots}$ (Peczalski)[45]
where $A, a, b, c, \ldots$ are constants.

48. $\left(\pi + 3\frac{e^{1-\theta}}{\phi^2}\right)(3\phi - 1) = 8\theta$ (Dalton)[46]

49. $\left(p + \frac{2a}{(v+b)^2}\right) = \frac{RT}{v}$ (Kam)[47]

50. $\left(p + \frac{a}{Tv(v-b)} - \frac{c}{T^2v^3}\right)(v - b) = RT$ (Wohl)[48]

51. $p = -\frac{RT}{2b}\log\frac{v-2b}{v}$ (Shaha and Basu;[49] Chapman and Appleby)[50]

where b is a constant (*cf.* (44) above).

52. $v = \frac{RT}{p} - \frac{c}{T} + \frac{kp^2}{T} + b$ (Schrieber)[51]

where c, k, R, and b are constants.

53. $\pi = \frac{\theta}{\phi - \frac{1}{4}}\, e^{\left(-\frac{a}{\phi} - \frac{b}{\phi^2} - \frac{c}{\phi^3}\right)}$ (Porter)[52]

where π is the reduced pressure, θ the reduced temperature, ϕ the reduced volume and a, b, c constants.

54. $\left(p + \frac{a}{v^2}\right)(v - b) = RT\left(1 + \frac{\psi^2}{T^2}\right)^{-1}$ (Boynton and Bramley)[53]

where ψ is a characteristic temperature.

55. $\left[p + \frac{\phi}{(v+n)^2}\right](v - b) = RT$ (Fouché)[54]

where $\phi = f(T)$, and n, b are constants; deduced for air.

56. For pure nitrogen, Smith and Taylor[55] suggest

$p = 2{\cdot}92855\,T/(v - \delta) - 1623{\cdot}63/(v + 0{\cdot}2954)^2$

where $\delta = 0{\cdot}18683 - 0{\cdot}3113/v$.

Note: A similar compilation is that by Otto (1929) and a longer one is that of Vukalovich & Novikov (1948).

[1]For this history, see Partington, p. 552, *1950*.

[2]*Phil. Trans.*, 144, 336, *1854.*

[3]*Ibid.*, 152, 579, *1862.*

[4]*Ann. Chim. et. Phys.*, 4, 426, *1865*; or *Bull. Acad. roy. Belg.*, 14, 46, *1887.*

[5]*Ann. Chim. et Phys.*, 11, 5, *1867*; *cf.* de Heen.

[6]*Ann. phys. Ergbd.*, p. 563, *1871.*

[7]*Over de Continuiteit van den Gas en Vloeistoftoestand*, Leyden, *1873*; tr. Threlfall and Adair, *1890*; *cf.* Jüptner, *Zeit. phys. Chem.*, 63, 579, *1908*; 73, 343, *1910.*

[8]*Ann. Phys.*, 9, 337, *1880*; 14, 279, *1881*; *cf.* Sarrau, *C. R.*, 101, 941, 994, 1145, *1885.*

[9]*Ann. Phys.*, 14, 692, *1881.*

[10]*Ibid.*, 12, 127, *1881.*

Table 1.2 *(continued)*

[11]*Ibid.*, 16, 500, *1882.*
[12]*C.R.*, 94, 847, *1882*; 118, 566, *1894.*
[13]*Ibid.*, 110, 880, *1890.*
[14]*Ann. Phys.*, 24, 467, *1885.*
[15]*Kinetische Theor. unvolk. Gase*, Dorpat, *1887*; *Ann. Phys.*, 33, 683, *1888*; *Arch. de Genève*, 28, 112, *1892.*
[16]*Phil. Mag.*, 24, 113, *1887*; 35, 211, *1893.*
[17]*Bull. Acad. roy. Belg.*, 16, 171, *1888.*
[18]*Rend. R. Acc. dei Lincei*, 4, 285, 316, 462, 513, *1888.*
[19]*C.R.*, 108, 896, *1889*; 110, 131, 1122, *1890*; 112, 284, 1892.
[20]*Proc. Roy. Soc. Edin.*, 16, 65, *1889*; 18, 265, *1891*; *Trans. Roy. Soc. Edin.*, 36, 257, *1892.*
[21]*Diss. Amsterdam, 1890*; *Beib.*, 15, 339, *1891.*
[22]*Ann. Phys.*, 40, 149, *1890.*
[23]*Wien. Ber.*, 99, 1028, 1892, *1891*; 101, 1675, *1892*; 105, 15, 97, *1896.*
[24]*Ann. Chim. et Phys.*, 25, 38, *1892.*
[25]*Journ. de Phys.*, 2, 113, *1893.*
[26]*Ann. Phys.*, 54, 544, *1895.*
[27]*Phil. Mag.*, 44, 76, *1897*; 45, 102, *1898.*
[28]*Gastheorie*, II, 153, *1898.*
[29]*Ann. Phys.*, 68, 350, *1899.*
[30]*C.R.*, 128, 538, 649, *1899.*
[31]*Diss. Göttingen, 1899*; *Arch. Néerl.*, 5, 574, *1900*; *Ann. Phys.*, 6, 533, *1901.*
[32]*Ann. Phys.*, 69, 685, *1899*; 5, 51, *1901.*
[33]*Ibid.*, 66, 826, *1898.*
[34]*Am. J. Sci.*, 7, 129, *1899.*
[35]*Arch. Néerl.*, 5, 417, *1900.*
[36]*Wien. Ber.*, 108, 1058, *1899*; 111, 524, *1902.*
[37]*Zeit. phys. Chem.*, 38, 257, *1901.*
[38]*Ann. Phys.*, 11, 657, *1903.*
[39]*Zeit. phys. Chem.*, 47, 471, *1904*; 49, 129, *1906.*
[40]Boltzmann, *Festschrift*, Leipzig, p. 626, *1904.*
[41]*Ann. Phys.*, 19, 310, *1906*; 21, 1001, *1906.*
[42]*Preuss. Akad. Wiss.*, p. 633, *1908.*
[43]*C.R.*, 148, 1670, *1909.*
[44]*Konink. Akad. Wetens*, p. 273, *1912*; *cf.* Jeans, *Dyn. Theory of Gases*, p. 164, *1916.*
[45]*C.R.*, 156, 1884, *1913.*
[46]*Trans. Roy. Soc. S. Africa*, pt. 2, 123, *1914.*
[47]*Phil. Mag.*, 37, 65, *1919.*
[48]*Zeit. phys. Chem.*, 87, 1, *1914.*
[49]*Phil. Mag.*, 36, 199, *1918.*
[50]*Phil. Mag.*, 40, 197, *1920.*
[51]*Phys. Zeit.*, 21, 430, *1920.*
[52]*Phil. Mag.*, 44, 1020, *1922.*
[53]*Phys. Rev.*, 20, 46, *1922.*
[54]*C.R.*, 169, 1089, *1919.*
[55]*J. Am. C.S.*, 45, 2107, *1923.*

several modifications it is widely used for the lighter hydrocarbons and inorganic gases, for liquid and vapor phases, and for mixtures. Some of the versions have as many as thirty constants, so they are usable only with computers.

Scores of modifications of the van der Waals equation have been made. A highly successful one is due to Redlich & Kwong (1949) and more recent ones to Soave (1972) and Peng & Robinson (1976). These are cubic equations whose parameters are basically expressible in terms of critical properties but include modifications for temperature and another property such as critical compressibility or acentric factor (see Section 1.3.1).

Most modifications of these equations have been empirical and arbitrary, with parameters that are adjustable to fit certain kinds of experimental data such as vapor pressures, densities, or enthalpies. Recently, however, some statistical mechanical theory has been applied toward improving cubic EOS. Comparatively simple results were obtained by Carnahan & Starling (1972), somewhat more complicated ones by Donohue & Prausnitz (1977). The work of Yokoyama et al. (1982) extends the method to polar substances.

Knowledge of the behavior of liquids is less quantitative than that of gases, although much work, both theoretical and correlative, is being done along these lines. A correlation of

Table 1.2A. Some Early and Recent Vapor Pressure Formulas

Formula	Source
1. $\ln P = (k_1 + k_2 T) \ln T$	Schmidt (1797)
2. $\ln P = k_1 \ln (k_2 + k_3 T)$	Young (1807)
3. $\ln P = k_1 T/(T + k_2)$	Wrede (1841)
4. $\ln P = k_1 - k_2/T - k_3 T^2$	Rankine (1849); Kirchhoff (1858)
5. $\ln P = k_1 - k_2/(T + k_3)$	Antoine (1888)
6. $\ln (P_c/P) = k(T_c/T - 1), k \simeq 3$	van der Waals (1899)
7. $P_r(V_g - V_L) = RT(1 - P_r)$	Nernst (1906)
8. $\ln P_r = (1 - 1/T_r)(k_1 - k_2 \ln T_r)$	Carbonelli (1919)
9. $\ln P = k_1 + k_2/T + k_3 T + k_4 T^2$	Cragoe (1928) (Int. Crit, Tables III, p. 228)
10. $\ln P = k(T_c/T - 1) + (T_c/T) \ln P_c$	Pollara (1942)
11. $\ln P = k_1 + k_2/T + k_3 \ln T + k_4 P/T^2$	Frost & Kalkwarf (1953) Harlacher & Braun (1970); Reid, Prausnitz, & Sherwood (1977, p. 188)
12. $\ln P = k_1 + k_2/T + k_3 T + k_4 T^3$	Miller (1964)
13. $T_r \ln P_r = k_1 \tau + k_2 \tau^{1.5} + k_3 \tau^3 + k_4 \tau^6$ $\tau = 1 - T/T_c$	Wagner (1973); Ambrose (1978)

saturated liquid densities by Hankinson & Thomson (1979) is typical of recent work. Many equations of state, notably the complex BWR (1940) and the simpler Peng-Robinson (1976) and Harmens-Knapp (1980), also have been designed for good representation of liquid densities. A particularly noteworthy achievement has been the prediction of the behavior of nonideal liquid mixtures in terms of their molecular structures by the ASOG and UNIFAC methods (see Chapter 4).

Equations of state remain an active field of research, primarily in three areas:

1. Highly accurate equations with many constants for important pure substances or mixtures such as air, water, ammonia, carbon dioxide, light hydrocarbons, and cryogenic fluids.
2. Complex equations or computer algorithms for mixtures encountered in the natural gas and petroleum industries.
3. Simpler equations, such as cubics, that may be adequate for making the repeated calculations of phase equilibrium and deviation functions for process design of multistage separation of mixtures, without placing extreme demands on computer time.

Continued emphasis is being devoted to electrolytes, polymers, coal liquids, and polar substances generally—a class that has proved largely intractable thus far.

1.2. NONIDEAL GASES

As a basis of comparison, the concept of the ideal gas is invaluable since all gases approach ideality at low pressures. Moreover, the methods of statistical mechanics are readily applicable to the calculation of properties of ideal gases, with results that are regarded as often superior to experimental data on nearly ideal gases. If that topic is of interest, reference may be made to a brief book by Rowlinson (1963).

1.2.1. Deviations from Ideality

Many properties of real substances can be expressed conveniently as deviations from ideality—for example, the residual volume, $\Delta V = V - RT/P$, whose difference from zero is a measure of nonideality, or the compressibility, $z = PV/RT$. The marked deviation of the latter quantity from unity is shown for some substances in Figure 1.2a, b. Other informative plots of the compressibility are shown in Figure 1.8. Several correlations of this property will be discussed later. Ideal gas isotherms are rectangular hyperbolas, but those of carbon dioxide and isopentane in Figure 1.5a, b deviate sharply from that shape, particularly in the vicinity of the critical temperature.

Another important distinction between real and ideal gases is their thermal behavior. Heat capacities of ideal gases are independent of pressure; but, as Figures 1.3 and 1.4 show, this is far from being the case with real substances. Many other properties of real substances also exhibit unusual behavior in the vicinity of the critical state that ideal gases do not—notably the properties of carbon dioxide, which are shown in Figure 1.28.

1.2.2. Intermolecular Forces

Sizes, shapes, and structures of molecules determine the forces between them and ultimately their *PVT* behavior. Attractive forces tend to keep molecules together, and repulsive ones prevent them from mutual annihilation. The former come into play at greater separations of the molecules; the latter are effective at close range. The effects of forces are describable in terms of potentials, quantities whose gradients are forces. Figure 1.24 and Table 1.23 show how the potentials are visualized to depend on the distance of separation of molecules. Some quantitative aspects of these concepts are considered in Section 1.7. Here it may be simply

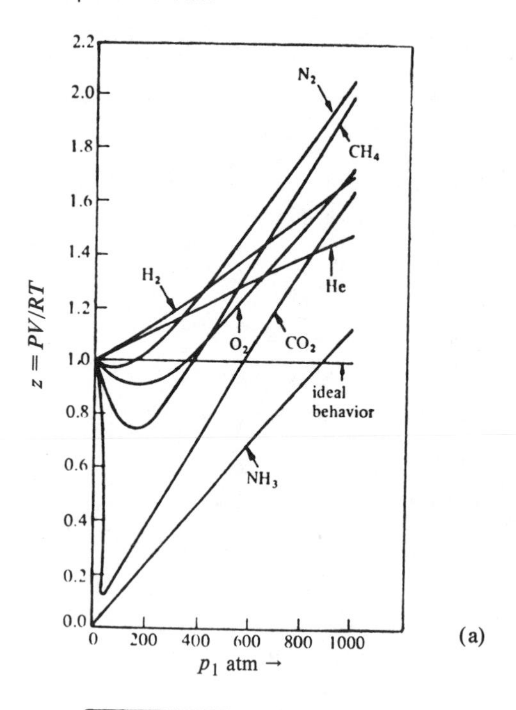

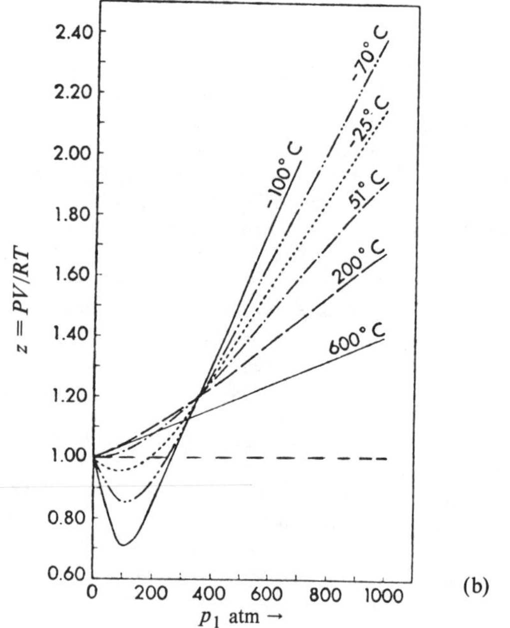

Figure 1.2 Effects of temperature and pressure on the compressibility, $z = PV/RT$, of some gases: (a) Compressibility of nitrogen at several temperatures over a range of pressures (Fried et al. 1977); (b) compressibilities of several gases at 0 C over a range of pressures (Fried et al. 1977).

stated that attractive forces may be regarded as resulting in a pressure that is greater than that due to the kinetic energies of the molecules, whereas repulsive forces reduce the effective volume that is available to molecular movement.

In terms of their electrical properties, molecules may be classified as

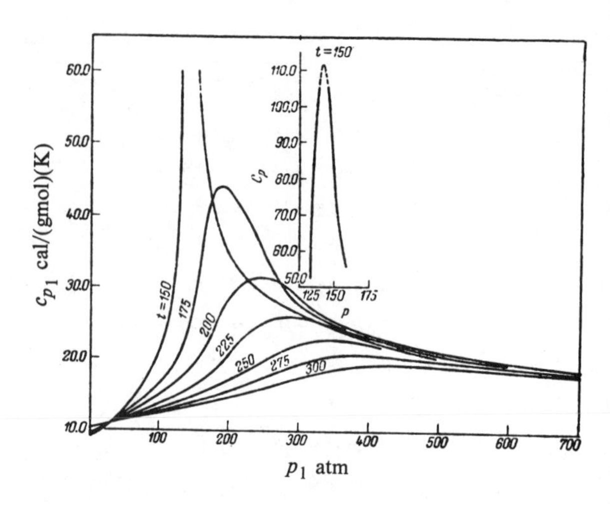

Figure 1.3 Nonideality as manifested by the effect of pressure on heat capacity: Data of ammonia. The critical properties are 111.3 atm, 405.6 K. (Karapetyants 1978).

1. Electrically neutral and symmetrical, and usually nonpolar.
2. Electrically neutral but unsymmetrical—that is, possessing dipole moments and called polar.
3. Those having residual valences that may result in association and hydrogen bonding.

Forces of repulsion and attraction are present in all molecules, but they appear in exaggerated forms in associating and polar molecules. Most success with *PVT* correlations has been achieved for nonpolar substances, which include the important category of hydrocarbons. Second virial coefficients even of polar substances, however, have been well correlated in terms of certain geometrical and electrical characteristics of the molecules. Some quantitative work on the effect of molecular association on *PVT* behavior has been done and is discussed in Section 1.3.3. Several theoretical and empirical relations for the effects of molecular attraction and repulsion have been incorporated in equations of state. Historically, rather more attention has been paid to modifying the attraction term, a/V^2 in the van der Waals equation, whereas the repulsion term has been largely accepted in the form proposed by him. Recently, however, there has been a revival of interest in the latter term; some useful developments are mentioned in Section 1.7.

1.2.3. The van der Waals Equation

Since the ideal gas equation was recognized early to be inadequate for the demands of science and technology, many other equations of varying merit were proposed over the years. Almost every one of these has been shown, or claimed, to be superior in some respects to earlier ones—because of a sound theoretical basis, or in some particular range of temperature and pressure, or for some particular substances, or for the evaluation of some particular thermodynamic property, or for being easier to use, or because the inventor had become interested in the topic. Most of these equations have not been accepted, not always because they were inferior, but simply because they were not superior.

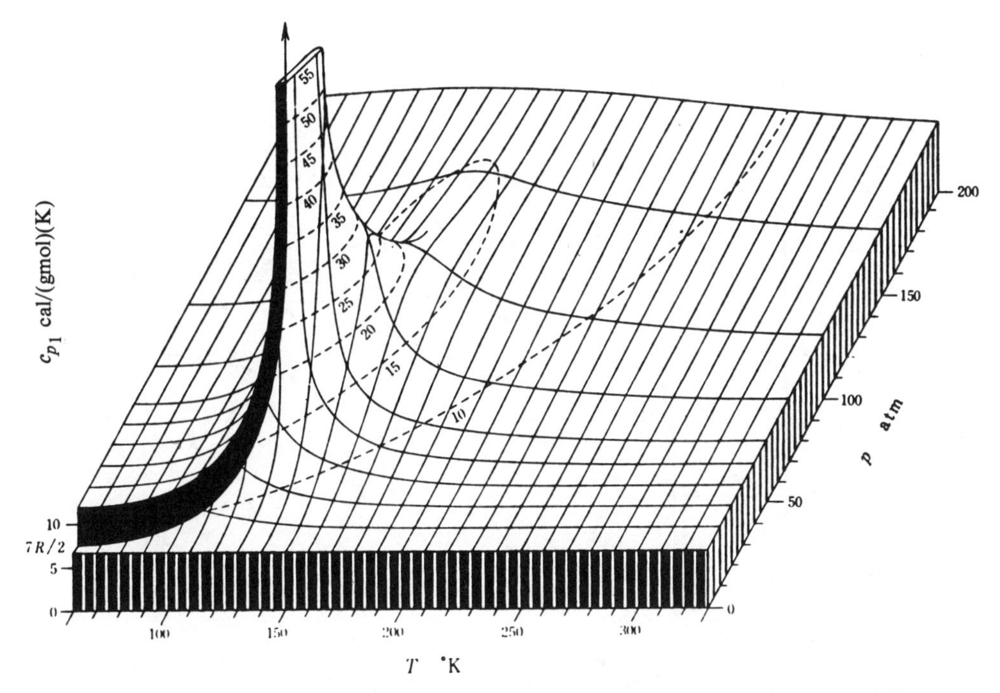

Figure 1.4 The $C_p - T - P$ surface of air in the vicinity of the critical point, 37.2 atm, 132.5 K (Kubo 1968).

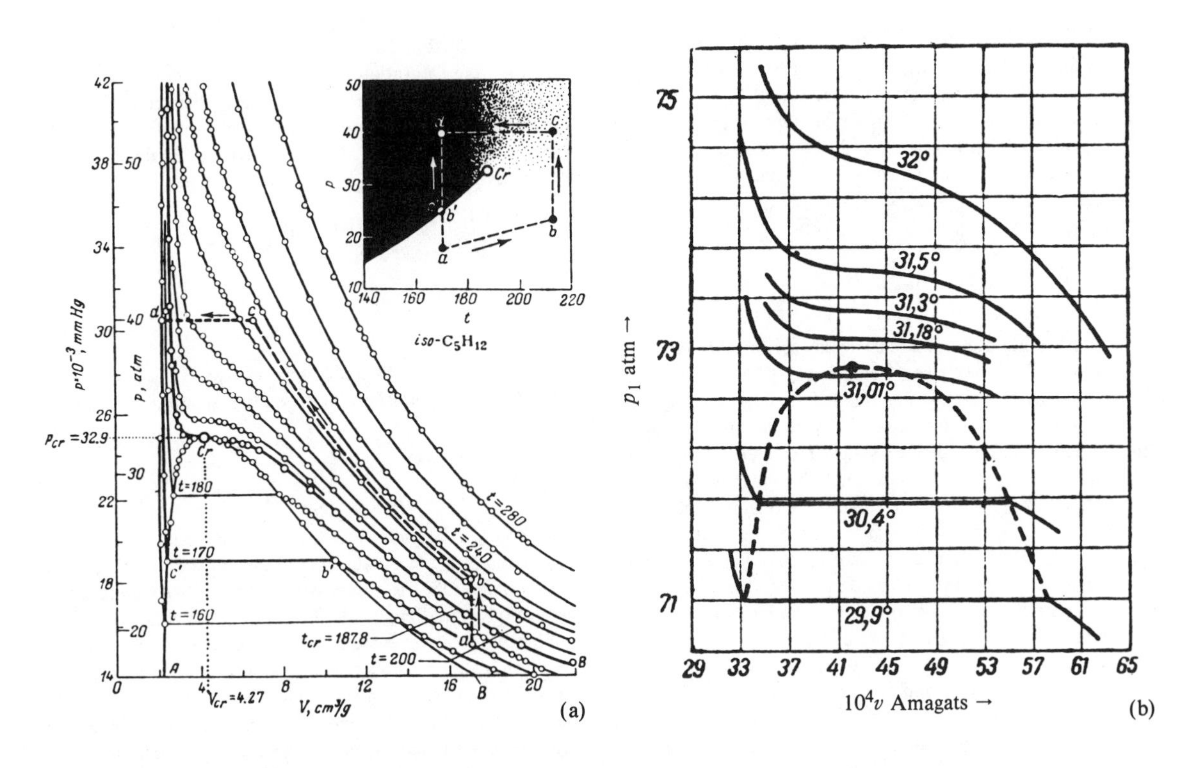

Figure 1.5 Experimental isotherms: (a) Phase diagram of isopentane in the two-phase and homogeneous regions. Near the critical point, the transition from a vapor to a liquid phase may be a heterogeneous path (*a-b'-c'-d*), or a homogeneous one (*a-b-c-d*) (Karapetyants 1978). (b) carbon dioxide in the vicinity of the critical point, 72.8 atm and 42.7 (E-4) Amagats or 94 ml/gmol (data of Michels et al., 1937).

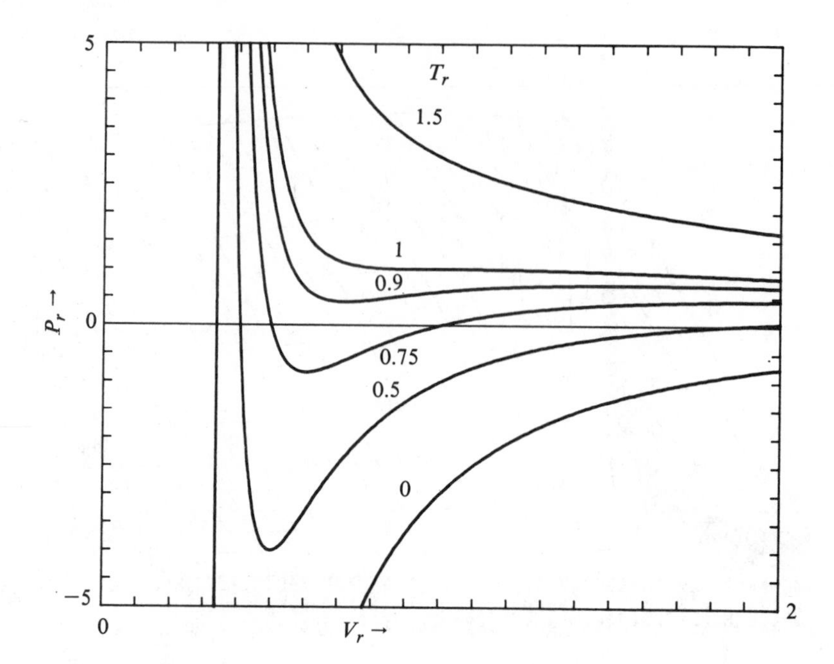

Figure 1.6 Isotherms of the reduced van der Waals equation of state, $P_r = 8T_r/(3V_r - 1) - 3/V_r^2$. All curves go to $\pm\infty$ at $V_r = 0$ and 1/3.

The most famous and one of the most fruitful equations of state is that of van der Waals (1873), whose dissertation was "on the continuity of the gas and liquid states." After 1873 he wrote more than a hundred papers, largely on the theory of liquids and mixtures. In a centennial paper, Rowlinson (1973) has shown how much of van der Waals's work is still embodied in currently accepted theory. In the van der Waals equation,

$$\left(P + \frac{a}{V^2}\right)\left(V - b\right) = RT, \tag{1.1}$$

a is called the attraction parameter and b the repulsion parameter. The latter also is called the *effective molecular volume,* which van der Waals theorized to be four times the actual volume of the molecules. The reasoning that led to this equation and its deduction are given conveniently by Partington (1950, p. 164), for example, where much historical material also is presented. A three-dimensional representation of the equation is shown as Figure 1.1d. Some isotherms are shown in Figure 1.6 and elsewhere in this chapter. Related mathematical material is collected in Table 1.3.

The S-shaped isotherm implying the continuity of the liquid and vapor phases was recognized by Thomson (1871) before van der Waals's publication. It is characteristic of all equations of cubic form and of some that are not cubic—for example the BWR equation diagram in Figure 1.21. It was pointed out by Maxwell (1875) that the reversible isothermal work of the cycle *BCDEFDB* of Figure 1.7 is zero, and consequently that the two areas between the curve above and below the horizontal *FDB* are equal. In accordance with this principle, the saturation pressure and the saturation volumes at a given temperature may be found by the intersection of a horizontal line with the two-phase envelope, located so that the two areas are equal. Mathematically, this condition may be written:

$$\text{Area} = \int_{V_L}^{V_g} P\,dV - P^{\text{sat}}(V_g - V_L)$$

$$= -\int_{P^{\text{sat}}}^{P^{\text{sat}}} V\,dP = 0. \tag{1.2}$$

Also, as shown in Chapter 3, this result follows from the fact that at phase equilibrium the fugacities of the two phases are the same:

$$f_g - f_L = \int_{P^{\text{sat}}}^{P^{\text{sat}}} V\,dP = 0. \tag{1.3}$$

In Figure 1.7 points B and F represent the specific volumes of gas and liquid in equilibrium at the corresponding pressure and temperature. In practice, however, the prediction of saturation conditions with the van der Waals equation is qualitative only, as evidenced by Figures 1.17 and 1.19 and Example 1.2, for instance. A check of the saturation condition with the Clausius equation, which is similar in form to the *vdW*, is asked for in Problem 1.2. Some other cubic equations have been designed to represent saturation more accurately (see Problem 3.26, for instance). Portions *BC* and *FE* of the curve of Figure 1.7 represent metastable conditions that have

Table 1.3. The van der Waals Equation of State

Standard forms of the equation are:

$$(P + a/V^2)(V - b) = RT, \tag{1}$$

$$z = PV/RT = V/(V - b) - a/RTV. \tag{2}$$

Polynomial forms are:

$$V^3 - \left(b + \frac{RT}{P}\right)V^2 + \frac{a}{P}V - \frac{ab}{P} = 0, \tag{3}$$

$$z^3 - \left(\frac{bP}{RT} + 1\right)z^2 + \frac{aP}{(RT)^2}z - \frac{abP^2}{(RT)^3} = 0. \tag{4}$$

Virial form:

$$z = 1 + \left(b - \frac{a}{RT}\right)\frac{1}{V} + \left(\frac{b}{V}\right)^2 + \left(\frac{b}{V}\right)^3 + \cdots. \tag{5}$$

Reduced form:

$$\left(P_r + \frac{3}{V_r^2}\right)(3V_r - 1) = 8T_r \tag{6}$$

The parameters in terms of critical properties:

$$a = 3P_cV_c^2 = 27R^2T_c^2/64P_c, \tag{7}$$

$$b = V_c/3 = RT_c/8P_c, \tag{8}$$

$$R = 8P_cV_c/3T_c, \tag{9}$$

$$z_c = 0.375. \tag{10}$$

The value of R given by Eq. 9 is not the same as the true gas constant, 8.314 joules/gmol-K; the true value should be used for evaluation of the parameters a and b from Eqs. 7 & 8.

Derivation of the formulas for a and b:

Method 1: At the critical condition, the first and second derivatives with respect to V at constant T are zero. Performing these differentiations,

$$\left(\frac{\partial P}{\partial V}\right)_T = -\frac{RT_c}{(V_c - b)^2} + \frac{2a}{V_c^3} = 0, \tag{11}$$

$$\left(\frac{\partial^2 P}{\partial V^2}\right)_T = \frac{2RT_c}{(V_c - b)^3} - \frac{6a}{V_c^4} = 0. \tag{12}$$

Also,

$$\left(P_c + \frac{a}{V_c^2}\right)(V_c - b) = RT_c. \tag{13}$$

Eqs. 11–13 are solved simultaneously to obtain the results of lines 7–9.

Method 2: At the critical point, the three roots of the polynomial, Eq. 3, are equal.

$$(V - V_c)^3 = V^3 - 3V_cV^2 + 3V_c^2V - V_c^3 = 0. \tag{14}$$

Comparison of coefficients of like powers of Eqs. 3 and 14 will lead to the same results for the parameters as by method 1.

For mixtures, the parameters are expressed in terms of the pure component parameters by the *combining rules:*

$$a = (\Sigma y_i\sqrt{a_i})^2 = \Sigma\Sigma y_iy_j\sqrt{a_ia_j} = \Sigma\Sigma y_iy_ja_{ij}; \tag{15}$$

$$b = \Sigma y_ib_i. \tag{16}$$

Fugacity equations are in tables 3.3 and 3.4. Residual property equations are in Table 11.3.

---

been realized experimentally, but portion *EDC* is not physically possible for a pure substance since it corresponds to changes of pressure and volume in the same direction at fixed temperature.

Much thought has been devoted to interpreting the meaning of the S-shaped isotherm at negative pressures. A survey of recent work is in the book of Temperley & Trevena (1978). The conclusion appears to be that liquids are definitely under tension under these conditions; for instance, a tensile strength of water of 277 atm at 10 C has been measured.

Since the van der Waals equation is of the third degree in volume, any subcritical isotherm has three real positive roots, whereas supercritical isotherms have only one real root. When there are three real roots, the smallest is interpreted as the specific volume of a liquid phase, the largest as that of the vapor phase, and the intermediate one as physically meaningless. Cardan's solution of a cubic equation is especially simple for real roots, but several other general root-finding methods are readily implemented on calculators. Although they are no longer of practical value, several chart methods of solution of cubic equations have been published, one by Lipka (1918) and another by Collatz (1955). A large numerical table prepared by Salzer et al. (1958) sometimes may be of interest.

The Reduced Form of the Equation

From the several plots of critical isotherms that have been shown in this chapter, it is clear that a point of inflection appears to exist just at the critical point. Mathematically the point of inflection may be found on equating the first and second derivatives to zero at the critical volume, thus:

$$\left(\frac{\partial P}{\partial V}\right)_T = \left(\frac{\partial^2 P}{\partial V^2}\right)_T = 0. \tag{1.4}$$

When these rules are applied to the van der Waals equation, the parameters a and b may be found in terms of the critical properties. The results are shown in Table 1.3. The same results also can be deduced from the observation that the three roots are equal at the critical point, by comparing the coefficients of the expansion of $(V - V_c)^3 = 0$ with those of the polynomial form of the EOS.

A conclusion of great significance, the Principle of Corresponding States, is suggested when parameters a and b are eliminated from the van der Waals equation in terms of their critical equivalents, with the result:

$$(P_r + 3/V_r^2)(3V_r - 1) = 8\,T_r. \tag{1.5}$$

The ratios $P_r = P/P_c$, and so on, are called *reduced properties*, and the equation is called a *reduced equation of state.* Clearly, it is applicable in principle to any substance; but the actual properties can be extracted from the equation only when the individual critical properties are known. Substances with the same reduced properties are said to be in *corresponding states.* This concept has widespread ramifications, some of which are discussed in Section 1.3.

The van der Waals analysis leads to an expression of the gas constant in terms of the critical properties as

$$R_{vdw} = 8\,P_c V_c / 3\,T_c,$$

which is numerically different for each substance. Only by accident in some case is it equal to the universal gas constant, $R = 8.314$ joules/(gmol) (K). Since it is desirable that the EOS reduce to the ideal gas form at low pressure, however, the universal value of R always is used to evaluate the parameters a and b. The values in Table 1.4 were obtained in this way. Since the vdW equation usually is not a perfect fit to data, R_{vdw} could be regarded as another parameter that varies from substance to substance, just as a and b vary. Thus there are cases in which the equation $(P + a/V^2)(V - b) = R_{vdw}T$ fits experimental data at least as well as $(P + a/V^2)(V - b) = RT$. Even tampering with the parameters a and b may result in an improved EOS. Examples are given later. A comparison of a and b determined from experimental data and from the critical properties is made in Problem 1.2.

Table 1.4. Some Critical Compressibilities and the Values Predicted by Some Equations of State

	Z_c
Equation	
Ideal	1
van der Waals	0.375
Redlich-Kwong	0.333
Soave	0.333
Third-order virial	0.333
Peng-Robinson	0.307
Berthelot	0.281
Dieterici	0.271
Substance	
He	0.3141
H_2	0.3049
CO_2	0.2869
SO_2	0.2774
Cl_2	0.2755
$i-C_5H_{12}$	0.2678
C_6H_6	0.2663
C_6H_5F	0.2634
$(C_2H_5)_2O$	0.2430
C_2H_5OH	0.2243
H_2O	0.229

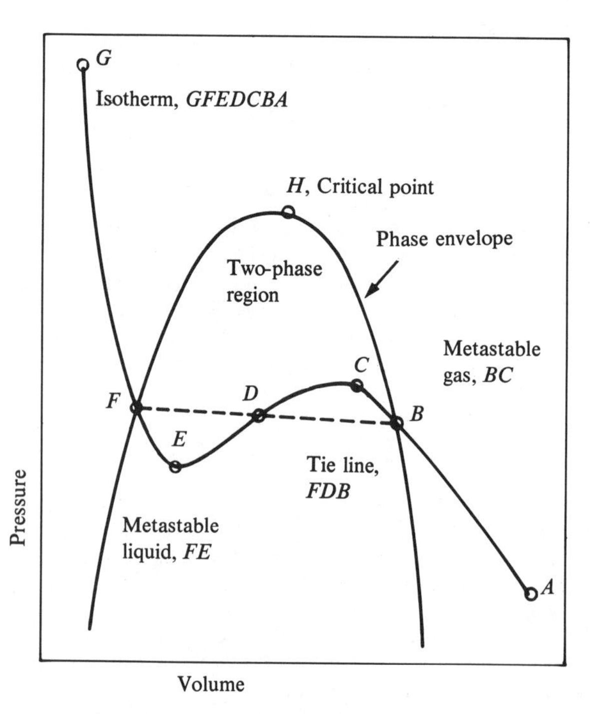

Figure 1.7 Diagram of a cubic equation of state in the two-phase region. Shows metastable regions and tie line connecting volumes of liquid and vapor phases in equilibrium. The areas *FEDF* and *DCBD* are equal (Maxwell's principle). The region *EDC* is physically impossible for a pure substance.

Since the vdW equation is seldom better than moderately accurate and is often quite poor, many efforts have been made to improve it by making its parameters dependent on some properties in addition to the critical ones—for instance, the normal boiling point or density or thermal expansion coefficient. Some of the early work of this kind is summarized by Partington (1950, p. 686). The parameters of other EOS similar in form to the vdW sometimes have been correlated in terms of temperature and acentric factor or critical compressibility (see Figure 1.8): this work is discussed later.

Mixtures

The utility of EOS is greatly increased when they can be made applicable to mixtures. Often this can be accomplished by expressing the parameters of the mixture in terms of the composition and the parameters of the pure components. For the van der Waals equation, the *combining rules* proposed by Lorentz (1881) and Berthelot (1898) are:

$$\sqrt{a} = \Sigma y_i \sqrt{a_i}, \tag{1.6}$$

$$b = \Sigma y_i b_i, \tag{1.7}$$

and are known as the Lorentz-Berthelot rules. Their use in the statistical mechanics of mixtures is illustrated by Rowlinson & Swinton (1982). Another approach to the problem is to evaluate the parameters of the mixture from its critical properties with the same equations that are applicable to pure substances. Since critical properties of random mixtures rarely are known from experiment and are not easy to estimate, however, it is fortunate that superior methods can be based on certain specially defined *pseudocritical properties,* M_{pc}, of which the simplest are mol fraction weighted sums of the

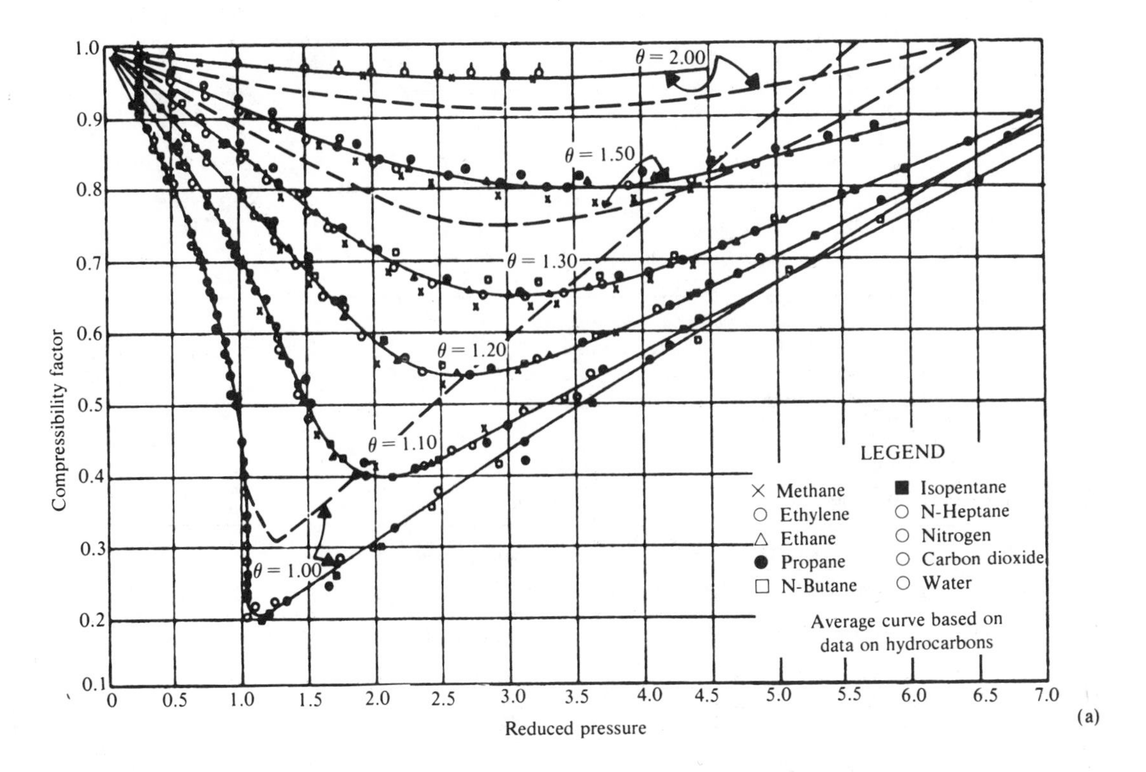

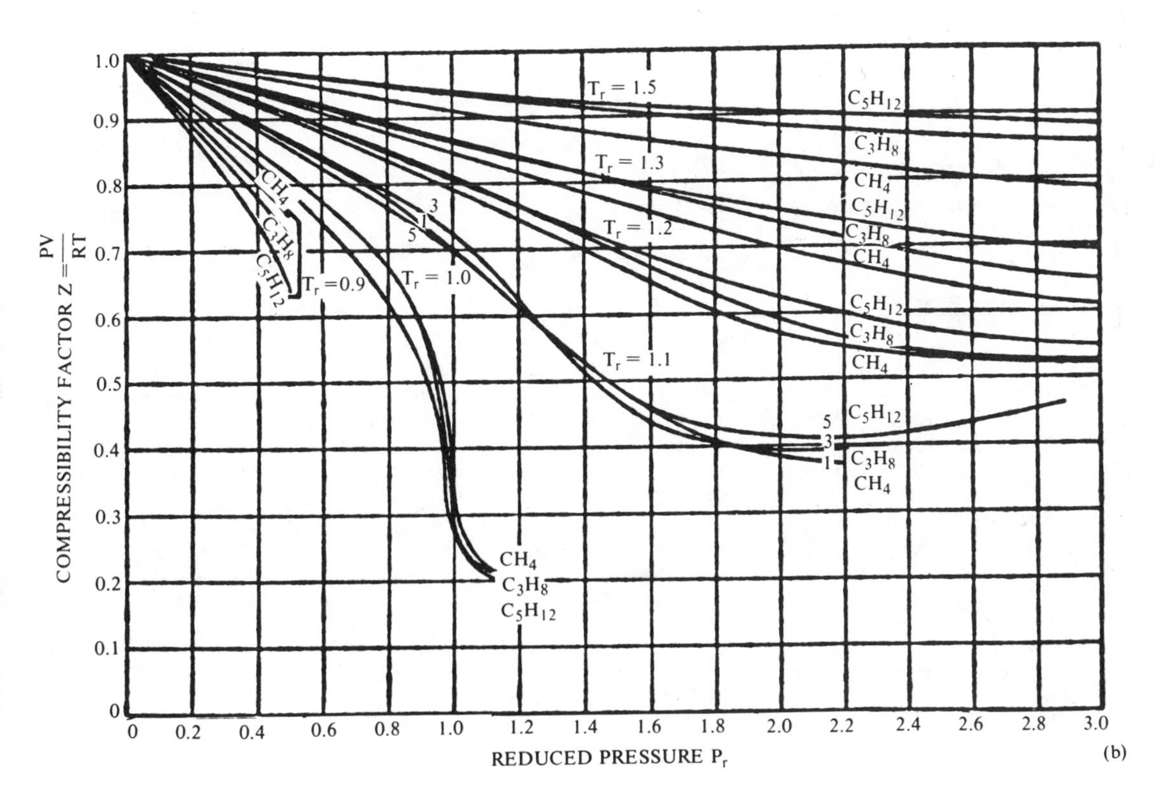

Figure 1.8 Corresponding states representations: (a) correlation of compressibilities, $z = PV/RT$, of common gases. The broken lines are based on van der Waals equation. $\theta = T/T_c$ (after Su, 1946). (Source: Copyright American Chemical Society; used with permission.) (b) compressibilities of methane, propane, and pentane, which do not conform to the principle of corresponding states over the full range of conditions (Brown et al. 1932). (Source: Copyright American Chemical Society; used with permission.)

Table 1.5. Some Combining Rules for Pseudocritical Properties

Pseudocritical Property	*Kay (1938)*	*Prausnitz & Gunn (1958)*	*Lorentz-Berthelot Type (LB)*
T_c	$\Sigma y_i T_{ci}$	$\Sigma y_i T_{ci}$	$(1-k_{ij})\sqrt{T_{ci}T_{cj}}$
V_c	$\Sigma y_i V_{ci}$	$\Sigma y_i V_{ci}$	$(V_{ci}^{1/3}+V_{cj}^{1/3})^3/8$
z_c	$\Sigma y_i z_{ci}$	$\Sigma y_i z_{ci}$	$0.5(z_{ci}+z_{cj})=0.291-0.080\omega$
ω	$\Sigma y_i \omega_i$	$\Sigma y_i \omega_i$	$0.5(\omega_i+\omega_j)$
P_c	$\Sigma y_i P_{ci}$	$z_c RT_c/V_c$	$z_c RT_c/V_c$

Note: Binary interaction parameters, k_{ij}, are given for some substances in Table E.3. The rules for T_c and V_c in the last column are known as the Lorentz-Berthelot Rules. They apply only to pairs of substances and are used to find the cross parameters a_{ij} of cubic equations or the B_{ij} of the virial equation.

Table 1.6. Constants of Some Equations of State of Some Substances

van der Waals: $(P+a/V^2)(V-b)=RT$, $a=27R^2T_c^2/64P_c$, $b=RT_c/8P_c$.

Berthelot: $(P+a/TV^2)(V-b)=RT$, $a=27R^2T_c^3/64P_c$, $b=9RT_c/128P_c$.

Dieterici: $P(V-b)=RT\exp(-a/RTV)$, $a=4R^2T_c^2/P_ce^2$, $b=RT_c/P_ce^2$, $e=2.718$.

Beattie-Bridgeman: $P=\dfrac{RT}{V}+\dfrac{\beta}{V^2}+\dfrac{\gamma}{V^3}+\dfrac{\delta}{V^4}$,

$$\beta=RTB_0-A_0-\frac{Rc}{T^2},$$

$$\gamma=RTB_0b+A_0a-\frac{RcB_0}{T^2},$$

$$\delta=\frac{RB_0bc}{T^2}.$$

Units: P, atm; V, liters/ gmol; T, °K; $R=0.08206$.

Gas	T_c	P_c	*van der Waals* a	*van der Waals* b	*Berthelot* a	*Berthelot* b	*Dieterici* a	*Dieterici* b
H_2	33.2	12.8	0.24463	0.02661	8.1217	0.01497	0.3139	0.02881
He	5.19	2.24	0.034161	0.023766	0.1773	0.013369	0.04383	0.025731
Ar	150.8	48.1	1.3431	0.032159	202.54	0.018089	1.7235	0.034818
N_2	126.2	33.5	1.3506	0.03864	170.45	0.02174	1.7331	0.04184
O_2	154.6	49.8	1.3634	0.03184	210.78	0.01791	1.7496	0.03448
CO_2	304.2	72.8	3.6111	0.04286	1098.5	0.02411	4.6337	0.04641
CH_4	190.6	45.4	2.2732	0.04306	433.27	0.02422	2.9169	0.04662
C_2H_5OH	516.2	63.0	12.016	0.08405	6202.7	0.04728	15.418	0.09100
C_6H_6	562.1	48.3	18.583	0.11937	10446.0	0.06715	23.845	0.12924

Beattie-Bridgeman

Gas	A_0	a	B_0	b	$10^{-4}c$
He	0.0216	0.05984	0.01400	0	0.004
Ne	0.2125	0.2196	0.02060	0	0.101
Ar	1.2907	0.02328	0.03931	0	5.99
H_2	0.1975	−0.00506	0.02096	−0.04359	0.050
N_2	1.3445	0.02617	0.05046	−0.00691	4.20
O_2	1.4911	0.02562	0.04624	0.004208	4.80
Air	1.3012	0.01931	0.04611	−0.01101	4.34
CO_2	5.0065	0.07132	0.10476	0.07235	66.00
CH_4	2.2769	0.01855	0.05587	−0.01587	12.83
$(C_2H_5)_2O$	31.278	0.12426	0.45446	0.11954	33.33

corresponding properties of the components of the mixture. Thus,

$$P_{pc} = \Sigma y_i P_{ci}, \quad T_{pc} = \Sigma y_i T_{ci}, \quad V_{pc} = \Sigma y_i V_{ci}. \quad (1.8)$$

Other combining rules are described later.

Direct combination of individual parameters on the basis of composition is usually preferred to the use of pseudocritical properties with many EOS, but both methods have their advocates. In Example 1.8, although one method predicts the mixture to be subcritical (three real roots of the EOS) and the other supercritical (one real root), they yield substantially the same value for the compressibility of this particular mixture. Since an EOS is of comparatively little value if it is not applicable to mixtures, much effort has been devoted to formulating effective combining rules for pseudocritical properties and mixture parameters from those of the components with at most binary interaction parameters (see Table 1.5). This topic is continued in Section 1.3.7.

Current Status of the van der Waals Equation

This equation has received much attention in the recent past. An exhaustive comparison with experimental data and with two other equations that were popular at the time (those of Berthelot and Dieterici) was made for nine gases at pressures up to 1,000 atm by Pickering (1925). A few comparisons also are made in this chapter. Although references to it continue in recent papers, the landmark *vdW* equation is now obsolete except as an example of a simple model that incorporates some corrections to the ideal gas law for attractions and repulsions between molecules. Other current equations of not much greater complexity are far superior. Parameters of the van der Waals and some other equations are shown for several substances in Table 1.6.

1.3. CORRESPONDING STATES

When van der Waals wrote his equation of state in terms of reduced properties,

$$(P_r + 3/V_r^2)(3V_r - 1) = 8T_r, \quad (1.9)$$

he made the first statement of the *principle of corresponding states* (PCS): Substances at equal reduced pressures and temperatures have equal reduced volumes. Other two-parameter EOS also may be cast into reduced forms—for example, the virial equation in Table 1.8, the Redlich-Kwong in Table 1.9, and the Clausius in Problem 1.4. Although reduced equations are in principle applicable to all substances, they are of course no more accurate than the original forms; and the several reduced EOS can differ widely from each other, as shown for example on Figure 1.18. Apparently, one of the difficulties is that some pertinent variables must have been overlooked in the development of these equations. It does appear, however, that experimental data often are more nearly in accord with the principle of corresponding states than with specific EOS. Thus Figure 1.8a demonstrates that the compressibilities of many gases correlate quite well with reduced properties, whereas the van der Waals lines show much poorer agreement. At higher P_r and T_r, however, even the PCS breaks down, as shown by the data for some hydrocarbons in Figure 1.8b.

Example 1.1. Saturation Volumes with the van der Waals Equation

For methyl chloride at 322 K, the vapor pressure is 10.49 atm, and the specific volumes of the vapor and liquid are 0.6710 and 0.0186 cu ft/lb. These data will be compared with those predicted by the van der Waals equation.

Since $P_c = 65.9$ atm, $T_c = 416.3$ K, and $R = 0.08206$, the constants of the equation are:

$$a = \frac{27}{64}\frac{R^2T_c^2}{P_c} = 7.4709,$$

$$b = RT_c/8P_c = 0.0648.$$

The polynomial equation,

$$V^3 - (b + RT/P)V^2 + \frac{a}{P}V - \frac{ab}{P} = 0,$$

becomes

$$V^3 - 2.5837\,V^2 + 0.7122\,V - 0.0461 = 0,$$

of which the roots are

$$V = 2.2802,\ 0.2047,\ 0.0988 \text{ liters/gmol}.$$

The largest and smallest of these values are equivalent to

$$V_g = 0.724 \text{ cu ft/lb},$$

$$V_L = 0.0314 \text{ cu ft/lb}.$$

The specific volume of the gas is checked within 8 percent of the experimental value, but that of the liquid only poorly.

1.3.1. Dimensional Similitude

The rule of dimensional similitude is that any relation between physical variables is rearrangeable as a relation between a limited number of dimensionless ratios. Thus an equation of state, $f_1(P, V, T, P_c, V_c, T_c) = 0$, is the equivalent of some other equation of state, $f_2(P_r, V_r, T_r) = 0$. The adequacy of any relation in describing a particular phenomenon depends on how completely the pertinent variables are identified. In the case of the equation of state, although the reduced variables undoubtedly are necessary, they are not sufficient. Differences in sizes and shapes of molecules, mechanical features such as moment of inertia or radius of gyration, and electrostatic parameters of polar molecules are other factors that may influence *PVT* behavior as well as chemical reactivity. Many efforts to improve equations of state have concentrated on finding easily evaluated additional parameters, a_i, thus making the general equation of state

$$P_r = f(T_r, V_r, a_1, a_2, \ldots) = 0. \quad (1.10)$$

Example 1.2. Vapor Pressure and Saturation Volumes from the vdW Equation

The vapor pressure and saturation volumes of methyl chloride at 322 K will be found and compared with the values given in Example 1.1.

Apply Maxwell's Principle:

$$-\int V\,dP = \int P\,dV - \Delta(PV) = 0, \quad (1)$$

$$\int_{V_L}^{V_g} \left(\frac{RT}{V-b} - \frac{a}{V^2}\right) dV - P^{\text{sat}}(V_g - V_L) = 0, \quad (2)$$

$$26.42 \ln \frac{V_g - b}{V_L - b} + a\left(\frac{1}{V_g} - \frac{1}{V_L}\right) - P^{\text{sat}}(V_g - V_L) = 0. \quad (3)$$

The values of a and b from Example 1.1 will be used. At any given pressure, the volumes are obtained from the cubic equation,

$$V^3 - \left(0.0648 + \frac{26.42}{P}\right)V^2 + \frac{7.4709}{P}V - \frac{0.4841}{P} = 0. \quad (4)$$

The roots and the LHS of Eq. 3 are found at several pressures. The last line of the table is interpolated. The result also is obtained graphically. Comparison with the data cited in Example 1.1 is quite poor.

P	V_g	V_x	V_L	*LHS Eq. 3*
10.49	2.2802	0.2047	0.0988	15.13
20	1.0528	0.2352	0.0977	1.413
25	0.7634	0.2611	0.0972	−2.596
21.8	0.949		0.0975	0

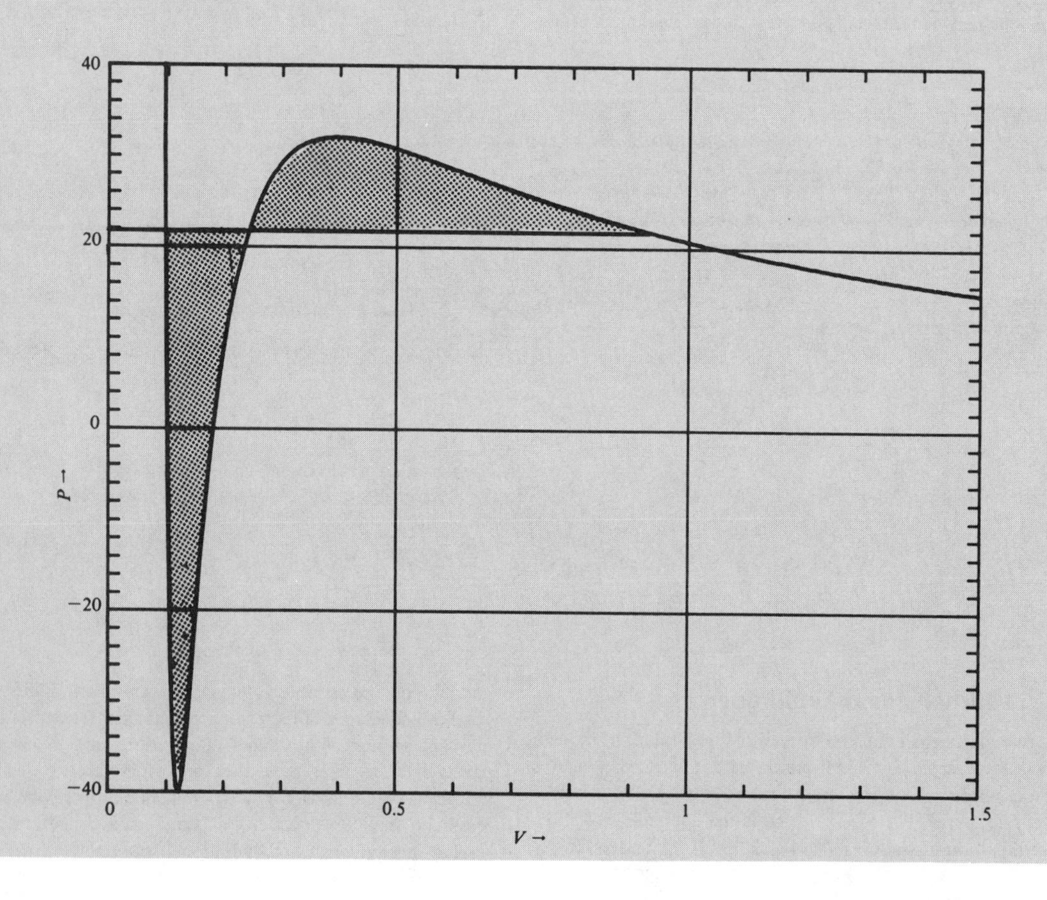

Example 1.3. Numerical Solution of Cubic Equations

Both real and complex roots may be found by Cardan's method. When only real roots are required, which is always the case with equations of state, the Newton-Raphson method is convenient. The BASIC progam for finding compressibilities is shown. This form of the equation of state is convenient to handle, since estimates of 1 and 0 usually lead to rapid convergence. Examples are shown of cases with one real root and with three real roots, with numerical and graphical solutions.

```
10  ! Real roots of the cubic equation, Z ^ 3+AZ ^ 2+
    BZ+C=0. With one real root, Zv=Z1.
20  INPUT A,B,C
30  Z=1 ! Trial value
40  GOSUB 210
50  H=F/(3*Z ^ 2+2*A*Z+B)
60  IF ABS(H/Z)<=.0001 THEN 90
70  Z=Z-H
80  GOTO 40
90  Z1=Z
100 Z=.01 ! Second root trial
110 GOSUB 210
120 H=F/(3*Z ^ 2+2*A*Z+B)
130 IF ABS(H/Z)<=.0001 THEN 160
140 Z=Z-H
150 GOTO 110
160 Z2=Z
170 Z3=-A-Z1-Z2
180 PRINT USING 190 ; Z1,Z2,Z3
190 IMAGE "Zv="D.DDDD,2X,"Z1="D.
    DDDD,2X,"Zx=" D.DDDD
200 END
210 F=Z ^ 3+A*Z ^ 2+B*Z+C
220 RETURN
```

Case with 3 real roots: A=−1, B=.089, C=−.0013
Zv= .9030   Z1= .0183   Zx= .0787

Case with 1 real root: A=−1,B=.4224, C=−.0608
Zv= .2697   Z1= .2697   Zx= .4606

In the latter case, Zx is not given by line 170.

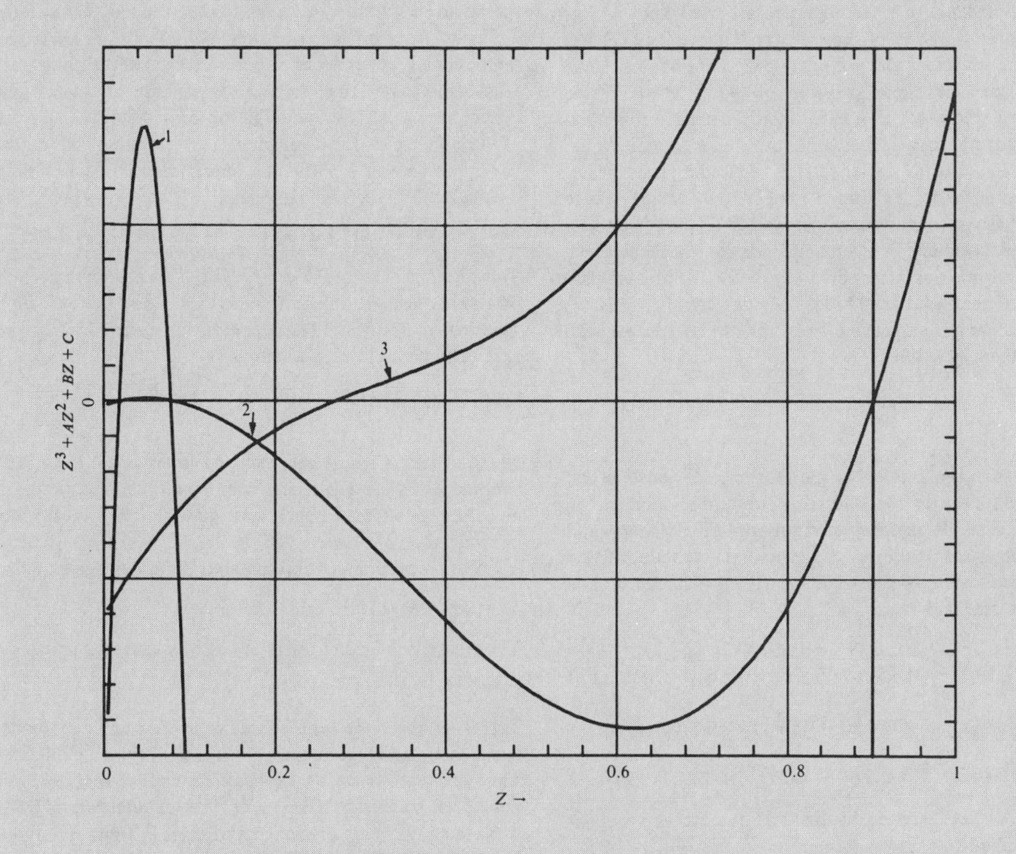

Curve 1 is the same as curve 2, but with expanded ordinate. Curve 2 is the plot of $z^3 - z^2 + 0.089z - 0.0013$, with three real roots. Curve 3 is the plot of $z^3 - z^2 + 0.4224z - 0.0608 = 0$.

Desirable features of additional parameters have been listed by Sterbacek et al. (1979, p. 21) as follows:

1. They should be related to the molecular structure and to the electrostatic properties of the molecule.
2. They should follow from a minimum of data.
3. Their values should not be directly affected by critical properties.
4. In their evaluation, *PVT* data should be avoided; otherwise the meaning of the reduced equation gets lost.
5. They should be defined functions of temperature, preferably of reduced temperature.

Among the simpler additional parameters that have been used are the critical compressibility and the slope of the vapor pressure curve at the critical point; but another quantity that is more easily measured, the acentric factor, has proved to be superior and therefore will be discussed at length. Correlation of the *PVT* behavior of mixtures also may be improved with experimentally determined binary interaction parameters that will be discussed in connection with combining rules.

1.3.2. Reference Substance

Another kind of normalization or reduction process is to relate the properties of a substance in question to those of some reference substance whose properties are well known. For instance, over a period of many years, Othmer (1947) has developed good linear plots of many properties against some property, such as vapor pressure of water at the same temperature. Pitzer et al. (1955–1958) evaluate the compressibility factor as a deviation from that of a "simple" fluid such as argon. Lee & Kesler (1975) relate the properties to those of two fluids, a "simple" one and *n*-octane whose properties also are very well known. Teja et al. (1981) employ methane and *n*-octane as reference fluids for some cases.

A generalization of this method was developed by Mollerup (1980) and Mentzer et al. (1981). They develop shape factors for temperature and volume which are the ratios of the reduced values of these quantities,

$$\theta_T = T_\alpha^R / T_0^R, \tag{1.11}$$

$$\theta_V = V_\alpha^R / V_0^R, \tag{1.12}$$

where the subscript α refers to the substance in question and subscript 0 refers to the reference substance—methane in Mentzer's case. When the substitutions $T_0^R = T_\alpha^R/\theta_T$ and $V_0^R = V_\alpha^R/\theta_V$ are made in the equation of state of the reference substance, the result is valid for substance α. The empirical shape factors are

$$\theta_T = 1 + (\omega_a - \omega_0)[0.0892 - 0.8493 \ln (T_a^R) + (0.3063 - 0.4506/T_a^R)(V_a^R - 0.5)], \tag{1.13}$$

$$\theta_V = \{1 + (\omega_a - \omega_0)[0.3903(V_a^R - 1.0177) - 0.9462(V_a^R - 0.7663) \ln (T_a^R)]\} \frac{Z_a^c}{Z_0^c} \tag{1.14}$$

where the ω_i are acentric factors and the Z_{ci} are critical compressibilities.

With or without added parameters, the principle of corresponding states underlies much of numerical thermodynamics, from equations of state on. van der Waals would have been pleased to know how important his discovery has become. In recent times the principle has been placed on a firm theoretical basis by the methods of statistical thermodynamics. Some of the work in this area is discussed at length by Rowlinson & Swinton (1982) and by Leland & Chappelear (1968). Later chapters bring out some of the applications to phase equilibria. An appreciation of the ramifications of the principle of corresponding states also may be obtained from the books by Bretsznajder (1971), Reid et al. (1977) and Sterbacek et al. (1979).

1.3.3. Vapor Pressure and the Acentric Factor

Application of the principle of corresponding states to vapor pressures does not lead to a unique relation for all substances. Thus, in Figure 1.9, selected substances do appear to have nearly the same reduced vapor pressure behavior, but in general the plots form a pencil of fairly straight lines. That differences in slope should exist is apparent from the Clapeyron equation,

$$d \ \ln P_r^{\text{sat}}/dT_r = \Delta H_v / RT_c \, \Delta z T_r^2, \tag{1.15}$$

and the approximate integrated form,

$$\ln P_r^{\text{sat}} = k - \left(\frac{\Delta H_V}{RT_c \Delta z} \right) \frac{1}{T_r}. \tag{1.16}$$

The coefficient of the term, $1/T_r$, depends on several properties of individual substances and can hardly be expected to be the same for all. Pitzer et al. (1955–1958) adopted the reduced pressure at a particular reduced temperature, $T/T_c = 0.7$, as an identification of the substance. At these temperatures, the reduced vapor pressures of the noble gases, whose molecules they termed *simple*, are about 0.1. This observation leads to the definition of a new parameter, the acentric factor ω, for describing the deviation of the reduced vapor pressure of a particular substance from that of simple molecules, by the defining equation,

$$\omega = (\log P_c / 10 \, P^{\text{sat}}) \ @ \ T_r = 0.7. \tag{1.17}$$

This factor is said by Pitzer (1977) to "measure the deviation of intermolecular potential functions from that of simple spherical molecules." The equation for reduced vapor pressure is assumed to be a series in ω,

$$\log P_r^{\text{sat}} = (\log P_r)^{(0)} + \omega (\log P_r)^{(1)} + \omega^2 (\log P_r)^{(2)} + \dots, \tag{1.18}$$

which usually is truncated after the linear term. Both superscripted items are evaluated from experimental data and are given in the original publications in tabular form as functions of reduced temperature.

Other properties are representable in the same general form,

$$M = M^{(0)} + \omega M^{(1)} + \omega^2 M^{(2)} + \dots. \tag{1.19}$$

The equation for compressibility also is usually truncated,

$$z = z^{(0)} + \omega z^{(1)}. \tag{1.20}$$

Tables of the superscripted items as functions of reduced temperature and pressure are given in the original publication. They are reproduced by Lewis & Randall et al. (1961) and have been graphed by Edmister (1958). Updated data by Lee & Kesler (1975) are presented as Tables E.1 and E.2 and are graphed in Figure 1.22. Empirical equations for the second virial coefficient and some other properties in terms of the Pitzer data are given in Figure 1.10. The compressibility can be found from the equation for the second virial coefficient by

$$z = 1 + (P_r/T_r)\,(BP_c/RT_c), \tag{1.21}$$

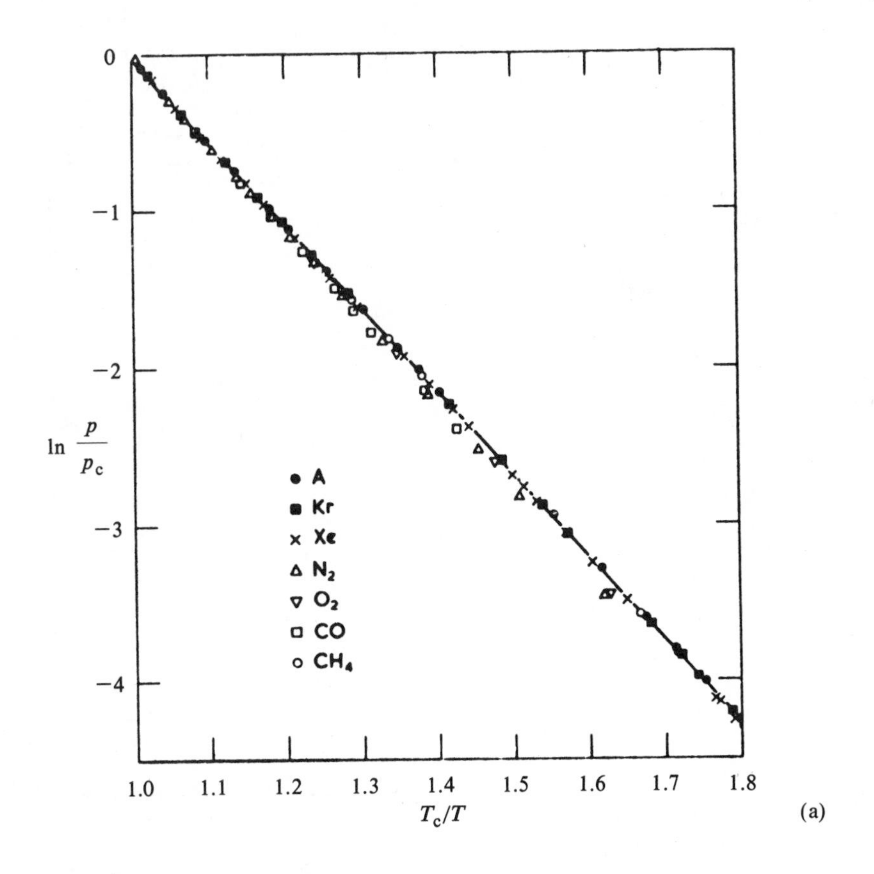

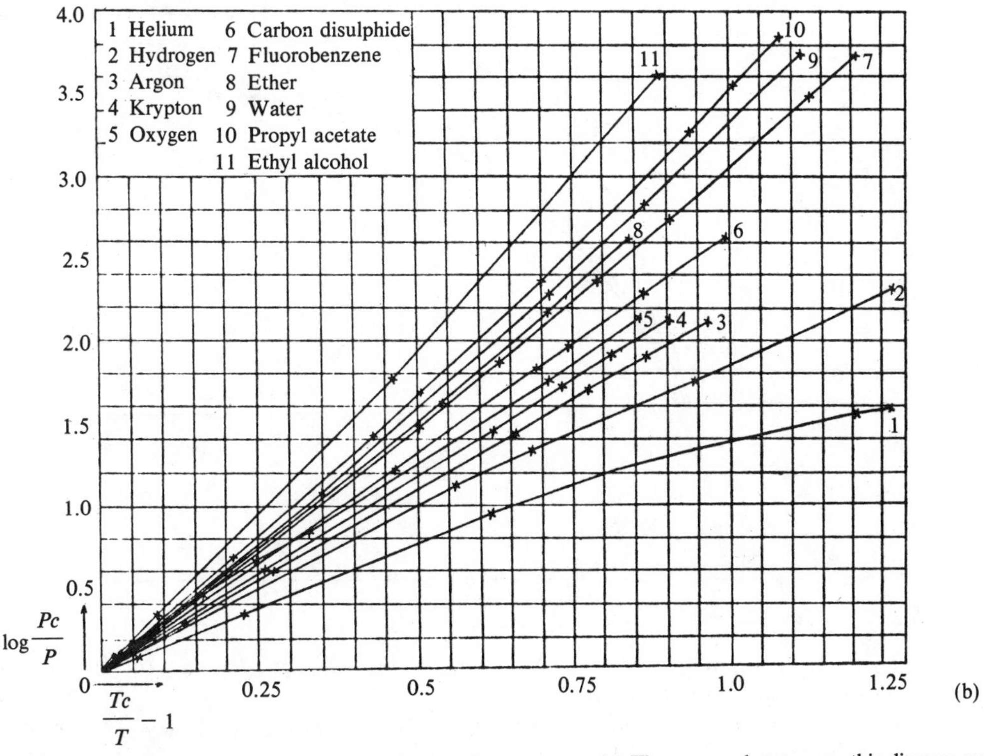

Figure 1.9 Reduced vapor pressures as functions of reduced temperature: (a) The seven substances on this diagram are fairly well correlated by a single line whose equation is $\ln(P/P_c) = 5.25(1 - T/T_c)$. Of these substances, only A, Kr and O_2 also appear in Figure 1.9(b) (Guggenheim 1977). (b) When a greater diversity of substances is taken into account, the plots of reduced vapor pressure against reduced temperature are not the same for all (Partington 1950).

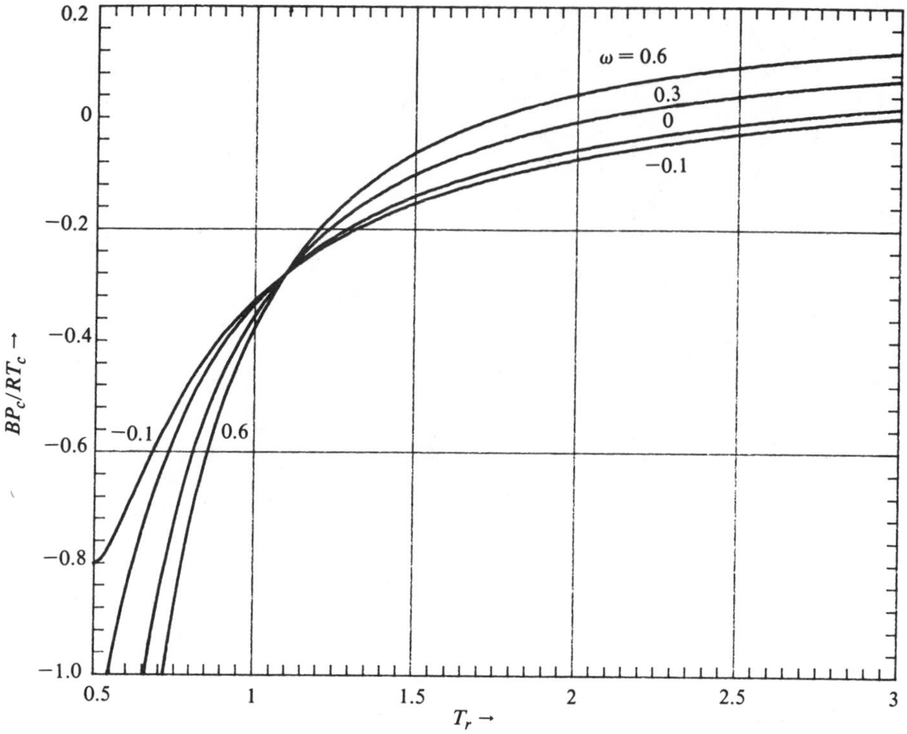

(a)

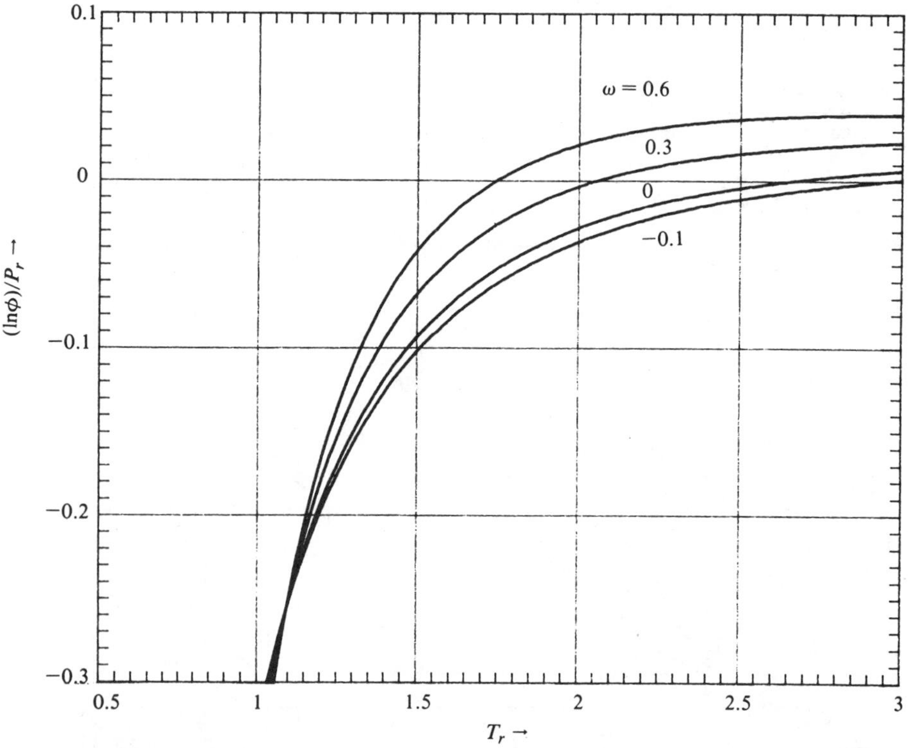

(b)

Figure 1.10 Correlations in terms of acentric factor (Pitzer & Curl, 1958; cited in Lewis & Randall et al. 1961): (a) The second virial coefficient is represented by the equation

$BP_c/RT_c = (0.1445 + 0.073\omega) - (0.330 - 0.46\omega)T_r^{-1} - (0.1385 + 0.50\omega)T_r^{-2} - (0.0121 + 0.097\omega)T_r^{-3} - 0.0073\omega T_r^{-8}$;

(b) the fugacity coefficient is represented by the equation

$$\log_{10}\phi = P_r/2.303\,[(0.1445 + 0.073\omega)T_r^{-1} - (0.330 - 0.46\omega)T_r^{-2} - (0.1385 + 0.50\omega)T_r^{-3} - (0.0121 + 0.097\omega)T_r^{-4} - 0.0073\omega T_r^{-9}];$$

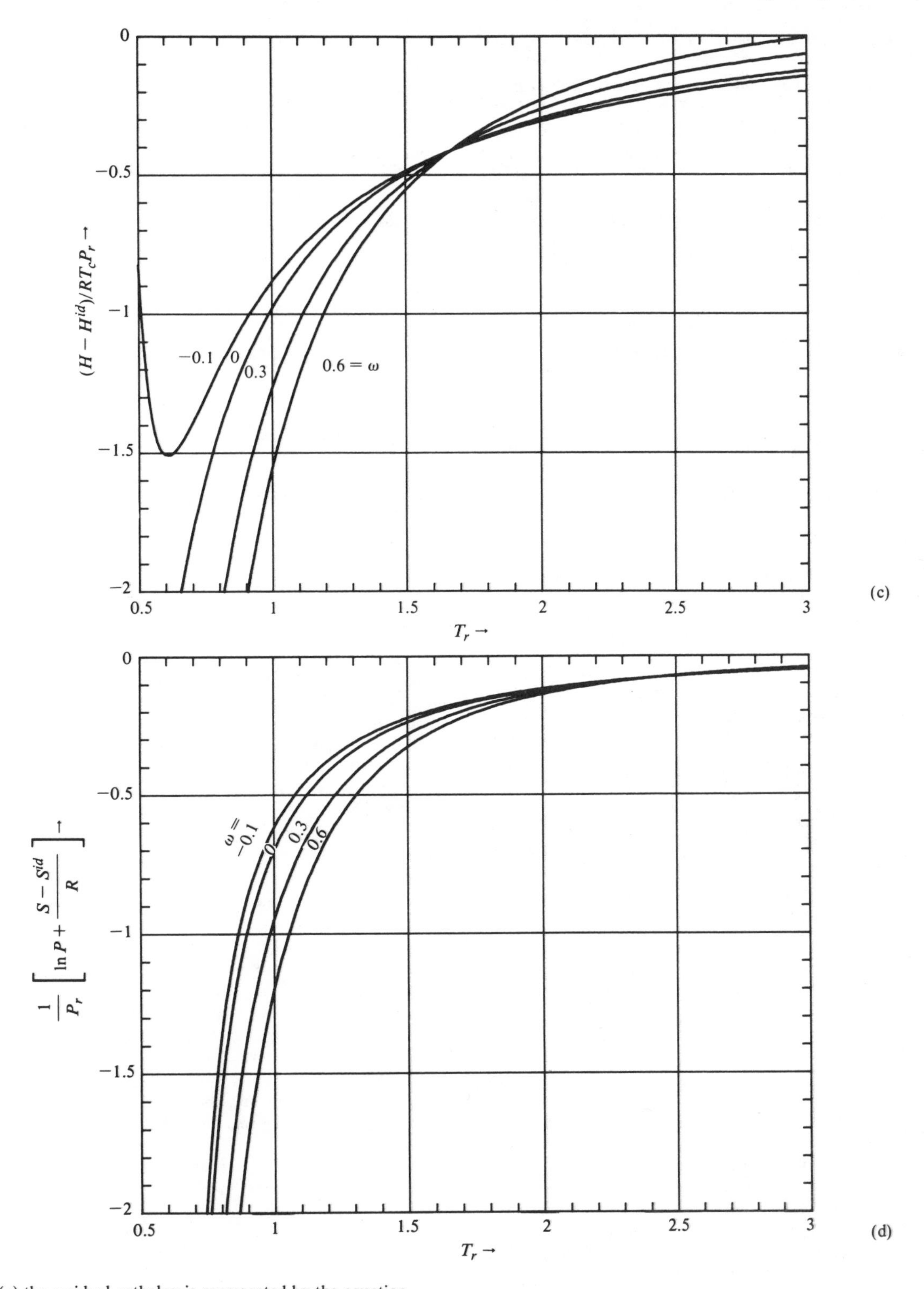

(c) the residual enthalpy is represented by the equation

$$(H - H^{id})/RT_c = P_r[(0.1445 + 0.073\omega) - (0.660 - 0.92\omega)T_r^{-1} - (0.4155 + 1.50\omega)T_r^{-2} - (0.0484 + 0.388\omega)T_r^{-3} - 0.0657\omega T_r^{-8}];$$

(d) the residual entropy is represented by the equation

$$(S - S^{id})/R = -\ln P + P_r[-(0.330 - 0.46\omega)T_r^{-2} - (0.2770 + 1.00\omega)T_r^{-3} - (0.0363 + 0.29\omega)T_r^{-4} - 0.0584T_r^{-9}].$$

but the result is not accurate over the full range of tabulated $z^{(0)}$ and $z^{(1)}$.

A recent account and some updating of the concept of acentric factor is given by Pitzer (1977). Some extensions of this concept have been advocated but have not been developed fully. For example, Stiel (1972) pointed out that the original definition is useful for nonpolar and slightly polar substances at $T_r > 0.7$, but for polar substances at lower temperatures it is proposed to represent the reduced vapor pressures by

$$\log P_r^{\text{sat}} = (\log P_r)^{(0)} + \omega(\log P_r)^{(1)} + \omega_2(\log P_r)^{(2)}, \quad (1.22)$$

with

$$\omega_2 = 1.552 + 1.7\omega + (\log P_r^{\text{sat}}) \ @\ T_r = 0.6, \quad (1.23)$$

and similar terms for quadripolar substances like carbon dioxide. These ideas have not been extended to compressibilities, fugacities, and excess properties.

Several generalized vapor pressure equations involving the acentric factor have been proposed. For hydrocarbons, primarily, Lee & Kesler (1975) give

$$\begin{aligned}\ln P_r^s = {} & 5.92714 - 6.09648/T_r - 1.28862 \ln T_r \\ & + 0.169347\, T_r^6 + \omega(15.2518 - 15.6875/T_r \\ & - 13.4721 \ln T_r + 0.43577\, T_r^6).\end{aligned} \quad (1.24)$$

When this equation is applied at the normal boiling point T_b, a general equation for the acentric factor in terms of $T_{br} = T_b/T_c$ and $P_{br}^s = P_b^{\text{sat}}/P_c$ is the result:

$$\omega = \frac{\ln P_{br}^s - 5.92714 + 6.09648/T_{br} + 1.28862 \ln T_{br} - 0.169347\, T_{br}^6}{15.2518 - 15.6875/T_{br} - 13.4721 \ln T_{br} + 0.43577\, T_{br}^6}. \quad (1.25)$$

Kesler et al. (1979) later developed an equation that covers also the vapor pressures of hydrocarbons that decompose well below the critical point, and has only one-half the error of the preceding one. The reduced vapor pressure behavior of some highly polar substances as well as hydrocarbons is correlated by Nath et al. (1976) as

$$2A \log P_r = -B + [B^2 - 4A(T_r^{-1} - C)]^{1/2}, \quad (1.26)$$

where

$$A = 0.0094 - 0.0144\,\omega^2, \quad (1.27)$$

$$B = 0.4506 - 0.4371\omega + 0.2127\omega^2 \quad (1.28)$$

and

$$C = 0.9827 + 0.0736\omega - 0.0210\omega^2. \quad (1.29)$$

Problem 1.17 asks for a comparison of this equation with the Antoine equation of pyridine, which has a dipole moment of 2.3 debye; for practical purposes the comparison is quite favorable, better at lower temperatures than the Lee-Kesler equation, which is not, however, intended for polar substances.

When only the normal boiling point and the critical pressure and temperature are known, the acentric factor may be estimated from the equation of Edmister (1958)

$$\omega = \frac{3}{7}\left[\frac{T_b}{T_c - T_b} \log P_c(\text{atm})\right] - 1. \quad (1.30)$$

The approximate relation, also derived by Edmister, between critical compressibility and acentric factor,

$$z_c = 0.291 - 0.080\omega, \quad (1.31)$$

often is substituted into equations that call for z_c since ω generally is known more accurately.

Besides the vapor pressure relations that have been given here, many others have been proposed, of which more than sixty have been collected by Partington (1951). Other formulas are cited by Ambrose (1972). A few of these relations are listed chronologically in Table 1.2A; those numbered 5, 9, 11, 12, and 13 are of current interest. In some cases the parameters are related to other properties that are more or less readily available, such as critical properties, normal boiling points, radii of gyration, and so on. A particularly interesting formula using a Chebyshev polynomial is adopted by the CONCEPT commercial simulation system,

$$T \ln P = k_0 + k_1 x + k_2(2x^2 - 1) + k_3(4x^3 - 3), \quad (1.32a)$$

where

$$x = \frac{2T - (T_{\max} + T_{\min})}{T_{\max} - T_{\min}}. \quad (1.32b)$$

$T_{\max}$ and $T_{\min}$ are temperatures that define the range of measured vapor pressures, and the k_i are characteristic of individual substances (Leesley & Heyen 1982; Ambrose, Counsell, & Davenport 1970).

1.3.4. Polarity

A dipole moment results when the electrical center of a bond does not coincide with its center of mass. Polarity is defined as the magnitude of a dipole moment expressed in debye units; as a rough rule, however, only molecules with dipole moments greater than unit debye actually are classified as polar, because certain behaviors regarded as polar are not manifest at the lower values. Some data are shown in Table 1.7 and in Appendix D.

Symmetrical molecules have zero dipole moments; but if the constituent elements of the dipole are far apart, polar properties may become evident. This may be true also of symmetrical molecules with several pairs of opposed dipoles—for instance, carbon dioxide, which has two opposed dipoles, called *quadrupoles*, and has strong polar characteristics. Polar molecules have greater viscosities and higher boiling and melting points than do nonpolar ones of about the same size. Toluene, $C_6H_5CH_3$, with $\mu = 0.4$ and $T_b = 383.8$ K, and aniline, $C_6H_5NH_2$, with $\mu = 1.6$ and $T_b = 457.5$ K, are examples. Also, polar molecules are more soluble in water, which has $\mu = 1.84$, and are only sparingly soluble in nonpolar substances. Particularly significant are the greater excess thermodynamic properties displayed by polar mixtures.

Polarity of a bond increases with increasing separation in the electronegativity series, which is H < C < O < Cl < Br < I. Groups of atoms such $-OH$, $-CN$, $-NH_2$, $-COOH$, and $-NO$ also contribute to polarity, whereas alkyl groups, for instance, have nonpolar characteristics. Unsymmetrical molecules containing polar atoms have strong dipole moments; for example, CH_3Cl has $\mu = 1.9$ debye, whereas symmetrical CCl_4 has $\mu = 0$. An example of a symmetrical molecule with zero dipole moment but nevertheless with decided polar properties because of the wide separation of the poles is that of p-dichlorbenzene; despite its $\mu = 0$, it has nearly the same boiling point as the meta isomer, with $\mu = 1.4$, or the ortho isomer, with $\mu = 2.3$. There are,

Table 1.7. Dipole Moments of Some Substances

Substance	μ*, debye*
Phthalic anhydride	5.3
Maleic anhydride	4.0
Propionitrile	3.7
Acetonitrile	3.5
Methyl ethyl ketone	3.3
Nitromethane	3.1
Acetic anhydride	3.0
Glycerol	3.0
Hydrazine	3.0
Acetone	2.9
o-dichlorbenzene	2.3
m-dichlorbenzene	1.4
p-dichlorbenzene	0
Formaldehyde	2.3
1-chlorobutane	2.0
Ethyl acetate	1.9
Water	1.8
Ethanol	1.7
Sulfur dioxide	1.6
Sulfur trioxide	0
Vinyl chloride	1.5
Ammonia	1.5
Acetic acid	1.3
Isobutylene	0.5
1,2-butadiene	0.4
1,3-butadiene	0
Carbon monoxide	0.1
Carbon dioxide	0

however, marked differences in the melting points of these substances.

Interactions between molecules are described in terms of *potential functions* such as the Lennard-Jones 6-12 (Section 1.7),

$$u(r) = 4\varepsilon[(\sigma/r)^{12} - (\sigma/r)^6] \tag{1.33}$$

where

$u(r)$ = potential.
r = distance between centers of molecules.
σ = diameter of the molecule.
ε/k = characteristic energy, °K.
k = Boltzmann constant.

The dimensionless group ε/kT is an energy parameter that should be employed in quantifying the behavior of fluids along with dipole moment, ionization potential, and other electrostatic and mechanical characteristics of molecules.

1.3.5. Molecular Association and Hydrogen Bonding

Because their hydrogen atoms behave as though they had a residual valence, the molecules of polar substances tend to form groups. This effect is called *hydrogen bonding,* and its tendency decreases with decreasing electronegativity of the constituent atoms. Fluorine is the most strongly negative ion, so hydrogen fluoride, for instance, associates strongly in both liquid and vapor phases, the formula in the gas phase being $(HF)_6$ under normal conditions. Acetic and formic acid vapors just above their atmospheric boiling points are bimolecular. Association may be detected spectroscopically. Chemical equilibrium constants for dimerization equilibria are well

Example 1.4 Calculation of the Acentric Factor

Calculate the acentric factor of methyl chloride by several methods. The value in Table D.2 is 0.156, and that in Henley & Seader (1981) is 0.1530. Physical properties are $P_c = 65.9$ atm, $T_c = 416.3$ K, $T_b = 248.9$ K, dipole moment = 1.9 debye, $A = 16.1052$, $B = 2077.97$, $C = -29.55$. The last three numbers are the coefficients of the Antoine equation with the vapor pressure in Torr.

a. When the vapor pressure equation $\log P^{\text{sat}} = A + B/T$ is applied at the normal boiling and critical points, it becomes

$$\log P^{\text{sat}} = \frac{T_b T_c}{T_c - T_b}\left(\frac{1}{T} - \frac{1}{T_b}\right)\log P_c.$$

When this is applied in the definition of ω, Edmister's equation (1958) results.

$$\omega = -1 + \frac{3}{7}\left[\frac{T_b}{T_c - T_b}\right]\log P_c$$

$$= -1 + \frac{3}{7}\left(\frac{248.9}{416.3 - 248.9}\right)\log 65.9 = 0.159.$$

b. In terms of the Antoine equation for the vapor pressure:

$$\omega = 0.4343\left[\ln\frac{P_c}{10P^{\text{sat}}}\right]_{@\,0.7T_c}$$

$$= 0.4343\left[\ln(0.1P_c) - \left(A - \frac{B}{0.7T_c + C}\right)\right]$$

$$= 0.4343\left[\ln[0.1(760)(65.9)] - 16.1052 + \frac{2077.97}{0.7(416.3) - 29.55}\right]$$

$= 0.152.$

c. With the Lee & Kesler equation of Section 1.3.3,

$T_{br} = T_b/T_c = 248.9/416.3 = 0.5979.$

Substitution into Eq. 1.25 gives the value

$\omega = 0.146.$

established. Such data are cited in Problem 1.11. Of the common substances, carboxy acids possess the strongest dimerization tendencies; but alcohols, esters, aldehydes, and other substances also display appreciable association at normal pressures and temperatures. The extent decreases as the pressure and concentration are lowered and as the temperature is raised. A classification of molecules with a tendency for association by hydrogen bonding is made by Ewell et al. (1944).

In the liquid phase association is quite common. The abnormal behavior of water in view of its low molecular weight is due largely to oligomer formation, and deviation of solutions of polar substances from ideality often is the result of association. A few examples of association in the liquid phase are shown in Figure 1.11, of which (a) shows complete dimerization being approached as the concentrations rise. Higher oligomers evidently form in some of the cases of (b), where it also appears that the nature of the solvent has a strong bearing on the extent of association. The strong effect of temperature is evident for phenol in carbon tetrachloride. These effects are, of course, more or less characteristic of chemical reactions in general. Figure 1.12 represents some dimerization equilibria.

A relation between the second virial coefficient and the chemical equilibrium constant for weak dimerization has long been known and has been developed into a working tool by Nothnagel et al. (1973). Some details of this work are given in Section 1.4.

To a certain extent, the deviation of some substances from ideal gas behavior is attributable to a reduction in the number of molecules present because of *compound formation*, with an apparent increase in compressibility of the parent substance. For example, if a molecule reacts according to a process

$$nA \rightleftharpoons A_n$$

with a fractional conversion ε, the apparent compressibility will be

$$z = \frac{\text{number of molecules at equilibrium}}{\text{number of original parent molecules}} = 1 - \varepsilon + \varepsilon/n, \tag{1.34}$$

where ε is given by the equilibrium constant of the polymerization,

$$K = \frac{\varepsilon}{n(1-\varepsilon)}\left[\frac{1-\varepsilon(1-1/n)}{P(1-\varepsilon)}\right]^{n-1}. \tag{1.35}$$

Similar relations can be developed when dimers, trimers, and so on, or intermolecular compounds, exist simultaneously, although equilibrium data for such cases are hard to come by. As a numerical example, if $K = 0.05$, $P = 2$, and $n = 3$, then $\varepsilon = 0.3117$ and $z = 0.7925$.

Values of z greater than unity possibly could be interpreted as due to dissociation, but in most instances there is no evidence for such a process and the explanation of such compressibility behavior must be sought elsewhere.

Many instances of self and complex associations have been discovered, to which recent references are made by Knobler (1978, p. 221). Equilibrium constants obtained mostly from second virial coefficients are tabulated for many substances and mixtures by Prausnitz et al. (1980).

The representation of multicomponent PVT behavior in terms of association equilibria is extremely cumbersome, and the data rarely are available. Therefore, such problems usually are handled with empirical equations of state that ignore chemical equilibria, except in some cases of dimerizations of oxygenated organic substances, where the equilibrium constants are known adequately.

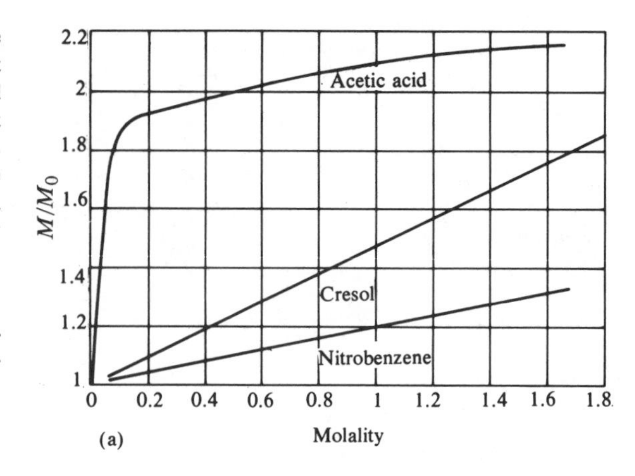

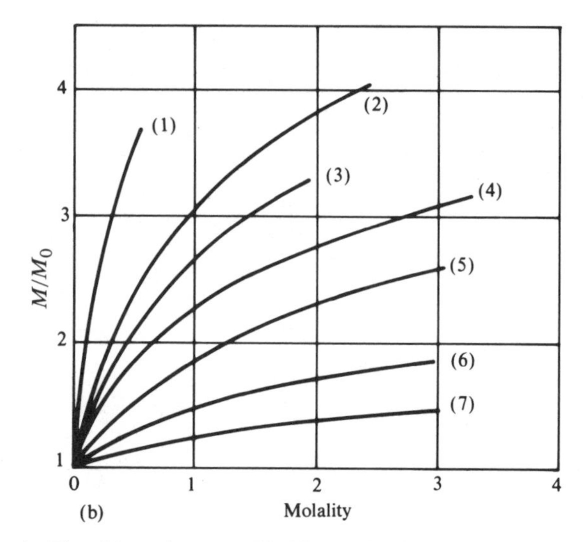

1. Phenol in carbon tetrachloride at −24 C.
2. Methanol in carbon tetrachloride at 20 C.
3. Ethanol in carbon tetrachloride at 20 C.
4. Phenol in carbon tetrachloride at 20 C.
5. *p*-chlorphenol in carbon tetrachloride at 20 C.
6. Phenol in chlorbenzene at 20 C.
7. Phenol in benzene at 20 C.

Figure 1.11 Molecular association in the liquid phase: (a) Relative molecular weights of several substances in benzene solution. M/M_0 is the ratio of the observed molecular weight in solution to the formula weight (Data collected by Glasstone, *Textbook of Physical Chemistry*, 1940); (b) degrees of association of several alcohols and phenols in various solvents and at various temperatures (data of Mecke 1948).

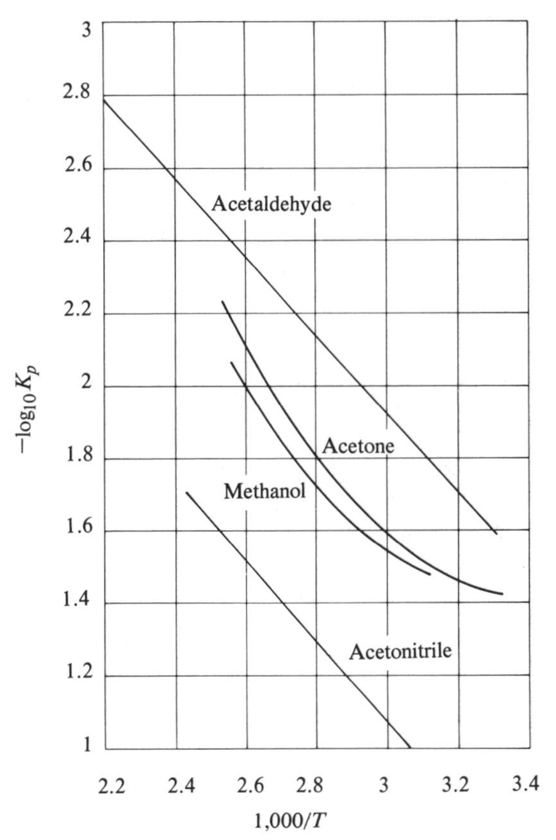

Figure 1.12 Some dimerization equilibrium constants, $K_p = P_{\text{dimer}}/P^2_{\text{monomer}}$ (data of Lambert et al., 1949).

1.3.6. Pseudocritical Properties

There is no reason to suppose that the principle of corresponding states cannot be applied to mixtures as well as to pure substances, and it has been found applicable with some restrictions. Although reduced equations of state prove not to be valid when based on the true critical properties of mixtures, it has been found possible to devise satisfactory pseudocritical properties in terms of the compositions and critical properties of the pure constituents. These pseudocritical properties usually differ quite markedly from the true values, as indicated in Figure 1.33, for instance. In fact, the true critical properties of mixtures often can be evaluated with equations of state based on pseudocritical properties, as described in Section 1.8.4.

One starting point for devising pseudocritical properties is the original combining rules of the van der Waals equation, which are

$$\sqrt{a} = \Sigma y_i \sqrt{a_i}, \tag{1.36}$$

$$b = \Sigma y_i b_i. \tag{1.37}$$

Since

$$a = 3P_c V_c^2 = \frac{27}{64} R^2 T_c^2 / P_c, \tag{1.38}$$

$$b = V_c/3 = RT_c/8P_c, \tag{1.39}$$

equations for pseudocritical properties could be given by

$$V_{pc} = \Sigma y_i V_{ci}, \tag{1.40}$$

$$T_{pc}/\sqrt{P_{pc}} = \Sigma y_i (T_{ci}/\sqrt{P_{ci}}), \tag{1.40a}$$

$$T_{pc}/P_{pc} = \Sigma y_i (T_{ci}/P_{ci}), \tag{1.40b}$$

and the parameters by

$$a = \frac{27}{64} R^2 [\Sigma y_i T_{ci}/\sqrt{P_{ci}}]^2, \tag{1.41}$$

$$b = \frac{R}{8} \Sigma y_i T_{ci}/P_{ci}. \tag{1.42}$$

These relations do lead to quite a satisfactory equation of state for some mixtures, but usually only slightly better than one based on a simpler method known as Kay's Rules (1936):

$$P_{pc} = \Sigma y_i P_{ci}, \tag{1.43}$$

$$T_{pc} = \Sigma y_i T_{ci}, \tag{1.44}$$

$$V_{pc} = \Sigma y_i V_{ci}. \tag{1.45}$$

An informative early study of the work of Kay and others on this topic, particularly as applied to petroleum fractions, is made by Edmister (1948).

A measure of self-consistency of pseudocritical properties was introduced by Prausnitz & Gunn (1958), who wrote,

$$P_{pc} = z_{pc} RT_{pc}/V_{pc}, \tag{1.46}$$

where the properties on the right are evaluated with Kay's Rule. In connection with evaluation of the cross-parameters, a_{ij} of cubic equations of state and B_{ij} of the virial equation, Prausnitz & Gunn also adopted these Lorentz-Berthelot-type rules for binary mixtures:

$$T_{cij} = (T_{ci} T_{cj})^{0.5}, \tag{1.47}$$

$$V_{cij}^{1/3} = 0.5(V_{ci}^{1/3} + V_{cj}^{1/3}). \tag{1.48}$$

Their complete system includes also

$$T_{cij} = (1 - k_{ij})(T_{ci} T_{cj})^{0.5}, \tag{1.49}$$

$$z_{cij} = 0.5(z_{ci} + z_{cj}) \tag{1.50}$$

$$= 0.291 - 0.04(\omega_i + \omega_j), \tag{1.51}$$

$$P_{cij} = z_{cij} RT_{cij}/V_{cij} \tag{1.52}$$

The binary interaction parameters k_{ij} are determined empirically. Some values are given in Table E.3; they were obtained mostly from experimental data on cross-second virial coefficients. The interaction parameters appear to be small for hydrocarbon pairs but are appreciable for pairs of dissimilar compounds. For some polar pairs they may be negative.

Many other combining rules for pseudocritical properties and equation parameters have been proposed, often superior to the rules already mentioned but usually more complicated. The subject is not closed. Some of the alternative methods are cited in connection with individual equations of state. A major investigation by Leland & Mueller evaluated the compressibilities of fifty-eight mixtures with Kay's Rule and with an improved method of their own. An extended summary of pseudocritical properties, combining rules and interaction parameters, may be found in the book by Sterbacek et al. (1979, p. 84). Discussions of theoretical bases for the formulation of pseudocritical properties are given by Leland & Chappelear (1968), Gunn (1972), and Mollerup (1980).

p	pv/RT	
atm	Volume Series	Pressure Series
1	$1-0.00064+0.00000+\ldots(+0.00000)$	$1-0.00064+0.00000+\ldots(+0.00000)$
10	$1-0.00648+0.00020+\ldots(-0.00007)$	$1-0.00644+0.00015+\ldots(-0.00006)$
100	$1-0.06754+0.02127+\ldots(-0.00036)$	$1-0.06439+0.01519+\ldots(+0.00257)$
1,000	$1-0.38404+0.68788+\ldots(+0.37272)$	$1-0.64387+1.51895+\ldots(-0.19852)$

(a)

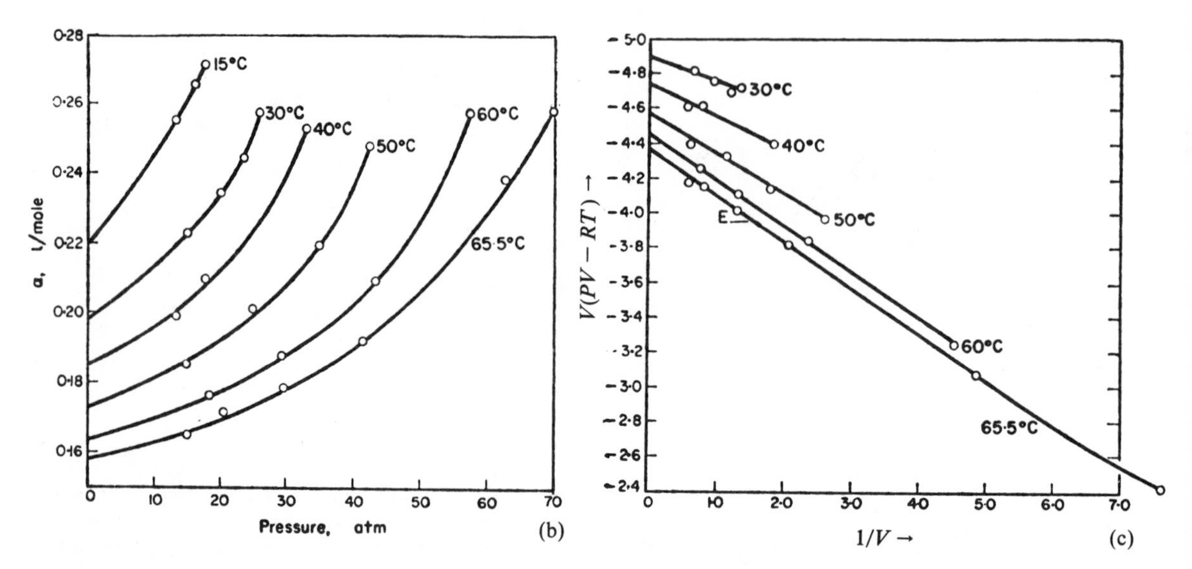

Figure 1.13 The volumetric and pressure forms of the virial equation: (a) Values of $z = PV/RT$ of argon at 25 C. The quantities in parentheses are the contributions of all terms in the series beyond the third one (Mason & Spurling 1969). (b) The residual volume, α, of an equimolal mixture of CO_2 and C_3H_6. If the equation were $z = 1 + BP/RT$, the plots would be horizontal; if the equation were $z = 1 + B'P + C'P^2$, the plots would be linear. If the equation were $z = 1 + BP/RT$, the plots would be horizontal; if the equation were $z = 1 + B'P + C'P^2$, the plots would be linear (King 1969). (c) The data of Figure 1.13(b) plotted as $V(PV - RT)$ against $1/V$. Since the plots are substantially linear, the C-truncated equation, $z = 1 + B/V + C/V^2$, is obeyed (King 1969).

Example 1.5. Pseudocritical Properties of Mixtures of Ethane (1) and *n*-Heptane (2)

Use the two methods,

(a) $P_{pc} = z_{pc}RT_{pc}/V_{pc}$,

(b) $P_{pc} = \Sigma y_i P_{ci}$,

with Kay's Rule for z_{pc}, T_{pc}, and V_{pc}. Pure component data are:

	Ethane	*n-heptane*	*units*
T_c	305.4	540.2	°K
P_c	48.2	27.0	atm
V_c	148	432	ml/gmol
z_c	0.285	0.263	

The results shown are compared with the true critical pressures in Figure 1.33.

	P_{pc}	
y_1	*a*	*b*
0.0	26.99	27.00
.1	27.86	29.12
.2	28.85	31.24
.3	29.97	33.36
.4	31.26	35.48
.5	32.78	37.60
.6	34.60	39.72
.7	36.82	41.84
.8	39.62	43.96
.9	43.27	46.08
1.0	48.26	48.20

Example 1.6. Comparison of Combining Rules for the Calculation of Compressibility of a Mixture with the Pitzer-Curl Correlation

The case is that of an equimolal mixture of carbon dioxide and propylene at 30 C and 25.5 atm. The experimental value of the compressibility is $z = 0.737$, and the experimental values of the critical properties of the mixture are $T_c = 319.8$ K, $P_c = 70.7$ atm.

Method 1 uses Kay's Rules for the pseudocritical properties. Method 2 uses $P_{pc} = z_{pc}RT_{pc}/V_{pc}$ for the pseudocritical pressure and Kay's Rules for the other pseudocritical properties. The properties of the pure components and those of the mixture are tabulated. Note that Kay's Rules check the experimental data most closely, whereas the value obtained from the true critical properties is much in error.

		T_c, °K	P_c, atm	V_c ml/gmol	z_c	ω
CO_2		304.2	72.9	94.1	0.274	0.225
C_3H_6		364.9	45.45	182.4	0.275	0.148
Mixture	Method 1	334.6	59.18	138.25	0.2745	0.1825
Mixture	Method 2	334.6	54.51	138.25	0.2745	0.1825
Mixture	True	319.8	70.7			0.1825
		T_r	P_r	$z^{(0)}$	$z^{(1)}$	z
Mixture	Method 1	0.906	0.431	0.761	−0.12	0.728
Mixture	Method 2	0.906	0.470	0.734	−0.15	0.693
Mixture	True	0.910	0.361	0.825	−0.05	0.811
Mixture	Experimental					0.737

Example 1.7. Comparison of Pseudocritical Rules with the van der Waals Equation

The problem of Example 1.6 is solved with the van der Waals equation. The parameters are calculated with pseudocritical properties obtained by the three methods of the previous example. Eq. 4 of Table 1.3 may be written

$$z^3 - \left(1 + \frac{P_r}{8T_r}\right)z^2 + \frac{27P_r}{64T_r^2}z - \frac{27P_r^2}{512T_r^3} = 0.$$

With the values of the reduced properties calculated previously, this equation becomes for the three cases:

Method 1: $z^3 - 1.0595z^2 + 0.2215z - 0.01317 = 0$,

$z = 0.8045$,

Method 2: $z^3 - 1.0501z^2 + 0.2416z - 0.01566 = 0$,

$z = 0.7813$,

True crit: $z^3 - 1.0501z^2 + 0.1880z - 0.00943 = 0$,

$z = 0.8396$.

The roots shown at the ends of the lines above were found with a calculator program for polynomial solutions. Clearly, none of the methods of obtaining the parameters of the van der Waals equation leads to compressibilities close to the experimental value $z = 0.737$. The method is clearly inferior to the Pitzer-Curl correlation with the same combining rules for pseudocritical properties.

Example 1.8 Compressibility of a Ternary Mixture

Find the compressibilities of the pure components and of a ternary mixture of methyl chloride (1), chloroform (2), and carbon tetrachloride (3) at 10 atm and 450 K with $y_1 = 0.2$ and $y_2 = 0.3$. Use the van der Waals equation.

The constants are taken from Lange's *Handbook of Chemistry* (1973).

	y	a	b	z	T_c	P_c
CH_3Cl	0.2	7.471	0.06483	0.9617	416.3	65.9
$CHCl_3$	0.3	15.17	0.1022	0.9091	536.4	54.0
CCl_4	0.5	20.39	0.1383	0.8738	556.4	45.0
Mix		15.784	0.1128	0.9073	522.38	51.88

First apply the Lorentz-Berthelot rules to evaluation of the parameters of the mixture—namely:

$$a = (\Sigma y_i \sqrt{a_i})^2 = 15.784.$$

$$b = \Sigma y_i b_i = 0.1128.$$

Thus the polynomial form of the vdW equation becomes

$$z^3 - 1.03055z^2 + 0.11575z - 0.003536 = 0,$$

of which the roots are

$$z = 0.9073,\ 0.0616 \pm 0.0095j.$$

The compressibilities of the pure components are found similarly and are shown in the preceding table.

Amagat's Law implies that

$$z = \Sigma y_i z_i = 0.9020,$$

which is about 6 percent higher than that given with the Lorentz-Berthelot Rules.

If Kay's Rules are used to evaluate the pseudocritical properties and thence the parameters of the equation, the results are:

$$P_c = \Sigma y_i P_{ci} = 51.88 \text{ atm},$$

$$T_c = \Sigma y_i T_{ci} = 522.38 \text{ K},$$

$$a = 27R^2T_c^2/64P_c = 14.9351,$$

$$b = RT_c/8P_c = 0.1033.$$

The polynomial becomes,

$$z^3 - 1.0280z^2 + 0.10958z - 0.003066 = 0,$$

of which the roots are $z = 0.9115,\ 0.0639,\ 0.0527$. The largest of these compares favorably with that obtained when the parameters of the mixture are evaluated directly from those of the pure components at the beginning of this exercise. The other difference between the two methods is that one of them finds only one real root and the other three such roots although the largest one is approximately the same as that found by the other methods.

Example 1.9 Comparison of Several Combining Rules with the Abbott-Virial Equation.

The specific volume of a mixture of 1/3 ethane + 2/3 *n*-pentane at 400 K and 20 atm will be found with the virial equation using Abbott's correlation for the second virial coefficient. The three sets of combining rules of Table 1.5 will be used. Kays Rule is $M_c = \Sigma y_i M_{ci}$ for all the critical properties. The Prausnitz-Gunn Rules are the same as the Kay Rules except that $P_c = z_c RT_c/V_c$. The results are summarized in the table that follows. Clearly, there are substantial differences in the predicted specific volumes of the mixture by the three methods. Experimental data are not available for comparison, but the Lorentz-Berthelot Rules are favored since they have some theoretical justification. The specific volumes are found by solution of the quadratic equation,

$$V^2 - \frac{RT}{P}V - \frac{BRT}{P} = 0.$$

ij		T_c	V_c	z_c	ω	P_c
11		305.4	148	0.285	0.098	48.2
22		469.6	363	0.262	0.251	33.3
12	Kay	414.9	291	0.270	0.200	31.5
12	PG	414.9	240	0.270	0.200	35.5
12	LB	378.7	240	0.274	0.175	38.3
		T_r	T_{r12}	B_{12}	B	V ml/gmol
ethane		1.3098			−95.1	1731.1
pentane		0.8518			−592.8	2103.4
mix	Kay	0.9642			−335.0	1926.4
mix	PG	0.9642			−406.9	1978.5
mix	LB		1.0562	−304.8	−409.5	1980.3

Examples 1.5–1.9 apply the rules for pseudocritical properties that have been mentioned.

1.3.7. Combining Rules

These have the purpose of representing a property of a mixture in terms of the composition and the properties of the pure components. Depending on the property, the composition may be in mol or volume or weight fractions. Some of the combining rules have a rational basis, but most are purely empirical. In many instances the prediction of a property can be improved by incorporating a limited amount of experimental data on the mixture or on component pairs of the mixture, since binary interactions have the major impact on the mixture behavior. The justifiable extent of elaboration of combining rules depends, of course, on the accuracy that is needed. Binary interaction parameters are covered in Section 1.3.8.

In connection with EOS, the main interest is in rules for pseudocritical properties, which are discussed in Section 1.3.6, and those for direct evaluation of the parameters. Such rules are stated with each equation that is cited in this chapter, but some general remarks may be made. Only the rules for the coefficients of the virial equation are on a rational basis—the second virial coefficient, for example, being represented by

$$B = \Sigma\Sigma y_i y_j B_{ij}. \tag{1.53}$$

In some instances the cross-parameter B_{ij}, when $i \neq j$, can be approximated by

$$B_{ij} \simeq (B_{ii}B_{jj})^{0.5}, \tag{1.54}$$

so that Eq. 1.53 simplifies to

$$B^{0.5} = \Sigma y_i B_i^{0.5}. \tag{1.55}$$

The particularly simple combining rules used for the parameters of the van der Waals EOS have been mentioned—namely,

$$\sqrt{a} = \Sigma y_i \sqrt{a_i}, \tag{1.56}$$

$$b = \Sigma y_i b_i. \tag{1.57}$$

For related cubic equations such as the Redlich-Kwong equation and its many modifications, $\sqrt{a_i a_j}$ is replaced by the cross-parameter a_{ij}; and binary interaction parameters k_{ij} may be introduced, as for the Soave equation:

$$a_{ij} = (1 - k_{ij})\sqrt{a_i a_j}. \tag{1.58}$$

A basis for the combining rules of equations of the van der Waals type is suggested by the theoretical combining rules for second virial coefficients. Since

$$B = \lim_{V \to \infty} V(z - 1), \tag{1.59}$$

relations between the virial coefficient and parameters of other equations of state become

$$B = b - a/RT \tag{1.60}$$

for the van der Waals EOS, and

$$B = b - a/RT^{1.5} \tag{1.60a}$$

for the Redlich-Kwong case. Take the van der Waals case. The virial coefficient of the mixture becomes

$$\begin{aligned} B &= \Sigma\Sigma y_i y_j B_{ij} \\ &= a - b/RT = \Sigma\Sigma y_i y_j b_{ij} - (1/RT)\,\Sigma\Sigma y_i y_j a_{ij}, \end{aligned} \tag{1.61}$$

so that the combining rules are

$$a = \Sigma\Sigma y_i y_j a_{ij}, \tag{1.61a}$$

$$b = \Sigma\Sigma y_i y_j b_{ij}. \tag{1.61b}$$

It has been found empirically, however, that Eq. 1.57 is better than Eq. 1.61b.

Combining equations for the parameters of some cubic equations of state were devised by Huron & Vidal (1979), who related the parameters to excess Gibbs energies. Their method requires experimental data on activity coefficients and is developed only for binary mixtures, being a correlative method rather than a predictive one.

Complex equations of state that are largely empirical in form likewise use rather arbitrary but empirically founded combining rules. The eight coefficients of the BWR equation (Table 1.16) applied to mixtures are given by the rules

$$A = (\Sigma y_i A_i^{1/m})^m, \tag{1.62}$$

with $m = 1$, 2, or 3, depending on the coefficient. The eleven constant Starling-Han modification employs similar rules and in addition has interaction parameters for four of the parameters, of the form $(1 - k_{12})^m$, with m as large as 5.

1.3.8. Binary Interaction Parameters

The behavior of mixtures naturally is affected by interactions of unlike molecules, particularly if some are polar. Interactions between triplets and higher combinations usually are less important than those between pairs of components. Some idea of the relative contributions of binary and ternary interactions may be obtained by examining virial equations with second and third virial coefficients that incorporate the effects of such interactions between unlike molecules. At modest pressures somewhat removed from the critical temperature, the contribution of the third virial coefficient may be quite small. For example, some data from Figures 1.14(c) and 1.14(d) are:

		Maximum error in z, %	
T_r	P_r	*With B*	*With B and C*
2	1	1	0.01
2	5	20	1

For many purposes an error of 1 percent in compressibility is acceptable, so the contribution of the third virial coefficient may be neglected at $T_r = 2$ and $P_r = 1 - 2$ for the substances covered by that figure. Equations of the form $z - 1 = BP + CP^2$ for mixtures of helium and nitrogen are compared in Figure 1.15(e). The effect of the third virial coefficient is quite modest even at 50 atm, with $CP^2/BP = 0.030$ for the upper curve and -0.022 for the lower one.

Higher-order interactions often are small and thus hidden by imperfections of practicable equations of state, so that incorporation of only binary data in addition to those of pure components generally leads to the major possible improvement in the accuracy of the equation of state. For an n-component mixture there are $n(n-1)/2$ possible binary interaction parameters.

Such parameters are commonly applied in one of two ways:

1. As an adjustment of a pseudocritical property of a pair: For the Redlich-Kwong equation, Chueh & Prausnitz (1967) write:

$$T_{cij} = (1 - k_{ij})\sqrt{T_{ci}T_{cj}}, \tag{1.63}$$

and

$$a_{ij} = 0.42748(1 - k_{ij})R^2(T_{ci}T_{cj})^{1.25}/P_{cij}. \tag{1.64}$$

k_{ij} usually is in the range from 0 to 0.2 or so. Values are given in Table E.3. Such cross-pseudocritical temperatures also are used in the evaluation of cross-second virial coefficients, as by the equations of Abbott and Tsonopoulos. Data of k_{ij} for hydrocarbons with 8–30 carbon atoms and nonpolar com-

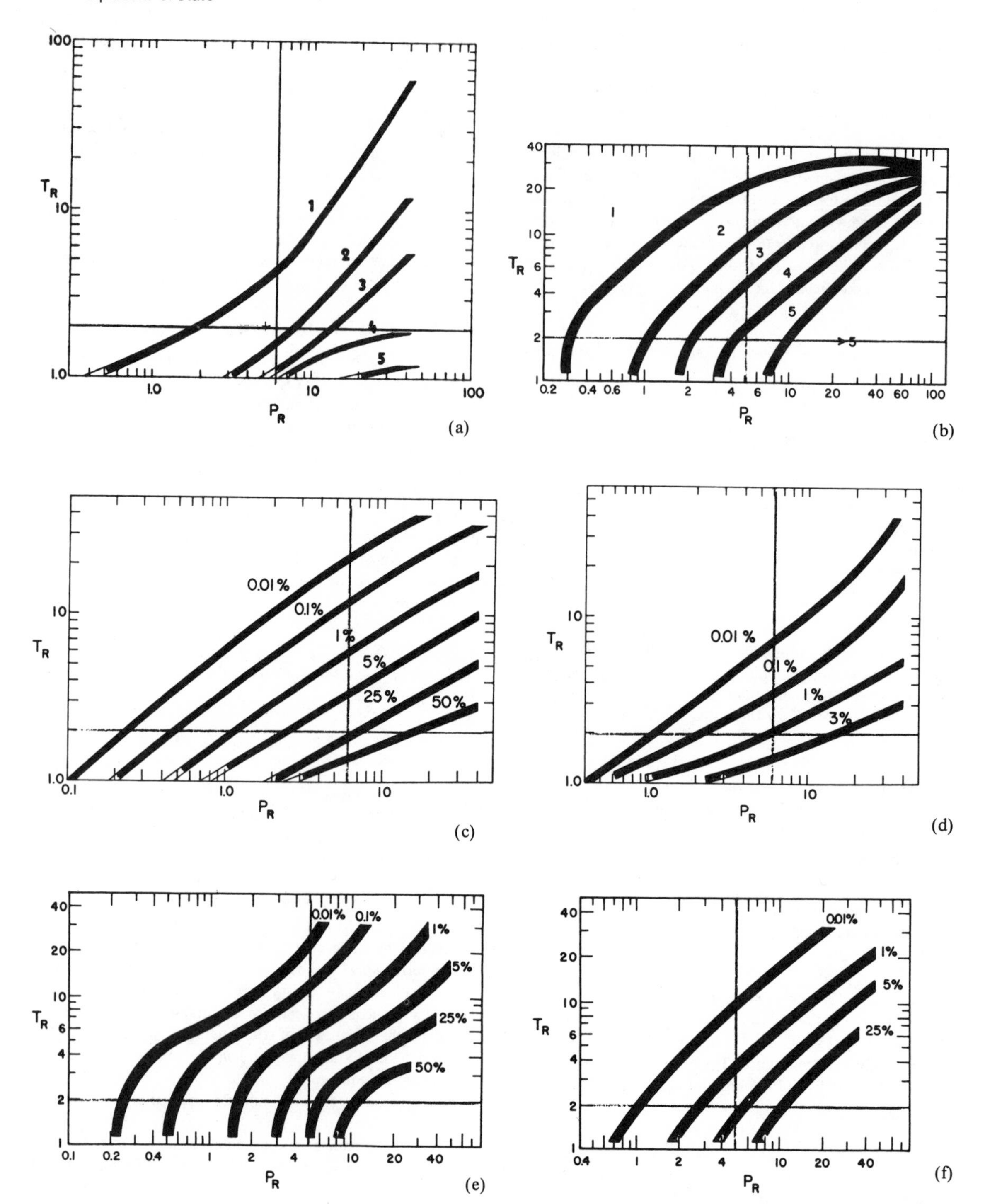

Figure 1.14 Truncation errors associated with the density and pressure forms of the virial equation (Lee, Eubank, & Hall 1978). Data for He, N_2, A, CH_4, C_2H_6, and their mixtures. Examples cited are for $T_r = 2$, $P_r = 5$. (a) Statistically optimum number of terms with the truncated density series. For the example, the optimum number of terms is two. (b) Statistically optimum number of terms with the truncated pressure series. For the example, the optimum number of terms is five. (c) Percentage errors from least-squares fit of $z = 1 + B\rho$. For the example, the error is 25%. (d) Percentage errors from least-squares fit of $z = 1 + B\rho + C\rho^2$. For the example, the error is 1 percent. (e) Percentage errors from least-squares fit of $z = 1 + B'P$. For the example, the error is 20 percent. (f) Percentage errors from least-squares fit of $z = 1 + B'P + C'P^2$. For the example, the error is 5 percent.

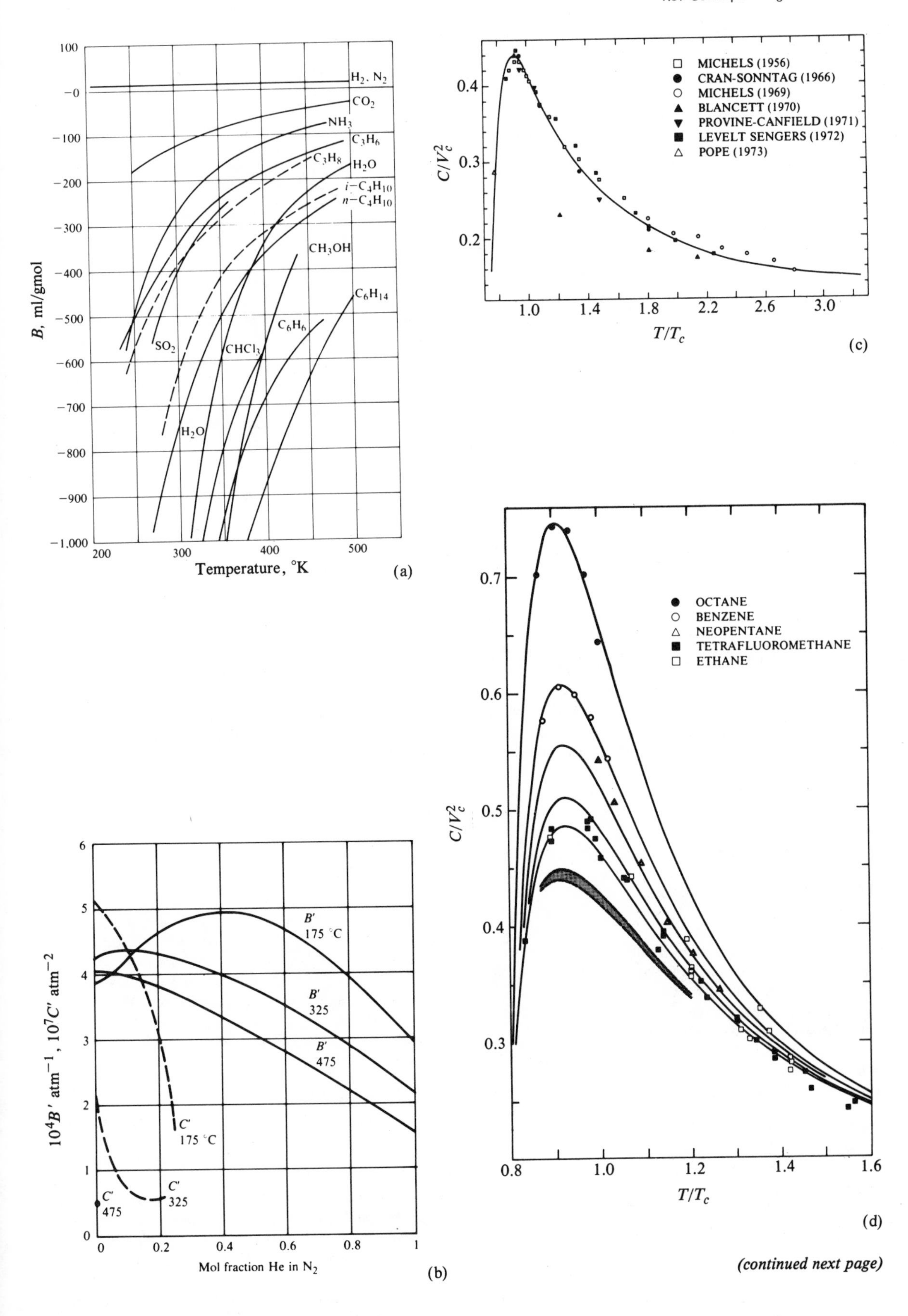

(continued next page)

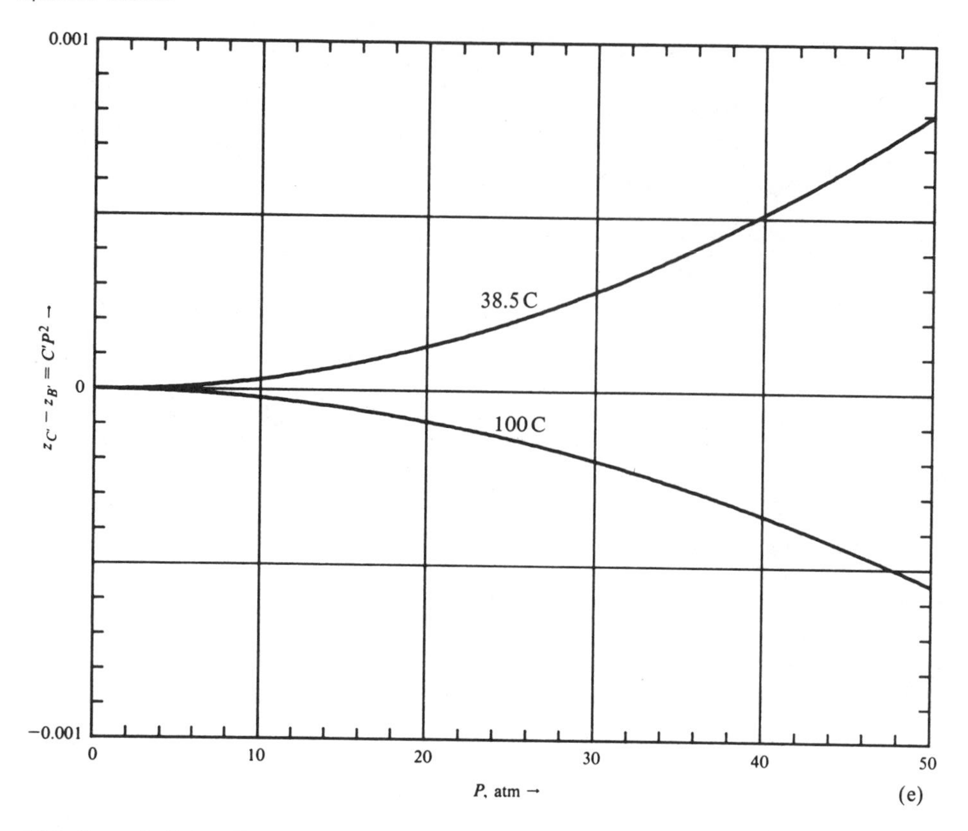

Figure 1.15 Data of virial coefficients: (a) Second virial coefficients of some substances as functions of temperature (from the compilation of Kogan 1967); (b) second and third virial coefficients, B' and C', of mixtures of helium and nitrogen. Cross-coefficients are $(10^4 B'_{12}, °C) = (6.234, 175), (4.429, 325)$, and $(3.313, 475)$ (data of Witonsky & Miller 1963); (c) reduced third virial coefficients of argon (De Santis & Grande 1979); (d) third virial coefficients of some nonpolar gases. The darkened region represents data of fluids such as Kr, Xe, CH_4, F_2, N_2, and O_2 (De Santis & Grande 1979); (e) the contribution of the third virial coefficient to the calculated compressibilities of mixtures of He and N_2. The upper curve is for 54.6% He at 38.5 C, and the lower one for 51.3% He at 100 C. The ordinate is the difference, $z_{C'} - z_{B'} = C'P^2$ (data of Ku & Dodge 1967).

pounds such as water and the lower oxygenated organics are given by Tsonopoulos (1974, 1979).

2. As a direct adjustment of some of the parameters of an equation of state, for example the cross-parameter of the Redlich-Kwong or Soave equations,

$$a_{ij} = (1 - c_{ij})\sqrt{a_i a_j}. \tag{1.65}$$

This scheme was proposed by Zudkevich & Joffe (1970). Values for the API version of the Soave equation are in Table 1.12. For hydrocarbon pairs in this correlation, the interaction parameters are zero, whereas for mixtures of hydrocarbons with H_2S, CO_2, and N_2, they were correlated with solubility parameters for each of these gases, but no general correlation was achieved. An improved correlation for CO_2 with hydrocarbons was obtained by Ezekwe (1982), namely

$$\log k_{ij} = -0.8849 - 0.2145\,\omega_i\,|\delta_i - \delta_j|. \tag{1.66}$$

Several of the eleven parameters of BWRS equation carry binary adjustments; in that system, interactions of hydrocarbon pairs also are significant.

For the Redlich-Kwong equation, Kato et al. (1976) point out that the two kinds of parameters are related by

$$1 - c_{ij} = \frac{V_{cij}}{Rz_{cij}}\sqrt{\frac{P_{ci}P_{cj}}{T_{ci}T_{cj}}}(1 - k_{ij})^{1.5}, \tag{1.67}$$

with

$$z_{cij} = 0.5(z_{ci} + z_{cj}) \tag{1.68}$$

$$V_{cij}^{1/3} = 0.5(V_{ci}^{1/3} + V_{cj}^{1/3}). \tag{1.69}$$

These investigators evaluated the k_{ij} for several systems and found only a small effect of temperature over a 100 C range. However, in equilibria involving an aqueous phase, the binary interaction parameters of water with CO_2 and H_2S were stated to be definitely temperature dependent (Peng & Robinson 1980).

The cross-virial coefficients were obtained by Hiza & Duncan (1970) from

$$B_{12} = (1 - c_{12})\sqrt{B_1 B_2}, \tag{1.70}$$

where the interaction parameters are correlated with ionization potentials of the pure components by the equation

$$c_{12} = 0.17(I_1 - I_2)^{0.5}\ln(I_1/I_2), \tag{1.71}$$

for methane, ethylene, ethane, and the light inorganic gases including hydrogen and helium. A tabulation of ionization potentials for about 1,000 substances is in the *CRC Handbook of Chemistry and Physics* (1969), but not in some later editions.

An outstanding success of the concept that binary interactions largely determine the deviations of multicomponent mixtures from ideality are the Wilson and related correlations for activity coefficients (Chapter 4). Since activity coefficients often are adequately computed from molecular structural contributions by the ASOG or UNIFAC methods, it may be possible that binary interaction parameters c_{ij} or k_{ij} also could be evaluated from pure component properties, but this has not yet been done in comprehensive form.

1.3.9. Correlation of *PVT* Behavior

Such behavior may be represented directly in terms of reduced properties in graphical or tabular form, without prior fitting an equation to *PVT* data. The compressibility, $z = PV/RT$, is shown in this way in Figure 1.8(a). Clearly all the substances represented here conform closely to the principle of corresponding states, much better in fact than to the van der Waals, Berthelot, or Dieterici equations, as shown by Ott et al. (1970). The broken lines of Figure 1.8(a) are an example. Further improvement may be achieved by including a third parameter in addition to T_c and P_c—for example, the critical compressibility z_c or the acentric factor ω, which is related to vapor pressure behavior.

The method of Pitzer et al. (1955–1958), already mentioned in Section 1.3.3, presents compressibility in the form

$$z = z^{(0)} + \omega z^{(1)}, \tag{1.72}$$

where $z^{(0)}$ and $z^{(1)}$ are empirical factors established as functions of P_r and T_r and developed originally in tabular form. A similar tabulation of compressibilities was prepared with the Lee-Kesler EOS for convenience of manual calculations because of the great complexity of that equation; the data are given as Table E.1 and E.2, and are generally considered more accurate than the Pitzer & Curl values. They are graphed in the API Data Book and in Figure 1.22.

Several of the correlations of the second virial coefficient, like those shown with Figure 1.14, may be used to find the compressibility from

$$z = 1 + BP/RT. \tag{1.73}$$

The critical compressibility z_c was adopted as a third correlating parameter by Lydersen et al. (1955). From data on some eighty-two compounds, they made up tables from which the compressibility could be obtained at specified values of P_r, T_r, and z_c. These tables are reproduced by Hougen et al. (1959), along with tables for liquid densities and other useful thermodynamic information. Because of the greater experimental difficulties in obtaining accurate values of z_c, however, the acentric factor instead has been adopted almost universally as the third correlating parameter. Lydersen et al. obtained the relation

$$z_c = 1/(3.41 + 1.28\omega), \tag{1.74}$$

and Edmister (1958) found the relation between the two factors to be

$$z_c = 0.291 - 0.080\,\omega. \tag{1.75}$$

These equations are compared with experimental data in the following, from which it can be seen that neither one is consistent enough for the purpose of predicting compressibilities from acentric factors, except for hydrocarbons, for which they are commonly adequate.

			Calculated z_c	
Substance	ω	z_c	*Lydersen*	*Edmister*
Acetone	0.309	0.232	0.263	0.266
Ammonia	0.250	0.242	0.268	0.271
Ethanol	0.635	0.248	0.237	0.240
Benzene	0.212	0.271	0.272	0.274
Hexane	0.296	0.260	0.264	0.267
Hydrogen sulfide	0.100	0.284	0.283	0.283
Methane	0.008	0.288	0.292	0.290

1.4. THE VIRIAL EQUATION

Three interrelated forms of the power series for the compressibility factor are

$$z = PV/RT = 1 + B/V + C/V^2 + \ldots \tag{1.76}$$

$$= 1 + B\rho + C\rho^2 + \ldots \tag{1.77}$$

$$= 1 + B'P + C'P^2 + \ldots. \tag{1.78}$$

These are called the volume (or density) and the pressure forms of the virial equation. Relations between the two sets of coefficients are given in Table 1.8.

The equation was originally proposed on a purely empirical basis by Thiesen (1885) and then developed thoroughly by Onnes (1901), but it later evolved naturally from a statistical-mechanical analysis of the forces between molecules, starting with Ursell (1927). If assumptions are made about the mathematical form of intermolecular potentials (Section 1.7), theoretical expressions can be developed for the several coefficients. The coefficient B corresponds to interaction between pairs of molecules, C to triplets, and so on. The unprimed terms, B, C, . . . , are called the second, third, and so on virial coefficients. In theory, for a given substance they are functions of temperature only. The application of virial theory goes beyond *PVT* relations: The same coefficients are involved in the description of other properties of gases such as viscosity, sonic velocity, and heat capacity. A particularly valuable consequence of the theory of the virial equation is an exact relation between the coefficients of a mixture and those of the pure components and the composition that appears in Table 1.8. An abbreviated statement of the theory may be found in the book of Mason & Spurling (1969).

As appears in Figure 1.18, isotherms of the virial equation do not have the S-shaped form of the van der Waals and related equations, so even with a large number of terms it cannot represent the liquid state and coexistence of liquid and vapor phases (see Problem 1.23). At high densities the infinite series diverges (Mason & Spurling 1969).

1.4.1. Truncated Forms

Since the infinite series is impractical for purposes of calculation, truncations at B or C are commonly used. For a given truncation the series in density is usually more accurate than the one in pressure. This is clearly true for the mixture of CO_2 and propylene represented in Figure 1.13, and also for argon, except at 1,000 atm, where both representations are poor.

Data for H_2, N_2, Ar, CH_4, C_2H_6, and their mixtures are represented in Figure 1.14. A particular statistical criterion was developed to establish the optimum number of parameters

Table 1.8. The Virial Equation

Basic forms:

$$z = PV/RT = 1 + B/V + C/V^2 + D/V^3 + \cdots \tag{1}$$

$$= 1 + B\rho + C\rho^2 + D\rho^3 + \cdots \tag{2}$$

$$= 1 + B'P + C'P^2 + D'P^3 + \cdots. \tag{3}$$

Common truncated form:

$$z = 1 + B/V = 1 + B'P = 1 + BP/RT. \tag{4}$$

According to Prausnitz (1957), the truncated equation is often adequate when

$$P \leq 0.5T(\Sigma y_i P_{ci})/(\Sigma y_i T_{ci}). \tag{5}$$

Relations between the coefficients:

$$B = RTB', \tag{6}$$

$$C = (RT)^2(C' + B'^2) \text{ or } C' = (C - B^2)/(RT)^2, \tag{7}$$

$$D = (RT)^3(D' + 3B'C' + B'^3) \text{ or } D' = (D - 3BC + 2B^3)/(RT)^3, \tag{8}$$

$$E = (RT)^4(E' + 4D'B' + 6C'B'^2 + 2C'^2 + B'^4) \tag{9}$$

$$\text{or } E' = (E - 4DB - 2C^2 + 10CB^2 - 5B^4)/(RT)^4. \tag{10}$$

Coefficients from limiting values of PVT data:

$$B = \lim_{1/V \to 0} \left(\frac{PV}{RT} - 1\right) V, \tag{11}$$

$$C = \lim_{1/V \to 0} \left(\left(\frac{PV}{RT} - 1\right) V - B\right) V, \tag{12}$$

$$D = \lim_{1/V \to 0} \left(\left(\frac{PV}{RT} - 1\right) V^2 - BV - C\right) V. \tag{13}$$

Other equations in virial form:

van der Waals: $(P - a/V^2)(V - b) = RT.$ (14)

virial form: $z = 1 + (b - a/RT)\rho + b^2\rho^2 + b^3\rho^3 + \cdots.$ (15)

Beattie-Bridgeman: $z = (1 - c/VT^2)(V + B_0 - bB_0/V)/V - A_0(1 - a/V).$ (16)

virial form:

$$z = 1 + (B_0 - A_0/RT - c/T^3)\rho + (-B_0 b + aA_0/RT - B_0 c/T^3)\rho^2 + (B_0 bc/T^3)\rho^3. \tag{17}$$

Benedict-Webb-Rubin:

$$z = 1 + (B_0 - A_0/RT - C_0/RT^3)\rho + (b - a/RT)\rho^2 + (a\alpha/RT)\rho^5$$
$$+ (c/RT^3)(1 + \gamma\rho^2)\rho^2 \exp(-\gamma\rho^2) \tag{18}$$

(the last term containing the exponential is a catch-all for the missing terms of the virial expansion).

Reduced form of the C-truncated equation.

$$P_r V_r = 3T_r \left(1 - \frac{1}{V_r} + \frac{1}{3V_r^2}\right). \tag{19}$$

The parameters are found in terms of critical properties by equating the first and second derivative of P with respect to V at constant T to zero at the critical condition. The parameter equations are:

$$B = -V_c, \tag{20}$$

$$C = V_c^2/3, \tag{21}$$

(continued on next page)

Table 1.8 *(continued)*

$$R = 3P_cV_c/T_c, \tag{22}$$

$$z_c = 1/3. \tag{23}$$

Combining rules for mixtures:

$$B = \Sigma\Sigma y_i y_j B_{ij} = y_1^2 B_{11} + y_2^2 B_{22} + 2y_1 y_2 B_{12}, \tag{24}$$

$$C = \Sigma\Sigma\Sigma y_i y_j y_k C_{ijk} = y_1^3 C_{111} + y_2^3 C_{222} + 3(y_1^2 y_2 C_{112} + y_1 y_2^2 C_{122}). \tag{25}$$

The expansions shown are for two components.

Correlations of virial coefficients: Data on second virial coefficients are the most abundant and several correlations have been developed, primarily for nonpolar gases. The Pitzer-Curl form is

$$B_{ij} = \left(\frac{RT_{cij}}{P_{cij}}\right)(B_{ij}^{(0)} + \omega B_{ij}^{(1)}), \tag{26}$$

where the superscripted items have been correlated in terms of reduced temperature. Figs. 1.10(a) and 1.16 present some of the available correlations. Pseudocritical properties for evaluation of cross parameters, B_{ij}, are found with the modified Lorentz-Berthelot rules, the last column of Table 1.5.

Third virial coefficients of nonpolar gases have been correlated analogously to the Pitzer model by Orbey & Vera (1983).

$$C = \left(\frac{RT_c}{Pc}\right)^2 (C^{(0)} + \omega C^{(1)}), \tag{27}$$

$$C^{(0)} = 0.01407 + 0.02432/T_r^{2.8} - 0.00313/T_r^{10.5}, \tag{28}$$

$$C^{(1)} = -0.02676 + 0.01770/T_r^{2.8} + 0.040/T_r^{3.0} - 0.003/T_r^{6.0} - 0.00228/T_r^{10.5}. \tag{29}$$

For mixtures the combining rules and binary interaction parameters of Tables 1.5 and E.3 are used.

Equations for fugacities are in Tables 3.3 and 3.4; those for residual properties are in Table 11.2.

in the virial equations to be used as a function of reduced temperatures and pressures. Here also it is quite clear that the density series usually is more accurate. For example, Figures 1.14(d) and 1.14(f) show that at $P_r = 5$ and $T_r = 2$ the expected errors are about 1 percent with the density and 5 percent with the pressure series.

1.4.2. Virial Coefficients

An exhaustive compilation of experimental data of second and third virial coefficients is by Dymond & Smith (1980), and briefer ones are by Kogan (1968) and Landolt-Börnstein (1970, Vol. II/1). Some data of pure substances are shown in Figure 1.15 and of mixtures in Problems 1.6 and 1.12. Theoretical methods for the evaluation of virial coefficients, particularly the second, from potential functions are mentioned in Section 1.7. The square-well potential was applied for obtaining second virial coefficients at low temperatures by Chang & Lu (1972), who express the constants of the potential function in terms of the acentric factor and give data for nineteen substances. Calculations with more complex potential functions were made by Gibbons (1974), and good checks of experimental data for polar substances including water were obtained. Experimental techniques in the determination of virial coefficients of He and CO_2 are described by Holste et al. (1980).

Some plots of second and third virial coefficients are presented in Figure 1.15. Accuracy of third virial coefficients generally is poor, as indicated in Figures 1.15(c) and 1.15(d). The scatter reflects both experimental uncertainty and the possibility that variables other than the reduced properties may be required for correlation. In the case of Figure 1.15(e), the third virial coefficient affects the compressibility factor only slightly.

Several correlations of second virial coefficients of nonpolar gases have been cast in the form

$$B = (RT_c/P_c)(B^{(0)} + \omega B^{(1)}), \tag{1.79}$$

following Pitzer et al. The superscripted items are correlated in terms of reduced temperature. Three correlations of this type are compared in Figure 1.16. Some extensions to polar substances have been made by including additional properties—for instance:

1. Hayden & O'Connell (1975) include dipole moments and radii of gyration in their elaborate procedure for second virial coefficients and equilibrium constants of dimerization. An extension of this work to mixtures of amines with methanol was made by Stein & Miller (1980).
2. Tarakad & Danner (1977) include the radius of gyration and a polarity factor based on the second virial coefficient at $T_r = 0.6$; they give data for more than 100 substances of all kinds.

3. Virial coefficients of some of the large hydrocarbon molecules that often are present in small quantities in gas streams at high pressures were obtained by Kaul & Prausnitz (1957).
4. Experimental results for perfluorohexanes was obtained by Taylor & Reed (1970).

Much less work has been done in correlating third virial coefficients, in part perhaps because the data are much fewer and of lesser accuracy. An equation is cited by Sterbacek (1979) without reference. DeSantis & Grande (1979) achieved a relatively simple correlation in terms of reduced temperature, acentric factor, molecular volume, and dipole polarizability, and checked some data on mixtures. A Pitzer-Curl type correlation by Orbey & Vera (1983) is given in Table 1.8.

1.4.3. Mixtures

Theory predicts that the second virial coefficients of mixtures are given in terms of the composition by the relation,

$$B = \Sigma\Sigma y_i y_j B_{ij} = y_1^2 B_{11} + y_2^2 B_{22} + \cdots + 2(y_1 y_2 B_{12} + y_1 y_3 B_{13} + \cdots + y_2 y_3 B_{23} + \cdots). \quad (1.80)$$

Analogous equations hold for the higher virial coefficients. The coefficients with repeated subscripts, B_{ii}, are those of pure components and correspond to interactions of pairs of like molecules, whereas the cross-coefficients correspond to interactions of pairs of unlike molecules. Theoretical methods for unlike interactions are no more complex than for like interactions, and the correlations that have been achieved are similar in form, but in terms of pseudocritical properties as well as binary interaction parameters in many cases. The cross-coefficient is given by

$$B_{ij} = (RT_{cij}/P_{cij})(B_{ij}^0 + \omega_{ij} B_{ij}^1), \quad (1.81)$$

where the superscripted items are determined from the same correlations as those for pure substances, but in terms of cross-pseudocritical properties. The forms of those that are currently accepted are:

$$\omega_{ij} = 0.5(\omega_i + \omega_j), \quad (1.82)$$

$$z_{cij} = 0.5(z_i + z_j), \quad (1.83)$$

$$V_{cij}^{1/3} = 0.5(V_{ci}^{1/3} + V_{cj}^{1/3}), \quad (1.84)$$

$$T_{cij} = (1 - k_{ij})(T_{ci} T_{cj})^{0.5}, \quad (1.85)$$

$$P_{cij} = z_{cij} RT_{cij}/V_{cij}. \quad (1.86)$$

The most sensitive mixing rule is that for T_{cij}. Much effort has been expended on correlation of binary interaction parameters, k_{ij}, but no universal one has been achieved. Some discussion is in Section 1.3.8. A review of recent work is by Tsonopoulos (1979).

For paraffin binaries, a moderately successful relation is

$$k_{ij} = 1 - 8\sqrt{V_{ci} V_{cj}}/(V_{ci}^{1/3} + V_{cj}^{1/3})^3. \quad (1.87)$$

Separate correlations also were developed by Tsonopoulos for hydrocarbons with methane, ethylene, ethane, and inorganic gases in terms of the number of hydrocarbon atoms. For methane-hydrocarbons, for instance,

$$k_{ij} = 0.0279(\ln n_{Cj})^2. \quad (1.88)$$

Polar-nonpolar binaries were not correlated as successfully, but many data are cited in the paper. The correlation of Hiza & Duncan (1970) in terms of ionization potential is cited in Section 1.3.8.

1.4.4. Other Equations of State in Virial Form

Several well-known equations of state may be recast into virial form, of which examples are shown in Table 1.8, Example 1.11, and Problem 1.22. These equations prescribe particular forms of the effect of temperature on the coefficients; and, probably because of these features, they often are more accurate than the virial equation truncated at B or C. Since the van der Waals equation in the virial expansion has third and higher coefficients independent of temperature, this may account in part for its inaccuracy compared with other cubic equations. The expansions are obtained with the aid of Eqs. 11–13 of Table 1.8 for the virial coefficients, as used in Example 1.11.

1.4.5. Dimerization

Negative deviation of PVT behavior from ideality sometimes is attributable to molecular associations, of which dimerization usually is most prevalent. A simple relation has been developed between the second virial coefficient and a dimerization equilibrium constant. For the reaction

$$2A \rightleftarrows A_2,$$

the equilibrium constant at pressure P is

$$K_p = p_{A_2}/p_A^2 = \frac{e}{2}\left(1 - \frac{e}{2}\right)\Big/(1-e)^2 P, \quad (1.89)$$

where e is the fractional dimerization of A. This may be solved for e,

$$e = 1 - \sqrt{1/(1+4PK_p)}. \quad (1.90)$$

At small degrees of association the square terms may be ignored, so approximately

$$e = 2PK_p. \quad (1.91)$$

The second virial coefficient is given by

$$z = n_t(1 + BP/RT) = (1 - e/2)(1 + BP/RT). \quad (1.92)$$

For small e this may be written

$$z = 1 - e/2 + BP/RT = 1 - PK_p + BP/RT = 1 + (B - RTK_p)P/RT = 1 + B_0 P/RT, \quad (1.93)$$

where the observed virial coefficient B_0 is related to the "true" one, B, by

$$B_0 = B - RTK_p. \quad (1.94)$$

If B is obtained from correlations on nonassociating molecules, the second virial coefficient of an associating system can be evaluated when the equilibrium constant is known. Some measurements of K_p by Lambert et al. (1949) are shown in Figure 1.12. Example 1.10 uses these data. The correlation method of Hayden & O'Connell (1975) may be used for obtaining the equilibrium constants. A comprehensive listing of such results is by Prausnitz et al. (1980). Oxygenated organics are the chief types with significant associations. Of course, not all negative deviations from ideality are caused by associations, and rarely are positive deviations due to dissociation.

Example 1.10 Compressibility of Acetone on the Basis of Dimerization Equilibria

The data of Lambert (1949) from Fig. 1.12 will be used. At 400 K and 2 atm, $K_p = 0.004$. Use the Abbott correlation for the virial coefficient of nonassociated acetone.

$T_c = 508.1, P_c = 46.4, \omega = 0.309,$

$T_r = 400/508.1 = 0.787,$

$$B = \frac{508.1\ R}{46.4}[-0.536 + 0.309(-0.331)] = -6.989\ R.$$

Equation 1.94:

$B_0 = B - RTK_p = -6.989\,R - 400(0.004)\,R = -8.589\,R.$

The compressibilities are:

Nonassociated:

$$z = 1 + BP/RT = 1 - 6.989(2)/(400) = 0.966.$$

Associated:

$$z = 1 + B_0P/RT = 1 - 8.589(2)/400 = 0.958.$$

The effect of association is slight in this case. A comparison of the approximate (Eq. 1.91) and exact expressions (Eq. 1.90) of the fractional dimerization is shown on the graph:

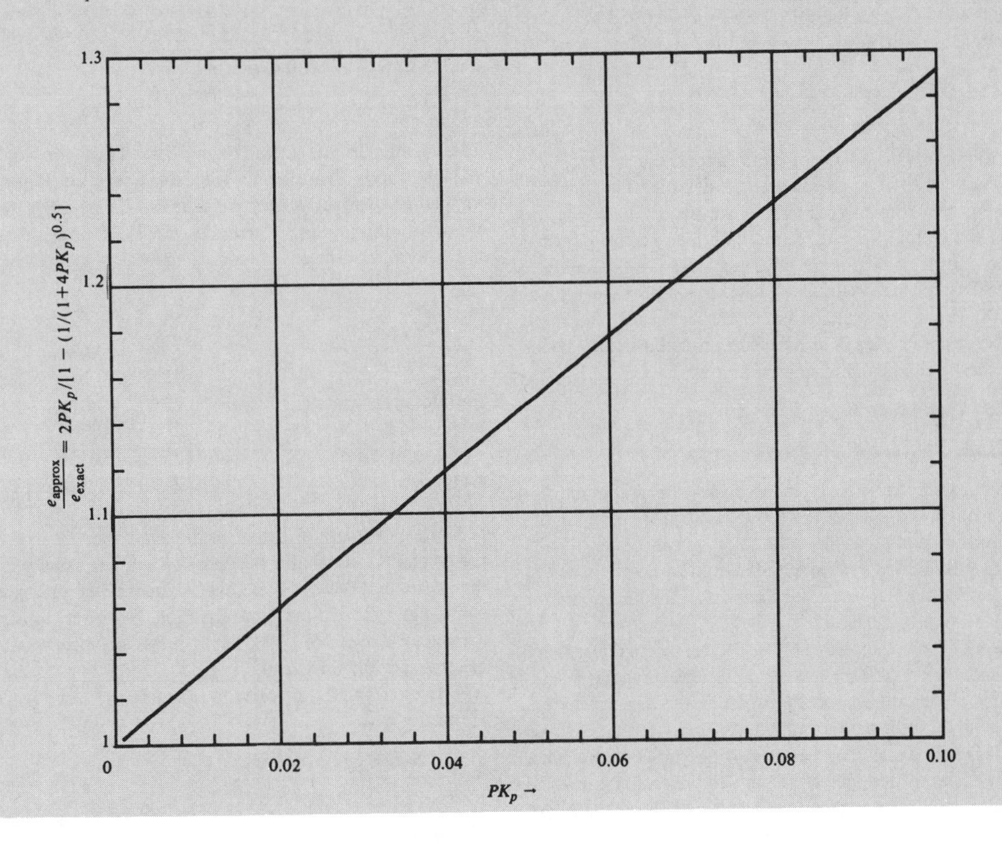

1.4.6. Applicability of the Virial Equation

Little is known about the higher virial coefficients, even the third, although a few data even on the fourth and fifth coefficients of argon have been obtained (Mason & Spurling 1969, p. 13), and theoretical expressions in terms of the square-well potential have been derived up to the eighth coefficient. Below the critical temperature the series is convergent up to the density of the saturated vapor, but divergent for liquid densities. Above the critical temperature the maximum useful density is about 0.5–0.75 times the critical density. For comparison, the BWR equation, which is somewhat related to the virial equation, applies to many liquids and vapors up to about 1.8 times the critical density. Criteria for accuracy of truncated equations also appear in Figure 1.14.

At pressures below a few atmospheres, the B-truncated equation is adequate for evaluating fugacity coefficients and deviation functions of gases, for which operations the equations are particularly simple and thus attractive with iterative procedures. These equations are summarized in Tables 3.3, 3.4, and 11.2.

For mixtures, the theoretical combining rules serve as the basis for the combining rules used with several other equations of state. The theoretical interest of the virial equation has not been exhausted, and much work remains to be done in figuring the second and higher coefficients for pure substances and mixtures with realistic potential functions.

1.5. CUBIC EQUATIONS

1.5.1. Introduction

Equations of state that are explicit in pressure and of the third degree in volume are among the most successful of the simpler forms. Examination of the collection of early equations by Partington & Shilling (Table 1.1) reveals many examples of this type, including the celebrated one of van der Waals; many more have been proposed in the half century since. They are solvable analytically for the volume, but numerical methods often are more convenient. The ambiguity of multiple roots usually is not a serious problem.

1.5.2. Evaluation of the Parameters

An equation of at least the third degree is required to satisfy the condition of criticality meaningfully,

$$\left(\frac{\partial P}{\partial V}\right)_T = \left(\frac{\partial^2 P}{\partial V^2}\right)_T = 0, \tag{1.95}$$

and to allow representation of liquid and vapor phases. The numerical solution of cubic equations is illustrated in Example 1.3. The largest root corresponds to the vapor phase and the smallest to the liquid phase; the intermediate has no physical meaning. At the critical point all three roots are the same.

Together with the original equation applied at the critical point, the criticality conditions permit evaluation of three independent parameters in terms of the critical properties. For the van der Waals equation the parameters are given by

$$a = 27R^2T_c^2/64P_c = \Omega_a R^2T_c^2/P_c, \tag{1.96}$$

$$b = RT_c/8P_c = \Omega_b RT_c/P_c. \tag{1.97}$$

The third parameter is R. In many cases, unfortunately, the critical properties are not completely adequate for defining the parameters of equations of state. Somewhat better results often can be obtained by letting the numerical coefficients Ω_a and Ω_b vary slightly from substance to substance, but of course with loss of generality.

Since equations of state can be used to represent densities of liquid and vapor phases, vapor pressures, fugacities, and deviations of thermal properties from ideality, the parameters can be adjusted to fit data of some but usually not all of these categories. Such a task may be easier with equations that have larger numbers of parameters. Thus cubic equations may be made to represent vapor pressures accurately, but usually not enthalpy deviations simultaneously.

Several useful cubic equations have only two basic parameters—for example, the Redlich-Kwong, Soave, and Peng-Robinson—but others with three or more parameters have been proposed for better representation of certain kinds of data. One that is based on critical compressibilities is said by Horvath & Lin (1977) to be particularly accurate in the saturation region of normal and polar substances. A modification of the three-parameter equation of Clausius (1880) was made by Elshayal & Lu (1973); they expressed the parameters as functions of temperature and correlated them with data on saturation volumes and the condition that fugacities of the two phases are equal at equilibrium.

1.5.3. General Forms of Cubic Equations

Several recent papers have analyzed generalized forms of cubic EOS of which some of the commonly used ones are special cases. Martin (1979) began with the form

$$P = \frac{RT}{V} - \frac{\alpha(T) - \delta(T)/V}{(V+\beta)(V+\gamma)} \tag{1.98}$$

and concluded that possibly the best form was

$$P = \frac{RT}{V-b} - \frac{a}{(V+c)^2} \tag{1.99}$$

or in terms of reduced properties,

$$P_r = \frac{T_r}{z_c V_r - B} - \frac{0.4219}{T_r^n(z_c V_r + 0.125 - B)^2}. \tag{1.100}$$

Two parameters, B and n, are involved. B is related to the critical compressibility. It is selected to give a good fit in the desired density range. At twice the critical density,

$$B = 0.857z_c - 0.1674. \tag{1.101}$$

At 1.5 times the critical density,

$$B = 0.752z_c - 0.1520. \tag{1.102}$$

The constant, n, is selected to give a good fit to the slope of the reduced vapor pressure curve at the critical point. The reduced vapor pressure curve may be estimated adequately from the normal boiling point. Martin describes the procedure. Some typical values are:

Substance	z_c	B	n
Argon	0.291	0.082	0.55
Ethylene	0.281	0.074	0.62
Perfluorocyclobutane	0.278	0.071	1.20
Isopentane	0.270	0.064	1.10
Trifluoromethane	0.259	0.055	1.00
Ammonia	0.242	0.040	1.10

Since more data are used in the evaluation of parameters, the fit of PVT data with the Martin equation usually is better than with the Soave or the other two parameter equations. For some binary mixtures this equation has been found quite satisfactory by Joffe (1981), but it has not been completely generalized for mixtures.

A five-parameter equation was studied by Abbott (1979). He began with

$$P = \frac{RT(V^2 + \alpha V + \beta)}{V^3 + \lambda V^2 + \mu V + \nu}, \tag{1.103}$$

and rearranged it to

$$P = \frac{RT}{V-b} - \frac{\theta(V-\eta)}{(V-b)(V^2+\delta V+\varepsilon)} \tag{1.104}$$

where θ is a function of temperature and possibly of the acentric factor; the term $V-b$ is included so that the steep gradient of the liquid density can be represented. By appropriate choice of the five parameters, most of the common cubic equations can be obtained, and those equations are commented on in this paper. Some of the features and limitations of cubic equations are brought out by a study of the critical isotherm.

Other three-parameter equations were investigated by Harmens & Knapp (1980) and by Schmidt & Wenzel (1980). Harmens & Knapp expressed the parameters of the equation

$$P = \frac{RT}{V-b} - \frac{a}{V^2 + bcV + b^2(c-1)}, \tag{1.105}$$

in terms of the critical parameters, including the critical compressibility and the acentric factor, for which the equations are shown in Table 1.14. For 831 data of pure substances, the H-K and S-W equations had about half the error of the API-Soave equation. Both groups of investigators say that application to mixtures is being worked out.

All the cubic equations cited so far retain the van der Waals repulsion term, $RT/(V-b)$, but use a different attraction term corresponding to van der Waals a/V^2. Recently a converse process has received attention: The attraction terms of individual equations are retained, but the term $RT/(V-b)$ is replaced with one deduced from a "perturbed hard sphere" model. The parameters of the resulting EOS still are basically expressed in terms of the critical properties. Some details of this work are presented in Section 1.7.5.

1.5.4. Roots of Cubic Equations

When a cubic EOS in z or V has three positive real roots, the largest one is that of the vapor, the smallest one that of the liquid, and the intermediate one without physical significance. During a series of calculations, as in a distillation problem, it is necessary to decide immediately to which phase a found root corresponds. This decision can be made with an empirical criterion devised by Poling et al. (1981) in terms of the isothermal compressibility coefficient. For example, for the equation

$$P = RT/(V-b) - a/\sqrt{T}\ V(V+b), \tag{1.106}$$

this coefficient is

$$\beta = -\frac{1}{V}\left(\frac{\partial V}{\partial P}\right)_T = \frac{V(V^2-b^2)^2}{RTV^2(V+b)^2 - a(2V+b)(V-b)^2/\sqrt{T}}. \tag{1.107}$$

With the pressure in atmospheres, the Poling criteria for deciding whether a calculated saturated volume is that of a liquid or vapor phase are:

for a liquid phase, $\beta < 0.005/\text{atm}$,

for a vapor phase, $0.9/P < \beta < 3/P$.

Although these rules are not rigorous, they are claimed to be valid over wide ranges of temperature and pressure. The results of Example 1.13 and Problem 1.24 show that they are valid except at one point near the critical condition.

1.5.5. The Redlich-Kwong Equation

At the time of its introduction (Redlich & Kwong 1949), this equation was a considerable improvement over current equations of relatively simple forms. To a large extent it has retained its popularity over the past three decades, with various modifications. The equation

$$\left(P + \frac{a}{T^{0.5}V(V+b)}\right)(V-b) = RT \tag{1.108}$$

clearly is modeled on that of van der Waals (1873) or that of Clausius (1880), following them by some seventy years. Recently Redlich stated in his book (1976) that they had had no particular theoretical basis for their equation, so it is to be regarded as an arbitrary but inspired empirical modification of its predecessors. Soon after van der Waals's publication, it had been recognized that the parameter a particularly was dependent on temperature, so inclusion of the term $T^{0.5}$ does have a precedent. Other quadratic terms in volume had been proposed earlier for the denominator of the attraction, but none exactly like Redlich & Kwong's. Comparisons of several such equations are made in Figures 1.17 and 1.18.

The Parameters

Various forms of the equation and related quantities are summarized in Table 1.9. By application of the criticality conditions, the parameters a and b are found in terms of the critical properties, as shown in Example 1.12. Thus:

$$a = \Omega_a R^2 T_c^{2.5}/P_c, \tag{1.109}$$

$$b = \Omega_b RT_c/P_c, \tag{1.110}$$

$$\Omega_a = 1/9(2^{1/3}-1) = 0.427480, \tag{1.111}$$

$$\Omega_b = (2^{1/3}-1)/3 = 0.086640. \tag{1.112}$$

Usually the fit of data to the equation of state is improved by allowing the coefficients Ω_a and Ω_b to vary from substance to substance. Example 1.15A shows the sensitivity of the equation to these coefficients.

In recent times these coefficients have been correlated in terms of reduced temperature and acentric factor. Perhaps the most widely used of such correlations is that of Soave, which is discussed in the next section. In other work, Djordjevic et al. (1977, 1980) evaluated second virial coefficients of some polar substances from the Redlich-Kwong equation modified in this way. Haman et al. (1977) represented Ω_a and Ω_b in terms of reduced temperatures by equations with four constants that are different for each of the thirteen substances investigated. Somewhat simpler equations were devised by Kato et al. (1976), but again the constants were specific to each substance.

The effect of temperature on Ω_b was noted by Medani & Hasan (1978) to be much greater than on Ω_a. Accordingly, they kept the latter constant and found Ω_b as a function of acentric factor and reduced temperature; but since they covered a highly limited data base, their results cannot be used generally. In another approach, Raimondi (1980) wrote the Redlich-Kwong term

$$A = \Omega_a P_r/T_r^{2.5} = a\ P/R^2T^{2.5} \tag{1.113}$$

as

$$A = \alpha\Omega_a P_r/T_r^2, \tag{1.114}$$

and found α as a function of reduced temperature and acentric factor. The same principle had been employed by Soave. Similar treatment by Simonet & Behar (1976) led to an equation that could represent hydrocarbon data to 700 atm over a reduced temperature range of 0.5–1.5. Interestingly, Redlich (1975) largely abandoned his earlier equation and introduced the critical compressibility into a generalized cubic

$$Pz_c = T/(V-b) - Q(T)/(V^2+fV+g). \tag{1.115}$$

Relations between b, f, g, and Q were developed in the paper.

Clearly, the Redlich-Kwong equation has received much attention, although it is being recognized that the process of modifying that equation has reached the point of diminishing returns, and new approaches to improving equations of state

Table 1.9 Redlich-Kwong Equation of State

Standard form:

$$P = \frac{RT}{V-b} - \frac{a}{\sqrt{T}\,V(V+b)}. \qquad (1)$$

Parameters: (see Example 1.12):

$$a = \Omega_a R^2 T_c^{2.5}/P_c = 0.42748\, R^2 T_c^{2.5}/P_c, \qquad (2)$$

$$b = \Omega_b R T_c/P_c = 0.08664 R T_c/P_c, \qquad (3)$$

$$A = aP/R^2T^{2.5} = 0.42748 P_r/T_r^{2.5}, \qquad (4)$$

$$B = bP/RT = 0.08664 P_r/T_r. \qquad (5)$$

Polynomial forms:

$$V^3 - \frac{RT}{P}V^2 + \frac{1}{P}\left(\frac{a}{\sqrt{T}} - bRT - Pb^2\right)V - \frac{ab}{P\sqrt{T}} = 0, \qquad (6)$$

$$z^3 - z^2 + (A - B - B^2)z - AB = 0, \qquad (7)$$

$$z^3 - z^2 + \frac{P_r}{T_r}\left[\frac{0.42748}{T_r^{1.5}} - 0.08664 - 0.007506\frac{P_r}{T_r}\right]z - 0.03704\frac{P_r^2}{T_r^{3.5}} = 0. \qquad (8)$$

Reduced form:

$$P_r = \frac{3T_r}{V_r - 3\Omega_b} - \frac{9\Omega_a}{T_r^{0.5}V_r(V_r + 3\Omega_b)}. \qquad (9)$$

Compressibility relations:

$$h = \frac{b}{V} = \frac{bP}{zRT} = \frac{0.08664RT_c}{VP_c} = \frac{0.08664\,P_r}{z\,T_r}, \qquad (10)$$

$$z = \frac{1}{1-h} - \frac{a}{bRT^{1.5}}\left(\frac{h}{1+h}\right) = \frac{1}{1-h} - \frac{4.934}{T_r^{1.5}}\left(\frac{h}{1+h}\right), \qquad (11)$$

$$z = \frac{V}{V-b} - \frac{a}{RT^{1.5}(V+b)}. \qquad (12)$$

Mixtures:

$$a = \Sigma\Sigma y_i y_j a_{ij} = y_1^2 a_{11} + y_2^2 a_{22} + \cdots + 2(y_1y_2a_{12} + y_1y_3a_{13} + \cdots + y_2y_3a_{23} + \cdots), \qquad (13)$$

$$b = \Sigma y_i b_i, \qquad (14)$$

$$A = \Sigma\Sigma y_i y_j A_{ij}, \qquad (15)$$

$$B = \Sigma y_i B_i. \qquad (16)$$

Cross-parameters:

$$a_{ij} = \sqrt{a_i a_j} \text{ (Redlich \& Kwong's original rule)}, \qquad (17)$$

$$a_{ij} = (1 - c_{ij})\sqrt{a_i a_j} \text{ (Zudkevich \& Joffe 1970)}, \qquad (18)$$

$$a_{ij} = \frac{\Omega_a R(V_{ci}^{1/3} + V_{cj}^{1/3})^3[(1-k_{ij})\sqrt{T_{ci}T_{cj}}]^{1.5}}{8[0.291 - 0.04(\omega_i + \omega_j)]} \text{ (Prausnitz \& Chueh 1968)}. \qquad (19)$$

Fugacity relations are in Tables 3.3 and 3.4. Residual function equations are in Table 11.3.

may need to be developed. A review of the literature on the R-K equation by Horvath (1974) cites 112 references, and many more have appeared since that publication date. A number of adjustments of the parameters of the R-K equation that have been reported is summarized in Table 1.10.

With all the changes that have been proposed over the years, there may be a question whether a given equation is properly called a modified R-K or a modified van der Waals or something new, but the term modified R-K seems to be applied to an equation with an attraction term of the form

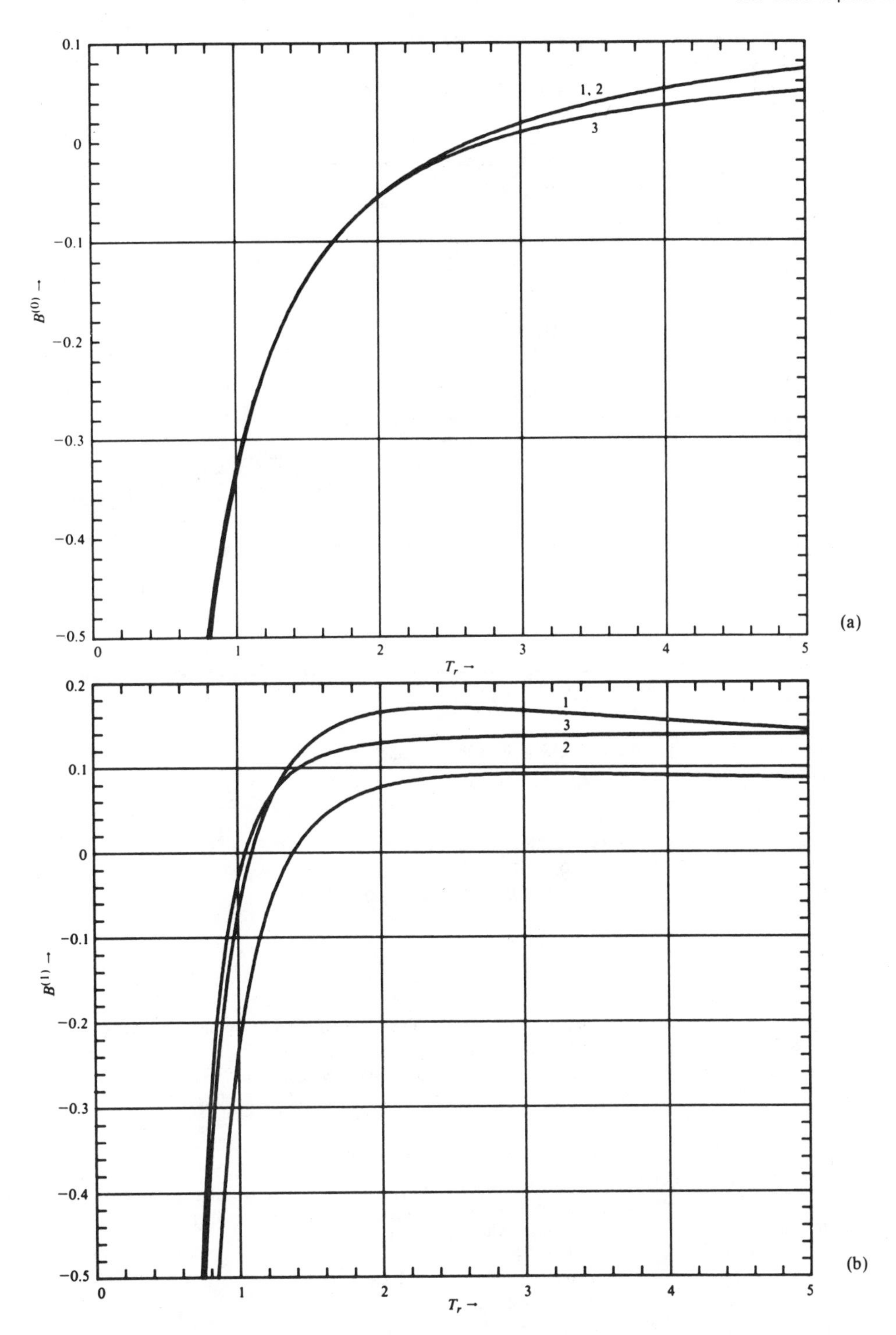

Figure 1.16 Three correlations of the second virial coefficients of nonpolar gases in the form $B = (RT_c/P_c)[B^{(0)} + \omega B^{(1)}]$: (a) $B^{(0)}$ as a function of T_r; (b) $B^{(1)}$ as a function of T_r.

1. Pitzer & Curl (1958):

$B^{(0)} = 0.1445 - 0.33/T_r - 0.1385/T_r^2 - 0.0121/T_r^3$,

$B^{(1)} = 0.073 + 0.46/T_r - 0.5/T_r^2 - 0.097/T_r^3 - 0.0073/T_r^8$.

2. Abbott, cited in Smith & Van Ness (1975):

$B^{(0)} = 0.083 - 0.422/T_r^{1.6}$, $B^{(1)} = 0.139 - 0.172/T_r^{4.2}$.

3. Tsonopoulos (1974):

$B^{(0)} = 0.1445 - 0.33/T_r - 0.1385/T_r^2 - 0.0121/T_r^3 - 0.000607/T_r^8$,

$B^{(1)} = 0.0637 + 0.331/T_r^2 - 0.423/T_r^3 - 0.008/T_r^8$.

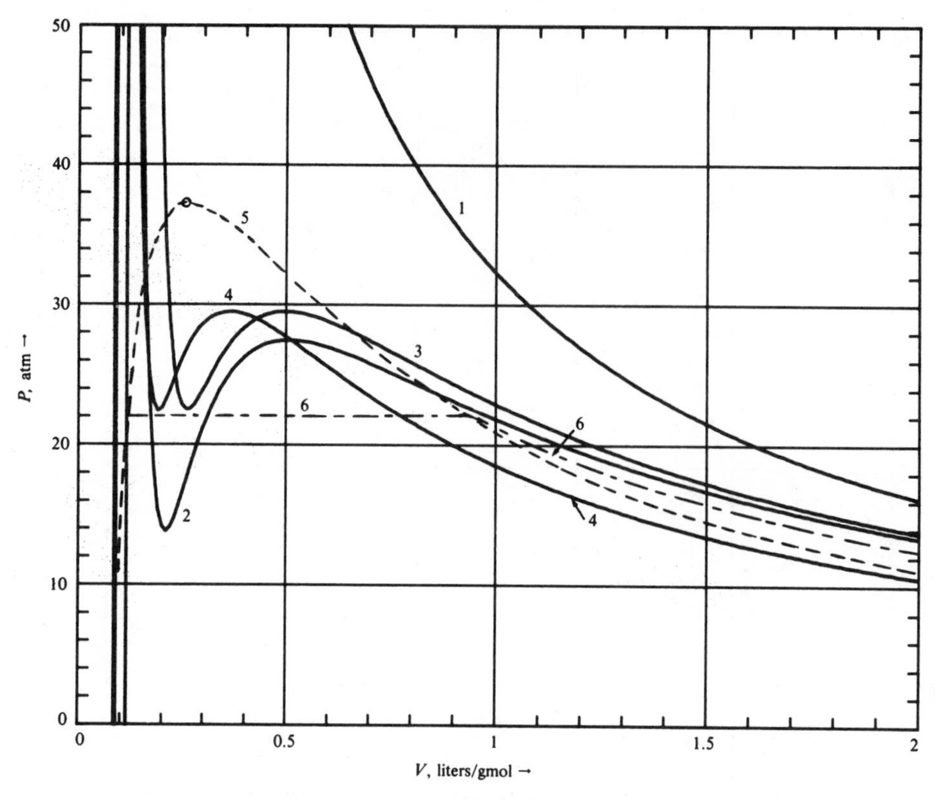

Figure 1.17 Saturation curve and 394.2 K isotherm of *n*-butane, and comparisons with predictions from the ideal, van der Waals and Redlich-Kwong equations of state.

Curve 1. Ideal gas: $P = 0.0825\ T/V$.
Curve 2. Redlich-Kwong: $P = 0.08205\ T/(V - 0.0806) - 286.1/[T^{0.5}V(V + 0.0806)]$.
Curve 3. van der Waals: $P = 0.08205\ T/(V - 0.1163) - 13.693/V^2$.
Curve 4. van der Waals with $R = 8P_cV_c/3T_c = 0.060, a = 7.322, b = 0.0850$.
Curve 5. Saturation curve.
Curve 6. True isotherm.

$a(T_r, \omega, \dots)/(V(V + b)$. Names are a matter of convenience, and since Redlich died in 1978, it will not hurt his feelings if we use other names when substantial changes in the form of the EOS are made.

Forms of the Equation

In addition to the pressure-explicit form of the equation, Table 1.9 also shows polynomials in volume and compressibility. The reduced form is useful for comparison with other equations. The roots of the polynomials may be found by the method of Example 1.3. The particular form of equation to be used depends on which pair of the three variables is specified. The pair of equations, 10 and 11 of Table 1.9, for h and z was proposed by Redlich & Kwong; direct iteration almost always works with them, or the Wegstein method can be used to force convergence. The roots of the polynomial forms are readily solved by the Newton-Raphson method, starting with unity for the vapor compressibility and zero for the liquid compressibility, or all three roots may be found by Cardan's method. For H-P calculators there is a program for finding the real and complex roots of polynomials. Example 1.13 utilizes this program for finding the roots of the equation for propylene over a range of saturation pressures.

The Redlich-Kwong equation is the basis for a series of graphs of compressibility factor and residual properties by Edmister (Sept.-Dec. 1968, Sept. 1971, and Jan. 1972) which may be convenient for spot checking.

Mixtures

The original combining rules for the R-K parameters of mixtures were the same as those for the van der Waals equation,

$$a = \Sigma\Sigma y_i y_j a_{ij}, \tag{1.116}$$

$$b = \Sigma y_i b_i, \tag{1.116a}$$

with the cross parameter taken to be

$$a_{ij} = (a_i a_j)^{0.5}. \tag{1.117}$$

Later two major improvements were made by the introduction of cross-pseudocritical properties of the Lorentz-Berthelot type for the evaluation of a_{ij} and the introduction of binary interaction parameters. These concepts were discussed in Sections 1.3.7 and 1.3.8. Thus a_{ij} is evaluated from the definition of a_i, but in terms of the following scheme:

Table 1.10. Proposed Adjustments of the Redlich-Kwong Parameters at $T_r < 1$ (Wichterle 1978c)

Author	*Parameters adjusted*	*Adjusting conditions*	*Quantities sacrificed*	*Note*
Wilson 1964, 1966	a	$f^G_{calc} = f^G_{exp}$	V^G, V^L	$\Omega_b = 0.0876$; f^G_{exp} from generalized fugacity charts; a is generalized
Soave 1972	α	$f^L_{calc} = f^G_{calc}$	V^G, V^L	$\Omega_b = 0.08664$; $\alpha = a/T^{0.5}$ is generalized
Prausnitz and Chueh 1968	Ω^L_a, Ω^L_b Ω^G_a, Ω^G_b	$V^L_{calc} = V^L_{exp}$ $V^G_{calc} = V^G_{exp}$	$f^L_{calc} = f^G_{calc}$	temperature independent parameters are optimized for saturated densities between normal boiling point and critical point.
Zudkevitch and Joffe, 1970	Ω_a, Ω_b	$V^L_{calc} = V^L_{exp}$ $f^L_{calc} = f^L_{exp}$	V^G	f^L_{exp} from generalized tables
Joffe et al. 1970	Ω_a, Ω_b	$V^L_{calc} = V^L_{exp}$	V^G	
Chang and Lu 1970		$f^G_{calc} = f^L_{calc}$		
Hirata et al. 1975				
Harmens 1975				
Hamam et al. 1977	Ω_a, Ω_b	$V^L_{calc} = V^L_{exp}$ $f^G_{calc} = f^L_{calc}$	V^G	Ω's are generalized
Kato, Chung and Lu 1976a	Ω_a, Ω_b	$V^L_{calc} = V^L_{exp}$ $f^L_{calc} = f^G_{calc}$	V^G	adjustment of apparent critical properties
Vogl and Hall 1970	Ω_a/Ω_a^{crit} Ω_b/Ω_b^{crit}	optimalization of the whole isotherm	V^G, V^L, $f^L = f^G$	reduced parameters are generalized
Chaudron et al. 1973	Ω_a, Ω_b	optimalization of the whole isotherm	V^G, V^L, $f^L = f^G$	excellent generalization
Simon et al. 1976				
Djordjević et al. 1977	Ω_a, Ω_b	optimalization of the whole	V^G, V^L, $f^L = f^G$	adjusted Ω's are not in any relation to reduced temperature
Hederer et al. 1976	A, B, α	$V^L_{calc} = V^L_{exp}$ $f^G_{calc} = f^L_{calc}$ $\Delta H^{vap}_{exp} = \Delta H^{vap}_{calc}$	V^G	A, b, α are known functions of normal boiling point and are independent of temperature; adjustment from arbitrary two vapor pressure points and liquid density
Wenzel and Peter 1971	a, b	$V^L_{calc} = V^L_{exp}$ $f^G_{calc} = f^L_{calc}$	V^G	parameters adjusted from liquid density and vapor pressure at the system temperature

$$z_{cij} = 0.5(z_{ci} + z_{cj}),\ V_{cij}^{1/3} = 0.5\ (V_{ci}^{1/3} + V_{cj}^{1/3}),$$
$$T_{cij} = (1 - k_{ij})\sqrt{T_{ci}T_{cj}},\ \ P_{cij} = z_{cij}RT_{cij}/V_{cij},$$
$$a_{ij} = \Omega_a RT_{cij}^{2.5}\ /P_{cij}. \qquad (1.118)$$

Values of k_{ij} for about 100 pairs of substances are given in Table E.3. Other data are cited in Section 1.3.8. Such a datum is used in Example 1.15.

Several methods of obtaining the parameters of a binary mixture are tried in Example 1.14, with the conclusion that the true critical properties are not the proper ones for evaluating the behavior of this mixture, a conclusion that was known early to be true generally for all mixtures.

An extensive application of the R-K equation to reservoir fluids was made by Yarborough (1979). He used a considerably modified equation, with Ω_a and Ω_b as functions of reduced temperature and acentric factor and with binary interaction parameters developed for the cases in hand.

Applicability of the Redlich-Kwong Equation

In one guise or another the Redlich-Kwong equation is still of interest, although for serious applications it has been largely replaced by other equations of the same type, such as the Soave or Peng-Robinson, even though they require more information about the components of the mixtures, such as acentric factor and binary interaction parameters for highest accuracy. The R-K equation is not at all satisfactory for the liquid phase, so it cannot be used by itself for calculating vapor-liquid equilibria; but in combination with separate liquid-phase correlations, it is part of the successful Chao-Seader procedure for evaluating vapor-liquid equilibria (Chapter 6).

A few comparisons with data and other EOS are shown in Figures 1.17–1.20(a). Since the arithmetic is simpler, the R-K equation continues in limited use and is roughly comparable with the B-truncated virial equation of state. Both are satisfactory for gas-phase fugacity calculations at reduced pressures less than about one-half the reduced temperatures, and for evaluation of the deviations of enthalpy and entropy and other properties from ideal behavior, particularly in situations where many such evaluations must be made, as in distillation calculations. Many comparisons of the R-K equation with other EOS, including some complex forms, appear in the literature, some of them quite favorable to the R-K form. Details of these comparisons are given at the end of the chapter.

Example 1.11 The Virial Expansion of the Cubic Equation of State, $P = RT/(V-b) - a/(V^2 + cV + d)$

Relations will be found between the virial coefficients and the parameters a, b, c, and d of the given equation. The equation is rearranged to

$$z - 1 = \frac{b}{V-b} - \frac{aV}{\mathrm{RT}(V^2 + cV + d)}.$$

Table 1.8 gives certain limiting expressions for the virial coefficients, B, C, D, ... Thus

$$B = \lim_{V\to\infty} (z-1)V$$

$$= \lim \frac{bV}{V-b} - \frac{aV^2}{RT(V^2 + cV + d)}$$

$$= b - \frac{a}{RT}.$$

Similarly,

$$C = \lim_{V\to\infty} [(z-1)V - B]V = b^2 + \frac{ac}{RT},$$

and

$$D = \lim_{V\to\infty} [(z-1)V^2 - BV - C]V = b^3 + \frac{a(d-c^2)}{RT}.$$

For the van der Waals equation, $c = d = 0$, so that $B = b - a/RT$, $C = b^2$ and $D = d^3$, etc.

Accordingly, the equivalent virial equation is

$$z = 1 + \frac{b}{V-b} - \frac{aV}{RT(V^2 + cV + d)}$$

$$= 1 + \left(b - \frac{a}{RT}\right)\frac{1}{V} + \left(b^2 + \frac{ac}{RT}\right)\frac{1}{V^2}$$

$$+ \left(b^3 + \frac{a(d-c)^2}{RT}\right)\frac{1}{V^3} + \ldots,$$

and that of the vdW

$$z = 1 + \frac{b}{V-b} - \frac{a}{RTV}$$

$$= 1 + \left(b - \frac{a}{RT}\right)\frac{1}{V} + \left(\frac{b}{V}\right)^2 + \left(\frac{c}{V}\right)^3 + \ldots.$$

Example 1.12 Derivation of Equations of the Parameters of the R-K Equation in Terms of the Critical Properties

The polynomial form of the R-K equation is

$$V^3 - \frac{RT}{P}V^2 + \left(\frac{a}{PT^{0.5}} - \frac{bRT}{P} - b^2\right)V - \frac{ab}{PT^{0.5}} = 0. \quad (1)$$

Relations between the parameters and the critical properties are found by comparing coefficients of this equation at the critical point with the form of the equation derived on the observation that the three roots are equal at the critical point, which is

$$(V - V_c)^3 = V^3 - 3V_cV^2 + 3V_c^2V - V_c^3 = 0. \quad (2)$$

Thus:

$$\frac{RT_c}{P_c} = 3V_c, \quad (3)$$

$$\frac{a}{PT_c^{0.5}} - \frac{bRT_c}{P_c} - b^2 = 3V_c^2, \quad (4)$$

$$\frac{a}{P_cT_c^{0.5}} = \frac{V_c^3}{b}. \quad (5)$$

The result of eliminating a between Eqs. 4 and 5 is

$$b^3 + 3V_cb^2 + 3V_c^2b - V_c^3 = 0. \quad (6)$$

The roots of this equation will be found analytically with the standard procedure reproduced here from *The CRC Handbook of Mathematical Sciences.* The relation between that notation and that of Eq. 6 is:

$$b = y, \quad (7)$$

$$p = 3V_c, \quad (8)$$

$$q = 3V_c^2, \quad (9)$$

$$r = -V_c^3. \quad (10)$$

Accordingly,

$$\alpha = (9V_c^2 - 9V_c^2)/3 = 0, \quad (11)$$

Example 1.12 *(continued)*

$$\beta = \frac{1}{27}[2(3V_c)^3 - 9(3V_c)(3V_c^2) + 27(-V_c^3)] = -2V_c^3, \tag{12}$$

$$D = \sqrt{\frac{\beta^2}{4} + \frac{\alpha^3}{27}} = V_c^3, \tag{13}$$

$$A = (V_c^3 + V_c^3)^{1/3} = 2^{1/3}V_c, \tag{14}$$

$$B = 0. \tag{15}$$

Therefore,

$$b = x - \frac{p}{3} = A + B - \frac{3V_c}{3} = (2^{1/3} - 1)V_c$$

$$= (2^{1/3} - 1)\frac{RT_c}{3P_c} = 0.086640\frac{RT_c}{P_c}. \tag{16}$$

To find a, eliminate V_c between Eq. 3 and Eq. 5; then substitute for b from Eq. 16:

$$a = \frac{P_cT_c^{0.5}}{b}V_c^3 = \frac{P_cT_c^{0.5}}{b}\left[\frac{RT_c}{3P_c}\right]^3 = \frac{R^2T_c^{2.5}}{9(2^{1/3} - 1)P_c}$$

$$= 0.427480\frac{R^2T_c^{2.5}}{P_c}. \tag{17}$$

Cubic Equations:

A cubic equation, $y^3 + py^2 + qy + r = 0$ may be reduced to the form

$$x^3 + \alpha x + \beta = 0$$

by substituting for y the value, $x - p/3$. Here

$$\alpha = \frac{1}{3}(3q - p^2) \text{ and } \beta = \frac{1}{27}(2p^3 - 9pq + 27r).$$

For solution, let

$$A = \sqrt[3]{-\frac{\beta}{2} + \sqrt{\frac{\beta^2}{4} + \frac{\alpha^3}{27}}}, \quad -B = \sqrt[3]{+\frac{\beta}{2} + \sqrt{\frac{\beta^2}{4} + \frac{\alpha^3}{27}}};$$

then the values of x will be given by,

$$x = A + B, \quad -\frac{A+B}{2} + \frac{A-B}{2}\sqrt{-3},$$

$$-\frac{A+B}{2} - \frac{A-B}{2}\sqrt{-3}.$$

If p, q, r are real, then:

If $\frac{\beta^2}{4} + \frac{\alpha^3}{27} > 0$, there will be one real root and two conjugate complex roots.

If $\frac{\beta^2}{4} + \frac{\alpha^3}{27} = 0$, there will be three real roots of which at least two are equal.

If $\frac{\beta^2}{4} + \frac{\alpha^3}{27} < 0$, there will be three real and unequal roots.

Example 1.13 Compressibilities of Propylene along the Saturation Curve from the Redlich-Kwong Equation

Although the Antoine equation is not strictly accurate over the entire range, it will be used nevertheless for illustrative purposes. The vapor pressure equation is

$$P = \frac{1}{760}\exp[15.7027 - 1807.53/(T - 26.15)].$$

Since $T_c = 365$ K and $P_c = 46.5$ atm, the parameters of the R-K equation are:

$$A = 0.42748(P/46.5)(365/T)^{2.5} = 23398.9P/T^{2.5},$$

$$B = 0.08664(P/46.5)(365/T) = 0.6801P/T.$$

The coefficients and the corresponding roots of the R-K polynomial are tabulated:

$$z^3 - z^2 + (A - B - B^2)z - AB = 0.$$

The true vapor pressure at 360 K is 41.5 atm instead of the predicted 38.68. Thus the prediction places the substance in the superheated region where the cubic equation has only one real root, as shown in the tabulation, so the vapor compressibility is missed. More accurate but more complicated vapor pressure equations could be used.

T °K	P atm	$A - B - B^2$	AB	z_v	z_x	z_L
300	11.81	0.14978	0.00475	0.8255	0.1304	0.0441
320	18.51	0.19553	0.00930	0.7583	0.1692	0.0725
340	27.39	0.24288	0.01647	0.6773	0.2027	0.1200
350	32.72	0.26645	0.02124	0.6312	0.2034	0.1655
360*	38.68*	0.28962	0.02690	0.5814	[0.2093 ± 0.0494j]	
370	45.27	0.31212	0.03347	0.5307	[0.2346 ± 0.0895j]	

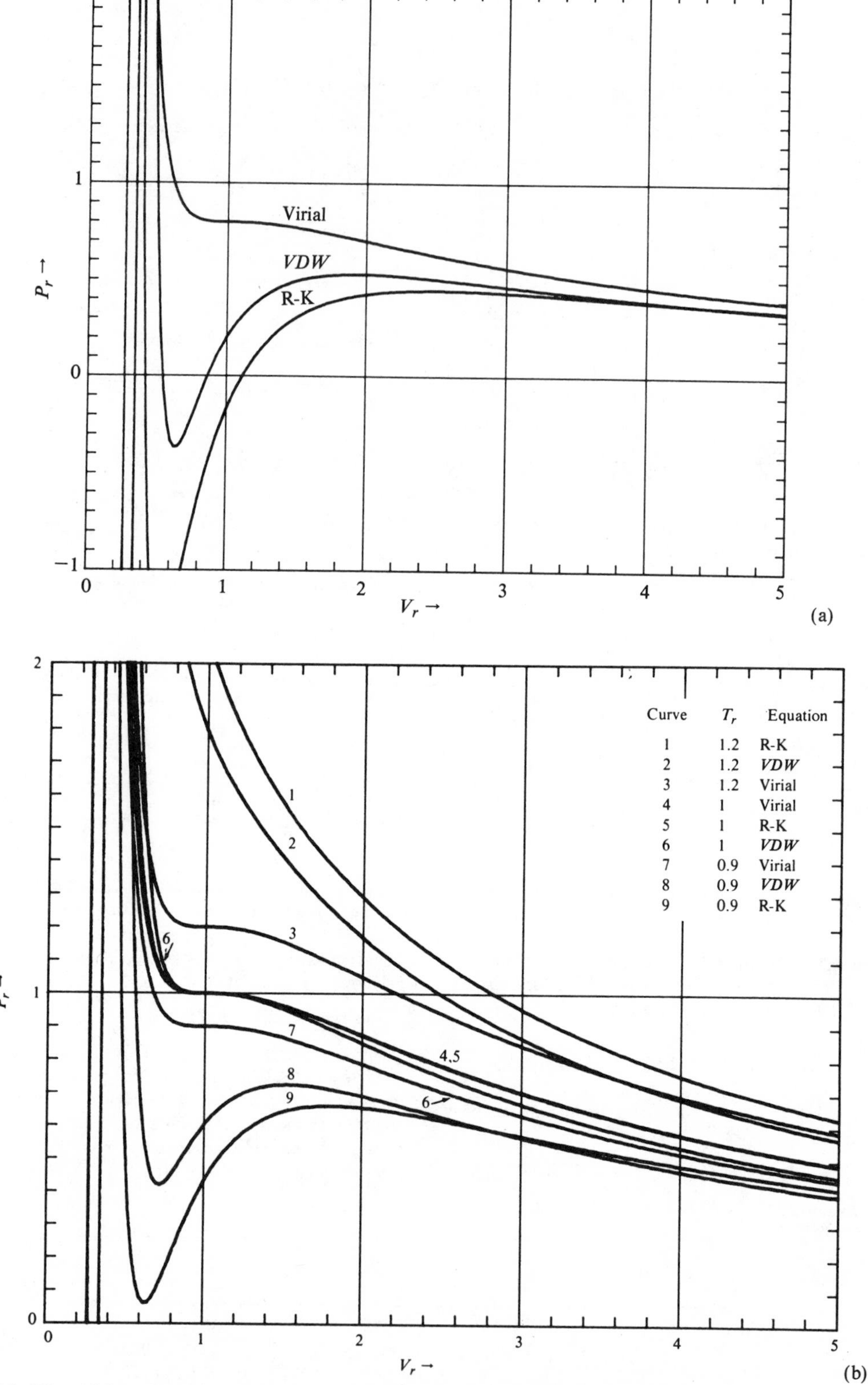

Figure 1.18 Plots of the reduced forms of the van der Waals, virial, and Redlich-Kwong equations of state: (a) $T_r = 0.8$:

van der Waals: $P_r = 8T_r/(3V_r - 1) - 3/V_r^2$.

Virial: $P_r = 3T_r(1/V_r - 1/V_r^2 + 1/3V_r^3)$. Redlich-Kwong:

$$P_r = \frac{3T_r}{V_r - 0.2599} - \frac{1}{0.2599T_r^{0.5}V_r(V_r + 0.2599)};$$

(b) $T_r = 0.9$, 1.0, and 1.2.

Example 1.14 Evaluation of the Compressibility of a Mixture Using Redlich-Kwong Parameters Obtained by Several Methods

For an equimolal mixture of carbon dioxide and propylene at 30 C and 25.5 atm, the experimental compressibility is 0.737. The true critical properties are 70.7 atm and 319.8 K. $R = 82.05$.

Method 1: Take $a_{ij} = \sqrt{a_1 a_2}$.

Method 2: Use the true critical properties to find a and b of the mixture.

Method 3: Use the pseudocritical properties figured with Kay's Rule to find a and b of the mixture.

Method 4: Evaluate the cross-parameter a_{ij} from the cross-critical properties by the rules:

$$z_{cij} = 0.5(z_{ci} + z_{cj}), \qquad V_{cij}^{1/3} = 0.5(V_{ci}^{1/3} + V_{cj}^{1/3}),$$

$$T_{cij} = \sqrt{T_{ci}T_{cj}}, \qquad P_{cij} = z_{cij}RT_{cij}/V_{cij}.$$

	T_c, °K	P_c, atm	V_c ml/gm	z_c	$10^{-6}a$	b
CO_2	304.2	72.9	94.1	0.274	63.72	29.664
C_3H_6	364.9	45.45	182.4	0.275	161.05	57.074

method	T_c mix or (T_{cij})	P_c mix or (P_{cij})	$10^{-6}a_{ij}$	$10^{-6}a$	b	z
1	—	—	101.3	106.84	43.36	0.7425
2	319.8	70.7	—	74.47	32.16	0.8388
3	334.55	59.18	—	99.58	40.19	0.7646
4	(333.17)	(56.24)	103.7	108.04	43.36	0.73746
Experiment:						0.737

z is obtained with Eqs. 10 and 11 of Table 1.9.

Clearly, the use of true critical properties is unsatisfactory; but figuring the cross-parameter with the Lorentz-Berthelot–type rules (method 4) gives a good check of experiment.

Example 1.15 The Effect of the Binary Interaction Parameter on the Calculated Compressibility of an Equimolal Mixture of CO_2 and C_3H_6

Calculations are made for the conditions of Example 1.14. From Table E.3, $k_{ij} = 0.1$. Use the combining rules of Table 1.5.

Method 3:

$$T_{cij} = (1 - k_{ij})(T_{ci}T_{cj})^{0.5} = 0.9(333.17) = 299.85.$$

Accordingly,

$$P_{cij} = 0.9(56.24) = 50.62,$$

and

$$a_{ij} = 88.54\ (\text{E}+6),$$

$$a = 100.46\ (\text{E}+6),$$

$b = 43.36$, as before.

Eq. 10 and 11 of Table 1.9 become

$$h = \frac{43.36(25.5)}{82.05(299.85)z} = \frac{0.0449}{z},$$

$$z = \frac{1}{1-h} - \frac{100.46(\text{E}+6)}{43.36(82.05)(299.85)^{1.5}} \frac{h}{1+h}$$

$$= \frac{z}{z - 0.0449} - \frac{5.4384}{1 + 22.251z} = 0.7588,$$

which compares with the experimental value, 0.737, cited in Example 1.14.

Method 4:

Apply the correction for binary interaction directly to the parameter.

$$a_{ij} = (1 - k_{ij})(a_i a_j)^{0.5} = 96.16\ (\text{E}+6).$$

Therefore,

$$a = 104.27(\text{E}+6),$$

$b = 43.36$, as before,

$$z = \frac{z}{z - 0.0449} - \frac{5.4384}{1 + 22.251z}\left(\frac{104.27}{100.46}\right)$$

$$= \frac{z}{z - 0.0449} - \frac{5.6447}{1 + 22.251z} = 0.7421,$$

which compares more favorably with the experimental 0.737.

The superiority of Method 4 over Method 3 is not always as clear-cut as in this case.

Example 1.15A Sensitivity of the Redlich-Kwong Equation to Numerical Values of the Coefficients Ω_a and Ω_b

The largest and smallest values of the coefficients in the listing of twenty-one substances by Prausnitz & Chueh (1968, p. 20) are:

	Ω_a	Ω_b
Methane	0.4278	0.0867
n-Octane	0.4760	0.0968
Theoretical	0.42748	0.08664

Plots of the reduced equation, Eq. 9 of Table 1.9, are drawn for the two substances at reduced temperatures of 0.9 and 1.1. The theoretical curve coincides with that of methane on this scale, but there are substantial differences between the two hydrocarbons.

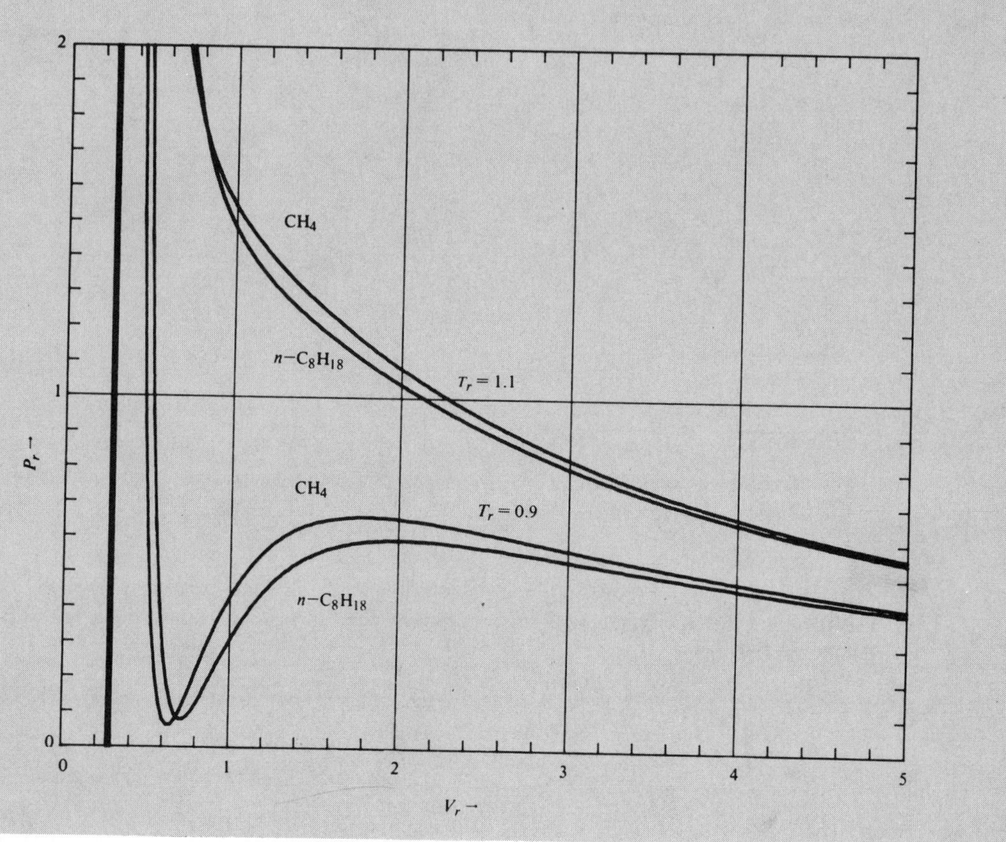

1.5.6. The Soave Equation

The temperature-dependent term $a/\sqrt{T}$ of the R-K equation was replaced by a function $a(T, \omega)$ involving the temperature and the acentric factor by Soave (1972) (Tables 1.11 and 1.12), so that the equation of state becomes

$$P = RT/(V-b) - a(T, \omega)/V(V+b). \tag{1.119}$$

The parameter $a(T, \omega)$ was formulated primarily to make the equation fit the vapor pressure data of hydrocarbons, with the result

$$a(T, \omega) = a\alpha = 0.42748(R^2T_c^2/P_c)\alpha, \tag{1.120}$$

where

$$\alpha^{0.5} = 1 + (1 - T_r^{0.5})(0.480 + 1.574\omega - 0.176\omega^2). \tag{1.121}$$

The coefficients of this term were modified by Graboski & Daubert (1978) to

$$\alpha^{0.5} = 1 + (1 - T_r^{0.5})(0.48508 + 1.55171\omega - 0.15613\omega^2). \tag{1.122}$$

For hydrogen, Graboski & Daubert (1979) wrote

$$\alpha^{0.5} = 1.096\exp(-0.15114\,T_r). \tag{1.123}$$

The effect of the acentric factor on the shape of reduced isotherms and a comparison with the Redlich-Kwong equation are shown in Example 1.16. Figure 1.19 is another comparison with other equations.

For some mixtures the mixing rules are the same as for the R-K equation, with the cross-parameter

$$a_{ij} = (1 - k_{ij})\sqrt{a_i a_j}. \tag{1.124}$$

For strongly polar substances such as water, alcohols, and others, further changes are proposed by Soave (1979b), making

$$\alpha = 1 + (1 - T_r)(m + nT_r), \tag{1.125}$$

Table 1.11. The Soave Equation (Graboski & Daubert coefficients)

Standard form:

$$P = \frac{RT}{V-b} - \frac{a\alpha}{V(V+b)} \quad (1)$$

Parameters:

$$a = 0.42747R^2T_c^2/P_c, \quad (2)$$

$$b = 0.08664RT_c/P_c, \quad (3)$$

$$\alpha = [1 + (0.48508 + 1.55171\omega - 0.15613\omega^2)(1 - T_r^{0.5})]^2, \quad (4)$$

$$\alpha = 1.202 \exp(-0.30288T_r)$$
for hydrogen (Graboski & Daubert 1979), (5)

$$A = a\alpha P/R^2T^2 = 0.42747\alpha P_r/T_r^2, \quad (6)$$

$$B = bP/RT = 0.08664P_r/T_r \quad (7)$$

Polynomial forms:

$$V^3 - \frac{RT}{P}V^2 + \frac{1}{P}(a\alpha - bRT - Pb^2)V - \frac{a\alpha b}{P} = 0, \quad (8)$$

$$z^3 - z^2 + (A - B - B^2)z - AB = 0. \quad (9)$$

Partly reduced form (see Example 1.16):

$$P_r = \frac{3T_r}{V_r - 0.2599} - \frac{3.8473\alpha}{V_r(V_r + 0.2599)}. \quad (10)$$

Mixtures:

$$a\alpha = \Sigma\Sigma y_i y_j (a\alpha)_{ij}, \quad (11)$$

$$b = \Sigma y_i b_i, \quad (12)$$

$$A = \Sigma\Sigma y_i y_j A_{ij}, \quad (13)$$

$$B = \Sigma y_i B_i. \quad (14)$$

Cross-parameters:

$$(a\alpha)_{ij} = (1 - k_{ij})\sqrt{(a\alpha)_i(a\alpha)_j}, \quad (15)$$

k_{ij} in Table 1.12; (16)

$k_{ij} = 0$ for hydrocarbon pairs and hydrogen. (17)

Fugacity coefficients are in Tables 3.3 and 3.4. Residual properties are in Table 11.3.

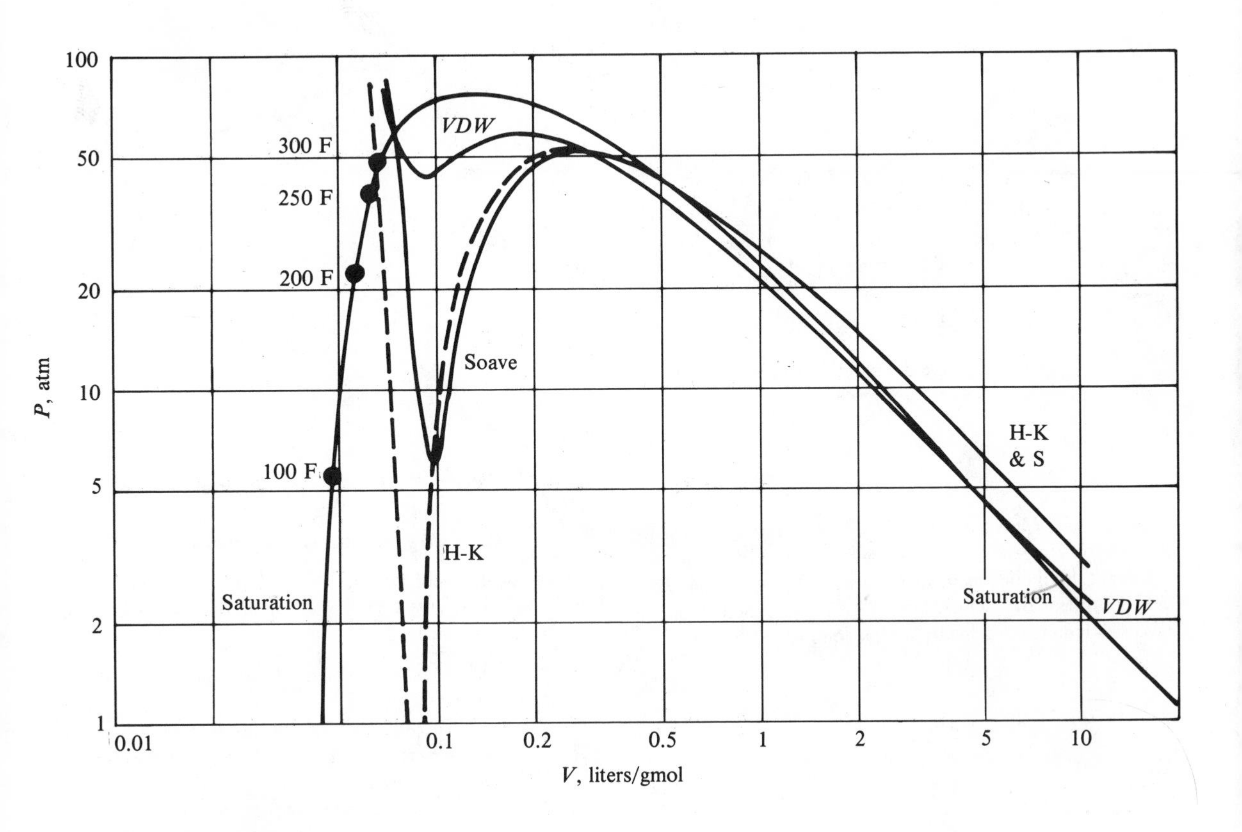

Figure 1.19 Plots of 250 F isotherms of sulfur dioxide with the Harmens-Knapp, Soave, and van der Waals equations, and comparison with the true saturation curve.

Note: The solution of Problem 3.26 gives these results for saturation pressures and volumes:

	P	V_{liq}	V_{vapor}
True values	41.1	0.063	0.516
van der Waals	54.6	0.0765	0.258
Soave	41.96	0.0756	0.5187
Harmens-Knapp	41.98	0.0597	0.4826

Table 1.12. Binary Interaction Parameters for the Soave Equation (Graboski & Daubert 1979)

1. Correlations in terms of absolute differences between solubility parameters of the hydrocarbon, δ_{HC}, and of the inorganic gas.

Gas	k_{ij}
H_2S	$0.0178 + 0.0244\lvert\delta_{HC} - 8.80\rvert$
CO_2	$0.1294 - 0.0292\lvert\delta_{HC} - 7.12\rvert - 0.0222\lvert\delta_{HC} - 7.12\rvert^2$
N_2	$-0.0836 + 0.1055\lvert\delta_{HC} - 4.44\rvert - 0.0100\lvert\delta_{HC} - 4.44\rvert^2$

For CO_2 and hydrocarbons, Ezekwe (1982) found

$\log k_{ij} = -0.8849 - 0.2145\omega_{HC}\lvert\delta_{HC} - 7.12\rvert$.

2. Direct values from vapor-liquid equilibrium measurements.

	H_2S	CO_2	N_2	CO
H_2S	—	0.102	0.140	—
CO_2	0.102	—	−0.022	−0.064
N_2	0.140	−0.022	—	0.046
CO	—	−0.064	0.046	—
Methane	0.0850	0.0973	0.0319	0.03
Ethane	0.0829	0.1346	0.0388	0.00
n-Propane	0.0831	0.1018	0.0807	0.02
2-Methylpropane	0.0523	0.1358	0.1357	—
n-Butane	0.0609	0.1474	0.1007	—
2-Methylbutane	—	0.1262	—	—
n-Pentane	0.0697	0.1278	—	—
n-Hexane	—	—	0.1444	—
n-Heptane	0.0737	0.1136	—	—
n-Octane	—	—	—	0.10
n-Nonane	0.0542	—	—	—
n-Decane	0.0464	0.1377	0.1293	—
Propylene	—	0.0914	—	—
Cyclohexane	—	0.1087	—	—
Isopropylcyclohexane	0.0562	—	—	0.01
Benzene	—	0.0810	0.2131	—
1,3,5-Trimethylbenzene	0.0282	—	—	—

where m and n are two adjustable parameters that must be derived from experimental vapor pressure data for each pure substance. Simplified methods for evaluating these parameters are described by Soave (1979b, 1980). Application to mixtures requires two binary interaction parameters:

$$a_{ij} = 0.5(a_i + a_j)(1 - C_{ij}), \qquad (1.126)$$

$$b_{ij} = 0.5(b_i + b_j)(1 - D_{ij}), \qquad (1.127)$$

$$a = \Sigma\Sigma y_i y_j a_i a_j, \qquad (1.128)$$

$$b = \Sigma\Sigma y_i y_j b_i b_j. \qquad (1.129)$$

No generalizations are made for these interaction parameters in this paper. However, some values were obtained for sour natural gas systems by Evelein & Moore (1979).

1.5.7. The Peng-Robinson Equation

Several goals were set by Peng & Robinson (1976) in developing a new two-parameter equation of state cubic in volume (see Table 1.13):

1. The parameters should be expressible in terms of P_c, T_c and acentric factor.
2. The model should result in improved performance in the vicinity of the critical point, particularly for calculations of z_c and liquid density.
3. The mixing rules should not employ more than one binary interaction parameter, and that should be independent of temperature, pressure, and composition.
4. The equation should be applicable to all calculations of all fluid properties in natural gas processes.

Example 1.16 Isotherms of the Reduced Redlich-Kwong, Soave, and Peng-Robinson Equations of State

The reduced form of the R-K equation is (from Table 1.9):

$$P_r = \frac{3T_r}{V_r - 0.2599} - \frac{3.8473}{T_r^{0.5} V_r (V_r + 0.2599)}. \tag{1}$$

For the Soave equation, substitution of

$$a = \Omega_a R^2 T_c^2 / P_c, \tag{2}$$

$$b = \Omega_b R T_c / P_c, \tag{3}$$

$$\alpha = [1 + (1 - T_r^{0.5})(0.48508 + 1.55171\omega - 0.15613\omega^2)]^2 \tag{4}$$

into

$$P = \frac{RT}{V - b} - \frac{a\alpha}{V(V + b)} \tag{5}$$

gives the result

$$P_r = \frac{3T_r}{V_r - 0.2599} - \frac{3.8473\alpha}{V_r (V_r + 0.2599)}. \tag{6}$$

This is not a completely reduced equation since α depends on the acentric factor. At $T_r = 1$, the isotherm is independent of ω, but the sensitivity of other isotherms to this factor is substantial, particularly below the critical temperature. At $\omega = 0$, the Soave and R-K equations practically coincide on the scale of the graph shown.

A similar development for the Peng-Robinson equation gives the result

$$P_r = \frac{3.2573\, T_r}{V_r - 0.2534} - \frac{4.8514\, \alpha}{V_r^2 + 0.5068 V_r - 0.0642}. \tag{7}$$

Substantial differences between the P-R and Soave equations in the vicinity of the critical point are evident on these graphs.

Example 1.16(a) Comparison of Soave and Redlich-Kwong Equations at Several Reduced Temperatures and Acentric Factors

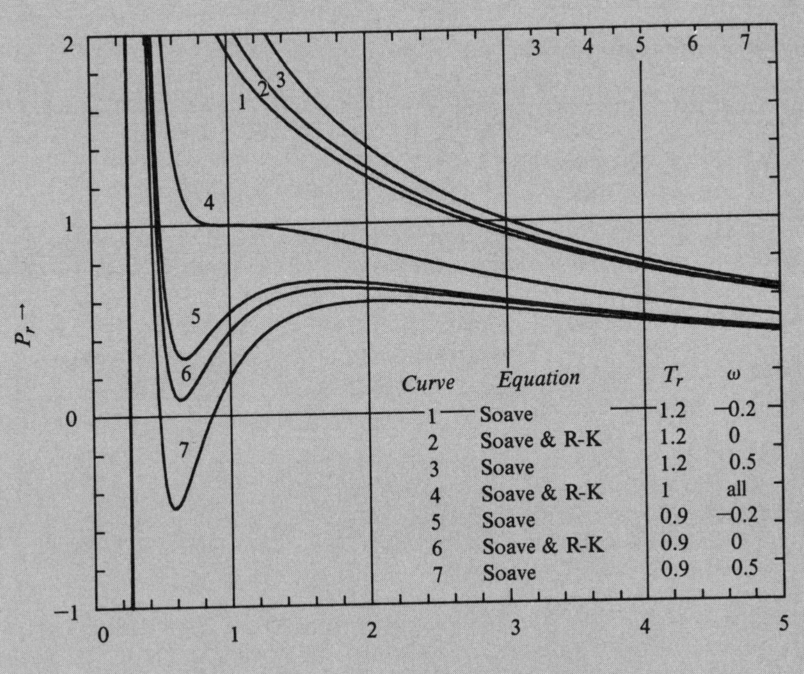

Example 1.16(b) Comparison of Soave and Peng-Robinson equations at $T_r = 0.9$ and several values of ω

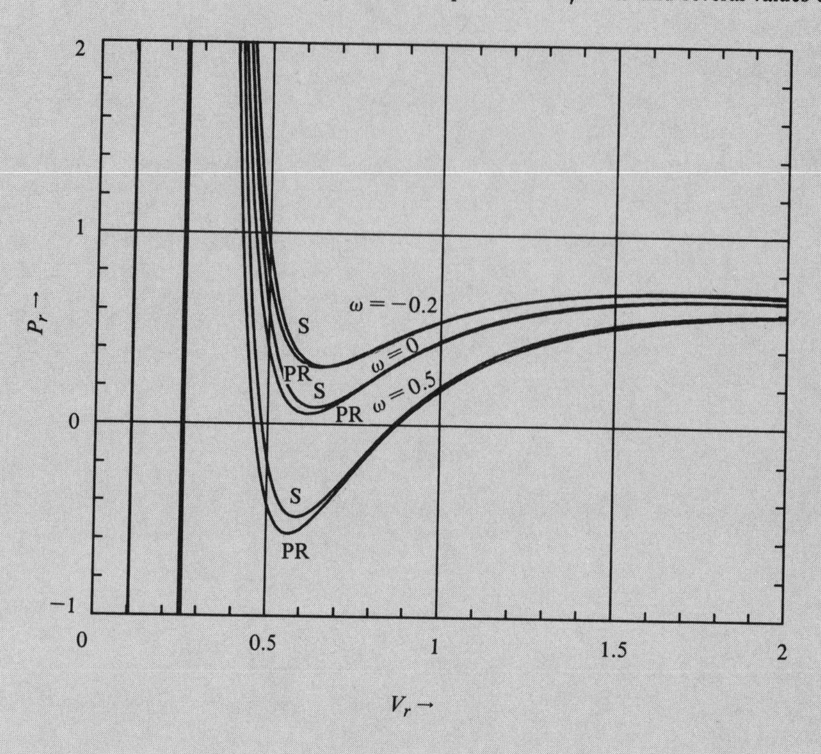

Example 1.16(c) Comparison of Soave and Peng-Robinson equations at $T_r = 1.2$ and several values of ω

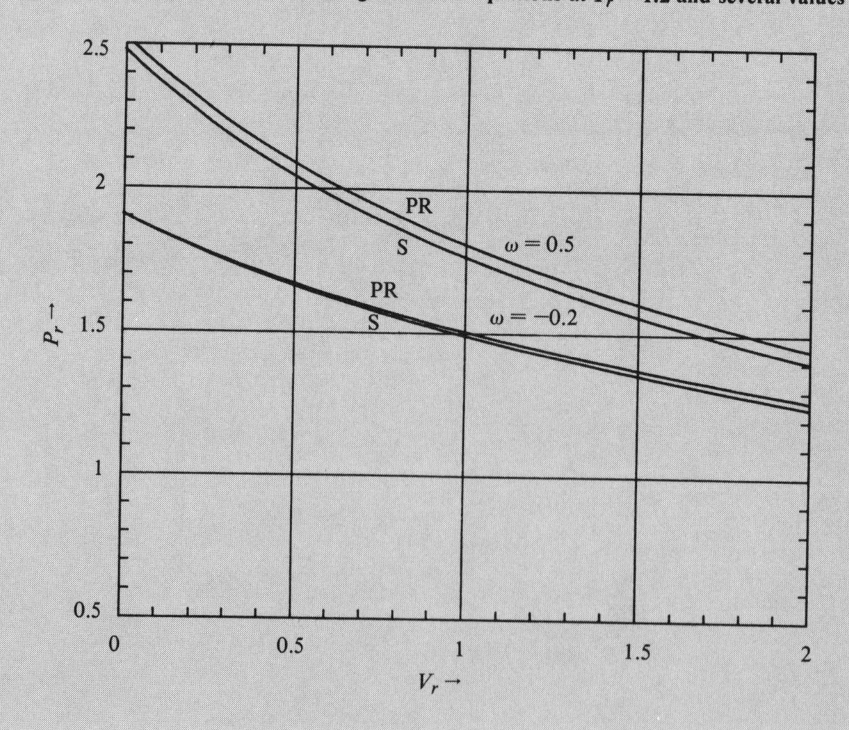

They proposed the equation:

$$P = \frac{RT}{V-b} - \frac{a(T)}{V(V+b)+b(V-b)}$$

$$= \frac{RT}{V-b} - \frac{a(T)}{V^2+2bV-b^2}. \qquad (1.130)$$

At the critical point,

$$a(T_c) = 0.45724R^2T_c^2/P_c, \qquad (1.131)$$

$$b(T_c) = 0.07780RT_c/P_c, \qquad (1.132)$$

$$z_c = 0.307. \qquad (1.133)$$

At other temperatures the parameter $a(T)$ is corrected to

$$a(T) = a(T_c)\,\alpha\,(T_r, \omega), \qquad (1.134)$$

similarly to the Soave treatment. Correlation in terms of the vapor pressure curve up to the critical point resulted in

$$\alpha^{0.5} = 1 + (1 - T_r^{0.5})(0.37464 + 1.5422\omega - 0.26992\omega^2). \qquad (1.135)$$

A range of applications of this equation of state is described by Robinson et al. (1977). A comparison of isotherms in the vicinity of the critical between this equation and that of Soave is shown in Example 1.16. According to the P-R equation the critical compressibility is $z_c = 0.307$. As indicated by the summary of Table 1.4, this value is nearer the true values of many substances, particularly nonpolar ones, than are the z_cs of some other two-parameter EOS. This is a partial explanation of the fact that the P-R equation is able to predict liquid densities more accurately that does the Soave equation, which is otherwise quite similar in performance. Improved prediction of liquid densities, however, also has been achieved by Fuller (1976), who expressed the parameters a and b of the Soave equation as functions of the temperature and used critical volumes and parachors in their evaluation. Even such polar molecules as water and ammonia are covered by this correlation. Liquid density predictions by the Soave and Fuller methods also are discussed by Chung et al. (1977).

Combining rules of the P-R system for mixtures are the usual ones of cubic equations, but binary interaction parameters for the cross-parameter are considered essential:

$$a_{ij} = (1 - k_{ij})\sqrt{a_i a_j}. \qquad (1.136)$$

They are found by optimizing bubblepoint pressures over an appropriate range of pressures and temperatures. Parameters for water with CO_2 and H_2S are temperature-dependent. No very extensive set of interaction parameters is in the open literature. A value of 0.13 is cited for carbon dioxide + n-butane; those of water with hydrocarbons range from 0.38 with 1-butene to 0.50 with methane and ethane. Other data are cited in Table 1.13.

Fugacity coefficients of both phases were found to be equally accurate with either the Soave or the P-R equation by Lin & Daubert (1980). Application to three-phase mixtures containing water was made by Peng & Robinson (1976), and a computational scheme was described. Further use of this equation in the calculation of multiphase equilibria is reviewed by Robinson & Peng (1980); systems with solid and liquid phases of carbon dioxide and of water with hydocarbons are mentioned. Routines for handling heavy-hydrocarbon fractions are referred to by Robinson et al. (1978). The first comprehensive program for the calculation of critical properties from an equation of state was developed for the P-R equation by Peng & Robinson (1977).

Computer programs for making various calculations with this system may be purchased from the Gas Processors Association in Tulsa, Oklahoma. These include routines for calculating vaporization equilibrium ratios, formation of hydrates, three-phase systems with water, critical properties, and handling heavy-hydrocarbon fractions.

Publication of the P-R equation was too late for a comprehensive evaluation by the API Committee, as reported by Daubert et al. (1978). Although the Soave and Peng-Robinson were judged more or less equally satisfactory for the calculation of vaporization ratios, the former equation—but with slightly modified coefficients and a comprehensive set of binary interaction parameters developed for the purpose (given in Table 1.12)—was adopted as the standard method of phase equilibrium calculations. For cryogenic conditions the BWRS equation still is preferred, and for enthalpy deviations the Lee-Kesler equation appears to be the best. The comparatively simple forms and acceptable accuracy of the Soave and P-R equations, however, make them favorites for inclusion in computer programs that call for repeated evaluations of properties.

1.5.8. Other Cubic Equations of State

A few equations of historical interest and additional ones of recent vintage will be discussed.

Berthelot (1899) took the parameter c of the Clausius equation (1880),

$$(P + a/T(V+c)^2)(V-b) = RT, \qquad (1.137)$$

to be zero and after some further approximations concluded that

$$PV/RT = 1 + (b - a/RT^2)(P/RT). \qquad (1.138)$$

Application of the criticality conditions gave for the parameters

$$a = \Omega_a P_c V_c^2 T_c, \qquad (1.139)$$

$$b = \Omega_b V_c, \qquad (1.140)$$

$$R = \Omega_R P_c V_c / T_c. \qquad (1.141)$$

By arbitrarily adjusting the coefficients Ω_i, a good fit to data is obtainable at moderate pressures near room temperature. The theoretical and Berthelot's adjusted values of the coefficients are

	Theoretical	*Adjusted*
Ω_a	3	16/3
Ω_b	1/3	1/4
Ω_R	8/3	32/9

With the adjusted coefficients the equation becomes in reduced form:

$$(P_r + 16/3T_rV_r^2)(V_r - 1/4) = 32T_r/9 = RT/P_cV_c. \qquad (1.142)$$

A convenient form of the equation is

$$z = 1 + \frac{9}{128}\frac{P_r}{T_r}(1 - 6/T_r^2). \qquad (1.143)$$

Table 1.13. The Peng-Robinson Equation of State (Peng & Robinson 1976)

Standard form:

$$P = \frac{RT}{V-b} - \frac{a\alpha}{V^2 + 2bV - b^2} \tag{1}$$

Parameters:

$$a = 0.45724R^2T_c^2/P_c, \tag{2}$$

$$b = 0.07780RT_c/P_c, \tag{3}$$

$$\alpha = [1 + (0.37464 + 1.54226\omega - 0.26992\omega^2)(1 - T_r^{0.5})]^2, \tag{4}$$

$$A = a\alpha P/R^2T^2 = 0.45724\alpha P_r/T_r^2, \tag{5}$$

$$B = bP/RT = 0.07780P_r/T_r. \tag{6}$$

Polynomial form:

$$z^3 - (1 - B)z^2 + (A - 3B^2 - 2B)z - (AB - B^2 - B^3) = 0. \tag{7}$$

Mixtures:

$$a\alpha = \Sigma\Sigma y_i y_j (a\alpha)_{ij}, \tag{8}$$

$$b = \Sigma y_i b_i, \tag{9}$$

$$(a\alpha)_{ij} = (1 - k_{ij})\sqrt{(a\alpha)_i(a\alpha)_j}, \tag{10}$$

$$A = \Sigma\Sigma y_i y_j A_{ij}, \tag{11}$$

$$B = \Sigma y_i B_i, \tag{12}$$

$$A_{ij} = (1 - k_{ij})(A_i A_j)^{0.5}, \tag{13}$$

$$k_{ii} = 0. \tag{14}$$

Data of Katz & Firoozabadi (1978):

System		k_{ij}
nitrogen + HC		0.12
CO_2 + HC		0.15
ethane + HC		0.01
propane + HC		0.01
methane +	ethane	0
	propane	0
	nC4	0.02
	nC5	0.02
	nC6	0.025
	nC7	0.025
	nC8	0.035
	nC9	0.035
	nC10	0.035
	nC20	0.054
	benzene	0.06
	cyclohexane	0.03

In older books these equations for heat capacities and entropy changes with pressure derived from the Berthelot equation sometimes appear and are applicable at moderate pressures near room temperature (Partington, 1950):

$$C_p - C_v \simeq R(1 + 27P_r/16T_r^3). \tag{1.144}$$

$$-\left(\frac{\delta S}{\partial P}\right)_T = \left(\frac{\partial V}{\partial T}\right)_P \simeq \frac{R}{P}\left[1 + \frac{27}{32}P_r/T_r^3\right]. \tag{1.145}$$

Two equations were developed by Dieterici (1899) on a semitheoretical basis, as described by Partington (1950). The first is

$$(P + a/V^{5/3})(V - b) = RT. \tag{1.146}$$

The expanded form,

$$V^{8/3} - (bP + RT/P)V^{5/3} + (a/P)V - ab/P = 0, \tag{1.147}$$

can be shown to have only three real roots in $V^{1/3}$.

Consideration of the variation of potential energy near the walls of containers led Dieterici (1899) to the unusual form

$$P = \frac{RT}{V-b}\exp(-a/RTV). \tag{1.148}$$

At low pressures or small values of the exponent the equation becomes

$$P = \frac{RT}{V-b} - \frac{a}{V(V-b)}. \tag{1.148a}$$

In terms of reduced variables the exponential form becomes

$$P_r = \frac{T_r}{2V_r - 1}\exp[2(1 - 1/T_rV_r)]. \tag{1.148b}$$

Derivatives useful for calculating deviation functions are:

$$\left(\frac{\partial P}{\partial V}\right)_T = \frac{P}{V-b}[1 - a(V-b)/RTV^2], \tag{1.149}$$

$$\left(\frac{\partial P}{\partial T}\right)_V = \frac{P}{T}(1 + a/RTV), \tag{1.150}$$

$$\left(\frac{\delta V}{\partial T}\right)_P = \frac{R}{P}\left[\frac{1 + a/RTV}{1 - a(V-b)/RTV^2}\right)\exp(-a/RTV). \tag{1.151}$$

The Berthelot and Dieterici equations are compared with data and with the van der Waals equation by Pickering (1925) at pressures to 1,000 atm. At room temperature or so, the Berthelot equation is generally superior in the range of 0–200 atm, except for ethylene and carbon dioxide, where the Dieterici is better. At higher pressures, the van der Waals and Dieterici usually give better results. For gases with critical temperatures above 300 K, the three equations are about equivalent. Reduced forms of the two equations are requested to be compared in Problem 1.45; the solution shows that the behavior in the vicinity of the critical point is unrealistic, particularly that of the Dieterici equation.

A recent example of a fairly involved cubic equation of state is that of Harmens & Knapp (1980), of which a summary is in Table 1.14. Their equation,

$$P = \frac{RT}{V-b} - \frac{a}{V^2 + bcV - b^2(c-1)}, \tag{1.152}$$

reduces to several common cubic equations of state at particular values of parameter c. Polynomial forms are

$$V^3 + \left(bc - b - \frac{RT}{P}\right)V^2 + \left[\frac{a}{P} + b^2(1-2c) - \frac{bcRT}{P}\right]V + b^3(c-1) + \frac{b^2(c-1)RT}{P} - \frac{ab}{P} = 0, \tag{1.153}$$

or

$$V^3 + AV^2 + BV + C = 0, \tag{1.154}$$

and

$$z^3 + \frac{AP}{RT}z^2 + B\left(\frac{P}{RT}\right)^2 z + C\left(\frac{P}{RT}\right)^3 = 0. \tag{1.155}$$

Auxiliary parameters are $\beta = b/V_c$ and $\zeta = (P_c/RT_c)V_c$. They were correlated in terms of acentric factor and reduced temperature by using critical isotherms and vapor pressure data of twenty substances ranging from argon to n-decane. In comparison with the Soave and Peng-Robinson equations, the new equation has superior performance for calculation of volumes along the critical isotherm, boiling temperatures, and saturated liquid volumes. The plots of Figure 1.19 for sulfur dioxide at 250 F shows that the HK equation predicts the saturation pressure and volumes most closely.

Though the equation is somewhat more complex than other cubics, there is some improvement in accuracy. Applications to mixtures and to vapor liquid equilibria are promised for later papers.

1.6. COMPLEX EQUATIONS OF STATE

1.6.1. Introduction

Every equation of state that has been proposed has more or less severe limitations with regard to the kinds of substances that it could represent, or the range of operating conditions, or the phases. Some equations are better for *PVT*, others for phase equilibria, and still others for enthalpy or entropy deviations. There is little hope that a universal equation of state of moderate complexity ever will be discovered.

Increasing the number of constants in the equation has often, but not always, served to improve coverage. There are instances where a two-parameter equation is superior to some with eight or more parameters; this cannot be depended on in general, however, so the multiparameter equations are used extensively, particularly since computers have made them tractable.

The question of equation fitting can be avoided by expressing the information in graphic or tabular form. For example, the Pitzer-Curl compressibility, $z = z^{(0)} + \omega z^{(1)}$, represents the superscripted items as functions of reduced properties in this way. Such an approach is not feasible when numerous calculations are to be made, however, because of the laboriousness of lookup and interpolation.

Four main types of equations of state may be recognized:

1. Equations that are specific for individual substances, such as water, or definite mixtures, such as air. Great accuracy over a wide range of conditions is required, so the equations have many constants.
2. Equations of a particular form with different numerical coefficients for different substances. Usually combining rules for the constants of pure components extend the equations to mixtures.
3. Equations with universal parameters that are evaluated in terms of the readily known properties of individual pure substances.
4. Equations of the preceding two types when applied to mixtures and incorporating binary interaction parameters obtained from experimental data on binary mixtures.

Cubic and other relatively simple equations of state likewise can be classified into the same groups. Examples will be considered of all these types. It should be borne in mind that for practical reasons a somewhat superior new equation need not always replace an older one. The original equation may be embedded in a comprehensive computer program in such a

Table 1.14. The Harmens-Knapp Three-Parameter Cubic Equation of State (Harmens & Knapp 1980)

Standard form:

$$P = \frac{RT}{V-b} - \frac{a}{v^2 + bcV - b^2(c-1)} \tag{1}$$

Parameters:

$$\beta = 0.10770 + 0.76405\zeta - 1.24282\zeta^2 + 0.96210\zeta^3 \tag{2}$$

$$\zeta = 0.3211 - 0.080\omega + 0.0384\omega^2 \tag{3}$$

$$\Omega_a = 1 - 3\zeta + 3\zeta^2 + \beta\zeta(3 - 6\zeta + \beta\zeta) \tag{4}$$

$$\Omega_b = \beta\zeta \tag{5}$$

$$\alpha(T_r) = \left\{1 + A(1 - \sqrt{T_r}) - B\left(1 - \frac{1}{T_r}\right)\right\}^2 \tag{6}$$

when $\omega \leq 0.2$,

$$A = 0.50 + 0.27767\omega + 2.17225\omega^2 \tag{7}$$

$$B = -0.022 + 0.338\omega - 0.845\omega^2 \tag{8}$$

when $\omega > 0.2$,

$$A = 0.41311 + 1.14657\omega \tag{9}$$

$$B = 0.0118 \tag{10}$$

when $T_r > 1.0$,

$$\alpha(T_r) = 1.0 - (0.6258 + 1.5227\omega)\ln T_r + (0.1533 + 0.41\omega)(\ln T_r)^2 \tag{11}$$

$$a = \alpha(T_r)\Omega_a \frac{R^2T_c^2}{P_c} \tag{12}$$

$$b = \Omega_b \frac{RT_c}{p_c} \tag{13}$$

$$c = 1 + \frac{1 - 3\zeta}{\beta\zeta} \tag{14}$$

Fugacity coefficient and residual properties:

$$\ln\phi = z - 1 + \ln\frac{RT}{P(V-b)} - \frac{a}{RT}L \tag{15}$$

$$H - H^{id} = PV - RT + \left(T\frac{da}{dT} - a\right)L \tag{16}$$

$$S - S^{id} = L\frac{da}{dT} - R\ln\frac{RT}{V-b} \tag{17}$$

(S^{id} is the entropy of ideal gas at 1 atm)

$$K = \sqrt{c^2 + 4c - 4} \tag{18}$$

$$L = \frac{1}{Kb}\ln\frac{2V + b(c+K)}{2V + b(c-K)} \tag{19}$$

way that replacing the equation of state may require changing too many other factors for replacement to be feasible. Many of the parts of a computer program may be mutually compensated, so that changing one part will require changing others.

1.6.2. The BWR Equation

The celebrated equation of Benedict, Webb, & Rubin (1940, 1942, 1951) was devised as an improvement on the Beattie-Bridgeman equation. Some steps in the evolution of these

Table 1.15. Evolution of the BWR Equation of State

Virial equation (1885–1901):

$$P = RT\rho(1 + B\rho + C\rho^2 + D\rho^3 + \cdots).$$

Beattie-Bridgeman equation (1927):

$$P = RT\rho + \left(B_0RT - A_0 - \frac{Rc}{T^2}\right)\rho^2 + \left(-B_0bRT + A_0a - \frac{RB_0c}{T^2}\right)\rho^3 + \frac{RB_0bc}{T^2}\rho^4.$$

Benedict-Webb-Rubin equation (1940):

$$P = RT\rho + \left(B_0RT - A_0 - \frac{C_0}{T^2}\right)\rho^2 + (bRT - a)\rho^3 + a\alpha\rho^6 + \frac{c}{T^2}\rho^3(1 + \gamma\rho^2)\exp(-\gamma\rho^2).$$

Starling (1970):

$$P = RT\rho + \left(B_0RT - A_0 - \frac{C_0}{T^2} + \frac{D_0}{T^3} - \frac{E_0}{T^4}\right)\rho^2 + \left(bRT - a - \frac{d}{T}\right)\rho^3 + \alpha\left(a + \frac{d}{T}\right)\rho^6 + \frac{c}{T^2}\rho^3(1 + \gamma\rho^2)\exp(-\gamma\rho^2).$$

Nishiumi (1980):

$$Z = 1 + \left(B_0^* - \frac{A_0^*}{T_r} - \frac{C_0^*}{T_r^3} - \frac{D_0^*}{T_r^4} - \frac{E_0^* + \psi_E}{T_r^5}\right)\rho_r + \left(b^* - \frac{a^*}{T_r} - \frac{d^*}{T_r^2} - \frac{e^*}{T_r^5} - \frac{f^*}{T_r^{24}}\right)\rho_r^2 + \alpha^*\left(\frac{a^*}{T_r} + \frac{d^*}{T_r^2} + \frac{e^*}{T_r^5} + \frac{f^*}{T_r^{24}}\right)\rho_r^5 + \left(\frac{c^*}{T_r^3} + \frac{g^*}{T_r^9} + \frac{h^*}{T_r^{18}} + y(T_r)\right)\rho_r^2(1 + \gamma^*\rho_r^2)e^{-\gamma^*\rho_r^2}.$$

The fifteen coefficients with asterisks are functions of the acentric factor. The quantities ψ_E and $y(T_r)$ express the effects of polarity on vapor and liquid properties, respectively.

equations are shown in Table 1.15. Details of the BWR equation and auxiliary matter are summarized in Table 1.16, and values of the parameters are given in Table E.5. Figures 1.20a and 1.20c show comparisons of the Beattie-Bridgeman equation with others and with data.

The equation defines the pressure or compressibility as a polynomial in terms of density with coefficients that are dependent on temperature, with an exponential term tacked on to compensate for the higher-power terms of the virial equation. The defect of the Beattie-Bridgeman equation that the BWR equation was intended to rectify was an inability to represent behavior of liquids and of gases above the critical density.

Data on PVT properties of gases, critical properties, and vapor pressures were employed in the evaluation of the eight parameters. For highest accuracy each substance requires its own parameters, but some generalizations of the parameters in terms of properties of the pure components have been made. Most emphasis has been placed on light hydrocarbons and the inorganic gases associated commonly with them.

Because of its high degree of nonlinearity, the BWR equation is more difficult to use than are cubic equations for which analytical solution for the volume or compressibility is possible. The Newton-Raphson method is generally satisfactory. Details of a calculation procedure are given by Johnson & Colver (1968), and a method that is always convergent though often slow has been devised by Tang (1970).

Plots of the equation of n-pentane are shown in Figure 1.21, where the isotherms display the familiar S-shapes and three

Table 1.16. The Benedict-Webb-Rubin Equation of State

Standard forms:

$$P = RT\rho + (B_0RT - A_0 - C_0T^{-2})\rho^2 + (bRT - a)\rho^3 + a\alpha\rho^6 + cT^{-2}\rho^3(1 + \gamma\rho^2)\ \exp(-\gamma\rho^2),$$

$$z = 1 + \left(B_0 - \frac{A_0}{RT} - \frac{C_0}{RT^3}\right)\rho + \left(b - \frac{a}{RT}\right)\rho^2 + \frac{a\alpha}{RT}\rho^5 + \frac{c}{RT^3}\rho^2(1 + \gamma\rho^2)\exp(-\gamma\rho^2).$$

Parameters: Numerical values are given in Table E.5. Below the normal point, C_0 often is expressed as a function of temperature; reference to many papers on this subject is made by Reid et al. (1977, p. 47).

Mixtures: The original combining rules are:

$B_0 = \Sigma_i x_i B_{0i}$ $\quad a = [\Sigma_i x_i (a_i)^{1/3}]^3$

$A_0 = [\Sigma_i x_i (A_{0i})^{1/2}]^2$ $\quad c = [\Sigma_i x_i (c_i)^{1/3}]^3$

$C_0 = [\Sigma_i x_i (C_{0i})^{1/2}]^2$ $\quad \gamma = [\Sigma_i x_i (\gamma_i)^{1/2}]^2$

$b = [\Sigma_i x_i (b_i)^{1/3}]^3$ $\quad a = [\Sigma_i x_i (a_i)^{1/3}]^3$

Binary interaction parameters were incorporated into the combining rules by Bishnoi & Robinson (1972a, 1972b) whose conclusions are:

$B_0 = \sum_{i=1}^{n}\sum_{j=1}^{n} x_i x_j B_{0ij}$ $\quad$ where $B_{0ij} = \sqrt{B_{0i}B_{0j}}$

$A_0 = \sum_{i=1}^{n}\sum_{j=1}^{n} x_i x_j A_{0ij}$ $\quad$ where $A_{0ij} = \sqrt{A_{0i}A_{0j}}(1 - k_{ij})$

$C_0 = \sum_{i=1}^{n}\sum_{j=1}^{n} x_i x_j C_{0ij}$ $\quad$ where $C_{0ij} = \sqrt{C_{0i}C_{0j}}(1 - k_{ij})^3$

$b = \sum_{i=1}^{n}\sum_{j=1}^{n}\sum_{k=1}^{n} x_i x_j x_k (b_{ij}\, b_{jk}\, b_{ik})^{1/3}$ $\quad$ where $b_{ij} = \sqrt{b_i b_j}$

$a = \sum_{i=1}^{n}\sum_{j=1}^{n}\sum_{k=1}^{n} x_i x_j x_k (a_{ij}\, a_{jk}\, a_{ik})^{1/3}$ $\quad$ where $a_{ij} = \sqrt{a_i a_j}(1 - k_{ij})$

$c = \sum_{i=1}^{n}\sum_{j=1}^{n}\sum_{k=1}^{n} x_i x_j x_k (C_{ij}\, C_{jk}\, C_{ik})^{1/3}$ $\quad$ where $c_{ij} = \sqrt{c_i c_j}(1 - k_{ij})^3$

$\alpha = \sum_{i=1}^{n}\sum_{j=1}^{n}\sum_{k=1}^{n} x_i x_j x_k (\alpha_{ij}\, \alpha_{jk}\, \alpha_{ik})^{1/3}$ $\quad$ where $\alpha_{ij} = \sqrt{\alpha_i \alpha_j}$

$\gamma = \sum_{i=1}^{n}\sum_{j=1}^{n} x_i x_j \gamma_{ij}$ $\quad$ where $\gamma_{ij} = \sqrt{\gamma_i \gamma_j}$

Values of the k_{ij} are as follows:

	CO_2	H_2S	N_2	CH_4	C_2H_6	C_3H_8	i-C_4H_{10}	n-C_4H_{10}	i-C_5H_{12}	n-C_5H_{12}	n-C_6H_{14}	n-C_7H_{14}
CO_2	0	0.057*	0.03*	0.03†	0.08†	0.11†	0.14	0.11†	0.166	0.166	0.188*	0.209*
H_2S		0	0.068*	0.09 ± 0.01*	0.085†	0.076*	0.09	0.09	0.102	0.102	0.116*	0.129*
N_2			0	0.03	0.06*	0.09	0.113*	0.113*	0.14*	0.14*	0.166*	0.193*
CH_4				0	0.01	0.022	0.034*	0.03*	0.047*	0.05*	0.06*	0.072*
C_2H_6					0	0	0	0	0.01	0.01	0.02	0.03
C_3H_8						0	0	0	0.006*	0.006	0.014	0.021
i-C_4H_{10}							0	0	0	0	0.006*	0.012*
n-C_4H_{10}								0	0	0	0.006*	0.012*
i-C_5H_{12}									0	0	0	0.003*
n-C_5H_{12}										0	0	0
n-C_6H_{14}											0	0
n-C_7H_{14}												0

Equations for fugacities and residual enthalpy and entropy are obtained from the corresponding equations based on the BWRS equation of Table 1.17 by setting $D_0 = E_0 = d = 0$.

† From second virial coefficients.
*Estimated.

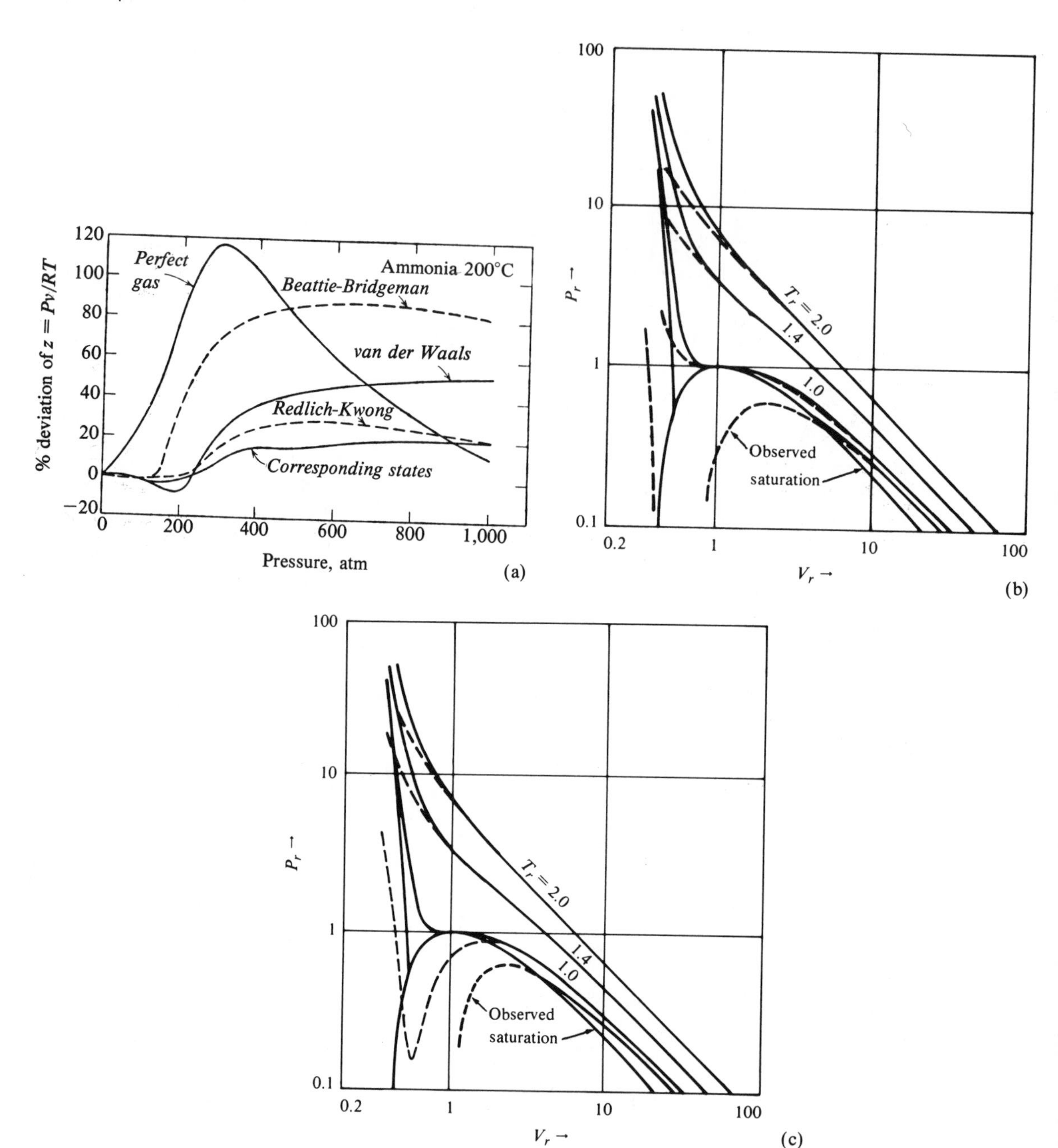

Figure 1.20 Observed and calculated *PVT* data: (a) Percentage deviation of calculated compressibilities from the true values of ammonia at 200 C (Comings 1956); (b) argon data and the Clausius equation:

$$\left(P + \frac{a}{T(V+c)^2}\right)(V-b) = RT.$$

Broken lines are experimental data (Shah & Thodos 1965); (c) argon data and the Beattie-Bridgeman equation:

$$p\,V^2 = RT\left[V + B_0\left(1 - \frac{b}{V}\right)\right]\left[1 - \frac{c}{VT^3}\right] - A_0\left(1 - \frac{a}{V}\right).$$

Broken lines are experimental data (Shah & Thodos 1965).

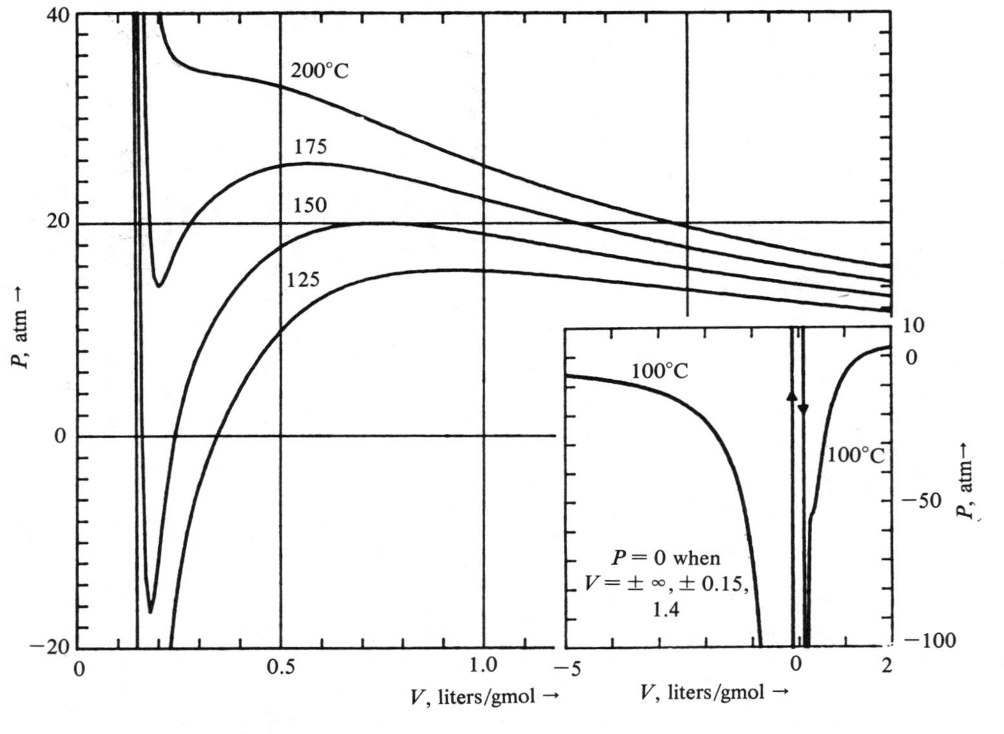

Figure 1.21 Isotherms of n-pentane with the BWR equation, at 100, 125, 150, 175, and 200 C, showing characteristic valleys

$$P = RT\rho + \left(B_0RT - A_0 - \frac{C_0}{T^2}\right)\rho^2 + (bRT - a)\rho^3 \quad + a\alpha\rho^6 + \frac{c\rho^3}{T^2}(1 + \gamma\rho^2)\exp(-\gamma\rho^2)$$

Range of application											
Vapor		*Liquid*									
Density, g-mol/l	*Temp., °C*	*Pressure, atm*	*Temp., °C*	*a, $(l/gmol)^3atm$*	*A_0, $(l/gmol)^2atm$*	*b, $(l/gmol)^2$*	*B_0, l/gmol*	*c, $(l/gmol)^3K^2atm$*	*C_0, $(l/gmol)^2K^2atm$*	*a, $(l/gmol)^3$*	*γ, $(l/gmol)^2$*
0.46	140	2.1	60	4.07480	1.21794E+1	6.68120E−2	1.56751E−1	8.24170E+5	2.12121E+6	1.81000E−5	4.75000E−2
4.8	280	25.5	180								

positive real roots appear to be possible. The behavior of the equation as the pressure, temperature, and composition are varied has been examined in some detail by Coward et al. (1978) and compared with that of the Redlich-Kwong equation. An improved algorithm for the evaluation of density has been obtained by Mills et al. (1980).

The Parameters

Many evaluations of the parameters for both pure substances and for generalized equations have been made—for example:

1. The first BWR paper of 1951 gives constants for twelve hydrocarbons, methane through n-heptane.
2. Cooper & Goldfrank (1967) have fifty-two sets of parameters covering thirty-eight substances. A generalization of a reduced equation is made by expressing the parameters as linear functions of the acentric factor, for a total of sixteen constants.
3. Coefficients of sixteen substances of interest to the natural gas industry were newly evaluated by Bishnoi et al. (1974).
4. Data for thirty-two compounds in thirty-eight sets were collected or determined by Novak et al. (1972) (cited by Holub & Vonka 1976). Hydrogen, acetylene, butadiene, and a few other hydrocarbons are unique to this list. They are given as Table E.5.
5. Ethanol is covered by Gomez-Nieto & Thodos (1976) and acetone by the same workers (1978).
6. Parameters valid at cryogenic temperatures were obtained for twenty substances by Orye (1969).
7. The parameter C_0 was made temperature-dependent by Sood & Hazelden (1970).
8. Extension to hydrocarbons heavier than n-decane was made possible by a generalization of Bishnoi et al. (1974) and by Nishiumi (1980a).
9. Water has not been successfully modeled by either the BWR or BWRS equation, according to Hopke (1977), but see the next item.
10. Extension to polar substances including water was made by Nishiumi (1980b). Three polar parameters are needed for each substance, and data are given for ninety-two substances.

Various expressions relating to the BWR equation are summarized in Table 1.16. A comprehensive treatment of all aspects of the BWR equation, including procedures for determining the coefficients from experimental data and evaluation of thermodynamic properties, is given by Opfell et al. (1959).

Mixtures

A strong feature of the BWR equation is its extension to modeling of mixtures. The combining rules originally proposed for the parameters M are:

$$M = (\Sigma y_i M_i^{1/r})^r, \qquad (1.56)$$

where $r = 1$ for B_0; $r = 2$ for A_0, C_0, and γ; and $r = 3$ for the other four parameters. Binary interaction parameters were incorporated for four of the parameters by Bishnoi & Robinson (1972a), and values of k_{ij} for twelve substances are given by Bishnoi & Robinson (1972b). The combining rules and the numerical values are given in Table 1.16.

Interaction parameters k_{ij} have been correlated in terms of the total number of carbon atoms by Labinov & Boiko (1967), primarily for n-alkanes, but a more comprehensive correlation in terms of the ratios of critical volumes was made by Nishiumi & Saito (1977), who checked the vapor-liquid equilibria of 125 binary systems.

Modified and Extended Equations

Two modifications have already been mentioned:

1. The parameter C_0 has been made a function of temperature.
2. The parameters have been generalized as linear functions of the acentric factor, making them universal parameters, applicable to all substances whose acentric factors and critical properties are known.

More drastic changes also have been developed. A generalized equation with sixteen parameters and a total of forty-four coefficients by Yamada (1973) gives good results at reduced densities as high as 2.8 and was checked for mixtures. A modification with thirty-three constants was applied to methane by McCarty (1974). An equation with fifteen coefficients, each correlated in terms of T_c, V_c, and ω, was applied to heavy hydrocarbons C_{10} to C_{20} by Nishiumi (1980).

The most widely accepted version is that of Starling (1973, and earlier), and has in turn been subject to much alteration (see Table 1.17). Three parameters have been added—D_0, E_0, and d—to improve the temperature dependence of the coefficients of ρ^2, ρ^3, and ρ^6. Binary interaction parameters are applied to four of the eleven parameters, and values are given for eighteen substances in this compilation. The equations are said to be capable of predicting properties at reduced temperatures as low as 0.3 and reduced densities as great as 3.0, thus being applicable to the cryogenic range. A generalization in terms of properties of pure substances is known by the name Han-Starling. The book cited has completely worked out tables and thermodynamic diagrams for fifteen substances, and also has the computer programs.

The BWRS equation was applied to mixtures of nitrogen with methane, ethane, and propane by Lin & Hopke (1974). In developing the values of the parameters, they used 21,000 experimental data points of density, enthalpy, vapor pressure and vaporization equilibriun ratios of these substances. The regression procedure is described briefly.

Application of This Family of Equations

The BWR and BWRS equations are widely accepted by industry, as stated by Adler et al. (1977) and Hopke (1977). They are superior to cubics, for instance, for all thermodynamic calculations, particularly for cryogenic systems, and have perhaps only the Lee-Kesler types as competitors for accurate calculations. However, their complex form hampers their use in distillation calculations, for example, where repeated determinations of fugacities, volumes and enthalpies need to be made, so that cubic equations often are employed for these purposes. This objection is not regarded as serious by some workers.

Table 1.17. The Benedict-Webb-Rubin-Starling Equation of State

Standard form:

$$P = \rho RT + \left(B_0 RT - A_0 - \frac{C_0}{T^2} + \frac{D_0}{T^3} - \frac{E_0}{T^4}\right)\rho^2 + \left(bRT - a - \frac{d}{T}\right)\rho^3$$

$$+ \alpha\left(a + \frac{d}{T}\right)\rho^6 + \frac{c\rho^3}{T^2}(1 + \gamma\rho^2)\exp(-\gamma\rho^2).$$

Parameters: Values of the eleven parameters are given for fifteen substances in the book by Starling (1973).

Mixtures: The combining rules include binary interaction parameters which also are given in Starling's book.

$$B_0 = \sum_i x_i B_{0i}$$

$$A_0 = \sum_i \sum_j x_i x_j A_{0i}^{1/2} A_{0j}^{1/2} (1 - k_{ij})$$

$$C_0 = \sum_i \sum_j x_i x_j C_{0i}^{1/2} C_{0j}^{1/2} (1 - k_{ij})^3$$

$$\gamma = \left[\sum_i x_i \gamma_i^{1/2}\right]^2$$

$$b = \left[\sum_i x_i b_i^{1/3}\right]^3$$

$$a = \left[\sum_i x_i a_i^{1/3}\right]^3$$

$$\alpha = \left[\sum_i x_i \alpha_i^{1/3}\right]^3$$

$$c = \left[\sum_i x_i c_i^{1/3}\right]^3$$

$$D_0 = \sum_i \sum_j x_i x_j D_{0i}^{1/2} D_{0j}^{1/2} (1 - k_{ij})^4$$

$$d = \left[\sum_i x_i d_i^{1/3}\right]^3$$

$$E_0 = \sum_i \sum_j x_i x_j E_{0i}^{1/2} E_{0j}^{1/2} (1 - k_{ij})^5$$

Generalized (Han-Starling) equation: The parameters are correlated in terms of the critical properties and acentric factors.

$$\rho_{ci} B_{0i} = A_1 + B_1 \omega_i$$

$$\frac{\rho_{ci} A_{0i}}{RT_{ci}} = A_2 + B_2 \omega_i$$

$$\frac{\rho_{ci} C_{0i}}{RT_{ci}^3} = A_3 + B_3 \omega_i$$

$$\rho_{ci}^2 \gamma_i = A_4 + B_4 \omega_i$$

$$\rho_{ci}^2 b_i = A_5 + B_5 \omega_i$$

$$\frac{\rho_{ci}^2 a_i}{RT_{ci}} = A_6 + B_6 \omega_i$$

Table 1.17. *(continued)*

$$\rho_{ci}^3\alpha_i = A_7 + B_7\omega_i$$

$$\frac{\rho_{ci}^2 c_i}{RT_{ci}^3} = A_8 + B_8\omega_i$$

$$\frac{\rho_{ci}D_{0i}}{RT_{ci}^4} = A_9 + B_9\omega_i$$

$$\frac{\rho_{ci}^2 d_i}{RT_{ci}^2} = A_{10} + B_{10}\omega_i$$

$$\frac{\rho_{ci}E_{0i}}{RT_{ci}^5} = A_{11} + B_{11}\omega_i \exp(-3.8\ \omega_i)$$

Parameter Subscript (j)	Parameter Value	
	A_j	B_j
1	0.443690	0.115449
2	1.28438	−0.920731
3	0.356306	1.70871
4	0.544979	−0.270896
5	0.528629	0.349261
6	0.484011	0.754130
7	0.0705233	−0.044448
8	0.504087	1.32245
9	0.0307452	0.179433
10	0.0732828	0.463492
11	0.006450	−0.022143

Partial fugacities: The result for the BWR equation is obtained from the following by making $D_0 = E_0 = d = 0$.

$$RT \ln(\hat{f}_i/x_i) = RT \ln(\rho RT) + \rho(B_0 + B_{0i})RT$$

$$+ 2\rho \sum_{j=1}^{n} x_j \left[-(A_{0j}A_{0i})^{1/2}(1 - k_{ij}) - \frac{(C_{0j}C_{0i})^{1/2}}{T^2} \right.$$

$$(1 - k_{ij})^3 + \frac{(D_{0j}D_{0i})^{1/2}}{T^3}(1 - k_{ij})^4 - \frac{(E_{0j}E_{0i})^{1/2}}{T^4}$$

$$\left.(1 - k_{ij})^5 \right] + \frac{\rho^2}{2}\left[3(b^2 b_i)^{1/3} RT - 3(a^2 a_i)^{1/3} \right.$$

$$\left. - \frac{3(d^2 d_i)^{1/3}}{T} \right] + \frac{\alpha\rho^5}{5}\left[3(a^2 a_i)^{1/3} + \frac{3(d^2 d_i)^{1/3}}{T}\right]$$

$$+ \frac{3\rho^5}{5}\left(a + \frac{d}{T}\right)(\alpha^2\alpha_i)^{1/3}$$

$$+ \frac{3(c^2 c_i)^{1/3}\rho^2}{T^2}\left[\frac{1 - exp(-\gamma\rho^2)}{\gamma\rho^2} - \frac{exp(-\gamma\rho^2)}{2}\right]$$

Table 1.17. *(continued)*

$$-\frac{2c}{\gamma T^2}\left(\frac{\gamma_i}{\gamma}\right)^{1/2}\left\{1-exp(-\gamma\rho^2)\left[1+\gamma\rho^2+\frac{1}{2}\gamma^2\rho^4\right]\right\}.$$

Residual enthalpy and entropy:

$$H - H^0 = \rho(B_0RT - A_0 - 4C_0/T^2 + 5D_0/T^3 - 6E_0/T^4)$$
$$+ 1/2\rho^2(2bRT - 3a - 4d/T) - 1/5p^5\alpha(6a + 7d/T)$$
$$+ c/\gamma T^2(3 - (3 + 1/2\gamma\rho^2 - \gamma^2\rho^4)\exp(-\gamma\rho^2)),$$

$$S - S^0 = -\Sigma_i x_i R\ln(RT\rho x_i) - \rho(B_0R + 2C_0/T^3 - 3D_0/T^4$$
$$+ 4E_0/T^5) - 1/2\rho^2(bR + d/T^2)$$
$$+ 1/5\rho^5\alpha d/T^2 + 2c/\gamma T^3(1 - [1 + 1/2\rho^2\gamma]\exp[-\gamma\rho^2]),$$

1.6.3. Corresponding States and the Lee-Kesler Equation

A successful class of *PVT* relations is based on correlations of deviations of these properties from those of particular reference substances. Modern schemes of this kind began with the correlations of Pitzer et al. (1955–1958) of the compressibility as a polynomial in acentric factor,

$$z = z^{(0)} + \omega z^{(1)} + \omega^2 z^{(2)} + \dots.$$

In practice this is usually truncated at the linear term. The term $z^{(0)}$ is the compressibility of a "simple" fluid such as argon, and the other items are corrections for deviation from simple behavior. Both $z^{(0)}$ and $z^{(1)}$ originally were prepared in tabular form as functions of the reduced temperature and pressure, but they were plotted by Edmister (1958) and are updated as Figure 1.22.

Critical compressibility is adopted as a correlating parameter by Lydersen et al. (1955). Tabulations are given of compressibility, liquid density, fugacity coefficients, and enthalpy and entropy deviations. This and the Pitzer system are limited to reduced temperatures above 0.7 or so, and their tabular or graphical form makes them difficult to implement on a computer. In any case, they have been superseded by the tabulations and equations of Lee & Kesler, to be discussed next.

A number of correlations of $z^{(0)}$ and $z^{(1)}$ in equation form have been made. Lee & Kesler (1975) adopted equations similar to the BWR for this purpose. They referred properties of all substances relatively to two substances, a "simple" fluid and a "reference" fluid. For $z^{(0)}$ they used mostly data on methane, argon, and krypton; as a reference fluid they used *n*-octane, the heaviest hydrocarbon for which there are extensive data. They write for the compressibility:

$$z = z^{(0)} + (\omega/0.3978)(z^{(r)} - z^{(0)}) = z^{(0)} + \omega z^{(1)}, \quad (1.157)$$

where 0.3978 is the acentric factor of *n*-octane. Clearly, Pitzer's $z^{(1)}$ is

$$z^{(1)} = (z^{(r)} - z^{(0)})/0.3978. \quad (1.158)$$

Both compressibilities are represented by equations of the same form but with different numerical coefficients. These equations and related ones are given in Table 1.18.

The solution procedure at specified T_r and P_r is to find $V_r^{(0)}$ and $V_r^{(1)}$ by trial from Eq. 3 of the Table, then to find z (0) and $z^{(1)}$ and to combine them into the compressibility. A convenient form for practical solution is obtained by replacing $1/V_r = P_r/zT_r$ throughout since the range of z usually is small, from approximately 1 to 0.05 or so. Note that while P_r and T_r are the conventional reduced properties, V_r is *not* the reduced volume, being

$$V_r = P_cV/RT_c.$$

Equations for fugacity and deviation functions are of analogous form, and may be evaluated after the V_r or z have been established. The equations have been solved and the results tabulated for the ranges $0.3 \le T_r \le 4$ and $0.01 \le P_r \le 10$ for convenience of manual operations, but T_r as high as 8.7 and P_r as high as 31 have been checked. These tables make those of Pitzer and Lydersen obsolete. Example 1.17 applies the L-K equation. Comparison with other equations is made on Figure 1.23.

Mixtures

The mixing rules are given in Table 1.18, but they have been altered by other people working in this area. The original paper gives equations for fugacity and enthalpy, entropy, and heat capacity deviations only for pure substances. Fugacities in mixtures were figured by Joffe (1976), who modified the combining rules and incorporated binary interaction factors. More extensive checking of vapor liquid equilibria of mixtures was done by Plöcker et al. (1978). They also modified the combining rules and provided binary interaction factors for about 140 pairs. Vapor liquid equilibria and enthalpy deviation comparisons with those figured from the BWRS equation

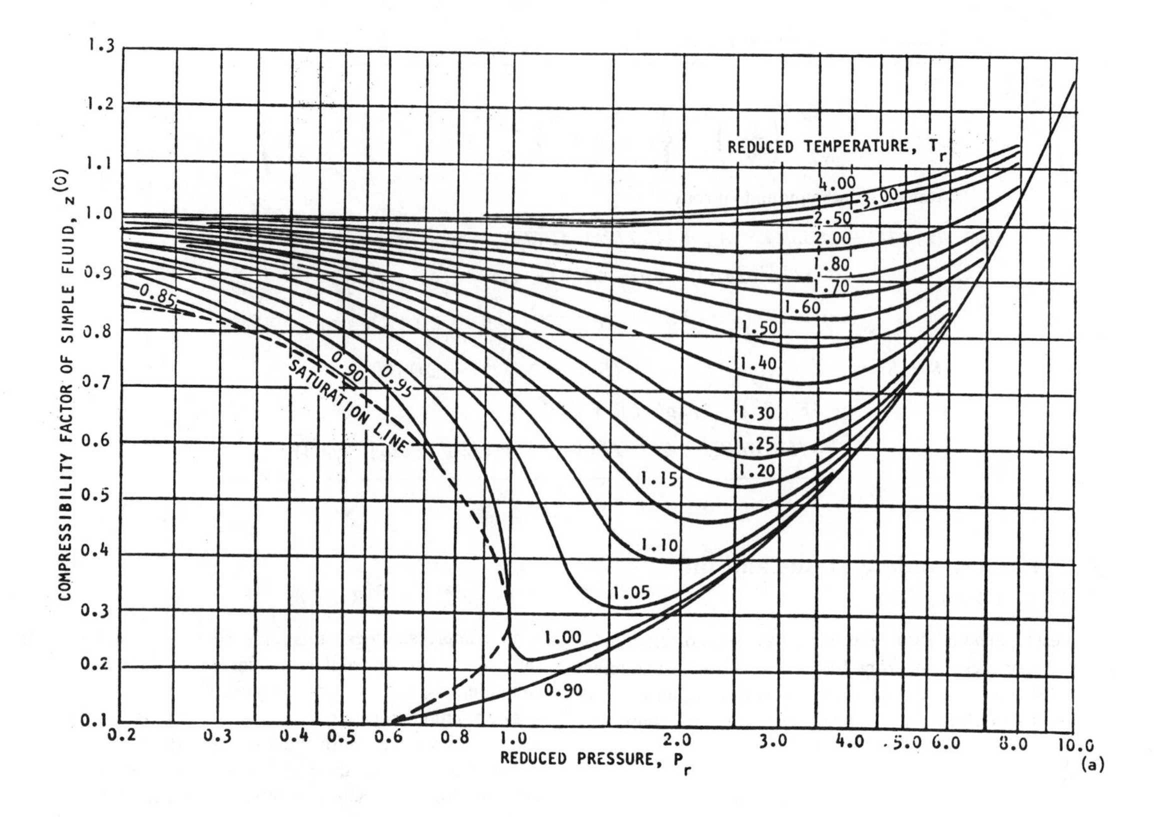

Figure 1.22 Compressibility factor, $z = z^{(0)} + \omega z^{(1)}$, based on the Lee-Kesler equation (Table 1.18); (a) factor $z^{(0)}$ for simple fluids; (b) correction factor $z^{(1)}$. See also Tables E.1 and E.2.

were highly favorable to this modified Lee-Kesler method. Very near the critical, calculations may encounter difficulties, but the authors conclude that no problems usually exist for T/T_c below 0.96.

Partial molal volumes were figured with the Lee-Kesler equation by Lin & Daubert (1979) who found the results generally within 15 percent of experimental values.

The "simple" fluid as a reference is eliminated by Teja et al. (1981), who employ two reference fluids that may be varied from case to case, and do not restrict the equation of state. They applied the method to figuring densities of liquid hydrocarbons with ethane and decane as reference fluids and with a modified Redlich-Kwong equation of state. Further work along this line is reported by Teja et al. (1980).

Conclusion

In the form used by Plöcker et al. (1978), the corresponding states Lee-Kesler method seems to be the most accurate general method available for nonpolar substances and mixtures. Extension to polar compounds could be made by adopting the polar factor form of the BWRS equation developed by Nishiumi (1980). Plöcker et al. (1978) have made computer programs available for carrying out calculations of compressibilities, fugacities and deviation functions.

1.6.4. Other General Equations

Recently attention has been focused on the repulsion term, $RT/(V-b)$, of the van der Waals equation, as described in Section 1.7.5. Along this line, Hlavaty (1974) took the compressibility factor to be made up of three parts: repulsion of hard spheres; attraction of the van der Waals type, a/V^2; and a small correction for which several empirical expressions were proposed. On this basis, Drahos et al. (1978) developed the equation

$$z = \frac{V^3 + bV^2 + b^2V}{(V-b)^3} - \frac{a}{V^2} - \frac{c(V-V_c)^2}{V^3} \tag{1.159}$$

for which the parameters are in Table 1.19. It is intended only for the vapor phase. For 1,046 data points, the density of the vapor phase was in error an average of 1.7 percent. In figuring high-pressure vapor liquid equilibria, the equation is applied to the vapor phase and combined with a modified solubility parameter method for the liquid phase, apparently quite successfully in comparison with other more complicated methods. Combining rules are given and tested for mixtures, and equations are given for fugacity and enthalpy deviations. More testing probably is required if this method is to compete with the BWRS or L-K methods.

Another combination of new types of repulsion and attraction terms is named BACK after its inventors: Boublik, Alder, Chen, and Kreglewski. It is described and applied to phase equilibria by Simnick et al. (1979). Thermodynamic properties of an azeotropic mixture of CH_3Cl_2 and CCl_2F_2 are developed with the aid of the BACK equation by Kudchaker (1979). The equation appears to be highly accurate in fitting

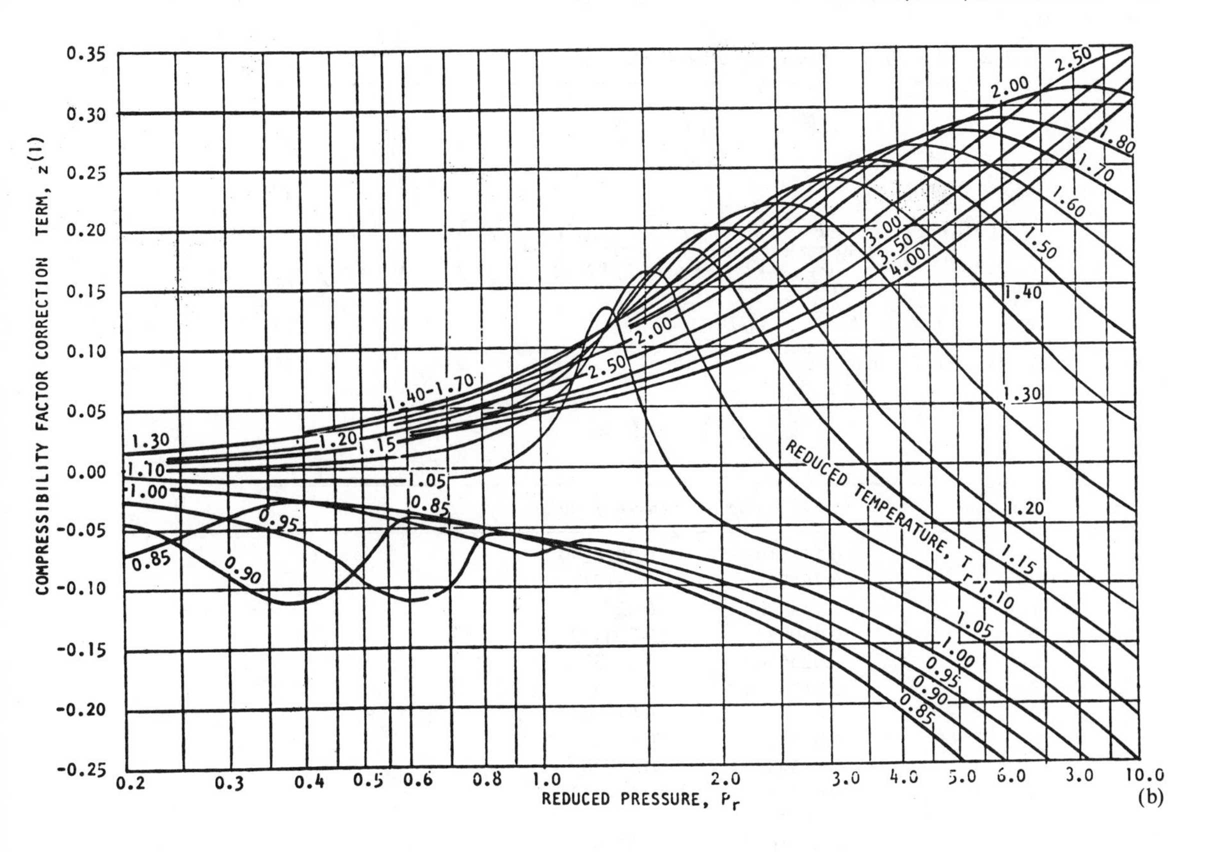

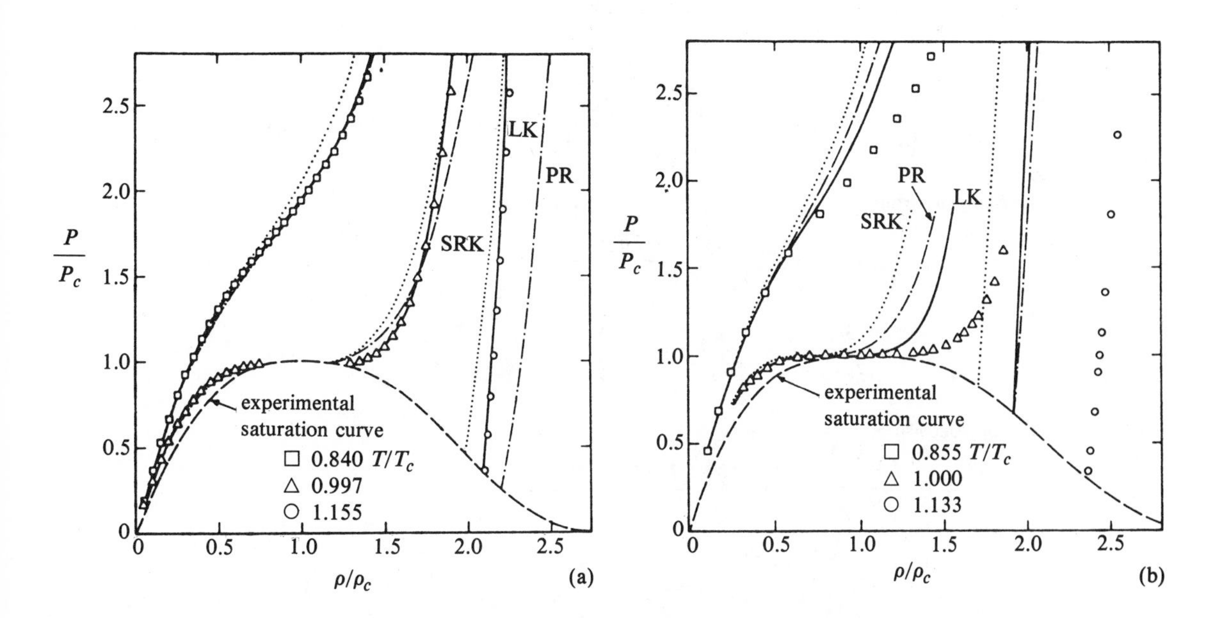

Figure 1.23 (a) Experimental isotherms of methane compared with calculations from the Soave-Redlich-Kwong (SRK), Peng-Robinson (PR), and Lee-Kesler (LK) equations. The Lee-Kesler predictions are best, but the SRK is acceptable (Prausnitz 1980). (b) Same as Figure 1.23(a) but for water. In this case none of the equations is satisfactory except at the lower temperatures over narrow ranges of density (Prausnitz, 1980).

Table 1.18. The Lee-Kesler Equation of State

$$Z = Z^{(0)} + \frac{\omega}{\omega^{(r)}}(Z^{(r)} - Z^{(0)}) = Z^{(0)} + \omega Z^{(1)}. \quad (1)$$

$$\omega^{(r)} = 0.3978. \quad (2)$$

$$Z = \left(\frac{P_r V_r}{T_r}\right) = 1 + \frac{B}{V_r} + \frac{C}{V_r^2} + \frac{D}{V_r^5} + \frac{c_4}{T_r^3 V_r^2}\left(\beta + \frac{\gamma}{V_r^2}\right)\exp\left(-\frac{\gamma}{V_r^2}\right). \quad (3)$$

$$B = b_1 - b_2/T_r - b_3/T_r^2 - b_4/T_r^3. \quad (4)$$

$$C = c_1 - c_2/T_r + c_3/T_r^3. \quad (5)$$

$$D = d_1 + d_2/T_r. \quad (6)$$

Constant	*Simple Fluids*	*Reference Fluids*
b_1	0.1181193	0.2026579
b_2	0.265728	0.331511
b_3	0.154790	0.027655
b_4	0.030323	0.203488
c_1	0.0236744	0.0313385
c_2	0.0186984	0.0503618
c_3	0.0	0.016901
c_4	0.042724	0.041577
$d_1 \times 10^4$	0.155488	0.48736
$d_2 \times 10^4$	0.623689	0.0740336
β	0.65392	1.226
γ	0.060167	0.03754

Fugacity coefficient:

$$\ln\left(\frac{f}{P}\right) = Z - 1 - \ln(Z) + \frac{B}{V_r} + \frac{C}{2V_r^2} + \frac{D}{5V_r^5} + E, \quad (7)$$

where

$$E = \frac{c_4}{2T_r^3\gamma}\left\{\beta + 1 - \left(\beta + 1 + \frac{\gamma}{V_r^2}\right)\exp\left(-\frac{\gamma}{V_r^2}\right)\right\}. \quad (8)$$

Enthalpy departure:

$$\frac{H - H^*}{RT_c} = T_r\left\{Z - 1 - \frac{b_2 + 2b_3/T_r + 3b_4/T_r^2}{T_r V_r} - \frac{c_2 - 3c_c/T_r^2}{2T_r V_r^2} + \frac{d_2}{5T_r V_r^5} + 3E\right\}. \quad (9)$$

Entropy departure:

$$\frac{S - S^*}{R} + \ln\left(\frac{P}{P^*}\right) = \ln(Z) - \frac{b_1 + b_3/T_r^2 + 2b_4/T_r^3}{V_r} - \frac{c_1 - 2c_3/T_r^3}{2V_r^2} - \frac{d_1}{5V_r^5} + 2E, \quad (10)$$

where $P^* = 1$ atm. for API Research Project 44 S^* data.

(continued on next page)

Table 1.18. *(continued)*

Pseudocritical properties:

$$V_{ci} = Z_{ci} R T_{ci} / P_{ci}. \quad (11)$$

$$Z_{ci} = 0.2905 - 0.085\omega_i. \quad (12)$$

$$V_c = \frac{1}{8} \sum_j \sum_k x_j x_k (V_{cj}^{1/3} + V_{ck}^{1/3})^3. \quad (13)$$

$$T_c = \frac{1}{8V_c} \sum_j \sum_k x_j x_k (V_{cj}^{1/3} + V_{ck}^{1/3})^3 \sqrt{T_{cj} T_{ck}}. \quad (14)$$

$$\omega = \sum_j x_j \omega_j. \quad (15)$$

$$P_c = Z_c R T_c / V_c = (0.2905 - 0.085\omega) R T_c / V_c. \quad (16)$$

Acentric factor:

$$\omega = \frac{\ln P_{br}^s - 5.92714 + 6.09648/T_{br} + 1.28862 \ln T_{br} - 0.169347\, T_{br}^6}{15.2518 - 15.6875/T_{br} - 13.4721 \ln T_{br} + 0.43577\, T_{br}^6} \quad (17)$$

Table 1.19. A Three-Parameter Equation of State (Hlavaty 1974; Drahos, Wichterle, & Hala 1978)

$$P = RT(V^3 + V^2 b + V b^2 - b^3)/V(V-b)^3 - a/V^2 - c(V - V_c)^2/V^4,$$

$$a = R^2 T_c^2 (a_1 + a_2 T_r + a_3/T_r)/P_c,$$

$$b = R T_c (b_1 + b_2 T_r + b_3/T_r)/P_c,$$

$$c = R^2 T_c^2 (c_1 + c_2 T_r)/P_c,$$

where

$$a_1 = -0.45307 + 5.8405\omega,$$

$$a_2 = 0.42448 - 3.5365\omega,$$

$$a_3 = 0.49780 - 2.3553\omega,$$

$$b_1 = -0.07788 + 0.60629\omega,$$

$$b_2 = 0.05548 - 0.35642\omega,$$

$$b_3 = 0.06603 - 0.25554\omega,$$

$$c_1 = 0.20570 + 0.38928\omega,$$

$$c_2 = -0.15299 - 0.28783\omega.$$

Mixture parameters:

$$a = \sum\sum y_i y_j a_{ij},$$

$$b = \sum\sum y_i y_j b_{ij},$$

$$c = a \sum y_i c_i / a_i,$$

$$V_c = b \sum y_i V_{ci} / b_i,$$

$$a_{ij} = (a_{ii} a_{jj})^{1/2},$$

$$b_{ij} = (b_{ii}^{1/3} + b_{jj}^{1/3})^3 / 8.$$

$$d_i^a = 2 \sum_j y_j a_{ij},$$

$$d_i^b = 2 \sum_j y_j b_{ij},$$

(continued on next page)

Table 1.19. *(continued)*

$$d_i^c = c_i/a_i,$$

$$d_i^{V_c} = V_{ci}/b_i.$$

Partial fugacity coefficients:

$$\ln \hat{\phi}_i^G = 3b^3 - 2Vbd_i^b - 5Vb^2 + 4V^2d_i^b)/(V-b)^3 - d_i^a/RTV$$
$$+ [c - cd_i^a/a - ad_i^c + (cV_cd_i^a/a + aV_cd_i^c + cV_cd_i^b/b$$
$$+ bcd_i^{V_c} - 3cV_c)/V + V_c(5cV_c - cV_cd_i^a/a - aV_cd_i^c$$
$$- 2cV_cd_i^b/b - 2cbd_i^{V_c})/3V^2]/RTV - \ln(PV/RT).$$

Example 1.17 Compressibility of *n*-Pentane by the Lee-Kesler Method

Several isotherms of P_r against $V_r^{(0)}$ or $V_r^{(r)}$ are constructed on the basis of Eq. 3 of Table 1.18. The two sets of curves are practically identical on the scale of these drawings.

Values of V_r at $P_r = 0.4$ are read off the plots and converted into compressibilities with the equation

$$z = \frac{P_r}{T_r}\left[V_r^{(0)} + \frac{0.251}{0.3978}(V_r^{(r)} - V_r^{(0)})\right],$$

where 0.251 is the acentric factor of *n*-pentane and 0.3978 that of the reference substance, *n*-octane. In the present instance, however, the last term essentially drops out so that

$$z = P_rV_r^{(0)}/T_r.$$

The Lee-Kesler Tables, E.1 and E.2, also are used to find the compressibilities with

$$z = z^{(0)} + 0.251z^{(1)}.$$

The results are tabulated.

At $T_r = 0.8$, the graphs identify both liquid and vapor phases, but the Lee-Kesler Table identifies only a liquid phase.

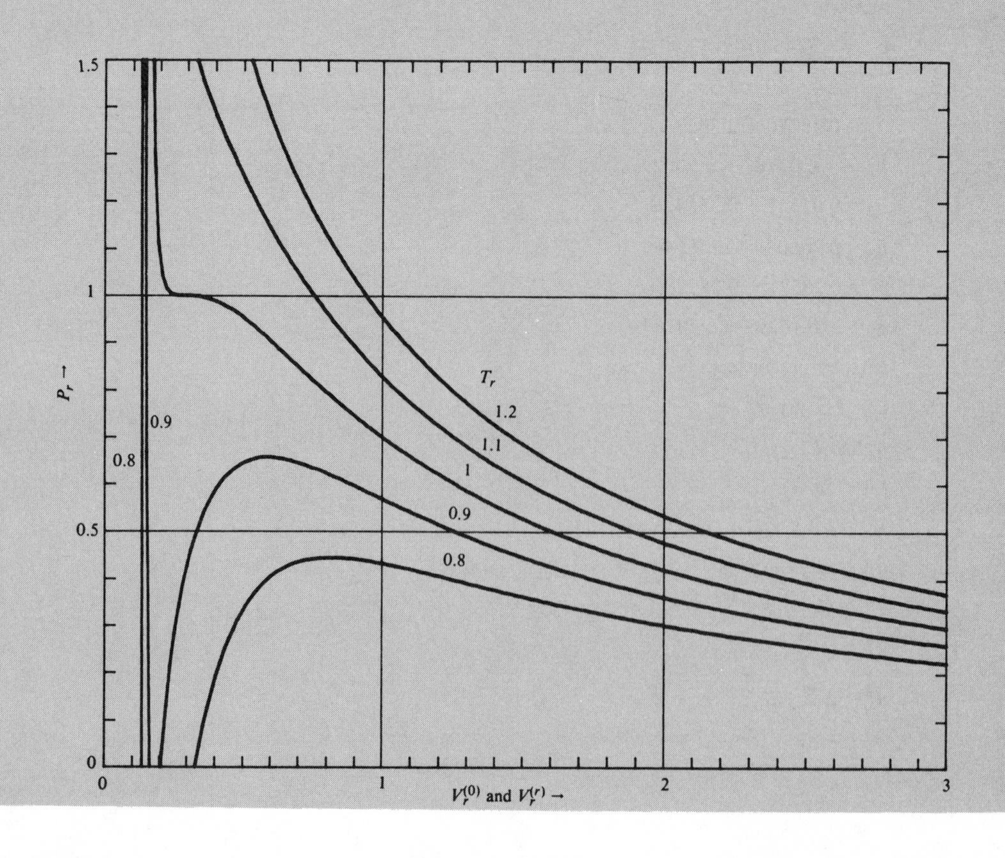

Example 1.17. *(continued)*

$P_r = 0.40$			L-K Tables		
T_r	$V_r^{(0)} = V_r^{(r)}$	z	$z^{(0)}$	$z^{(1)}$	z
0.8	1.27(0.12)	0.63(0.06)	0.0661	−0.0272	0.0593
0.9	1.78	0.756	0.7800	−0.1118	0.7549
1.0	2.18	0.844	0.8509	−0.0285	0.8437
1.1	2.48	0.891	0.8930	0.0038	0.8940
1.2	2.82	0.923	0.9205	0.0190	0.9253

Example 1.18 Pseudocritical Properties for Use with the Lee-Kesler Method

This is an exercise in the use of the rather complex rules for evaluation of pseudocritical properties of mixtures with the Lee-Kesler system. The pertinent properties of the pure components and the results for the mixture are tabulated. The formulas are Eqs. 11–16 of Table 1.18.

$$z_{ci} = 0.2905 - 0.085\omega_i,$$

$$\omega = \Sigma x_i \omega_i,$$

$$z_c = 0.2905 - 0.085\Sigma x_i \omega_i,$$

$$V_{ci} = 82.05\, z_{ci} T_{ci}/P_{ci},$$

$$\begin{aligned} V_c &= x_1^2 V_{c1} + x_2^2 V_{c2} + x_3^2 V_{c3} + 0.25[x_1x_2(V_{c1}^{1/3} + V_{c2}^{1/3})^3 \\ &\quad + x_1x_3(V_{c1}^{1/3} + V_{c3}^{1/3})^3 + x_2x_3(V_{c2}^{1/3} + V_{c3}^{1/3})^3] \\ &= 0.04(269.29) + 0.09(175.38) + 0.25(157.21) \\ &\quad + 0.25[0.06(6.4576 + 5.5975)^3 \\ &\quad + 0.1(6.4576 + 5.3971)^3 + 0.15(5.5975 + 5.3971)^3] \\ &= 183.63, \end{aligned}$$

$$\begin{aligned} V_cT_c &= x_1^2 V_{c1}T_{c1} + x_2^2 V_{c2}T_{c2} + x_3^2 V_{c3}T_{c3} \\ &\quad + 0.25[x_1x_2(T_{c1}T_{c2})^{0.5}(V_{c1}^{1/3} + V_{c2}^{1/3})^3 \\ &\quad + x_1x_3(T_{c1}T_{c3})^{0.5}(V_{c1}^{1/3} + V_{c3}^{1/3})^3 \\ &\quad + x_2x_3(T_{c2}T_{c3})^{0.5}(V_{c2}^{1/3} + V_{c3}^{1/3})^3] \\ &= 87{,}162, \end{aligned}$$

$$T_c = 87{,}162/183.63 = 474.67,$$

$$P_c = z_cRT_c/V_c = 0.2514(82.05)(474.67)/183.63 = 53.32.$$

The calculated pseudocritical properties are somewhat intermediate to the critical properties of the pure components, as expected.

i	x	T_c	P_c	ω	z_c	V_c	$V_c^{1/3}$
1	0.2	600	50	0.2	0.2735	269.29	6.4576
2	0.3	500	60	0.4	0.2565	175.38	5.5975
3	0.5	400	50	0.6	0.2395	157.21	5.3971
Mixture		474.67	53.32	0.46	0.2514	183.63	

the *PVT* data of many substances and its parameters depend only on properties of pure substances. The entire subject of equations of state that are based on the theory of repulsions of hard spheres is reviewed recently by Leland (1980). The statistical mechanical background is reviewed by Boublik (1977).

1.6.5. Liquids

Equations such as the Soave, Peng-Robinson, and Lee-Kesler usually are not precise enough for densities of liquids. Many special equations for that purpose have been proposed. A corresponding states equation for saturated liquid densities of pure substances and mixtures is due to Hankinson & Thomson (1979, 1982). It was tested against 6,500 data points, with superior results. The saturated specific volume, V_s, is given by

$$V_s/V^* = V_R^{(0)}(1 - \omega V_R^{(\delta)}). \tag{1.160}$$

Here V^* is a "characteristic volume" that has been obtained for several hundred substances and has been correlated in terms of the acentric factor for other substances. $V_R^{(0)}$ is a characteristic of spherical molecules, and $V_R^{(\delta)}$ is a correction factor, both of which are correlated in terms of reduced temperature. Some details are given in Table 1.20.

Densities of compressed liquids are related to those of saturated liquids by the Tait equation by Thomson et al. (1982). The formula is

$$V/V_s = 1 - C \ln \frac{B+P}{B+P_s}. \tag{1.160a}$$

C is represented as a linear function of the acentric function, and B as a more complex function of both acentric factor and reduced temperature. The formulas are given in Table 1.20. The methods are extended to mixtures with appropriate mixing rules.

Another density correction method of interest is due to Rea et al. (1973), who write

$$V/V_R = K_R/K, \tag{1.161}$$

where K is a third-degree polynomial in T_r of which the coefficients are fourth-degree polynomials in P_r, and K_R is the same function evaluated at the temperature and pressure at which the reference-specific volume V_R is known. The twenty numerical coefficients given in Table 1.21 were evaluated from a set of 2,600 data points.

A comparatively simple set of relations was proposed by Yamada & Gunn (1973):

$$V_s = V' z_c^{(1+(1-T_r)^{2/7})} \exp(-(1-T_r')^{2/7}), \tag{1.161a}$$

$$z_c = 0.29056 - 0.08775\,\omega. \tag{1.161b}$$

The critical compressibility often is expressed in terms of the acentric factor because accurate experimental values of the former are not always available. The V' and T_r' are corresponding values of the specific volume and the reduced temperature. For mixtures, several combining rules were studied by Spencer & Danner (1973). One of the better equations for the saturated specific volume is

$$V_s = R(\Sigma x_i T_{ci}/P_{ci})\, z_c^{(1+(1-T_r)^{2/7})}, \tag{1.161c}$$

$$z_c = \Sigma x_i z_{ci}. \tag{1.161d}$$

For quite another approach to correlating the densities of saturated fluids, reference may be made to Thompson (1979). The corresponding states method of Teja et al. (1981) was cited in Section 1.6.3. A relatively simple method that can be used without computers relates densities at two temperatures, T and T^*, by

$$\Delta\rho = \rho_L - \rho_V = (\Delta\rho)^* \left[\frac{T_c - T}{T_c - T^*}\right]^k.$$

Values of k are given as 0.25 for alcohols and water, 0.29 for hydrocarbons and ethers, 0.31 for other organic compounds, and 0.33 for inorganic compounds by Fishtine (1963). A less generalized correlation of liquid densities was developed by Shah & Yaws (1976) as the equation

$$\rho = k_1/k_2^{(1-T/T_c)^{2/7}}. \tag{1.162}$$

Values of the constants are given for sixty-two liquids.

Densities at other pressures also can be related to those at saturation by this form of the Tait equation,

$$V/V_s = 1 - k_1 \ln(1 + P/k_2), \tag{1.163}$$

with constants that are specific to each substance; or the similar Eq. 1.160a with generalized constants may be adequate. Many compressibility data are given in reference works—for example, in Gray (1972) (*American Institute of Physics Handbook*) and in Bolz & Tuve (1973) (*Handbook of Tables for Applied Engineering Science*).

Some earlier work on liquid densities is described in Bretsznajder (1962), Reid et al. (1977), and Sterbacek et al. (1979). Several relations of early historical interest concerning thermal expansion and compressibility of liquids are discussed by Partington (1951)—for instance:

$$V = V_0/(1 - kT) \quad \text{(Mendeléev 1860)}, \tag{1.164}$$

$$V = k_1 - k_2 \ln(T_c - T) \quad \text{(Avenarius 1876)}, \tag{1.165}$$

$$\left(\frac{\delta V}{\delta P}\right)_T = k_1 \ln(1 + k_2 P) \quad \text{(Tait 1888)}. \tag{1.166}$$

The preceding equation has been used at pressures to 25,000 atm.

$$\left(\frac{\partial V}{\partial P}\right)_T = k_1/(k_2 + k_3 C + P), \quad \text{(Tait 1888)} \tag{1.167}$$

where C is the concentration of salt solution. This is an early observation that adding foreign molecules to a liquid may be numerically equivalent to raising the pressure, in its effect on the density.

$$V = V^*(kT_b - T^*)/(kT_b - T) \quad \text{(Guye \& Jordan 1896)}, \tag{1.168}$$

where T_b is the atmospheric boiling point.

$$V = k_1 + k_2/(1 + k_3 P) \quad \text{(Tumlirz 1909)}, \tag{1.169}$$

$$1/V - 1/V_c = k_1(T - T_c)^{k_2} \quad \text{(Herz 1919)}, \tag{1.170a}$$

$$V = k(1 - T/T_c)^{k_2} \quad \text{Herz (1932)}. \tag{1.170b}$$

1.6.6. Individual Substances

The importance of certain substances or mixtures requires accuracy that may not be provided by generalized equations of state. Successful equations of special forms have been developed for many substances. A prime example is the fundamental equation of water, which gives the Helmholtz energy in terms of density and temperature, and is shown in Table 1.22. As given by Keenan et al. (1978), the equation has fifty-eight numerical coefficients and is adequate for deriving all the data in the steam tables. Equations for the other properties in terms of the fundamental equation are cited in the table.

Parameters of the BWR and related equations are specific for individual substances as well as of a generalized character. Extended BWR forms of high accuracy have been developed by Bender (1975) for ethylene and propylene in both liquid and vapor phases, along with similar equations for other

Table 1.20. Densities of Saturated and Compressed Liquids and Mixtures (Hankinson & Thomson 1979; Hankinson, Coker, & Thomson 1982; Thomson, Brobst, & Hankinson 1982)

The saturated density is represented in the range, $0.25 < T_r < 0.95$, by the equations:

$$\frac{V_s}{V^*} = V_R^{(0)}[1 - \omega_{\text{SRK}} V_R^{(\delta)}],$$

$$V_R^{(0)} = 1 + a(1 - T_R)^{1/3} + b(1 - T_R)^{2/3} + c(1 - T_R) + d(1 - T_R)^{4/3},$$

$$V_R^{(\delta)} = [e + fT_R + gT_R^2 + hT_R^3]/(T_R - 1.00001).$$

$a = -1.52816$
$b = 1.43907$
$c = -0.81446$
$d = 0.190454$
$e = -0.296123$
$f = 0.386914$
$g = -0.0427258$
$h = -0.0480645$

Values of the characteristic volume, V^*, are given in the original for several hundred substances, but they have been correlated also in terms of the acentric factor for nine groups of substances as

$$V^* = \frac{RT_c}{P_c}(a + b\omega_{\text{SRK}} + c\omega_{\text{SRK}}^2),$$

where the constants of the nine groups are:

	Paraffins	*Olefins and Diolefins*	*Cycloparaffins*	*Aromatics*	*All Hydrocarbons*
a	0.2905331	0.3070619	0.6564296	0.2717636	0.2851686
b	−0.08057958	−0.2368581	−3.391715	−0.05759377	−0.06379110
c	0.02276965	0.2834693	7.442388	0.05527757	0.01379173
Avg abs. % error	1.23%	1.43%	1.00%	0.58%	1.89%

	Sulfur Compounds	*Fluorocarbons*	*Cryogenic Liquids*	*Condensable Gases*
a	0.3053426	0.5218098	0.2960998	0.2828447
b	−0.1703247	−2.346916	−0.05468500	−0.1183987
c	0.1753972	5.407302	−0.1901563	0.1050570
Avg abs. % error	1.98%	0.82%	0.85%	3.65%

The acentric factor, ω_{SRK}, is given adequately by the equation of Lee & Kesler (1975), where $P_{br}^s = 1/P_c$ and $T_{br} = T_b/T_c$ is the reduced atmospheric boiling point and P_c is in atm:

$$\omega = \frac{\ln P_{br}^s - 5.92714 + 6.09648/T_{br} + 1.28862 \ln T_{br} - 0.169347\, T_{br}^6}{15.2518 - 15.6875/T_{br} - 13.4721 \ln T_{br} + 0.43577\, T_{br}^6}.$$

The ratio of compressed and saturated specific volumes is given in terms of the pressure, the saturation pressure P_s, the acentric factor, and the reduced temperature by the set of relations and parameters:

$$V = V_s\left(1 - C \ln \frac{B + P}{B + P_s}\right),$$

$$B/P_c = -1 + a(1 - T_R)^{1/3} + b(1 - T_R)^{2/3} + d(1 - T_R) + e(1 - T_R)^{4/3},$$

$$e = \exp(f + g\,\omega_{SRK} + h\,\omega_{SRK}^2),$$

$$C = j + k\omega_{SRK}.$$

$a = -9.070217$, $g = 0.250047$,
$b = 62.45326$, $h = 1.14188$,
$d = -135.1102$, $j = 0.0861488$,
$f = 4.79594$, $k = 0.0344483$.

The original papers present combining rules for applying these correlations to mixtures.

Table 1.21. Adjustment of a Known Liquid Density for Temperature and Pressure (Rea, Spencer, & Danner 1973)

$\rho_T = (K_T/K_r)\rho_r$.

K_r is evaluated from the given equation at the temperature and pressure at which the density, ρ_r, is known. K_T is evaluated from the same equation at the desired temperature and pressure:

$K = A_0 + A_1T_r + A_2T_r^2 + A_3T_r^3$,

where A_i is given by

$A_i = B_{0,i} + B_{1,i}P_r + B_{2,i}P_r^2 + B_{3,i}P_r^3 + B_{4,i}P_r^4$.

i	$B_{0,i}$	$B_{1,i}$	$B_{2,i}$	$B_{3,i}$	$B_{4,i}$
0	1.6368	−0.04615	2.1138(10⁻³)	−0.7845(10⁻⁵)	−0.6923(10⁻⁶)
1	−1.9693	0.21874	−8.0028(10⁻³)	−8.2823(10⁻⁵)	5.2604(10⁻⁶)
2	2.4638	−0.36461	12.8763(10⁻³)	14.8059(10⁻⁵)	−8.6895(10⁻⁶)
3	−1.5841	0.25136	−11.3805(10⁻³)	9.5672(10⁻⁵)	2.1812(10⁻⁶)

substances. Experimental data and a criticism of equations of state of neon at very low temperatures are given by Gibbons (1969). Many data of second and some higher virial coefficients are available and have been referred to in Section 1.4. Precise measurement of the second virial coefficient of ethylene at low temperatures is reported by Waxman & Davis (1979). An equation of basic structure similar to the Beattie-Bridgeman was developed with parameters specific to propane by Goodwin (1979).

A comprehensive discussion of the recent literature of experimental and correlation work for many pure substances such as carbon dioxide, inert gases, air, ammonia, water, hydrogen, light hydrocarbons, and some mixtures is given by Douslin (1975). This article has a long table of references to experimental investigations on mixtures, classifying them with respect to thirteen thermodynamic properties. A list of about two hundred "most important reports" of *PVT* data is given by Brielles et al. (1975). Pressure and temperature ranges, estimated precision, and the kind of experimental equipment are shown.

1.7. INTERMOLECULAR FORCES AND EQUATIONS OF STATE

Ultimately, bulk properties of substances are a result of the properties of individual molecules and interactions between them. This molecular viewpoint is the subject of statistical thermodynamics. Some results pertinent to equations of state that have been achieved by this discipline will be cited in this section.

1.7.1. Energy Relations

The only forms of energy possessed by ideal gases are those of translation of the constituent molecules. Real substances also have rotational, vibrational, and potential energies corresponding to separations of the atoms in the molecules. If these forms of energy are designated by ε_i, the sum

$$Q = \Sigma g_i \exp(-\varepsilon_i/kT)$$

is called the molecular partition function. The meanings of the terms are:

ε_i = a quantized energy of any of the types mentioned.

g_i = number of quantized states having the same or nearly the same energy, ε_i.

k = Boltzmann constant, 1.3805 ($E-16$) erg/°K.

The energies of the molecules of an ensemble cover a wide range of magnitudes whose distribution is expressed by the Maxwell-Boltzmann Law:

$$p_i = N_i/N = \frac{g_i\exp(-\varepsilon_i/kT)}{\Sigma g_i\exp(-\varepsilon_i/kT)} = \frac{g_i\exp(-\varepsilon_i/kT)}{Q}, \tag{1.171}$$

where p_i = fraction of the molecules having the quantized energy, ε_i. Translational, rotational, and vibrational energies, which are deducible with great accuracy from spectral observations, are the major contributors to molecular energies. Potential energies corresponding to separations of atoms in complex molecules also are significant quantities, but most of this theory has been applied only to simpler molecules.

Major thermodynamic properties are expressible in terms of the partition function Q. Thus the internal energy of the ensemble is the sum of all the energy levels of all the molecules, or

$$U = \Sigma N_i\varepsilon_i = \frac{N\Sigma\varepsilon_i g_i\exp(-\varepsilon_i/kT)}{\Sigma g_i\exp(-\varepsilon_i/kT)} \tag{1.172}$$

$$= \frac{N}{Q}\Sigma\varepsilon_i g_i\exp(-\varepsilon_i/kT) = RT^2\left(\frac{\partial \ln Q}{\partial T}\right)_V. \tag{1.173}$$

The deduction for the other key property, the entropy, results in

$$S = R\left[T\left(\frac{\partial \ln Q}{\partial T}\right)_V + \ln Q\right]. \tag{1.174}$$

Table 1.22. The Fundamental Equation of Water and Thermodynamic Functions Derived from It (Keenan et al. 1978)

The Helmholtz energy is

$$\psi = \psi_0(T) + RT[\ln \rho + \rho Q(\rho, \tau)], \quad (1)$$

where

$$\psi_0 = \sum_{i=1}^{6} C_i/\tau^{i-1} + C_7 \ln T + C_8 \ln T/\tau, \quad (2)$$

and

$$Q = (\tau - \tau_c) \sum_{j=1}^{7} (\tau - \tau_{aj})^{j-2} \left[\sum_{i=1}^{8} A_{ij}(\rho - \rho_{aj})^{i-1} + e^{-E\rho} \sum_{i=9}^{10} A_{ij}\rho^{i-9} \right]. \quad (3)$$

In (1), (2), and (3) T denotes temperature on the Kelvin scale, τ denotes $1{,}000/T$, ρ denotes density in g/cm^3, $R = 4.6151$ bar cm^3/g°K or 0.46151 J/g°K, $\tau_c \equiv 1{,}000/T_{\text{crit}} = 1.544912$, $E = 4.8$, and

$$\tau_{aj} = \tau_c (j = 1) \qquad \rho_{aj} = 0.634 (j = 1)$$
$$= 2.5 (j > 1), \qquad = 1.0 (j > 1).$$

The coefficients for ψ_0 in joules per gram are given as follows:

$C_1 = 1857.065$ $\quad C_4 = 36.6649$ $\quad C_7 = 46.$

$C_2 = 3229.12$ $\quad C_5 = -20.5516$ $\quad C_8 = -1011.249$

$C_3 = -419.465$ $\quad C_6 = 4.85233$

Values of the 50 numerical coefficients A_{ij} are given in the reference. Derived properties are:

Pressure:

$$p = \rho RT \left[1 + \rho Q + \rho^2 \left(\frac{\partial Q}{\partial \rho} \right)_\tau \right]. \quad (4)$$

Internal energy:

$$u = RT \rho\tau \left(\frac{\partial Q}{\partial \tau} \right)_\rho + \frac{d(\psi_0 \tau)}{d\tau}. \quad (5)$$

Entropy:

$$s = -R \left[\ln \rho + \rho Q - \rho\tau \left(\frac{\partial Q}{\delta \tau} \right)_\rho \right] - \frac{d\psi_0}{dT}. \quad (6)$$

Enthalpy:

$$h = RT \left[\rho\tau \left(\frac{\partial Q}{\partial \tau} \right)_\rho + 1 + \rho Q + \rho^2 \left(\frac{\partial Q}{\partial \rho} \right)_\tau \right] + \frac{d(\psi_0 \tau)}{d\tau}. \quad (7)$$

Second virial coefficient:

$$B \equiv \left\{ \left[\frac{\partial (p/\rho RT)}{\partial \rho} \right]_T \right\}_{\rho=0} = Q_{\rho=0}. \quad (8)$$

Fugacity:

$$\ln(f/P) = \frac{\psi - \psi_0}{RT} + \ln \frac{V}{V^0} - \ln z + z - 1. \quad (9)$$

Compressibility:

$$z = P/\rho RT. \quad (10)$$

Other properties follow from these two with familiar thermodynamic manipulations—for example:

$$A = U - TS = -RT \ln Q, \quad (1.175)$$

$$P = -\left(\frac{\partial A}{\partial V} \right)_T = RT \left(\frac{\partial \ln Q}{\partial V} \right)_T, \quad (1.176)$$

$$H = U + PV = RT \left[T \left(\frac{\partial \ln Q}{\partial T} \right)_V + V \left(\frac{\partial \ln Q}{\partial V} \right)_T \right], \quad (1.177)$$

$$G = RT \left[V \left(\frac{\partial \ln Q}{\partial V} \right)_T - \ln Q \right], \quad (1.178)$$

$$C_p = \left(\frac{\partial H}{\partial T} \right)_P, \quad (1.179)$$

$$C_v = \left(\frac{\partial U}{\partial T} \right)_V, \quad (1.180)$$

$$\mu_i = -RT \left(\frac{\partial \ln Q}{\partial n_i} \right)_{TVn_j}. \quad (1.181)$$

1.7.2. Real Substances

Forces of attraction keep molecules from flying off into space; those of repulsion keep them from mutual destruction. The mathematical formulation of such interactions is expressed conveniently in terms of potential functions whose gradients are related to the forces and which are functions of the distances of separation of the molecules. The relation is

$$u(r) = -\frac{dF}{dr} \quad (1.182)$$

for a one-dimensional case. Energy changes are found by integration over distances.

Molecules are considered to interact as pairs, triplets, and so on. These interactions result in the deviations from ideality that correspond to the successive virial coefficients of the equation of state,

$$z = 1 + B\rho + C\rho^2 + D\rho^3 + \ldots. \quad (1.183)$$

Equations for the coefficients in terms of the potential function are known, that for the second virial coefficient being

$$B = 2\pi N_a \int_0^\infty (1 - \exp(-u/kT)) r^2 dr, \quad (1.184)$$

where N_a is the Avogadro number. A derivation of this relation is made, for example, by Beattie & Stockmayer (1951). Equations for the higher virial coefficients are more complicated but are still based on the potential function. Accordingly, all thermodynamic properties that are derivable from an equation of state are ultimately derivable from the potential function.

Table 1.23. Some Commonly Used Potential Functions

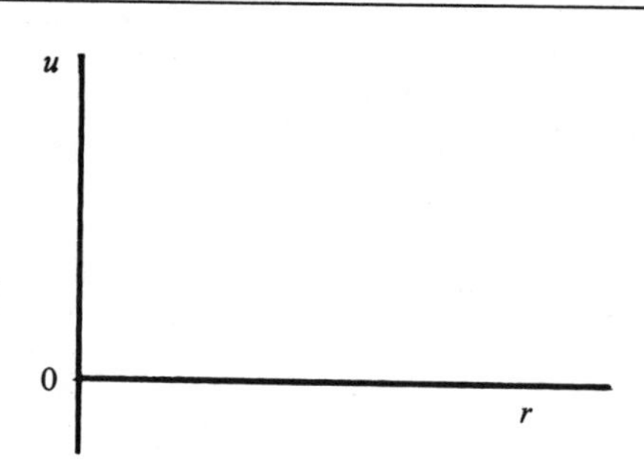

a. Ideal gas, no intermolecular forces:

$u = 0.$

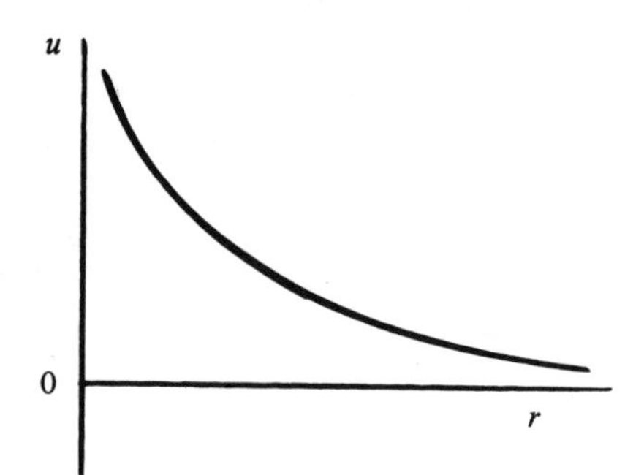

b. Point of repulsion is at the centers of the molecules:

$u = ar^{-\alpha}.$

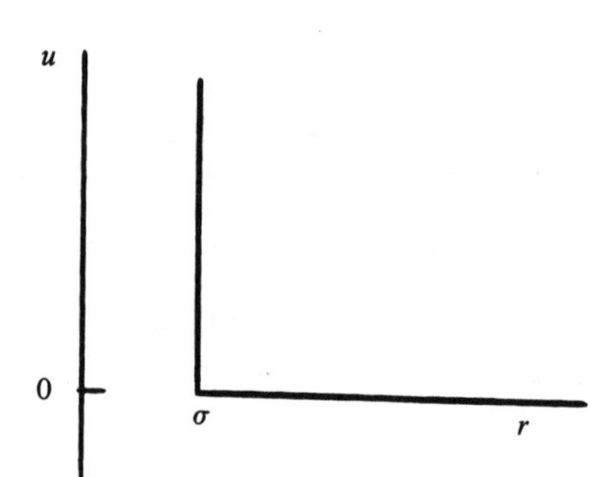

c. Hard-sphere potential. Point of repulsion is at surfaces of the molecules:

$$u = \begin{cases} \infty, & r < \sigma. \\ 0, & r > \sigma. \end{cases}$$

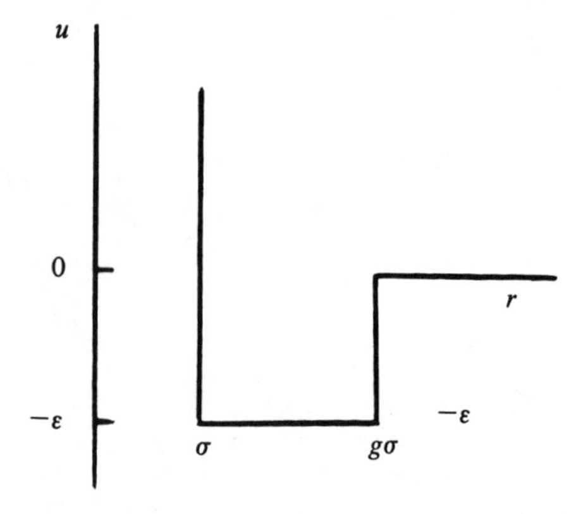

d. Square-well potential. Point of repulsion at surface, with constant attraction over a limited distance:

$$u = \begin{cases} \infty, & r < \sigma. \\ -\varepsilon, & \sigma < r < g\sigma. \\ 0 & r > g\sigma. \end{cases}$$

Table 1.23. *(continued)*

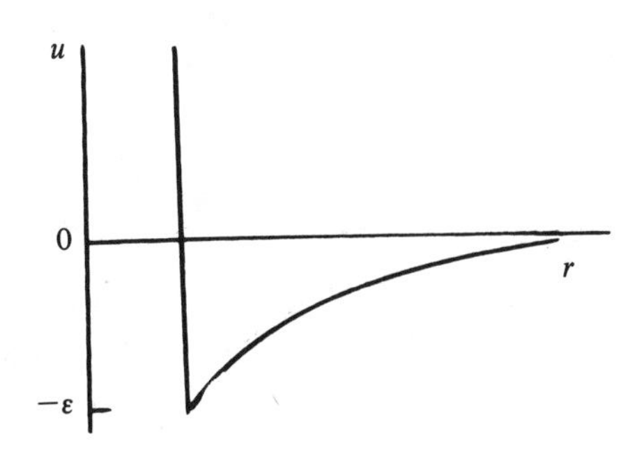

e. Sutherland potential. Point of repulsion at surface, attraction varies with distance:

$$u = \begin{cases} \infty, & r < \sigma. \\ -\varepsilon \left(\dfrac{\sigma}{r}\right)^{\alpha}, & r > \sigma. \end{cases}$$

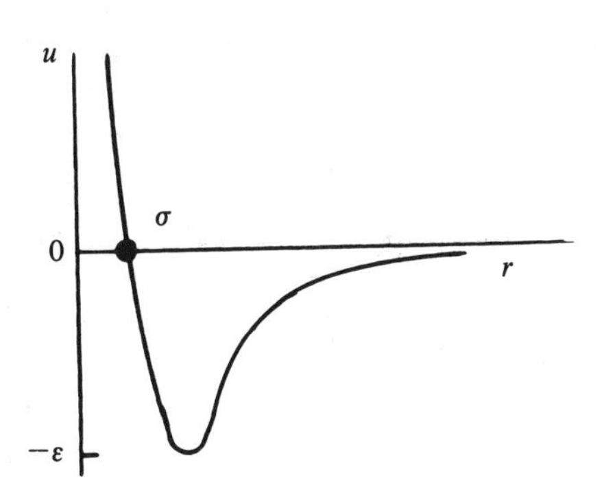

f. Bireciprocal, of which the Lennard-Jones is a special case.

$$u = a\left(\frac{\sigma}{r}\right)^{\alpha} - b\left(\frac{\sigma}{r}\right)^{\beta}.$$

$$u_{\mathrm{LJ}} = 4\varepsilon\left[\left(\frac{\sigma}{r}\right)^{12} - \left(\frac{\sigma}{r}\right)^{6}\right].$$

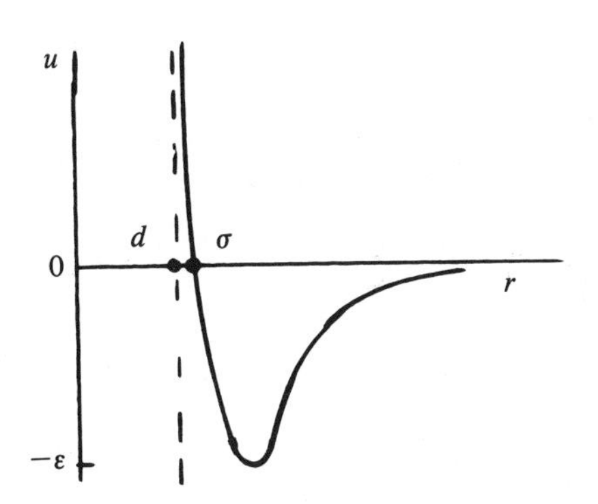

g. Kihara potential. Point of repulsion is a hard core of diameter d, which is smaller than the molecular diameter σ.

$$u = \begin{cases} \infty, & r < d. \\ 4\varepsilon\left[\left(\dfrac{\sigma - d}{r - d}\right)^{12} - \left(\dfrac{\sigma - d}{r - d}\right)^{6}\right], & r \geq d. \end{cases}$$

Note: Some detail and literature references are given in Mason & Spurling (1969).

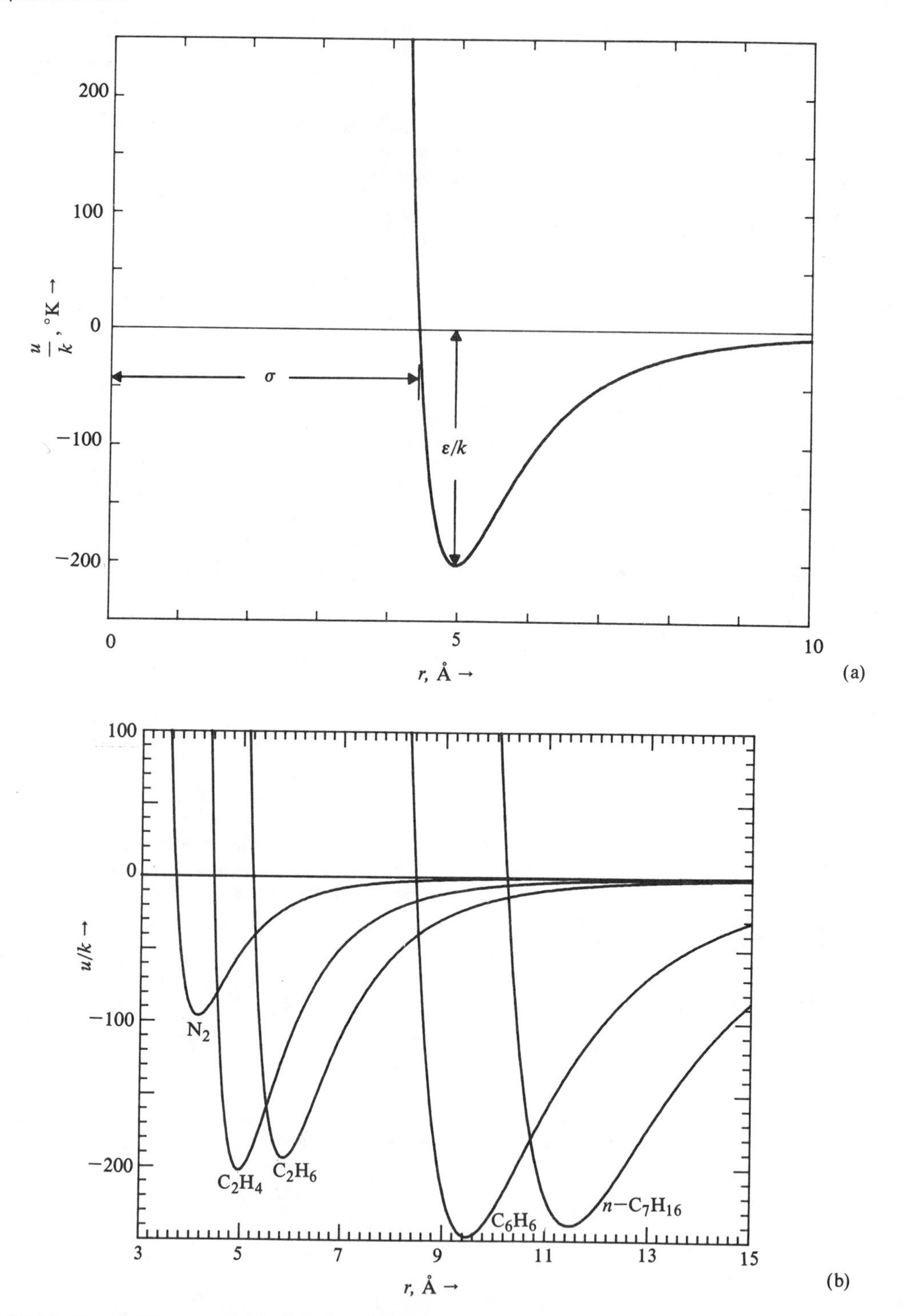

Figure 1.24 (a) Lennard-Jones potential of ethylene derived from second virial coefficients. The parameters are $\sigma = 4.433$ Å, $\varepsilon/k = 202.52$ K. (b) Lennard-Jones potentials of several molecules with ε/k and σ derived from measurements of second virial coefficients. The parameters are from a collection by Tee, Gotoh, & Stewart (1966).

1.7.3. Potential Functions

Potential functions (see Table 1.23) are empirical. They correspond to forces of attraction and repulsion. There is a theoretical basis, due to London, that attraction potentials vary inversely as the sixth power of the separation, whereas some higher but not specific power is involved with repulsion potentials. A widely used function incorporating these observations is due to Mie (1903) and is called bireciprocal:

$$u = ar^{-n} - br^{-6}. \tag{1.185}$$

Infinite series solutions for the virial coefficients in terms of any $n - m$ potential have been obtained. A special case with $n = 12$ was introduced by Lennard-Jones (1924) and has been widely employed. The meanings of the parameters are indicated in Figure 1.24(a) which is a plot for ethylene.

Besides virial coefficients, some properties that involve potential functions are viscosity and molecular beam scattering, so such data can be used to find the parameters of potential functions. Hirschfelder et al. (1964) record some comparisons of the parameters from virial coefficients and viscosity. Lennard-Jones potentials derived from virial coefficients are shown for several molecules in Figure 1.24(b). Since the Lennard-Jones potential is consistent with the principle of corresponding states, the parameters should be expressible in terms of the critical constants. When no better information is available, the parameters can be estimated from the approximate equations,

$$\varepsilon/k = 0.775T_c, \tag{1.185a}$$

$$\sigma^3 = 0.983kT_c/P_c. \tag{1.185b}$$

Many other potential functions have been used; a few of the simpler ones are shown in Table 1.23. Some were developed on rational grounds, others partly because they lead to integrable relations that are convenient for theoretical studies. Those shown have 0, 1, 2, or 3 parameters that can be fitted to data. Besides the Lennard-Jones, two potentials are especially useful in theoretical work:

1. The hard-sphere or rigid sphere, which assumes that attractive forces are absent and that repulsive forces are infinite when the molecules touch and zero at finite separation. There is only one parameter.
2. Square-well potential. A hard-sphere repulsion term is retained, and a constant attraction potential over a finite distance is included. Since there are three parameters, ε, σ, and g, use of the square-well potential can result in quite flexible correlations.

1.7.4. Virial Coefficients

Integration for the second virial coefficient with these potential functions is straightforward, as shown in Example 1.19. For the Lennard-Jones, a numerical integration is convenient since the integrand diminishes rapidly with r. Tabulations of integrals with bireciprocal functions are available for $n = 8$ to 150 (Reed & Gubbins 1973, p. 452).

Quite different shapes of potential functions can give rise to essentially the same virial coefficient. The square-well and Lennard-Jones potentials corresponding to the second virial coefficients of argon are shown in Figure 1.25(a); Figure 1.25(b) illustrates how closely second virial coefficients can match a substantial portion of a square-well correlation.

Third and higher virial coefficients have been evaluated from several potential functions. A bibliography of such calculations is presented by Reed & Gubbins (1973, p. 460). The hard-sphere potential has been applied up to the seventh coefficient, with the results shown in Table 1.24. The "exact" hard-sphere virial equation is shown as the top line of Table 1.25, where other expansions and polynomial equivalents also are presented.

1.7.5. Perturbed Hard-Sphere Equations of State

Consideration of the forces of repulsion between rigid spheres leads to a virial equation for the repulsion pressure P_R (Ree & Hoover 1967):

$$P_R = \frac{RT}{V}(1 + 4y + 10y^2 + 18.36y^3 + 28.3y^4 + \ldots)$$

$$= \frac{RT}{V}\phi(y), \tag{1.186}$$

where

$$y = b/4V, \tag{1.187}$$

and b is a measure of the volumes occupied by the molecules but may be regarded as an empirical parameter. Several approximations of the infinite series in finite form have been devised and are shown in Table 1.25.

For real gases the pressure is made up of contributions of forces of attraction as well as repulsion. For instance, in the van der Waals equation,

$$P = \frac{RT}{V - b} - a/V^2, \tag{1.188}$$

the second term on the right is regarded as the attraction term, and $RT/(V - b)$, which is common to all the usual cubic EOS, is the repulsion term and may be replaced with P_R. The result should be an improvement over the original vdW. In terms of the Carnahan & Starling (1972), Table 1.25, approximation of $\phi(y)$, the modified vdW equation becomes

$$P = \frac{RT}{V}\left(\frac{1 + y + y^2 - y^3}{(1 - y)^3}\right) - a/V^2. \tag{1.189a}$$

Similarly, the modified Redlich-Kwong equation becomes

$$P = \frac{RT}{V}\left(\frac{1 + y + y^2 - y^3}{(1 - y)^3}\right) - \frac{a}{T^{0.5}V(V + b)}. \tag{1.189b}$$

The parameters a and b in these equations may be related to the critical properties by application of the mathematical conditions at the critical point, as in Example 1.12. Carnahan & Starling obtain these results:

	vdW		*RK*	
	Original	*Modified*	*Original*	*Modified*
a/RT_cV_c	9/8	1.3824	$1.283T_c^{0.5}$	$1.463T_c^{0.5}$
b/V_c	1/3	0.5216	0.2599	0.3326
z_c	3/8	0.3590	1/3	0.3157

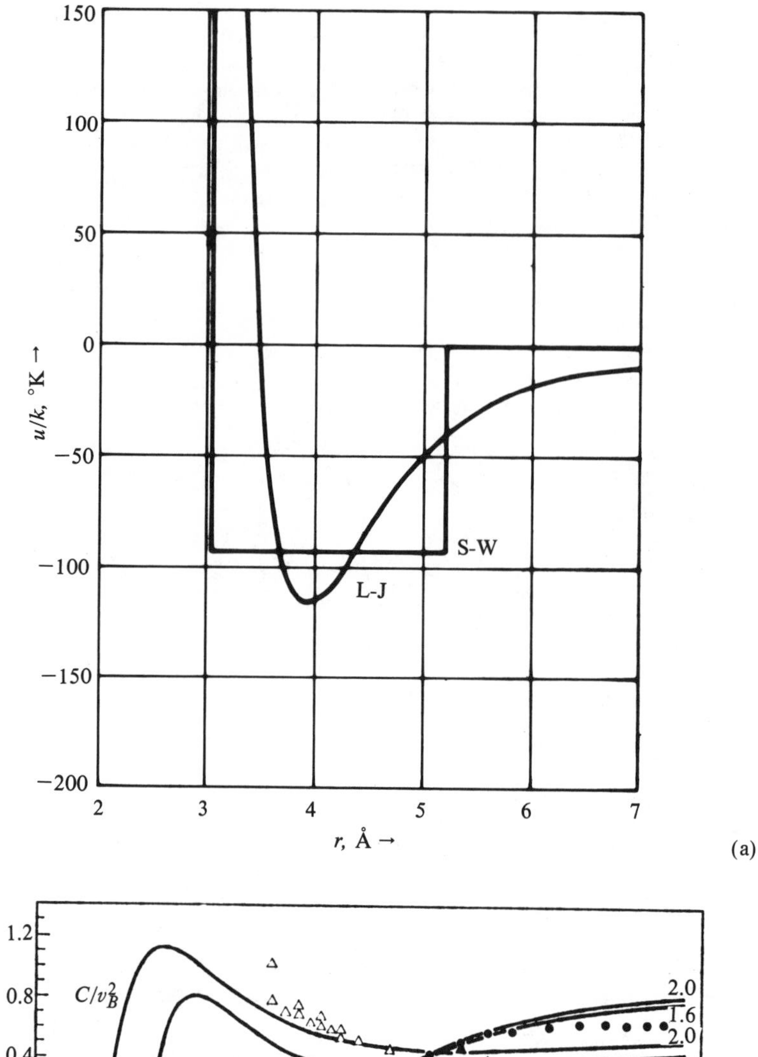

(a)

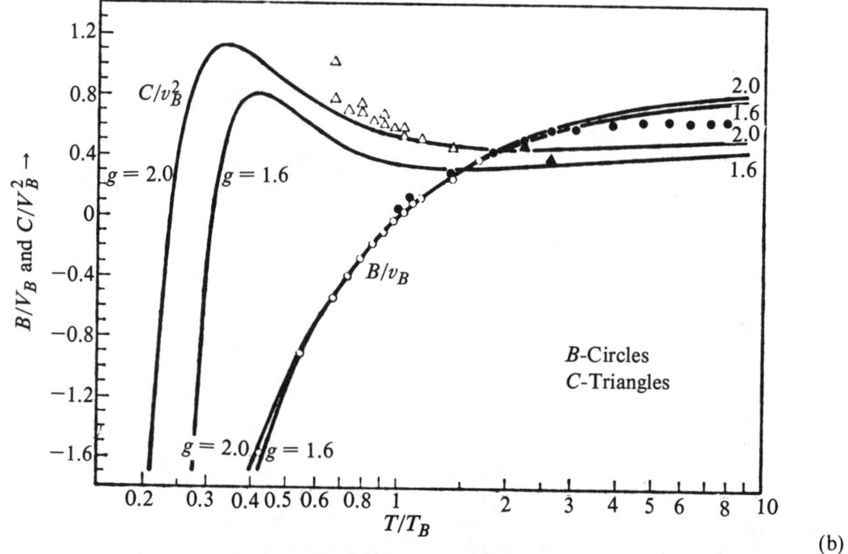

(b)

Figure 1.25 Virial coefficients and potential functions: (a) Square-well and Lennard-Jones potentials corresponding to second virial coefficients of argon (Prausnitz 1969); (b) second and third virial coefficients for the square-well potential of argon (○) and neon (●). $V_B = T(dB/dt)_{T=T_B}$. T_B is the temperature at which $B = 0$. (Mason & Spurling 1969)

Table 1.24. Equations for Virial Coefficients Based on the Hard-Sphere Potential (Mason & Spurling 1969, p. 186)

$$B = (2\pi N_0/3)\sigma^3 \equiv b_0,$$

$$C/b_0^2 = 5/8 = 0.625,$$

$$D/b_0^3 = \frac{1283}{8960} + \frac{3}{2}\left[\frac{73\sqrt{2} + (81)(17)(\arctan\sqrt{2} - \pi/4)}{(32)(35\pi)}\right],$$

$$= 0.286949,$$

$$E/b_0^4 = 0.11040 \pm 0.00006,$$

$$F/b_0^5 = 0.0386 \pm 0.0004,$$

$$G/b_0^6 = 0.0138 \pm 0.0004,$$

$$H/b_0^7 \approx 0.005.$$

Proof of these results is requested in Problem 1.36.

Several studies based as perturbed hard-sphere equations of state may be cited.

1. Carnahan & Starling (1972) worked with modified van der Waals and Redlich-Kwong equations. The latter was found to give superior performance for densities and enthalpy deviations of pure substances and some hydrocarbon binaries.

2. A unique attraction term was developed by Beret & Prausnitz (1975) and combined with a hard-sphere repulsion to form a three-parameter equation of state that appears to be valid even for large structurally complex molecules. This work was continued by Donohue & Prausnitz (1977, 1978). Parameters were developed for all substances commonly encountered in the natural gas industry, and a computer program for utilizing the complex system was prepared. The results are applicable at high pressures to mixtures of components differing widely in size and shape. Predictions are compared with data but not with other procedures used in this industry, such as the API-Soave or the Peng-Robinson systems.

3. De Santis et al. (1976) adopted the Clausius attraction term with the result

$$P = \frac{RT}{V}\left(\frac{1 + y + y^2 + y^3}{(1 - y)^3}\right) - \frac{a}{V(V + b)}. \qquad (1.190)$$

The parameters a and b were correlated as linear functions of the temperature for twenty-one substances.

4. Nagata & Yasuda (1977) applied the Carnahan-Starling-Redlich-Kwong equation to saturated vapors.

5. Nakamura, Breedveld, & Prausnitz (1976) studied the hard-sphere equation in the form

$$P = \frac{RT}{V}\left[\frac{1 + y + y^2 + y^3}{(1 - y)^3}\right] - \frac{a}{V(V + c)}, \qquad (1.191)$$

where

$$a = \alpha + \beta/T, \qquad (1.192)$$

$$y = \frac{1}{4V}\exp\left(\frac{1}{\gamma + \delta T}\right), \qquad (1.193)$$

Table 1.25. The "Exact" and Approximate Forms of the Compressibility Factor Based on the Hard-Sphere Potential Obtained by Several Investigators (Scott 1971, p. 74)

$(PV/RT)_{hs} = \phi(y)$.

$y = b/4V$.

$\phi(y)$	*Expansion in Powers of y*
Exact	$1 + 4y + 10y^2 + 18.365y^3 + 28.24y^4 + 39.5y^5 + 56.5y^6 + \cdots$
I (van der Waals) $1/(1-4y)$	$1 + 4y + 16y^2 + 64y^3 + 256y^4 + 1024y^5 + 4096y^6 + \cdots$
II (PY-pressure) $(1 + 2y + 3y^2)/(1 - y)^2$	$1 + 4y + 10y^2 + 16y^3 + 22y^4 + 28y^5 + 34y^6 + \cdots$
III [scaled particle (PY-compressibility)] $(1 + y + y^2)/(1 - y)^3$	$1 + 4y + 10y^2 + 19y^3 + 31y^4 + 46y^5 + 64y^6 + \cdots$
IV (Guggenheim) $1/(1 - y)^4$	$1 + 4y + 10y^2 + 20y^3 + 35y^4 + 56y^5 + 84y^6 + \cdots$
V $(1 + 2y)/(1 - 2y)$	$1 + 4y + 8y^2 + 16y^3 + 32y^4 + 64y^5 + 128y^6 + \cdots$
VI $(1 + 3y + 4y^2)/(1 - 2y)(1 + y)$	$1 + 4y + 10y^2 + 18y^3 + 38y^4 + 74y^5 + 150y^6 + \cdots$
VII (Flory) $1/(1 - y^{1/3})$	—
VIII. Carnahan & Starling $\dfrac{1 + y + y^2 - y^3}{(1 - y)^3}$	$1 + 4y + 10y^2 + 18y^3 + 28y^4 + 40y^5 + 54y^6 + \cdots$ $= 1 + \sum_2^\infty (n^2 + n - 2)\, y^{n-1}$.

Example 1.19. Second Virial Coefficients in Terms of Several Different Potential Functions

It is shown in books on statistical thermodynamics that the second virial coefficient is related to a potential function, $u(r)$, by the equation

$$B = 2\pi N_0 \int_0^\infty r^2(1 - \exp(-u/kT))dr, \tag{1}$$

where N_0 is the Avogadro number and k the Boltzmann number. Usually a numerical integration is called for, but the Lennard-Jones potential

$$u = 4\varepsilon[(\sigma/r)^{12} - (\sigma/r)^6] \tag{2}$$

lends itself to analytical solution—namely,

$$B = -\frac{2\pi}{3} N_0\sigma^3 \Sigma \frac{2^{n+0.5}}{4n!} (\varepsilon/kT)^{0.5n+0.25} \Gamma\left(\frac{2n-1}{4}\right). \tag{3}$$

Tables for the numerical evaluation of this equation are available, for instance in the book of Reed & Gubbins (1973, Appendix H), where a bibliography of evaluations of virial coefficients from various potential functions also is given (p. 456). In terms of the hard-sphere potential,

$$u = \begin{cases} \infty, & r < \sigma, \\ 0, & r \geq \sigma, \end{cases} \tag{4}$$

$$B = 2\pi N_0 \int_0^\sigma r^2 dr = \frac{2\pi}{3} N_0 \sigma^3, \tag{5}$$

and in terms of the square-well function,

$$u = \begin{cases} \infty, & r < \sigma \\ -\varepsilon, & \sigma < r < g\sigma, \\ 0 & r \geq g\sigma \end{cases} \tag{6}$$

$$B = 2\pi N_0 \left[\int_0^\sigma r^2 dr + \int_\sigma^{g\sigma} \left(1 - \exp\frac{\varepsilon}{RT}\right) r^2 dr\right]$$

$$= \frac{2\pi}{3} N_0 \sigma^3 \left[1 + (g^3 - 1)\left(1 - \exp\frac{\varepsilon}{kT}\right)\right]. \tag{7}$$

Plots of virial coefficients from these equations with several values of the parameters are shown. At least in some range of temperature, any experimental virial coefficient data can be represented closely by appropriate choice of the parameters σ, ε, and g.

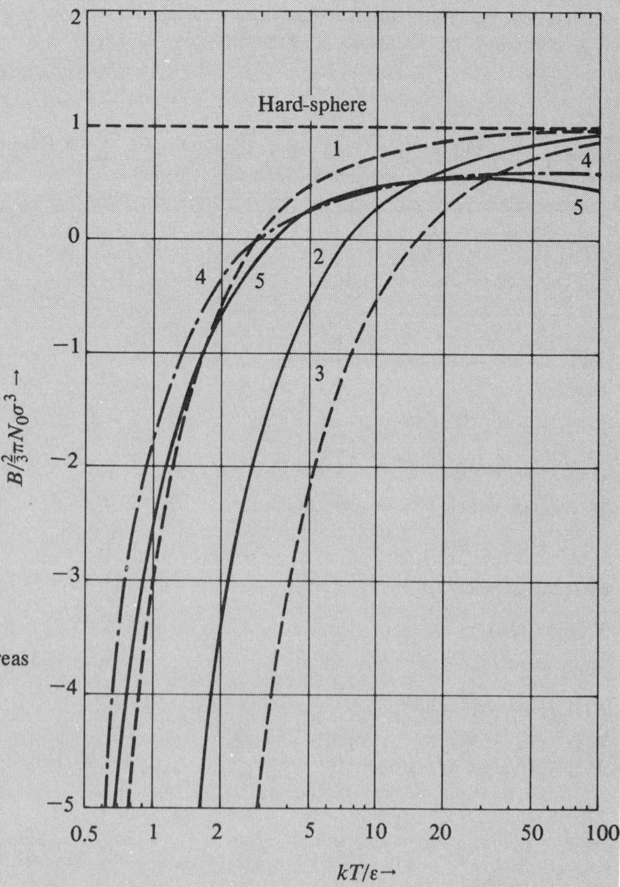

Curves 1, 2, 3, and 5 have the same value of σ^3, whereas that of Curve 4 is greater by a factor of 5/3.

Curve 1: Square-well, $g = 1.5$, $\sigma = \sigma_{LJ}$.
Curve 2: Square-well, $g = 2$, $\sigma = \sigma_{LJ}$.
Curve 3: Square-well, $g = 2.5$, $\sigma = \sigma_{LJ}$.
Curve 4: Lennard-Jones, $\sigma = \sigma_{LJ}$.
Curve 5: Square-well, $g = 1.5$, $\sigma^3 = 1.66\sigma_{LJ}^3$.

Example 1.20. Evaluation of the Parameters of a Square-Well Potential from Data of Second Virial Coefficients

The data of second virial coefficients are:

T, °K	360	450	540
$10^6 B$, m³/gmol	−25.45	−17.68	−12.96

Eq. 7 of Example 1.19,

$$B = \frac{2}{3}\pi N_0 \sigma^3 \left[1 + (g^3 - 1)\left(1 - \exp\frac{\varepsilon}{kT}\right)\right] \quad (1)$$

may be shorthanded as

$$B = x[1 + y(1 - \exp(z/T))]. \quad (2)$$

Use two of the data to eliminate x,

$$1 + [y(1 - \exp(z/T_1))] = \frac{B_1}{B_2}[1 + y(1 - \exp(z/T_2))]$$

$$= \frac{B_1}{B_3}[1 + y(1 - \exp(z/T_3))]. \quad (3),(4)$$

Solve for y,

$$y = \frac{B_1/B_2 - 1}{1 - \exp(z/T_1) - [1 - \exp(z/T_2))]B_1/B_2} \quad (5)$$

$$= \frac{B_1/B_3 - 1}{1 - \exp(z/T_1) - [1 - \exp(z/T_3)]B_1/B_3}. \quad (6)$$

After substitution of the data, the last equation is solved for $z = \varepsilon/k$ by trial, then the parameters g and σ are found directly.

$z = \varepsilon/k = 200.25$ K,

$y = 7.0097$,

$g = (1 + y)^{1/3} = 2.0008$,

$x = 6.0365(E{-}6)$,

$$\sigma = \left[\frac{x}{2\pi(6.02)(E23)/3}\right]^{1/3} = 1.685(E{-}10)\text{ m}, 1.682 \text{ Å},$$

$\varepsilon = 200.25\,k = 2.76(E - 21)$ joules.

Figures 1.24(a), 1.24(b) are results of such calculations.

and parameter c is independent of temperature but is characteristic of the substance, being zero for nonpolars and slightly positive for the few polar substances tested. For mixtures, binary interaction parameters are required for evaluation of the αs. Constants for fourteen substances are given.

6. A simpler form of the hard-sphere repulsion term was adopted by Ishiwaka, Chung, & Lu (1980), who wrote

$$P = \frac{RT}{V}\left[\frac{2V + b}{2V - b}\right] - \frac{a}{\sqrt{T}\,V(V + b)}, \quad (1.194)$$

with

$$a = \Omega_a R^2 T_c^{2.5}/P_c, \quad (1.195)$$

$$b = \Omega_b R T_c/P_c \quad (1.196)$$

and the coefficients Ω_a and Ω_b were evaluated from vapor pressures and saturated liquid densities for twenty-two substances. This system was claimed as somewhat superior to the Soave and Peng-Robinson equations.

The statistical-mechanical equations of state that have been described generally require much more information about individual substance than the critical properties and the acentric factor and are more demanding of computer time. Except for the work of Donohue & Prausnitz, they have not been tested as exhaustively as the Soave or BWR, for instance, which have been compared with thousands of data points. Moreover, they are not on completely rational bases, since the hard-sphere and square-well potentials are strictly empirical and probably often unrealistic. Nevertheless, this change in point of view and attention to the repulsion term after so many years of neglect may lead to valuable developments in the near future.

1.8. THE CRITICAL STATE AND CRITICAL PROPERTIES

1.8.1. Introduction

The earliest observations of critical behavior were made by heating fluids in sealed glass tubes. If the average density of the fluid in the tube is appropriate, the interface between the coexisting liquid and vapor phases will disappear at the critical condition. Inspection of a critical isotherm like those of Figure 1.5 reveals that it is flat near the critical and that the density need not be known precisely for this experiment to be successful. Photographs of the behavior of a mixture of aniline and cyclohexane as the temperature is lowered past the critical are shown by Ferrell (1968). Other graphic illustrations are given by Sengers & Levelt-Sengers (1968), who also present much historical material. Early work in this field is well described and thoroughly referenced by Partington (1950).

As the critical state is approached, all properties of the two phases become the same, and many change abruptly, often by several orders of magnitude. A selection of properties of carbon dioxide that exhibit anomalies in the vicinity of the critical is given in Figure 1.28. A common method for identifying the critical state is by a series of measurements of densities of the two phases. Two methods of plotting data for this purpose are shown in Figure 1.26.

Power laws: The anomalous or asymptotic behavior of physical properties near the critical point often is closely represented by power laws and critical exponents, on both

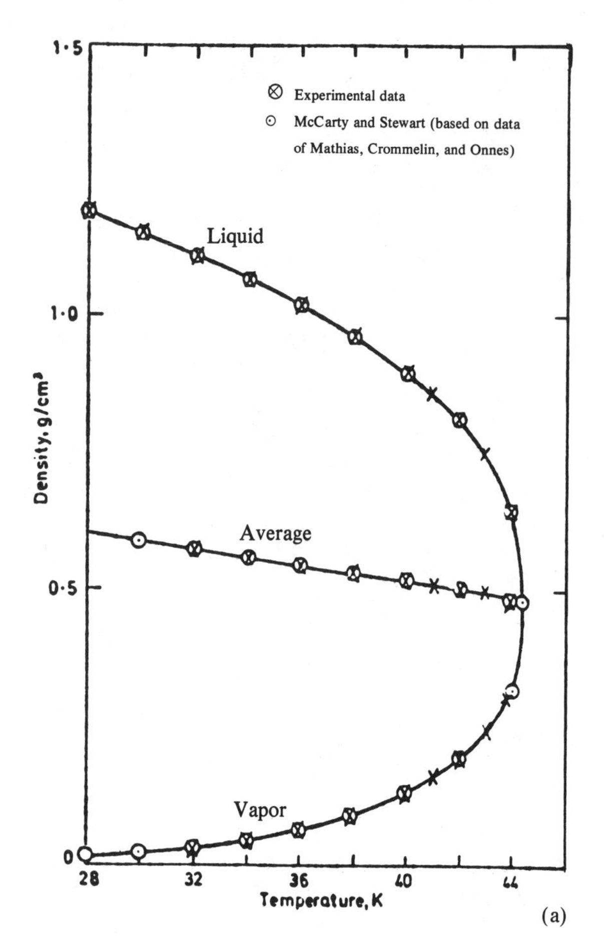

theoretical and experimental grounds. This is equivalent to saying that there exists a linear relation between the logarithms of two properties, say M and N:

$$\log M = \log k + \beta \log N, \tag{1.197}$$

or

$$M = kN^{\beta}. \tag{1.198}$$

Figure 1.27 shows this for the saturation density function

$$(\rho/\rho_c - 1) = k(T/T_c - 1)^{\beta}, \tag{1.199}$$

where the exponents are very nearly the same, 0.35, for all three substances. The historical background of such relations is presented by Levelt Sengers (1976), and the subject itself is treated elsewhere by the same author (1977).

Near the critical condition, latent heats approach zero, heat capacities increase sharply, and vaporization equilibrium ratios become unity. Accordingly, the performance of equipment depending on these properties, such as fractionators, flash drums, and reboilers, will be inadequate if, through ignorance, the operating conditions prove to be too close to the critical. Clearly, it is very important to know or to be able to predict the critical properties of process streams, be they pure substances or mixtures. Many correlation methods have been developed and are reviewed, for instance, by Bretsznajder (1962), Reid et al. (1977), Sterbacek et al. (1979), and Spencer et al. (1973). Calculation from equations of state will be discussed later in this section.

For pure substances the critical temperature and pressure are the highest at which liquid and vapor phases can coexist. This is clearly indicated by isotherms of carbon dioxide and isopentane of Figure 1.5. Except in the case of azeotropes,

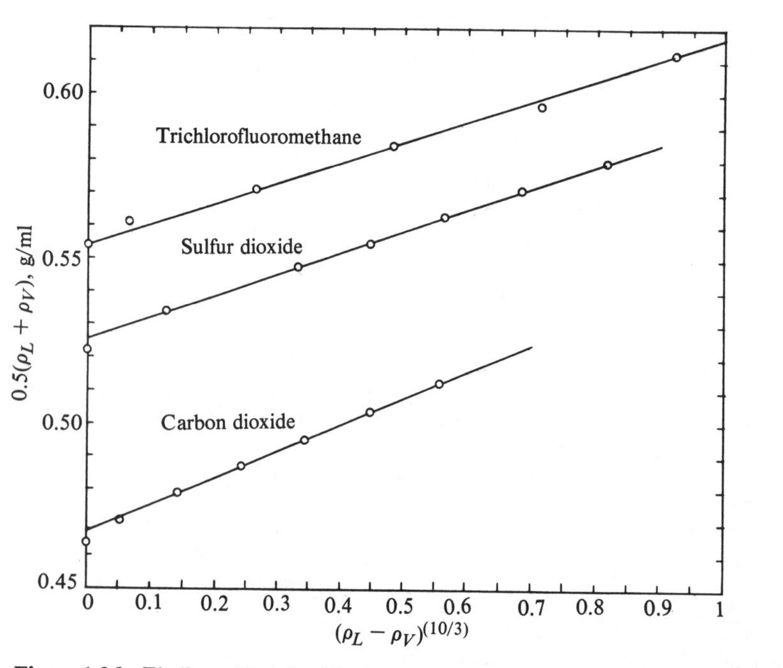

Figure 1.26 Finding critical densities by extrapolation: (a) Saturated molar densities of vapor and liquid densities and their average values for neon (Gibbons 1969). (b) The method of Hakala assumes that $0.5\ (\rho_L + \rho_V) = \rho_c + K(\rho_L - \rho_V)^{(10/3)}$. In these examples the results are only approximate. The method is cited by Fried, Hameka, & Blukis (1977).

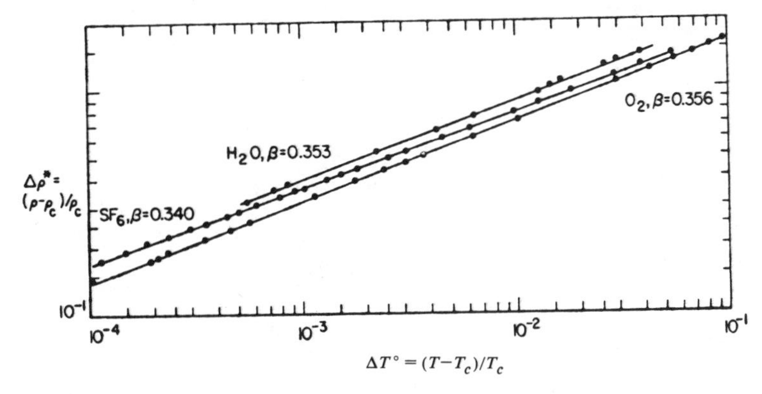

Figure 1.27 Power-law behavior of the saturated densities of H_2O, SF_6, and O_2. The relation is $\rho/\rho_c = \alpha(T/T_c - 1)^\beta$, where β is nearly the same for all these substances (Levelt Sengers 1977).

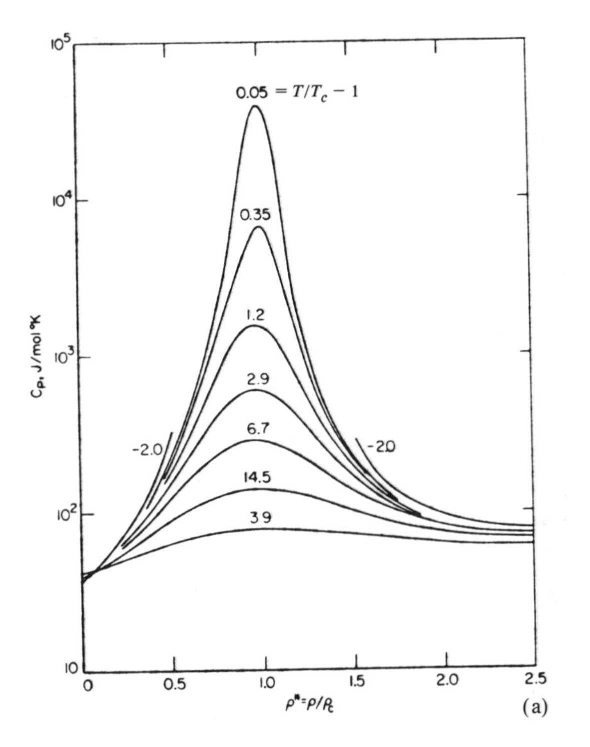

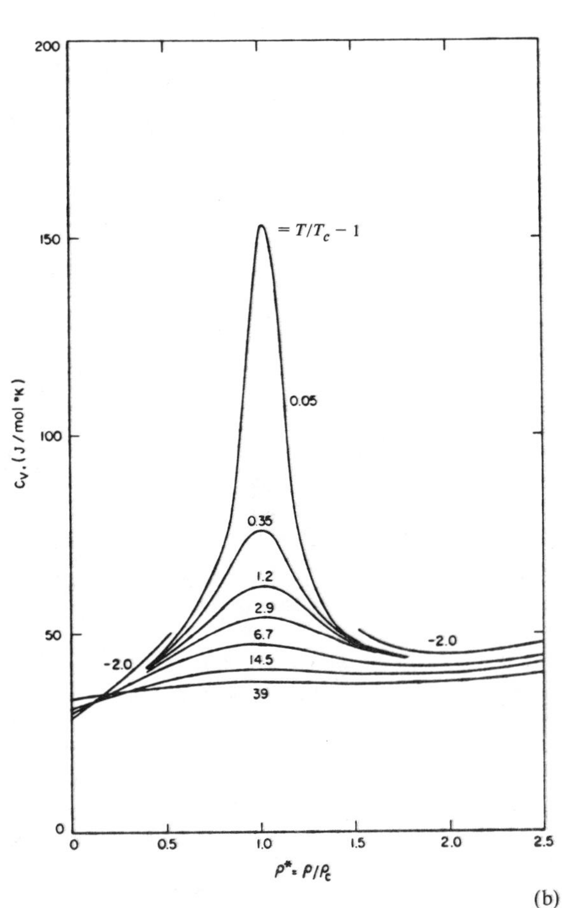

Figure 1.28 Properties of CO_2 near the critical point (Levelt Sengers 1977): (a) The heat capacity C_p becomes very large; (b) the heat capacity C_v becomes very large; (c) the thermal conductivity becomes very large; (d) the heat of vaporization vanishes; (e) the thermal expansion coefficient becomes very large; (f) the isothermal compressibility becomes very large; (g) the Prandtl number becomes large; (h) the sonic velocity diminishes greatly.

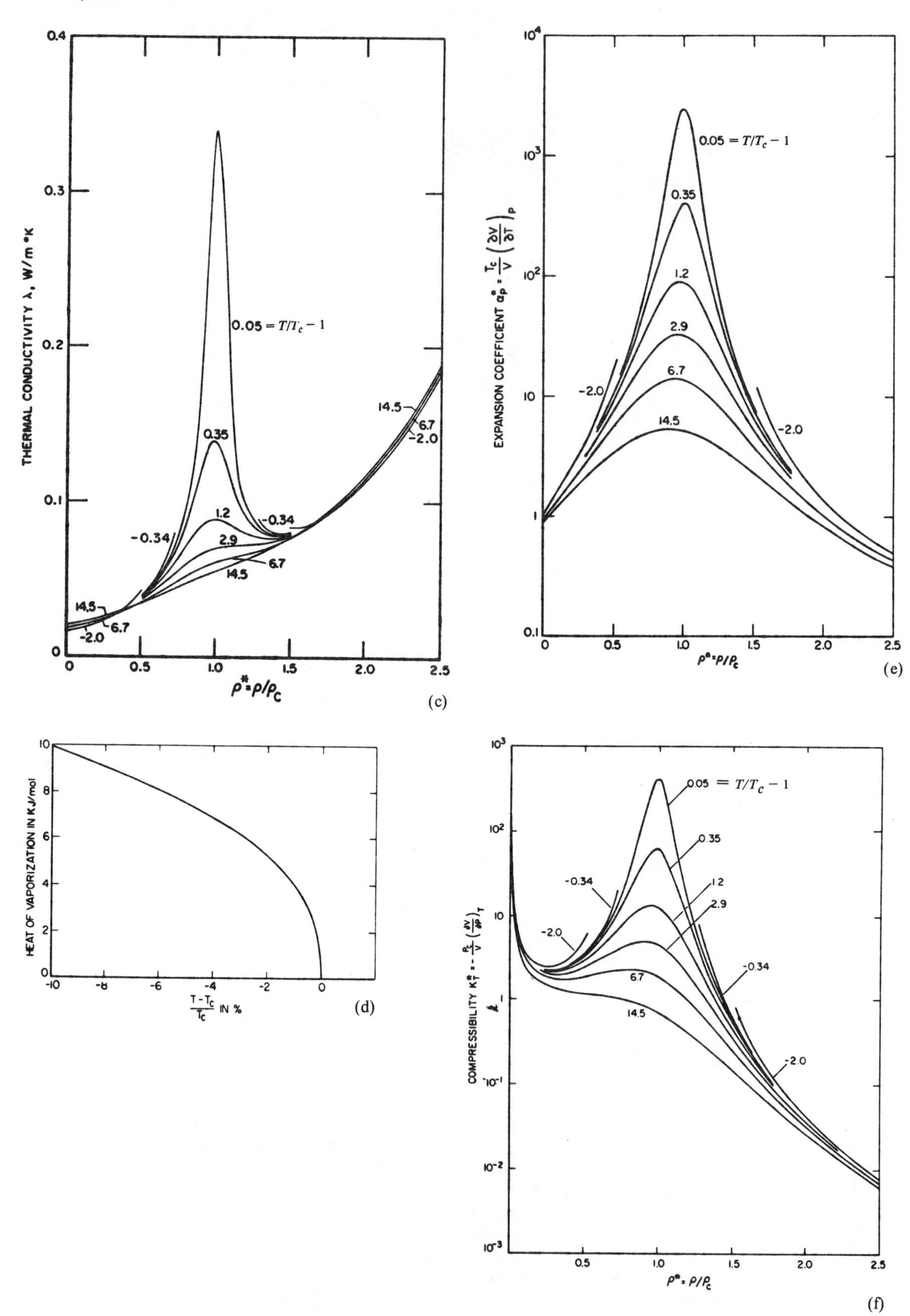
THERMAL CONDUCTIVITY λ, W/m °K
0.05 = T/Tc − 1
0.35
1.2
2.9
6.7
14.5
−0.34
−2.0
ρ* = ρ/ρc
EXPANSION COEFFICIENT
HEAT OF VAPORIZATION IN KJ/mol
(T − Tc)/Tc IN %
COMPRESSIBILITY
(c)

(e)

(d)

(f)

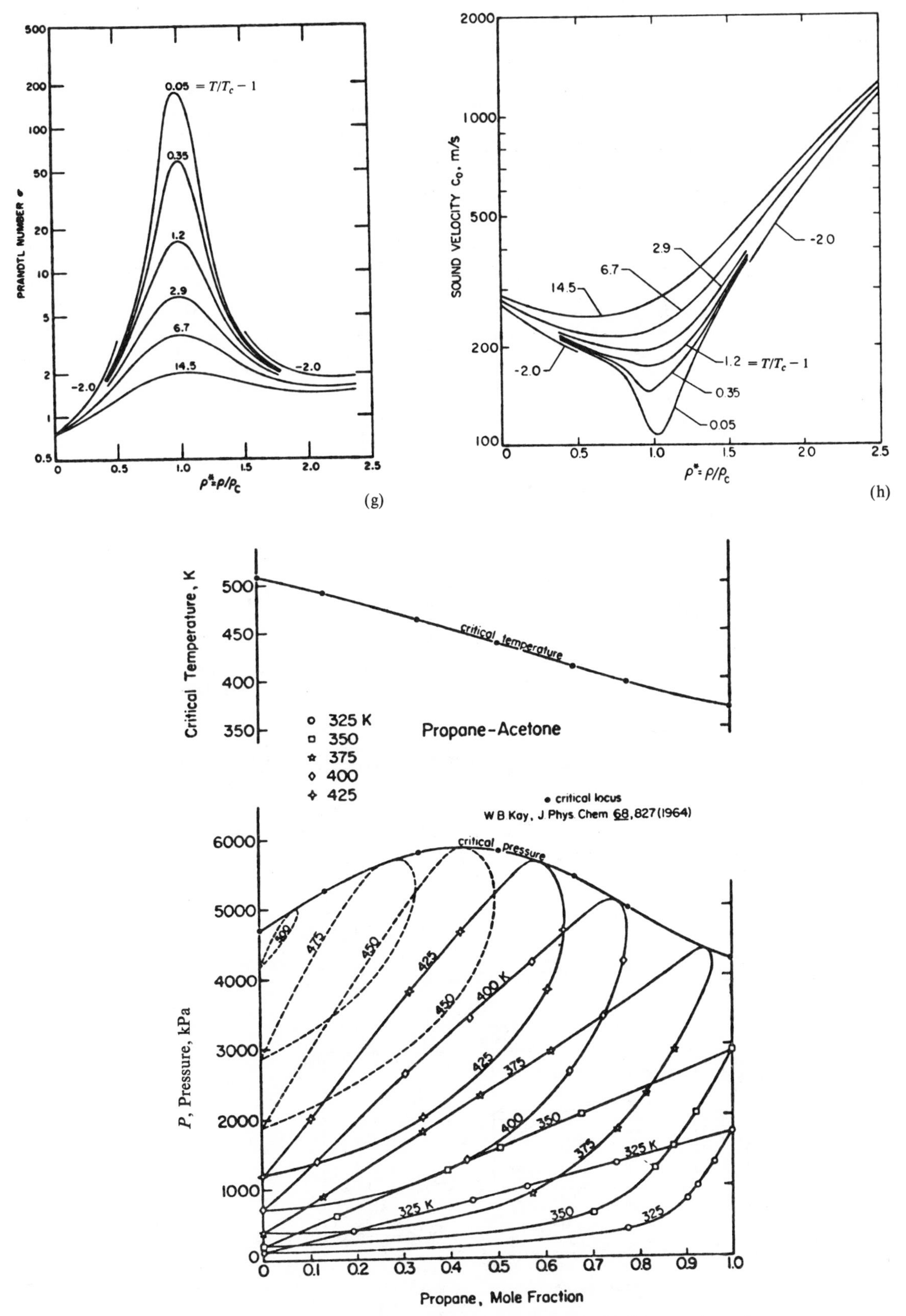

Figure 1.29 Isotherms and loci of critical pressures and critical temperatures of mixtures of propane and acetone (Gomez-Nieto & Thodos 1978).

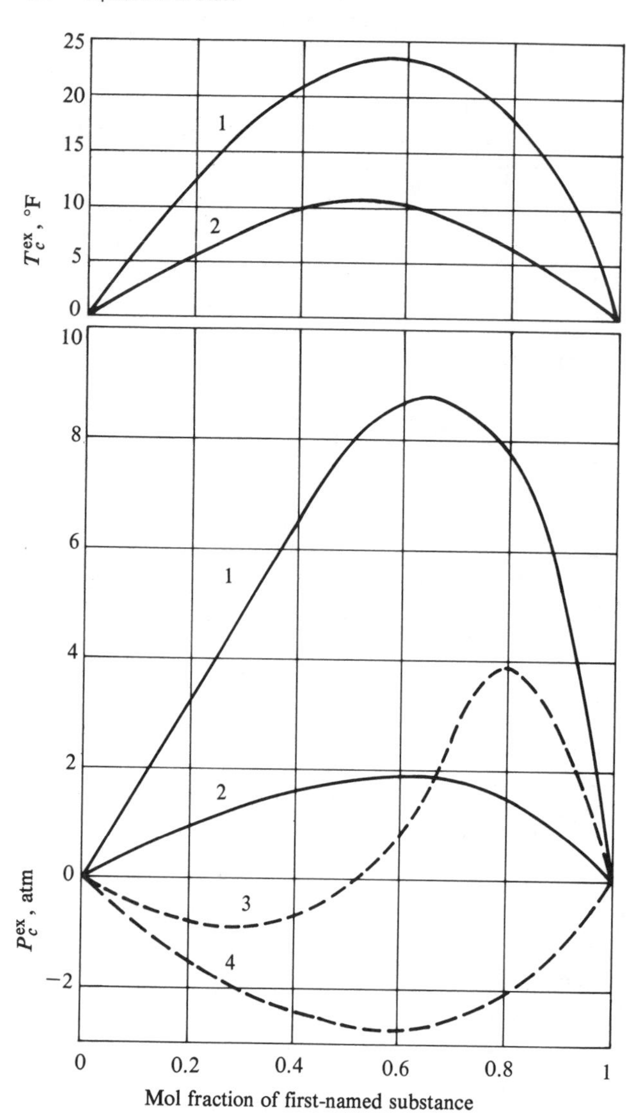

$P_c^{ex} = P_c - \Sigma y_i P_{ci}$,

$T_c^{ex} = T_c - \Sigma y_i T_{ci}$.

System 1. *n*-butane-*n*-octane.
System 2. Nonane-tridecane.
System 3. Benzene-tridecane.
System 4. Benzene-heptane.

Figure 1.30 Excess critical pressures and temperatures of some hydrocarbon binaries (data of *API Technical Data Book* 1974).

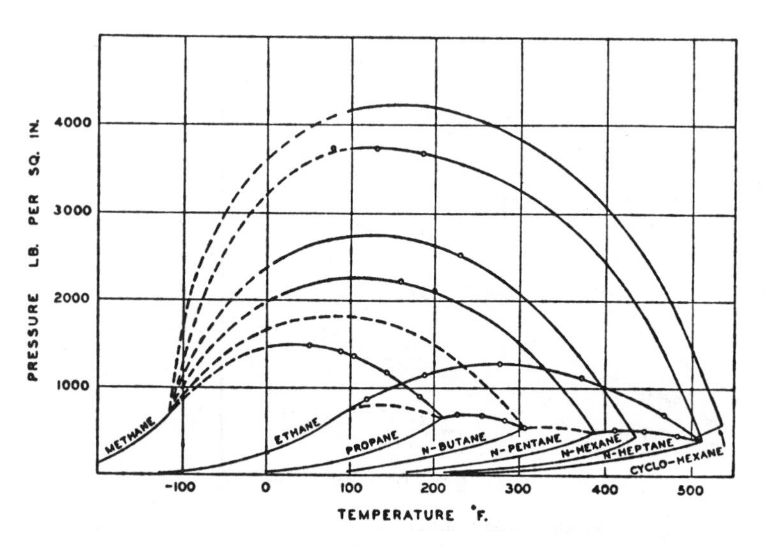

Figure 1.31 Loci of the critical point of several hydrocarbon mixtures (Sage & Lacey 1939). See also Figure 6.2. Each combination of P and T corresponds to a different composition.

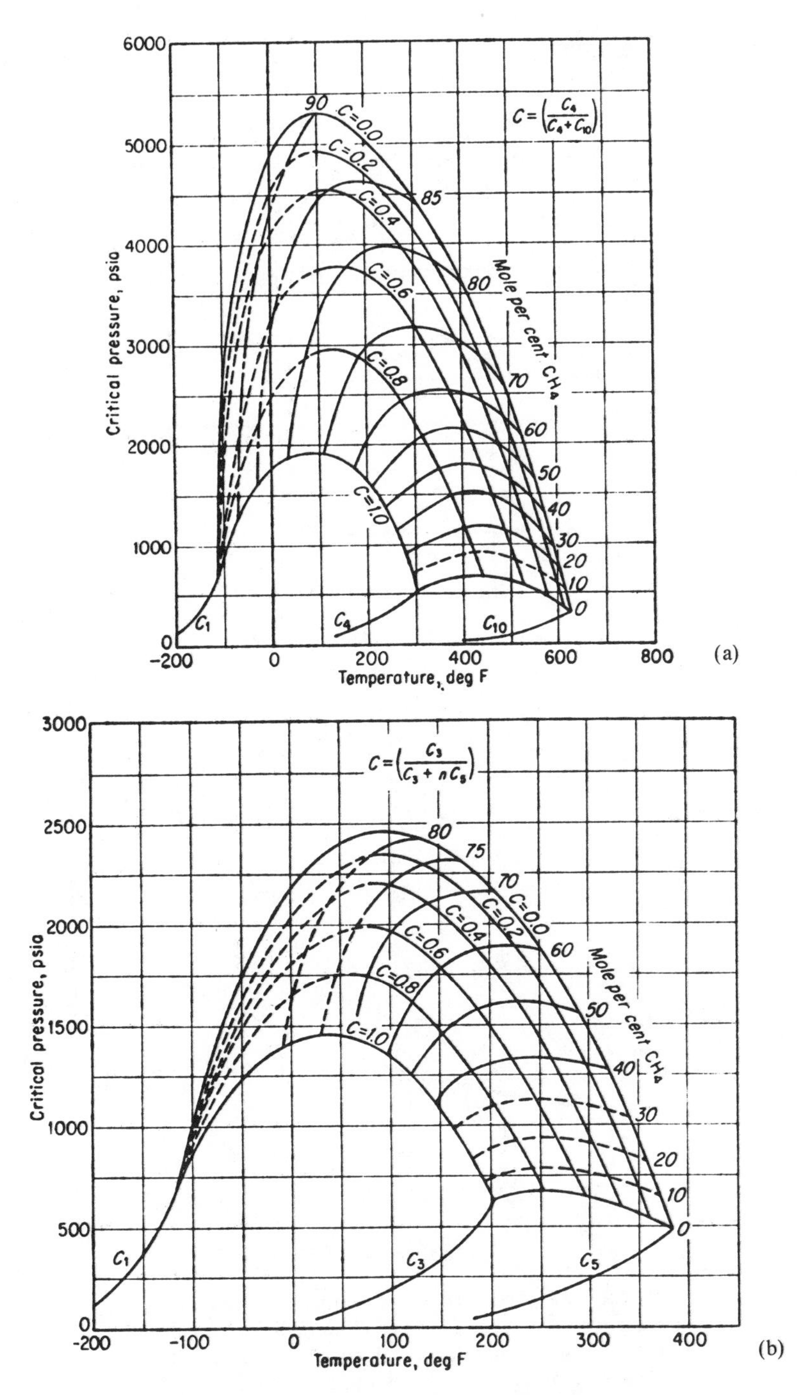

Figure 1.32 Critical properties of ternary mixtures: (a) Mixtures of methane, *n*-butane, and decane (Katz et al. 1959; data of Sage & Lacey); (b) mixtures of methane, propane, and pentane (Katz et al. 1959; data of Sage & Lacey).

however, this is not generally true of mixtures, and two phases sometimes can exist above either the critical temperature or the critical pressure of the mixture. Such behavior, which was called *retrograde* by Kuenen, is described in detail later.

1.8.2. Pure Substances

Examination of any family of isotherms on a *PV* diagram reveals that the critical isotherm has a point of inflexion at the critical. Mathematically, this is described by saying that

$$\left(\frac{\partial P}{\partial V}\right)_T = \left(\frac{\partial^2 P}{\partial V^2}\right)_T = 0 \tag{1.200}$$

at the critical point. This conclusion is very useful in that it allows the evaluation of two parameters of an equation of state in terms of critical properties, and led van der Waals to the concept of corresponding states. A third parameter also can be

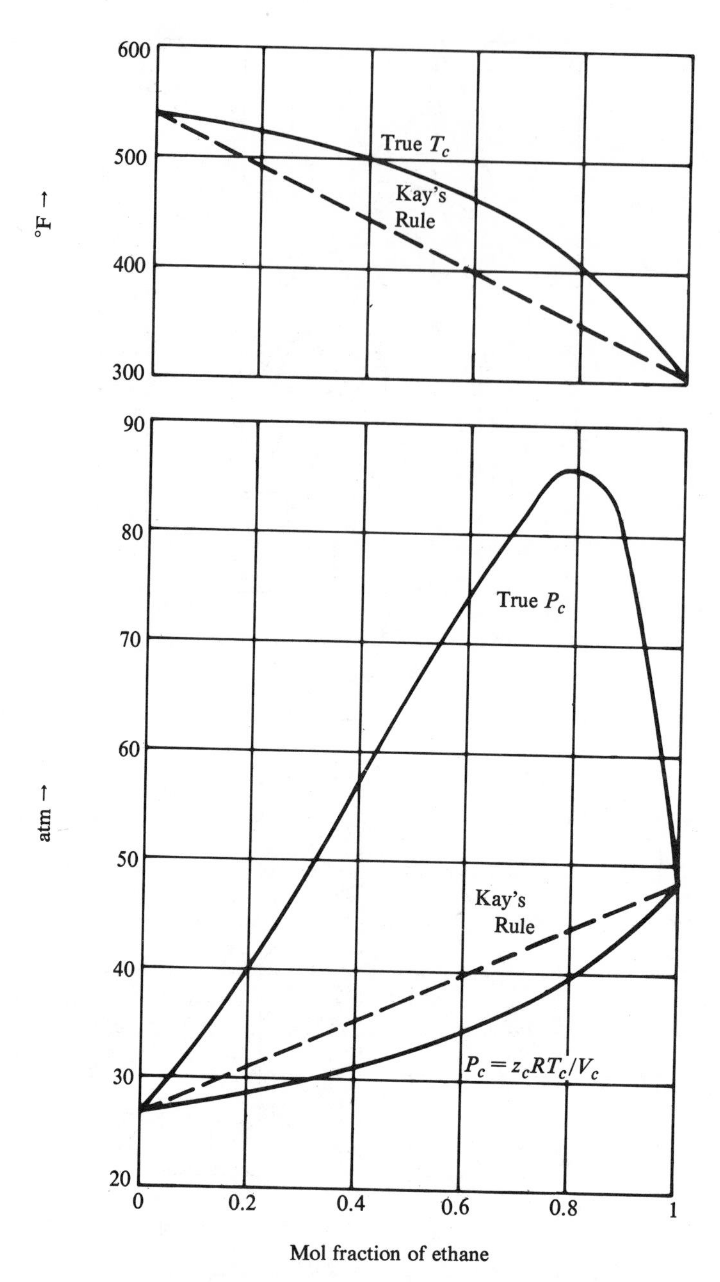

Figure 1.33 True critical properties and pseudocritical properties calculated with Kay's Rules of mixtures of ethane and n-heptane. The bottom curve is of $P_c = z_c RT_c/V_c$ with z_c, T_c, and V_c obtained with Kay's Rules (data of Kay 1938).

evaluated if the equation of state is assumed to apply at the critical.

Experimental critical data are moderately abundant. Some estimation methods are based on structural contributions. Pure component data are of course required for estimation of critical properties of mixtures, but the relations are quite complex. Critical compressibilities often are estimated adequately from vapor pressure data by way of the acentric factor, as mentioned in Section 1.3.1.

1.8.3. Mixtures

Critical states of mixtures, like those of pure substances, are identified as the condition at which the properties of coexisting liquid and vapor phases become indistinguishable. The *PVT* and composition behaviors of several groups of binary mixtures in the vicinity of the critical are shown in Figures 1.29–1.33. For the propane-acetone mixtures, the variation of critical temperature with composition is fairly linear, but not

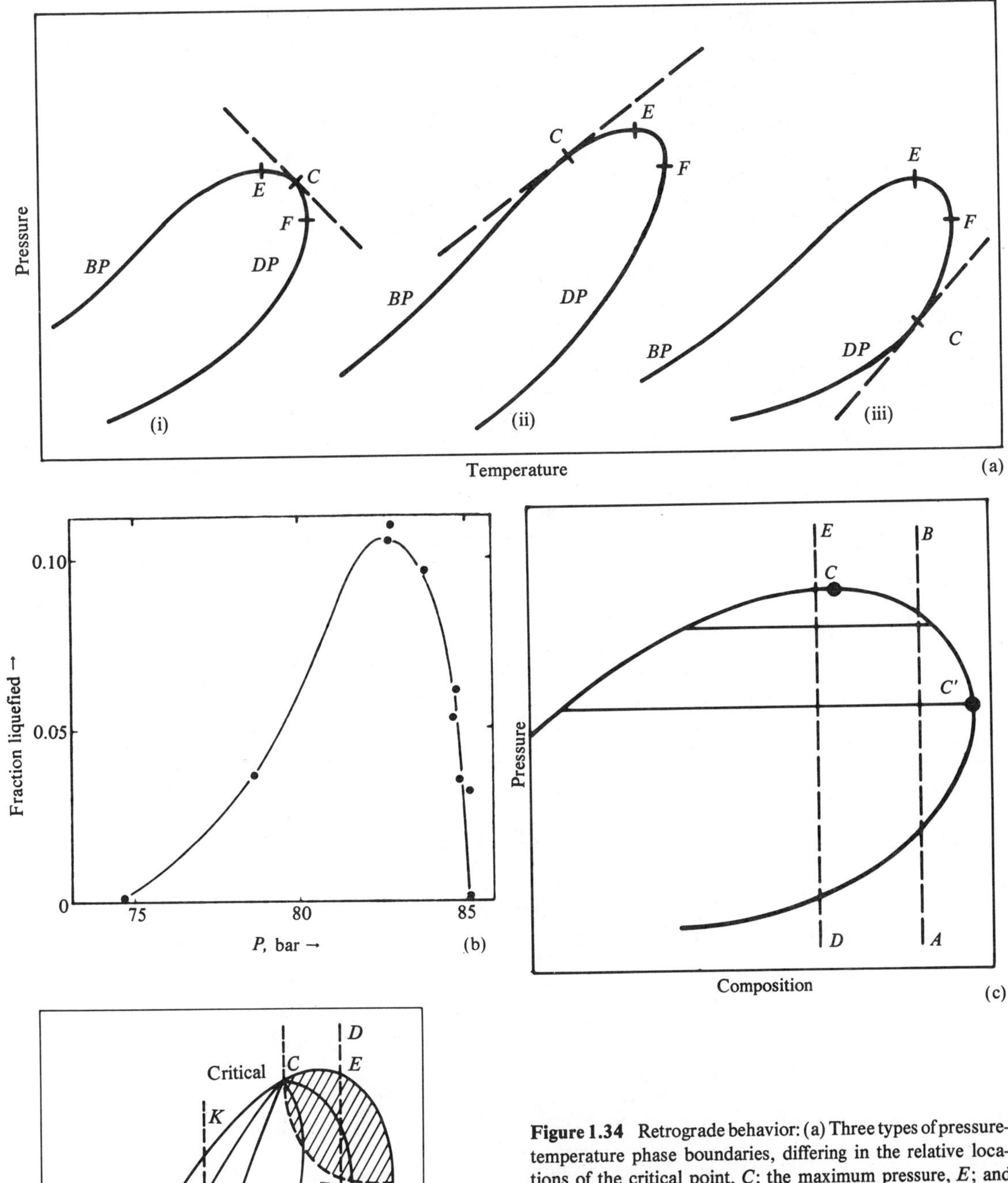

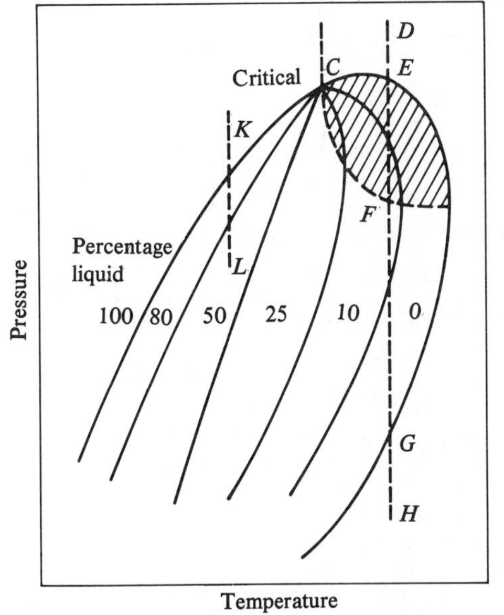

Figure 1.34 Retrograde behavior: (a) Three types of pressure-temperature phase boundaries, differing in the relative locations of the critical point, C; the maximum pressure, E; and the maximum temperature, F. BP is the bubblepoint curve, DP is the dewpoint. (b) Retrograde condensation of a 41/59 mixture of CO_2 and CH_3Cl at 378.2 K between the two dewpoint pressures, 74.5 and 85.0 bar; type (b) loop (data of Kuenen; Rowlinson & Swinton 1982). (c) Retrograde condensation on a pressure-composition diagram. Along path AB condensation reaches a maximum amount at C'. Along path DE the condensation behavior is normal; that is, it increases continually with rise in pressure. (d) retrograde condensation on a pressure-temperature diagram. The cross-hatched area is retrograde. As the pressure is lowered along line DH, maximum condensation of 14 percent is attained at point F. Along line KL, the amount of condensation decreases normally with decreasing pressure (after Katz et al. 1959).

so for the ethane-heptane mixtures; the variation of the critical pressures is highly nonlinear for both. For the latter system it is clear that there are temperatures and pressures at which two phases exist in excess of the critical values. Curves 3 and 4 of Figure 1.30 illustrate how extreme the critical pressure behavior of mixtures of dissimilar substances can be. As indicated in Figure 1.31, the further apart are the critical pressures of the pure components, the greater the excess critical pressure of the mixture. This effect also is shown by the ternary mixtures of Figure 1.32.

Two kinds of pseudocritical pressures are compared with true criticals of binary mixtures in Figure 1.33. Clearly, Kay's rule and the other one used there are no way to calculate true criticals, but it is pointed out in Section 1.3.8 that the pseudocriticals have a utility of their own, in that they are the better terms to use for developing equations of state of mixtures when using the usual combining rules for parameters of those equations.

The name *retrograde* is given to phase behavior above the critical temperature and pressure when liquid and vapor coexist because the amount of condensation of vaporization changes with pressure and temperature in the opposite direction to normal behavior. Figure 1.34(a) shows three relative locations of critical properties and maximum temperature and pressure. Part (b) shows retrograde condensation of a mixture of carbon dioxide and methyl chloride. At first the amount of liquefaction increases with pressure, but above about 81 atm it *decreases,* becoming zero at 83.3 atm. Usually the region in which retrograde behavior occurs is fairly narrow, as indicated in Figure 1.34(d), for instance. Knowledge of this phenomenon may be very important in understanding the behavior of petroleum reservoirs, for example. Figure 1.34(c) illustrates retrograde behavior on a pressure-composition diagram.

Even more complex phase separation behavior can exist beyond the critical point; some coverage of such phenomena is in Chapter 9.

1.8.4. Critical Properties from Equations of State

The critical point corresponds to incipient separation into liquid and vapor phases. The mathematics of all incipient phase separations is the same, whether liquid-vapor, liquid-liquid, or other. The stability conditions for liquid-liquid separation are treated in Chapter 7 from the standpoint of Gibbs energy behavior. Similarly, the condition for the liquid-vapor critical point of multicomponent mixtures is expressed in terms of derivatives of the chemical potentials, or partial molal Gibbs energies,

$$\mu_i = \bar{G}_i = (\partial G/\partial n_i)_{T,P,n_j}. \qquad (1.201)$$

The mathematical conditions for the critical point were first stated by Gibbs; developed by Prigogine & Defay (1954, pp. 226, 250); and generalized more recently—for example, by Reid & Beegle (1977). At the critical point, two Jacobians are equal to zero—namely,

$$J_1 = \frac{\partial(\mu_1, \mu_2, \ldots, \mu_{n-1})}{\partial(x_1, x_2, \ldots, x_{n-1})} = 0, \qquad (1.202)$$

$$J_2 = \frac{\partial(J_1, \mu_2, \mu_3, \ldots, \mu_{n-1})}{\partial(x_1, x_2, \ldots, x_{n-1})} = 0. \qquad (1.203)$$

In these equations the chemical potentials μ_i may be replaced by the logarithms of the partial fugacities, $\ln f_i$. For binary mixtures these equations simplify to

$$\left(\frac{\partial \mu_1}{\partial x_2}\right) = \left(\frac{\partial^2 \mu_1}{\partial x_2^2}\right) = 0, \qquad (1.204)$$

or

$$\left(\frac{\partial^2 G}{\partial x_2^2}\right) = \left(\frac{\partial^3 G}{\partial x_2^3}\right) = 0, \qquad (1.205)$$

or

$$\left(\frac{\partial \ln \hat{f}_1}{\partial x_2}\right) = \left(\frac{\partial^2 \ln \hat{f}_1}{\partial x_2^2}\right) = 0, \qquad (1.206)$$

or

$$\left(\frac{\partial \ln \hat{\phi}_1}{\partial x_2}\right) = \left(\frac{\partial^2 \ln \hat{\phi}_1}{\partial x_2^2}\right) = 0. \qquad (1.207)$$

After the derivatives of the fugacities have been obtained from a suitable equation of state, the critical temperature and pressure or volume can be evaluated by simultaneous solution of the pair of equations at a specified composition, x_2. Since the equations are highly nonlinear, a computer solution usually is called for even when only two components are present. Example 1.21 develops the equations for a binary mixture on the basis of the Redlich-Kwong equation of state.

For a ternary mixture, the Jacobian relations are

$$J_1 = \begin{vmatrix} \left(\frac{\partial^2 G}{\partial x_2^2}\right)_{T,P,x_1} & \left(\frac{\partial^2 G}{\partial x_2 \partial x_3}\right)_{T,P} \\ \left(\frac{\partial^2 G}{\partial x_2 \partial x_3}\right)_{T,P} & \left(\frac{\partial^2 G}{\partial x_3^2}\right)_{T,P,x_2} \end{vmatrix} = 0, \qquad (1.208)$$

$$J_2 = \begin{vmatrix} \left(\frac{\partial J_1}{\partial x_2}\right)_{T,P,x_3} & \left(\frac{\partial J_1}{\partial x_3}\right)_{T,P,x_1} \\ \left(\frac{\partial^2 G}{\partial x_2 \partial x_3}\right)_{T,P} & \left(\frac{\partial^2 G}{\partial x_3^2}\right)_{T,P,x_2} \end{vmatrix} = 0. \qquad (1.209)$$

The expansion of the second Jacobian is

$$J_2 = \left(\frac{\partial^2 G}{\partial x_3^2}\right)^2_{T,P,x_2} \left(\frac{\partial^3 G}{\partial x_2^3}\right)_{T,P,x_3} - \left(\frac{\partial^2 G}{\partial x_2 \partial x_3}\right)_{T,P} \times \left[3 \left(\frac{\partial^2 G}{\partial x_3^2}\right)_{T,P,x_2} \left(\frac{\partial^3 G}{\partial x_2^2 \partial x_3}\right)_{T,P} + \left(\frac{\partial^2 G}{\partial x_2^2}\right)_{T,P,x_3} \left(\frac{\partial^3 G}{\partial x_3^3}\right)_{T,P,x_3}\right]$$

$$+\left(\frac{\partial^3 G}{\partial x_2 \partial x_3^2}\right)_{T,P}$$

$$\left[\left(\frac{\partial^2 G}{\partial x_2^2}\right)_{T,P,x_3}\left(\frac{\partial^2 G}{\partial x_3^2}\right)_{T,P,x_2}+2\left(\frac{\partial^2 G}{\partial x_2 \partial x_3}\right)_{T,P}^2\right]. \tag{1.210}$$

Readers wishing to check their algebra will find these equations written out in terms of the Redlich-Kwong equation by Domnina & Zakharov (1978).

Some recent utilization of these conditions for finding critical properties of mixtures may be cited. Joffe & Zudkevitch (1967) employed the Redlich-Kwong equation in the study of several binary mixtures, and found it necessary to include binary interaction parameters for good prediction of the critical properties. Spear et al. (1969) likewise employed a corrected Redlich-Kwong equation of state for binaries and were able to computerize the solution. The BWR equation was used by Sarashina et al. (1974) and the Soave equation by Huron (1976).

Ternary systems were studied by Spear et al. (1971) and by Domnina & Zakharov (1978). Both groups used the Redlich-Kwong equation with interaction parameters.

Critical data for mixtures containing up to twelve components were obtained by Peng & Robinson (1977) with their equation of state, and by Heidemann & Khalil (1980), who used the Soave equation. The latter were able to develop a procedure that markedly reduced computation time.

As a result of the work that has been cited and of other important investigators, the problem of identifying a critical state has been largely solved for many important cases. Case studies of difficult industrial problems involving phase equilibria in the critical region are cited by Gray (1980).

1.9. COMPARISONS AND TENTATIVE RECOMMENDATIONS

1.9.1. Introduction

The existence of a large number of recent review articles on equations of state and phase equilibria perhaps attests to the

Example 1.21 Equations for the Critical Properties of a Binary Mixture from the Redlich-Kwong EOS

The condition is Eq. 1.207:

$$\frac{\partial \ln \hat{\phi}_1}{\partial y_2}=\frac{\partial^2 \ln \hat{\phi}_1}{\partial y_2^2}=0. \tag{1, 2}$$

From Table 3.4, the equation of the fugacity coefficient is

$$\ln \hat{\phi}_i=\frac{b_i}{b}(z-1)-\ln\left[z\left(1-\frac{b}{V}\right)\right]$$

$$+\frac{1}{bRT^{1.5}}\left[\frac{ab_i}{b}-2\sqrt{aa_i}\right]\ln\left(1+\frac{b}{V}\right), \tag{3}$$

$$z=\frac{V}{V-b}-\frac{a}{RT^{1.5}(V+b)}, \tag{4}$$

$$a=(1-y_2)^2a_1+y_2^2a_2+2y_2(1-y_2)(a_1a_2)^{0.5}, \tag{5}$$

$$b=(1-y_2)b_1+y_2b_2. \tag{6}$$

Intermediate derivatives are

$$\frac{dz}{dy_2}=\frac{V}{(V-b)^2}\frac{db}{dy_2}$$

$$-\frac{1}{RT^{1.5}(V+b)^2}\left[(V+b)\frac{da}{dy_2}+a\frac{db}{dy_2}\right], \tag{7}$$

$$\frac{da}{dy_2}=2(-2y_1a_1+y_2a_2+(1-2y_2)(a_1a_2)^{0.5}, \tag{8}$$

$$\frac{db}{dy_2}=-b_1+b_2. \tag{9}$$

The derivative of $\ln \hat{\phi}_1$ is

$$\frac{\partial \ln \hat{\phi}_1}{\partial y_2}=\frac{b_1}{b}\frac{dz}{dy_2}-\frac{(z-1)}{b}\frac{db}{dy_2}$$

$$-\frac{1}{z}\frac{dz}{dy_2}+\frac{1}{V-b}\frac{db}{dy_2}$$

$$+\frac{1}{RT^{1.5}}\left\{\frac{1}{b(V+b)}\left(\frac{ab_i}{b}-2\sqrt{aa_i}\right)\frac{db}{dy_2}\right.$$

$$+\ln\left(1+\frac{b}{V}\right)\left[\frac{1}{b}\left(\frac{b_i}{b}-\sqrt{\frac{a_i}{a}}\right)\frac{da}{dy_2}\right.$$

$$\left.\left.+\frac{2}{b^2}\left(\sqrt{a_ia}-\frac{ab_i}{b}\right)\frac{db}{dy_2}\right]\right\}=0. \tag{10}$$

With the exercise of a little patience, the second derivative also can be formulated. The unknown T and V in the resulting two equations are the critical properties. The quantities a, b, and z are defined in terms of T_c and P_c in Table 1.9. The two equations can be expressed in terms of T_c and P_c as unknowns by eliminating V_c with the relation for the R-K equation,

$$V_c=\frac{RT_c}{3P_c}. \tag{11}$$

The resulting highly nonlinear pair of equations then may be solved for P_c and T_c.

Table 1.26. Recommended Equations for Estimating Gas Phase Compressibility and Fugacity for Several Types of Systems in Particular *PVT* Ranges. (Tarakad, Spencer, & Adler 1979)[a]

Type of System	*Subcritical Temp. Away from Critical Region*	*Critical Region*	*Supercritical Temp. Away from Critical Region*
Pure compounds:			
Nonpolar, mildly polar.	RK, RKS, JBA, virial	RK, RKS, JBA	RK, RKS, virial
Highly polar.	Virial	RK	Virial, RK
Mixtures:			
Only nonpolar and mildly polar components are present.	RK, RKS, JBA, virial	RK, RKS	RK, RKS, virial, NBP
One highly polar component present.	RK, virial	RK	RK, virial, NBP
More than one polar component present.	Virial	RK	RK, virial, NBP
Aqueous-gas mixtures:			
Water-hydrocarbon and water–carbon dioxide.	RK, virial	RKD, RK	NBP, RKD, virial, RK
Systems containing ammonia and water.	RKG	RKG	RKG, RK
Natural-gas mixtures		JBA, RKC, RK	JBA, RKC, RK

Note: RK = Redlich-Kwong, RKC = Chueh-Redlich-Kwong, JBA = Joffe-Barner-Adler, RKS = Soave-Redlich-Kwong, NBP = Nakamura-Breedveld-Prausnitz, RKD = de Santis-Redlich-Kwong, RKG = Guerreri-Prausnitz-Redlich-Kwong.

[a](1) In the critical region, errors in density are generally large, and errors in fugacity coefficients could be significant. (2) Whenever the virial equation is used, it should be remembered that there is an upper bound on the pressure, above which the two-term virial equation is either poor or completely fails. For most systems, the limit may be taken as $P_r < (T_r - 0.65)$ for $T_r < 1$; $P_r < (1.05T_r - 0.35)$ for $T_r > 1$, where $P_r = P/\Sigma y_i P_{c_i}$; $T_r = T/\Sigma y_i T_{c_i}$. These limits were established by plotting a large number of predicted compressibility factor deviations, and then isolating that region where the deviation was less than 2% for pure components, 3% for nonpolar or mildly polar mixtures, and 5% for highly polar mixtures.

uncertainties that exist in this area. A universally applicable equation does not exist and is not likely to be developed soon. Even computerized equations with thirty or forty constants have significant limitations. Attempts at generalization of equations to cover broader classes of substances usually lead to a decrease of accuracy in particular instances. Thus the Han-Starling equation with its generalized coefficients is substantially less accurate than the corresponding BWRS equation with coefficients specific to individual substances.

A great deal is desired from an equation of state. Besides the usual prediction of *PVT* behavior, there are vapor pressures; liquid densities; fugacities; phase equilibrium ratios of vapors, liquids, and solids; and enthalpy and entropy deviations from ideality to be taken care of. Further, conditions in the cryogenic and critical ranges may need to be covered. When such data are available in limited amount, they may be used for the reverse process of developing a general equation of state. One problem in this connection is an insufficiency of data. The greatest abundance of data is of light hydrocarbons below *n*-nonane, the inert and common gases, some of their binary mixtures, and a very few ternary and higher mixtures. Of polar substances, only for water and methanol are there data adequate for developing broadly applicable correlations.

Still, the field has advanced far beyond the laws of ideal gases and Raoult and Henry, and from an industrial point of view many important situations can be handled with fair accuracy. In this section something will be said about the utility of the five main categories of equations of state in solving engineering problems. Recommendations of equations also are summarized in Table 1.26.

1.9.2 Recommendations

The main classes of equations of state that are developed sufficiently for practical applications are the virial type, cubics in volume or compressibility, complex equations of the BWR type, corresponding states equations patterned after the Pitzer approach, and the most recent types based on hard-sphere repulsion terms. A promising group contribution approach by Wilson (1977) has not been developed much as yet.

The Virial Equation

Its simple form, the availability of a large amount of data—particularly of second virial coefficients, and a rational method of handling mixtures maintain the usefulness of the virial equation even in comparison with much more complex equations. At pressures corresponding to $P/P_c \leq 0.5(T/T_c)$ or to $\rho/\rho_c \leq 0.75$, the *B*-truncated form usually is acceptable for evaluation of vapor phase fugacities. The equation is not applicable to highly compressed gases or to liquids.

Cubic Equations

Although new treatments of this category still appear frequently, the API-Soave and the Peng-Robinson forms appear to be the standards. Both have strong proponents, and applications are being made constantly. Both equations need binary interaction parameters, which are available in the open literature only for the Soave. A comprehensive study, sponsored by the API and reported by Daubert et al. (1979), of these equations and other systems for figuring vapor-liquid equilibria concludes that they are about equivalent for this purpose and are generally superior to the more complex methods also investigated. Representation of *PVT* data in the vicinity of the critical is better with the Peng-Robinson. Both equations are inferior to the Lee-Kesler for enthalpy deviations, but they are more simply incorporated into distillation calculation programs, for example, computer time being perhaps one-fifth that of the BWRS and similar equations. Figure 1.23 compares the Soave and P-R equations with data of methane and water, and thus gives some indication of their applicability.

The BWR Type

A limitation of the original equation is that coefficients are known for fewer than fifty substances, mostly hydrocarbons. The Starling modification, the BWRS equation, is an industry standard; but even fewer sets of coefficients are known, and binary interaction parameters are available only for substances of interest to the light-hydrocarbon industry. The generalized version of Yamada (1973) with forty-four constants is said to be applicable at reduced densities up to 2.8. The generalized versions of Han-Starling, Yamada, and Nishiumi, however, are not always superior to the cubic equations. The Barner-Adler modification is intended for evaluation of liquid enthalpies and equilibria in the cryogenic range of natural gas systems. The Nishiumi development, which covers polar compounds, is stated to be acceptable everywhere except in the $1 \leq T_r \leq 1.3$ and $1 \leq P_r \leq 3$ region.

Corresponding States Equations

These are typified by the Lee-Kesler form, particularly with the combining rules of Plöcker et al. and thus known as the L-K-P equation, which is perhaps the most accurate generalized equation now available—except that polar compounds are not covered—for calculation of vaporization equilibrium ratios, enthalpies, and other thermodynamic properties. The use of the system is completely computerized. The range of validity of the L-K equation is claimed as $0.3 \leq T_r \leq 4$ and $0 \leq P_r \leq 10$, but Plöcker et al. recognize some difficulties in the range $0.93 \leq T_r \leq 1$ or so. For manual calculations, tabulations of compressibility and other properties as functions of reduced temperature and pressure have been prepared by Lee & Kesler and graphed in the API Data Book.

Perturbed-Hard-Sphere Equations

These are a comparatively recent and highly promising development, and work is continuing. Pertinent reviews of these theories have been made by Boublik (1977) and in the book of Rowlinson & Swinton (1982). The work sponsored by the Gas Producers Association and reported by Donohue & Prausnitz (1978) is in a usable form. That procedure can handle most mixtures encountered in petroleum refining and natural gas processing, including mixtures whose components differ considerably in size, shape, and potential energy distribution. Computer programs are available from the sponsoring organization. The comparisons were made with experimental data but not with the other widely used methods for these computations.

1.9.3. Review Literature

The flood of original literature on equations of state and phase equilibria has inspired an abundant review literature intended to provide some guidance to the user. The contents of several more or less recent articles will be indicated briefly. Since many of these are voluminous, it is not possible to do them justice; but perhaps the statements made here will be more informative than quotations of the titles. The chronological order is: 23, 1, 2a, 2, 20, 26, 28, 22, 21, 14, 9, 4, 5, 16, 24, 7, 11, 15, 18, 29, 3, 12, 19, 27, 10, 13, 17, 25.

The articles will be described in the alphabetical order given at the end of this section, except for the books by Otto, Partington, and Vukalovich & Novikov, which are given first. References follow the descriptions.

1. *Otto*. Experimental methods and results of *PVT* behavior of gases are covered at length. A list of fifty-six EOS published before 1927 is compiled, some of which are not in Table 1.2. A few of the equations receive particular attention, especially the work of Onnes on the direct and reduced forms of the virial equation of which the coefficients are correlated in terms of reduced temperatures.

2. *Partington.* Section 7, covering some 300 pages of this work, is concerned with properties of gases, emphasizing the historical and experimental aspects of the subject. Critical phenomena, *PVT* relations and equations, densities, specific heats, viscosities, conductivities, and diffusion are covered. Of characteristic equations, the main emphasis is on the original van der Waals equation; but several dozen others, including the virial and Beattie-Bridgeman, also are discussed, though only briefly. This reference constitutes indispensable background material.

2a. *Vukalovich & Novikov.* A noteworthy feature is the collection of 150 equations of state—the latest of which is dated 1944—and including several by East European authors. The body of this 300 page book is devoted largely to the van der Waals and virial equations and to the association theory of nonideality of *PVT* behavior.

3. *Abbott.* It is shown that many popular equations of state are special cases of a general equation that is a cubic in volume. The background and limitations of the Redlich-Kwong, Soave, and Peng-Robinson forms are mentioned. Such equations strike a practical balance between accuracy, simplicity, and generality.

4. *Adler et al.* The industrial viewpoint is presented. For those substances for which the constants are known, the BWRS equation is preferred by major industrial users in spite of its fivefold greater demands on computer time in comparison with equations like the API-Soave. The former is particularly superior in the cryogenic range. This long article presents several case studies employing the two equations mentioned and the Chao-Seader method, including multiple bubblepoints, a flash calculation of C_2-C_5 at 1,015 psia, and solubilities of solids. The limitations of existing equations of state include lack of generality of binary interaction parameters, and poor applicability to polar substances, petroleum fractions, and other complex mixtures that are not readily characterized by critical properties.

5. *Boublik* is concerned with recent progress in the statistical thermodynamic description of systems. Perturbation theories proposed for pure and mixed fluids are summarized with stress laid on relations for describing their thermodynamic behavior. Work on nonspherical and polar molecules

is active but incomplete. Few numerical comparisons of theory and experiment are mentioned. Of the 200 or so references, those later than 1970 are preponderant.

6. *Coward et al.* For the R-K and BWR equations, numerical examples are worked out to show the changing behaviors of these equations as pressure, temperature, and composition are varied. Binary mixtures of methane and propane are taken as the example, for which phase envelopes and phenomena such as retrograde behavior are analyzed. The problem of finding the correct root (of the three that may be present) is addressed.

7. *Cox & Lawrenson.* Several equations of state are mentioned, but their relative merits or accuracy are not considered in detail. Useful parts of this review are descriptions of *PVT* measurements and references to many such done recently. The principle of corresponding states, second virial coefficients and association in the vapor phase are covered. They mention what might be considered a challenge by Robertson & Babb (1969) that no known equation adequately represents their precise results.

8. *Daubert et al.* The 273 pages of this report are concerned primarily with checks of experimental data on vapor-liquid equilibria of pure and mixed fluids by several graphical methods and nine equation of state methods. Several thousand data points were utilized. The work led to improvements of the coefficients of the Soave equation, a correlation extending it to hydrogen-containing mixtures and data on binary interaction parameters, with their correlation in terms of acentric factors. The Soave and Peng-Robinson proved generally to be the best, but the former was adopted for the API because it is slightly simpler. For calculation of enthalpy deviations, the Lee-Kesler equation was tentatively judged the best. This document does emphasize that nowadays a vast amount of data must be analyzed to prove the merit or superiority of a correlation in the area of phase equilibria.

9. *Douslin* references the recent literature on the critical region, intermolecular potentials, virial coefficients, *PVT*, and thermodynamic properties and vapor pressures for many substances, primarily of experimental work. A large table classifies some of the references by substance and by the kind of data obtained on each, including excess properties, *PVTx*, equations of state, and so on.

10. *Fredenslund et al.* Literature citations are given to recent work, including perturbed-hard-sphere equations, polar compounds, and equations whose parameters are obtainable from group contributions, a method that promises to be useful for mixtures such as coal liquids whose critical properties are not known and cannot be determined.

11. *Gibbons et al.* The paper makes an analysis, based on 2,500 experimental *K*-values, of the Soave; Peng-Robinson; BWRS; and a corresponding states equation, MCSP, due to McCarty, which has thirty-three constants. Vapor and liquid densities, *K*-values, and dewpoints are compared. For liquid densities the MCSP and BWRS equations are definitely superior to the simpler ones, but the advantage is less marked for the other properties mentioned.

12. *Gray* appraises the strengths and weaknesses of a modification of the R-K equation by Joffe & Zudkevich, the RKJZ equation, whose performance is similar to those of Soave and Peng-Robinson. He points out that, surprisingly, water-hydrocarbon and carbon dioxide–hydrocarbon systems are represented quite well by these equations, but that mixtures of widely different molecular sizes cannot be represented within experimental accuracy. The equations are not widely extrapolatable, and it is implied that the parameters could be adjusted for different ranges. The article closes with a list of eight features that might well be incorporated in a future equation of state.

13. *Grenier & Cabre* describe in a general way their programs for distillation and flash calculations and thermodynamic and physical properties that use a modification (not identified) of the Redlich-Kwong equation. Good accuracy is obtained for common gases such as hydrogen, nitrogen, oxygen, carbon monoxide, and hydrocarbons.

14. *Horvath's* paper is a bibliography of the literature on the original, unmodified form of the Redlich-Kwong equation, comprising 112 references complete with titles. Those items giving comparisons with experimental data or other equations of state are identified. An appendix collects equations for fugacity, enthalpy deviations, and so on, of pure and mixed fluids based on the R-K equations.

15. *Knobler.* This chapter in a series largely devoted to experimental aspects of chemical thermodynamics describes the literature of investigations of the volumetric properties of gaseous mixtures. It covers the virial equation of state, experimental methods for determining virial coefficients, calculation of cross-virial coefficients, vapor phase association, third virial coefficients, and excess functions. The references supplement the bibliography of the earlier book by Mason & Spurling (1969).

16. *Krolikowski.* Computer programs for multicomponent distillation calculations and process simulation are described in a general way. Ten equations of state are listed as being in the program at that time, ranging from the ideal, through the Soave and Hayden-O'Connell virial, to the BWRS. The author states that new models are being added constantly, but old models are never discarded. Once a process has been designed successfully with a certain model, that model is made available for future process calculations of similar processes, since the effort to redo an acceptable old model cannot be justified economically in such a situation. The case of a mixture of hydrogen, light hydrocarbons, and several oxygenated organics is mentioned, for which the vapor model is a modified R-K and the liquid model the Wilson equation. Because of his position in the chemical industry as distinct from the petroleum industry, the author looks for models to be applicable to a wide variety of compounds over a wide range of temperature and pressure, and which are simple in form and converge rapidly—nice for everybody.

17. *Leland.* The capabilities of the van der Waals and BWR types of models and combining rules for mixtures are discussed. Equations with repulsion terms of the hard-sphere type (the Carnahan-Starling, for instance) are the most promising and a number of references is given to these developments. For mixtures, the McCarty corresponding states model, MCSL, which embodies such repulsion terms, is favored. Extension to polar mixtures is indicated, but developments are not far advanced.

18. *Luna & Castro.* Several modifications of the R-K equation were compared with the Han-Starling and Chao-Seader models for figuring thermodynamic properties. For mixtures of natural gas components, the Soave equation with modified mixing rules came out very well in this study.

19. *Martin.* This is the latest of a number of perceptive papers on equations of state by this author over a period of more than twenty years. A general form of cubic equations is formulated, of which many common ones are special cases; the problem of which form is the best is addressed; and a

successful four-parameter equation is developed. Later a Soave type of temperature function is introduced into Martin's equation by Joffe (1980), making it more accurate than the Soave in predicting liquid and vapor densities, but not better in predicting vapor-liquid equilibria.

20. *Opfell et al.* This work is devoted primarily to the BWR equation, with a few pages on its predecessor, the Beattie-Bridgeman equation. A general bibliography of equations of state, with titles, of 102 items is provided.

21. *Ott et al.* This low-key study compares reduced forms of the classical equations of van der Waals, Berthelot, and Dieterici with a corresponding states plot of experimental compressibilities. None of these comparisons would be considered favorable nowadays.

22. *Otto.* Nine equations ranging from the van der Waals to the BWR are compared with experimental data on substances such as argon and butane (and in some instances with each other) and presented in thirteen graphs taken mostly from Shah & Thodos (1965). Included are tabulations of Beattie-Bridgeman and BWR constants, and a comprehensive tabulation of experimental fugacity coefficients.

23. *Pickering.* This extensive comparison of the equations of van der Waals, Berthelot, and Dieterici at pressures up to 1,000 atm is of interest historically, both for the collection of experimental data and for the bibliography of early literature. A summary is made (p. 34) of the light gases and the pressure range in which each equation is most satisfactory.

24. *Prausnitz.* This innovator in the study of phase equilibria covers many topics in this and the next review. The conditions under which vaporization equilibrium ratios, or only vapor fugacities, can be figured are considered. Several comparisons of equations of state with experimental data are given. Although the inference may be drawn that each equation has its successes and failures, the examples shown may provide some indication of what may be expected in the way of reliability. A larger proportion of space is devoted to activity coefficients of the liquid phase.

25. *Prausnitz.* The detail of this paper is somewhat less than that of the preceding one, but naturally much the same ground is covered. Some examples are given for the Soave, Peng-Robinson, and Lee-Kesler equations in which they all come out fairly well at moderate pressures. For improved equations of state, perturbation theory and association equilibria are regarded as good possibilities. The author also believes that more attention should be devoted to improving combining rules for mixtures and to accounting for the effects of differences in molecular sizes.

26. *Shah & Thodos.* The Redlich-Kwong equation of state was checked with eleven others against data for argon and with the Beattie-Bridgeman and the BWR equations against data for *n*-butane. Constants of the BWR equation were not available for argon. The R-K equation was found at least as good generally as the B-B and BWR, superior to the BWR in the vicinity of the critical, and substantially superior to all the other equations examined. Those included the van der Waals, Berthelot, Clausius, Dieterici, Wohl, and six others invented for this project. Vigorous response to this criticism of the BWR equation, however, was voiced by Barner & Adler (1968), and it appears that the Shah & Thodos calculations were wrong.

27. *Tarakad et al.* The eight equations studied are the virial, the original Redlich-Kwong, the Soave, two other modifications of the R-K, and two others. The BWR was not included because it was considered usually reliable when the necessary constants and interaction parameters were known. Pure components, mixtures, and polar systems including water-gas mixtures are covered. For figuring gas compressibilities, the original R-K was found as good as the modified equations, except for the carbon dioxide–propane system. At low to moderate pressures nonpolar systems are adequately represented by the virial equation. At high pressures none of the equations is in general reliable. Table 1.26 summarizes recommendations of equations to use at various conditions for several types of systems.

28. *Tsonopoulos & Prausnitz.* This study is addressed primarily to equations that are applicable in the cryogenic range. The *C*-truncated virial equation generally is adequate up to 0.75 times the critical density. Several equations for figuring liquid densities are mentioned. The Beattie-Bridgeman equation is applicable up to nearly the critical density, the BWR up to 1.8 times the critical density. An extension of the BWR by Strobridge was applied up to three times the critical density. A modified van der Waals equation with twenty-one constants applied up to about twice the critical density. Several modifications of the Redlich-Kwong equation with binary interaction parameters are found to be adequate for enthalpies and vaporization equilibrium ratios in the cryogenic range. Since this article was published, BWR-type equations with 30–40 constants by Bender and by Goodwin and a corresponding states equation by McCarty, also with a large number of constants, have been used to represent cryogenic fluids.

29. *Wichterle.* The fourth paper in this series is concerned with equations of state for figuring vaporization equilibrium ratios. Several modifications of the Redlich-Kwong equation are referred to. Table 1.10 summarizes the various adjustments of the R-K parameters that have been proposed in the literature. Modification of combining rules and binary interaction parameters are reviewed. The R-K equation is recommended for liquid densities when the parameters are evaluated from liquid densities (as in the case of the P-R equation), and is applicable to polar compounds that do not form strong hydrogen bonds and to azeotropic systems. The review of BWR literature mentions a half-dozen or so modifications. Since the R-K and BWR types are so satisfactory, only a few other equations (mostly with three parameters) are mentioned. Corresponding states methods, for instance the Lee-Kesler-Plöcker, and the perturbed-hard-sphere repulsion system of Donohue and Prausnitz also are mentioned favorably.

1.10. REVIEW LITERATURE ON EQUATIONS OF STATE

1. Otto, in Wien-Harms, "Handbuch der Experimentalphysik," Bd. 8, Teil 2, 1–246 (1929).
2. Partington, "An Advanced Treatise on Physical Chemistry," Vol. 1, Sect. 7 (1950).
2a. Vukalovich & Novikov, "Equations of State of Real Gases" (in Russian), Gosenergoizdat, Moscow (1948). The Library of Congress has a copy.

A selection of recent articles is:

3. Abbott, "Cubic equations, an interpretive review," *ACS Advances in Chemistry Series 182,* 47–70 (1979).
4. Adler, Spencer, Ozkardesh, & Kuo, "Industrial uses of equations of state: a state of the art review," *ACS Symposium Series 60,* 150–199 (1977).

5. Boublik, "Progress in statistical thermodynamics applied to fluid phase," *Fluid Phase Equilibria 1*, 37–87 (1977).
6. Coward, Gale, & Webb, "Process engineering calculations with equations of state," *Trans IChemE 56*, 19–27 (1978).
7. Cox & Lawrenson, "The PVT behavior of single gases," in McGlashan (ed.), *Chemical Thermodynamics*, Vol. 1, The Chemical Society, 162–203 (1973).
8. Daubert, Graboski, & Danner, "Documentation of the basis for selection of the contents of Chapter 8—vapor-liquid K-values in Technical Data Book—Petroleum Refining," API (1979).
9. Douslin, "The pressure, volume and temperature properties of fluids," in Skinner (ed.) *International Review of Science, Physical Chemistry*, Vol. 2, Vol. 10 (1975).
10. Fredenslund, Rasmussen, & Michelsen, "Recent progress in the computation of equilibrium ratios," *Chem. Eng. Commun. 4*, 485–500 (1980).
11. Gibbons, Coulthurst, Farrel, Gough, & Gillett, "Industrial uses of thermodynamic data," *Proceedings of the NPL Conference, National Physical Laboratory, Chemical Thermodynamic Data*, 8–25 (1978).
12. Gray, "Industrial experience in applying the Redlich-Kwong equation to vapor-liquid equilibria," *ACS Advances in Chemistry Series 182* (1979).
13. Grenier & Cabre, "Equation of state for process calculations, L'Air Liquide experience in this field," *Phase Equilibrium and Fluid Properties in Chemical Industry*, EFCE 2nd International Conference, 389–398 (1980).
14. Horvath, "Redlich-Kwong equation of state: Review for chemical engineering calculations," *Chem. Eng. Science 29*, 1334–1340 (1974).
15. Knobler, "Volumetric properties of gaseous mixtures," in McGlashan (ed.), *Chemical Thermodynamics*, Vol. 2, The Chemical Society, 199–237 (1978).
16. Krolikowski, "Industrial view of the state-of-the-art in phase equilibria," *ACS Symposium Series 60*, 62–86 (1977).
17. Leland, "Equations of state for phase equilibrium computations: Present capabilities and future trends," EFCE 2nd International Conference, 283–334 (1980).
18. Luna & Castro, "Evaluations of various modifications of the Redlich-Kwong equation," *International Chemical Engineering 18*, 611–626 (1978).
19. Martin, "Cubic equations of state—which?" *IEC Fund 18*, 81–97 (1979); *19*, 130–131 (1980).
20. Opfell, Pings, & Sage, "Equations of state for hydrocarbons," *API Monograph*, 184 pp. (1959).
21. Ott, Coates, & Hall, "Comparisons of equations of state in effectively describing *PVT* relations," *J. Chem. Education 48*, 515–517 (1971).
22. Otto, "Zustandsgleichungen," in Landolt-Börnstein *II/1* 298–309 (1970).
23. Pickering, "Relations between the temperatures, pressures and densities of gases," *Circular of the Bureau of Standards*, No. 279, 85 pp. (1925).
24. Prausnitz, "State of the art review of phase equilibria," *ACS Symposium Series 60*, 11–61 (1977).
25. Prausnitz, "State of the art review of phase equilibria," *Phase Equilibria and Fluid Properties in the Chemical Industry*, EFCE Second International Conference, 231–282 (1980).
26. Shah & Thodos, "A comparison of equations of state," *Ind. Eng. Chem. 57* (3), 30–37 (1965).
27. Tarakad, Spencer & Adler, "A Comparison of eight equations of state to predict gas-phase fugacity, density and fugacity," *IECPDD 18*, 726–729 (1979).
28. Tsonopoulos & Prausnitz, "Equations of state, a review for engineering applications," *Cryogenics* 315–327, October (1969).
29. Wichterle, "High pressure vapor-liquid equilibrium," *Fluid Phase Equilibria 1*, 161–172, 225–245, 305–316 (1978); *2*, 59–78, 143–159, 293–295 (1979).

1.11. PROBLEMS

1.1. Using the Clausius (1880) equation of state in the form

$$\left[P + \frac{11{,}000}{T(V+0.14)^2}\right][V - 0.14] = 0.08206T,$$

find the saturation pressure and the specific volumes of liquid and vapor phases in equilibrium at 350 K. Apply the Maxwell rule that in the S-shaped region of the *P-V* plot, the area above the saturation pressure P_s equals that below. The equivalent statement is

$$\int_{V_1}^{V_g} V dP = P_s(V_g - V_1) - \int_{V_1}^{V_g} P dV = 0.$$

P_s is expected to be in the vicinity of 8 atm. See Figure 1.7.

1.2. The densities of saturated methanol vapor at two conditions are:

T,°C	*P, atm*	*g/ml*
140	10.84	0.01216
230	68.04	0.1187

Find numerical values of the van der Waals constants from these data, and compare with those figured from the critical properties, which are

$T_c = 512.6$ K, $P_c = 79.9$ atm, $V_c = 118$ ml/gmol.

1.3. Compare compressibility factors of saturated ethylene vapor at 40 F and 654.56 psia, using

a. The Redlich-Kwong equation.
b. The *B*-truncated virial equation with the Pitzer-Curl correlation for B.
c. The Lydersen tables.

For comparison, the experimental specific volume is 0.139 cu ft/lb.

1.4. Find reduced forms of the Lorentz (1881) and of the two versions of the Clausius (1880) equations:

Lorentz: $(P + a/V^2)V = RT(1 + b/V)$.

Clausius 1: $[(P + a/T(V + b)^2](V - b) = RT$.

Clausius 2: $[(P + a/T(V + c)^2](V - b) = RT$.

Note: In the last case only a partially reduced form is

obtainable. Letting $V_r^* = (V + c)/(V_c + c)$, the final equation is

$(P_r + 3/T_r V_r^{*2})(3V_r^* - 1) = 8T_r$.

1.5. Find the specific volume of an equimolal ternary mixture at 40 atm and 450 K with the Soave equation, neglecting binary interaction parameters. Pure component properties are:

	P_c, *atm*	T_c, °*K*	V_c *ml/gmol*	z_c	ω
Methanol	79.9	512.6	118	0.224	0.559
Ethanol	63.0	516.2	167	0.248	0.635
Acetone	46.4	508.1	209	0.232	0.309

1.6. For mixtures of helium and nitrogen at 30 C, containing the percentages of helium of the following table, compare compressibilities by these three methods:

a. Use the virial coefficient data cited in Landolt-Börnstein II/1 (1970, p. 260), namely,

A—B	Mol-% *A*	$B' \cdot 10^6$ atm^{-1}	$C' \cdot 10^6$ atm^{-2}
He–N_3	21.608	2.099	1.505
	50.747	5.066	0.597
	80.349	5.590	0.141

b. The truncated virial equation with Abbott's formulas for *B* and the Lorentz-Berthelot combining rules (Table 1.5).

c. The truncated virial equation with Tsonopoulos's formula for *B* and the Lorentz Berthelot combining rules (Table 1.5).

1.7. Second virial coefficient data of an equimolal mixture of propane, butane and methyl bromide are cited in Landolt Börnstein II/1 (1970, p. 265) as follows:

T	*B in cm³/mol*		
°*K*	*Propane*	*Butane*	*Methylbromide*
244.0	−610	−1230	−1040
273.0	−477	−932	−718
297.0	−394	−758	−567
321.0	−340	−635	−451

T	B_{12} *in cm³/mol*	
°*K*	C_3H_3—CH_3Br	C_4H_{10}—CH_3Br
244.0	−617	−915
273.0	−474	−661
297.0	−411	−546
321.0	−356	−459

Compare these data with calculations from the formulas of Abbott and of Tsonopoulos, using the Lorentz-Berthelot combining rules (Table 1.5).

1.8. Find formulas for the acentric factor in terms of the constants of the Antoine and the Wagner (1973) vapor pressure equations and the critical constants:

Antoine: $\log P^0 = A - B/(T + C)$ and $\log P^0 = A - B/T$.

Wagner: $\log P°/P_c = [(A(1 - T_r) + B(1 - T_r)^{1.5} + C(1 - T_r)^3 + D(1 - T_r)^6)]/T_r$.

1.9. A mixture consisting of 2/3 propylene and 1/3 sulfur dioxide is at 425 K and 60 atm. What density is predicted for this mixture by the van der Waals equation, taking $R = 0.08205$(liter)(atm)/(gmol)(*K*)?

1.10. An equimolal mixture of methanol and nitrogen at 100 atm and 600 K is expanded in an isothermal turbine at the rate of 1 gmol/sec. The Soave equation applies. How much work is being obtained? Exit $P = 5$ atm.

1.11. Equilibrium constants for the dimerization of carboxy acids are given by Gmehling & Onken (1977, Vol. 1, Pt. 1, p. 687) in the form

$\log K = \log (P_{dimer}/P^2_{monomer}) = A + B/T$, atm, °K,

where

Acid	*A*	*B*
Formic	−10.743	3083
Acetic	−10.421	3166
Propionic	−10.843	3316
Butyric	−10.100	3040

a. Assuming that deviation from ideality is due entirely to dimerization, find the compressibilities at 323, 373, and 423 K, and 1, 5, and 10 atm, for all four acids.

b. For acetic acid, compare with the compressibilities from the Redlich-Kwong equation, neglecting dimerization and condensation.

1.12. Check these experimental data of the second virial coefficients of acetone (1) and butane (2) with the Abbott, Pitzer, & Curl and Tsonopoulos formulas, and the Lorentz-Berthelot combining rules.

°*K*	$-B_{11}$	$-B_{22}$	$-B_{12}$, *ml/gmol*
282.3	2733	862	805
297.0	2268	758	656
312.0	1876	674	569
321.0	1680	635	504

(Kappallo et al. 1963)

1.13. Draw isotherms of the reduced Soave equation for $T_r = 0.9$, 1.0, and 1.2 at $\omega = -0.2$, 0, and 0.5.

1.14. The third virial coefficient is correlated by DeSantis & Grande (1979) by the equations:

$$\frac{C}{v_c^2} = C^0(T_R) + dC'(T_R) + d^2C''(T_R),$$

$$C^0 = \frac{0.1961}{T_R^{1/4}} + \frac{0.3972}{T_R^5} + \left(0.06684T_R^4 - \frac{0.5428}{T_R^6}\right)\exp(-T_R^2),$$

$$C' = \frac{64.5}{T_R^9}[1 - 2.085 \exp(-T_R^2)],$$

$$C'' = \frac{801.7}{T_R^7}.$$

Use these together with the Tsonopoulos equations for the second virial coefficient to figure the compressibility of benzene at 24.15 atm and reduced temperatures in the range 0.8–2.5. The parameter $d = 0.0069$. Other properties are $T_c = 562.1$ K, $P_c = 48.3$ atm, $V_c = 259$ ml/gmol, and $\omega = 0.215$.

1.15. Show that the second virial coefficient is given by

$$B' = \lim_{P \to 0} \left(\frac{\partial z}{\partial P}\right)_T.$$

Hence, find expressions for the second virial coefficient in terms of the parameters of the van der Waals and Redlich-Kwong equations of state.

1.16. For an equimolal mixture of propane (1), sulfur dioxide (2), and carbon dioxide (3), apply the appropriate combining rules to find the parameters of

a. The van der Waals equation, given the data

i	a	b
1	8.664	0.08445
2	6.714	0.05636
3	3.592	0.04267

b. The Redlich-Kwong equation, given that

ij	P_c	T_c
11	41.9	369.8
22	77.8	430.8
33	72.8	304.2
12	56.54	399.1
13	53.97	335.4
23	74.99	362.0

c. The BWR equation, using the data of this chapter.

1.17. For pyridine, plot vapor pressures figured from the Antoine equation,

$$\ln P^s = 16.091 - 3095.13/(T - 61.15),$$

and from the Nath et al. and from the Lee & Kesler equations of Section 1.3.3, over the range $\ln P_r(0, -40)$, $1/T_r(1, 5)$. Properties are $\omega = 0.245$, $T_c = 620$ K, $P_c = 55.6$ atm. The units of P^s are Torr and those of T are °K.

1.18. At the Boyle point of a gas, the derivative $(\partial(PV)/\partial P)_T = 0$. Find equations relating the temperature to either the volume or pressure at this condition in terms of (1) the van der Waals equation and (2) the Redlich-Kwong equation. For ethylene, Partington (1950, p. 555) cites data of Amagat for the Boyle point as 164 atm at 100 C and 105 atm at 40 C. Compare these values with calculated ones, given that for van der Waals,

$$a = 4.5574,$$

$$b = 0.0583,$$

and for Redlich-Kwong,

$$a = 77.6030,$$

$$b = 0.0404,$$

in the units atm, liters/gmol, °K.

1.19. Find equations for the following in terms of the van der Waals equation:

Thermal expansion coefficient: $\frac{1}{V}\left(\frac{\partial V}{\partial T}\right)_P$.

Isothermal compressibility: $-\frac{1}{V}\left(\frac{\partial V}{\partial P}\right)_T$.

Joule-Thomson coefficient: $\left(\frac{\partial T}{\partial P}\right)_H$.

Pressure coefficient of enthalpy: $\left(\frac{\partial H}{\partial P}\right)_T$.

1.20. Entropies of ideal gases are given in the *API Data Book* as functions of temperature, $S = f(T)$. Find a relation for the partition function, $\ln Q$, and then the equation for internal energy, U in terms of such functions of temperature.

1.21. Find the values of the Lennard-Jones parameters ε and σ for a substance that has second virial coefficients $25(10^{-6})$ m^3/gmol at 360 K and $11.9(10^{-6})$ at 540 K. Some values of the integral, $B/N\sigma^3$, from the compilation of Reed & Gubbins (1973, p. 452) are:

Tk/ε	$-B/N\sigma^3$
0.3	27.89
0.4	13.79
0.45	10.76
0.5	8.720
0.6	6.198
0.75	4.176
0.9	2.775
1	2.538
1.5	1.200
2	0.628
3	0.115

1.22. Use the results of Example 1.11 to find the virial expansions of the Redlich-Kwong equation $P = RT/(V - b) - a/\sqrt{T}\,V(V + b)$ and of the Peng-Robinson equation, $P = RT/(V - b) - a/[V(V + b) + b(V - b)]$.

1.23. For the reduced form of the truncated virial equation,

$$V_r^3 - 3(T_r/P_r)V_r^2 + 3(T_r/P_r)V_r - T_r/P_r = 0,$$

apply Cardan's method for the solution of a cubic to show that the equation has only one real root. (See, for example, *CRC Handbook of Mathematical Sciences*, p. 40, 1978).

1.24. Apply the criteria of Poling et al. cited in Section 1.5.4 to the vapor and liquid phase roots found in Example 1.13.

1.25. Virial coefficients and critical properties of $n - C_6F_{14}$ (and other perfluorohexanes) were determined by Taylor & Reed (1970) (*AIChE Journal* 16 738–741). Some of the results are:

°K	B liters/gmol	C (liters/gmol)^2
395.56	−1.051	0.25
415.48	−0.920	0.23
432.68	−0.818	0.21
451.55	−0.725	0.20

$P_c = 18.1$ atm, $T_c = 451$, $V_c = 0.555$ liters/gmol, $\omega = 0.483$.

a. What are the compressibilities when both B and C are taken into account?
b. What are the compressibilities when only B is taken into account?
c. Find the second virial coefficients and the corresponding compressibilities, from the Tsonopoulos correlation.

The pressure is 10 atm.

1.26. Taking the van der Waals equation of state, find equations for

a. The Joule effect: $\left(\dfrac{\partial T}{\partial V}\right)_V$.

b. The heat capacity difference: $C_p - C_v$.

1.27. Develop an equation for the relation between T and V for an adiabatic reversible process of a van der Waals gas, finding first $(\partial T/\partial V)_S$ as a function of T and V.

1.28. Find $\mu C_p/V_c$ as a function of P_r and T_r with the reduced form of the van der Waals equation over a practicable range of the variables.

1.29. The Joule-Thomson inversion point is the temperature at which a cross over from heating to cooling occurs, for a throttling process. Experimental data are correlated by Miller (1970) by the equation

$P_r = 24.21 - 18.54/T_r - 0.825\, T_r^2$.

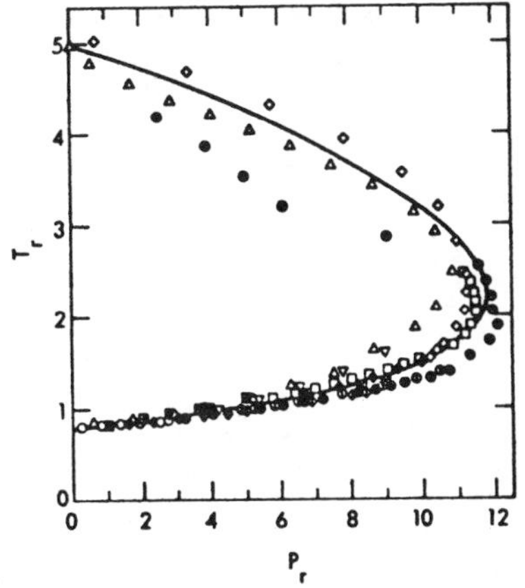

The data are shown on the sketch. Over the range of conditions shown, find the (P_r, T_r) relations at the inversion point from

a. The C-truncated virial equation.
b. The Redlich-Kwong equation.
c. The Soave equation with acentric factor = 0.5 and any other values that appear interesting.

1.30. Throttling experiments at 20 atm and two temperatures gave these results for the Joule-Thomson coefficients:

T, °K	V, liters/gmol	μC_p
600	2.11	0.98
700	2.60	0.72

Assuming the gas to conform to the van der Waals equation, find the constants a and b for this substance.

1.31. The B-truncated virial equation often is written $z = 1 + BP/RT$. However, make the proper substitution, $V = zRT/P$, into the standard form, $z = 1 + B/V$; then compare correct values of zs with those figured from the first equation, over a range of BP/RT ranging from −0.25 to 0.25.

1.32. Densities of ethane at 25 C and several pressures are:

P, atm	10	20	30	40
ρ, gmol/1	0.4447	0.9893	1.7227	3.0817

Find the second and third virial coefficients.

1.33. The second virial coefficient is given as a function of temperature by Guggenheim's equation:

$B/V_c = 1.51 - 1.60 \exp(0.936T_c/T)$.

For ethane, $V_c = 142$ ml/gmol, and $T_c = 305.7$ K. At a pressure of 100 atm, at what temperature is the specific volume one-half that at 75 atm and 400 K?

1.34. Find the work required to compress a gas from $P_r = 0.2$ to $P_r = 10$ at $T_r = 1.2$, using the Soave equation with acentric factors of 0, 0.5, and 1.0. Express the results as W/P_cV_c. The form of Soave equation derived in Example 1.16 may be useful.

1.35. The EOS of Wohl (1914, 1921) (*Z Physik Chem* *87*, 1; *99*, 207),

$$P = \frac{RT}{V-b} - \frac{a}{TV(V-b)} + \frac{c}{T^{4/3}V^3},$$

has the polynomial form

$V^4 - (b + RT/P)V^3 + aV^2/PT - cV/PT^{4/3} + bc/PT^{4/3} = 0.$

Derive these relations for the parameters in terms of the critical properties:

$a = 6V_c^2P_cT_c$; $b = V_c/4 = RT_c/15P_c$;

$c = 4V_c^3P_cT_c^{4/3}$; $RT_c/P_cV_c = 15/4 = 3.75$.

Hence find the reduced form of this EOS.

1.36. The van der Waals equation modified with a perturbed-rigid-sphere repulsion term is

$$P = \frac{RT}{V}\left(\frac{1+y+y^2-y^3}{(1-y)^3}\right) - a/V^2,$$

where $y = b/4V$. Show that the parameters are given in terms of the critical properties by

$$a/RT_cV_c = 1.38(9/8),$$

$$b/V_c = 0.5216(1/3),$$

$$z_c = 0.359(3/8).$$

For comparison, the quantities in parentheses are those for the original *VDW* equation.

1.37. Data for hydrogen sulfide taken from Starling, "*Fluid Thermodynamic Properties*" (1973), are

P, atm	40.82	27.21
V, 1/gmol	0.2583	0.5864
T, °K	355.4	456.4

a. Find the parameters of the van der Waals equation.
b. Find the parameters of the perturbed van der Waals equation given in Problem 1.36.

1.38. Hydrogen fluoride in the gas phase forms dimer, hexamer, and octamer. The equilibrium constants (K_p) are:

$$K_2 = \exp(6429.73/T - 24.171),$$

$$K_6 = \exp(21011.97/T - 69.7327),$$

$$K_8 = \exp(25225.72/T - 83.4731).$$

(Schotte 1980). Find the apparent compressibility of the parent substance at 20 atm and 400 K.

1.39. Equations for the second virial coefficients of propylene (1) and carbon dioxide (2) are cited by Kogan (1967).

$$B_{11} = -\frac{38{,}360}{T} - \frac{2.3184(E9)\log T}{T^2},$$

$$B_{22} = 1{,}000\left(\frac{-291.61 + 102.87\log T}{T}\right),$$

$$B_{12} = \frac{4{,}810}{T} + \frac{1.393(E7)}{T^2},$$

in ml/gmol. van der Waals parameters in the units atm, liters/gmol, °K are $a_1 = 8.379$, $a_2 = 3.592$, $b_1 = 0.08272$, $b_2 = 0.04267$. Find the compressibilities at 20 atm and 400 and 600 K, of mixtures for which $x_1 = 0, 0.5$, and 1, with the virial and van der Waals equations.

1.40. Second virial coefficients of propane, butane, and methyl bromide are given by Kappallo et al. (1963). Find compressibilities of equimolal mixtures at the several temperatures.

T	*B in cm³/mol*		
°*K*	C_3H_8	C_4H_{10}	CH_3Br
244.0	−610	−1,230	−1,040
273.0	−477	−932	−718
297.0	−394	−758	−567
321.0	−340	−635	−451

T	B_{12} *in cm³/mol*	
°*K*	C_3H_8—CH_3Br	C_4H_{10}—CH_3Br
244.0	−617	−915
273.0	−474	−661
297.0	−411	−546
321.0	−356	−459

1.41. The second virial coefficient of carbon tetrachloride is represented by the equation

$$B = 575 - 2E8/T^2, \text{ ml/gmol}$$

over the range 315–345 K. (Kogan 1967, p. 400.) Find the parameters of a square-well potential function that fits these data as closely as possible. Note Example 1.19.

1.42. The second virial coefficient of nitromethane is represented by the empirical equation

$$B = 100 - 4.19(E5)/T - 1.91(E23)/T^2, \text{ ml/gmol}$$

over the range 315–345 K. (Kogan 1967, p. 400.) Find the parameters of Lennard-Jones and square-well potential functions that will fit the experimental virial coefficients as closely as possible. Note Example 1.19.

1.43. Experimental pressures of methane (Keyes & Burks 1927) are

V, ml/gmol	320	320	320	640	640	640
T, °C	0	100	200	0	100	200
P, atm	60.13	92.24	121.81	32.297	46.474	60.486

Find the best values of the constants of the van der Waals and Redlich-Kwong equations and compare them with the values obtained from the critical properties.

1.44. Inspection of Figure 1.19 indicates that the isotherms of the *C*-truncated virial equation do not possess valleys. Show that $V_r = 1$ and $P_r = T_r$ when $dP_r/dV_r = 0$. Furthermore, show that this is a point of inflection and not a minimax.

1.45. Make plots of isotherms at $T_r = 0.9$, 1.0, and 1.2 of the reduced Berthelot and Dieterici equations of state.

$$P_r = \frac{32T_r}{9(V_r - 0.25)} - \frac{16}{3T_rV_r^2},$$

$$P_r = \frac{T_r}{2V_r - 1}\exp\left[2\left(1 - \frac{1}{T_rV_r}\right)\right].$$

Cover the range of P_r from −2 to +4 and V_r from 0 to 2.

Comment on any peculiarities that the plots may have in the vicinity of the critical point.

1.46. Refer to Figure 1.7 for identification of regions of metastability. The van der Waals equation of acetic acid at 500 K is $(P + 17.59/V^2)(V - 0.1068) = 0.08205T$. Find the limits of metastability with this equation. Note that there are three points of minimax according to the solution of $(\partial P/\partial V)_T = 0$, one of which is not realizable physically. Trace the complete isotherm to show all the minimax.

1.47. A gas phase mixture of four substances has the composition: 60% A, 20% B, and 10% each of C and D. Equilibrium constants of dimerization are assumed to be like those of the acids in Problem 1.11 in the same order. Find the compressibility z on the assumption that the deviation from ideality is due entirely to the dimerization process. Do this at 0.5 atm and temperatures 300 and 325 K.

1.48. Specific volumes of isobutane at several pressures are given in the *Chemical Engineer's Handbook* (1973) as follows at 220 F:

P, psia	50	100	150	200	250	300
V, cuft/lb	2.379	1.1278	0.7091	0.4977	0.3649	0.2696

Compare these values with those calculated from

a. The Redlich-Kwong equation with a and b from Table 1.9.
b. Eq. 1.189b with $a = 1.463RT_c^{1.5}V_c$, $b = 0.3326V_c$, and $V_c = 0.263$ liters/gmol.
c. Eq. 1.189b as in part b, but with $V_c = 0.3157RT_c/P_c$.
d. The Soave equation.

Express the results as ratios of calculated to true pressures at each volume.

2 Thermodynamic Functions and Equilibrium

This chapter reviews those aspects of thermodynamics that are especially pertinent to the study of phase and chemical equilibria. Mainly this comprises the evaluation of those properties that are significant in defining equilibrium states and the evaluation of energy changes accompanying transition from one equilibrium state to another. It is beyond the scope of this book to derive all the useful equations in detail. For convenience, basic relations are extended in summary tables and in Appendix A.

2.1. ENERGY FUNCTIONS

Magnitudes of changes in stored energy functions are determined by net heat, work, and mass transfers. The basic property of this kind is the internal energy U, which is related to other properties in the combined statement of the first and second laws of thermodynamics:

$$d\underline{U} = d(nU) = dQ - dW + \Sigma(\partial \underline{U}/\partial n_i)_{\underline{S}\underline{V}n_j} dn_i \quad (2.1)$$

$$= Td\underline{S} - Pd\underline{V} + \Sigma \mu_i dn_i, \quad (2.2)$$

where μ_i is the change in internal energy of the system per mol of substance i transferred at constant entropy and volume. Other equivalents are given later. The internal energy is a function of the independent variables $S, V, n_1, n_2, \ldots$, which may be written in functional notation:

$$\underline{U} = \underline{U}(\underline{S}, \underline{V}, n_1, n_2, \ldots). \quad (2.3)$$

Several important results are obtained by manipulation of Eq. 2.2. With the substitutions $\underline{U} = nU$, $\underline{S} = nS$, $\underline{V} = nV$, and $n_i = nx_i$, the expanded and rearranged result is

$$d(nU) - Td(nS) + Pd(nV) - \Sigma \mu_i d(nx_i)$$
$$= (dU - TdS + PdV - \Sigma \mu_i dx_i)n$$
$$+ (U - TS + PV - \Sigma x_i \mu_i)dn = 0. \quad (2.4)$$

Since n and dn are arbitrary, each quantity in parentheses is separately zero so that

$$dU = TdS - PdV + \Sigma \mu_i dx_i, \quad (2.5)$$

$$U = TS - PV + \Sigma x_i \mu_i. \quad (2.6)$$

When the composition does not change, Eq. 2.5 becomes the familiar

$$dU = TdS - PdV. \quad (2.7)$$

After differentiation, Eq. 2.6 may be rearranged to

$$dU - TdS + PdV - \Sigma \mu_i dx_i - (SdT - VdP + \Sigma x_i d\mu_i) = 0. \quad (2.8)$$

In view of Eq. 2.5, the quantity in parentheses likewise is zero and rearranges to

$$\Sigma x_i d\mu_i = -SdT + VdP = (\partial G/\partial T)_P dT + (\partial G/\partial P)_T dP \quad (2.9)$$

This relation is one form of the Gibbs-Duhem equation; it has many applications.

Other useful energy functions are the enthalpy, H; the Helmholtz energy, A; and the Gibbs energy, G, whose relations to the internal energy are shown in Table A.3 and developed further later. It may be shown that the various energy functions are related quite simply to work transfers under specific conditions. The energy balance of a closed system is

$$dU = dQ - dW = TdS - PdV, \quad (2.10)$$

and of a steady-flow open system,

$$dH = dQ - dW_s = TdS + VdP. \quad (2.11)$$

Accordingly, the work transfers of reversible processes at either isothermal or adiabatic conditions may be derived and summarized as:

System Type	*Condition*	*Work Transfer*	
Closed	Adiabatic	$-dU$	(2.12)
Closed	Isothermal	$-dA$	(2.13)
Flow	Adiabatic	$-dH$	(2.14)
Flow	Isothermal	$-dG$	(2.15)

A certain historical significance attaches to these work equations, for example the term *maximum work* for A and the term *free energy* (for work performance) for G, but all the energy functions have many applications not directly related to work and heat transfers.

The integral and differential forms of H, A, and G are derived from Eqs. 2.5 and 2.6 by substituting $H = U + PV$, $A = U - TS$, and $G = U + PV - TS$, with the results

$$H = TS + \Sigma x_i \mu_i, \quad (2.16)$$

$$A = -PV + \Sigma x_i \mu_i, \quad (2.17)$$

$$G = \Sigma x_i \mu_i, \quad (2.18)$$

$$dH = TdS + VdP + \Sigma \mu_i dx_i, \quad (2.19)$$

$$dA = -SdT - PdV + \Sigma \mu_i dx_i, \quad (2.20)$$

$$dG = -SdT + VdP + \Sigma \mu_i dx_i. \quad (2.21)$$

Thermodynamic relations are closely interrelated and may be used to evaluate quantities that are difficult to measure in terms of some that are easily measured. Table A.4 lists some of these readily measurable quantities and their formulas, including compressibility; heat capacity; the Joule-Thomson coefficient; and, by implication, pressure, temperature, volume, and composition.

Although few substances approach ideal-gas behavior over ranges of practical interest, such behavior is of interest as a limiting case. The main thermodynamic characteristics of ideal gases are collected in Table A.5; Table A.6 summarizes the changes in their properties in some common processes. Other limiting relations, particularly those of mixtures, such as the laws of Henry, Raoult, and Lewis & Randall, are used in later chapters. Table A.2 lists a few named relations in thermodynamics.

2.2. FUNDAMENTAL EQUATIONS

The functional relationship for a system of constant composition,

$$U = U(S, V), \quad (2.22)$$

is known as a *fundamental equation*, since all thermodynamic information about such a system is derivable from it, as will be shown shortly. The mathematical form of such an equation is not prescribed and must be developed empirically, a process that has been accomplished for very few substances. The methods of statistical mechanics (for example, Section 1.7) do lead to theoretical equations in terms of partition functions,

Example 2.1 An Open System with Work and Heat Transfers

When work and heat transfers occur, an open system is analyzed most simply with the internal energy balance in the form,

$$\underline{dU} = d(nU) = ndU + Udn$$
$$= dQ - dW + \Sigma H_{\text{in}} dn_{\text{in}} - \Sigma H_{\text{out}} dn_{\text{out}} \qquad (1)$$

rather than Eq. 2.2. In this equation the energy associated with each mol, n_i, crossing the system boundary is made up of its flow work, P_iV_i, plus its internal energy, U_i, which total its enthalpy, H_i.

In case only one stream crosses the boundary and that is inward, the equation becomes

$$ndU = dQ - dW + (H_{\text{in}} - U)dn. \qquad (2)$$

Introducing

$$dU = C_v dT, \qquad (3)$$
$$dH = C_p dT, \qquad (4)$$

then

$$H - U = P_0V_0 + \int_{T_0}^{T_{\text{in}}} C_p dT - \int_{T_0}^{T} C_v dT, \qquad (5)$$

where the subscript 0 refers to an arbitrary reference condition. Several conditions need to be imposed on the energy balance before Eq. 2 can be integrated, say for a relation between n and T. Problem 2.18 is a specific example.

but the process of developing fundamental equations in this way is far from being a working tool. Nevertheless, the concept of fundamental equations is illuminating. Equivalent relations between other groups of variables are:

$$H = H(S, P), \qquad (2.23)$$
$$A = A(T, V), \qquad (2.24)$$
$$G = G(T, P). \qquad (2.25)$$

These groups of variables have evolved naturally, but they are deducible mathematically from the original fundamental equation, Eq. 2.22, with the Legendre Transformation (e.g., Modell & Reid, 1982). Another group of variables from which all other thermodynamic data are deducible is

$$f(P, V, T, C_p^{id}) = 0, \qquad (2.26)$$

which comprises the PVT equation of state and the heat capacity at zero pressure or in the ideal-gas state. Since the quantities of Eq. 2.26 are the most easily measured ones, that version is the most used of the fundamental relations, usually not as a single equation but as a group of relations between these variables.

Although it is not a formal proof of the principle, the entries of Table 2.1 do indicate that all thermodynamic properties may be represented in terms of each of the four groups of three variables each: (U, S, V); (H, S, P); (A, T, V); or (G, T, P). Reading from the table, for example:

$$G = U - S(\partial U/\partial S)_V - V(\partial U/\partial V)_S \qquad (2.27)$$
$$= H - S(\partial H/\partial S)_P \qquad (2.28)$$
$$= A - V(\partial A/\partial V)_T \qquad (2.29)$$
$$= G(T, P). \qquad (2.30)$$

Such representations also may be made in terms of (P, V, T, C_p^{id}). For instance, Example 2.2 develops $H - H^{id}$ in this way, whereas Example 2.3 develops expressions for $H - H^{ia}$ and $S - S^{id}$ and indicates how other properties may be obtained from them.

Although P and V are interrelated by an EOS, one or the other may be a preferred variable for representing changes in the thermodynamic properties, depending on which of them is solvable for explicitly from the EOS. This topic is considered in Section 2.7. In Example 2.2 the particular EOS may be solved explicitly for either P or V, so there is no advantage of one system over the other; but in most other cases there is a substantial difference in the ease with which a useful relation can be developed.

Interrelations of the properties usually involve derivatives, of which those involving P, V, and T are the most easily obtained. If Eq. 2.26 is valid, all derivatives should be expressible in terms of those variables and the ideal heat capacity, and several convenient methods actually have been worked out for making such transformations. The easiest to use and the most popular is that of Bridgman (1925). As given in Table A.7, all derivatives are based on T, P, S, and the three derivatives most susceptible of measurement: $(\partial V/\partial T)_P$, $(\partial V/\partial P)_T$, and $C_p = (\partial H/\partial T)_P$. Second derivatives also were covered by Bridgman. Another, perhaps more versatile system, is due to Shaw (1935), which is also explained by Sherwood & Reed (1939) and simplified by Carroll (1965). A system presented by Modell & Reid (1974) involves the compressibility, $z = PV/RT$, and is convenient for use with corresponding state correlations or equations of state explicit in z.

As an example, the Bridgman scheme may be applied to finding the isothermal derivatives of U and H with respect to S. The derivatives are ratios of two terms listed in Table A.7:

$$\left(\frac{\partial U}{\partial S}\right)_T = \frac{(\partial U)_T}{(\partial S)_T}$$
$$= \left[T\left(\frac{\partial V}{\partial T}\right)_P + P\left(\frac{\partial V}{\partial P}\right)_T\right] \Big/ \left(\frac{\partial V}{\partial T}\right)_P$$
$$= T + P\left(\frac{\partial V}{\partial P}\right)_T \left(\frac{\partial T}{\partial V}\right)_P = T - P\left(\frac{\partial T}{\partial P}\right)_V,$$

$$\left(\frac{\partial H}{\partial S}\right)_T = \frac{(\partial H)_T}{(\partial S)_T}$$
$$= \left[-V + T\left(\frac{\partial V}{\partial T}\right)_P\right]\left(\frac{\partial T}{\partial V}\right)_P$$
$$= T - V\left(\frac{\partial T}{\partial V}\right)_P.$$

Table 2.1. Thermodynamic Properties Expressed in Terms of the Variables of Each of the Four Fundamental Equations

Fundamental Equation	P	V	T	S	U	H
$U(S, V)$	$-\left(\frac{\partial U}{\partial V}\right)_S$	—	$\left(\frac{\partial U}{\partial S}\right)_V$	—	—	$U - V\left(\frac{\partial U}{\partial V}\right)_S$
$H(S, P)$	—	$\left(\frac{\partial H}{\partial P}\right)_S$	$\left(\frac{\partial H}{\partial S}\right)_P$	—	$H - P\left(\frac{\partial H}{\partial P}\right)_S$	—
$A(T, V)$	$-\left(\frac{\partial A}{\partial V}\right)_T$	—	—	$-\left(\frac{\partial A}{\partial T}\right)_V$	$A - T\left(\frac{\partial A}{\partial T}\right)_V$	$A - T\left(\frac{\partial A}{\partial T}\right)_V - V\left(\frac{\partial A}{\partial V}\right)_T$
$G(T, P)$	—	$\left(\frac{\partial G}{\partial P}\right)_T$	—	$-\left(\frac{\partial G}{\partial T}\right)_P$	$G - T\left(\frac{\partial G}{\partial T}\right)_P - P\left(\frac{\partial G}{\partial P}\right)_T$	$G - T\left(\frac{\partial G}{\partial T}\right)_P$

	A	G	C_p	C_v
$U(S, V)$	$U - S\left(\frac{\partial U}{\partial S}\right)_V$	$U - S\left(\frac{\partial U}{\partial S}\right)_V - V\left(\frac{\partial U}{\partial V}\right)_S$	$\left[\partial\frac{[U - V(\partial U/\partial V)_S]}{\partial(\partial U/\partial S)_V}\right]_{(\partial U/\partial V)_S}$	$\left[\frac{\partial U}{\partial(\partial U/\partial S)_V}\right]_V$
$H(S, P)$	$H - P\left(\frac{\partial H}{\partial P}\right)_S - S\left(\frac{\partial H}{\partial S}\right)_P$	$H - S\left(\frac{\partial H}{\partial S}\right)_P$	$\left[\frac{\partial H}{\partial(\partial H/\partial S)_P}\right]_P$	$\left[\frac{\partial[H - P(\partial H/\partial P)_S]}{\partial(\partial H/\partial S)_P}\right]_{(\partial H/\partial P)_S}$
$A(T, V)$	—	$A - V\left(\frac{\partial A}{\partial V}\right)_T$	$-T\left(\frac{\partial^2 A}{\partial T^2}\right)_{V, (\partial A/\partial V)_T}$	$-T\left(\frac{\partial^2 A}{\partial T^2}\right)_V$
$G(T, P)$	$G - P\left(\frac{\partial G}{\partial P}\right)_T$	—	$-T\left(\frac{\partial^2 G}{\partial T^2}\right)_P$	$T\left[-\left(\frac{\partial^2 G}{\partial T^2}\right)_P + \left(\frac{\partial^2 G}{\partial P \partial T}\right)_V^2 \Big/ \left(\frac{\partial^2 G}{\partial P^2}\right)_T\right]$

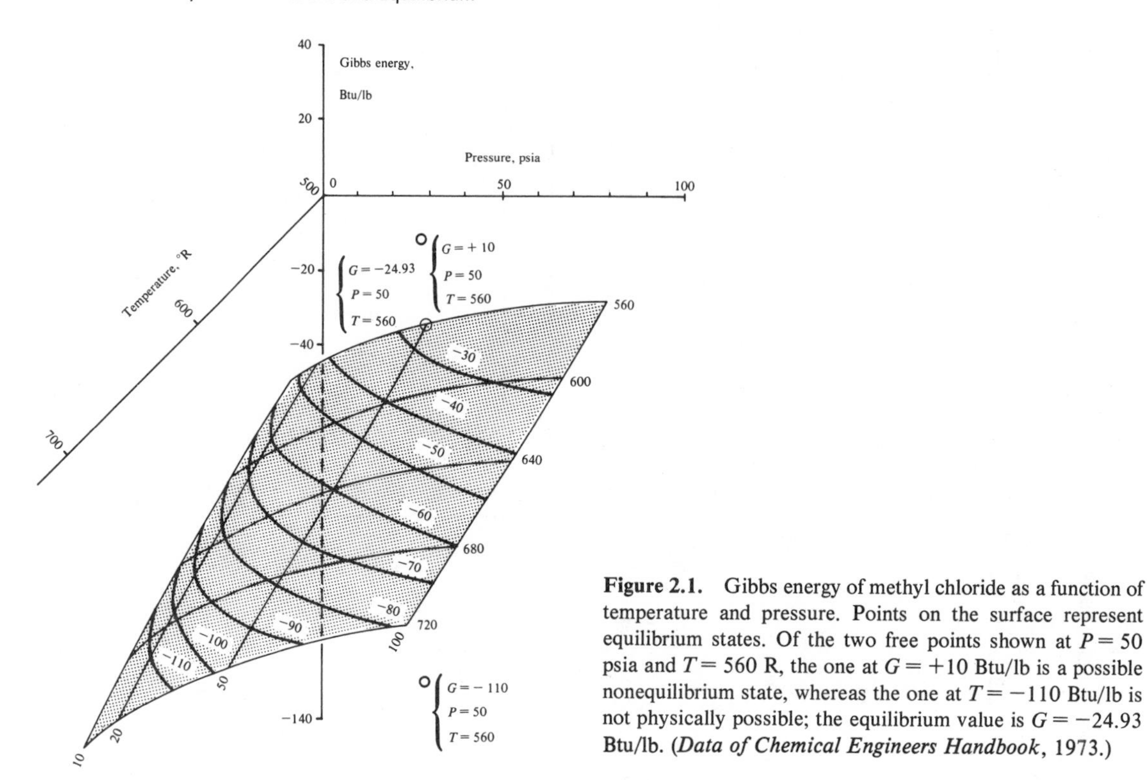

Figure 2.1. Gibbs energy of methyl chloride as a function of temperature and pressure. Points on the surface represent equilibrium states. Of the two free points shown at $P = 50$ psia and $T = 560$ R, the one at $G = +10$ Btu/lb is a possible nonequilibrium state, whereas the one at $T = -110$ Btu/lb is not physically possible; the equilibrium value is $G = -24.93$ Btu/lb. (*Data of Chemical Engineers Handbook*, 1973.)

The usefulness of the Table in handling some uncommon combinations of variables is illustrated with problems at the end of the chapter. Several "rules for partial differentiation" listed in Appendix B also may be helpful in devising relations between derivatives.

Example 2.2 Enthalpy Change Resulting from Changes in Temperature and Pressure when the Equation of State and the Heat Capacity in the Ideal Gas State Are Known

The path to be followed from condition (H_1, T_1, P_1) to condition (H_2, T_2, P_2) is represented on the sketch. It comprises an isothermal reduction of pressure to zero, heating at zero pressure, then isothermal compression to the final value. Equations will be derived for both pressure-explicit and volume-explicit equations of state, and illustrated with the B-truncated virial equation.

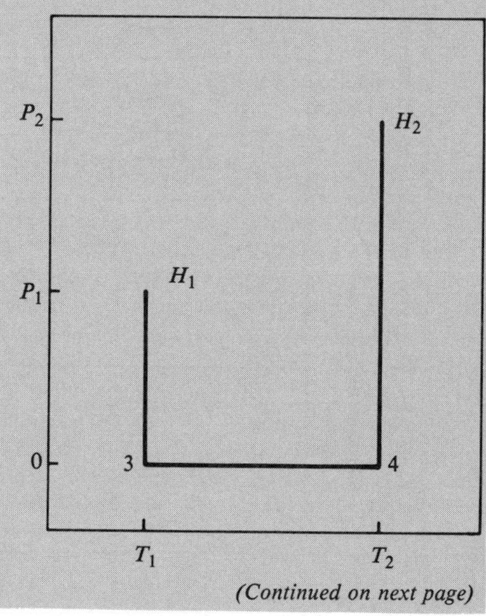

Start with the differential,

$$dH = C_p dT + \left[V - T\left(\frac{\partial V}{\partial T}\right)_P \right] dP. \qquad (1)$$

For the path described, the integral is

$$H_2 - H_1 = -\int_0^{P_1} \left[V - T\left(\frac{\partial V}{\partial T}\right)_P \right]_{@T_1} dP + \int_{T_1}^{T_2} C_p^{\text{id}} dT$$
$$+ \int_0^{P_2} \left[V - T\left(\frac{\partial V}{\partial T}\right)_P \right]_{@T_2} dP. \qquad (2)$$

(Continued on next page)

Example 2.2 *(continued)*

With a pressure-explicit EOS, integration with respect to volume is necessary. Start with,

$$dH = d(PV) + dU$$

$$= d(PV) + C_v dT + \left[T\left(\frac{\partial P}{\partial T}\right)_V - P\right] dV, \qquad (3)$$

of which the integral is

$$H_2 - H_1 = P_2V_2 - P_1V_1 + \int_{T_1}^{T_2} (C_p^{\mathrm{id}} - R)dT - \int_{\infty}^{V_1}\left[T\left(\frac{\partial P}{\partial T}\right)_V - P\right]_{@T_1} dV + \int_{\infty}^{V_2}\left[T\left(\frac{\partial P}{\partial T}\right)_V - P\right]_{@T_2} dV. \qquad (4)$$

The B-truncated virial equation can be written in either the pressure-explicit

$$P = \frac{RT}{V - B}, \qquad (5)$$

or volume-explicit form,

$$V = \frac{RT}{P} + B. \qquad (6)$$

The required derivatives are

$$\left(\frac{\partial V}{\partial T}\right)_P = \frac{R}{P} + \frac{dB}{dT} \qquad (7)$$

$$\left(\frac{\partial P}{\partial T}\right)_V = \frac{R}{V - B} + \frac{RT}{(V - B)^2}\frac{dB}{dT}. \qquad (8)$$

For the volume-explicit form, the integral is

$$\int_0^P \left[V - T\left(\frac{\partial V}{\partial T}\right)_P\right] dP = \left(B - T\frac{dB}{dT}\right)P, \qquad (9)$$

so that the enthalpy change is

$$H_2 - H_1 = \int_{T_1}^{T_2} C_p^{\mathrm{id}}\, dT - \left(B - T\frac{dB}{dT}\right)_{@T_1} P_1 + \left(B - T\frac{dB}{dT}\right)_{@T_2} P_2. \qquad (10)$$

With the pressure-explicit form,

$$\int_{\infty}^{V}\left[T\left(\frac{\partial P}{\partial T}\right)_V - P\right] dV = \int_{\infty}^{V} \frac{RT^2}{(V - B)^2}\frac{dB}{dT} dV = -\frac{RT^2}{V - B}\frac{dB}{dT} = -PT\frac{dB}{dT} \qquad (11)$$

and the enthalpy change is

$$H_2 - H_1 = P_2V_2 - P_1V_1 + \int_{T_1}^{T_2} (C_p^{\mathrm{id}} - R)\, dT + \left(PT\frac{dB}{dT}\right)_1 - \left(PT\frac{dB}{dT}\right)_2 \qquad (12)$$

$$= P_2V_2 - RT_2 - P_1V_1 + RT_1 + \int_{T_1}^{T_2} C_p^{\mathrm{id}}\, dT + \left(PT\frac{dB}{dT}\right)_1 - \left(PT\frac{dB}{dT}\right)_2 \qquad (13)$$

$$= \int_{T_1}^{T_2} C_p^{\mathrm{id}}\, dT - \left(B - T\frac{dB}{dT}\right)_1 P_1 + \left(B - T\frac{dB}{dT}\right)_2 P_2, \qquad (14)$$

as before.

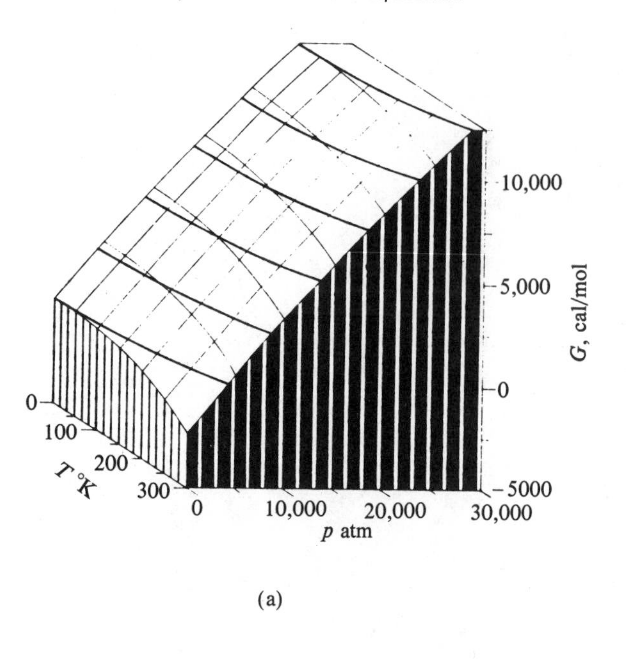

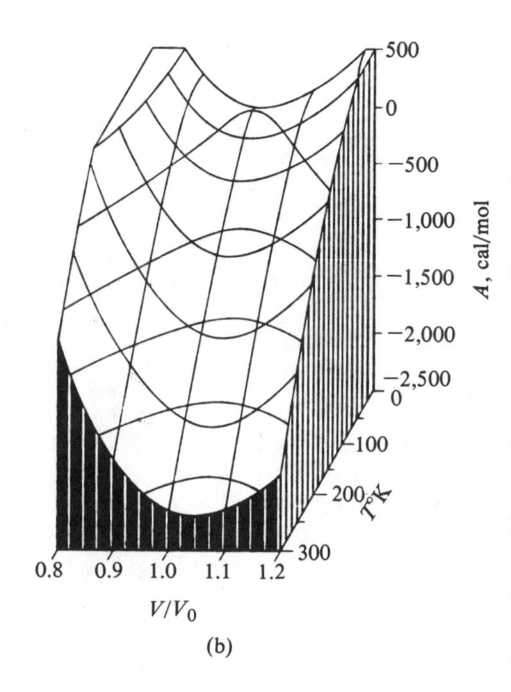

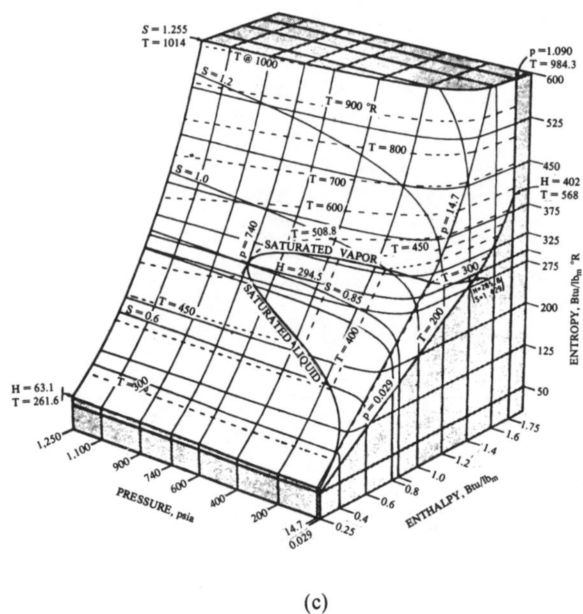

Figure 2.2. Fundamental thermodynamic surfaces: (a) Gibbs energy–pressure–temperature surface of sodium (Kubo, 1968); (b) Helmholtz energy–temperature–volume surface of sodium (Kubo, 1968); (c) Enthalpy–entropy–pressure surface of ethylene based on the Redlich-Kwong equation of state and a polynomial equation for the heat capacity (K. R. Jolls, Iowa State University, 1982, submitted to *J. Chem. Education*).

Example 2.3 Thermodynamic Properties from Data on P, V, T and C_p^*, an Equivalent Fundamental Equation

Basis: a reference temperature T_0 and a pressure p_0 at which ideal gas behavior holds. With an EOS explicit in volume,

$$V = f(P, T), \tag{1}$$

$$H = H_T^* + \int_0^P \left(\frac{\partial H}{\partial P}\right)_T dP = H_{T_0}^* + \int_{T_0}^T C_p^* \, dT + \int_0^P \left[V - T\left(\frac{\partial V}{\partial T}\right)_P\right] dP, \tag{2}$$

$$S = S(T_0, P_0) + \int_{T_0}^T \frac{C_p^*}{T} dT + \int_{P_0}^P \left(\frac{\partial S}{\partial P}\right)_T dP \tag{3}$$

$$= S(T_0, P_0) + \int_{T_0}^T \frac{C_p^*}{T} dT - \int_{P_0}^P \left(\frac{\partial V}{\partial T}\right)_P dP. \tag{4}$$

The reference condition commonly is taken as 298.15 K and 1 atm, but it cancels out when differences are taken.

Other properties are derived from H and S. Thus,

$$U = H - PV, \tag{5}$$

$$A = U - TS = H - PV - TS, \tag{6}$$

$$G = H - TS. \tag{7}$$

The heat capacities are given by

$$C_p = C_p^* - T \int_0^P \left(\frac{\partial^2 V}{\partial T^2}\right)_P dP, \tag{8}$$

$$C_v = C_p - T\left(\frac{\partial P}{\partial T}\right)_V \left(\frac{\partial V}{\partial T}\right)_P. \tag{9}$$

All integrands are evaluated at the final system temperature. When the equation of state is explicit in pressure, $P = f(T, V)$, it is more convenient first to find internal energy and entropy, starting with

$$dU = (C_p^* - R)dT + \left[T\left(\frac{\partial P}{\partial T}\right)_V - P\right] dV, \tag{10}$$

$$dS = \frac{C_p^* - R}{T} dT - \left(\frac{\partial P}{\partial T}\right)_V dV. \tag{11}$$

After integrations, continue with Eqs. 5–7.

2.3. TOTAL AND SPECIFIC PROPERTIES

With processes involving variation in the mass or composition of a system or its individual phases, it is necessary to distinguish between a total property, designated $\underline{M}$, and its magnitude per unit mass, M. This distinction was made in Section 2.1 but will be considered at greater length now.

Two letters can be used to identify the total property, but a single letter with understrike often is convenient. For a pure substance,

$$\underline{M} = nM, \tag{2.31}$$

and in differential form,

$$d\underline{M} = n\, dM + M\, dn. \tag{2.32}$$

When the property M is a function of two variables X and Y as well as the numbers of mols of the several substances,

$$\underline{M} = \underline{M}(X, Y, n_1, n_2, \ldots), \tag{2.33}$$

the differential is

$$d\underline{M} = d(nM) = \left(\frac{\partial(nM)}{\partial X}\right)_{Y_n} dX + \left(\frac{\partial(nM)}{\partial Y}\right)_{X_n} dY + \Sigma \left(\frac{\partial(nM)}{\partial n_i}\right)_{XY_{n_j}} dn_i. \tag{2.34}$$

The derivatives of the energy functions U, H, A, and G with respect to the number of mols are given special symbols,

$$\mu_i = \left(\frac{\partial \underline{M}}{\partial n_i}\right)_{XY_{n_j}} \tag{2.35}$$

and are called *chemical potentials*. They are also partial molal properties, when the independent properties X and Y are T and P. The particular forms of Eq. 2.34 corresponding to the four fundamental equations are

$$d\underline{U} = T d\underline{S} - P d\underline{V} + \Sigma \mu_i dn_i, \tag{2.36}$$

$$d\underline{H} = T d\underline{S} + \underline{V} dP + \Sigma \mu_i dn_i, \tag{2.37}$$

$$d\underline{A} = -\underline{S} dT - P d\underline{V} + \Sigma \mu_i dn_i, \tag{2.38}$$

$$d\underline{G} = -\underline{S} dT + \underline{V} dP + \Sigma \mu_i dn_i. \tag{2.39}$$

When these equations are divided through by n, they become Eqs. 2.5 and 2.19–2.21.

The chemical potential may be expressed in terms of any of the four fundamental groups of properties,

$$\mu_i = \left(\frac{\partial \underline{U}}{\partial n_i}\right)_{\underline{S}\underline{V}n_j} = \left(\frac{\partial \underline{H}}{\partial n_i}\right)_{\underline{S}Pn_j} = \left(\frac{\partial \underline{A}}{\partial n_i}\right)_{T\underline{V}n_j} = \left(\frac{\partial \underline{G}}{\partial n_i}\right)_{TPn_j}. \tag{2.40}$$

Because of the importance of T and P as independent properties, the chemical potential is most commonly thought of as the derivative of Gibbs energy with respect to the number of mols, which is also the partial molal Gibbs energy. The names and symbols are used interchangeably,

$$\mu_i = \bar{G}_i = (\partial \underline{G}/\partial n_i)_{TPn_j}. \tag{2.41}$$

Eqs. 2.9 and 2.21 are important relations between Gibbs energy and chemical potentials.

Examination of Eq. 2.36 reveals a simple interpretation of the physical meaning of the derivatives with respect to mols. For a system of constant composition, energy transfers by reversible processes are identified as

$$dW = PdV = -dU \text{ (when } dQ = dn = 0), \tag{2.42}$$

$$dQ = TdS = dU \quad \text{(when } dW = dn = 0), \tag{2.43}$$

$$dU = \mu\, dn \quad \text{(when } dQ = dW = 0). \tag{2.44}$$

Since pressure is the potential function for work transfer and temperature that for heat transfer, μ may be regarded as the potential function for internal energy transfer accompanying the transfer of mass across the boundary of the system.

2.4. PARTIAL MOLAL PROPERTIES

In view of the importance of the properties T and P, which are most readily measured and controlled, the fundamental equation of the Gibbs energy,

$$\underline{G} = G(T, P, n_1, n_2, \ldots) \tag{2.45}$$

is particularly useful. Its differential is

$$d\underline{G} = \left(\frac{\partial \underline{G}}{\partial T}\right)_{Pn} dT + \left(\frac{\partial \underline{G}}{\partial P}\right)_{Tn} dP + \Sigma \left(\frac{\partial \underline{G}}{\partial n_i}\right)_{TPn_j} dn_i \tag{2.45a}$$

$$= -\underline{S}\, dT + \underline{V}\, dP + \Sigma \bar{G}_i dn_i. \tag{2.46}$$

As shown in the last equation, the derivative with respect to mols at constant T and P is given a special symbol, $\bar{G}_i$. It is called the partial molal Gibbs energy, but it also is the chemical potential, μ_i, of Eq. 2.40. In terms of this symbol, Eq. 2.18 and 2.21 become

$$G = \Sigma x_i \bar{G}_i, \tag{2.47}$$

$$dG = \Sigma \bar{G}_i dx_i \tag{2.48}$$

which state that the Gibbs energy of a mixture is the mol fraction weighted sum of the partial molal Gibbs energies.

Other mol derivatives at *constant temperature and pressure* likewise are called partial molal properties. The general definition is

$$\bar{M}_i = \left(\frac{\partial (nM)}{\partial n_i}\right)_{TPn_j}, \tag{2.49}$$

The properties that are useful in this form include S, V, z, U, H, A, G, C_p, C_v $\ln f$, and $\ln \phi$. Except for the fugacity related properties, they conform to the relation,

$$M = \Sigma x_i \bar{M}_i. \tag{2.50}$$

In the case of fugacity, the partial molal relation is

$$\ln(\hat{f}_i/x_i) = (\partial(n \ln f)/\partial n_i)_{TPn_j}, \tag{2.51}$$

which is implied by Eq. 3.15. The logarithmic forms belong in this group by reason of their simple relation to the Gibbs energy,

$$dG = RT \ln f + \Lambda(T, x_1, x_2, \ldots), \tag{3.8}$$

$$d\bar{G}_i = RT \ln \hat{f}_i + \lambda(T), \tag{3.9}$$

as explained in Section 3.1. Applying Eq. 2.50,

$$\ln f = \Sigma x_i \ln (\hat{f}_i/x_i), \tag{2.51a}$$

or

$$f = \Pi(\hat{f}_i/x_i)^{x_i}. \tag{2.51b}$$

A parallel exists between the M and the $\bar{M}_i$; any relation between properties is convertible to one of the same form in partial molal properties by a one-to-one replacement of extensive properties—for instance:

$$G = H - TS \quad \text{and} \quad \bar{G}_i = \bar{H}_i - T\bar{S}_i,$$

$$dG = VdP \quad \text{and} \quad d\bar{G}_i = \bar{V}_i dP.$$

A more nearly complete list is in Table 2.2.

A derivation of a partial molal compressibility, $\bar{z}_i$, is made in Example 2.4. Derivatives of properties other than those in the list just before Eq. 2.49 with respect to n_i are sometimes needed; but for convenience if they are not made at constant T and P, they are not called partial molal properties. For instance, $(\partial P/\partial n_i)_{TVn_{j\neq i}}$ is evaluated in Problem 2.19 and in Example 3.4 for particular cases.

For many purposes it is more convenient to work with compositions expressed in mol fractions rather than mols of each constituent present. This topic is covered in Section 2.5, where mention also is made of methods for evaluating partial molal properties from experimental data.

2.5. DERIVATIVES WITH RESPECT TO MOL FRACTIONS

When compositions of mixtures are given in terms of mol fractions, the derivatives with respect to mol fractions may be obtained directly. Relating such derivatives to those with respect to the numbers of mols of individual components, however, as in the formulation of partial molal properties, requires a little care. Individual mol numbers may be independent, but the mol fractions are subject to the restriction $\Sigma x_i = 1$, so that at least two mol fractions must vary simultaneously when a partial derivative is taken.

For a binary mixture, note that

$$n_1 = nx_1 = (n_1 + n_2)x_1 \tag{2.52}$$

and

$$dn_1 = ndx_1 + x_1 dn_1 = \frac{ndx_1}{1 - x_1} \tag{2.53}$$

so that

$$\bar{M}_1 = (\partial nM/\partial n_1)_{TPn_2} = M + n(\partial M/\partial n_1)$$

$$= M + (1 - x_1)(\partial M/\partial x_1)_{TP}. \tag{2.54}$$

Similarly,

$$\bar{M}_2 = M + x_1(\partial M/\partial x_2)_{TP} = M - x_1(\partial M/\partial x_1)_{TP}. \tag{2.55}$$

Table 2.2. Analogies between Partial Molal and Corresponding Parent Properties

	Property	*Partial Molal Property*
1.	$M = \Sigma \bar{M}_i x_i$	$\bar{M}_i = \left(\dfrac{\partial nM}{\partial n_i}\right)_{TPn_j}$
2.	U_i	$\bar{U}_i$
3.	S_i	$\bar{S}_i$
4.	$H = U + PV$	$\bar{H}_i = \bar{U}_i + P\bar{V}_i$
5.	$A = U - TS$	$\bar{A}_i = \bar{U}_i - T\bar{S}_i$
6.	$G = U + PV - TS$	$\bar{G}_i = \bar{U}_i + P\bar{V}_i - T\bar{S}_i$
7.	$z = \dfrac{P}{RT} V$	$\bar{z}_i = \dfrac{P}{RT} \bar{V}_i$
8.	$dU = TdS - PdV$	$d\bar{U}_i = Td\bar{S}_i - Pd\bar{V}_i$
9.	$dH = TdS + VdP$	$d\bar{H}_i = Td\bar{S}_i + \bar{V}_i dP$
10.	$dA = -SdT - PdV$	$d\bar{A}_i = -\bar{S}_i dT - Pd\bar{V}_i$
11.	$dG = -SdT + VdP$	$d\bar{G}_i = -\bar{S}_i dT + \bar{V}_i dP$
12.	$\left(\dfrac{\partial G}{\partial P}\right)_T = V$	$\left(\dfrac{\partial \bar{G}_i}{\partial P}\right)_{Tn} = \bar{V}_i$
13.	$\left(\dfrac{\partial G}{\partial T}\right)_P = -S$	$\left(\dfrac{\partial \bar{G}_i}{\partial T}\right)_{Pn} = -\bar{S}_i$
14.	$\left(\dfrac{\partial (G/T)}{\partial T}\right)_P = -\dfrac{H}{T^2}$	$\left(\dfrac{\partial (\bar{G}_i/T)}{\partial T}\right)_{Pn} = -\dfrac{\bar{H}_i}{T^2}$
15.	$\left(\dfrac{\partial \ln f}{\partial P}\right)_T = \dfrac{V}{RT}$	$\left(\dfrac{\partial \ln (\hat{f}_i/x_i)}{\partial P}\right)_{Tx} = \left(\dfrac{\partial \ln \hat{f}_i}{\partial P}\right)_{Tx} = \dfrac{\bar{V}_i}{RT}$
16.	$\left(\dfrac{\partial \ln f}{\partial T}\right)_P = -\dfrac{H - H^{id}}{RT^2}$	$\left(\dfrac{\partial \ln (\hat{f}_i/x_i)}{\partial T}\right)_{Px} = \left(\dfrac{\partial \ln \hat{f}_i}{\partial T}\right)_{Px} = -\dfrac{\bar{H}_i - H^{id}}{RT^2}$
17.	$\left(\dfrac{\partial \ln \phi}{\partial P}\right)_T = \dfrac{V - V^{id}}{RT}$	$\left(\dfrac{\partial \ln \hat{\phi}_i}{\partial P}\right)_{Tx} = \dfrac{\bar{V}_i - V_i^{id}}{RT}$
18.	$\left(\dfrac{\partial \ln \phi}{\partial T}\right)_P = -\dfrac{H - H^{id}}{RT^2}$	$\left(\dfrac{\partial \ln \hat{\phi}_i}{T}\right)_{Px} = -\dfrac{\bar{H}_i - H^{id}}{RT^2}$
19.		$\left(\dfrac{\partial \ln \gamma_i}{\partial T}\right)_{Px} = -\dfrac{\bar{H}_i - H_i^{id}}{RT^2}$
20.	$G = RT \ln f + \Lambda(T)$	$\bar{G}_i = RT \ln \hat{f}_i + \lambda(T, x)$
21.	$\lim_{P \to 0} (f/P) = 1$	$\lim_{P \to 0} (\hat{f}_i/x_i P) = 1$

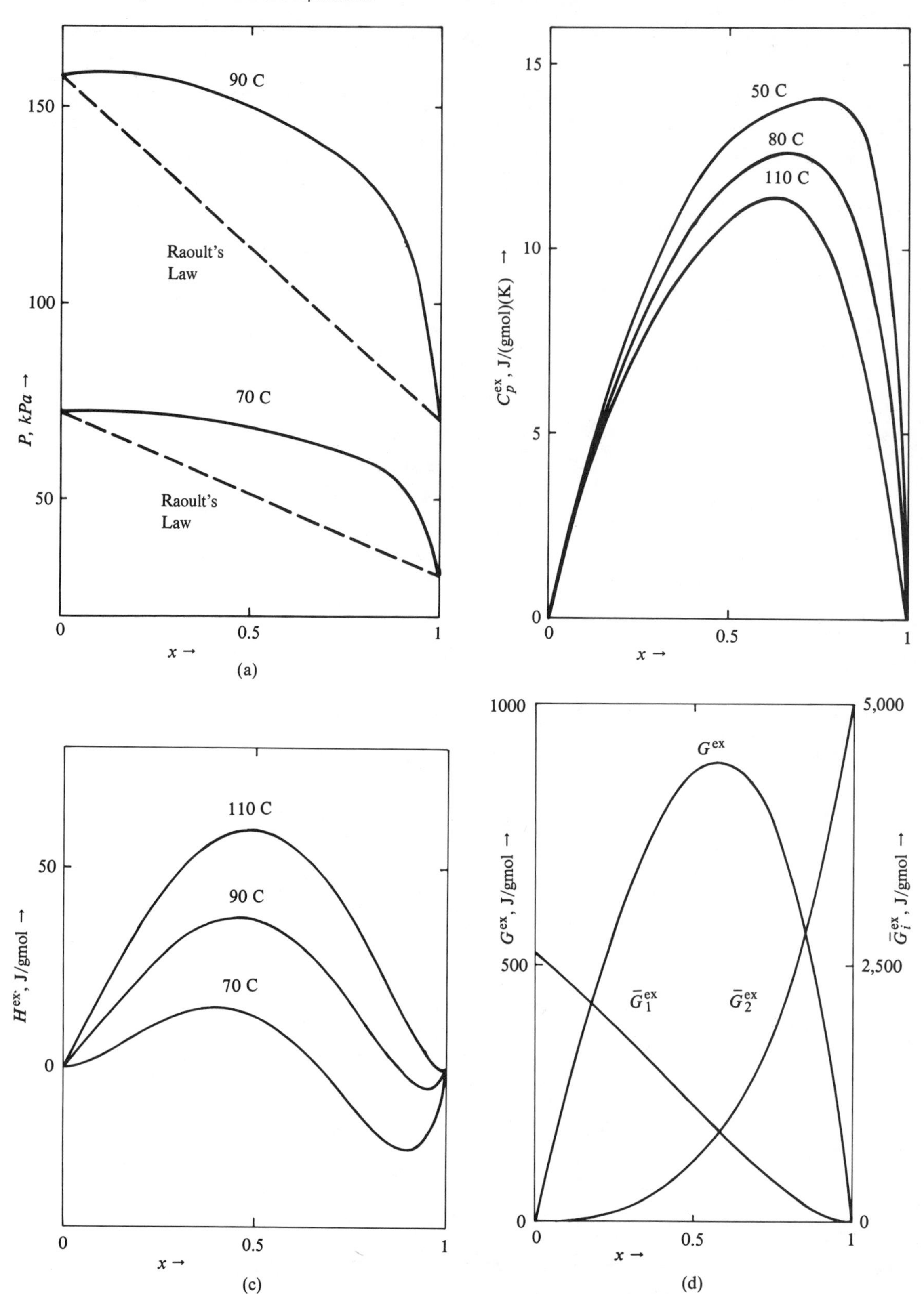

Figure 2.3. Excess properties of mixtures of water + ethanol and of water + methanol (Figure 2.3(f) only) (from *International Data Series, Selected Data on Mixtures,* 1978): (a) total pressure; (b) excess heat capacity; (c) excess enthalpy; (d) excess Gibbs energy at 70 C; (e) excess entropy at 70 C; (f) excess volumes of water + methanol mixtures at 25 C.

(Continued next page)

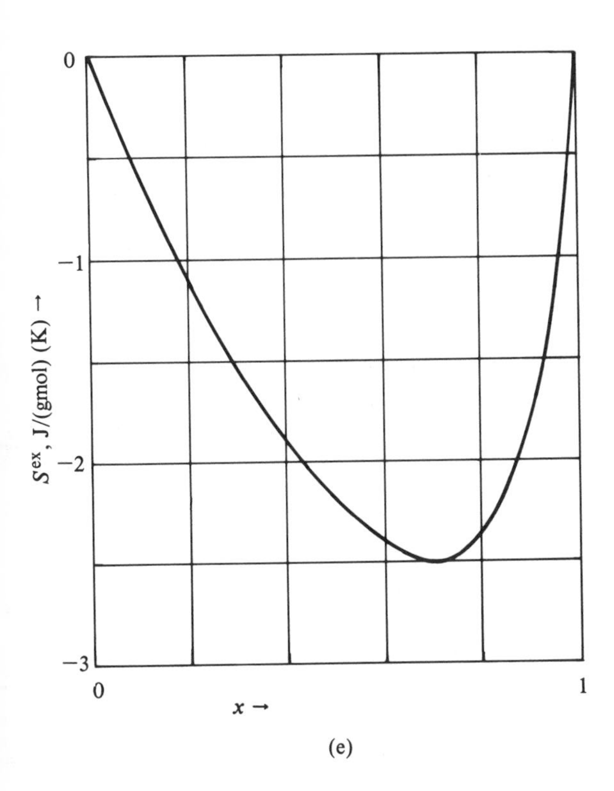

(e)

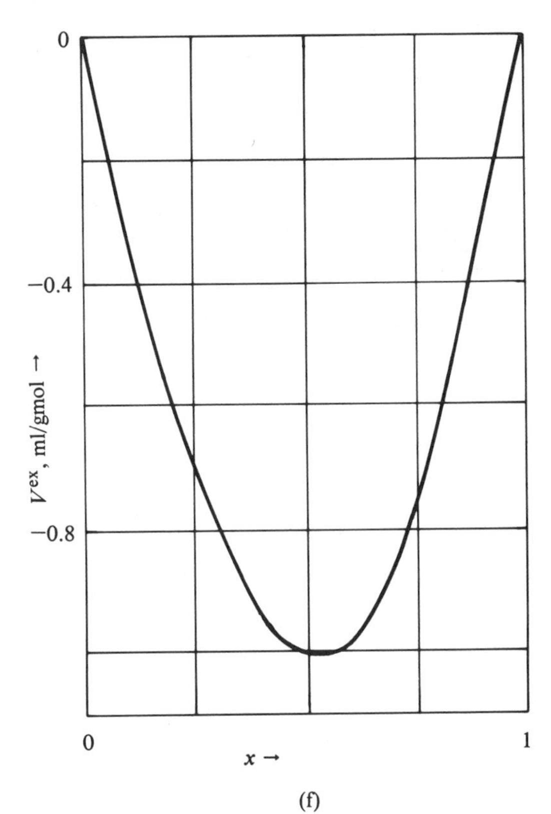

(f)

Figure 2.3 *(continued)*

In the general case the partial molal property is

$$\bar{M}_i = M - \sum_{k\neq i} x_k(\partial M/\partial x_k)_{TPx_{j\neq i,k}} \tag{2.56}$$

The simplest way to keep track of which xs are to be allowed to vary at any one time when differentiating is:

1. When forming $\bar{M}_i$, first eliminate x_i from the defining equation with the substitution $x_i = 1 - \Sigma_{j\neq i} x_j$.
2. Perform this operation for each $\bar{M}_i$ that is to be found.

Example 2.4 applies this procedure to a ternary mixture and Example 3.5 does it for the partial fugacities of a four component mixture.

Eq. 2.56 is derived, for example, by Smith & Van Ness (1975, p. 605) and an equivalent one by Modell & Reid (1983, p. 183). A simple derivation of an equivalent equation is based on elimination of the mol fraction x_1 in favor of the total mols, n. Then the functional relation becomes

$$M = M(n_1, n_2, \ldots, n_r) = nM(n, x_2, x_3, \ldots, x_r)$$

The derivative is

$$\left(\frac{\partial nM}{\partial n_i}\right)_{n_k} = M + n\left(\frac{\partial M}{\partial n}\right)_{x_j}\left(\frac{\partial n}{\partial n_i}\right)_{n_k} + n\sum_{j=2}^{r}\left(\frac{\partial M}{\partial x_j}\right)_{n,x_k}\left(\frac{\partial x_j}{\partial n_i}\right)_{n_k}. \tag{2.56a}$$

Since

$$(\partial M/\partial n)_{n_k} = 0,$$

$$(\partial n/\partial n_i)_{n_k} = 1$$

and

$$\left(\frac{\partial x_j}{\partial n_i}\right)_{n_k} = \frac{\delta_{ij} - x_j}{n},$$

$$\delta_{ij} \begin{cases} 1, & \text{when } i = j \\ 0, & \text{when } i \neq j \end{cases} \tag{2.56b}$$

the desired result becomes

$$\bar{M}_i = \left(\frac{\partial nM}{\partial n_i}\right)_{n_k} = M + \sum_{j=2}^{r}(\delta_{ij} - x_j)\left(\frac{\partial M}{\partial x_j}\right)_{n,x_k} \tag{2.56c}$$

For a ternary mixture the equations are:

$$\bar{M}_1 = M - x_2\left(\frac{\partial M}{\partial x_2}\right)_{x_3} - x_3\left(\frac{\partial M}{\partial x_3}\right)_{x_2}, \tag{2.56d}$$

$$\bar{M}_2 = M + (1 - x_2)\left(\frac{\partial M}{\partial x_2}\right)_{x_3} - x_3\left(\frac{\partial M}{\partial x_3}\right)_{x_2}, \tag{2.56e}$$

Example 2.4 Partial Molal Properties of Multicomponent Mixtures When the Property of the Mixture Is Known in Terms of Mol Fractions

The property in question is represented by the equation,

$$M = ax_1^2 + bx_2^2 + cx_3^2 + dx_1x_2 + ex_1x_3 + fx_2x_3. \quad (1)$$

Rewrite this equation three times by eliminating one of the mol fractions at a time with $x_i = 1 - \Sigma_{j\neq i}x_j$; thus

$$M = a(1 - x_2 - x_3)^2 + bx_2^2 + cx_3^2 + d(1 - x_2 - x_3)x_2 + e(1 - x_2 - x_3)x_3 + fx_2x_3 \quad (2)$$

$$= ax_1^2 + b(1 - x_1 - x_3)^2 + cx_3^2 + (dx_1 + fx_3)(1 - x_1 - x_3) + ex_1x_3 \quad (3)$$

$$= ax_1^2 + bx_2^2 + c(1 - x_1 - x_2)^2 + (ex_1 + fx_2)(1 - x_1 - x_2) + dx_1x_2. \quad (4)$$

Equations for the three partial molal properties are found with Eq. 2.56:

$$\bar{M}_1 = M - x_2\left(\frac{\partial M}{\partial x_2}\right)_{x_3} - x_3\left(\frac{\partial M}{\partial x_3}\right)_{x_2}, \quad (5)$$

$$\bar{M}_2 = M - x_1\left(\frac{\partial M}{\partial x_1}\right)_{x_3} - x_3\left(\frac{\partial M}{\partial x_3}\right)_{x_1}, \quad (6)$$

$$\bar{M}_3 = M - x_1\left(\frac{\partial M}{\partial x_1}\right)_{x_2} - x_2\left(\frac{\partial M}{\partial x_2}\right)_{x_1}. \quad (7)$$

Six derivatives are required:

$$\left(\frac{\partial M}{\partial x_1}\right)_{x_2} = (2a - e)x_1 + (d - 2b)x_2 + (e - f)x_3, \quad (8)$$

$$\left(\frac{\partial M}{\partial x_1}\right)_{x_3} = (2a - d)x_1 + (d - 2b)x_2 + (e - f)x_3, \quad (9)$$

$$\left(\frac{\partial M}{\partial x_2}\right)_{x_1} = (d - e)x_1 + (2b - f)x_2 + (f - 2c)x_3, \quad (10)$$

$$\left(\frac{\partial M}{\partial x_2}\right)_{x_3} = (d - 2a)x_1 + (2b - d)x_2 + (f - e)x_3, \quad (11)$$

$$\left(\frac{\partial M}{\partial x_3}\right)_{x_1} = (e - d)x_1 + (f - 2b)x_2 + (2c - f)x_3, \quad (12)$$

$$\left(\frac{\partial M}{\partial x_3}\right)_{x_2} = (e - 2a)x_1 + (f - d)x_2 + (2c - e)x_3. \quad (13)$$

Eqs. 8–13 are substituted into Eqs. 5–7 to obtain the desired partial molal properties. Example 3.5 evaluates partial molal fugacities in a similar manner. Equations 2.56(d)–(f) also can be applied to this problem; in that case the mol fraction x_1 will not be involved.

$$\bar{M}_3 = M - x_2\left(\frac{\partial M}{\partial x_2}\right)_{x_3} + (1 - x_3)\left(\frac{\partial M}{\partial x_3}\right)_{x_2}, \quad (2.56\text{f})$$

Comparatively few data of mixtures higher than binary are complete enough to require this operation very often. A special case is that of ternary mixtures, which often are made up by adding varying amounts of a third component to fixed proportions of the other two. In a mathematical sense this is really equivalent to a two-component system. Represent the ratio of two of the mol fractions by r,

$$r = x_1/x_2 \quad (2.57)$$

and note that

$$x_1 + x_2 + x_3 = 1. \quad (2.58)$$

The differential of a property, M, at constant temperature and pressure is

$$dM = \bar{M}_1\,dx_1 + \bar{M}_2\,dx_2 + \bar{M}_3\,dx_3, \quad (2.59)$$

which may be written

$$\frac{\partial M}{\partial x_3} = \bar{M}_1\frac{\partial x_1}{\partial x_3} + \bar{M}_2\frac{\partial x_2}{\partial x_3} + \bar{M}_3, \quad (2.60)$$

also

$$M = \bar{M}_1 x_1 + \bar{M}_2 x_2 + \bar{M}_3 x_3. \quad (2.61)$$

Manipulation of these equations gives

$$dx_1 = \frac{-r}{r+1}dx_3 = \frac{-x_1}{1 - x_3}dx_3 \quad (2.62)$$

and

$$dx_2 = -\frac{1}{r+1}dx_3 = -\frac{x_2}{1 - x_3}dx_3. \quad (2.63)$$

These relations are combined into the final expression:

$$\bar{M}_3 = M + (1 - x_3)(\partial M/\partial x_3)_{TPr} \quad (2.64)$$

which is of the same form as Eq. 2.54.

The binary equations lend themselves to graphical evaluation of partial molal properties, when the property of the mixture is known as a function of the mol fractions, by what is known as the *method of tangent intercepts* and which is illustrated in Fig. 3.4. On such a diagram, the partial molal quantities at a certain mol fraction, x_1, are the intercepts of the tangent at that composition with the ordinates at 0 and 1. Graphical methods are of low accuracy, however, so it may be more accurate and probably more convenient to perform the required differentiations numerically as in Example 2.6. Partial molal volumes are found by analytical differentiation of equations of state in Examples 2.5 and 2.7. Example 2.4 finds partial molal properties when the compositions are expressed in mol fractions.

Absolute values of V, z, C_p, C_v, $\ln f$, and $\ln \phi$ can be measured, but not ordinarily absolute values of H, A, G, U, and S where the experimenter is restricted to finding differences, for instance, $\Delta H = H - x_1H_1 - x_2H_2$, which can be found calorimetrically. Clearly, however, Eq. 2.54 may be

Example 2.5 Partial Molal Compressibilities in a Binary Gas Mixture with the *B*-Truncated Equation of State

The equation is

$$z = \frac{PV}{RT} = 1 + \frac{BP}{RT}. \tag{1}$$

The partial molal compressibility is

$$\bar{z}_1 = \frac{P}{RT}\left(\frac{\partial nV}{\partial n_1}\right)_{TPn_2} = \frac{P}{RT}\left[V + n\left(\frac{\partial V}{\partial n_1}\right)\right]. \tag{2}$$

Since

$$V = RT/P + B, \tag{3}$$

and

$$B = y_1^2B_1 + y_2^2B_2 + 2y_1y_2B_{12}$$

$$= \frac{1}{n^2}(n_1^2B_1 + n_2^2B_2 + 2n_1n_2B_{12}), \tag{4}$$

the derivative is

$$\left(\frac{\partial V}{\partial n_1}\right)_{TPn_2} = \left(\frac{\partial B}{\partial n_1}\right)_{TPn_2} = -\frac{2}{n} + \frac{2}{n^2}(n_1B_1 + n_2B_{12}), \tag{5}$$

$$= \frac{2}{n}(y_1B_1 + y_2B_{12} - B). \tag{6}$$

Substitute Eq. 6 into Eq. 2,

$$\bar{z}_1 = \frac{P}{RT}\left[\frac{RT}{P} + B + n\left(\frac{\partial B}{\partial n_1}\right)\right]$$

$$= 1 + \frac{P}{RT}(2y_1B_1 + 2y_2B_{12} - B). \tag{7}$$

Similarly,

$$\bar{z}_2 = 1 + \frac{P}{RT}(2y_2B_2 + 2y_1B_{12} - B). \tag{8}$$

As a check, from Eqs. 7 and 8,

$$z = y_1\bar{z}_1 + y_2\bar{z}_2 = 1 + \frac{BP}{RT}. \tag{9}$$

Example 2.6 Partial Molal Volumes in Mixtures of Ethyl Iodide (1) and Ethyl Acetate (2)

Measurements of the specific volumes of these mixtures are cited by Lewis & Randall et al. (1961, pp. 209, 221). The units of those data have been transformed and are recorded in the first three columns of the table following. The derivatives with respect to mol fraction are calculated with the Lagrange differentiation formulas (Appendix B) and are recorded in column 4. The partial molal volumes then are figured from the equations,

$$\bar{V}_1 = V + (1 - x_1)\left(\frac{\partial V}{\partial x_1}\right),$$

$$\bar{V}_2 = V - x_1\left(\frac{\partial V}{\partial x_1}\right),$$

and shown in the last two columns. The limiting values at infinite dilution of the excess molal volumes are $\bar{V}_1^{ex} = 3.49$ and $\bar{V}_2^{ex} = 3.36$ ml/gmol. The BASIC program for the differentiation is given.

X_{iodide}	V	$V - \Sigma x_iV_i$	$\partial V/\partial x_1$	$\bar{V}_1$	$\bar{V}_2$
.0000	102.0895	.0000	−14.9730	87.12	102.09
.1176	100.2749	.3550	−15.8880	86.26	102.14
.2333	98.3846	.6006	−16.7880	85.51	102.30
.3565	96.2870	.7776	−17.7310	84.88	102.61
.4560	94.5025	.8289	−18.1380	84.64	102.77
.5516	92.7498	.8401	−18.5290	84.44	102.97
.6235	91.3477	.7654	−18.8900	84.24	103.13
.7310	89.2606	.6625	−19.9400	83.90	103.84
.8219	87.4077	.4864	−20.8280	83.70	104.53
.9143	85.4777	.2615	−21.2110	83.66	104.87
1.0000	83.6342	.0000	−21.8110	83.63	105.45

```
10  3-point Lagrange differentiation
20  Input X0, X1, X2, Y0, Y1, Y2
30  Input X
40  A=(2*X−X1−X2)*Y0/(X0−X1)/(X0−X2)
50  B=(2*X−X0−X2)*Y1/(X1−X0)/(X1−X2)
60  C=(2*X−X0−X1)*Y2/(X2−X0)/(X2−X1)
70  D=A+B+C
80  Print using 90; X, D
90  Image D.DDDD, 3X, DDD.DDDD
100 GOTO 30
110 End
```

Example 2.7 Partial Molal Volumes in a Binary Mixture Represented by the Redlich-Kwong Equation

The case is mixtures of carbon dioxide (1) and propylene (2) at 400 K and 20 atm. The values of the parameters are $A_1 = 0.0592, A_2 = 0.1491, B_1 = 0.0181, B_2 = 0.0347$. The equation for the compressibility is from Table 1.9. Since

$$V = \frac{zRT}{P}, \tag{1}$$

the derivative is

$$\left(\frac{\partial V}{\partial y_1}\right) = \frac{RT}{P}\left(\frac{\partial z}{\partial y_1}\right), \tag{2}$$

and the equations for the partial molal volumes are

$$\bar{V}_1 = \frac{RT}{P}\left[z + y_2\left(\frac{\partial z}{\partial y_1}\right)\right], \tag{3}$$

$$\bar{V}_2 = \frac{RT}{P}\left[z - y_1\left(\frac{\partial z}{\partial y_1}\right)\right]. \tag{4}$$

The required derivatives are found as follows:

$$A = y_1^2 A_1 + (1 - y_1)^2 A_2 + 2y_1(1 - y_1)\sqrt{A_1 A_2}, \tag{5}$$

$$B = y_1 B_1 + (1 - y_1) B_2, \tag{6}$$

$$\left(\frac{\partial A}{\partial y_1}\right) = 2y_1 A_1 - 2(1 - y_1)A_2 + 2(1 - 2y_1)\sqrt{A_1 A_2}, \tag{7}$$

$$\left(\frac{\partial B}{\partial y_1}\right) = B_1 - B_2, \tag{8}$$

$$z^3 - z^2 + (A - B - B^2)z - AB = 0, \tag{9}$$

$$(3z^2 - 2z + A - B - B^2)\left(\frac{\partial z}{\partial y_1}\right) + z\left[\left(\frac{\partial A}{\partial y_1}\right) - \left(\frac{\partial B}{\partial y_1}\right) - 2B\left(\frac{\partial B}{\partial y_1}\right)\right] - A\left(\frac{\partial B}{\partial y_1}\right) - B\left(\frac{\partial A}{\partial y_1}\right) = 0. \tag{10}$$

Therefore,

$$\left(\frac{\partial z}{\partial y_1}\right) = \frac{(B - z)\left(\frac{\partial A}{\partial y_1}\right) + (A + z + 2Bz)\left(\frac{\partial B}{\partial y_1}\right)}{3z^2 - 2z + A - B - B^2}. \tag{11}$$

Substitution of Eqs. 11, 7, and 8 into Eqs. 3 and 4 will give the desired equation. The work is carried out with the given BASIC program. The results of the computation are shown tabularly and graphically. There are no experimental data to check the computation.

y_{CO_2}	Z	V	$\bar{V}_{CO_2}$	$\bar{V}_{C_3H_6}$
0.00	.878	1.441	1.624	1.441
.05	.883	1.450	1.618	1.441
.10	.889	1.458	1.613	1.441
.15	.894	1.467	1.608	1.442
.20	.899	1.475	1.604	1.443
.25	.904	1.483	1.599	1.444
.30	.908	1.491	1.596	1.445
.35	.913	1.498	1.592	1.447
.40	.917	1.505	1.589	1.449
.45	.921	1.512	1.586	1.451
.50	.925	1.519	1.584	1.453
.55	.929	1.525	1.582	1.456
.60	.933	1.531	1.580	1.458
.65	.937	1.537	1.578	1.461
.70	.940	1.543	1.577	1.464
.75	.944	1.549	1.576	1.467
.80	.947	1.554	1.575	1.470
.85	.950	1.559	1.574	1.473
.90	.953	1.564	1.574	1.477
.95	.956	1.569	1.573	1.480
1.00	.959	1.573	1.573	1.484

```
10  ! PARTIAL MOLAL VOLUMES OF CO2 and
    C3H6 FROM THE R-K EQN AT 400 K and 20
    ATM
20  READ A1, B1, A2, B2, R, T, P
30  DATA .0592, .0181, .1491, .0347, .08206, 400, 20
35  Z=1
40  Y=0
50  A=Y^2*A1+(1-Y)^2*A2+2*Y*(1-Y)*
    (A1*A2)^.5
60  B=Y*B1+(1-Y)*B2
70  A3=2*Y*A1-2*(1-Y)*A2+2*(1-2*Y)*
    (A1*A2)^.5
80  B3=B1-B2
100 GOSUB 300
110 F1=F
120 Z=1.001*Z
130 GOSUB 300
140 F2=F
150 H=.001*Z*F1/(F2-F1)
160 Z=Z/1.001-H
170 IF ABS(H/Z)<=.001 THEN 190
180 GOTO 100
190 Z1=((B-Z)*A3+(A+Z+2*B*Z)*B3)/
    (3*Z^2-2*Z+A-B-B^2)
200 M1=R*T/P*(Z+(1-Y)*Z1)
210 M2=R*T/P*(Z-Y*Z1)
220 V=Z*R*T/P
230 PRINT USING 240; Y, Z, V, M1, M2
240 IMAGE D.DD, 2X, D.DDD, 2X, D.DDD, 2X,
    D.DDD, 2X, D.DDD
250 IF Y<=.95 THEN 260 ELSE 280
260 Y=Y+.05
270 GOTO 50
280 END
300 F=Z^3-Z^2+(A-B-B^2)*Z-A*B
310 RETURN
```

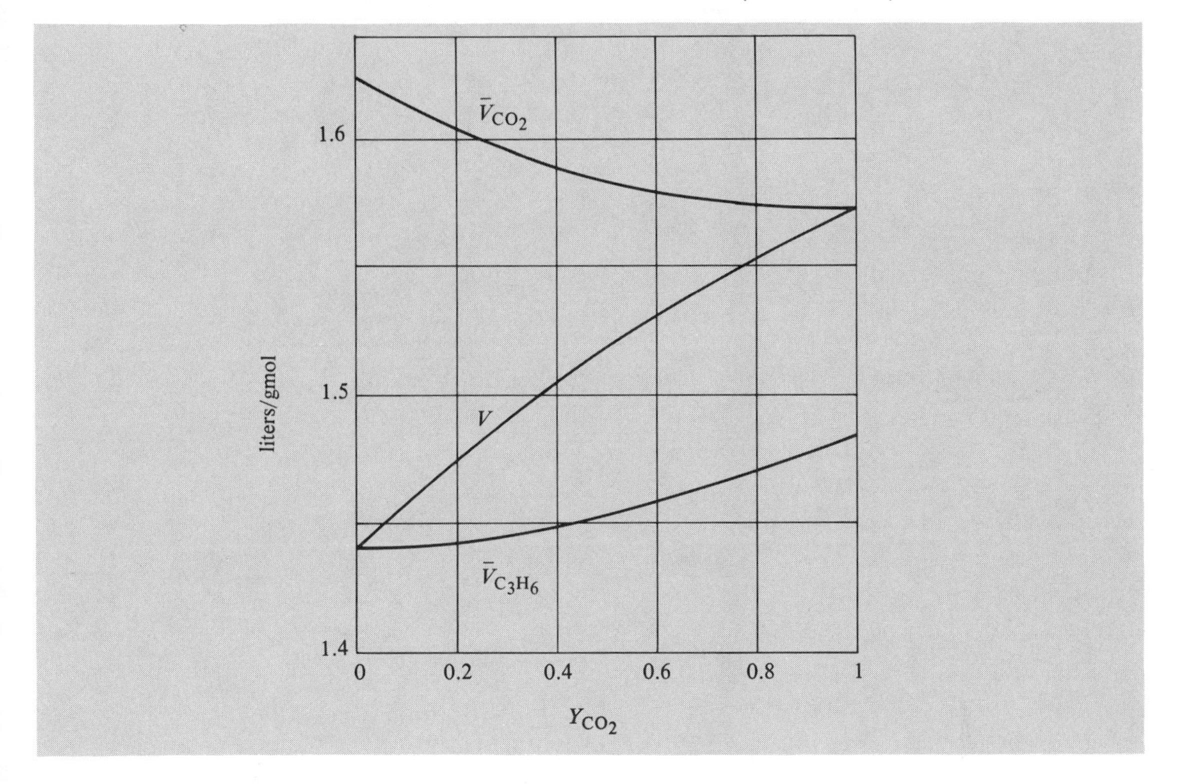

adapted to the partial molal ΔH as

$$\overline{\Delta H_1} = \Delta H + (1 - x_1)(\partial \Delta H/\partial x_1)_{TP}, \quad (2.65)$$

so there is no complication in principle by working with differences of properties rather than with their absolute values. Mixing and excess properties are considered at length in a later section.

2.6. HOMOGENEOUS FUNCTIONS

Thermodynamic properties whose magnitudes are proportional to the mass, which includes all the common properties except z, T, and P, are a mathematical class called homogeneous functions of the first degree. The functional relationship is

$$\underline{M} = \underline{M}(a, b, \ldots, nX, nY, nZ, \ldots) = nM(a, b, \ldots, X, Y, Z, \ldots), \quad (2.66)$$

where $\underline{M}$, X, Y, and Z are the variables proportional to mass and a, b, ... are independent of mass. An important theorem due to Euler relates such a function and its derivatives in this manner:

$$\underline{M} = X(\partial M/\partial X)_{YZ} + Y(\partial M/\partial Y)_{XZ} + Z(\partial M/\partial Z)_{XY} + \ldots \quad (2.67)$$

where only derivatives with respect to mass proportional properties are included.

Taking, for example, the enthalpy, $\underline{H} = \underline{H}(\underline{S}, P, n_1, n_2, \ldots)$, the variables $\underline{S}$ and n_i are extensive ones, but P is not. Accordingly,

$$\underline{H} = \underline{S}(\partial \underline{H}/\partial \underline{S})_{Pn} + \Sigma n_i(\partial \underline{H}/\partial n_i)_{SPn_j} = T\underline{S} + \Sigma n_i \mu_i, \quad (2.68)$$

which is a familiar result. Similar results for the other energy functions are obtained in the same way. The variables are identified with those of the type equation, Eq. 2.67, in this table:

Function	a	b	$\underline{X}$	$\underline{Y}$	$\underline{Z}$
$\underline{U}(\underline{S}, \underline{V}, n)$	—	—	$\underline{S}$	$\underline{V}$	n
$\underline{H}(\underline{S}, P, n)$	—	P	$\underline{S}$	—	n
$\underline{A}(T, \underline{V}, n)$	T	—	—	$\underline{V}$	n
$\underline{G}(T, P, n)$	T	P	—	—	n

Application of the Euler theorem gives these results:

$$U = TS + PV + \Sigma x_i \mu_i, \quad (2.6)$$

$$H = TS + \Sigma x_i \mu_i, \quad (2.16)$$

$$A = -PV + \Sigma x_i \mu_i, \quad (2.17)$$

$$G = \Sigma x_i \mu_i. \quad (2.18)$$

At constant compositions the summations are replaced by μ—for example,

$$G = \mu.$$

Some of these relations were obtained in a different manner in Section 2.1, but it is interesting to know that they are part of a general pattern. Example 2.8 is devoted to some less familiar results obtained by application of this theorem.

2.7. VARIATION OF PROPERTIES WITH TEMPERATURE AND PRESSURE

The effects of temperature and pressure (or volume) on various thermodynamic properties are covered in detail in Chapter 11, but the basic relations will be presented here.

Example 2.8 Further Applications of the Homogeneous Function Theorem

a. The volume function, in which only the n_i are extensive properties,

$$\underline{V} = \underline{V}(T, P, n_1, n_2, \ldots) = \Sigma n_i \left(\frac{\partial \underline{V}}{\partial n_i}\right)_{TPn_j}. \quad (1)$$

Dividing by n gives the familiar relation:

$$\underline{V} = \Sigma n_i \bar{V}_i. \quad (2)$$

b. The Massieu function:

$$\underline{\psi} = \underline{A}/T = \underline{\psi}(T, \underline{V}, n_1, n_2, \ldots)$$

$$= \underline{V}\left(\frac{\partial \underline{\psi}}{\partial \underline{V}}\right)_{Tn} + \Sigma n_i \left(\frac{\partial \underline{\psi}}{\partial n_i}\right)_{T\underline{V}n_j} \quad (3)$$

$$= \frac{\underline{V}}{T}\left(\frac{\partial \underline{A}}{\partial \underline{V}}\right) + \frac{1}{T}\Sigma n_i \mu_i. \quad (4)$$

c. The Planck function:

$$\underline{\Phi} = \underline{G}/T = \underline{\Phi}(T, P, n_1, n_2, \ldots) = \Sigma n_i \left(\frac{\partial \underline{\Phi}}{\partial n_i}\right)_{TPn_j}$$

$$= \frac{1}{T}\Sigma n_i \mu_i. \quad (5)$$

d. Internal energy of a mixture:

$$\underline{U} = \underline{U}(T, \underline{V}, n_1, n_2, \ldots) = \underline{V}\left(\frac{\partial \underline{U}}{\partial \underline{V}}\right)_{Tn}$$

$$+ \Sigma n_i \left(\frac{\partial \underline{U}}{\partial n_i}\right)_{T\underline{V}n_j}. \quad (6)$$

The first of these derivatives is readily convertible into known functions with the Bridgman Table, for instance. The derivatives $(\partial \underline{U}/\partial n_i)_{T\underline{V}n_j}$ are not the chemical potentials, however; those are $(\partial \underline{U}/\partial n_i)_{S\underline{V}n_j}$, and there is no simple way to evaluate them in terms of other properties.

Since heat capacities usually are known only at low (zero) pressures—that is, in the ideal-gas state—other properties are conveniently referred to zero pressure as a basis. Thus, in order to find the effect of going from state (P_1T_1) to state (P_2T_2), a three-step process is employed, as diagramed for enthalpy in Example 2.2:

1. Reduce the pressure of the system from P_1 to zero isothermally at T_1.
2. Evaluate the effect of changing the temperature from T_1 to T_2 in terms of ideal-gas heat capacities.
3. Raise the pressure to P_2 isothermally at T_2.

The difference between the values of a property in the real and ideal-gas states is called a residual and is represented as

$$\Delta M = M - M^{id}. \quad (2.69)$$

Since the internal energy and enthalpy of an ideal gas are independent of pressure, their residuals may be written:

$$\Delta U = U - U^{id} = U - U_0 \quad (2.70)$$

$$\Delta H = H - H^{id} = H - H_0 \quad (2.71)$$

where the subscript 0 refers to zero pressure. Since the entropy, and hence the Helmholtz and Gibbs energies, of ideal gases does depend on the pressure, the quantities $S - S^{id}$, $A - A^{id}$, and $G - G^{id}$ remain the differences between the real and ideal-gas properties but at a *given pressure,* although, as will be seen, they also involve integration of particular expressions over the interval from zero to P.

For convenience, $H - H^{id} = H - H_0$ and $S - S^{id}$ frequently have been evaluated from several PVT correlations and tabulated or graphed over practical ranges of reduced temperatures and pressures; some examples are in Chapter 11. Such information is used for evaluating changes in H and S when the process is $(P_1T_1) \rightarrow (P_2T_2)$ in accordance with the three-step process described earlier. The procedure for enthalpy change is outlined in Example 2.2, whereas that for entropy consists of Eq. 2.72a–d.

$$\Delta S_1 = -(S - S^{id})_{P_1T_1} - R\ln(P_0/P_1), \quad (2.72a)$$

$$\Delta S_2 = \int_{T_1}^{T_2} (C_p^{id}/T)dT + \Delta H_v/T_v, \quad (2.72b)$$

$$\Delta S_3 = (S - S^{id})_{P_2T_2} - R\ln(P_2/P_0). \quad (2.72c)$$

Overall,

$$S_2 - S_1 = (S - S^{id})_{P_2T_2} - (S - S^{id})_{P_1T_1} - R\ln(P_2/P_1)$$
$$+ \int_{T_1}^{T_2} (C_p^{id}/T)dT + \Delta H_v/T_v. \quad (2.72d)$$

where ΔH_v is the enthalpy of phase change at temperature T_v when that is between T_1 and T_2. Reference pressure Po goes to zero in the limit, but cancels out of Eq. 2.72d.

In Table 11.1 may be found the expressions for evaluating $M - M^{id}$ from either pressure-explicit or volume-explicit equations of state. The basic relations are for enthalpy and entropy. Thus:

$$dH = (\partial H/\partial T)_P dT + (\partial H/\partial P)_T dP = C_p dT$$
$$- (V - T(\partial V/\partial T)_P)dP, \quad (2.73)$$

$$dH = d(PV) + dU = d(PV) + (\partial U/\partial T)_V dT$$
$$+ (\partial U/\partial V)_T dV \quad (2.74)$$

$$= d(PV) + C_v dT + (T(\partial P/\partial T)_V - P)dV, \quad (2.75)$$

and for entropy:

$$dS = (\partial S/\partial T)_P dT + (\partial S/\partial P)_T dP$$
$$= (C_p/T)dT - (\partial V/\partial T)_P dP. \quad (2.76)$$

$$dS = (\partial S/\partial T)_V dT + (\partial S/\partial V)_T dV$$
$$= (C_v/T)dT + (\partial P/\partial T)_V dV. \quad (2.77)$$

Also, when only the pressure is varied,

$$d(S - S^{id}) = (R/P - (\partial V/\partial T)_P)dP \quad (2.78)$$
$$= ((\partial P/\partial T)_V - R/V)dV. \quad (2.79)$$

Note that $S - S^{id} = 0$ when $P = 0$.

When the composition remains constant, the preceding equations of this section apply to mixtures. For variable composition, relations between partial molal quantities are analogous. Those for partial molal entropy, for instance, are

$$d\bar{S}_i = \left(\frac{\partial \bar{S}_i}{\partial T}\right)_P dT + \left(\frac{\partial \bar{S}_i}{\partial P}\right)_T dP$$
$$= \left(\frac{\bar{C}_{pi}}{T}\right) dT - \left(\frac{\partial^2 V}{\partial n_i \partial T}\right)_P dP, \quad (2.80)$$

$$d\bar{S}_i = \left(\frac{\partial \bar{S}_i}{\partial T}\right)_V dT + \left(\frac{\partial \bar{S}_i}{\partial V}\right)_T dV$$
$$= \left(\frac{\bar{C}_{vi}}{T}\right) dT + \left(\frac{\partial^2 P}{\partial n_i \partial T}\right)_V dV. \quad (2.81)$$

When the summation $dS = \Sigma x_i d\bar{S}_i$ is made, Eqs. 2.76 and 2.77 result.

2.8. MIXING AND EXCESS FUNCTIONS

Many data are more informative and easier to handle when expressed relative to some kind of ideal behavior, the obvious example being the difference between the volume of a real gas and that of one conforming to the ideal-gas law at the same T and P. Two kinds of differences are used widely for mixtures:

1. The mixing property

$$M^{mix} = M - \Sigma x_i M_i \quad (2.82)$$

is the difference between the property of the mixture and the weighted sum of the properties of the pure constituents at the same conditions.

2. The excess property

$$M^{ex} = M - M^{id} \quad (2.83)$$

is the difference between the magnitude of the actual property and the value it would have at the same conditions if it were ideal. In this definition, the ideal mixture property is related to the weighted sum of the pure constituent properties by

$$M^{id} = \Sigma x_i M_i + \Sigma x_i m_i = \Sigma x_i (M_i + m_i). \quad (2.84)$$

For all properties except the entropy and those defined in terms of entropy, the $m_i = 0$. For the others, the change in entropy accompanying the formation of an ideal solution is taken into account:

$$\Delta S^{id} = -R\Sigma x_i \ln x_i. \quad (2.85)$$

The derivation of this equation is obtained in Example 2.9. Accordingly, the values of m_i of various properties may be summarized as:

Extensive Property	m_i	M^{id}	
H, U, V, C_p, C_v	0	$\Sigma x_i M_i$	(2.86)
S	$-R \ln x_i$	$\Sigma x_i (S_i - R \ln x_i)$	(2.87)
A	$RT \ln x_i$	$\Sigma x_i (A_i + RT \ln x_i)$	(2.88)
G	$RT \ln x_i$	$\Sigma x_i (G_i + RT \ln x_i)$	(2.89)
$A/RT = \psi$	$\ln x_i$	$\Sigma x_i (A_i/RT + \ln x_i)$	(2.90)
$G/RT = \Phi$	$\ln x_i$	$\Sigma x_i (G_i/RT + \ln x_i)$	(2.91)

Relations of identical form hold between original properties, the excess and the excess partial molal properties. Some of the basic analogies are collected in Table 2.4, and those involving derivatives and integrals are in Table 2.5. Nomenclature for properties of mixtures is summarized in Table 2.3.

In terms of the several contributions, a property, M, is given by

$$M = M^{ex} + M^{id} = M^{ex} + \Sigma x_i (M_i + m_i) \quad (2.92)$$
$$= \Sigma x_i (\bar{M}_i^{ex} + M_i + m_i), \quad (2.93)$$

Example 2.9 Derivation of the Equation for the Entropy Change of Mixing of Ideal Gases at Constant T and P.

The mixing process is visualized on the sketch. Two gases in the amounts n_a and n_b at the same pressure P are separated by a partition in a closed vessel. When the partition is removed, the gases mix and their partial pressures become $p_i = y_i P$. Since, at constant temperature,

$$dS = -\frac{V}{T} dP = -R\, d \ln P, \quad (1)$$

the total entropy change for the process is

$$(n_a + n_b)\Delta S = -R[n_a \ln(p_a/P) + n_b \ln(p_b/P)], \quad (2)$$

or

$$\Delta S = R[y_a \ln(1/y_a) + y_b \ln(1/y_b)]. \quad (3)$$

Clearly the result is generalizable to ideal solutions of any number of components as

$$\Delta S = R\Sigma x_i \ln(1/x_i), \quad (4)$$

which is Eq. 2.85.

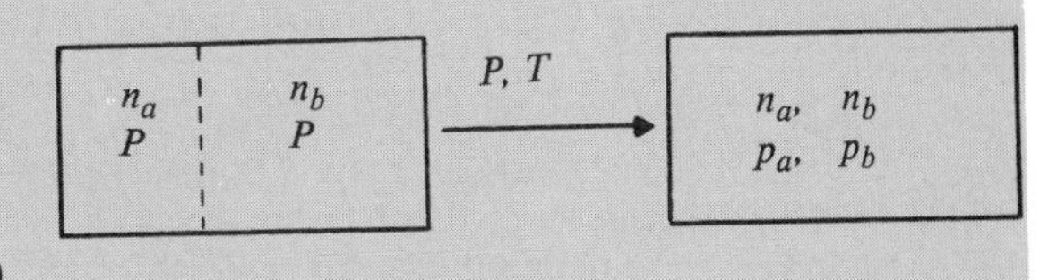

Table 2.3. Nomenclature for Properties of Mixtures

M = a general property of a mixture: $U, H, V, C_p, C_v, S, A, G$
$= \Sigma x_i \bar{M}_i$.
M_i = property of substance i.
M_i' = property of ideal gas i.
M^{id} = property of an ideal solution
$= \Sigma x_i(M_i + m_i)$,

$m_i = 0$ except for S, A, G,

$s_i = -R \ln x_i$,

$a_i = RT \ln x_i$,

$g_i = RT \ln x_i$.

$\bar{M}_i$ = partial molal property

$$= \left(\frac{\partial nM}{\partial n_i}\right)_{TPn_j},$$

all properties listed under M plus $\ln(\hat{f}_i/x_i)$, $\ln \hat{\phi}_i$, $\ln(a_i/x_i)$, $\ln \gamma_i$.

$\Delta M'$ = residual property = (property with components in the ideal gas state) − (true property). The notation $\Delta M' = M^{id} - M$ sometimes is used.
$= M' - M = \Sigma x_i[M_i' + m_i - \bar{M}_i]$

M^{mix} = the difference between the property of the mixture and the weighted sum of the properties of the components
$\equiv \Delta M$ in some books
$= M - \Sigma x_i M_i = \Sigma x_i(\bar{M}_i - M_i)$
$= M^{ex} + \Sigma x_i m_i$.

M^{ex} = the difference between the property of the real mixture and that of an ideal mixture.
$= M - M^{id} = \Sigma x_i(\bar{M}_i - M_i - m_i)$
$= M^{mix} - \Sigma x_i m_i$

m_i is defined in Section 2.8 for various properties. For those properties having the $m_i = 0$, for the gaseous state $\Delta M' = -M^{ex}$.

where the relation

$$M^{ex} = \Sigma x_i \bar{M}_i^{ex} \tag{2.94}$$

has been incorporated. The Gibbs energy has received particular attention,

$$G = G^{ex} + G^{id} = \Sigma x_i(\bar{G}_i^{ex} + G_i + RT \ln x_i), \tag{2.95}$$

and special terms have been invented for convenient description of this behavior. These are covered in detail in Chapters 3 and 4 but may be mentioned here also. The terms in question are the fugacity and the activity and the related fugacity and activity coefficients.

At constant temperature, for an ideal gas,

$$dG^{id} = VdP = RT\, d \ln P. \tag{2.96}$$

The fugacity, f, is defined by

$$dG = RT\, d \ln f. \tag{2.96a}$$

Integrating from a lower limit of $P = 0$ at which $G = G^{id}$ and $f/P = 1$,

$$G^{ex} = G - G^{id} = RT \ln \frac{f}{P} = RT \ln \phi. \tag{2.97}$$

Similarly, the excess partial Gibbs energy is

$$\bar{G}^{ex} = \bar{G}_i - G^{id} = RT \ln(\hat{f}_i/x_i) - RT \ln P = RT \ln(\hat{f}_i/x_i P)$$
$$= RT \ln \hat{\phi}_i. \tag{2.98}$$

Condensed phases are not naturally referred to the ideal-gas state, but to the pure substance at the system condition. Accordingly,

$$G^{ex} = G - G^{ref}, \tag{2.99}$$

and

$$\bar{G}_i^{ex} = \bar{G}_i - G^{ref} = RT \ln(\hat{f}_i/x_i) - RT \ln f = RT \ln(\hat{f}_i/x_i f)$$
$$= RT \ln \gamma_i, \tag{2.100}$$

where

$$\gamma_i = \hat{f}_i/x_i f \tag{2.101}$$

is the activity coefficient. The excess Gibbs energy becomes

Table 2.4. Analogous Relations of Excess Properties and Partial Molal Excess Properties

Property	*Excess Property*	*Partial Molal Excess Property*	
V	$V^{ex} = V - \Sigma x_i V_i$	$\bar{V}_i^{ex} = \bar{V}_i - V_i$	(1)
U	$U^{ex} = U - \Sigma x_i U_i$	$\bar{U}_i^{ex} = \bar{U}_i - U_i$	(2)
$H = U + PV$	$H^{ex} = U^{ex} + PV^{ex}$	$\bar{H}_i^{ex} = \bar{U}_i^{ex} + P\bar{V}_i^{ex}$	(3)
$A = U - TS$	$A^{ex} = U^{ex} - TS^{ex}$	$\bar{A}^{ex} = \bar{U}^{ex} - T\bar{S}^{ex}$	(4)
$G = H - TS$	$G^{ex} = H^{ex} - TS^{ex}$	$\bar{G}_i^{ex} = \bar{H}_i^{ex} - T\bar{S}_i^{ex}$	(5)
$G/T = H/T - S$	$G^{ex}/T = H^{ex}/T - S^{ex}$	$\bar{G}_i^{ex}/T = \bar{H}_i^{ex}/T - \bar{S}_i^{ex}$	(6)

Note: V_i could be written V_i^{ref} and U_i could be U_i^{ref}, but these superscripts usually are omitted for convenience.

Table 2.5. Some Parallel Integral and Derivative Relations of Excess and Partial Molal Excess Properties

	Excess Property	*Partial Molal Excess Property*
1.	$G^{ex} = G - \Sigma x_i(G_i + RT \ln x_i)$	$\bar{G}_i^{ex} = \bar{G}_i - (G_i + RT \ln x_i)$
2.	$= RT\, \Sigma x_i \ln \gamma_i$	$= RT \ln \gamma_i$
3.	$= \int_{P^*}^{P} V^{ex}\, dP$	$= \int_{P^*}^{P} \bar{V}_i^{ex}\, dP$
4.	$V^{ex} = V - \Sigma x_i V_i$	$\bar{V}_i^{ex} = \bar{V}_i - V_i$
5.	$= \left(\dfrac{\partial G^{ex}}{\partial P}\right)_{Tx}$	$= \left(\dfrac{\partial \bar{G}_i^{ex}}{\partial P}\right)_{Tx}$
6.	$H^{ex} = \left(\dfrac{\partial (G^{ex}/T)}{\partial (1/T)}\right)_{Px}$	$\bar{H}_i^{ex} = \left(\dfrac{\partial (\bar{G}_i^{ex}/T)}{\partial (1/T)}\right)_{Px}$
7.	$= -T^2 \int_{P^*}^{P} \left(\dfrac{\partial (V^{ex}/T)}{\partial T}\right)_{Px} dP$	$= -T^2 \int_{P^*}^{P} \left(\dfrac{\partial (\bar{V}_i^{ex}/T)}{\partial T}\right)_{Px} dP$
8.	$S^{ex} = -\left(\dfrac{\partial G^{ex}}{\partial T}\right)_{Px}$	$\bar{S}_i^{ex} = -\left(\dfrac{\partial \bar{G}_i^{ex}}{\partial T}\right)_{Px}$
9.	$= -\left(\dfrac{\partial A^{ex}}{\partial T}\right)_{Vx}$	$= -\left(\dfrac{\partial \bar{A}_i^{ex}}{\partial T}\right)_{Vx}$
10.	$= -\int_{P^*}^{P} \left(\dfrac{\partial V^{ex}}{\partial T}\right)_{Px} dP$	$= -\int_{P^*}^{P} \left(\dfrac{\partial \bar{V}_i^{ex}}{\partial T}\right)_{Px} dP$
11.	$= (H^{ex} - G^{ex})/T$	$= (\bar{H}_i^{ex} - \bar{G}_i^{ex})/T$
12.	$C_p^{ex} = \left(\dfrac{\partial H^{ex}}{\partial T}\right)_{Px}$	$\bar{C}_{pi}^{ex} = \left(\dfrac{\partial \bar{H}_i^{ex}}{\partial T}\right)_{Px}$
13.	$= T\left(\dfrac{\partial S^{ex}}{\partial T}\right)_{Px}$	$= T\left(\dfrac{\partial \bar{S}_i^{ex}}{\partial T}\right)_{Px}$

Note: P^* is a low pressure at which gas behavior becomes ideal; contributions of this term cancel out when differences are taken.

$$G^{ex} = \Sigma x_i \bar{G}_i^{ex} = RT\Sigma x_i \ln\gamma_i, \tag{2.102}$$

which is a fundamental relation in the thermodynamics of solutions. The fugacity coefficient and the activity coefficient are determined experimentally and are functions of composition, temperature, and pressure. Some experimental excess function data are in Figure 2.3 and more are in Chapter 4. The literature is quite extensive; the bibliography of Wisniak & Tamir (1978) is 1,500 pages long. An extensive compilation of heats of mixing is in the book by Christensen et al. (1982). Experimental techniques and literature data on mixing volumes of two liquids are reviewed by Handa & Benson (1978). A pair of recent papers may be mentioned to emphasize current activity in this area: Low-temperature techniques are used by Miller & Hiza (1978) for measuring volumes of natural gas liquids. A concise paper by Nakanishi et al. (1975) reports on the effects of ring structure and aromaticity on excess functions.

Binary excess functions of course vanish at $x_1 = 0$ and $x_2 = 0$. Accordingly if it is desired to fit an equation in terms of the composition, the form

$$M^{ex} = x_1 x_2 f(x_1, x_2) \tag{2.103}$$

satisfies these requirements. The form of $f(x_1, x_2)$ is empirical and a variety has been used, particularly for Gibbs energy-related functions. The *IDS Selected Data on Mixtures* (Kehiaian, 1971-date) commonly uses

$$M^{ex} = x_1(1 - x_1) \sum_{0}^{n} a_i(1 - 2x_i)^i \tag{2.104}$$

where the summation usually extends to four or five terms. A scheme based on the more complex but flexible Legendre polynomials is adopted for consistency analysis by the DECHEMA *Vapor Liquid Equilibrium Data Collection,* Vol. 1, Pt. 1 (1979).

Experimental methods for determining mixing and excess properties are quite varied, depending on the accuracy required, the nature of the system, and the range of pressures and temperatures to be covered. Excess volumes of liquids are measured in some kind of dilatometer. Calorimetric methods of varying degrees of elaborateness are used for excess enthalpies. Phase equilibrium data leading directly to evaluation of activity or fugacity coefficients are the usual method for finding excess Gibbs energies. Excess entropies then are calculated from $S^{ex} = (H^{ex} - G^{ex})/T$. Chemical potentials are known when the activity coefficients are known. Other partial molal properties are obtained by differentiation of the parent properties. Details of experimental techniques may be tracked down by consulting recent issues of *Journal of Chemical Thermodynamics* or other periodicals. In book form, the 1,300 pages of the IUPAC sponsored work edited by LeNeindre & Vodar (1975) is the most comprehensive reference. The book of McGlashan (1979) stresses the experimental aspects throughout.

2.9. THE GIBBS-DUHEM EQUATION AND THERMODYNAMIC CONSISTENCY

In Section 2.1, the differential of the extensive internal energy

$$\underline{U} = \underline{U}(\underline{S}, \underline{V}, n_1, n_2, \ldots)$$

was rearranged as Eq. 2.4, which is

$$(dU - TdS + PdV - \Sigma\mu_i dx_i)\,n + (U - TS + PV - \Sigma x_i\mu_i)\,dn = 0. \quad (2.105$$

The homogeneous function theorem will be used to introduce a variation on the analysis given there. According to that theorem,

$$\underline{U} = (\partial U/\partial S)_{Vn}\underline{S} + (\partial U/\partial V)_{Sn}\underline{V} + \Sigma\mu_i n_i \quad (2.106)$$

$$= T\underline{S} - P\underline{V} + \Sigma\,\mu_i n_i, \quad (2.107)$$

or

$$U - TS + PV - \Sigma\mu_i x_i = 0. \quad (2.108)$$

Accordingly, the first part of Eq. 2.105 likewise is zero, that is

$$dU - TdS + PdV - \Sigma\mu_i dx_i = 0. \quad (2.109)$$

The differential of Eq. 2.108 in turn is

$$dU - TdS - SdT + PdV + VdP - \Sigma\mu_i dx_i - \Sigma x_i d\mu_i = 0. \quad (2.110)$$

On comparing Eqs. 2.109 and 2.110, the conclusion is drawn that

$$-SdT + VdP - \Sigma x_i d\mu_i = 0, \quad (2.111)$$

or

$$(\partial G/\partial T)_{Px}dT + (\partial G/\partial P)_{Tx}dP - \Sigma x_i d\bar{G}_i = 0. \quad (2.112)$$

Similar equations can be written for any energy function, for which the generic equation is

$$(\partial M/\partial T)_{Px}dT + (\partial M/\partial P)_{Tx}dP = \Sigma x_i d(\partial nM/\partial n_i)_{TPn_j}. \quad (2.113)$$

This general form of the Gibbs-Duhem equation relates any property that is a function of T, P, and the composition to the chemical potentials. Since $G = G(T, P)$ is a fundamental equation, the Gibbs-Duhem equation is most widely applied with functions closely related to the Gibbs energy. The basic form is

$$-SdT + VdP = \Sigma x_i d\bar{G}_i = \Sigma x_i d\mu_i. \quad (2.114)$$

Versions in terms of the Planck function, $\Phi = G/RT$, and its mixing and excess forms are collected in Table 2.6. Use will be made of these in later chapters.

At constant T and P, Eq. 2.114 becomes simply

$$\Sigma x_i\, d\mu_i = 0. \quad (2.115)$$

It is particularly useful for binary mixtures in the arrangement,

$$d\,\mu_2 = -\,x_1 d\mu_1/(1 - x_1), \quad (2.116)$$

or the integrated form,

$$\mu_2 = G_2 - \int_0^{x_1} \frac{x_1}{1 - x_1}\,d\mu_1, \quad (2.117)$$

which allows evaluation of one chemical potential or partial molal property of a pair when the other is known as a function of composition. Eq. 2.116 also is useful in the form,

$$x_1\frac{\partial\mu_1}{\partial x_1} + x_2\frac{\partial\mu_2}{\partial x_1} = 0. \quad (2.118)$$

Three characteristics of Eq. 2.118 may be employed to check the thermodynamic consistency of experimental data—namely:

1. The equation predicts that the slopes, $d\mu_i/dx_1$, are of opposite signs for the two components at the same composition.
2. The slope of μ_1 is zero at $x_1 = 1$ and that of μ_2 is zero at $x_1 = 0$.
3. The area test developed from this equation in Example 2.10 must be satisfied, subject to some uncertainty if the data are not truly isothermal and isobaric, though the principal effect is that of temperature since pressure normally has little effect on the behavior of condensed phases.

Evaluation of the thermodynamic consistency of data has received much attention in the literature. The area test is a necessary but insufficient condition, since individual data may be off in ways that compensate each other. Schemes that examine the data point-by-point have been worked out and are referred to in Chapter 4.

Although it is a laborious procedure and requires many data, the consistency test has been applied to mixtures of three and more components, for example by Herington (1951).

2.10. CONDITIONS OF EQUILIBRIUM

Thermodynamics is concerned largely with relations between various properties of systems at equilibrium, and the differences between these properties in distinct equilibrium states. For instance, the PVT equation of state gives the equilibrium

Table 2.6. Some Forms of the Gibbs-Duhem Equation

Property	*Gibbs-Duhem Equation*	
M	$\left(\frac{\partial M}{\partial T}\right)_{Px} dT + \left(\frac{\partial M}{\partial P}\right)_{Tx} dP = \Sigma x_i d\left(\frac{\partial (nM)}{\partial n_j}\right)_{TPn_j}$	(1)
	$= \Sigma x_i d\bar{M}_i$	
$M^{mix} = M - \Sigma x_i M_i$	$\left(\frac{\partial M^{mix}}{\partial T}\right)_{Px} dT + \left(\frac{\partial M^{mix}}{\partial P}\right)_{Tx} dP = \Sigma x_i d(\bar{M}_i - M_i)$	(2)
G	$-SdT + VdP = \Sigma x_i d\bar{G}_i = \Sigma x_i d\mu_i$	(3)
G/RT	$-\frac{H}{RT^2} dT + \frac{V}{RT} dP = \Sigma x_i d(\mu_i/RT)$	(4)
$\frac{\Delta G'}{RT} = \frac{G - G^{id}}{RT} = \ln\frac{f}{P}$	$-\frac{H^{ex}}{RT^2} dT + \frac{V^{ex}}{RT} dP = \Sigma x_i d \ln(\hat{f}_i/x_i)$	(5)
	$= \Sigma x_i d \ln \hat{f}_i$	(6)
	$= \Sigma x_i d \ln \hat{\phi}_i$	(7)
$\frac{G^{ex}}{RT} = \frac{G - G^{ref}}{RT}$	$-\frac{H^{ex}}{RT^2} dT + \frac{V^{ex}}{RT} dP = \Sigma x_i d \ln(\hat{f}_i/x_i f_i^{ref})$	(8)
$= \ln\frac{f}{f^{ref}}$	$= \Sigma x_i d \ln \gamma_i$	(9)
	$= \Sigma x_i d \ln(\hat{f}_i/f_i^{ref})$	(10)
	$= \Sigma x_i d \ln a_i$	(11)

Note: In connection with numbers 5 and 8, note that $\Delta H' = H - H' = H^{ex} = H - H^{id}$ since $H' = H^{id}$.

pressure of an ideal gas at specified T and V, although in practice higher or lower pressures could exist momentarily or even indefinitely under metastable conditions. All the equations in this book are valid only at equilibrium. The rate of attainment of equilibrium is of practical importance, but that is another story.

The state of a pure substance in a single phase is determined when two of its thermodynamic properties are specified. As described in Section 2.2, complete information about such a system is embodied in a fundamental equation of state. An equilibrium state is characterized as having a maximum entropy or a minimum energy function, at specified values of the two other properties of the particular fundamental equation. The possible extrema at equilibrium are identified in the following tabulation:

Independent Variables	*Property*	
	Maximum	*Minimum*
U, V	S	—
S, V	—	U
P, H	S	—
P, S	—	H
T, V	—	A
P, T	—	G

Equilibrium and other states may be identified on spatial diagrams like those of Figures 2.1 and 2.2. On the diagram of methyl chloride, for instance, at $P = 50$ and $T = 560$ the equilibrium Gibbs energy is -24.93, which is the smallest value possible. Only states above the surface are physically possible; thus the point at $G = -110$ at the cited T and P is impossible to attain, whereas that at $G = 10$ is possible though unstable. On the Helmholtz diagram of sodium, only points above the surface are possible. On the ethylene diagram, in the direction of the entropy axis only points on and in front of the surface are possible, whereas in the direction of the enthalpy axis only points on and above the surface are possible.

2.10.1. Mixtures

The numbers of mols of individual species also are variables in the fundamental equation of a mixture—for instance:

$$\underline{G} = \underline{G}(T, P, n_1, n_2, \ldots), \tag{2.119}$$

and

$$d\underline{G} = -\underline{S}dT + \underline{V}dP + \Sigma\mu_i dn_i. \tag{2.120}$$

Thus, in addition to T and P, the chemical potentials are involved in the specification of an equilibrium state. There are

Table 2.7. Excess Properties of Some Equimolal Binary Mixtures

Mixture	°C	G^{ex}	H^{ex}	S^{ex}
2-propanone + decane	65	1,000	1,980	−2.90
methyl acetate + cyclohexane	30	959	1,770	−2.68
dioxane + heptane	40	838	1,639	−2.56
dioxane + heptane	45	838	1,632	−2.50
2-butanone + dodecane	25	934	1,647	−2.39
benzene + heptane	50	297	867	−1.77
2-butanone + heptane	25	860	1,338	−1.60
2-butanone + hexane	25	830	1,252	−1.42
cyclohexane + dioxane	25	1,069	1,445	−1.24
acetone + hexane	−5	1,167	1,385	−0.81
carbon tetrachloride + furan	30	190	323	−0.44
methyl acetate + benzene	25	237	317	−0.27
cyclohexane + 2,3-dimethylbutane	25	87	156	−0.23
cycloheptane + 2,3-dimethylbutane	25	135	163	−0.09
dichloromethane + furan	30	−6.7	16.9	−0.08
cycloheptane + cyclopentane	25	−4.5	3.9	−0.03
cycloheptane cyclohexane	25	8.6	6.0	0.01
cycloheptane + cyclooctane	45	2.7	−2.0	0.02
cyclopentane + 2,3-dimethylbutane	25	12.7	−1.8	0.05
1,2-dichloroethane + methanol	45	1,114	957	0.49
carbon tetrachloride + cyclopentane	25	34	80	0.15
tetrahydrofuran + cyclopentane	25	226	369	0.48
dichloroethane + methanol	45	1,114	957	0.49
dichloromethane + acetone	30	−404	−887	1.59
1-propanol + heptane	30	1,291	660	2.08
dimethylsulfoxide + dibromomethane	35	−208	−889	2.21
dimethylsulfoxide + dibromomethane	25	−157	−959	2.69
water + ethanol	50	821	−121	2.92
water + ethanol	90	901	378	1.44

Data from *Selected Data on Mixtures: International Data Series,* Thermodynamics Research Center, Texas A&M University, 1973–1981. G^{ex} and H^{ex} in J/gmol, S^{ex} in J/(gmol) (K). Some data are interpolated.

some practical differences in the handling of phase equilibria and chemical equilibria, so they will be discussed separately.

2.10.2. Phase Equilibria

Equality of temperature of phases in contact is required for thermal equilibrium and equality of pressure for hydrostatic equilibrium. The uniformity of chemical potentials for diffusive equilibrium can be shown readily. In osmotic processes where an interface is permeable to one of the substances, a difference in pressure is required to maintain diffusive equilibrium when the concentrations of the diffusable substance are different in the two phases. Here hydrostatic equilibrium is sacrificed to prevent changes in concentration or chemical potential.

For a transfer of dn_i mols of a substance between two phases at the same T and P, the change in Gibbs energy is

$$d\underline{G} = (\mu_i^{(2)} - \mu_i^{(1)})\, dn_i. \tag{2.121}$$

Since G is a minimum at equilibrium, its derivative is zero:

$$(\partial \underline{G}/\partial n_i)_{TPn_j} = 0, \tag{2.122}$$

so that from Eq. 2.121,

$$\mu_i^{(1)} = \mu_i^{(2)}. \tag{2.123}$$

When the transfer of more than one substance between more than two phases occurs, equality of chemical potentials clearly extends to all phases and all substances:

$$\mu_i^{(1)} = \mu_i^{(2)} = \ldots = \mu_i^{(k)}; \qquad i = 1, 2, \ldots, n. \tag{2.124}$$

The equilibrium condition may be found either by direct minimization of the Gibbs energy or by utilizing the principle of equality of chemical potentials. In vapor-liquid equilibria, for instance, the condition of minimum Gibbs energy may be written

$$\underline{G}/RT = \Sigma\, l_i[G_{i0}^{(L)} + \ln x_i] + \Sigma v_i[G_{i0}^{(V)} + \ln y_i] \rightarrow \text{minimum} \tag{2.125}$$

where l_i and v_i are the mols of individual constituents in the liquid and vapor phases. For given overall amounts of the

Example 2.10 The Gibbs-Duhem Consistency Test for a Binary Mixture at Constant Temperature and Pressure

The differential of the property

$$M = x_1\bar{M}_1 + x_2\bar{M}_2 = x_1\bar{M}_1 + (1 - x_1)\bar{M}_2 \quad (1)$$

is

$$dM = (x_1 d\bar{M}_1 + x_2 d\bar{M}_2) + (\bar{M}_1 dx_1 - \bar{M}_2 dx_1). \quad (2)$$

By the Gibbs-Duhem equation the first group in parentheses is zero, which leaves

$$dM = (\bar{M}_1 - \bar{M}_2)\, dx_1. \quad (3)$$

Terminal conditions are $M = \bar{M}_1$ when $x_1 = 1$ and $M = \bar{M}_2$ when $x_1 = 0$. The integral of Eq. 3 is

$$\int_{x_1=0}^{x_1=1} dM = M_1 - M_2 = \int_0^1 (\bar{M}_1 - \bar{M}_2)\, dx_1, \quad (4)$$

which can be written as

$$\int_0^1 (\bar{M}_1 - M_1) dx_1 - \int_0^1 (\bar{M}_2 - M_2) dx_1 = 0, \quad (5)$$

or

$$\int_0^1 (\bar{M}_1^{ex} - \bar{M}_2^{ex}) dx_1 = 0. \quad (6)$$

This theorem will be checked with the data of the excess partial molal Gibbs energies for water + ethanol mixtures at 343.15, which are the basis of Figure 2.3. With Simpson's Rule, the integrals are

$$\int_0^1 \mu_1^{ex}\, dx_1 - \int_0^1 \mu_2^{ex} dx_1$$

$$= 1{,}179.57 - 1{,}180.33 = -0.76. \quad (7)$$

This is considered an excellent check of the consistency of the data. Some other consistency checks are shown in Figure 4.7. Units are J/gmol.

x_1	G^{ex}	μ_1^{ex}	μ_2^{ex}
0.000	0.0	2640	0
0.050	128.5	2499	4
0.100	249.9	2375	14
0.150	364.2	2238	34
0.200	469.9	2086	66
0.250	565.8	1926	112
0.300	651.0	1766	173
0.350	724.9	1607	250
0.400	786.5	1448	345
0.450	834.8	1289	463
0.500	868.5	1129	608
0.550	886.5	972	782
0.600	887.1	819	989
0.650	868.9	672	1235
0.700	830.1	531	1528
0.750	768.3	397	1881
0.800	680.6	273	2310
0.850	563.3	163	2829
0.900	412.7	76	3445
0.950	225.5	19	4146
1.000	0.0	0	4920

individual substances the number of independent l_i or v_i equals the number of substances present, so finding the minimum Gibbs energy as a function of distribution between the phases is a fairly tedious numerical problem. The work is simplified, however, by using the principle of equality of chemical potentials in the two phases or, equivalently, the equality of fugacities as explained in Chapter 6. Representation of the Gibbs energy of binary liquid phase systems is easier and direct minimization is feasible; several examples are shown in Chapter 7.

2.10.3. Chemical Equilibria

For a chemical reaction between chemical species designated by M_i and stoichiometric coefficients by ν_i

$$\nu_i M_i = 0, \quad (2.126)$$

the total Gibbs energy is

$$\underline{G} = \Sigma n_i (G_{i0} + RT \ln x_i). \quad (2.127)$$

If a single reaction is taking place, the amounts of all substances present are related by the stoichiometric equation, so there is only one independent variable—say, the fractional conversion of one of the components. Thus finding the minimum Gibbs energy and hence the equilibrium conversion is simpler than finding an equilibrium phase distribution. For instance, chemical equilibria in the oxidation of sulfur dioxide at several temperatures and pressures are found as minima on the Gibbs energy plots of Example 2.11. For multiple reactions with many participants, calculation methods are available that are simpler sometimes than direct Gibbs energy minimization; examples are given in Chapter 10.

2.10.4. Evaluation of Chemical Potentials

Because of their fundamental character in the identification of equilibrium states, numerical values of chemical potentials are needed in practical work. They are functions of T and P as well as composition. Their evaluation from measured excess properties such as G^{ex}, H^{ex}, and V^{ex}, is simply by mathematical differentiation of these data. The excess Gibbs energy, for instance, usually is found from vapor-liquid and other phase equilibrium data as explained in Chapter 4. The chemical potential is

$$\mu_i^{ex} = \bar{G}_i = (\partial n G^{ex} / \partial n_i)_{TPn_j}, \quad (2.128)$$

or, in binary systems,

Example 2.11 Plots of the Gibbs Energies of Reacting Mixtures, Showing Equilibrium Conversion as the Minima of Those Plots

The example is of the oxidation of sulfur dioxide with stoichiometric proportions of the reactants at several temperatures and pressures. The amounts of the participants present are expressed in terms of the fraction, e, of SO_2 converted.

$$SO_2 + 0.5O_2 \rightleftharpoons SO_3, \tag{1}$$

$$y_{SO_2} = 2(1-e)/(3-e), \tag{2}$$

$$y_{O_2} = (1-e)/(3-e), \tag{3}$$

$$y_{SO_3} = 2e/(3-e). \tag{4}$$

The Gibbs energy of the mixture is a function of e.

$$\frac{G}{RT} = \frac{1}{RT}\Sigma n_i\mu_i = \Sigma n_i\left[\frac{\mu_{i0}}{RT} + \ln(y_iP)\right] \tag{5}$$

$$= (1-e)\left[\frac{\mu_{10}}{RT} + \ln\frac{2(1-e)P}{3-e}\right] + 0.5(1-e)\ln\frac{(1-e)P}{3-e} + e\left[\frac{\mu_{30}}{RT} + \ln\frac{2eP}{3-e}\right]. \tag{6}$$

The μ_{i0} are the Gibbs energies of formation which are tabulated following.

	$\mu_{i0}/RT = -\ln K_f$	
°F	SO_2	SO_3
1,100	−41.467	−43.827
1,200	−38.437	−39.998
1,300	−35.752	−36.606

The Gibbs energies are plotted against the fractional conversion at several combinations of temperature and pressure. The low points on the curves correspond to the equilibrium conversions. The set of large curves is plotted with amplified ordinates of different range in each case for accurate location of the minima.

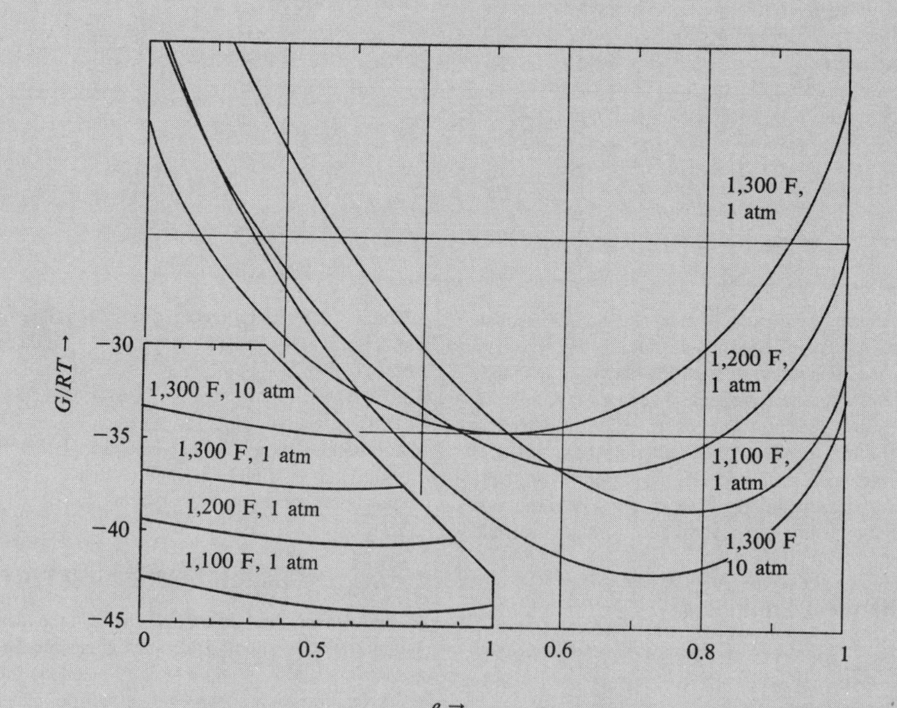

$$\mu_i^{ex} = G^{ex} + (1 - x_i)(\partial G^{ex}/\partial x_i)_{TP}, \quad (2.129)$$

as explained in Section 2.5. The special functions, fugacity and activity, have been invented to simplify representation of G^{ex}, as explained in Chapters 3 and 4.

The chemical potentials also are related to other partial molal properties, for example

$$(\partial \mu_i/\partial P)_{Tn_j} = \bar{V}_i, \quad (2.130)$$

$$(\partial(\mu_i/T)/\partial T)_{Pn_j} = -\bar{H}_i/T^2. \quad (2.131)$$

Since these relations involve differentiations with respect to compositions over wide ranges of temperature and pressure, followed by integrations, they are not often used. The calorimetric data, particularly, are more difficult to measure accurately than are vapor-liquid equilibria from which activity coefficients and chemical potentials are derived most readily.

2.11. PROBLEMS

2.1. Construct a thermodynamic diagram of entropy, S, vertically and internal energy, U, and log(specific volume) horizontally for refrigerant 22, using the data given in Bolz & Tuve, *Handbook of Tables for Applied Engineering Science* (1973), p. 77.

2.2. The fundamental equation of a particular substance is given as a second-order response surface, namely:

$$U = 200 + 50\,S - 0.3\,V + 0.8\,S^2 - 0.004\,V^2 - 0.03\,S\,V.$$

a. Determine the equations for the properties T, P, H, A, and G in terms of the variables U, S, and V.
b. Plot the surface $U(S, V)$ for the ranges $10 \le S \le 30$ and $100 \le V \le 300$.

2.3. Find the partial molal volumes in an equimolal mixture of carbon dioxide (1) and propylene (2) with the Redlich-Kwong equation for which these data hold at 400 K and 20 atm:

	A	B
CO_2	0.05914	0.0181
C_3H_6	0.1495	0.0348
Mixture	0.0992	0.0265

It is convenient to start with the polynomial form

$$z^3 - z^2 + (A - B - B^2)z - AB = 0.$$

The value of z at the specified conditions is 0.9253.

2.4. Use the virial equation $V = B + RT/P$ to find equations for these departures from ideality at a specified temperature:

$A - A^{id}, S - S^{id}, H - H^{id}, U - U^{id}, G - G^{id}$, and $C_p - C_p^{id}$

Define a reference condition (P^*, V^*) at which the substance behaves as an ideal gas. Start with the relations:

$$-(A - A^{id}) = \int_{V^*}^{V} P\,dV$$

$$= \int_{\infty}^{V}\left(P - \frac{RT}{V}\right)dV + \int_{V^*}^{V}\frac{RT}{V}\,dV$$

$$= \int_{\infty}^{V}\left(P - \frac{RT}{V}\right)dV + RT\ln\frac{P^*V}{RT},$$

$$S - S^{id} = -\left(\frac{\partial(A - A^{id})}{\partial T}\right)_V.$$

2.5. Find equations for the partial molal volumes in binary mixtures in terms of the virial equation, $V = B + RT/P$, and the van der Waals equation, starting with the polynomial form

$$V^3 - \left(b + \frac{RT}{P}\right)V^2 + \frac{a}{P}V - \frac{ab}{P} = 0.$$

2.6. The fundamental equation $A = A(T, V)$ or $\psi = \psi(T, \rho)$ is given for water by Keenan et al. (1969). Verify the equations also given there for P, U, S, H, and G in terms of the variables of the fundamental equation. (See Table 1.22.)

2.7. Find equivalents for the following derivatives at constant enthalpy in terms of $PVTC_p$ data: $(\partial U/\partial V)_H$, $(\partial T/\partial P)_H$, and $(\partial G/\partial T)_H$.

2.8. Find a relation between the Joule-Thomson coefficient, $(\partial T/\partial P)_H$, the heat capacity at constant pressure, and the derivative $(\partial H/\partial P)_T$.

2.9. Transform $(\partial H/\partial P)_T$ into groups of derivatives in which the two independent variables are selected in turn from the group P, V, T, and S. The method of Shaw (1935) will be convenient. It is presented, for example, by Sherwood & Reed, *Applied Mathematics in Chemical Engineering* (1939, p. 180).

2.10. A differential amount dn_e of a substance with properties u_e, p_e, v_e, and T_e enters a tank already containing an amount n of the same material with properties U and T. Heat and work transfers are dQ and dW.

a. Show that the energy balance may be written

$$d(nU) = n\,dU + U\,dn = dQ - dW + (u_e + p_e v_e)dn_e.$$

b. Integrate the equation under the assumptions that $U = C_v T$, $H = C_p T$, heat capacities are constant, and $(dQ - dW)/dn = \alpha$, a constant. This will give a relation between the temperature and the amount n of material in the tank.

2.11. Given the fundamental equation

$$U = a\,SV + b\,S^2 + c\,V^2,$$

find G and C_v in terms of these variables.

2.12. Tabulate or plot the partial molal volumes of mixtures of carbon dioxide (1) and propylene at 400 K and 20 atm from the van der Waals equation. The values of the parameters are

	a	b
CO_2	3.592	0.04267
C_3H_6	8.379	0.08272

in the units atm, °K, and liters/gmol.

2.13. Find the Joule coefficient, $\eta = (\partial T/\partial V)_U$, from the van der Waals equation with the aid of the Bridgman Table.

2.14. Express the derivative $(\partial G/\partial T)_H$ in terms of $(\partial H/\partial P)_T$, C_p and PVT derivatives.

2.15. Find $(\partial U/\partial V)_H$ for (a) an ideal gas; (b) a gas with equation of state $P(V - b) = RT$.

2.16. Find the derivative $(\partial G/\partial P)_H$ of the Joule-Thomson experiment in terms of the van der Waals equation.

2.17. Apply the procedure of Example 2.2 with the B-truncated virial equation in the explicit forms

$$P = (1 + B/V)RT/V,$$

or

$$V = (RT/2P)(1 + \sqrt{4BP/RT}),$$

and also with the approximation

$$V = (RT/P)(1 + BP/RT).$$

2.18. Complete the solution of Example 2.1 when the flow is outwards, $(dQ - dW)/dn =$ constant, the gas is ideal, and the heat capacities are constant.

2.19. Show that for the equation of state,

$$P = \frac{RT}{V - y_1 b_1 - y_2 b_2},$$

the derivative is

$$\left(\frac{\partial P}{\partial n_1}\right)_{TVn_2} = \frac{P^2(b_1 - b_2)}{nRT} y_2.$$

2.20. Show that for $U(S, V)$ the "change of variable held constant" rule of Appendix B gives

$$\left(\frac{\partial U}{\partial S}\right)_T = \left(\frac{\partial U}{\partial S}\right)_V + \left(\frac{\partial U}{\partial V}\right)_S \left(\frac{\partial V}{\partial S}\right)_T.$$

Verify that this is correct with the Bridgman Table. Also find $(\partial G/\partial T)_S$ both ways.

2.21. The graph shows excess property data of mixtures of methanol and piperidine at 298.15 K by Nakanishi, Wada, & Touhara (1975).

a. Find $\bar{G}_i^{ex}$, $\bar{H}_i^{ex}$ and $T\bar{S}_i^{ex}$ and $x_{MeOH} = 0.4$.
b. With your values check $G^{ex} = \Sigma x_i \bar{G}_i^{ex}$ and the other properties.
c. Also compare the $\bar{H}_i^{ex} - \bar{G}_i^{ex}$ with the values of $T\bar{S}_i^{ex}$ that were calculated.

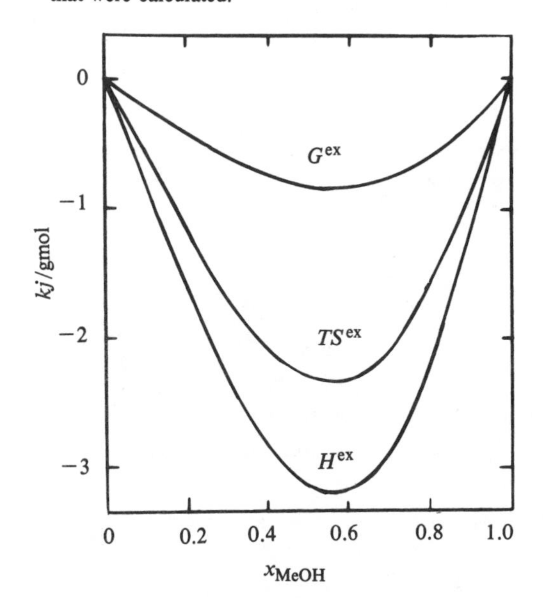

2.22. Excess volumetric data of water-methanol mixtures, ml/gmol at 25° C, *are given. Fit a polynomial of the second* degree in $(1 - 2x_1)$ to these data (*IDS Selected Data on Mixtures*).

x_1	V^{ex}	V^{ex}/x_1x_2	$1 - 2x_1$
.0970	−324	−3,699	.8060
.1843	−573	−3,812	.6314
.3952	−939	−3,929	.2096
.4706	−1,004	−4,030	.0588
.6260	−947	−4,045	−.2520
.6938	−854	−4,020	−.3876
.7279	−799	−4,034	−.4558
.7670	−706	−3,951	−.5340
.8092	−600	−3,886	−.6184
.8464	−492	−3,749	−.6892
.8892	−348	−3,532	−.7784
.9146	−262	−3,354	−.8292
.9526	−137	−3,034	−.9052
.9623	−107	−2,949	−.9246
.9714	−68	−2,448	−.9428

2.23. The excess Gibbs energy of mixtures of ethanol (1) and *n*-heptane (2) is represented by

$$G^{ex}/RT = -x_1 \ln(x_1 + Ax_2) - x_2 \ln(Bx_1 + x_2),$$

where

$$A = 1.8857 \exp(-1055.11/T),$$

$$B = 0.5303 \exp(-310.57/T).$$

The data are taken from Table E.7. Some data of excess enthalpies of equimolal mixtures are taken from the *IDS Selected Data on Mixtures,* as follows:

T, K	283	318	333	348
H^{ex}, J/gmol	1,450	1,250	850	450

Evaluate the excess enthalpies from the given equation for G^{ex} and compare with the quoted values.

2.24. Excess enthalpy data of mixtures of *n*-hexane and *n*-hexadecane are given by the equation

$$H^{ex} = x_1x_2(a_0 + a_1(1 - 2x_1)) \text{ Joules/gmol},$$

where the coefficients are functions of temperature:

°C	20	30	40	50
a_0	517.7	379.9	254.1	140.3
a_1	44.0	35.0	26.0	17.0

Find the change in excess Gibbs energy resulting from a change of temperature from 20 C to 50 C with $x_1 = 0.25$.

2.25. Excess properties of a mixture containing 25 percent water in ethanol are, in the units J/gmol:

°C	G^{ex}	H^{ex}
30	497.5	−211.0
50	536.2	−48.3
70	565.8	111.7
90	584.0	289.7

Do these data satisfy the Gibbs-Helmholtz equation?

$T(dG^{ex}/dT) + H^{ex} = G^{ex}$

2.26. Data of Joule-Thomson coefficient, μK/atm, heat capacity, C_p cal/gmol-K and second virial coefficient, B ml/gmol, of nitrogen are given.

a. Derive the relation, $\mu C_p = -d(B/T)/d(1/T)$.
b. Assuming the value at 273 to be correct, find values of the virial coefficient at other temperatures from the data of μ and C_p.

T	μ	C_p	B
273	0.2655	6.596	−10.3
298	0.2216	6.958	
323	0.1854	6.960	−0.3
348	0.1555	6.966	
373	0.1291	6.974	6.1
398	0.1069	6.986	
423	0.0868	7.000	11.5

2.27. Precise measurements of the specific volumes of aqueous H_2O_2 at 25 C are fitted by the equation

V, ml/$g = 0.0030(1 - w) + 0.6935w - 0.0360w(1 - w)$,

where w is the mass fraction of H_2O_2. Find the partial molal volumes over the entire range of concentrations.

2.28. Enthalpies of mixing of carbon tetrachloride and cyclohexane at 25 C are given. The solutions are believed to be "regular" so that the symmetrical equation,

$\Delta H^{mix} = kx_1x_2$

may be applicable. How nearly is this true? Suggest a better curve fit to these data:

x_i	0.8529	0.7213	0.5955
ΔH^m, J/gmol	76.6	118.6	141.6

x_1	0.4745	0.3495	0.2335	0.0657
ΔH^m, J/gmol	146.7	134.9	107.9	37.8

2.29. Find an integrable relation between temperature and volume in the isentropic expansion of a gas that follows the virial equation, $V = RT/P + B$, with $B = a + bT$ and $C_p = c + dT$.

2.30. The molal volumes of a binary mixture are correlated by

$V = 3 + 2(x_1 - x_2) + (x_1 - x_2)^2$.

Find the partial molal excess volumes.

2.31. The Gibbs energy of a mixture is represented by

$G = x_1 \ln \gamma_1 x_1 + x_2 \ln \gamma_2 x_2$,

with $\ln \gamma_1 = 0.5x_2^2$ and $\ln \gamma_2 = 3.0x_1^2$. What is the equilibrium composition?

2.32. The excess enthalpy of mixtures of water (1) and ethanediol (2) is represented by the equation

$$H^{ex} = x_1x_2[-2{,}431.9 + 1{,}925(x_1 - x_2) - 1{,}467.6(x_1 - x_2)^2 + 385.9(x_1 - x_2)^3],$$

J/gmol at 25° C. Derive equations for the partial molal enthalpies. (IDS Series B, No. 67, 1978).

3 Fugacity and Fugacity Coefficient

3.1. DEFINITIONS AND BASIC RELATIONS

The concept of fugacity arises from a consideration of the change in Gibbs energy that results from changes in pressure and temperature—namely:

$$dG = -SdT + VdP. \tag{3.1}$$

For an ideal gas at constant temperature,

$$dG = VdP = RT\,d\ln P, \tag{3.2}$$

and for a component i of an ideal mixture in which the partial pressure is $P_i = x_iP$,

$$d\bar{G}_i = RT\,d\ln P_i = RT\,d\ln Px_i = \bar{V}_i\,dP. \tag{3.3}$$

The simple forms of these relations may be preserved for other substances and mixtures by defining terms called *fugacity* and *partial fugacity* at a fixed temperature as

$$dG = RT\,d\ln f = V\,dP;\ \text{constant } T; \tag{3.4}$$

$$d\bar{G}_i = RT\,d\ln\hat{f}_i = \bar{V}_i\,dP;\ \text{constant } T. \tag{3.5}$$

In view of the following auxiliary conditions, the fugacity and partial fugacity reduce to the pressure and partial pressure at low values.

$$\lim_{P\to 0} f/P = \lim_{P\to 0} \phi = 1, \tag{3.6}$$

$$\lim_{P\to 0} \hat{f}_i/x_iP = \lim_{P\to 0} \hat{\phi}_i = 1. \tag{3.7}$$

The ratios $\phi = f/P$ and $\hat{\phi}_i = \hat{f}_i/x_iP$ are called fugacity and partial fugacity coefficients.

Integration at constant temperature generates integration constants that are functions of temperature.

$$G = RT\ln f + \Lambda(T), \tag{3.8}$$

$$\bar{G}_i = RT\ln\hat{f}_i + \lambda_i(T). \tag{3.9}$$

When only changes in Gibbs energies are required, the integration constants drop out. There is, however, a definite relation between the integration constants, which is derived as follows:

Since

$$G = \Sigma x_i\bar{G}_i, \tag{3.10}$$

substitution in terms of fugacities gives

$$RT\ln f + \Lambda(T) = \Sigma x_i[RT\ln\hat{f}_i + \lambda_i(T)]. \tag{3.11}$$

At a low pressure P^* at which ideal gas behavior exists, the last equation may be written

$$RT\ln P^* + \Lambda(T) = RT\,\Sigma x_i\ln x_iP^* + \Sigma x_i\lambda_i(T). \tag{3.12}$$

The pressure cancels out so that

$$\Lambda(T) = RT\,\Sigma x_i\ln x_i + \Sigma x_i\lambda_i(T), \tag{3.13}$$

which is the desired relation. Use of this result gives a simple relation between the fugacity of the mixture, f, and the partial fugacities—namely:

$$RT\ln f + RT\,\Sigma x_i\ln x_i + \Sigma x_i\lambda_i(T) = RT\,\Sigma x_i\ln\hat{f}_i + \Sigma x_i\lambda_i(T), \tag{3.14}$$

or

$$\ln f = \Sigma x_i\ln(\hat{f}_i/x_i). \tag{3.15}$$

Similarly,

$$\ln\phi = \Sigma x_i\ln\hat{\phi}_i. \tag{3.16}$$

From these results also follow the interesting conclusions,

$$f = \Pi(\hat{f}_i/x_i)^{x_i}, \tag{3.15a}$$

$$\phi = \Pi(\hat{\phi}_i)^{x_i}. \tag{3.16a}$$

Individual partial fugacity coefficients are obtained by the method for partial molal quantities discussed in Section 2.5.

Table 3.1. Fugacity and Fugacity Coefficient Relations of Pure Substances

Fugacity	*Fugacity Coefficient*
	1b. $\phi = f/P$
2a. $dG = RT\,d\ln f = V\,dP$	2b. $d(G - G^{\text{id}}) = RT\,d\ln\phi = \left(V - \frac{RT}{P}\right)dP$
3a. $RT\ln f_2/f_1 = \int_{P_1}^{P_2} V\,dP = \int_{V_1}^{V_2} V\left(\frac{\partial P}{\partial V}\right)_T dV$	3b. $RT\ln\phi = \int_0^P \left(V - \frac{RT}{P}\right)dP = \int_\infty^V \left(V - \frac{RT}{P}\right)\left(\frac{\partial P}{\partial V}\right)_T dV$
4a. $\left(\frac{\partial\ln f}{\partial T}\right)_P = -\frac{H - H^{\text{id}}}{RT^2}$	4b. $\left(\frac{\partial\ln\phi}{\partial T}\right)_P = -\frac{H - H^{\text{id}}}{RT^2}$
5a. $\left(\frac{\partial\ln f}{\partial P}\right)_T = \frac{V}{RT}$	5b. $\left(\frac{\partial\ln\phi}{\partial P}\right)_T = \frac{V - V^{\text{id}}}{RT}$
6a. $d\ln f = -\left(\frac{H - H^{\text{id}}}{RT^2}\right)dT + \frac{V}{RT}dP$	6b. $d\ln\phi = -\left(\frac{H - H^{\text{id}}}{RT^2}\right)dT + \left(\frac{V - V^{\text{id}}}{RT}\right)dP$

For a binary mixture, for example,

$$\ln \hat{\phi}_1 = \ln \phi + (1 - x_1)\left(\frac{\partial \ln \phi}{\partial x_1}\right) \tag{3.17}$$

$$= \left(\frac{\partial (n \ln \phi)}{\partial n_1}\right)_{TPn_2}, \tag{3.18}$$

$$\ln \hat{\phi}_2 = \ln \phi - x_1\left(\frac{\partial \ln \phi}{\partial x_1}\right), \tag{3.19}$$

$$\ln(\hat{f}_1/x_1) = \ln f + (1 - x_1)\left(\frac{\partial \ln f}{\partial x_1}\right), \tag{3.20}$$

$$\ln(\hat{f}_2/x_2) = \ln f - x_1\left(\frac{\partial \ln f}{\partial x_1}\right). \tag{3.21}$$

Other relations involving fugacities and fugacity coefficients are summarized in Tables 3.1 and 3.2.

3.2. PURE SUBSTANCES

When the volume is known as a function of the pressure, either from direct measurement or from an empirical equation of state, changes in fugacity may be found by integration; thus,

$$\ln \frac{f_2}{f_1} = \frac{1}{RT}\int_{P_1}^{P_2} V\,dP. \tag{3.22}$$

Since $\phi = 1$ when $P = 0$, absolute values of the fugacity coefficient are determinable; thus,

$$\ln \phi = \ln \frac{f}{P} = \frac{1}{RT}\int_0^P \left(V - \frac{RT}{P}\right) dP = \int_0^P \frac{z-1}{P}\,dP. \tag{3.23}$$

Clearly the integrands represent deviations from ideal-gas behavior and vanish at $P = 0$.

Table 3.2. Fugacity and Fugacity Coefficient Relations of Mixtures

Fugacity	*Fugacity Coefficient*
	1b. $\hat{\phi}_i = \hat{f}_i / y_i P$
2a. $d\bar{G}_i = RT\,d\ln \hat{f}_i = \bar{V}_i\,dP$	2b. $d\bar{G}_i^{ex} = RT\,d\ln \hat{\phi}_i = \bar{V}_i^{ex}\,dP$
3a. $RT\ln(\hat{f}_{i2}/\hat{f}_{i1}) = \bar{G}_{i2} - \bar{G}_{i1} = \int_{P_1}^{P_2} \bar{V}_i\,dP$	3b. $RT(\ln \hat{\phi}_i - \ln \phi_i) = \bar{G}_i^{ex} = \int_0^P \left(\bar{V}_i - \frac{RT}{P}\right) dP$
	4b. $= \int_V^\infty \left[\left(\frac{\partial P}{\partial n_i}\right)_{TVn_j} - \frac{RT}{V}\right] dV - \ln \frac{PV}{RT}$
5a. $\ln f = \Sigma y_i \ln(\hat{f}_i / y_i)$	5b. $\ln \phi = \Sigma y_i \ln \hat{\phi}_i$
6a. $\ln(\hat{f}_i/y_i) = \ln f - \sum_{k \neq i} y_k \left(\frac{\partial \ln f}{\partial y_k}\right)_{y_{j \neq i,k}}$	6b. $\ln \hat{\phi}_i = \ln \phi - \sum_{k \neq i} y_k \left(\frac{\partial \ln \phi}{\partial y_k}\right)_{y_{j \neq i,k}}$
7a. $\left(\frac{\partial \ln \hat{f}_i}{\partial T}\right)_{Py} = -\frac{\bar{H}_i - H_i}{RT^2} = -\frac{\bar{H}_i^{ex}}{RT^2}$	7b. $\left(\frac{\partial \ln \hat{\phi}_i}{\partial T}\right)_{Py} = -\frac{\bar{H}_i - H_i}{RT^2} = -\frac{\bar{H}_i^{ex}}{RT^2}$
8a. $\left(\frac{\partial \ln \hat{f}_i}{\partial P}\right)_{Ty} = \frac{\bar{V}_i}{RT}$	8b. $\left(\frac{\partial \ln \hat{\phi}_i}{\partial P}\right)_{Ty} = \frac{\bar{V}_i - V_i^{id}}{RT} = \frac{\bar{V}_i^{ex}}{RT}$
9a. $\left(\frac{\partial \ln(\hat{f}_i/y_i)}{\partial y_k}\right)_{TPy_{j \neq i,k}} = P\left(\frac{\partial \ln \hat{\phi}_i}{\partial y_k}\right)_{TPy_{j \neq i,k}}$	9b. $\left(\frac{\partial \ln \hat{\phi}_i}{\partial y_k}\right)_{TPy_{j \neq i,k}} = \frac{\partial}{\partial y_k}\left[\int_0^P \frac{\bar{V}_i^{ex}}{RT}\,dP\right]_{TPy_{j \neq i,k}}$
10a. $d\ln \hat{f}_i = -\frac{\bar{H}_i^{ex}}{RT^2}\,dT + \frac{\bar{V}_i}{RT}\,dP$	10b. $d\ln \hat{\phi}_i = -\frac{\bar{H}_i^{ex}}{RT^2}\,dT + \frac{\bar{V}_i^{ex}}{RT}\,dP$
11a. Gibbs-Duhem equation $\Sigma y_i\,d\ln \hat{f}_i = -\frac{H^{ex}}{RT^2}\,dT + \frac{V}{RT}\,dP$	11b. Gibbs-Duhem equation $\Sigma y_i\,d\ln \hat{\phi}_i = -\frac{H^{ex}}{RT^2}\,dT + \frac{V^{ex}}{RT}\,dP$
12a. Gibbs-Duhem equation at constant *T, P* $\Sigma y_i\,d\ln \hat{f}_i = 0$	12b. Gibbs-Duhem equation at constant *T, P* $\Sigma y_i\,d\ln \hat{\phi}_i = 0$

Equations of state that are explicit in pressure are handled more conveniently by replacing VdP with its equivalent,

$$\int_{P_0}^{P} VdP = \int_{P_0}^{P} d(PV) - \int_{V_0}^{V} PdV$$

$$= PV - P_0V_0 - \int_{V_0}^{V} PdV. \tag{3.24}$$

Accordingly, the fugacity coefficient may be evaluated as follows:

$$RT \ln \phi = PV - RT - \int_{V_0}^{V} PdV - RT \ln \frac{P}{P_0} \tag{3.25}$$

$$= PV - RT - \int_{V_0}^{V} \left(P - \frac{RT}{V}\right) dV - \int_{V_0}^{V} \left(\frac{RT}{V}\right) dV - RT \ln \frac{P}{P_0}, \tag{3.26}$$

or

$$\ln \phi = z - 1 - \ln z - \frac{1}{RT} \int_{\infty}^{V} \left(P - \frac{RT}{V}\right) dV, \tag{3.27}$$

when the lower limit of pressure is made zero.

Integrations of several of the more common equations of state are presented in Tables 3.3 and 3.4 and others are shown in Chapter 1 under particular equations of state. The fugacity coefficient derived from the Redlich-Kwong equation was plotted in terms of the parameters by Edmister (1968). Several correlations in terms of reduced properties and the acentric factor have been developed. That of Pitzer & Curl in equation form is plotted as Figure 1.10b. The improved correlation of Lee & Kesler (1975) is in the form

$$\log(f/P) = (\log(f/P))^{(0)} + \omega(\log(f/P))^{(1)}, \tag{3.28}$$

where the logarithmic terms in parentheses are functions of the reduced temperature and pressure; some values are given in Tables E.6 and E.7. Example 3.1 provides a comparison of fugacities derived from experimental compressibilities with those estimated from some equations of state. Two forms of the B-truncated virial equation are applied in Example 3.2.

Derivatives are obtained by starting with the defining Eq. 3.8

$$G = RT \ln f + \Lambda(T), \tag{3.29}$$

and the Gibbs-Helmholtz equation (Table A.3),

$$\left(\frac{\partial(G/T)}{\partial T}\right)_P = -\frac{1}{T^2}\left(\frac{\partial(G/T)}{\partial(1/T)}\right)_P = -\frac{H}{T^2}. \tag{3.30}$$

The integration constant $\Lambda(T)$ is eliminated by differencing between the real and ideal gas states as

$$\ln f - \ln P = G/RT - G^{id}/RT. \tag{3.31}$$

Since the pressure is held constant when the temperature derivatives are formed, the results will be the same for both fugacity and fugacity coefficient,

$$\left(\frac{\partial \ln f}{\partial T}\right)_P = \left(\frac{\partial \ln (f/P)}{\partial T}\right)_P = \left(\frac{\partial \ln \phi}{\partial T}\right)_P = -\frac{H - H^{id}}{RT^2}. \tag{3.32}$$

The pressure derivative comes from rearrangement of Eq. 3.4 with the result

$$\left(\frac{\partial \ln f}{\partial P}\right)_T = \frac{V}{RT}, \tag{3.33}$$

and

$$\left(\frac{\partial \ln \phi}{\partial P}\right)_T = \left(\frac{\partial \ln (f/P)}{\partial P}\right)_T = \frac{V}{RT} - \frac{1}{P} = \frac{V - V^{id}}{RT}. \tag{3.34}$$

The combined effects of changes in pressure and temperature are represented by

$$d \ln f = \frac{V}{RT} dP - \frac{H - H^{id}}{RT^2} dT, \tag{3.35}$$

$$d \ln \phi = \frac{V - V^{id}}{RT} dP - \frac{H - H^{id}}{RT^2} dT. \tag{3.36}$$

For illustration, these results may be applied to the B-truncated virial equation, $V = RT/P + B$, with these results:

$$\ln \phi = \int_0^P \left(\frac{z-1}{P}\right) dP = \frac{BP}{RT}, \tag{3.37}$$

$$\left(\frac{\partial \ln \phi}{\partial P}\right)_T = \frac{1}{RT}\left(\frac{RT}{P} + B - \frac{RT}{P}\right) = \frac{B}{RT}. \tag{3.38}$$

Eq. 3.38 is obtained either by substitution of $V = RT/P$ into Eq. 3.34 or by direct differentiation of Eq. 3.37. In order to find the temperature derivative, the right-hand side of Eq. 3.32 may be evaluated with the aid of

$$H - H^{id} = \int_0^P \left[V - T\left(\frac{\partial V}{\partial T}\right)_P\right] dP = \left(B - T\frac{dB}{dT}\right) \tag{3.39}$$

Combining with Eq. 3.32 gives the result,

$$\left(\frac{\partial \ln \phi}{\partial T}\right)_P = \left(T\frac{dB}{dT} - B\right)\Big/RT^2. \tag{3.39a}$$

3.3. MIXTURES

In a homogeneous mixture at a fixed temperature, the fugacity of a particular component of a mixture is defined by Eqs. 3.5 and 3.7. Additional formulas are summarized in Table 3.2. A change in partial fugacity with pressure is evaluated by integration of the partial molal volume at constant composition,

$$\ln(\hat{f}_{i2}/\hat{f}_{i1}) = \frac{1}{RT} \int_{P_1}^{P_2} \bar{V}_i \, dP. \tag{3.40}$$

Absolute values of the partial fugacity coefficient are obtainable by taking advantage of Eq. 3.7, which makes $\ln \phi_i = 0$ when $P = 0$. The pressure integrals have several useful equivalents:

Table 3.3. Fugacity Coefficients of Pure Substances from Some Equations of State

van der Waals (Table 1.3):

$$P = \frac{RT}{V-b} - \frac{a}{V^2} \tag{1}$$

$$\ln\phi = z - 1 - \frac{a}{RTV} - \ln\left[z\left(1-\frac{b}{V}\right)\right] \tag{2}$$

$$= \frac{b}{V-b} - \frac{2a}{RTV} - \ln\left[z\left(1-\frac{b}{V}\right)\right] \tag{3}$$

Virial (Table 1.8):

$$z = 1 + \frac{B}{V} + \frac{C}{V^2} \tag{4}$$

$$\ln\phi = \frac{2B}{V} + \frac{1.5C}{V^2} - \ln z \tag{5}$$

$$z = 1 + B\left(\frac{P}{RT}\right) + (C - B^2)\left(\frac{P}{RT}\right)^2 + \ldots \tag{6}$$

$$\ln\phi = B\left(\frac{P}{RT}\right) + \frac{(C-B^2)}{2}\left(\frac{P}{RT}\right)^2 + \frac{(D - 3BC + 2B^3)}{3}\left(\frac{P}{RT}\right)^3 + \ldots \tag{7}$$

Redlich-Kwong (Table 1.9):

$$P = \frac{RT}{V-b} - \frac{a}{\sqrt{T}V(V+b)} \tag{8}$$

$$\ln\phi = z - 1 - \ln\left[z\left(1-\frac{b}{V}\right)\right] - \frac{a}{bRT^{1.5}}\ln\left(1+\frac{b}{V}\right) \tag{9}$$

$$= z - 1 - \ln(z - B) - \frac{A}{B}\ln\left(1+\frac{B}{z}\right) \tag{10}$$

Soave (Table 1.11):

$$P = \frac{RT}{V-b} - \frac{a\alpha}{V(V+b)} \tag{11}$$

$$\ln\phi = z - 1 - \ln\left[z\left(1-\frac{b}{V}\right)\right] - \frac{a\alpha}{bRT}\ln\left(1+\frac{b}{V}\right) \tag{12}$$

$$= z - 1 - \ln(z - B) - \frac{A}{B}\ln\left(1+\frac{B}{z}\right) \tag{13}$$

Peng-Robinson (Table 1.13):

$$P = \frac{RT}{V-b} - \frac{a\alpha}{V^2 + 2bV - b^2} \tag{14}$$

$$\ln\phi = z - 1 - \ln(z - B) - \frac{A}{2\sqrt{2}B}\ln\left(\frac{z + 2.414B}{z - 0.414B}\right) \tag{15}$$

BWR, BWRS, and *Lee-Kesler* equations are in Tables 1.16, 1.17, and 1.18

Table 3.4. Partial Fugacity Coefficients from Some Equations of State

van der Waals (Table 1.3):

$$P = \frac{RT}{V-b} - \frac{a}{V^2} \quad (1)$$

$$a = (\Sigma y_i \sqrt{a_i})^2 \quad (2)$$

$$b = \Sigma y_i b_i \quad (3)$$

$$\ln \hat{\phi}_i = \frac{b_i}{V-b} - \ln\left[z\left(1-\frac{b}{V}\right)\right] - \frac{2\sqrt{a\,a_i}}{RTV} \quad (4)$$

Virial (Table 1.8):

$$z = 1 + \frac{B}{V} + \frac{C}{V^2} \quad (5)$$

$$\ln \hat{\phi}_i = \frac{2}{V}\sum_k y_k B_{ki} + \frac{1.5}{V^2}\sum_k \sum_1 y_k y_1 C_{k1i} - \ln z \quad (6)$$

$$z = 1 + \frac{BP}{RT} \quad (7)$$

$$\ln \hat{\phi}_i = \frac{P}{RT}\left\{B_{ii} + 0.5\left[\sum_j \sum_k y_j y_k (2\delta_{ji} - \delta_{jk})\right]\right\} \quad (8)$$

$$B = \Sigma\Sigma y_i y_j B_{ij} \quad (9)$$

$$\delta_{ji} = 2B_{ji} - B_{jj} - B_{ii} \quad (10)$$

$$\delta_{jk} = 2B_{jk} - B_{jj} - B_{kk} \quad (11)$$

For two components,

$$\ln \hat{\phi}_1 = \frac{P}{RT}[B_1 + y_2^2(2B_{12} - B_1 - B_2)] \quad (12)$$

$$\ln \hat{\phi}_2 = \frac{P}{RT}[B_2 + y_i^2(2B_{12} - B_1 - B_2)] \quad (13)$$

Redlich-Kwong (Table 1.9):

$$P = \frac{RT}{V-b} - \frac{a}{\sqrt{T}V(V+b)} \quad (14)$$

a and b of the mixture by Eqs. 2 and 3:

$$A = (\Sigma y_i \sqrt{A_i})^2 \quad (15)$$

$$B = \Sigma y_i B_i \quad (16)$$

$$\ln \hat{\phi}_i = \frac{b_i}{b}(z-1) - \ln\left[z\left(1-\frac{b}{V}\right)\right] + \frac{1}{bRT^{1.5}}\left[\frac{ab_i}{b} - 2\sqrt{a\,a_i}\right]\ln\left(1+\frac{b}{V}\right) \quad (17)$$

$$\ln \hat{\phi}_i = \frac{B_i}{B}(z-1) - \ln(z-B) + \frac{A}{B}\left(\frac{B_i}{B} - 2\sqrt{\frac{A_i}{A}}\right)\ln\left(1+\frac{B}{z}\right) \quad (18)$$

When the combining rule is $a = \Sigma\Sigma y_i y_k a_{ik}$ instead of Eq. 2, replace $\sqrt{aa_i}$ in Eq. 17 by

$$\sqrt{a\,a_i} - \sum_k y_k a_{ik} = y_1 a_{i1} + y_2 a_{i2} + y_3 a_{i3} + \ldots \quad (19)$$

Soave (see Table 1.11):

$$(a\alpha)_{ij} = (1 - k_{ij})\sqrt{(a\alpha)_{ii}(a\alpha)_{jj}} \quad (20)$$

(Continued next page)

Table 3.4. *(continued)*

$$k_{ii} = 0 \tag{21}$$

$$a\alpha = \Sigma\Sigma y_i y_j (a\alpha)_{ij} \tag{22}$$

$$b = \Sigma y_i b_i \tag{23}$$

$$A = (a\alpha)P/(RT)^2 \tag{24}$$

$$B = bP/RT \tag{25}$$

$$B_i = b_i P/RT \tag{26}$$

$$\ln \hat{\phi}_i = \frac{b_i}{b}(z-1) - \ln\left[z\left(1-\frac{b}{V}\right)\right] + \frac{a\alpha}{bRT}\left[\frac{b_i}{b} - \frac{2}{a\alpha}\sum_j y_j (a\alpha)_{ij}\right]\ln\left(1+\frac{b}{V}\right) \tag{27}$$

$$= \frac{B_i}{B}(z-1) - \ln(z-B) + \frac{A}{B}\left[\frac{B_i}{B} - \frac{2}{a\alpha}\sum_j y_j (a\alpha)_{ij}\right]\ln\left(1+\frac{B}{z}\right) \tag{28}$$

Peng-Robinson (see Table 1.13):

The symbols have the same meanings as for the Soave equation, although the numerical coefficients in α are slightly different.

$$\ln \hat{\phi}_i = \frac{B_i}{B}(z-1) - \ln(z-B) + \frac{A}{4.828B}\left[\frac{B_i}{B} - \frac{2}{a\alpha}\sum_j y_j (a\alpha)_{ij}\right]\ln\left[\frac{z+2.414B}{z-0.414B}\right] \tag{29}$$

Fugacity coefficients based on the BWR, BWRS, and Lee-Kesler EOS are in Tables 1.16, 1.18, and 1.19.

Example 3.1. Fugacity of Ammonia from Experimental Data, and Comparisons with Predictions from Several Correlations

Data: $T_c = 405.6$ K, $P_c = 111.5$ atm, $\omega = 0.25$.

Compressibilities at 100 C and 200 C over a range of pressures are given in Gmelin [*Handbuch der anorganischen Chemie,* 4 (1935):426]. Use the equation

$$\ln \phi = \int_0^P \frac{z-1}{P}\, dP.$$

The Pitzer & Curl equation in terms of T_r, P_r, and ω is given with Figure 1.10(b).

The Lee-Kesler data are given in Tables E6 and E7 from which

$$\log \phi = (\log \phi)^{(0)} + \omega(\log \phi)^{(1)}.$$

Double interpolation is required in these tables, so only a few values are determined.

The Redlich-Kwong expression for ϕ is given in Table 3.3. The numerical values were found by computer.

Calculations from the experimental zs, Pitzer-Curl, Lee-Kesler, and Redlich-Kwong are summarized as follows. The Lee-Kesler and Redlich-Kwong methods are roughly comparable in checking the z-calculation, but the Pitzer-Curl is quite poor, particularly above 400 atm.

	φ @ 100 C from				*φ @ 200 C from*			
P	*exp.*	*PC*	*LK*	*RK*	*exp.*	*PC*	*LK*	*RK*
100	0.5464	0.6560		0.5268		0.8322		0.8157
200	0.2896	0.4298		0.3126	0.6676	0.6926		0.6538
300	0.2124	0.2818	0.2106	0.2392	0.5466	0.5764	0.5595	0.5402
400	0.1756	0.1847		0.2057	0.4081	0.4797		0.4760
600	0.1426	0.0794	0.1508	0.1781	0.3888	0.3322	0.4237	0.4166
800	0.1299	0.0341		0.1713	0.3547	0.2301		0.3980
1000	0.1254	0.0147	0.1441	0.1742	0.3404	0.1593	0.3997	0.3985

Experimental compressibilities of ammonia and calculated fugacity coefficients as function of pressure at 100 C and 200 C. These data are excerpted for the preceding table.

Example 3.1. *(continued)*

100 C				200 C			
P, atm	z	$\frac{-10^4(z-1)}{P}$	ϕ	*P*, atm	z	$\frac{-10^4(z-1)}{P}$	ϕ
0	1	(0)	1	0	1	(0)	1
1.374	0.9928	52	0.9964	1.747	0.9964	21	0.9982
3.537	0.9830	48	0.9857	4.519	0.9913	19	0.9929
5.832	0.9728	47		7.489	0.9856	19	
8.632	0.9599	46		11.153	0.9786	19	
11.352	0.9468	47		14.766	0.9716	19	
14.567	0.9315	47	0.9362	19.098	0.9635	19	
19.109	0.9085	48		25.361	0.9513	19	0.9544
22.84	0.8890	49		30.64	0.9409	19	
26.12	0.8714	49		35.40	0.9318	19	
30.47	0.8471	50	0.8660	41.89	0.9188	19	
33.21	0.8310	51		46.12	0.9104	19	
36.47	0.8111	52		51.29	0.9000	19	0.9085
40.41	0.7864	53		57.74	0.8865	20	
45.19	0.7538	54	0.8016	66.05	0.8693	20	
51.09	0.7102	57		77.13	0.8460	20	
58.28	0.6481	60	0.7440	92.57	0.8122	20	0.8365
100	0.1158	88	0.5464	100			
200	0.2221	39	0.2896	200	0.5506	22	0.6676
300	0.3202	23	0.2124	300	0.4615	18	0.5466
400	0.4145	15	0.1756	400	0.4949	13	0.4681
500	0.5060	10.0	0.1550	500	0.5567	8.9	0.4195
600	0.5955	6.7	0.1426	600	0.6211	6.3	0.3888
700	0.6828	4.5	0.1348	700	0.6826	4.5	0.3684
800	0.7684	2.9	0.1299	800	0.7546	3.1	0.3547
900	0.8507	1.7	0.1269	900	0.8235	2.0	0.3457
1,000	0.9333	0.7	0.1254	1,000	0.8915	1.1	0.3404
1,100	1.0140	−0.1	0.1251	1,100	0.9590	0.4	0.3379

$$\ln \hat{\phi}_i = \frac{1}{RT}\int_0^P \left(\bar{V}_i - \frac{RT}{P}\right) dP = \frac{1}{RT}\int_0^P \bar{V}_i^{\text{ex}}\, dP \tag{3.41}$$

$$= \int_0^P \left(\frac{\bar{z}-1}{P}\right) dP. \tag{3.42}$$

All integrands vanish at $P = 0$.

When measurements of the volume or compressibility of the mixture are available, the partial molal quantities, $\bar{V}_i$ or $\bar{z}_i$, may be calculated with Eq. 2.56. For a binary mixture, for instance,

$$\bar{z}_1 = z + (1 - x_1)\left(\frac{\partial z}{\partial x_1}\right), \tag{3.43}$$

$$\bar{z}_2 = z - x_1\left(\frac{\partial z}{\partial x_1}\right). \tag{3.44}$$

These formulas are used in Example 2.5, and an application of Eq. 2.56 is made in Problem 3.20.

Individual partial fugacities and fugacity coefficients are derivable from those of the mixture by the usual rules for partial molal quantities:

$$\ln \frac{\hat{f}_i}{y_i} = \left(\frac{\partial\, n \ln f}{\partial n_i}\right)_{TPn_{j\neq i}}, \tag{3.45}$$

$$\ln \hat{\phi}_i = \left(\frac{\partial n \ln \phi}{\partial n_i}\right)_{TPn_{j\neq i}}. \tag{3.46}$$

With some equations of state, it may be easier to find an expression for the fugacity coefficient of the mixture first, and then to find the partial fugacity coefficients by application of Eq. 3.45 or 3.46.

Since pressure-explicit EOS are the more common types, a useful relation for finding partial fugacity coefficients is

$$RT \ln \hat{\phi}_i = \int_V^\infty \left[\frac{\partial P}{\partial n_i} - \frac{RT}{V}\right] dV - RT \ln z, \tag{3.47}$$

which is derived in Example 3.3 and applied with the Redlich-Kwong equation in Example 3.4.

Example 3.2. Total and Partial Fugacity Coefficients of Binary Mixtures with the Two Forms of the *B*-Truncated Virial Equation

The two forms of the virial equation are

$$z = \frac{PV}{RT} = 1 + BP/RT \quad (1)$$

$$= 1 + B/V \quad (2)$$

From Eq. 1,

$$\ln \phi = \frac{1}{RT}\int_0^P \frac{z-1}{P}\,dP = \frac{BP}{RT} \quad (3)$$

For Eq. 2, start with

$$RT\,d\ln f = V\,dP = d(PV) - P\,dV. \quad (4)$$

Subtract $RT\,d\ln P$ from each side and rearrange,

$$d\ln\phi = dz - \frac{P}{RT}dV - d\ln P. \quad (5)$$

Substitute

$$\frac{P}{RT} = \frac{1}{V} + \frac{B}{V^2}, \quad (6)$$

and integrate between the limits P and 0, with the result

$$\ln\phi = z - 1 - \ln z + \frac{B}{V} = \frac{2B}{V} - \ln\left(1 + \frac{B}{V}\right)$$

$$= 2(z-1) - \ln z. \quad (7)$$

The partial fugacity coefficients are obtainable from Eqs. 3 and 7 with

$$\ln\hat{\phi}_i = \ln\phi + (1 - y_i)\left(\frac{\partial \ln\phi}{\partial y_i}\right), \quad i = 1, 2. \quad (8)$$

With the second virial coefficient dependence on the composition,

$$B = y_1B_{11} + y_2B_{22} + y_1y_2\delta_{12}, \quad (9)$$

$$\delta_{12} = 2B_{12} - B_{11} - B_{22}, \quad (10)$$

the resulting partial fugacity coefficients are:
From Eq. 1:

$$\ln\hat{\phi}_1 = \frac{P}{RT}(B_{11} + y_2^2\delta_{12}), \quad (11)$$

$$\ln\hat{\phi}_2 = \frac{P}{RT}(B_{22} + y_1^2\delta_{12}). \quad (11a)$$

From Eq. 2:

$$\ln\hat{\phi}_1 = \frac{2}{V}(y_1B_{11} + y_2B_{12}) - \ln z, \quad (12)$$

$$\ln\hat{\phi}_2 = \frac{2}{V}(y_1B_{12} + y_2B_{22}) - \ln z, \quad (12a)$$

Plots of Eqs. 11 and 12 are shown. They differ appreciably at higher pressures. On the plots the values of the parameters are $B_{11} = 0.5$, $B_{22} = 0.1$, and $B_{12} = 0.4$. Extensions of these results to multicomponent mixtures are shown in Tables 3.3 and 3.4. Empirical correlations of second virial coefficients B usually are based on the *B*-truncated form, Eq. 1, so Eq. 11 is the preferred one for the fugacity coefficients.

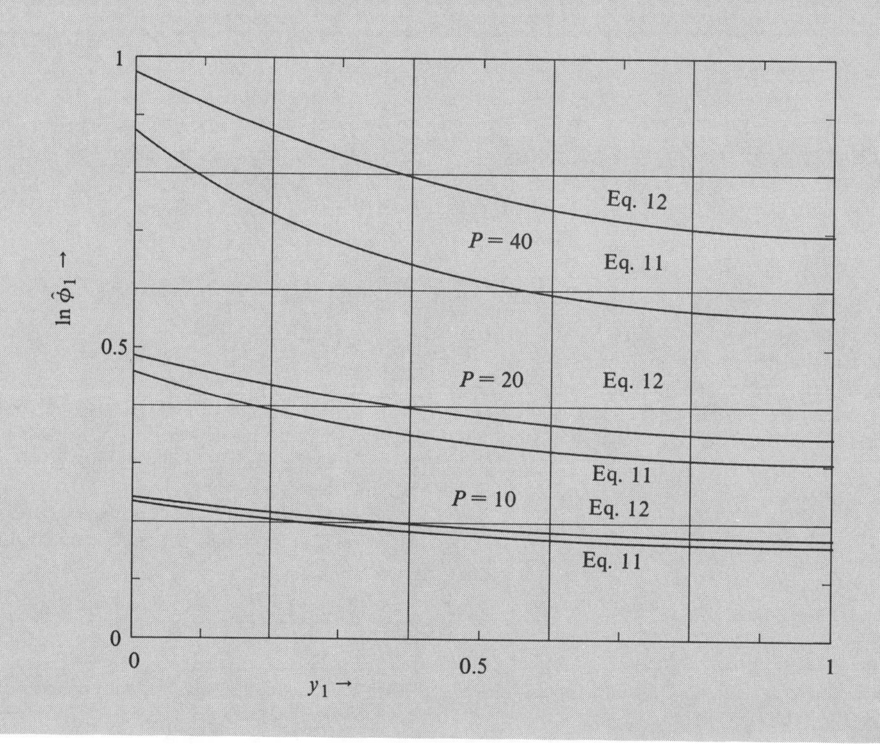

When the composition is expressed in terms of mol fractions, Eq. 2.56 assumes these forms for fugacity and fugacity coefficients,

$$\ln \frac{\hat{f}_i}{x_i} = \ln f - \sum_k x_k \left(\frac{\partial \ln f}{\partial x_k}\right)_{TPx_{j\neq i,k}}, \quad (3.48)$$

$$\ln \phi_i = \ln \phi - \sum_k x_k \left(\frac{\partial \ln \phi}{\partial x_k}\right)_{TPx_{j\neq i,k}} \quad (3.49)$$

When evaluating the properties of component i, the work is simplified by replacing $x_i = 1 - \Sigma x_{j\neq i}$ before taking derivatives. This rule is applied in Example 3.5.

Application of Eqs. 3.48 and 3.49 to a binary mixture leads to the result,

$$\left(\frac{\partial \ln \phi}{\partial x_1}\right) = \ln \hat{\phi}_1 - \ln \hat{\phi}_2, \quad (3.50)$$

of which the graphical solution is known as the *method of tangent intercepts,* and is illustrated in Figure 3.4. When the tangents can be drawn accurately, this is a rapid method of finding individual fugacity coefficients from that of the mixture.

3.3.1. Derivatives

Equations for the temperature and pressure derivatives of partial fugacity and fugacity coefficients are analogous to Eqs. 3.32–3.34. Thus,

$$\left(\frac{\partial \ln \hat{f}_i}{\partial T}\right)_{Px} = -\frac{\bar{H}_i - H_i}{RT^2} = -\frac{\bar{H}_i^{ex}}{RT^2}, \quad (3.51)$$

$$\left(\frac{\partial \ln \hat{\phi}_i}{\partial T}\right)_{Px} = -\frac{\bar{H}_i^{ex}}{RT^2}, \quad (3.52)$$

$$\left(\frac{\partial \ln \hat{f}_i}{\partial P}\right)_{Tx} = \frac{\bar{V}_i}{RT}, \quad (3.53)$$

$$\left(\frac{\partial \ln \hat{\phi}_i}{\partial P}\right)_{Tx} = \frac{\bar{V}_i^{ex}}{RT}. \quad (3.54)$$

The value of the integral of Eq. 3.42 depends implicitly on the composition as well as on the temperature and pressure, so the variation of the fugacity coefficient with composition is obtainable by differentiating the result of that integration. For the virial equation, the result of the integration is given as Eq. 11 of Example 3.2, so the corresponding derivative is

$$\left(\frac{\partial \ln \hat{\phi}_1}{\partial y_1}\right) = -2y_2\delta_{12}\frac{P}{RT}. \quad (3.55)$$

An explicit derivative from the Redlich-Kwong equation is much more difficult to derive. In Eq. 14 of Example 3.4, for instance, the variation of z and V with composition also must be taken into account. Numerical derivatives can be obtained readily, however. For example, in Problem 3.14 these results are obtained numerically:

y_{NH_3}	$\hat{\phi}_{NH_3}$	$10^6(\partial\hat{\phi}_1/\partial y_1)$
0.4	0.845856	−14.72
0.5	0.845877	443.11
0.6	0.845942	880.94

Formally, the effects of changes in temperature, pressure, and composition on a general property, M, are

$$d(nM) = n\left(\frac{\partial M}{\partial T}\right)_{Pn} dT + n\left(\frac{\partial M}{\partial P}\right)_{Tn} dP + \Sigma \bar{M}_i\, dn_i \quad (3.56)$$

On combining these two equations, the result is

$$\Sigma n_i d\bar{M}_i = n\left(\frac{\partial M}{\partial T}\right)_{Pn} dT + n\left(\frac{\partial M}{\partial P}\right)_{Tn} dP, \quad (3.58)$$

or

$$\Sigma x_i d\bar{M}_i = \left(\frac{\partial M}{\partial T}\right)_{Pn} dT + \left(\frac{\partial M}{\partial P}\right)_{Tn} dP, \quad (3.59)$$

which is the Gibbs-Duhem equation. Substitution of Eqs. 3.51–3.54 gives

$$\Sigma x_i d \ln(\hat{f}_i/x_i) = -\frac{H^{ex}}{RT^2} dT + \frac{V}{RT} dP \quad (3.60)$$

$$\Sigma x_i d \ln \phi_i = -\frac{H^{ex}}{RT^2} dT + \frac{V^{ex}}{RT} dP. \quad (3.61)$$

At constant temperature and pressure, the preceding equations assume the useful forms

$$\Sigma x_i\, d \ln(\hat{f}_i/x_i) = 0, \quad (3.62)$$

$$\Sigma x_i\, d \ln \phi_i = 0, \quad (3.63)$$

which are commonly applied to binary systems for calculating one of the fugacity coefficients when the other is known as a function of composition:

$$\ln \hat{\phi}_2 = \ln \phi_2 - \int_{x_1=0}^{x_1} \frac{x_1}{1-x_1} d \ln \hat{\phi}_1. \quad (3.64)$$

3.3.2. Experimental Data

One of the principal uses of equations of state is for the calculation of fugacities, so much of the evaluation of these equations has been with experimental fugacity data. Some recent work may be mentioned. Antezana & Cheh (1975, 1976) measured fugacities in mixtures of hydrogen, ammonia, and propane with the aid of a semipermeable membrane. Of the equations that they checked, the BWR, a third-order virial, and the Redlich-Kwong had substantially the same accuracy, within 2–3%. An analysis of a wide range of literature data on hydrocarbons by Lin & Daubert (1978) demonstrated that the Soave and Lee-Kesler methods were generally superior for fugacity predictions, usually within 1–2% of experiment. The literature on this subject is abundant; many other references are cited in Chapters 1 and 6.

Example 3.3. Derivation of Eq. 3.47 for the Partial Fugacity Coefficient in Terms of Integration of an EOS with Respect to Volume

In the definition of the partial fugacity,

$$RT\, d\ln \hat{f}_i = \bar{V}_i\, dP = \left(\frac{\partial nV}{\partial n_i}\right) dP, \tag{1}$$

the differential of pressure can be eliminated with the aid of the chain rule for partial differentiation (Appendix B),

$$\left(\frac{\partial nV}{\partial n_i}\right)\left(\frac{\partial n_i}{\partial P}\right)\left(\frac{\partial P}{\partial nV}\right) = -1, \tag{2}$$

from which

$$\left(\frac{\partial nV}{\partial n_i}\right) dP = \bar{V}_i\, dP = -\left(\frac{\partial P}{\partial n_i}\right) d(nV). \tag{3}$$

Substituting into Eq. 1 and adding $RT\, d\ln(V/RT)$ to each side,

$$RT\, d\ln\frac{\hat{f}_i V}{RT} = -\left(\frac{\partial P}{\partial n_i}\right) d(nV) + RT\, d\ln\frac{V}{RT} \tag{4}$$

$$= \left[-\left(\frac{\partial P}{\partial n_i}\right) + \frac{RT}{nV}\right] d(nV). \tag{5}$$

Note that

$$\lim_{V\to\infty} \ln\frac{\hat{f}_i V}{RT} = \lim_{P\to 0} \ln\frac{\hat{f}_i}{P} = \ln y_i. \tag{6}$$

Integrating Eq. 5,

$$RT\ln\frac{\hat{f}_i}{y_i} = \int_V^\infty \left(\frac{\partial P}{\partial n_i} - \frac{RT}{nV}\right) d(nV) - RT\ln\frac{V}{RT}. \tag{7}$$

Subtract $RT\ln P$ from each side, with the final result:

$$RT\ln\hat{\phi}_i = RT\ln\frac{\hat{f}_i}{y_i P}$$

$$= \int_V^\infty \left(\frac{\partial P}{\partial n_i} - \frac{RT}{nV}\right) d(nV) - RT\ln z. \tag{8}$$

Another useful form may be obtained by noting that $P = zRT/V$ and

$$\left(\frac{\partial P}{\partial n_i}\right)_{T,nV,n_j} = \frac{P}{nz}\left(\frac{\partial (nz)}{\partial n_i}\right)_{T,nV,n_j}. \tag{9}$$

Substituting this expression into Eq. 8 and rearranging leads to the form

$$\ln\hat{\phi}_i = \int_V^\infty \left(\frac{\partial (nz)}{\partial n_i} - 1\right)\frac{dV}{V} - \ln z. \tag{10}$$

Eq. 8 is applied in Example 3.4.

Example 3.4. Partial Fugacity Coefficients from the Redlich-Kwong Equation of State

Eq. 8 of Example 3.3 will be applied.

$$RT\ln\hat{\phi}_i = \int_V^\infty \left[\frac{\partial P}{\partial n_i} - \frac{RT}{\underline{V}}\right] d\underline{V} - RT\ln z. \tag{1}$$

In terms of the total volume, $\underline{V} = nV$, the R-K equation is

$$P = \frac{nRT}{\underline{V} - nb} - \frac{n^2 a}{\sqrt{T}\,\underline{V}(\underline{V} + nb)}. \tag{2}$$

The required derivative is

$$\left(\frac{\partial P}{\partial n_i}\right)_{T\underline{V}n_j} = \frac{RT}{\underline{V} - nb} + \frac{nRT}{(\underline{V} - nb)^2}\frac{\partial (nb)}{\partial n_i} - \frac{1}{\sqrt{T}\,\underline{V}(\underline{V} + nb)}\frac{\partial (n^2 a)}{\partial n_i} + \frac{n^2 a}{\sqrt{T}\,\underline{V}(\underline{V} + nb)^2}\frac{\partial (nb)}{\partial n_i}. \tag{3}$$

Since the derivatives on the RHS of Eq. 3 are independent of $\underline{V}$, the integration of Eq. 1 may proceed directly.

$$RT\ln\hat{\phi}_i = -RT\ln\frac{\underline{V} - nb}{\underline{V}} + \left[\frac{nRT}{\underline{V} - nb} - \frac{na}{b\sqrt{T}(\underline{V} + nb)} + \frac{a}{b^2\sqrt{T}}\ln\frac{\underline{V} + nb}{\underline{V}}\right]\frac{\partial (nb)}{\partial n_i} - \frac{1}{nb\sqrt{T}}\ln\left(\frac{\underline{V} + nb}{\underline{V}}\right)\frac{\partial (n^2 a)}{\partial n_i} - RT\ln z. \tag{4}$$

At the upper limit the integral was zero. Take the following combining rules:

$$a = (\Sigma y_i\sqrt{a_i})^2 = \frac{1}{n^2}(\Sigma n_i\sqrt{a_i})^2, \tag{5}$$

$$b = \Sigma y_i b_i = \frac{1}{n}\Sigma n_i b_i. \tag{6}$$

The various derivatives are,

$$\frac{\partial (n^2 a)}{\partial n_i} = 2\sqrt{a_i}\,\Sigma n_i\sqrt{a_i} = 2n\sqrt{a\,a_i}, \tag{7}$$

(Continued next page)

Example 3.4. *(continued)*

$$\frac{\partial(nb)}{\partial n_i} = b_i. \quad (8)$$

On substituting for the derivatives into Eq. 4 and canceling out n,

$$RT \ln \hat{\phi}_i = -RT \ln \frac{V-b}{V} + \left[\frac{RT}{V-b} - \frac{a}{b\sqrt{T}(V+b)} + \frac{a}{b^2\sqrt{T}} \ln \frac{V+b}{V}\right] b_i - \frac{2\sqrt{a\,a_i}}{b\sqrt{T}} \ln \frac{V+b}{V} - RT \ln z. \quad (9)$$

The substitution

$$\frac{b}{V-b} - \frac{a}{RT^{1.5}(V+b)} = z - 1 \quad (10)$$

and a final rearrangement give

$$\ln \hat{\phi}_i = -\ln\left[z\left(1 - \frac{b}{V}\right)\right] + \frac{b_i}{b}(z-1) + \frac{1}{bRT^{1.5}}\left(\frac{ab_i}{b} - 2\sqrt{a\,a_i}\right) \ln\left(1 + \frac{b}{V}\right). \quad (11)$$

If $a_{ij} \neq \sqrt{a_i a_j}$, the combining rule of Eq. 5 becomes

$$a = \frac{1}{n^2} \Sigma\Sigma n_i n_j a_{ij}. \quad (12)$$

The derivative is

$$\frac{\partial(n^2 a)}{\partial n_i} = 2n \sum_k y_i a_{ik}. \quad (13)$$

The fugacity coefficient becomes

$$\ln \hat{\phi}_i = -\ln\left[z\left(1 - \frac{b}{V}\right)\right] + \frac{b_i}{b}(z-1) + \frac{1}{bRT^{1.5}}\left(\frac{ab_i}{b} - 2\sum_j y_i a_{ij}\right) \ln\left(1 + \frac{b}{V}\right). \quad (14)$$

To obtain the equation for a pure substance, place $b_i = b$ and $a_i = a$:

$$\ln \phi = -\ln\left[z\left(1 - \frac{b}{V}\right)\right] + z - 1 - \frac{a}{bRT^{1.5}} \ln\left(1 + \frac{b}{V}\right). \quad (15)$$

Example 3.5. Equations for the Partial Fugacities of a Four-Component Mixture, Given an Equation for the Total Fugacity as a Function of Composition

The method of Section 2.6 will be applied. To evaluate $\hat{f}_i$, the mol fraction x_i is eliminated from the equation for f before making the required differentiations, with

$$x_i = 1 - \sum_{j \neq i} x_j.$$

The assumed equation of the fugacity of the mixture and the several required equivalents are:

$$\begin{aligned}
\ln f &= ax_1^2 + bx_2^2 + cx_3^2 + dx_4^2 \\
&= a(1 - x_2 - x_3 - x_4)^2 + bx_2^2 + cx_3^2 + dx_4^2 \\
&= ax_1^2 + b(1 - x_1 - x_3 - x_4)^2 + cx_3^2 + dx_4^2 \\
&= ax_1^2 + bx_2^2 + c(1 - x_1 - x_2 - x_4)^2 + dx_4^2 \\
&= ax_1^2 + bx_2^2 + cx_3^2 + d(1 - x_1 - x_2 - x_3)^2.
\end{aligned}$$

The equations for the partial fugacities are:

$$\ln \frac{\hat{f}_1}{x_1} = \ln f - x_2\left(\frac{\partial f}{\partial x_2}\right)_{x_3x_4} - x_3\left(\frac{\partial f}{\partial x_3}\right)_{x_2x_4} - x_4\left(\frac{\partial f}{\partial x_4}\right)_{x_2x_3} = \ln f - 2x_2(-ax_1 + bx_2) - 2x_3(-ax_1 + cx_3) - 2x_4(-ax_1 + dx_4),$$

$$\ln \frac{\hat{f}_2}{x_2} = \ln f - 2x_1(ax_1 - bx_2) - 2x_3(-bx_2 + cx_3) - 2x_4(-bx_2 + dx_4),$$

$$\ln \frac{\hat{f}_3}{x_3} = \ln f - 2x_1(ax_1 - cx_3) - 2x_2(bx_2 - cx_3) - 2x_4(-cx_3 + dx_4),$$

$$\ln \frac{\hat{f}_4}{x_4} = \ln f - 2x_1(ax_1 - dx_4) - 2x_2(bx_2 - dx_4) - 2x_3(cx_3 - dx_4).$$

Example 3.6. Calculation of Fugacity Coefficients in Mixtures of Carbon Dioxide and Propane

The conditions are 100 F and 200 psia. The virial coefficients are $B_{11}=-1.8$, $B_{22}=-5.87$, $B_{12}=-3.36$. The equations of Tables 3.3 and 3.4 will be used.

$$\delta_{12}=2B_{12}-B_{11}-B_{22}=0.95,$$

$$B=-1.8y_1-5.87y_2+0.95y_1y_2,$$

$$P/RT=200/10.73(559.6)=0.0333,$$

$$\ln\phi=BP/RT=0.0333B,$$

$$\ln\hat{\phi}_1=\frac{P}{RT}(B_{11}+y_2^2\delta_{12}),$$

$$\ln\hat{\phi}_2=\frac{P}{RT}(B_{22}+y_1^2\delta_{12}).$$

The calculations are summarized in the table. The coefficients depend quite substantially on the compositions, and the Lewis-Randall Rule, $\phi_i/\hat{\phi}_i=1$, certainly is not valid in this case.

y_{CO_2}	ϕ	$\hat{\phi}_{CO_2}$	$\hat{\phi}_{C_3H_8}$
0.0	.8224	.6958	.8224
.1	.8360	.7234	.8496
.2	.8493	.7506	.8760
.3	.8623	.7773	.9015
.4	.8749	.8035	.9260
.5	.8871	.8290	.9493
.6	.8989	.8536	.9714
.7	.9103	.8774	.9921
.8	.9213	.9001	1.0113
.9	.9318	.9216	1.0290
1.0	.9418	.9418	1.0449

```
10  FOR Y=0 TO 1 STEP .1
20  B=-1.8*Y^2-6.72*Y*(1-Y)-5.87*(1-Y)^2
30  B1=-2*(1.8*Y+3.36*(1-2*Y)-5.87*(1-Y))
40  F=EXP(.0333*B)
50  F1=EXP(.0333*(B-(1-Y)*B1))
60  F2=EXP(.0333*(B+Y*B1))
70  PRINT USING 80 ; Y,F,F1,F2
80  IMAGE D.D,2X,D.DDDD,2X,D.DDDD,2X,
    D.DDDD
90  NEXT Y
100 END
```

3.4. IDEAL MIXTURES

An ideal mixture is one whose properties are derivable from those of the pure components in the same way as those of a mixture of ideal gases. This is not to say that individual components are necessarily ideal gases since even liquid mixtures and solid solutions may behave ideally in the sense used here. In terms of the concept of excess properties described in Section 2.9, an ideal mixture is one whose excess properties are zero. The ideal relations for volume and enthalpy, for example, are:

$$V^{ex}=V-\Sigma x_iV_i=\Sigma x_i(\bar{V}_i-V_i)=0, \tag{3.65}$$

$$\bar{V}_i=V_i, \tag{3.66}$$

$$H^{ex}=H-\Sigma x_iH_i=\Sigma x_i(\bar{H}_i-H_i)=0, \tag{3.67}$$

$$\bar{H}_i=H_i, \tag{3.68}$$

but, as explained in Section 2.8, the ideal relations for entropy and related quantities are

$$S^{ex}=S-\Sigma x_iS_i+R\Sigma x_i\ln x_i=0, \tag{3.69}$$

$$\bar{S}_i=S_i+R\ln x_i \tag{3.70}$$

$$G^{ex}=G-\Sigma x_iG_i-RT\Sigma x_i\ln x_i=0, \tag{3.71}$$

$$\bar{G}_i=G_i-RT\ln x_i. \tag{3.72}$$

Plots of entropy and Gibbs energy of an ideal mixture are shown as Figure 3.1. Excess properties of real mixtures often are substantially different from zero, either positive or negative, depending on the concentration range and the chemical nature; a few examples are given in Chapter 4. Ideal behavior of partial pressures is represented by Raoult's law; some deviations from that ideality are illustrated by Figure 3.2.

From Eq. 3.66 and the definitions, Eqs. 3.4 and 3.5, it follows for ideal mixtures that

$$\hat{\phi}_i=\phi_i, \tag{3.73}$$

and

$$\hat{f}_i=x_if_i \tag{3.74}$$

These relations are called the Lewis-Randall Rule, which in effect states that the partial fugacity is related to that of the pure component in the same way as the partial pressure is to the vapor pressure. Since it is easier to figure the pure component fugacity than the partial fugacity, the L-R Rule is often used.

No general statement can be made about possible errors when taking $\hat{\phi}_i=\phi_i$, except that they can be large even at modest pressures. For instance, in the CO_2-propane system of Example 3.6, the errors approach 25%. The requirements for ideality can be interpreted, in part, by the van der Waals equation, for which the fugacity coefficients are (from Tables 3.3 and 3.4).

$$\ln\phi=\frac{b}{V-b}-\frac{2a}{RTV}-\ln\left[z\left(1-\frac{b}{V}\right)\right], \tag{3.74a}$$

$$\ln\hat{\phi}_i=\frac{b_i}{V-b}-\frac{2\sqrt{aa_i}}{RTV}-\ln\left[z\left(1-\frac{b}{V}\right)\right]. \tag{3.74b}$$

At a given temperature and pressure, $\phi=\hat{\phi}_i$ generally only when $b_i=b$ and $a_i=a$—that is, when all molecules of the mixture are of about the same size and are chemically similar.

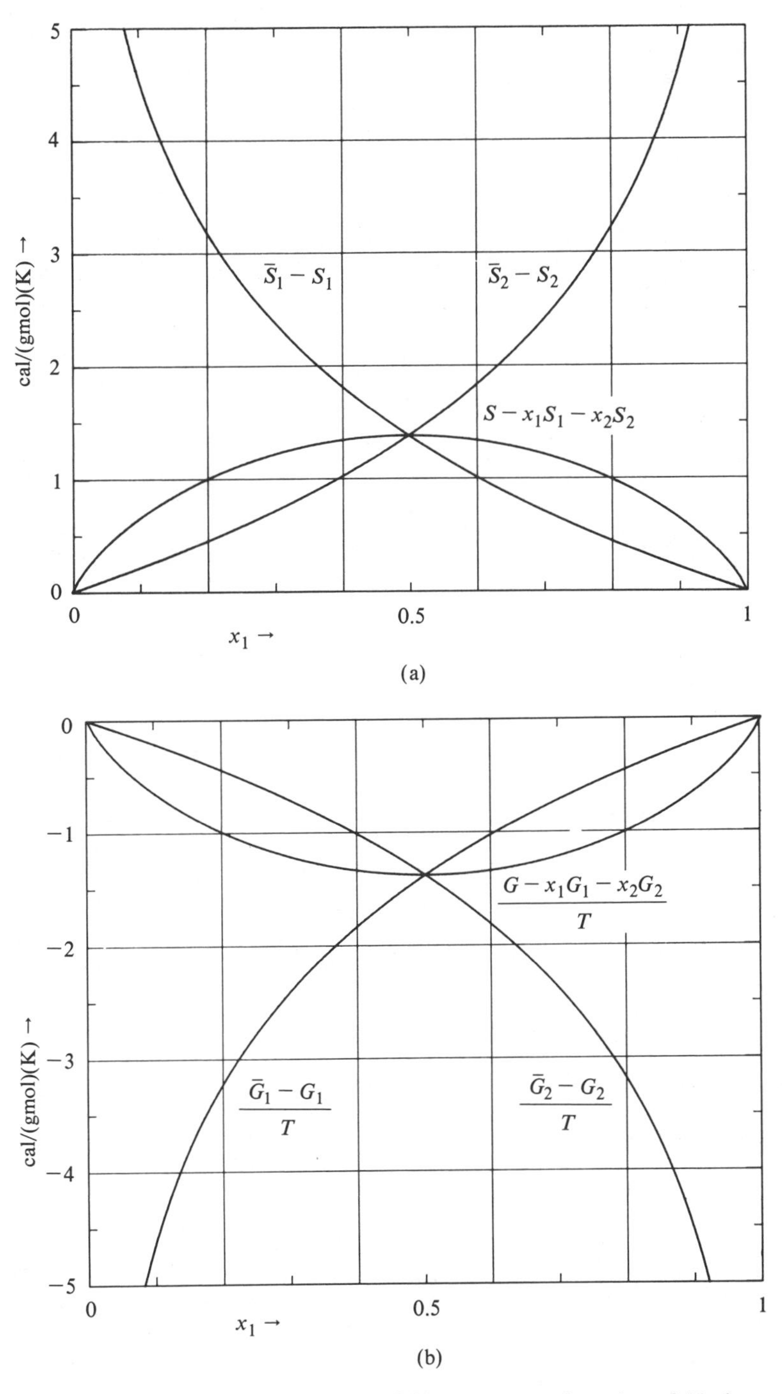

Figure 3.1 Changes in entropy and Gibbs energy on formation of ideal solutions. For the mixture, $S - \Sigma x_iS_i = R\Sigma x_i \ln x_i$ and $G - \Sigma x_iG_i = -RT\Sigma x_i \ln x_i$. For individual components, $\bar{S}_i - S_i = R \ln x_i$ and $\bar{G}_i - G_i = -RT \ln x_i$. Since $H = \Sigma x_iH_i$ for ideal solutions, $(G - \Sigma x_iG_i)/T = -(S - \Sigma x_iS_i)$.

Also, a_i and b_i approach the mixture values for those components present in large concentration, say y_i in the range of 0.9 or so. Finally, at low pressures the terms with V in them disappear and $\phi \rightarrow \hat{\phi} \rightarrow 1/z$. Except for these three limiting cases, there is rarely justification for use of the L-R Rule, although it is useful for starting iterative calculations that need an estimate of ϕ_i.

Categories other than the most general case of ideality of mixtures are often recognized. When the excess enthalpy is zero, the solution is called *athermal.* Such behavior is

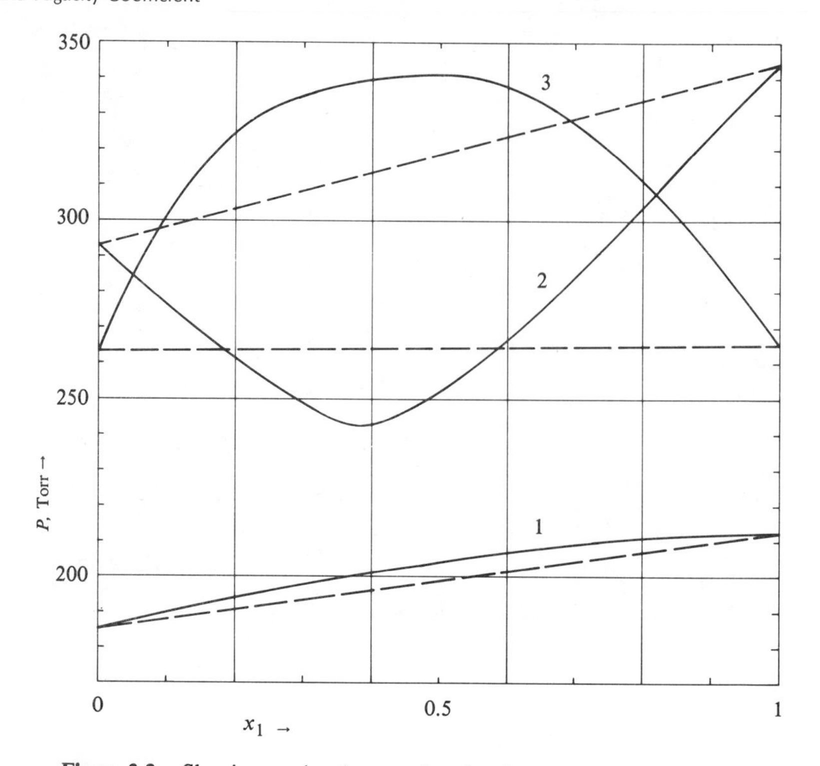

Figure 3.2. Showing varying degrees of conformity to Raoult's Law, which is represented by the broken straight lines.

Curve 1. Moderate agreement. Carbon tetrachloride + cyclohexane at 40 C.
Curve 2. Marked negative deviation. Acetone + chloroform at 35.3 C.
Curve 3. Marked positive deviation. Methylal + carbon disulfide at 16.5 C.

approximated by chemically similar substances usually of much the same size, such as some classes of hydrocarbons. There are theoretical methods for finding the excess entropies of some types of mixtures (see for example King 1969, p. 509); then the excess Gibbs energy of an athermal mixture is found from $G^{ex} = TS^{ex}$.

The class of solutions of which the excess entropy is zero is called *regular*. Substances forming regular solutions generally are free of strong associative forces, such as strong polarity, hydrogen bonding, and chemical association of like and unlike molecules. The important simplification in such cases is that the excess Gibbs energy equals the enthalpy of mixing, a quantity that is readily determinable by standard calorimetric measurements.

$$G^{ex} = H^{ex} - TS^{ex} \rightarrow H^{ex} = H - \Sigma x_i H_i, \tag{3.75}$$

$$\bar{G}^{ex} = \bar{H}^{ex} - T\bar{S}^{ex} \rightarrow \bar{H}^{ex} = \bar{H}_i - H_i. \tag{3.76}$$

Theoretical deductions for regular solutions are made in Chapter 4. A few values of excess properties are shown in Table 2.7, where several of the solutions could be considered regular, and other cases are cited in Chapter 4. A continuing compilation of excess thermodynamic data is edited by Kehiaian & Zwolinski (*Selected Data on Mixtures,* Thermodynamics Research Center, Texas A&M University). A complete literature survey of this subject is that of Wisniak & Tamir (1978).

For mixtures of molecules that differ considerably in size, such as a solvent and a polymer of high molecular weight, a modified excess entropy that sometimes is nearly zero may be defined in terms of the molar volume fraction rather than mol fraction. That is,

$$\Delta S_{mixing} = R\Sigma x_i \ln x_i \tag{3.77}$$

is replaced by

$$\Delta S_{mixing} = R\Sigma x_i \ln \Phi_i, \tag{3.78}$$

where

$$\Phi_i = x_i V_i / \Sigma x_i V_i. \tag{3.79}$$

Moreover, in some work, interactions between molecules are expressed in terms of surface-area fractions, for instance in the UNIQUAC equation and the related UNIFAC procedure.

3.5. LIQUID AND SOLID PHASES

The convenience of a reference or standard state is manifested in connection with evaluation of Gibbs energy changes or of phase equilibria. The condition of equality of fugacities of individual components—in all phases 1, 2, 3, . . . , at equilibrium

$$\hat{f}_i^{(1)} = \hat{f}_i^{(2)} = \ldots, \tag{3.80}$$

or

$$\ln \hat{f}_i^{(1)} = \ln \hat{f}_i^{(2)} = \ldots. \tag{3.81}$$

Table 3.5. Triple and Melting Points of Some Organic Substances

The triple point data are taken from Zernike (1955) and the melting points from Lange's *Handbook of Chemistry* (1973); the two sources are not necessarily consistent.

		Triple Point	
Substance	*Melting Temperature*	*Temperature, C*	*Pressure, Torr*
Naphthalene	80.2	80	7.4
Carbon tetrachloride	−22.96	−22.6	7.9
Acetic acid	16.6	16.7	9.7
Benzene[a]	5.5	5.47	36.1
Ammonia	−77.74	−77.7	43.4
Methane	−182.5	−183.2	70
Carbon monoxide	−207	−205.2	115.1
Camphor	179.5	175	354
Cyanogen	−27	−27.92	555
Acetylene	−81.5	−81	950
Carbon dioxide	−56	−56.4	3,906

[a]The fusion temperature of benzene is 32.5 C @ 1,000 bar, 114.6 C @ 5,000 bar, and 190.5 C @ 10,000 bar (Bridgman).

may be written to involve the fugacity of a reference state, f_i^0:

$$\ln(\hat{f}_i^{(1)}/f_i^0) = \ln(\hat{f}_i^{(2)}/f_i^0) = \ldots \tag{3.82}$$

For a change of state,

$$\Delta\bar{G}_i = RT\ln[(\hat{f}_i)_2/(\hat{f}_i)_1] = RT\ln\left[\frac{(\hat{f}_i)_2/f_i^0}{(\hat{f}_i)_1/f_i^0}\right]. \tag{3.83}$$

As long as the same reference state is used for a particular substance in a series of interrelated changes of state, all thermodynamic relations remain unchanged when the fugacities are replaced by their corresponding ratios to standard-state fugacities.

Although the choice is arbitrary and may vary from problem to problem, certain reference states have been found especially convenient and are widely adopted. Specification of a standard state includes the pressure and the physical state of the substance, usually the one that is stable at the system temperature, but not the temperature itself, so the standard state does depend on the temperature. For gases the natural reference state is unit fugacity at the temperature of the system, which in most cases is very nearly 1 atm. The relation is sketched in Figure 3.3a.

For condensed phases, the standard state naturally is chosen as one at which the fugacity is readily calculable. When the vapor pressure is known, the fugacity of the condensed phase may be taken as that of the vapor phase in equilibrium with it at the temperature of the system. Adjustment of that fugacity to the system pressure is given by

$$\ln\frac{f}{f^{sat}} = \int_{P^{sat}}^{P}\frac{V}{RT}\,dP. \tag{3.84}$$

Use of Eq. 3.84 is made in Examples 3.7 and 3.10.

For solids, vapor pressure data are not always available; but usually the melting point and sometimes the triple point are known and are nearly the same. Some data are given in Table 3.5. The pressure at these conditions is the vapor pressure, which can be obtained by extrapolation of a vapor-pressure equation below the freezing point. Subsequent adjustment of the fugacity of the solid to the system pressure and temperature is made with the equation,

$$\ln\frac{f}{f_{tp}} = \int_{P_{tp}}^{P}\frac{V}{RT}\,dP + \int_{T_{tP}}^{T}\frac{H^{id}-H}{RT^2}\,dT. \tag{3.85}$$

At moderate pressures particularly, $H^{id} - H$ may be taken as the heat of sublimation. When this has not been measured, it may be approximated as the sum of heats of fusion and vaporization, or it may be figured from vapor-pressure data of the solid with the Clausius-Clapeyron equation. A table of Antoine constants of many solids from which this calculation can be made is the *CRC Handbook of Chemistry & Physics* (pp. C722-3, 1979), for instance. Example 3.8 finds the fugacity of a liquid over a range of temperature and Example 3.9 is an application of Eq. 3.85.

3.5.1. Liquid and Solid Solutions

There are two commonly used standard states:

1. The standard state is the pure liquid or solid component at the temperature and pressure of the system; alternatively at a pressure of 1 atm; alternatively at the vapor pressure of the pure component.

2. A distinction is drawn between the solvent (the component present in larger amount) and the solute (the component present in smaller amount). The standard state of the solvent is the pure component, as in convention 1, whereas the standard state of the solute is based on Henry's Law in the modified form,

$$\hat{f}_i = k_H x_i. \tag{3.86}$$

The constant is the limiting value of the tangent to the plot of $\hat{f}_i$ against x_i at zero concentration so that the fugacity at the standard state is numerically equal to the Henry constant:

Example 3.7. The Effect of Pressure on the Fugacity of Liquid Chloroform

At 20 C and 1 atm, the specific volume is 80.17 ml/gmol, and the vapor pressure is 0.079 atm. The effect of pressure on the specific volume is represented by the formula (Lange 1973):

$$V = 80.17[1 - 94.9(E-6)(P-1)]$$

At 20 C the fugacity may be taken equal to the vapor pressure, 0.079 atm. At other pressures,

$$f = P^{\text{sat}}\exp \int_{P^{\text{sat}}}^{P} (V/RT)dP$$

$$= 0.079 \exp\left[\frac{80.17}{82.05(293.16)} \times [0.99991\,P - 47.45(E-6)\,P^2]\Big]_{0.079}^{P}\right].$$

The fugacity from this equation and also on the assumption that the liquid is incompressible are tabulated. The effect of pressure is significant even as low as 10 atm.

P, atm	*f*	f/P^{sat}	f/P^{sat} *(incompressible)*
0.079	0.079	1	1
1	0.0792	1.0031	1.0031
10	0.0817	1.0336	1.0336
100	0.1100	1.3929	1.3952

Example 3.8. The Fugacity Coefficient of Saturated Liquid Benzene between the Melting and Critical Temperatures

Note that the fugacity of the saturated liquid equals that of the saturated vapor. The *B*-truncated virial equation with Abbott's formulas will be used:

$$B = \frac{RT_c}{P_c}[0.083 - 0.422/T_r^{1.6} + \omega(0.139 - 0.172/T_r^{4.2})].$$

The critical properties are $T_c = 562.1$ K, $P_c = 48.3$ atm, $\omega = 0.212$, and the vapor pressure is

$$\ln(760P) = 15.9008 - 2788.51/(T - 52.36).$$

At the critical temperature this equation predicts 44.6 atm instead of the correct value of 48.3, but this inaccuracy is ignored here. The fugacity coefficient is found from

$$\phi = \exp(BP/RT).$$

P and ϕ are plotted against temperature from the melting point, 278.7 K, to the critical temperature, 562.1 K. There is a substantial variation of ϕ in this range.

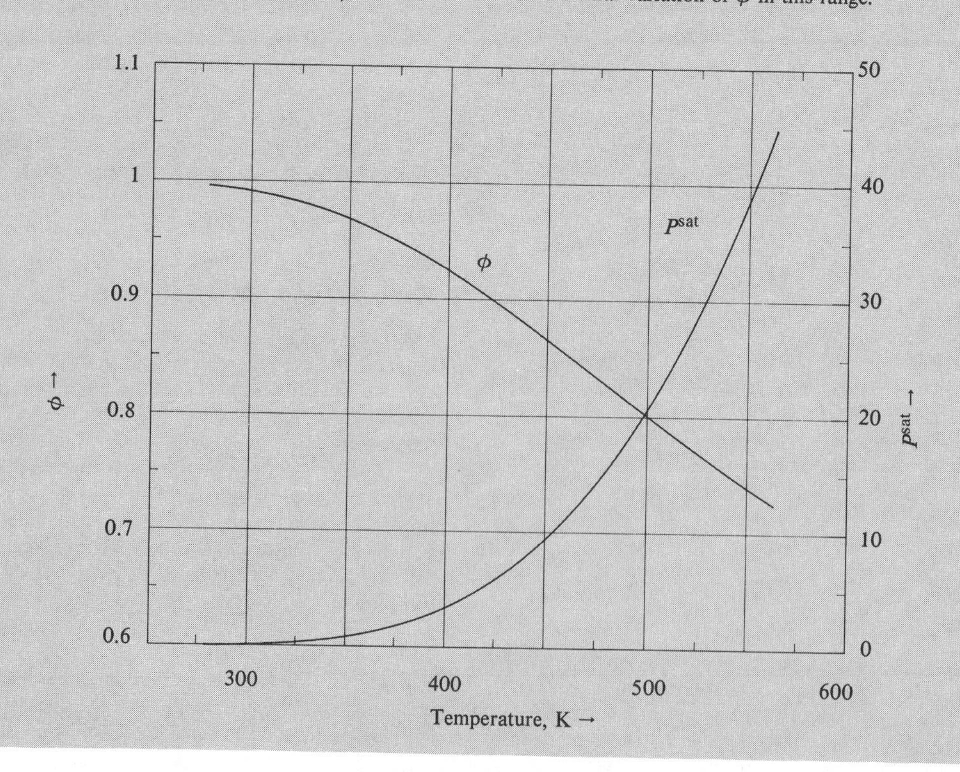

Example 3.9. The Fugacity of Solid Benzene at −30 C and 10 ATM.

Three problems will be treated:

1. The fugacity of the solid will be found from the vapor pressure data of the solid.
2. The fugacity of the solid will be found by adjusting the fugacity at the triple point for pressure and temperature.
3. The ratio of the fugacities of the solid and the subcooled liquid will be found.

Data: The melting point is 5.5 C, assumed to be the same as the triple point. Vapor pressure at this temperature is 35.7 Torr. At −30 C, the vapor pressure of the solid is 1.96 Torr, and that of the subcooled liquid is 3.62 Torr. Specific volumes of the liquid and solid are the same, 87 ml/gmol. The heat of sublimation is 10,566 cal/gmol and the heat of fusion is 2,378 cal/gmol.

1. At −30 C, raise the pressure from 1.96 Torr to 10 atm. At the low pressure the fugacity equals the pressure, 1.96 Torr:

$$f_{\text{solid}} = 1.96 \exp\left[\frac{87}{82.05(243.2)}\left(10 - \frac{1.96}{760}\right)\right]$$

$$= 2.047 \text{ Torr.}$$

2. Adjust the fugacity, 35.7 Torr, at the triple point to 10 atm and −30 C:

$$\ln \frac{f_{\text{solid}}}{35.7} = -\int_{278.7}^{243.2} \frac{\Delta H_{\text{sub}}}{RT^2}\, dT$$

$$+ \frac{87}{82.05(243.3)}\left(10 - \frac{37.5}{760}\right)$$

$$= -\frac{10{,}566}{1.987}\left(\frac{1}{243.2} - \frac{1}{278.7}\right) + 0.0434,$$

$$f_{\text{solid}} = 35.7 \exp(-2.7417) = 2.30 \text{ Torr}$$

3. Since the molal volumes of the two phases are nearly equal, Eq. 3.92 reduces to

$$\ln \frac{f_{\text{solid}}}{f_{\text{liquid}}} = -\frac{\Delta H_{\text{fusion}}}{R}\left(\frac{1}{T} - \frac{1}{T_{tp}}\right)$$

$$= -\frac{2{,}378}{1.987}\left(\frac{1}{243.2} - \frac{1}{278.7}\right) = -0.6136$$

$$\frac{f_{\text{solid}}}{f_{\text{liquid}}} = \exp(-0.6136) = 0.5414$$

It is interesting to note that 0.5414 also is the ratio of the vapor pressure of the solid to the extrapolated vapor pressure of the liquid,

$$1.96/3.62 = 0.5414.$$

In this case extrapolation from 5.5 C to −30 C apparently is fairly safe.

Example 3.10. The Effect of Temperature on the Decomposition Pressure of $CaCO_3$, Illustrating the Effect of Pressure on Fugacities of the Solid Phases.

The equilibrium constant of the reaction $CaCO_3 \rightleftharpoons CO_2 + CaO$ is given by Hougen et al. [*Thermodynamics* (1959), p. 1026].

$$\Delta G^0 = -RT \ln(K_a) = 45037 - 58.14T + 2.83T \ln(T) - 0.7415(10^{-3})T^2 + 0.4125(10^{-6})T^3 - 99{,}800/T, \text{ cal/gmol.} \quad (1)$$

The virial coefficient is given by Haselden et al. [*Proc Roy Soc,* A240, no. 1220 (1957):1] for the temperature range 273–873 K, but will be extrapolated here:

$$B = (-291.61 + 44.67 \ln T)(1000/T) \text{ ml/gmol.} \quad (2)$$

Specific volumes of the $CaCO_3$ and CaO are 36.93 and 21.4 ml/gmol:

$$K_a = a_{CO_2} a_{CaO}/a_{CaCO_3}, \quad (3)$$

$$a_{CO_2} = f = P\phi = P\exp(BP/82.05T) \quad (4)$$

$$a_{CaO} = \exp(V_{CaO}(P-1)/82.05T), \quad (5)$$

$$a_{CaCO_3} = \exp(V_{CaCO_3}(P-1)/82.05T). \quad (6)$$

In these equations the standard state of the gas is unit fugacity, and those of the solids are the pure substances at 1 atm. The equilibrium constant becomes

$$K_a = \exp(-\Delta G^0/1.987T)$$

$$= P \exp[[BP + (P-1)(V_{CaO} - V_{CaCO_3})]/82.05T]. \quad (7)$$

Solve for P at specified T. Four cases are evaluated, showing the independent and combined effects of CO_2 nonideality and pressure on the fugacity ratio of the solid phases. The results may not be truly quantitative since the virial equation is not accurate at the extreme T and P considered.

	P, atm			
K	*1*	*2*	*3*	*4*
1,169	1.00	1.00	1.00	1.00
1,200	1.58	1.58	1.58	1.58
1,400	18.1	18.1	18.0	18.0
1,600	106.9	108.1	104.8	106.1
1,800	406.3	423.5	381.8	397.1
2,000	1,130.2	1,263.9	979.5	1,069.7
1,954	—	1,000	—	—
1,973	1,000	—	—	—
1,985	—	—	—	1,000
2,005	—	—	1,000	—

Case 1: Ideal gas, no P correction
Case 2: Ideal gas, with P correction
Case 3: Virial gas, no P correction
Case 4: Virial gas, with P correction

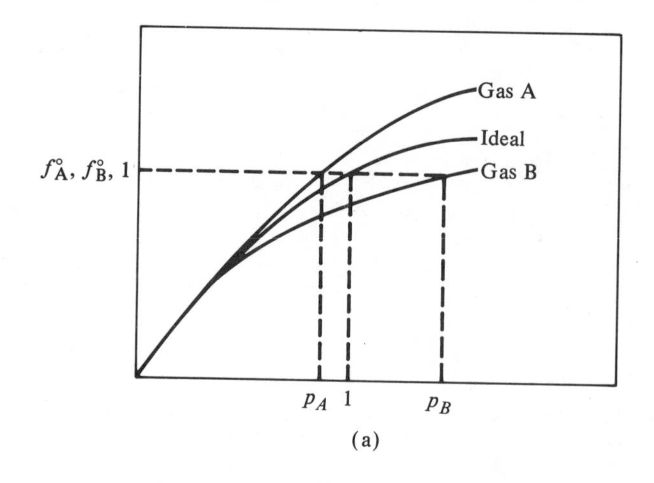

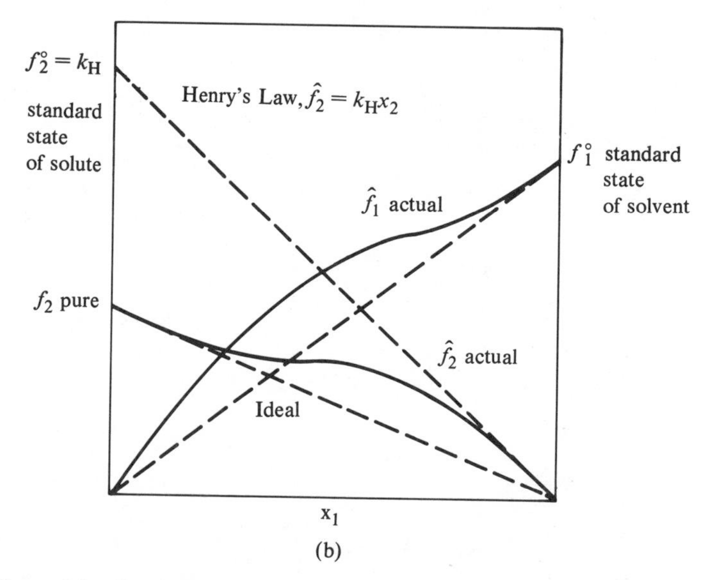

Figure 3.3. Standard states of gases, solvents, and solutes: (a) The two real gases, A and B, deviate from ideality to different extents but have the same standard state; (b) the standard state of the substance present in larger concentration, the solvent (1), is the pure substance; that of the solute (2) is the hypothetical liquid state represented by the intercept at $x_2 = 1$ of the tangent at zero concentration or infinite dilution.

$$f^\circ_i = k_H = \left(\frac{\partial f_i}{\partial x_i}\right)_{x_i=0} \tag{3.87}$$

These conventions for the reference state are illustrated in Figure 3.3b.

In binary systems, k_H is unambiguous, and infinite dilution is a practicable reference when the data are known; but in mixed solvents the concept is of limited value because k_H depends on the composition. Such dependence can be analyzed readily for a special case of ternary mixtures in which a third component is dissolved to various extents in a binary solvent of constant composition. Eq. 2.64 is applied to the partial fugacity coefficient,

$$\ln \hat{\phi}_3 = \ln \phi + (1 - x_3)\left(\frac{\partial \ln \phi}{\partial x_3}\right)_{TPr} \tag{3.88}$$

at constant $r = x_1/x_2$. From the definition, $\phi_i = \hat{f}_i x_i P$ it follows that

$$\frac{\partial \hat{f}_3}{\partial x_3} = P\left(\frac{\partial \hat{\phi}_3 x_3}{\partial x_3}\right) = P\left[x_3\left(\frac{\partial \hat{\phi}_3}{\partial x_3}\right) + \hat{\phi}_3\right]. \tag{3.89}$$

Accordingly, the Henry constant becomes

$$(k_H)_3/P = \lim_{x_3 \to 0}\left(\frac{\partial \hat{f}_3}{\partial x_3}\right)_{TPr}$$

$$= \lim_{x_3 \to 0} \exp\left[\ln\phi + \left(\frac{\partial \ln \phi}{x_3}\right)_{TPr}\right]. \tag{3.90}$$

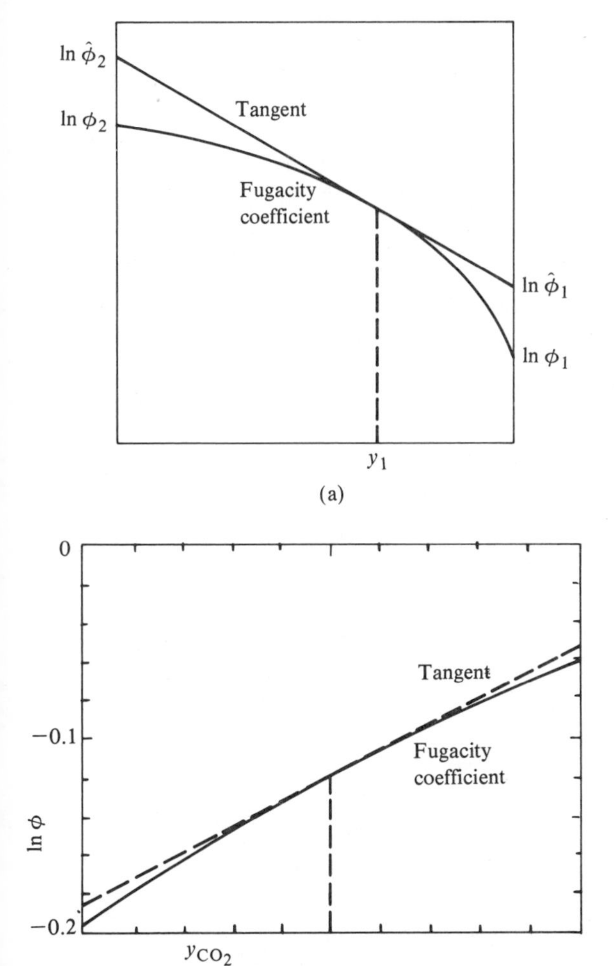

Figure 3.4. Partial fugacity coefficients by the method of tangent intercepts: (a) At a particular composition, the partial fugacities are determined as the intersections of the tangent to the plot of the fugacity coefficient of the mixture with the terminal ordinates; (b) Fugacity coefficients of carbon dioxide and propane mixtures. This typical graph points out the difficulty of obtaining accurate results by this method when the fugacity plot has little curvature. This case is treated in Example 3.6.

The considerable dependence of k_H on composition in some cases is indicated by some results of Problem 3.14:

r	0	2	10
k_3/P	0.135	0.801	0.508

Figure 3.5 shows schematically how the intercepts on the axes can depend on the shape of the surface. Tangents to the surface at $x_3 = 0$ and passing through the axis at point 3 do not intersect that axis at the same elevation.

3.5.2. Subcooled Liquid as a Reference State

The location of the temperature and pressure of a subcooled liquid state is shown in Figure 5.3. Such a reference state is particularly convenient when it is desired to calculate solubilities of solids in liquids. When Eq. 3.33 is applied to the two phases separately and the results are combined,

$$d \ln \frac{f_{\text{solid}}}{f_{\text{liquid}}} = \frac{V_s - V_L}{RT} dP - \frac{H_S - H_L}{RT^2} dT. \quad (3.91)$$

Assuming that the volume and enthalpy changes on fusion are constant, and noting that the fugacity ratio is unity at the triple point, integration from the triple point gives

$$RT \ln(f_{\text{solid}}/f_{\text{liquid}}) = \Delta V_{\text{fusion}}(P-P_{tp}) - \Delta H_{\text{fusion}}(1-T/T_{tp}). \quad (3.92)$$

Furthermore, when equilibrium exists between the two phases, the fugacity of the pure solid equals the partial fugacity of the dissolved material,

$$f_{i,\text{solid}} = \hat{f}_i = \gamma_i x_i f_{i,\text{liquid}} \quad (3.93)$$

where γ_i is an empirical activity coefficient (Chapter 4), often approximately unity. When the last two equations are combined, the reference fugacity of the pure subcooled liquid cancels out, and the solubility, x_i, of the solid is found directly from

$$RT \ln(\gamma_i x_i) = \Delta V_{\text{fusion}}(P - P_{tp}) - \Delta H_{\text{fusion}}(1 - T/T_{tp}). \quad (3.94)$$

Further discussion of reference states is in Chapter 4.

3.6. APPLICATIONS

Because of the many interrelations of thermodynamic functions, fugacity is involved in a variety of thermodynamic

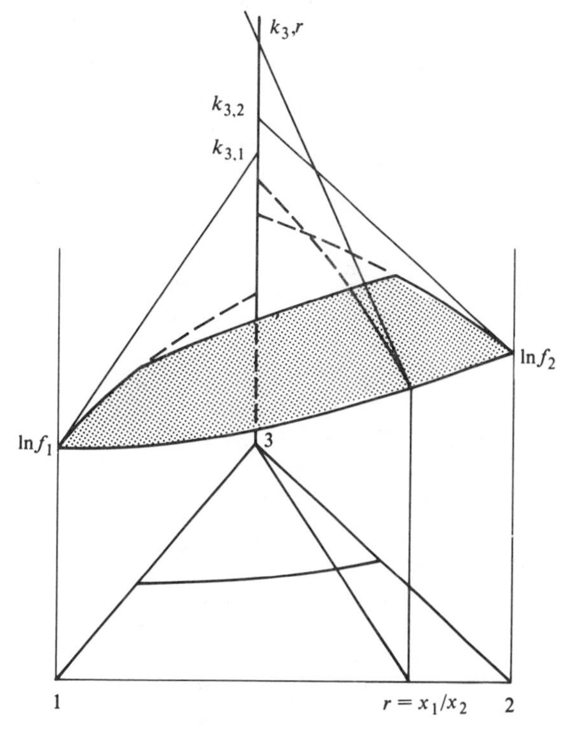

Figure 3.5. Showing that the Henry's Law constants of multicomponent mixtures depend on the composition. Note: Tangents to the surface at $x_3 = 0$ and passing through the axis at point 3 do not intersect that axis at the same elevation.

processes, of which a few of the more important ones may be mentioned.

1. The change in the Gibbs energy is a measure of the isothermal flow work that can be obtained. Thus,

$$-dW_s = dG = RT\,d\ln f, \tag{3.95}$$

and

$$-W_s = RT\ln(f_2/f_1) = RT\ln(\phi_2 P_2/\phi_1 P_1). \tag{3.96}$$

2. The vapor pressure can be determined from subcritical PV isotherms. At saturation the fugacities of the liquid and vapor phases are equal, so that

$$RT\ln(f_{\text{vapor}}/f_{\text{liquid}}) = 0 = \int V\,dP = P^{\text{sat}}V^{\text{sat}} - \int_{V_{\text{liquid}}}^{V_{\text{vapor}}} P\,dV. \tag{3.97}$$

The graphical significance of these relations is shown in Figure 1.7. Example 3.11 is an application of the principle of Eq. 3.97.

3. The general case of phase equilibrium is defined by equality of the partial fugacities of any particular component in all phases:

$$\hat{f}_i^{(a)} = \hat{f}_i^{(b)} = \hat{f}_1^{(c)} = \ldots, \text{ in phases } a, b, c, \ldots. \tag{3.98}$$

This is the main application in this book.

4. Chemical equilibrium is expressed in terms of a Gibbs energy change at standard conditions and the activities, $a_i = \hat{f}_i/f_i^0$, where the f_i^0 are fugacities at the standard conditions. For a reaction with this stoichiometric equation, for example,

$$aA + bB + \ldots \rightleftarrows rR + sS + \ldots, \tag{3.99}$$

the equilibrium constant is

$$K_a = (a_R)^r(a_S)^s \ldots/(a_A)^a(a_B)^b \ldots = K_{\hat{f}}/K_{f^0}, \tag{3.100}$$

and the relation to the Gibbs energy change is

$$\Delta G^0 = -RT\ln K_a. \tag{3.101}$$

Special forms of this relationship are described in Chapter 10.

Example 3.11. The Saturation Pressure of *n*-Pentane at 100 C with the Redlich-Kwong and Soave Equations.

At the saturation pressures the fugacities of the two phases are equal. Data are $T_c = 469.7$ K, $P_c = 33.25$ atm, $\omega = 0.251$. The experimental vapor pressure is 5.86 atm. Redlich-Kwong equations are in Table 1.9, the Soave in Table 1.11, and the fugacity equations in Table 3.3.

Redlich-Kwong equation

$$A = 0.42747\left(\frac{P}{33.25}\right)\left[\frac{469.7}{373.2}\right]^{2.5} = 0.022852\,P,$$

$$B = 0.08664\left(\frac{P}{33.25}\right)\left(\frac{469.7}{373.2}\right) = 0.003280\,P,$$

$$z^3 - z^2 + (A - B - B^2)z - AB = 0,$$

$$z^3 - z^2 + P[0.01957 - 10.76(E{-}6)P]z - 74.95(E{-}6)P^2 = 0$$

$$\ln f = z - 1 - \ln\frac{z - B}{P} - \frac{A}{B}\ln\left(1 + \frac{B}{z}\right)$$

$$= z - 1 - \ln\frac{z - 0.00328\,P}{P} - 6.967\ln\left(1 + \frac{0.00328\,P}{z}\right).$$

The compressibilities and the fugacities are tabulated. By interpolation, the predicted vapor pressure is 7.79 atm.

	z		f		
P	*Vapor*	*Liquid*	*Vapor*	*Liquid*	*Ratio*
4	0.9162	0.0206	3.9300	6.4823	
5	0.8931	0.0259	4.5148	6.5155	
6	0.8689	0.0310	5.3019	6.5493	
7	0.8434	0.0361	6.0502	6.5832	0.9190
8	0.8162	0.0412	6.7596	6.6172	1.0215
7.79 (interpolated)					1.0000

Soave equation

$$\alpha = [1 + (1 - \sqrt{T_r})(0.48508 + 1.55171\omega - 0.15163\,\omega^2)]^2 = 1.1968,$$

$$A = 0.42747\,\alpha\left(\frac{P}{P_c}\right)\left[\frac{T_c}{T}\right]^2 = 0.024377\,P,$$

$$B = 0.08644\left(\frac{P}{P_c}\right)\left[\frac{T_c}{T}\right]^2 = 0.00328\,P,$$

$$z^3 - z^2 + P[0.021097 - 10.76(E-6)P]z - 79.957(E{-}6)P^2 = 0,$$

$$\ln f = z - 1 - \ln\frac{z - B}{P} - \frac{A}{B}\ln\left(1 + \frac{B}{z}\right)$$

$$= z - 1 - \ln\frac{z - 0.00328\,P}{P} - 7.43238\ln\left(1 + \frac{0.00328\,P}{z}\right).$$

Interpolation in the following table gives the vapor pressure as 6.01, which compares more favorably than does the Redlich-Kwong value with the experimental value of 5.86 atm. The Soave equation generally fits liquid data better.

Example 3.11 *(continued)*

	z		f		
P	Vapor	Liquid	Vapor	Liquid	Ratio
4	0.8943	0.0184	3.6650	5.1960	
5	0.8637	0.0229	4.4774	5.2287	0.8563
6	0.8310	0.0275	5.2482	5.2512	0.9994
7	0.7955	0.03210	5.9775	5.2807	1.1320
6.01 (interpolated)					1.0000

3.7. PROBLEMS

3.1. Find the fugacity coefficient in terms of the Berthelot EOS, $PV = RT + (b - a/RT^2)P$.

3.2. Find the equation for the fugacity coefficient of a substance following the Dieterici EOS,

$$P = \left(\frac{RT}{V-b}\right)\exp\left(-\frac{a}{RTV}\right).$$

3.3. These data are for hydrogen at 0 C. Find the fugacity coefficient at 500 and 1,000 atm.

P	$\frac{Pv}{RT}$	P	$\frac{Pv}{RT}$
100	1.069	600	1.431
200	1.138	700	1.504
300	1.209	800	1.577
400	1.283	900	1.649
500	1.356	1000	1.720

3.4. The compressibility of a gas may be represented by

$$PV/RT = A + BP + CP^2 + DP^3,$$

where A, B, C, D are functions of the temperature; hence, derive an expression for the fugacity as a function of the pressure at a given temperature. For nitrogen at 0 C, A is 1.000, B is -5.314×10^{-4}, C is 4.276×10^{-6}, and D is -3.292×10^{-9} with P in atm up to 400 atm. [Bartlett, *J Am Chem Soc*, 49 (1927):687] Evaluate the fugacity of the gas at 300 atm pressure.

3.5. Find these derivatives in terms of the variables of the Bridgman Table A.7:

a. $(\partial \ln f/\partial T)_S$.
b. $(\partial \ln f/\partial S)_U$.
c. $(\partial \ln f/\partial S)_H$.

3.6. From the given data of fugacity coefficients against pressure at 0 C, find the power required to compress 1 kgmol/sec of the gas isothermally from 200 to 800 atm.

P	ϕ	P	ϕ
200	1.1219	600	1.4873
300	1.2030	700	1.5984
400	1.2898	800	1.7177
500	1.3851		

3.7. In Eq. 3.23, replace the differential dP with $(\partial P/\partial V)_T dV$. Apply the resulting equation

$$RT \ln \phi = \int_\infty^V \left(V - \frac{RT}{P}\right)\left(\frac{\partial P}{\partial V}\right)_T dV$$

to find the fugacity coefficient from

a. the equation of state $P(V - b) = RT$.
b. the van der Waals equation.

3.8. Find the fugacity coefficient from the two forms of the B-truncated virial equation:

a. $z = PV/RT = 1 + B/V = 1 + BP/zRT$,

$$\ln \phi = \int_\infty^V \left(\frac{z-1}{P}\right)\left(\frac{\partial P}{\partial V}\right)_T dV.$$

b. $z = 1 + BP/RT$,

$$\ln \phi = \int_0^P \left(\frac{z-1}{P}\right) dP.$$

Compare the two values of $\ln \phi$ over a range of values of BP/RT from -0.25 to 0.5 and the corresponding values of the compressibility.

3.9. Excess volumetric data of mixtures of water and tetrahydrofuran were obtained by Signer et al. [*Hely Chim Acta*, 52 (1969):2347]. Values read off a small graph are tabulated. Pure component volumes are 0.018 for water and 0.06403 liters/gmol for THF. Find the partial molal volumes at 20% intervals of the concentration, and relate them to fugacity ratios.

x_{THF}	0.1	0.2	0.4	0.6	0.8
$-V^{ex}$, ml/gmol	0.50	0.77	0.89	0.75	0.45

3.10. Volumetric data for the components A and B of a mixture are $V_a = 25 + 3x_a$ and $V_b = 15$ ml/gmol. An equimolal mixture is charged isothermally at 2 gmol/min to a 1 liter vessel that has an initial content of 10 gmol of B. How long does it take to fill the vessel?

3.11. The following data were obtained for the molar volumes V_1 and V_2 in ml. for pure hydrogen and nitrogen, respectively, at various pressures P and 0° C. On the other hand, $\bar{V}_1$ and $\bar{V}_2$ are the partial molar volumes in a mixture containing 0.6 mole hydrogen to 0.4 mole nitrogen, where P is now the total pressure.

P	V_1	V_2	$\bar{V}_1$	$\bar{V}_2$
50 atm.	464.1	441.1 ml.	466.4	447.5 ml.
100	239.4	220.6	241.3	226.7
200	127.8	116.4	129.1	120.3
300	90.5	85.0	91.1	86.9
400	72.0	70.5	72.5	71.8

Determine by the graphical method the fugacity of nitrogen in the mixture at the various total pressures, and compare the results with those obtained for the pure gases; hence, test the Lewis-Randall rule for the fugacity of a gas in a mixture [cf. Merz and Whittaker, *J Am Chem Soc* 50 (1928):1522].

3.12. The fugacity of a mixture is given by $\ln \phi = 0.2x_1^2 + 0.5x_2^2$. Plot $\ln \phi_1$ and $\ln \hat{\phi}_2$ against x_1.

3.13. The fugacity of a ternary mixture is

$$\ln \phi = 0.2x_1^2 + 0.5x_2^2 + 0.6x_3^2.$$

Use the results of Section 2.6 to plot $\ln \hat{\phi}_3$ against x_3 at values of $r = x_1/x_2 = 0, 0.2, 2, \infty$. Note that the value of $\ln \phi_3$ at any given value of x_3 is a maximum at about $r = 2$.

3.14. For the case of a ternary mixture in which the concentrations of two of the components are kept in a constant ratio, $x_1/x_2 = r$, find an expression for the Henry Law constant

$$k_3 = \lim_{x_3 \to 0} \left(\frac{\partial f_3}{\partial x_3}\right)_r = \lim_{x_3 \to 0} P \left(\frac{\partial x_3 \phi_3}{\partial x_3}\right)_r$$

as a function of r. When the fugacity coefficient of the mixture is represented by

$$\ln \phi = x_1 x_2 x_3 + x_1^3 + x_2^3 + x_3^3,$$

find numerical values of k_3/P over a suitable range of r.

3.15. Find equations for $\ln(\hat{f}_2/f_2)$ given these two relations for $\ln \hat{f}_1$:

a. $\ln \hat{f}_1 = Ay_2^2$.

b. $\ln \hat{f}_1 = Ay_1^2 + By_2^2 + Cy_1 y_2$.

3.16. Follow the procedure of Example 3.4 to find the partial fugacity coefficients from the van der Waals equation.

3.17. The molal volumes of a binary gas mixture are represented by the equation $V = RT/P + 0.2 + 0.1y_1 - 0.2y_1^2$, liters/gmol. The temperature is 300 K, and the pressure is 25 atm. Find the partial fugacity coefficients of an equimolal mixture.

3.18. Partial molal volumes of nitrogen in a nitrogen-hydrogen mixture containing 40% N_2 at 0 C are:

P, atm	50	100	150	200
$\bar{V}_{N_2}$, ml/gmol	447.5	226.7	154.9	120.3

a. Find the partial fugacity coefficients of nitrogen at 100 atm and 200 atm.

b. Find the change in the partial molal Gibbs energy of the nitrogen on compression of the mixture from 100 to 200 atm.

3.19. The partial fugacity coefficient of component 1 of a binary mixture is given by

$$\ln \hat{\phi}_1 = 0.18(1 - 2x_1).$$

The fugacity coefficient of pure component 2 is $\phi_2 = \exp(-0.1)$. Find the equation for the fugacity coefficient ϕ_m of the mixture as a function of the composition.

3.20. The fugacity coefficient of a ternary mixture is given by

$$\ln \phi_m = 0.2x_1 x_2 - 0.3x_1 x_3 + 0.15x_2 x_3.$$

Find each of the partial fugacity coefficients in an equimolar mixture.

3.21. Find the fugacity coefficient and the partial fugacity coefficients of an equimolal mixture of ethylene and ammonia at 350 K and 30 atm, using equations based on these equations of state:

a. van der Waals.

b. Redlich-Kwong.

c. Soave.

d. Tsonopoulos virial.

Pertinent properties of ethylene are $T_c = 282.4$, $P_c = 49.7$, $w = 0.085$; and of ammonia, $T_c = 405.6$, $P_c = 111.3$, $\omega = 0.25$.

3.22. For a pure van der Waals gas, the excess enthalpy is (Chapter 11),

$$H^{ex} = H - H^{id} = -[2a/V - bRT/(v - b)].$$

What is the ratio of fugacity coefficients of ammonia at 30 atm at 350 K and 450 K?

3.23. Find the values of $\hat{\phi}_{NH_3}$ in mixtures with ethylene at 350 K, 30 atm, and $x_{NH_3} = 0.4$, 0.5, and 0.6. Hence, estimate the values of the derivatives $\partial\hat{\phi}_{NH_3}/\partial x_{NH_3}$ at the three compositions. Use the van der Waals equation.

3.24. An ammonia reactor operates at 150 atm. Feed composition and partial fugacity coefficients are

	x_i	$\hat{\phi}_i$
H_2	0.72	1.05
N_2	0.24	0.95
NH_3	0.01	0.50
CH_4	0.03	0.80

The equilibrium constant is $K_f = \hat{f}_{NH_3}^2/\hat{f}_{N_2}\hat{f}_{H_2}^3 = 22.7(10^{-6})$. Find the fractional conversion.

3.25. Apply the Soave-Redlich-Kwong equation to find the saturation pressure and vapor and liquid specific volumes of n-pentane at 100 C. Properties are $P_c = 33.25$ atm, $T_c = 469.7$ K, and $\omega = 0.251$. Some roots of the equation in z-form are:

P, atm	z_G	z_X	z_L
4	0.8943	0.0873	0.0184
5	0.8637	0.1134	0.0229
6	0.8310	0.1415	0.0275
7	0.7955	0.1725	0.0320

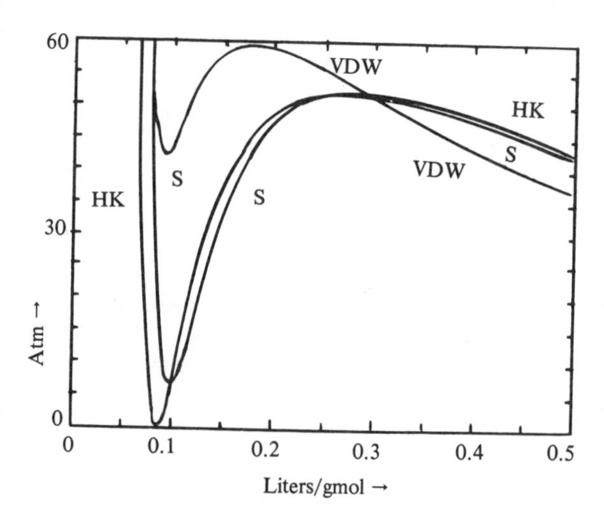

The experimental saturation pressure is 5.86 atm.

3.26. *PVT* data for saturated sulfur dioxide are given.

a. Construct the two-phase envelope, P against V.
b. Draw the 250 F isotherms in the vicinity of the saturation curve for the equations of state of van der Waals, Soave, and Harmens & Knapp.
c. Find the vapor pressures and saturation densities with each of these equations, and compare with the numbers corresponding to the intersections of the isotherms with the two-phase envelope.

Other properties of sulfur dioxide are: $P_c = 77.8$ atm, $T_c = 430.8$ K, $V_c = 0.122\ k$/gmol, $z_c = 0.268$, $w = 0.251$. A sample plot of the isotherms is given as a guide to estimating roots of the equations. (See page 158. The curves are identified with the initials of the originators of the respective equations.)

Temp., °F.	*Abs. pressure, lb./sq. in.*	*Volume cu. ft./lb.*	
		Liquid	*Vapor*
−100	0.294	0.009856	204.7
−90	.465	.009954	132.9
−80	.710	.010054	89.3
−70	1.056	.01015	61.0
−60	1.550	.01025	42.6
−50	2.225	.01034	30.4
−40	3.12	.01044	22.2
−30	4.33	.01053	16.5
−20	5.88	.01062	12.5
−10	7.83	.01072	9.48
0	10.26	.01082	7.35
10	13.3	.01092	5.77
20	16.9	.01103	4.59
30	21.3	.01114	3.70
40	26.6	.01125	3.02
50	32.9	.01137	2.48
60	40.3	.01149	2.05
70	49.1	.01162	1.70
80	59.3	.01175	1.42
90	71.0	.01189	1.20
100	84.1	.01204	1.02
110	99.1	.01219	0.868
120	116.3	.01235	.746
130	135.8	.01251	.646
140	157.7	.01269	.554
150	182	.01288	.481
160	209	.01309	.418
170	238	.01330	.364
180	272	.01350	.319
190	307	.01371	.281
200	347	.01396	.246
210	390	.01422	.217
220	437	.01453	.191
230	487	.01487	.168
240	543	.01527	.147
250	604	.01574	.129
260	669	.01628	.113
270	739	.01690	.0987
280	816	.01767	.0848
290	898	.01861	.0723
300	987	.02002	.0599
315.4	1143	.03070	.0307

3.27. Naphthalene forms an ideal solution in benzene. Its melting point is 80 C, and its heat of fusion is 35.6 cal/gm.

a. What is the solubility at 25 C?
b. At what temperature is the solubility 85 gm/mol benzene?

3.28. Show that $dG = \Sigma x_i d\bar{G}_i = RT\Sigma x_i d \ln \hat{f}_i$ also may be written $dG = RT\Sigma x d \ln(\hat{f}_i/x_i)$ by proving that $\Sigma x_i d \ln x_i = 0$. Note that logarithms of partial molal fugacities are partial molal quantities; that is,

$$\ln \hat{\phi}_i = \left(\frac{\partial n \ln \phi}{\partial n_i}\right)_{TPn_j}$$

3.29. The fugacity of a substance is correlated in terms of the equation $f = a + bP + cP^2$.

a. Find the equation of state corresponding to this result.
b. Find the equations for $H - H^{id}$ and $V - V^{id}$ corresponding to this result.

3.30. The solubilities of stannic iodide in carbon disulfide are given at several temperatures. Estimate the melting point and heat of fusion of stannic iodide.

°C	10	25	40
SnI_4, g/100 g solution	49.01	58.53	67.56

The experimental values are 143.4°C and 4330 cal/gmol.

3.31. The excess Gibbs energy of a ternary mixture is represented by

$$G^{ex}/RT = A_{12}x_1x_2 + A_{13}x_1x_3 + A_{23}x_2x_3.$$

Show that the activity coefficients are represented by

$$RT \ln \gamma_1 = A_{12}x_2^2 + A_{13}x_3^2 + (A_{12} + A_{13} - A_{23})x_2x_3,$$

$$RT \ln \gamma_2 = A_{12}x_1^2 + A_{23}x_3^2 + (A_{12} + A_{23} - A_{13})x_1x_3,$$

$$RT \ln \gamma_3 = A_{13}x_1^2 + A_{23}x_2^2 + (A_{13} + A_{23} - A_{12})x_1x_2.$$

3.32. Find the fugacity coefficient from the equation of state, $z = 1 + B/V$, with Eq. 3.27; and from the equation of state, $z = 1 + BP/RT$, with Eq. 3.23. What is the range of z for which the equations agree within 10%?

4 Activity Coefficients

Processes are characterized by changes in energy properties such as ΔS, ΔH, ΔG, $\Delta \ln f$, and so on. Although absolute values of such properties can be evaluated, at least with respect to pure crystalline states at absolute zero on the basis of the Third Law, usually it is not necessary to go that far back. The convenience of absolute values can be retained by arbitrary definition of a reference state that can be made the same for a wide class of problems.

4.1. ACTIVITY AND ACTIVITY COEFFICIENTS

Certain choices of reference or standard states have been found convenient and are widely accepted. The ones to be discussed here are those for fugacity, which are especially important because of the relation of fugacity to Gibbs energy. For a process from state 1 to state 2,

$$\Delta G = G_2 - G_1 = RT(\ln f_2 - \ln f_1)$$
$$= RT\,\Delta \ln f = RT \ln(f_2/f_1). \quad (4.1)$$

In terms of a standard state, designated by a superscript o, for a pure substance,

$$G = G^0 + RT \ln(f/f^0), \quad (4.2)$$

and for a component of a mixture,

$$\bar{G}_i = G_i^0 + RT \ln(\hat{f}_i/f_i^0) = G_i^0 + RT \ln a_i, \quad (4.3)$$

where

$$a_i = \hat{f}_i/f_i^0, \quad (4.4)$$

is called the relative fugacity or activity. A quantity called the activity coefficient,

$$\gamma_i = \hat{f}_i/x_i f_i^0 \quad (4.5)$$

is analogous to the fugacity coefficient

$$\hat{\phi}_i = \hat{f}_i/x_i P. \quad (4.6)$$

When rearranged as

$$\hat{f}_i = \gamma_i x_i f_i^0, \quad (4.7)$$

the definition resembles certain familiar limiting or ideal conditions. When $\gamma_i = 1$,

$$\hat{f}_i = x_i f_i^0, \quad (4.8)$$

which is of the same form as Raoult's Law, $P_i = x_i P_i^0$; and in the vicinity of zero concentration or infinite dilution,

$$\hat{f}_i = \gamma_i^{\infty} x_i f_i^0 = k_H x_i, \quad (4.9)$$

which is analogous to Henry's Law. These relations point out that in some respects the fugacity is an effective pressure.

4.2. STANDARD STATES

Standard or reference states have been discussed in Chapter 3, but because of the importance of this topic in connection with the liquid phases, which are the main concern of this chapter, the discussion will be repeated and extended somewhat here.

For gases the standard state is always taken as that of unit fugacity, so the activity is numerically the same as the fugacity,

$$a_i = \hat{f}_i^{(V)}/f_i^0 = \hat{f}_i^{(V)}. \quad (4.10)$$

For liquids: In order to take advantage of the requirement of equality of fugacities of individual components in all phases at equilibrium, the standard states of condensed and vapor phases must be consistent. The factor in common is the vapor pressure, at which the fugacities of vapor and condensed phases are the same, thus

$$f_i^{\text{sat}} = \phi_i^{\text{sat}} P_i^{\text{sat}}. \quad (4.11)$$

The standard state is defined as that of the pure substance at the temperature and pressure of the system. Taking the vapor pressure at the temperature of the system and adjusting the fugacity given by Eq. 4.11 to the system pressure by Eq. 3.33 gives for the fugacity of the standard state the equation

$$f_i^0 = f_i^{\text{sat}} \exp\left[\int_{P^{\text{sat}}}^{P} (\bar{V}^{(L)}/RT)\, dP\right] \quad (4.12)$$

$$= \phi_i^{\text{sat}} P_i^{\text{sat}} (PF)_i, \quad (4.13)$$

where

$$(PF)_i = \exp\left[\int_{P^{\text{sat}}}^{P} (\bar{V}^{(L)}/RT)\, dP\right]. \quad (4.14)$$

The exponential term of Eq. 4.14 is called the Poynting Factor; at moderate pressures it is usually little different from unity.

In all the material of this chapter the activity coefficient is related to other properties by the equation

$$y_i \hat{\phi}_i^{(V)} P = x_i \gamma_i \phi_i^{\text{sat}} P_i^{\text{sat}} (PF)_i, \quad (4.15)$$

which is based on the equilibrium relation

$$\hat{f}_i^{(V)} = \hat{f}_i^{(L)}. \quad (4.16)$$

The fugacity coefficients $\hat{\phi}_i^{(V)}$ and ϕ_i^{sat} are obtained from the same correlation or equation of state. The activity coefficients are calculated with Eq. 4.15 directly from experimental equilibrium data or from structural contributions as described later in this chapter.

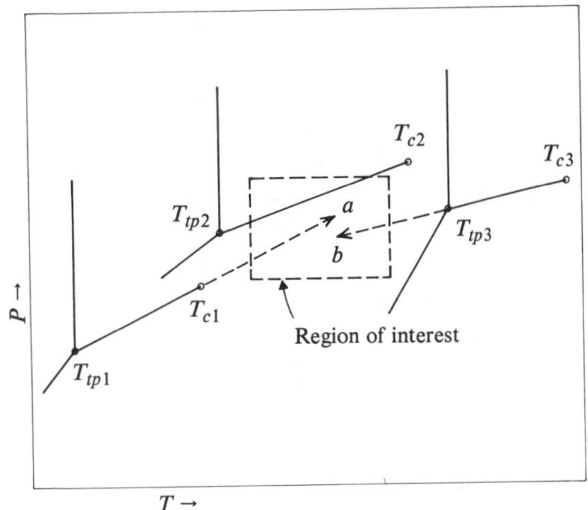

Figure 4.1. Showing extrapolation of vapor-pressure curves into superheated, point a, and subcooled, point b, liquid regions. Component 1 is above its critical point, component 2 is between its triple and critical points, and component 3 is below its triple point.

Solids and supercritical gases: Vapor pressure curves of such substances can be extrapolated to the system temperature, as indicated in Figure 4.1. The resulting hypothetical liquid states often serve satisfactorily as reference states when brought to the system pressure with the Poynting Factor. In special cases other standard states may be advantageous, for instance in describing the solubilities of gases and solids in liquids, as done in other sections of this book. They are discussed at some length by Van Ness & Abbott (1982, pp. 231–261) and Prausnitz (1980, pp. 56–61).

Solutions are discussed in Section 3.5.

4.3. THERMODYNAMIC RELATIONS INVOLVING ACTIVITY COEFFICIENTS

Activity coefficients are closely related to various excess properties, of which the great variety that exists is suggested by the gallery of Figures 4.2–4.6. Formulas representing these interrelations can be written down almost by inspection from corresponding ones for fugacity coefficients given in Table 3.2, starting with the definitions,

$$\bar{G} = RT \ln \hat{f}_i + \lambda_i(T), \quad (3.9, 4.17)$$

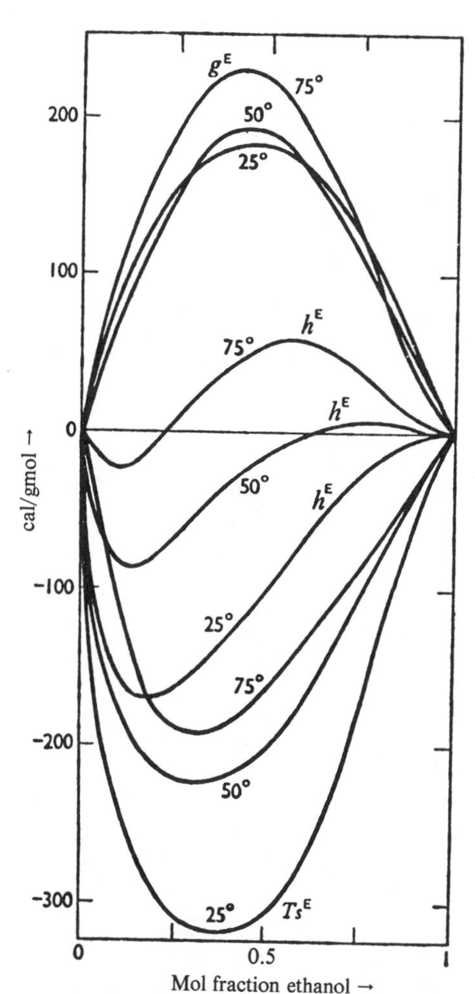

Figure 4.2. Excess properties of ethanol + water at several temperatures (Prigogine & Defay, 1954).

and

$$\hat{f}_i^{id} = x_i f_i^0. \quad (4.18)$$

Accordingly,

$$\bar{G}_i^{ex} = \bar{G}_i - \bar{G}_i^{id} = RT \ln \hat{f}_i - RT \ln x_i f_i^0 = RT \ln(a_i/x_i) = RT \ln \gamma_i, \quad (4.19)$$

since the $\lambda_i(T)$ cancel out. For the mixture as a whole

$$G^{ex} = RT\Sigma x_i \ln \gamma_i. \quad (4.20)$$

Clearly $\ln \gamma_i$ is a partial molal property and can be represented by Eq. 2.56 as

$$RT \ln \gamma_i = G^{ex} - \sum_{k \neq i} \left(\frac{\partial G^{ex}}{\partial x_k}\right)_{TPx_{j \neq ik}}, \quad (4.21)$$

and for component 1 of a binary mixture

$$RT \ln \gamma_1 = G^{ex} + (1 - x_1)\left(\frac{\partial G^{ex}}{\partial x_1}\right). \quad (4.22)$$

Eq. 4.22 is a linear differential equation whose solution for the excess Gibbs energy is

$$G^{ex} = (1 - x_1)\int_0^{x_1} \frac{RT \ln \gamma_1}{(1 - x_1)^2} dx_1. \quad (4.23)$$

For instance, if $RT \ln \gamma_1 = a(1 - x_1)^n$, the solution is

$$G^{ex} = \frac{a}{n-1} x_2(1 - x_2^{n-1}). \quad (4.24)$$

What is usually done, however, is to measure both activity coefficients and to calculate the excess Gibbs energy from them, rather than to perform the reverse operation of finding the activity coefficients from the excess Gibbs energies with Eq. 4.22.

Clearly, activity coefficients depend on T, P, and composition. These dependencies are found with the aid of these equations from Table 2.6:

$$(\partial(\bar{G}^{ex}/T)/\partial T)_P = -\bar{H}^{ex}/T^2, \quad (4.25)$$

$$(\partial \bar{G}^{ex}/\partial P)_T = \bar{V}^{ex}, \quad (4.26)$$

which lead to the desired relations, namely:

$$\left(\frac{\partial \ln \gamma_i}{\partial T}\right)_{Px} = -\frac{\bar{H}^{ex}}{RT^2}, \quad (4.27)$$

$$\left(\frac{\partial \ln \gamma_i}{\partial P}\right)_{Tx} = \frac{\bar{V}^{ex}}{RT}, \quad (4.28)$$

$$\left(\frac{\partial \ln \gamma_i}{\partial x_k}\right)_{PTx_{j \neq ik}} = \frac{1}{RT}\left(\frac{\partial \bar{G}^{ex}}{\partial x_k}\right)_{PTx_{j \neq ik}} = \frac{1}{RT}\int_0^P \left(\frac{\partial \bar{V}^{ex}}{\partial x_k}\right)_{PTx_{j \neq ik}} dP. \quad (4.29)$$

Since activity coefficients usually are measured as functions of the composition, the derivative is obtainable directly from those data, so the expression in terms of the excess volume rarely is useful.

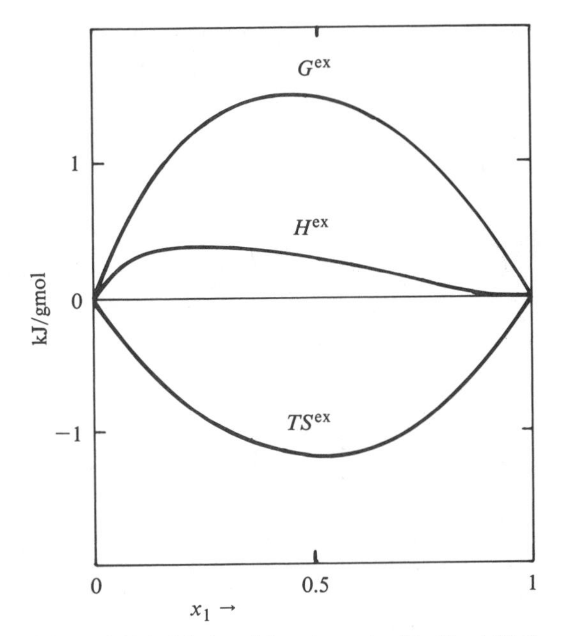

Figure 4.3(a). Methanol + carbon tetrachloride at 35 C.

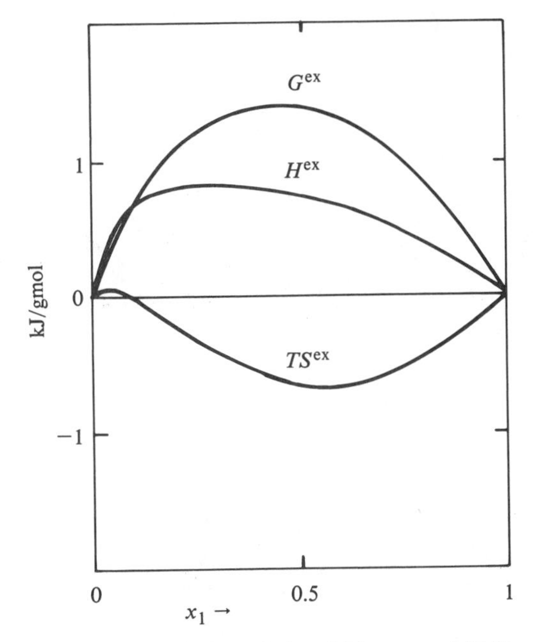

Figure 4.3(b). Methanol + benzene at 35 C.

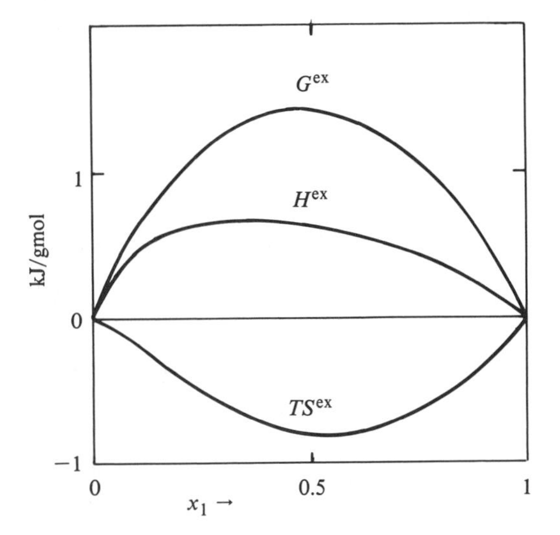

Figure 4.3(c). Ethanol + iso-octane at 25 C.

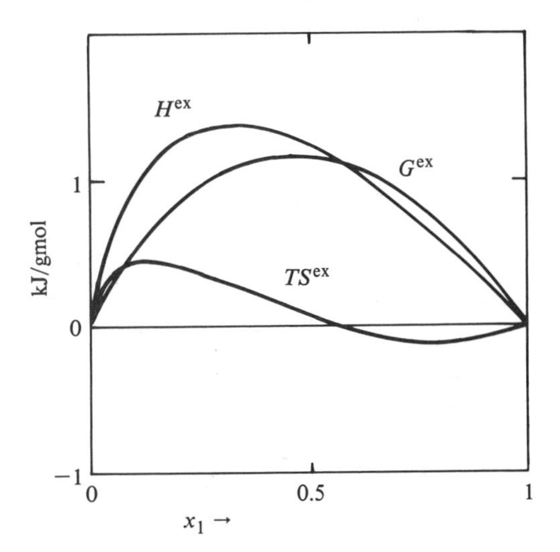

Figure 4.3(d). Ethanol + benzene at 45 C.

Figure 4.3. Experimental values of excess Gibbs energies, excess enthalpies, and excess entropies. The abscissa is the mol fraction of the first-named substance in each case. Data compiled by Rowlinson (1969 edition).

The activity coefficient in terms of both T and P is

$$d \ln \gamma_i = -\frac{\bar{H}^{ex}}{RT^2}\, dT + \frac{\bar{V}^{ex}}{RT}\, dP, \tag{4.30}$$

and the result of summing up for all the components is

$$\Sigma x_i \, d \ln \gamma_i = -\frac{H^{ex}}{RT^2}\, dT + \frac{V^{ex}}{RT}\, dP. \tag{4.31}$$

Eq. 4.31 is one form of the Gibbs-Duhem equation of which several other forms are listed in Table 2.7. In the binary form,

$$x_1 \, d \ln \gamma_1 + x_2 \, d \ln \gamma_2 = -\frac{H^{ex}}{RT^2}\, dT + \frac{V^{ex}}{RT}\, dP, \tag{4.32}$$

and especially at constant T and P,

$$x_1 \, d \ln \gamma_1 + x_2 \, d \ln \gamma_2 = 0, \tag{4.33}$$

it is the basis of a widely used test of the consistency of experimental data, which is derived as follows. In the equation of the differential of the Gibbs energy,

$$\begin{aligned} d(G^{ex}/RT) &= d(x_1 \ln \gamma_1 + x_2 \ln \gamma_2) \\ &= (x_1 \, d \ln \gamma_1 + x_2 \, d \ln \gamma_2) \\ &\quad + \ln \gamma_1 \, dx_1 - \ln \gamma_1 \, dx_1, \end{aligned} \tag{4.34}$$

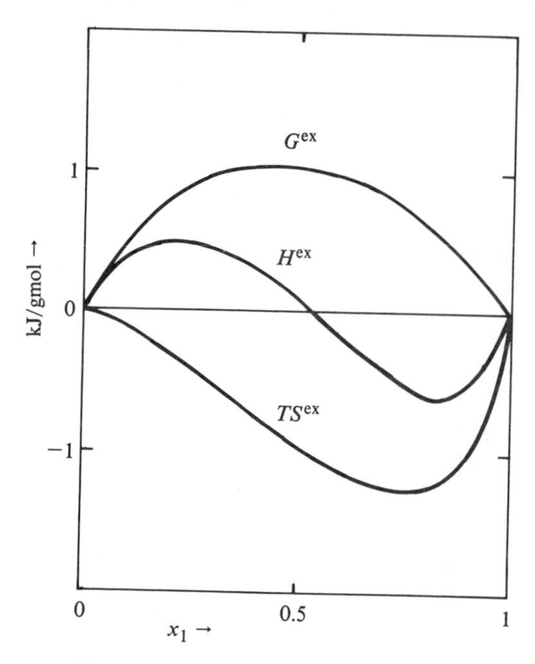

Figure 4.3(e). Water + dioxane at 25 C.

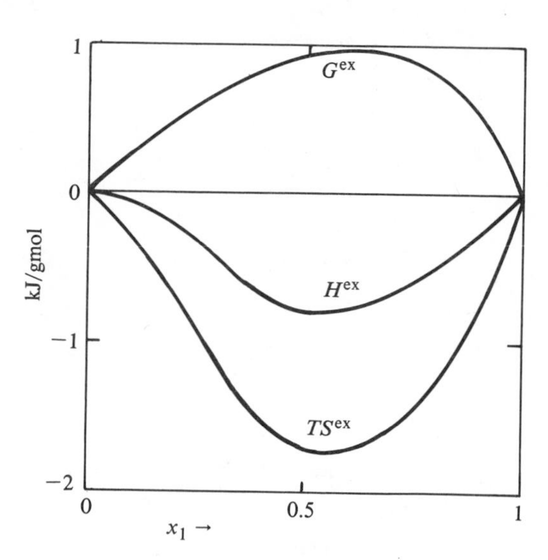

Figure 4.3(g). Water + pyridine at 80 C.

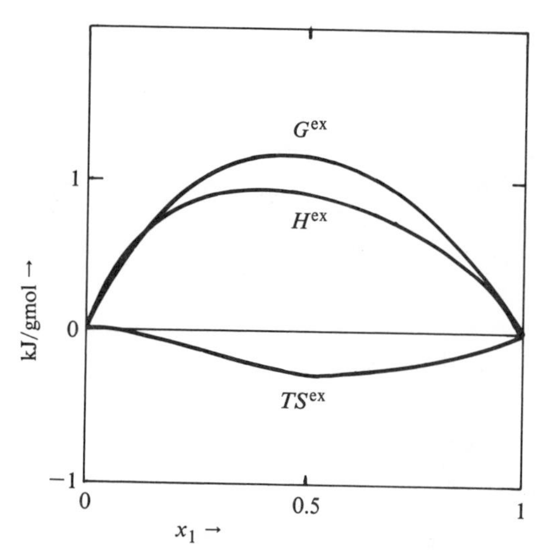

Figure 4.3(f). Acetonitrile + carbon tetrachloride at 25 C.

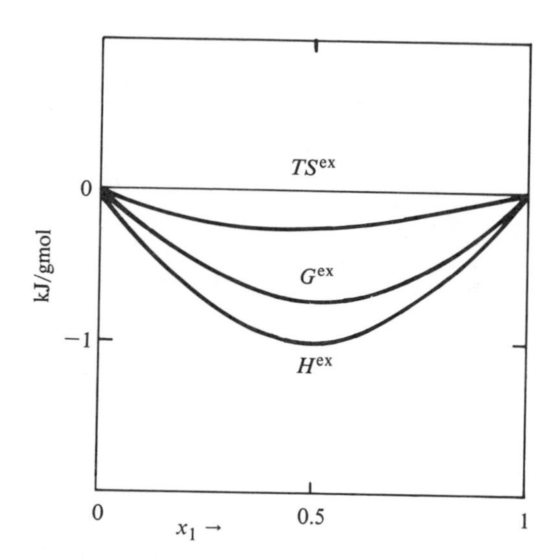

Figure 4.3(h). Water + hydrogen peroxide at 25 C.

the group in parentheses is zero by Eq. 4.33. Since the excess Gibbs energy vanishes at the terminals, integration of the remaining part of Eq. 4.34 results in

$$\int_0^1 d(G^{ex}/RT) = \int_0^1 \ln(\gamma_1/\gamma_2)\, dx_1 = 0. \tag{4.35}$$

Application of this equation is made in Example 4.2, and other instances are shown in Figure 4.7. Further details of consistency testing are provided, for example, by Van Ness & Abbott (1982, Chap. 6). When the effects of T and P are retained, the result is

$$\int_0^1 \ln\frac{\gamma_1}{\gamma_2}\, dx_1 - \int_{T_0}^{T_1} \frac{H^{ex}}{RT^2}\, dT + \int_{P_0}^{P_1} \frac{V^{ex}}{RT}\, dP = 0. \tag{4.36}$$

Usually, however, only the effect of temperature needs to be taken into account. In the absence of enthalpy of mixing data, approximations have been developed by Herington (1951), which are an improvement over neglecting the effect of temperature. That paper also considers the testing of consistency of multicomponent equilibrium measurements. Improved consistency testing examines the data point by point to avoid mutual cancellation of individual errors.

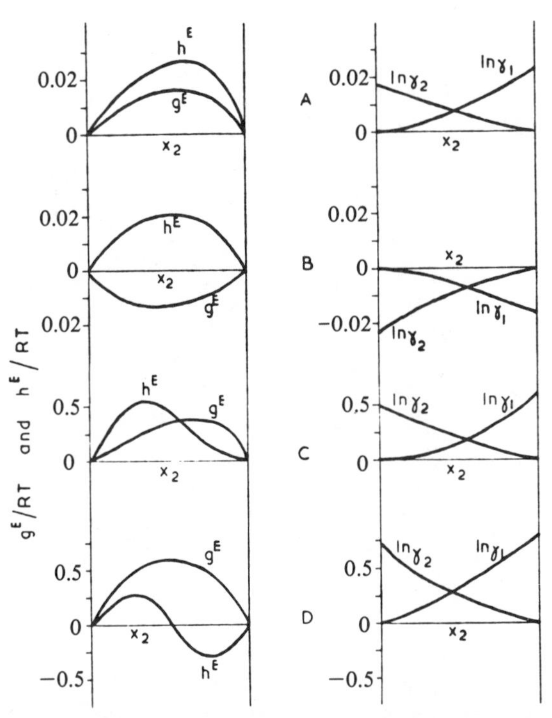

Figure 4.4. Excess Gibbs energies and enthalpies and activity coefficients of four systems [Bett, Rowlinson, & Saville, *Thermodynamics for Chemical Engineers* (1975)].

A. A(1) + O_2(2) at 86 K.

B. n-C_6H_{14}(1) + n-$C_{12}H_{26}$(2) at 25 C.

C. C_6H_6(1) + C_2H_5OH(2) at 45 C.

D. C_4H_8O (dioxane)(1) + H_2O(2) at 25 C.

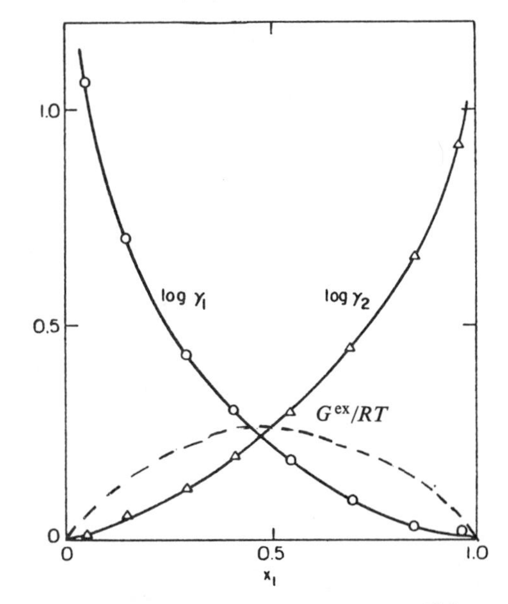

4.5(a) Ethanol + methlycyclohexane (Redlich, 1976).

Figure 4.5. Three systems that have very nearly symmetric activity-coefficient behavior.

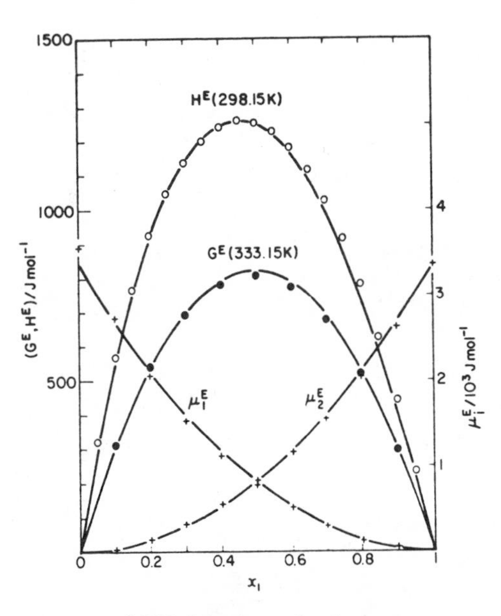

4.5(b) 2-butanone + n-hexane

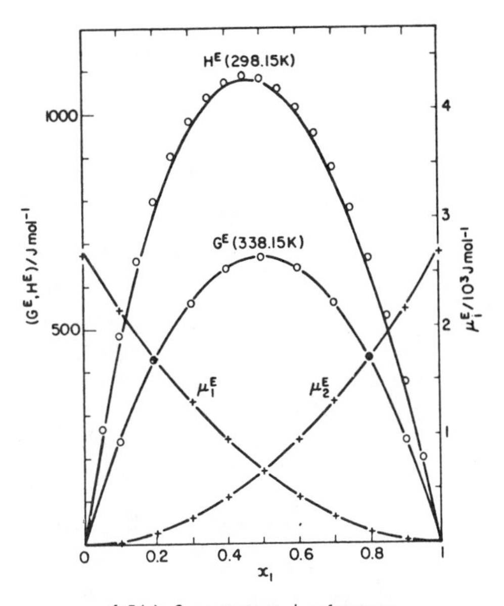

4.5(c) 3-pentanone + n-heptane

(b and c from Kehiaian et al., 1981).

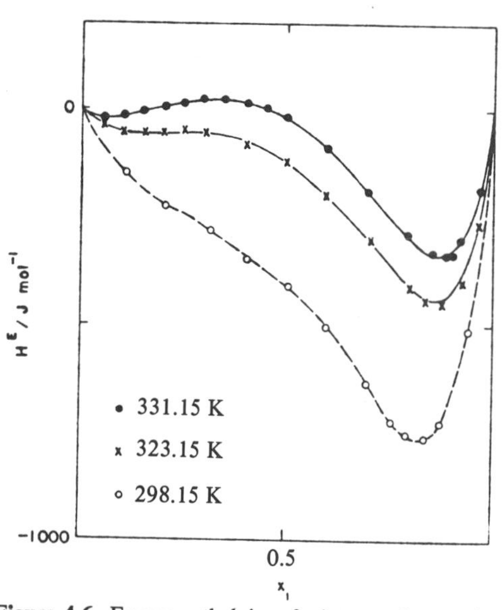

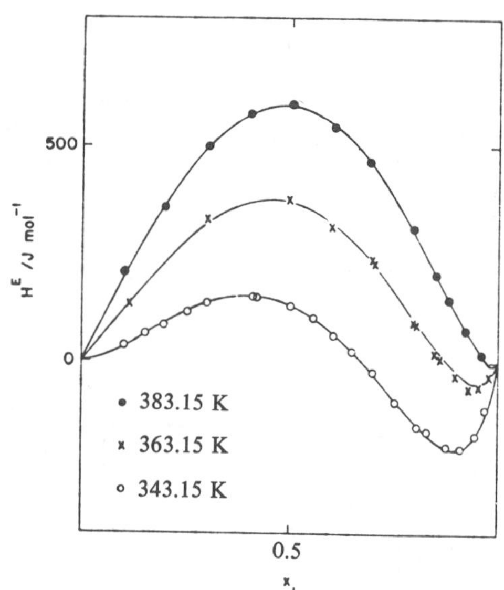

Figure 4.6. Excess enthalpies of mixtures of water + ethanol at several temperatures [Larkin, in International Data Series B, *Thermodynamic Properties of Aqueous Organic Systems*, Nos. 56 and 57 (1978)].

Example 4.1. Excess Properties Derived from Measurements of Activity Coefficients and Excess Volumes

The data from which other properties are to be obtained are stated in lines 1, 2, and 3. The conditions are 300 K, 25 atm, and $x_1 = 0.3$. The value of the gas constant is 1.987 cal/gmol-K or 82.07 ml-atm/gmol-K.

1. $\ln \gamma_1 = 0.4T^{0.2}(1-x_1)^2 = 0.6133.$
2. $\ln \gamma_2 = 0.4T^{0.2}x_1^2 = 0.1126.$
3. $V^{ex} = 25x_1x_2(1+0.2x_1^2)\left(\dfrac{50}{P+1}\right)^{0.02}$

   $= 5.4149$ ml/gmol.
4. $G^{ex} = RT(\Sigma x_i \ln \gamma_i) = 0.4RT^{1.2}x_1x_2$

   $= 156.68$ cal/gmol.
5. $(\partial V^{ex}/\partial x_1) = (1 - 2x_1 + 0.6x_1^2 - 0.08x_1^3)\left(\dfrac{50}{P+1}\right)^{0.02}$

   $= 0.4381.$
6. $\bar{V}_1^{ex} = V^{ex} + (1-x_1)\left(\dfrac{\partial V^{ex}}{\partial x_1}\right)$

   $= 5.4149 + 0.7(0.4324)(50/26)^{0.02}$

   $= 5.7216$ ml/gmol.
7. $\bar{V}_2^{ex} = V^{ex} - x_1\left(\dfrac{\partial V^{ex}}{\partial x_1}\right)$

   $= 5.4149 - 0.3(0.4324)(50/26)^{0.02}$

   $= 5.2835$ ml/gmol.
8. $\left(\dfrac{\partial \bar{V}_1^{ex}}{\partial x_1}\right) = (1-x_1)\left(\dfrac{\partial^2 V^{ex}}{\partial x_1^2}\right)$

   $= (1-x_1)(-2+1.2x_1-2.4x_1^2)(50/26)^{0.02}$

   $= -1.3163$ ml/gmol.
9. $\dfrac{\partial \ln \gamma_1}{\partial T} = \dfrac{0.2}{T}\ln \gamma_1 = 0.000409/K.$
10. $\dfrac{\partial \ln \gamma_2}{\partial T} = \dfrac{0.2}{T}\ln \gamma_2 = 0.000075/K.$
11. $\bar{H}_1^{ex} = -RT^2\left(\dfrac{\partial \ln \gamma_1}{\partial T}\right) = 73.12$ cal/gmol.
12. $\bar{H}_2^{ex} = -RT^2\left(\dfrac{\partial \ln \gamma_2}{\partial T}\right) = 13.42$ cal/gmol.
13. $H^{ex} = \Sigma x_i \bar{H}_i^{ex} = 31.33$ cal/gmol.
14. $H = H^{id} + H^{ex} = x_1H_1 + x_2H_2 + 31.33.$
15. $\dfrac{\partial \ln \gamma_1}{\partial P} = \bar{V}_1^{ex}/RT = 5.7216/(300)(1.987)$

    $= 0.00023$/atm.
16. $\dfrac{\partial \ln \gamma_1}{\partial x_1} = \dfrac{1}{RT}\displaystyle\int_0^P \dfrac{\partial \bar{V}_1^{ex}}{\partial x_1}\,dP$

    $= -\dfrac{1.2992}{RT}\displaystyle\int_0^P \left(\dfrac{50}{P+1}\right)^{0.02} dP.$

    $= -\dfrac{1.2992(50)^{0.02}}{300(0.98)(82.07)}(26^{0.98}-1)$

    $= -0.0014.$

Example 4.2. Thermodynamic Consistency of Vapor-Liquid Equilibrium Data of Water and Thiazole at 90 C (Metzger & Disteldorf, *J. Chim. Phys* 50, no. 3, (1953):156; Kogan #316)

The Gibbs-Duhem requirement is that $\int_0^1 \log_{10}(\gamma_1/\gamma_2)\, dx_1 = 0.$

Integrations with Simpson's Rule give:

Area above $\gamma_1/\gamma_2 = 1$ is 0.1917.

Area below $\gamma_1/\gamma_2 = 1$ is 0.1987.

Difference/sum = 0.0070/0.3904 = 0.018, within 2%, a good check.

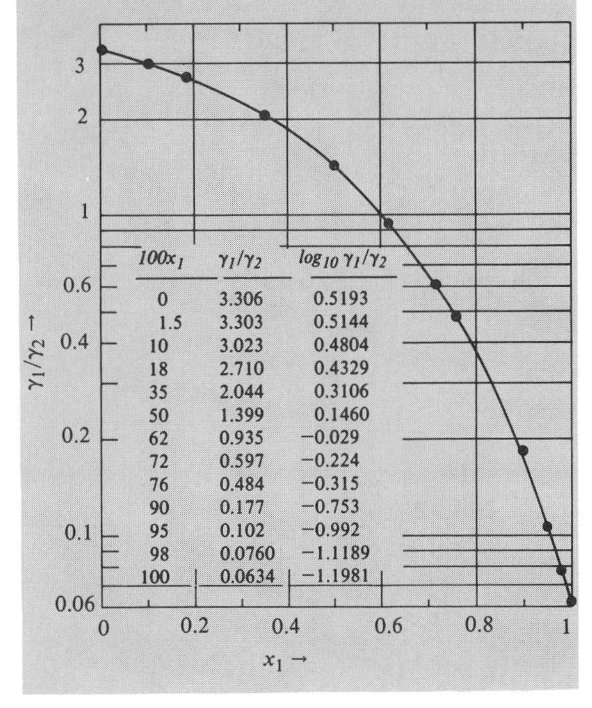

$100x_1$	γ_1/γ_2	$\log_{10}\gamma_1/\gamma_2$
0	3.306	0.5193
1.5	3.303	0.5144
10	3.023	0.4804
18	2.710	0.4329
35	2.044	0.3106
50	1.399	0.1460
62	0.935	−0.029
72	0.597	−0.224
76	0.484	−0.315
90	0.177	−0.753
95	0.102	−0.992
98	0.0760	−1.1189
100	0.0634	−1.1981

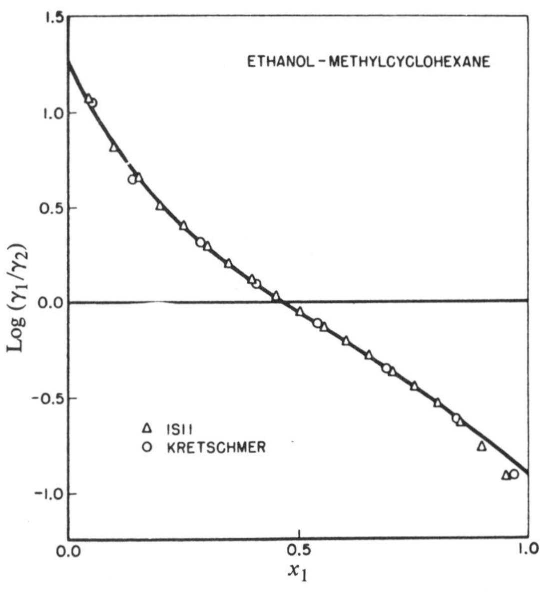

(a) Ethanol + methylcyclohexane. The data are consistent.

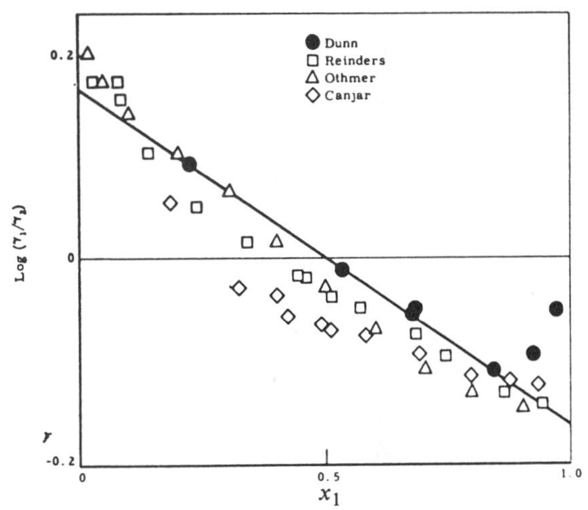

(b) Acetone + benzene. Only the fourth set of data is consistent.

Figure 4.7. Check of consistency of activity coefficients (Redlich, 1976).

4.4. DETERMINATION OF ACTIVITY COEFFICIENTS

Phase-equilibrium measurements are the principal kind from which activity coefficients are determined, usually vapor-liquid or liquid-liquid. Those techniques are described later in the chapter, but some mention may be made now of other methods of more limited scope that are useful in special cases, particularly of solutions of nonvolatile substances. These methods depend on measurements of

1. Osmotic pressure.
2. Freezing-point depression.
3. Boiling-point elevation.
4. Vapor-pressure lowering.

Activities in electrolyte solutions also can be determined from EMF measurements with galvanic cells, but this topic is beyond the present scope. Equations 4.15, 4.27, and 4.28 are the ones directly applicable to evaluations of activity coefficients.

4.4.1. Osmotic Pressure

The physical situation in an osmosis experiment is represented in Figure 4.8. A pure solvent A at a certain T and P and a solution of concentration x_a at the same temperature and initially at the same pressure are separated by a membrane that is permeable only to the solvent. Pressure on the solution side is raised by an amount $\Delta P_{\text{osmotic}}$ sufficient to establish equilibrium across the membrane. At that time the chemical potential balance becomes

$$\mu\,(T, P, x_a = 1) = \mu\,(T,\ P + \Delta P_{\text{osm}}, x_a). \qquad (4.37)$$

The change in chemical potential of the solvent by reason of the increase in pressure on the solution side is:

Table 4.1. Activity Coefficient Relations

1. $\gamma_i = \hat{f}_i / x_i f_i = a_i / x_i$.
2. $RT \ln \gamma_i = \bar{G}_i^{ex} = \bar{G}_i - (G_i + RT \ln x_i)$.
3. $RT \ln \gamma_i x_i = \bar{G}_i - G_i$.
4. $RT \Sigma x_i \ln \gamma_i = G^{ex} = G - G^{id} = G - \Sigma x_i (G_i + RT \ln x_i)$.
5. $RT \Sigma x_i \ln \gamma_i x_i = G - \Sigma x_i G_i$.
6. $\left(\dfrac{\partial \ln \gamma_i}{\partial T}\right)_{Px} = -\dfrac{\bar{H}_i - H_i}{RT^2} = -\dfrac{\bar{H}_i^{ex}}{RT^2}$.
7. $\left(\dfrac{\partial \ln \gamma_i}{\partial P}\right)_{Tx} = \dfrac{\bar{V}_i - V_i}{RT} = \dfrac{\bar{V}_i^{ex}}{RT}$.
8. $\left(\dfrac{\partial \ln \gamma_i}{\partial x_k}\right)_{TP} = \dfrac{1}{RT}\left(\dfrac{\partial \bar{G}_i^{ex}}{\partial x_k}\right) = \dfrac{1}{RT}\displaystyle\int_0^P \left(\dfrac{\partial \bar{V}_i^{ex}}{\partial x_k}\right) dP$.
9. $d \ln \gamma_i = -\dfrac{\bar{H}_i^{ex}}{RT^2} dT + \dfrac{\bar{V}_i^{ex}}{RT} dP$.
10. Gibbs-Duhem equation: $\Sigma x_i \, d \ln \gamma_i = -\dfrac{H^{ex}}{RT^2} dT + \dfrac{V^{ex}}{RT} dP$.
11. Gibbs-Duhem equation at constant T and P: $\Sigma x_i \, d \ln \gamma_i = 0$.
12. For a binary mixture, $\ln \gamma_2 = -\displaystyle\int_{x_1=0}^{x_1} \dfrac{x_1}{1-x_1} d \ln \gamma_1$.

$$\begin{aligned}\Delta\mu &= \mu(T, P + \Delta P_{osm}, x_a) - \mu(T, P, x_a) \\ &= \frac{1}{RT}\int_{P_o}^{P_o+\Delta P_{osm}} \bar{V}_a \, dP \\ &= \mu(T, P, x_a = 1) - \mu(T, P, x_a) \\ &= RT(\ln f_a^{pure} - \ln \hat{f}_a) = -RT \ln a_a = -RT \ln \gamma_a x_a.\end{aligned} \tag{4.38}$$

Finally,

$$\ln a_a = \ln \gamma_a x_a = -\frac{1}{RT}\int_{P_o}^{P_o+\Delta P_{osm}} \bar{V}_a \, dP, \tag{4.39}$$

from which the activity coefficient may be found when the concentration of the solvent and the osmotic pressure, as well as the pressure variation of the partial molal volume of the solvent, are known. With dilute solutions particularly, $\bar{V}$ may be taken constant and equal to the specific volume of the pure solvent.

4.4.2. Freezing-Point Depression

The process is the transfer of solvent from a solution of concentration x_a to the pure solid state at a temperature T that is below the freezing temperature T_0 of the pure solvent. At the freezing point T_0 of the pure solvent,

$$\mu_1(T_0, \text{liquid}, x_a = 1) = \mu_2(T_0, \text{pure solid } A), \tag{4.40}$$

while at the freezing point T of a solution,

$$\mu_3(T, \text{ liquid}, x_a) = \mu_4(T, \text{ solid } A). \tag{4.41}$$

Accordingly, the difference between the chemical potentials of frozen and pure liquid solvents at temperature T becomes

$$\begin{aligned}\Delta\mu &= \mu_4(T, \text{ solid } A) - \mu_5(T, \text{ pure liquid}) \\ &= \mu_3(T, \text{ liquid}, x_a) - \mu_5(T, \text{ pure liquid}) \\ &= RT \ln a_a \\ &= (\mu_1 - \mu_5) - (\mu_2 - \mu_4) = -\int_{T_0}^{T} \frac{H(\text{liquid})}{T} dT \\ &\quad + \int_{T_0}^{T} \frac{H(\text{solid})}{T} dT,\end{aligned} \tag{4.42}$$

which may be written

$$RT \, d \ln a_a = \int_{T_0}^{T} \frac{\Delta H(\text{fusion})}{T} dT, \tag{4.43}$$

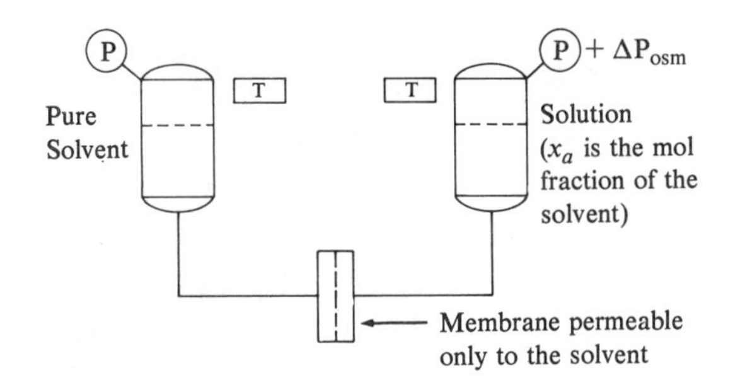

(a) Osmotic-pressure behavior of solutions. ΔP_{osm} is the excess pressure on the solution required to stop flow of solvent through the semipermeable membrane.

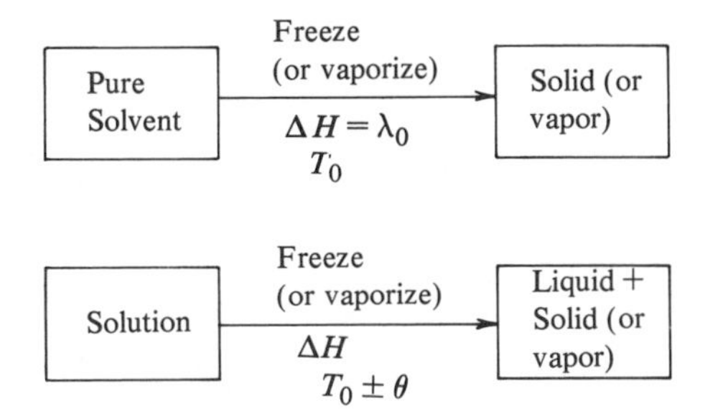

(b) Freezing (or vaporization) behavior of solutions compared with that of a pure liquid. θ is either the freezing-point depression or the boiling-point rise, relative to those of the pure liquid.

Figure 4.8. Activities and changes in colligative properties of solutions, those depending primarily on the numbers of molecules, not their nature.

and

$$\ln a_a = \ln \gamma_a x_a = \int_{T_0}^{T} \frac{\Delta H_f}{RT^2}\, dT. \tag{4.44}$$

Integration of this equation may be performed under several assumptions:

1. When ΔH_f is substantially constant and equal to the enthalpy of freezing, λ_0, of the pure liquid, and the depression of the freezing point is $\theta = T_0 - T$, the integral becomes

$$\ln a_a = -\lambda_0 \theta / R T_0 (T_0 - \theta). \tag{4.45}$$

2. When ΔH_f is expanded in a truncated Taylor series,

$$\Delta H_f = \lambda_0 + (\partial \Delta H_f / \partial T)(T - T_0)$$
$$= \lambda_0 + (C_{p(\text{liquid})} - C_{p(\text{solid})})(T - T_0), \tag{4.46}$$

and the result is integrated,

$$\ln a_a = \frac{(\lambda_0 - T_0 \Delta C_p)\,\theta}{RT_0(T_0 - \theta)} + \frac{\Delta C_p}{R} \ln \frac{T_0 - \theta}{\theta}. \tag{4.47}$$

When $\Delta C_p = 0$, this reduces to Eq. 4.45 as expected.

3. When the integrand $\Delta H_f / T^2$ is expanded similarly,

$$\frac{\Delta H_f}{T^2} = \frac{\lambda_0}{T_0^2} + \frac{1}{T_0^2}\left(\Delta C_p - \frac{2\lambda_0}{T_0}\right)(T - T_0), \tag{4.48}$$

and the integral is

$$\ln a_a = -\frac{\lambda_0 \theta (T_0 + \theta)}{RT_0^3} + \frac{\Delta C_p}{2RT_0^2}\theta^2. \tag{4.49}$$

Eq. 4.49 is easier to solve for θ if that is desired than is Eq. 4.47, but it does not reduce to Eq. 4.45 when $\Delta C_p = 0$. When that is the case, however, the ratio of the right-hand sides of Eqs. 4.45 and 4.49 is

$$1/[1 - (\theta/T_0)^2],$$

which differs by less than 1 percent from unity when $\theta/T_0 < 0.1$; in the case of water this occurs when the freezing-point depression is less than 27 K or so.

Integration of the equation also is accomplished readily when the enthalpy of crystallization is known as a polynomial in T; such an equation but for enthalpy of vaporization is cited in Problem 4.8.

4.4.3. Boiling-Point Elevation

When the solute is nonvolatile, the development for the boiling-point rise, $\theta = T - T_0$, is similar to that for the freezing-point depression. Thus:

$$\ln a_a = \ln \gamma_a x_a = -\int_{T_0}^{T} \frac{H_v - H_L}{RT^2}\, dT$$

$$= -\int_{T_0}^{T_0+\theta} \frac{\Delta H_{\text{vaporization}}}{RT^2}\, dT. \tag{4.50}$$

At constant $\Delta H_v = \lambda_0$, this becomes

$$\ln a_a = -\lambda_0 \theta // R T_0 (T_0 + \theta), \tag{4.51}$$

and when ΔH_v is expanded in the Taylor series,

$$\ln a_a = -\frac{1}{R}\left[\frac{(\lambda_0 - T_0 \Delta C_p)\theta}{T_0(T_0 + \theta)} + \Delta C_p \ln \frac{T_0 + \theta}{T_0}\right]. \tag{4.52}$$

Very dilute solutions have activity coefficients that are substantially unity; then the activity is identical with the mol fraction x_a. In such cases the equations of this section directly relate the mol fraction of the solute, $1 - x_a$, to the osmotic pressure, the freezing-point depression, and the boiling-point elevation. When independent means can be used to find the activity coefficients, the effect of composition on those three properties can be evaluated even for concentrated solutions.

4.4.4. Vapor Pressure Lowering by Nonvolatile Solutes

In the rearrangement of Eq. 4.15,

$$\gamma_i = \left[\frac{y_i \hat{\phi}_i}{\phi_i^{\text{sat}}(PF)_i}\right] \frac{P}{x_i P_i^{\text{sat}}} \longrightarrow \frac{P}{x_i P_i^{\text{sat}}}, \tag{4.53}$$

y_i is unity when the solute is nonvolatile. Moreover, the entire group of terms in brackets also is substantially unity at pressures not far removed from atmospheric. Consequently, the activity coefficient is obtained simply from the measured

Example 4.3. Activity Coefficients and Vapor Pressures of Aqueous Glycerol Solutions from Freezing-Point Depressions

Data of mol fractions of glycerol, freezing temperature, °C, and vapor pressures of pure water, Torr, are given in the table following. The enthalpy of fusion is 1,436 cal/gmol. Vapor pressure of glycerol is substantially zero.

Applying Eq. 4.45 for the first point,

$$\ln \gamma_w x_w = -\frac{1436\,(1.918)}{1.987(273.15)(273.15-1.918)} = -0.0187$$

$$\gamma_w = \frac{\exp(-0.0187)}{1-0.0176} = 0.9991$$

Applying Eq. 4.53

$$P = \gamma_w\, x_w\, P_w^{sat}$$
$$= 0.9991\,(1-0.0176)(3.98) = 3.906 \text{ Torr.}$$

Solutions for all the compositions are tabulated.

$x_{glycerol}$	T_f °C	P_ω^{sat}	γ_ω	P
0.0176	−1.918	3.98	0.9991	3.906
0.0346	−3.932	3.43	0.9966	3.302
0.0822	−10.58	2.14	0.9795	1.924

pressure, P, the mol fraction of solvent, x_i, and the vapor pressure, P_i^{sat}, of pure solvent.

4.5. ACTIVITY COEFFICIENTS AND EQUATIONS OF STATE

Activity coefficients are properties of the liquid state and they are calculable, by Eq. 4.59, with the aid of equations of state that are applicable to both liquid and vapor phases. Currently, however, only select substances are covered by such equations of state, notably the Soave and the BWR family for light hydrocarbons and a few other gases. Phase distribution ratios $K_i = y_i/x_i$ are expressible as ratios of partial fugacity coefficients,

$$K_i = \hat{\phi}_i^{(L)}/\hat{\phi}_i^{(V)}, \tag{4.54}$$

or partly in terms of activity coefficients,

$$K_i = \gamma_i \phi_i^{sat} P_i^{sat}/\hat{\phi}_i^{(V)} P. \tag{4.54a}$$

When a mixture contains any substances that are not represented by a liquid-vapor equation of state, all activity coefficients must be found from phase equilibrium measurements.

The condition of equality of partial fugacities at equilibrium is expressible in several ways:

$$\hat{f}_i^{(V)} = \hat{f}_i^{(L)}, \tag{4.55}$$

$$y_i \hat{\phi}_i^{(V)} P = x_i \hat{\phi}_i^{(L)} P, \tag{4.56}$$

$$y_i \hat{\phi}_i^{(V)} P = \gamma_i x_i f_i^0 \tag{4.57}$$

$$= \gamma_i x_i \phi_i^{sat} P_i^{sat} (PF)_i. \tag{4.58}$$

Accordingly, the activity coefficient is given in terms of the properties of the liquid phase as

$$\gamma_i = \frac{\hat{\phi}_i^{(L)} P}{\phi_i^{sat} P_i^{sat} (PF)_i}, \tag{4.59}$$

where the fugacity coefficients are obtained from the universal equation of state.

In Example 4.4 the activity coefficients obtained with Eq. 4.54 and the Soave equation are compared with experimental data, with only mediocre results but somewhere in range. In many other cases the check is much better; the Soave equation really is not satisfactory in general for highly unsaturated substances like the isoprene of that example.

Example 4.4. Activity Coefficients with the Soave Equation

Measurements of Mervart [*Coll Czech Chem Communications* 24, no. 12 (1959):4034] on vapor-liquid equilibria in the system isobutene (1)-isoprene (2) gave as one result:

$x_1 = 0.288, y_1 = 0.555, T = 72$ C, $P = 5$ atm.

Use the Soave equation to calculate the activity coefficients of both phases with

1. The liquid-phase partial fugacity coefficients.
2. The vapor-phase partial fugacity coefficients.

	Isobutene	Isoprene	Liquid Mixture	Vapor Mixture
T_c	417.9	484		
P_c	39.5	38.0		
ω	0.190	0.164		
A	15.7528	15.8548		
B	2125.75	2467.4		
C	−33.15	−39.64		
P^o	10.04	3.14		
T_r	0.8259	0.7131		
P_r^o	0.2542	0.0826		
m	0.7743	0.7354		
$a\alpha$	14.588	22.037	19.735	17.714
b	0.0752	0.0906	0.0862	0.0821
A	0.1826	0.0863	0.1230	0.1104
B	0.0267	0.0100	0.0152	0.0145
b_i/b(liq)	0.8724	1.0510		
b_i/b(vap)	0.9160	1.1035		

Example 4.4 *(continued)*

$$a\alpha = (0.288)^2(14.588) + (0.712)^2(22.037) + (0.288)(0.712)\sqrt{14.588(22.037)} = 19.735$$

$$b = \Sigma x_i b_i$$

$$A = aP/R^2T^2$$

$$B = bP/RT$$

Liquid mixture, $x_2 = 0.288$:

$$z_m^3 - z_m^2 + 0.1076 z_m - 0.001871 = 0$$

$$z_m = \underline{0.0217}, 0.0982, 0.8802$$

Pure component 1.

$$z_1^3 - z_1^2 + 0.1552 z_1 - 0.004869 = 0$$

$$z_1 = 0.0425, 0.1400, \underline{0.8174}$$

Pure component 2.

$$z_2^3 - z_2^2 + 0.07606 z_2 - 0.000865 = 0$$

$$z_2 = 0.0139, 0.0679, \underline{0.9182}$$

For the vapor phase, $y_1 = 0.555$

$$z_{mv}^3 - z_{mv}^2 + 0.0957 z_{mv} - 0.001601 = 0$$

$$z_{mv} = 0.0214, 0.0835, \underline{0.8951}$$

$$\ln \phi_1^{sat} = 0.8174 - 1 - \ln(0.8174 - 0.0267) - \frac{0.1826}{0.0267} \ln\left(1 + \frac{0.0267}{0.8174}\right)$$

$$\phi_1^{sat} = 0.8457$$

$$\ln \phi_2^{sat} = 0.9182 - 1 - \ln(0.9182 - 0.01) - \frac{0.0863}{0.010} \ln\left(1 + \frac{0.01}{0.9182}\right)$$

$$\phi_2^{sat} = 0.9240$$

$$\ln \hat{\phi}_1^v = 0.9160(0.8951 - 1) - \ln(0.8951 - 0.0145) + \frac{0.1104}{0.0145}\left(0.9160 - \frac{2[0.555(14.588) + 0.445(17.93)]}{17.714}\right] \times \ln\left(1 + \frac{0.0145}{0.8951}\right)$$

$$\hat{\phi}_1^v = 0.9241$$

$$\ln \hat{\phi}_2^v = 1.1035\,(0.8951 - 1) - \ln(0.8951 - 0.10145) + \frac{0.1104}{0.0145}\left[1.1035 - \frac{2[0.555(17.93) + 0.445(22.04)]}{17.714}\right] \times \ln\left(1 + \frac{0.0145}{0.8951}\right)$$

$$\hat{\phi}_2^v = 0.8812$$

$$\ln \hat{\phi}_1^L = 0.8724(0.0217 - 1) - \ln(0.0217 - 0.0152) + \frac{0.1230}{0.0152}\left[0.8724 - \frac{2[0.288(14.588) + 0.712(17.93)]}{19.735}\right] \times \ln\left(1 + \frac{0.0152}{0.0217}\right)$$

$$\hat{\phi}_1^L = 1.7214$$

$$\ln \hat{\phi}_2^L = 1.0510(0.0217 - 1) - \ln(0.0217 - 0.0152) + \frac{0.1230}{0.0152}\left[1.0510 - \frac{2[0.288(17.93) + 0.712(22.037)]}{19.735}\right] \times \ln\left(1 + \frac{0.0152}{0.0217}\right)$$

$$\hat{\phi}_2^L = 0.5732$$

$$(PF)_1 = \exp\frac{94.46\,(5 - 10.04)}{82.06(345.16)} = 0.9833$$

$$(PF)_2 = \exp\frac{100.03\,(5 - 3.14)}{82.06(345.16)} = 1.0066$$

Activity coefficients are figured from the two equations

$$\gamma_i = \frac{\hat{\phi}_i^L P}{\phi_i^{sat} P_i^{sat}(PF)_i} \qquad (1)(\text{Eq. } 4.59)$$

$$\gamma_i = \frac{\hat{\phi}_i^v P}{\phi_i^{sat} P_i^{sat}(PF)_i}\left(\frac{y_i}{x_i}\right) \qquad (2)(\text{Eqs. } 4.15, 4.58)$$

	γ		
Component	*Eq. 1*	*Eq. 2*	*Mervart*
Isobutene	1.0309	1.0665	1.17
Isoprene	0.9811	0.9429	1.06

The values calculated from the liquid and vapor phases are roughly comparable. The calculation method and vapor-pressure data used by Mervart have not been checked, but the comparison with the present calculation is not good.

4.6. CORRELATION OF DATA

Many equations have been proposed for correlating activity coefficients with composition and to a lesser extent with temperature, some on more or less rational grounds, others purely empirically but with intuition. Usually the composition is expressed in mol fractions, x_i, but the use of volume fractions ϕ_i or molecular surface fractions may be preferable when the molecules differ substantially in size or chemical nature.

At the present time, six more-or-less different kinds of correlations of activity coefficients are finding favor, so most of the attention in this chapter will be devoted to them. Since the superiority of one method over the others is not always clear cut, practice still must rely on experience and analogy. The most comprehensive comparison of five of the methods is made in the DECHEMA Vapor-Liquid Data Collection (1979–date). In that work the best fit of experimental data by these formulas is identified in every case. The summary of Table 4.10 covers 3,563 binary systems, which may be more than half the data that will eventually appear. From this purely statistical analysis, the Wilson equation comes out the best, and the van Laar and UNIQUAC tie for last; but there are also marked differences for particular classes of substances, the NRTL, for instance, coming out markedly the best for aqueous systems.

The equations for binary mixtures are listed in Table 4.4, and the simpler forms applicable to infinite dilution are in Table 4.5. These equations have two parameters each, except the symmetrical Marqules equation, which has one parameter, and the NRTL, which has three. Four of these equations are applicable to multicomponent mixtures without any parameters beyond those of the constituent pairs. These multicomponent formulas are listed in Table 4.6.

From a developmental point of view, activity coefficients derive from excess Gibbs energies but in practice the process is reversed and G^{ex} is evaluated from knowledge of activity coefficients. The basic relations are

$$G^{ex}/RT = \Sigma x_i \ln \gamma_i, \tag{4.60}$$

and

$$\ln \gamma_i = \frac{G^{ex}}{RT} - \sum_{k \neq i} x_k \left(\frac{\partial(G^{ex}/RT)}{\partial x_k}\right)_{TPx_{j \neq ik}} \tag{4.61}$$

and for binary mixtures,

$$\frac{G^{ex}}{RT} = x_1 \ln \gamma_1 + x_2 \ln \gamma_2, \tag{4.62}$$

$$RT \ln \gamma_1 = G^{ex} + (1 - x_1)\left(\frac{\partial G^{ex}}{\partial x_1}\right), \tag{4.63}$$

$$RT \ln \gamma_2 = G^{ex} - x_1 \left(\frac{\partial G^{ex}}{\partial x_1}\right). \tag{4.64}$$

Correlations of excess Gibbs energies are empirical in form or based on semirational grounds. For binary mixtures the following general types of formulas are useful:

$$G^{ex} = x_1 x_2 f(x_1), \tag{4.65}$$

$$G^{ex} = \phi_i \phi_2 f(\phi_1), \tag{4.66}$$

in terms of mol fractions and volume fractions, respectively.

Table 4.2. Excess Gibbs Energies of Binary Mixtures (Nomenclature is in Table 4.4)

Equation	G^{ex}/RT
Symmetrical	Ax_1x_2
Margules	$x_1x_2(A_{21}x_1 + A_{12}x_2)$
van Laar	$1/(1/A_{12}x_1 + 1/A_{21}x_2)$
Wilson	$-x_1 \ln(x_1 + \Lambda_{12}x_2) - x_2 \ln(\Lambda_{21}x_1 + x_2)$
T-K-Wilson	$x_1 \ln \dfrac{x_1 + V_2x_2/V_1}{x_1 + \Lambda_{12}x_2} + x_2 \ln \dfrac{V_1x_1/V_2 + x_2}{\Lambda_{21}x_1 + x_2}$
NRTL	$x_1x_2 \left[\dfrac{\tau_{21}G_{21}}{x_1 + G_{21}x_2} + \dfrac{\tau_{12}G_{12}}{G_{12}x_1 + x_2}\right]$
UNIQUAC	$x_1 \left[\ln \dfrac{\phi_1}{x_1} + \dfrac{q_1 z}{2} \ln \dfrac{\theta_1}{\phi_1} - q_1 \ln(\theta_1 + \theta_2\tau_{21})\right]$
	$+ x_2 \left[\ln \dfrac{\phi_2}{x_2} + \dfrac{q_2 z}{2} \ln \dfrac{\theta_2}{\phi_2} - q_2 \ln(\theta_1\tau_{12} + \theta_2)\right]$
Scatchard-Hildebrand	$\dfrac{(\delta_1 - \delta_2)^2}{RT\left(\dfrac{1}{V_1x_1} + \dfrac{1}{V_2x_2}\right)}$

Table 4.3. Excess Gibbs Energies of Multicomponent Mixtures (Nomenclature is in Table 4.4)

Equation	G^{ex}/RT
Wilson	$-\sum_i x_i \ln\left(\sum_j x_j \Lambda_{ij}\right)$
T-K-Wilson	$\sum_i x_i \left[\ln\left(\sum_j x_j V_j/V_i\right) - \ln\left(\sum_j x_j \Lambda_{ij}\right)\right]$
NRTL	$\sum_i x_i \left[\sum_j \tau_{ji} G_{ji} x_j \Big/ \sum_k G_{ki} x_k\right]$
UNIQUAC	$\sum_i x_i \ln\frac{\phi_i}{x_i} + \frac{z}{2}\sum_i q_i x_i \ln\frac{\theta_i}{\phi_i} - \sum_i q_i x_i \ln\left(\sum_j \theta_j \tau_{ji}\right)$
Scatchard-Hildebrand	$\frac{1}{RT}\sum_i x_i V_i (\delta_i - \bar{\delta})^2$

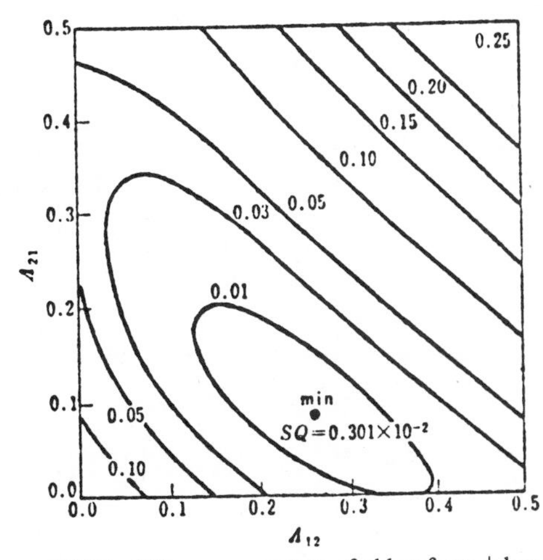

Figure 4.9. Contour plots showing the allowable ranges of multiple parameters for representing activity-coefficient data with specified magnitudes of objective functions, O.F.

Figure 4.9(b). Wilson parameters of chloroform + benzene. O.F. $= [(\Sigma x_i \ln \gamma_i)_{\text{calc}} - (\Sigma x_i \ln \gamma_i)_{\text{exp}}]^2$ (Hirata et al., 1976).

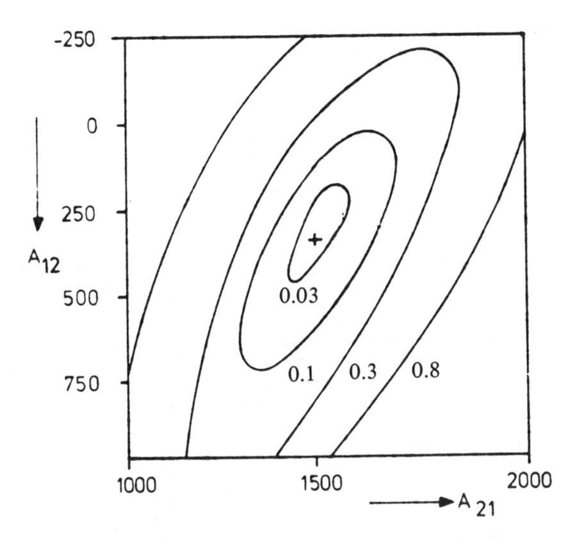

Figure 4.9(a). Wilson parameters for the system acetone + water at one atm. The objective function is

$$\text{O.F.} = \sum_n \sum_i (1 - \gamma_{\text{calc}}/\gamma_{\text{exp}})^2,$$

where the summations are over components, n, and data points, i. At the cross, O.F. = 0.0197 (DECHEMA, VLE Data Collection, I/1, 1979).

Figure 4.9(c). Van Laar parameters for acetone + methanol. The objective function takes into account differences between measured and calculated P, T, x_i, and y_i. The values of the parameters can be expected to lie within the ellipses at confidence levels of 0.87 or 0.99, as shown (Prausnitz et al., 1980, p. 104).

Table 4.4. Equations of Activity Coefficients of Binary Mixtures

Name	*Parameters*	*ln γ_1 and γ_2*	
Symmetrical	A	Ax_2^2	(1)
		Ax_1^2	(2)
Scatchard-Hildebrand	δ_1, δ_2	$\frac{V_1}{RT}(1-\phi_1)^2(\delta_1-\delta_2)^2$	(3)
		$\frac{V_2}{RT}\phi_1^2(\delta_1-\delta_2)^2$	(4)
	$\phi_1 = V_1x_1/(V_1x_1+V_2x_2)$		(5)
Margules	A_{12}	$[A_{12}+2(A_{21}-A_{12})x_1]\,x_2^2$	(6)
	A_{21}	$[A_{21}+2(A_{12}-A_{21})x_2]\,x_1^2$	(7)
van Laar	A_{12}	$A_{12}\left(\frac{A_{21}x_2}{A_{12}x_1+A_{21}x_2}\right)^2$	(8)
	A_{21}	$A_{21}\left(\frac{A_{12}x_1}{A_{12}x_1+A_{21}x_2}\right)^2$	(9)
Wilson	λ_{12}	$-\ln(x_1+\Lambda_{12}x_2)+x_2\left(\frac{\Lambda_{12}}{x_1+\Lambda_{12}x_2}-\frac{\Lambda_{21}}{\Lambda_{21}x_1+x_2}\right)$	(10)
	$\lambda_{21}-\lambda_{22}$	$-\ln(x_2+\Lambda_{21}x_1)-x_1\left(\frac{\Lambda_{12}}{x_1+\Lambda_{12}x_2}-\frac{\Lambda_{21}}{\Lambda_{21}x_1+x_2}\right)$	(11)
	$\Lambda_{12}=\frac{V_2^L}{V_1^L}\exp\left(-\frac{\lambda_{12}}{RT}\right)$	$\Lambda_{21}=\frac{V_1^L}{V_2^L}\exp\left(-\frac{\lambda_{21}}{RT}\right)$	(12, 13)
	V_i^L molar volume of pure liquid component i.		
T-K-Wilson (Tsuboka-Katayama-Wilson)	$a_{12}-a_{11}=\lambda_{12}$ $a_{21}-a_{22}=\lambda_{21}$	$\ln\frac{x_1+V_2x_2/V_1}{x_1+\Lambda_{12}x_2}+(\beta-\beta_v)x_2$	(14)
		$\ln\frac{V_1x_1/V_2+x_2}{\Lambda_{21}x_1+x_2}-(\beta-\beta_v)x_1$	(15)
	$\beta_v=\frac{V_2/V_1}{x_1+V_2x_2/V_1}-\frac{V_1/V_2}{V_1x_1/V_2+x_2}$,	$\beta=\frac{\Lambda_{12}}{x_1+\Lambda_{12}x_2}-\frac{\Lambda_{21}}{\Lambda_{21}x_1+x_2}$	(16)
	Λ_{12} and Λ_{21} as for the Wilson equation.		
NRTL	$g_{12}-g_{22}$ $g_{21}-g_{11}$	$x_2^2\left[\tau_{21}\left(\frac{G_{21}}{x_1+x_2G_{21}}\right)^2+\left(\frac{\tau_{12}G_{12}}{(x_2+x_1G_{12})^2}\right)\right]$	(17)
	α_{12}	$x_1^2\left[\tau_{12}\left(\frac{G_{12}}{x_2+x_1G_{12}}\right)^2+\left(\frac{\tau_{21}G_{21}}{(x_1+x_2G_{21})^2}\right)\right]$	(18)
	$\tau_{12}=\frac{g_{12}-g_{22}}{RT}$	$\tau_{21}=\frac{g_{21}-g_{11}}{RT}$	(19, 20)
	$G_{12}=\exp(-\alpha_{12}\tau_{12})$	$G_{21}=\exp(-\alpha_{21}\tau_{21})$	(21, 22)

Table 4.4. *(continued)*

Name	*Parameters*	*ln γ_1 and γ_2*	
	g_{ij} parameter for interaction between components i and j; $g_{ij} = g_{ji}$		
	α_{ij} nonrandomness parameter; $\alpha_{ij} = \alpha_{ji}$		
UNIQUAC	$u_{12} - u_{22}$	$\ln \gamma_1^C + \ln \gamma_1^R$ [3]	(23)
	$u_{21} - u_{11}$	$\ln \gamma_2^C + \ln \gamma_2^R$	(24)

$$\ln \gamma_1 = \ln \gamma_1^C + \ln \gamma_1^R \tag{25}$$

$$\ln \gamma_1^C = \ln \frac{\varphi_1}{x_1} + \frac{z}{2} q_1 \ln \frac{\vartheta_1}{\varphi_1} + \varphi_2 \left(l_1 - \frac{r_1}{r_2} l_2 \right) \tag{26}$$

$$\ln \gamma_1^R = -q_1 \ln(\vartheta_1 + \theta\tau_{21}) + \vartheta_2 q_2 \left(\frac{\tau_{21}}{\vartheta_1 + \vartheta_2\tau_{21}} - \frac{\tau_{12}}{\vartheta_1\tau_{12} + \vartheta_2} \right) \tag{27}$$

$$\ln \gamma_2 = \ln \gamma_2^C + \ln \gamma_2^R \tag{28}$$

$$\ln \gamma_2^C = \ln \frac{\varphi_2}{x_2} + \frac{z}{2} q_2 \ln \frac{\vartheta_2}{\varphi_2} + \varphi_1 \left(l_2 - \frac{r_2}{r_1} l_1 \right) \tag{29}$$

$$\ln \gamma_2^R = -q_2 \ln(\vartheta_1\tau_{12} + \vartheta_2) + \vartheta_1 q_2 \left(\frac{\tau_{12}}{\vartheta_1\tau_{12} + \vartheta_2} - \frac{\tau_{21}}{\vartheta_1 + \vartheta_2\tau_{21}} \right) \tag{30}$$

$$l_i = \frac{z}{2}(r_i - q_i) - (r_i - 1) \qquad z = 10 \tag{31}$$

q_i area parameter of component i

r_i volume parameter of component i

u_{ij} parameter of interaction between components i and j; $u_{ij} = u_{ji}$

z coordination number

γ_i^C combinatorial part of activity coefficient of component i

γ_i^R residual part of activity coefficient of component i

$$\vartheta_i = \frac{q_i x_i}{\sum_j q_j x_j} \text{ area fraction of component } i \tag{32}$$

$$\phi_i = \frac{r_i x_i}{\sum_j r_j x_j} \text{ volume fraction of component } i \tag{33}$$

$$\tau_{ji} = \exp\left(-\frac{u_{ji} - u_{ii}}{RT} \right) \tag{34}$$

The functional forms may be based on some theoretical mode of interactions between molecules, or may be purely empirical. A common statistical mechanical basis of many of the usual formulas has been shown to exist, for example, by Mollerup (1981). The prefatory terms, x_1x_2 or $\phi_1\phi_2$, clearly belong since G^{ex} must vanish for pure substances. Representative plots of activity coefficients and the corresponding G^{ex} are shown in Figure 4.4.

When the function $f(x_1)$ is taken constant, the resulting equation is called the symmetrical equation or the Margules symmetrical form,

$$\frac{G^{ex}}{RT} = Ax_1x_2. \tag{4.67}$$

On application of Eq. 4.22, the activity coefficients are found to be

$$\ln \gamma_1 = Ax_1x_2 + A(1 - x_1)(1 - 2x_1) = Ax_2^2, \tag{4.68}$$

$$\ln \gamma_2 = Ax_1^2. \tag{4.69}$$

The constant is equal to the limiting values of the logarithms of the activity coefficients, at zero concentration or at infinite dilution, the latter term being more widely used,

Table 4.5. Binary Activity-Coefficient Correlations at Infinite Dilution

van Laar & Margules

$$A_{12} = \ln \gamma_1^\infty \tag{1}$$

$$A_{21} = \ln \gamma_2^\infty \tag{2}$$

Scatchard-Hildebrand

$$\ln \gamma_1^\infty = \frac{V_1}{RT}(\delta_1 - \delta_2)^2 \tag{3}$$

$$\ln \gamma_2^\infty = \frac{V_2}{RT}(\delta_1 - \delta_2)^2 \tag{4}$$

Scatchard-Hildebrand-Flory-Huggins

$$\ln \gamma_i^\infty = \frac{V_1}{RT}(\delta_1 - \delta_2)^2 + \ln \frac{V_1}{V_2} + 1 - \frac{V_1}{V_2} \tag{5}$$

$$\ln \gamma_2^\infty = \frac{V_2}{RT}(\delta_1 - \delta_2)^2 + \ln \frac{V_2}{V_1} + 1 - \frac{V_2}{V_1} \tag{6}$$

Wilson

$$\ln \Lambda_{12} + \Lambda_{21} = 1 - \ln \gamma_1^\infty = k_1 \tag{7}$$

$$\Lambda_{12} + \ln \Lambda_{21} = 1 - \ln \gamma_2^\infty = k_2 \tag{8}$$

$$\Lambda_{12} = \exp(k_1 - \exp(k_2 - \Lambda_{12})) \tag{9}$$

$$\Lambda_{21} = \exp(k_2 - \Lambda_{12}) \tag{10}$$

T-K-Wilson

$$\ln \Lambda_{12} + \Lambda_{21} = \ln \frac{V_2}{V_1 \gamma_1^\infty} + \frac{V_1}{V_2} = k_3 \tag{11}$$

$$\Lambda_{12} + \ln \Lambda_{21} = \ln \frac{V_1}{V_2 \gamma_2^\infty} + \frac{V_2}{V_1} = k_4 \tag{12}$$

$$\Lambda_{12} = \exp[k_3 - \exp(k_4 - \Lambda_{12})] \tag{13}$$

$$\Lambda_{21} = \exp(k_4 - \Lambda_{12}) \tag{14}$$

NRTL

$$\tau_{12}\exp(-\alpha_{12}\tau_{12}) + \tau_{21} = \ln \gamma_1^\infty = k_5 \tag{15}$$

$$\tau_{12} + \tau_{21}\exp(-\alpha_{12}\tau_{21}) = \ln \gamma_2^\infty = k_6 \tag{16}$$

$$\tau_{21} = k_5 - \tau_{12}\exp(-\alpha_{12}\tau_{12}) \tag{17}$$

$$\tau_{12} = k_6 - \tau_{21}\exp(-\alpha_{12}\tau_{21}) \tag{18}$$

UNIQUAC

$$k_7 = \frac{1}{q_1}\left[\ln \frac{r_1}{r_2} - \ln \gamma_1^\infty + 5q_1 \ln \frac{q_1 r_2}{q_2 r_1} - \frac{r_1 l_2}{r_2} + q_1 + l_1\right] + 1 \tag{19}$$

$$k_8 = \frac{1}{q_2}\left[\ln \frac{r_2}{r_1} - \ln \gamma_2^\infty + 5q_2 \ln \frac{q_2 r_1}{q_1 r_2} - \frac{r_2 l_1}{r_1} + q_2 + l_2\right] + 1 \tag{20}$$

$$\tau_{12} = \exp[k_8 - \exp(k_7 - \tau_{12})] \tag{21}$$

$$\tau_{21} = \exp(k_7 - \tau_{12}) \tag{22}$$

Table 4.6. Equations of Activity Coefficients of Multicomponent Mixtures

Equation	*Parameters*	*ln* γ_i
Scatchard-Hildebrand	δ_i	$\frac{V_i}{RT}\left[\delta_i - \sum_j \frac{x_j V_j \delta_j}{\sum_k x_k V_k}\right]^2$
Wilson	$\Lambda_{ij} = \frac{V_j^L}{V_i^L} \exp\left(-\frac{\lambda_{ij}}{RT}\right)$ $\Lambda_{ii} = \Lambda_{jj} = 1$	$- \ln\left(\sum_{j=1}^{m} x_j \Lambda_{ij}\right) + 1 - \sum_{k=1}^{m} \frac{x_k \Lambda_{ki}}{\sum_{j=1}^{m} x_j \Lambda_{kj}}$
NRTL	$\tau_{ji} = \frac{(g_{ji} - g_{ii})}{RT}$ $G_{ji} = \exp(-\alpha_{ji}\tau_{ji})$ $\tau_{ii} = \tau_{jj} = 0$ $G_{ii} = G_{jj} = 1$	$\frac{\sum_{j=1}^{m} \tau_{ji} G_{ji} x_j}{\sum_{l=1}^{m} G_{li} x_l} + \sum_{j=1}^{m} \frac{x_j G_{ij}}{\sum_{l=1}^{m} G_{lj} x_l}\left(\tau_{ij} - \frac{\sum_{n=1}^{m} x_n \tau_{nj} G_{nj}}{\sum_{l=1}^{m} G_{lj} x_l}\right)$
UNIQUAC	$\tau_{ji} = \exp\left(-\frac{u_{ji} - u_{ii}}{RT}\right)$ $\tau_{ii} = \tau_{jj} = 1$ $l_i = \frac{z}{2}(r_i - q_i) - (r_i - 1)$	$\ln \gamma_i^C + \ln \gamma_i^R$ $\ln \gamma_i^C = \ln \frac{\varphi_i}{x_i} + \frac{z}{2} q_i \ln \frac{\vartheta_i}{\varphi_i} + l_i - \frac{\varphi_i}{x_i} \sum_j x_j l_j$ $\ln \gamma_i^R = q_i \left[1 - \ln\left(\sum_{j=1}^{m} \vartheta_j \tau_{ji}\right) - \sum_{j=1}^{m} \frac{\vartheta_j \tau_{ij}}{\sum_{k=1}^{m} \vartheta_k \tau_{kj}}\right]$ $z = 10$
T-K-Wilson	$\Lambda_{ij}, \Lambda_{ji}$	$- \ln\left(\sum_{j=1}^{m} x_j \Lambda_{ij}\right) - \sum_{k=1}^{m} \frac{x_k \Lambda_{ki}}{\sum_{j=1}^{m} x_j \Lambda_{kj}}$ $+ \ln\left(\sum_j x_j V_j / V_i\right) + \sum_k x_k \left(\frac{V_i / V_k}{\sum_j V_j x_j / V_k}\right)$

See Table 4.4 for nomenclature.

$$A = \ln \gamma_1^\infty = \ln \gamma_2^\infty. \quad (4.70)$$

Only mixtures of chemically similar molecules conform to this simple kind of correlation. As an example, Redlich (1976, p. 110) cites the six binary mixtures of ethylbenzene and the isomers of xylene; in those cases, however, the constants range only from 0.0035 to 0.0003, indicating substantially ideal behavior. The mixtures of Figure 4.5 also exhibit nearly symmetrical behavior, but their deviations from ideality are appreciable.

Somewhat more complex formulas are needed to represent activity coefficient data with accuracy of the same order of magnitude as the accuracy of the measurements. The chief formulas that are now widely used will be discussed in this chapter. A detailed treatment of older formulas is provided by Hala et al. (1967). Finding the best values of parameters of the empirical equations from experimental data usually is based on linear or nonlinear least squares, with appropriate constraints or so-called objective functions. Much thought has been devoted to this subject by statistically informed investigators. Typical procedures in the area of interest are described by Hirata et al. (1976), by Prausnitz et al. (1980), and in the prefatory parts of the DECHEMA VLE and LLE Collections (1979 ff.). Some comments on this topic are made in Section 4.9. One point can be emphasized now: When a multiparameter equation is involved, within limits many combinations of values of the parameters can provide comparable accuracy. Figure 4.9 shows several examples of this kind.

For purposes of comparison at this time, the listings of parameters for typical mixtures of the van Laar, Margules, and Wilson equations in Tables E-8, E-9, and E-10 may be of interest.

4.7. THE MARGULES AND RELATED EQUATIONS

The oldest of the formulas still in common use is that of Margules (1895). Its frequently superior performance, as attested by Table 4.10, is perhaps disquieting to believers in progress since his time. Margules's work was before fugacity and activity coefficients were invented, but essentially his proposals are equivalent to these power expansions in composition:

$$\ln \gamma_1 = a_1 x_2^2 + b_1 x_2^3 + \ldots, \tag{4.71}$$

$$\ln \gamma_2 = a_2 x_1^2 + b_2 x_1^3 + \ldots \tag{4.72}$$

for binary mixtures. Generally they are used in the truncated rearrangements of Carlson & Colburn (1942):

$$\ln \gamma_1 = (A + 2(B - A)x_1)x_2^2, \tag{4.73}$$

$$\ln \gamma_2 = (B + 2(A - B)x_2)x_1^2, \tag{4.74}$$

in which the parameters are simply related to the activity coefficients at infinite dilution

$$A = \ln \gamma_1^\infty, \tag{4.75}$$

$$B = \ln \gamma_2^\infty. \tag{4.76}$$

The excess Gibbs energy corresponding to these equations is

$$G^{ex}/RT = x_1 x_2 (A x_2 + B x_1). \tag{4.77}$$

Somewhat generalized equations with more parameters also have been used; thus,

$$G^{ex}/RT = x_1 x_2 (A x_2 + B x_1 + C x_1 x_2 + \ldots) \tag{4.78}$$

The parameters of Eqs. 4.73, 4.74 can be isolated:

$$A = \frac{x_2 - x_1}{x_2^2} \ln \gamma_1 + \frac{2 \ln \gamma_2}{x_1}, \tag{4.79}$$

$$B = \frac{x_1 - x_2}{x_1^2} \ln \gamma_2 + \frac{2 \ln \gamma_1}{x_2}. \tag{4.80}$$

The parameters can be found from a single set of activity coefficients. When more data are available, however, the linearized form can be used:

$$\frac{1}{RT}\left(\frac{\ln \gamma_1}{1 - x_1} + \frac{\ln \gamma_2}{x_1}\right) = A + (B - A)x_1. \tag{4.81}$$

It is applied in Example 4.5.

An expansion similar in form to Eq. 4.78 is that of Redlich & Kister (1948):

$$G^{ex}/RT = x_1 x_2 [B + C(x_1 - x_2) + D(x_1 - x_2)^2 + \ldots], \tag{4.82}$$

from which three-parameter forms of activity-coefficient equations are:

$$\ln \gamma_1 = x_2^2 [B + C(3x_1 - x_2) + D(x_1 - x_2)(5x_1 - x_2)] \tag{4.83}$$

$$\ln \gamma_2 = x_1^2 [B + C(x_1 - 3x_2) + D(x_1 - x_2)(x_1 - 5x_2)]. \tag{4.84}$$

The parameters are evaluated readily from experimental data, as shown in Example 4.6.

An equation due to Scatchard & Hamer (1935) uses volume fractions as a measure of composition; that is,

$$\phi_i = V_i x_i / \Sigma V_j x_j. \tag{4.85}$$

The derivation starts out with

$$G^{ex}/RT = \phi_1 \phi_2 (a + b\phi_1 + c\phi_2), \tag{4.86}$$

from which the activity coefficients are obtained as

$$\ln \gamma_1 = \left[A + 2\left(\frac{BV_1}{V_2} - A\right)\phi_1\right]\phi_2^2, \tag{4.87}$$

$$\ln \gamma_2 = \left[B + 2\left(\frac{AV_2}{V_1} - B\right)\phi_2\right]\phi_1^2. \tag{4.88}$$

This system takes into account an important difference between molecules and should possibly have a role to play in the representation of data; but because it is slightly more

Example 4.5. Linear Plots of the Margules and van Laar Equations

Water + ethanol data of Pemberton & Mash (1978) at 30 C.

Margules: The data are plotted as $G^{ex}/x_w x_a$ against x_a. The regression line is

$$G^{ex}/x_w x_a = 3{,}652.5 - 1{,}353.0\, x_a. \tag{1}$$

Comparing with the standard form

$$G^{ex}/RTx_w x_a = B + (A - B)x_a, \tag{2}$$

the values of the parameters are

$A = 0.9146,$

$B = 1.4514.$

van Laar: The regression line is

$$\sqrt{\ln\gamma_w} = 0.968 - 0.795\sqrt{\ln\gamma_a}. \tag{3}$$

Comparing with

$$\sqrt{\ln\gamma_w} = \sqrt{A} - \sqrt{A/B}\sqrt{\ln\gamma_a}, \tag{4}$$

the values of the parameters are

$A = 0.937,$

$B = 1.4826.$

Extrapolation of the experimental data to infinite dilution with the Lagrange three-point formula gives

$\ln \gamma_w^\infty = 0.906,$

$\ln \gamma_a^\infty = 1.407,$

which check the Margules and van Laar parameters only moderately well.

Example 4.5 *(continued)*

x_a	y_a	p/kPa	$\delta p/kPa$	$ln\ \gamma_w$	$ln\ \gamma_a$	$G^E/J\ mol^{-1}$
0.00435	0.0412	4.413	0.001	0.00004	1.391	15.4
0.01524	0.1280	4.803	0.003	0.00039	1.355	53.0
0.02727	0.2043	5.203	0.002	0.00113	1.320	93.5
0.04633	0.2975	5.781	−0.007	0.00295	1.273	155.7
0.06783	0.3753	6.386	0.003	0.00608	1.221	223.0
0.10991	0.4743	7.329	0.001	0.01709	1.109	345.6
0.17111	0.5479	8.189	0.002	0.04813	0.9205	497.5
0.24688	0.5907	8.723	−0.007	0.1082	0.6926	636.4
0.32385	0.6194	9.085	0.011	0.1820	0.5069	723.9
0.38655	0.6406	9.303	−0.003	0.2471	0.3885	760.5
0.41758	0.6508	9.403	−0.005	0.2811	0.3378	768.2
0.50492	0.6797	9.663	0.001	0.3835	0.2179	755.8
0.58087	0.7087	9.869	0.005	0.4758	0.1398	707.4
0.63434	0.7329	9.999	−0.001	0.5391	0.09889	655.0
0.72455	0.7810	10.199	−0.003	0.6437	0.04932	537.0
0.80840	0.8337	10.341	0.001	0.7447	0.01844	397.2
0.85785	0.8705	10.394	−0.003	0.7987	0.00754	302.5
0.89064	0.8979	10.427	0.003	0.8255	0.00365	235.7
0.89934	0.9056	10.435	0.004	0.8309	0.00301	217.6
0.92444	0.9284	10.445	−0.001	0.8432	0.00181	164.8
0.95370	0.9556	10.457	−0.004	0.8554	0.00103	102.3
0.97315	0.9739	10.467	−0.002	0.8691	0.00052	60.1
0.98153	0.9819	10.473	0.003	0.8788	0.00030	41.6

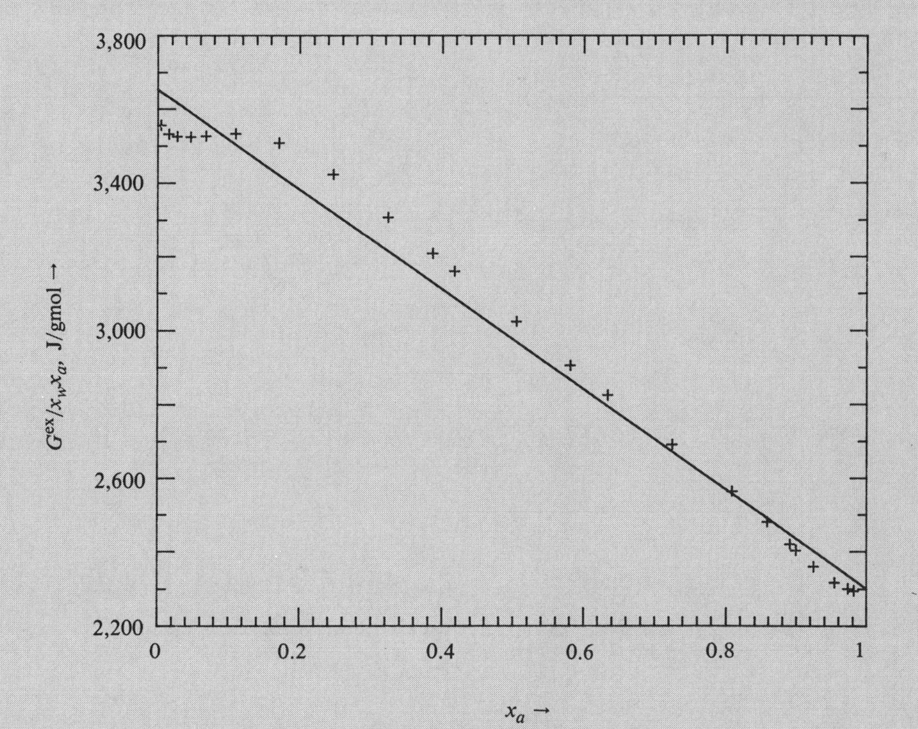

Example 4.5(a). Margules plot.

Example 4.5 *(continued)*

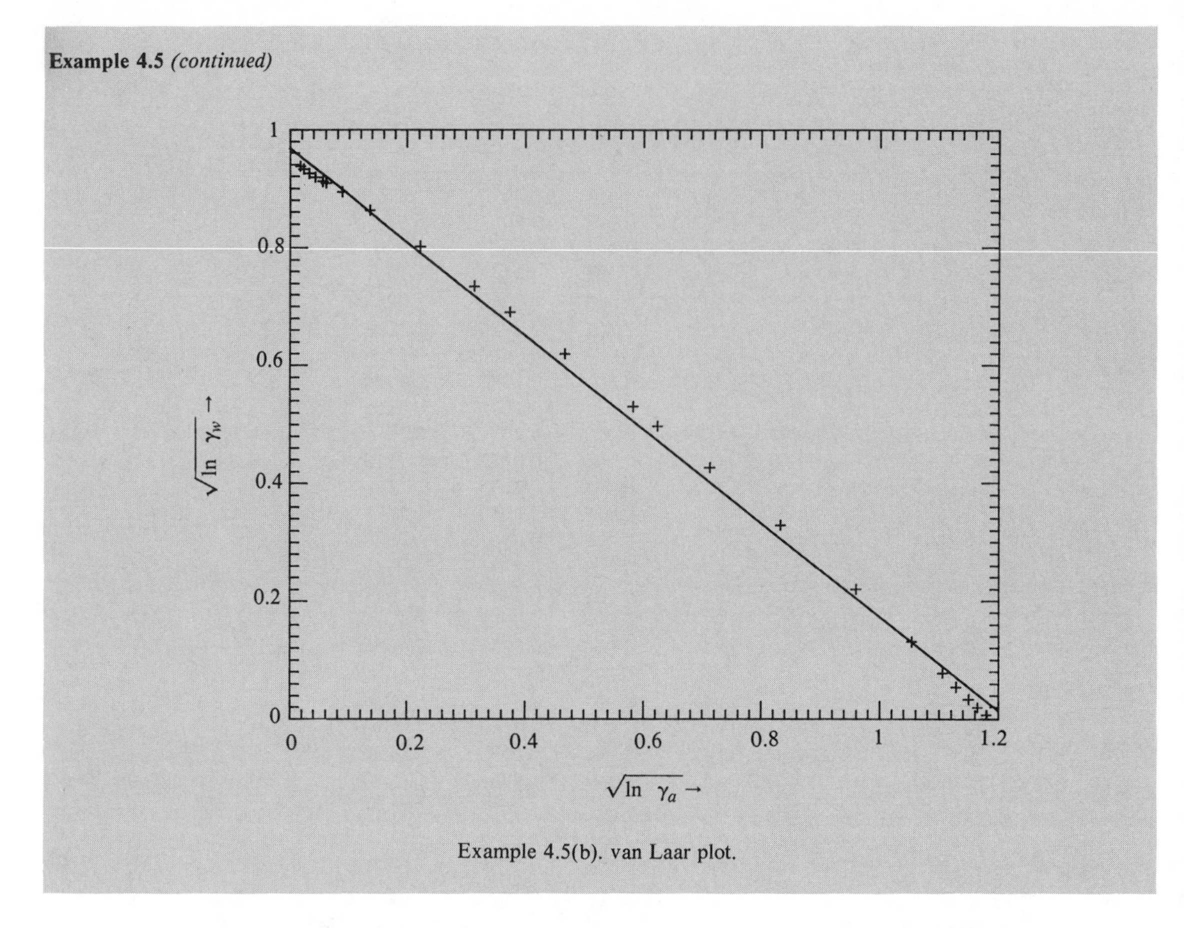

Example 4.5(b). van Laar plot.

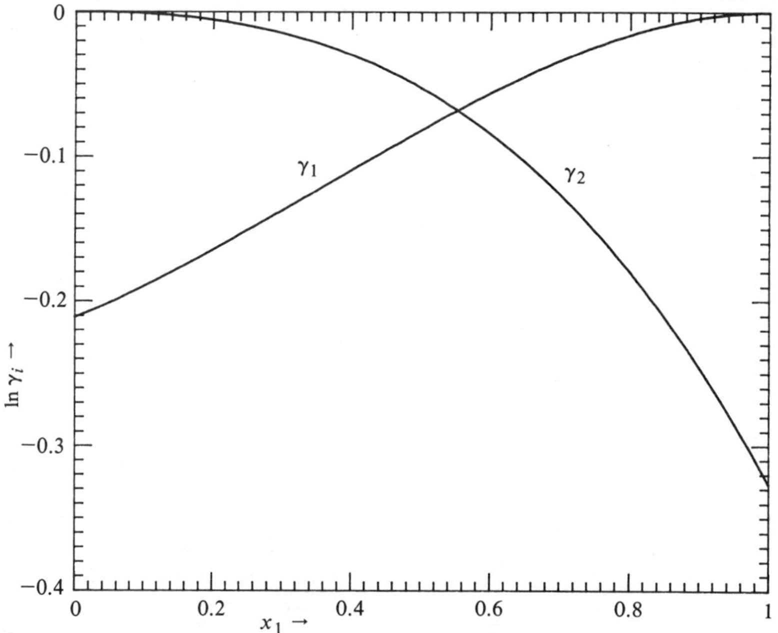

Figure 4.10(a). Uncommon activity-coefficient behavior represented by the Margules equation: Negative values of the parameters. $A = -0.2112$, $B = -0.3270$. Chloroform + methyl acetate. Parameters from Table E.9.

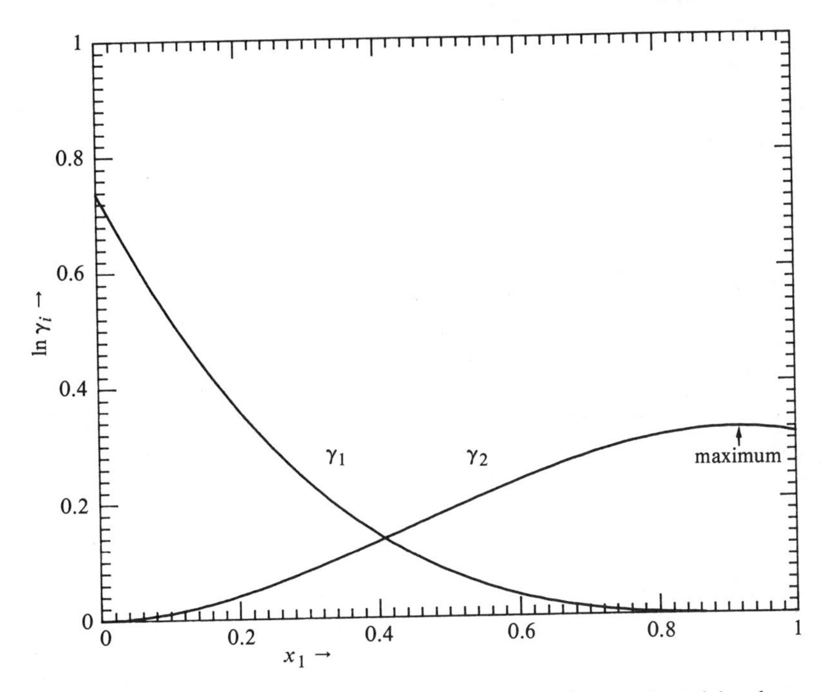

Figure 4.10(b). Uncommon activity-coefficient behavior represented by the Margules equation: A maximum value of an activity coefficient. $A = 0.7358$, $B = 0.3105$. Hexylene glycol + ethylbenzene. Parameters from Table E.9.

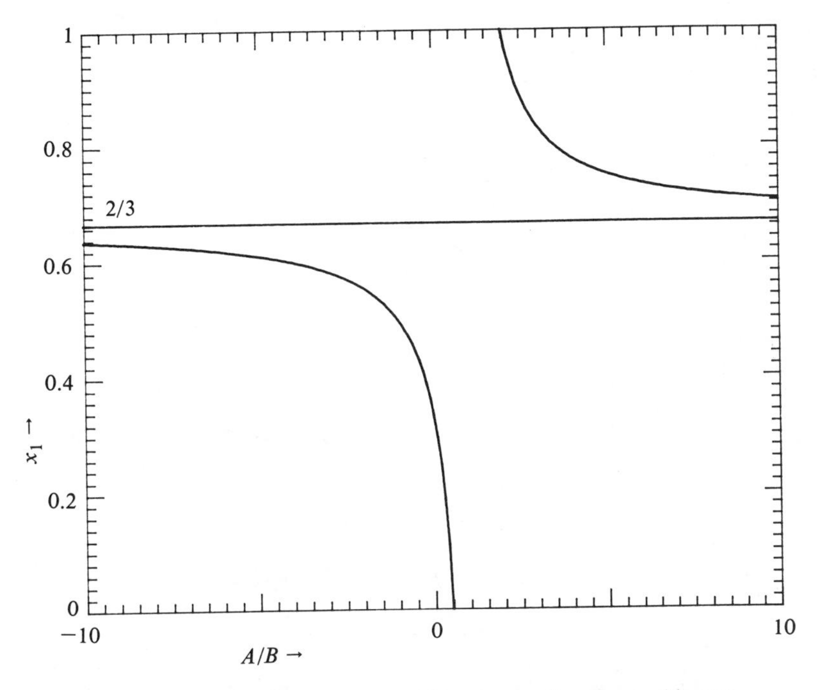

Figure 4.11(a). Extrema represented by the Margules equation: Compositions at which extrema of the activity coefficient occur, as a function of the ratio A/B of the parameters (Eq. 4.89).

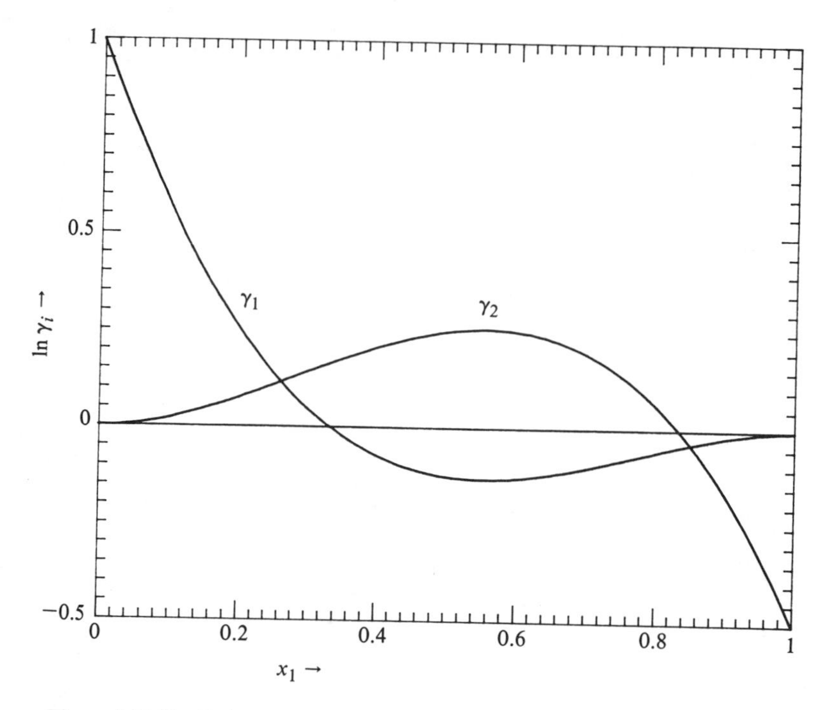

Figure 4.11(b). Extrema represented by the Margules equation: Both activity coefficients have extrema. $A = 1$, $B = -0.5$.

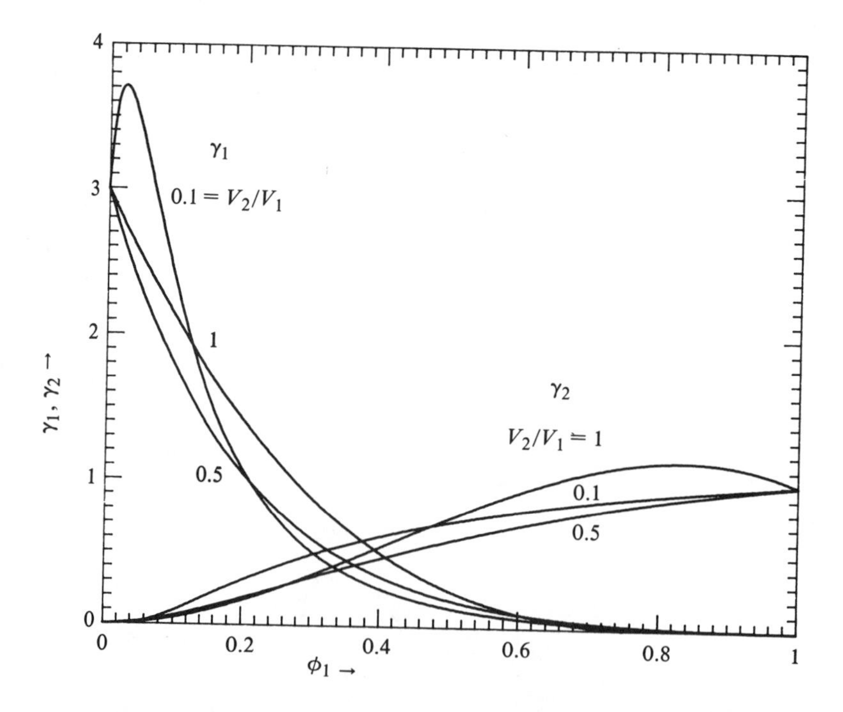

Figure 4.12. Sensitivity of the activity coefficients predicted by the Scatchard-Hamer formulas, Eq. 4.87 & 4.88, to the molal volume fraction. $\phi_1 = V_1x_1/(V_1x_1 + V_2x_2)$. $A = 3$, $B = 1$. When $V_2/V_1 = 1$, the formulas become those of Margules.

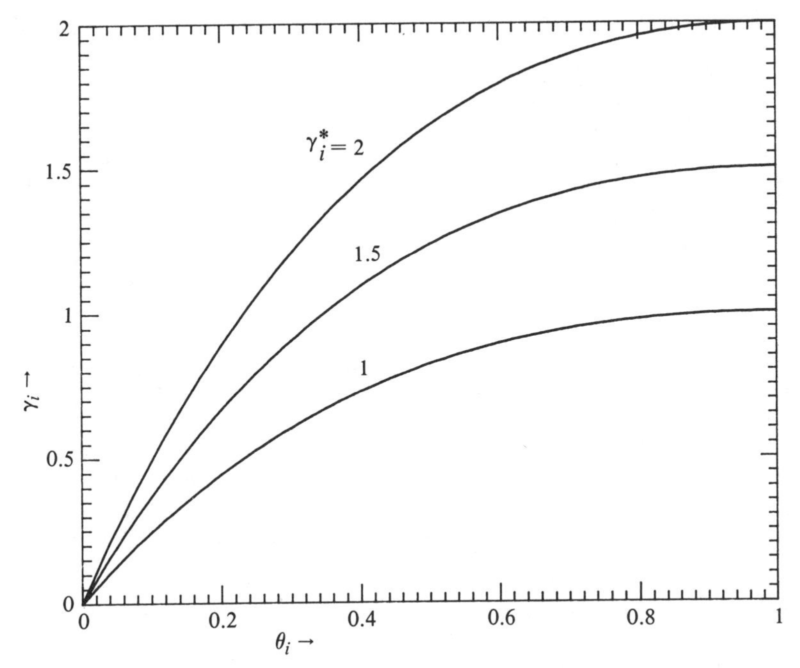

Figure 4.12A. The effect of the volume ratio on the prediction of activity coefficients by the Scatchard-Hildebrand-Flory-Huggins equation.

$\ln \gamma_i = \ln \gamma_i^* + \ln \theta_i - \theta_i + 1.$ SHFH equation

$\ln \gamma_i^* = V_i(\delta_i - \bar{\delta})^2/RT.$ SH equation

$\theta_i = V_i/\Sigma x_j V_j.$

complex than other methods, it does not receive much attention nowadays. The sensitivity of the activity-coefficient profiles to the molal volume fraction is shown in Figure 4.12. For equal molal volumes, the equation becomes that of Margules. Figure 4.17(b) is an example.

Besides the easier procedures for evaluating the parameters, polynomial correlations have two advantages over other equations:

1. One or both of the parameters may be negative. Both van Laar parameters must be of the same sign if data over the full concentration range are to be represented, and both parameters must be positive in equations that involve logarithms of the composition.

2. Data that exhibit a maximum or minimum activity coefficient can be represented. The Margules equation, for instance, has an extremum at the value of composition given by

$$x_1 = \frac{1 - 2A/B}{3(1 - A/B)}, \tag{4.89}$$

which is plotted as Figure 4.11.

4.8. THE VAN LAAR EQUATION

Originally the van Laar (1910, 1913) development was based on the van der Waals EOS; but since the fit of activity-coefficient data with van der Waals parameters is poor, the van Laar equation now is regarded as a purely empirical one. The excess Gibbs energy is related to the mol fractions in reciprocal form

$$\frac{RT}{G^{ex}} = \frac{1}{Ax_1} + \frac{1}{Bx_2}, \tag{4.90}$$

from which the equations of the activity coefficients are

$$\ln \gamma_1 = A\left[\frac{Bx_2}{Ax_1 + Bx_2}\right]^2, \tag{4.91}$$

$$\ln \gamma_2 = B\left[\frac{Ax_1}{Ax_1 + Bx_2}\right]^2. \tag{4.92}$$

The parameters can be calculated from a single set of activity coefficient data with

$$A = \ln \gamma_1^\infty = \ln \gamma_1 \left[1 + \frac{x_2 \ln \gamma_2}{x_1 \ln \gamma_1}\right]^2, \tag{4.93}$$

$$B = \ln \gamma_2^\infty = \ln \gamma_2 \left[1 + \frac{x_1 \ln \gamma_1}{x_2 \ln \gamma_2}\right]^2 \tag{4.94}$$

A linear form of van Laar's equation is due to Black (1958),

$$\sqrt{\ln \gamma_1} = \sqrt{A} - \sqrt{\frac{A}{B}}\sqrt{\ln \gamma_2}. \tag{4.95}$$

This linear form and that of the Margules are applied to

experimental data in Example 4.5, where the representations are seen to be fair.

Extrema of activity coefficients cannot be represented by van Laar's equations, and both parameters must have the same sign if the data are to be representable over the full concentration range. An attempt to remove the latter difficulty was proposed by Null (1970), whose equation uses absolute values of the parameters and reduces to the standard form when the parameters are of the same sign. He wrote:

$$\frac{RT}{G^{ex}} = \frac{\left[\frac{|B|}{ABx_1} + \frac{|A|}{ABx_2}\right]^2}{\frac{1}{Ax_1} + \frac{1}{Bx_2}}. \tag{4.96}$$

Consequently, when the parameters are of opposite signs, the equations for the activity coefficients become

$$\ln \gamma_1 = A(1 - Z)^2[1 + 2Z(AB/|AB| - 1)], \tag{4.97}$$

$$\ln \gamma_2 = BZ^2[1 + 2(1 - Z(AB/|AB| - 1)], \tag{4.98}$$

where

$$Z = \frac{|A|x_1}{|A|x_1 + |B|x_2} \tag{4.99}$$

No evidence was shown in support of this proposal. The comparisons in Figure 4.13 with the Margules equation, which admits of having parameters of opposite signs, certainly are arguments against the Null equations.

The case is that of water + 3-hydroxy-2-butanone, for which the Margules parameters are given in the DECHEMA VLE Collection I/1, 401, as $A = -0.2997$ and $B = 1.5637$; this is the best fit of the five correlations used in that work. The same parameters will hold for the van Laar equation if the infinite-dilution activity coefficients are taken the same. The plots of Eqs. 4.96–4.98 are compared with the Margules plots and clearly disagree with them very markedly, so that the Null modification is not valid in this case. Figure 4.13(b) is a purely artificial example with $A = 1$ and $B = -0.5$, in which considerable disagreement between the Margules and Null equations also is apparent.

Several three-parameter extensions of the van Laar equation have been made, usually with some improvement in accuracy of representation. Black (1958) adds a term to the original excess Gibbs energy, making it

$$G^{ex}/RT = 1/(1/Ax_1 - 1/Bx_2) + Cx_1x_2(x_1 - x_2)^2. \tag{4.100}$$

The corresponding expressions for the activity coefficients are cited in Problem 4.46. The parameters are related to the infinite dilution activity coefficients by

$$\ln \gamma_1^\infty = A + C, \tag{4.101}$$

$$\ln \gamma_2^\infty = B + C. \tag{4.102}$$

One additional measurement suffices to establish all three parameters. The effect of C on the shapes of the curves can be quite marked, as indicated in Figure 4.14.

The parameters of all activity-coefficient equations vary with temperature, but not according to any simple pattern. In Example 4.20 both van Laar parameters decrease with temperature, whereas in Example 4.19, A decreases and B increases.

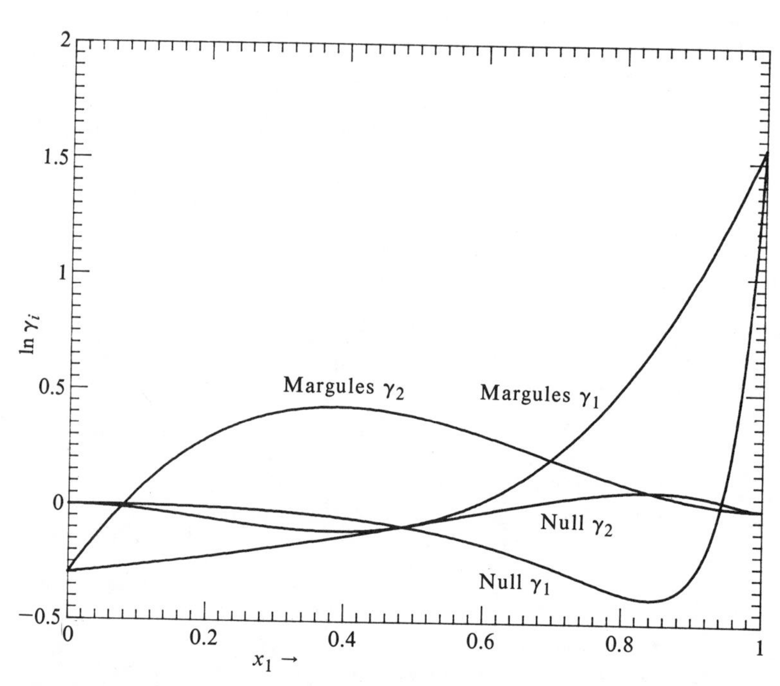

Figure 4.13(a). Margules and Null plots for water + 3-hydroxy-2-butanone, with parameters $A = -0.2977$ and $B = 1.5637$.

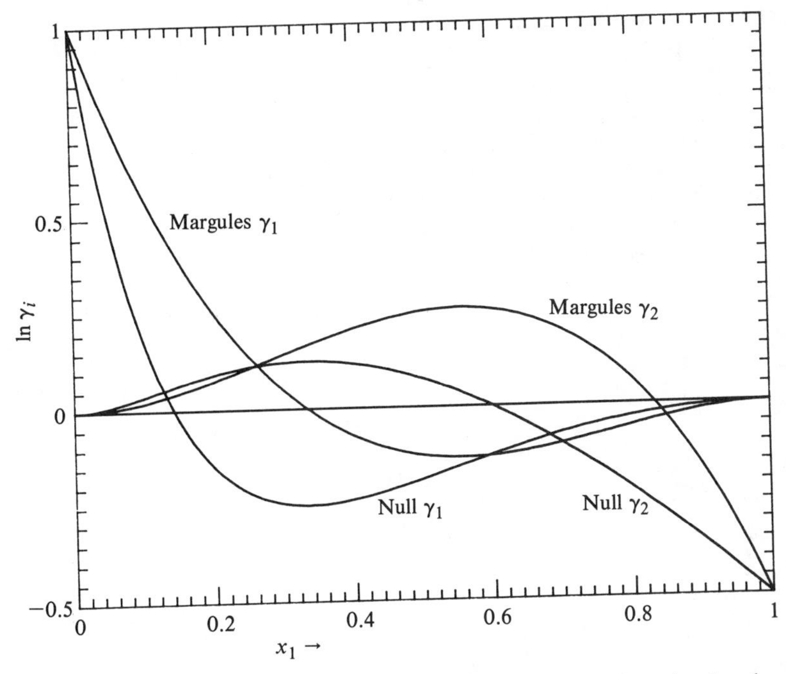

Figure 4.13(b). Margules and Null plots when the parameters are $A = 1$ and $B = -0.5$.

Figure 4.13. Check of Null's modification of the van Laar equations when the parameters are of opposite signs.

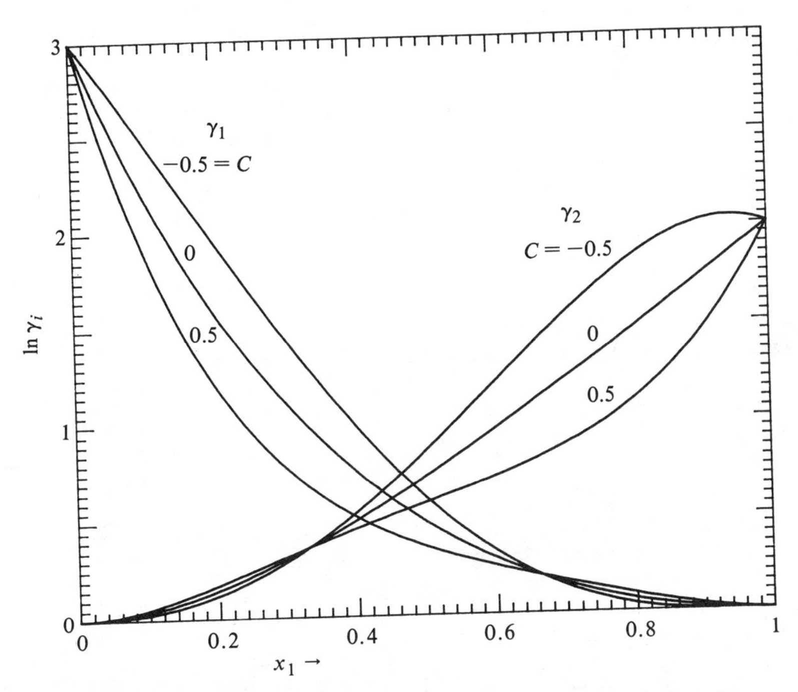

Figure 4.14. The effect of the third parameter, C, in the extended van Laar equation, Eq. 4.100. The equations of the activity coefficients are cited in Problem 4.46. The graphs are drawn for $\ln \gamma_1^\infty = 3$, $\ln \gamma_2^\infty = 2$.

$G^{ex}/RT = 1/(1/Ax_1 + 1/Bx_2) + Cx_1x_2(x_1 - x_2)^2.$

4.9. THE WILSON EQUATION AND SOME MODIFICATIONS

In the development of this equation, Wilson (1964) conceived that interactions between molecules depend primarily on "local concentrations," which he expressed as volume fractions. These concentrations are defined in probabilistic terms, the Boltzmann distribution of energies. Figure 4.15 represents a binary mixture that consists of a mixture of molecules of types 1 and 2. In the vicinity of a molecule of type 1, the probability of finding other molecules of the same type is p_{11} and the probability of finding the other type is p_{12}. Accordingly the ratios of the two probabilities are:

$$\frac{p_{11}}{p_{21}} = \frac{x_1 \exp(-a_{11}/RT)}{x_2 \exp(-a_{21}/RT)}, \tag{4.103}$$

$$\frac{p_{12}}{p_{22}} = \frac{x_1 \exp(-a_{12}/RT)}{x_2 \exp(-a_{22}/RT)}. \tag{4.104}$$

The volume fractions in terms of these probabilities are:

$$z_1 = \frac{p_{11}V_1}{p_{11}V_1 + p_{21}V_2} = \frac{1}{1 + \dfrac{x_2V_2}{x_1V_1}\exp[-(a_{21} - a_{11})/RT]} \tag{4.105}$$

$$= \frac{x_1}{x_1 + \Lambda_{12}x_2}. \tag{4.106}$$

$$z_2 = \frac{1}{1 + \dfrac{x_1V_1}{x_2V_2}\exp[-(a_{12} - a_{22})/RT]} = \frac{x_2}{\Lambda_{21}x_1 + x_2}, \tag{4.107}$$

where

V_i = molal volume

$$\Lambda_{12} = \frac{V_2}{V_1}\exp[-(a_{21} - a_{11})/RT] = \frac{V_2}{V_1}\exp(-\lambda_{12}/RT), \tag{4.108}$$

$$\Lambda_{21} = \frac{V_1}{V_2}\exp[-(a_{12} - a_{22})/RT] = \frac{V_1}{V_2}\exp(-\lambda_{21}/RT), \tag{4.109}$$

$$\lambda_{12} = a_{21} - a_{11}, \tag{4.110}$$

$$\lambda_{21} = a_{12} - a_{22}, \tag{4.111}$$

$$a_{12} = a_{21}. \tag{4.112}$$

Next, the assumption is made that the Gibbs energy depends on $\ln z_i$ in the same way that the ideal value, G^{id}, depends on $\ln x_i$, so that the excess Gibbs energy is represented by

$$G^{ex}/RT = (G - G^{id})/RT = \Sigma x_i \ln(z_i/x_i). \tag{4.113}$$

On substituting for the z_i,

$$G^{ex}/RT = -x_1 \ln(x_1 + \Lambda_{12}x_2) - x_2 \ln(\Lambda_{21}x_1 + x_2). \tag{4.114}$$

Activity coefficients are found by applying Eq. 4.21:

$$\ln \gamma_1 = -\ln(x_1 + \Lambda_{12}x_2) + \beta x_2 \tag{4.115}$$

$$\ln \gamma_2 = -\ln(\Lambda_{21}x_1 + x_2) - \beta x_1, \tag{4.116}$$

$$\beta = \frac{\Lambda_{12}}{x_1 + \Lambda_{12}x_2} - \frac{\Lambda_{21}}{\Lambda_{21}x_1 + x_2}. \tag{4.117}$$

The generalization of these results to any number of components is shown in Table 4.6. As seen there, only binary parameters are required for representing multicomponent mixtures, which makes the Wilson equation much superior to the equations considered earlier in this chapter. Another attractive feature is the built-in effect of temperature, Eqs. 4.108, 4.109. At least to a first approximation the parameters λ_{ij} are independent of temperature. In most collections of Wilson parameters, the λ_{ij} are the ones recorded.

Several attempts have been made to put the Wilson derivation on a more rigorous basis and to improve the equation in some way. Mollerup (1981) derives it as well as several other equations from a local composition version of the

Example 4.6. Finding the Constants of the Redlich-Kister Equation for the System Water + Thiazole of Example 4.2.

Eqns. 4.83 and 4.84 are combined as

$$\ln \gamma_1/\gamma_2 = B(x_2 - x_1) + C(6x_1x_2 - 1) + D(x_2 - x_1)(1 - 8x_1x_2). \tag{1}$$

This result assumes especially simple forms for particular values of the compositions. These are tabulated and the corresponding values of $\ln \gamma_1/\gamma_2$ are read off the figure of Example 4.2.

Condition	x_1	$\ln \gamma_1/\gamma_2$
$x_1 = x_2$	0.5	0.3365
$x_1x_2 = 1/6$	0.2113, 0.7887	1.0438
$x_1x_2 = 1/8$	0.1464, 0.8536	0.9555

Substituting into equation (1),

$$0.3365 = [6(0.25) - 1]C, \tag{2}$$

$$1.0438 = (0.7887 - 0.2113)\left[B + \left(1 - \frac{8}{6}\right)D\right], \tag{3}$$

$$0.9555 = (0.8536 - 0.1464)\left[B + \left(\frac{6}{8}\right)D\right], \tag{4}$$

whence $B = 2.1732$, $C = 0.6730$, and $D = 1.0960$.

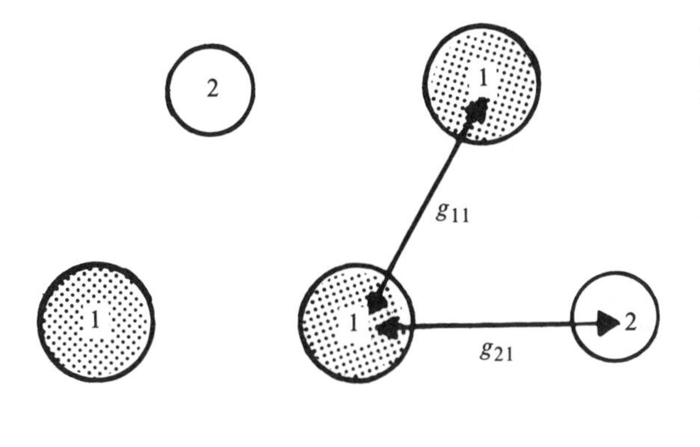

(a). Type 1 molecule at the center of the cluster.

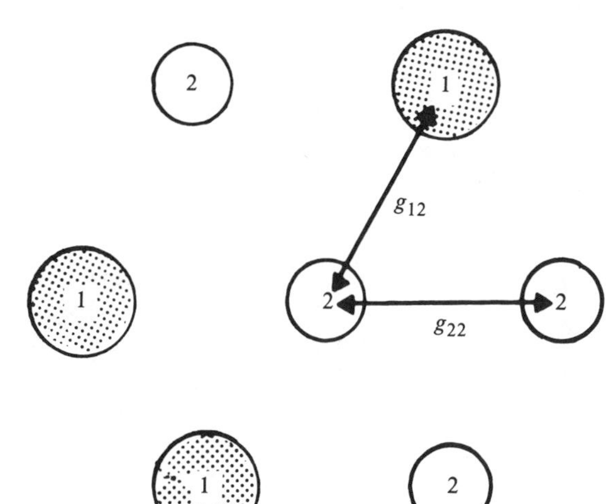

(b). Type 2 molecule at the center of the cluster.

Figure 4.15. Two types of molecular clusters or cells according to Scott's theory of binary liquid mixtures. Each molecule of type 1 or of type 2 is regarded as being surrounded by molecules of both types in proportions determined by the Gibbs energies of interaction, g_{ij}, which are indicated on the diagram.

van der Waals equation, with somewhat different assumptions for each of the cases. Nitta & Katayama (1974) interpret the Wilson equation as a result of an associated solution theory. Tsuboka & Katayama (1975) show it as a special case of their development when the molal volumes of the components are equal. Some of the modifications of Wilson's concepts are discussed later in this chapter.

4.9.1. Evaluation of the Parameters

The Wilson equations are a pair of transcendental equations that are solvable only numerically for the parameters when experimental activity coefficients are known. Given a number of measurements, finding the most reliable values of the parameters is the problem. A criterion called the objective function and designated O.F. must be selected and used as a control for the computation. The O.F. is a less-than deviation of some predicted property from the corresponding measured one. The appropriate objective function to use depends on the kind of data that are available, which may be measurements of T, P, x, or y, or a derived quantity such as $G^{ex}/RT = \Sigma x_i \ln \gamma_i$.

The objective function used in the establishment of the data collection of Hirata et al. (1976) is

Table 4.7. The Wilson and T-K-Wilson Equations Expanded (See also Table C.11)

Wilson

$$\ln \gamma_i = 1 - \ln(x_1\Lambda_{i1} + x_2\Lambda_{i2} + x_3\Lambda_{i3}) - \frac{x_1\Lambda_{1i}}{x_1 + x_2\Lambda_{12} + x_3\Lambda_{13}} - \frac{x_2\Lambda_{2i}}{x_1\Lambda_{21} + x_2 + x_3\Lambda_{23}} - \frac{x_3\Lambda_{3i}}{x_1\Lambda_{31} + x_2\Lambda_{32} + x_3}$$

$\Lambda_{ii} = 1$

T-K-Wilson

$$\ln \gamma_i = \ln \frac{x_1V_{i1} + x_2V_{i2} + x_3V_{i3}}{x_1\Lambda_{i1} + x_2\Lambda_{i2} + x_3\Lambda_{i3}}$$

$$+ x_1\left[\frac{V_{1i}}{x_1 + x_2V_{12} + x_3V_{13}} - \frac{\Lambda_{1i}}{x_1 + x_2\Lambda_{12} + x_3\Lambda_{13}}\right]$$

$$+ x_2\left[\frac{V_{2i}}{x_1V_{21} + x_2 + x_3V_{23}} - \frac{\Lambda_{2i}}{x_1\Lambda_{21} + x_2 + x_3\Lambda_{23}}\right]$$

$$+ x_3\left[\frac{V_{3i}}{x_1V_{31} + x_2V_{32} + x_3} - \frac{\Lambda_{3i}}{x_1\Lambda_{31} + x_2\Lambda_{32} + x_3}\right]$$

$V_{jk} = V_k/V_j$

$\Lambda_{ii} = 1$

$$\text{O.F.} = \Sigma[(G^{ex}/RT)_{calc} - (G^{ex}/RT)_{exp}] \le \varepsilon_G, \quad (4.118)$$

whereas that used by Gmehling & Onken (1979, Part 1) for the DECHEMA collection is either

$$\text{O.F.} = \Sigma(1 - \gamma_{calc}/\gamma_{exp})^2 \le \varepsilon_\gamma, \quad (4.119)$$

or

$$\text{O.F.} = \Sigma(P_{calc} - P_{exp})^2 \le \varepsilon_P. \quad (4.120)$$

When estimates can be made of standard deviations of the several measured properties, Prausnitz et al. (1980, Chap. 6) recommend the use of

$$\text{O.F.} = \sum \left[\frac{(\Delta P)^2}{\sigma_P^2} + \frac{(\Delta T)^2}{\sigma_T^2} + \frac{(\Delta x)^2}{\sigma_x^2} + \frac{(\Delta y)^2}{\sigma_y^2} \right] \leq \varepsilon, \tag{4.121}$$

where the Δs are differences between calculated and experimental values. Typical magnitudes of the standard deviations found by Anderson & Prausnitz (1978) are:

$\sigma_P = 1$ Torr, $\sigma_T = 0.05\ C$, $\sigma_x = 0.001$, and $\sigma_y = 0.0003$.

Pairs of parameters corresponding to a specified maximum of some objective function are not unique. For example, in Figure 4.9 any pair of values within one of the ellipses will correspond to less than the specified value of the objective function. The actual pair found by a particular calculation depends on the algorithm and the starting values of the parameters. Of the many methods that have been used for evaluating parameters, in the area of phase equilibria the methods that have found most favor are nonlinear least squares, gradient search, and simplex pattern search. Hirata et al. (1976) use the first of these; Gmehling & Onken (1979, Part 1) use the last method.

For many purposes the unsophisticated methods about to be described are quite adequate, particularly when the data are so few that a statistical analysis cannot be made properly. The parameters of most activity-coefficient equations, including the Wilson, are comparatively simply related to infinite-dilution activity coefficients. The Wilson equations reduce to

$$\ln \gamma_1^\infty = -\ln \Lambda_{12} + 1 - \Lambda_{21}, \tag{4.122}$$

$$\ln \gamma_2^\infty = -\ln \Lambda_{21} + 1 - \Lambda_{12}. \tag{4.123}$$

These equations can be reduced to an equation in a single variable that is readily solved, namely,

$$\Lambda_{12} = \frac{1}{\gamma_1^\infty} \exp\left[1 - \frac{1}{\gamma_2^\infty} \exp(1 - \Lambda_{12})\right]. \tag{4.124}$$

Then

$$\Lambda_{21} = 1 - \ln(\Lambda_{12}\gamma_1^\infty). \tag{4.125}$$

A starting value for the trial solution of Eq. 4.124 can be read off the nomogram, Fig. 4.16. The practical range of Wilson parameters is quite large. In the collection of about 90 systems in Table E.8, the λ_{ij} range from about −500 to 3,500, which corresponds roughly to a range of Λ_{ij} of 2.5 to 0.005. As is evident from the nomogram, more than one set of parameters is mathematically possible when the infinite-dilution activity coefficients are less than unity. Such possibilities occur with other activity-coefficient equations; some attention will be devoted to multiple roots later in this chapter.

Another easy way to find the parameters holds for equimolal mixtures, in which case also the equations reduce to one in a single variable. The variables β and Λ_{21} are first isolated:

$$\beta = \frac{2\Lambda_{12}}{1 + \Lambda_{12}} - \frac{2\Lambda_{21}}{1 + \Lambda_{21}}, \tag{4.126}$$

$$\Lambda_{21} = \frac{4}{\gamma_1\gamma_2(1 + \Lambda_{12})} - 1, \tag{4.127}$$

then substituted into

$$\Lambda_{12} = -1 + 2\exp(0.5\beta - \ln\gamma_1). \tag{4.128}$$

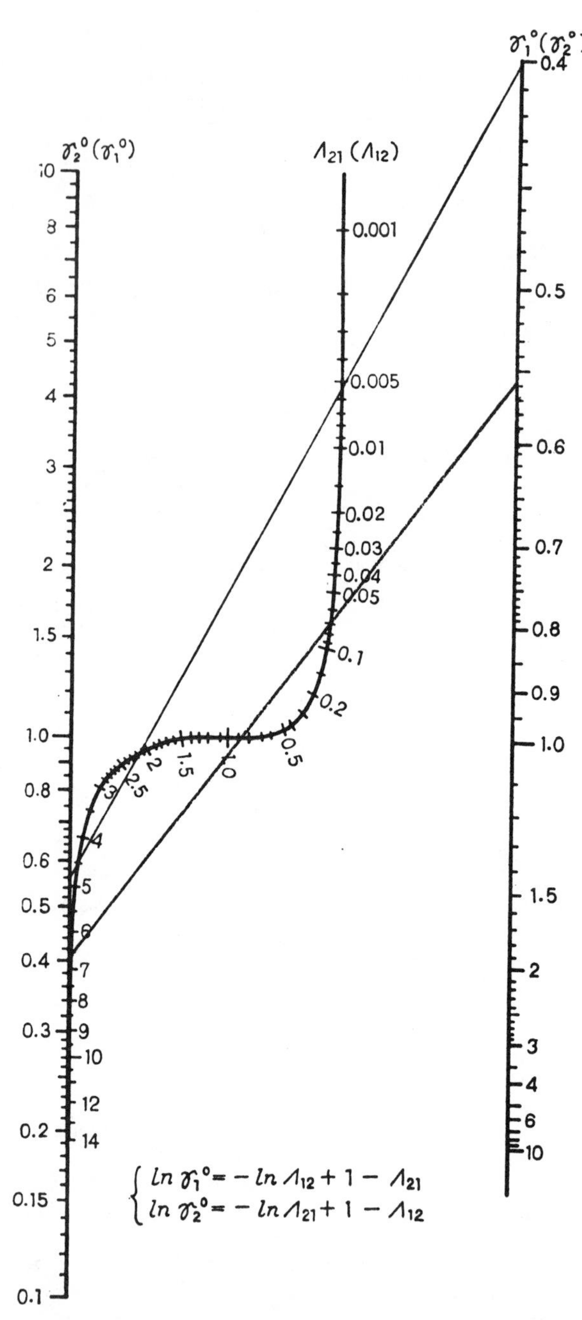

Figure 4.16. Nomogram for Wilson parameters in terms of infinite dilution coefficients (Miyahara et al., 1970). The construction lines show that when $\gamma_1^\infty = 0.4$ and $\gamma_2^\infty = 0.56$, there are three sets of values,

$(\Lambda_{12}, \Lambda_{21}) = (0.08, 4.6), (0.83, 2.15)$ and $(6.7, 0.005)$

It is not possible to tell from the diagram what the proper pairs are but they were established with Eqs. 4.122, 4.123.

Eq. 4.128 contains only one unknown and is readily solvable by trial.

Utilization of any other single pair of γ_i to find the parameters requires the solution of simultaneous nonlinear equations. Example 4.7 adopts the Newton-Raphson method

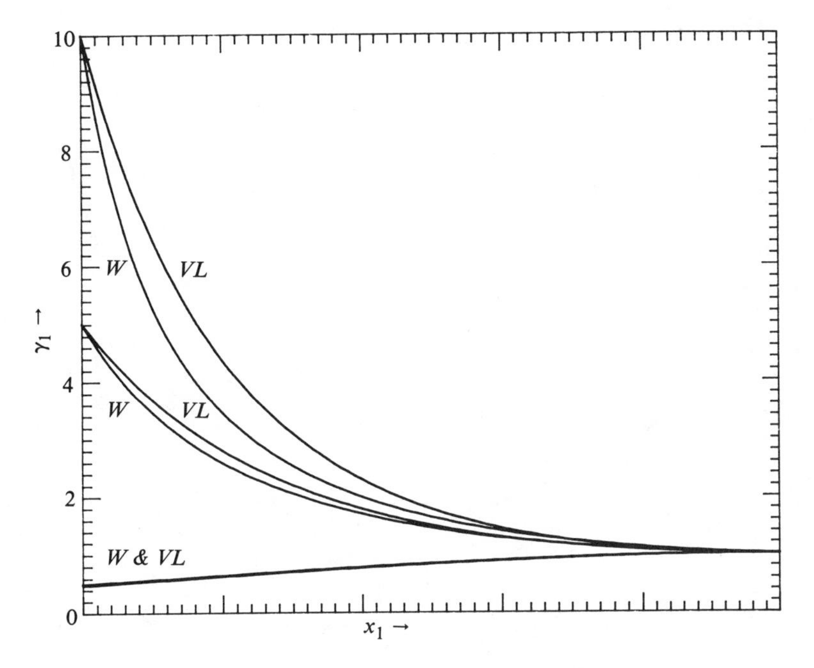

Figure 4.17(a). Comparison of activity-coefficient correlations: Symmetrical Wilson and van Laar equations for the same value of infinite-dilution activity coefficients. The plots of the γ_2 are mirror images of the given γ_1 plots.

for this purpose. A word of caution is in order when using a single pair of γ_i to find the parameters: Negative values may be obtained, which are, of course, unacceptable if the data are to be represented over the entire composition range. An example is when $\gamma_1 = 1.5$ and $\gamma_2 = 2$ at $x_1 = 0.5$; then the parameters come out $\Lambda_{12} = -0.0485$ and $\Lambda_{21} = 0.4013$. On the other hand, it is clear from Eqs. 4.124, 125 that negative values are never obtained with infinite dilution γ's; in fact

$$0 \leq \gamma_i^\infty \Lambda_{ij} \leq 2.303. \tag{4.129}$$

Other arguments in favor of infinite dilution activity coefficients are in Section 4.15.

4.9.2. Pros and Cons

The outstanding features of the Wilson equations are the generally superior representation of activity-coefficient behavior of both polar and nonpolar mixtures, and the capability of representing multicomponent behavior with only binary parameters. They do have a few drawbacks, however:

1. The presence of multiple roots when the γ_i are below unity necessitates making a choice that is not easy to incorporate in an automatic computer program.
2. Negative values of the parameters are not allowable if the data are to be represented over the full composition range.
3. The equations cannot handle liquid-liquid immiscibility. The topic of liquid-liquid equilibria is covered in Chapter 7, but some of the modifications of the Wilson equation that have been made primarily with the object of dealing with this problem have applications to other equilibria, so they will be discussed briefly here.

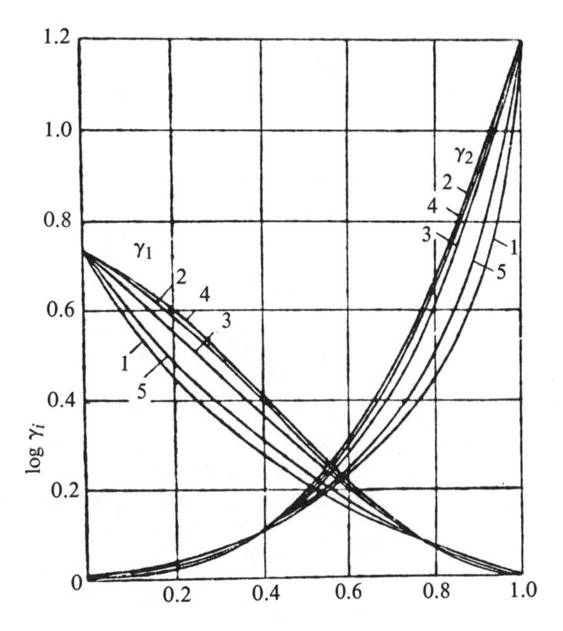

Figure 4.17(b). Comparison of activity-coefficient correlations: Activity coefficients of 2-methylbutene-2 + n-methylpyrollidone. Comparison of experimental values with those obtained from several equations whose parameters are found from the infinite-dilution activity coefficients. (1) Experimental data. (2) Margules equation. (3) van Laar equation. (4) Scatchard-Hamer equation. (5) Wilson equation. (Kogan, 1967, p. 206).

(continued next page)

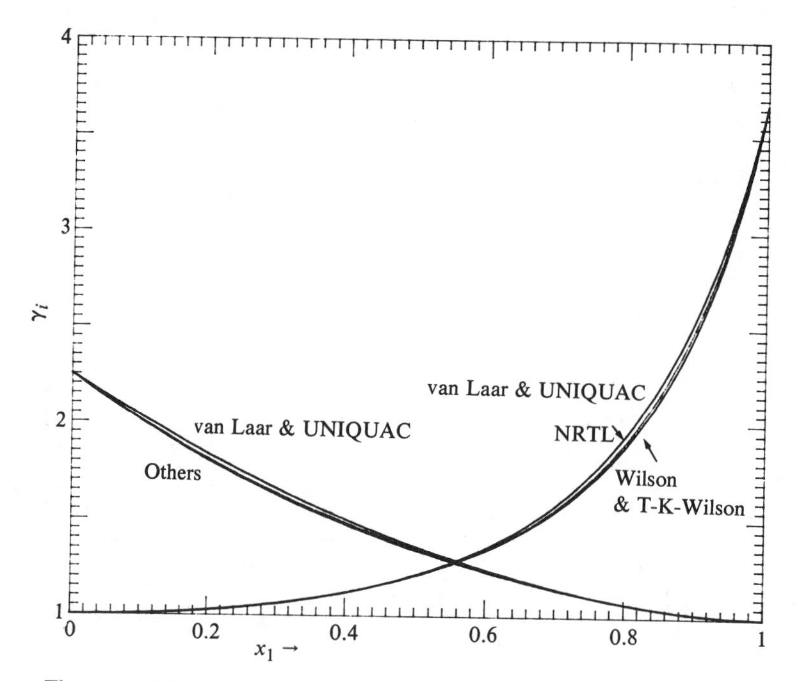

Figure 4.17(c). Correlations of activity coefficients from the infinite-dilution values. Mixtures of n-hexane (1) and diethylketone (2) at 65 C. $\gamma_1^\infty = 2.25$, $\gamma_2^\infty = 3.67$, $V_1/V_2 = 1.236$.

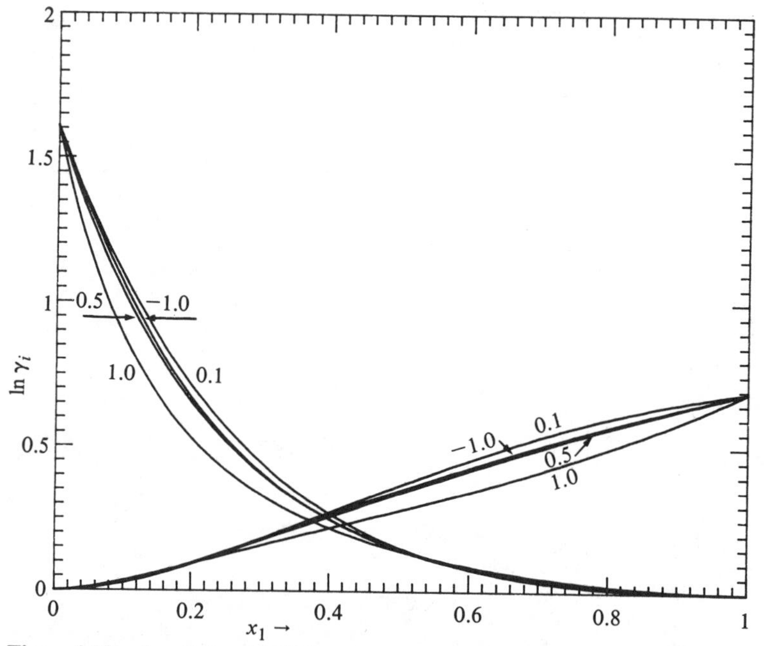

Figure 4.18. Sensitivity of NRTL activity coefficients to values of the parameter α_{12} with $\gamma_1^\infty = 5$ and $\gamma_2^\infty = 2$. The numbers on the graph are the values of this parameter.

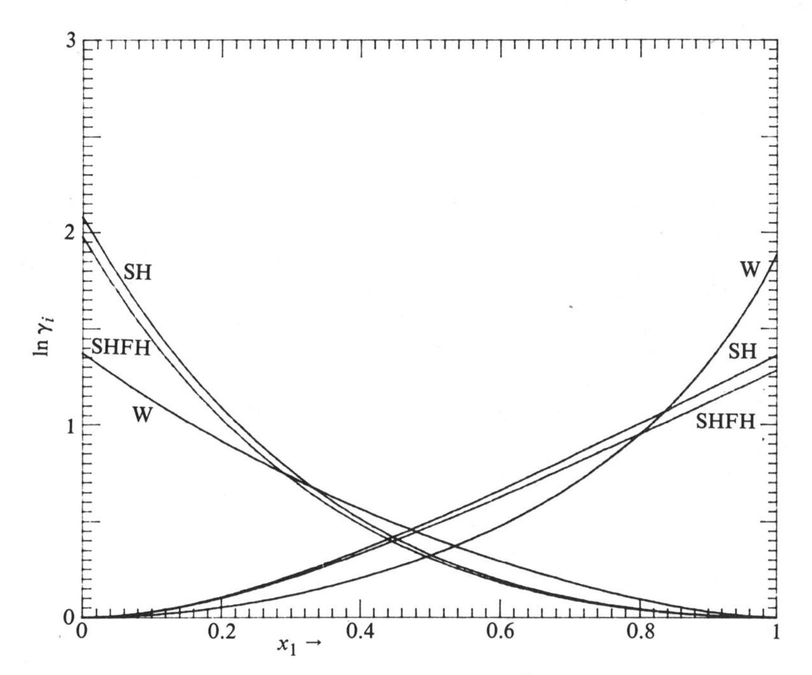

Figure 4.19(a). Mixtures of benzene and ethanol.

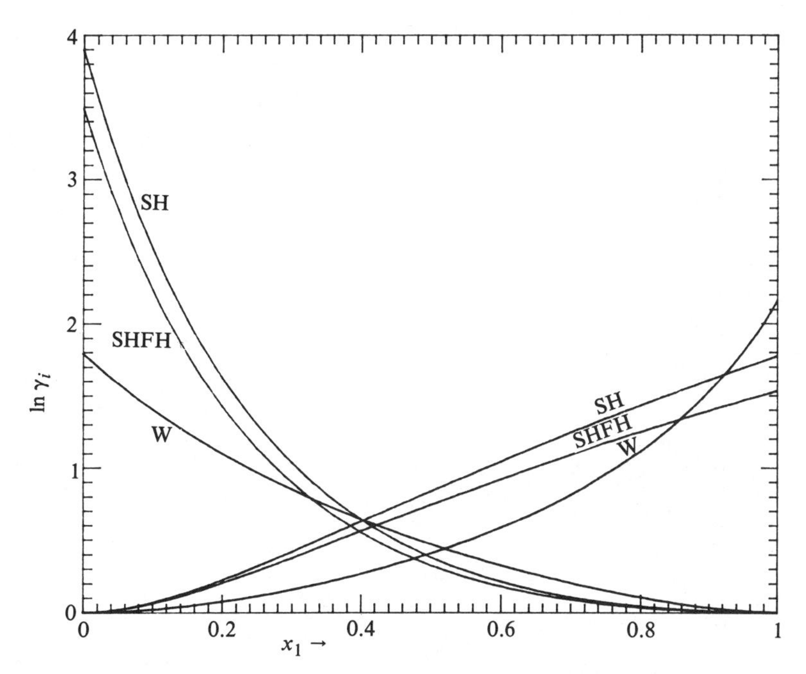

Figure 4.19(b). Mixtures of benzene and methanol.

Figure 4.19. Comparison of activity coefficients obtained from the Wilson equation with parameters from Table E.8, and from the Scatchard-Hildebrand equation with solubility parameters from Appendix D and from the S-H-F-H equation. See also Table 4.12.

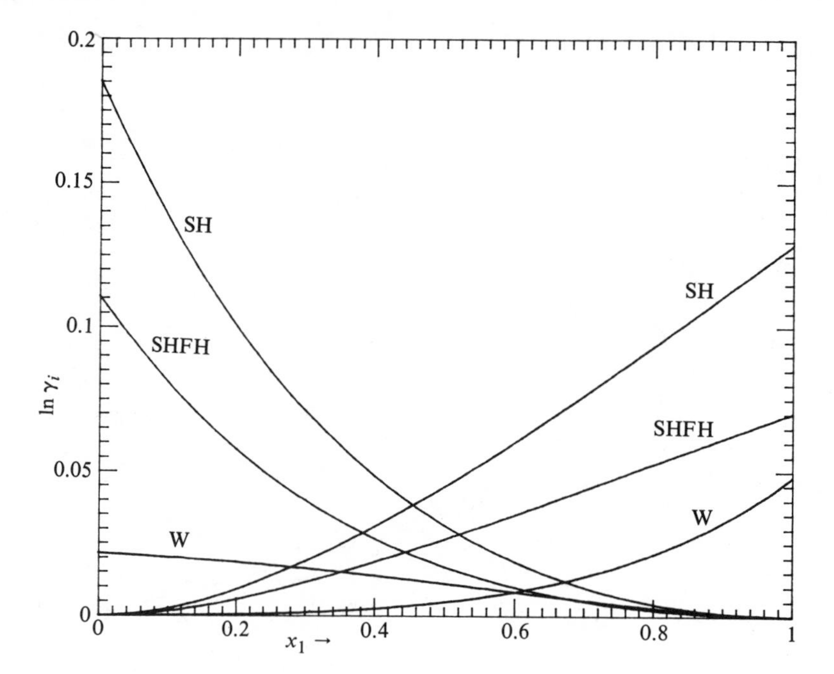

Figure 4.19(c). Mixtures of ethanol and methanol.

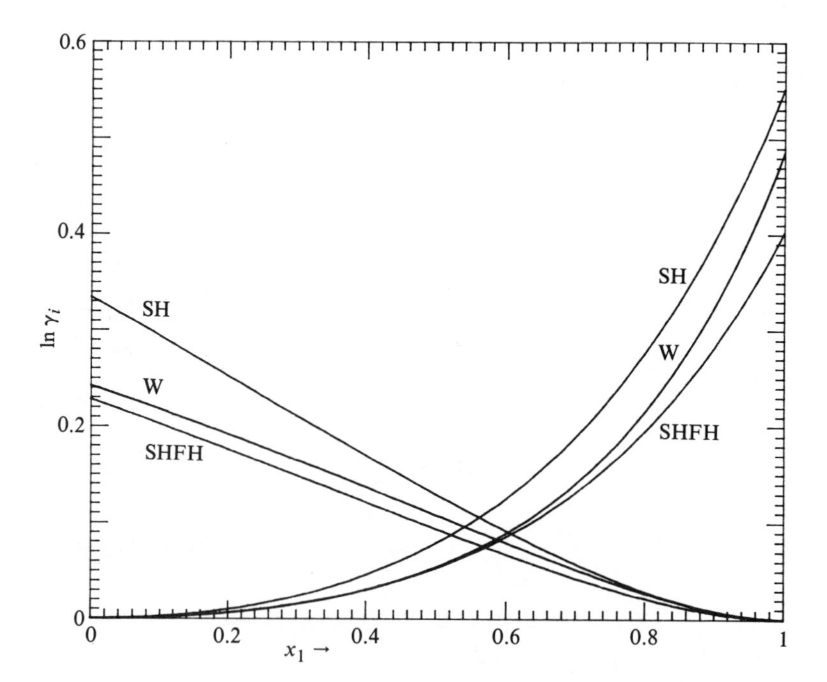

Figure 4.19(d). Mixtures of benzene and n-heptane.

Example 4.7. Wilson Parameters by the Newton-Raphson Method

The Wilson parameters for the mixture of methyl acetate (1) and methanol (2) will be found, given that at 5,907 Torr and $x_1 = 0.736$ the activity coefficients are $\gamma_1 = 1.0224$ and $\gamma_2 = 1.3983$. The Wilson equations are written in the forms,

$$f = \ln\gamma_1 + \ln(x_1 + Ax_2) - x_2\left(\frac{A}{x_1 + Ax_2} - \frac{B}{Bx_1 + x_2}\right) = 0, \quad (1)$$

$$g = \ln\gamma_2 + \ln(Bx_1 + x_2) + x_1\left(\frac{A}{x_1 + Ax_2} - \frac{B}{Bx_1 + x_2}\right) = 0. \quad (2)$$

The derivatives are

$$\frac{\partial f}{\partial A} = A\left(\frac{x_2}{x_1 + Ax_2}\right)^2, \quad (3)$$

$$\frac{\partial f}{\partial B} = \left(\frac{x_2}{Bx_1 + x_2}\right)^2, \quad (4)$$

$$\frac{\partial g}{\partial A} = \left(\frac{x_1}{x_1 + Ax_2}\right)^2, \quad (5)$$

$$\frac{\partial g}{\partial B} = B\left(\frac{x_1}{Bx_1 + x_2}\right)^2. \quad (6)$$

The N-R equations for the corrections, h and k, to initial estimates of A and B are

$$f_0 + h\left(\frac{\partial f}{\partial A}\right)_0 + k\left(\frac{\partial f}{\partial B}\right)_0 = 0, \quad (7)$$

$$g_0 + h\left(\frac{\partial g}{\partial A}\right)_0 + k\left(\frac{\partial g}{\partial B}\right)_0 = 0. \quad (8)$$

Starting with initial estimates, $A_0 = B_0 = 0.5$, and requiring the values to be correct to 0.0001, the successive iterations are

A	B
.5	.5
.57151	.83994
.31395	1.1717
.29528	1.2602
.29236	1.2671

The computer program is in Appendix C.5.

The parameters of the Tsuboka & Katayama modification of the Wilson equation will be obtained similarly. The ratio of volumes is $V_1/V_2 = 0.5107$. The equations for f and g are different, but the derivatives are the same.

$$f = \ln\gamma_1 + \ln\frac{x_1 + Ax_2}{x_1 + 0.5107x_2} - x_2(\beta - \beta_v) = 0, \quad (9)$$

$$g = \ln\gamma_2 + \ln\frac{Bx_1 + x_2}{x_1/0.5107 + x_2} + x_1(\beta - \beta_v) = 0, \quad (10)$$

$$\beta = \frac{A}{x_1 + Ax_2} - \frac{B}{Bx_1 + x_2}, \quad (11)$$

$$\beta_v = \frac{1}{x_1/0.5107 + x_2} - \frac{1}{x_1 + 0.5107x_2}. \quad (12)$$

Six iterations are required for the solution.

A	B
.5	.5
.72203	.85748
.36679	1.291
.33135	1.4531
.32109	1.4782
.32104	1.4785

Example 4.7A. Negative Values of the Wilson Parameters

Negative values of the Wilson parameters may develop when single data sets are used to evaluate them. Thus measurements of the equilibrium of n-heptane (1) + toluene (2) gave the results

$$x_1 = 0.45, y_1 = 0.54, \gamma_1 = 1.0914, \gamma_2 = 2.0893.$$

With the program of Table C.5, the Wilson parameters are

$\Lambda_{12} = -0.2475$,

$\Lambda_{21} = 1.3331$,

and with Eqs. 4.79, 4.80, the Margules parameters are

$A_{12} = 3.3037$,

$A_{21} = -0.0458$.

Wilson parameters corresponding to these Margules infinite-dilution activity coefficients are

$\Lambda_{12} = 0.0059$,

$\Lambda_{21} = 2.8289$.

Activity coefficients from both sets of Wilson parameters are compared with those from the Margules equation on the figures. With the negative Λ_{12}, $(x_1 + \Lambda_{12}x_{12})$ becomes zero at $x_1 = 0.1984$. At some distance removed from this composition, the Wilson and Margules activity coefficients are roughly the same, so perhaps the Wilson could be used to make calculations in such a composition range. It is not safe, however, to assume that even approximately correct values always will be given with negative parameters, so in general they should be avoided. How safe the parameters obtained from Margules infinite-dilution activity coefficients are in any particular case requires an exercise of judgment. The justification for evaluating Wilson parameters from Margules γ_i^∞ is that they are usable for multicomponent equilibria, whereas the Margules are not.

(continued next page)

Example 4.7A *(continued)*

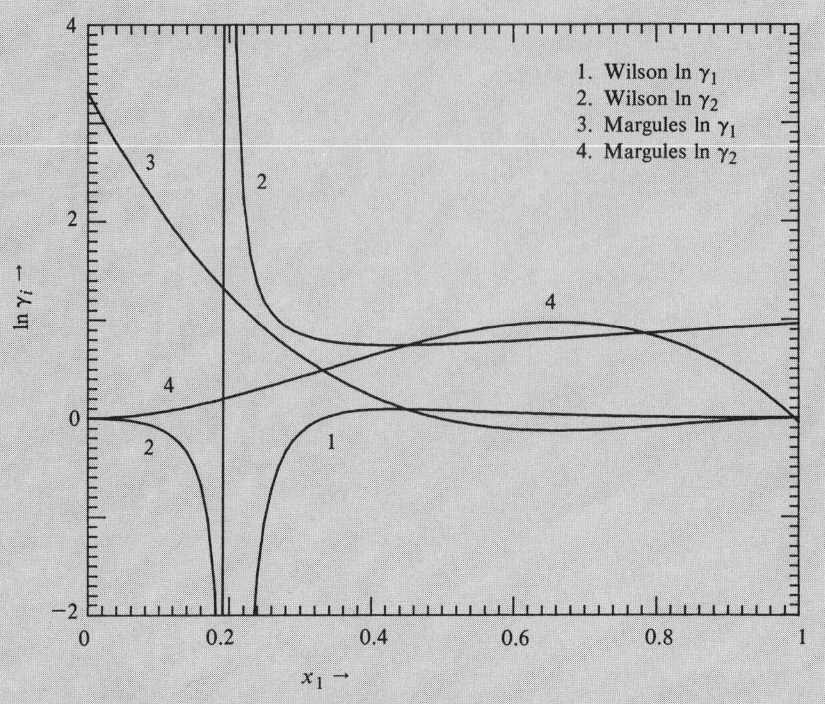

Example 4.7A(a). Comparison of the Margules equation with the Wilson equation possessing a negative parameter.

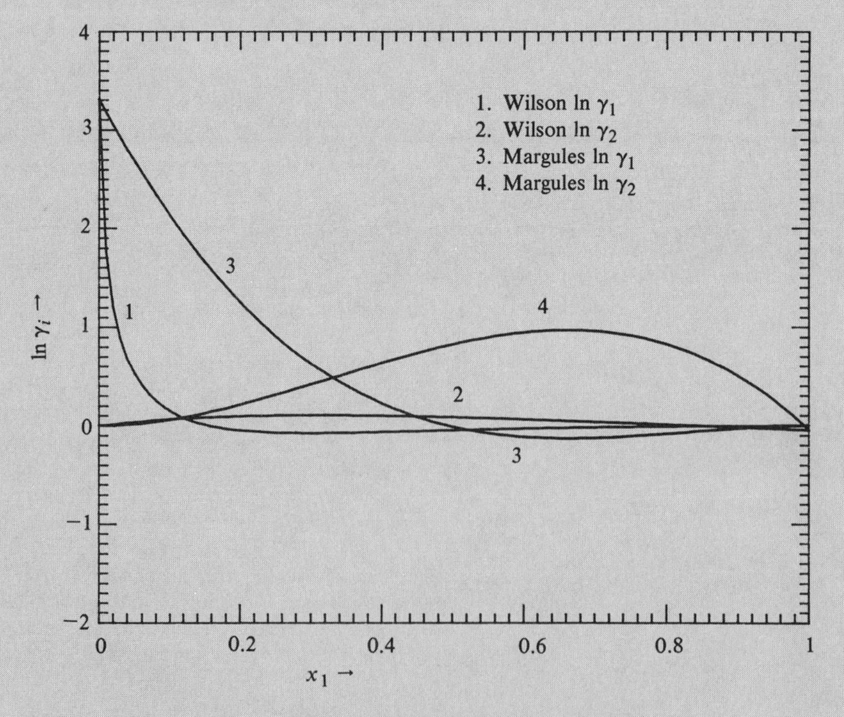

Example 4.7A(b). Comparison of Margules and Wilson equations with parameters based on the same infinite-dilution activity coefficients.

4.9.3. Modifications of the Equation

Simply incorporating a multiplier C_{12} into

$$\ln \gamma_1 = C_{12}[-\ln(x_1 + \Lambda_{12}x_2) + \beta x_2] \quad (4.130)$$

permits representation of liquid-liquid equilibria, but it is not clear that the C_{ij} can be carried over to multicomponent mixtures. Adding a term with a third parameter seems to be on a sounder theoretical basis. Hiranuma (1974, 1975) developed quite a complicated equation with a theoretically based third parameter. This work was continued by Schulte et al. (1980). Nagata et al. (1975) added a term,

$$(\delta_1 - \delta_2)^2\phi_1\phi_2(V_1x_1 + V_2x_2), \quad (4.131)$$

to the excess Gibbs energy; the δ_i are solubility parameters, the ϕ_i are volume fractions, and the V_i are the molal volumes. Since the solubility parameters are properties of pure substances, Nagata's equations remain two-parameter ones.

Perhaps the simplest successful modification is that of Tsuboka & Katayama (1975), which will be identified in this book as the T-K-Wilson equation. For the excess Gibbs energy they write,

$$\frac{G^{ex}}{RT} = x_1 \ln \frac{x_1 + V_{12}x_2}{x_1 + \Lambda_{12}x_2} + x_2 \ln \frac{V_{21}x_1 + x_2}{\Lambda_{21}x_1 + x_2}, \quad (4.132)$$

where the $V_{ij} = V_j/V_i$ are the ratios of the molal volumes; when these are unity, the Wilson equation results. The formulas for activity coefficients of binary mixtures then become

$$\ln \gamma_1 = \ln \frac{x_1 + V_{12}x_2}{x_1 + \Lambda_{12}x_2} + (\beta - \beta_v)x_2, \quad (4.133)$$

$$\ln \gamma_2 = \ln \frac{V_{21}x_1 + x_2}{\Lambda_{21}x_1 + x_2} - (\beta - \beta_v)x_1, \quad (4.134)$$

$$\beta_v = \frac{V_{12}}{x_1 + V_{12}x_2} - \frac{V_{21}}{V_{21}x_1 + x_2}, \quad (4.135)$$

$$\beta = \frac{\Lambda_{12}}{x_1 + \Lambda_{12}x_2} - \frac{\Lambda_{21}}{\Lambda_{21}x_1 + x_2}, \quad (4.136)$$

$$V_{12} = V_2/V_1, \quad (4.137)$$

$$V_{21} = V_1/V_2. \quad (4.138)$$

The infinite dilution version of the T-K-Wilson equations can be given the same mathematical form as the Wilson equations, Eqs. 4.122, 4.123, by introducing pseudo-infinite-dilution activity coefficients:

$$(\ln \gamma_1^\infty)^* = \ln \gamma_1^\infty + \ln(V_1/V_2) + 1 - V_1/V_2 = 1 - \ln \Lambda_{12} - \Lambda_{21}, \quad (4.139)$$

$$(\ln \gamma_2^\infty)^* = \ln \gamma_2^\infty + \ln(V_2/V_1) + 1 - V_2/V_1 = 1 - \ln \Lambda_{21} - \Lambda_{12}. \quad (4.140)$$

Although only a few T-K-Wilson sets of parameters have been recorded, the extensive collections of Wilson parameters by Holmes & Van Winkle (1970), Hirata et al. (1976), and DECHEMA (1979 ff.) can be used to find them. The primary justification for such an effort is that the T-K-Wilson equation is applicable to liquid-liquid equilibria, whereas the Wilson is not. One way to find the TKW parameters is first to find the infinite dilution activity coefficients with the Wilson parameters, then to substitute those values into Eqs. 4.139, 4.140, and finally into Eqs. 4.124 and 4.125.

Such T-K-W parameters may be satisfactory approximations for analysis of processes where both vapor-liquid and liquid-equilibria occur. Liquid-liquid equilibria, however, are quite sensitive to the values of the parameters; even parameters evaluated from a whole range of activity coefficients based on vapor-liquid equilibria sometimes are not good enough. More discussion of this matter is in Chapter 7.

Figure 4.17c compares several equations whose parameters are based on the same γ_i^∞; the Wilson and T-K-Wilson plots are practically identical on this scale, perhaps because the volume ratio, 1.236, is not much different from unity. Less agreement between the different methods is shown by the plots of Figures 4.17 and 4.19.

4.10. THE NRTL (RENON) EQUATION

The model for the derivation of the NRTL (nonrandom two-liquid) equation of the excess Gibbs energy is a two-cell theory represented in Figure 4.15. Here it is assumed that the liquid has a structure made up of cells of molecules of types 1 and 2 in a binary mixture, each surrounded by assortments of the same molecules, with each of the surrounding molecules in turn surrounded in a similar manner, and so on. Gibbs energies of interaction between molecules are identified by g_{ij}, where subscript j refers to the central molecule, and mol fractions in the surrounding regions, x_{ij}, are identified in the same way. Gibbs energies of the two kinds of cells are

$$g^{(1)} = x_{11}g_{11} + x_{21}g_{21}, \quad (4.141)$$

$$g^{(2)} = x_{12}g_{12} + x_{22}g_{22}, \quad (4.142)$$

where g_{11} and g_{22} are the Gibbs energies of the pure substances and the assumption is made that $g_{12} = g_{21}$. The excess Gibbs energy of the assemblage of cells then becomes

$$g^{ex} = x_1x_{21}(g_{21} - g_{11}) + x_2x_{12}(g_{12} - g_{22}). \quad (4.143)$$

the local mol fractions, x_{ij}, are given by equations similar to Wilson's Eqs. 4.102, 103:

$$\frac{x_{21}}{x_{11}} = \frac{x_2 \exp(-\alpha_{12}g_{21}/RT)}{x_1 \exp(-\alpha_{12}g_{11}/RT)}, \quad (4.144)$$

$$\frac{x_{12}}{x_{22}} = \frac{x_1 \exp(-\alpha_{12}g_{12}/RT)}{x_2 \exp(-\alpha_{12}g_{22}/RT)}, \quad (4.145)$$

where α_{12} is a constant that is assumed to be characteristic of the nonrandomness of the mixture. Since

$$x_{21} + x_{11} = 1, \quad (4.146)$$

$$x_{12} + x_{22} = 1, \quad (4.147)$$

the local mol fractions may be solved for as

$$x_{21} = \frac{x_2 \exp(-\alpha_{12}(g_{21} - g_{11})/RT)}{x_1 + x_2 \exp(-\alpha_{12}(g_{21} - g_{11})/RT)}, \quad (4.148)$$

$$x_{12} = \frac{x_1 \exp(-\alpha_{12}(g_{12} - g_{22})/RT)}{x_2 + x_1 \exp(-\alpha_{12}(g_{12} - g_{22})/RT)}. \quad (4.149)$$

When these are substituted into Eq. 4.143, the final equation for the excess Gibbs energy becomes

$$\frac{G^{ex}}{RT}=x_1x_2\left[\frac{\tau_{21}G_{21}}{x_1+x_2G_{21}}+\frac{\tau_{12}G_{12}}{G_{12}x_1+x_2}\right], \quad (4.150)$$

$$\tau_{12}=(g_{12}-g_{22})/RT, \quad (4.151)$$

$$\tau_{21}=(g_{12}-g_{11})/RT, \quad (4.152)$$

$$G_{12}=\exp(-\alpha_{12}\tau_{12}), \quad (4.153)$$

$$G_{21}=\exp(-\alpha_{12}\tau_{21}). \quad (4.154)$$

The activity coefficients are obtained by differentiation as

$$\ln\gamma_1=x_2^2\left[\tau_{21}\left(\frac{G_{21}}{x_1+x_2G_{21}}\right)^2+\left(\frac{\tau_{12}G_{12}}{(x_2+x_1G_{21})^2}\right)\right], \quad (4.155)$$

$$\ln\gamma_2=x_1^2\left[\tau_{12}\left(\frac{G_{12}}{x_2+x_1G_{12}}\right)^2+\left(\frac{\tau_{21}G_{21}}{(x_1+x_2G_{21})^2}\right)\right]. \quad (4.156)$$

These equations have three independent parameters: α_{12}, τ_{12}, and τ_{21}. They are shown generalized to multicomponent mixtures in Table 4.6. Like the Wilson equations, the NRTL multicomponent equations are expressed entirely in terms of binary parameters.

4.10.1. Evaluation of the Parameters

Although the third parameter, α_{12}, was vaguely related by Renon & Prausnitz (1968) to the reciprocal of the coordination number, which is the number of molecules just touching a reference molecule, the range of numerical values found experimentally shows that it is strictly an empirical factor not clearly related to any mechanism. In the original paper, the authors suggest a range of 0.2–0.47, depending on the chemical natures of the constituents. On the other hand, Marina & Tassios (1973) found that $\alpha_{12}=-1$ gives excellent representation of both miscible and partially immiscible binaries.

When an estimate of α_{12} can be made satisfactorily, charts prepared by Renon & Prausnitz (1969) enable figuring the other two parameters in terms of infinite-dilution activity coefficients. A thorough treatment of the NRTL equation is given by Renon et al. (1971), including computer programs for finding parameters by Newton-Raphson least squares and for making distillation and extraction calculations.

At infinite dilution, and when α_{12} can be estimated, the activity-coefficient equations can be reduced to a single variable equation that can be solved readily by trial. When τ_{21} from

Table 4.8. The NRTL and UNIQUAC Equations Expanded

NRTL

$$\ln\gamma_i=\frac{\tau_{1i}G_{1i}x_1+\tau_{2i}G_{2i}x_2+\tau_{3i}G_{3i}x_3}{G_{1i}x_1+G_{2i}x_2+G_{3i}x_3}$$

$$+\frac{x_1G_{i1}}{x_1+G_{12}x_2+G_{13}x_3}\left[\tau_{i1}-\frac{x_2\tau_{21}G_{21}+x_3\tau_{31}G_{31}}{x_1+x_2G_{21}+x_3G_{31}}\right]$$

$$+\frac{x_2G_{i2}}{G_{12}x_1+x_2+G_{32}x_3}\left[\tau_{i2}-\frac{x_1\tau_{12}G_{12}+x_3\tau_{32}G_{32}}{x_1G_{12}+x_2+x_3G_{32}}\right]$$

$$+\frac{x_3G_{i3}}{G_{13}x_1+G_{23}x_2+x_3}\left[\tau_{i3}-\frac{x_1\tau_{13}G_{13}+x_2\tau_{23}G_{23}}{G_{13}x_1+G_{23}x_2+x_3}\right]$$

$$\tau_{ii}=0$$

$$G_{ii}=1$$

UNIQUAC

$$\ln\gamma_i=\ln\frac{\phi_i}{x_i}+5q_i\ln\frac{\theta_i}{\phi_i}+l_i-\frac{\phi_i}{x_i}(x_1l_1+x_2l_2+x_3l_3)+q_i[1-\ln(\theta_1\tau_{1i}+\theta_2\tau_{2i}+\theta_3\tau_{3i})]$$

$$-\frac{\theta_1\tau_{i1}}{\theta_1+\theta_2\tau_{21}+\theta_3\tau_{31}}-\frac{\theta_2\tau_{i2}}{\theta_1\tau_{12}+\theta_2+\theta_3\tau_{32}}-\frac{\theta_3\tau_{i3}}{\theta_1\tau_{13}+\theta_2\tau_{23}+\theta_3}$$

$$\tau_{ii}=1$$

$$\phi_i=\frac{r_ix_i}{r_1x_1+r_2x_2+r_3x_3}$$

$$\theta_i=\frac{q_ix_i}{q_1x_1+q_2x_2+q_3x_3}$$

$$l_i=5(r_i-q_i)-r_i+1$$

$$\tau_{21} = \ln \gamma_1^\infty - \tau_{12} \exp(-\alpha_{12}\tau_{12}) \quad (4.157)$$

is substituted into

$$\tau_{12} = \ln \gamma_2^\infty - \tau_{21} \exp(-\alpha_{12}\tau_{21}), \quad (4.158)$$

the desired single-variable equation results.

4.10.2. Numerical Values of α_{12}

The relative insensitivity of activity coefficients to values of α_{12} in the range 0.1–0.5 is brought out in Figure 4.18, where it also appears that $\alpha_{12} = -1$ has very nearly the same effect as $\alpha_{12} = 0.3$. The values obtained by correlations of experimental data in the DECHEMA VLE Collection (1979), however, cover a wide range of positive values, from $\alpha_{12} = 0.01$ to 100 or more; presumably the simplex method was programmed to give only the best positive values. In the hope of finding guidelines for estimating α_{12} for particular classes of mixtures, the data of the DECHEMA VLE Collection were examined in detail with, however, only rather indefinite conclusions.

Part 6A: C_4–C_6 hydrocarbons with hydrocarbons and other organics except oxygenated ones. For 588 data sets, the mean $\alpha_{12} = 0.336$ with standard deviation 0.096; 16 data sets were excluded, however, because they were far out of the range, varying from 0.0167 to 3.53.

Part 6B: C_7–C_{18} hydrocarbons with hydrocarbons and other organics except oxygenated ones. For 419 data sets, the mean $\alpha_{12} = 0.316$ with standard deviation 0.063; but 34 data sets were excluded, with α_{12} varying from 0.0207 to 125.57, mostly above 1.0.

Parts 3 + 4: Aldehydes, ketones, and ethers: For 478 data sets, the mean $\alpha_{12} = 0.310$ with standard deviation 0.056. The data included in the average are in the range 0.1–0.8; 40 sets were excluded from the average and were in the range 0.008–7.86. In many instances mixtures with parameter values much different from the mean also had some values close to the mean. For instance, acetone + heptane data were 0.0724, 0.2892, 0.2911, 0.2924, 0.2943; diethylether + chloroform data were 0.2988, 0.3089, 0.3140, 1.9233; chloroform + dipropylether data were 0.3054, 0.3027, 0.3156, 2.5188, 2.5550.

Part 1: Aqueous organic systems. For 456 systems the mean $\alpha_{12} = 0.388$ with standard deviation 0.136. Perhaps half were grouped around 0.3 and the other half in the range 0.56–0.6. Forty-eight systems were below 0.1 and above 0.9 or so and were left out of the average. Here also systems with some values considerably out of range had some values near the mean. Of 42 data sets of water + ethanol, 32 had a mean $\alpha_{12} = 0.296$ with standard deviation 0.004, but the other 10 varied from 0.0204 to 1.1614. Some other ranges were: water + 2,3-methylbutanediol, 0.0050, 0.2926, 0.3154, 0.3225, and 0.5553; 12 sets for acetaldehyde ranged from 0.0154 to 0.9817; 14 sets for ethylenediamine ranged from 0.3764 to 4.4456.

In conclusion, if an estimate of α_{12} has to be made, it should be about 0.3 for nonaqueous mixtures and about 0.4 for aqueous organic mixtures. There may be explanations for the sizable divergences of 5–6 percent of the cases from these mean values, but they have not been investigated. The somewhat greater variation of aqueous organic systems is not unexpected since water often is a nonconformist. Here the third parameter of the NRTL equation apparently does help in correlating data in comparison with the four two-parameter equations evaluated in the DECHEMA Collection, as shown in Table 4.9. All data in the DECHEMA LLE Collection (1978–79) are correlated with $\alpha_{12} = 0.2$.

4.10.3. Pros and Cons

The NRTL equation usually represents binary-equilibrium data quite well with its three parameters. It is superior to the Margules and van Laar in that it is applicable to multicomponent mixtures with only binary parameters, and superior to the Wilson in that it can represent liquid-liquid equilibria, although the T-K-Wilson also is competent in this area. In particular cases some one of the correlations may be best for vapor-liquid equilibria, but as a general rule the requirement of three parameters is the NRTL's main handicap in comparison with the Wilson and UNIQUAC. Even this is not really a handicap when more than two data points are available and the computer program is set up. The Wilson and UNIQUAC equations are the bases for activity-coefficient prediction by group contributions, Wilson by ASOG and UNIQUAC by UNIFAC, but it is possible to take the outputs of those programs and to fit NRTL parameters to them.

4.10.4. Effect of Temperature

Part of this effect is built into Eqs. 4.151, 152 since Δg_{ij} may be taken as approximately independent of temperature in

Table 4.9. Frequencies of Best Fits of Five Activity Coefficient Correlations of the DECHEMA Vapor-Liquid Equilibrium Data Collection

Part of Collection	*Number of Data*	*Margules*	*van Laar*	*Wilson*	*NRTL*	*UNIQUAC*
1 Aqueous organics	504	0.143	0.071	0.240	0.403*	0.143
2A Alcohols	574	0.166	0.085	0.395*	0.223	0.131
2B Alcohols and phenols	480	0.213	0.119	0.342*	0.225	0.102
3/4 Alcohols, ketones, ethers	490	0.280*	0.167	0.243	0.155	0.155
6A C_4-C_6 hydrocarbons	587	0.172	0.133	0.365*	0.232	0.099
6B C_7-C_{18} hydrocarbons	435	0.225	0.170	0.260*	0.209	0.136
7 Aromatics	493	0.260*	0.187	0.225	0.160	0.172
Total of 7 parts	3563	0.206	0.131	0.300*	0.230	0.133

*Identifies the most frequent best fit in each category.

Example 4.8. Finding the Three NRTL Parameters from Experimental Activity Coefficients

Data are given for carbon disulfide + acetonitrile (Kogan, No. 507). The parameters will be found by using $\gamma_1^\infty = 8.9$, $\gamma_2^\infty = 19.5$, and $\gamma_1 = 2.019$ at $x_1 = 0.475$.

The procedure is to assume a value of α_{12}, then calculate τ_{12} and τ_{21} from Eqs. 4.157, 4.158, and finally γ_1 at $x_1 = 0.475$ from Eq. 4.155. From the following table, the parameters appear to be

$$\alpha_{12} = 0.30,\ \tau_{12} = 2.913,\ \text{and}\ \tau_{21} = 0.9704.$$

α_{12}	τ_{12}	τ_{21}	γ_1 @ $x_1 = 0.475$
.10	3.0339	−.0541	2.219
.15	2.8412	.3307	2.169
.20	2.8590	.5720	2.146
.25	2.8889	.7829	2.094
.30	2.9130	.9704	2.019
.35	2.9296	1.1353	1.930
.40	2.9407	1.2791	1.837
.45	2.9480	1.4038	1.745
.50	2.9530	1.5115	1.657

$CS_2 - C_2H_3N$

x	y	t	P	γ_1	γ_2
0.0	0.0	20.5	70.6	8.9	1.000
12.8	79.4		318.0	6.508	1.064
17.5	81.7		355.2	5.470	1.115
47.5	81.7		355.2	2.019	1.753
71.6	81.7		355.0	1.336	3.241
73.6	81.7		356.0	1.300	3.487
87.1	81.7		355.2	1.099	7.136
100.0	100.0		303.0	1.000	19.50

Example 4.9. The Effect of Temperature on Activity Coefficients Represented by the NRTL Equation

The NRTL parameters of n-hexane + acetone are given as a function of temperature by Renon et al. (1971) as follows:

$$\alpha_{12} = 0.4539 + 0.001937\,(T - 273.15),$$

$$\Delta g_{12} = 758 - 2.67\,(T - 273.15),\ \text{cal/gmol},$$

$$\Delta g_{21} = 754 - 0.66\,(T - 273.15),\ \text{cal/gmol}.$$

The range of temperature covered is 183.15–373.15 K. The activity coefficients are evaluated at 300 and 350 K and are plotted; they exhibit a substantial effect of temperature in this interval.

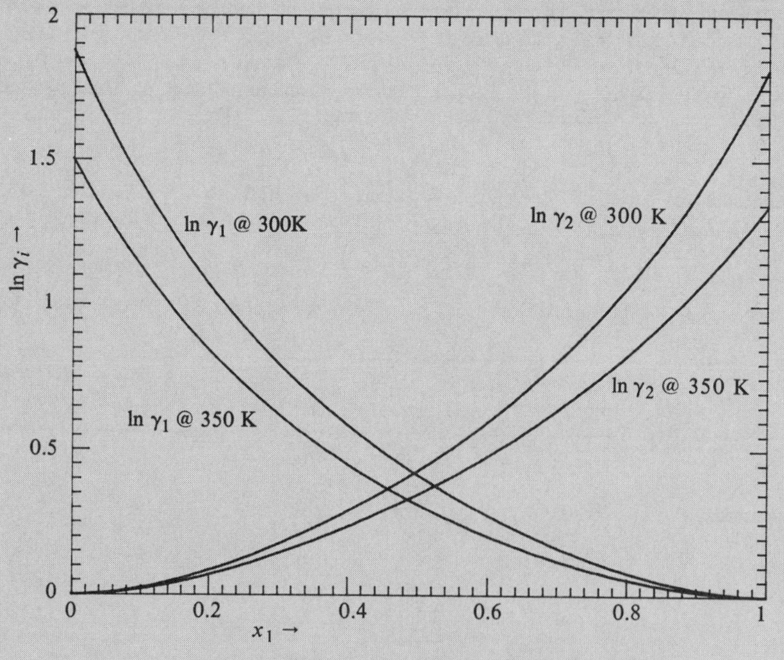

Example 4.9. n-hexane (1) + acetone (2)

some cases. Renon et al. (1971), however, represent α_{12}, Δg_{12}, and Δg_{21} as linear functions of the temperature and give the six constants for 82 mixtures. The effect found in Example 4.9 is quite substantial.

4.11. THE UNIQUAC (UNIVERSAL QUASI-CHEMICAL) EQUATION

The two-liquid model and the concept of local compositions were adopted by Abrams & Prausnitz (1975) in the semi-theoretical development of this equation. Their model also stipulates that the excess Gibbs energy is made up of two parts:

1. A contribution due to differences in sizes and shapes of the molecules (configurational or combinatorial part).
2. A contribution due to energetic interactions between them (residual part).

In binary form the resulting equations are given in Table 4.4, the multicomponent generalization in Table 4.6, and the ones at infinite dilution in Table 4.5. The volume and surface factors, r_i and q_i, are obtained basically from crystallographic measurements. Values for many substances are given by Prausnitz et al. (1980) and the DECHEMA VLE Data Collection (1979 ff., Appendix of each volume). Also, these factors may be obtained from group contributions; tables are given by Reid et al. (1977), Fredenslund et al. (1977), Skjold-Jorgensen et al. (1979), and Zakarian et al. (1979). Some data are given in Tables E.11, E.13. Special factors are proposed by Anderson & Prausnitz (1978) for water and alcohols.

A simpler derivation is made by Maurer & Prausnitz (1978), who also derive a three-parameter version that, however, is not generally translatable to multicomponent mixtures. An equation similar to UNIQUAC is presented by Nagata & Katoh (1980), who also found that a third common parameter is needed for accurate representation of ternary liquid-liquid equilibria. Correlation of the behavior of mixtures of aliphatic hydrocarbons is improved by modifying the combinatorial part of the UNIQUAC by Kikic et al. (1980). It is shown by Abrams & Prausnitz (1975) that the van Laar, Margules, Wilson, or NRTL equations are special cases of the UNIQUAC when suitable values are assigned to the r_i and q_i. Similar family relations are demonstrated by Tsuboka & Katayama (1975b), who include their own modification of the Wilson equation in this group.

Like the Wilson and NRTL, the UNIQUAC equations at infinite dilution are transformable so that they can be solved for one parameter at a time, thus

$$\tau_{12} = \exp[k_8 - \exp(k_7 - \tau_{12})], \tag{4.159}$$

$$\tau_{21} = k_8 - \ln \tau_{12}, \tag{4.160}$$

where k_7 and k_8 are constants defined in Table 4.5. The parameters Δu_{12} and Δu_{21} are the ones listed in the DECHEMA Collection; they are related simply to the τ_{ij} in Table 4.4.

4.11.1. Effect of Temperature

All the factors in the equation originally were taken independent of temperature. Some improvements in the simultaneous representation of excess Gibbs energies and enthalpies, however, were obtained with temperature-dependent interaction parameters u_{ij} by Skjold-Jorgensen et al. (1980). More marked improvements were achieved by making the coordination number, z, a function of temperature:

$$z = 35.2 - 0.1272T + 0.00014T^2. \tag{4.161}$$

The coordination number is the number of nearest touching neighbors possessed by any particular molecule. For many liquids at normal conditions the number is 10. The preceding equation gives 9.6 at 300 K and 6.7 at 400 K. Although the value $z = 10$ has been more or less accepted, Krumins et al. (1980), for example, found cases where $z = 6$ gave better correlation.

4.11.2 One Parameter Form

A one-parameter form of the equation is developed by placing

$$u_{11} = -\Delta u_1^{\text{vap}}/q_1, \tag{4.162}$$

$$u_{22} = -\Delta u_2^{\text{vap}}/q_2, \tag{4.163}$$

where Δu_i^{vap} is the internal energy change of vaporization of component i, which is presumed to be known. The further assumption

$$u_{12} = u_{21} = (1 - c_{12})\sqrt{u_{11}u_{22}} \tag{4.164}$$

leads to an equation in which the only parameter to be determined experimentally is c_{12}. This form of equation is useful when only very limited data are available, such as a single infinite-dilution activity coefficient determined by chromatographic or differential ebulliometric means. There is much recent interest in those experimental techniques, for example Eckert et al. (1981), Santacesaria et al. (1978) and Tassios (1972). A comparison of several one-parameter equations was made by Krumins et al. (1980), but no outstanding one was discovered.

4.11.3. Pros and Cons

Major characteristics of the UNIQUAC equation are:

1. Applicability to multicomponent mixtures in terms of binary parameters only.
2. Applicability to liquid-liquid equilibria.
3. A built-in temperature dependence valid over at least a moderate range.
4. Possibly superior representation of mixtures of widely different molecular sizes.
5. It is the basis of a group contribution method for obtaining activity coefficients, the UNIFAC method, from properties of the pure components. A great deal of effort is being expended in this direction.

Perhaps the main disadvantages of UNIQUAC are its somewhat greater algebraic complexity and the fact that (as suggested by Table 4.10) the representation of data often is poorer than by some simpler equations. Another comparison of activity-coefficient plots from infinite dilution data with the several equations is in Figure 4.18.

4.12. MULTICOMPONENT MIXTURES

As the number of components in a mixture goes to three and beyond, the experimental effort increases sharply and is rarely

Example 4.10. Parameters of the UNIQUAC Equation for Mixtures of Acetone (1) and Chloroform (2)

After making the replacements,

$\gamma_1^\infty \rightarrow \exp(1-k_8)$,

$\gamma_2^\infty \rightarrow \exp(1-k_7)$,

where k_7 and k_8 are as defined in Table 4.5, the parameters may be found by the technique used for the Wilson equation in Section 4.9. There are three sets of roots.

	γ^∞	r	q
Acetone	0.40	2.57	2.34
Chloroform	0.56	2.70	2.34

$r_1/r_2 = 0.9519$

$q_1/q_2 = 1$

$l_1 = -0.4200$

$l_2 = 0.1000$

$k_7 = 1.3968$

$k_8 = 1.2537$

Use the procedure for the Wilson equation to find the UNIQUAC parameters, after making the replacements

$\gamma_1^\infty \rightarrow \exp(1-k_8) = 0.7759,$

$\gamma_2^\infty \rightarrow \exp(1-k_7) = 0.6725.$

The three sets of parameters are:

Case	τ_{12}	τ_{21}
I	0.08581	3.7097
II	1.8203	0.6549
III	2.5757	0.3076

attempted beyond three components in any comprehensive fashion. Sometimes it is justifiable to assume that the excess Gibbs energy of the mixture is the sum of those of all pairs of components plus additional terms for interactions of triplets and other combinations. One of the simpler expressions of this type for a ternary mixture is

$$G^{ex} = G_{12}^{ex} + G_{13}^{ex} + G_{23}^{ex} + (C_0 + C_1x_1 + C_2x_2 + C_3x_3)x_1x_2x_3, \quad (4.165)$$

where the C_i are the ternary parameters. Several of the older equations with only one ternary parameter are given, for example by Hala (1967), of which two are:

Symmetrical:
$$G^{ex}/RT = x_1x_2(x_1+x_2)A_{12} + x_1x_3(x_1+x_3)A_{13} + x_2x_3(x_2+x_3)A_{23} + x_1x_2x_3(A_{12}+A_{13}+A_{23}-C). \quad (4.166)$$

Table 4.10. Some Values of Parameters

Substances	T	P	γ_i^∞	γ_2^∞	*Wilson* λ_{12}	*Wilson* λ_{21}
1-hexane + toluene	30		1.27	1.34	−78.07	236.02
acetonitrile + water		300	8.22	14.39	142.91	1961.72
acetonitrile + water		760	32.50	6.06	1391.96	1356.93
water + diacetone alcohol		100	3.14	11.23	1684.67	−3.88
water + pyridine		760	2.20	29.19	1020.97	1486.51
1-octene + ethylbenzene		760	1.28	1.09	532.50	−211.19
octane + butyronitrile	90		2.78	4.10	−157.95	1299.67
heptane + pyridine	67.8		7.94	3.58	747.06	717.91
heptane + dimethylformamide	5.0		28.71	37.36	1188.98	2426.56
isoprene + acetonitrile		760	4.86	8.14	163.51	1226.52
hexane + octane		149.2	0.68	0.82	−464.70	301.61
perfluorohexane + hexane	35		15.71	8.18	1133.06	1051.54
hexane + heptane		92.3	0.93	1.43	−514.84	1244.18
ethanol + water	10		3.36	2.89	−189.77	959.16
ethanol + water	70		7.24	2.62	471.04	883.75

Notes: (1) γ_1^∞ and γ_2^∞ are given for the equation that is the best fit to the data.
(2) Units are cal/gmol for the Wilson, UNIQUAC and NRTL parameters.
(3) Pressures are Torr, temperatures °C.
(4) Data excerpted from *DECHEMA Vapor-Liquid Equilibrium Data Collection.*

$$\text{Margules: } G^{ex}/RT = x_1x_2(A_{21}x_1 + A_{12}x_2) + x_1x_3(A_{31}x_1 + A_{13}x_3) + x_2x_3(A_{32}x_2 + A_{23}x_3) + x_1x_2x_3(A_{21} + A_{13} + A_{32} - C). \tag{4.167}$$

The ternary parameter sometimes can be approximated; for instance, Colburn (1950) used

$$C = 0.5(A_{21} + A_{13} + A_{32} - A_{31} - A_{23}). \tag{4.168}$$

Abbott et al. (1975) found that neither

$$C = 0 \tag{4.168a}$$

nor

$$C = 0.5\Sigma\Sigma A_{ij} \tag{4.168b}$$

represented their data adequately, but that representation with the full Eq. 4.168 was well within experimental uncertainty. In general, with Margules, van Laar, and other older equations, some experimental data beyond those for binary mixtures should be available.

A considerable advance in the representation of excess Gibbs energies of multicomponent mixtures came with the Wilson equation, which is theoretically applicable with only binary parameters. That equation is not a simple sum of the excess Gibbs energies of all pairs but is expressed in terms of binary contributions,

$$G^{ex}/RT = -\sum_i \left[x_i \ln \left(\sum_j x_j \Lambda_{ij} \right) \right]. \tag{4.169}$$

Many systems were checked against this equation in an early paper by Schreiber & Eckert (1971). Many ternary systems are correlated in the DECHEMA VLE Collection.

The NRTL equation similarly is applicable to multicomponent systems; it was tested against seventeen vapor-liquid ternaries and several liquid-liquid ternaries by Renon (1971).

Anderson & Prausnitz (1978) checked the UNIQUAC equation with nine ternary vapor-liquid systems, and with ten ternary and one quaternary liquid-liquid systems. They found that ternary liquid-liquid correlation is much improved if at least one ternary tie-line datum is used in the evaluation of the parameters, a conclusion that is undoubtedly true of the NRTL and T-K-Wilson equations as well.

The modified Wilson equation of Tsuboka & Katayama (1975), called here the T-K-Wilson equation, was tested on seven vapor-liquid and four liquid-liquid equilibria; the latter were predicted more closely than with the NRTL equation.

Predictions of ternary liquid-liquid equilibria of fifty systems were made with varying degrees of success by Fredenslund et al. (1977). A general discussion of the use of the UNIQUAC equation for representing ternary and quaternary LLE data was made by Fredenslund et al. (1980a). The most extensive compilation of LLE data, experimental and correlated with the NRTL and UNIQUAC equations, is that of Sorensen & Arlt (1979) (DECHEMA LLE Data Collection). For vapor-liquid equilibrium of some ternary systems, the DECHEMA VLE Data Collection provides Wilson, NRTL, and UNIQUAC parameters.

Although in theory binary parameters are sufficient for representation of multicomponent data, better results for specific mixtures are obtained when the parameters are based on multicomponent equilibrium data. Many comparisons are made in the DECHEMA LLE Data Collection of ternary equilibria with binary parameters and with the six parameters fitted to ternary data. It has not been demonstrated, however, that equilibrium of quaternary and higher mixtures is represented better with parameters of pairs found from ternary data than with those found from binary measurements.

NRTL			*UNIQUAC*		*Margules*		*van Laar*	
Δg_{12}	Δg_{21}	α_{12}	Δu_{12}	Δu_{21}	A_{12}	A_{21}	A_{12}	A_{21}
171.21	−13.78	0.3001	235.06	−175.58	0.2378	0.2604	0.2377	0.2613
1301.23	1000.06	0.5352	1088.43	−124.55	2.0094	2.3275	1.9868	2.3643
1259.00	2085.68	0.5960	−1.76	852.83	2.5858	1.6373	2.9128	1.5513
1214.83	691.22	0.6205	−364.57	1245.78	1.1210	2.2439	1.2119	2.3600
2273.97	10.91	0.4855	−448.24	434.50	−0.1893	2.6832	0.7014	3.1029
−512.93	817.96	0.3382	−322.41	423.29	0.2072	0.0728	0.2499	0.0917
914.60	137.07	0.2982	758.50	−289.27	1.0224	1.4104	1.0526	1.4399
85.43	1200.21	0.2886	385.03	−28.85	1.7562	1.1417	1.9010	1.1691
1302.97	737.97	0.0693	714.48	38.71	3.3358	3.5318	3.3363	3.5321
929.99	695.80	0.6034	771.15	−76.82	1.4987	1.7361	1.4973	1.7588
901.90	−348.10	1.8262	688.45	−503.33	−0.2503	−0.8665	−0.4106	−0.9466
764.85	1237.73	0.4269	293.04	9.12	2.5122	1.8896	2.5666	1.8980
1224.51	−723.33	0.3160	668.92	−420.24	−0.1619	0.1246	−0.1587	−0.0102
223.43	488.82	0.2978	580.53	−242.22	1.2103	1.0529	1.2132	1.0596
−121.27	1337.86	0.2974	−30.19	337.00	1.6346	0.8563	1.7966	0.9238

Table 4.11. Sample Page from the DECHEMA Collection of Vapor-Liquid Equilibrium Data (1979, Vol I/3&4 p. 228)

(1) ACETONE C3H60

(2) HEXANE C6H14

***** ANTOINE CONSTANTS				REGION *****			CONSISTENCY
(1)	7.11714	1210.595	229.664	-13-	55 C	METHOD 1	*
(2)	6.91058	1189.640	226.280	-30-	170 C	METHOD 2	*

TEMPERATURE= 20.00 DEGREE C

LIT: RALL W.,SCHAEFER K.,Z.ELECTROCHEM.63,1019(1959).

CONSTANTS:	A12	A21	ALPHA12
MARGULES	1.7448	1.8012	
VAN LAAR	1.7416	1.8044	
WILSON	1077.8013	375.5248	
NRTL	632.4249	583.8331	0.2913
UNIQUAC	−41.9959	512.3937	

WILSON

$\gamma_1^\infty = 7.24$

$\gamma_2^\infty = 6.96$

$Y_1 \rightarrow$ 0.00 0.20 0.40 0.60 0.80 1.00

$X_1 \rightarrow$ 0.00 0.20 0.40 0.60 0.80 1.00

EXPERIMENTAL DATA			*MARGULES*		*VAN LAAR*		*WILSON*		*NRTL*		*UNIQUAC*	
P MM HG	*X1*	*Y1*	*DIFF P*	*DIFF Y1*	*DIFF P*	*DIFF Y1*	*DIFF P*	*DIFF Y1*	*DIFF P*	*DIFF Y1*	*DIFF P*	*DIFF Y1*
119.60	0.0	0.0	−0.67	0.0	−0.67	0.0	−0.67	0.0	−0.67	0.0	−0.67	0.0
187.20	0.0913	0.3966	4.24	0.0024	4.35	0.0028	−1.26	−0.0110	2.90	−0.0011	3.60	0.0008
226.70	0.2563	0.5421	−0.19	−0.0166	−0.13	−0.0166	1.52	−0.0023	0.07	−0.0137	−0.01	−0.0151
232.30	0.3019	0.5595	0.85	−0.0161	0.89	−0.0162	3.13	−0.0007	1.28	−0.0128	1.10	−0.0144
232.40	0.3543	0.5737	−2.36	−0.0154	−2.33	−0.0155	−0.02	−0.0009	−1.89	−0.0121	−2.09	−0.0137
237.00	0.4035	0.5827	0.37	−0.0153	0.38	−0.0155	2.40	−0.0032	0.76	−0.0125	0.58	−0.0139
238.80	0.5325	0.6092	−0.02	−0.0043	−0.03	−0.0046	0.92	−0.0020	0.10	−0.0039	0.04	−0.0043
237.70	0.6609	0.6362	−1.47	0.0045	−1.49	0.0042	−0.89	−0.0028	−1.44	0.0027	−1.46	0.0033
239.30	0.7309	0.6564	1.22	0.0065	1.20	0.0063	1.76	−0.0033	1.24	0.0042	1.22	0.0052
237.90	0.7679	0.6722	1.21	0.0081	1.19	0.0080	1.62	−0.0017	1.20	0.0060	1.19	0.0069
234.30	0.7862	0.6825	−1.38	0.0097	−1.41	0.0097	−1.10	0.0004	−1.43	0.0078	−1.43	0.0086
234.10	0.8219	0.6975	1.18	0.0038	1.14	0.0038	1.08	−0.0037	1.02	0.0024	1.07	0.0030
230.30	0.8528	0.7202	0.88	0.0028	0.83	0.0028	0.29	−0.0021	0.60	0.0021	0.70	0.0024
220.60	0.9105	0.7778	1.70	−0.0051	1.64	−0.0049	−0.07	−0.0026	1.18	−0.0041	1.38	−0.0044
202.90	0.9619	0.8739	−0.14	−0.0071	−0.18	−0.0069	−2.24	0.0009	−0.66	−0.0050	−0.44	−0.0059
181.50	1.0000	1.0000	−3.96	0.0	−3.96	0.0	−3.96	0.0	−3.96	0.0	−3.96	0.0
MEAN DEVIATION:			1.23	0.0084	1.23	0.0084	1.31	0.0027	1.13	0.0064	1.17	0.0073
MAX. DEVIATION:			4.24	0.0166	4.35	0.0166	3.13	0.0110	2.90	0.0137	3.60	0.0151

4.13. COMPARISON OF EQUATIONS

The merits of the individual activity-coefficient correlation methods have been cited locally, and a comparison of a sort is in Table 4.9. Those conclusions may be resummarized.

1. The Margules, van Laar, and related algebraic forms have the merit of mathematical simplicity, ease of evaluation of the parameters from activity-coefficient data, and often adequate representation of even fairly nonideal binary mixtures, including partially miscible liquid systems. They are not applicable to multicomponent systems without ternary or higher interaction parameters.

2. The Wilson equation represents vapor-liquid equilibria of binary and multicomponent mixtures very well with only binary parameters. Because of its greater simplicity it may be preferable to the NRTL and UNIQUAC equations for this purpose. Although it is not directly applicable to liquid-liquid equilibria, the equally simple modified T-K-Wilson equation probably is satisfactory, though it has not been tested as thoroughly as some others. The Wilson equation is the basis of the ASOG group contribution method for activity coefficients.

3. The NRTL equation represents vapor-liquid and liquid-liquid equilibria of binary and multicomponent systems quite well, and is often superior to the others for aqueous systems. It is simpler in form than the UNIQUAC method but has the disadvantage of involving three parameters for each pair of constituents. The third parameter α_{12} often can be estimated from the chemical natures of the components, and a strong claim has been made for a universal value $\alpha_{12} = -1$. The value $\alpha_{12} = 0.2$ has been adopted for all mixtures in the DECHEMA LLE Data Collection.

4. Although it employs only two parameters per pair of components, the UNIQUAC equation is algebraically the most complex one. It utilizes knowledge of molecular surfaces and volumes of the pure components, which can be estimated from structural contributions, and for this reason the method may be particularly applicable to mixtures of widely different molecular sizes. It is applicable to vapor-liquid and liquid-liquid equilibria of multicomponent mixtures with binary parameters and pure component data only. UNIQUAC is the basis of the UNIFAC group contribution method for activity coefficients from structure and receives much support in the literature.

Results of an examination of the frequency of best fits to data in the DECHEMA Collection by the various equations are shown in Table 4.9. Representative values of parameters are shown in Table 4.10. Table 4.11 is a sample page from the DECHEMA Collection. A comparison between the Wilson and Scatchard-Hildebrand equations is shown in Table 4.12 and indicates little agreement. A review of the applicability of the Wilson, NRTL, and UNIQUAC models was made recently by Eckert, Thomas, & Johnston (1980).

4.14. MULTIPLE ROOTS

Not surprisingly, equations of the complexity of those used for prediction of activity coefficients may possess more than one set of parameters corresponding to particular sets of activity coefficients, either at infinite dilution or throughout. The Wilson equation has been most investigated for this aspect. The infinite dilution form

$$\Lambda_{12} = \frac{1}{\gamma_1^\infty} \exp\left(1 - \frac{1}{\gamma_2^\infty} \exp(1 - \Lambda_{12})\right), \tag{4.170}$$

Table 4.12. Comparison of Activity Coefficients at Infinite Dilution Obtained with the Wilson Parameters of Table E.8 and Those Obtained with the Scatchard-Hildebrand Equation, Using Solubility Parameters of Hoy (1970), Hansen (1971), and Henley & Seader (1981)

	van Winkle		*Hoy*		*Hansen*		*Henley & Seader*	
Substance	γ_1^∞	γ_2^∞	γ_1^∞	γ_2^∞	γ_1^∞	γ_2^∞	γ_1^∞	γ_2^∞
1-Butanol	2.398	3.887	1.984	2.185	1.797	1.944	1.824	1.975
Carbon tetrachloride	1.084	1.121	1.049	1.053	1.016	1.017	1.004	1.004
Chloroform	0.846	0.884	1.000	1.000	1.006	1.005	1.001	1.001
Cyclohexane	1.401	1.459	1.127	1.157	1.097	1.120	1.124	1.153
Cyclopentane	1.464	1.383	1.170	1.164	1.000	1.000	1.198	1.190
Ethanol	3.935	6.599	5.355	2.997	7.061	3.575	6.061	3.221
n-Heptane	1.264	1.610	1.396	1.786	1.340	1.663	1.612	2.205
n-Heptane	1.281	1.724	1.428	1.800	1.367	1.674	1.664	2.229
n-Hexane	1.400	1.631	1.600	1.956	1.500	1.783	1.597	1.949
Methanol	6.035	8.699	44.649	5.247	50.118	5.509	45.126	5.152
Methyl acetate	1.355	1.250	1.012	1.010	1.001	1.001	1.003	1.002
Methylcyclohexane	1.242	1.542	1.000	1.000	1.203	1.325	1.236	1.381
Methylcyclopentane	1.367	1.454	1.251	1.319	1.000	1.000	1.250	1.319
1-Propanol	2.470	6.371	3.029	2.660	2.866	2.534	2.766	2.430
1-Propanol	2.906	6.235	3.251	2.695	3.065	2.565	2.951	2.459
2-Propanol	3.008	4.964	1.931	1.770	2.133	1.926	2.092	1.887
2-Propanol	3.697	6.369	1.970	1.809	2.182	1.975	2.138	1.933

Mixtures of benzene with the substances identified.

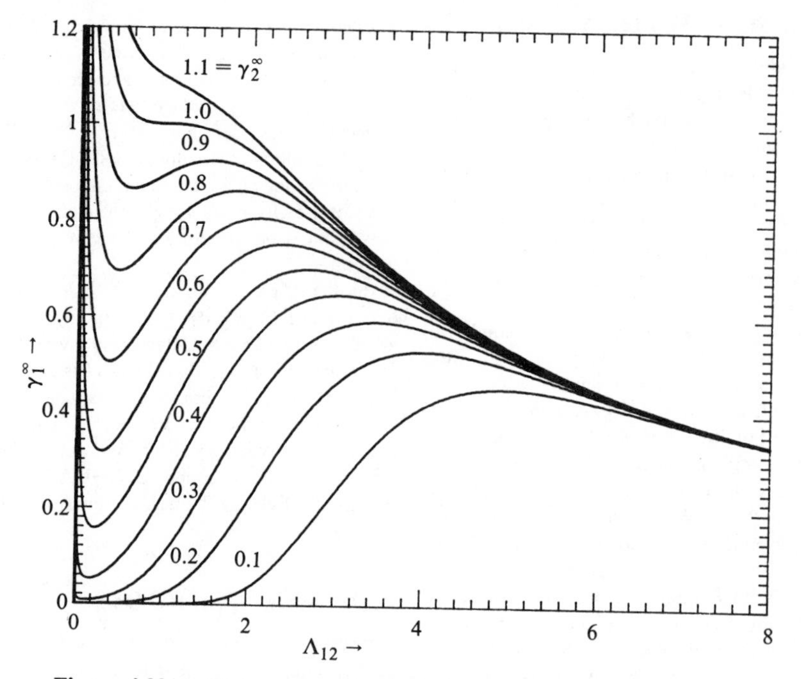

Figure 4.20(a). Plots of the infinite-dilution form of the Wilson equation showing the region in which multiple roots are present. The other parameter is given by $\Lambda_{21} = 1 - \ln(\Lambda_{12}\gamma_1^\infty)$.

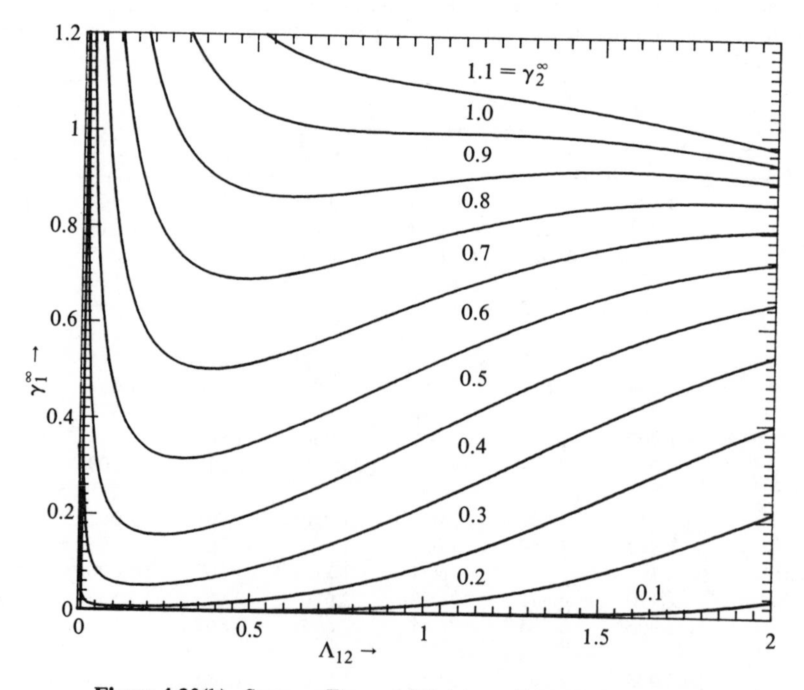

Figure 4.20(b). Same as Figure 4.20(a) but with expanded abscissa.

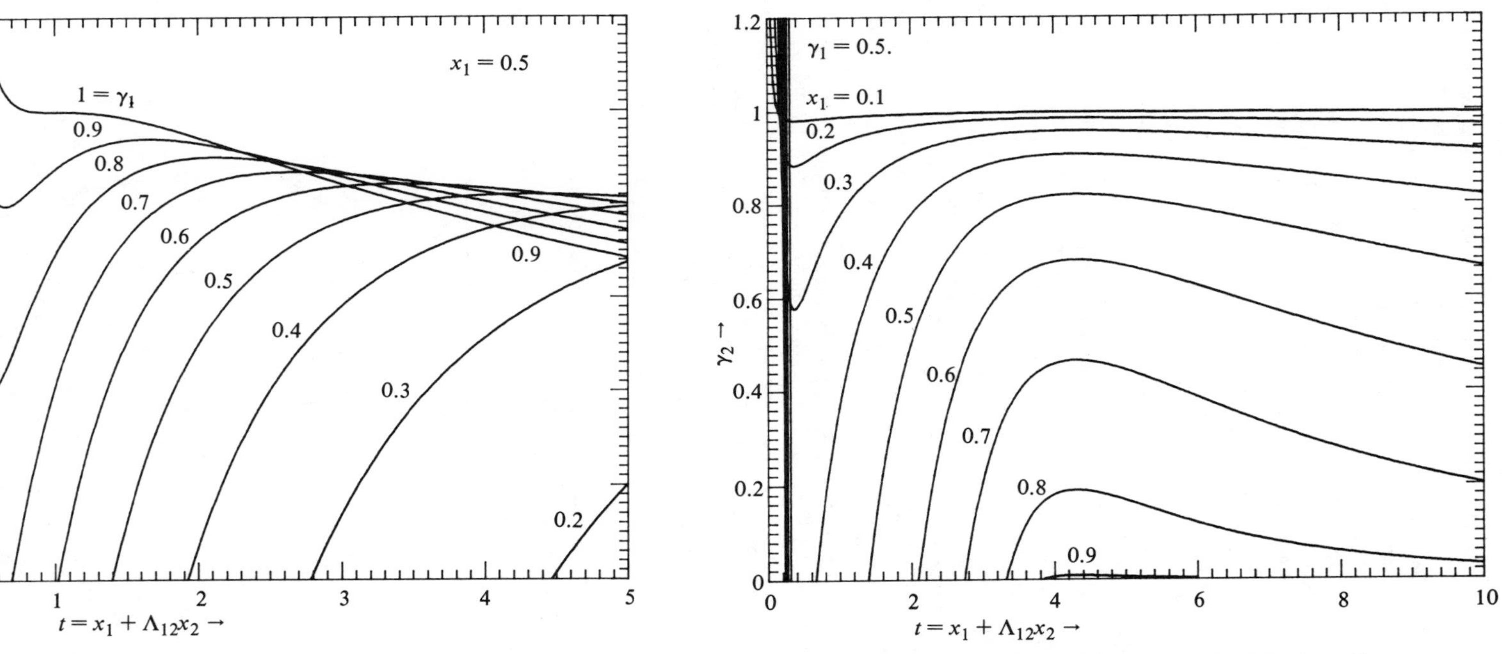

Figure 4.21(a). Plots with $x_1 = 0.5$ and indicated values of γ_1.

Figure 4.21(b). Plots with $\gamma_1 = 0.5$ and indicated values of x_1.

Figure 4.21. Plots of a rearranged Wilson equation showing multiple roots in some ranges of activity coefficients

The Wilson equation is put in the following form by Silverman & Tassios (1977):

$$\gamma_2 = \frac{(\gamma_1 t)^{(-x_1/x_2)}}{x_2}\left[1 + \frac{x_1}{x_2}\left(\ln(\gamma_1 t) - 1 + \frac{x_1}{t}\right)\right],$$

$$t = x_1 + \Lambda_{12} x_2.$$

From the plots, for example, when $x_1 = 0.5$, $\gamma_1 = 0.5$, and $\gamma_2 = 0.3$, two values of t are 0.22 and 1.70, and the corresponding values of the parameter Λ_{12} are -0.56 and 2.40.

$$\Lambda_{21} = 1 - \ln(\gamma_1^\infty \Lambda_{12}), \qquad (4.171)$$

is plotted in Fig. 4.20. Clearly, only one set of parameters is obtained when both γ_i^∞ exceed unity, but as many as three sets may be obtained when either or both γ_i^∞ are less than unity. Taking as an example the case of the acetone-chloroform system for which $\gamma_1^\infty = 0.4$ and $\gamma_2^\infty = 0.5$, the three sets of parameters and the corresponding activity-coefficient plots are shown in Example 4.11. Only one of the sets of parameters fits the experimental data closely.

The nomogram of Figure 4.16 shows clearly the ranges in which multiple roots of the infinite dilution form of the Wilson equation occur.

For compositions other than infinite dilution, also, it is evident from Fig. 4.21 that in some ranges of activity coefficients multiple values of Λ_{12} can occur.

Identifying the physically correct choice from multiple sets of parameters is a problem when no data are available beyond the few that may have been used for determination of the parameters. Activity coefficient curves that display pronounced maxima or minima, as in some of the examples of this section, usually may be discounted. A rule proposed by Ladurelli et al. (1975) is that for mixtures of molecules of similar sizes and shapes, the right solution is likely to be the one for which the product, $\Lambda_{12}\Lambda_{21} > 1$; this is true of the acetone-chloroform data just cited. Eight cases of multiple roots were studied by Silverman & Tassios who expressed their conclusions in terms of the exponential parameters,

$$\lambda_{ij} = RT \ln(\Lambda_{ij} V_i / V_j). \qquad (4.172)$$

In all their cases the best fit of vapor compositions was obtained when the sum of the absolute values, $|\lambda_{12}| + |\lambda_{21}|$, was the smallest. That paper goes on to explore the impact of multiple roots on multicomponent vapor-liquid equilibria calculations.

Parameters of the UNIQUAC equation are found by the same procedures as those of the Wilson, after replacing the activity coefficients at infinite dilution in the latter equation by

$$\ln \gamma_1^\infty \rightarrow 1 - k_8, \qquad (4.173)$$

$$\ln \gamma_2^\infty \rightarrow 1 - k_7, \qquad (4.174)$$

where k_7 and k_8 are as defined in Table 4.5. Multiple sets of parameters τ_{12} and τ_{21} will occur when $(1 - k_7)$ and $(1 - k_8)$ are less than unity. Example 4.10 is concerned with this problem.

Because of the third parameter α_{12}, the root situation of the NRTL is more complex. For positive values of α_{12}, three roots exist for values of the infinite-dilution activity coefficients less than unity, and only one root at larger values of the coefficients. For $\alpha_{12} = -1$, the situation is reversed, multiple roots obtaining at γ_1^∞ and γ_2^∞ greater than unity, and only one root at smaller values of the activity coefficients. Figures 4.22 and 4.23 illustrate these conclusions, and Example 4.12 is of a case with three sets of parameters. The NRTL equation was studied by Tassios (1978), who concluded that the best fit is obtained when the absolute values of the parameters τ_{12} and τ_{21} sum to a minimum.

Figure 4.22. Multiple sets of parameters τ_{12} and τ_{21} of the NRTL equation with some positive values of α_{12}. The values of τ_{21} are obtained from

$$\tau_{21} = \ln \gamma_1^\infty - \tau_{12} \exp(-\alpha_{12}\tau_{12}).$$

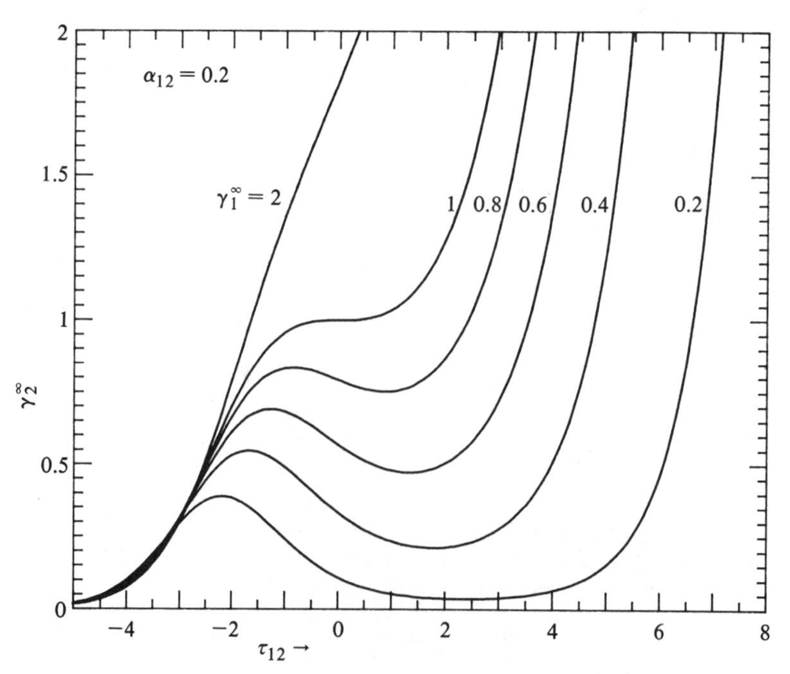

Figure 4.22(a). Plots with $\alpha_{12} = 0.2$ and indicated values of γ_1^∞.

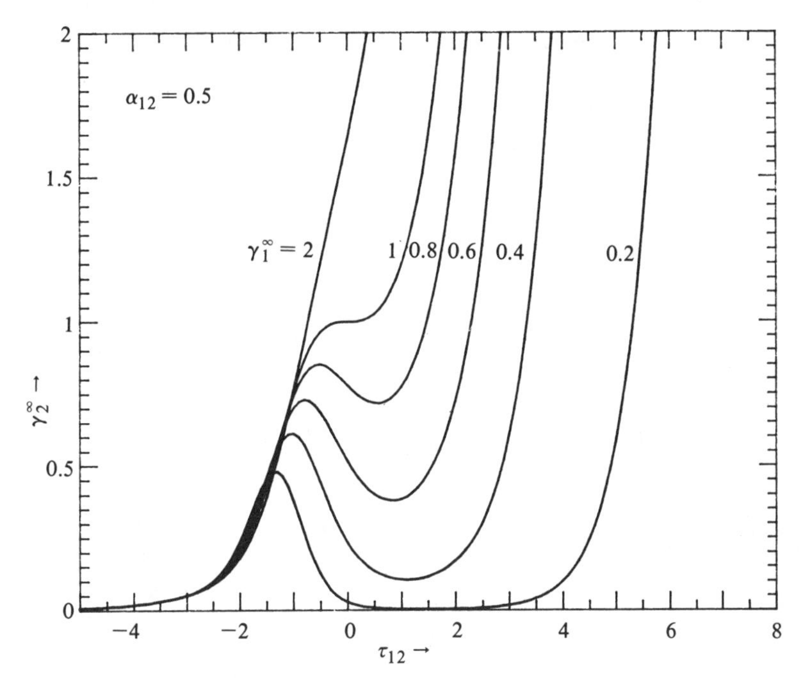

Figure 4.22(b). Plots with $\alpha_{12} = 0.5$ and indicated values of γ_1^∞.

Example 4.11. Plots of Activity Coefficients of the System Acetone (1) + Chloroform (2) for which the Infinite-Dilution Values Are $\gamma_1^\infty = 0.40$ and $\gamma_2^\infty = 0.56$.

The three sets of Wilson parameters are:

Case	Λ_{12}	Λ_{21}
I	0.0754	4.5017
II	0.8454	2.0842
III	6.7576	0.00564

Case II fits the experimental data closely. Note that the product of the parameters for this case is greater than unity, in accordance with the Ladurelli (1975) rule.

Number II fits the experimental data best. This case conforms to the Tassios rule that the right solution is the one for which $|\tau_{12} + \tau_{21}| \rightarrow$ minimum.

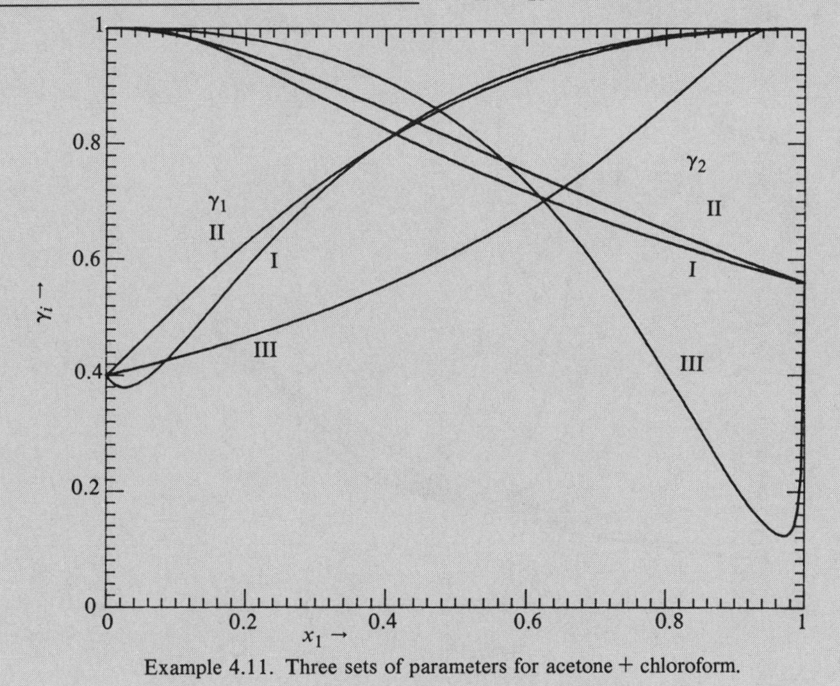

Example 4.11. Three sets of parameters for acetone + chloroform.

Figure 4.23. Multiple sets of parameters τ_{12} and τ_{21} of the NRTL equation with $\alpha_{12} = -1$. The values of τ_{21} are obtained with

$\tau_{21} = \ln \gamma_1^\infty - \tau_{12} \exp(-\alpha_{12}\tau_{12})$.

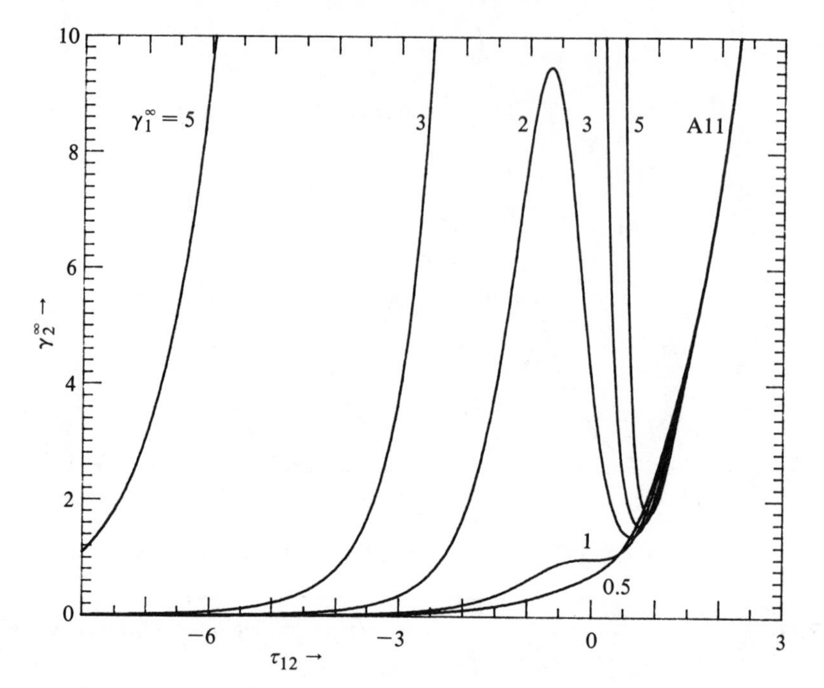

Figure 4.23(a). Plots with wide range of τ_{12} at indicated values of γ_1^∞.

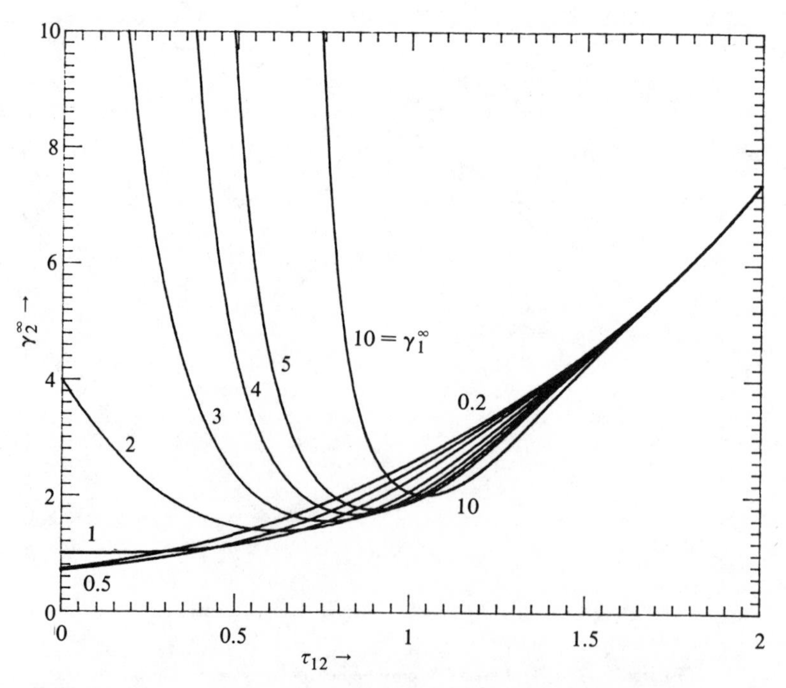

Figure 4.23(b). Plots with narrow range of τ_{12} at indicated values of γ_1^∞.

Example 4.12. Multiple Roots in the NRTL Representation of Equilibria in the System *n*-Hexane (1)- Diethyl Ketone (2)

The infinite dilution activity coefficients are $\gamma_1^\infty = 2.25$ and $\gamma_2^\infty = 3.67$. Taking $\alpha_{12} = -1$, the three sets of parameters are:

Case	τ_{12}	τ_{21}
I	−1.94000	1.08972
II	0.16703	0.61353
III	1.3532	−4.4257

The activity-coefficient plots for the three cases are shown. Number II fits the experimental data best. This case conforms to the Tassios rule that the right solution is the one for which $|\tau_{12}| + |\tau_{21}| \rightarrow$ minimum.

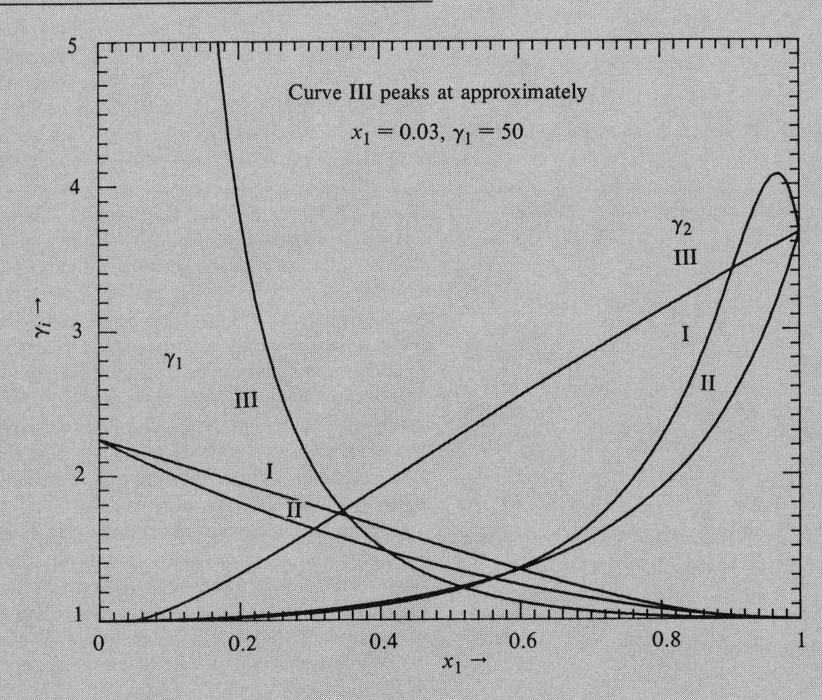

Example 4.12. Three sets of parameters for *n*-hexane + diethylketone.

In the absence of more data than are used in the evaluation of the parameters, the choice between multiple sets of parameters that satisfy the activity-coefficient correlations still may not be clearcut. Since there is no ambiguity with either the van Laar or Margules equations, and those correlations usually are fairly close to the Wilson or NRTL or UNIQUAC, comparison with their plots may be employed sometimes to discriminate between multiple sets of parameters of the other equations. Otherwise, in a practical design situation the most conservative result is advisable.

4.15. INFINITE DILUTION

The utility of the limiting values of activity coefficients at infinite dilution in evaluating the parameters of correlations such as the van Laar, Wilson, etc., has been amply illustrated. Moreover, such coefficients are of particular importance in the evaluation of stage requirements for separation processes since the greatest demands for equipment are in the dilute ranges. With newly improved techniques, accurate measurements at low concentrations can be made with much less effort and cost than those of conventional vapor-liquid measurements, and with greater overall accuracy since extrapolation of infinite-dilution activity coefficients to middle concentrations generally is safer than extrapolation from middle concentrations to low values.

The most common measurements are of boiling temperatures of known compositions at a known pressure, or measurements of total pressure as a function of known composition and fixed temperature. The classical method of measuring both liquid and vapor compositions in equilibrium is the most difficult but the best when it can be done accurately. Several extrapolation techniques are applicable. If vapor-phase nonidealities are small, the activity coefficient can be written

$$\gamma_1 = \frac{P - \gamma_2 x_2 P_2^0}{x_1 P_1^0}. \tag{4.175}$$

Since $\gamma_2 \to 1$ when $x_1 \to 0$,

$$\gamma_1^\infty = \lim_{x_1 \to 0} \frac{P - x_2 P_2^0}{x_1 P_1^0}, \tag{4.176}$$

and similarly for the other component. Example 4.15 applies this method. The infinite-dilution activity coefficient also is related simply to the limiting slope of the x-y diagram:

$$\gamma_1 = \frac{P y_1}{P_1^0 x_1} = \frac{P(y_1 - 0)}{P_1^0 (x_1 - 0)}, \tag{4.177}$$

so that

$$\gamma_1^\infty = \lim_{x_1 \to 0} \frac{P_2^0}{P_1^0} \frac{dy_1}{dx_1}, \tag{4.178}$$

since the system pressure becomes the vapor pressure of the solvent in the limit.

Calculations of γ_1^∞ from isothermal measurements can be made with the aid of equations derived by Gautreaux & Coates (1955),

$$\gamma_1^\infty = \frac{P_2^0 \phi_{1P_2^0}}{P_1^0 \phi_{1P_1^0}} \left[1 + \left(\frac{dP}{dx_1} \right)_{x_1=0}^{\infty} \left(\frac{V_2^v - V_2^L}{RT} \right) \right] \exp\left(- \int_{P_1^0}^{P_2^0} \frac{V_1^L}{RT} \, dP \right), \tag{4.179}$$

$$\gamma_2^\infty = \frac{P_1^0 \phi_{2P_1^0}}{P_2^0 \phi_{2P_2^0}} \left[1 - \left(\frac{dP}{dx_1} \right)_{x_1=1.0}^{\infty} \left(\frac{V_1^v - V_1^L}{RT} \right) \right] \exp\left(- \int_{P_2^0}^{P_1^0} \frac{V_2^L}{RT} \, dP \right). \tag{4.180}$$

The key quantities are the slopes, $(dP/dx_1)_{x_1=0}$ and $(dP/dx_1)_{x_1=1}$. Their determination can be made more reliable with the aid of an auxiliary function introduced by Ellis & Jonah (1962):

$$P_D = P - P_2^0 - (P_1^0 - P_2^0) x_1, \tag{4.181}$$

which has the useful limiting properties

$$\lim_{x_1 \to 0} \left(\frac{P_D}{x_1 x_2} \right) = \frac{0}{0} = \lim_{x_1 \to 0} \frac{dP/dx_1}{1 - 2x_1} = \left(\frac{dP}{dx_1} \right)_{x_1=0}, \tag{4.182}$$

$$\lim_{x_2 \to 0} \left(\frac{P_D}{x_1 x_2} \right) = \left(\frac{dP_D}{dx_2} \right)_{x_2=0} = - \left(\frac{dP_D}{dx_1} \right)_{x_1=1}. \tag{4.183}$$

Therefore,

$$\left(\frac{dP}{dx_1} \right)_{x_1=0} = \left(\frac{P_D}{x_1 x_2} \right)_{x_1=0} + P_1^0 - P_2^0, \tag{4.184}$$

$$\left(\frac{dP}{dx_1} \right)_{x_1=1} = - \left(\frac{P_D}{x_1 x_2} \right)_{x_1=1} + P_1^0 - P_2^0, \tag{4.185}$$

thus obviating the need for numerical differentiation or graphical slope determinations. According to Maher & Buford (1979), P_D/x_1x_2 varies linearly with x_1 in many instances.

The problem is to make accurate measurements in the low concentration ranges so that only limited extrapolation to zero will be necessary. Eckert et al. (1981, 1982) got around the difficulties of accurate ebulliometric measurements at low concentrations by employing a differential technique. They measured the difference in boiling temperatures of a pure solvent and a dilute solution with instruments capable of detecting differences of 0.001 K. Concentrations well under 0.005 mol fraction could be tested adequately. A precise ebulliometer for concentrations below 3 percent or so also was used by Tochigi & Kojima (1976), as part of a program for rapid evaluation of group contributions to activity coefficients for the ASOG method that they have developed at length.

A favorite method for dealing with low concentrations, gas chromatography, was proposed for activity-coefficient determination as early as 1952, but only recently has it been developed into a rapid and accurate technique. Two basic methods are commonly employed, both using a solvent of low volatility occluded on an inert support such as glass beads. In the *retention-time method*, a pulse of solute of 0.001 ml or so is injected into helium carrier that flows through the column. The activity coefficient of the solute is related mathematically to the retention time of the solute in the column and other properties of the pure components.

In the *stripping method*, the solute to the extent of 0.01 ml or so in 20 ml or so of solvent is charged to the chromatograph. The variation with time of the concentration of solute in a stripping gas such as helium is measured. This also is related mathematically to the activity coefficient. Thermal conductivity or flame ionization detectors are used, depending on the volatility of the solvent.

Chromatography was used by Tassios (1972) for evaluating the merits of extractive distillation solvents. Pairs of solutes in equal volume proportions were injected into a column that had been precharged with a solvent on an inert support. Presumably, the effluent concentrations in an elutriating gas were monitored. From these measurements, the relative volatilities of close-boiling pairs of substances were obtained in the presence of several possible extractive distillation solvents.

The retention-time and stripping methods were checked within 10 percent or so of each other by Santacesaria et al. (1979), who tested six hydrocarbons in phenol, furfural, and dimethylformamide, which are common extractive distillation solvents. Some improvements in technique were made by Eckert et al. (1981), who were able to measure activity coefficients even in solvents that were slightly more volatile than the solutes. Equilibration was rapid enough to allow the measurement of several dozen points in one day. Data on about a dozen solutes are reported in solvents such as benzene, 1-2-dichlor ethane, and ethanol. The chromatographic results compared favorably with some vapor-liquid equilibrium and differential ebulliometric data.

Application of chromatographic techniques to measurements of vapor-liquid equilibria is reviewed in detail by Letcher (1978). A serious disadvantage of these methods is that they usually cannot measure both infinite-dilution activity coefficients of a binary mixture, but only that of the more volatile component. This information is sufficient for determining the single parameter of one-parameter versions of the Margules, Wilson, or UNIQUAC equations. Those equations may be of value in some exploratory studies. Perhaps the major use of the chromatographic technique is for the evaluation of relative volatilities of mixtures in the presence of a less volatile solvent, as for extractive distillation processes. Such data can be extended, however, by the Gibbs-Duhem equation or Eq. 4.23.

Differential ebulliometry, on the other hand, does not have these limitations and is a highly promising technique, even though extrapolation to the middle concentrations is subject to serious error in some cases. The fact that mixtures may have essentially the same infinite-dilution activity coefficients does not guarantee that intermediate values likewise will be the same, a point that is made in Figures 4.17 and 4.18, for example. Accurate intermediate values of activities still may be desirable for appraisal of azeotropy and liquid-phase splitting.

4.16. ACTIVITY COEFFICIENTS FROM PURE COMPONENT PROPERTIES

If the accuracy is sufficient, the most desirable way to evaluate activity coefficients is from the properties of the pure components without recourse to experiments on mixtures. Second best is to deduce them from measurements on binary mixtures alone with the aid of suitable theory—for example, the Wilson, NRTL and UNIQUAC equations. The pure-component approach has been developed into a working tool in three ways that will be described.

4.16.1. Regular Solutions and the Solubility Parameter

Regular solutions are those without molecular interactions such as dipole interactions, association, or chemical effects generally. Thermodynamically they are characterized by having excess entropy, $S^{\text{ex}} = 0$. If, moreover, the excess volume likewise is zero, these relations follow:

$$G^{\text{ex}} = H^{\text{ex}} = U^{\text{ex}} = RT\,\Sigma x_i \ln \gamma_i. \tag{4.186}$$

Some data are shown in Table 2.7.

In this theory the volatility and miscibility behavior of liquid mixtures is visualized to depend on the relative ease with which molecules are separable from each other. Plausibly, this effect is measured by the energy of vaporization per unit volume. For reasons that will be clear shortly, this energy is given a special symbol,

$$\delta^2 = \Delta U/V, \tag{4.187}$$

where δ is called the solubility parameter; it turns out that the more nearly equal the solubility parameters of two substances are, the greater their miscibility. Some further assumptions led Scatchard (1931) and Hildebrand (1933) to quantitative results for the excess Gibbs energy and hence the activity coefficients. More detailed development of this theory may be found in books by Hildebrand et al. (1970, Chap. 6) and Prausnitz (1969, Chap. 7).

First, the combining rules for the second virial coefficient are assumed to apply to the ratio, internal energy/volume, on a volume-fraction basis,

$$\frac{\Delta U}{V} = [\Sigma\phi_i(\Delta U/V)_i^{0.5}]^2, \tag{4.188}$$

or

$$\delta^2 = (\Sigma\phi_i\delta_i)^2, \tag{4.189}$$

where

$$\phi_i = x_i V_i/\Sigma x_j V_j. \tag{4.190}$$

The further assumptions of zero excess volume and entropy and that

$$\delta_{12} = (\delta_1\delta_2)^{0.5} \tag{4.191}$$

ultimately lead to these equations for binary mixtures:

$$\ln \gamma_1 = \frac{V_1}{RT}(1-\phi_1)^2(\delta_1-\delta_2)^2, \tag{4.192}$$

$$\ln \gamma_2 = \frac{V_2}{RT}\phi_1^2(\delta_1-\delta_2)^2, \tag{4.193}$$

and

$$G^{\text{ex}} = x_1x_2\left[\frac{V_1V_2}{x_1V_1+x_2V_2}\right](\delta_1-\delta_2)^2. \tag{4.194}$$

For component i of a multicomponent mixture,

$$\ln \gamma_i = \frac{V_i}{RT}(\delta_i-\bar{\delta})^2, \tag{4.195}$$

where the weighted mean-solubility parameter of the mixture is

$$\bar{\delta} = \Sigma\phi_i\delta_i = \frac{\Sigma x_iV_i\delta_i}{\Sigma x_iV_i}. \tag{4.196}$$

Eq. 4.195 is called the Scatchard-Hildebrand or regular-solution activity-coefficient equation, abbreviated SH. It always predicts values of activity coefficient in excess of unity. The modification by Flory and Huggins is called the extended SH or the SHFH equation,

$$\ln \gamma_i = V_i(\delta_i-\bar{\delta})^2/RT + \ln\theta_i + 1 - \theta_i, \tag{4.197}$$

where

$$\theta_i = V_i/\Sigma x_iV_i = \phi_i/x_i \tag{4.197a}$$

is the ratio of the molal volume of component i to the weighted molal volume of the mixture. The effect of θ_i on the activity coefficient is shown graphically in Figure 4.12A and appears always to reduce the value given by the original SH equation.

The accuracy of the solubility parameter method often is quite poor. For instance, Table 4.17 is a comparison of infinite-dilution activity coefficients figured with the Wilson parameters of Holmes & van Winkle (1970) and with those figured with three different sets of solubility parameters from the literature. Even for hydrocarbon pairs, the agreement often is quite poor. Several attempts have been made to remedy this situation. Chao & Seader (1961, and also Chap. 6) retained the form of Eq. 4.195 but modified some of the solubility parameters arbitrarily in order to make the equation a better fit to experimental equilibrium data. Maffiolo et al. (1975) retained some aspects of the Chao-Seader method but used the SHFH equation as part of the system. That procedure was found by Daubert et al. (1978), in their comprehensive study, to be superior to the original method for naphthenic and aromatic mixtures and for heavy hydrocarbons at low pressures.

The Daubert team found the solubility parameter useful as a correlating variable for binary interaction coefficients in the Soave procedure (Table 1.12). Other properties also have been related to the solubility parameter, for example van der Waals parameters, surface tension, refractive index, and viscosity; literature is given by Barton (1975, 1983), together with a large compilation of values of δ.

The concept has been extended to nonpolar mixtures that do not generally conform to regular-solution theory, by assuming the solubility parameter to be made up of several contributions that can be evaluated separately. When the energy of vaporization is written as

$$\Delta U = \Delta U_{np} + \Delta U_p + \Delta U_H, \tag{4.198}$$

the solubility parameter becomes

$$\delta = \sqrt{\delta_{np}^2 + \delta_p^2 + \delta_H^2}, \tag{4.199}$$

where the meanings of the subscripts are: np = nonpolar, p = polar, and H = hydrogen bonding. The nonpolar component is taken to be that of a hydrocarbon of approximately the same size, shape, and structure, called a *homomorph* of the polar molecular in question. The other components are evaluated by methods described, for example, by Hansen & Beerbower (1971) and Barton (1975); also they may be obtained from group contributions as shown in these same references. Values of solubility parameters for about 200 substances are given in these articles; a more extensive compilation based on vapor pressure data has been made by Hoy (1970). A typical application of this theory is to the selection of solvents for a solute with known components δ_i. Suitable solvents are liquids whose components are within the ellipsoid with axes δ_{np}, δ_p, and δ_H of the solute. A specific example is described by Barton (1975, p. 743). Solubility parameters with only one or two components also are useful for describing solvent power and some other properties.

4.16.2. Effect of Temperature

The difference between the solubility parameters of two substances often is only slightly affected by temperature, so it is common practice to evaluate the group $V_i(\delta_i - \bar{\delta})^2$ at 25 C and to assume it invariant with temperature. Individual parameters do vary appreciably with temperature, however; for example, Hildebrand (1970) correlates the function $\delta/P_c^{0.5}$ with the reduced temperature and the acentric factor. The temperature correlations of Hansen are cited in Problem 4.50.

4.16.3. Applications

The theory is applied to figuring activity coefficients, solubilities of gases, and mutual solubilities of liquids. The last topic has received the most attention in the literature, the principle being that the extent of miscibility depends on the nearness of the solubility parameters of the components of the mixture to each other. Liquid miscibilities also are discussed in Chapter 7 and solubilities of gases in Chapter 6.

The use of Eq. 4.195 in representing vaporization equilibrium ratios by Chao & Seader (1961) has been mentioned. Their solubility parameters are given in Table 6.2. A study of C_3-C_5 vapor-liquid equilibria using solubility parameters was made by Barton et al. (1974); they also used "apparent" solubility parameters to improve their predictions of equilibria. A comparison between activity coefficients from the Wilson equation, which may be regarded as the best values in this instance, with those from the Scatchard-Hildebrand equations and the Flory-Huggins modification is shown in Figure 4.19. In all cases the SHFH values agree more closely with the Wilson values, though except for the benzene + heptane system the agreement still is poor.

A fair correlation of infinite dilution activity coefficients was made by Helpinstill & van Winkle (1968) in terms of a two-component solubility parameter,

$$\Delta U/V = \lambda^2 + \tau^2, \tag{4.200}$$

where λ is the solubility parameter of the homomorph and τ is the nonpolar contribution. For the activity coefficient they found

$$RT \ln \gamma_2^\infty = RT\left[\ln\frac{V_2}{V_1} + 1 - \frac{V_2}{V_1}\right] + V_2[(\lambda_1 - \lambda_2)^2 + (\tau_1 - \tau_2)^2(1-k)], \tag{4.201}$$

where

$$k = \begin{cases} 0.798 & \text{saturated hydrocarbons} \quad \pm 11.6\% \\ 0.776 & \text{unsaturated hydrocarbons} \quad \pm 8.5\% \\ 0.894 & \text{aromatic hydrocarbons} \quad \pm 13.5\% \end{cases}$$

The temperature range is 0–125 C. The percentages shown are the extents of agreement with experimental data.

No really thorough evaluation of activity coefficients from solubility parameters appears to have been published, but the results in Table 4.17 are discouraging. Pure component methods based directly on group contributions, ASOG and UNIFAC, have received all the attention. Eq. 4.195 has the drawback that only activity coefficients greater than unity can be represented, but it does have the advantage of a built-in temperature dependence, particularly when the effect of temperature on solubility parameters also is taken into account. The inverse relation between T and $\ln \gamma_i$ also is a useful approximation with the van Laar and Margules equations.

4.16.4. Structural Contributions: The ASOG Model

Many properties of complex molecules can be evaluated, at least approximately, on the basis that a smaller group of atoms within the molecule contributes in a fixed way to that property, independently of the natures of other groups that may be present. This was called the principle of independent action by Langmuir (1925). Applications of this principle to heat capacities, enthalpies of formation, and other thermodynamic properties are well established. The pattern of current methods of figuring activity coefficients in this way was set by Wilson & Deal (1962) and Derr & Deal (1969), and has been elaborated into a working tool by Kojima & Tochigi (1979). The method comprises several steps and some special concepts.

1. The activity coefficient is assumed to be made up of two contributions, one associated with differences in size and the other with interactions of specific groups of atoms making up the molecule, thus for molecule i,

$$\ln \gamma_i = \ln \gamma_i^S + \ln \gamma_i^G. \tag{4.202}$$

2. For the size contribution, the composition is expressed in terms of the number, ν_j, of atoms other than hydrogen in molecule j, as

$$r_i = \frac{\nu_i}{\Sigma x_j \nu_j} = \frac{\nu_i}{x_1\nu_1 + x_2\nu_2 + x_3\nu_3 + \ldots}, \tag{4.203}$$

where x_j is the mol fraction of molecule j and the summation is over all the types of molecules present in the mixture. The size contribution to the activity coefficient of substance i is a Flory-Huggins term

$$\ln \gamma_i^S = 1 + \ln r_i - r_i. \tag{4.204}$$

3. When figuring the contributions from interaction of groups, the composition of the system is expressed in terms of the groups and not in terms of the mol fractions of the components. These terms are used:

ν_{ki} = the number of atoms other than H in group k in molecule i, but with these special values: $\nu_{H_2O} = 1.6$, $\nu_{CH} = 0.8$, $\nu_C = 0.5$.

$$\begin{aligned} S &= \sum_i \left(x_i \sum_k \nu_{ki} \right) \\ &= x_1(\nu_{11} + \nu_{21} + \nu_{31} + \ldots) \\ &\quad + x_2(\nu_{12} + \nu_{22} + \nu_{32} + \ldots) \\ &\quad + \ldots + x_i(\nu_{1i} + \nu_{2i} + \nu_{3i} + \ldots). \end{aligned} \tag{4.205}$$

When groups H_2O, CH, and C are absent, $\nu_i = \Sigma \nu_{ki}$. The fraction of group L in the mixture is

$$X_L = \frac{1}{S} \sum_i x_i \nu_{Li} = \frac{1}{S}(x_1\nu_{L1} + x_2\nu_{L2} + x_3\nu_{L3} + \ldots). \tag{4.206}$$

The fraction of group L in an individual molecule i is obtained by making $x_i = 1$ in Eq. 4.206, so that

$$X_L^{(i)} = \nu_{Li} / \sum_i \nu_{ki} = \frac{Li}{\nu_{1i} + \nu_{2i} + \nu_{3i} + \ldots}. \tag{4.207}$$

4. Interactions of groups k and L in the mixture are represented by empirical parameters a_{kL} which are functions of temperature according to the equation

$$a_{kL} = \exp(m_{kL} + n_{kL}/T). \tag{4.208}$$

Note that $a_{kL} \neq a_{Lk}$ and that $a_{kk} = 1$. In a later paper, Tochigi et al. (1981) propose an improved form of temperature dependence but do not provide coefficients for all groups. Two kinds of sums involving these parameters are required. For group k,

$$D_k = X_1 a_{k1} + X_2 a_{k2} + X_3 a_{k3} + \ldots, \tag{4.209}$$

and

$$C_k = X_1 a_{1k}/D_1 + X_2 a_{2k}/D_2 + X_3 a_{3k}/D_3 + \ldots, \tag{4.210}$$

where the sums are over all the groups. The corresponding expressions for pure components, $D_k^{(i)}$ and $C_k^{(i)}$, are obtained from the preceding by replacing X_k with $X_k^{(i)}$.

5. The contribution of group k to the activity coefficient is given by a Wilson-type equation

$$\ln \Gamma_k = 1 - \ln D_k - C_k. \tag{4.211}$$

Similarly, the contribution of group k in each pure substance is

$$\ln \Gamma_k^{(i)} = 1 - \ln D_k^{(i)} - C_k^{(i)}. \tag{4.212}$$

When only one kind of group is present in the molecule, $\ln \Gamma_k^{(i)} = 0$.

The total group contribution to the activity coefficient of molecule i is

$$\ln \gamma_i^G = \sum_k \nu_{ki}(\ln \Gamma_k - \ln \Gamma_k^{(i)}). \tag{4.213}$$

Substitution of Eq. 4.204, 214 into 202 will give the activity coefficient of component i in the mixture. The method is used in Example 4.13. A BASIC program for the ASCG method is in Appendix C; the parameters a_{kL} are part of that program listing, steps 270 to 3080.

4.16.5. The UNIFAC Method (UNIQUAC Functional Group Activity Coefficients) (Fredenslund et al. 1975)

In this method the size and group contributions to the activity coefficient are called *configurational* (C) and *residual* (R):

$$\ln \gamma_i = \ln \gamma_i^C + \ln \gamma_i^R. \tag{4.214}$$

Both parts are based on the UNIQUAC equation. In addition to group interaction parameters, a_{mn} and a_{nm}, group volume parameters, R_k, and surface parameters, Q_k, are involved. The configurational part is the same as that of the UNIQUAC equation of Table 4.4. The residual part is found as in the ASOG method, but with

$$\psi_{mn} = \exp(-a_{mn}/T), \tag{4.215}$$

instead of the interaction term a_{kL} of Eq. 4.208. The fraction of group L in the mixture, X_L, is defined by Eqs. 4.206, 4.207. The group surface-area fraction is

$$\theta_m = X_m Q_m / \sum_n X_n Q_n. \tag{4.216}$$

The weighted interaction parameter of group k is

$$E_k = \theta_1 \psi_{1k} + \theta_2 \psi_{2k} + \theta_3 \psi_{3k} + \ldots, \tag{4.217}$$

and an auxiliary function F_k is defined as

Example 4.13. Activity Coefficients of Ethanol + Benzene by the ASOG Method

The temperature is 345 K, the composition is $x_1 = 0.2$.

Ethanol has 2 type CH_2 and 1 type OH groups; benzene has 6 type ArCH groups. Interaction parameters from the Tochigi & Kojima table are:

	Group l	1 CH_2		3 ArCH		6 OH	
No.	Group k	m_{kl}	n_{kl}	m_{kl}	n_{kl}	m_{kl}	n_{kl}
1	CH_2	0	0	−0.7457	146.0	−41.2053	7686.4
3	ArCH	0.7297	−176.8	0	0	2.2682	−1111.5
6	OH	4.7125	−3,060.0	−0.5859	−939.1	0	0

ν_{ki}

i	Group k / i Compound	1 CH_2	3 ArCH	6 OH	$\Sigma\nu_{ki}$	ν_i^S
1	Ethanol	2 ν_{11}	0 ν_{31}	1 ν_{61}	3	3
2	Benzene	0 ν_{12}	6 ν_{32}	0 ν_{62}	6	6

For convenience, the groups CH_2, ArCH, and OH will be renumbered 1, 2, and 3 in the following.

$a_{kl} = \exp(m_{kl} + n_{kl}/T)$.

k	l: 1	2	3
1	1 (a_{11})	0.7243 (a_{12})	0 (a_{13})
2	1.2426 (a_{21})	1 (a_{22})	0.3854 (a_{23})
3	0.0157 (a_{31})	0.0366 (a_{32})	1 (a_{33})

$\Sigma x_j \nu_j^s = 0.2(3) + 0.8(6) = 5.4,$

$\Sigma \nu_{k1} = 3,$

$\Sigma \nu_{k2} = 6,$

$\Sigma x_i(\Sigma \nu_{ki}) = 0.2(3) + 0.8(6) = 5.4,$

$X_k = (x_1 \nu_{k1} + x_2 \nu_{k2})/5.4 = (0.2\,\nu_{k1} + 0.8\,\nu_{k2})/5.4,$

$X_1 = (0.2(2) + 0.8(0))/5.4 = 0.0741,$

$X_2 = (0.2(0) + 0.8(6))/5.4 = 0.8889,$

$X_3 = (0.2(1) + 0.8(0))/5.4 = 0.0370.$

For compound 1,

$X_1^{(1)} = 2/3, X_2^{(1)} = 0, X_3^{(1)} = 1/3.$

For compound 2,

$X_1^{(2)} = 0, X_2^{(2)} = 1, X_3^{(2)} = 0.$

Summary:

	Mixture	Cpd 1	Cpd 2
X_1	0.0741	2/3	0
X_2	0.8889	0	1
X_3	0.0370	1/3	0

Size contribution

Ethanol, $r_1 = \nu_1^s/\Sigma x_j \nu_j^s = 3/5.4 = 0.5556,$

Benzene, $r_2 = 6/5.4 = 1.1112,$

$\ln \gamma_1^s = 1 - 0.5556 + \ln 0.5556 = -0.1433,$

$\ln \gamma_2^s = 1 - 1.1112 + 1.1112 = -0.0058,$

$D_k = X_1 a_{k1} + X_2 a_{k2} + X_3 a_{k3},$

$D_1 = X_1 a_{11} + X_2 a_{12} + X_3 a_{13},$

$D_2 = X_1 a_{21} + X_2 a_{22} + X_3 a_{23},$

$D_3 = X_1 a_{31} + X_2 a_{32} + X_3 a_{33},$

$C_k = X_1 a_{1k}/D_1 + X_2 a_{2k}/D_2 + X_3 a_{3k}/D_3,$

$C_1 = X_1 a_{11}/D_1 + X_2 a_{21}/D_2 + X_3 a_{31}/D_3,$

$C_2 = X_1 a_{12}/D_1 + X_2 a_{22}/D_2 + X_3 a_{32}/D_3,$

$C_3 = X_1 a_{13}/D_1 + X_2 a_{23}/D_2 + X_3 a_{33}/D_3,$

$\ln \Gamma_k = 1 - \ln D - C.$

Summary of numerical results:

	D	C	$\ln\Gamma_k$
Mixture			
k = 1	0.7179	1.2212	0.1101
k = 2	0.9953	0.9870	0.0176
k = 3	0.0707	0.8675	0.7819
Ethanol			
k = 1	0.6667	1.0152	0.3893
k = 2	0.3438	0.9695	1.0983
Benzene			
k = 3	1	1	0

(continued next page)

Example 4.13 *(continued)*

$$\ln \gamma_1^G = 2(0.1101 - 0.3893 - 0) + 1(2.7819 - 0 - 1.0983) = 1.1252,$$

$$\ln \gamma_2^G = 6(0.0176) = 0.1056,$$

$$\ln \gamma_1 = 1.1252 - 0.1433 = 0.9819,$$

$$\gamma_1 = 2.6695,$$

$$\ln \gamma_2 = 0.1056 - 0.0058 = 0.0998,$$

$$\gamma_2 = 1.1049.$$

Over the full range of concentrations, the computer solution with Program C.7 is:

x_1	γ_1	γ_2
0	7.4666	1
0.2	2.6626	1.1059
0.4	1.6234	1.3600
0.6	1.2258	1.7961
0.8	1.0518	2.5644
1	1	4.0501

$$F_k = \frac{\theta_1 \psi_{k1}}{E_1} + \frac{\theta_2 \psi_{k2}}{E_2} + \frac{\theta_3 \psi_{k3}}{E_3} \cdots \tag{4.218}$$

The quantities θ_k, E_k, and F_k are evaluated both for the mixture and the pure components. For the mixture,

$$\ln \Gamma_k = Q_k(1 - \ln E_k - F_k), \tag{4.219}$$

and for pure component i,

$$\ln \Gamma_k^{(i)} = Q_k^{(i)}(1 - \ln E_k^{(i)} - F_k^{(i)}). \tag{4.220}$$

When there is only one kind of group in the molecule, $\ln \Gamma_k^{(i)} = 0$. The residual part of the activity coefficient becomes

$$\ln \gamma_i^R = \sum_k Q_k (\ln \Gamma_k - \ln \Gamma_k^{(i)}), \tag{4.221}$$

and the complete expression for the activity coefficient of component i is

$$\ln \gamma_i = \ln \gamma_i^C + \sum_k Q_k (\ln \Gamma_k - \ln \Gamma_k^{(i)}). \tag{4.222}$$

The method is applied in Example 4.14 on the same problem to which the ASOG method was applied in Example 4.13.

4.16.6. Comparison of ASOG and UNIFAC

Both methods have been tested amply; many examples are given in the books of Tochigi & Kojima (1979) and Fredenslund et al. (1977). From the two examples worked out here, it is clear that computer solution is desirable. Both books give FORTRAN programs for these calculations. The methods seem to be of about equal reliability, although no exhaustive comparison appears to have been published. Somewhat more development effort is being expended on the UNIFAC system and the contributions of more groups have been analyzed for it. The latest in a series of compilations is that by Gmehling et al. (1982) (Tables E11, E12). A special set of parameters also has been developed for the calculation of liquid-liquid equilibria by Magnussen et al. (1981) (Tables E13, E14). These two compilations are based on the exhaustive DECHEMA collections of equilibrium data.

Example 4.14. Activity Coefficients of Ethanol + Benzene by the UNIFAC Method

The temperature is 345 K and the composition is $x_1 = 0.2$.

The group breakup is a little different from the ASOG method: Ethanol has 1 type CH_3, 1 type CH_2, and 1 type OH groups, while benzene has 6 type ArCH groups. Parameter data are from Tables E.11, E.12.

	ν			
	Eth.	Benz.	R_k	Q_k
CH_3	1	0	0.9011	0.848
CH_2	1	0	0.6744	0.540
OH	1	0	1.0000	1.2000
ArCH	0	6	0.5313	0.4000

$$a_{12} = 0 \qquad a_{21} = 0$$

$$a_{23} = a_{13} = 986.5 \qquad a_{32} = a_{31} = 156.4$$

$$a_{24} = a_{14} = 61.13 \qquad a_{42} = a_{41} = -11.12$$

$$a_{34} = 89.60 \qquad a_{43} = 636.1$$

$$\psi_{12} = 1 \qquad \psi_{21} = 1$$

$$\psi_{23} = \psi_{13} = 0.0573 \qquad \psi_{32} = \psi_{31} = 0.6355$$

$$\psi_{24} = \psi_{14} = 0.8376 \qquad \psi_{42} = \psi_{41} = 1.0328$$

$$\psi_{34} = 0.7713 \qquad \psi_{43} = 0.1582$$

$$r_1 = 0.9011 + 0.6744 + 1 = 2.5755$$

$$r_2 = 6(0.5313) = 3.1878$$

$$q_1 = 0.848 + 0.540 + 1.200 = 2.588$$

$$q_2 = 6(0.4) = 2.4$$

$$\phi_1 = \frac{0.2(2.5755)}{0.2(2.5755) + 0.8(3.1878)} = 0.1680$$

$$\phi_2 = 0.8320$$

$$\theta_1 = \frac{0.2(2.588)}{0.2(2.588) + 0.8(2.4)} = 0.2123$$

(continued next page)

Example 4.14 *(continued)*

$\theta_2 = 0.7877$

$l_1 = 5(2.5755 - 2.588) + 1 - 2.5755 = -1.6380$

$l_2 = 5(3.1878 - 2.4) + 1 - 3.1878 = +1.7512$

$$\ln \gamma_1^S = \ln \frac{0.168}{0.2} + 5(2.588) \ln \frac{0.2123}{0.168} + 0.832 \left(-1.638 - \frac{2.5755(1.7512)}{3.1878}\right) = 0.3141$$

$$\ln \gamma_2^S = \ln \frac{0.832}{0.8} + 5(2.4) \ln \frac{0.7877}{0.832} + 0.168 \left(1.7512 - \frac{3.1878(-1.638)}{2.5755}\right) = 0.0175$$

$$\bar{X}_m = \frac{0.2\, \nu_m^{(1)} + 0.8\, \nu_m^{(2)}}{0.2(3) + 0.8(6)}$$

		Mix	Eth.	Benz.
CH_3	$\bar{X}_1$	0.037	1/3	0
CH_2	$\bar{X}_2$	0.037	1/3	0
OH	$\bar{X}_3$	0.037	1/3	0
ArCH	$\bar{X}_4$	0.8889	0	1

$$\theta_m = \frac{\bar{X}_m Q_m}{0.037(0.848 + 0.54 + 1.2) + 0.8889(0.4)} = \frac{\bar{X}_m Q_m}{0.4513}$$

	Mix	Eth.	Benz.
θ_1	0.0695	0.3277	—
θ_2	0.0443	0.2087	—
θ_3	0.0984	0.4637	—
θ_4	0.7879	—	1

$$E_k = \theta_1 \phi_{1k} + \theta_2 \phi_{2k} + \theta_3 \psi_{3k} + \theta_4 \psi_{4k}$$

	Mix	Eth.	Benz.
E_1	0.9901	0.8311	0
E_2	0.9901	0.8311	0
E_3	0.2296	0.4944	0
E_4	0.9591	—	1

$$F_k = \frac{\theta_1 \psi_{k1}}{E_1} + \frac{\theta_2 \psi_{k2}}{E_2} + \frac{\theta_3 \psi_{k3}}{E_3} + \frac{\theta_4 \psi_{k4}}{E_4}$$

	Mix	Eth.	Benz.
F_1	0.8276	0.6992	—
F_2	0.8276	0.6992	—
F_3	1.1352	1.3481	—
F_4	1.0158	—	1

$$\ln \Gamma_k = Q_k(1 - \ln E_k - F_k)$$

	Mix	Eth.	Benz.
$\ln \Gamma_1$	0.1546	0.4120	—
$\ln \Gamma_2$	0.0985	0.2623	—
$\ln \Gamma_3$	1.6035	0.4276	—
$\ln \Gamma_4$	0.0104	—	1

$$\ln \gamma_1 = 0.3141 + 1(0.1546 - 0.4120) + 1(0.0985 - 0.2623) + 1(1.6035 - 0.4276) = 1.0688$$

$\gamma_1 = 2.9119$

$\ln \gamma_2 = 0.0175 + 6(0.0104) = 0.0799$

$\gamma_2 = 1.0832$

Comparison with the ASOG results, $\gamma_1 = 2.67$ and $\gamma_2 = 1.10$, is only fair.

4.17. ACTIVITY COEFFICIENTS FROM MEASUREMENTS OF EQUILIBRIUM

4.17.1. Basic Considerations

Carefully obtained experimental data generally are preferable to calculated results such as those from equations of state or group contributions. In any event, the scope and validity of those methods and of the correlating equations can be established only with experimental work. The intent of this section is to note the various kinds of data and the numerical procedures suitable for these purposes, and thus for the interpolation and extrapolation of limited data.

In the most general case, the problem of finding the parameters of equations is one of nonlinear regression for which the Newton-Raphson method often is suitable; this is given as Program C.6 in the Appendix, as well as by Hirata et al. (1976). The DECHEMA work utilizes the simplex method of Nelder & Mead (1965). The subject of appropriate objective functions is mentioned in Section 4.9. Theoretically, a number of data equal to the number of parameters in the equation is sufficient to determine those parameters, a procedure that has been explained in the section on infinite-dilution activity coefficients; but more data over the entire range of concentrations are of course statistically desirable.

Equality of partial fugacities of individual components in all phases in contact is a basic condition for phase equilibrium. For a substance, i, in liquid and vapor phases,

$$\hat{f}_i^V = \hat{f}_i^L, \tag{4.223}$$

or

$$\hat{\phi}_i P y_i = \gamma_i x_i f_i^0 = \gamma_i x_i \phi_i^{sat} P_i^{sat} \exp\left(\int_{P_i^{sat}}^{P} \frac{\bar{V}^L}{RT} dP\right), \quad (4.224)$$

from which

$$\gamma_i = \frac{\hat{\phi}_i}{\phi_i^{sat}} \frac{P}{P_i^{sat}} (PF)_i \frac{y_i}{x_i}. \quad (4.225)$$

At pressures of only a few atmospheres the Poynting Factor is little different from unity and is usually ignored. Likewise, at moderate pressure the ratio of the fugacity coefficients may be near unity, so the expression for the activity coefficient that is often adequate is simply

$$\gamma_i \simeq \frac{\hat{\phi}_i}{\phi_i^{sat}} \frac{P}{P_i^{sat}} \frac{y_i}{x_i} \simeq \frac{P}{P_i^{sat}} \frac{y_i}{x_i}. \quad (4.226)$$

Since binary parameters usually are applicable to vapor-liquid equilibria of multicomponent mixtures with modern equations, and somewhat less accurately to liquid-liquid equilibria, binary mixtures primarily will be considered here. Vapor-phase nonidealities, represented by the ratio of fugacity coefficients, $\hat{\phi}_i/\phi_i^{sat}$, will be neglected. Examples of how they are taken account of when necessary are in Chapter 6.

Example 4.15. Extrapolation of Isobaric and Isothermal Bubblepoint Data to Find Infinite-Dilution Activity Coefficients

Data of methanol + ethylacetate at 1 atm and of methanol + dichlorethane at 60 C are available. Rearranging of

$$P = \gamma_1 x_1 P_1^0 + \gamma_2 x_2 P_2^0, \quad (1)$$

gives the activity coefficient formulas

$$\gamma_1 = (P - \gamma_2 x_2 P_2^0)/x_1 P_1^0, \quad (2)$$

$$\gamma_2 = (P - \gamma_1 x_1 P_1^0)/x_2 P_2^0, \quad (3)$$

The infinite-dilution activity coefficients are the limits of "apparent" activity coefficients * as the concentrations go to zero,

$$\gamma_1^* = (P - x_2 P_2^0)/x_1 P_1^0, \quad (4)$$

$$\gamma_2^* = (P - x_1 P_1^0)/x_2 P_2^0, \quad (5)$$

$$\gamma_1^\infty = \lim_{x_1=0} \gamma_1^*, \quad (6)$$

$$\gamma_2^\infty = \lim_{x_2=0} \gamma_2^*. \quad (7)$$

The plots of experimental and "apparent" activity coefficients are shown. With the exercise of a little imagination, it is clear that the extrapolations to zero concentrations agree. The infinite-dilution activity coefficients are read off the figures as:

Methanol + ethylacetate: $\gamma_1^\infty = 3.05$, $\gamma_2^\infty = 2.9$,

Methanol + dichlorethane: $\gamma_1^\infty = 6.3$, $\gamma_2^\infty = 3.9$.

Although in these two examples the vapor compositions have been determined, allowing calculation of the activity coefficients, the "apparent" activity-coefficient plots do not require this information, and accordingly allow finding infinite-dilution activity coefficients with a knowledge of only liquid compositions and either the total pressures at constant temperature or bubblepoint temperatures at constant pressure.

(a). Extrapolation of Data on Bubble Pressures of Liquids of Known Composition to Find Infinite-Dilution Activity Coefficients

The system is methanol + dichlorethane at 60 C, data of McKetta & Katz [*IEC* 40 (1948):853] The "apparent" activity coefficients of Section 4.15 are extrapolated to infinite dilution. The true activity coefficients are available for this system and are shown on plot (a) for comparison.

(b) Extrapolation of Bubble Temperatures of Liquids of Known Compositions to Find Infinite-Dilution Activity Coefficients

The system is methanol + ethyl acetate at 1 atm, data of Miller & Bliss [IEC 32 (1940):123]. The true activity coefficients are known for this system and are plotted for comparison with the "apparent" activity coefficients on Figure (b).

x_1	y_1	t	γ_1	γ_2	P_1^{sat}	P_2^{sat}	γ_1^*	γ_2^*
2.8	12.0	74.4	3.020	1.002	1108.4	694.0	2.75	1.08
3.7	13.3	74.0	2.533	0.997	1092.1	684.8	2.49	1.09
7.3	22.0	71.5	2.329	1.016	994.8	629.9	2.43	1.18
12.3	31.0	69.3	2.115	1.028	915.2	584.4	2.20	1.26
21.1	42.0	66.4	1.867	1.086	818.4	528.6	1.99	1.41
23.6	44.2	66.0	1.784	1.076	805.7	521.2	1.90	1.43
23.9	44.0	65.8	1.767	1.092	799.4	517.6	1.92	1.44
26.5	46.8	65.3	1.729	1.094	783.9	508.6	1.86	1.48
35.2	52.6	64.0	1.539	1.157	744.8	485.7	1.70	1.58
40.8	55.8	63.7	1.425	1.193	736.0	480.5	1.58	1.62

(Continued next page)

Example 4.15 *(continued)*

x_1	y_1	t	γ_1	γ_2	P_1^{sat}	P_2^{sat}	γ_1^*	γ_2^*
44.0	57.3	63.6	1.363	1.222	733.1	478.8	1.52	1.63
53.3	62.0	63.1	1.221	1.363	718.7	470.3	1.41	1.72
58.5	64.7	62.9	1.189	1.432	713.0	466.9	1.36	1.77
66.4	68.7	62.4	1.135	1.560	698.9	458.6	1.31	1.92
70.8	71.1	62.4	1.102	1.658	698.9	458.6	1.27	1.98
74.8	73.7	62.4	1.081	1.748	698.9	458.6	1.23	2.05
79.3	76.8	62.4	1.062	1.877	698.9	458.6	1.20	2.17
82.2	79.0	62.5	1.050	1.969	701.7	460.3	1.18	2.24
88.3	84.2	62.8	1.030	2.230	710.2	465.3	1.13	2.44
96.1	93.4	64.0	1.001	2.667	744.8	485.7	1.04	2.43

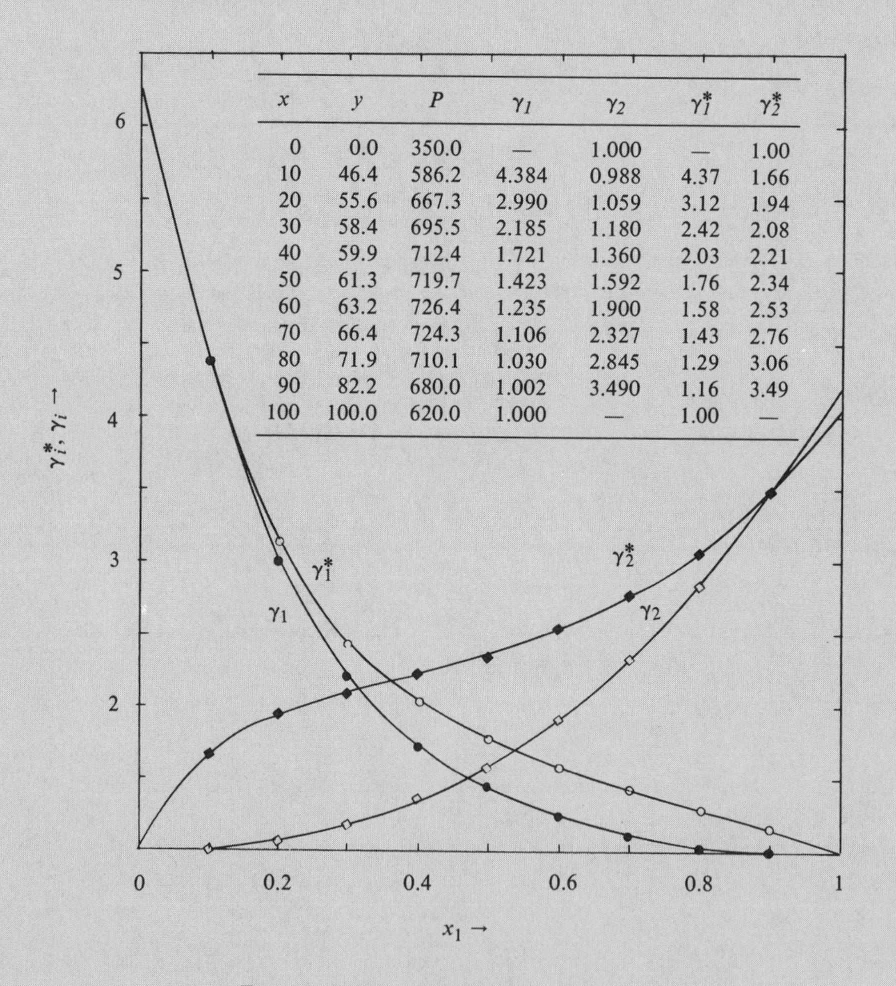

x	y	P	γ_1	γ_2	γ_1^*	γ_2^*
0	0.0	350.0	—	1.000	—	1.00
10	46.4	586.2	4.384	0.988	4.37	1.66
20	55.6	667.3	2.990	1.059	3.12	1.94
30	58.4	695.5	2.185	1.180	2.42	2.08
40	59.9	712.4	1.721	1.360	2.03	2.21
50	61.3	719.7	1.423	1.592	1.76	2.34
60	63.2	726.4	1.235	1.900	1.58	2.53
70	66.4	724.3	1.106	2.327	1.43	2.76
80	71.9	710.1	1.030	2.845	1.29	3.06
90	82.2	680.0	1.002	3.490	1.16	3.49
100	100.0	620.0	1.000	—	1.00	—

Example 4.15 (a). Extrapolation of bubble pressures.

4.17.2. Azeotropy

For many binary systems, and some ternary and higher ones, the only phase-equilibrium data known are the azeotropic compositions, temperature, and pressure since these are relatively easy to find. Several thousand such data are in the books of Lecat (1949) and Horsley (1952–1973). Sin $x_1 = y_1$, by definition, at this condition, the activity coef cients become

$$\gamma_i = \hat{\phi}_i P/\phi_i^{sat} P_i^{sat}, \qquad (4.22$$

or

Example 4.15 *(continued)*

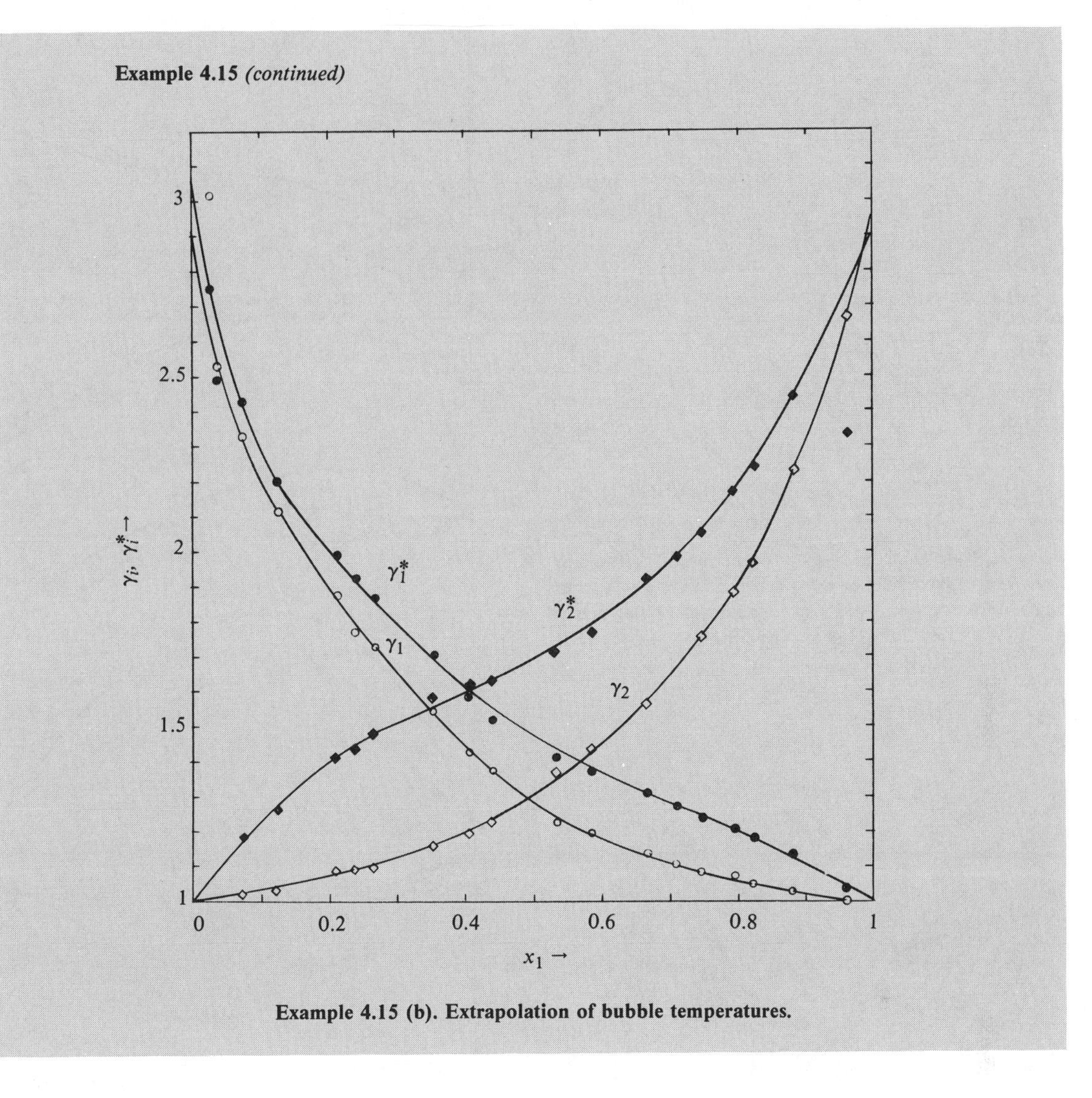

Example 4.15 (b). Extrapolation of bubble temperatures.

$$\gamma_i \simeq P/P_i^{sat}, \tag{4.228}$$

when the fugacity ratio is near unity. Once these activity coefficients have been evaluated, the finding of the parameters of a two-parameter equation is straightforward, and has been explained for several common equations in earlier sections.

The reverse process of finding the azeotropic conditions when the activity-coefficient correlation is known is of particular interest, especially for multicomponent mixtures since experimentation with them is highly time-consuming. At the azeotropic point each relative volatility is unity, so a suitable criterion for locating the azeotrope is to find a minimum at or near zero of the sum

$$\sum_{i=1}^{k-1} |\alpha_{ik} - 1| = \text{minimum, zero} \tag{4.229}$$

At moderate pressures, the simplified form is adequate:

$$\alpha_{ik} = \gamma_i P_i^o / \gamma_k P_k^o \rightarrow 1. \tag{4.230}$$

This criterion was applied by Aristovich & Stepanova (1970), who used the Wilson equation in their calculations. Any one of several search methods for the minimum may be used—for example, the method of steepest descent. They give data on nineteen ternary and one quaternary mixture obtained in this way. An interesting system is that of benzene + cyclohexane + isopropanol that forms the three binary and a ternary azeotropes at 1 atm. The mol fractions, temperatures, and normal boiling points are:

	Benzene	*Cyclohexane*	*Isopropanol*	*T, °C*
	0.667	—	0.333	71.9
	0.502	0.498	—	77.4
	—	0.589	0.411	69.1
	0.164	0.454	0.382	69.0
NBP	80.2	80.8	82.5	

Some cases have been discovered that exhibit more than one local maximum or minimum of the temperature-composition or pressure composition diagram. Figure 4.24(a) is of a rare instance of a binary mixture with both maximum and minimum boiling azeotropes. Local dimples or pimples on ternary-mixture surfaces are more common. The first case that

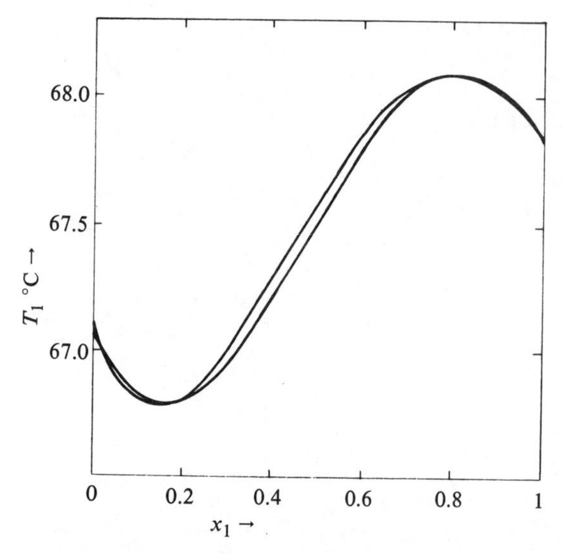

Figure 4.24(a). Unusual cases of formation of azeotropes: The system of hexafluorobenzene + benzene has a minimum boiling azeotrope at $x_1 = 0.19$ and a maximum boiling one at $x_1 = 0.79$, at 500 Torr [Gaw & Swinton, Nature 212 (Oct. 15, 1966):283–284].

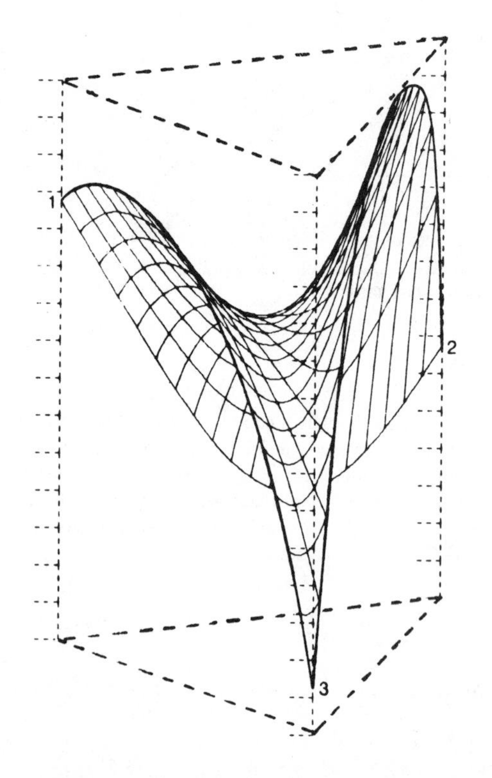

Figure 4.24(b). Unusual cases of formation of azeotropes: The system of acetone (1) + chloroform (2) + methanol (3) has a saddle-shaped pressure-composition surface—that is, a positive azeotrope in one direction and a negative one in another direction. Discovered by Ewell & Welch (1945); drawn by Van Ness & Abbott (1982, p. 347).

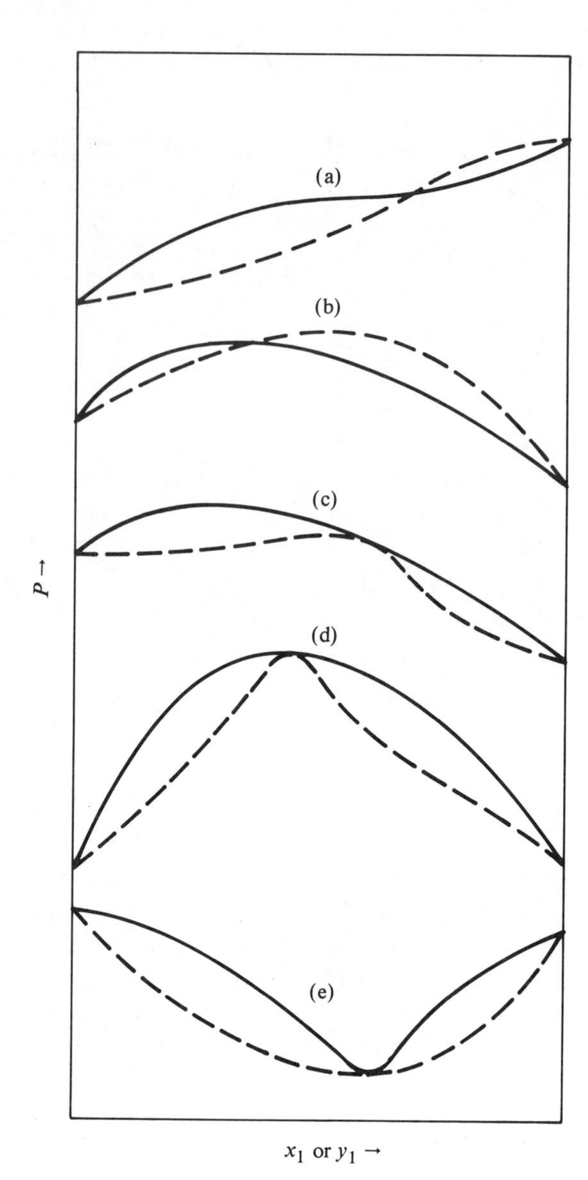

Figure 4.25. Possible and impossible types of isothermal binary boiling diagrams:

(a) Composite boiling curves resulting with formation of intermolecular compound having two parts of component 1 and one part of component 2.

(b) Thermodynamically it is not possible for the liquidus and vaporus to cross.

(c) Thermodynamically it is not possible for the liquidus and vaporus to touch except at maximum or minimum.

(d) A maximum pressure (minimum temperature) azeotrope formation.

(e) A minimum pressure (maximum temperature) azeotrope formation. Proof of statements (b) and (c) is requested in Problem 4.49.

Example 4.16. Azeotropic Compositions of Mixtures of Methyl Acetate (1) and Methanol (2) at Several Pressures

The Wilson parameters are given by

$\Lambda_{12} = 0.5108 \exp(54.9958/T)$,

$\Lambda_{21} = 1.9578 \exp(-467.79/T)$.

Antoine equations with pressure in Torr and temperature in °K:

$\ln P_1^0 = 16.5835 - 2{,}838.7/(T - 45.16)$,

$\ln P_2^0 = 18.1412 - 3{,}391.96/(T - 43.16)$.

At the azeotropic condition, the relative volatility becomes unity:

$\alpha_{12} = \gamma_1 P_1^0/\gamma_2 P_2^0 \rightarrow 1$.

Neglecting the slight vapor-phase nonideality, the bubblepoint temperature is found from

$P = \gamma_1 P_1^0 x_1 + \gamma_2 P_2^0 (1 - x_1)$.

Plots of relative volatility at the bubblepoint against the mol fractiom of methyl acetate are shown at 0.1, 1, and 5 atm. The azeotropic composition at one atm, $x_1 = 0.67$, compares favorably with the experimental data, $x_1 = 0.675$, of Balashov et al. [*Zh Fiz Khim* 41 (1937):739] (Hirata et al. (1976)).

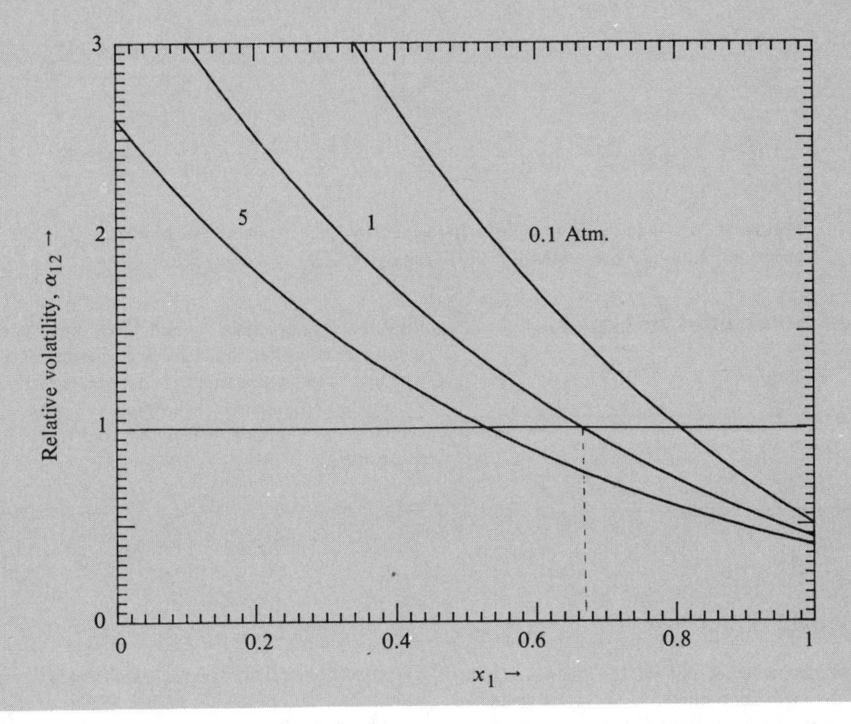

was discovered of a surface with a ridge possessing a depression, called a saddlepoint, is shown in Figure 4.24(b). Variety of equilibrium surface behavior is to be expected—the saddlepoint was predicted almost a hundred years ago by Ostwald—and the possible occurrence of local extrema must be borne in mind during mathematical searches for azeotropes. Some of what is possible and what is not possible in isothermal boiling diagrams is shown in Figure 4.25.

Some idea of the effects of activity coefficients and vapor pressures on azeotropic compositions may be obtained in terms of the symmetrical Margules equations,

$$RT \ln \gamma_1 = A(1 - x_1)^2, \qquad (4.231)$$

$$RT \ln \gamma_2 = A x_1^2. \qquad (4.232)$$

On making the relative volatility unity, the azeotropic composition becomes

$$x_1 = 0.5[1 - (RT/A)\ln(P_2^0/P_1^0)]. \qquad (4.233)$$

The equation is plotted in Figure 4.26. According to this approximation, the content of the less volatile component in the azeotrope decreases with increasing vapor pressure ratio at positive values of the Margules parameter.

Another problem of interest is finding the effect of pressure on the azeotropic composition. For a binary system this is done in Example 4.16 with the Wilson equation and in Problem 4.38 with the van Laar. The effect of temperature at constant pressure can be explored in a similar manner. Fig. 5.18 shows additional azeotropic studies.

Except by way of activity-coefficient correlations and the search procedure mentioned, there is no simple way to predict the occurrence of multicomponent azeotropes. If several of the pairs of components form azeotropes, there is likelihood of a multicomponent azeotrope; but this is not certain. For instance, in the system of acetone-ethanol-hexane the pairs with hexane form azeotropes, but the ternary mixture does not, at least at atmospheric pressure. Some of the theory of binary azeotropy is discussed by King (1969), and of ternary along with binary by Kogan (1967). Many aspects of this subject also are treated by Malesinski (1965).

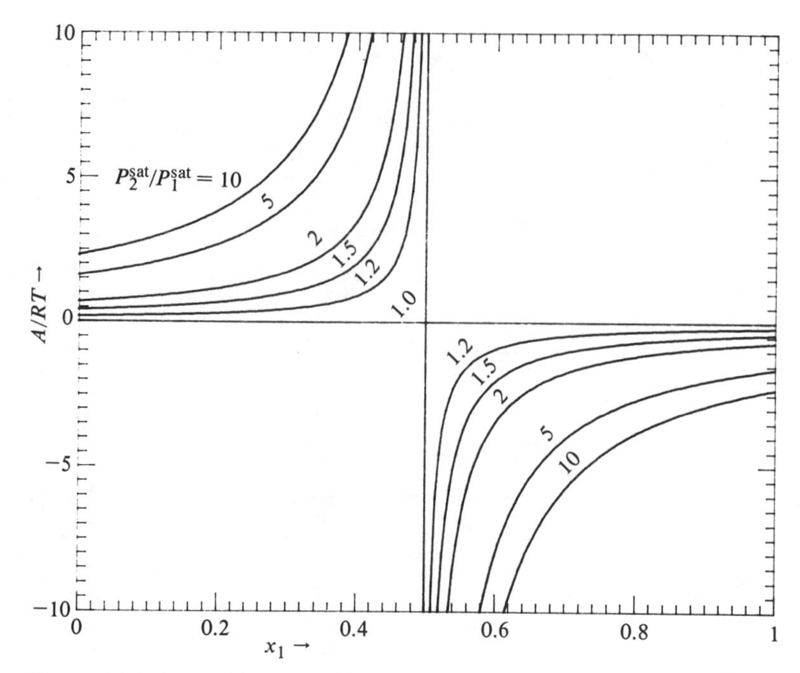

Figure 4.26. Azeotropic compositions given by the symmetrical Margules equations. Eq. 4.233. $x_1 = 0.5[1 - RT/A \ln(P_2^{sat}/P_1^{sat})]$.

4.17.3. Mutual Solubilities of Liquids

Equating the partial fugacities in two liquids, for every component,

$$\hat{f}_i = \gamma_i x_i f_i^\circ, \tag{4.234}$$

$$\hat{f}_i^* = \gamma_i^* x_i^* f_i^\circ, \tag{4.235}$$

leads to the equilibrium relationship

$$\gamma_i x_i = \gamma_i^* x_i^*, \tag{4.236}$$

or

$$\ln \gamma_i - \ln \gamma_i^* = \ln(x_i^*/x_i), \tag{4.236a}$$

where the asterisk distinguishes one of the phases. These composition-activity-coefficient relations are considered in detail in Chapter 7 but a brief treatment belongs here, since they provide a method of finding activity coefficients experimentally.

A useful application of Eq. 4.236 is to cases of very low mutual solubilities, such as those of many organic substances and water. Let the subscript 1 identify the solute and the superscript * the essentially pure solute phase. Then Eq. 4.236 becomes

$$\gamma_1 x_1 = \gamma_1^* x_1^* = 1, \tag{4.237}$$

since both x_1^* and γ_1^* of the pure solute are unity. Accordingly,

$$\gamma_1 = 1/x_1. \tag{4.237a}$$

When the solubility x_1 is very small, the activity coefficient is essentially the infinite-dilution value.

For a binary mixture

$$\ln \gamma_1 - \ln \gamma_1^* = \ln(x_1^*/x_1), \tag{4.238}$$

$$\ln \gamma_2 - \ln \gamma_2^* = \ln[(1 - x_1^*)/(1 - x_1)]. \tag{4.239}$$

When the four activity coefficients are related to the two measured compositions x_1 and x_1^* by some appropriate two-parameter equation, the resulting two equations can be solved first for the two parameters, then for the activity coefficients if those are still desired. The simplest case is that of the symmetrical equation, which has only one parameter. On substituting

$$\ln \gamma_2 = A x_1^2 \tag{4.240}$$

into Eq. 4.239 and solving for the parameter,

$$A = \frac{1}{x_1^2 - x_1^{*2}} \ln \frac{1 - x_1^*}{1 - x_1}. \tag{4.241}$$

The condition of incipient phase splitting is $x_1 = x_1^* = 0.5$, for which the value of $A = 2$. Larger values of A correspond to more limited miscibilities.

With the Margules relation, the miscibility equations become

$$\ln x_1^*/x_1 = \ln \gamma_i - \ln \gamma_1^* = A(x_2^2 - x_2^{*2}) + 2(B - A)(x_2 x_2^2 - x_1^* x_2^{*2}), \tag{4.242}$$

$$\ln \frac{1 - x_1^*}{1 - x_1} = \ln \gamma_2 - \ln \gamma_2^* = B(x_1^2 - x_1^{*2}) - 2(B - A)(x_2 x_1^2 - x_2^* x_1^{*2}). \tag{4.243}$$

An explicit solution of these equations for A and B is shown in Section 7.4. The van Laar equation also leads to explicit solutions, but other common equations require solution by approximation. The Wilson equation is limited in that it cannot represent equilibria between liquid phases.

Graphs of the limits of liquid-liquid miscibilities with several activity coefficient equations are given in Chapter 7; there infinite dilution activity coefficients are used as parameters. Location within the (x, x^*) network determines the γ_i^∞ that may be used subsequently to find the usual equation parameters (A_{12}, A_{21}), (τ_{12}, τ_{21}), and so on. These graphs reveal that the several equations do not predict the same phase splits in all ranges. To a more limited extent, of course, the prediction of VLE with the same infinite-dilution activity coefficients also is not the same with all equations. Experience with particular systems may be a guide as to which equation to use in a particular situation, but most attention is being paid to the NRTL with $\alpha_{12} = 0.2$ and the UNIQUAC equations, particularly since they are readily applicable to multicomponent mixtures also.

In general, activity coefficients fitted to LLE data do not

Example 4.17. Comparison of van Laar Parameters Obtained from LLE and From VLE Data

The mutual solubilities of water and n-butanol at 110 C are $x_1 = 0.9788$, $x_1^* = 0.6759$ [Butler et al., *J Chem Soc* (London) 674 (1933)]. The van Laar parameters are obtained on substitution into Eq. 7.39:

$$\frac{A}{B} = \frac{\left(\dfrac{0.9788}{0.0212} + \dfrac{0.6759}{0.3241}\right)\dfrac{\log(0.6759/0.9788)}{\log(0.0212/0.3241)} - 2}{\dfrac{0.9788}{0.0212} + \dfrac{0.6759}{0.3241} - \dfrac{2 \times 0.9788 \times 0.6759 \log(0.6759/0.9788)}{0.0212 \times 0.3241 \log(0.212/0.3241)}} = 0.2059.$$

$$A = \frac{-0.1608}{(1 + 0.2059 \times 46.17)^{-2} - (1 + 0.2059 \times 2.085)^{-2}} = 0.335.$$

$$B = \frac{0.335}{0.2059} = 1.627.$$

The best van Laar representation of atmospheric (93–116 C) VLE data in the DECHEMA VLE Data Collection, part 1, p. 409 is

$A = 1.1295$,

$B = 3.9571$.

Plots of the activity coefficients with the two sets of parameters show little resemblance, even in the range of mutual solubilities. Predicted miscibilities with the VLE van Laar parameters, however, are obtained from Figure 7.8 as $x_1 = 0.98$, $x_1^* = 0.50$, and do not differ astronomically from the measured values. Difference in temperature may account for some of the disagreement.

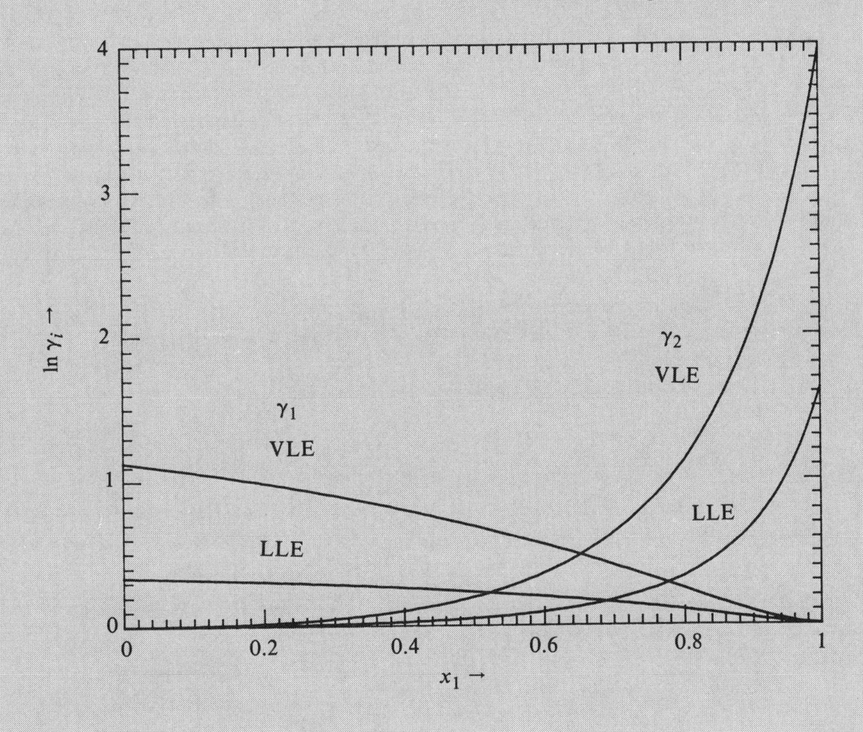

represent VLE particularly well, nor do VLE equations represent LLE well, as shown in Example 4.17 for one case. More about this topic is in Chapter 7.

4.17.4. $P-x$ or $T-x$ Data

Extrapolation of data of liquid composition and bubblepoint pressure or temperature to find infinite-dilution activity coefficients was explained in Section 4.14. Such data likewise can be used to find the parameters of an activity-coefficient equation directly and thus the activity coefficients over the full range of concentrations. By this means the difficult experimental problems of sampling and analyzing the vapor phases can be avoided, but it is necessary to assume the validities of specific forms of the equation of state of the vapor phase and of the activity-coefficient correlation method of the liquid phase. Although any number of data sets can be regressed, the procedure will be explained with only two data sets of a binary mixture, using the B-truncated virial and the van Laar equations.

The starting point is the total pressure expression; when the Poynting Factor can be neglected, this is

$$P = \gamma_1 x_1 P_1^\circ \phi_1^\circ / \hat{\phi}_1 + \gamma_2 x_2 P_2^\circ \phi_2^\circ / \hat{\phi}_2. \tag{4.244}$$

These steps can be followed when two isothermal sets (xq_1, P) are known:

1. Since the temperature is known, the vapor pressures, P_i° are known and the pure-component fugacity coefficients, ϕ_i°, are found from the virial equation.

2. Since the vapor composition is unknown, as a first trial assume that each partial fugacity coefficient is unity, $\hat{\phi}_i = 1$, or that the ratio, $\phi_i^\circ/\hat{\phi}_i = 1$, if preferred. After a trial value of the vapor composition has been established, the partial fugacity coefficients are figured with the virial equation.

3. Replace the activity coefficients in the total pressure equation with their van Laar equivalents, resulting in

$$P = x_1 P_1^\circ \frac{\phi_1^\circ}{\hat{\phi}_1} \exp\left[A\left(\frac{Bx_2}{Ax_1 + Bx_2}\right)^2\right] + x_2 P_2^\circ \frac{\phi_2^\circ}{\hat{\phi}_2} \exp\left[B\left(\frac{Ax_1}{Ax_1 + Bx_2}\right)^2\right]. \tag{4.245}$$

4. After substituting two sets of pressure-composition data into the equation of step 3, solve the simultaneous equations for the parameters A and B, and ultimately for the activity coefficients at the two liquid compositions.

5. Find the approximate vapor compositions from

$$y_i = \gamma_i x_i \frac{P_i^\circ}{P} \frac{\phi_i^\circ}{\hat{\phi}_i}. \tag{4.246}$$

If the mol fractions do not sum to unity, they should be normalized before proceeding.

6. The partial fugacity coefficients are obtained with these equations from Table 3.4:

$$\ln \hat{\phi}_1 = \frac{P}{RT}[B_1 + y_2^2(2B_{12} - B_{11} - B_{22})], \tag{4.247}$$

$$\ln \hat{\phi}_2 = \frac{P}{RT}[B_2 + y_1^2(2B_{12} - B_{11} - B_{22})]. \tag{4.248}$$

7. Return to step 3. Recycle between steps 3 and 7 until the values of A and B or the values of the vapor compositions become steady.

When more than two sets of data are available, all can be used to establish the van Laar parameters by a least-squares method—that is, by solving the pair of equations

$$\frac{\partial}{\partial A}[\Sigma(P_i - RHS_i)^2] = 0, \tag{4.249}$$

$$\frac{\partial}{\partial B}[\Sigma(P_i - RHS_i)^2] = 0. \tag{4.250}$$

Isobaric data can be treated by a similar procedure, but if the temperature range is very wide an activity coefficient equation with a built-in temperature dependence, such as the Wilson, should be used.

Essentially this method of finding activity coefficients was proposed by Barker (1953), who used the virial equation for the vapor phase and the Scatchard-Raymond (1938) for the activity coefficients. Swift and co-workers have used the Barner-Adler and the Wilson or T-K-Wilson equations. At moderate pressures—that is, at pseudoreduced pressures less than one-half the pseudoreduced temperatures—the virial equation is adequate and is particularly easy to use because it is not complicated by multiple roots in contrast to cubic equations of state. Prausnitz et al. (1980) have standardized on the B-truncated virial and the UNIQUAC equations. The

Example 4.18. Activity Coefficients from T-x Data with the Symmetrical and Margules Equations

The system is methyl acetate + methanol at 5,907 Torr. Data are:

x_1	T, °C	P_1^0	P_2^0
0.294	123.5	4,951.9	5,145.4
0.736	125.6	5,195.2	5,445.4

Assume that $\hat{\phi}_i/\phi_i^\circ = 1$ and that the activity coefficients are independent of temperature. The symmetrical equations are:

$$\gamma_1 = \exp Ax_2^2,$$

$$\gamma_2 = \exp Ax_1^2.$$

With the first set of data,

$$P = \gamma_1 x_1 P_1^{sat} + \gamma_2 x_2 P_2^{sat},$$

$$5{,}907 = 0.294(4{,}951.9)\exp[A(1-0.294)^2] + 0.706(5{,}145.4)\exp[A(0.294)^2].$$

By trial,

$A = 0.688$ and $\gamma_1 = 1.41$, $\gamma_2 = 1.06$.

With the second set of data,

$$0.6473\exp(0.0697A) + 0.2434\exp(0.5417A) - 1 = 0$$

(continued next page)

Example 4.18 *(continued)*

from which

$A = 0.548$ and $\gamma_1 = 1.04$, $\gamma_2 = 1.35$

Since the two values of A differ appreciably, apparently the symmetrical equation does not represent these data well. With the full Margules equations the equations to be solved are,

$$0.2456 \exp\left[(A + 2(B - A)x_1)x_2^2\right] + 0.6150 \exp\left[(B + 2(A - B)x_2)x_1^2\right] - 1 = 0, \quad (1)$$

$$0.6473 \exp\left[(A + 2(B - A)x_1)x_2^2\right] + 0.2434 \exp\left[(B + 2(A - B)x_{21}^2\right] - 1 = 0. \quad (2)$$

Substitute the mol fractions.

$$0.2465 \exp[0.4984(A + 0.588(B - A))] + 0.6150 \exp[0.0864(B + 0.1412(A - B))] - 1 = 0$$

$$0.6473 \exp[0.0697(A + 0.1472(B - A))] + 0.2434 \exp[0.5417(B + 0.5280(A - B))] - 1 = 0$$

The last two equations can be solved by the Newton-Raphson method as in Example 4.7, but in this instance they are solved graphically. The graphs are drawn point by point, by assuming values of A and finding corresponding values of B by single-variable Newton-Raphson. The intersection of the curves is at $A = 0.83$, $B = 0.42$. As shown in the following table, agreement between calculated and measured vapor compositions is fair.

x_1	γ_1	γ_2	y_1 Calc.	y_1 Meas.
0.294	1.34	1.09	0.330	0.336
0.736	1.02	1.42	0.660	0.678

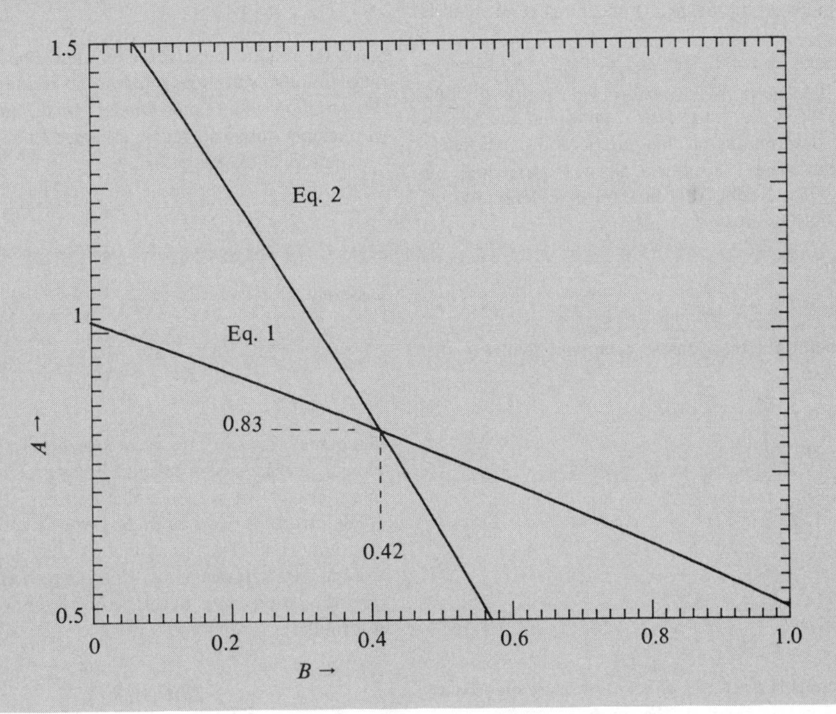

correlations of the DECHEMA collection, however, which include data up to 15 bar, do not take account of vapor-phase nonidealities except of monocarboxy acids, which are strongly associating.

4.18. THE EFFECT OF TEMPERATURE AND EXCESS ENTHALPY

Several of the common activity coefficient equations have built-in temperature dependence, but this compensation is only partial and data fit usually is improved by making the parameters temperature-dependent. Comments on specific equations are summarized in Table 4.13, and the dependence of Wilson and Margules parameters on temperature is examined with some experimental data in Tables 4.14, 4.15. Those data include all the systems in the cited reference for which at least four temperatures have been observed. A general pattern of temperature dependence of the Wilson parameters is not readily discernible in these results, but it is clear that the variation of infinite-dilution activity coefficients figured with the given Wilson parameters with temperature is more regular. This is tied in with the fact that pairs of quite dissimilar parameters can represent particular data with the same accuracy (Section 4.9, Figure 4.9). The very carefully obtained data of ethanol + water of Example 4.19 do show simple regularities with temperature. Those of Example 4.20

Table 4.13. Temperature Dependence of the Parameters of Activity-Coefficient Equations

Equation	*Parameter*	*Comment*
Scatchard-Hildebrand	$\ln \gamma_i = \frac{V_i}{RT}(\delta_i - \bar{\delta})^2$	The difference, $\delta_i - \bar{\delta}$, often is only slightly dependent on temperature. Individual δ_i are functions of ω, P_c and T_r.
Wilson	$\Lambda_{ij} = (V_j/V_i)\exp(-\lambda_{ij}/RT)$	λ_{ij} often is assumed independent of temperature, but see Table 4.14.
NRTL	$\tau_{ij} = C_{ij}/RT$ $G_{ij} = \exp(-\alpha_{ij}\tau_{ij})$	Both C_{ij} and α_{ij} are given as linear functions of temperature by Renon et al. (1971, p. 168).
UNIQUAC	$\tau_{ij} = \exp(-\Delta u_{ij}/RT)$ $z = 35.2 - 0.1272T + 0.00014T^2$	Δu_{ij} usually is taken independent of temperature. The coordination number, z, has been correlated in terms of the temperature by Skjold-Jorgensen et al. (1980).
Margules & van Laar	A_{12}, A_{21}	Approximately linear functions of $1/T$, sometimes increasing, sometimes decreasing, sometimes in the same direction, sometimes in opposite directions. See Table 4.15.

exhibit rather more erratic behavior, but the general trend is clear.

Besides the activity coefficient and the associated excess Gibbs energy, the excess properties that are most easily and commonly measured are the excess volume and the excess enthalpy. Brief descriptions of dilatometric and calorimetric techniques in these applications are given, for example, by McGlashan (1979). Besides the fundamental definitions in terms of excess Gibbs energy,

$$\bar{G}_i^{ex} = RT \ln \gamma_i, \tag{4.251}$$

$$G^{ex} = RT\, \Sigma x_i \ln \gamma_i, \tag{4.252}$$

immediately useful thermodynamic relations involving the activity coefficient are

$$\bar{V}_i^{ex} = RT\left(\frac{\partial \ln \gamma_i}{\partial P}\right)_{Tx}, \tag{4.253}$$

$$\bar{H}_i^{ex} = -RT^2\left(\frac{\partial \ln \gamma_i}{\partial T}\right)_{Px}, \tag{4.254}$$

$$H^{ex} = -RT^2 \Sigma x_i\left(\frac{\partial \ln \gamma_i}{\partial T}\right)_{Px}. \tag{4.255}$$

Since V^{ex} of liquids is small, the effect of pressure on activity coefficients usually is ignored, but the temperature effect given by

$$\ln(\gamma_i/\gamma_{io}) = -\int_{T_o}^{T} \frac{\bar{H}^{ex}}{RT^2}\, dT \tag{4.256}$$

is of course significant. Since the partial excess enthalpy, for example,

$$\bar{H}_i^{ex} = H^{ex} + (1 - x_1)(\partial H^{ex}/\partial x_1), \tag{4.257}$$

is a fairly complex function of temperature whose mathematical form usually is not known, analytical integration in general is not possible. The variety of curve shapes in Figures 4.3–4.6 points out the difficulty of correlating excess enthalpy with temperature and composition in a simple way. The general form,

$$H^{ex} = x_1 x_2 f(T, x_1) \tag{4.258}$$

meets the requirements at the terminal compositions. In some ranges of concentration, often in the vicinity of equimolality, the variation with composition is small, and if the variation with temperature also can be assumed small, the integral with a symmetrical excess enthalpy,

$$H^{ex} = A_H x_1 x_2, \tag{4.259}$$

$$\bar{H}_1^{ex} = A_H(1 - x_1)^2 \tag{4.260}$$

is simply

$$\ln \gamma_1/(\gamma_1)_0 = \frac{A_H(1 - x_1)^2}{R}\left(\frac{1}{T_0} - \frac{1}{T}\right), \tag{4.261}$$

which may be satisfactory sometimes for extrapolation of activity coefficients over small ranges of temperature when values are known at two temperatures.

The inverse process of finding excess enthalpies from data of activity coefficients as a function of temperature sometimes is of interest. In terms of the Wilson equation, for example, the needed relationship is developed with the following sequence of formulas:

$$H^{ex} = -RT^2\left(\frac{\partial(G^{ex}/RT)}{\partial T}\right), \tag{4.262}$$

$$G^{ex}/RT = -x_1 \ln(x_1 + \Lambda_{12}x_2) - x_2 \ln(\Lambda_{21}x_1 + x_2), \tag{4.263}$$

$$\Lambda_{12} = \frac{V_2}{V_1}\exp(-\lambda_{12}/RT), \tag{4.264}$$

$$\Lambda_{21} = \frac{V_1}{V_2}\exp(-\lambda_{21}/RT), \tag{4.265}$$

$$\frac{\partial \Lambda_{12}}{\partial T} = \frac{\lambda_{12}\Lambda_{12}}{RT^2}, \tag{4.266}$$

Table 4.14. Wilson Parameters and Corresponding Infinite-Dilution Activity Coefficients at Several Temperatures. Parameters from the DECHEMA VLE DATA Collection, Part 3/4. All data of each system obtained by the same investigators.

System	T	λ_{12}	λ_{21}	γ_1^∞	γ_2^∞	V_1/V_2
Acetone + chloroform	0	−290.9	−81.4	0.503	0.396	0.917
	10	−395.2	−103.8	0.410	0.273	
	20	−307.9	−164.4	0.436	0.352	
	30	−343.6	−52.1	0.517	0.395	
	50	−472.7	194.1	0.606	0.412	
Carbon disulfide + acetone	0	715.8	689.7	6.50	8.66	0.801
	10	661.3	796.4	5.81	9.51	
	30	537.2	961.1	4.51	10.03	
	35	604.8	783.2	4.68	7.66	
Acetone + hexane	−20	1184.8	502.5	13.09	11.09	0.562
	−5	1160.6	487.5	10.77	9.87	
	20	1077.8	375.5	7.24	6.96	
	45	962.8	341.4	5.05	5.63	
Methyl ethyl ketone + ethylbenzene	25	568.1	−210.6	1.83	1.54	0.732
	55	−96.4	495.8	1.22	1.63	
	65	162.8	134.8	1.39	1.55	
	75	−65.3	670.6	1.37	2.18	
1,4-dioxane + nitromethane	30	−483.0	590.5	1.07	1.12	1.590
	46	−487.3	623.4	1.11	1.18	
	80	−419.0	593.5	1.20	1.27	
	98	−407.6	580.7	1.21	1.26	
1,4-dioxane + N,N-dimethylformamide	30	−242.8	642.4	1.71	1.93	1.667
	40	−260.5	668.6	1.69	1.92	
	50	−275.5	685.2	1.66	1.89	
	60	−293.5	691.4	1.62	1.82	
Diethylether + chloroform	0	6437.1	−583.2	11163	0.72	1.297
	10	6401.8	−6471.1	26950	1.12	
	20	−674.0	+674.7	0.74	0.57	
	30	4051.9	−634.8	71.2	0.73	

Note: The Wilson parameters for diethylether + chloroform seem to be wrong; the data are well fitted by the Margules equation, which predicts infinite-dilution activity coefficients in the range 0.6–0.9 (see Table 4.15).

$$\frac{\partial \Lambda_{21}}{\partial T} = \frac{\lambda_{21}\Lambda_{21}}{RT^2}, \tag{4.267}$$

so that

$$H^{ex} = x_1 x_2 \left[\frac{\lambda_{12}\Lambda_{12}}{x_1 + \Lambda_{12}x_2} + \frac{\lambda_{21}\Lambda_{21}}{\Lambda_{21}x_1 + x_2}\right], \tag{4.268}$$

which assumes that the λ_{ij} are constant. Eq. 4.268 is tested against three sets of data in Example 4.21, but with somewhat disappointing results.

Since equilibrium relations and enthalpy balances commonly are involved together in many calculations, it would be convenient to have them expressed in terms of the same parameters. Unfortunately, at least with the local composition models (Wilson, NRTL, UNIQUAC), the simultaneous correlation of VLE and H^{ex} usually leads to a worsening of the fit to at least one of the data types. Making the parameters temperature-dependent often helps, but this procedure at least doubles the number of parameters, which becomes a serious matter with multicomponent systems. Some success was achieved by Skjold-Jorgensen et al. (1980), who expressed only the coordination number, z, of the UNIQUAC equation as a general function of temperature (quoted in Table 4.13). When this function is used, UNIQUAC/UNIFAC parameters are obtained by regression of both VLE and H^{ex} data. No general set of UNIFAC parameters of this kind has been prepared, and more exhaustive testing of this temperature correlation probably needs to be made.

Meanwhile, in most applications it is necessary to rely on enthalpy correlations that are independent of activity-coefficient correlations.

Table 4.15. Margules Parameters at Several Temperatures

System	*T*	A_{12}	A_{21}
Acetone + chloroform	0	−0.6632	−0.9089
	10	−0.8380	−1.2639
	20	−0.8050	−1.0303
	30	−0.5948	−0.9382
	50	−0.3922	−0.8919
Carbon disulfide	0	1.7119	1.8948
+ acetone	10	1.6133	1.9497
	30	1.3842	1.9494
	35	1.4354	1.8091
Acetone + hexane	−20	2.1671	2.1376
	−5	2.0155	2.1024
	20	1.7448	1.8012
	45	1.5061	1.6352
Methyl ethyl ketone	25	0.5482	0.4348
+ ethylbenzene	55	0.1121	0.4893
	65	0.3242	0.4370
	75	0.3069	0.6176
1,4-dioxane	30	0.0509	0.1288
+ nitromethane	46	0.0899	0.1687
	80	0.1815	0.2372
	98	0.1837	0.2314
1,4-dioxane	30	0.2755	0.5793
+ N,N-dimethyl-	40	0.2701	0.5554
formamide	50	0.2596	0.5351
	60	0.2271	0.5058
Diethylether	0	−0.0556	−0.5893
+ chloroform	10	−0.2485	−0.6292
	20	−0.2429	−0.5252
	30	−0.3380	−0.5353

The systems are the same as those in Table 4.14.

Example 4.19. Effect of Temperature on van Laar and Wilson Parameters for Mixtures of Water and Ethanol (Data of Pemberton & Mash, *J Chem Thermodynamics* 10 (1978):867–88).

The parameters have been calculated by least-squares regression and are:

	van Laar		*Wilson*			
K	*A*	*B*	Λ_{12}	Λ_{21}	λ_{12}/R	λ_{21}/R
303.15	0.9428	1.4769	0.78925	0.23896	428.50	77.198
323.15	0.9409	1.6463	0.83185	0.19486	439.78	148.22
343.15	0.9312	1.7427	0.86504	0.17273	453.57	198.76
363.15	0.9044	1.7890	0.90524	0.15985	463.51	238.49

The ordinate scales are different for each line. The plots indicate roughly linear variation of AT and BT and of the four Wilson parameters.

(continued next page)

Example 4.19 *(continued)*

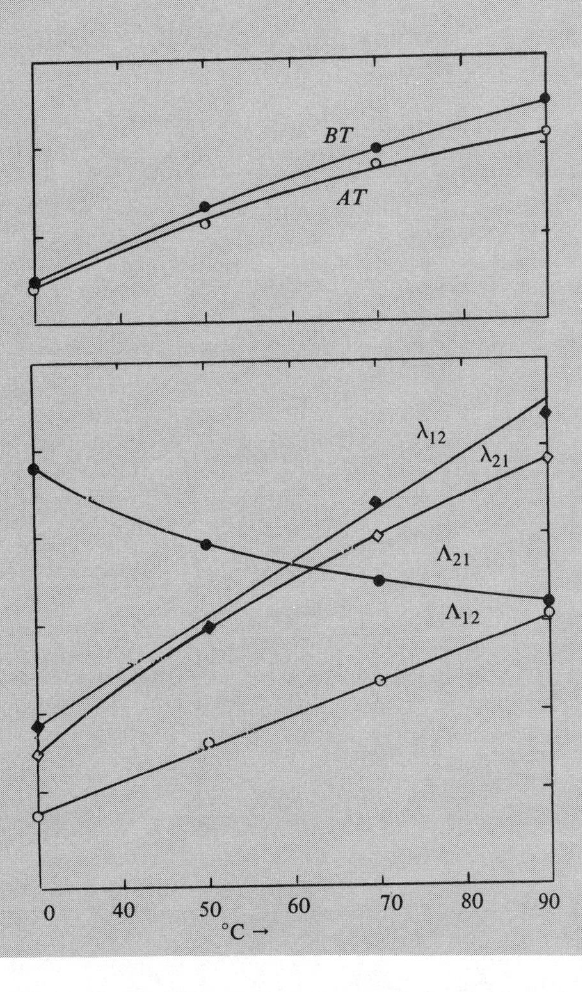

Example 4.20. The Effect of Temperature on van Laar Parameters

Activity coefficients of mixtures of diethylamine and triethylamine at several temperatures were measured by Bittrich & Kauer (*Z Phys Chem Neue Folge* 219, nos. 3, 4 (1962):151). The data were linearly regressed with the van Laar equation, with the results shown in the table and on the graph. The small values of the coefficients of determination, r^2, indicate rather poor correlations of the data, but the parameters do vary fairly smoothly with temperature.

T	A	B	r^2
50	0.7664	0.2447	0.6157
55	0.7260	0.1978	0.4720
60	0.6259	0.1971	0.5644
65	0.5961	0.1761	0.5763
70	0.5763	0.1620	0.6235

Original data are reproduced for only two of the temperatures.

t	x	y	γ_1	γ_2
50	0	0.00	—	1.000
	2	8.13	2.090	1.007
	5	18.01	1.700	1.014
	10	29.37	1.380	1.027
	20	45.71	1.148	1.042
	30	57.99	1.094	1.057
	40	67.92	1.007	1.072
	50	75.69	1.057	1.089
	60	82.35	1.045	1.104
	70	87.73	1.035	1.122
	80	92.44	1.023	1.140
	90	96.40	1.014	1.191
	95	98.24	1.007	1.282
	98	99.17	1.005	1.849
	100	100.00	1.000	—
70	0	0.00	—	1.000
	2	6.96	1.778	1.005

(continued)

Example 4.20 *(continued)*

t	x	y	γ_1	γ_2
	5	15.31	1.496	1.012
	10	26.13	1.227	1.018
	20	43.07	1.102	1.032
	30	55.87	1.064	1.040
	40	66.11	1.045	1.050
	50	74.38	1.034	1.052
	60	81.25	1.030	1.072
	70	86.96	1.021	1.084
	80	92.01	1.014	1.104
	90	96.21	1.009	1.143
	95	98.16	1.005	1.178
	98	90.18	1.002	1.496
	100	100.00	1.000	—

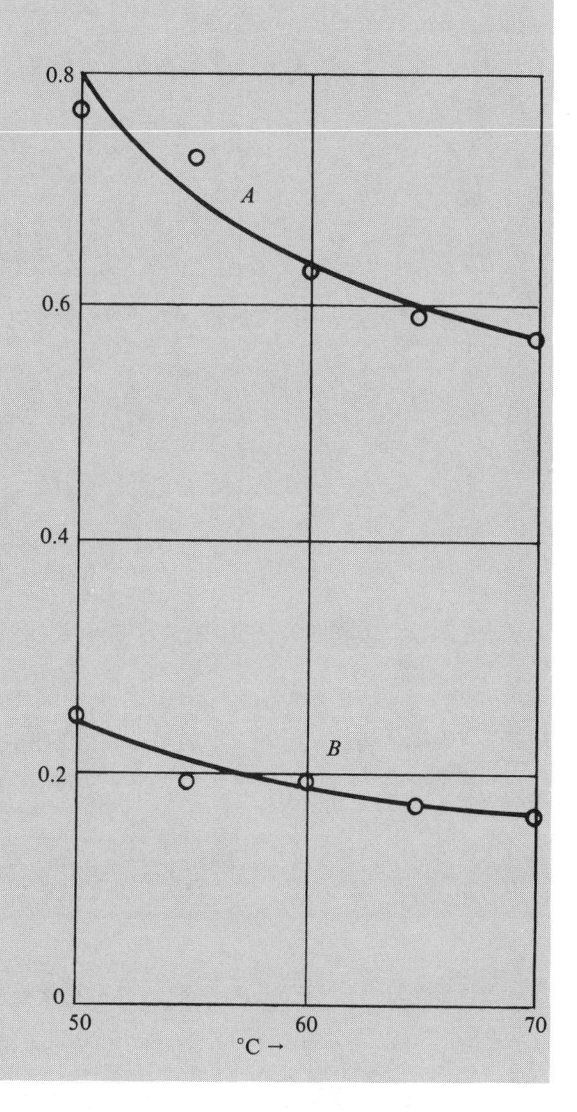

Example 4.21. Excess Enthalpies in Terms of Wilson Parameters

Wilson parameters from Table E.8 and volume ratios of the three systems at the respective temperatures are:

	V_1/V_2	λ_{12}	λ_{21}	Λ_{12}	Λ_{21}
Cyclohexane + benzene	1.224	80.02	187.23	0.7265	0.9301
Chloroform + acetone	1.092	−332.23	−72.2	1.6403	1.2335
Benzene + methanol	2.179	153.86	1620.36	0.3570	0.1546

Experimental data of excess properties are summarized on the graphs (Malesinski, 1965). Excess enthalpy may be calculated with Eq. 4.268 and compared with these data. For $x_1 = 0.5$ this equation becomes

$$H^{ex} = 0.5\left[\frac{\lambda_{12}\Lambda_{12}}{1+\Lambda_{12}} + \frac{\lambda_{21}\lambda_{21}}{1+\Lambda_{21}}\right]$$

The calculated and graphical values are tabulated and clearly do not compare favorably.

	Calculated	*Graph*
Cyclohexane + benzene	62.0	170
Chloroform + acetone	−122.3	−450
Benzene + methanol	128.7	160

Example 4.21 *(continued)*

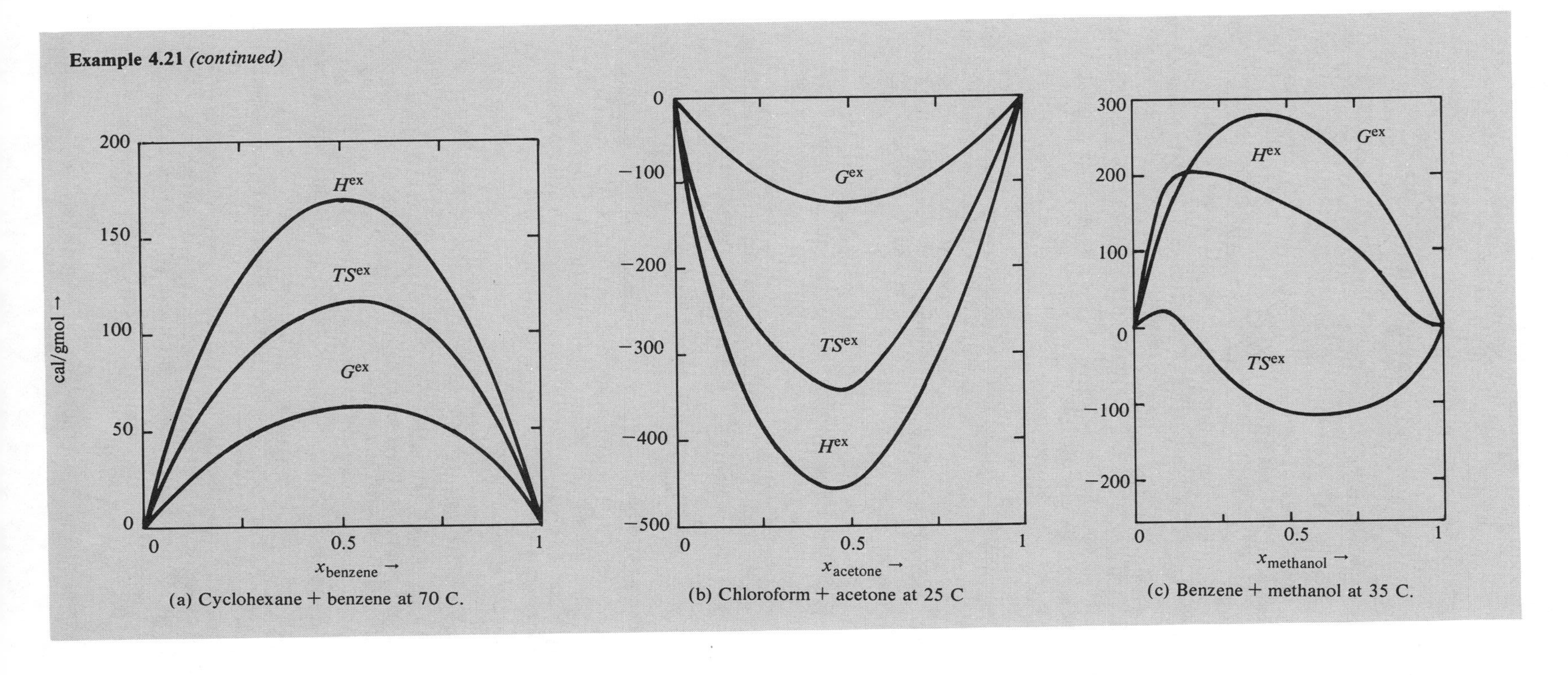

(a) Cyclohexane + benzene at 70 C.

(b) Chloroform + acetone at 25 C

(c) Benzene + methanol at 35 C.

4.19 PROBLEMS

4.1. The volume of a binary liquid mixture is represented by the equation

$$V = k(T/P)^{0.1}(1 + x_2^3).$$

Find equations for V^{ex}, $\bar{V}_2$, γ_1, and γ_2.

4.2. The excess enthalpy of a binary mixture is expressed as a function of composition and temperature (K) by

$$H^{ex} = H - H^{id} = x_1x_2(1{,}200 + 200x_1 + 1.5x_1T) \text{ cal/gmol}.$$

At 300 K, infinite-dilution activity coefficients are $\gamma_1^\infty = 6.00$, $\gamma_2^\infty = 4.4$. At 400 K, vapor pressures are $P_1^0 = 4{,}008$ Torr, $P_2^0 = 2{,}643$ Torr.

For an equimolal mixture, find

a. the partial molal enthalpies, $\bar{H}_1^{ex}$ and $\bar{H}_2^{ex}$ as functions of T.
b. the activity coefficients as functions of T, assuming the van Laar equation to hold at 300 K.
c. the bubblepoint pressure at 400 K, given that the partial fugacity coefficients of the vapors are $\hat{\phi}_i^v = 0.95$ and those of the pure liquids at their vapor pressures are $\phi_i^0 = 0.98$.

The excess volume of the liquid phase is substantially zero.

4.3. These measurements were made for a binary mixture at 350 C:

x_1	H	γ_1	γ_2
0	500	15.00	1
0.2	1,500	5.66	2
0.4	2,500	2.65	3
0.6	4,000	1.54	5
0.8	2,000	1.11	12
1.0	0	1.00	30

Find the excess entropy as a function of composition.

4.4. An aqueous solution containing 1.00 g of antipyrine ($C_{11}H_{12}N_2O$) in 100 ml has an osmotic pressure of 1.18 atm at 0 C. Calculate the molecular weight of the compound from these data, and compare with the formula weight.

4.5. The average osmotic pressure of human blood is 7.7 atm at 40 C. (a) What is the mol fraction of various solutes in the blood? (b) What is the freezing point of blood?

4.6. The osmotic pressure of an aqueous solution containing 0.001 mol fraction of a solute B is 0.139 MPa at 300 K when the solvent A is at 0.100 MPa. Taking $V_a = 0.018$ l/gmol and independent of pressure, calculate the value of

$$\mu(A,\ 300\text{ K},\ 0.1\text{ MPa},\ x_b = 0.001) - \mu(A,\ 300\text{ K},\ 0.1\text{MPa},\ x_b = 0).$$

4.7. Osmotic pressures of aqueous solutions of sucrose are given. For this system the excess volumes are substantially zero over the entire range of concentrations. Prepare isotherms of the activity coefficients of water against the mol fraction of sucrose.

m (*moles/1,000 g H_2O*)	*Osmotic Pressure (atm)* 0°	20°	40°	60°	80°
0.100	2.46	2.59	2.66	2.72	
0.200	4.72	5.06	5.16	5.44	
0.300	7.09	7.61	7.84	8.14	
0.400	9.44	10.14	10.60	10.87	
0.500	11.90	12.75	13.36	13.67	
0.600	14.38	15.39	16.15	16.54	
0.700	16.89	18.13	18.93	19.40	
0.800	19.48	20.91	21.80	22.33	23.06
0.900	22.12	23.72	24.74	25.27	25.92
1.000	24.83	26.64	27.70	28.37	28.00

4.8. The enthalpy of vaporization of water is given by the equation

$$\Delta H_v = 13{,}425 - 9.91T + 7.5(E-5)T^2 + 4.46(E-7)T^3 \text{ cal/gmol}.$$

Find an equation for the activity of water in aqueous solutions as a function of boiling-point elevation.

4.9. The freezing temperature of seawater is 271.240 K. Given that the freezing temperature of pure water is 273.150 K, the molar volume of water is 18 ml/gmol and the molar enthalpy of melting of ice is 6,000 J/gmol, find the osmotic pressure of seawater at about 273 K.

4.10. Calculate the fugacity ratio

$$f(H_2O,\text{ liquid},\ m = 0)/\hat{f}(H_2O,\text{ liquid},\ m = 0.10459)$$

where m is the molality of HCl, gmol/kg, at 273.15 K and atmospheric pressure. The freezing temperature is depressed 0.3684 K. At normal freezing of water, $\Delta H_f = 6{,}008$ J/gmol and $\Delta C_p = 38.09$ J $K^{-1}gmol^{-1}$. How much difference would it make if ΔC_p were neglected?

4.11. The freezing-point depressions of solutions of urea in water are given. The heat of fusion of water at 273.16 K is 1,436.4 cal/gmol. Find the activity coefficients of water. The units of m are gmols urea/1,000 g water.

m	ΔT_f
0.000538	0.001002
0.004235	0.007846
0.007645	0.01413
0.012918	0.02393
0.01887	0.03496
0.03084	0.05696
0.04248	0.07850

4.12. Data on the freezing points of aqueous brines are given. With the additional aid of the given heat-capacity data, find the activity coefficients of water over a range of concentrations to the eutectic. [Data are from Lange, (*Handbook of Chemistry*, 1973) and CRC (*Handbook of Chemistry and Physics* 1979).]

Freezing Point of Sodium Chloride Brines

% NaCl by weight	*Spec. Grav.* 15C. (59F.)	*Freezing Point*	
		C.	F.
0	1.000	0.00	32.0
1	1.007	−0.58	31.0
2	1.014	−1.13	30.0
3	1.021	−1.72	28.9
4	1.028	−2.35	27.8
5	1.036	−2.97	26.7
6	1.043	−3.63	25.5
7	1.051	−4.32	24.2
8	1.059	−5.03	22.9
9	1.067	−5.77	21.6
10	1.074	−6.54	20.2
11	1.082	−7.34	18.8
12	1.089	−8.17	17.3
13	1.097	−9.03	15.7
14	1.104	−9.94	14.1
15	1.112	−10.88	12.4
16	1.119	−11.90	10.6
17	1.127	−12.93	8.7
18	1.135	−14.03	6.7
19	1.143	−15.21	4.6
20	1.152	−16.46	2.4
21	1.159	−17.78	+0.0
22	1.168	−19.19	−2.5
23	1.176	−20.69	−5.2
23.3 (E)	1.179	−21.13	−6.0
24	1.184	−17.0*	+1.4*
25	1.193	−10.4*	13.3*
26	1.201	−2.3*	27.9*
26.3	1.203	0.0*	32.0*

*Saturation temperatures of sodium chloride dihydrate; at these temperatures $NaCl \cdot 2H_2O$ separates leaving the brine of the eutectic composition (E).

Specific Heat of Ice—Cal./g/°C

Temp. °C.	*Specific Heat*	*Observer*
−31.8	.4454	Dickinson-Osborne, 1915
−23.7	.4599	Dickinson-Osborne, 1915
−24.5	.4605	Dickinson-Osborne, 1915
−20.8	.4668	Dickinson-Osborne, 1915
−14.8	.4782	Dickinson-Osborne, 1915
−14.6	.4779	Dickinson-Osborne, 1915
−11.0	.4861	Dickinson-Osborne, 1915
− 8.1	.4896	Dickinson-Osborne, 1915
− 4.3	.4989	Dickinson-Osborne, 1915
− 4.5	.4984	Dickinson-Osborne, 1915
− 4.9	.4932	Dickinson-Osborne, 1915
− 2.6	.5003	Dickinson-Osborne, 1915
− 2.2	.5018	Dickinson-Osborne, 1915
Water Below 0°C		
−6	1.0119	Martinetti, 1890
−5	1.0155	Barnes, 1902
−5	1.0113	Martinetti, 1890
−4	1.0105	Martinetti, 1890
−3	1.0102	Martinetti, 1890
−2	1.0097	Martinetti, 1890
−1	1.0092	Martinetti, 1890

4.13. The boiling points of aqueous solutions of NaOH at atmospheric pressure are:

Wt% NaOH	20	30	40	50	60
Boiling point, °F	221	233	249	288	305

In this range the enthalpy of liquid water may be assumed independent of pressure. Find the activity coefficients of water at the various temperatures.

4.14. Mutual solubilities of furfural (A) and water (B) at several temperatures are (Stephen et al., 1979):

Mutual Solubility Wt.%			*Mutual Solubility Wt.%*			*Mutual Solubility Wt.%*		
A	*B*	°C	*A*	*B*	°C	*A*	*B*	°C
7.9	92.1	10	14.8	85.2	80	91.4	8.6	60
8.3	91.7	20	16.6	83.4	90	92.4	7.6	50
8.8	91.2	30	18.4	81.6	97.9	93.3	6.7	40
9.5	90.5	40	84.1	15.9	97.9	94.2	5.8	30
10.4	89.6	50	86.7	13.3	90	95.2	4.8	20
11.7	88.3	60	88.7	11.3	80	96.1	3.9	10
13.2	86.8	70	90.3	9.7	70			

a. Find the activity coefficients for each equilibrium condition.
b. Fit empirical equations to plots of the van Laar parameters against the temperature.

4.15. Vapor pressures of aqueous $KHCO_3$ at 100 C are given, in atm, as a function of mol fraction of the salt.

a. Find the activity coefficient of water at the several concentrations.
b. Find the ratio, γ/γ^{∞}, of the salt as a function of temperature. See Lewis & Randall (1961, p. 263) for a method of extrapolation to zero concentration.

x_{salt}	P_w^0	x_{salt}	P_w^0
0.0089	0.9847	0.0822	0.8263
0.0176	0.9689	0.0970	0.7895
0.0346	0.9224	0.1253	0.7237
0.0510	0.8979	0.1519	0.6392
0.0668	0.8629		

4.16. The van Laar parameters of ethanol-water mixtures are $A = 0.7292$, $B = 0.4104$. Approximate the osmotic pressure as a function of the mol fraction of ethanol in the solution.

4.17. Find excess enthalpies of all equimolal binaries and the equimolal ternary at 300 K, given the following Wilson parameters, cal/gmol:

	ij	a_{ij}	a_{ji}
methanol-chloroform	12	1,703.68	−373.30
methanol-ethyl acetate	13	985.69	−200.36
chloroform-ethyl acetate	23	−367.50	−92.50

4.18. A substance, A, is to be recovered from a solution in substance B by diffusion through a semipermeable membrane. The downstream side contains solvent C, which is maintained pure by replenishment. On the upstream side, $P_a^o = 5.7$ atm, $x_{ao} = 0.9$, $\gamma_{ao} = \gamma_{bo} = 1.25$. The van Laar equation applies. Find the concentration, x_a, at which the rate of mass transfer across the membrane becomes 10 percent of the initial value.

4.19. The following data on the partial pressures (in mm Hg) of toluene and of acetic acid at 69.94°C have been taken from the International Critical Tables, Volume III, pages 217, 223, and 288. For the purposes of this exercise, assume that the partial pressure of each component is identical with its fugacity.

N_1 (toluene)	N_2 (acetic acid)	p_1 (toluene)	p_2 (acetic acid)
0.0000	1.0000	0	136
.1250	.8750	54.8	120.5
.2310	.7690	84.8	110.8
.3121	.6879	101.9	103.0
.4019	.5981	117.8	95.7
.4860	.5140	130.7	88.2
.5349	.4651	137.6	83.7
.5912	.4088	145.2	78.2
.6620	.3380	155.7	69.3
.7597	.2403	167.3	57.8
.8289	.1711	176.2	46.5
.9058	.0942	186.1	30.5
.9565	.0435	193.5	17.2
1.0000	.0000	202	0

a. Draw a graph of f_1 vs. N_1. Indicate Raoult's law by a dotted line.

b. Draw a graph of f_2 vs. N_2. Indicate Raoult's law and Henry's law, each by a dotted line.

c. Find the constant in Henry's law for acetic acid in toluene solutions by extrapolating a graph of f_2/N_2 vs. N_2 to infinite dilution.

d. Calculate the activities and activity coefficients of acetic acid on the basis of an f_2^0 established from Henry's law. Plot these values vs. N_2.

e. Calculate the activities and activity coefficients of acetic acid on the basis of an f_2^0 established from Henry's law. Plot these values on the same graph as in part d.

f. Calculate the activities and activity coefficients of toluene, the solvent in these solutions.

4.20. The activity of a component of a binary mixture is given by the equation $\ln \gamma_1 = a/(1 + bx_1/x_2)^2$. Find the equation for the excess Gibbs energy using Eq. 4.22.

4.21. Equilibrium data were obtained in a recirculating still for carbon tetrachloride + chloroform at 40 C. Compositions, vapor pressure, total pressure, second virial coefficients, and liquid molal volumes are known. Find the activity coefficients and the excess Gibbs energy.

$x_2 = 0.5242$, $y_2 = 0.6456$, $P_1^{sat} = 28.443$ kPa,

$P_2^{sat} = 48.064$, $P = 40.242$, $B_{11} = -1{,}465$ ml/gmol,

$B_{22} = -1{,}040$, $\delta_{12} = 0$, $V_1^{(L)} = 99$ ml/gmol, $V_2^{(L)} = 82$.

4.22. Boiling temperatures of ethanol + benzene at 750 Torr are given. By extrapolation of the plots of "apparent activity coefficients",

$$\gamma_1^* = (P - x_2P_2^0)/x_1P_1^0$$

$$\gamma_2^* = (P - x_1P_1^0)/x_2P_2^0$$

find the infinite-dilution activity coefficients. Compare plots of the apparent activity coefficients with those calculated from the van Laar equation with these γ_1^∞ and γ_2^∞.

x_1	T	P_1^{sat}	P_2^{sat}
0.0	79.7	804	750
0.04	75.2	671	648
0.11	70.8	560	562
0.28	68.3	507	518
0.43	67.8	497	509
0.61	68.3	507	518
0.80	70.1	545	549
0.89	72.4	598	592
0.94	74.4	650	632
1.00	78.1	750	711

4.23. The total pressure of a mixture of C_6H_6 (1) and $CCl_2F\cdot CClF_2$ (2) at 25 C with $x_1 = 0.26$ is 291.9 Torr. Other data are

	V^L, ml/gmol	P^{sat}
1	89	95.2
2	120	333.1

a. Find the difference in solubility parameters, $\delta_1 - \delta_2$.

b. Find the composition of the vapor.

4.24. Formulas for the effect of temperature on solubility parameters are given by Hansen & Beerbower (1971) in terms of the coefficient of thermal expansion, α. These formulas are:

$d \ln \delta_D/dT = -1.25\alpha$,

$d \ln \delta_P/dT = -0.5\alpha$,

$d \ln \delta_H/dT = -(0.00122 + 0.5\alpha)$.

For ethanol at 298.2 K, $\delta_D = 7.7$, $\delta_P = 4.3$, $\delta_H = 9.5$, and $\alpha = 0.0011$. Find the solubility parameter at 75 C.

4.25. Evaluate infinite dilution activity coefficients with the Wilson equation and with Hoy's solubility parameters for selected mixtures in the collection of Table E.8.

4.26. Equilibrium data and vapor pressures of methyl acetate + methanol at 5907 Torr are:

x	y	T°C	P_1^0	P_2^0
0.294	0.336	123.5	4,951.9	5,145.4
0.471	0.471	123.2	4,917.8	5,103.6
0.736	0.678	125.6	5,195.2	5,445.4

Assuming that $\phi_i^{sat}/\phi_i = 1$, find the activity coefficients and the Margules and van Laar parameters.

4.27. Equilibrium compositions and total pressures at −5 C of acetone + hexane are:

x	y	P
0.2618	0.5688	70.29
0.6039	0.6161	73.26
0.8535	0.6850	70.87

The vapor pressures are $P_1^0 = 53.54$ and $P_2^0 = 34.23$ Torr. Find the activity coefficients and the Margules and van Laar parameters. Regression of complete data for this system is given in the DECHEMA VLE Data Collection part 3–4, 227: Margules, $A = 2.016$, $B = 2.102$; van Laar, $A = 2.012$, $B = 2.106$.

4.28. Apply the equations of Ellis & Jonah (Section 4.15) to find the infinite-dilution activity coefficients from the data given in Problem 4.35 for methanol + dichlorethane at 40 C. Assume ideal-gas behavior and that the liquid specific volumes are negligible in comparison with those of the vapors.

4.29. Find the excess enthalpies of an equimolal mixture of ethanol + water at 25, 50, and 75°C. The data off Figure 4.2 are

T, C	25	50	75
H^{ex}, cal/gmol	−100	−10	50

Wilson parameters are known at 50 C and may be assumed constant (DECHEMA VLE Data Collection, Part 1, p. 191): $\lambda_{12} = 101.81, \lambda_{21} = 916.49$ cal/gmol, $V_1/V_2 = 0.308$. Note: Agreement may be found quite poor.

4.30. These data are known for a binary mixture: $\ln \alpha_i^\infty = 3$, $\ln \gamma_2^\infty = 2$, $V_1/V_2 = 2.5$, NRTL $\alpha_{12} = 0.3$, UNIQUAC $k_7 = -2.2$, $k_8 = -0.8$ (Table 4.5). Find these activity-coefficient equation parameters: Wilson λ_{12}, Λ_{21}; T-K-Wilson Λ_{12}, Λ_{21}; NRTL τ_{12}, τ_{21}; UNIQUAC τ_{12}, τ_{21}. Make plots of the $\ln \gamma_i$ with the Wilson and NRTL equations.

4.31. The melting point of δ is 1808 K and the enthalpy of melting is 1.536×10^4 J mol^{-1}. The heat capacity of the liquid iron exceeds that of the solid by about 1.3 J K^{-1} mol^{-1}. Show that the difference of free energy of liquid and solid iron as a function of temperature at constant pressure is given in J mol^{-1} by

$$\Delta G^0 = 1.3 \times 10^4 - 1.3T \ln T + 2.6T.$$

At 1,673 K a liquid mixture of iron and iron sulphide containing 0.870 mole fraction of iron is in equilibrium with almost pure solid δ iron. Estimate the activity coefficient of the iron in this liquid melt. State clearly the reference basis. (Denbigh)

4.32. Data for benzene (1) + cyclopentane (2) at 35 C were obtained by Hermensen & Prausnitz (1963). Compare values of the activity coefficients calculated with the Scatchard-Hildebrand equation and with the SH equation plus the Flory-Huggins entropy correction with the experimental data.

x_1	γ_1	γ_2
0.0	1.5541	1.0000
0.1	1.4218	1.0047
0.2	1.3150	1.0186
0.3	1.2294	1.0416
0.4	1.1611	1.0742
0.5	1.1075	1.1165
0.6	1.0664	1.1693
0.7	1.0362	1.2333
0.8	1.0157	1.3096
0.9	1.0038	1.3993
1.0	1.0000	1.5044

4.33. Liquid-liquid-vapor equilibrium data of butane + water are given as mol % butane as follows (Kogan, no. 391):

x_1	x_1^*	y_1	°C	P	P_1^{sat}	P_2^{sat}
0.0044	99.909	98.57	37.9	3.550	3.78	0.065
0.0059	99.662	96.49	71.1	8.493	8.63	0.323
0.0105	99.075	94.70	104.6	17.741	16.90	1.178
0.0186	97.20	92.49	137.8	33.429	29.24	3.355

Find the activity coefficients of butane and water in the two liquid phases,

a. From the mutual solubilities, using the van Laar equation.
b. From the vapor-liquid equilibrium data, using the given vapor pressures.
c. From the simple rule for slightly soluble substances, $\gamma_i = 1/x_i$.

4.34. From the Redlich-Kister expansion for the excess Gibbs energy,

$$G^{ex}/RT = x_1x_2[A + B(x_1 - x_2) + C(x_1 - x_2)^2 + D(x_1 - x_2)^3 + \ldots],$$

derive these equations for the activity coefficients,

$$\ln \gamma_1 = a^{(1)}x_2^2 + b^{(1)}x_2^3 + c^{(1)}x_2^4 + d^{(1)}x_2^5 + \ldots,$$

$$\ln \gamma_2 = a^{(2)}x_1^2 + b^{(2)}x_1^3 + c^{(2)}x_1^4 + d^{(2)}x_1^5 + \ldots,$$

where:

$a^{(1)} = A + 3B + 5C + 7D$, $a^{(2)} = A - 3B + 5C - 7D$,

$b^{(1)} = -4(B + 4C + 9D)$, $b^{(2)} = 4(B - 4C + 9D)$,

$c^{(1)} = 12(C + 5D) \qquad c^{(2)} = 12(C - 5D),$

$d^{(1)} = -32D \qquad d^{(2)} = 32D.$

4.35. For the accompanying four sets of activity coefficients, find the parameters of the Redlich-Kister equations of the second, third, fourth, and fifth degrees. Apply the statistical F-test to decide which is the most complex equation that is justified by the data (data from Kogan et al. 1966).

Nitric acid + water

x	y	t	P	γ_1	γ_2
6.1	0.36	104.0	760	0.0324	0.9288
9.6	0.95	106.4		0.0188	0.8884
11.7	1.50	107.8		0.0633	0.8592
13.9	2.11	109.4		0.0703	0.8420
17.5	4.23	111.8		0.1046	0.8065
18.3	5.10	112.3		0.1212	0.7525
22.5	8.95	114.8		0.1588	0.7080
26.6	13.60	116.8		0.1955	0.6937
27.7	15.99	117.5		0.2151	0.6447
34.1	25.90	119.4		0.2660	0.5919
37.4	36.50	119.9		0.3430	0.5408
38.3	37.45	120.0		0.3420	0.5372
48.5	73.00	116.1		0.5720	0.3110
52.1	81.08	113.4		0.6393	0.2517
54.7	85.00	110.8		0.7063	0.2316
65.1	94.20	102.9		0.8434	0.1501
71.9	97.20	96.1		0.9425	0.1153
76.5	98.75	92.0		—	—
81.6	99.30	88.4		—	—
100.0	100.00	83.4		—	—

Heptane + furfural

x	y	t	P	γ_1	γ_2
10	78.4	114.0	760	5.262	1.019
20	86.8	102.3		3.960	1.082
30	88.6	98.7		2.939	1.201
40	89.1	98.3		2.236	1.393
50	88.8	98.7		1.762	1.690
60	88.5	99.2		1.445	2.158
70	88.4	99.2		1.235	2.897
80	89.2	98.9		1.100	4.103
90	92.1	98.4		1.024	6.152

Methylethylketone + cyclohexane

x	y	t	P	γ_1	γ_2
6.0	16.4	77.0	760	2.9678	0.9947
13.1	25.3	74.7		2.2586	1.0326
17.5	29.9	73.7		2.0644	1.0532
24.7	35.6	72.6		1.8055	1.0976
36.9	42.0	71.8		1.4640	1.2101
48.0	46.5	71.5		1.2585	1.3675
64.5	57.4	72.1		1.1334	1.5647
73.6	63.4	73.0		1.0650	1.7568
80.4	69.8	74.0		1.0387	1.8919
87.6	77.7	75.3		1.0173	2.1202

Methanol + dichlorethane

x	y	t	P	γ_1	γ_2
0	0.0	40	150.0	—	1.000
10	47.9		265.4	4.800	1.024
20	56.9		303.5	3.261	1.093
30	59.2		319.9	2.380	1.239
40	60.3		325.0	1.848	1.435
50	60.5		326.8	1.493	1.720
60	61.4		327.9	1.278	2.110
70	64.0		327.4	1.130	2.619
80	69.7		320.4	1.053	3.241
90	81.0		301.7	1.012	4.020
100	100.0		265.0	1.000	—

4.36. Consider the mixture of n-hexane (1) and diethyl ketone (2) of Problem 4.50. Use only one of the infinite-dilution activity coefficients at one time to find c_{12} of the one-parameter UNIQUAC equation and hence τ_{12} and τ_{21} corresponding to each of these γ_i^∞. The internal energies of vaporization are figured with the aid of the latent enthalpies of vaporization from Pitzer's correlation (RPS, p. 200) with the results

$\Delta u_1^{vap} = 6161$ cal/gmol,

$\Delta u_2^{vap} = 7{,}222$ cal/gmol.

4.37. For acetone (1) + chloroform (2) at 1 atm the infinite-dilution activity coefficients are 0.37 and 0.46, respectively. The mixture forms an azeotrope with $x_1 = 0.345$ at 64.5 C. Find the activity coefficients at $x_1 = 0.25$, 0.50, and 0.75 from both sets of data with

a. the Margules equation.
b. the Wilson equation.
c. the NRTL equation with $\alpha_{12} = 0.30$.

4.38. van Laar parameters of the system methyl acetate (1)-methanol(2) from Table E.10 are $A = 0.4262$ and $B = 0.4394$ based on data of Bushmakin & Kish (1957) (*Zh Prikl Khim* **30** 200). Find the azeotropic composition at 1 atm and compare with the results of Example 4.16, which incidentally are based on other experimental data. The parameters are for 1 atmosphere.

4.39. The activity coefficients of the components of certain binary mixtures are given by the relations

$RT \ln \gamma_1 = \alpha x_2^2,$

$RT \ln \gamma_2 = \alpha x_1^2,$

where α is a function of pressure only. Obtain expressions for the increase in the Gibbs function and enthalpy in the process of mixing the pure components at constant temperature and pressure.

In a steady-flow process an equimolal mixture, for which $\alpha = 418$ J mol^{-1}, is separated into the pure components by a process of distillation. The inflow and outflow are at 20 °C and 1 atm. The only source of energy is a heat reservoir maintained at a steady temperature of 100 °C. Calculate the amount of heat which must be removed by cooling water at 20 °C, per mole of the mixture distilled, if the energy of the reservoir is used at maximum efficiency (Denbigh).

4.40. At atmospheric pressure ethyl acetate and ethyl alcohol form an azeotropic mixture containing 53.9 mole % of the former component and boiling at 71.8 C.

Estimate: (a) the values of the constants A and B in the empirical equations of van Laar:

$$\ln \gamma_1 = \frac{A}{\left(1 + \dfrac{Ax_1}{Bx_2}\right)^2},$$

$$\ln \gamma_2 = \frac{B}{\left(1 + \dfrac{Bx_2}{Ax_1}\right)^2},$$

and (b) if A and B remain unchanged, the azeotropic composition and the corresponding total pressure for boiling at 56.3 C.

The vapor pressures, in mm Hg, of the pure liquids are as follows:

	71.8 °C	*56.3 °C*
Ethyl alcohol	587	298
Ethyl acetate	636	360

(Denbigh).

4.41. Check the possible existence of extrema in the activity-coefficient curves represented by

a. the van Laar equation.
b. the Wilson equation.

4.42. Explore the values of the Wilson parameters that will give extrema in the plots of $\ln \gamma_1$ against x_1. Take the special case for which $\Lambda_{12} = 0$ and find the values of Λ_{21} that will give extrema at values of x_1 in the range 0.1–0.9.

4.43. Excess Gibbs energies of methane-propane mixtures are fitted by the equation

$$G^E = (1 - x)x \sum_{i=0}^{n} a_i(1 - 2x)^i.$$

	J mol⁻¹		
T, K	a_0	a_1	a_2
90	756.8	277.4	19.8
95	755.1	287.0	40.0
100	759.6	297.7	45.8
105	766.8	308.7	43.5
110	774.2	318.9	54.0

where x is the mol fraction of propane (*Selected Data on Nonelectrolyte Mixtures,* 1973).

a. Find corresponding equations for the activity coefficients.
b. Check the assumption sometimes made that $T \ln \gamma_i$ = constant.

4.44. For a mixture with $\gamma_1^\infty = 3$ and $\gamma_2^\infty = 2$, make plots of the Wilson and Margules equations. The Wilson parameters are $\Lambda_{12} = 0.3465$, $\Lambda_{21} = 0.9612$.

4.45. For the integral used for evaluation of thermodynamic consistency, $\int_0^{x_1} \ln(\gamma_1/\gamma_2)\, dx_1$, find only the value above $\gamma_1/\gamma_2 = 1$, using

a. The symmetrical equation for which $G^{ex}/RT = Ax_1x_2$.
b. The van Laar equations.

4.46. Show that the excess Gibbs energy function

$$G^{ex}/RT = \frac{1}{\dfrac{1}{Ax_1} + \dfrac{1}{Bx_2}} + Cx_1x_2(x_1 - x_2)^2$$

leads to Black's equations (1958):

$$\ln \gamma_1 = A\left(\frac{Bx_2}{Ax_1 + Bx_2}\right)^2 + Cx_2^2(2x_2 - 1)(6x_2 - 5)$$

$$\ln \gamma_2 = B\left(\frac{Ax_1}{Ax_1 + Bx_2}\right)^2 + Cx_1^2(2x_1 - 1)(6x_1 - 5)$$

Find values of the parameters given that $\ln \gamma_1^\infty = 3$, $\ln \gamma_2^\infty = 2$ and $\ln \gamma_1 = 0.6$ when $x_1 = 0.5$.

4.47. In the system acetone + hexane at 55 C, find the compositions of vapors in equilibrium with $x_1 = 0.1846$ and $x_1 = 0.8021$ with the Margules, Wilson, and NRTL equations. Vapor pressures are 731.5 and 483.0 Torr and the specific volume ratio is $V_1/V_2 = 0.562$. The parameters are:

Margules: $A = 1.4449$, $B = 1.6488$.

Wilson: $\lambda_{12} = 921.52$ cal/gmol, $\lambda_{21} = 420.36$.

NRTL: $RT\tau_{12} = 746.25$ cal/gmol, $RT\tau_{21} = 424.42$,

$\alpha_{12} = 0.292$.

The experimental values are $y_1 = 0.476$ and $y_1 = 0.700$ (DECHEMA VLE Data Collection part 3/4, p. 224).

4.48. Experimental data and Wilson parameters are given for vapor-liquid equilibrium of a ternary mixture. For the given liquid composition, compare the calculated vapor composition with the experimental one.

		x	y	P^{sat}, *atm*
Acetone	1	0.3130	0.5240	1.190
Tetrachlormethane	2	0.2440	0.2100	0.666
Benzene	3			0.596
$\lambda_{12} = 744.085$		$\lambda_{21} = -74.704$	cal/gmol	
$\lambda_{13} = 442.011$		$\lambda_{13} = -152.877$	cal/gmol	
$\lambda_{23} = -26.525$		$\lambda_{32} = 142.519$	cal/gmol	
$T = 64.2$C				

4.49. With the Gibbs-Duhem equation in the form

$$-\frac{H^{ex}}{RT^2}\frac{dT}{dx_1} + \frac{V^{ex}}{RT}\frac{dP}{dx_1} = x_1\frac{d\ln\gamma_1}{dx_1} + x_2\frac{d\ln\gamma_2}{dx_1},$$

show that at the azeotropic condition

a. At constant temperature, $dP/dx_1 = dP/dy_1 = 0$,
b. At constant pressure, $dT/dx_1 = dT/dy_1 = 0$,

which mean that the liquidus and vaporus cannot intersect or touch except at the extrema on the $P - x_1$ or $T - x_1$ diagrams (see Figure 4.25).

4.50. Infinite-dilution activity coefficients of n-hexane (1) and diethyl ketone (2) at 65 C from the data of Maripuri & Ratcliff (1972) (*J Appl Chem Biotechnol* 22, no. 899) are $\gamma_1^\infty = 2.25$ and $\gamma_2^\infty = 3.67$. Vapor-pressure data are represented by

$$P_1^0 = \exp[15.8366 - 2{,}607.55/(T - 48.78)], \text{ Torr, °K}$$

$$P_2^0 = \exp[16.8138 - 3{,}410.51/(T - 40.15)].$$

Specific gravities are 0.659 and 0.814, respectively.

a. Find the parameters of these correlation equations for activity coefficients:

   1. Van Laar.
   2. Wilson.
   3. T-K-Wilson.
   4. UNIQUAC
   5. NRTL with $\alpha_{12} = -1$.

b. With each of these correlations and also for ideal mixtures find theoretical plate requirements for separating an equimolal mixture into an overhead with $y_1 = 0.95$ and a bottom with $x_1 = 0.02$ when using a reflux ratio 20% greater than the minimum. Constant molal overflow may be assumed for the distillation, and the feed may be taken to be saturated liquid. Pressure is 1 atm.

4.51. Show that the Margules Eqs. 4.73 and 4.74 are consistent with the Gibbs-Duhem equation in the form

$$x_1 \, d \ln \gamma_1 + x_2 \, d \ln \gamma_2 = 0.$$

Also show that when $\ln \gamma_1 = Ax_2^2$ and $\ln \gamma_2 = Bx_1^2$, A must equal B.

4.52. With aid of the program of Table C.11, find the composition of vapor in equilibrium with a liquid of composition: methanol, 0.1; ethanol, 0.1; isopropanol, 0.18; water, 0.62. Vapor pressures are 3.50, 2.23, 1.11, and 1.00 in order. Wilson parameters Λ_{ij} are given by the matrix:

	j			
i	1	2	3	4
1	1	2.3357	2.7385	0.4180
2	0.1924	1	1.6500	0.1108
3	0.2419	0.5343	1	0.0465
4	0.9699	0.9560	0.7795	1

4.53. Write a computer program for the NRTL equation similar to that of Table C.11 for the Wilson equation.

4.54. For mixtures of methyl acetate (1) and methanol (2) at 40 C, Wilson parameters are given by Hirata et al. (1976, No. 378) as $\Lambda_{12} = 0.45159$ and $\Lambda_{21} = 0.57116$. Construct

a. The x-y diagram at 760 Torr,
b. Plots of G^{ex}, H^{ex}, and TS^{ex} against x_1 at 40°C.

5 Phase Diagrams

In this chapter the intent is to present classifications of some types of equilibria and to offer specific examples of many of them, chiefly of those involving organic chemicals. The correlation and prediction of quantitative phase behavior and the use of such data are treated in subsequent chapters.

A phase diagram is a record of the effects of temperature, pressure, and composition on the kinds and numbers of phases that can exist in equilibrium with each other. The number of phases is given in terms of other variables by the Gibbs phase rule (Section 5.2). The kinds of phases that can exist at any particular condition are characteristic of the chemical natures of the components. In comparison with phase-equilibrium information in the form of a numerical tabulation, geometric representation has the advantages of readily grasped interrelationships between all of the variables as well as ease of interpolation and extrapolation. Several types of diagrams are useful, depending on which variables are emphasized. Planar diagrams are easiest to construct but are of course limited to continuous variation of only two variables. In order to show the effects of other variables, a series of planar diagrams may be drawn, each at constant values of one or more of those variables, such as isotherms, isobars, or isopleths. Spatial three-dimensional diagrams are often worthwhile. A few spatial models have been made, a famous one by Roozeboom (Vogel, 1959, p. 139), a pioneer in the systematization of phase-equilibrium data. Excellent stereoscopic views to scale, eighty in number, have been prepared by Tamas & Pal (1970).

Apparently a great variety of systems exists. The more complicated diagrams, however, often may be regarded as made up of combinations of simpler types, so the apparent diversity is not overwhelming. A practical classification is that of Nyvlt (1977) which is shown here as Table 5.1, and similar ones are used in the comprehensive books of Ricci (1951) and Vogel (1959). Many types of phase behavior also are identified in Figures 5.16 and 5.21.

The primary grouping is according to the number of components, then the kinds and numbers of phases. The occurrence of intermolecular compounds and polymorphism of the solid state also are taken into account. The number of possible

Table 5.1. Classification of Solid-Liquid Systems (J. Nyvlt, 1977)

Number of Components	*Class*	*Group*		*Properties*
1	I			A single modification
		a		Triple point below atmospheric pressure
		b		Triple point above atmospheric pressure
	II			Several modifications
		a		All modifications stable
		b		One modification unstable
2	I			Components immiscible in the solid phase
		a		Components do not form a stoichiometric compound
			a_1	Components completely miscible in the liquid phase
			a_2	Components partially miscible in the liquid phase
			a_3	Components immiscible in the liquid phase
		b		Components form a stoichiometric compound
			b_1	A compound stable up to its m.p.
			b_2	A compound unstable at its m.p.
	II			Components completely miscible in the solid phase
		a		Components do not form a solid compound
			a_1	Melting point curve without a maximum or minimum
			a_2	Melting point curve exhibits a maximum or minimum
		b		Components form a solid compound
	III			Components partially miscible in the solid phase
		a		Close melting points of the components
		b		Melting points of the components far apart
3	I			Components immiscible in the solid phase
		a		Components do not form a stoichiometric compound
		b		Components form a stoichiometric compound
			b_1	Binary compounds formed
			b_2	Ternary compounds formed
	II			Components miscible in the solid phase
		a		Components completely miscible in the solid phase
		b		Components partially miscible in the solid phase
4	I			Four different components
	II			Reciprocal salt pairs

Table 5.2. Main Characteristics Identifying or Describing Phase Equilibrium Behavior

I. Unary systems, for which the phase rule becomes $Ph = 3 - F$
  1. Vapor-solid equilibria
  2. Vapor-liquid equilibria
  3. Liquid-solid equilibria
  4. Polymorphism of the crystalline state
  5. Critical and triple points

II. Binary systems. Ph = 4 − F
  1. Vapor-liquid $T - x$ and $P - x$ behavior
    a. One liquid phase
      1. Monotonic vapor pressure behavior
      2. Minimax vapor pressure behavior
  2. Liquid-solid $T - x$ behavior
    a. Pure solid phases
    b. Solid solutions
    c. Minimum melting (eutectic) and maximum melting solids
    d. Peritectic (two solid solutions and one liquid)
  3. Intermolecular compounds in the solid phase
    a. Congruent melting (without decomposition)
    b. Incongruent melting (with decomposition)

III. Ternary systems. $Ph = 5 - F$
  1. Vapor and one liquid phase
  2. Multiple liquid phases
  3. Liquid and solid phases
    a. Pure solid phases
    b. Solid solutions
  4. Intermolecular compounds in the solid phase

IV. Quaternary and higher order systems. $Ph = 2 + C - F$

combinations increases sharply with the number of components, so the experimental coverage of cases is small with more than three components. In some areas the phase behavior of multicomponent systems can be estimated from data on simpler systems involving the same components in fewer combinations.

Not all combinations of variables are of frequent occurrence or practical interest. The most common are systems having only vapor and liquid phases, or two or three liquid phases, or liquid and solid phases, each group with two or three components. To some extent, data on one kind of system, say vapor-liquid, are transferable to another, say liquid-liquid or liquid-solid, for the same components; the common concept is partial fugacities or chemical potentials. The effects of temperature and pressure (particularly with vapor-liquid systems), or of temperature alone for condensed systems, are the most important features studied in addition to compositions. The bulk of research on phase equilibria of condensed systems has been done with ceramics, metals and aqueous salt solutions. Similar behavior, however, occurs in organic systems, which are the main concern of this text. A comprehensive bibliography has been prepared by Wisniak & Tamir (1981). A comprehensive older compilation is Landolt-Börnstein (II/2a, b, c; II/3; IV/4c;NS IV/3). Other references are given in specific chapters.

The elucidation of phase behavior in multicomponent systems, like some of those in metallurgy, has required the labor of many investigators over a period of years. Specific complex organic systems rarely have had the economic importance of metallurgical or ceramic systems, so much less work of this kind has been done with organics. Behavior of organic systems sometimes is described adequately on the assumption of ideal behavior, or by measurements on binary mixtures alone, which may be combined in some cases to express the behavior of the multicomponent systems made up of the pairs. A brief review of experimental methods for phase equilibria is presented in Chapter 12.

5.1. GEOMETRIC REPRESENTATION

The variables that may need to be represented on a phase diagram of a system of n substances are T, P, and $n - 1$ mol fractions. Three variables need representation on spatial diagrams, but equivalent representations that are easier to draw and use are series of planar diagrams at several fixed values of the third variable, or by contours of the third variable on a single diagram. Mol fractions of ternary mixtures are represented on a planar triangular diagram, and mol fractions of a quaternary mixture are represented on a spatial diagram of a regular tetrahedron. The meaning of various regions of phase diagrams is not always unique (insofar as the number of phases is concerned) and may depend on the number of components of the system. The relation between the number of components, number of phases, and number of controllable variables is discussed in the next section.

5.1.1. One Component

On a unary diagram such as Figure 5.1, single phases exist in the open areas, solid above line AB and to the left of line BD. Along a line such as AB, two phases coexist in equilibrium, solid and vapor in this case. At the intersection of the three lines, point B, three phases coexist in equilibrium, this being the greatest number of phases that can coexist in equilibrium in a unary system. As exemplified later, a pure solid can have several stable crystalline forms so the solid region ABD is not necessarily homogeneous. Similarly, liquid crystals can form with certain classes of substances, so the region DBC also may not be homogeneous throughout. Point C on the diagram is a critical point at which the properties of the liquid and vapor phases become the same. There are no critical points corresponding to the other pairs of phases. The dashed lines represent metastable subcooled liquid and superheated vapor.

5.1.2. Two Components

The effects of both T and P in addition to the composition of binary mixtures are represented on spatial diagrams such as Figures 5.2(a) and (b). Essentially the same information is represented in a more convenient fashion by the isobaric sections (or in some cases isothermal sections) shown on these diagrams, or by projection onto a base, as shown with Figure 5.2(b). Two-phase regions are shown shaded. Figure 5.3 has three common kinds of binary-phase diagrams, drawn at constant pressure and showing the relation between temperature and composition. Along the isotherm a-f of Figure 5.3(c), single phases of varying compositions exist between x_a and x_b, between x_c and x_d and between x_e and x_f. A two-phase mixture of varying proportions of phases of compositions x_b and x_c, and a two-phase mixture of varying proportions of phases of compositions x_d and x_e occupy the other composition ranges. At this temperature the overall composition

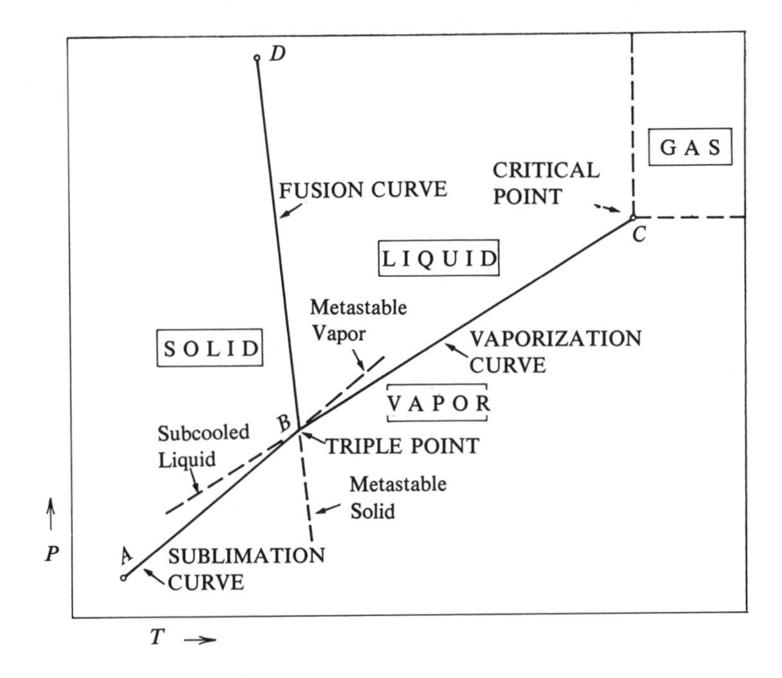

Figure 5.1. Geometric representation of phase equilibrium of a unary system.

determines the phase makeup. In the two-phase regions the compositions of the phases are fixed, but their relative amounts vary with the overall composition and do not affect the equilibrium conditions.

As the temperature is lowered along an isopleth such as g-k, the crossing of any line indicates that some change in phase condition is taking place. Between points h and i, two phases of varying compositions and proportions are formed, whereas between points i and j a single solid phase exists, and beyond point j a mixture of two solid phases in fixed proportions is present.

5.1.3. Lever Rule

The amounts and compositions of two phases in equilibrium are related by the overall composition of the mixture and may be found by material balances. In Figure 5.4 the amounts and compositions of the two phases are designated by prime(′) and double prime (″) signs. The material balances are

$$mx = m'x' + m''x'' \tag{5.1}$$

$$= m'x' + (m - m')x'', \tag{5.2}$$

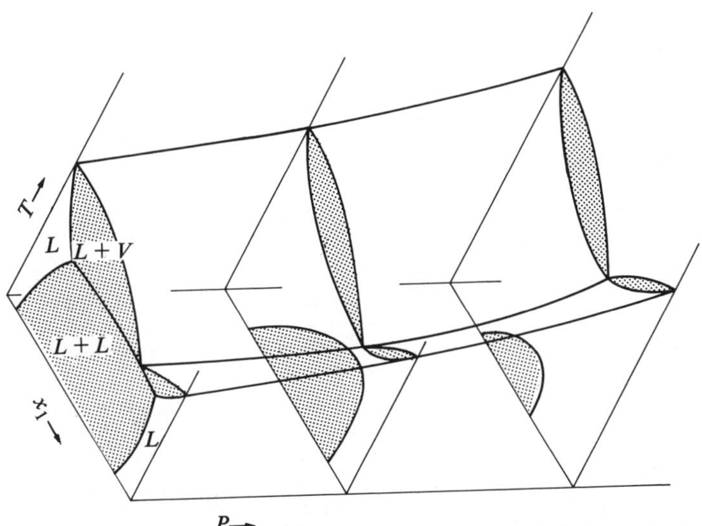

Figure 5.2. Spatial and corresponding planar diagrams, showing the effects of temperature and pressure on mixtures. **(a)** Composition-temperature-pressure diagram of a binary mixture showing isobaric sections. The shaded regions are two-phase.

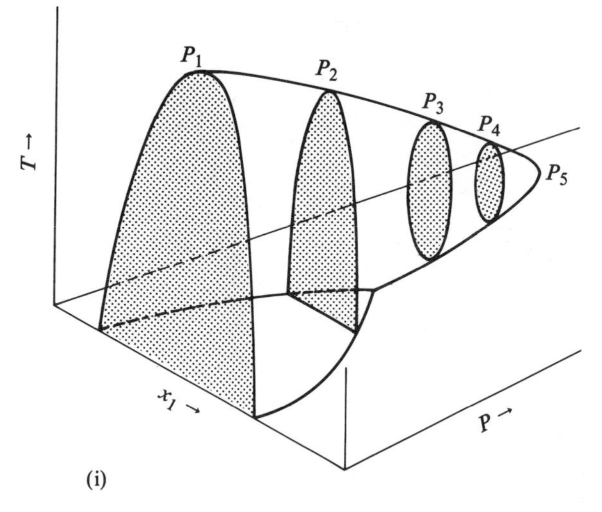

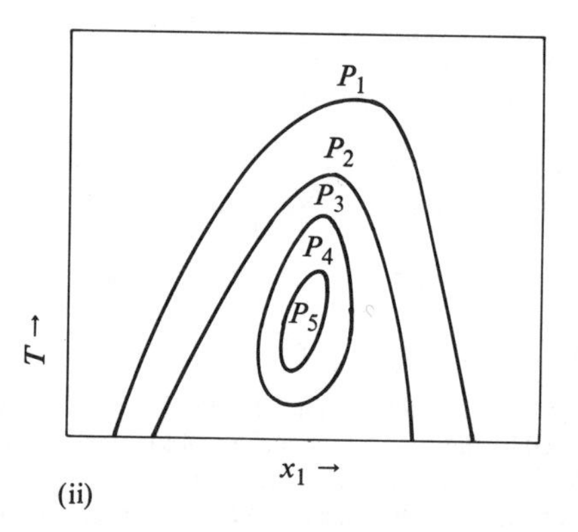

Figure 5.2(b). Variation of the miscibility range of a binary mixture with temperature and pressure; the planar figure shows the isobars indicated on the spatial diagram.

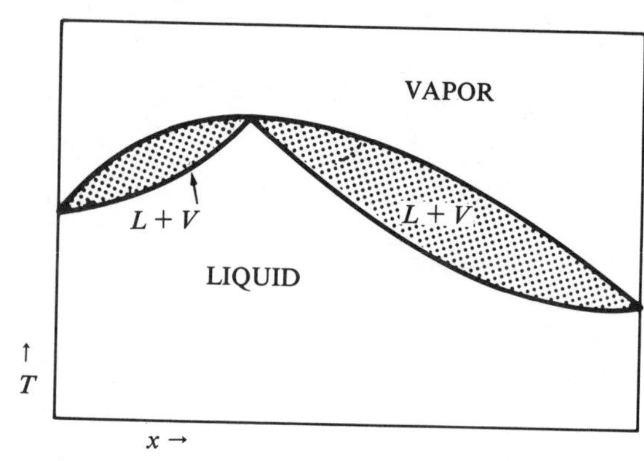

a. Showing vapor-liquid equilibrium of a binary system, temperature against composition.

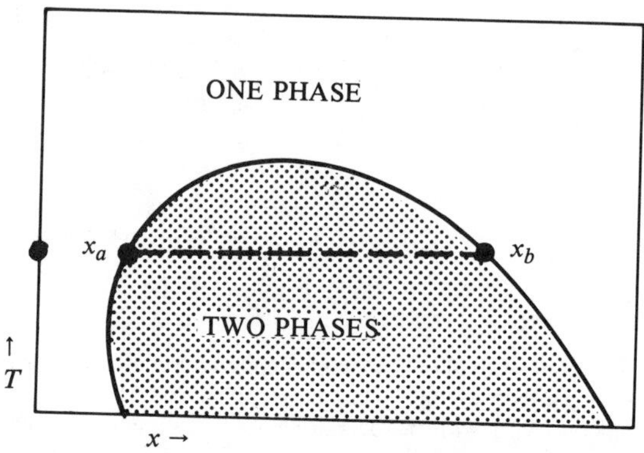

b. Showing partial miscibility of two liquids, temperature against composition. x_a and x_b are the compositions of the phases in equilibrium at the indicated temperature.

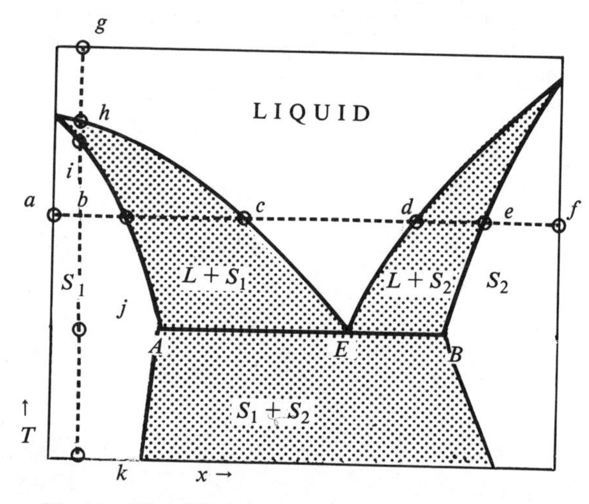

c. Liquid-solid equilibria, with points A and B as the solubility limits in the solid phases. At the eutectic point, E, liquid is in equilibrium with solid solutions of compositions x_A and x_B.

Figure 5.3. Geometric representation of phase equilibria of binary systems at fixed pressure.

from which several pertinent ratios are obtained:

$$m'/m'' = (x'' - x)/(x - x') = DB/AD, \qquad (5.3)$$

$$m'/m = (x'' - x)/(x'' - x') = DB, \qquad (5.4)$$

$$m''/m = (x - x')/(x'' - x') = AD. \qquad (5.5)$$

The name *lever rule* for these relations is suggested by the fact that they are derivable by taking moments, $m\,\Delta x$, around points x, x', or x'' in the order of the given equations. These ratios also can be read off as projections on the x axis as indicated on the diagram and with the equations preceding.

5.1.4. Three Components

Compositions of ternary mixtures are represented by points in the interiors of equilateral or isosceles triangles of unit heights. The equilateral triangle has the property that the sum of the perpendiculars from any point to the sides equals the height of the triangle. Although the matter is not of direct interest to phase representation, the sum of the perpendiculars from any external point to the extended sides of the triangle likewise

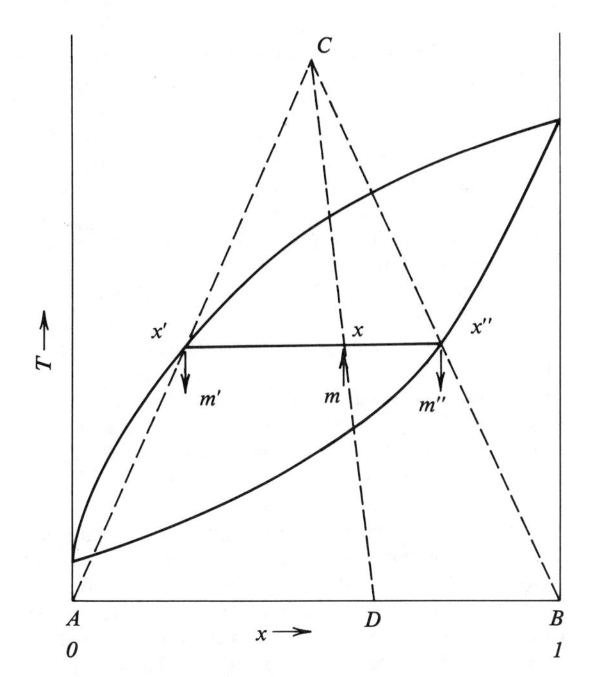

Figure 5.4. Illustrating the "lever rule." When the overall composition of the mixture is x, the fraction of the phase whose composition is x' is the distance DB and that of the other phase, whose composition is x'', is the distance AD. The m's represent masses.

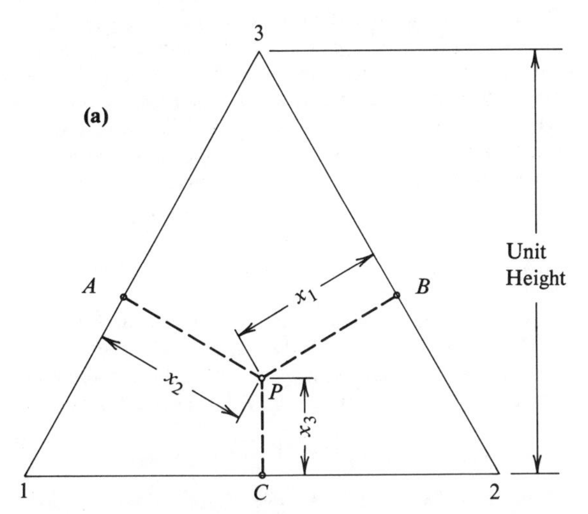

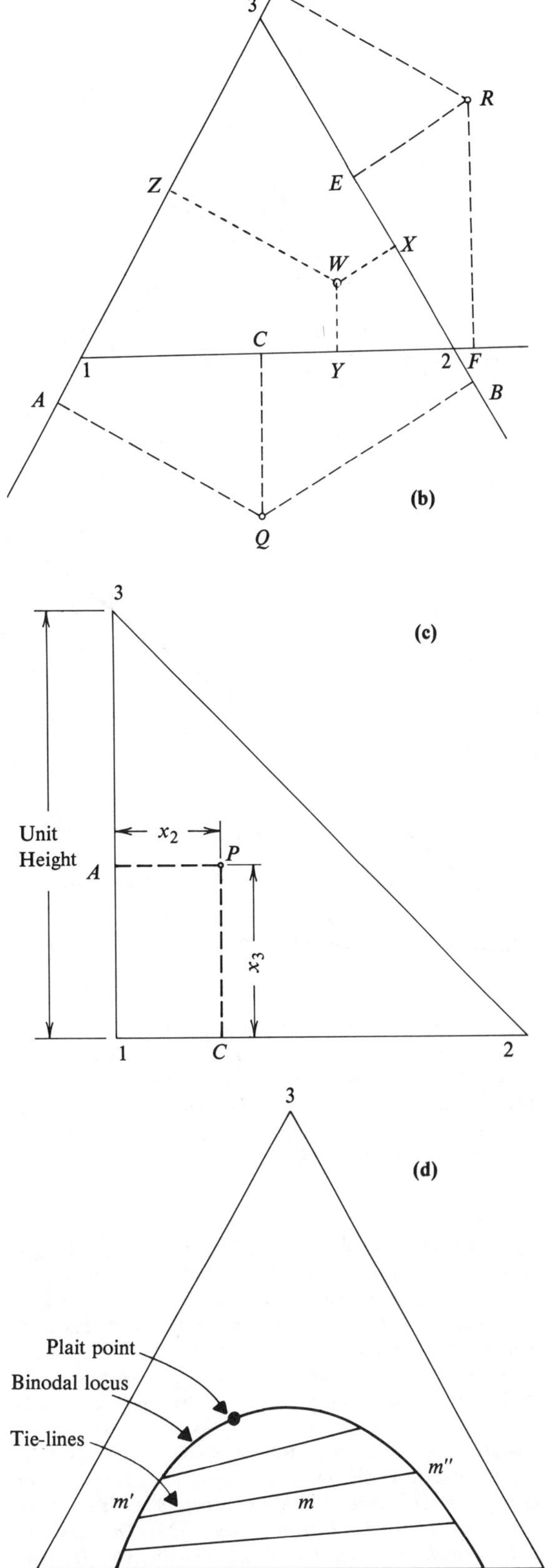

Figure 5.5. Geometric representation of ternary mixtures. (a) Representation of ternary compositions on an equilateral triangle. $PA + PB + PC =$ height of triangle. (b) Generalized use of triangular (or homogeneous or Möbius) coordinates, either inside or outside the triangle. Distances QC and RE are negative since they are in the opposite directions to WY and WX, whereas QA, QB, RD, and RF are positive since they are in the same directions as the corresponding lines WZ, WX, WZ, and WY. (c) Representation of ternary compositions on an isosceles right triangle of unit height. The composition of the remaining component is found by difference, $x_1 = 1 - x_2 - x_3$. (d) A ternary mixture with a two-phase region, showing a binodal curve, tie-lines, and the plait point.

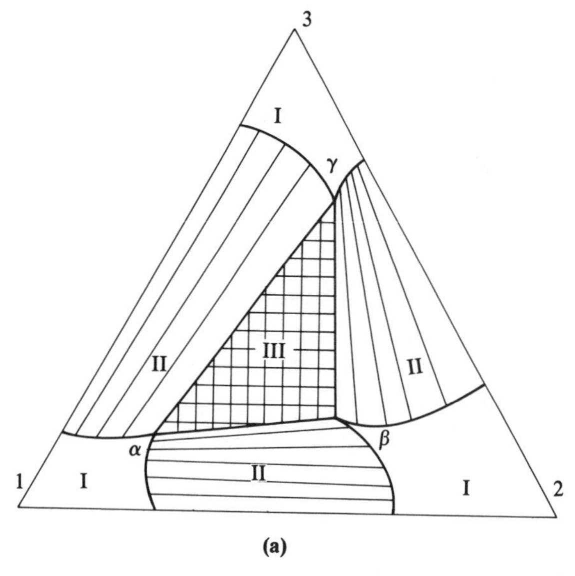

(a)

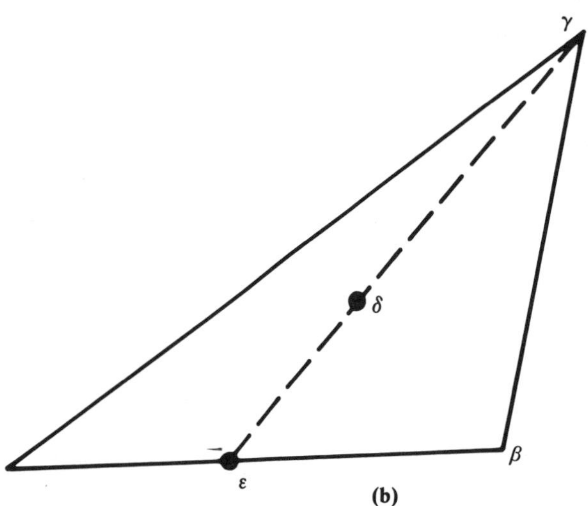

(b)

Figure 5.6. A ternary mixture with multiphase regions. (a) A ternary mixture with one three-phase region (III) and three each of two-phase (II) and one-phase (I) regions. (b) Diagram for determination of the relative amounts of three phases. Section 5.1 develops the formulas in terms of distances $\alpha\varepsilon$, $\varepsilon\beta$, $\gamma\delta$, and $\delta\varepsilon$.

equals the height of the triangle, if proper identification of positive and negative directions is made. Thus, in Figure 5.5(b), if the perpendicular from an external point (line *RD*, for example) is in the same direction as from an internal point to the same side, the sign is positive; and if in the opposite direction (line *RE*, for example), the sign is negative. As is well known, these principles are used in the graphical analysis of stage requirements for separating the components of ternary mixtures. The geometry of triangular coordinates has been developed extensively and also goes under the names of geometry of homogeneous or Möbius coordinates. Two useful treatments of this subject are by Loney (1923) (*The Elements of Coordinate Geometry*, Part II) and Durell (1961) (*Homogeneous Coordinates*).

Although several diagrams sometimes may be superimposed for convenience, they usually are drawn to show composition relations at a particular temperature and pressure. An area represents a composition range in which one, two, or three phases may exist in equilibrium. Since all substances are in principle at least slightly soluble, any region that has one or more corners of the triangle as part of its boundary is single-phase. Practically, the one-phase region may disappear when the mutual solubilities are very low, as for example at the upper corner of Figure 5.37(a) or the right-hand corner of Figure 5.38(b).

A line on a diagram separates two regions that differ by unity in the number of phases. At any point that is common to three regions, three phases coexist. Any region in which three phases coexist is necessarily bounded by a straight-sided triangle, as $\alpha\beta\gamma$ in Figure 5.6(a) or (b); here line $\alpha\beta$ is a tie-line, which must be straight, connecting all mixtures of phases α and β, and similarly for the sides $\beta\gamma$ and $\gamma\alpha$.

The number of phases that can be present at a particular overall composition is not predicted by the phase rule but depends on the chemical natures of the substances present. Thus in Figure 5.39(a) there are three three-phase regions, each made up of varying proportions of phases of fixed but different compositions in the three regions. The phase rule, however, does state that at the fixed T and P of this diagram, when there are three phases, the compositions are fixed, although there is no statement of what those compositions are. The same diagram also has several one- and two-phase regions whose existence is very sensitive to temperature.

Compositions of phases in equilibrium may be connected with tie-lines, also called *connodals*, along which may be found all possible proportions of the two phases. The relative amounts of the two phases are determined by material balances on individual components. When the compositions are truly along a tie-line, as $m'mm''$ in Figure 5.5(d), the relative amounts of the two phases are related to the mol fractions of a particular component in the same way as with binary mixtures, Eqs. 5.1–5.5, for instance,

$$m'/m = (x_i - x_i'')/(x_i' - x_i'') \tag{5.6}$$

where subscript i may be 1, 2, or 3.

In a three-phase region, such as that bounded by triangle $\alpha\beta\gamma$ of Figure 5.6(a) or (b), the fractions $\phi_i = m_i/m$ of the three phases that are present when the overall composition is represented by point δ are given by the material balances,

$$(x_{1\alpha} - x_{1\gamma})\phi_\alpha + (x_{1\beta} - x_{1\gamma})\phi_\beta = x_{1\delta} - x_{1\gamma}, \tag{5.7}$$

$$(x_{2\alpha} - x_{2\gamma})\phi_\alpha + (x_{2\beta} - x_{2\gamma})\phi_\beta = x_{2\delta} - x_{2\gamma}, \tag{5.8}$$

$$\phi_\gamma = 1 - \phi_\alpha - \phi_\beta. \tag{5.9}$$

The linear equations are readily solvable for the ϕ_i by determinants. Alternatively, the phase compositions may be found by measuring off distances on the phase diagram. With the notation of Figure 5.6(b),

$$\frac{\phi\gamma}{\phi_\alpha + \phi_\beta} = \frac{1}{\phi_\alpha + \phi_\beta} - 1 = \overline{\delta\varepsilon}/\overline{\gamma\delta}, \tag{5.10}$$

and

$$\phi_\beta/\phi_\alpha = \overline{\alpha\varepsilon}/\overline{\beta\varepsilon}, \tag{5.11}$$

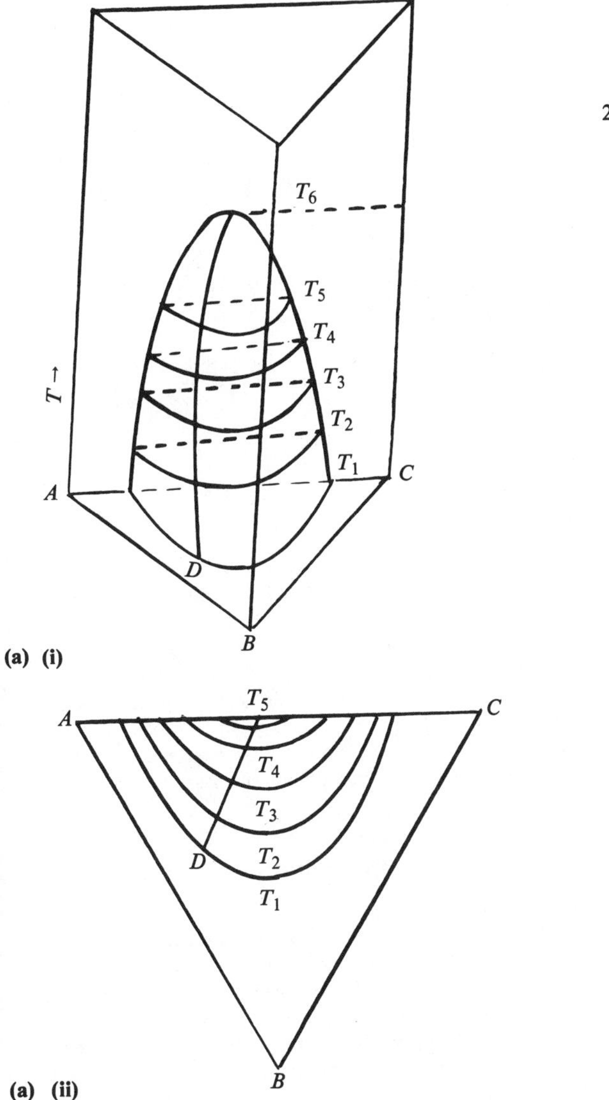

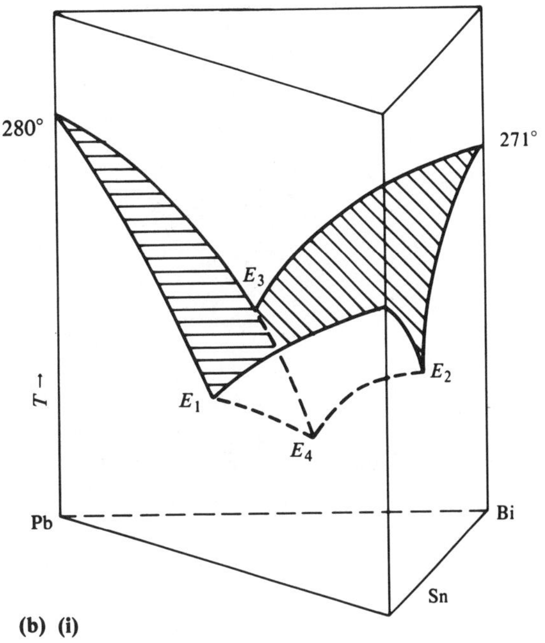

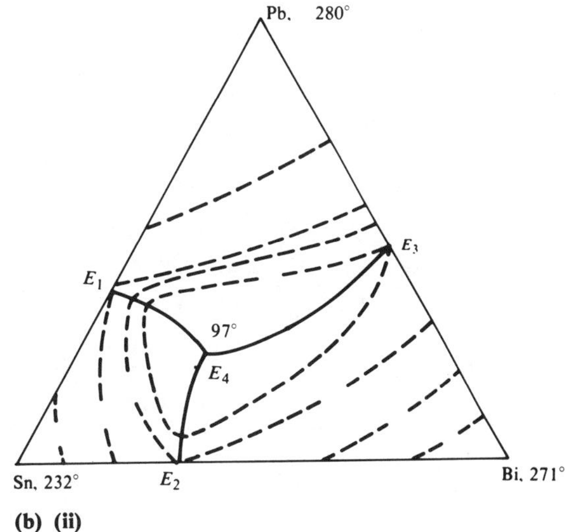

Figure 5.7. Spatial and corresponding planar diagrams, showing the effects of temperature and pressure on ternary mixtures: (a) (i) Variation of the miscibility range of a ternary mixture with temperature at a particular pressure; (ii) the planar diagram shows the isotherms indicated on the spatial diagram. (b) Diagrams of mixtures of bismuth, lead, and tin that have binary and a ternary eutectic, $E4$, at which one liquid and three solid phases are in equilibrium at 97 C, with composition Bi = 0.51, Pb = 0.33, and Sn = 0.16. The projection onto the base also locates the eutectics and isotherms.

whence

$$\phi_\alpha = \left(\frac{\overline{\beta\varepsilon}}{\overline{\alpha\beta}}\right)\left(\frac{\overline{\gamma\delta}}{\overline{\gamma\varepsilon}}\right), \tag{5.12}$$

$$\phi_\beta = \left(\frac{\overline{\alpha\varepsilon}}{\overline{\alpha\beta}}\right)\left(\frac{\overline{\gamma\delta}}{\overline{\gamma\varepsilon}}\right), \tag{5.13}$$

$$\phi_\gamma = 1 - \phi_\alpha - \phi_\beta. \tag{5.9}$$

Tie-line correlations are discussed in detail in other chapters. With binary mixtures the simple relation

$$x_1'' = K_1 x_1' \tag{5.14}$$

may hold over a limited range of compositions with a constant value of the distribution coefficient, K_1; there is usually a strong effect of temperature, however. Relations between compositions of vapor and liquid phases are approximated by

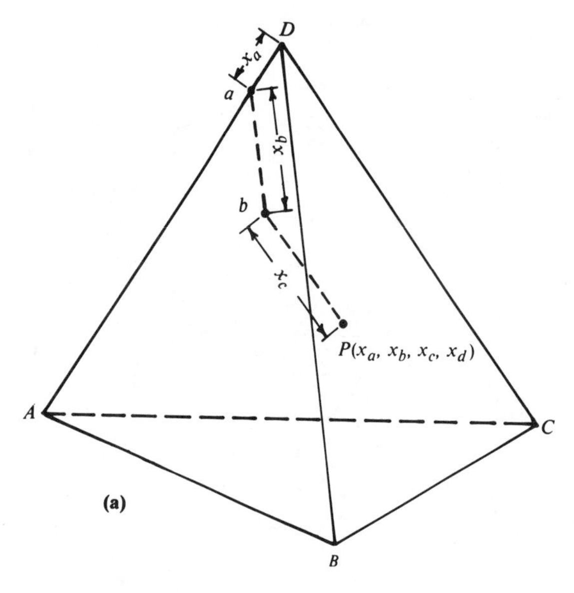

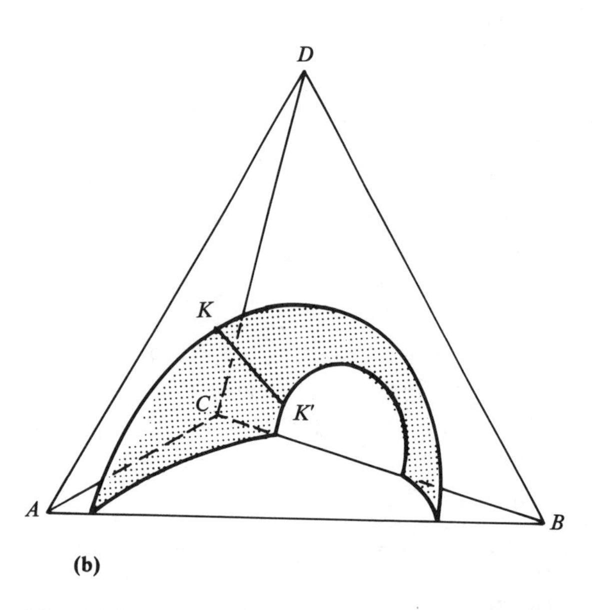

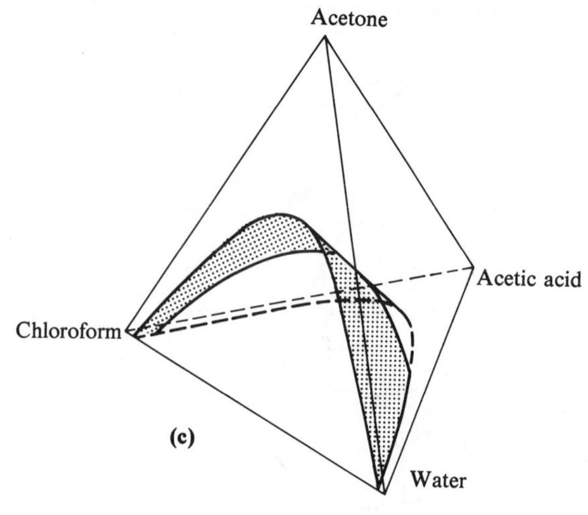

Figure 5.8. Quaternary composition tetrahedra. (a) Composition tetrahedron with unit edge. In order to locate a point, $P(x_a, x_b, x_c, x_d)$: Make line $Da = x_a$, draw line $ab = xb$ parallel to edge BD, and draw line $bP = x_c$ parallel to edge DC. (b) The system naphtha (A) + furfural (B) + isobutane (C) + butadiene (D) at 20 F. KK' is the locus of the plait points (data of A.S. Smith & T.B. Braun, *Ind. Eng. Chem. 37* 1047 (1945)). Copyright American Chemical Society; used with permission. (c) The system acetone + water + chloroform + acetic acid at 25 C (Brancker et al., 1940). Copyright by American Chemical Society; used with permission.

Henry's and Raoult's laws, but more precise representation is made in terms of activity and fugacity coefficients, as explained elsewhere in the text.

The effects of temperature (or pressure) can be represented on a triangular prism. Figure 5.7(a) shows a ternary system whose isothermal sections exhibit limited liquid miscibility, and Figure 5.7(b) represents a system in which liquid and solid phases are present. Both diagrams are accompanied by isothermal contours projected onto the bases.

5.1.5. Multicomponents

Compositions of quaternary systems can be represented by points within a regular tetrahedron—for example, Figures 5.8(a), (b), and (c). One such diagram is needed for each pair of T and P represented. The tetrahedron has the property that the sum of the perpendiculars from an internal point to the sides equals the height. It is more convenient, however, to use a tetrahedron of unit edge rather than unit height. On Figure 5.8(a) a composition is located readily by proceeding parallel to the several edges, as indicated. Since the spatial diagram is awkward to make and to use, several methods have been devised to use suitable groups of planar projections to embody the information of the solid model, some of which are described by Ricci (1951, p. 392) and Vogel (1959, p. 670). One obvious scheme is to use a triangular diagram to represent the compositions of A, B, and C on a D-free basis, for several values of D, either by a series of such diagrams or by a single diagram with contour lines corresponding to the several values of D.

Only a few quaternary systems of organic substances have been analyzed. Two simple types involving only liquid phases are shown somewhat idealized in Figures 5.8(b) and (c). Multicomponent phase relations can be approximated by assuming ideal-solution behavior. More accurate quaternary diagrams can be constructed if the binary activity-coefficient behavior is known, and even better ones if the four ternary diagrams also are known. The activity-coefficient method is applicable to any number of components and phases, in the absence of intermolecular compound formation, but has rarely been applied except to liquid-liquid and vapor-liquid equilibria.

Table 5.3. Components (C), Phases (Ph), and Variances (F) at Phase Equilibrium

				Free variables, F		
Order	*Type*	C	Ph	*General* $F=C+2-Ph$	*Fixed P* $F=C+1-Ph$	*Fixed T&P* $F=C-Ph$
Unary	Bivariant	1	1	T&P	T	None
	Univariant	1	2	T or P	None	—
	Invariant	1	3	None	—	—
Binary	Trivariant	2	1	T, P, x	T or x	x
	Bivariant	2	2	any 2	x	None
	Univariant	2	3	any 1	None	—
	Invariant	2	4	None	—	—
Ternary	Tetravariant	3	1	T, P, x_1, x_2	T, x_1, x_2	x_1, x_2
	Trivariant	3	2	any 3	x_1, x_2	x_1
	Bivariant	3	3	any 2	x_1	None
	Univariant	3	4	any 1	None	—
	Invariant	3	5	None	—	—

5.2. THE PHASE RULE

The Gibbs phase rule relates the number of phases, Ph, present at equilibrium to the number of components, C, at specified composition, temperature, and pressure. Gravitational and electrical variables and the like are considered not pertinent here. The composition is expressible by giving $C-1$ mol or weight fractions per phase. Together with temperature and pressure, therefore, the total number of variables defining all phases of the system is $Ph(C-1)+2$. Since the chemical potential, $\mu_i^{(j)}$ or $\bar{G}_i^{(j)}$ of each component i is the same in each phase j at equilibrium, the number of these restraints is $(Ph-1)C$. The difference between the number of condition variables and the number of chemical potential restraints is the variance or the number of degrees of freedom, F, of the system, which becomes

$$F = Ph(C-1) + 2 - (Ph-1)C = C + 2 - Ph.$$

When using planar diagrams, reduced forms of the phase rule are convenient. With binary systems, only $T-x$ and $P-x$ planar diagrams are possible, so the rule becomes $Ph = 3 - F_{net}$, where F_{net} may have a maximum value of 2, say x_1 and either P or T. A planar diagram of a ternary system is drawn at constant T and P, so the rule for this case also becomes $Ph = 3 - F_{net}$, where F_{net} again has a maximum value of 2 (two of the three mol fractions). These results are extended in Table 5.3.

The meaning of C in the phase rule is that it is the minimum number of chemical species required to define the chemical nature of the system. When the chemical species are *inert*, C is simply their number, as a mixture of oxygen and nitrogen at atmospheric conditions has $C=2$. When the species are *reactive*, each *independent* reaction that is not inhibited from reaching equilibrium reduces by unity the number of independent components. How to find the number of independent reactions is explained in Chapter 10. For reactive systems, accordingly, the number of components is expressed by

$$C = C_e - C_p + C_i,$$

where

C_e = number of chemical elements present in the system.

C_p = number of reactions resulting in intermolecular compounds or isomers or polymers, in each of which the chemical elements are present in the same proportions on the left-hand and right-hand sides of the stoichiometric equations.

C_i = number of reactions inhibited from reaching equilibrium.

Several examples of these cases may be given. For the dimerization of butadiene,

$$2\ C_4H_6 \rightleftharpoons (C_4H_6)_2,$$

$$C = 2 - 1 + 0 = 1.$$

When H_2 and N_2 react to form ammonia,

$$C = 2 - 0 + 0 = 2,$$

and when they are inhibited from reacting, as at low temperature without catalysts,

$$C = 2 - 0 + 1 = 3.$$

When propane forms a hydrate,

$$C_3H_8 + n\ H_2O \rightleftharpoons C_3H_8 \cdot (H_2O)_n,$$

$$C = 3 - 1 + 0 = 2.$$

The rule about intermolecular compounds is particularly important since they occur frequently on phase diagrams. Although the number of chemical species changes when intermolecular compounds are formed, a stoichiometric relation is generated for each one, and the net effect on the "components" in the phase rule is zero.

Systems are known according to the number of components as unary, binary, ternary, and so on, and according to the number of degrees of freedom as invariant, univariant, and so on. These identifications are summarized in Table 5.3. For example, the table states that a ternary system can have five phases at some particular temperature, pressure, and composition—that is, when $F=0$. On a ternary composition triangle at fixed T and P, however, a maximum of three phases can be shown. On a spatial diagram such as Figure 5.7(b) the coexistence of four phases (three solid and one liquid) is identified as ternary eutectic E_4. According to the phase rule,

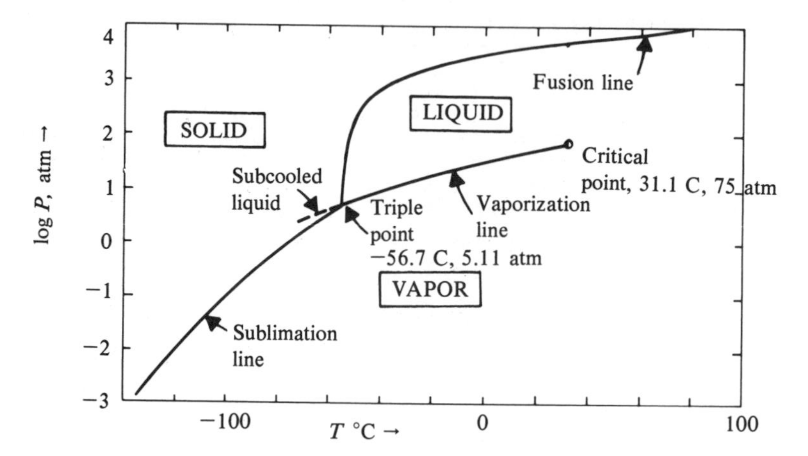

Figure 5.9. Phase diagram of carbon dioxide (after Gmehling, Handbuch der anorganischen Chemie: Kohlenstoff, T3/L1 (1970)).

this point is univariant; but practically, pressure has little effect on the condensed system, so the ternary eutectic is essentially invariant. At some very low pressure, however, a vapor phase could form and be in equilibrium with the three solid phases and one liquid phase of the ternary eutectic, making a total of five phases, as stated by the phase rule.

At fixed temperature and pressure, for which many multicomponent data are obtained, the maximum possible number of phases at equilibrium equals the number of components. Problem 5.15 cites an example of a six-component system with six liquid phases at essentially atmospheric conditions.

5.3 UNARY SYSTEMS

For a one-component system the phase rule becomes $Ph = 3 - F$, which means that one, two, or three phases may be present at equilibrium. The carbon dioxide diagram, Figure 5.9, is typical. Distinguishing characteristics are areas in which single phases exist, lines along which two phases coexist, a critical point at which the properties of the liquid and vapor phases become identical, and a triple point at which three phases coexist in equilibrium. There are no corresponding points at which the properties of the other pairs of phases become the same.

At high pressures many substances display solid polymorphism—that is, more than one crystalline form. The examples of phosphorus, sulfur, tin, and water are commonly cited in chemistry books. Figures 5.10(a), (b) are particularly simple examples of such behavior. Acetic acid possesses two crystalline forms and urethane has three. There are other far more complicated behaviors of organic substances—for example, camphor, which has eleven crystalline phases.

The water system of condensed phases is especially interesting. Nine solid phases have been identified. *PT* diagrams, triple-point data, and a schematic *PVT* diagram are shown in Figure 5.11. Ice IV seems to be a metastable form. Other metastabilities and some uncertainties also are recorded on the diagram. It is interesting that the forms of ice that can exist directly in equilibrium with liquid have melting points that increase with pressure, except for ordinary ice I. At 40,000 kg/sq cm, ice melts at 200 C. A claim has been made that there is a form of ice more dense than the liquid.

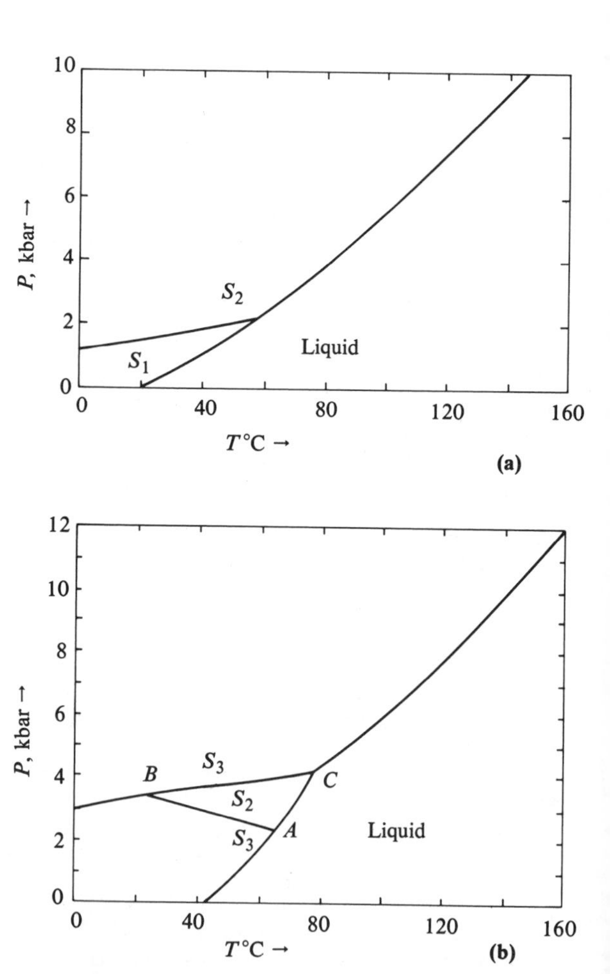

Figure 5.10. Liquid-solid equilibria of individual substances (after Bridgman, *Proc. Am. Acad. Arts & Sciences 52,* 91 (1916); *51,* 53 (1915)). (a) Phase diagram of acetic acid. The triple point is at 55.7 C, 2,100 bar. (b) Phase diagram of urethane. The triple points are A(66.2 C, 2,350 bar), B(25.5 C, 3,400 bar), C(76.8 C, 4,230 bar).

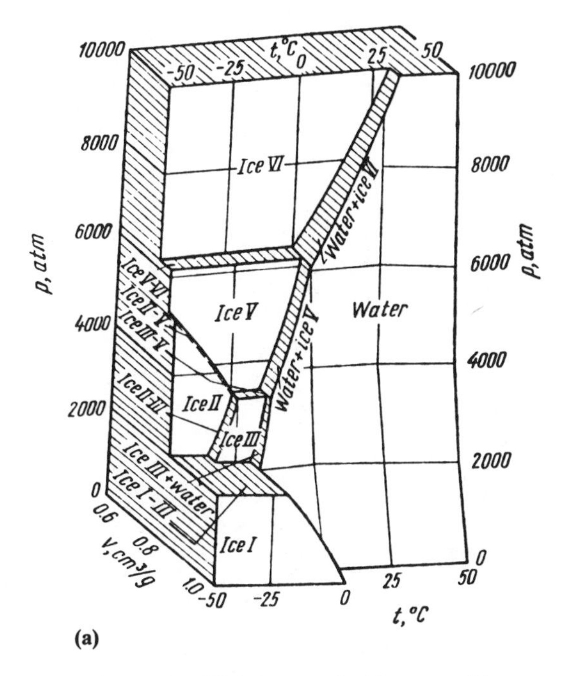

(a)

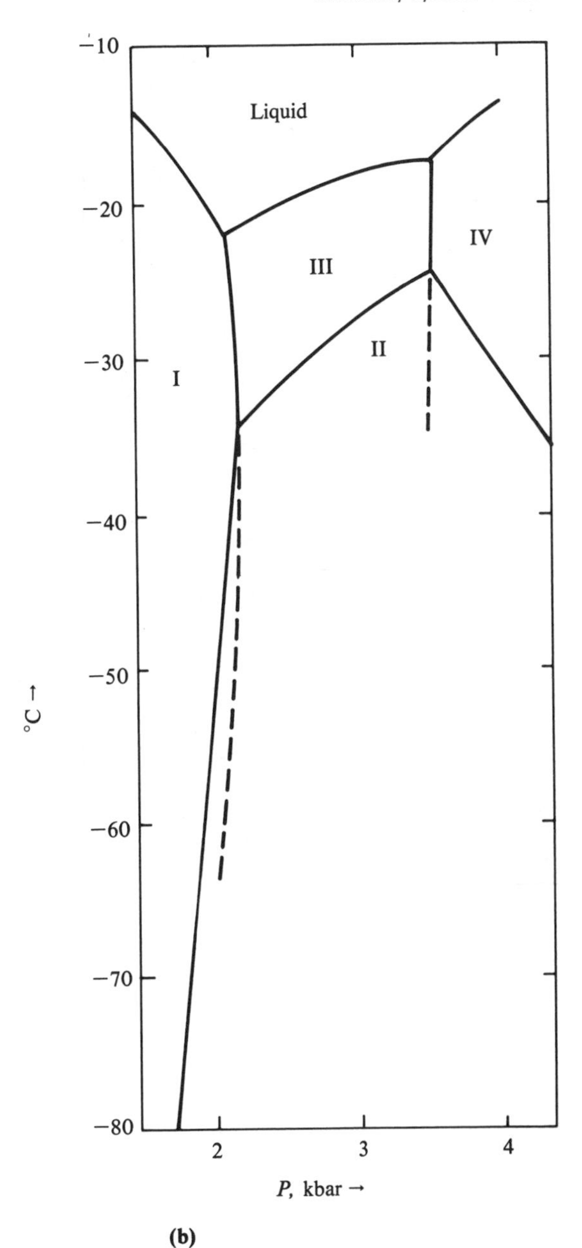

(b)

Figure 5.11. Equilibria between liquid water and its several crystalline phases. (a) *PVT* diagram to 10,000 atm (Gerasimov, 1974). (b) Details of the temperature-pressure diagram (Gmehling, *Handbuch*).

Although the investigations emobodied in these diagrams have extended over a period of sixty years or so, there still remain uncertainties.

Transition points of crystalline forms are measured in thousands of atmospheres, so such phenomena are mostly out of the range of normal chemical processing.

A *PVT* diagram of propane is shown as Figure 5.12. Figure 5.14 shows that solid propane exists below −300 F.

5.4. BINARY SYSTEMS

Phase behavior of binary systems usually is more complex than that of most pure substances. Ordinarily it is most convenient to use planar diagrams, which means either a $T-x$ or a $P-x$ diagram, although a limited use is made of $P-T$ diagrams at constant composition (isopleths). For convenience, diagrams usually are restricted to showing the relations between limited numbers of phases such as vapor-liquid, liquid-liquid, liquid-solid, and so on. The compositions of phases in equilibrium may be indicated by tie-lines, which are horizontal lines on $P-x$ or $T-x$ diagrams, or they may be shown separately as plots of the composition of one phase against the composition of the other; these are called distribution diagrams or $x-y$ diagrams; and may be either isothermal or isobaric.

Three examples may be given of phase diagrams with vapor, liquid, and solid phases. In the methanol + water system at 1 atmosphere, Figure 5.13, the vapor-liquid behavior is almost ideal, but the freezing behavior is complicated by the formation of a solid monohydrate. The mutual solubilities of propane and water are limited, so two liquid phases are present over most of the composition range, Figure 5.14; the three solid phases present are ice, propane, and a hydrate or clathrate with 7.5 mols of water per mol of hydrocarbon. In the ammonia + water system of Figure 5.15 there are four solid phases: ice, ammonia, hemihydrate, and monohydrate. Vapor-liquid equilibria are represented on these diagrams, but not vapor-solid equilibria since those occur only at low pressures and cannot be drawn on the scale of these diagrams.

For utilitarian purposes, only limited ranges of T and P usually are needed, and the number of phases involved also is limited. In Landolt-Börnstein, separate listings are made of liquid-vapor, liquid-liquid and liquid-solid diagrams, but all three kinds of diagrams are not always known for the same pairs of substances.

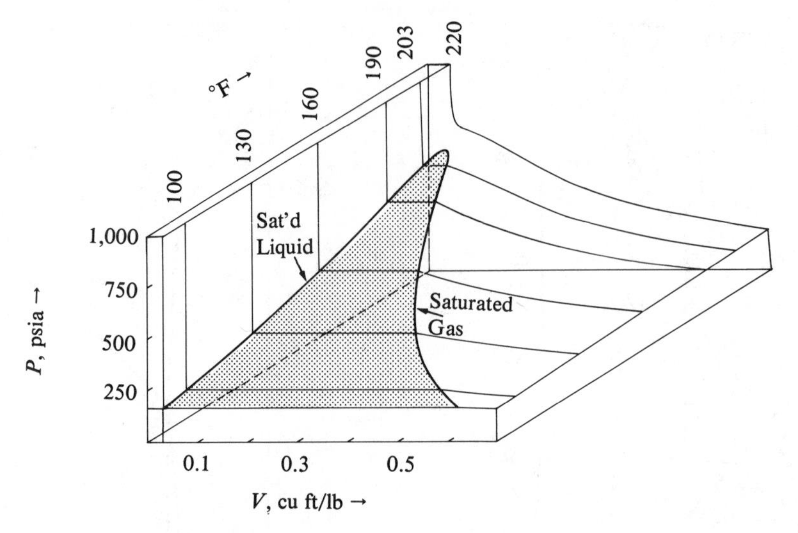

Figure 5.12. *PVT* diagram of propane. The shaded area is two-phase (Sage & Lacey, 1939).

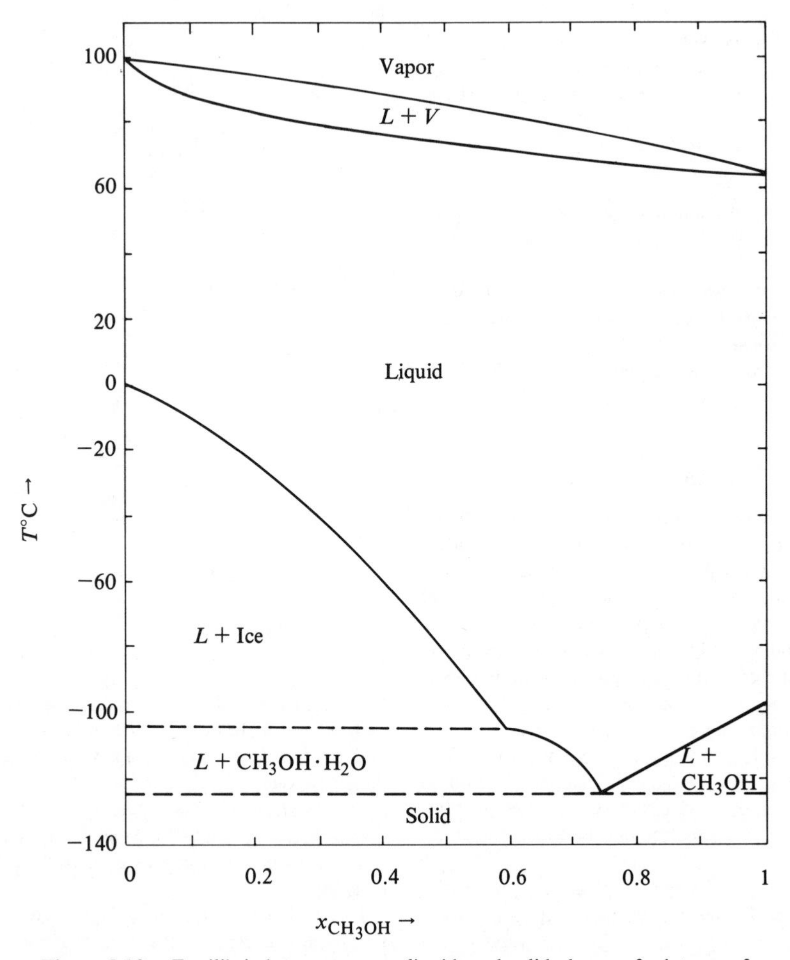

Figure 5.13. Equilibria between vapor, liquid, and solid phases of mixtures of methanol and water. The solid monohydrate, $CH_3OH \cdot H_2O$, is stable in the range from −104.5 to −125 C.

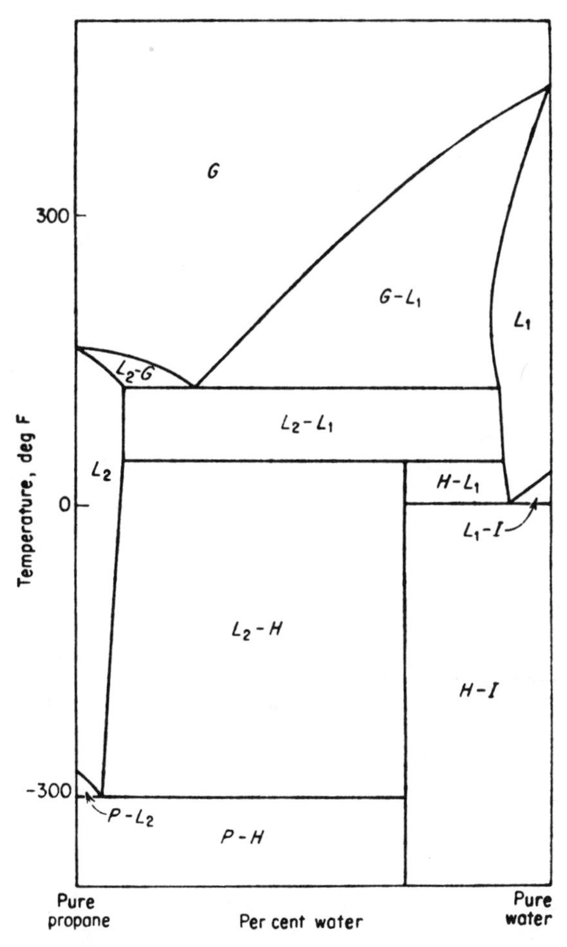

Figure 5.14. Schematic phase diagram of the propane-water system below the critical pressure of propane. H = hydrate, I = ice, P = solid propane (Kobayashi, 1951, in Katz et al., 1959).

5.4.1. Vapor-Liquid Equilibria

The main types of vapor-liquid behaviors are summarized in Figure 5.16. The popular kinds of diagrams are $T-x$, $P-x$, and $x-y$. In this classification type I is that for which all compositions have boiling points between those of the pure substances, which includes systems conforming to Raoult's Law. An azeotropic condition is one for which the compositions of the liquid and vapor phases are the same, or for which the boiling temperature and vapor composition are constant over a range of liquid compositions. On the $x-y$ diagram, the curve of an azeotropic system crosses the 45° line. Types II and III are of homogeneous azeotropes. Minimum boiling azeotropes are quite common, the maximum boiling ones less so. A system having two azeotropes is shown in Figure 4.24(a). Types IV and V involve liquid phases that are partially miscible or immiscible, respectively, and exhibit heterogeneous azeotropy. In the less common systems of type VI, immiscible liquid phases exist at temperatures intermediate to the condensation temperatures of the pure substances; here the two liquid-phase compositions remain constant over a wide range of vapor compositions. If a system formed intermolecular compounds that were stable at their boiling points, the phase diagrams would look like type VII. No specific examples come to mind, but similar diagrams of liquid-solid systems are given in some parts of Figure 5.22.

Marked changes in the shapes of phase diagrams often are caused by variation of temperature and pressure. Figure 5.17 illustrates this for mixtures of ethane and n-heptane. The isopleths, Figure 5.17(a), show the critical points as well as the maximum temperatures and pressures on the two-phase envelopes. The range over which two phases can coexist narrows sharply as the pressure goes up. An important effect of pressure is on the azeotropic composition. For the ethanol + water system, for example, the corresponding pressures and azeotropic compositions are: (Torr, wt % water) = (94.9, 0.5), (200, 2.7), (760, 4.4), (1,500, 4.87). In most cases, as in this one, the azeotrope can be broken by operating at a low enough pressure. Similar effects appear on the $x-y$ diagrams of methyl acetate + methanol, consisting of several isobars and isotherms (Figure 5.18). These curves are of data smoothed with the Wilson equation, but agreement with experimental data is quite good.

Although a few more $x-y$ diagrams are shown in Figure 5.19, it is not profitable to cite any extended number of liquid-vapor data since several thousand sets have been compiled in readily available works by Gmehling et al. (1978 ff.); Hirata et al. (1976); Kojima & Tochigi (1979); Kogan, Fridman, & Kafarov (1966); and Landolt-Börnstein. The last two are largely in tabular form; the others have tables, graphs, and parameters of correlating equations.

5.4.2. Liquid-Liquid Systems

Phase diagrams of binary systems restricted to liquid phases are comparatively simple in form: convex, concave, or closed in shape as functions of temperature. Some examples of $T-x$ diagrams are in Figure 5.20. The effect of pressure on equilibria of condensed phases is significant only at high pressures or near the critical point. The directional effect is predictable by Le Chatelier's principle. Usually the size of the two-phase region is reduced by pressure, as indicated schematically in Figures 5.1 and 5.2, and shown for methyl ethyl ketone + water in Figure 5.20(a), where raising the pressure 150 atm changes the two-phase region from a convex to a closed shape.

Other parts of Figure 5.20 display the varied effects of temperature on several systems. In most cases the mutual solubilities of condensed phases increase with temperature. The temperature at which complete miscibility is attained is called the critical solution temperature, CST. Figure 5.20(e) has an upper CST, Figure 5.20(d) a lower CST, and Figures 5.20(a) and (c) have both upper and lower CSTs. Many systems do show a tendency to form a closed two-phase region if a wide enough temperature range is explored—for example, Figure 5.20(c), which covers a 230°C range in temperature. Figure 5.20(e) also seems to be looping in. Vaporization may occur before an upper CST or freezing before a lower CST can be obtained, however. The most unusual behavior is that of sulfur + benzene, Figures 5.20(b), which has an upper CST that is lower than a lower CST.

Several comprehensive collections of liquid-liquid equilibrium data are available. Stephen, Stephen, & Silcock (1979) is in seven volumes. More than 2,000 binary systems are represented on 321 graphs in Landolt-Börnstein (1964). Each graph has a common solute and sometimes more than a dozen

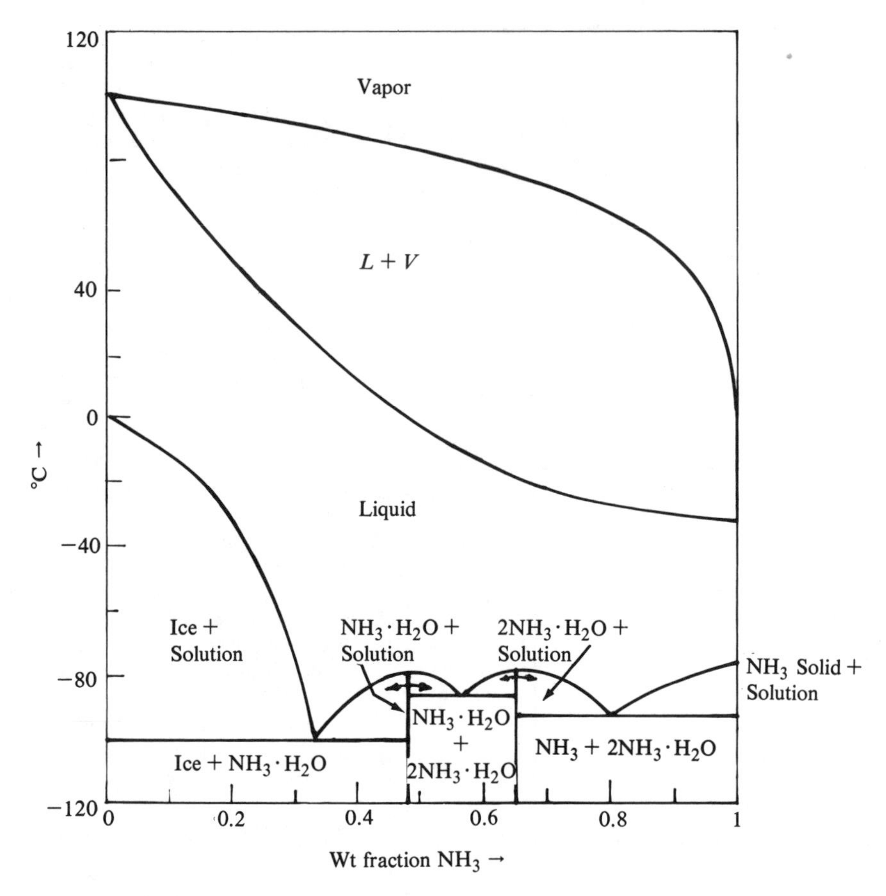

Figure 5.15. The ammonia + water system at 1 atmosphere (after Landolt-Börnstein, II/2a, p 377, 1960).

Figure 5.16. Types of binary $T-x$, $P-x$ and $x-y$ phase diagrams showing vapor and liquid phases

I. Intermediate-boiling systems, including Raoult's Law behavior

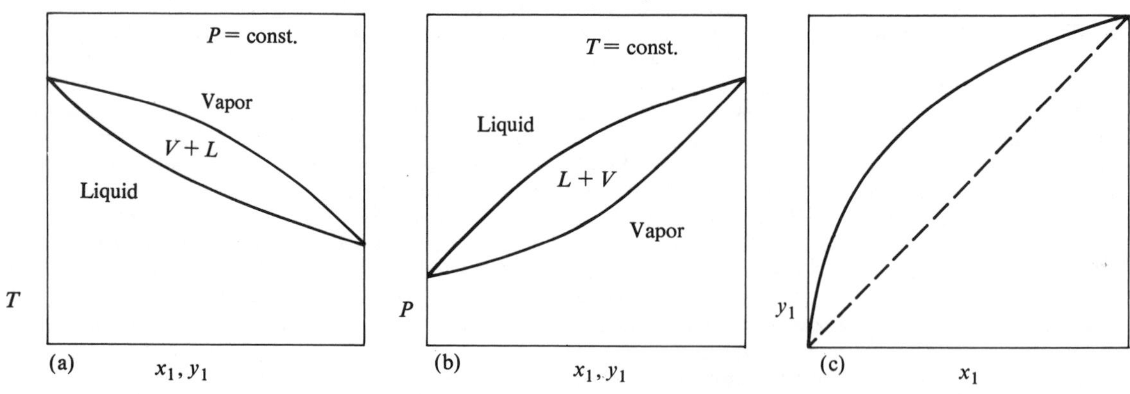

Figure 5.16 (*continued next page*)

II. Systems having a minimum boiling azeotrope

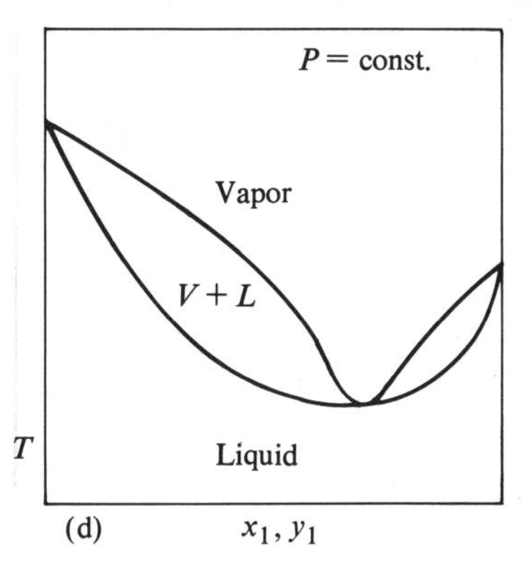

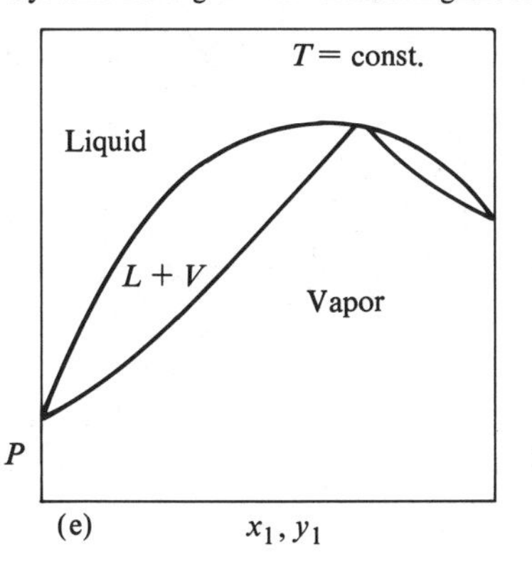

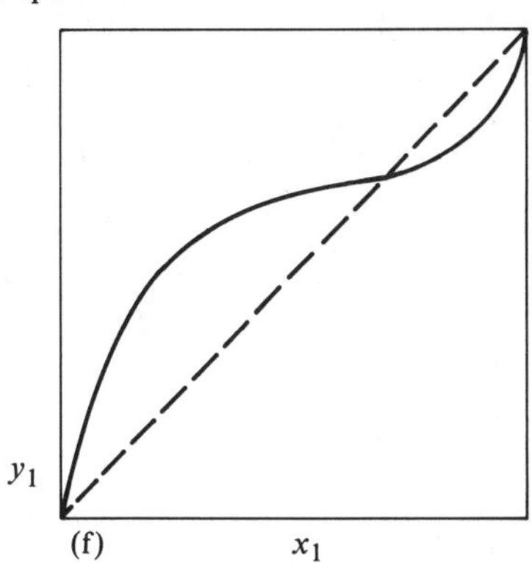

III. Systems having a maximum boiling azeotrope

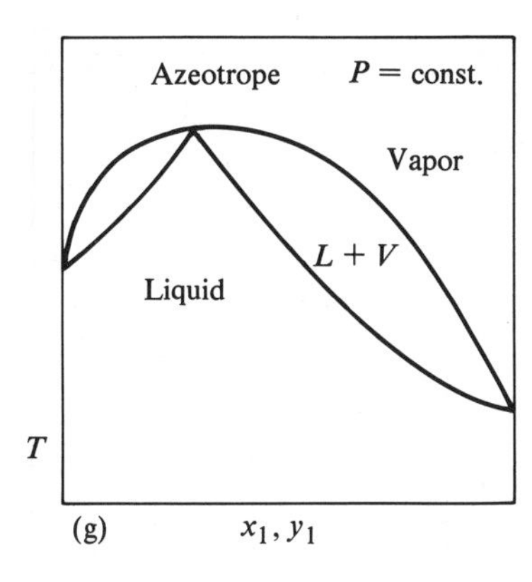

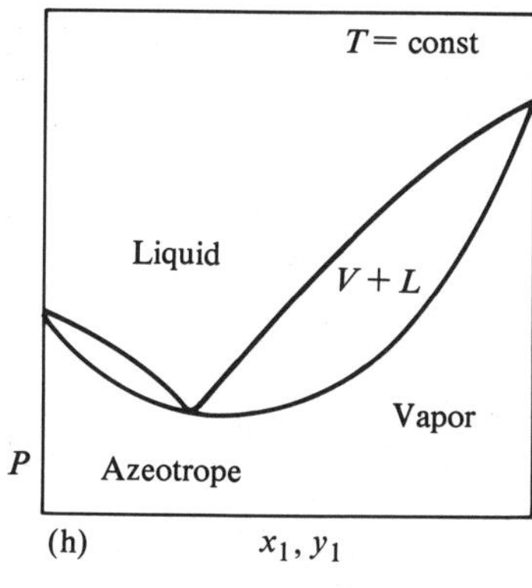

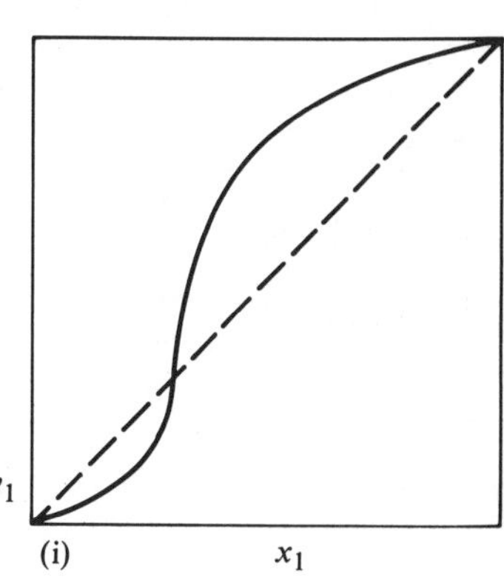

IV. Systems having immiscible liquid phases

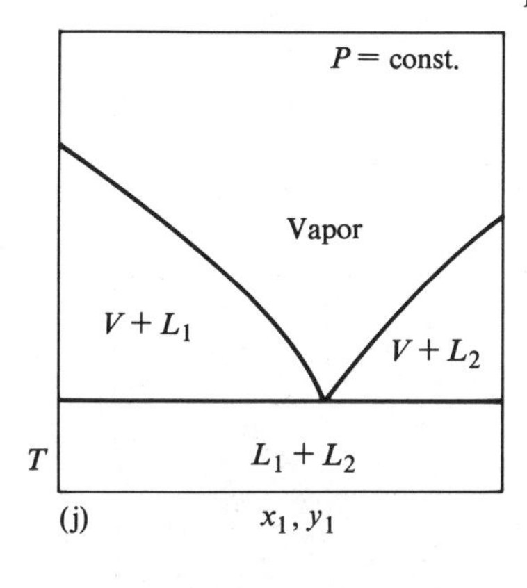

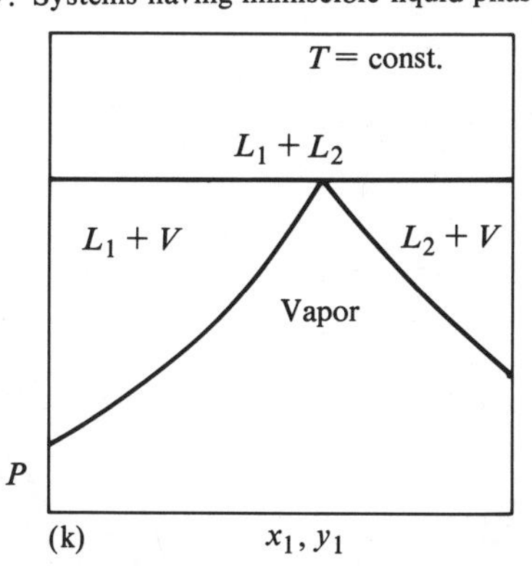

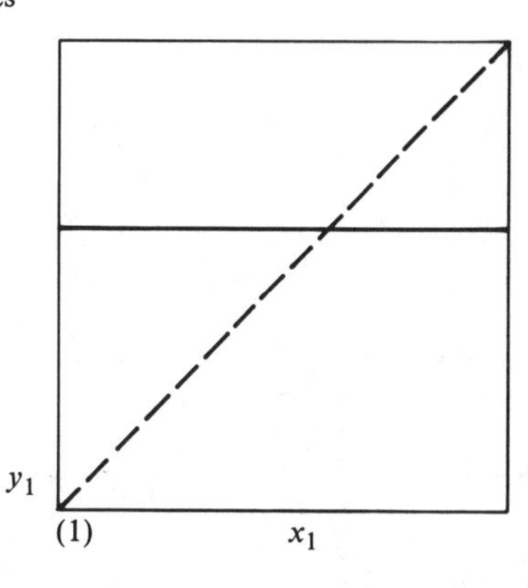

V. Systems having partially miscible liquid phases

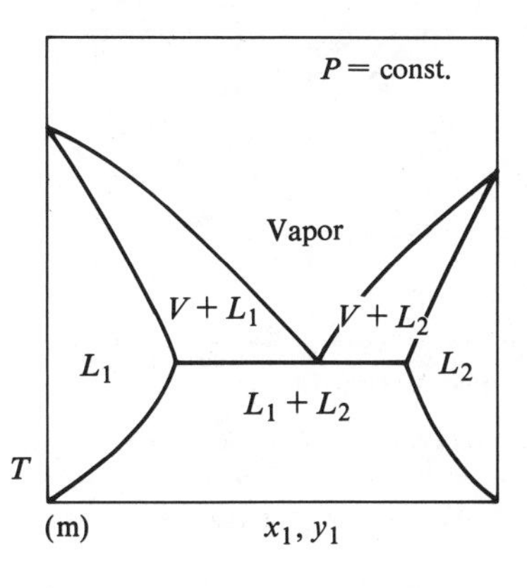

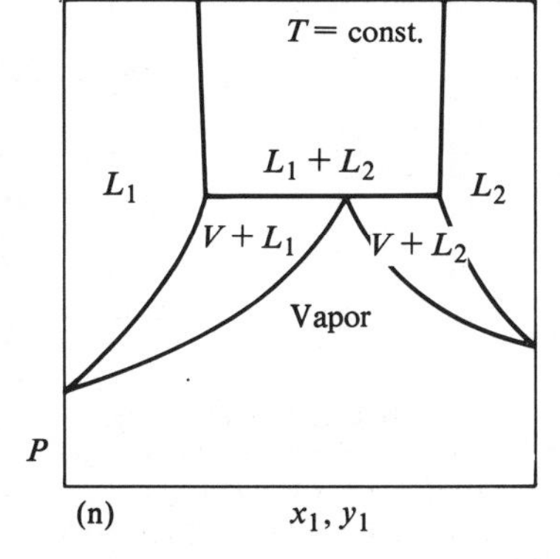

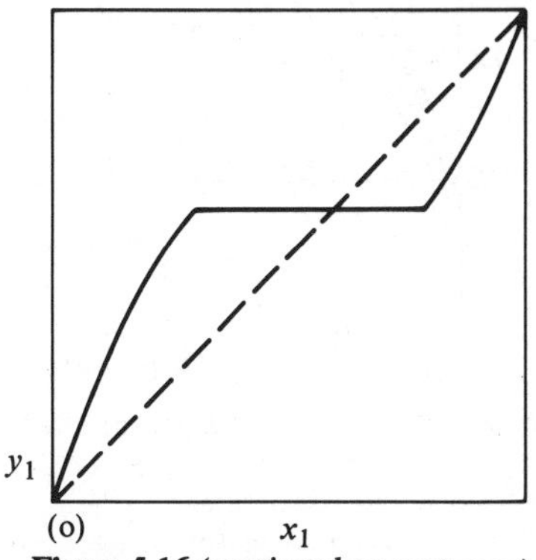

Figure 5.16 (*continued on next page*)

VI. Systems with immiscible liquid phases at temperatures between the condensation temperatures of the pure substances.

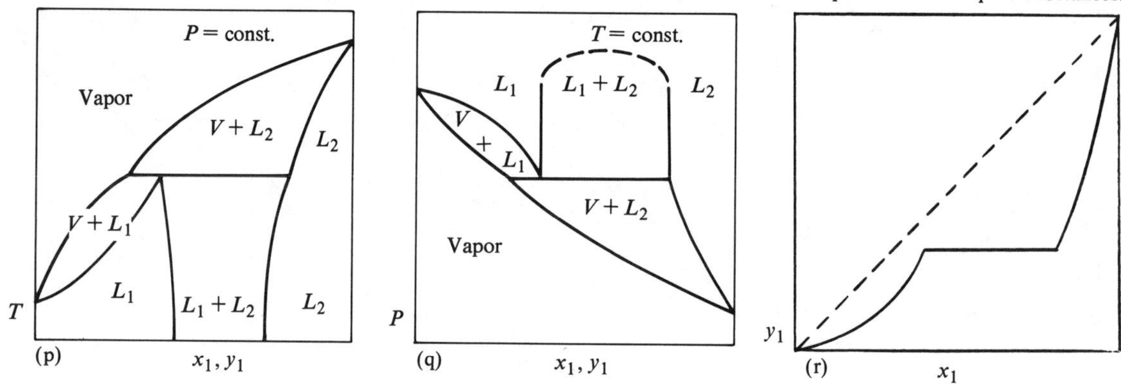

VII. System forming an intermolecular compound with boiling point between those of the pure components.

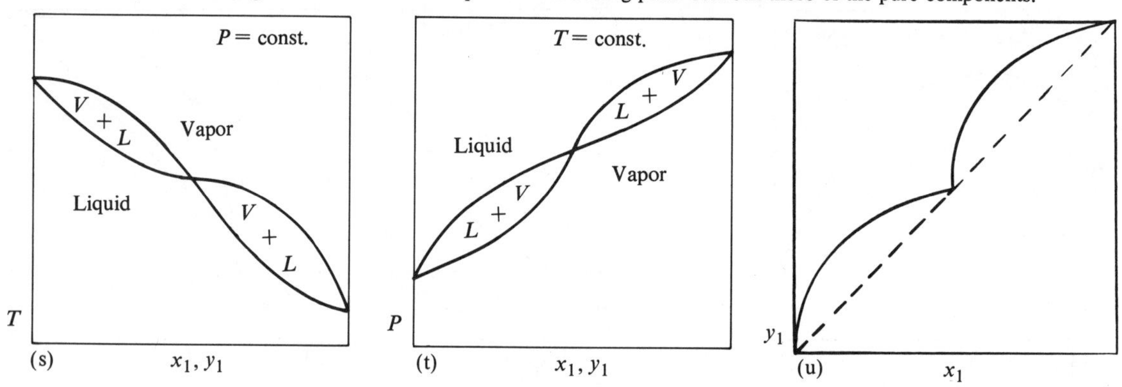

Figure 5.16 *(concluded)*

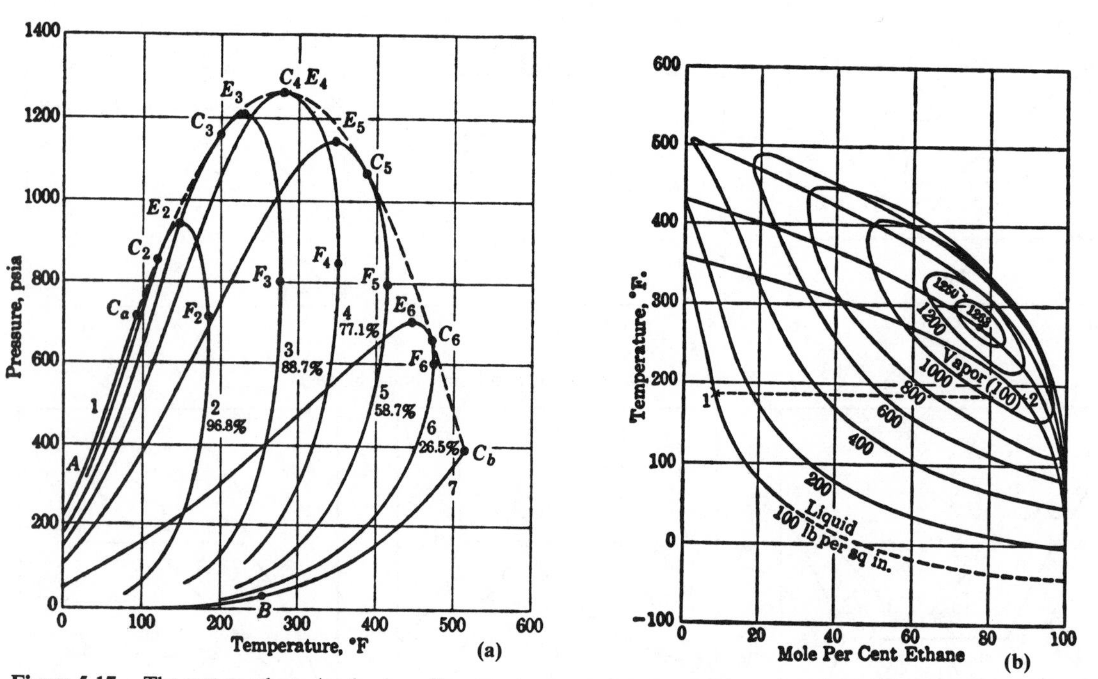

Figure 5.17. The system ethane + n-heptane: Pressure-temperature-composition behavior (Kay, 1938); redrawn by Hougen et al., "Thermodynamics," (1959). Copyright American Chemical Society; used with permission. (a) Pressure-temperature lines at several compositions. C_i designate the critical points, while E_i and F_i designate the maxima of temperature and pressure at a given composition. (b) Temperature-composition curves at several pressures.

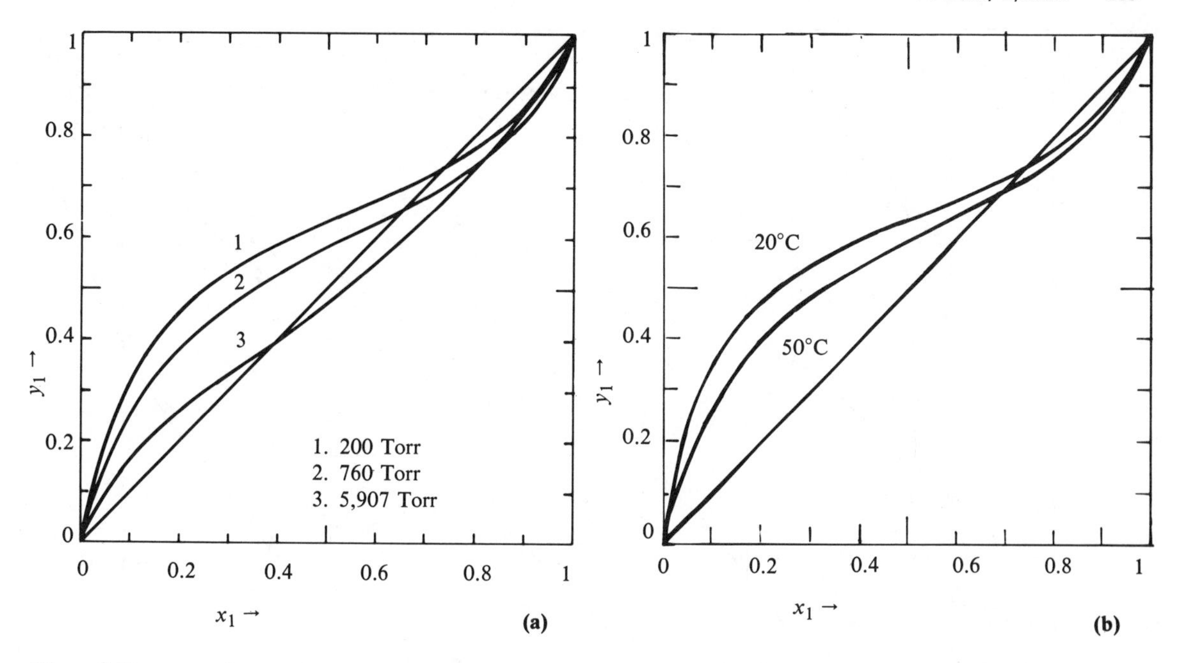

Figure 5.18. $x - y$ diagrams at constant pressure and at constant temperature of methylacetate + methanol (after Hirata et al., 1976). (a) Vapor-liquid equilibria at constant pressures. (b) Vapor-liquid equilibria at constant temperatures.

solvents, usually chemically related, over temperature ranges that may extend into the freezing range. It is illuminating to study this compilation, but also frustrating in the absence of any correlating or predictive theory. The most comprehensive collection of liquid-liquid equilibria in graphical as well as tabular form is that of Sorensen & Arlt (DECHEMA Chem Data Series V/1, 2, 3, 1979–1980). NRTL and UNIQUAC parameters are given at each temperature but are not correlated with temperature; in some cases the variation is approximately linear. A bibliography of liquid-liquid equilibrium data has been prepared by Wisniak & Tamir (1980).

5.4.3. Liquid-Solid Melt Equilibria

The principal features of phase diagrams of melt equilibria are identified in the classification of Table 5.1. The behavior usually is more complex than that of liquid-liquid systems since it may be necessary to take account of

1. Liquid-solid equilibria.
2. Limited liquid-liquid miscibility.
3. Limited miscibility of solid solutions.
4. Polymorphism of the solid state.
5. Formation of intermolecular compounds more or less stable in the solid state.

The discussion will be arranged according to the behaviors of the solid phases. Figure 5.21 illustrates the main types of binary diagrams.

5.4.3.1. Miscible Solid Phases

When the pure substances are completely miscible in both the liquid and solid phases, the diagrams resemble those of vapors and miscible liquids, that is, types I, II, and III of Figure 5.16. Examples of intermediate, minimum, and maximum melting solid phases are shown in Figure 5.22(a), (b), (c). Combinations of these simple types also occur. The iodine-bromine diagram, Figure 5.22(d), is made up of two type (a) diagrams, corresponding to equilibria of the 1:1 intermolecular compound with each of the elements. In Figure 5.22(e) the 1:1 compound of carbon tetrachloride and cyclohexanone forms a minimum melting diagram with each of the pure substances, while the 2:3 compound of Figure 5.22(f) forms a normal type of melting diagram with 2,2-dimethylbutane and a minimum melting one with 2,3-dimethylbutane. Nitrogen + carbon monoxide, in Figure 5.22(g), exhibit equilibria between two miscible solid forms as well as between a liquid and a solid.

5.4.3.2. Immiscible Solid Phases

The principal types of diagrams involving partially miscible or completely immiscible solid phases are shown in the various parts of Figure 5.23. The characteristic behavior of binary systems is that each of the substances lowers the melting point of the other. On a $T - x$ diagram the two melting or freezing curves meet at a minimum point called the eutectic ("easy melting"). Some of the characteristics of such systems are shown on Figure 5.23(a). A specific example is that of pyridine + formamide shown on Figure 5.23(b). Perhaps a majority of common organic substances crystallize out of melts as pure substances, but in metallic systems partial miscibility is quite common. In appearance, a eutectic is an intimate mixture of fine crystals, whereas the crystals that are formed above the eutectic temperature are coarse. Binary eutectics are a two-phase system in spite of their being mixed intimately.

Intermolecular compounds that are stable in the solid state are of fairly common occurrence. In Figure 5.23(c) the 1:1 compound of *p*-toluidine and phenol is seen to form simple eutectics with each of the pure substances. The solidification behavior of the partially miscible liquids phenol + water includes two eutectics, Figure 5.23(d).

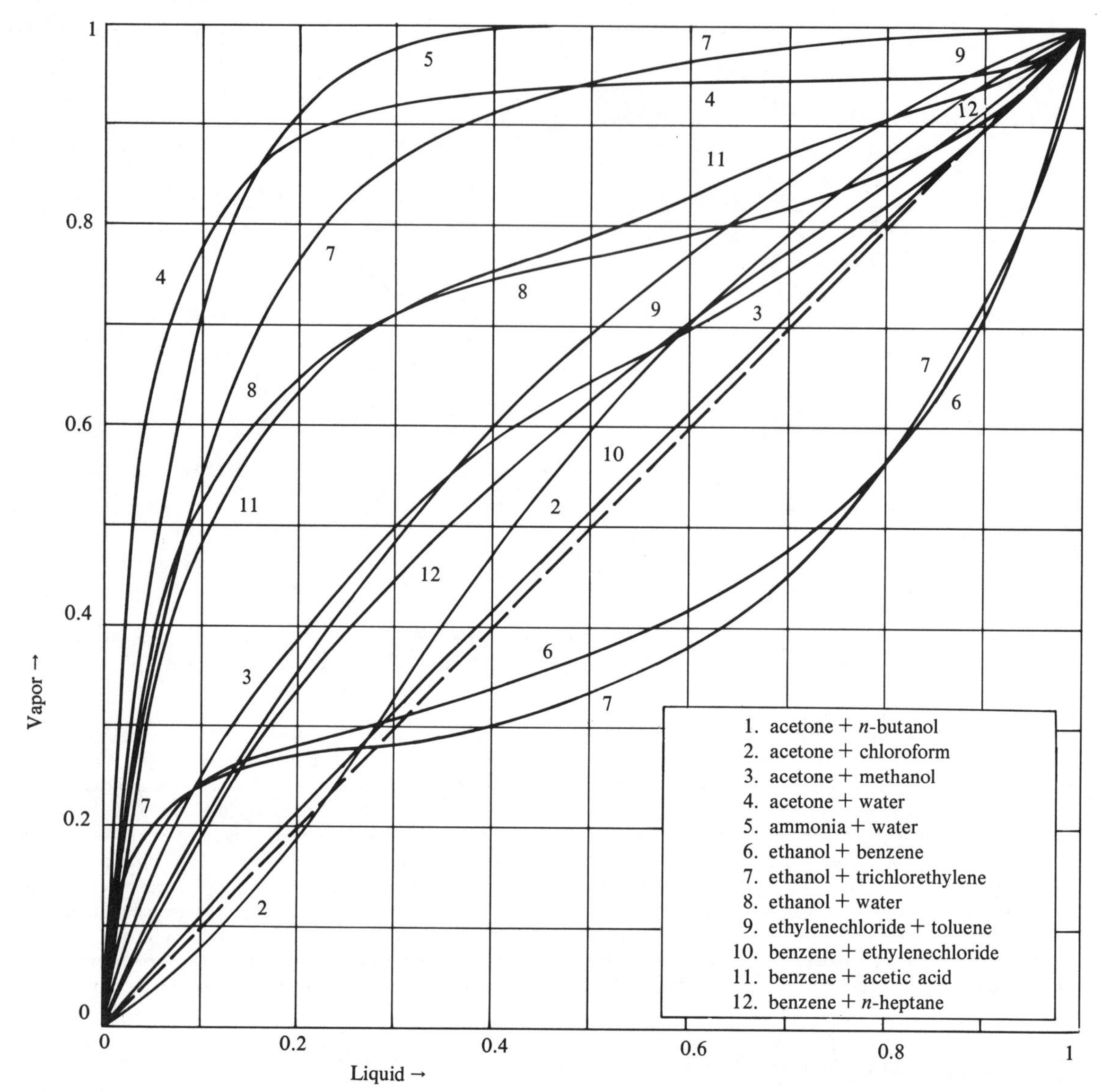

Figure 5.19. Some vapor-liquid composition diagrams at essentially atmospheric pressure. This is one of four such diagrams in the original reference (Kirschbaum, "Destillier und Rektifiziertechnik," 1969). Compositions are in weight fractions of the first-named.

A less common kind of situation is depicted in Figure 5.23(f), which involves substances that each have two distinct liquid forms, an amorphous one and an optically isotropic one called a liquid crystal. Some details about such substances are given in Chapter 9. In this example, equilibrium between the liquid phases is represented by the common lens-shaped diagram, whereas the freezing behavior of the liquid crystals is simple eutectic.

Some melt diagrams show kinks in the arms, as at point D in Figure 5.24(a). These kinks correspond to changes that occur at constant temperature and may be either chemical transformations or changes in the lattice structure. The dashed curve DCE is a hypothetical freezing line in the presence of a compound that decomposes below its expected or metastable melting point; the hidden maximum, point C, is that metastable melting point. Systems with a break in the melting curve known to be due to chemical transformation in the solid phase are called *meritectic* (part melting; Bowden, 1938); but the general process covering both chemical and allotropic transformations in the solid phase is called *peritectic.* In equation form the process is,

$$\text{Liquid} + \text{Solid}\, A \rightleftharpoons \text{Solid}\, B,$$

where A and B may or may not be chemically identical.

A compound with a stable melting point is called *congruent melting* while one with an unstable melting point is called *incongruent melting*. Examples of meritectic behavior are systems of picric acid with benzene and anthracene, Figure 5.24(c), (d), but the chemically similar system with naphthalene is congruent, Figure (b). Peritectic systems are quite common, particularly in the field of alloys. The genesis of

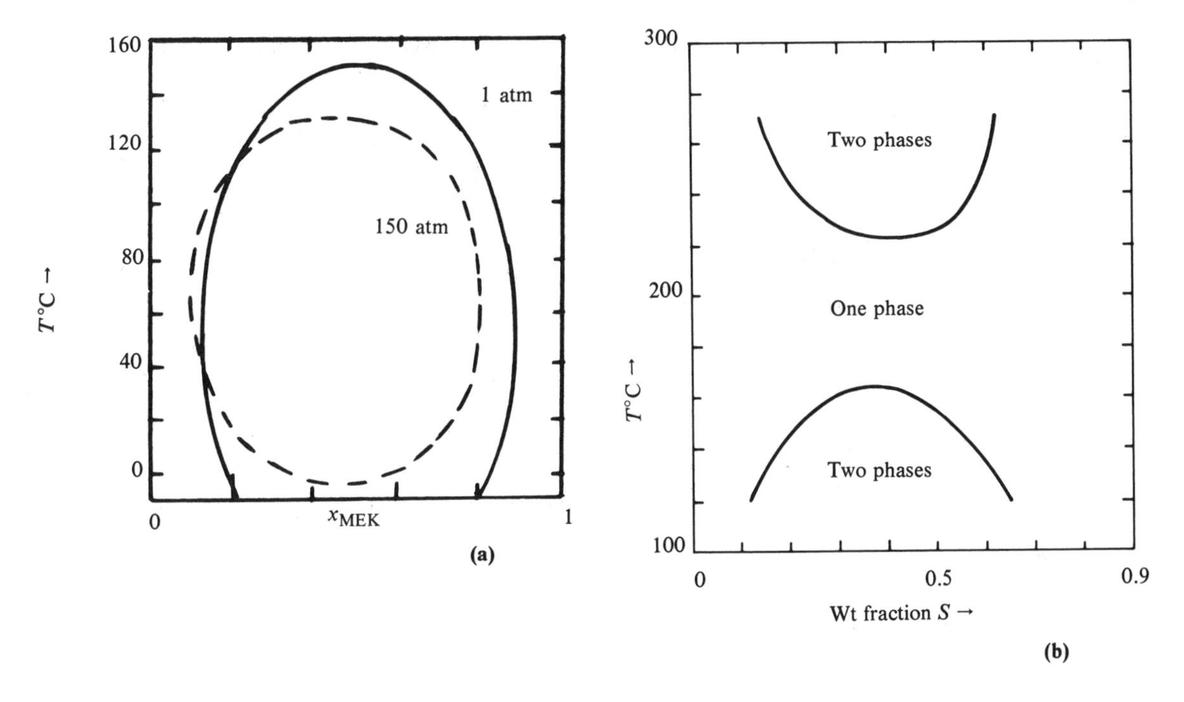

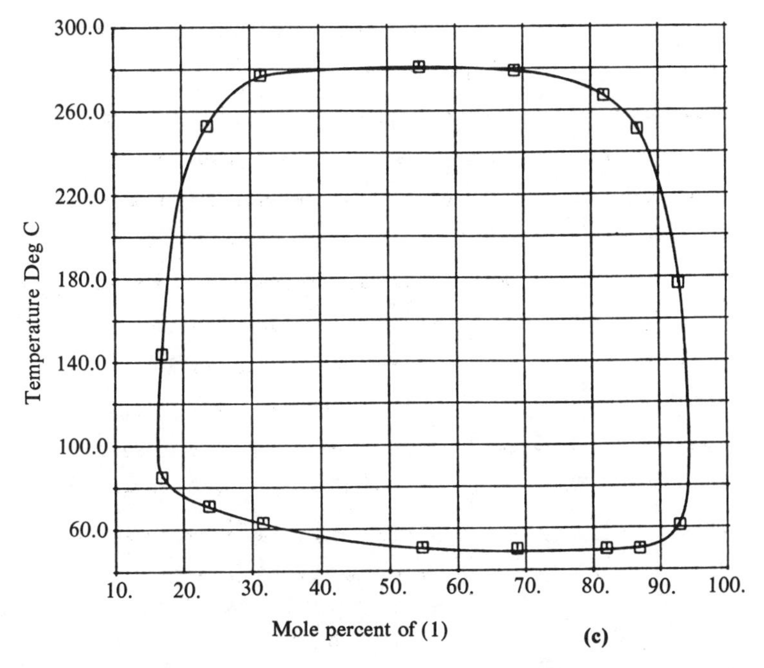

Figure 5.20. Examples of binary liquid-liquid equilibria. (a) Water + methylethylketone at 1 atm and at 150 atm. (b) The sulfur + benzene system, a rare type in which a lower critical-solution temperature is greater than an upper one. (c) The system glycerol + benzyl ethyl amine, with both upper and lower critical-solution temperatures (DECHEMA LLE Data Collection, V/1, 1979). (d) The system 3-ethyl, 4-methyl pyridine + water. Though there is a tendency to forming an upper critical-solution temperature, vaporization interferes with its attainment (DECHEMA LLE Data Collection, V/1, 1979). (e) The system 2-butanone + water. Though there is an approach to a lower critical solution temperature, solidification sets in before it can occur (DECHEMA LLE Data Collection, V/1, 1979).

Figure 5.20 (*continued on next page*)

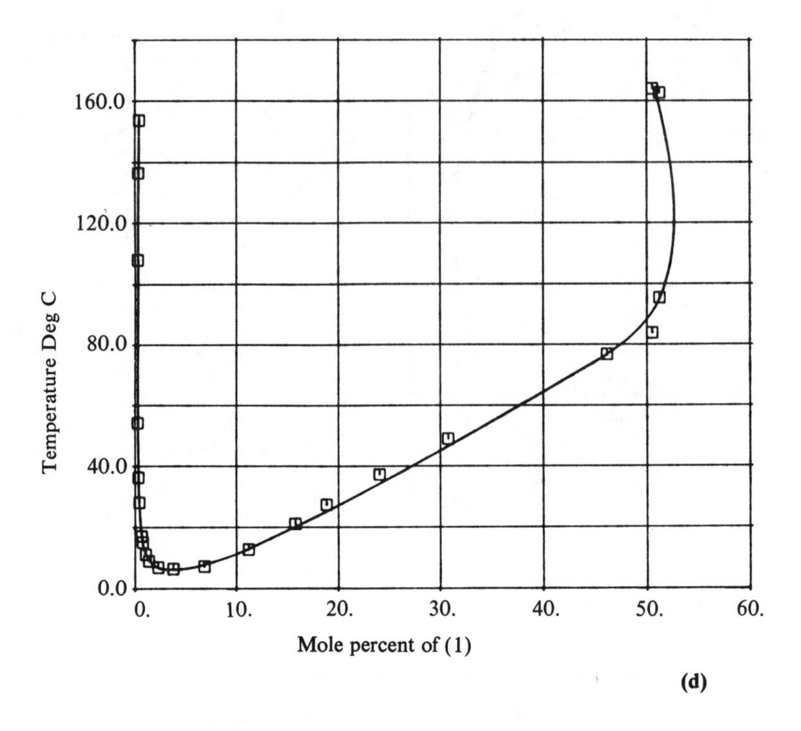

(d)

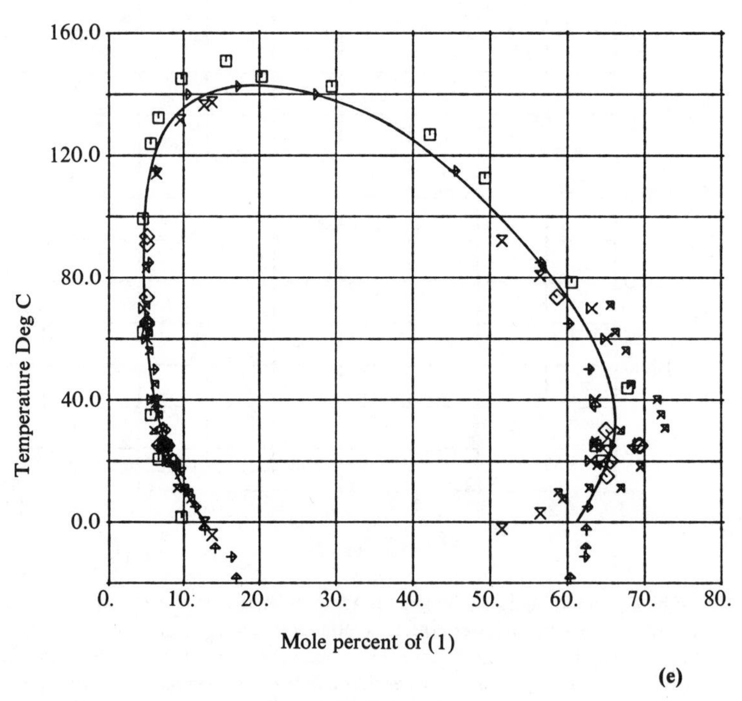

(e)

Figure 5.20 (*concluded*)

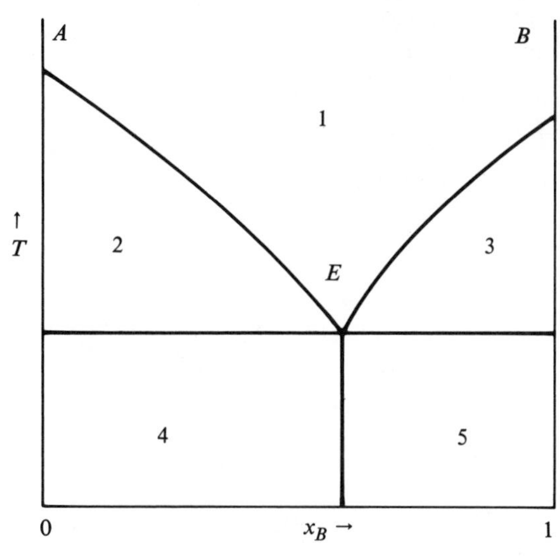

a. Simple eutectic system
1. homogeneous liquid
2. liquid + A-crystals
3. liquid + B-crystals
4. A-crystals + eutectic mixture
5. eutectic mixture + B-crystals

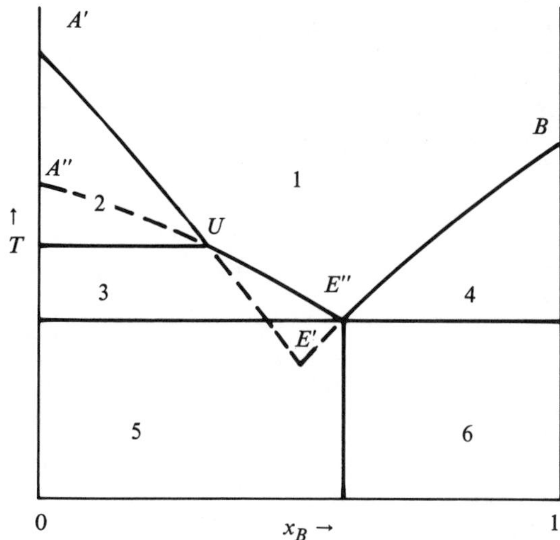

b. A has two crystalline forms.
The melt line $A''UE'E''$ is unstable.
1. homogeneous liquid
2. liquid + A'-crystals
3. liquid + A''-crystals
4. liquid + B-crystals
5. A''-crystals + eutectic mixture
6. eutectic mixture + B-crystals

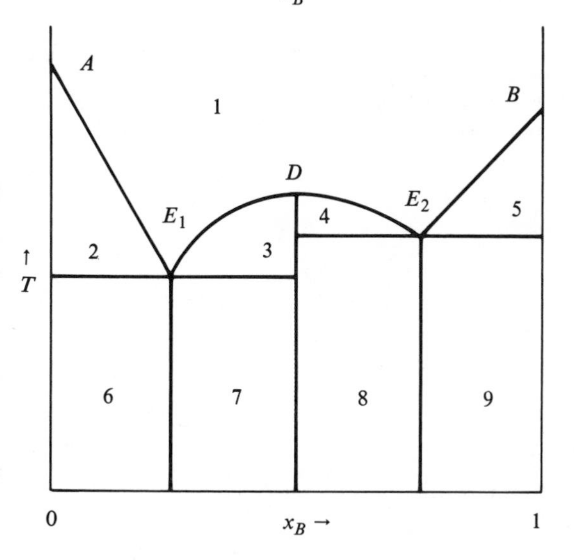

c. Molecular compound A_aB_b identified by point D.
1. homogeneous liquid
2. liquid + A crystals
3. liquid + A_aB_b crystals
4. liquid + A_aB_b crystals
5. B crystals + liquid
6. A crystals + eutectic mixture
7. eutectic mixture + D crystals
8. eutectic E_2 + D crystals
9. eutectic E_2 + B crystals

Figure 5.21. Types of binary solid-liquid equilibrium diagrams, with identification of their regions. A and B represent the components, E the eutectic, D the compound A_aB_b, P the peritectic, S a solubility limit, and K a critical-solution point. (after Mauser, in Ullmann Enzyklopädie der technischen Chemie, II/1, 638–664 1972).

Figure 5.21 *(continued on next page)*

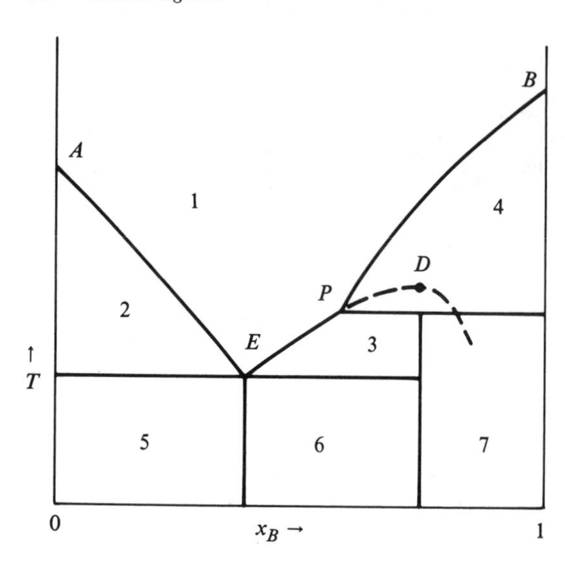

d. Molecular compound A_aB_b decomposes below its melting point (incongruent melting, point D).
1. homogeneous liquid
2. liquid + A crystals
3. liquid + A_aB_b crystals
4. liquid + B crystals
5. A crystals + eutectic mixture
6. A_aB_b crystals + eutectic mixture
7. A_aB_b crystals + B crystals

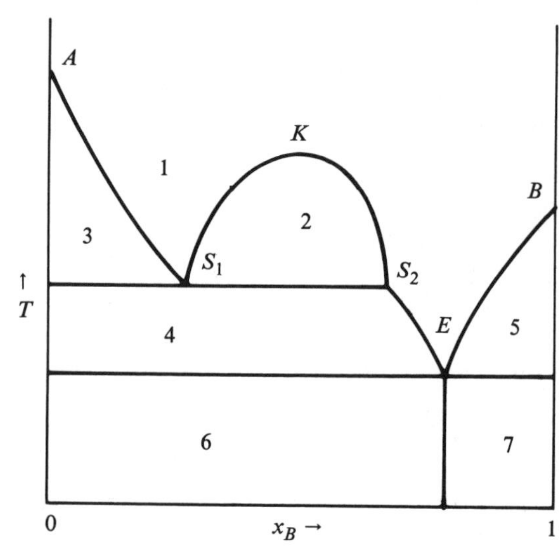

e. Limited miscibility of the liquid phase.
1. homogeneous liquid
2. mixture of two liquid phases
3. A crystals + liquid
4. A crystals + A-poor liquid
5. B crystals + liquid
6. A crystals + eutectic mixture
7. eutectic mixture + B crystals

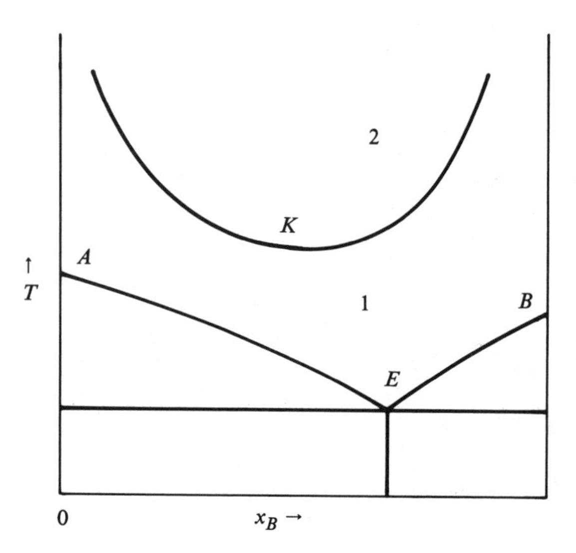

f. Region of limited liquid miscibility separate from the melt lines. K is at the lower CST.
1. homogeneous liquid
2. two liquid phases

Figure 5.21 *(continued on next page)*

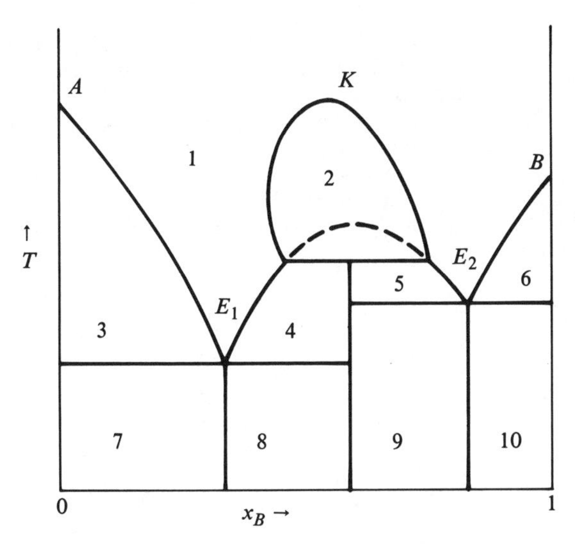

g. A two-liquid phase region superimposed on the melt lines of a molecular compound.
1. homogeneous liquid
2. two liquid phases
3. A crystals + liquid
4. A_aB_b crystals + liquid 1
5. A_aB_b crystals + liquid 2
6. B crystals + liquid
7. A crystals + eutectic mixture
8. A_aB_b crystals + eutectic mix 1
9. A_aB_b crystals + eutectic mix 2
10. B crystals + eutectic mix 2

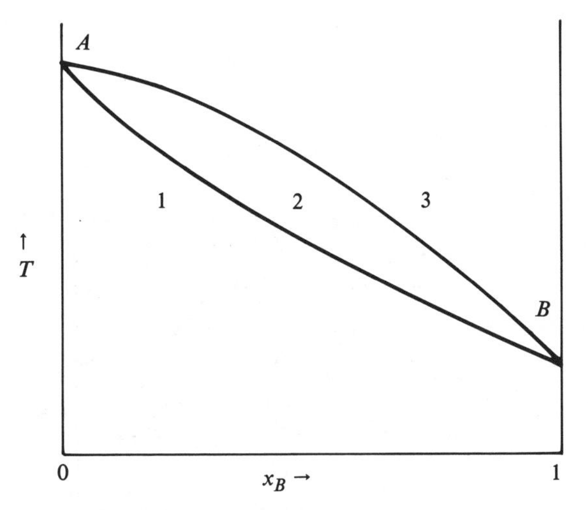

h. Complete miscibility of liquid and of solid phases.
1. homogeneous solid solution
2. liquid + mixed crystals
3. homogeneous liquid

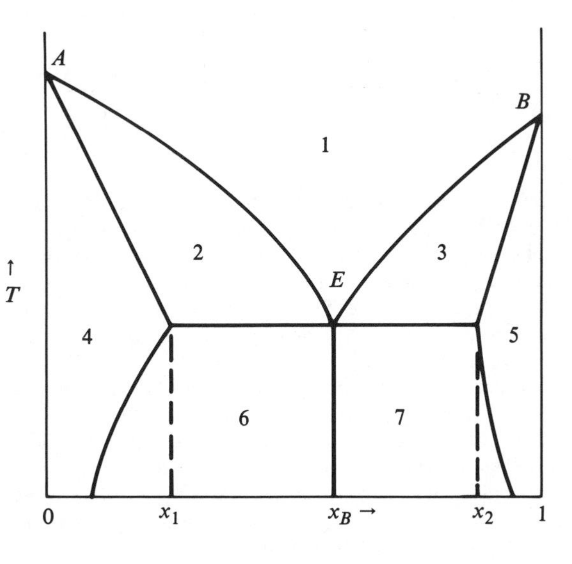

i. Limited miscibilities of solid phases. Eutectic E is in equilibrium with solid solutions of compositions x_1 and x_2.
1. homogeneous liquid
2. liquid + A-rich crystals
3. liquid + B-rich crystals
4. A-rich solid solution
5. B-rich solid solution
6. A-rich solid solution + eutectic
7. B-rich solid solution + eutectic

Figure 5.21 *(continued on next page)*

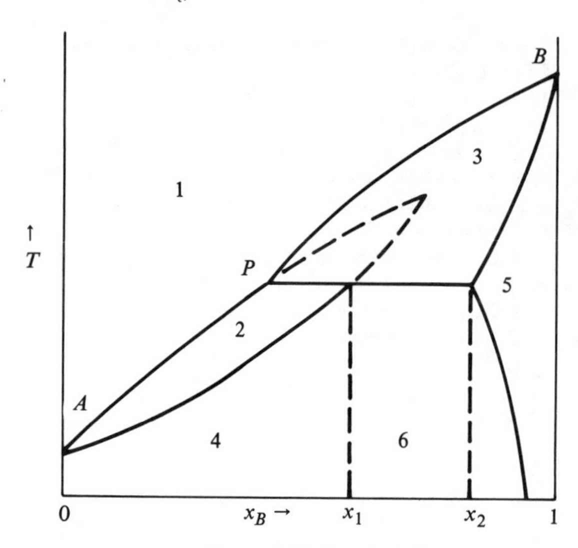

Figure 5.21 *(concluded)*

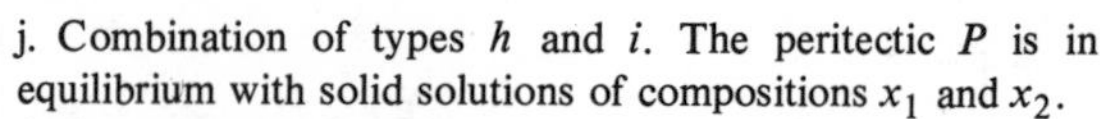

j. Combination of types h and i. The peritectic P is in equilibrium with solid solutions of compositions x_1 and x_2.
1. homogeneous liquid
2. liquid + A-rich crystals
3. liquid + B-rich crystals
4. A-rich solid solution
5. B-rich solid solution
6. A-rich solid solution + B-rich solid solution.

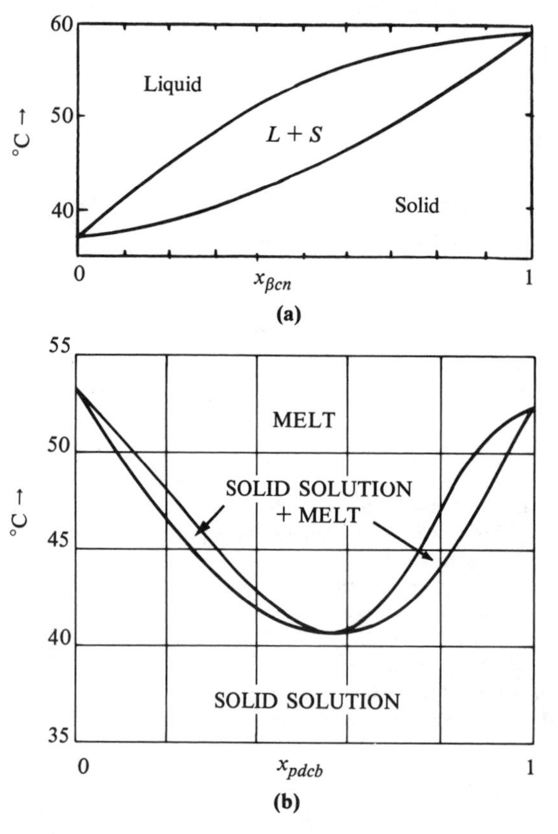

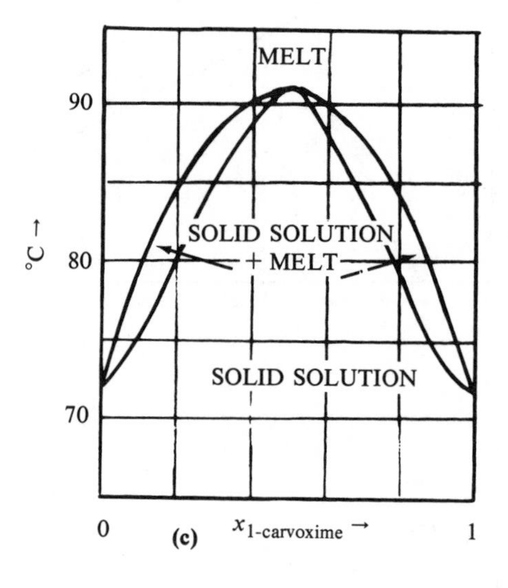

Figure 5.22. Real systems that are completely miscible in the liquid and solid phases. (a) The system β-methylnaphthalene + β-chloronaphthalene [Data of Grimm, Gunther, & Titus, Z physik Chem (B) 14, 169 (1931)]. (b) The system p-dichlorobenzene + p-chloroiodobenzene has a minimum melting point. (c) The system d-carvoxime + 1-carvoxime has a maximum melting point. [Adriani, Z physik Chem 33, 469 (1900)]. (d) The system bromine + iodine forms a compound, IBr, and has two lozenge-shaped two-phase regions. [Meerum-Tervogt, Z anorg Chem 47, 209 (1905)]. (e) The system carbon tetrachloride + cyclohexanone forms an intermolecular compound melting at −39.6 C. The compound forms minimum melting solutions with each of the pure components. [Pariaud, Bull Soc. Chim France (5) 17, 1239 (1950)]. (f) The system 2,2-dimethylbutane + 2,3-dimethylbutane forms an intermolecular compound that melts at −102.6 C. It forms a minimum melting mixture with the second substance. [Data of Fink, Ciner, Frey, & Aston, JACS 69, 1501 (1947)]. Copyright American Chemical Society, used with permission. (g) The system nitrogen + carbon monoxide, showing equilibria between liquid and solid phases and between two solid phases. The solid lines are curves for ideal solutions. (Hildebrand & Scott, "Solubility of Non-Electrolytes," 1950).

Figure 5.22 *(continued on next page)*

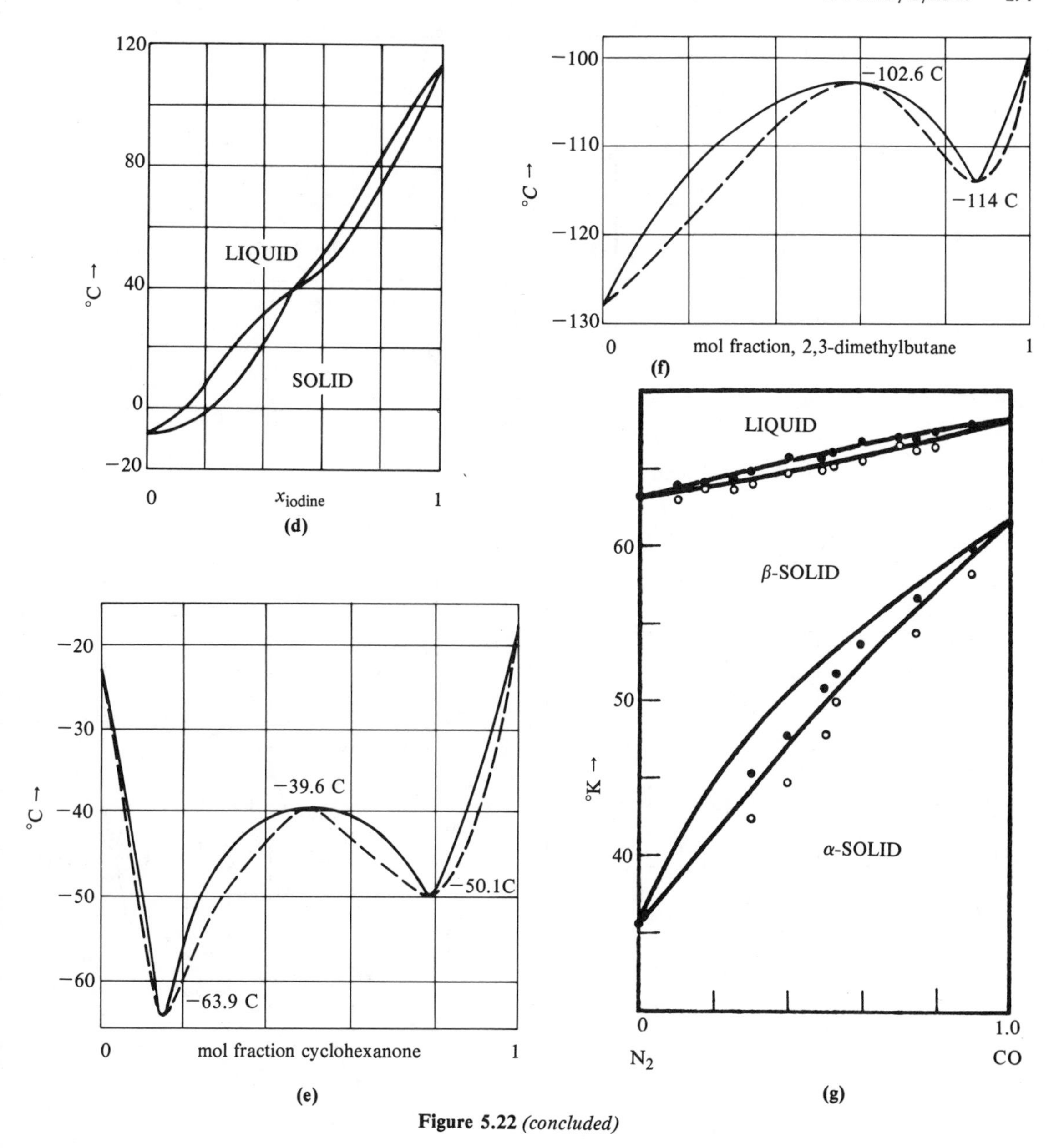

Figure 5.22 *(concluded)*

some such systems is shown schematically in Figures 5.25 (b) and (c).

5.4.3.3. Partially Miscible Solid Phases

Examples of solid-solution behaviors similar to those of the liquid-liquid systems represented in Figure 5.21 are plentiful. When phase diagrams attempt to record all liquid and solid interactions they may become fairly complex. The liquid-solid and the solid-solid zones may remain distinct or they may merge partially, as indicated on the schematics of Figures 5.25(a), (b), and (c). When an isopleth crosses the line AV in Figures (b) and (c), a peritectic process occurs:

Solid solution $\alpha \rightleftarrows$ Solid solution $\beta +$ Liquid.

Solid solutions also may undergo transformations, simply because of temperature changes or because the crystalline forms may change as the temperature is lowered. In Figure 5.25(d), both cerium and lanthanum are polymorphic; when the temperature is lowered further after solidification has occurred, equilibrium develops between the two sets of crystalline forms, in the form of a lens-shaped diagram. Transformation of a solid solution may be of the eutectic type, as in Figure 5.25(e). Such a eutectic is known in the iron-carbon system at 721 °C and 0.9 wt% C.

5.4.3.4. Compound Diagrams

When several compounds can form and when transformations continue in the solid states, very complicated diagrams can result but may be regarded as made up of several simple diagrams of individual substances and their compounds taken just two at a time. In Figure 5.23(h) the diagram of water and sulfuric acid is made up of four simple eutectics with

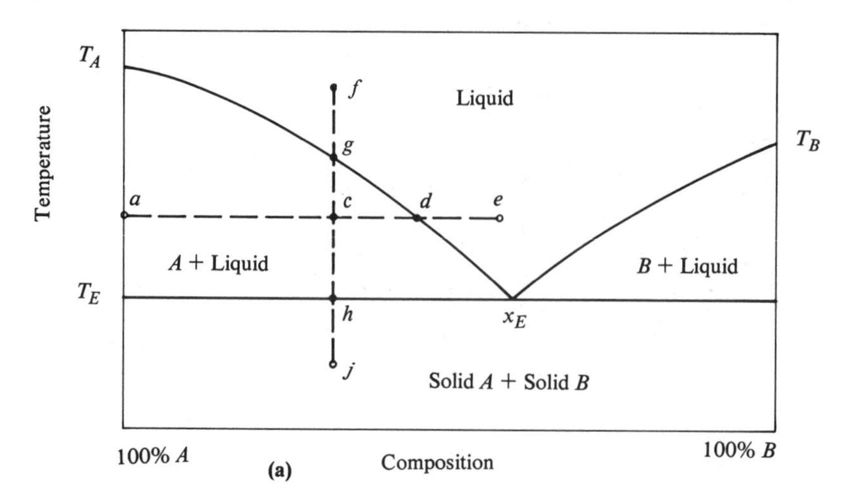

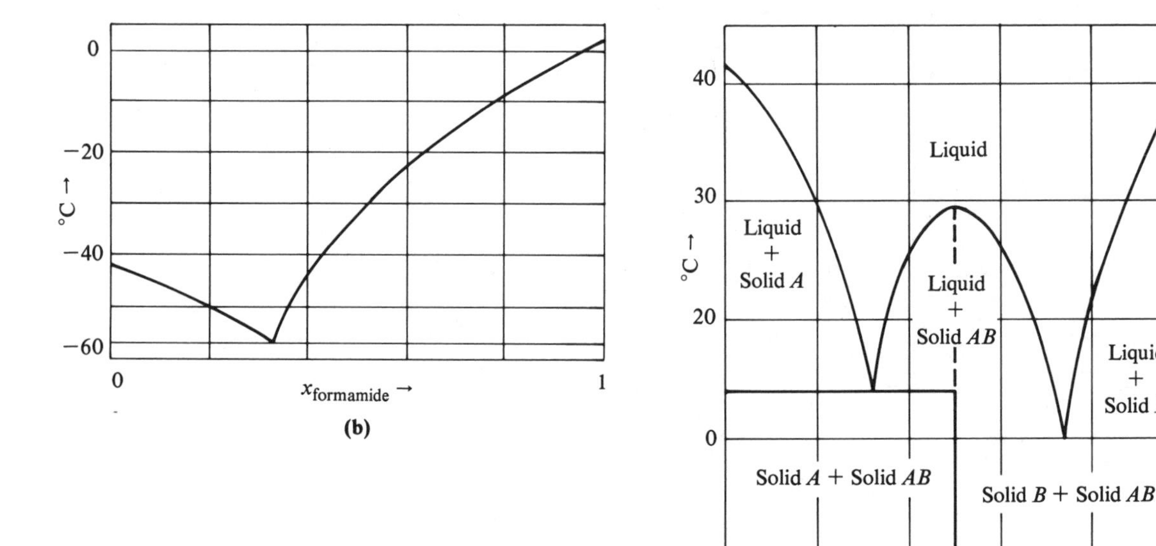

Figure 5.23. Liquid-solid equilibria involving immiscible solid phases. (a) Schematic of a system that forms a simple eutectic. Along an isopleth, *fj*, the mixture begins to solidify at point *g* and is completely solid at point *h*. At point *c* the amounts of liquid phase and solid *A* are in the ratio of the line segments, *ac*/*cd*; the composition of the liquid corresponds to point *d*. The melting temperatures of the substances are labeled T_A and T_B; that of the eutectic is labeled T_E. (b) The system pyridine + formamide has a simple eutectic at 32.3 mol% formamide and −56.7 C. [Stephanou, Vanderwerf, & Sisler, JACS 70, 264 (1948)]. Copyright American Chemical Society; used by permission. (c) The system *p*-toluidine (*A*) + phenol (*B*) forms a 1:1 intermolecular compound that forms simple eutectics with each of the components. (d) The system phenol + water forms partially miscible liquid phases and immiscible ice and solid phenol. (e) The system triphenylmethane + *m*-phenylenediamine forms immiscible liquid phases and forms a 1:1 intermolecular compound that decomposes below its melting point. (f). Equilibria in the system of *p,p'*-azoxyphenetole + *p*-methoxycinnamic acid, showing equilibria between liquid crystals and amorphous liquid and the freezing behavior of the liquid crystals [Prins, Z physik Chem 67; 718 (1909)]. (g) Potassium nitrate and thallium nitrate form a eutectic made up of two solid solutions. Point *K* is a hypothetical critical-solution temperature of the solid solutions S_1 and S_2. (Bowden, "The Phase Rule and Phase Reactions," 1938). (h) The system of water (*W*) and sulfuric acid (*S*) involves intermolecular compounds, the monohydrate (*M*), the dihydrate (*D*), and the trihydrate (*T*). The melting points of these compounds are +7 °C, −40 °C, and −25 °C.

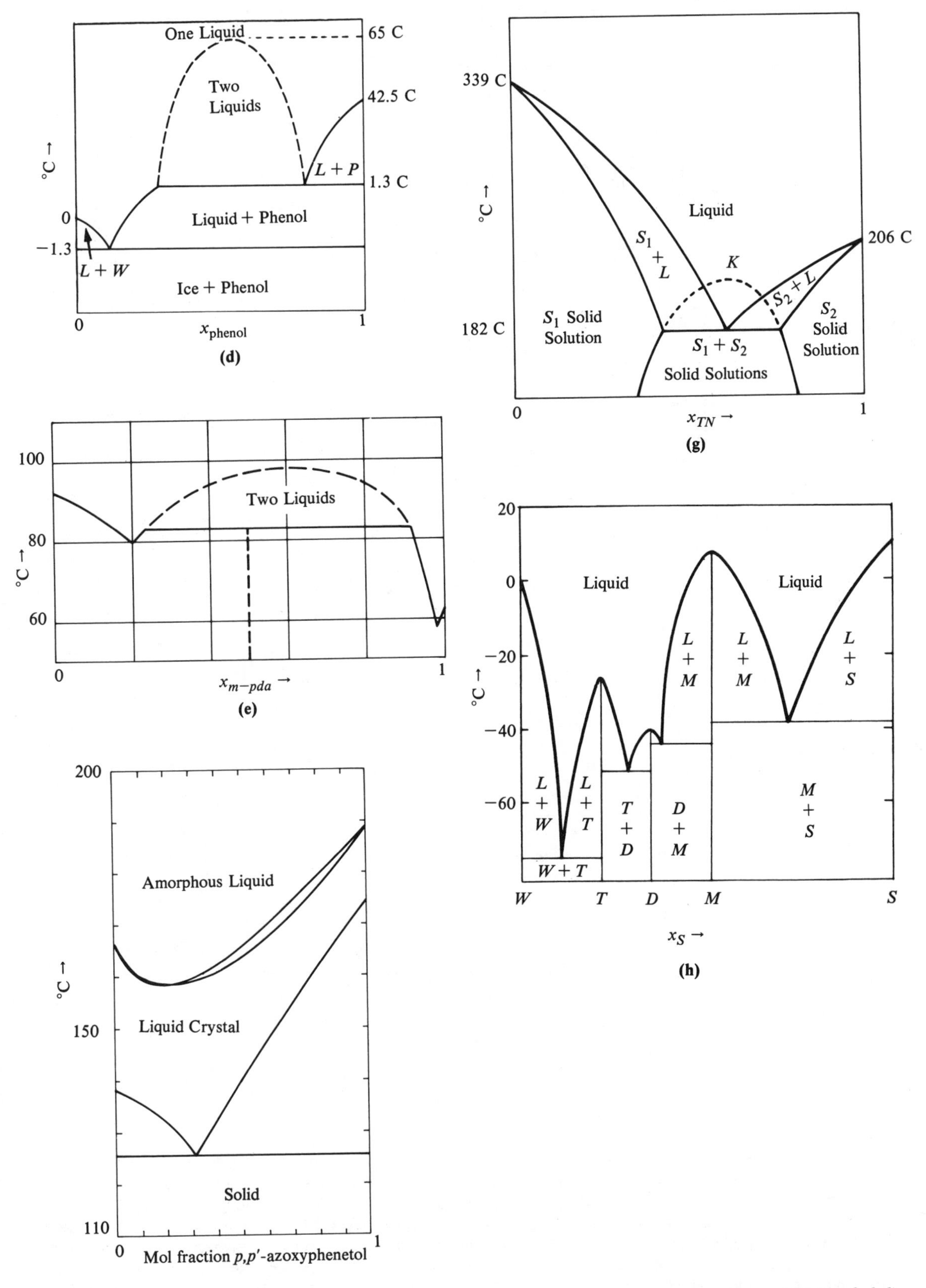

Figure 5.23 *(concluded)*

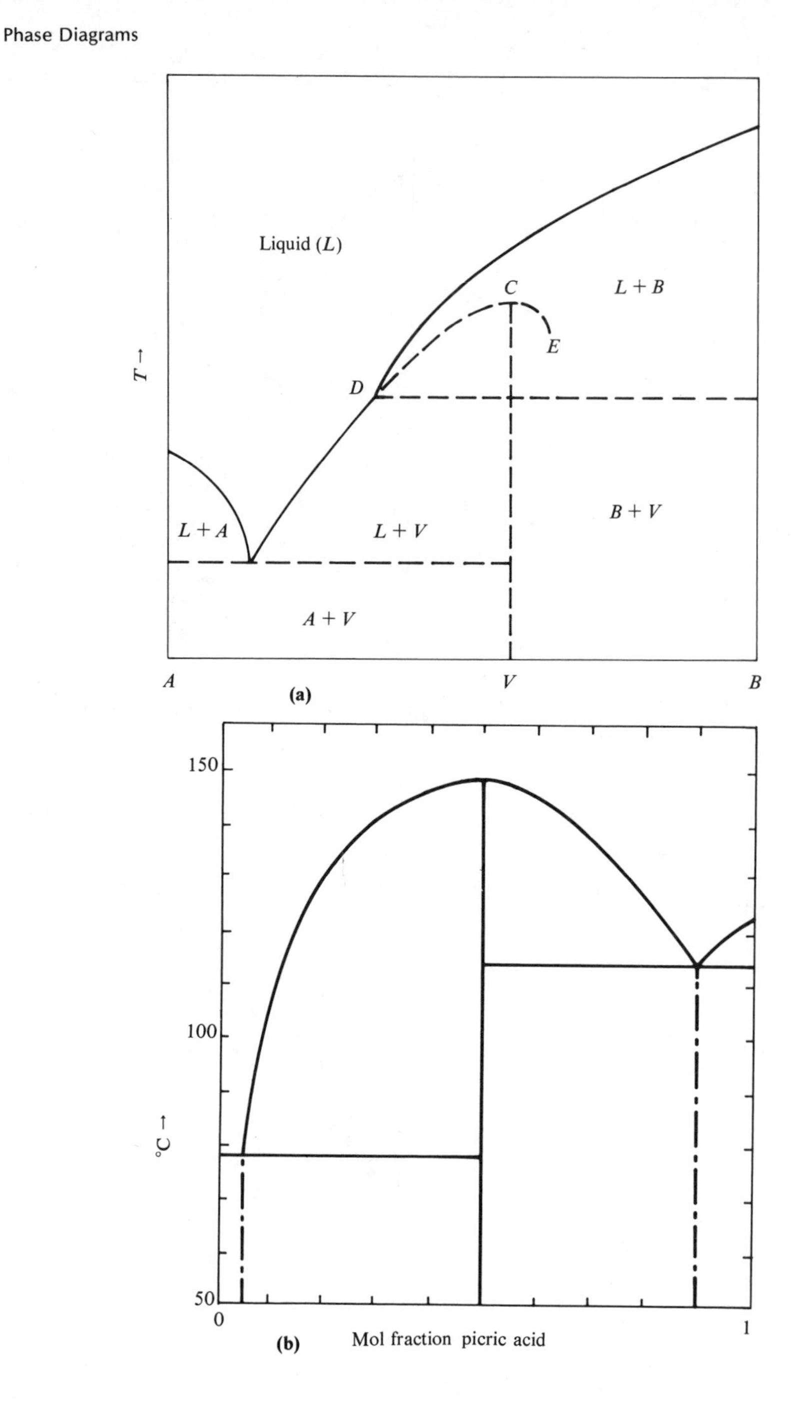

Figure 5.24. Intermolecular compounds in systems with picric acid (Mindovich, 1956). (a) Schematic diagram of a system which forms an intermolecular compound, *V*, that decomposes (at point *D*) before it can reach its melting point. Point *C* identifies its unstable melting point, and line *DCE* is a hypothetical freezing curve with *V* as the solid phase. (b) Picric acid and naphthalene form a 1:1 intermolecular compound that is stable at its melting point of 148 C and forms two eutectics, containing about 5% and 90% picric acid. (c) Picric acid and benzene form an intermolecular compound that decomposes slightly below its stable melting point. The broken line *VA* represents an unstable continuation of the freezing curve with the compound. (d) Picric acid and anthracene form an intermolecular compound that decomposes slightly below its expected melting point. The left branch of the melting curve of part (b) is almost horizontal.

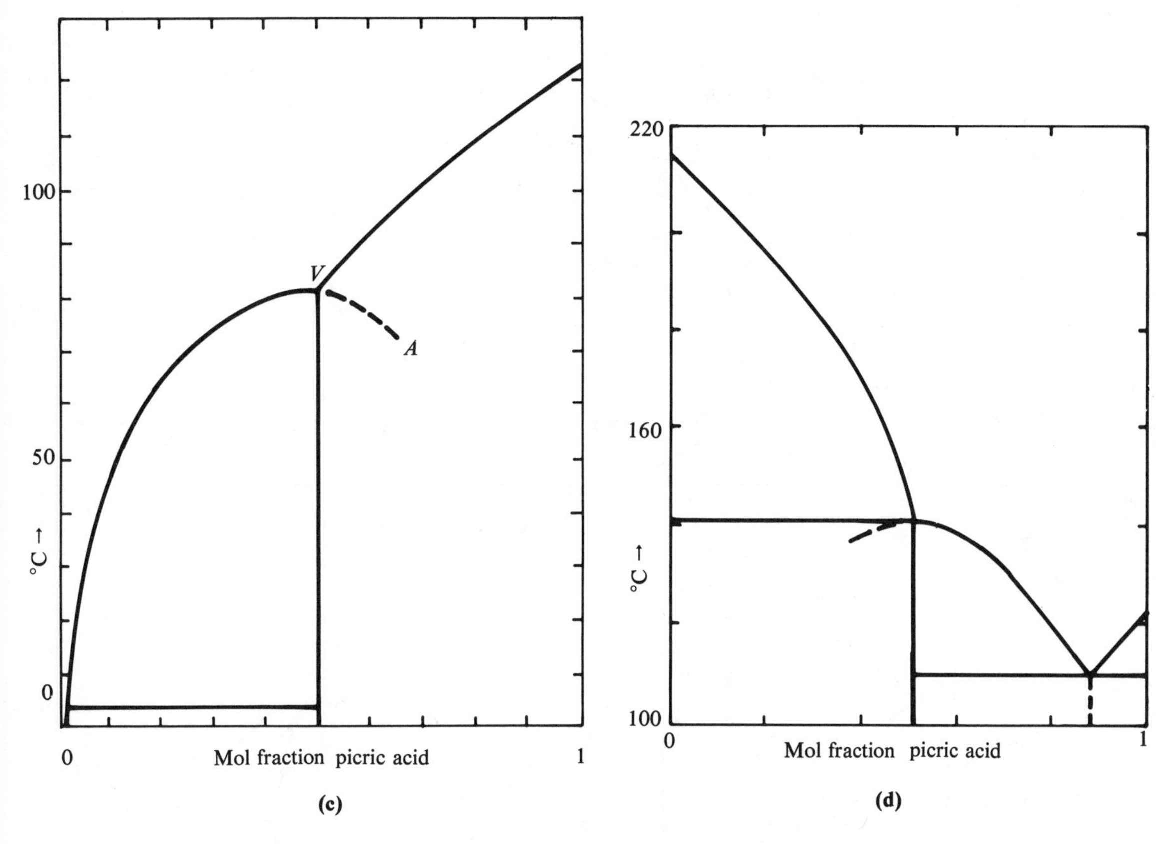

Figure 5.24 *(concluded)*

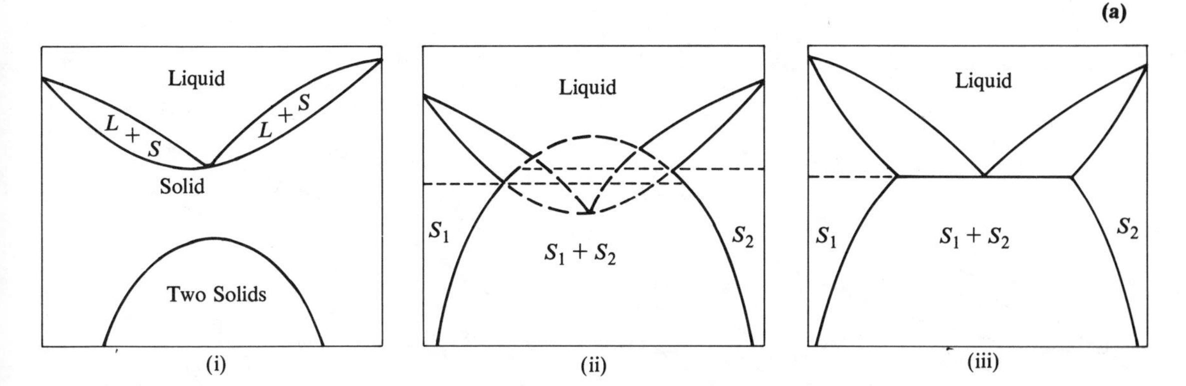

Figure 5.25. Systems with limited miscibilities of the solid phases and with transformations in the solid phases. Ordinates are temperatures and abscissas are compositions. (a) Genesis of a eutectic system of solid solutions, visualized to result from expansion and intersection of simpler two-phase regions. To a limited extent such transformations can be effected by changes of pressure, but usually differences in diagrams are associated with differences in chemical natures of the components. (b) Genesis of a peritectic (unstable melting point) type of system, visualized to result from expansion and intersection of simpler two-phase regions. The peritectic point is V. For the significance of line VA, see Section 5.4.3.3. (c) Genesis of a peritectic system with a minimum melting point. For the significance of line AV, see Section 5.4.3.3. (d) Cerium and lanthanum exhibit polymorphism of the solid state. (e) A type of system in which the solid solution has limited miscibility with solid forms of the pure components and forms a solid-phase eutectic with them. The iron-carbon system forms such a eutectic at 721 C and 0.9 wt% C. (f) Azoxybenzene and azobenzene form solid solutions (Hildebrand & Scott, "Solubility of Non-Electrolytes," 1964) (g) *p*-chloroiodobenzene and *p*-diodobenzene form solid solutions and have a peritectic point (Hildebrand & Scott, 1964). (h) Equilibrium of mixtures of cyclohexane and methylcyclopentane. Cyclohexane has two stable crystalline forms in this temperature range. [Dale, "Encyclopedia of Chemical Processing and Design" 13, 471 (1981)].

Figure 5.25 *(continued on next page)*

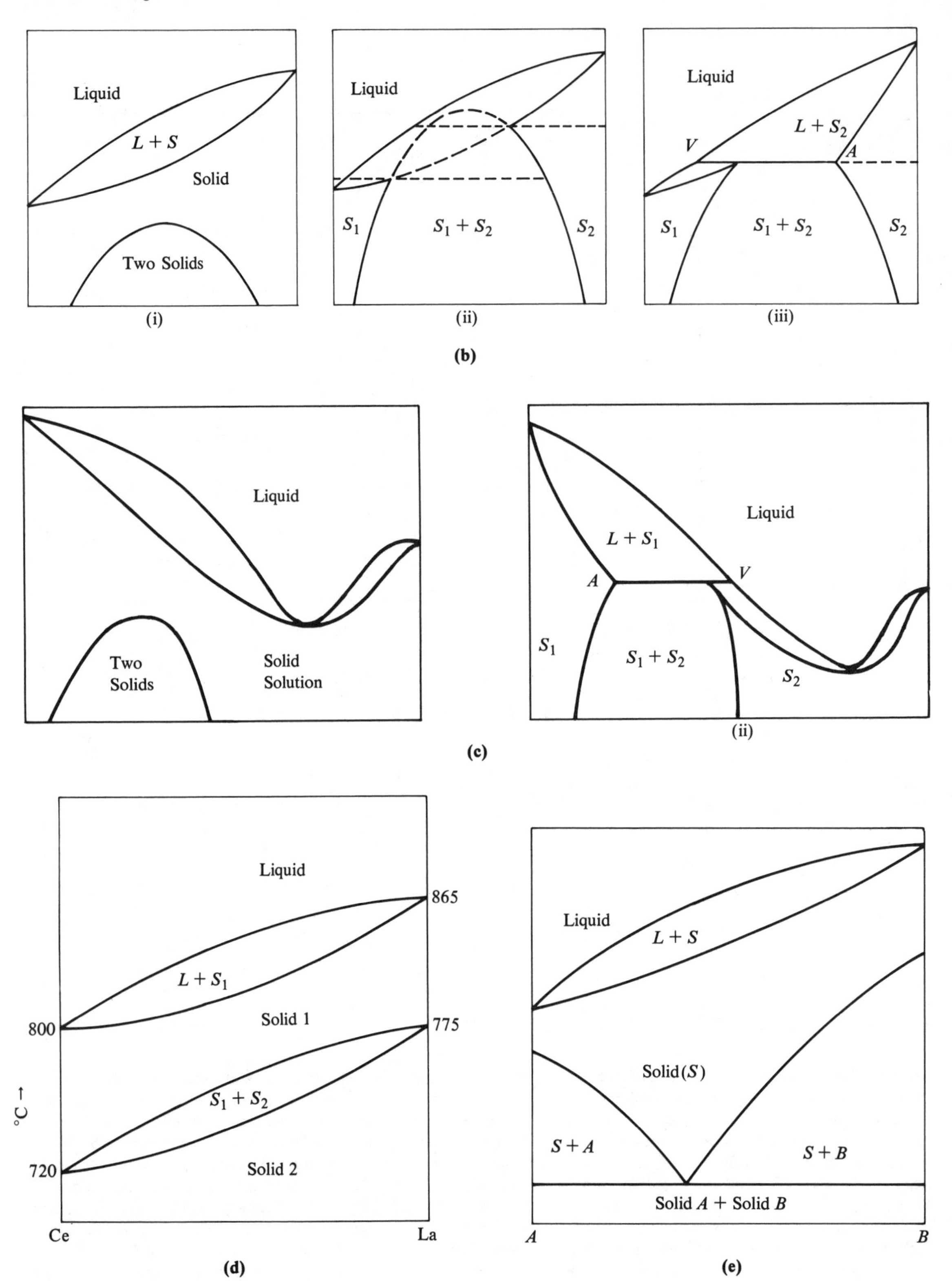

Figure 5.25 *(continued)*

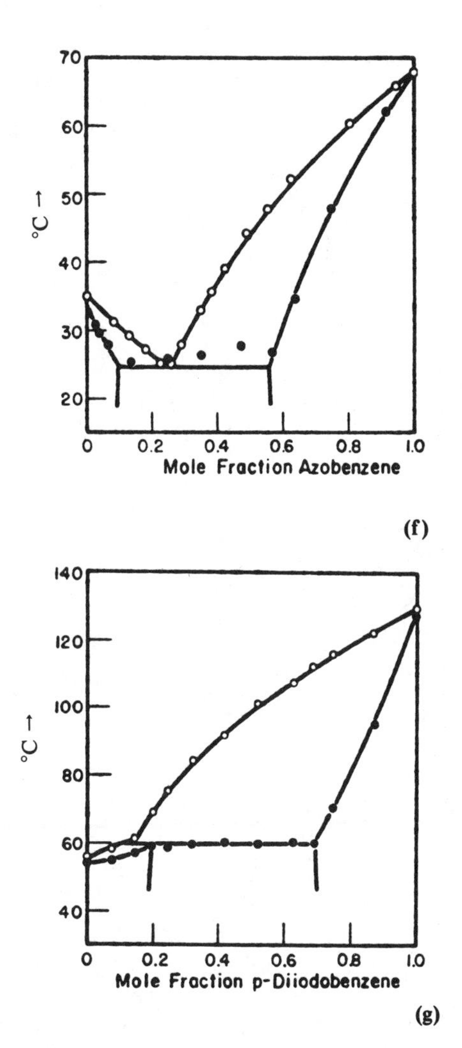

(f)

(g)

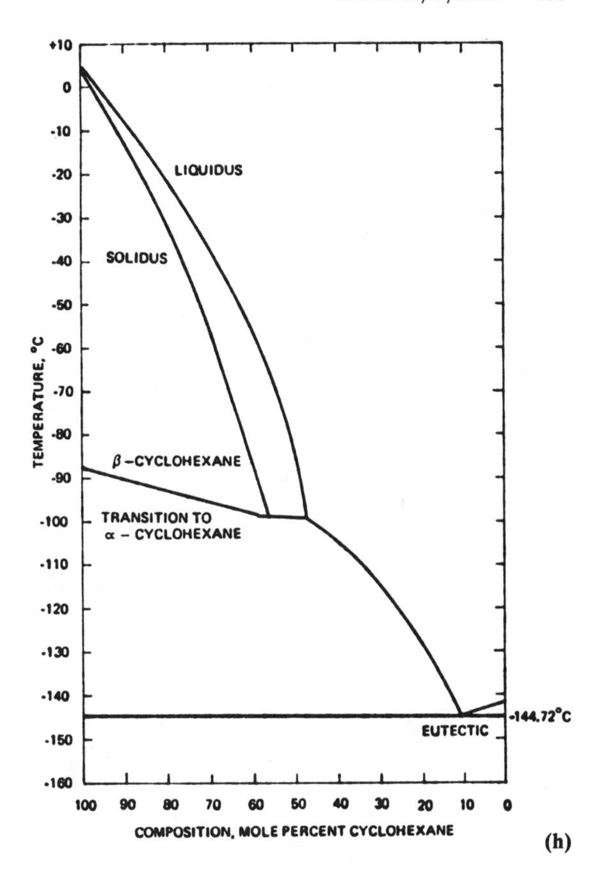

(h)

Figure 5.25 *(concluded)*

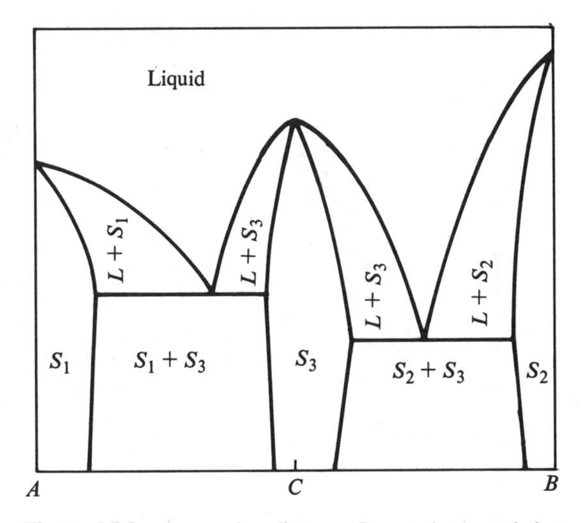

Figure 5.26. A complex diagram. It may be regarded as made of two simpler diagrams: of substance A and intermolecular compound C, and of intermolecular compound C with substance B. All substances have limited miscibilities in the solid phase. Several items of earlier figures also are similarly complex.

immiscible solid phases. Figure 5.26 is a schematic of compound formation and limited miscibilities of the solid phases.

5.4.4. Equilibria between Solid and Supercritical Fluid

Some gases are well known to be soluble in solids, for example hydrogen in metals. On the other hand, under certain conditions a gas phase may hold much higher concentrations of liquid and solid substances than would be expected from their normal vapor pressures. In some cases of scientific and technological interest, the pressures employed are above the critical point of the gas, so such equilibrium is between a condensed phase and a supercritical fluid. A P-x diagram of such a system is shown in Figure 5.27(a). The deviation from ideality indicated here often is quite marked in the supercritical region, generally leading to a greatly enhanced content of the normally condensed material in the fluid phase. In the schematic diagram, Figure 5.27(a), a fluid phase having the composition of point a deposits some solid phase and becomes a fluid phase at point b when the pressure is reduced. Real systems are shown in parts (b), (c), and (d) of this figure. Naphthalene has a minimum solubility in ethylene near the critical point of the solvent, which then increases sharply with pressure. Figure 5.27(c) shows similar behavior for another solid in ethylene and also shows an effect of temperature. Figure 5.27(d) compares real and ideal behavior of benzoic acid in carbon dioxide at elevated pressures and several temperatures. These effects of pressure are well predicted by

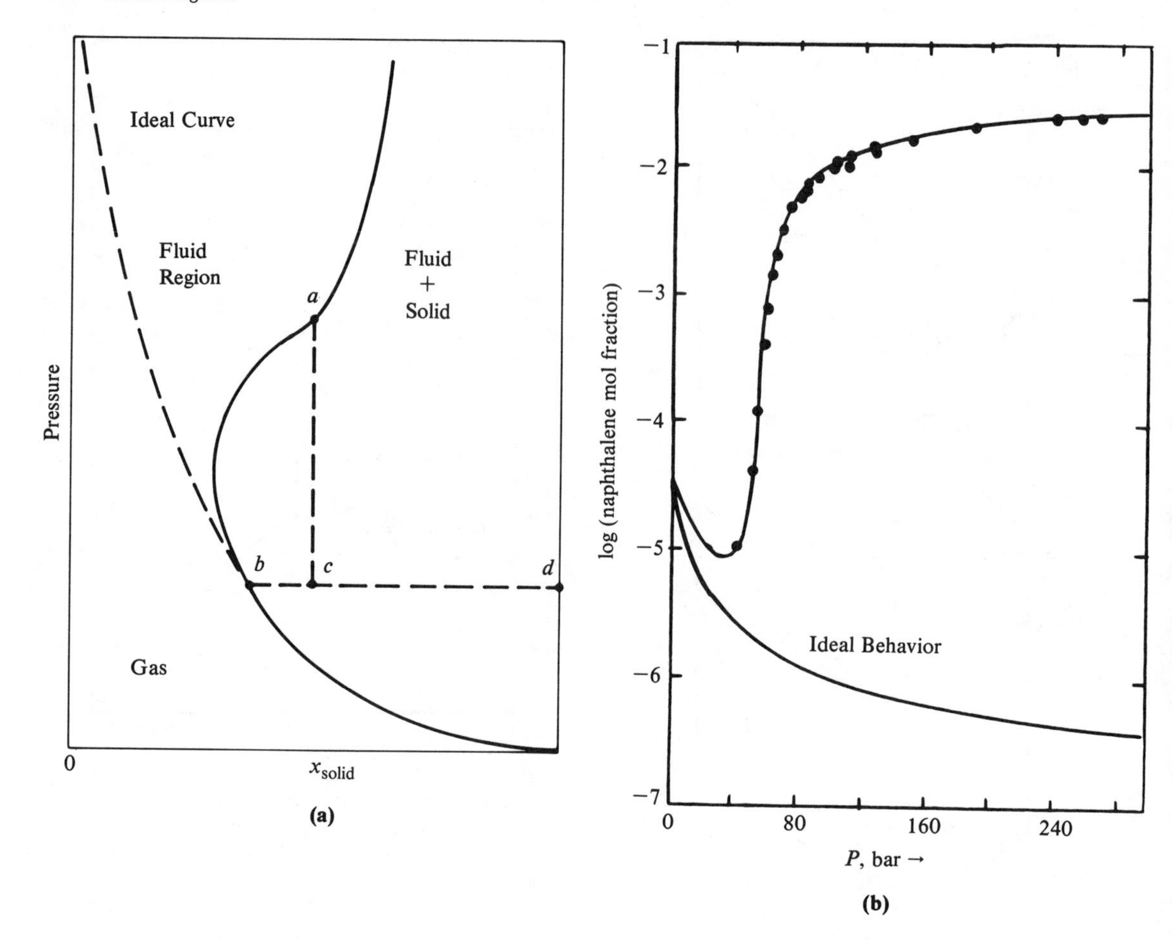

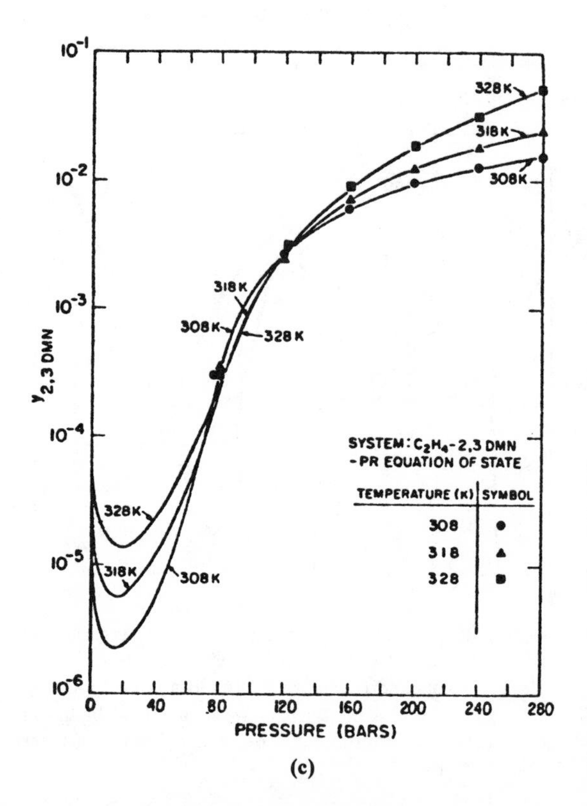

Figure 5.27. Binary solid-fluid equilibria at supercritical conditions of the fluid substance. Related data are in Figure 8.16 and in Chapter 9. (a) Equilibrium between a solid phase (S) and a fluid (F) above its critical point. A solution at point a decomposes into a solution of composition b and the pure solid d when the pressure is lowered to point c. (b) Solubility of naphthalene in ethylene at 25 C at high pressures. For example, at about 100 bar, naphthalene is present in the fluid phase to the extent of $x = 0.01$; the solubility is a minimum, $x = 10^{-5}$, at about 40 bar and is $4(10^{-5})$ at atmospheric pressure. [Data of Tsekanskaya et al., Zh Fiz Khim 38, 2166 (1964)]. (c) Solubility of 2,3-dimethylnaphthalene in ethylene at elevated pressures. The curves are predictions with the Peng-Robinson equation (Kurnik, Holla, & Reid, 1981). Copyright American Chemical Society; used with permission. (d) Solubility of benzoic acid in carbon dioxide at elevated pressures. The solid lines are predicted with the Peng-Robinson equation. (Kurnik, Holla, & Reid, 1981). Copyright American Chemical Society; used with permission.

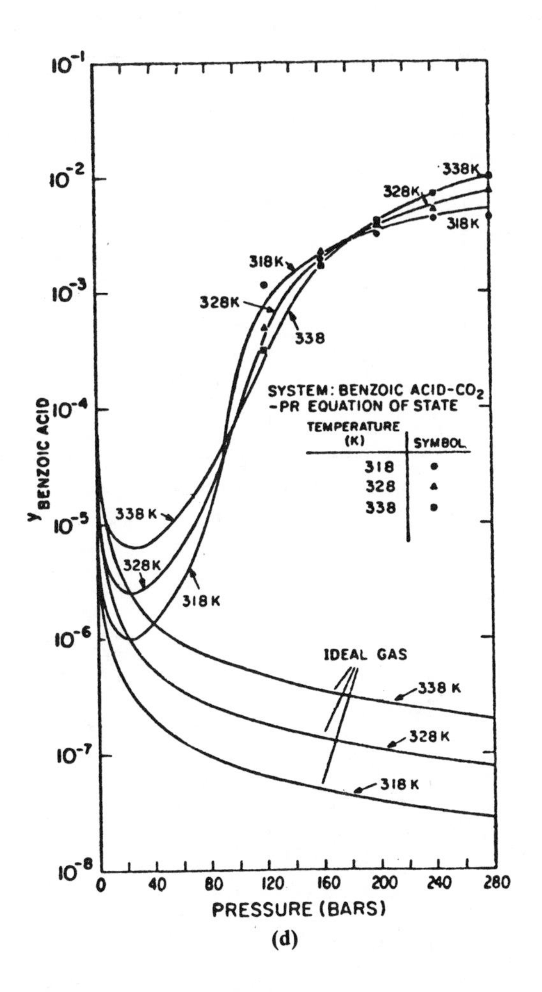

(d)

the Soave and Peng-Robinson equations. Some of the industrial applications of supercritical gas extraction are mentioned in Chapter 9. A comprehensive collection of papers on this subject has been published recently (Paulaitis et al., 1983).

5.4.5. Effect of Pressure on Melt Equilibria

The directional effects of pressure on the melting and solubilities of solids are predictable with the LeChatelier principle. Generally the effect is to raise the melting point and to increase the solubility. An example of the liquid phase is Figure 5.20(a), where the mutual solubilities are changed significantly by pressure. Figure 5.28 reports several examples of the effect of pressure on melt equilibria; in part (a) the two-phase region is not changed much in shape, although it is shifted upward. In parts (b) and (d) the eutectic temperatures are raised somewhat but there is little change in their compositions; finally in part (c) both temperature and composition of the eutectic are changed greatly. The pressures employed in these and the few other investigations of this type that have been made are somewhat out of the range of ordinary processing of condensed phases, but it is nevertheless interesting to have the information.

5.4.6. Equilibrium and Reality

A phase diagram represents conditions of equilibrium. Transformations in the solid phase at lower temperatures may be particularly slow and unless some way can be found to speed up attainment of equilibrium, the investigation may not be practicable. Figures 5.25(d) and (e) are two of the few cases that have been investigated of solid-phase equilibria. Temperatures should be made as high as possible and the necessary time allowed. Some solid-phase transformations can be accelerated by introducing a melting step. For instance, with a system like that of Figure 5.23(g), the solid solution containing 25% thallium nitrate could not be prepared by an impatient individual by mixing the ingredients at room temperature in the proper proportions. If the mixture were melted at 300 C, however, and then cooled slowly below 285 C, the solid solution would be obtained. Below eutectic temperatures, unfortunately, such tricks are not applicable. Also, since rates of solid transformations are predominantly diffusion-controlled, it is unlikely that they can be catalyzed by traces of foreign materials. Thus time and temperature are the only controls available.

5.5. TERNARY SYSTEMS

Planar diagrams of ternary systems at constant T and P are composition diagrams showing regions, lines, and points at which different phases exist. Compositions of phases in equilibrium may be connected with tie-lines (connodals), but for interpolation purposes and to reduce the clutter on a small diagram some kind of continuous tie-line correlation is preferred. On Figure 5.29(i), the dashed tie-line curve is the locus of intersections of horizontals through the right ends of the tie-lines and verticals through the left ends, but any consistent set of lines through the tie-line terminals may be used. A separate distribution diagram may be constructed as indicated in Figure 5.29(ii). Here x_{CA} is the fraction of component C in the A-rich (left-hand) phase and x_{CB} that in the B-rich (right-hand) phase. Distribution diagrams also are useful for representing three-phase compositions, as shown later. Some modes of plotting have been devised that give linear plots in some cases, which is especially convenient for interpolation or extrapolation. Examples are given in Chapter 7.

Complete diagrammatic representation of the effects of T, P, and composition in ternary systems is not always simple. For LV and LLV systems, algebraic representation in terms of activity and fugacity coefficients and vapor pressures may be preferred and is adequate when complications such as compound formation do not occur. With certain classes of compounds, oxygenated organics for instance, the effect of compound formation on phase equilibria can be taken into account because equilibrium constants are known.

Certain restrictions apply to the numbers of phases that may be present at equilibrium of ternary systems.

1. Two phases may exist over a range of compositions in equilibrium at fixed T and P.
2. Three phases have unique compositions at specified T and P, but a range of compositions is possible when only T or P is specified. The proportions of the three phases are determined by the overall composition.
3. Four phases are unique when only the temperature or pressure is specified. When only condensed phases are present, pressure usually has little effect so it is not regarded as a pertinent variable and the four phase condition is invariant.

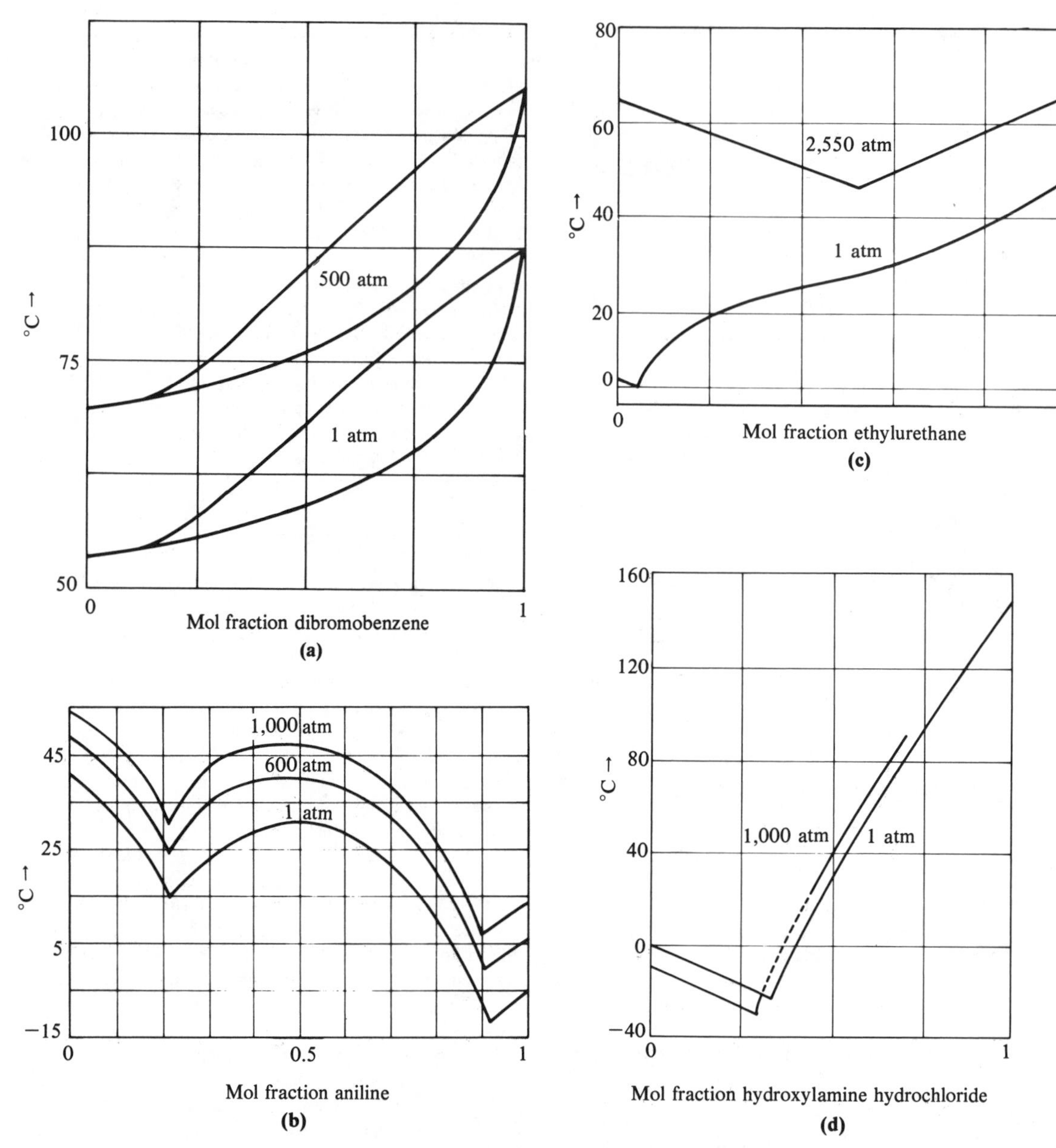

Figure 5.28. The effect of pressure on some binary liquid-solid equilibria. (a) The system 1,4-dichlorbenzene + 1,4-dibromobenzene at 1 atm and 500 atm. [Data of Deffet, Bull Soc Chim Belgique 47, 461 (1930)]. (b) The system phenol + aniline at 1 atm, 600 atm, and 1,000 atm, displaying intermolecular compound formation and two eutectics. [Data of Deffet, Bull Soc Chim Belgique 47, 461 (1938)]. (c) The system benzene + ethylurethane at 1 atm and 2,550 atm, showing a great effect on the eutectic composition. [Data of Puschin, Z physik Chem 118, 447 (1925)]. (d) The effect of pressure on the freezing curve of aqueous solutions of hydroxylamine hydrochloride, showing only a slight effect on the eutectic point. [Data of Mathieu, Bull Soc Chim Belge 58, 112 (1949)].

4. Five phases constitute an invariant condition when the pressure can affect the behavior.

The ensuing discussion of ternary systems is arranged according to the numbers and kinds of phases that are present and whenever possible representative examples of organic systems are given. Since systems comprised of all solid phases are of less importance in the organic field and comparatively little is known about gas-solid equilibria little attention will be devoted to these topics here, but see Chapter 9. Some properties of triangular diagrams used to represent ternary compositions are illustrated with Figures 5.5 and 5.6.

5.5.1. Vapor-Liquid Systems

Some representative kinds of diagrams are shown in Figure 5.30 for the system acetone-methanol-water:

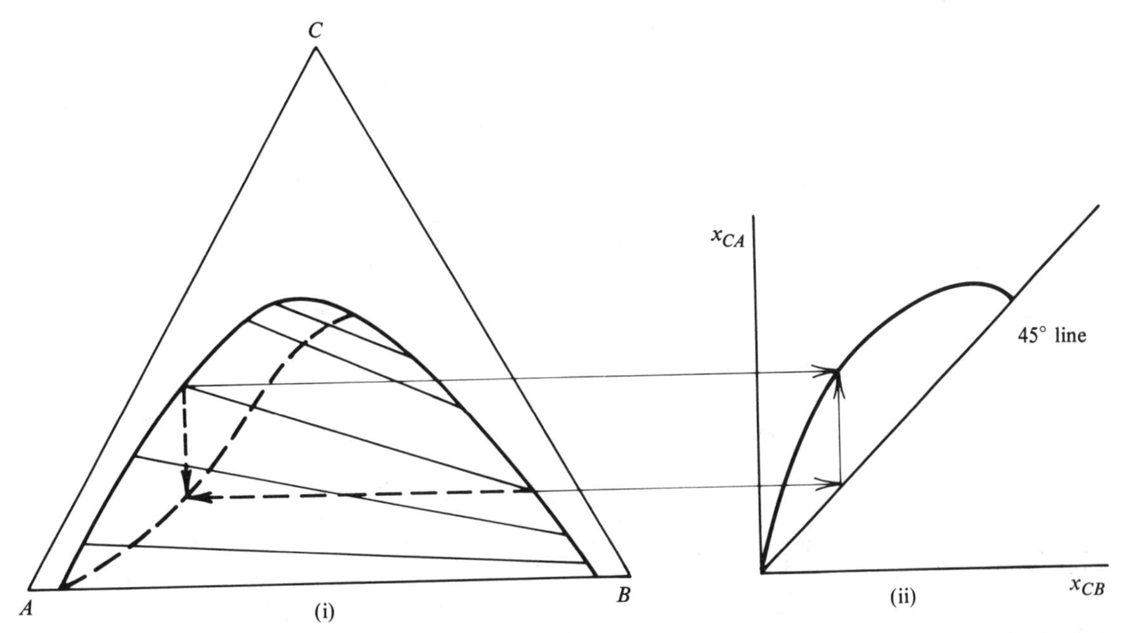

Figure 5.29. Representation of the equilibrium compositions of two phases of a ternary mixture: (i) The broken line is the tie-line correlation; it is constructed as the locus of intersections of horizontals through the right-hand ends with verticals through the left-hand ends of the tie-lines; (ii) the distribution curve is constructed as the locus of intersections of reflections of horizontals from the right-hand ends of the tie-lines off the 45° line with horizontals from the left-hand ends of the tie-lines.

a. This shows the composition of the liquid phase at 100 C for a number of bubblepoint pressures. For example, a mixture containing 30% each of water and acetone has a bubblepoint of 3.46 atm.
b. The composition for a number of bubblepoint temperatures at one atm is shown. For example, a mixture containing 30% each of water and acetone has a bubblepoint of about 61.7 C. at 1 atm.
c. This gives the composition of the vapor phase in equilibrium with a given liquid composition at one atm. For example, the liquid mixture containing 30% each of water and acetone is in equilibrium with a vapor containing 10% water and 57% acetone.

On Figure 5.30(d) and (e), representing hydrogen + nitrogen + carbon monoxide at a fixed T and P, compositions of phases in equilibrium are connected with tie-lines. For example, a liquid containing 10% H_2 and 43% CO is in equilibrium with a vapor containing 10% CO and 3% N_2 at −195 C and 150 atm. Vapor-liquid compositions are shown at three pressures for the system methane-n-butane-decane of Figure 5.30(h). Another way of recording vapor-liquid equilibria is that shown for acetone-chloroform-4-methylpetanone(4) in Figure 5.31. For completeness, the bubblepoint temperatures also could have been drawn on the triangular chart, but on this scale the clutter would have been too great and a separate diagram would be preferable.

5.5.2. Vapor and Two Liquid Phases

Several ways of representing the compositions of the three phases in equilibrium are given in Figure 5.32. In Figure 5.32(a) the liquid-liquid equilibrium compositions are identified with tie-lines, whereas on the dashed vapor line compositions are identified by the same numbers as on the corresponding tie-lines. Distribution curves shown in Figure 5.32(b) are with mol fraction of acetone in the water-rich phase as abscissa, the liquid-liquid distribution line is constructed as in Figure 5.29 and the mol fractions of acetone and of chloroform in the vapor phase are plotted as ordinates. Scattered $x - y$ data of the homogeneous liquid region of Figure 5.33 are indicated by lines connecting the proper compositions; the scheme of Figure 5.30 could be used for this purpose if sufficient data were available for crossplotting.

5.5.3. Liquid Phases

Ternary systems may have a multiplicity of one-phase, two-phase, and three-phase regions at particular temperatures and pressures, although the majority of systems that have been investigated have only one or two each of one-phase and two-phase regions. The shapes of such regions may differ markedly from system to system and often are sensitive to temperature. The spatial diagram of Figure 5.34 illustrates one kind of temperature effect, where both upper and lower critical-solution temperatures are present. The most frequent kind of behavior is increase of miscibility range with temperature. In the contour drawing of Figure 5.35, the effect of increasing temperature is to develop an upper critical-solution point with a ternary composition. Figure 5.7 shows a ternary system that has a binary critical-solution point. Other kinds of temperature effects are shown on several other figures of this section.

A simple kind of classification of diagrams is with respect to the numbers of one-, two-, or three-phase regions that may be present. Some of the simpler types are shown in Figure 5.36. The single two-phase regions of Figure 5.36(a) or (b) are by far the most common, particularly Figure 5.36(a) that has a

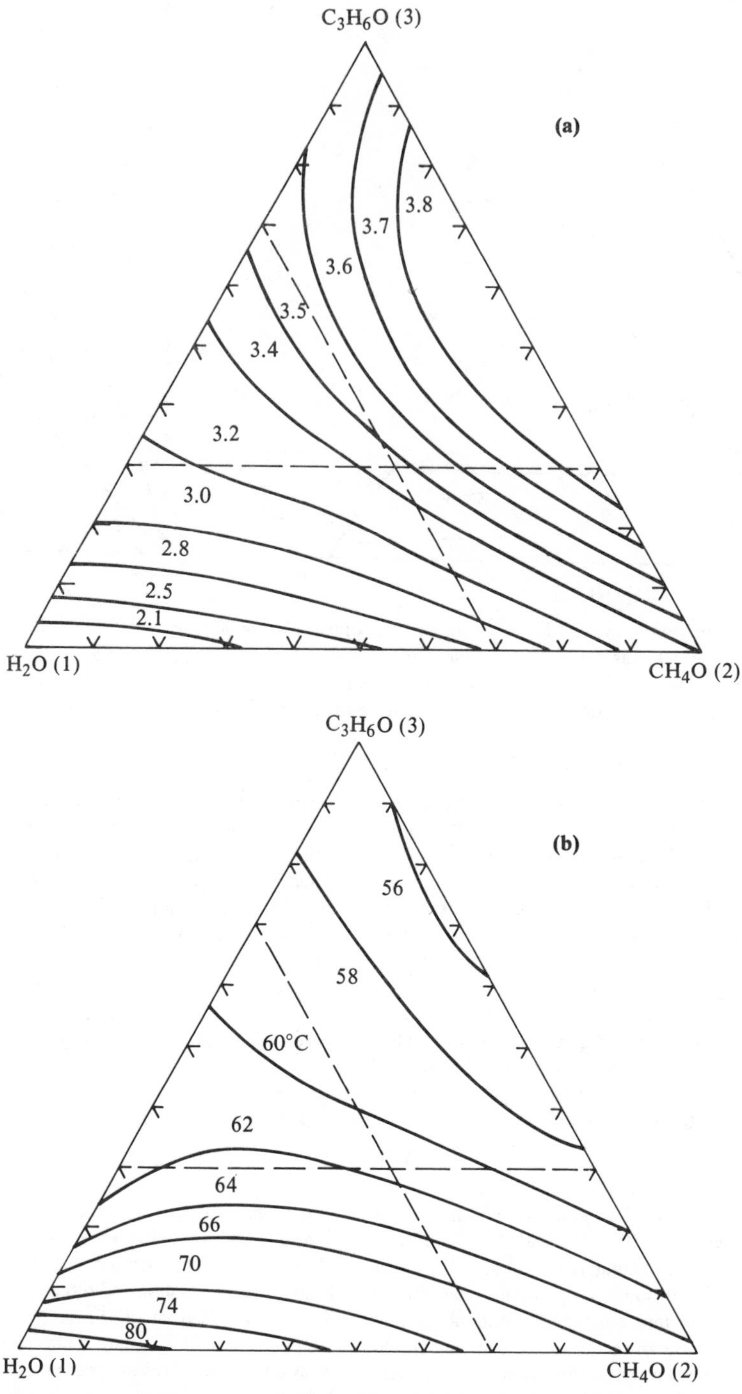

Figure 5.30. Isotherms, isobars, and liquid-vapor equilibrium compositions in some ternary systems. (a) Isobars of mixtures of water, methanol, and acetone at 100 C. The construction shows that a mixture containing 30% each of water and acetone has a pressure of 3.45 atm [Griswold & Wong, CEP Symposium Series 3, 18 (1952)]. (b) Isotherms of mixtures of water, methanol, and acetone at 1 atm. The construction shows that a mixture containing 30% each of water and acetone has a saturation temperature of 61.5 C at 1 atm [Griswold & Wong, CEP Symposium Series 3, 18 (1952)]. (c) Vapor-liquid equilibrium compositions of mixtures of water (1), methanol (2), and acetone (3) at 1 atm. Full lines are labeled y_1 and broken lines are labeled y_3. The construction shows that a liquid phase containing 30% each of water and acetone is in equilibrium with a vapor having $y_1 = 0.1$ and $y_3 = 0.57$. [Griswold & Wong, CEP Symposium Series 3,

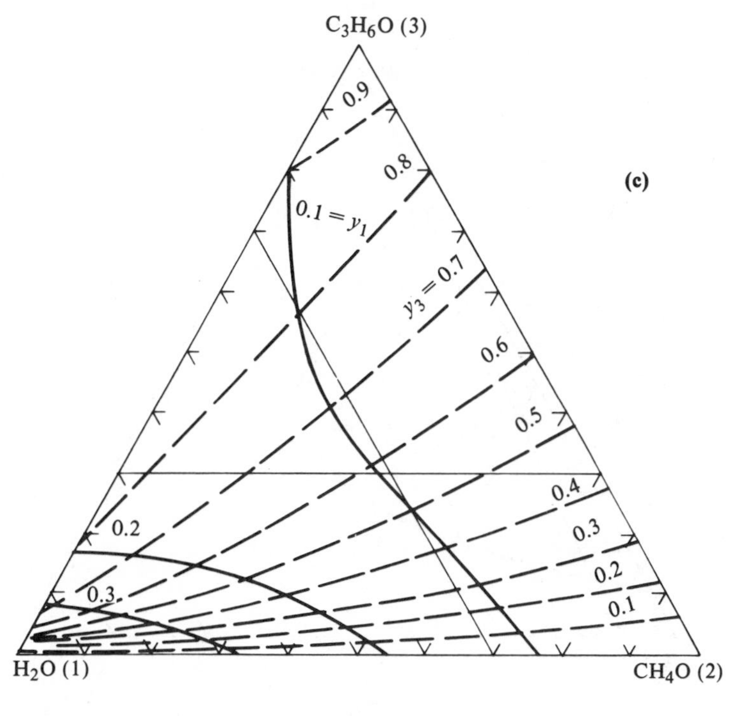

(c)

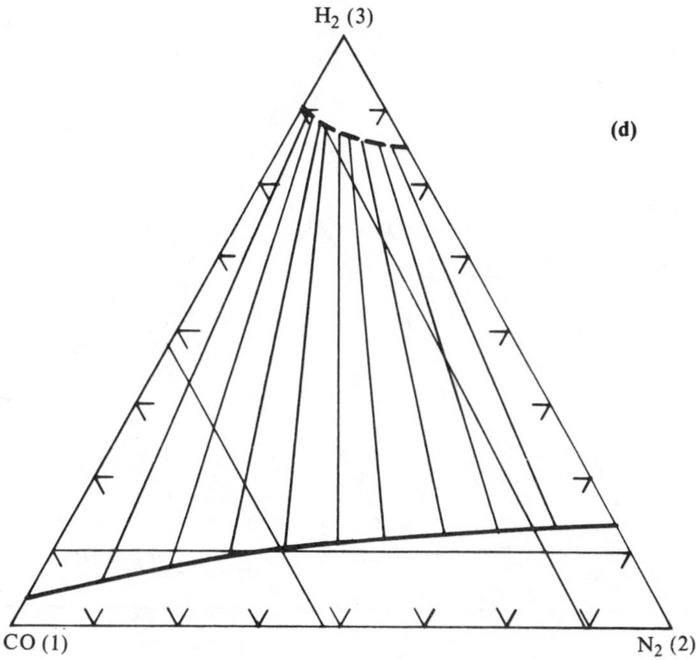

(d)

18 (1952)]. (d) Equilibrium compositions of liquid and vapor phases in the system of hydrogen, nitrogen, and carbon monoxide at 150 atm and −195C. Vapor compositions are on the broken line [Ruhemann & Zinn, Zh Fiz Khim 12, 389 (1937)]. (e) The same system as in Figure 5.30(d) but at 150 atm and −185 C. Note the sharp change in the shape of the two-phase region. (f). Compositions of coexisting phases in the system methane + *n*-butane + decane at 280 F and several pressures. The lower ends of the tie-lines represent liquid compositions and the upper ends represent vapor compositions. [Riemer, Sage, & Lacey, Ind Eng Chem 43, 1436 (1951)]. Copyright American Chemical Society; used by permission.

Figure 5.30 *(continued on next page)*

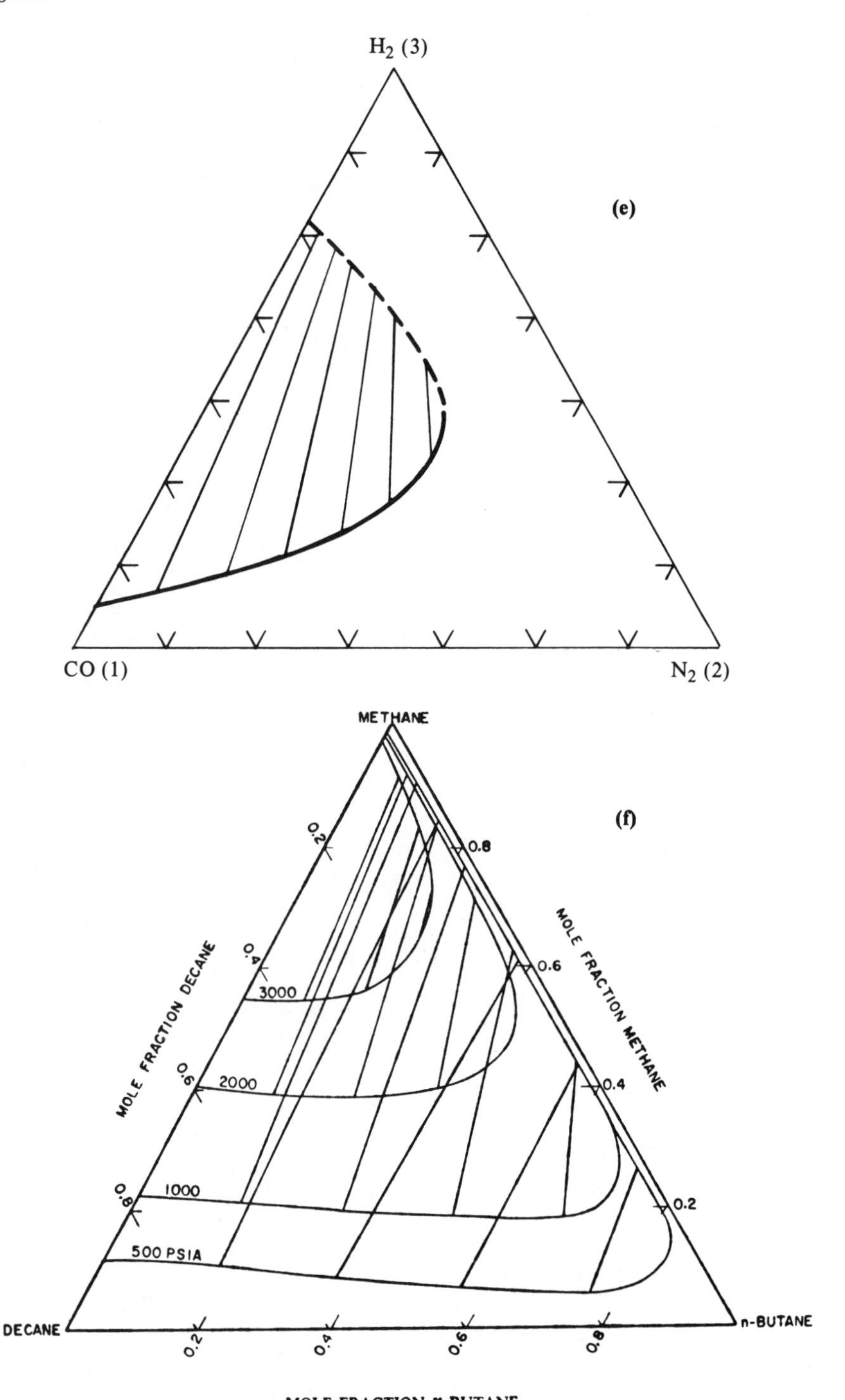

Figure 5.30 *(concluded)*

single one-phase region also. The pair of two-phase regions of Figure 5.36(c) may be visualized to expand and merge, thus dividing the one-phase region in two parts, like Figure 5.36(b). Similarly, the three heterogeneous regions of Figure 5.36(d) may be visualized to expand until they touch or merge. If they just touch, they will form a three-phase region that is necessarily bounded by a straight-sided triangle, as in Figure 5.6 or some of the later figures of this section.

In view of the availability of several comprehensive collections of phase diagrams, notably in Landolt-Börnstein, it is not feasible to reproduce many here. A particularly illuminating set of 464 ternary diagrams with carbon dioxide as one of the components was prepared by Francis (1954) and displays a great variety of behaviors. The three sets of diagrams shown here as Figures 5.37, 5.38, and 5.39 demonstrate primarily the effects of temperature. In the case

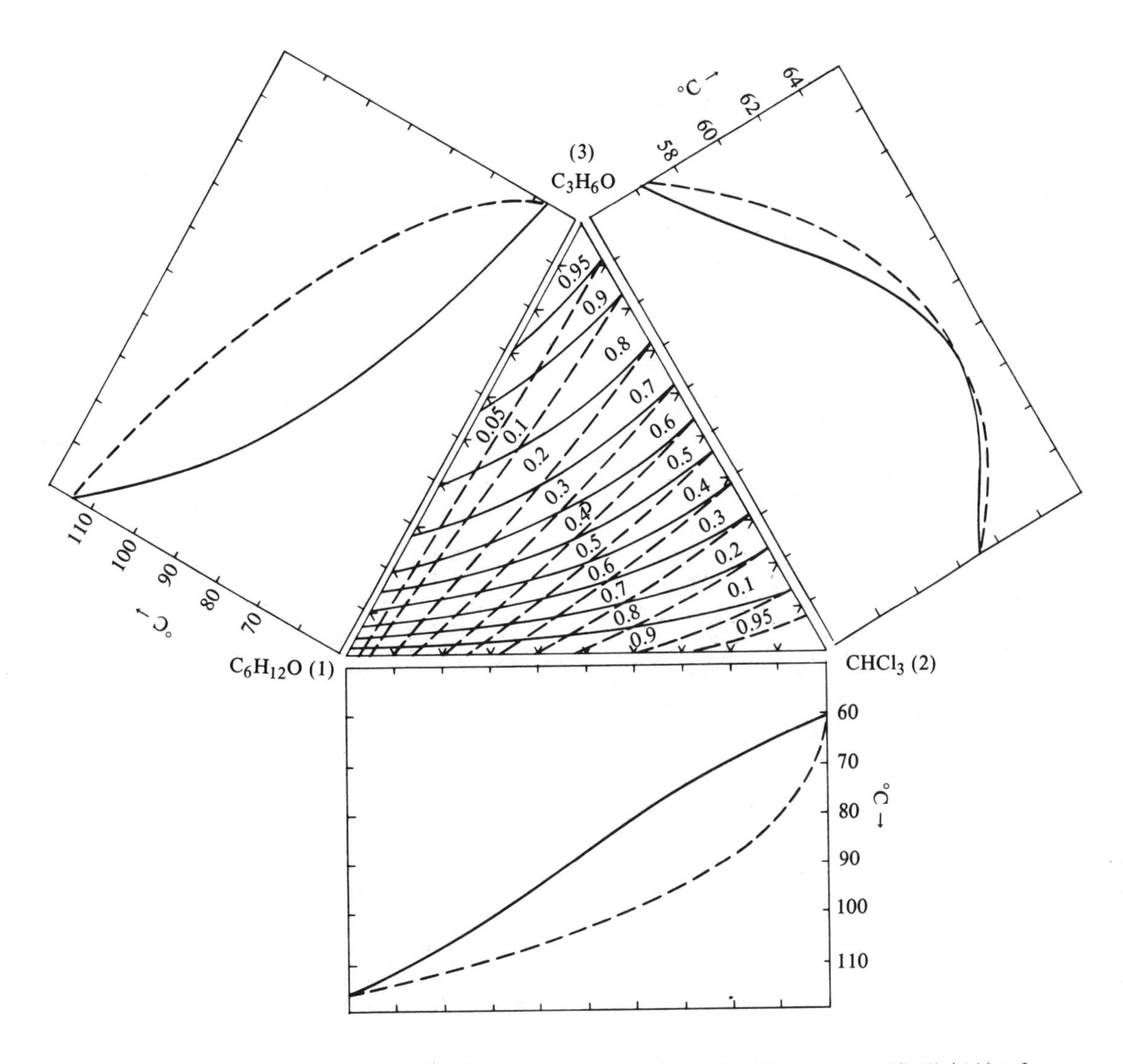

Figure 5.31. Ternary $x - y$ diagram and binary $T - x - y$ diagrams of the system 2-methylpentanone-(4) (1) + chloroform (2) + acetone (3) at one atm. On the ternary diagram, the broken lines represent y_2 and the full ones y_3. For example, an equimolal liquid is in equilibrium with a vapor composition $y_2 = 0.35$ and $y_3 = 0.57$ (based on data in Landolt-Börnstein, II/2a 527, 608, 609, and 756).

of water + triethylamine + phenol, raising the temperature solubilizes the phenol but makes the water and triethylamine less miscible. In Figure 5.38 the six regions present at 25 C become only two at the higher temperature. Only a 10 degree rise in temperature is sufficient to eliminate the three three-phase regions from the system of Figure 5.39 and to reduce the 12 regions to four simpler ones. Micellar systems display particularly complex behavior, for example in Figure 9.12.

The few cases that have been cited may suggest the magnitude of the challenge to theoretical methods of correlation or prediction of multicomponent phase behavior. Thus far, correlations with the NRTL or UNIQUAC methods have been applied primarily to two-phase systems that are of practical importance and occasionally to three-phase cases. In theory the methods are extendable indefinitely with binary parameters, but practically the situation is complicated by higher-order interactions and the effect of temperature, which has not been systematized.

5.5.4. Liquid-Solid Melt Equilibria

The only entirely satisfactory diagrams for the understanding of the full range of phase behavior are three-dimensional models such as the schematics of Figures 5.7 and 5.8 or stereoscopic views like those of Tamas & Pal (1970). Unfortunately, many spatial diagrams in the literature are drawn with so many lines and without adequate shading or silk-screening that they are difficult to understand completely

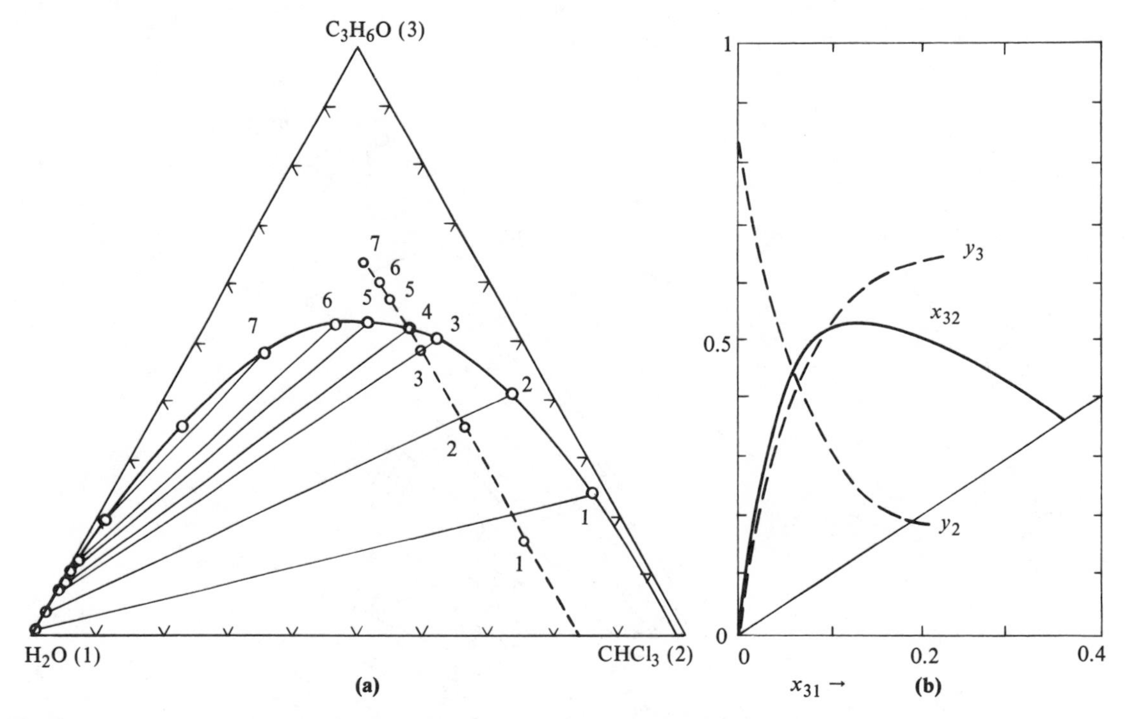

Figure 5.32. The system acetone + chloroform + water at 1 atm: (a) Liquid-liquid equilibrium tie-lines are shown. The broken line shows the composition of the vapor in equilibrium with the liquid-liquid mixture having a tie-line with the same number; (b) distribution curves: x_{31} is the mol fraction of acetone (3) in the water-rich (1) phase; x_{32} that of acetone in the chloroform-rich (2) phase; y_2 and y_3 are the mol fractions of chloroform and acetone in the vapor phase. Data from Landolt-Börnstein, II/2 727.

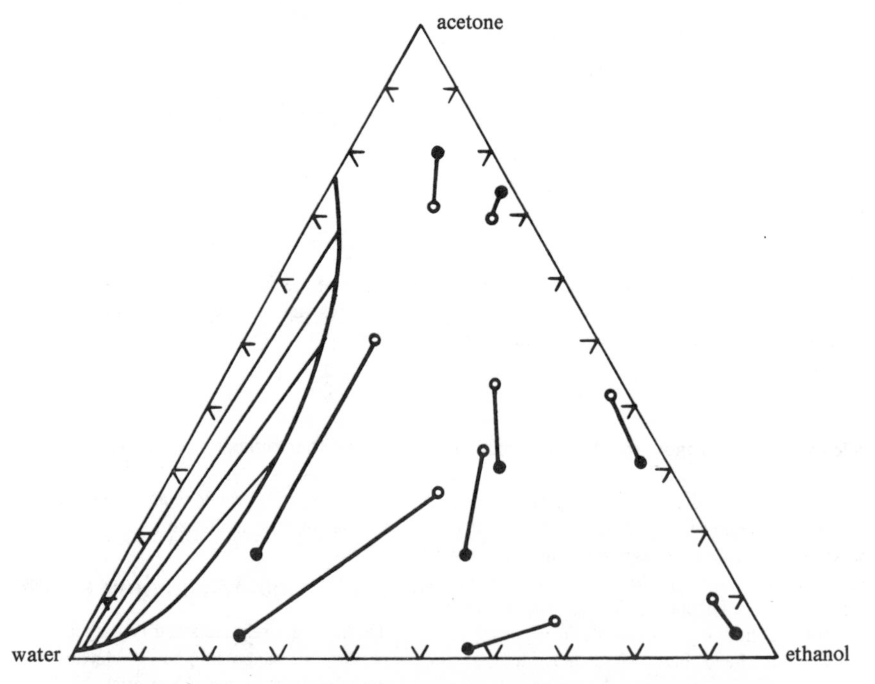

Figure 5.33. A diagram showing liquid-liquid equilibria and some vapor-liquid equilibria in the homogeneous liquid region of the system water + ethanol + ethyl acetate at 70°C. Open circles represent vapor compositions, and solid circles liquid compositions on the same tie-line [Mertl, Coll Czech Chem Commun 37, 366 (1972)].

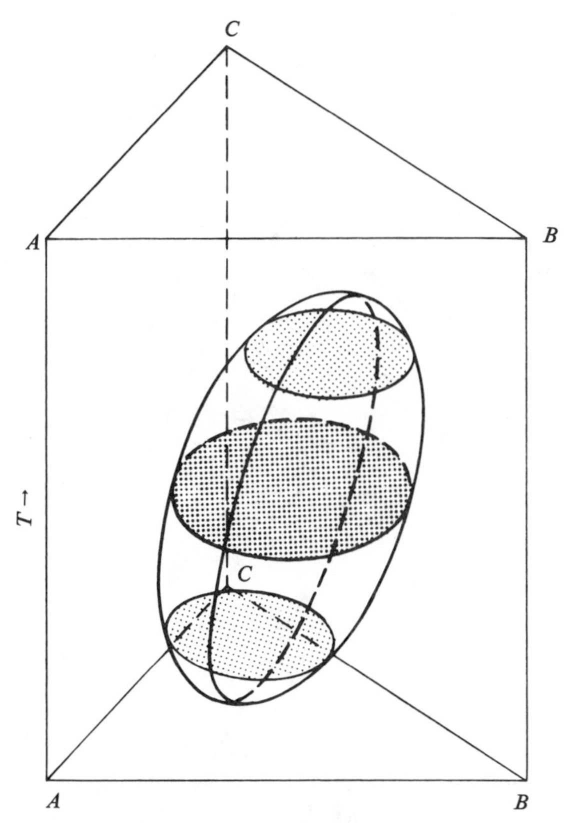

Figure 5.34. The effect of temperature on the size and shape of the two-phase region of a ternary mixture: a system with upper and lower critical solution temperatures; see also Figures 5.2(a) and 5.2(b).

except by the originators. The reader should take the opportunity to examine the stereos of Tamas & Pal, of which some of the particularly illuminating ones may be listed as follows:

No. 14: The system $H_2S_2O_7$ + HSO_3Cl + H_2SO_4, showing two binary compounds.
No. 18: A system having a congruent ternary compound.
No. 20: A system having a peritectic ternary compound.
No. 37: Path of equilibrium crystallization.
No. 38: The system $C_6H_4Cl_2 + C_6H_4Br_2 + C_6H_4ClBr$.
No. 59: Showing liquid immiscibility.
No. 11: System of diphenyl + benzophenone + diphenylamine with a congruent binary compound.

For a number of reasons that will be stated, melt equilibria often are more complex than just liquid-liquid or vapor-liquid equilibria. Fortunately pressure is not a significant variable with condensed systems, which permits often adequate representation of a system by a series of isothermal diagrams, assisted in some cases by an occasional isopleth.

Among the factors that contribute to the complexity of ternary diagrams are:

1. Occurrence of solid solutions of substances of varying degrees of miscibility ranging from zero to complete. Such solutions form with substances that have the same crystalline form and nearly the same molecular volumes. This is very often the case with organic substances that it may be desired to separate, for example, by crystallization from a solvent.

2. Formation of molecular compounds that may be stable (congruent) or unstable (incongruent) at their melting points. The effects of intermolecular compounds have been noted for binary systems in Section 5.4. Compounds of pairs of substances are naturally fairly common in ternary organic systems, but no examples of ternary organic compounds come to mind. They occur often, however, in the ceramic field, for example in the system CaO-Al_2O_3-SiO_2 (Grieg, 1927).

A few ternary melt diagrams are shown here. Figure 5.40(a) is of the simplest type with three binary eutectics and a single ternary eutectic. An intermolecular compound is formed in the system of Figure 5.40(b) and two ternary eutectics are

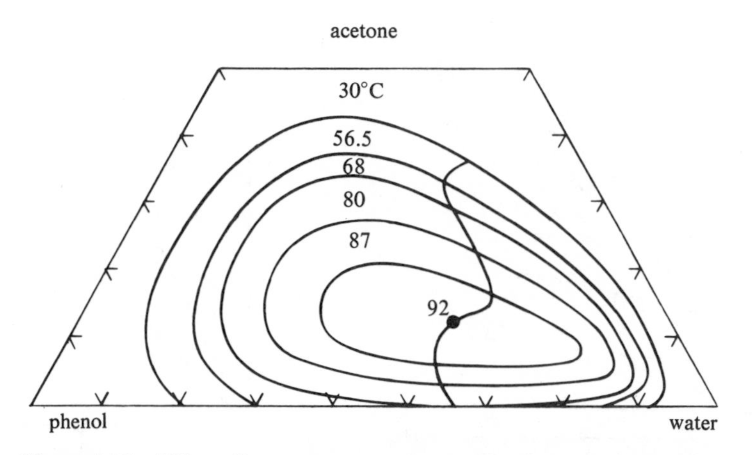

Figure 5.35. Effect of temperature on the two-phase region of the system acetone + water + phenol. The upper critical-solution temperature is 92 C. The wavy line is the locus of the plait points [data of Schreinemakers, Z physik Chem 33, 75 (1900)].

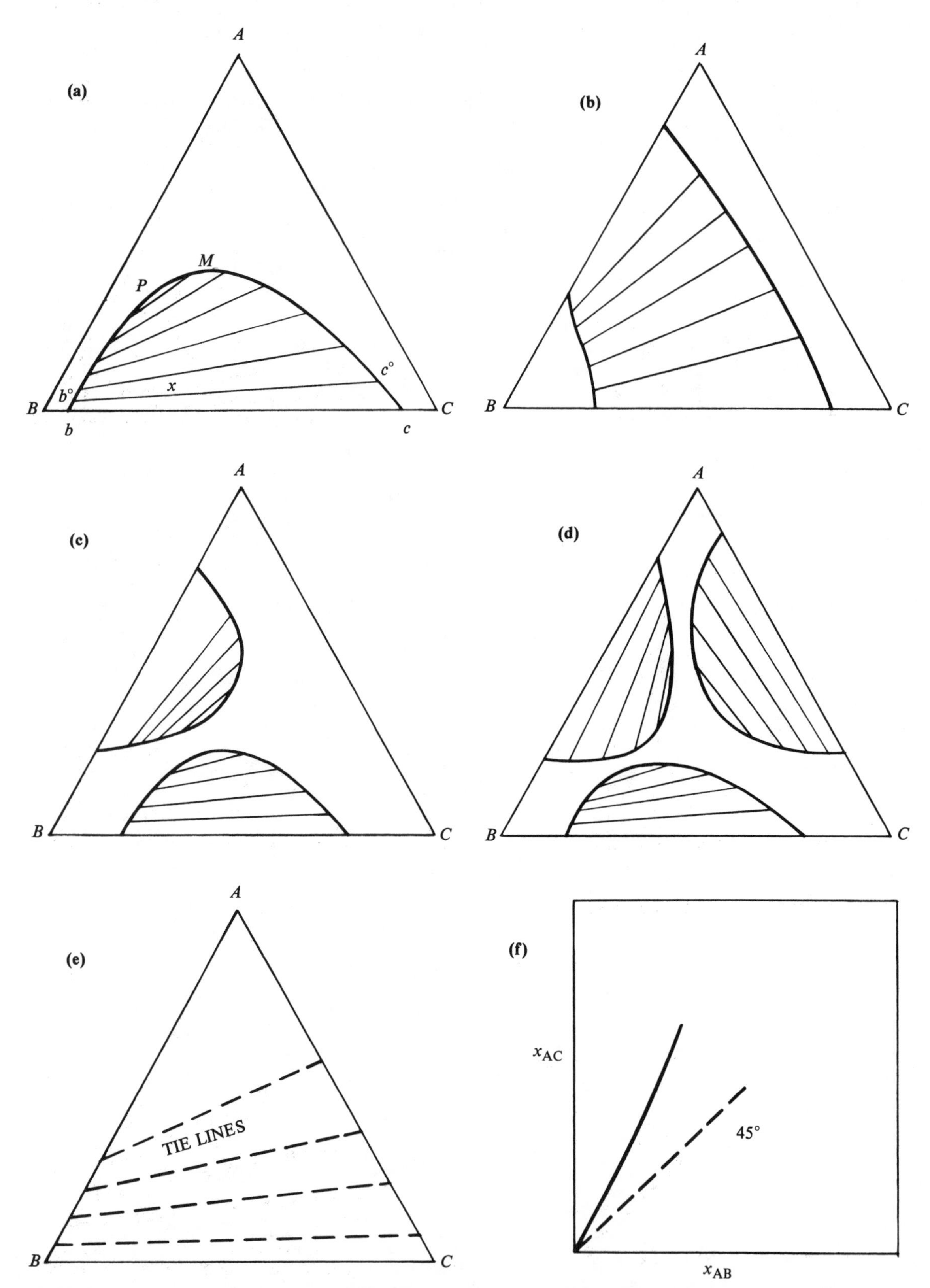

Figure 5.36. Common types of ternary diagrams with one-phase and two-phase regions, showing tie-lines connecting phase compositions in equilibrium. (a) One each of one-phase and two-phase regions. (b) Binodal curves forming a band and making two one-phase regions. (c) A pair of two-phase regions. (d) Three two-phase regions. (e) A two-phase region with only two components in each phase. (f) The distribution curve for Figure (e) constructed as in Figure 5.29.

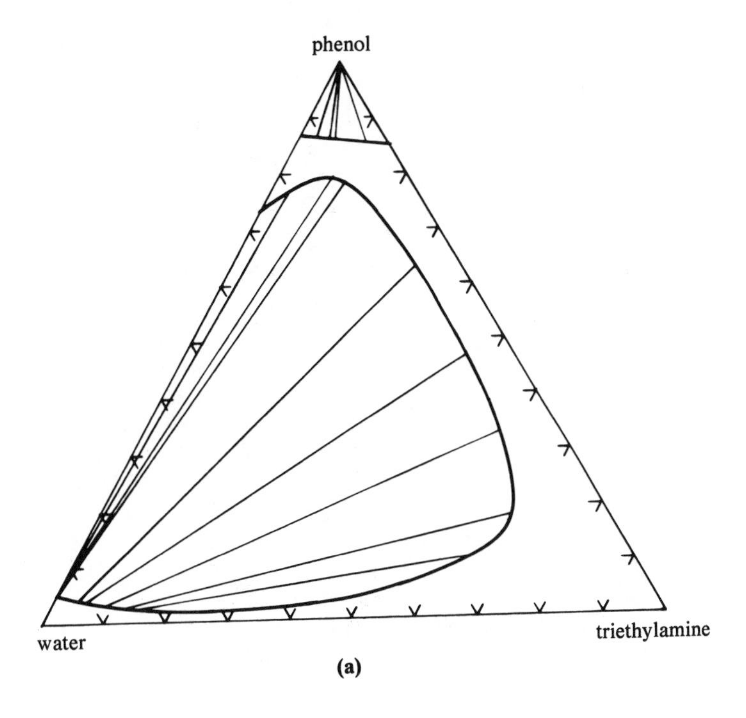

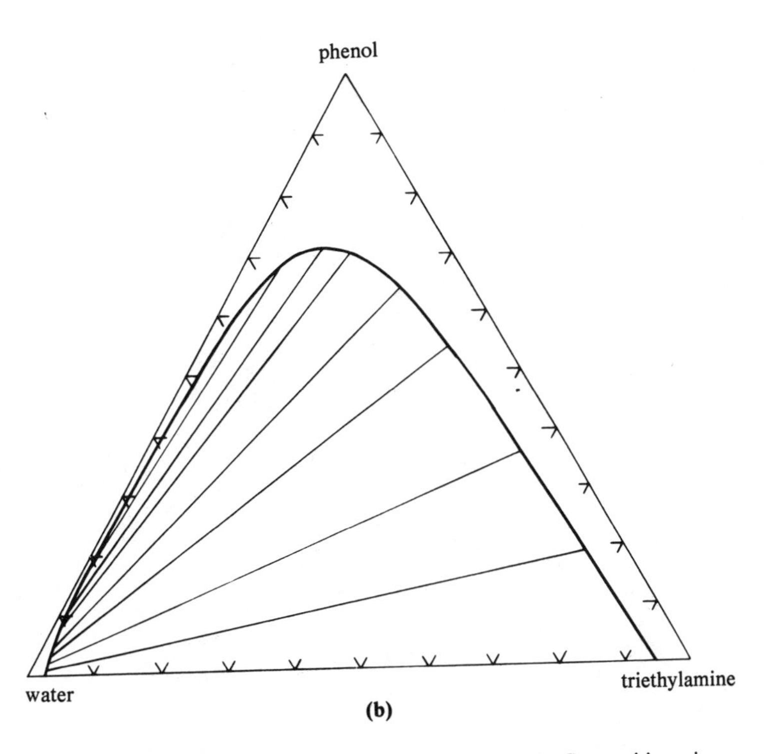

Figure 5.37. The system water + triethylamine + phenol. Compositions in weight fractions [Meerburg, Z physik Chem 40, 642 (1902)]. (a) At 10 C the system has two two-phase regions. (b) At 75 C, two pairs of components become completely miscible, and water + triethylamine become only partially miscible.

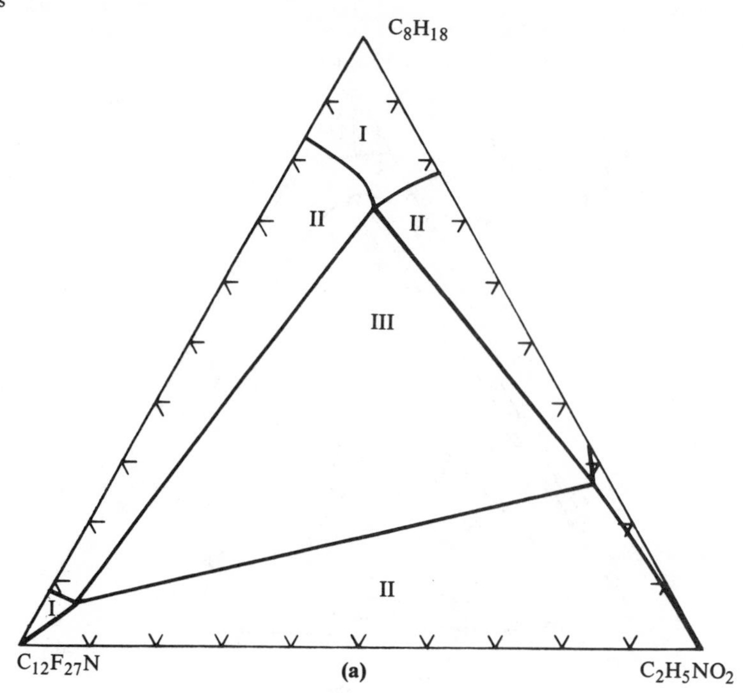

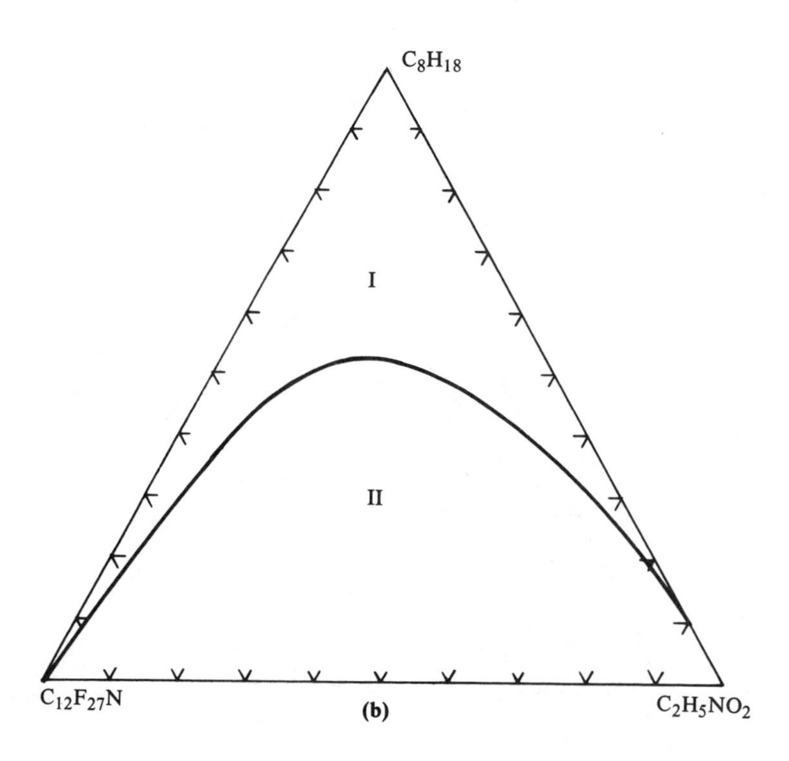

Figure 5.38. The system 2,2,4-trimethylpentane + nitroethane + perfluorobutylamine, showing multiple-phase regions at several temperatures. The numbers of phases present are designated I, II, and III. Compositions in weight fractions [data of Vreeland & Dunlap, J Phys Chem 61, 329 (1957)]. Copyright American Chemical Society; used by permission. (a) At 25 C the system has six different regions, including a three-phase region. (b) At 51.3 C only one multiple-phase region remains.

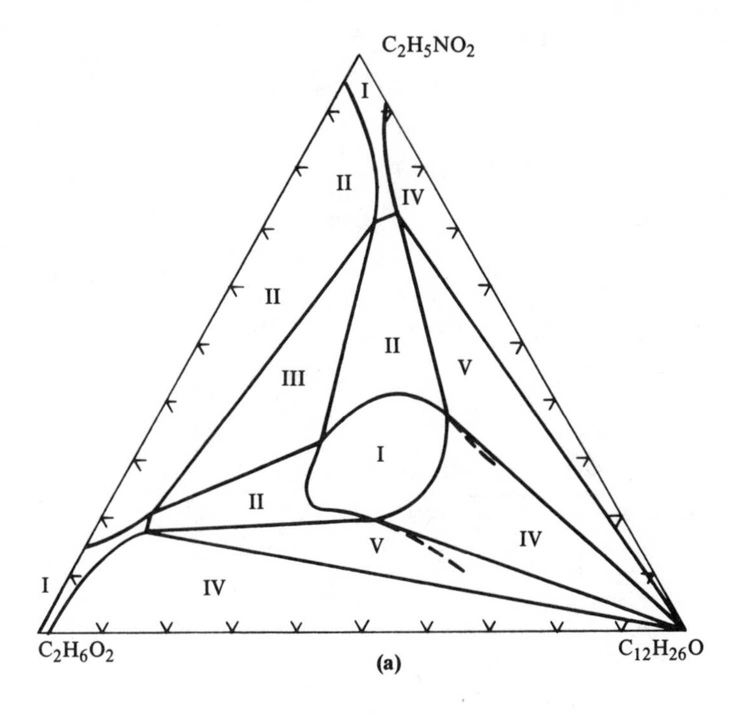

(a)

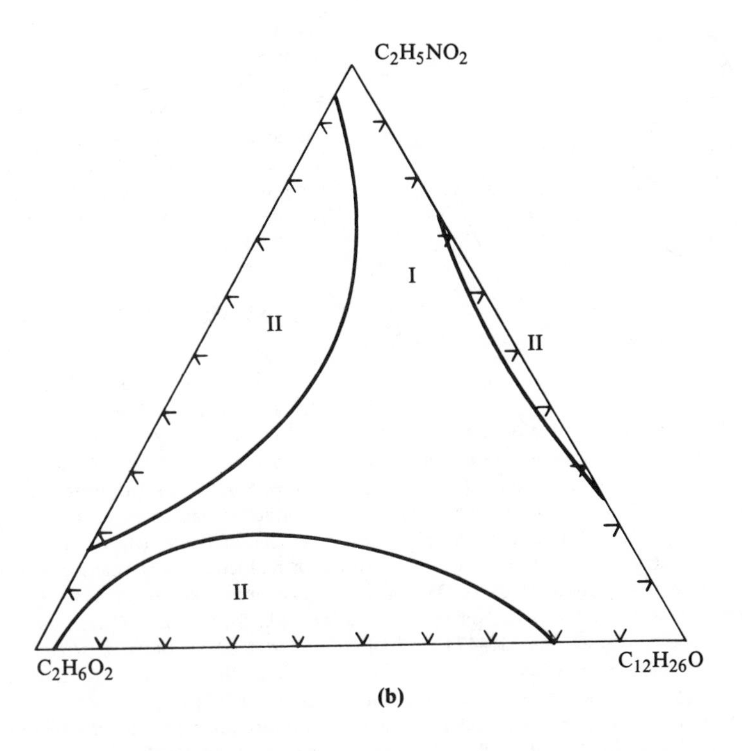

(b)

Figure 5.39. The system glycol + dodecanol + 1-nitroethane, showing a multiplicity of homogeneous and heterogeneous regions as a function of temperature. Compositions in weight fractions. *Key:* I = 1 liquid phase; II = 2 liquid phases; III = 3 liquid phases; IV = one liquid and one solid phase; V = two liquid phases and one solid phase (Francis, 1956). Copyright American Chemical Society; used by permission. (a) Conditions at 14 C and 1 atm. (b) Conditions at 24 C and 1 atm.

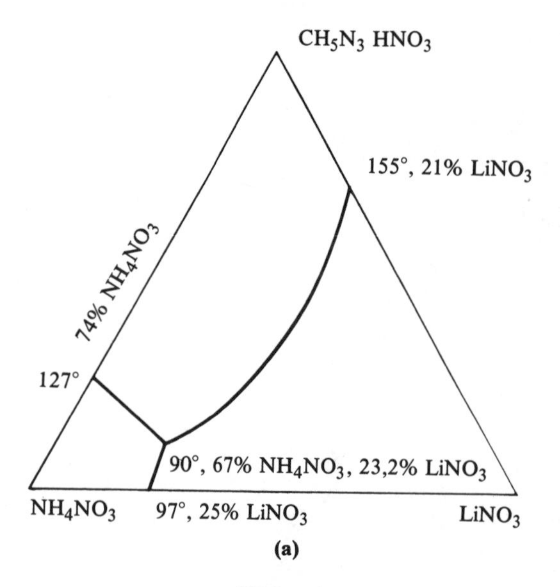

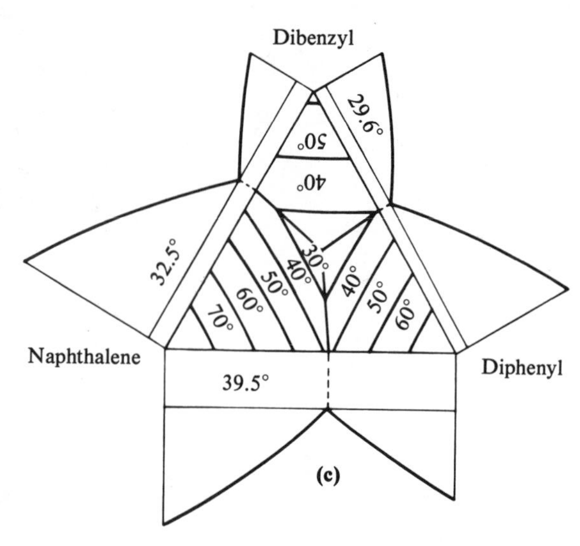

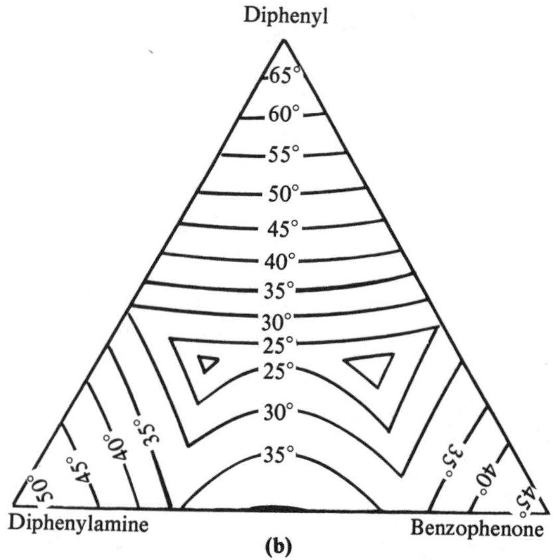

Figure 5.40. Melt equilibria of some ternary mixtures. (a) The system ammonium nitrate + lithium nitrate + guanidine nitrate forms simple binary and ternary eutectics (Clark et al., J Phys & Colloid Chem 53, 225 (1949)]. Compositions in weight %. (b) The system diphenyl + benzophenone + diphenylamine has a 1:1 compound of the last two substances. The ternary eutectics are at 18.8°, 0.205 benzophenone, 0.49 diphenylamine; 16.5°, 0.52 benzophenone, 0.18 diphenylamine [Lee & Warner, JACS 4474 (1933)]. Copyright American Chemical Society; used by permission. (c) Binary and ternary melt diagrams of naphthalene + diphenyl + dibenzyl. The ternary eutectic is at 17.4° and 0.27 naphthalene, 0.338 diphenyl, 0.392 dibenzyl mol fractions [Data of Lee & Warner, JACS 57, 318 (1935)]. Copyright American Chemical Society; used by permission.

formed. Binary and ternary diagrams are shown for the system of naphthalene + diphenyl + dibenzyl in Figure 5.40(c); it behaves almost ideally.

Several aspects of the system phenol + urea + urethane are shown on the several parts of Figure 5.41. Urea and urethane and urea and phenol form intermolecular compounds which decompose before melting; the peritectic points are shown on the spatial diagram and on the projection, Figure 5.41(b). Binary sections at constant amounts of the third component are shown as Figures 5.41(c) and (d) and display a number of phase equilibria both in the liquid-solid and solid-solid regions, between the pure substances and the compounds. Development of these diagrams was a substantial achievement.

5.6. MORE THAN THREE COMPONENTS

A system of four components requires three dimensions for recording its composition alone on a regular tetrahedron at constant temperature and pressure. Examples are Figures 5.8(b) and (c). For practical work, binary diagrams at constant values of two of the mol fractions, analogous to Figures 5.41(c) and (d), or ternary diagrams at one constant mol fraction can be used. Constant water content would be practical for the system of Figure 5.8(c). Much of the work with multicomponent solution and melt equilibria has been with aqueous salt mixtures and is described by Teeple (1929) and Blasdale (1927). Quaternary and quinary systems are beyond the scope of the present discussion; they are treated at length in the books of Ricci (1951), Vogel (1959), and Zernike (1955).

Multicomponent systems with immiscible solid phases and without intermolecular compounds can be analyzed with the Schröder equation when binary activity-coefficient correlations are available, or with unit activity coefficients in the ideal case. Examples of such calculations are in Chapter 7. For the most part, complex diagrams cannot be synthesized in the present state of knowledge, but some progress is being made and is reported, for example, in the book of Kaufman & Bernstein (1970) and in the article of Rudman (1970). The abundant literature is concerned primarily with experimental phase relations and with their classification.

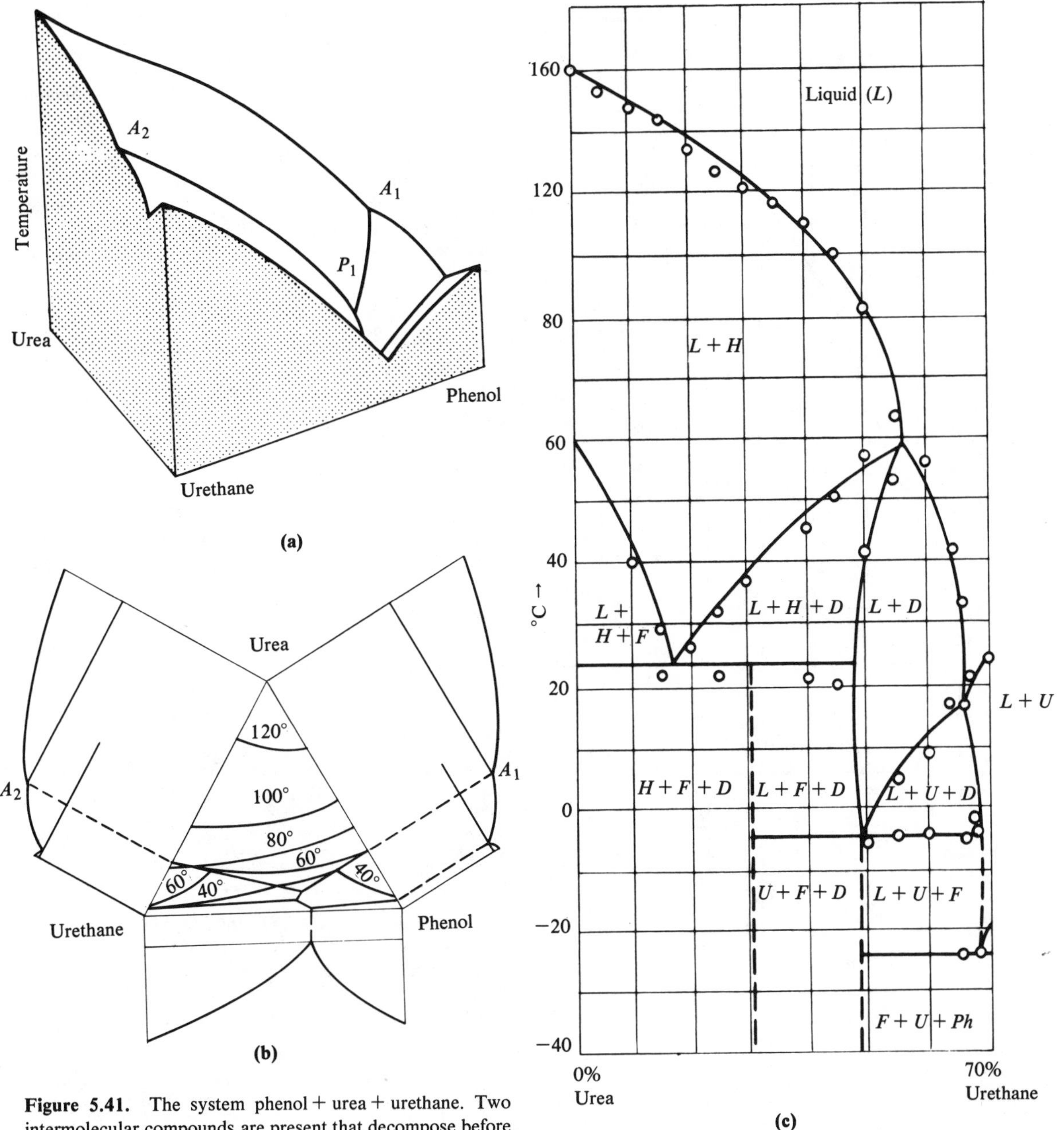

Figure 5.41. The system phenol + urea + urethane. Two intermolecular compounds are present that decompose before melting [Hrynakowski & Szmit, Z physik Chem 175A, 83 (1933)]. (a) A spatial model of the system. A_1 and A_2 are peritectics, and P_1 is the ternary eutectic involving the two metastable compounds. (b) Projections of the ternary and binary systems. A_1 and A_2 are peritectic points. The three ternary eutectics are at 23.2°, −3.5°, and −24°. (c) Vertical section at 30 wt% phenol. Code: D = compound of urea and urethane; F = compound of urea and phenol; H = urea; L = liquid; Ph = phenol; U = urethane. (d) Vertical section at 20 wt% urethane. Code is the same as for Fig. 5.41(c).

Figure 5.41 *(continued on next page)*

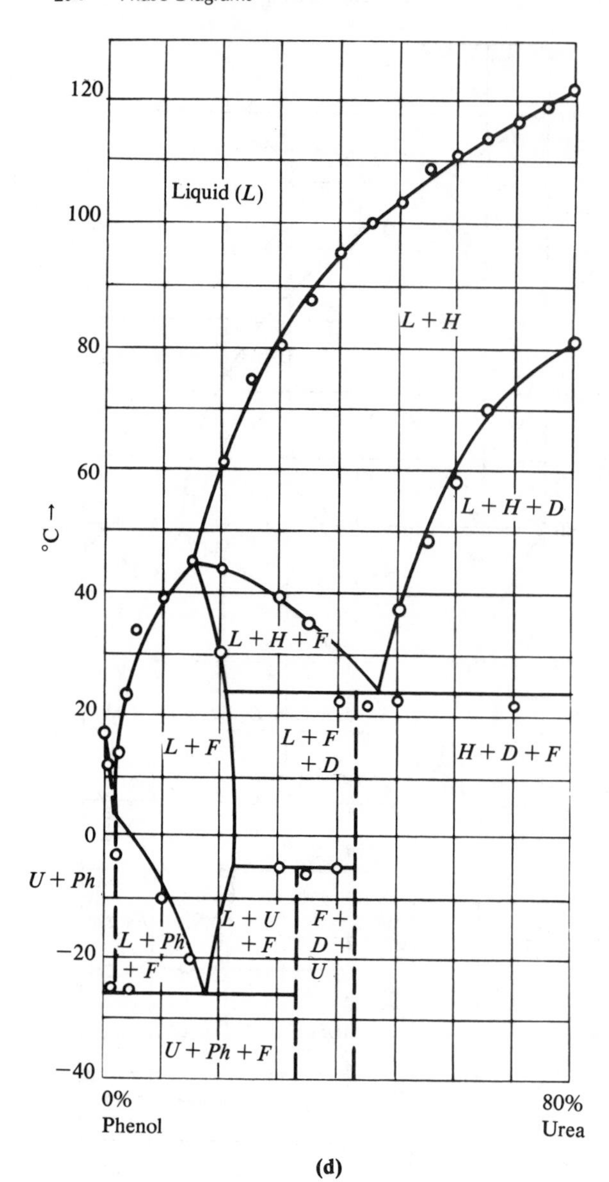

(d)

Figure 5.41 *(concluded)*

5.7. SELECTED BOOKS ON PHASE DIAGRAMS

1. S. T. Bowden, *The Phase Rule and Phase Reactions* (London: Macmillan, 1938).
2. F. D. Ferguson & T. K. Jones, *The Phase Rule* (Woburn: Butterworths, 1966).
3. R. Haase & H. Schönert, *Solid-Liquid Equilibrium* (New York: Pergamon, 1969).
4. L. Kaufman & H. Bernstein, *Computer Calculation of Phase Diagrams* (New York: Academic Press, 1970).
5. Landolt-Börnstein, *Zahlenwerte und Funktionen*, II/2a (1960), II/2b (1962), II/2c (1964), II/3 (1956), *Neue Serie* IV/3 (New York: Springer, 1975).
6. J. Nyvlt, *Solid-Liquid Phase Equilibrium* (New York: Elsevier, 1977).
7. A. Prince, *Alloy Phase Equilibria* (New York: Elsevier, 1966).
8. A. Reisman, *Phase Equilibrium* (New York: Academic, 1970).
9. J. Ricci, *The Phase Rule and Heterogeneous Equilibrium* (New York: Van Nostrand, 1951).
10. F. Tamas & I. Pal, *Phase Equilibrium Spatial Diagrams* (Woburn, MA: Butterworths, 1970).
11. R. Treybal, *Liquid Extraction* (New York: McGraw-Hill, 1963).
12. R. Vogel, *Die Heterogenen Gleichgewichte* (Leipzig: Akademische, 1959).
13. D. R. F. West, *Ternary Equilibrium Diagrams* (London: Chapman & Hall, 1982).
14. J. Zernike, *Chemical Phase Theory* (Deventer: Kluwer, 1955).

5.8. PROBLEMS

5.1. The mol fractions of propionic acid in the water and octane phases are

x_w	0.030	0.081	0.167	0.274
x_o	0.057	0.124	0.181	0.252

Fit an equation for the distribution coefficient as a function of the concentration of propionic acid in one of the phases.

5.2. For the system of Figure 5.38 at 25 C, the coordinates of the three-phase points are, in wt fractions:

g_2	0.042	0.712	0.159
g_3	0.075	0.269	0.722

Find the relative amounts of the three phases for a mixture having equal weight fractions overall.

5.3. For the system hexadecane (1)-aniline (2)-benzene (3), equilibrium data at one atm and 25 C are given following. Construct a tie-line correlating curve and estimate the plait point composition by extrapolation. (Hunter & Brown, 1947, Ind Eng Chem 39, 1343). Locus of midpoints is approximately linear.

C_6H_7N-poor Phase		*C_6H_7N-rich Phase*	
x_2	x_3	x_2	x_3
0.141	0.000	0.990	0.000
0.144	0.147	0.896	0.095
0.147	0.180	0.879	0.111
0.152	0.227	0.852	0.137
0.162	0.303	0.804	0.183
0.165	0.323	—	—
0.181	0.370	0.744	0.237
—	—	0.742	0.239
0.194	0.396	—	—
—	—	0.671	0.299
0.249	0.449	—	—
—	—	0.562	0.378
0.340	0.456	—	—
—	—	0.453	0.426

5.4.a. The system SO_3–H_2O exhibits congruent melting points at compositions by weight of 68.96%, 81.63%, and 89.89% SO_3. What are the formulas of the corresponding compounds?

b. The system *n*-heptane and 2,2,4-trimethyl pentane exhibits a simple eutectic point at −114.4°C corresponding to 24 mole per cent of *n*-heptane [Smittenberg, Hoog, and Henkes, *J. Am. Chem. Soc.* 60, no. 17 (1938)]. Determine analytically the maximum mole per cents of *n*-heptane that can be recovered by crystallization from mixtures of the two compounds containing 80, 90, and 95 mole per cent of *n*-heptane.

5.5. Sketch the phase diagram of the HF-KF system from these facts [G.H. Cady, *J. Am. Chem. Soc.* 56, 1431 (1934)]:

a. The transition temperature for α-KHF_2 to β-KHF_2 occurs at 195°C.
b. α-KHF_2, β-KHF_2, and liquid HF are in equilibrium at 195°C when the overall mole percentage of HF is 53.82.
c. Compounds and melting points: KF, 880°C; β-KF·HF, 239.0°C; KF·2 HF, 71.7°C; 2 KF·5 HF, 64.3°C; KF·3 HF, 65.8°C; KF·4 HF, 72.0°C; HF, −83.7°C.
d. Eutectic temperatures and mole percentages of HF: −97°C at 93.11%; 63.6°C at 77.1%; 62.4°C at 72.7%; 61.8°C at 69.69%; 68.3°C at 64.9%; 229.5°C at 48.60%. Describe what happens at 65.0°C as KF is gradually added to liquid HF until the mixture is mostly KF.

5.6. Construct the constant-pressure phase diagram of CCl_4 (A) and dioxane-1,4 (B) from these melting points [S.M.S. Kennard and P.A. McCusker, *J. Am. Chem. Soc.* 70, 3375 (1948)]:

X_B	0.000	0.035	0.049	0.060	0.090
T(°C)	−22.7	−24.0	−24.6	−24.2	−23.0
X_B	0.312	0.331	0.349	0.370	0.411
T(°C)	−18.4	−18.3	−18.3	−18.4	−18.8
X_B	0.645	0.713	0.755	0.802	0.846
T(°C)	−7.3	−3.7	−1.6	+0.9	+3.4

X_B	0.109	0.141	0.205	0.250	0.284
T(°C)	−22.4	−21.2	−19.2	−18.6	−18.4
X_B	0.472	0.489	0.514	0.556	0.604
T(°C)	−19.8	−20.0	−17.6	−13.1	−9.9
X_B	0.912	0.967	1.000		
T(°C)	+7.0	+10.1	+11.8		

What are the compositions and melting temperatures of any eutectics and compounds?

5.7. The System $CaCl_2 + CaF_2$ is represented on the diagram due to Plato [Z physik Chem 58, 363 (1907)]. The peritectic at 737 C corresponds to the decomposition of a 1:1 intermolecular compound. Identify the species within each region. Describe the changes that occur when the system is heated slowly along the lines at 50% and 55% $CaCl_2$.

5.8. A series of mixtures of acetamide and acetic acid (x = molar fraction of acetic acid) was prepared, and the freezing point, T°C, measured: |100*x*, *T*°C |0 80.2|11.5 72.5 |20.9 63.0 |30.0 51.9 |40.2 35.6 |51.5 8.0 |51.9 −0.2 |53.6 −0.5 |55.8 −0.6 |57.9 −1.8 |60.3 −2.6 | 62.5 −3.9 |68.2 −9.2 |69.9 −11.7 |70.0 −11.1 |78.9 0.0 |89.0

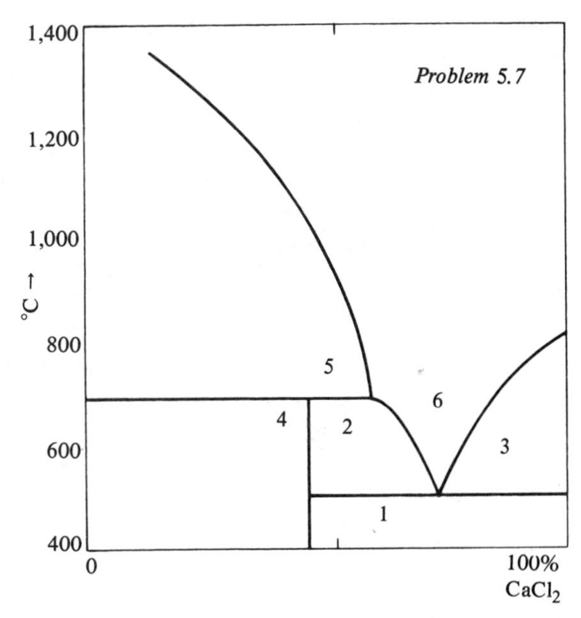

8.9 | 100.0 16.6 |. Under what conditions can three phases be in equilibrium in this temperature range? [Sisler, Davidson, Stoenner, and Lyon, JACS 66, 1888 (1944)].

5.9. The diagram shows equilibria of a mixture of *A* and *B*.

a. Identify the phases present in each area.
b. Draw cooling curves for the compositions at *a, b,* and *c*.
c. Find the maximum weight of pure *B* obtainable by crystallization from 100 kg of a melt having an initial composition of 90 wt% B.

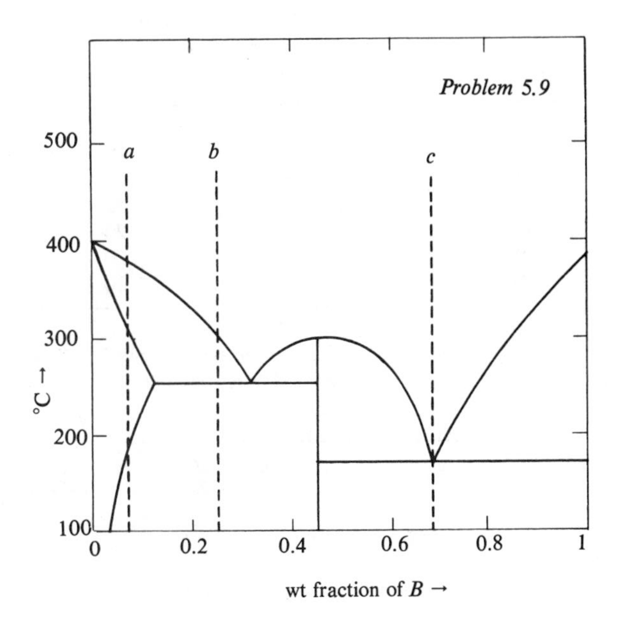

5.10. Vapor-liquid-liquid equilibria of substances A and B are shown on the diagram.

a. At what temperature would a mixture containing 25 wt% A begin to boil?

b. If the mixture of part a is distilled until the temperature has risen 10°, what is the composition of the residue?

c. What is the composition of the distillate?

d. If the distillate of part b is redistilled, what are the initial boiling point and composition?

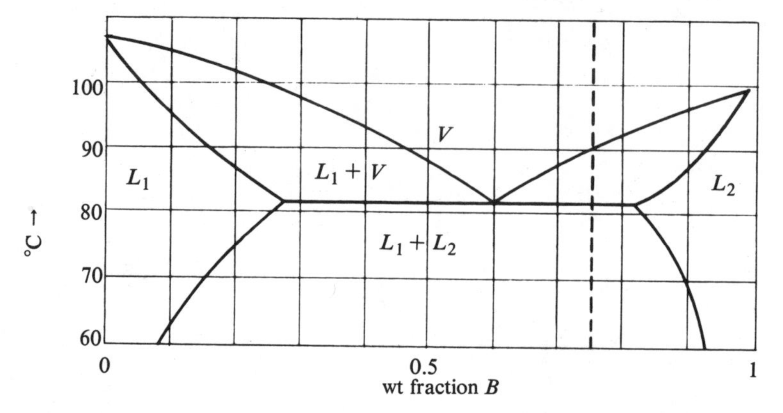

5.11. Hudson [*Z. Physik. Chem.* 47, 113 (1904)] made a phase study of the liquid region of the system nicotine-water.

Weight percent nicotine in the total system	*Temperature, °C at which second phase appears*	*Temperature, °C at which the system becomes homogeneous again*
6.8	94	95
7.8	89	155
10	75	—
14.8	65	200
32.2	61	210
49.0	64	205
66.8	72	190
80.2	87	170
82.0	129	130

Draw the temperature-composition phase diagram of this system. Label all areas and indicate the composition(s) of all phase(s) in equilibrium with one another under the following conditions:

	Weight percent nicotine in total system	*Temperature, °C*
(a)	20	45
(b)	20	120
(c)	20	220
(d)	90	120

5.12. The following data are given by Andrews [*J. Phys. Chem.* 29, 1041 (1925)] for the system *o*-dinitrobenzene-*p*-dinitrobenzene:

Mol % of Para Compound	*Initial Melting Point, °C*
100	173.5
90	167.7
80	161.2
70	154.5
60	146.1
50	136.6
40	125.2
30	111.7
20	104.0
10	110.6
0	116.9

Construct a temperature-composition diagram for the system, and determine therefrom the eutectic temperature and composition.

Find the maximum percentage of *p*-dinitrobenzene which can be recovered pure by crystallization from mixtures of the two compounds containing originally 95%, 75%, and 45% of the para compound.

5.13. For each of the following systems determine the number of components.

a. $NH_4Cl(c)$, $NH_4^+(aq)$, $Cl^-(aq)$, $H_2O(liq)$, $H_3O^+(aq)$, $H_2O(g)$, $NH_3(g)$, $OH^-(aq)$, $NH_4OH(aq)$.

b. $NH_4Cl(c)$, $NH_3(g)$, $HCl(g)$, where the partial pressure of NH_3 is always equal to the partial pressure of HCl as in the case where all the gas is formed by the sublimation of $NH_4Cl(c)$.

c. $NH_4Cl(c)$, $NH_3(g)$, $HCl(g)$, where the partial pressure of NH_3 is not necessarily equal to the partial pressure of HCl.

d. $CH_3COONH_4(c)$, $CH_3COO^-(aq)$, $NH_4^+(aq)$, $H_3O^+(aq)$, $NH_3(aq)$, $OH^-(aq)$, $CH_3COOH(aq)$, $H_2O(liq)$, $H_2O(g)$, taking hydrolysis into account.

e. $NaCl(c)$, $KBr(c)$, $K^+(aq)$, $Na^+(aq)$, $Cl^-(aq)$, $Br^-(aq)$, $H_2O(liq)$, $H_2O(g)$.

f. $NaCl(c)$, $KCl(c)$, $Na^+(aq)$, $Cl^-(aq)$, $H_2O(liq)$, $H_2O(g)$.

g. $CaCl_2 \cdot 6H_2O(c)$, $Ca^{2+}(aq)$, $Cl^-(aq)$, $H_2O(liq)$, $H_2O(g)$.

h. $CaCO_3(c)$, $CaO(c)$, $CO_2(g)$, where all the CaO and CO_2 in the system are formed by the decomposition of $CaCO_3(c)$

[Labowitz & Arents, *Physical Chemistry Problems and Solutions* (1969)].

5.14. Give for each of the following systems the number C of components (propose a suitable set of components), a list of the internal variables, and the number F of degrees of freedom (propose a suitable set of independent internal variables):

i. (a) $CaCO_3$ in equilibrium with CaO and CO_2. (b) NiO and Ni in equilibrium with H_2O, H_2, CO_2, and CO.
ii. (a) Hydrogen and oxygen are in a gas phase and dissolved in water at 25°C. (b) Water vapor is heated to 2,000°C, at which temperature it also contains H_2, O_2, OH, O, and H. (c) An arbitrary mixture of hydrogen and oxygen is heated to 2,000°C, at which temperature H_2O, OH, etc., are formed.
iii. (a) Liquid N_2O_4 is in equilibrium with a gas, containing NO, N_2O_3, NO_2, and N_2O_4. (b) Nitrogen, oxygen, and NO are mixed at 25°C; by reaction between NO and oxygen, NO_2 and other oxides of nitrogen are formed. (c) The mixture in (b) is heated to 3,000°C.
iv. (a) A mixture of solid NaCl, KCl, $NaNO_3$, and KNO_3 is shaken to equilibrium with water. (b) A mixture of NaCl and KNO_3 is shaken to equilibrium with water. (c) An aqueous solution contains a different cations, b different anions, and c different complexes between them (e.g., Na^+, H^+, K^+, Hg^{2+}; SO_4^{2-}, Cl^-; HSO_4^-, $HgSO_4$, $HgCl^+$, $HgCl_2$)

[Sillen et al., *Problems in Physical Chemistry* (1952)].

5.15. A vessel in the shape of a truncated cone contains equal weights of the six immiscible liquids mentioned in Bowden's "Phase Rule". These are hexane ($\rho = 0.65$), water ($\rho = 0.99$), aniline ($\rho = 1.02$), phosphorus ($\rho = 1.83$), gallium ($\rho = 6.1$), and mercury ($\rho = 13.6$). One end of the vessel is 1 cm dia, and the other is 3.5 cm dia. The composite liquid is balanced against a column of brine with density $\rho = 1.15$. What is the relative height of the brine column?

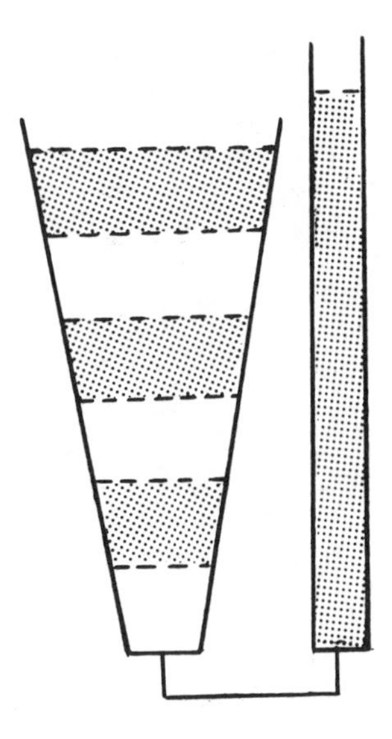

5.16. In Figure 5.6(b) the compositions of the three phases are $\alpha(0.5, 0.2, 0.3)$; $\beta(0.1, 0.7, 0.2)$; and $\gamma(0.2, 0.1, 0.7)$. Find the proportions of the three phases as a function of the composition x_1 when $x_3 = 0.35$.

6 Vapor-Liquid Equilibrium

Coexistence of vapor and liquid phases occurs so frequently and is of such great importance that equilibria of this kind have been studied intensively and are perhaps better understood than any other kind.

6.1. VAPOR AND LIQUID PHASES

This chapter is concerned with finding the state of equilibrium between liquid and vapor phases when particular conditions or properties are specified. The properties in question may be the compositions and relative amounts of the phases, the temperature, pressure, enthalpy, entropy, or some other thermodynamic property. Equilibria are expressed most generally in terms of fugacity and activity coefficients.

Any case of equilibrium between phases requires equality of T, P, and partial fugacities of individual components throughout the system, that is,

$$T_{(1)} = T_{(2)} = \dots \quad \text{for all phases}_{(j)} \tag{6.1}$$

$$P_{(1)} = P_{(2)} = \dots \quad \text{for all phases} \tag{6.2}$$

$$\hat{f}_{i(1)} = \hat{f}_{i(2)} = \dots \quad \text{for all phases} \quad (i = 1, 2, \dots, n \text{ components}) \tag{6.3}$$

(but see Section 2.10.2 for an exception to the pressure rule). It is convenient to distinguish between systems having a vapor phase in contact with one liquid phase or with several liquid phases.

6.1.1. One Liquid Phase

The condition of equilibrium between a vapor and a liquid,

$$\hat{f}_{iV} = \hat{f}_{iL} \tag{6.4}$$

becomes, in terms of fugacity coefficients,

$$y_i \hat{\phi}_{iV} P = x_i \hat{\phi}_{iL} P, \tag{6.5}$$

where the partial fugacity coefficients may be derivable from an equation of state. When this is not possible for the liquid phase, activity coefficients are introduced:

$$y_i \hat{\phi}_{iV} P = x_i \gamma_i f_{iL} \tag{6.6}$$

$$= x_i \gamma_i \phi_i^s P_i^s \, (PF)_i, \tag{6.7}$$

where the Poynting Factor,

$$(PF)_i = \exp \int_{P_i^s}^{P} (\bar{V}_{iL}/RT)\, dP, \tag{6.8}$$

provides correction of the liquid-phase fugacity from the vapor pressure to the system pressure.

Eq. 6.7 may be rearranged into an apparently direct proportionality between the phase compositions:

$$y_i = \frac{\gamma_i \, \phi_i^s P_i^s (PF)_i}{\hat{\phi}_i \, P} x_i. \tag{6.9}$$

Since, however, the partial fugacities and activity coefficients depend on the compositions of their respective phases as well as on the temperature and pressure, equilibrium compositions under particular conditions must be found by successive approximation, as will be explained later. Occasionally, some simplifications of Eq. 6.9 are justifiable: (1) At relatively low pressures the $(PF)_i$ approach unity; (2) the ratio of fugacity coefficients may be nearly unity. Several combinations of equations for fugacity and activity coefficients will be discussed later.

6.1.2. Several Liquid Phases

When more than one liquid phase is present, Eq. 6.4 is simply extended to all of them. Taking the case of two liquid phases and identifying the second one by asterisks, Eqs. 6.5–6.7 become:

$$y_i \hat{\phi}_{iV} P = x_i \hat{\phi}_{iL} P = x_i^* \hat{\phi}_{iL}^* P \tag{6.10}$$

$$= x_i \gamma_i f_{iL} = x_i^* \gamma_i^* f_{iL} \tag{6.11}$$

$$= x_i \gamma_i \phi_i^s P_i^s (PF)_i = x_i^* \gamma_i^* \phi_i^s P_i^s (PF)_i. \tag{6.12}$$

In cases of practical interest more than two or three liquid phases rarely are present. The most common multiphase liquid situation in organic systems occurs when water is present. In a typical case, the amounts and compositions of all phases in equilibrium are to be determined. Problems involving immiscible liquids will be covered in this chapter, but consideration of more complex liquid-liquid-vapor equilibria is deferred to Chapter 7.

6.1.3. Vapors above Their Critical Temperatures

In the equation for the partial fugacity in the liquid phase,

$$\hat{f}_{iL} = x_i \gamma_i f_{iL}, \tag{6.13}$$

the liquid state at which the pure component fugacity is evaluated is a hypothetical one when the substance is a vapor above its critical temperature. Several ways of evaluating the behavior in the liquid are being used:

1. The question of supercriticality is avoided by employing Eq. 6.5 with the same equation of state for both phases. This is often possible for light hydrocarbons and other nonpolar substances.

2. Values of f_{iL} obtained from

$$f_{iL} = \frac{\hat{\phi}_{iV} P}{\gamma_i} \frac{y_i}{x_i} \tag{6.14}$$

with experimental data in the measurable range are correlated in terms of suitable variables and are considered extrapolatable. One such correlation is described in Section 6.5.

3. A correlation of f_{iL} is made in terms of acentric factor and reduced temperature and pressure, for example, by Chao & Seader (Section 6.2.3).

4. Extrapolation of an accurate vapor pressure equation (not the Antoine) is used to evaluate the liquid fugacity from

$$f_{iL} = P_i^s \phi_i^s (PF)_{i0}, \tag{6.15}$$

and correlated as

$$\ln f_{iL} = C_{1i} + C_{2i}/T + C_{3i}T + C_{4i} \ln T + C_{5i} T^2 \tag{6.16}$$

by Prausnitz et al. (1980, Appendix B). Values of the coefficients are given for 92 substances. The Poynting Factor, $(PF)_{i0}$, is the pressure correction from the saturation pressure, P_i^s, to 0.

5. The partial fugacity coefficient of the liquid may be expressed in terms of a Henry's Law type coefficient,

$$\hat{f}_{iL} = k_{Hi} x_i. \tag{6.17}$$

The dependence of k_{Hi} on the composition of the mixture and other properties is discussed in Section 3.5 and in greater detail by Van Ness & Abbott (1982, pp. 248–261) and Prausnitz et al. (1980, p. 55).

6. King et al. (1977) employ Eq. 6.7 with the SHFH and Antoine equations to achieve a successful representation of data of substances whose critical temperatures are not more than 50 C or so lower than the temperature of the system. This work is discussed in Section 6.5.

Methods (1) and (3) particularly are covered in some detail in this chapter.

6.2. VAPORIZATION EQUILIBRIUM RATIO

The distribution coefficient,

$$K_i = y_i/x_i, \tag{6.18}$$

also called the vaporization equilibrium ratio, abbreviated VER, is a key quantity in the analysis of vapor-liquid equilibria. Many relations are expressible compactly in terms of the K_i, but because they depend on T, P, and the compositions of the two phases, solutions of typical problems in terms of K_i must be accomplished by successive approximation. Ideal values independent of composition are readily evaluated, however, and serve as starting values for more accurate solutions.

By utilizing Eqs. 6.5–6.7, the VER may be given several forms:

$$K_i = y_i/x_i = \hat{\phi}_{iL}/\hat{\phi}_{iV} \tag{6.19}$$

$$= \gamma_i f_{iL}/\hat{\phi}_{iV}P \tag{6.20}$$

$$= \gamma_i \phi_{iL}/\hat{\phi}_{iV} \tag{6.21}$$

$$= \frac{\gamma_i \phi_i^s P_i^s (PF)_i}{\hat{\phi}_{iV}P}. \tag{6.22}$$

In some cases the group of terms, $\phi_i^s(PF)_i/\hat{\phi}_{iV}$, is little different from unity (see Example 6.1) so that

$$K_i \sim \gamma_i P_i^s/P, \tag{6.23}$$

which is often taken to be adequate below pressures of 5-6 atm or so. Ideal values of the VER are given by

$$(K_i)_{\text{ideal}} = P_i^s/P. \tag{6.24}$$

The several ways of evaluating K_i by Eqs. 6.19–6.22 will be taken up separately. An exhaustive appraisal of many different ways was made by Daubert et al. (1979) for the API; the conclusions are summarized in Section 6.2.6. Implementation of the K_i will be made in calculations of dewpoints, bubblepoints, and flashes.

6.2.1. Equations of State

Some of the established equations of state that are suitable under some conditions for calculating fugacity coefficients of both phases are the Soave, Peng-Robinson, BWRS, and Plöcker-Lee-Kesler. The first of these is adopted here for illustrative purposes. Pertinent formulas are in Tables 1.11, 3.3, and 3.4. The polynomial form of the equation,

$$z^3 - z^2 + (A - B - B^2)z - AB = 0 \tag{6.25}$$

is solved for the largest root at a given vapor composition and for the smallest root at a given liquid composition. The two individual partial fugacity coefficients then are found from the Soave equation,

$$\ln \hat{\phi}_i = -\ln\left[z\left(1 - \frac{b}{V}\right)\right] + \frac{b_i}{b}(z-1) + \frac{a\alpha}{bRT}\left[\frac{b_i}{b} - \frac{2\sqrt{(a\alpha)_i}}{\sqrt{a\alpha}}\right]\ln\left(1 + \frac{b}{V}\right), \tag{6.26}$$

after substitution of the appropriate sets of compositions and mixture compressibilities.

The BWR equation has long been a standard for figuring the K_i of both phases, but its complexity makes it unattractive to some workers for repetitive calculations such as those in distillation problems, so simpler methods have been developed. For example, a multicomponent distillation program by Christiansen et al. (1979) incorporates the Soave equation. Results obtained by a general algorithm with emphasis on the critical and high-pressure regions are illustrated with the Soave equation by Asselineau et al. (1979). A distillation scheme using either the Soave or Peng-Robinson equation for figuring the K_i in cryogenic applications was found superior to the Chao-Seader method by Shah & Bishnoi (1978). Sim & Daubert (1980) found the Soave method the most reliable for flash calculations of petroleum mixtures. They divided the mixture into cuts with a spread of about 25 C in boiling range, and related the average boiling point, T_b, and the specific gravity, S, to the molecular weight, M, and the critical properties that are required in the Soave equation. These empirical relations are

$$M = 5.805(10^{-5})\,T_b^{2.3776}/S^{0.9371}, \tag{6.27}$$

$$P_c = 6.1483(10^{12})\,S^{2.4853}/T^{2.3177},\ \text{Pa}, \tag{6.28}$$

$$T_c = 0.5776 \exp(4.2009\,T_b^{0.08615}S^{0.04614}),\ {}^\circ K. \tag{6.29}$$

6.2.2. Vapor-Pressure Method

In Eq. 6.9 both fugacity coefficients pertain to the vapor phase, so any suitable equation of state applicable to the gas phase only may be used. When using Eq. 6.26, for example, the fugacity of the saturated vapor is found after making $a_i = a$ and $b_i = b$. Any suitable equation may be used for the activity coefficient; but the Scatchard-Hildebrand equation is convenient because it depends on properties of pure components only. With an accurate equation, a modest extrapolation of the vapor pressure above the critical temperatures of some of the components may be permissible. Other methods of handling supercritical gases are mentioned in Sections 6.1.3 and 6.5.

6.2.3. Correlation of Pure-Liquid Fugacity Coefficients

The longest established method of this type is that of Chao & Seader (1961). Other papers devoted to the method include: Grayson & Streed (1963); Robinson & Chao (1971); Lee et al. (1973); Maffiolo et al. (1975). Here the basic Chao-Seader relations with the Grayson-Streed numerical values of the coefficients and the extended Scatchard-Hildebrand equation proposed by Robinson & Chao and by Maffiolo et al. will be adopted. Example 6.7 uses this method.

Formulas, numerical coefficients, and restrictions on the use of the method are summarized in Table 6.1. The restrictions are those proposed in the original paper and

Table 6.1. The Chao-Seader Method

1. $K_i = \gamma_i \phi_{iL} / \hat{\phi}_i$.
2. $\gamma_i = \exp[V_i(\delta - \bar{\delta})^2/RT + \ln \theta_i + 1 - \theta_i]$, $\theta_i = \dfrac{V_i}{\Sigma x_i V_i}$.
3. $\hat{\phi}_i$ from the Redlich-Kwong equation, Table 3.4.
4. $\log \phi_{iL} = \log \phi_{iL}^{(0)} + \omega \log \phi_{iL}^{(1)}$.
5. $\log \phi_{iL}^{(0)} = A_0 + \dfrac{A_1}{T_{r_i}} + A_2 T_{r_i} + A_3 T_{r_i}^2 + A_4 T_{r_i}^3 + (A_5 + A_6 T_{r_i} + A_7 T_{r_i}^2) P_{r_i} + (A_8 + A_9 T_{r_i}) P_{r_i}^2 - \log P_{r_i}$.
6. $\log \phi_{iL}^{(1)} = A_{10} + A_{11} T_{r_i} + \dfrac{A_{12}}{T_{r_i}} + A_{13} T_{r_i}^3 + A_{14}(P_{r_i} - 0.6)$.
7. Values of the coefficients:

	Simple Fluid, ω = 0	*Methane*	*Hydrogen*
A_0	2.05135	1.36822	1.50709
A_1	− 2.10899	− 1.54831	2.74283
A_2	0	0	− 0.02110
A_3	− 0.19396	0.02889	0.00011
A_4	0.02282	− 0.01076	0
A_5	0.08852	0.10486	0.008585
A_6	0	− 0.02529	0
A_7	− 0.00872	0	0
A_8	− 0.00353	0	0
A_9	0.00203	0	0

$A_{10} = -4.28393$, $A_{12} = -1.22060$, $A_{14} = -0.025$
$A_{11} = 8.65808$, $A_{13} = -3.15224$

8. Restrictions:
   a. $T < 500°F$ (260 C), $T > 0°F$ (−18°C)
   b. $P < 1{,}000$ psia (6.89 MPa).
   c. For hydrocarbons (except methane), $0.5 < T_{r_i} < 1.3$ and mixture reduced critical pressure < 0.8.
   d. For systems containing methane and/or hydrogen, molal average $T_r < 0.93$, and methane mole fraction < 0.3. Mole fraction of other dissolved gases < 0.2.
   e. When predicting K-values of paraffins or olefins, liquid-phase aromatic mole fraction should be < 0.5. Conversely, when predicting K-values of aromatics, liquid-phase aromatic mole fraction should be > 0.5.
9. *Caution:* The logarithms in these equations are to the base 10!

Table 6.2. Pure Component Properties for Use with the Chao-Seader Method. [Additional data are given by Henley & Seader (1981) taken mostly from the FLOWTRAN Data Bank.]

	Modified ω	δ, *(cal./ml.)*$^{1/2}$	*V, ml./g.-mole*
Hydrogen		3.25	31
Paraffins			
Methane		5.68	52
Ethane	0.1064	6.05	68
Propane	0.1538	6.40	84
i-Butane	0.1825	6.73	105.5
n-Butane	0.1953	6.73	101.4
i-Pentane	0.2104	7.02	117.4
n-Pentane	0.2387	7.02	116.1
neo-Pentane	(0.195)	7.02	123.3
n-Hexane	0.2927	7.27	131.6
n-Heptane	0.3403	7.430	147.5
n-Octane	0.3992	7.551	163.5
n-Nonane	0.4439	7.65	179.6
n-Decane	0.4869	7.72	196.0
n-Undecane	0.5210	7.79	212.2
n-Dodecane	0.5610	7.84	228.6
n-Tridecane	0.6002	7.89	244.9
n-Tetradecane	0.6399	7.92	261.3
n-Pentadecane	0.6743	7.96	277.8
n-Hexadecane	0.7078	7.99	294.1
n-Heptadecane	0.7327	8.03	310.4
Olefins			
Ethylene	0.0949	6.08	61
Propylene	0.1451	6.43	79
1-Butene	0.2085	6.76	95.3
cis-2-Butene	0.2575	6.76	91.2
trans-2-Butene	0.2230	6.76	93.8
i-Butene	0.1975	6.76	95.4
1,3-Butadiene	0.2028	6.94	88.0
1-Pentene	0.2198	7.05	110.4
cis-2-Pentene	(0.206)	7.05	107.8
trans-2-Pentene	(0.209)	7.05	109.0
2-Methyl-1-Butene	(0.200)	7.05	108.7
3-Methyl-1-Butene	(0.149)	7.05	112.8
2-Methyl-2-Butene	(0.212)	7.05	106.7
1-Hexane	0.2463	(7.40)	125.8
Naphthenes			
Cyclopentane	0.2051	8.11	94.7
Methylcyclopentane	0.2346	7.85	113.1
Cyclohexane	0.2032	8.20	108.7
Methylcyclohexane	0.2421	7.83	128.3
Aromatics			
Benzene	0.2130	9.16	89.4
Toluene	0.2591	8.92	106.8
o-Xylene	0.2904	8.99	121.2
m-Xylene	0.3045	8.82	123.5
p-Xylene	0.2969	8.77	124.0
Ethylbenzene	0.2936	8.79	123.1

extended by Lenoir & Koppany (1967). In Eq. 6.21 the partial fugacity coefficient is obtained with the Redlich-Kwong equation, and the fugacity coefficient of the pure liquid is correlated by a Pitzer-Curl type relation in terms of the acentric factor:

$$\ln \phi_{iL} = \ln \phi_L^{(0)} + \omega \ln \phi_{iL}^{(1)}. \quad (6.30)$$

The logarithmic terms on the right are given in terms of T_r and P_r. Originally the activity coefficient was given by the Scatchard-Hildebrand equation,

$$\ln \gamma_i = V_i(\delta_i - \bar{\delta})^2/RT, \quad (6.31)$$

but the extended form given in Table 6.1 is superior; in that equation the molal volume fractions are represented by θ_i.

Values of the solubility parameters, acentric factors, and molal volumes on which this correlation is based are in Table 6.2. All the values are assumed independent of temperature. For some of the lighter substances, δ_i are not true values but are adjusted to fit vapor pressure and experimental K_i data; some 2,000 data points were used in developing this correlation. The modification of this system by Robinson & Chao makes the method more complex, but it incorporates methane into the general pattern and is claimed to be generally better at temperatures down to −45 C.

For substances with critical temperatures above that of the system, the modified method of Maffiolo et al. (1975) is significantly more accurate. They use the extended SH equation for the activity coefficient (Eq. 2 of Table 6.1), and for the liquid fugacity they write

$$f_{iL} = \phi_i^s P_i^s (PF)_i, \quad (6.32)$$

where the fugacity of the saturated vapor is obtained with the Redlich-Kwong equation. Below two atmospheres or so the vapor pressure is represented by Antoine's equation; above that pressure and up to the critical, they use a five-coefficient equation. They retain the original Chao-Seader correlation for supercritical vapors.

According to an analysis by Coward & Webb (1978), the use of the Redlich-Kwong equation in the Chao-Seader method can give rise to nonunique dew and bubblepoints. They propose some guidelines for appraising the physical realism of particular solutions.

6.2.4. Methods with the Virial Equation

The B-truncated virial equation is satisfactory for representing the vapor phase at pressures given by

$$P_r \le 0.5\, T_c/\Sigma y_i T_{ci}, \quad (6.33)$$

or correspondingly for

$$\rho/\rho_c \le 0.5. \quad (6.34)$$

It has the important numerical advantages of having only a single root and a relatively simple form of the equation for the fugacity coefficient—namely:

$$\ln \hat{\phi}_i = \frac{P}{RT}\left(2 \sum_j y_i B_{ij} - B\right), \quad (6.35)$$

where

$$B = \Sigma\Sigma y_i y_j B_{ij}, \quad (6.36)$$

and for pure substances

$$\ln \phi_i = B_i P/RT. \quad (6.37)$$

Many data and correlations of second virial coefficients in terms of pure component properties are available. In the range of the majority of distillation problems, for instance, the virial equation is quite adequate.

When the virial and SH equations are combined to form the VER, it assumes the forms:

$$K_i = \frac{\gamma_i \phi_i^s P_i^s (PF)_i}{\hat{\phi}_{iV} P} \quad (6.22)$$

$$= \frac{P_i^s (PF)_i \exp\left[\frac{V_i}{RT}(\delta_i - \bar{\delta})^2\right] \exp\frac{B_i P_i^s}{RT}}{P \exp\left[\frac{P}{RT}\left(2\sum_j y_j B_{ij} - B\right)\right]} \quad (6.38)$$

$$= \frac{P_i^s (PF)_i}{P} \exp\left\{\frac{1}{RT}\left[(V_i(\delta_i - \bar{\delta})^2 + B_i P_i^s + BP - 2P\sum_j y_j B_{ij}\right]\right\}. \quad (6.39)$$

Only pure component properties are involved in this equation. The extended SHFH equation could be incorporated instead of the original SH.

When the data are available, other activity-coefficient equations, such as the Wilson, may be preferable. The system of Prausnitz et al. (1980) combines the virial and UNIQUAC equations, which is also done in the book by Fredenslund et al. (1977). In another paper, Fredenslund et al. (1977) describe a distillation program based on UNIFAC activity coefficients and the Hayden-O'Connell (1975) second virial coefficients. A new EOS is combined with a "surface regular solution" model of the activity coefficient by Drahos et al. (1977); this method requires one binary parameter per pair and is recommended for the range of temperatures 127 K–500 K at pressures up to 0.9 times the convergence pressure of the mixture. The NRTL equation is highly favored for the liquid phase, and many data are available.

6.2.5. Chart Methods

A number of nomograms and network charts representing data on vaporization equilibrium ratios as functions of T, P, and some composition parameter has been devised. At present their utility is more for spot checking and occasional manual calculation, since at least equivalent analytical methods suitable for computer use are available. These methods still have some value, however, for high-boiling petroleum mixtures and for understanding or modification of older plant designs that may have been made with their aid. Two of the currently accepted systems of charts will be described briefly.

6.2.5.1. Convergence pressure

At the critical point of a mixture the compositions of vapor and liquid phases become identical; an equivalent statement is that for every component the vaporization equilibrium ratio becomes unity, $K_i = 1$. For example, the plots of critical

Example 6.1. Fugacity Coefficient Ratio, $\hat{\phi}_i/\phi_i$, and Poynting Factor at Several Pressures with the *B*-Virial Equation

This will be found for equimolal methane (1) + pentane (2) at 100 F. Experimental virial coefficients are available (Kogan, 1967):

$B_{11} = -38$ ml/gmol,

$B_{12} = -196$,

$B_{22} = -1{,}220$.

Liquid molal volumes are

$V_1 = 52$ ml/gmol,

$V_2 = 116.1$.

Equations for the fugacity coefficients are in Table 3.4. The Poynting factors are calculated for pressures in excess of 1 atm.

$$\ln(\hat{\phi}_i/\phi_i) = \frac{P}{RT}(1 - y_i)^2(2B_{12} - B_{11} - B_{22})$$

$$= \frac{0.25(-392 + 38 + 1220)P}{82.05\,(310.9)} = 0.0085\,P,$$

$$(PF)_i = \exp\left[\frac{V_{iL}}{RT}(P - 1)\right]$$

	$P = 5$		$P = 10$		$P = 25$	
	$\hat{\phi}/\phi$	PF	$\hat{\phi}/\phi$	PF	$\hat{\phi}/\phi$	PF
CH_4	1.0433	1.0082	1.0886	1.0185	1.2364	1.0501
C_5H_{12}	1.0433	1.0184	1.0886	1.0412	1.2364	1.1154

In this case, the ratio $(PF)_i/(\hat{\phi}_i/\phi_i)$ is appreciably different from unity even at at 5 atm. However, normally the Poynting Factor is figured at pressures in excess of the vapor pressure and may be more nearly unity in such cases than with $P = 1$ as the lower limit of integration.

pressures of binary mixtures of Figures 1.30 and 1.31 show the wide range of combinations of temperature and composition at which $K_1 = K_2 = 1$. The behavior of multicomponent mixtures is similar. The pressure at which all VERs become unity is called the *convergence pressure*. Clearly it is also a measure of the composition of the mixture at a specified temperature, and that is the use to which it is put in correlating the VERs in complex mixtures.

As an example, data of mixtures of propane and benzene are plotted in Figure 6.11 as K_i against the pressure at two different temperatures; the rapid change of the K of the heavier component shown in the vicinity of the critical point is representative. Figure 1.32 represents one of the few sets of ternary data that have been recorded; temperatures and compositions at which $K_1 = K_2 = K_3 = 1$ can be deduced from such data.

The true critical pressure of a multicomponent mixture can be computed with some equations of state, but this process is too time consuming for everyday use with K-charts. In practical work, estimates of convergence pressures are used instead. At a particular temperature the convergence pressure is taken to be that of an equivalent binary mixture devised according to some definite rules. In both the API and NGPSA systems these rules are:

1. Estimate the composition of the liquid phase,
2. As the lighter component of the equivalent binary, choose the lightest substance other than hydrogen that is present to at least 0.1 mol%,
3. The heavier component is assumed to have the weight average critical temperature and pressure of all the remaining components of the liquid phase except the heaviest 2 mol%.
4. Locate the convergence pressure on the critical locus of a binary diagram like Figure 6.2; expanded diagrams are given in the references.
5. Use this convergence pressure, P_{cv}, with the appropriate charts for the K_i of all the substances in the mixture.
6. Calculate the phase composition by the bubblepoint method. If this does not match the estimate of step 1, recycle the computation.

Other schemes for dividing up a mixture into two equivalent components have been used; one of these simply divides the mixture into two groups and figures their weight average boiling points. Some historical background of the concept of convergence pressure is given by Edmister (1949).

6.2.5.2. *The NGPSA system (Natural Gasoline Processors Suppliers Association)*

Charts like the sample of Figure 6.4 are provided in this reference for methane through decane, some inorganic gases but not hydrogen, and petroleum fractions characterized by average boiling points to 800 F, at discrete convergence pressures from 800 to 10,000 psia. Special charts are available for several binaries containing methane and one for ethane + nitrogen. Over limited ranges, interpolation for temperature or pressure can be made with such typical equations as

$$\ln K = A - B/(T + C), \tag{6.40}$$

$$(K/T)^{1/3} = a + bT + cT^2, \tag{6.41}$$

$$\ln K = a + b/T + c \ln P. \tag{6.41a}$$

6.2.5.3. The API system (American Petroleum Institute)

This includes several nomograms for different temperature ranges, for mixtures of hydrocarbons with nonhydrocarbons, and for hydrogen. The identity of the substance, the temperature and a "grid" pressure determine the K_i. The grid pressure is determined by the system pressure and the convergence pressure, as in Figure 6.3. At high convergence pressures, or low system pressures, the system and grid pressures become the same.

As they stand the API and NGPSA systems are not usable for computer solutions of equilibrium problems. What is usually done is to input several values from the charts in the range of the problem at hand and to provide an interpolation routine, by equations such as 6.40 and 6.41 or a more complex form:

$$\ln KP = a + b/T + c \ln T + dT + e\,T^2. \tag{6.42}$$

Several proprietary computerizations of these schemes apparently exist. One is mentioned by Dowling & Todd (1973) but no details are given. Curve-fits with the objective of extending the API system to high-boiling petroleum fractions have been made by Zhvanetskii & Platonov (1978) and possibly could be used for specific light hydrocarbons. They point out rightly that at the pressures at which crude oil fractions are usually handled the convergence pressure correction is unnecessary. As usual, the TBP of the mixture is divided up into 25 C spans which are characterized by the average boiling points. The K of a cut is related to those of ethane, K_e, and heptane, K_h, by the equation

$$K = K_h^{\beta+1}/K_e^{\beta}. \tag{6.43}$$

The Ks of the reference substances are functions of the temperature, °C, and the pressure, bars,

$$\ln K = a_0 + a_1/T + a_2/T^2 + a_3/T^3 + a_4/P + a_5/P^2 + a_6/TP + a_7/T^2P + a_8/T^3P, \tag{6.44}$$

$$32 < T < 473 \text{ C}, \; 1 < P < 10 \text{ atm}$$

for which the coefficients are given in the following table. The volatility index, β, is a function of the average boiling point, T°C, of a petroleum cut:

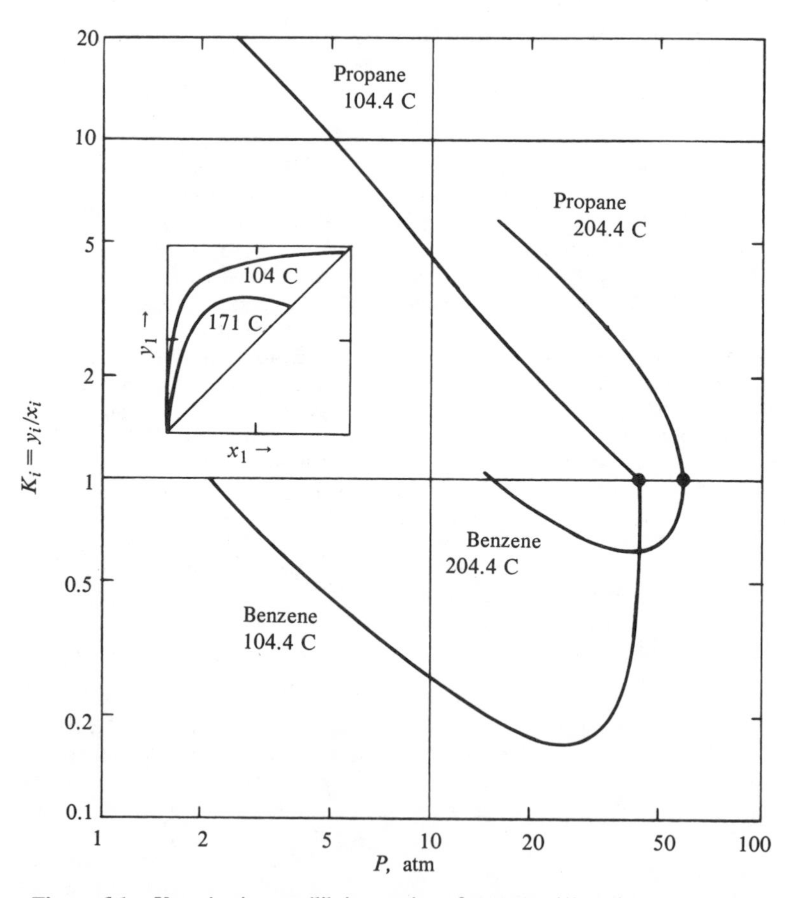

Figure 6.1. Vaporization equilibrium ratios of propane (1) and benzene (2) [data of Glanville, Sage, & Lacey, Ind Eng Chem 42, 508 (1952)].

$$\beta = -0.723 + 0.692(10^{-2})T - 0.291(10^{-5})T^2 + 0.263(10^{-7})T^3. \quad (6.45)$$

Deviations of values of K_i calculated with these equations from those on the chart averaged about 7%.

	Ethane	*Heptane*
a_0	$+1.750$	$+2.708$
a_1	$+3.608 \cdot 10^2$	$-6.377 \cdot 10^2$
a_2	$-6.260 \cdot 10^4$	$+4.032 \cdot 10^3$
a_3	$+2.702 \cdot 10^6$	$+8.807 \cdot 10^5$
a_4	$+4.750$	$+4.782$
a_5	-2.156	$+2.198$
a_6	$-1.386 \cdot 10^2$	$-1.723 \cdot 10^2$
a_7	$+2.192 \cdot 10^4$	$+2.328 \cdot 10^4$
a_8	$+1.060 \cdot 10^6$	$-8.993 \cdot 10^5$

Two formulas are recommended as particularly suitable for interpolation over limited ranges of T and P by Leesley & Heyen (1982, pp. 357–366):

$$\ln KP = k_1 + k_2 \ln P^{\text{sat}}, \quad (6.45\text{a})$$

which is good in the range 50–100 C below 3 atm or so, and

$$\ln K = k_1 + k_2 \ln P^{\text{sat}} + k_3 \ln P, \quad (6.45\text{b})$$

which is good at higher pressures. The effect of temperature is represented by the vapor-pressure term.

6.2.6. Comparison of Methods

The relative merits of various equations of state were brought out throughout Chapter 1, especially Section 1.9. Those that are suitable in some cases for both vapor and liquid phases have been identified, primarily the Soave, Peng-Robinson, BWRS, and Plöcker-Lee-Kesler. Activity-coefficient equations have been compared in Chapter 4, particularly Section 4.13. The advantage of the Scatchard-Hildebrand, UNIFAC, and ASOG methods is that they require only pure-component data for representation of multicomponent behavior, although the SH equation is applicable to a smaller variety of substances than the other two are. When the data are available, the Wilson or related equations may be preferred for the liquid phase.

A detailed comparison of nine analytical methods was made by Daubert et al. (1979) in connection with the API *Technical Data Book.* The methods considered were:

1. Original Chao-Seader and modifications by Grayson-Streed, Cavett, Lee-Edmister-Erbar, and Maffiolo et al.
2. Lee-Kesler.
3. Han-Starling BWR.
4. Soave.
5. Peng-Robinson.

A data bank of more than 4,000 experimental data was used for the evaluation. On the whole, the Soave and Peng-Robinson were the best for VER, the Peng-Robinson somewhat better for liquid densities, and the Lee-Kesler the best for enthalpy data. The Soave was adopted for the API *Data Book.* Binary-interaction parameters were developed for H_2S, CO_2, N_2, and CO with hydrocarbons. Hydrogen and petroleum fractions characterized by T_c, P_c, and ω were accommodated within the system. The Peng-Robinson equation was not evaluated as thoroughly as the others since it appeared while the work was in progress. Computer programs for computing VER and other relations with this equation, however, are available from the Gas Processors Association, Tulsa, at a price.

The convergence pressure calculation methods now are the same in the NGPSA and API systems. Data for hydrocarbons with H_2, H_2O, NH_3, SO_2, HCl, and a few other substances are in the API system but not in the other. No comparisons of chart and analytical methods were made in the API survey.

6.3. EQUILIBRIA WITH *K*s INDEPENDENT OF COMPOSITION

Problems such as the determination of bubble- and dewpoints and flashes are fomulated in terms of vaporization-equilibrium ratios. When those can be approximated as independent of composition, the solutions become relatively straightforward. In any event, such solutions can be good starting conditions for obtaining correct ones by successive approximation that take into account the effect of originally unknown compositions. In the definitions,

$$K_i = \frac{\hat{\phi}_{iL}}{\hat{\phi}_{iV}} = \frac{\gamma_i \phi_i^{\text{sat}} P_i^{\text{sat}} (PF)_i}{\hat{\phi}_{iV} P}, \quad (6.46)$$

the partial fugacity and activity coefficients are composition-dependent. In some cases the effect of vapor composition is relatively small (see Example 6.1) but usually this should be verified.

As a basis for discussion in this section, it will be assumed that the VER are obtained from the API or NGPSA charts. For calculator and computer procedures, it is necessary to have equations for the effects of temperature and pressure. Over the narrow ranges of particular cases, these equations with two or three coefficients may be used:

$$\ln K_i = A_i - B_i/(T + C_i), \quad (6.47)$$

$$\ln K_i = A_i - B_i/(T + 18 - 0.19T_b), \quad (6.48)$$

$$K_i = a_i P^{b_i}, \quad (6.49)$$

where T_b is the normal boiling point, °K. The coefficients are evaluated with two or three readings off the *K*-charts.

In the applications of this section the derivatives of these equations will be needed. They are:

$$\frac{\partial K_i}{\partial T} = \frac{B_i K_i}{(T + C_i)^2}, \quad (6.50)$$

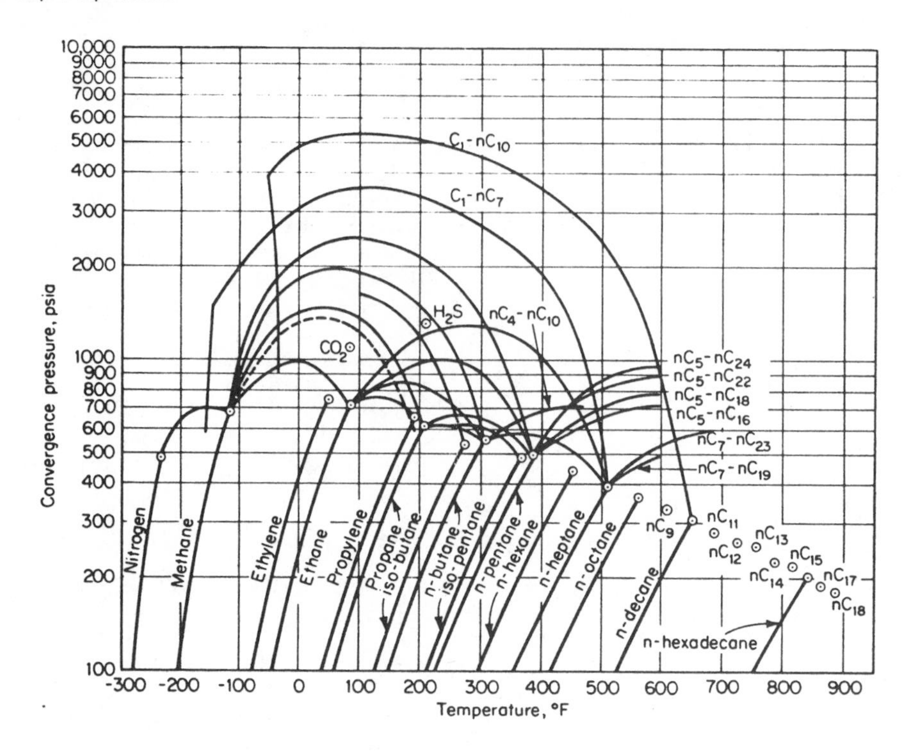

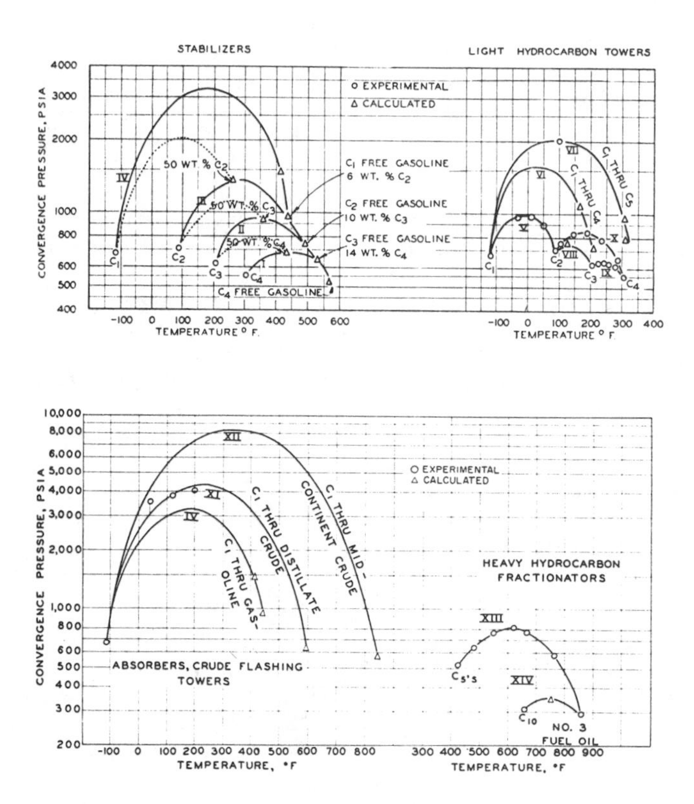

Figure 6.2. Convergence pressures of hydrocarbon mixtures. (a) Binary mixtures (*NGPSA Engineering Data Book*, 1976). (b) Typical refinery mixtures (Grayson & Streed, 1961).

$$\frac{\partial^2 K_i}{\partial T^2} = \frac{B_i[B_i - 2(T + C_i)]K_i}{(T + C_i)^4}, \tag{6.51}$$

$$\frac{\partial K_i}{\partial P} = b_i K_i / P, \tag{6.52}$$

$$\frac{\partial^2 K_i}{\partial P^2} = b_i(b_i - 1)K_i/P^2. \tag{6.53}$$

For solving the equations that crop up in this chapter, the Newton-Raphson method usually is satisfactory. For example, when an estimate $x^{(r)}$ of the root of the equation

$$f(x) = 0 \tag{6.54}$$

is known, an improved value of the root is

$$x^{(r+1)} = x^{(r)} - f(x^{(r)})/f'(x^{(r)}). \tag{6.54a}$$

When many calculations of this kind need to be made, as in distillation problems, the efficiency of a root-finding process becomes important. An often superior method, attributable to Richmond (1944), was applied by Jelinek & Hlavacek (1971) to figuring bubblepoints. The algorithm is

$$x^{(r+1)} = x^{(r)} - 2/(2f'/f - f''/f'). \tag{6.55}$$

The derivatives are easy to form when equations like Eq. 6.47 are involved. A derivation of this equation from Taylor's theorem is made in the book by Lapidus (1962).

6.3.1. Bubblepoint

The temperature at which a liquid of known composition first begins to boil is found from the equation

$$f(T) = \Sigma K_i x_i - 1 = 0, \tag{6.56}$$

where the K_i are known functions of the temperature. In terms of Eq. 6.47, the Newton-Raphson algorithm is

$$T = T - \frac{-1 + \Sigma K_i x_i}{\Sigma\left[\dfrac{B_i K_i x_i}{(T + C_i)^2}\right]} \tag{6.57}$$

The corresponding Richmond algorithm is quoted in Example 6.3. Similarly, when Eq. 6.49 represents the effect of pressure, the bubblepoint pressure is found with the N-R algorithm,

$$f(P) = \Sigma K_i x_i - 1 = 0, \tag{6.58}$$

$$P = P - \frac{-1 + \Sigma a_i P^{b_i} x_i}{\Sigma a_i b_i P_i^{b_i - 1} x_i}. \tag{6.59}$$

6.3.2. Dewpoint

The temperature or pressure at which a vapor of known composition first begins to condense is given by solution of the appropriate equation,

$$f(T) = \Sigma y_i/K_i - 1 = 0, \tag{6.60}$$

$$f(P) = \Sigma y_i/K_i - 1 = 0. \tag{6.61}$$

In terms of Eqs. 6.47 and 6.49, the N-R algorithms are

$$T = T + \frac{-1 + \Sigma y_i/K_i}{\Sigma\left[\dfrac{y_i}{K_i^2}\dfrac{\partial K_i}{\partial T}\right]} = T + \frac{-1 + \Sigma y_i/K_i}{\Sigma\left[\dfrac{B_i y_i}{K_i(T + C_i)^2}\right]}, \tag{6.62}$$

$$P = P + \frac{-1 + \Sigma y_i/K_i}{\Sigma\left[\dfrac{y_i}{K_i^2}\dfrac{\partial K_i}{\partial P}\right]} = P + \frac{(-1 + \Sigma y_i/K_i)P}{\Sigma(b_i y_i/K_i)}. \tag{6.63}$$

6.3.3. Flash at Fixed *T* and *P*

At temperatures and pressures between those of the bubblepoint and dewpoint, a mixture of two phases exists whose amounts and compositions depend on the conditions that are imposed on the system. The most common sets of such conditions are fixed *T* and *P*, or fixed *H* and *P*, or fixed *S* and *P*; they are represented in Figure 6.6 and will be considered in turn, the (*T, P*) condition now.

For each component the material balances and equilibria are:

$$Fz_i = Lx_i + Vy_i, \tag{6.64}$$

$$y_i = K_i x_i. \tag{6.65}$$

On combining these equations and introducing $\beta = V/F$, the fraction vaporized, the flash condition becomes

$$f(\beta) = -1 + \Sigma K_i x_i = -1 + \Sigma \frac{z_i}{1 + \beta(K_i - 1)} = 0, \tag{6.66}$$

and the corresponding N-R algorithm is

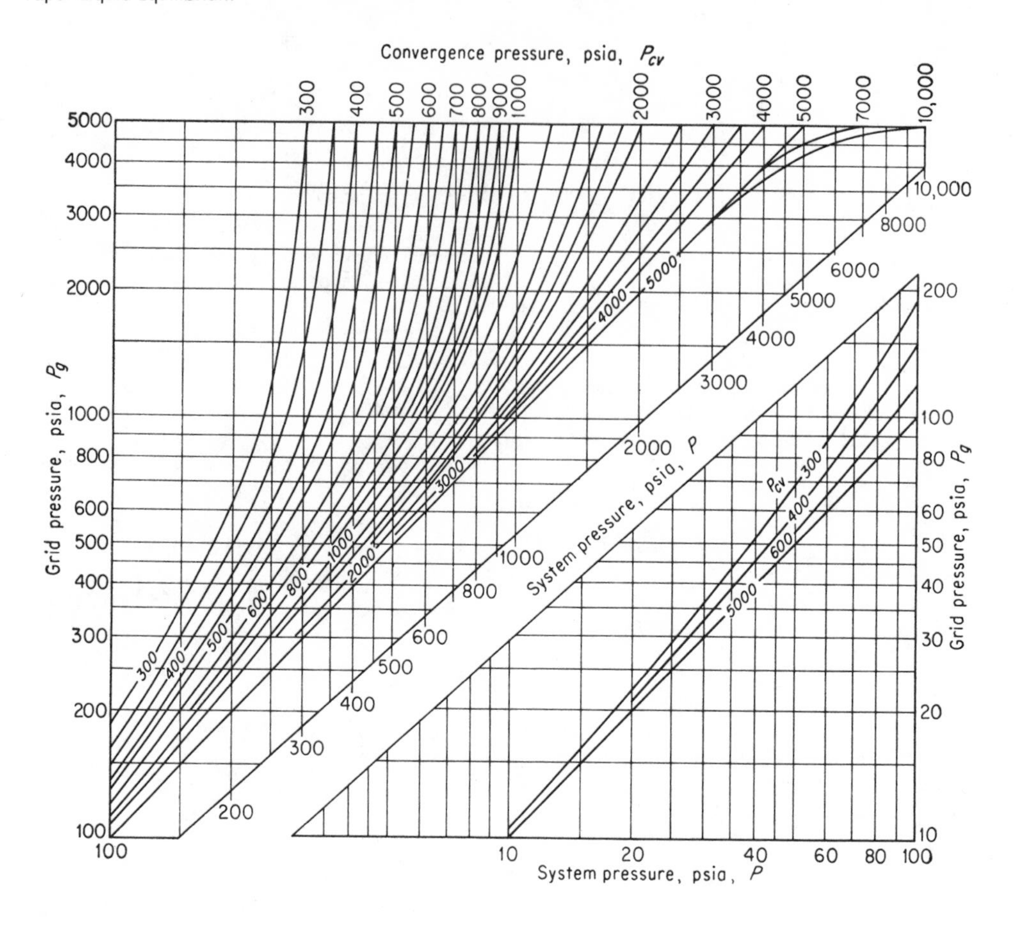

Figure 6.3. Grid pressure as a function of the system and convergence pressures for use with Figure 6.5. (Hadden & Grayson, 1961) [reproduced in Edmister & Lee, "Applied Hydrocarbon Thermodynamics," vol. 1, 1983]. Copyright Gulf Publishing Company.

$$\beta = \beta + \frac{-1 + \Sigma \dfrac{z_i}{1 + \beta(K_i - 1)}}{\Sigma \dfrac{(K_i - 1)z_i}{[1 + \beta(K_i - 1)]^2}}. \tag{6.67}$$

After β has been found by successive approximation, the phase compositions are obtained with

$$x_i = \frac{z_i}{1 + \beta(K_i - 1)}, \tag{6.68}$$

$$y_i = K_i x_i. \tag{6.65}$$

A starting value of $\beta = 1$ always leads to a converged solution by this method.

An objective function that often leads to more rapid convergence is that of Rachford & Rice (1952), who use

$$f(\beta) = \Sigma y_i - \Sigma x_i = 0. \tag{6.69}$$

The flash equation is

$$f(\beta) = \Sigma \frac{(K_i - 1)z_i}{1 + \beta(K_i - 1)} = 0, \tag{6.70}$$

and its N-R algorithm is

$$\beta = \beta + \frac{\Sigma \dfrac{(K_i - 1)z_i}{1 + \beta(K_i - 1)}}{\Sigma \left[\dfrac{K_i - 1}{1 + \beta(K_i - 1)}\right]^2 z_i}. \tag{6.71}$$

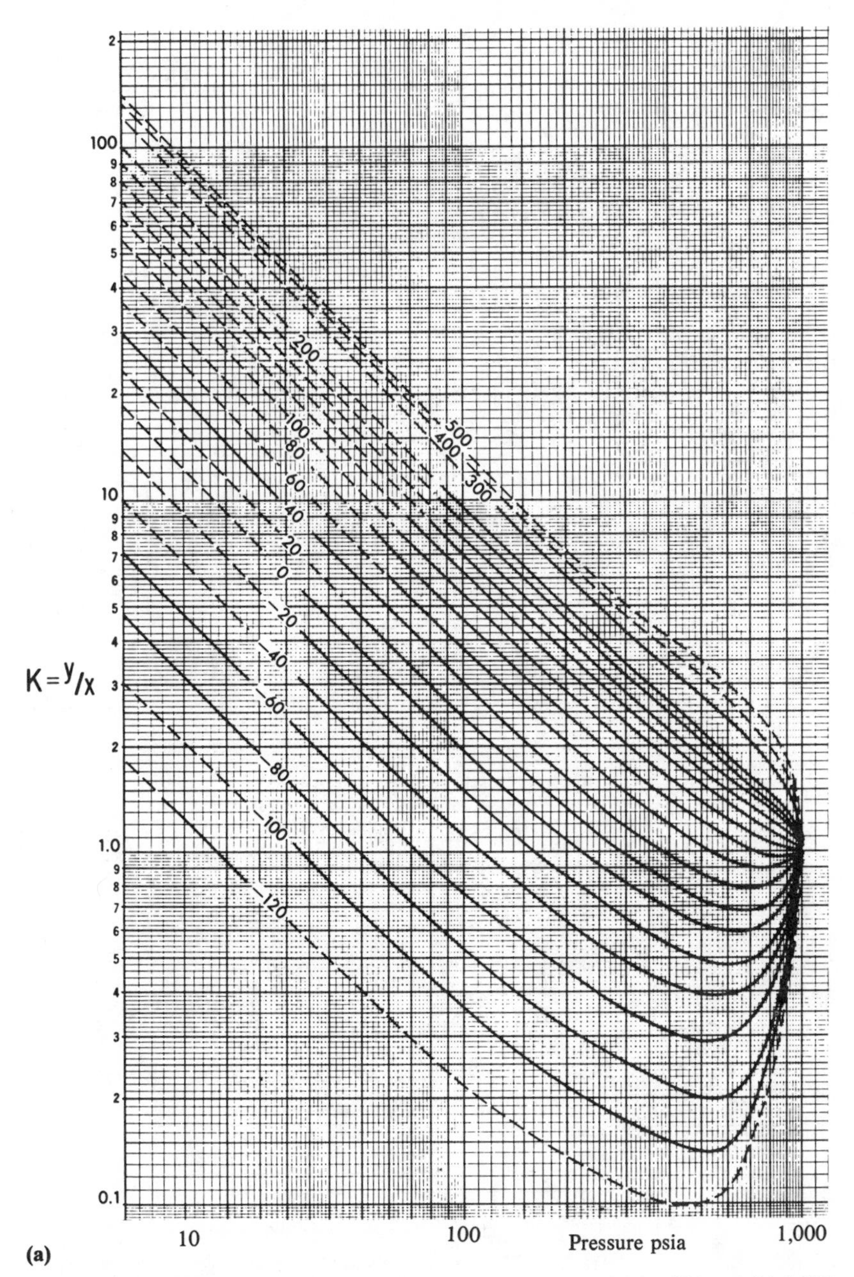

Figure 6.4. NGPSA charts of ethane at convergence pressures of (a) 1,000 and (b) 5,000 psia. (**Figure 6.4** *continued on next page*)

6.3.4. Flash at Fixed Enthalpy

The problem will be formulated for a specified final pressure and enthalpy, and under the assumption that the enthalpies are additive (that is, with zero enthalpy of mixing) and are known functions of temperature at the given pressure. The enthalpy balance is

$$H_F = (1-\beta)\Sigma x_i H_{iL} + \beta \Sigma y_i H_{iV}$$

$$= (1-\beta)\Sigma \frac{z_i H_{iL}}{1+\beta(K_i-1)} + \beta\Sigma \frac{K_i z_i H_{iV}}{1+\beta(K_i-1)}. \quad (6.72)$$

This equation and the flash Eq. 6.66 constitute a set,

$$f(\beta, T) = -1 + \Sigma \frac{z_i}{1+\beta(K_i-1)} = 0, \quad (6.73)$$

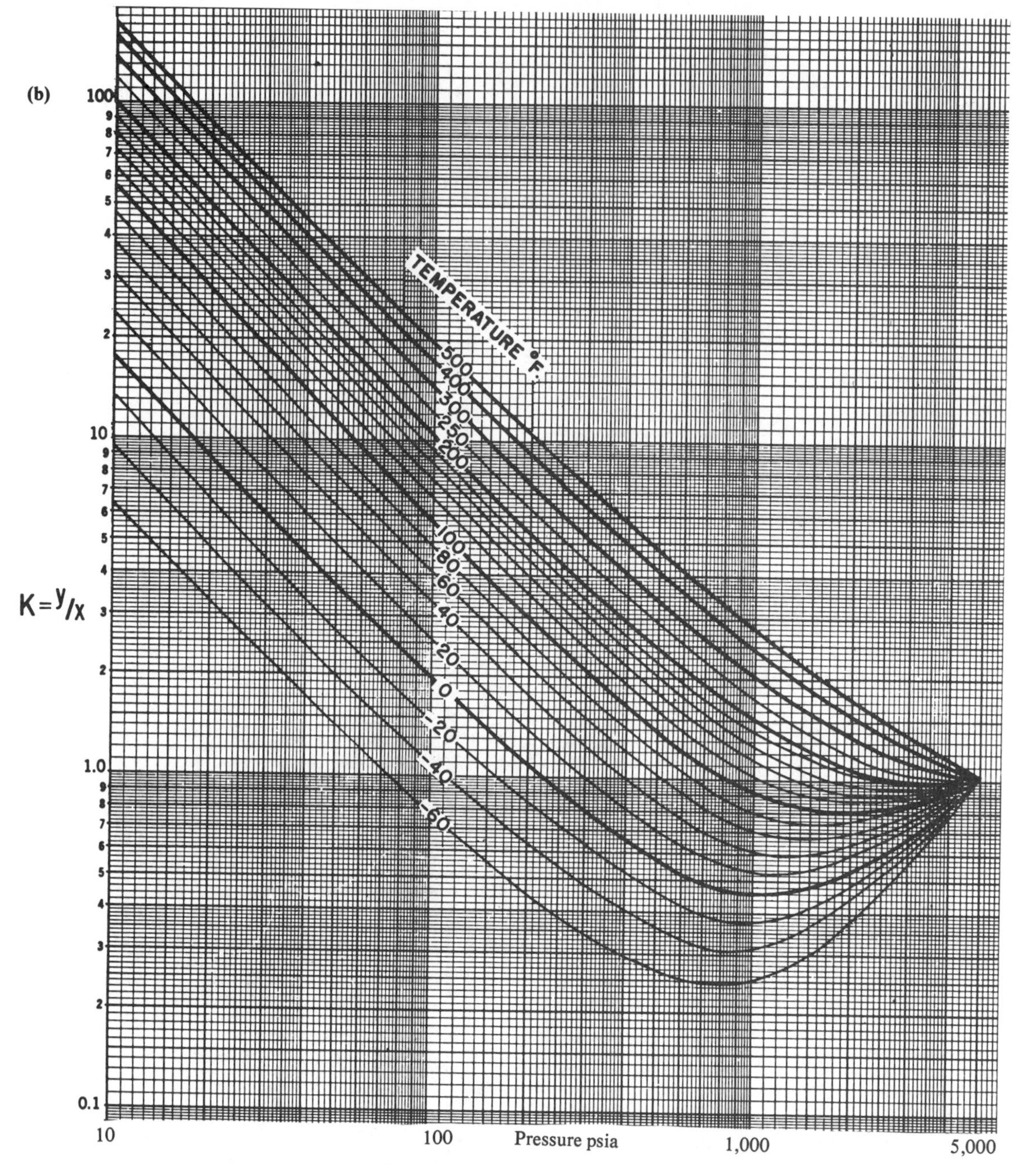

Figure 6.4(b) *continued*

$$g(\beta, T) = H_F - (1-\beta)\,\Sigma \frac{z_i H_{iL}}{1+\beta(K_i - 1)} - \beta \Sigma \frac{K_i z_i H_{iV}}{1+\beta(K_i - 1)} = 0, \tag{6.74}$$

from which the phase split, β, and temperature can be found when the enthalpies and the vaporization equilibrium ratios are known functions of temperature. The N-R method applied to Eqs. 6.73 and 6.74 finds corrections to initial estimates of β and T by solving the linear equations

$$h \frac{\partial f}{\partial \beta} + k \frac{\partial f}{\partial T} + f = 0, \tag{6.75}$$

$$h \frac{\partial g}{\partial \beta} + k \frac{\partial g}{\partial T} + g = 0, \tag{6.76}$$

where all terms are evaluated at the assumed values (β_0, T_0) of the two unknowns. The corrected values, suitable for the next trial if that is necessary, are

$$\beta = \beta_0 + h, \tag{6.77}$$

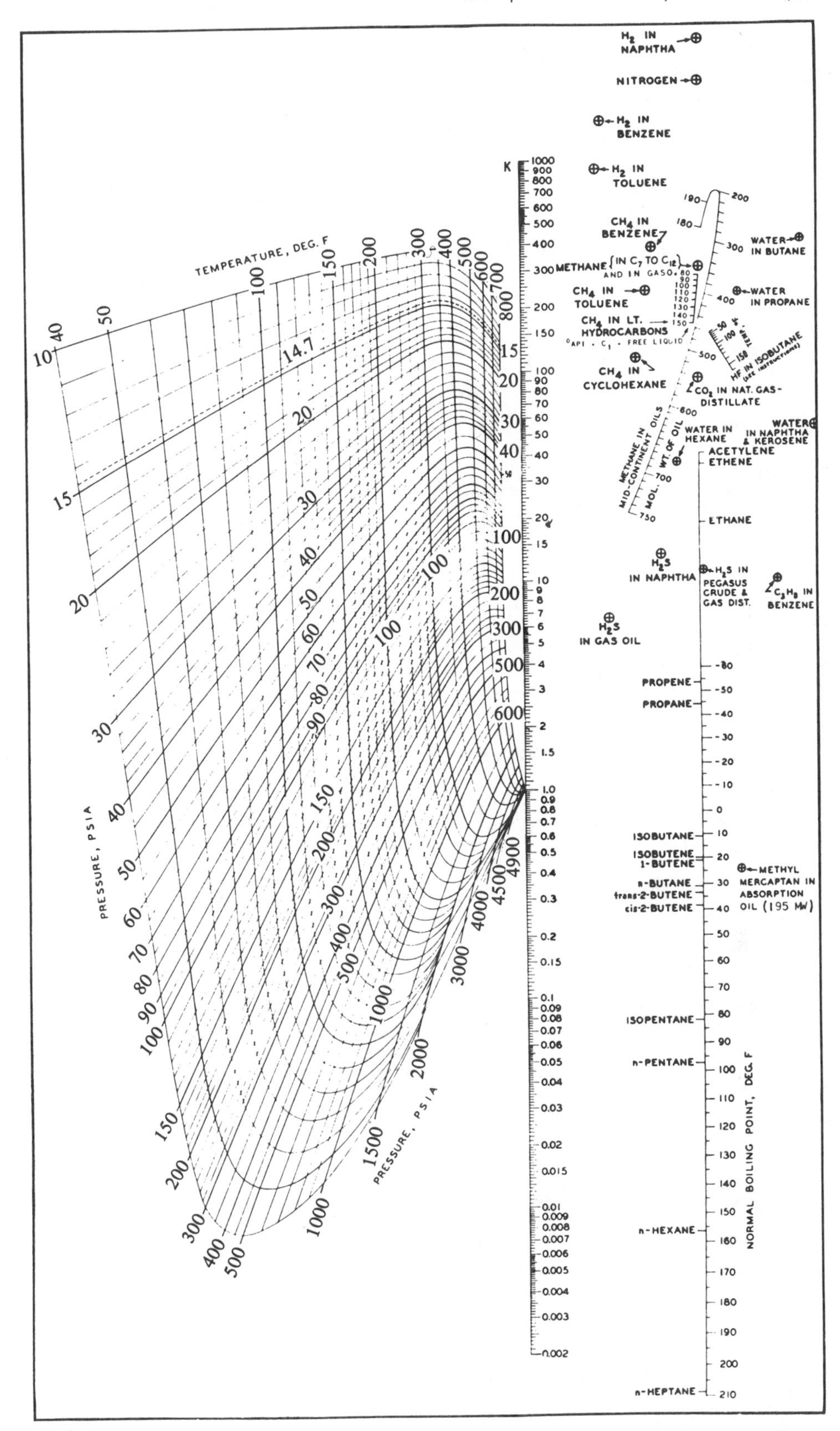

Figure 6.5. Nomogram for vaporization equilibrium ratios (Hadden & Grayson, 1961) [reproduced in Edmister & Lee, "Applied Hydrocarbon Thermodynamics," vol. 1, 1983]. Copyright Gulf Publishing Company.

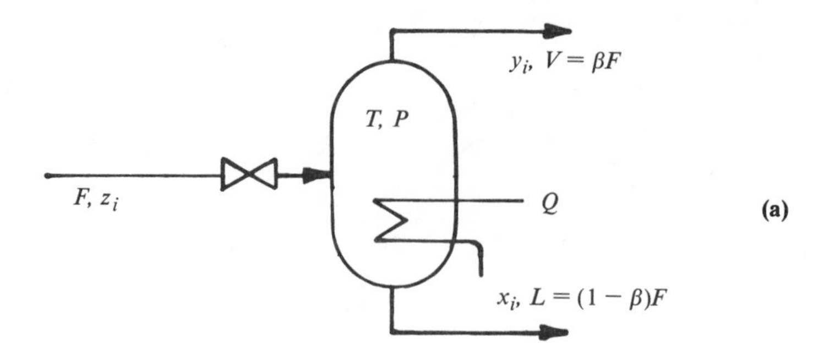

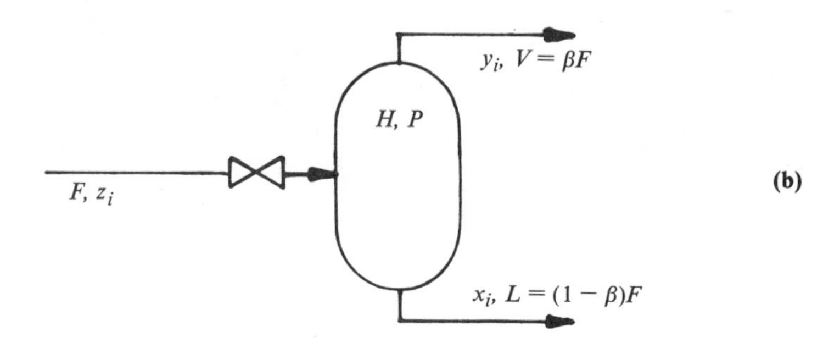

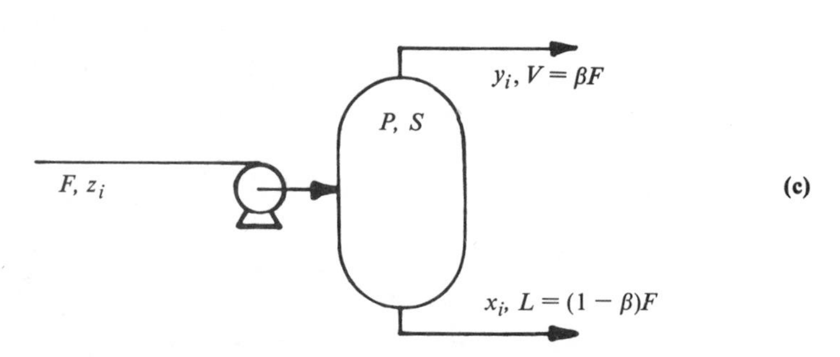

Figure 6.6. Equilibrium flashes at specified (T, P) or (H, P) or (S, P). (a) The flash drum is at specified temperature and pressure. (b) Throttling to specified enthalpy and pressure. (c) Expansion or compression to specified entropy and pressure.

$$T = T_0 + k. \tag{6.78}$$

Another way of finding β and T, used in Example 6.3, starts with an assumed value of T, then finds β from Eq. 6.73 by single variable N-R algorithm. Correctness of these values of the unknowns then is checked by substitution into Eq. 6.74; several trial Ts may be needed.

6.3.5. Isentropic Flash

The first step in figuring the expansion or compression of a fluid is to find what happens under reversible adiabatic, or isentropic, conditions. Subsequent analysis to find what happens with a particular isentropic efficiency will not be dealt with here.

Figure 6.6(c) represents the process to be studied. The reference condition is the liquid state at reference conditions T_f and P_f at which the latent heats of vaporization are λ_i. The vapor at T_f and P_f will expand isentropically to a final pressure P at which the fractional vapor content will be $\beta = V/F$. The entropy balance is:

$$\Sigma z_i \left(\frac{\lambda_i}{T_f} - R \ln z_i\right) = (1 - \beta)\, \Sigma x_i \left[\int_{T_f}^{T} \frac{dH_{iL}}{T} - R \ln \frac{P}{P_f} - R \ln x_i\right] + \beta \Sigma y_i \left[\frac{\lambda_i}{T_f} + \int_{T_f}^{T} \frac{dH_{iV}}{T} - R \ln \frac{P}{P_f} - R \ln y_i\right], \tag{6.79}$$

and may be rearranged to,

$$f(T, \beta, x_i, y_i) = R \ln \frac{P}{P_f} + \Sigma z_i \left(\frac{\lambda_i}{T_f} - R \ln z_i \right) - (1-\beta)\Sigma x_i \left[\int_{T_f}^{T} \frac{dH_{iL}}{T} - R \ln x_i \right] - \beta \Sigma y_i \left[\frac{\lambda_i}{T_f} + \int_{T_f}^{T} \frac{dH_{iV}}{T} - R \ln y_i \right] = 0. \quad (6.80)$$

After elimination of x_i and y_i with Eqs. 6.68 and 6.65, the resulting equation can be solved simultaneously with the flash Eq. 6.73 to find β and T and ultimately the phase compositions, again when the enthalpies and K_i are known functions of T. Either of the equation-solving procedures described in Section 6.3.4 is applicable to the present problem.

Determination of isentropic flash conditions may be done with the algorithm of Figure 6.12, except that in box 10 the given condition is replaced by: "Is Eq. 6.80 Satisfied?"

Example 6.2. Use of the API System to Find Bubble- and Dewpoints, with and without the Convergence Pressure Correction

The pressure is 300 psia. Composition and properties of the mixture are:

	x	T_c, °K	P_c, atm
C_2H_6	0.05		
C_3H_8	0.20	370.0	42.0
nC_4	0.30	425.3	37.4
nC_5	0.20	470.1	32.6
nC_6	0.25	508.1	29.4
Weight Avg.		453 K 355.8 F	32.8 atm 482 psia

The lightest component having a content in excess of 0.1 mol % is ethane, so this is the equivalent light component; the equivalent heavy component is characterized by $T_c = 355.8$ F, $P_c = 482$ psia. From Figure 6.2 the convergence pressure is 600 psia and roughly independent of temperature. From Figure 6.3, the grid pressure at a system pressure of 300 psia is 530 psia. The bubble- and dewpoint temperatures are found with the aid of Figure 6.5 at a grid pressure of 530 psia, and also for comparison at 300 psia, which neglects the convergence-pressure correction.

530 psia	300 F K	300 F Kx	300 F x/K	400 F K	400 F Kx	400 F x/K
C_2H_6	3.3			4.0		
C_3	1.8			2.5		
C_4	1.05			1.65		
C_5	0.54			1.02		
C_6	0.30			0.62		
Sum		1.020	1.616		1.554	0.874

300 psia	200 F K	200 F Kx	200 F x/K	300 F K	300 F Kx	300 F x/K
C_2	3.8			5.2		
C_3	1.65			2.7		
C_4	0.70			1.43		
C_5	0.30			0.75		
C_6	0.14			0.40		
Sum		0.83	2.76		1.48	1.19

By interpolation: At 530 psia, $BP = 296$ F, $DP = 383$ F
At 300 psia, $BP = 226$ F, $DP = 312$ F

Clearly, the convergence pressure correction is very important in this case.

Example 6.3. Vaporization and Condensation of a Ternary Mixture

The pressure is 100 psia. The overall composition, the normal boiling points, K_i at two temperatures and enthalpies of pure liquid and vapor at two temperatures are known. The bubble- and dewpoint temperatures, a flash at 100 F and 100 psia, and an adiabatic flash at 100 psia of a mixture that is initially liquid at 100 F will be determined.

	z	T_b °F	K 100 F	K 40 F	Coefficients A	Coefficients B	Coefficients C
C_2	0.3	−128	5.4	3.2	5.7799	2167.12	−30.6
nC_4	0.3	31	0.53	0.21	6.1418	3382.90	−60.8
nC_5	0.4	97	0.18	0.057	6.4610	3978.36	−73.4

(Continued next page)

Example 6.3. *(continued)*

The coefficients of the K-equation are

$$\ln K_i = A_i - B_i/(T + C_i) \quad (1)$$

$$C = 32.4 - 0.19T_b\,(°R) \quad (2)$$

$$B = \frac{\ln(K_2/K_1)}{\dfrac{1}{T_1 + C} - \dfrac{1}{T_2 + C}} \quad (3)$$

$$A = \ln K_1 + B/(T_1 + C) \quad (4)$$

Numerical values are tabulated in the preceding. Bubblepoint algorithms are

Newton-Raphson: $$T = T - \frac{-1 + \Sigma K_i x_i}{\Sigma \dfrac{B_i K_i x_i}{(T + C_i)^2}} \quad (5)$$

Richmond: $$F = -1 + \Sigma K_i x_i \quad (6)$$

$$F' = \Sigma \frac{B_i K_i x_i}{(T + C_i)^2} \quad (7)$$

$$F'' = \Sigma \frac{B_i[B_i - 2(T + C_i)]K_i x_i}{(T + C_i)^2} \quad (8)$$

$$T = T - \frac{2}{\dfrac{2F'}{F} - \dfrac{F''}{F'}} \quad (9)$$

The N-R dewpoint algorithm is

$$T = T + \frac{-1 + \Sigma y_i/K_i}{\Sigma \left[\dfrac{B_i y_i}{K_i(T + C_i)^2}\right]} \quad (10)$$

Results of the bubblepoint and dewpoint calculation are shown. The Richmond method requires slightly fewer iterations to convergence.

Bubblepoint		
N-R	*Richmond*	*Dewpoint*
1,000.0000	1,000.0000	700.0000
695.1614	620.3217	597.8363
560.1387	504.4799	625.9790
506.5023	495.8020	635.3072
496.1742	495.7963	636.0697
495.7968		636.0743
495.7963		636.0743

The conventional flash algorithm is

$$\beta = V/F = \beta + \frac{-1 + \Sigma \dfrac{z_i}{1 + \beta(K_i - 1)}}{\Sigma \dfrac{B_i y_i}{K_i(T + C_i)^2}}, \quad (11)$$

and the Rachford-Rice algorithm is

$$\beta = \beta + \frac{\Sigma \dfrac{(K_i - 1)z_i}{1 + \beta(K_i - 1)}}{\Sigma \left\{\left[\dfrac{K_i - 1}{1 + \beta(K_i - 1)}\right]^2 z_i\right\}}. \quad (12)$$

The compositions are

$$x_i = \frac{z_i}{1 + \beta(K_i - 1)}, \quad (13)$$

$$y_i = K_i x_i. \quad (14)$$

Successive iterations by the two methods are listed.

Conventional	*Rachford-Rice*
1.0000	1.0000
.8257	.7890
.5964	.4610
.3986	.2720
.3038	.2818
.2830	.2819
.2819	

The vapor and liquid compositions are:

	z_i	x_i	y_i
C_2	0.3	0.1339	0.7231
nC_4	0.3	0.3458	0.1833
nC_5	0.4	0.5203	0.0936

Adiabatic flash: Liquid and vapor enthalpies off charts in the API data book are fitted with linear equations,

$$h = a + bT\,(°F), \quad (15)$$

$$H = c + dT\,(°F). \quad (16)$$

The enthalpies and the corresponding coefficients are tabulated following. The inlet material to the flash drum is liquid at 100 F, with $H_0 = 8.575.8$ Btu/lbmol. The flash Eq. 11 applies to this part of the example. The enthalpy balance is

$$H_0 = 8{,}575.8 \quad (17)$$

$$= (1 - \beta)\,\Sigma M_i x_i h_i + \beta \Sigma M_i y_i H_i$$

$$= (1 - \beta)\,\Sigma \frac{M_i z_i h_i}{1 + \beta(K_i - 1)} + \beta \Sigma \frac{K_i M_i z_i H_i}{1 + \beta(K_i - 1)} \quad (18)$$

The procedure consists of the steps,

1. Assume T.
2. Find the K_i, h_i, and H_i.
3. Find β from the flash equation, Eq. 11.
4. Evaluate the enthalpy of the mixture and compare with H_0, Eq. 18.

The results of several trials are shown.

(Continued next page)

Example 6.3. *(continued)*

$T°R$	β	H	
530.00	.1601	8,475.70	
532.00	.1681	8,585.46	
531.82	.1674	8,575.58	~ 8575.8, Check.

The final VERs and the liquid and vapor compositions are:

	K	x	y
C_2	4.2897	0.1935	0.8299
nC_4	0.3534	0.3364	0.1189
nC_5	0.1089	0.4701	0.0512

Coefficients of the enthalpy correlations are:

		h, Btu/lb		*H*					
	M	*0 F*	*100 F*	*0 F*	*100 F*	*a*	*b*	*c*	*d*
C_2	30	122	195	290	335	122	0.73	290	0.45
nC_4	58	96	152	267	301	96	0.56	267	0.34
nC_5	72	90	145	260	300	90	0.55	260	0.40

Bubblepoint

```
10  T=600
20  READ A1,A2,A3,B1,B2,B3,C1,C2,C3,X1,X2,X3
30  DATA 5.7799,6.1418,6.461,216,7.12,3382.9,
    3978.36,30.6,60.8,73.4,.3,.3,.4
40  GOSUB 150
50  H=F/F1
60  H1=2/(2*F1/F-F2/F1)
70  PRINT USING 75 ; T
75  IMAGE DDDD.DDDD
80  T=T-H
90  IF ABS(H/T)<=.0000000001 THEN 110
100 GOTO 40
110 END
150 K1=EXP(A1-B1/(T-C1))
160 K2=EXP(A2-B2/(T-C2)
170 K3=EXP(A3-B3/(T-C3))
180 F=-1+X1*K1+X2*K2+X3*K3
190 F1=B1*X1*K1/(T-C1)^2+B2*X2*K2/(T-C2)
    ^2+B3*X3*K3/(T-C3)^2
195 G=(T-C3)^4
200 F2=B1*(B1-2*(T-C1))*K1*X1/(T-C1)^4+B2*
    (B2-2*(T-C2))*K2*X2/(T-C2) ^ 4+B3*(B3-2*
    (T-C3))*K3*X3/G
210 RETURN
220 END
```

Flash
Including Rachford-Rice

```
10  B=1
20  READ A1,A2,A3,B1,B2,B3,C1,C2,C3,X1,X2,X3,T
30  DATA 5.7799,6.1418,6.461,2167.12,3382.9,
    3978.36,30.6,60.8,73.4,.3,.3,.4,560
40  GOSUB 150
50  H=F3/F4
70  PRINT USING 75 ; B
75  IMAGE D.DDDD
80  B=B+H
90  IF ABS(H/B)<=.0001 THEN 110
100 GOTO 40
110 END
150 K1=EXP(A1-B1/(T-C1))
160 K2=EXP(A2-B2/(T-C2))
170 K3=EXP(A3-B3/(T-C3))
180 D1=1+B*(K1-1)
190 D2=1+B*(K2-1)
200 D3=1+B*(K3-1)
210 F=-1+X1/D1+X2/D2+X3/D3
215 F1=(K1-1)*X1/D1^2+(K2-1)*X2/D2^2+
    (K3-1)*X3/D3^2
220 F3=(K1-1)*X1/D1+(K2-1)*X2/D2+(K3-1)*
    X3/D3
230 F4=((K1-1)/D1)^2*X1+((K2-1)/D2)^2*X2+
    ((K3-1)/D3)^2*X3
240 RETURN
```

```
5   ! ADIABATIC FLASH
10  INPUT T
20  B=1
30  READ A1,A2,A3,B1,B2,B3,C1,C2,C3,X1,X2,X3
40  DATA 5.7799,6.1418,6.461,2167.12,3382.9,
    3978.36,30.6,60.8,73.4,.3,.3,.4
50  GOSUB 200
60  H=F/F1
70  B=B+H
80  IF ABS(H/B)<=.0001 THEN 100
90  GOTO 50
100 H1=122+.73*(T-459.6)
110 H2=96+.56*(T-459.6)
120 H3=90+.55*(T-459.6)
130 H4=290+.45*(T-459.6)
140 H5=267+.34*(T-459.6)
150 H6=260+.4*(T-459.6)
160 H=(1-B)*(9*H1/D1+17.4*H2/D2+28.8*H3/D3)+
    B*(9*H4*K1/D1+17.4*H5*K2/D2+28.8*H6*
    K3/D3)
170 PRINT USING 180 ; T,B,H
180 IMAGE DDD.DD,3X,.DDDD,3X,DDDDD.DD
190 END
200 K1=EXP(A1-B1/(T-C1))
210 K2=EXP(A2-B2/(T-C2))
220 K3=EXP(A3-B3/(T-C3))
230 D1=1+B*(K1-1)
240 D2=1+B*(K2-1)
250 D3=1+B*(K3-1)
260 F=-1+X1/D1+X2/D2+X3/D3
270 F1=(K1-1)*X1/D1^2+(K2-1)*X2/D2^2+
    (K3-1)*X3/D3^2
280 RETURN
```

6.3.6. Vapor and Immiscible Liquid Phases

Although traces of hydrocarbons in water and of water in hydrocarbons have significance in some applications, the performance of condensers and vaporizers of such mixtures is adequately analyzed on the assumption that the liquids are immiscible. Taking the case of condensation, the mixture will have two distinct dewpoints, one at which the hydrocarbon first begins to condense and the other when the water first does so. The computation of the condensing curves, temperature against amounts of liquids and vapor, depends on which dewpoint is first. One procedure that can be followed is this:

1. Lower the temperature of the vapor gradually and keep track of $\Sigma_{HC} y_i/K_i$ of the hydrocarbons and of the partial pressure, P^s_w/P of the water.
2. Note which condition of the following occurs first,

a. $\Sigma_{HC} y_i/K_i = 1$, (6.81)

b. $P^s_w/P = y_{wo}$, the initial mol fraction of the water in the vapor. (6.82)

3a. If the hydrocarbon dewpoint is higher, continue with flash calculations at lower temperatures, meanwhile keeping track of when condensation of water first occurs. Continue cooling until everything is condensed.

3b. If the water dewpoint is higher, replace the hydrocarbon mol fractions by

$$z_i = z_i(1 - P^s_w/P), \tag{6.83}$$

and find the hydrocarbon dewpoint, which is the temperature at which

$$(1 - P^s_w/P) \sum_{HC} K_i z_i = 1. \tag{6.84}$$

Beyond the dewpoint of the hydrocarbon, continue cooling until everything is condensed.

This procedure is followed in Example 6.4. The equilibrium vaporization and condensation curves of a given mixture are identical.

In keeping with the accuracy of the rest of this section, the vaporization equilibrium ratio of water is taken in the simplified form,

$$K_w = \frac{\phi^s_w P^s_w (PF)_w}{\hat{\phi}_w P} \sim P^s_w/P. \tag{6.85}$$

Bubblepoint: This temperature is the one at which vaporization first begins, and is also the one at which condensation is complete. When the liquid phases are immiscible the bubblepoint condition:

$$\Sigma y_i = 1 \tag{6.86}$$

becomes

$$K_w + \sum_{HC} K_i x_i = 1, \tag{6.87}$$

or

$$P^s_w/P + \sum_{HC} K_i x_i = 1. \tag{6.88}$$

Dewpoint: The introductory material of this section explains how the dewpoints of the water and of the hydrocarbon are found.

Flash at constant pressure: The material balances are

$$F = V + L + W, \tag{6.89}$$

$$W = Fy_{wo} - Vy_w, \tag{6.90}$$

$$W/F = y_{wo} - \beta y_w = \text{fraction of the material that is liquid water}, \tag{6.91}$$

$$L/F = 1 - y_{wo} - \beta(1 - y_w) \tag{6.92}$$

= fraction of the mixture that is liquid hydrocarbon.

For each hydrocarbon component,

$$Fz_i = Lx_i + Vy_i = (L/K_i + V)y_i. \tag{6.93}$$

Since

$$\sum_{HC} y_i = 1 - y_w = \sum_{HC} K_i z_i/(L/F + \beta K_i), \tag{6.94}$$

the flash equation becomes

$$f(\beta) = -1 + y_w + \sum_{HC} \frac{K_i z_i}{1 - y_{wo} + \beta(K_i - 1 + y_w)} = 0. \tag{6.95}$$

This equation applies below the hydrocarbon dewpoint, which may be above or below that of the water. Example 6.4 develops the condensation curves for both cases.

6.4. EQUILIBRIA WITH *K*s DEPENDENT ON COMPOSITION

In the formulas for vaporization equilibrium ratios,

$$K_i = \gamma_i \phi^s_i P^s_i (PF)_i / \hat{\phi}_i P, \tag{6.96}$$

$$K_i = \hat{\phi}_{iL}/\hat{\phi}_{iV}, \tag{6.97}$$

the fugacity and activity coefficients are the factors that depend strongly on composition, particularly at extreme T and P or in the critical region and of mixtures containing strongly polar and associating substances. Since the Poynting Factor makes only a minor contribution, the variation of partial molal volume with composition usually may be neglected. The effect of the sensitivity of partial molal volume to pressure, however, may be significant; this is discussed in the book of Prausnitz (1969, Chap. 10).

When the variation of the VERs with composition must be taken into account, solutions of problems are obtained by successive approximation. Since the composition of either or both phases is unknown at the start, a trial calculation is made by assuming ideal behavior—that is, with unit fugacity and activity coefficients—whereby estimates of the composition are obtained from which corrected values of the coefficients can be made for the next trial.

Routines for the familiar types of processes are depicted on the flowcharts, Figures 6.7–6.12, and will be discussed briefly in turn. They are phrased in terms of Eq. 6.96, but procedures only slightly modified will do with Eq. 6.97. No specific equations for γ_i and ϕ_i are identified here; any of those mentioned in Section 6.2 and elsewhere in the book may be of course applicable. Example 6.5 employs the Virial and Wilson equations, whereas Example 6.6 uses the Soave for both phases.

Example 6.4. Condensation Curves of a Mixture of Water and Hydrocarbons

The hydrocarbon mixture of Example 6.3 with

1. 5 mol% water vapor,
2. 10 mol% water vapor,

is to be condensed at 100 psia. Below the water dewpoint, the mol fraction of water remaining in the vapor is given in terms of the vapor pressure by the Antoine equation as

$$y_w = P_w^s/P = \frac{1}{5{,}170}\exp\left(18.3036 - \frac{6{,}869.6}{T - 83.03}\right).$$

Equations for the K_i of the hydrocarbons are taken from the previous example. Below the hydrocarbon dewpoint, Eq. 6.95, step 300 of the program, is solved by Newton-Raphson. The fractions of the mixture remaining as vapor and produced as liquid hydrocarbons and water and the ratio $K_w = y_w = P_w^s/P$ are tabulated and plotted as functions of the temperature.

1. In the first case, $\Sigma_{HC} y_i/K_i = 1$ at 631.77 R while $P_w^s/P = 0.0629$, well above the value $y_{w0} = 0.05$. At 626.23 R, condensation of water begins, while total condensation is complete at 495.71 R.
2. In the second case, $P_w^s/P = 1$ at 652.89 R, at which temperature $\Sigma_{HC} y_i/K_i \ll 1$. Hydrocarbon begins to condense at 630.76 R, while total condensation again is complete at 495.71 R.

L = fraction of the original mixture that is liquid hydrocarbon
V = fraction of the original mixture that remains vapor
W = fraction of the original mixture that is liquid water

Output and program for case with 5 mol % water. To find the hydrocarbon dewpoint, replace $K4 = P_w^s/P$ with $z4 = 0.05$ in steps 220, 230, and 300. S in step 100 then is unity.

T	β	L	W	K_w
631.77	1.0000	.	.	.0629
631.76	.9998	.0002	.0000	.0629
630.00	.9924	.0176	−.	.0604
626.23	.9022	.0978	.0000	.0554
626.22	.9020	.0980	.0001	.0554
625.00	.8756	.1215	.0029	.0538
620.00	.7798	.2075	.0127	.0478
610.00	.6319	.3418	.0263	.0375
600.00	.5236	.4417	.0347	.0292
590.00	.4408	.5191	.0401	.0224
580.00	.3750	.5814	.0436	.0171
570.00	.3207	.6334	.0459	.0129
560.00	.2740	.6787	.0474	.0096
550.00	.2320	.7196	.0484	.0070
540.00	.1926	.7584	.0490	.0051
530.00	.1539	.7966	.0494	.0036
520.00	.1141	.8362	.0497	.0026
510.00	.0713	.8789	.0499	.0018
500.00	.0231	.9269	.0500	.0012
496.00	.0017	.9483	.0500	.0010
495.71	.0000	.9500	.0500	.0010

```
20  READ A1,A2,A3,A4,B1,B2,B3,B4,C1,C2,C3,C4,
    Z1,Z2,Z3,Z4
30  DATA 5.7799,6.1418,6.461,18.3036,2167.12,
    3382.9,3978.36,6869.6,30.6,60.8,73.4,83.03
40  DATA .285,.285,.38,.05
50  INPUT T
60  K1=EXP(A1-B1/(T-C1))
70  K2=EXP(A2-B2/(T-C2))
80  K3=EXP(A3-B3/(T-C3))
90  K4=EXP(A4-B4/(T-C4))/5170
100 S=.3*(1-Z4)/K1+.3*(1-Z4)/K2+.4*(1-Z4)/K3
110 B=.1
120 GOSUB 300
130 F1=F
140 B=1.0001*B
150 GOSUB 300
160 F2=F
170 H=.0001*B*F1/(F2-F1)
180 B=B/1.0001
190 IF ABS(H/B)<=.0001 THEN 220
200 B=B-H
210 GOTO 120
220 L=.95-B*(1-K4)
230 W=.05-B*K4
240 PRINT USING 245 ; T,B,L,W,K4
245 IMAGE DDD.DD, X,D.DDDD, X,.DDDD, X,
    .DDDD, X,D.DDDD
250 GOTO 50
260 END
300 F=-1+K4+K1*Z1/(.95+B*(K1-1+K4))+K2*
    Z2/(.95+B*(K2-1+K4))+K3*Z3/(.95+B*(K3-1+
    K4))
310 RETURN
```

Output and program for case with 10 mol % water. Water dewpoint is higher than the hydrocarbon dewpoint.

T	β	L	W	K_w
652.85	.9999	.0001	.0000	.1000
630.80	.9590	.0000	.0410	.0615
630.76	.9589	.0001	.0410	.0615
.630.00	.9402	.0166	.0432	.0604
620.00	.7388	.1966	.0647	.0478
610.00	.5987	.3238	.0775	.0375
600.00	.4960	.4185	.0855	.0292
590.00	.4175	.4918	.0906	.0224
580.00	.3553	.5508	.0939	.0171
570.00	.3038	.6001	.0961	.0129
560.00	.2595	.6430	.0975	.0096
550.00	.2198	.6817	.0985	.0070
540.00	.1825	.7184	.0991	.0051
530.00	.1458	.7547	.0995	.0036
520.00	.1081	.7922	.0997	.0026
510.00	.0675	.8326	.0999	.0018
500.00	.0219	.8782	.1000	.0012
495.71	.0000	.9000	.1000	.0010

(Continued next page)

Example 6.4. *(continued)*

```
20  READ A1,A2,A3,A4,B1,B2,B3,B4,C1,C2,C3,C4,
    Z1,Z2,Z3,Z4
30  DATA 5.7799,6.1418,6.461,18.3036,2167.12,
    3382.9,3978.36,6869.6,30.6,60.8,73.4,83.03
40  DATA .27,.27,.36,.1
50  INPUT T
60  K1=EXP(A1-B1/(T-C1))
70  K2=EXP(A2-B2/(T-C2))
80  K3=EXP(A3-B3/(T-C3))
90  K4=EXP(A4-B4/(T-C4))/5170
100 S=.3*(1-K4)/K1+.3*(1-K4)/K2+.4*(1-K4)/K3
110 B=.1
120 GOSUB 300
130 F1=F
140 B=1.0001*B
150 GOSUB 300
160 F2=F
170 H=.0001*B*F1/(F2-F1)
180 B=B/1.0001
190 IF ABS(H/B)<=.0001 THEN 220
200 B=B-H
210 GOTO 120
220 L=.9-B*(1-K4)
230 W=.1-B*K4
240 PRINT USING 245 ; T,B,L,W,K4
245 IMAGE DDD.DD, X,D.DDDD, X,.DDDD,
    X.DDDD,X,.DDDD,X,D.DDDD
250 GOTO 50
260 END
300 F=-1+K4+K1*Z1/(.9+B*(K1-1+K4))+K2*Z2/
    (.9+B*(K2-1+K4))+K3*Z3/(.9+B*(K3-1+K4))
310 RETURN
```

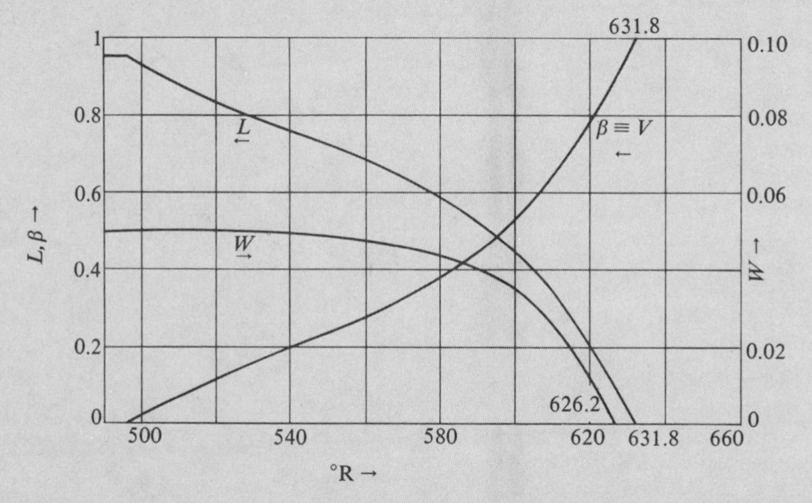

Example 6.4(a). With $y_{w0} = 0.05$, the hydrocarbon dewpoint is higher than that of water.

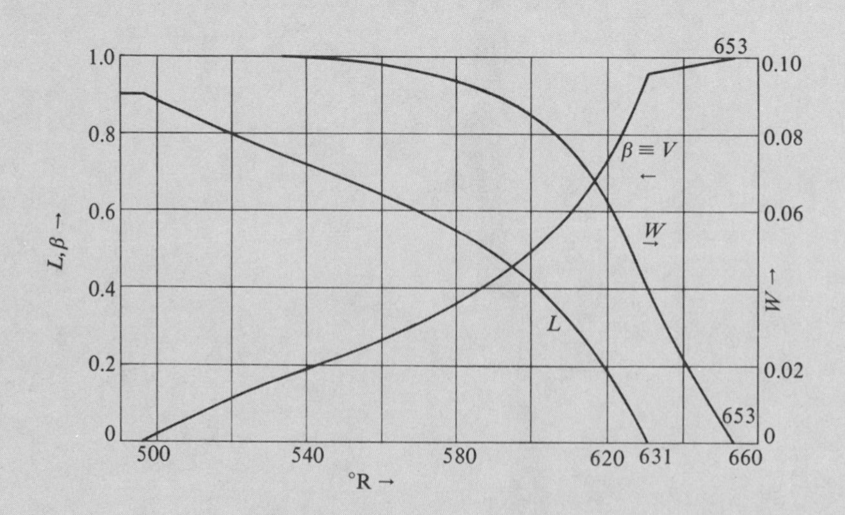

Example 6.4(b). With $y_{w0} = 0.10$, the water dewpoint is higher than that of the hydrocarbon.

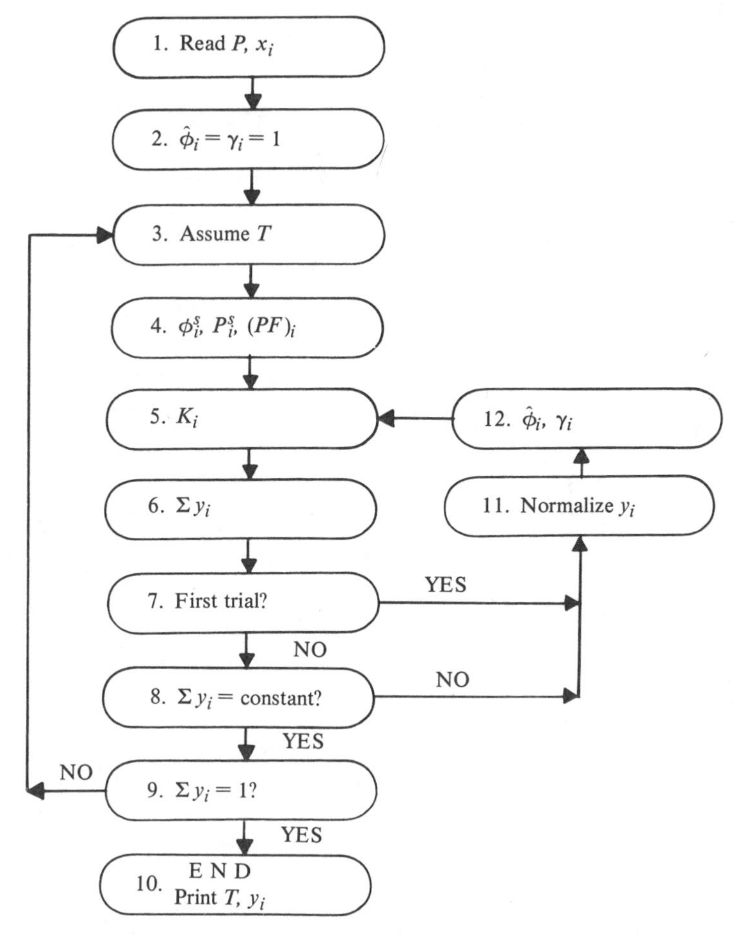

Figure 6.7. Bubblepoint temperature.

6.4.1. Bubblepoints (Figures 6.7, 6.8)

The liquid composition is known but that of the vapor is not. For temperature determination, the key starting assumptions are in box 2, although it may be preferable to start with $\phi_i^s/\hat{\phi}_i = 1$, which is more nearly true than $\hat{\phi}_i = 1$. The method iterates at a particular temperature until $S = \Sigma y_i$ becomes constant. If Σy_i is greater than unity, the temperature should be reduced for the next series of iterations. After two values of the temperature have been tried, an improved trial value can be found by linear interpolation; thus,

$$T = T_2 + (1 - S_2)\left[\frac{T_2 - T_1}{S_2 - S_1}\right]. \tag{6.98}$$

It may be desirable to adopt a more complex interpolation routine as the information becomes available. A saving in the total number of iterations may be achievable by not iterating to constant S at each trial temperature, but instead proceeding directly to another temperature after a single trial, until the correct temperature is approached closely.

Selections of variables for the next trial of bubblepoint pressure and dewpoint temperature and pressure are automated similarly. For bubblepoint pressure determination, only the starting assumption $\hat{\phi}_i = 1$ need be made, since the activity coefficients are essentially independent of pressure.

6.4.2. Dewpoints (Figures 6.9, 6.10)

Only the starting assumption $\gamma_i = 1$ need be made for either the temperature or the pressure determination. The flowcharts and calculation methods are much the same for all four bubble- and dewpoint determinations.

6.4.3. Flash at Fixed T and P (Figure 6.11)

Since compositions of both phases are unknown, the starting assumption is $\hat{\phi}_i = \gamma_i = 1$. The convergence criterion is constancy of the compositions of the phases, which is expressed in box 7 in terms of successive calculations of the vapor mol fractions. The sums of the mol fractions automatically are unity by the flash evaluation of the phase split, β, which is done by the Newton-Raphson or a similar method.

6.4.4. Flash at Fixed H and P (Figure 6.12)

The starting assumption in box 2 is the same as in the previous case. The calculation is started with an assumed temperature and converged to a steady composition. If the enthalpy balance is not satisfied, another temperature is assumed and the cycle is repeated. After two temperatures have been tried

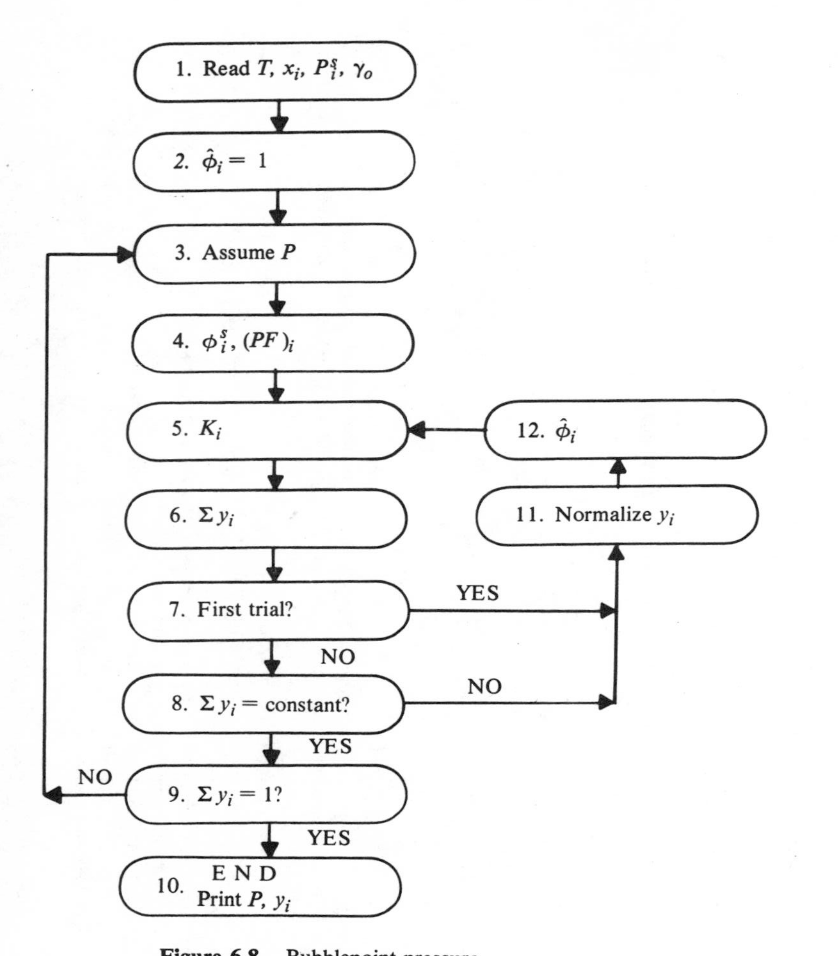

Figure 6.8. Bubblepoint pressure.

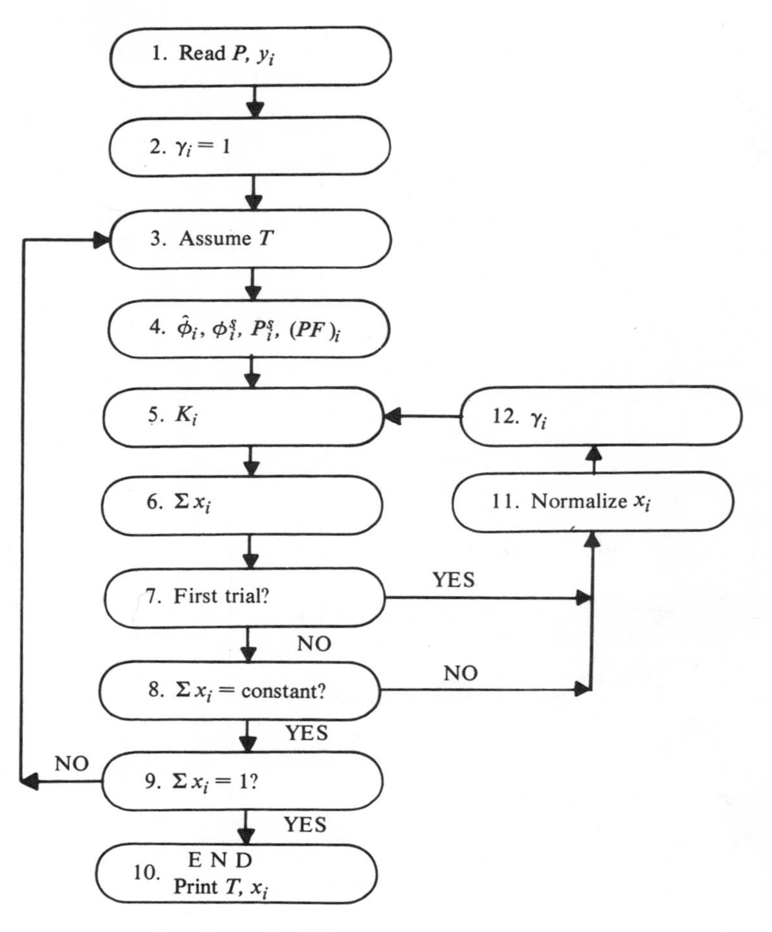

Figure 6.9. Dewpoint temperature.

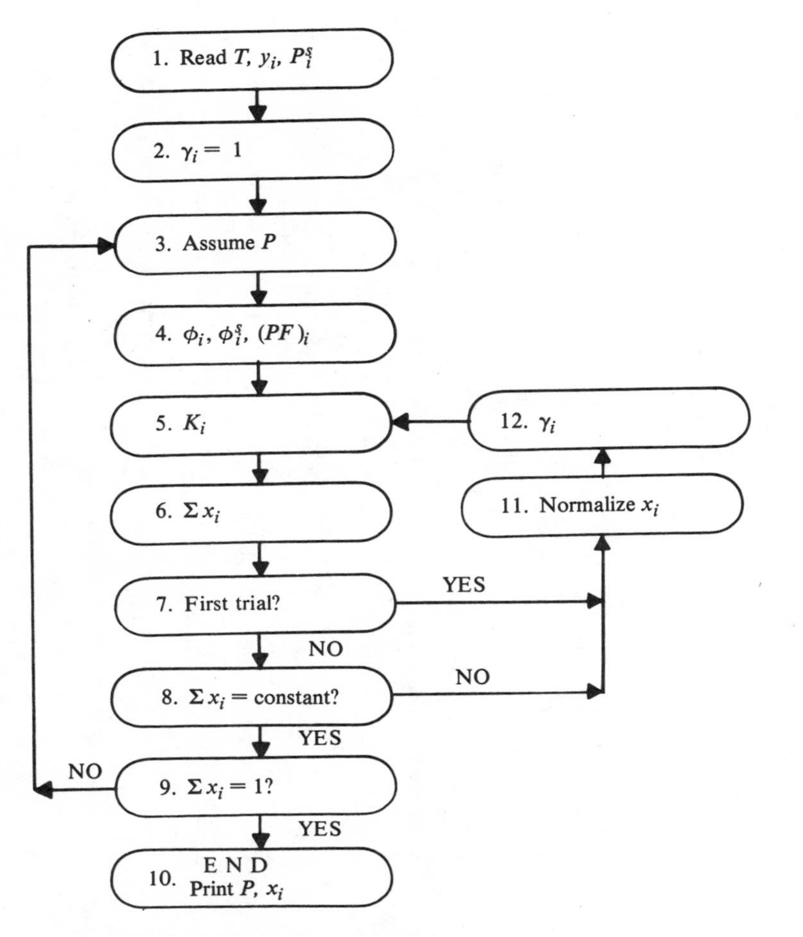

Figure 6.10. Dewpoint pressure.

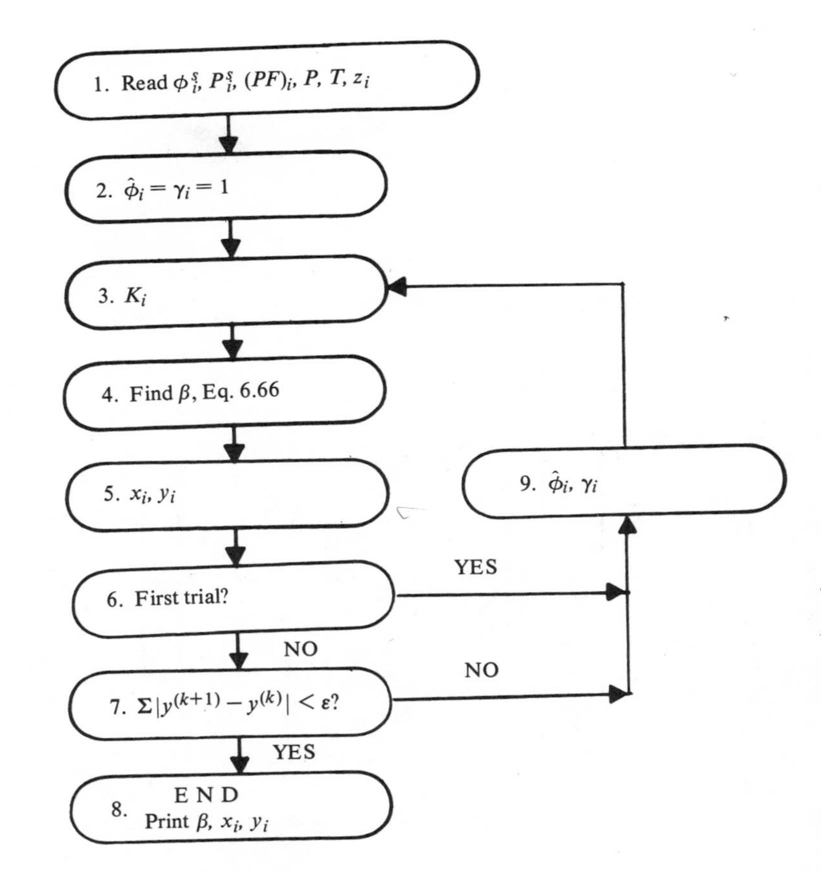

Figure 6.11. Flash at fixed temperature and pressure.

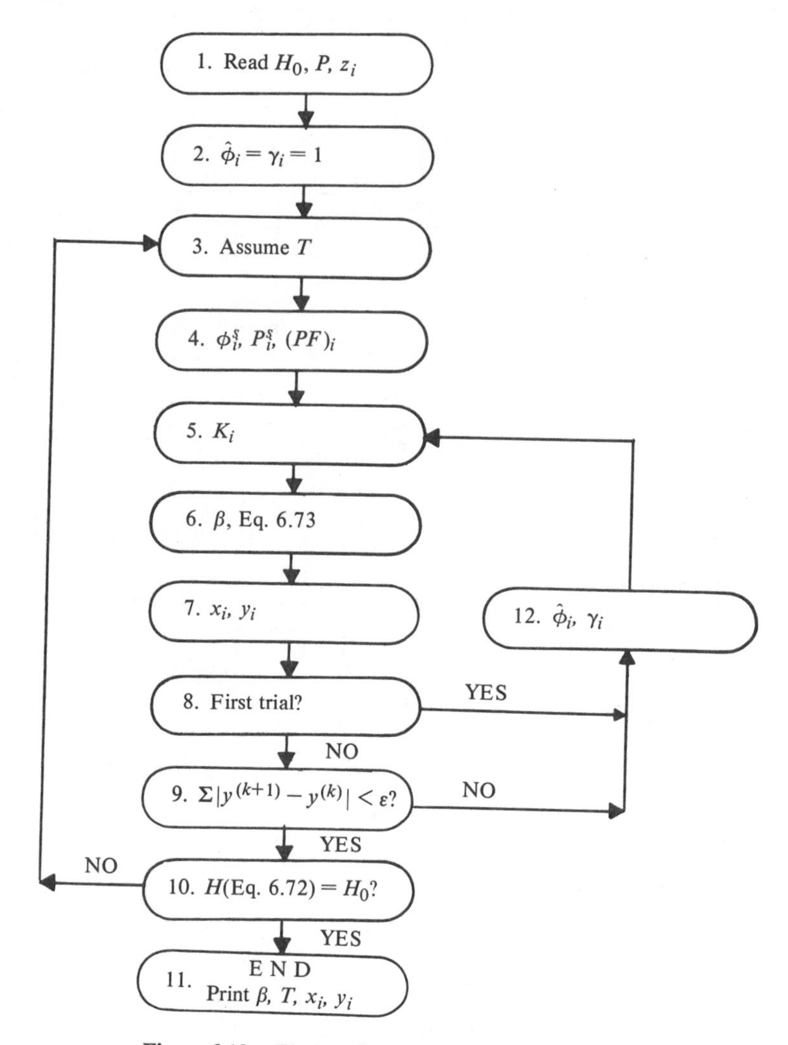

Figure 6.12. Flash at fixed enthalpy and pressure.

and two values of the mixture enthalpy have been found, a new trial temperature can be deduced by linear interpolation,

$$T = T_2 + (H_{\text{mix}} - H_2)(T_2 - T_1)/(H_2 - H_1). \tag{6.99}$$

6.4.5. The Method of Chao & Seader (1961)

In this system, the vaporization equilibrium is represented by the familiar combination of three terms,

$$K_i = \frac{\gamma_{iL}\,\phi_{iL}}{\hat{\phi}_{iV}}, \tag{6.100}$$

but with special means of evaluating the individual terms. Complete equations are summarized in Tables 6.1 and 6.2. Some details also are given in Section 6.2.3.

In the current version of this method, the activity coefficient is derived from the SHFH equation,

$$\gamma_{iL} = \exp\left[\frac{V_{iL}(\delta_i - \bar{\delta})^2}{RT} + \ln(V_{iL}/V_L) + 1 - V_{iL}/V_L\right]. \tag{6.101}$$

The partial fugacity coefficient, $\hat{\phi}_{iV}$, is found with the Redlich-Kwong equation. The fugacity coefficient, ϕ_{iL}, of the pure liquid is an empirical function of the acentric factor and the reduced temperature and pressure. The currently accepted coefficients are due to Grayson & Streed (1963). A detailed study of this method was made by Lenoir & Koppany (1967); the list of restrictions shown in Table 6.1 is due largely to them. Example 6.7 applies this calculation method for obtaining a dewpoint. Other examples may be found in the book of Henley & Seader (1981). The Chao-Seader method was one of those investigated in the API survey by Daubert et al. (1978). There it was found generally inferior to the use of the Soave equation for both phases but it is still used widely for mixtures of light hydrocarbons as in depropanizers and debutanizers, and with the Grayson-Streed coefficients it is useful in the simulation of synthetic fuel systems.

6.4.6. Enthalpy as a Function of *T*, *P*, and Composition

For a proper evaluation of enthalpy balances, the partial molal enthalpies should be known as functions of composition as

Example 6.5. Bubblepoint of a Ternary Mixture with the Virial and Wilson Equations

A mixture of acetone (1) + 2-butanone (2) + ethylacetate (3) is at 20 atm and has the composition $x_1 = x_2 = 0.3$, $x_3 = 0.4$. The Abbott formulas (Fig. 1.16) will be used for the virial coefficients, with cross-critical properties figured by the Lorentz-Berthelot rules (Table 1.5). Fugacity-coefficient formulas are in Tables 3.3 and 3.4, and activity-coefficient formulas are in Table 4.7. Wilson exponential parameters are from DECHEMA, Vol. 1, parts 3 & 4, p. 389, and are listed as step 160 of the program. Since the system pressure is near the vapor pressures of all the components, the Poynting factors are essentially unity. Accordingly, the VERs are computed from

$$K_i = \gamma_i \phi_i^s P_i^s / \hat{\phi}_i P.$$

The procedure of Figure 6.7 is followed.

The printout shows the bubblepoint temperature to be 468.7 K and the vapor composition (0.3779, 0.2951, 0.3270). As shown by the runs at several temperatures, in this case step 8 of Figure 6.7 did not appear to be necessary, since going directly to 468.7 gave the same final results in fewer iterations than starting at 500 and going down 10° at a time while steadying off at each temperature. However, the $\Sigma y_i = 4.1$ at 500, and the comparison of the ratio of fugacity coefficients at the two temperatures also shows considerable differences:

i	$\phi_i^s/\hat{\phi}_i$ @ 500	@ 468.7
1	1.0003	1.0001
2	1.1058	1.0018
3	1.2411	1.0078

Final values of vapor pressures, ideal VERs, partial fugacity coefficients, pure-component fugacity coefficients, ratio of fugacity coefficients, and activity coefficients at 468.7 K.		Showing converged values of Σy_i at each temperature. At 468.7 K the sum has converged to unity.	
		500 K	
25.1151633414 (1)	P_i^s	1.8044	$S = \Sigma y_i$
14.4904039759 (2)		.3317	y_1
15.4532211081 (3)		.2955	y_2
		.3728	y_3
		500 K	
1.25575816707	$P_i^s/P = (K_i)$ideal	1.7844	
.724520198795		.3351	
.772661055405		.2993	
		.3657	
		500 K	
1.25991956422	K_i	1.7815	
.984045189215		.3356	
.817621023225		.2993	
		.3651	
		470 K	
.843532547419	$\hat{\phi}_i$	1.0362	
.790714279772		.3724	
.785362161074		.2973	
		.3304	
		470 K	
.843632255947	ϕ_i^s	1.026	
.792192034686		.3760	
.791524258278		.2957	
		.3284	
		470 K	
1.00011820353	$\phi_i^s/\hat{\phi}_i$	1.0253	
1.00186888608		.3762	
1.0078461855		.2954	
		.3283	
		468.7K	
1.00319527112	γ_i	1.0007	
1.35566896974		.3777	
1.04995036875		.2952	
		.3270	
		468.7 K	
		1.0000	Final
		.3779	Final
		.2951	Final
		.3270	Final

BUBBLEPOINT TEMPERATURE PROGRAM WITH THE HP-85 COMPUTER

```
10  ! "FLASH" BUBBLEPOINT T
20  SHORT S
30  OPTION BASE 1
40  READ P,R
50  DATA 20,.08205
60  DIM B(3,3),D(3),F(3),F1(3),G(3),K(3),L(3,3),P(3),
    Y(3)
70  ! MATRICES ENTERED IN ORDER: T11,T12,
    T13,T21,T22,T23,T31,T32,T33
80  ! T1(I,J) CRIT TEMP;W(I,J) A CENT FACTOR;
    P1(I,J) CRIT PRESS
90  DIM T1(3,3),W(3,3),P1(3,3)
100 DATA 508.1,521.7,515.6,521.7,535.6,529.4,515.6,
    529.4,523.2
110 DATA .309,.319,.336,.319,.329,.346,.336,.346,.363
120 DATA 46.4,43.5,40,43.5,41,37.8,40,37.8,37.8
130 MAT READ T1,W,P1
140 ! A(I,J) WILSON EXPONENTIAL
    PARAMETERS
150 DIM A(3,3)
160 DATA 0,1371.31,−292.975,−650.152,0,−405.21,
    644.481,2704.427,0
170 MAT READ A
180 ! V(I) MOLAL VOLUMES
190 DIM V(3)
200 DATA 73.52,89.57,97.79
210 MAT READ V
220 ! A1,B1,C1, ANTOINE COEFFS
230 DIM A1(3),B1(3),C1(3)
240 DATA 16.6513,16.5986,16.1516,2940.46,3150.42,
    2790.5,−35.93,−36.65,−57.15
```

(Continued next page)

Example 6.5. *(continued)*

```
250  MAT READ A1,B1,C1
260  INPUT T
270  ! B(I,J) VIRIAL COEFFS
280  ! L(I,J) WILSON PARAMS
290  J=1
300  FOR I=1 TO 3
310  B(I,J)=R*T1(I,J)/P1(I,J)*(.083-.422*(T1(I,J)/T) ^
     1.6+W(I,J)*(.139-.172*(T1(I,J)/T)^4.2))
320  L(I,J)=V(J)/V(I)*EXP(-(A(I,J)/1.987/T))
330  NEXT I
340  J=J+1
350  IF J<=3 THEN 300
360  DIM X(3)
370  DATA .3,.3,.4
380  MAT READ X
390  FOR J=1 TO 3
400  D(J)=X(1)*L(J,1)+X(2)*L(J,2)+X(3)*L(J,3)
410  P(J)=1/760*EXP(A1(J)-B1(J)/(T-C1(J)))
420  Y(J)=P(J)*X(J)/P
430  NEXT J
440  S=Y(1)+Y(2)+Y(3)
450  DISP S
460  FOR J=1 TO 3
470  Y(J)=Y(J)/S
480  NEXT J
490  MAT DISP Y
500  ! G(I) ACTIVITY COEFFS
510  FOR I=1 TO 3
520  G(I)=EXP(1-LOG(D(I))-X(1)*L(1,I)/D(1)-X(2)
     *L(2,I)/D(2)-X(3)*L(3,I)/D(3))
530  NEXT I
540  ! B MIXTURE VIRIAL COEFF
550  B2=Y(1) ^2*B(1,1)+Y(2) ^2*B(2,2)+Y(3) ^2*B
     (3,3)
560  B=B2+2*(Y(1)*Y(2)*B(1,2)+Y(1)*Y(3)*B(1,3)+Y
     (2)*Y(3)*B(2,3))
570  ! F1(I) PURE COMP FUG COEFF
580  ! F(I) PARTIAL FUG COEFF
590  ! P(I) VAPOR PRESS
600  ! K(I) VAP EQUILIB RATIO
610  FOR I=1 TO 3
620  F1(I)=EXP(B(I,I)*P/R/T)
630  F(I)=EXP(P/R/T*(2*(Y(1)*B(I,1)+Y(2)*B(I,2)+Y
     (3)*B(I,3))-B))
640  F2(I)=F1(I)/F(I)
650  P(I)=1/760*EXP(A1(I)-B1(I)/(T+C1(I)))
660  K(I)=G(I)*F1(I)*P(I)/F(I)/P
670  Y(I)=K(I)*X(I)
680  NEXT I
690  S=Y(1)+Y(2)+Y(3)
700  PRINT T
710  PRINT S
720  FOR I=1 TO 3
730  Y(I)=Y(I)/S
740  NEXT I
750  MAT PRINT USING "2X,.DDDD" ; Y
760  INPUT T
770  GOTO 510
780  END
```

Example 6.6. VLE When the Soave Equation Applies to Both Phases

The *x*-*y* data at 20 atm of propylene(1) + isobutane(2) will be found over the entire concentration range with the formulas in Tables 1.11 and 3.4. The basic properties are:

ij	*Substance*	T_c	P_c	V_c	z_c	ω	a	b
11	C_3H_6	365	45.6	181	0.275	0.148	8.4078	0.0569
22	i-C_4H_{10}	408.1	36.0	263	0.283	0.176	13.3125	0.0806

$$\alpha_1 = \left[1 + 0.7113\,(1 - \sqrt{T/365})\right]^2,$$

$$\alpha_2 = \left[1 + 0.7533\,(1 - \sqrt{T/408.\ 1})\right]^2,$$

$$a\alpha = \left(y_1\sqrt{a_1\alpha_1} + y_2\sqrt{a_2\alpha_2}\right)^2,$$

$$b = y_1 b_1 + y_2 b_2,$$

$$\ln\hat{\phi}_i = -\ln[z(1 - b/V)] + \frac{b_i}{b}(z - 1) + \frac{a\alpha}{bRT}\left(b_i/b - \frac{2\sqrt{(a\alpha)_i}}{\sqrt{a\alpha}}\right)\ln(1 + b/V).$$

Procedure: Equation 9 of Table 1.11 is solved for the largest and smallest roots, which become the compressibilities of the vapor and liquid phases. For a given liquid composition, a trial estimate of the vapor compositions is needed to start the solution. After the zs have been obtained, the partial fugacity coefficients are found with the given equation; this is Eq. 27 with the combining rules, Eq. 2, both of Table 3.4. The VERs are $K_i = \phi_i^L/\phi_i^V$. The bubblepoint is checked. For subsequent trials, the calculated $y_i = K_i x_i$ are normalized and employed to refigure the vapor fugacity coefficients. The results are tabulated and the computer program is provided.

(Continued next page)

Example 6.6. *(continued)*

x_1	T	y_1	z_V	z_L	$\hat{\phi}_1^V$	$\hat{\phi}_2^V$	$\hat{\phi}_1^L$	$\hat{\phi}_2^L$
0	373.42	0	0.6636	0.1026	0.9005	0.7538	1.5060	0.7538
0.1	367.00	0.1606	0.6793	0.0993	0.8823	0.7403	1.4165	0.6904
0.2	360.84	0.3067	0.6924	0.0963	0.8664	0.7281	1.3286	0.6311
0.3	354.93	0.4376	0.7024	0.0935	0.8525	0.7168	1.2434	0.5759
0.4	349.34	0.5535	0.7100	0.0909	0.8401	0.7063	1.1626	0.5256
0.5	344.05	0.6553	0.7154	0.0885	0.8290	0.6962	1.0866	0.4799
0.6	339.06	0.7445	0.7190	0.0862	0.8187	0.6865	1.0157	0.4385
0.7	334.39	0.8221	0.7211	0.0839	0.8093	0.6772	0.9504	0.4016
0.8	330.00	0.8898	0.7246	0.0817	0.8003	0.6687	0.8901	0.3683
0.9	325.89	0.9486	0.7215	0.0796	0.7920	0.6590	0.8348	0.3387
1.0	322.02	1	0.7202	0.0775	0.7840	0.6502	0.7839	0.3121

```
10  ! X-Y DATA WITH "SOAVE" EQN
20  OPTION BASE 1
30  DIM X(2),Y(2),Z(2),F(2),G(2),H(2)
40  DATA .8,.2,.9,.1
50  MAT READ X,Y
60  READ P,R
70  DATA 20,.08205
72  DIM B(2),A(2)
74  DATA .0569,.0806
80  MAT READ B
90  INPUT T
100 A(1)=8.4078*(1+.7113*(1-(T/365) ^.5)) ^2
110 A(2)=13.3125*(1+.7533*(1-(T/408.4) ^.5)) ^2
120 FOR I=1 TO 2
130 Z(I)=X(I)
140 NEXT I
150 GOSUB 460
160 Z=.05
170 GOSUB 500
180 DISP "ZL=";Z
190 GOSUB 600
200 FOR I=1 TO 2
210 G(I)=H(I)!φ̂_i^L
220 Z(I)=Y(I)
230 NEXT I
240 GOSUB 460
250 Z=1
260 GOSUB 500
270 DISP "ZV=";Z
280 GOSUB 600
290 FOR I=1 TO 2
300 F(I)=H(I)!φ̂_i^V
310 K(I)=G(I)/F(I)
320 Y(I)=K(I)*X(I)
330 NEXT I
340 S=Y(1)+Y(2)
350 DISP "S=";S
360 FOR I=1 TO 2
370 Y(I)=Y(I)/S
380 NEXT I
390 DISP USING 400 ; X(1),Y(1),T,F(1),F(2),G(1),
    G(2)
400 IMAGE D.D,X,D.DDD,X,DDD.D,X,D.DDD,X,
    D.DDD,X,D.DDD,X,D.DDD
405 IF ABS (S-1)<=0.0001 THEN 420
410 GOTO 90
420 END
460 ! SR FOR SOAVE COEFFS
470 A=(Z(1)*A(1) ^.5+Z(2)*A(2) ^.5) ^2
480 B=Z(1)*B(1)+Z(2)*B(2)
490 RETURN
500 ! SR FOR COMPRESSIBILITY Z
510 C=A*P/R ^2/T ^2
520 D=B*P/R/T
530 F1=Z ^3-Z ^2+(C-D-D ^2)*Z-C*D
540 F2=3*Z ^-2*Z+C-D-D ^2
550 H1=F1/F2
560 IF ABS(H1/Z)<=.0001 THEN 590
570 Z=Z-H1
580 GOTO 530
590 RETURN
600 ! SR FOR FUGACITY COEFFS
610 FOR I=1 TO 2
620 V=Z*R*T/P
630 H(I)=-LOG(Z*(1-B/V))+B(I)/B*(Z-1)
640 H(I)=H(I)+LOG(1+B/V)/B/R/T*(A*B(I)/B-2*
    (Z(1)*(A(I)*A(1)) ^.5+Z(2)*(A(I)*A(2)) ^.5))
650 H(I)=EXP(H(I))
660 NEXT I
670 RETURN
680 END
```

well as of temperature and pressure. These properties can be expressed in terms of fugacity and activity coefficients, which are derivable from equations of state or from generalized correlations. Such relations are discussed in detail in Chapter 11, and only a brief mention of suitable procedures or formulas is given here.

A general expression for the enthalpy of a mixture relative to zero pressure and temperature T_0 is

$$H = H_0^* + \int_{T_0}^{T} C_p^*\,dT + \int_0^P \left(\frac{\partial H}{\partial P}\right)_T dP \tag{6.102}$$

Example 6.7. The Dewpoint by the Chao-Seader Method

The dewpoint temperature of a mixture of propylene + isobutane with $y_1 = 0.607$ will be found at 20 atm.

Compressibilities are found with the Redlich-Kwong equation, Table 1.9, by trial from

$$z^3 - z^2 + (A - B - B^2)z - AB = 0 \quad \text{(Eq. 7)},$$

$$A = aP/R^2T^{2.5},$$

$$B = bP/RT,$$

$$R = 0.08205.$$

The partial fugacity coefficients are found with Eq. 18 of Table 3.4,

$$\ln \hat{\phi}_i = -\ln(z - B) + B_i(z - 1)/B + (A/B)(B_i/B - 2\sqrt{A_i/A}) \ln(1 + B/z).$$

Fugacities of the liquids and the activities of the liquid components are found with the formulas of Table 6.1. The dewpoint is found by trial by the method of Figure 6.9. This involves specifying a temperature, estimating a composition on the assumption that the activity coefficients are unity, then proceeding with improved values of the activity coefficients to convergence.

In this particular case the activity coefficients actually are near unity, so convergence is very rapid.

The result is $T = 346.4$, $x_1 = 0.541$. In Example 6.6, the corresponding result is $T = 346.6$, $x_1 = 0.451$.

Using the API chart, Figure 6.5, the $\Sigma y_i/K_i$ are

350 K, $\quad 0.607/1.50 + 0.393/0.68 = 0.9826,$

340 K, $\quad 0.607/1.30 + 0.393/0.56 = 1.1687.$

The interpolated value is $T = 349$, $x_1 = 0.41$.

The substantial disagreement between the three methods is disquieting. On a statistical basis, the API studies came out in favor of the Soave method of Example 6.6, as explained in Section 1.9. In a particular case, judgment may need to be exercised in favor of either a conservative or an optimistic design.

	T_c	P_c	ω	δ	V	a	b
C_3H_6	365	45.6	0.1451	6.43	79	160.64	0.0569
iC_4	408.1	36	0.1825	6.73	105.5	268.96	0.0806
mix						199.90	0.0662

T	Z	$\hat{\phi}_1$	$\hat{\phi}_2$	ϕ_{1L}	ϕ_{2L}	$\phi_{1L}/\hat{\phi}_1$	$\phi_{2L}/\hat{\phi}_2$
320	0.5827	0.7976	0.6277	0.6145	0.3565	0.7704	0.5679
330	0.6535	0.8121	0.6656	0.7310	0.4441	0.9001	0.6672
340	0.7000	0.8266	0.6957	0.8564	0.5441	1.0361	0.7821
350	0.7352	0.8401	0.7212	0.9886	0.6565	1.1768	0.9103

At 340 K, for example, the trial value $x_1 = 0.5383$ was based on unit activity coefficients. The VERs are $K_i = \gamma_i \phi_{iL}/\hat{\phi}_i$.

γ_1	γ_2	K_1	K_2	Σx_i	x_1
T=346					
.993	.992	1.111	.850	1.0085	.542
.993	.992	1.111	.850	1.0085	.542
.993	.992	1.111	.850	1.0085	.542
T=346.5					
.993	.992	1.120	.855	1.0015	.541
.993	.992	1.120	.855	1.0015	.541
.993	.992	1.120	.855	1.0015	.541
T=347					
.993	.992	1.124	.863	.9954	.542
.993	.992	1.124	.863	.9954	.542
.993	.992	1.124	.863	.9954	.542

```
10  ! DEWPOINT BY FIG. 6.9
20  READ T,L1,L2
30  DATA 347,1.132,.87
40  INPUT X
50  F1=79*X+105.5*(1-X)
60  F=79*X/F1
70  T1=79/F1
80  T2=105.5/F1
90  D3=6.43*F+6.73*(1-F)
100 G1=EXP(79*(6.43-D3)^2/1.987/T+LOG(T1)+
    1-T2)
110 G2=EXP(105.5*(6.73-D3)^2/1.987/T+LOG(T2)
    +1-T2)
120 K1=G1*L1
130 K2=G2*L2
140 X1=.607/K1
150 X2=.393/K2
160 S=X1+X2
170 X=X1/S
180 PRINT USING 190 ; G1,G2,K1,K2,S,X
190 IMAGE .DDD,X,.DDD,X,D.DDD,X,.DDD,X,
    D.DDDD,X,.DDD
200 GOTO 50
210 END
```

$$= \Sigma y_i \left[H_{i0}^* + \int_{T_0}^{T} C_{pi0}^* dT \right] + \int_0^P \left(\frac{\partial H}{\partial P} \right)_T dP \tag{6.103}$$

where the asterisk designates the ideal-gas state. The second integral may be evaluated with an equation of state of the mixture. With the B-truncated virial, for instance,

$$H = \Sigma y_i \left[H_{i0}^* + \int_{T_0}^{T} C_{pi0}^* dT \right] + \int_0^P \left(B - T \left(\frac{\partial B}{\partial T} \right)_P \right) dP \Bigg], \tag{6.104}$$

where the virial coefficient, B, of the mixture depends on the composition as given in Table 1.8.

Partial molal enthalpies may be evaluated from the equations,

$$\bar{H}_i = H_i - RT^2 \left(\frac{\partial \ln \hat{f}_i}{\partial T} \right)_{PN_i} = H_i - RT^2 \left(\frac{\partial \ln \hat{\phi}_i}{\partial T} \right)_{PN_i}, \tag{6.105}$$

for liquid phases as well as vapor. For some substances, the Soave and BWR equations of state, for example, are applicable to both phases. Equations for the partial fugacity coefficients are given in Table 3.4. A wider range of mixtures can be handled in terms of activity-coefficient correlations when these are known. The basic relation is

$$\hat{f}_i = \gamma_i x_i P_i^s \phi_i^s (PF)_i. \tag{6.106}$$

The coefficient, ϕ_i^s, of the pure saturated vapor is obtained from a suitable equation of state. In terms of the virial equation, for instance,

$$\hat{f}_i = \gamma_i x_i P_i^s \exp \left[\frac{B_i P_i^s + V_{Li}(P - P_i^s)}{RT} \right]. \tag{6.107}$$

Enthalpy relations derived from these equations are given in Chapter 11. Because of their complexity, computer demands for their use may be quite severe. As always, judgment must be exercised in choosing between the best available method and simpler methods that may be good enough. The most common assumption is that the partial molal enthalpies are the same as enthalpies of the pure components, and only the effects of temperature and pressure need be taken into account.

6.5. SOLUBILITY OF GASES IN LIQUIDS

Solutions are homogeneous mixtures. It is usual to designate the components present in significantly larger concentrations the *solvent*, and the other components the *solutes*. Gases and solids dissolved in liquids usually are called solutes. In the special case considered here of gases dissolved in liquids, the solvent is assumed to have negligible vapor pressure and to remain entirely in the liquid phase. From another viewpoint, the components of the solvent, if it is a mixed one, all have zero VERs. If the gas is below its critical temperature, the solubility can be represented or predicted by the earlier methods of this chapter. For the contrary case of gases above

Table 6.3. Solubilities of Gases, 10^4x Mol fraction of Solute, at 25 C and 1 Atm Partial Pressure (data from Battino & Clever, in Dack, ed., *Solutions and Solubilities,* pt. 1, 1976). Solubility parameters and molal volumes are given. Problem 6.1 requests a check of the CH_4 data against Eq. 6.109.

	10^4x Gas									
Solvent	*He*	*Ne*	*Ar*	H_2	N_2	O_2	CO_2	CH_4	δ	V
Water	0.068	0.082	0.254	0.142	0.119	0.231	—	0.248	23.53	18.1
n-Hexane	2.604	3.699	25.12	6.315	14.02	19.3	—	50.37	7.27	131.6
n-Octane	2.397	3.626	24.26	6.845	13.04	20.83	—	29.27	7.54	163.4
n-Tetradecane	2.249	3.340	26.50	—	—	—	—	—	—	—
Isooctane	3.083	4.593	29.21	7.832	15.39	28.14	138.7	29.66	—	—
Cyclohexane	1.217	1.792	14.80	4.142	7.61	12.48	76.0	32.75	8.19	108.7
Benzene	0.771	1.118	8.815	2.580	4.461	8.165	97.30	20.77	9.16	89.4
Toluene	0.974	1.402	10.86	3.171	5.74	9.09	101.3	24.14	8.93	106.8
m-Xylene	1.121	1.619	—	4.153	—	—	—	—	8.88	123.4
n-Perfluoroheptane	8.862	—	53.22	14.03	38.80	55.08	208.2	82.56	—	—
Hexafluorobenzene	2.137	3.455	23.98	—	17.95	24.18	220	38.42	—	—
Carbon tetrachloride	—	—	13.51	3.349	6.480	12.01	105.3	28.70	8.55	97.0
Chlorobenzene	0.691	0.979	8.609	2.609	4.377	7.910	98.06	20.47	9.67	102.2
Nitrobenzene	0.350	0.436	4.448	—	—	4.95	99.80	—	10.8	102.7
Methanol	0.595	0.814	4.491	—	2.747	4.147	55.78	8.695	14.5	40.7
Ethanol	0.769	1.081	6.231	2.067	3.593	5.841	63.66	12.80	12.78	58.7
Acetone	1.081	1.577	9.067	2.996	5.395	8.383	185.3	18.35	9.62	74.0
Dimethyl sulfoxide	0.284	0.368	1.54	0.761	0.833	1.57	90.8	3.86	12.0	71.3
δ	—	—	5.33	0	4.44	4.0	7.12	5.68		
V	—	—	—	31.0	53.0	28.4	44.0	52.0		

Example 6.8. The Solubilities of CO_2 and H_2S in *n*-Decane at 1 atm

Experimental values are given by King et al. (1977). These will be checked with calculations from the solubility parameter equation, using superheated liquid fugacities, (1) by extrapolation of Antoine vapor pressures; (2) by the corresponding states correlation of Prausnitz & Shair (1961). Raw and calculated data are in Table 2, and the comparison of solubilities is in Table 1. The procedure is:

1. Obtain vapor pressures with the Antoine equation.
2. Estimate the fugacity coefficients of the superheated liquid with the Lee-Kesler tables.
3. Figure the Poynting factor from $PF = \exp(V_2(1 - P^{sat})/RT)$.
4. Evaluate the fugacity of the superheated liquid as

   $$f^0_{2L} = P^{sat}_2 \phi^{sat}_2 (PF)_2,$$

   which is called the *extrapolated value.*
5. Also figure this quantity from the correlation,

   $$f^0_{2L} = P_c \exp(7.81 - 8.06/T_r - \ln T_r).$$

   which is Eq. 6.113.
6. Obtain the activity coefficients of the solutes from the simplified equation

   $$\gamma_2 = \exp(V_2(7.72 - \delta_2)^2/RT,$$

   where the effect of small values of x_2 on the RHS has been neglected.
7. Finally, obtain the solubilities from

   $$x_2 = f_{2V}/\gamma_2 f^0_L = 1/\gamma_2 f^0_{2L},$$

   where the fugacity has been assumed equal to the pressure at one atm.

Comparison: The extrapolation method agrees better with experiment for CO_2, but for H_2S the methods are roughly equivalent. With the extended Scatchard-Hildebrand equation, King et al. (1977) achieve correlations within 1% of experiment.

Table 1. Summary of Solubilities

	CO_2		H_2S	
	20 C	50 C	20 C	50 C
Experimental	0.0133	0.0099	0.0502	0.0325
P & S Correlation	0.021	0.013	0.060	0.031
Extrapolated vapor pressure	0.011	0.0082	0.052	0.031

Table 2. Raw and Calculated Data

	$C_{10}H_{22}$	CO_2 20 C	CO_2 50 C	H_2S 20 C	H_2S 50 C
P_c		72.8		88.2	
T_c		304.2		373.2	
ω		0.225		0.1	
δ	7.72	7.12		8.8	
V	196	44		34.3	
P^{sat}		214	572	17.3	33.7
ϕ^{sat}		0.6	0.53	0.91	0.88
Poynting factor		0.68	0.39	0.98	0.96
f^0_L (extrapolation)		87	118	15.5	28.5
f^0_L (correlation)		46.7	76.1	15.5	30.1
γ		1.0276	1.0250	1.0711	1.0643

Example 6.9. Isobaric and Isothermal *x-y* Diagrams with the One-Parameter Margules Equation

1. Isothermal *x-y* diagrams are constructed with vapor pressures $P^0_1 = 1.5$ and $P^0_2 = 1$. Activity coefficients are given by

$$\ln \gamma_1 = Ax_2^2,$$

$$\ln \gamma_2 = Ax_1^2,$$

with values of $A = 3$, 2, 1, or 0. Vapor compositions are obtained as in Example 6.10. Liquid-phase separation occurs with $A = 3$. The broken portion of that curve is not physically realizable.

2. For the isobaric condition, temperature-dependent forms of the activity coefficient equations are assumed to be

$$\ln \gamma_1 = \frac{M}{T} x_2^2,$$

$$\ln \gamma_2 = \frac{M}{T} x_1^2.$$

Several values of M are tried: 1,000, 500, and 0. Vapor pressures are given by

$$\ln P^0_1 = 12.5 - 4{,}300/T,$$

$$\ln P^0_2 = 12 - 4{,}500/T.$$

With the pressure set at $P = 0.5$, the bubblepoint temperatures at each liquid composition are determined with the condition

$$P = \gamma_1 x_1 P^\circ_1 + \gamma_2 x_2 P^\circ_2,$$

using the Newton-Raphson method. Vapor compositions then are found with

$$y_1 = 1/(1 + \gamma_2 x_2 P^\circ_2/\gamma_1 x_1 P^\circ_1).$$

At $M = 1{,}000$ on Figure B, the dashed portion of the curve embraces the two-liquid-phase region.

3. Temperature-composition plots are made with the aid of a portion of the program for part (b). For $M = 500$, there is a minimum temperature of 326° at $x_1 = 0.86$. The two-liquid mixture with $M = 1{,}000$ boils at 319°.

Example 6.9. *(continued)*

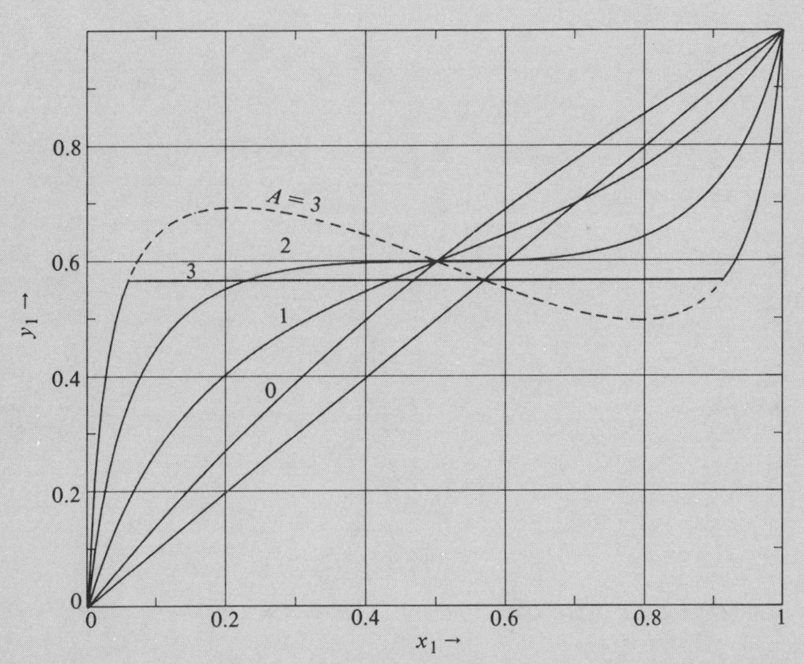

Example 6.9(a). x-y diagram at constant temperature. In the region where the line is broken, the physical situation is represented by the horizontal line.

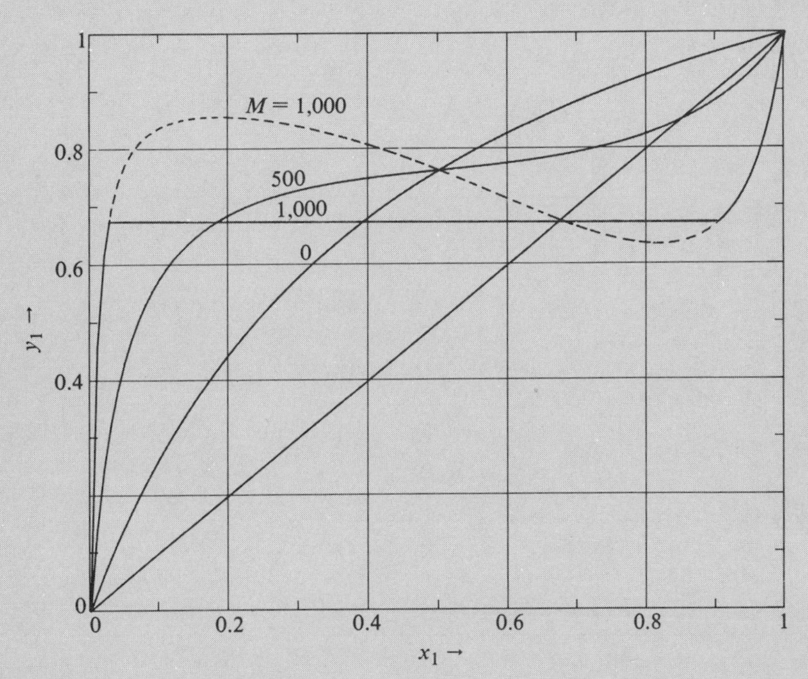

Example 6.9(b). x-y diagram at constant pressure. In the region where the line is broken, the physical situation is represented by the horizontal line.

(Continued next page)

Example 6.9. *(continued)*

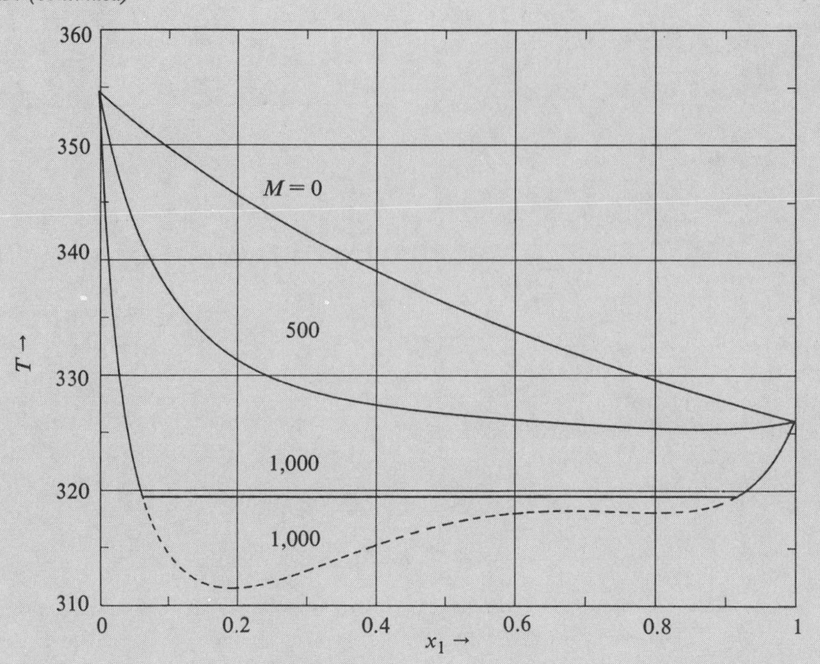

Example 6.9(c). Temperature against composition. In the region where the line is broken, the physical situation is represented by the horizontal line.

Example 6.10. Isothermal VLE with the Margules Equation

The effect of several combinations of Margules parameters on the shapes of x-y, P-x, and γ-x diagrams is examined. Vapor pressures are $P_1^0 = 3$ and $P_2^0 = 1$. Margules $A = 1$ in all cases, but B assumes the values 4, 2, −2, or −6. The activity coefficients are given by

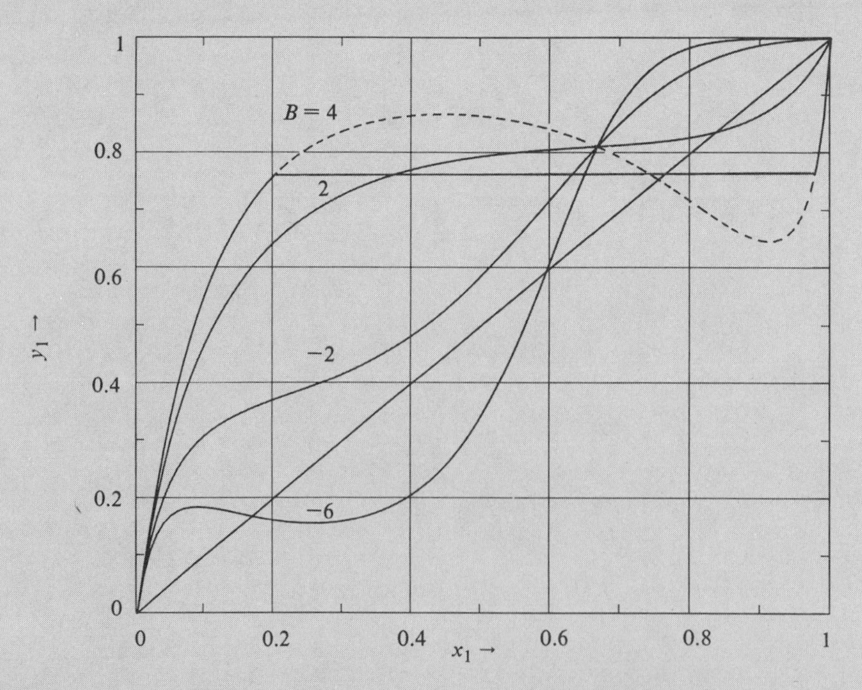

Example 6.10(a). The x-y diagram. The broken line is given by the mathematical equation, but the physical situation is represented by the horizontal line in this interval.

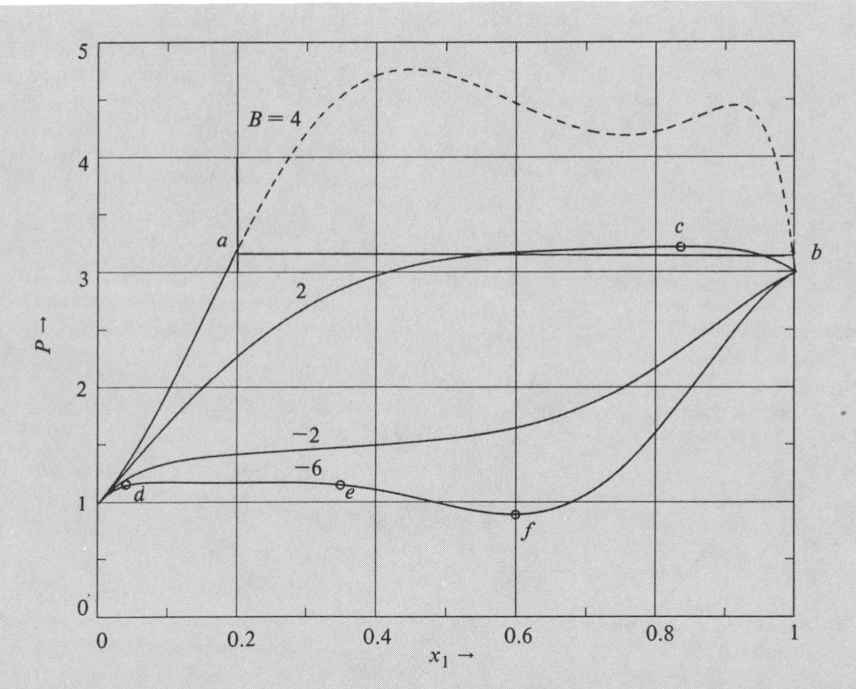

Example 6.10(b). Total pressure against composition, *P-x*. In the region where the line is broken, the physical situation is represented by the horizontal line ab.

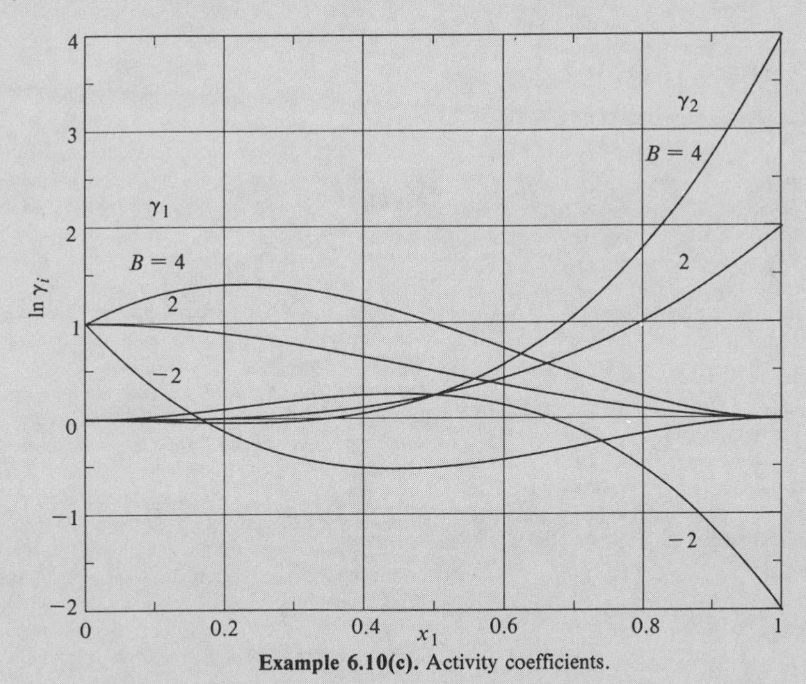

Example 6.10(c). Activity coefficients.

$$\ln \gamma_1 = [A + 2(B - A)x_1]x_2^2,$$

$$\ln \gamma_2 = [B + 2(A - B)x_2]x_1^2.$$

a. The vapor compositions are figured from

$$y_1 = 1/(1 + \gamma_2 x_2 P_2^0/\gamma_1 x_1 P_1^0).$$

With (A, B) equal to (1, 4) and (1, −6), two liquid phases form. In addition, at (1, −6) a homogeneous azeotrope is formed, as well as at (1, 2).

b. Total pressures are obtained from

$$P = \gamma_1 x_1 P_1^0 + \gamma_2 x_2 P_2^0.$$

The two-liquid-phase regions are limited by points a and b and points d and e on Figure B. A maximum-pressure azeotrope is identified at point c and a minimum-pressure one at point f.

c. Activity coefficients are plotted directly from the Margules equations. Noteworthy features of the plots of Figure c are the maximum and minimum values.

their critical temperatures, special methods have been developed.

From Henry's Law on, a number of relations of value has been developed. A recent example of a rather precise scheme is that of de Ligny et al. (1976) who use an equation of the form

$$x_2 = G_1S_1 + G_2S_2,$$

where the Gs are properties of the solute and the Ss are those of the solvent. The many kinds of units in which gas solubilities are commonly expressed are summarized by Kruis (1976), where a review of theory and many data also are given. Extended treatments of theory may be found in books by Shinoda (1978) and Dack (1976).

For a step beyond ideality, the concept of solubility parameter has proved useful in quantifying the solubility of gases and other phases. In the fugacity balance, where the subscript 2 identifies the gas,

$$f_{2G} = \hat{f}_{2L} = \gamma_2 x_2 f^0_{2L}, \tag{6.108}$$

the fugacity f^0_{2L} of pure solute in the liquid state is hypothetical when the gas is above its critical temperature, but can be estimated, as will be shown.

Solving Eq. 6.108 for the solubility, x_2, and introducing the Scatchard-Hildebrand equation,

$$x_2 = \frac{f_{2G}}{\gamma_2 f^0_{2L}} = \frac{f_{2G}}{f^0_{2L}} \exp\left[-\frac{V_{2L}}{RT}(\delta_1 - \delta_2)^2\phi_1^2\right]. \tag{6.109}$$

When the behavior of the gas phase is substantially ideal, pressures can be substituted for fugacities, so that

$$x_2 = \frac{P}{P_2^{sat}} \exp\left[-\frac{V_{2L}}{RT}(\delta_1 - \delta_2)^2\phi_1^2\right], \tag{6.110}$$

but this is not satisfactory above the critical temperature of the solute. The fugacity, f_{2G}, of the pure gas may be found with a suitable equation of state, which also may be used to estimate the fugacity of saturated liquid at the system temperature, and thereby used to make an estimate of f^0_{2L} at the system temperature and pressure by application of the Poynting factor, thus,

$$f^0_{2L} = f^s_{2L} \exp(V_{2L}(P - P^s_2)/RT). \tag{6.111}$$

This scheme was adopted by King et al. (1977) for representing the solubility of the gases CO_2, H_2S, and C_3H_8 in several alkanes. Since these gases have critical temperatures above 25 C, extrapolation of vapor pressures to the 70 C maximum worked with is modest. Even so, it was found better to use the Flory-Huggins extension of the S-H equation with interaction parameters:

$$\ln \gamma_2 = \frac{V_{2L}}{RT}\phi_1^2[(\delta_1 - \delta_2)^2 + 2\,l_{12}\delta_1\delta_2 + 1 - \phi_1/x_1 + \ln(\phi_1/x_1)]. \tag{6.112}$$

Normally the interaction coefficient l_{12} is taken to be zero, but in the King work it is a parameter; for light hydrocarbons it was in the range of 0.02 and for CO_2 and H_2S it was in the range of 0.1. Further improvement, to within 1% of experiment, was obtained by adjusting the solubility parameter away from the theoretical value; in the extreme case of H_2S, the adjusted value of δ was 5.6, compared with the true value 8.36.

When experimental solubility data are available and the molal volumes and solubility parameters are known, the effective fugacity of the superheated liquid can be back-calculated from Eq. 6.109. This was done by Prausnitz & Shair (1961) and Prausnitz (1962). Their graphical correlation in terms of the reduced temperature and the critical pressure can be represented by the equation

$$\ln(f^0_{2L}/P_{c2}) = 7.81 - 8.06/T_r - 2.94 \ln T_r, \quad 0.7 \le T_r \le 2.5. \tag{6.113}$$

A similar correlation is extended to multicomponent mixtures by Prausnitz et al. (1980, p. 55), who write for the solubility

$$x_i = f_{2V}/f^0_{2i}\gamma_i. \tag{6.114}$$

Contributions of individual components to the reference fugacity are given by

$$f^0_{2i} = 7.224 - 7.534(T_{ci}/T) - 2.598 \ln(T/T_{ci}), \tag{6.115}$$

and pair contributions by

$$\gamma_{ij} = \exp(a_{ij} + b_{ij}/T), \tag{6.116}$$

where the coefficients a_{ij} and b_{ij} are specific to the system; empirical values are given for four gases and two liquids, but no generalization is made. Combining rules for the overall properties, f^0_{2i} and γ_i, also are stated in the reference.

When the gas being dissolved is a mixture, the solubility of each component is expressible in terms of the multicomponent form of the SH equation, Eq. 4.195, so that, for example, Eq. 6.110 becomes for each gas:

$$x_i = \frac{P}{P_i^{sat}} \exp(-V_i(\delta_i - \bar{\delta})^2/RT) \tag{6.117}$$

A general procedure for handling mixtures with some supercritical components is developed by Van Ness & Abbott (1979). The data required are of partial fugacity coefficients of all pairs of components as functions of a limited range of compositions and the assumption of a specific form of the dependence of activity coefficients on composition, for example, the Wilson equation. The data are regressed so the parameters of the Wilson equation and extrapolated values of pure component reference fugacities are found simultaneously, for one pair of mixtures at a time. The merit of this method is that the reference states of all components are of the same type, pure liquid in the case of subcritical components and hypothetical pure liquid in the cases of the others. Example 6.12 outlines the method and applies the results to a bubblepoint calculation.

As indicated by Example 6.8, the results obtained by some of the methods of this section are not unreasonable when critical temperatures are not greatly different from that of the system, but less satisfactory at other conditions. How to take account of supercritical components is a problem that has not been resolved entirely. Research in this field is currently active. Useful entries into that literature are papers by O'Connell (1980) and Brandani & Prausnitz (1981).

6.6. BINARY VLE DIAGRAMS

Although binary mixtures are of infrequent occurrence industrially, quantitative knowledge of their behavior is of importance, since methods have been devised recently for representing the behavior of multicomponent mixtures in terms of

Example 6.11. Equilibrium as the Condition of Minimum Gibbs Energy

Equilibrium mixtures of cyclohexane + *m*-xylene at 25 C are well represented by $G^{ex}/RT = 0.48x_1x_2$ (DECHEMA I/6a, p. 314). Vapor pressures are $P_1^s = 97.6$ Torr, $P_2^s = 8.3$ Torr. The equilibrium composition at a system pressure of 50 Torr will be found by locating the minimum Gibbs energy.

At this low pressure, it is legitimate to assume $\hat{\phi}_i = \phi_i^s = (PF)_i = 1$, so that the vapor compositions may be eliminated from Eq. 6.128 with

$$y_i = (\gamma_i P_i^s/P)\,x_i.$$

Before substituting into that equation, however, it is necessary to normalize these vapor compositions; this procedure assures that the minimum Gibbs energy is correct, but not necessarily that the values at other compositions are exactly right. The plots show that the composition at the minimum Gibbs energy is independent of the overall composition, z_i, as it should be. The value, $x_1 = 0.379$, found graphically, compares favorably with the result of a flash calculation: $x_1 = 0.3786$, $y_1 = 0.8895$, $\beta = 0.2376$ when $z_1 = 0.5$.

```
      5 EXAMPLE 6.11
     10 SCALE 0,1,-4.5,-3.5
     20 XAXIS -4.5,.1
     22 XAXIS -3.5,.1
     24 YAXIS 1,.1
     30 YAXIS 0,.1
     40 READ Z,P,P1,P2
     50 DATA .2,.0658,.1284,.0109
     60 FOR X=.001 TO .999 STEP .01
γ1   70 G1=EXP(.48*(1-X) ^2)
γ2   80 G2=EXP(.48*X ^2)
     90 Y1=G1*P1*X/(G1*P1*X+G2*P2*(1-X))
    100 Y2=1-Y1
β   110 B=(Z-X)/(Y1-X)
γ   120 Y=B*(Y1*LOG(P*Y1)+Y2*LOG(P*Y2))
        +(1-B)*(X*LOG(G1*P1*X)+(1-X)
        *LOG(G2*P2*(1-X)))
    130 DRAW X,Y
    140 NEXT X
    150 MOVE 0,0
    160 BEEP
    170 END
```

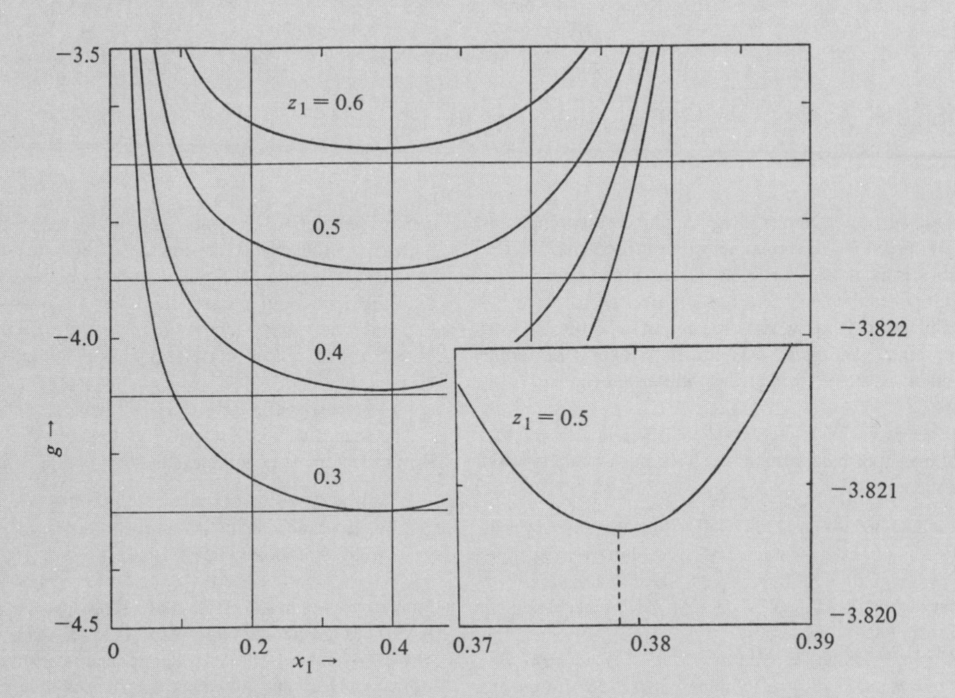

Example 6.11. Plot of g (from Eq. 6.128) against x_1. Minimum is at $x_1 = 0.379$.

Example 6.12. Activity Coefficients in a Ternary Mixture with a Supercritical Component and Application to a Bubblepoint Calculation

The supercritical component is designated number 1 and the others as 2 and 3. In the binary mixture of components 1 and 2, designate the extrapolated pure component fugacity by f°_{12}. At equilibrium,

$$\hat{f}_1^V = \hat{f}_1^L \tag{1}$$

Assume the Wilson equation for the excess Gibbs energy. Accordingly,

$$g^{ex} = G^{ex}/RT$$
$$= \Sigma x_i \ln \gamma_i = \Sigma x_i \ln(\hat{f}_i^V/x_i f_i^0) = \Sigma x_i \ln(\hat{\phi}_i^V P/f_i^0) \tag{2}$$
$$= x_1 \ln \hat{\phi}_1^V + x_2 \ln \hat{\phi}_2^V + \ln P - x_1 \ln f_{12}^0 - x_2 \ln f_2^0 \tag{3}$$
$$= -x_1 \ln(x_1 + \Lambda_{12} x_2) - x_2 \ln(\Lambda_{21} x_1 + x_2). \tag{4}$$

The unknown quantities in (3) and (4) are the Wilson parameters and the extrapolated pure component fugacity f_{12}^0 of the supercritical component 1 in the presence of subcritical component 2. These can be found by regression when at least three sets of vapor-phase fugacities $\hat{\phi}_1^V$ and $\hat{\phi}_2^V$ have been evaluated from an appropriate equation of state at the liquid phase composition x_1, x_2. The hypothetical liquid fugacity f_{13}^0 is found similarly.

For the mixture, the reference fugacity of component 1 is assumed to be

$$f_1^0 = [(f_{12}^0)^{x_2}(f_{13}^0)^{x_3}]^{\left(\frac{1}{x_2+x_3}\right)} \tag{5}$$

and equations for the activity coefficients are:

$$\ln \gamma_1 = \ln \gamma_1^{(Wilson)}, \tag{6}$$

$$\ln \gamma_2 = \ln \gamma_2^{(Wilson)} + \frac{x_1 x_3}{(x_2+x_3)^2} \ln(f_2^0/f_{13}^0), \tag{7}$$

$$\ln \gamma_3 = \ln \gamma_3^{(Wilson)} - \frac{x_1 x_2}{(x_2+x_3)^2} \ln(f_{12}^0/f_{13}^0), \tag{8}$$

where $\gamma_i^{(Wilson)}$ is the activity coefficient calculated from the Wilson equation of Table 4.6 with the parameters found here.

As an application, the equations for a bubblepoint calculation will be stated. The condition is

$$\Sigma y_i = \Sigma K_i x_i = 1. \tag{9}$$

For the supercritical component,

$$K_1 = \gamma_1 f_1^0/\hat{\phi}_1^V P \tag{10}$$

with γ_1 by Eq. 6 and f_1^0 by Eq. 5. For the other components

$$K_i = \gamma_i \hat{\phi}_i^{sat} P_i^{sat} (PF)_i/\hat{\phi}_i^V P \tag{11}$$

where the γ_i are obtained by Eqs. 7 and 8 and the other terms in the usual fashion.

Extension to complex mixtures containing more than one supercritical component is given by Van Ness & Abbott (1979).

data obtained from binaries alone. Such representations are not overly precise, and sometimes quite inadequate, but they are all that is available short of actual experimentation, which is ordinarily economically feasible only over narrow ranges of compositions. In fact few data are reported in the literature even on ternary and quaternary mixtures, except on ternary liquid-liquid systems, which are important in extraction processes.

In recent years, several comprehensive collections of VLE data, primarily of binary mixtures, have been made of which the main ones are,

1. Landolt-Börnstein: II2a (1960) 661 sets of binary data, 49 sets of ternary data, about 3,000 binary azeotropes; IV4b (1972) 138 binary sets, 29 ternary sets, New Series Group IV, vol. 3 (1975). 557 binary sets, 86 ternary sets, 4 quaternary.
2. Kogan, Fridman, & Kafarov, *Vapor-Liquid Equilibria* (in Russian) (1966): 1,765 binary sets, 358 ternary sets, 44 quaternaries.
3. Hirata, Ohe, & Nagahama, *Computer-Aided Data Book of Vapor-Liquid Equilibria* (1976): 800 binary sets, 133 high-pressure binary sets.
4. Maczynski et al., *Verified Vapor-Liquid Equilibrium Data,* 4 volumes (1977–1979).
5. Gmehling, Onken, & Arlt, *DECHEMA Vapor-Liquid Equilibrium Data Collection* (1977): About twelve volumes containing 8,000–9,000 data sets, mostly binary, some ternary and quaternary. A separate volume is devoted to high-pressure equilibria.

The correlation of VLE data in terms of fugacity and activity coefficients is covered in earlier chapters. In most instances deviations of liquid phases from ideality are greater than those of vapor phases; the latter are often neglected at moderate pressures, for instance in the DECHEMA and Hirata books. Those correlations that do not take explicit account of vapor-phase nonidealities really call the entire group of terms, $\gamma_i \phi^s_i (PF)_i/\hat{\phi}_i$, an effective activity coefficient.

The shapes of *x-y, T-x,* and *P-x* diagrams are quite varied. Perhaps the most useful classification is into systems

1. Without azeotropes.
2. With homogeneous azeotropes.
3. With two liquid phases.

Numerous examples of all these types can be found in the references given, although VLE data of immiscible liquid-phase systems are the least common. For instance, Landolt-Börnstein II/2a has only these diagrams with miscibility gaps: Nos. 60, 62, 65, 68, 90, 152, 378, 634, and 635.

Activity coefficients are of course a prime factor in determining the shapes of equilibrium diagrams. Examples of the effect of differences in activity coefficients on systems with otherwise the same properties are shown in Examples 6.9, 6.10. These use the simplified equation

$$y_i = \gamma_i P_i^s x_i/P.$$

Isotherms and isobars are shown with the symmetrical and Margules equations. One of the systems has a double azeotrope, a rare occurrence in actuality (see Figures 4.24a,

5.22f). The calculated x-y curves of systems with miscibility gaps always have a maximum and a minimum, but the portions of the curve in the vicinities of the extrema correspond to instability. To find the true equilibrium curve, which is a horizontal line in the two-phase region, it is necessary to make a liquid-liquid equilibrium calculation, as done in Example 7.12, where unstable regions of both the P-x and P-y curves also are shown.

6.7. GIBBS ENERGY MINIMIZATION

Although it is not always the easiest way in practice, a generally applicable method of finding the equilibrium condition is by finding when the Gibbs energy of the mixture is a minimum. Applications of this principle are made in Chapters 7 and 10; here the procedure will be developed for a vapor and a single liquid phase.

For every component of the vapor phase, the partial Gibbs energy may be written

$$\bar{G}_{iV} = G_{iV} + RT \ln (\hat{f}_{iV}/f_{iV}), \tag{6.118}$$

and of the liquid phase,

$$\bar{G}_{iL} = G_{iL} + RT \ln (\hat{f}_{iL}/f_{iL}) \tag{6.119}$$

$$= G_{iV} + RT \ln (\hat{f}_{iL}/f_{iV}). \tag{6.120}$$

When the standard state is chosen as gas at unit fugacity, the pure-component Gibbs energies become those of the standard state, G^0_{iv}. Moreover, if the Gibbs energies of the elements are taken arbitrarily as zero, these Gibbs energies of the components of the mixture become the Gibbs energies of formation. Accordingly, Eqs. 6.118, 6.120 may be used in the forms

$$\bar{G}_{iV} = G^0_{iV} + RT \ln \hat{f}_{iV}, \tag{6.121}$$

$$\bar{G}_{iL} = G^0_{iV} + RT \ln \hat{f}_{iL}. \tag{6.122}$$

Forms of the equations of the partial fugacities in terms of readily evaluated quantities are

$$\hat{f}_{iV} = \hat{\phi}_i P y_i, \tag{6.123}$$

$$\hat{f}_{iL} = \gamma_i \phi_i^s P_i^s (PF)_i. \tag{6.124}$$

A mixture containing n_i mols overall of component i, v_i mols in the vapor phase, l_i in the liquid phase and with a fraction

$$\beta = \Sigma\, v_i / \Sigma\, n_i = V/F \tag{6.125}$$

in the vapor phase has the total Gibbs energy given by:

$$\frac{nG}{RT} = \beta \Sigma\, v_i \left(\frac{G_{iV}}{RT} + \ln \hat{\phi}_i^v P y_i \right)$$

$$+ (1-\beta)\, \Sigma\, l_i \left(\frac{G_{iV}}{RT} + \ln (\gamma_i \phi_i^s P_i^s (PF)_i x_i) \right), \tag{6.126}$$

which may be written,

$$g = \frac{G - \Sigma\, z_i G^0_{iV}}{RT} \tag{6.127}$$

$$= \beta\, \Sigma\, y_i \ln \hat{\phi}_i^v P y_i + (1-\beta)\, \Sigma\, x_i \ln (\gamma_i \phi_i^s P_i^s (PF)_i x_i), \tag{6.128}$$

where $z_i = n_i / \Sigma\, n_i$ is the overall mol fraction of component i in the mixture. For a given overall composition, $\Sigma\, z_i G^0_{iV}$ is a constant.

Vapor compositions can be eliminated from this equation with

$$y_i = K_i x_i = \frac{\gamma_i \phi_i^s P_i^s (PF)_i}{\hat{\phi}_i P} x_i, \tag{6.129}$$

although this may need to be done by successive approximation whenever the ratio $\phi_i^s / \hat{\phi}_i$ is appreciably different from unity, since $\hat{\phi}_i$ does depend on the vapor composition. The phase split is eliminated by the material balance,

$$\beta = (z_i - x_i)/(y_i - x_i). \tag{6.130}$$

Which component to use for this substitution is arbitrary; preferably it is one for which the denominator is appreciably different from zero. One more constraint remains,

$$\Sigma\, x_i - 1 = 0. \tag{6.131}$$

Upon introducing the Lagrange multiplier, λ, for this constraint, the minimization of Eq. 6.128 subject to Eq. 6.131 is equivalent to minimization of ψ,

$$\psi = g + \lambda (\Sigma\, x_i - 1) \rightarrow \text{minimum}. \tag{6.132}$$

The derivatives of ψ with respect to the liquid mol fractions together with Eq. 6.131 constitute a system from which λ and the composition can be found, for example, by the Newton-Raphson or similar methods, or a direct search for the minimum value of ψ can be instituted.

Finding the equilibrium condition by minimization of Gibbs energy is, in many cases, a cumbersome procedure in comparison with the other methods shown in this chapter for flash calculations. For binary mixtures it is not necessary to employ the Lagrange multiplier since the Gibbs energy of the mixture is readily expressible in terms of a single mol fraction. Example 6.11 takes advantage of this fact. Detailed examples of the Lagrange multiplier technique are described in Chapter 10.

6.8. PROBLEMS

6.1. The bubblepoint pressures of mixtures of carbon disulfide and cyclohexane at 19.8 C are 150.8 Torr when $x_1 = 0.25$ and 249.5 Torr when $x_1 = 0.727$. The Redlich-Kwong equation applies to the vapor phase and the van Laar to the liquid phase. Find the vapor compositions and the van Laar constants.

6.2. An equimolal mixture of ethanol and benzene is liquid at 200 C. When its pressure is reduced adiabatically to 12.6 atm, some vaporization occurs. What will be the temperature? Activity coefficients at 179 C are $\gamma_1^\infty = 1.03$ and $\gamma_2^\infty = 3.30$. The virial equation with the Pitzer & Curl correlation for B is applicable to the vapor phase. The enthalpies may be assumed independent of pressure. Liquid heat capacities are 0.7 cal/(g)(C) for benzene and 1.4 for ethanol. Other physical properties may be taken from Table D.2.

6.3. An equimolal mixture at 600 R is half liquid and half vapor. Activity coefficients are represented by $\ln \gamma_i = 1.5(1 - x_i)^2$. Virial coefficients are $B_{11} = -1.8$, $B_{12} = -6.0$, $B_{22} = -3.4$ cu ft/lb mol. Vapor pressures are $P_1^\circ = 30$, $P_2^\circ = 20$ atm. Find the system pressure.

6.4. An equimolal mixture of acetone (1), ethanol (2), and ethyl ether (3) is at 55 C. Liquid densities are 0.790, 0.789, and 0.713 g/ml. Vapor-phase behavior may be taken as ideal. Wilson coefficients at the indicated temperatures are:

Mixture	*T*, °C	Λ_{ij}	Λ_{ji}
12	55	0.30771	1.20101
13	40	0.49880	0.86494
23	30	0.29207	0.77045

Find the bubblepoint and dewpoint pressures and compositions.

6.5. A liquid mixture of 1/3 methanol (1) and 2/3 ethyl acetate (2) is at 150 C. What are the bubblepoint pressure and vapor composition?

The Poynting factors may be taken as unity.
Vapor pressures are $P_1^0 = 13.85$ atm, $P_2^0 = 6.65$ atm.
Van Laar parameters are $A_{12} = 0.4227$, $A_{21} = 0.4470$.
Virial coefficients are $B_{11} = -0.3306$, $B_{22} = -0.7569$, $B_{12} = -0.4739$ liters/gmol.

6.6. The TBP (true boiling point) distillation of a kerosene-naphtha blend is

Mol % *distilled*	°*C*
0	5
10	38
20	69
30	100
40	130
50	156
60	182
70	212
80	224
90	282
100	320

Use the method of Zhvanetskii & Platonov (Section 6.2.5) to find

1. The bubblepoint at 5 atm.
2. The temperature at which 50 mol % of the mixture is volatilized at atmospheric pressure.

6.7. For the mixture of Example 6.3, prepare the condensation curve at 100 psia, a plot of $\beta = V/F$ against the temperature.

6.8. Find the solubility of hydrogen sulfide in benzene at 400 K and 12 atm with Abbott's formula (Fig. 1.16) for the second virial coefficient and Eq. 6.110 for the fugacity of hypothetical hydrogen sulfide liquid. Data are:

	δ	V_L	T_c	P_c	ω
H_2S	8.80	43.1	373.2	88.2	0.100
C_6H_6	9.16	89.4			

Units are cal, ml, gmol, °K, atm.

6.9. Obtain the *x-y* diagram by each of the five equations for each of the three systems for which the parameters are given. The meanings of the parameters, in cal/gmol, of the Wilson, NRTL, and UNIQUAC equations are:

	A 12	*A 21*
Wilson	$(\lambda_{12} - \lambda_{11})$	$(\lambda_{12} - \lambda_{22})$
NRTL	$(g_{12} - g_{22})$	$(g_{21} - g_{11})$
UNIQUAC	$(u_{12} - u_{22})$	$(u_{21} - u_{11})$

Acetone + cyclohexane, 25 C, DECHEMA, pt 3 & 4, 216

Constants	*A 12*	*A 21*	*ALPHA12*
Margules	2.0522	1.7201	
Van Laar	2.0684	1.7174	
Wilson	1114.6929	448.6899	
NRTL	693.7917	881.9712	0.4841
UNIQUAC	−55.5703	587.6497	

Acetone + toluene, 35 C, DECHEMA, pt 3 & 4, 232

Constants	*A 12*	*A 21*	*ALPHA12*
Margules	0.6277	0.5020	
Van Laar	0.6480	0.5017	
Wilson	604.4247	−199.5906	
NRTL	−50.4587	446.0895	0.3017
UNIQUAC	−248.2777	454.3634	

Acetone + heptane, 65 C, DECHEMA, pt 3 & 4, 239

Constants	*A 12*	*A 21*	*ALPHA12*
Margules	1.4370	1.4455	
Van Laar	1.4370	1.4455	
Wilson	1092.9664	157.4597	
NRTL	553.7388	545.2523	0.2911
UNIQUAC	−146.9628	628.9213	

Vapor pressures in Torr are: Acetone (25 C, 230.9); (35 C, 349.2); (65 C, 1,020.3); cyclohexane (25 C, 97.6); toluene (35 C, 46.8); heptane (65 C, 253.8). Molal volumes and UNIQUAC *R* and *Q* are:

	V	*R*	*Q*
Acetone	74.05	2.5735	2.3360
Cyclohexane	108.75	4.0464	3.2400
Toluene	106.85	3.9228	2.9680
Heptane	147.47	5.1742	4.3960

6.10. An equimolal mixture of propane and hexane is at its dewpoint at 20 atm. It is expanded isentropically to 5 atm. Heat capacities may be assumed independent of pressure and the Soave otherwise applies to both phases. Find the final temperature, phase split, and phase compositions.

6.11. Use the Chao-Seader method to construct the *x-y* and *T-x* diagrams of mixtures of cyclohexane and toluene at 5 atm.

6.12. Given the following values of the Wilson parameters, in cal/gmol,

	λ_{ij}	λ_{ji}
ethanol + benzene	1,297.90	131.47
ethanol + methanol	−511.39	598.44
methanol + benzene	1,620.36	153.86

find the phase compositions and temperature of a mixture whose overall composition is equimolal and which is at 2 atm and halfway between the bubble- and dewpoint temperatures. Vapor-phase nonidealities may be neglected.

6.13. A mixture with the overall composition: $C_2 = 0.10$, $C_3 = 0.14$, $iC_4 = 0.19$, $nC_4 = 0.54$, $iC_5 = 0.03$ is one-fourth liquid at 100 F. What are the pressure and the phase compositions? Figure 6.5 with grid pressure equal to the system pressure may be used.

6.14. The split achieved in a distillation column is shown. The outlet of the condenser is at 120 F and 5psi less than the pressure at the top of the column. The pressure at the reboiler is 10 psi greater than that at the top of the column. Find the following:

1. The pressure at the top.
2. The temperature of the top plate.
3. The temperature at the reboiler.

Use the Chao-Seader method. Take $\hat{\phi}_i = \phi_i$ in the vapor.

	F	D	B
C_2	10	10	0
C_3	14	13	1
iC_4	19	10	9
nC_4	54	4	50
iC_5	3	0	3

6.15. The thermocompressor distillation system shown makes an overhead containing 90% ethylene + 10% propylene and a bottoms with 10% ethylene and 90% propylene. Feed is equimolal. The pressure of the overhead is raised until its bubblepoint is 30 F greater than the dewpoint of the bottoms. The pressure P into the isentropic compressor is 20 atm, that at the reboiler is 1 atm greater, and that at the outlet of the compressor is to be figured. Enthalpies of the components of the mixture may be taken additive; data may be found, for example, in Starling (1973). Find

1. The pressure at the outlet of the compressor
2. The temperatures into and out of the reboiler.
3. The relation between the pressure at the outlet of the compressor and the amount of condensation in the reboiler when the reflux ratio at the top is 2:1. Total condensation is desirable.

6.16. The VERs of the components of a mixture are represented by equations of the form

$$\ln K_i P = A_i - B_i / T°\text{K}.$$

The overall composition and the values of these coefficients are:

	z	A	B
C_6H_{14}	0.1	11.15	3,811
C_6H_6	0.3	11.00	3,885
C_7H_{16}	0.4	10.85	4,030
C_8H_{18}	0.2	11.22	4,471

At a pressure of 3 atm find the bubble- and dewpoint temperatures and the phase compositions halfway between those two temperatures.

6.17. VERs and enthalpies in equation form at 120 psia and 264.7 psia are given by Holland (Multicomponent Distillation (1963)). The VER's are given as ratios to that of isopentane as

$$\alpha_i = K_i / K_{iC_5}.$$

A direct formula is given for iC_5. An equimolal mixture of the four substances is on hand.

1. Find the bubble and dew temperatures at 264.7 psia.
2. Find the temperature, phase splits, and phase compositions when the pressure on a saturated liquid at 264.7 psia is released adiabatically to 120 psia.
3. When saturated vapor at 264.7 psia is expanded isentropically to 120 psia, find the complete outlet conditions.

$$K\ (\text{for } i\text{-}C_5) = 0.37088 - 0.55786\,(T/100) + 0.44841(T/100)^2 - 0.03704\,(T/100)^3, \quad (T \text{ in °F}).$$

$$\alpha_i = a_{1i} + a_{2i}(T/100) + a_{3i}(T/100)^2, (T \text{ in °F}).$$

(Tables follow on p. 334.)

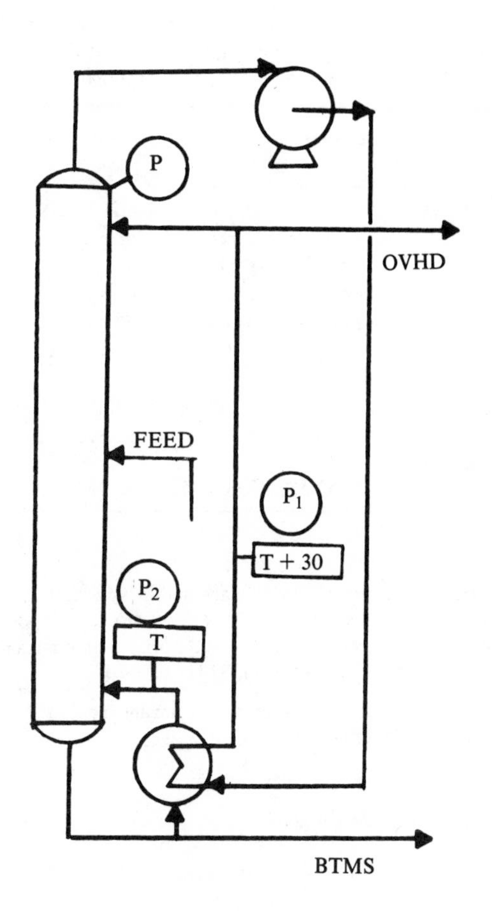

Problem 6.15

Component	a_1	a_2	a_3
120 psia			
C_3H_8	11.06095	−5.20067	0.92489
i-C_4	4.69290	−1.82431	0.31755
n-C_4	3.07033	−0.83565	0.12144
n-C_5	0.73827	0.05246	0.00189
264.7 psia			
C_3H_8	3.64797	−1.06605	0.16197
i-C_4	1.46077	−0.18097	0.02721
n-C_4	1.00000	0.00000	0.00000
n-C_5	0.17185	0.22019	−0.02826

$h_i = c_{1i} + c_{2i}(T/100) + c_{3i}(T/100)^2$, ($T$ in °F).

$H_i = e_{1i} + e_{2i}(T/100) + e_{3i}(T/100)^2$, ($T$ in °F).

Component	c_1	c_2	c_3
120 psia			
C_3H_8	0.00000	2,521.92	175.417
i-C_4	0.00000	3,345.00	150.000
n-C_4	0.00000	2,960.00	400.000
n-C_5	0.00000	3,845.00	250.000
264.7 psia			
C_3H_8	4,425.16	2,968.64	48.868
i-C_4	5,320.76	3,525.03	90.076
n-C_4	5,553.24	3,641.27	90.076
n-C_5	6,663.21	4,161.81	160.848

Component	e_1	e_2	e_3
120 psia			
C_3H_3	7,488.65	1,750.51	79.273
i»C_4	9,592.01	1,843.56	221.433
n-C_4	8,002.89	4,382.70	−401.587
n-C_5	12,004.88	3,168.66	67.456
264.7 psia			
C_3H_8	11,806.71	1,797.38	112.6239
i-C_4	13,726.02	2,261.20	167.3186
n-C_4	14,851.06	2,312.00	155.9965
n-C_5	18,757.52	2,903.45	233.0578

6.18. Mixtures of cyclohexane and *m*-xylene at 25 C are quite well represented by $G^{ex}/RT = 0.48x_1x_2$. Vapor pressures are $P_1^s = 97.6$ Torr, $P_2^s = 8.3$ Torr. Prepare *x-y* and *P-x* diagrams (DECHEMA I/6a, 314).

6.19. Mixtures of acetone and hexane at 760 Torr are represented by the Wilson equation with

$\lambda_{12} = 1{,}156.5822$ cal/gmol,

$\lambda_{21} = 262.7197$ cal/gmol.

Vapor pressures are represented by

$\log_{10} P_1^s = 7.11714 - 1{,}210.595/(229.664 + T°\text{C})$,

$\log_{10} P_2^s = 6.91058 - 1{,}189.640/(226.280 + T°\text{C})$,

in Torr. Molal volumes are 73.52 and 130.77 ml/gmol. Make the *x-y* and *T-x* plots (DECHEMA I, 3 & 4, 226).

6.20. Activity coefficients at a particular temperature are represented by $\ln \gamma_1 = Ax_2^2$ and $\ln \gamma_2 = Bx_1^2$. Vapor pressures are $P_1^s = 1.2$ and $P_2^s = 1$. Construct the *x-y*, *P-x*, and *P-y* diagrams for two cases:

1. $A = B = 1.5$
2. $A = 2$, $B = 1$.

In the latter case, interpret why the azeotropic composition does not correspond to the maximum pressure of the system. Does case (2) conform to the Gibbs-Duhem equation at constant T and P?

6.21. Check the solubility data in Table 6.3 against the Scatchard-Hildebrand correlation, Eqs. 6.109 and 6.113.

6.22. A binary vapor-liquid equilibrium can be represented by either of two one-parameter relations: (1) in terms of relative volatility, $y = \alpha x/[1 + (\alpha - 1)x]$; or (2) in terms of the Margules symmetric activity-coefficient equations, with the result

$$P_i^0 x_1 \exp A(1 - x_1)^2 + P_2^0(1 - x_1) \exp x_1^2 = P.$$

Plot the quantity, $\alpha P_2^0/P_1^0$, against x_1 at several values of A from 0 to 5. It will be found that a constant value of relative volatility over a range of concentrations does not correspond to a constant value of A.

6.23. Vaporization equilibrium ratios and vapor pressures are given at several conditions for propane and isobutane.

P \ T	*Propane* 100	*Propane* 150	*Isobutane* 100	*Isobutane* 150
150	1.2*	2.05	0.52*	0.94
200	1.0	1.55*	0.42	0.72*
P^{sat}	186.7	333.5	71.8	139.7

The temperatures above are in °F and the pressures in psia. Antoine coefficients (Torr, °K) are:

	A	*B*	*C*
Propane	15.726	1,872.46	−23.16
Isobutane	15.5381	2,032.73	−33.15

1. Use the two sets of starred K_is to find the constants of the formulas

$\ln K_i P = k_{1i} + k_{2i} \ln P_i^{sat}$.

2. Evaluate the K_is in the unstarred positions with the resulting formulas and compare with the given values.
3. Find the bubblepoint of an equimolal mixture at 160 psia.

6.24. For the mixture of Example 6.7, find the bubblepoint temperature at 20 atm for $x_1 = 0.5$.

6.25. An equimolal binary mixture is flashed to 50 mol % vapor. What are the compositions of the phases?

Data:

$$G^{ex}/RT = 1.8x_1x_2,$$

$$\ln \hat{\phi}_1 = 0.5 - 0.15y_1^2,$$

$$\phi_2 = 0.8,$$

$$\phi_1^s P_1^s/P = 0.4$$

$$\phi_2^s P_2^s/P = 1.2$$

Note that as an intermediate step it is necessary to find an equation for $\hat{\phi}_2$ with the Gibbs-Duhem equation.

6.26. Because of surface tension the pressure at equilibrium inside a liquid droplet is greater than that outside. The precise relation is

$$P_{\text{inside}} - P_{\text{outside}} = 4\sigma/d$$

where σ is the surface tension and d is the droplet diameter. As a consequence, the relative volatility of two substances depends on the diameter of the drop. Find this effect for mixtures of (1) methanol and (2) water, given the following data at 25 C for $x_1 = 0.3973$. Explore d down to 1 Angstrom.

$$\gamma_1 = 1.186, \gamma_2 = 1.100 \quad \text{(Kogan, 251)}$$

$$P_1^{sat} = 127.2 \text{ Torr}, P_2^{sat} = 22.9 \text{ Torr}$$

$$\sigma = 34.9 \text{ (E-3)N/m} \quad \text{(CRC Handbook, F44, 60th edition)}$$

$$V^{ex} = x_1x_2[-4041 - 428(x_1 - x_2) + 644(x_1 - x_2)^2 + 990(x_1 - x_2)^3]$$

in the units $(\text{mm})^3$/gmol. (Int Data Series, Series B, 4, 1978).

6.27. In terms of an equation of state applicable to both phases, the VER is given by Eq. 6.19, $K_i = \phi_{iL}/\phi_{iV}$. In such cases it is not possible to start a flash calculation by assuming unity fugacity coefficients by analogy to box 2 of Figure 6.11. Trial values of K_i may be taken, however, off Figure 6.5 or other API nomograms or ideal values $K_i = P_i^{sat}/P$ may be used. Modify Figure 6.11 for such cases.

7 Liquid-Liquid Equilibria

7.1. LIQUID MIXTURES

Unless they are members of some homologous series, liquids as a rule have only limited mutual solubilities. This phenomenon is a manifestation of nonideality, so equilibrium between liquid phases should be of the same thermodynamic pattern as that between a liquid and a vapor phase in that both are conditions of minimum Gibbs energy. Although there is no thermodynamic distinction between the two types of equilibria, there are some practical differences:

1. Liquid-liquid equilibria usually are much easier to determine experimentally, especially near room temperature, which usually is the case of greatest interest. Of published data, the great majority have been obtained within 10 C or so of room temperature. Data often are obtained only for the need at hand and may not be of such quantity and scope that correlations can be developed.
2. Equations of state of liquids usually are not available for calculating fugacities or activities.
3. The effect of temperature on liquid-liquid equilibria is more pronounced. Vapor-liquid equilibria are determined at boiling temperatures, which are usually somewhat removed from room temperature, the condition at which liquid-liquid equilibria normally are desired.

Three main aspects of liquid-liquid equilibria are of interest:

1. How to predict data from properties of pure components or pairs of components.
2. How to correlate limited data so they can be interpolated or extrapolated or combined into a representation of multicomponent behavior.
3. How to obtain data experimentally.

The first two topics are the principal concern of this chapter. They also are treated at length in a series of review articles by Sorensen et al. (1979–1980). Their first paper discusses the content of the *DECHEMA Liquid-Liquid Equilibrium Data Collection* (1979, 3 volumes), which now comprises 800–900 sets each of binary and ternary data, plus a few quaternary ones.

Experimental methods in this area are mostly simple in concept and are primarily of two types:

1. Solubility limits are found by gradual addition of known amounts of a substance to a solution of known composition until the onset of turbidity or some other measure of saturation.
2. Equilibrium compositions are found by first thoroughly agitating a mixture of known overall composition, then separating the phases and sampling and analyzing them by appropriate means. For precise work, the techniques can be quite sophisticated; typical details may be found in recent papers, such as those that appear in *The Journal of Chemical Thermodynamics* or related research publications.

7.2. PHASE DIAGRAMS

Examples of diagrams of binary and ternary mixtures are given in Chapter 5, Figures 5.20 and 5.30ff. Composition and temperature normally are the only variables considered since the effect of pressure on condensed phases is small; Figure 5.28, for instance, indicates the magnitudes of pressure required to affect solubility behavior—in this case of solids in liquids, but that of liquids in liquids is expected to be similar.

An analysis of some 900 binary data sets in the *DECHEMA Collection* reveals that about 41% have an upper CST (critical solution temperature), whereas 53% have neither upper nor lower CSTs. The few cases of lower CST are attributable to molecular association and dipole interaction, which become greater as the temperature is lowered. Of a similar number of ternary data sets, 75% have single one-phase and two-phase regions, as in Figure 5.36a, and 20% have banded diagrams with two one-phase and one two-phase regions, as in Figure 5.36b. Although they are infrequent, there are important industrial examples of the other types; thus the mixtures of water + 2-butanol + 2-butanone and of water + nicotine + 2-butanone are like Figure 5.36c. A few ternary systems have been studied over a wide enough temperature range to detect an upper CST, as in Figure 5.35. Almost 60% of the binaries and 80% of the ternaries had water as one of the components, which reflects the importance of water as solvent and possibly its availability and ease of handling for experimental work.

Some nomenclature associated with ternary diagrams is given in Figure 5.5d. Composition-distribution curves of liquid-liquid systems may be of the same types as those of Figures 5.29 and 5.32 for liquid-vapor systems, but several other kinds of plotting often are convenient, some of which are described, for example, by Treybal (1963). A scheme developed by Mapstone (1970) should appeal to persons interested in both tie-line interpolation and nomograms. Binodal curves of ternary systems having one partially miscible pair were correlated by a special form of empirical equation with three or more constants by Hlavaty (1972). The old but still popular correlation methods of Hand (1930) and Othmer & Tobias (1943) were checked against 110 systems by Carniti et al. (1978), who found that they are inadequate by current standards. The Hand method is cited in Problem 7.13. A fundamental and greatly superior method of correlating or predicting LLE is in terms of activity-coefficient correlations, which will be studied in this chapter.

7.3. MISCIBILITY AND THERMODYNAMIC STABILITY

Binary mixtures display a variety of vapor-pressure behavior with composition, some of which is shown in Figure 7.1. Isotherm T_1 in this figure enters the dotted domed region in which two liquid phases form and are in equilibrium.

A state of equilibrium is characterized as having a minimum Gibbs energy at a given T, P, and composition. Thus in Figure 7.2, point d is less stable than points e or f, the latter of which will be shown shortly to be an equilibrium condition. There are Gibbs energies smaller than G_f on the curve from a to h, but they are not all at the reference composition.

Continuing with Figure 7.2, the dashed curve G_2kG_1 represents unstable states of two completely immiscible liquids whose equilibrium states are along the straight line G_2mG_1. Curve G_2nG_1 is that of a completely miscible mixture, whereas line G_2mG_1 represents its Gibbs energy before the components have dissolved each other.

As will be explained shortly, portion adh of the intermediate curve represents unstable states whose equilibrium states

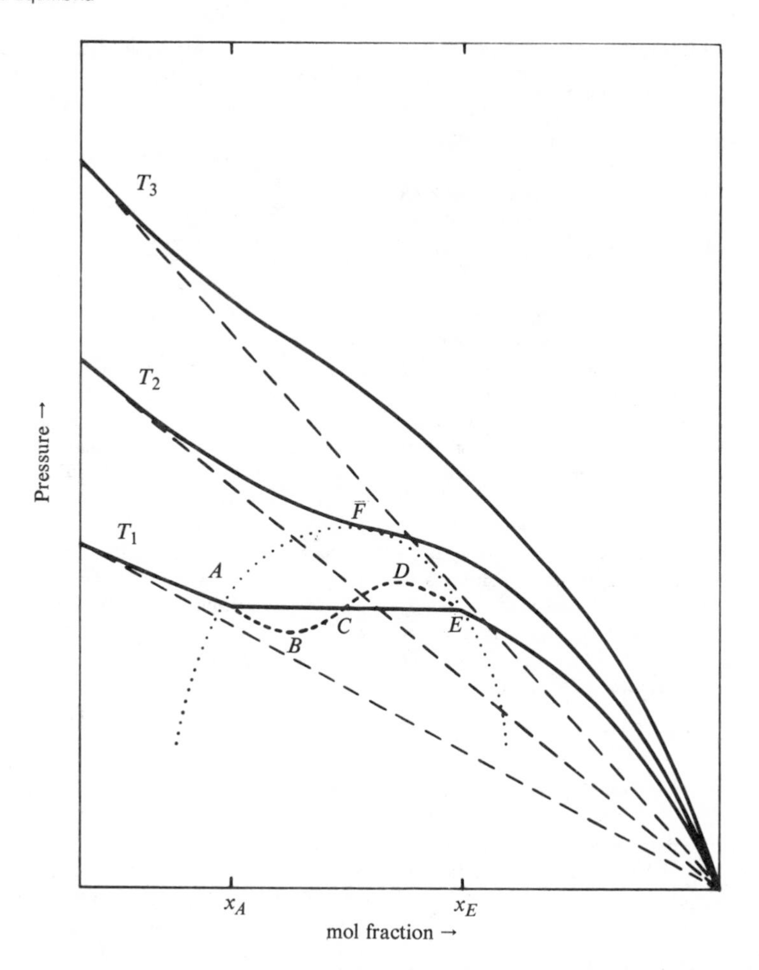

Figure 7.1. Vapor pressures of a binary mixture. At temperature T_1, the broken portion of the curve $ABCDE$ represents an unstable condition; in this composition range two liquid phases are present, with compositions x_A and x_E, and the vapor pressure is constant along the horizontal line ACE. Two liquid phases exist anywhere within the dotted dome. Point F on the T_2 isotherm is the point of incipient immiscibility, and T_2 itself is called the *critical solution temperature.* The dashed straight lines represent Raoult's Law behavior.

are on the line afh. Such a curve always has at least one local maximum and two or more local minima, and is described as having convex portions. The mathematical condition for convexity is that

$$\partial G/\partial x = 0,$$

$$\partial^2 G/\partial x^2 < 0.$$

If there are values of composition that satisfy these conditions, immiscibility will occur. The compositions at which the second derivatives are zero are the inflection points, c and g in the figure. The regions a-c and h-g, the terminals of which are at the points of inflection, are metastable. This behavior is analyzed by Modell & Reid (Chapter 9, 1983) and particularly by Lupis (1983). The region c-d-g is truly unstable. Local extrema, found by equating the first derivative to zero, are at points b, d and j. The compositions of phases in equilibrium are found as follows:

The intercepts of the double tangent afh with the axes are the partial molal properties, $\bar{G}_1$ and $\bar{G}_2$, as explained in Section 3.3. From the following relations,

$$G = \Sigma x_i \bar{G}_i = G^{ex} + G^{ideal} \tag{7.1}$$

$$= RT\ \Sigma x_i \ln(\gamma_i x_i) = RT\ \Sigma x_i \ln a_i$$

$$= RT\ \Sigma x_i \ln(\hat{f}_i / f_i^0), \tag{7.2}$$

it is clear that the two points of double tangency have the same partial molal Gibbs energies and the same partial fugacities, and consequently that they are in equilibrium with each other. As a matter of fact, all overall compositions between x_a and x_h have the same partial fugacities and differ only in the proportions of the two phases having these compositions. Inflection points, local minima, and points of double tangency are investigated in Example 7.1. Problem 7.26 asks for a spinodal curve, which is the locus of metastable composition limits or points of inflection. See also Example 7.14.

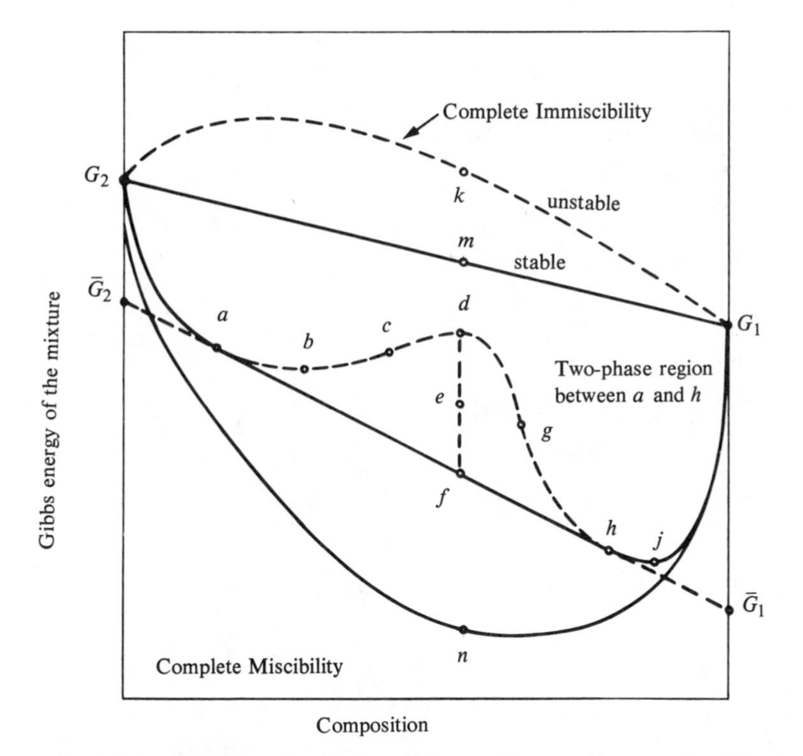

Figure 7.2. Gibbs energies of binary mixtures as functions of composition. The portion *adh* of the curve represents an unstable region of which the stable Gibbs energy is along the straight line *afh*: *a* and *h* are points of double tangency identifying the compositions of phases in equilibrium; *b* and *j* are local minima of the Gibbs energy; *c* and *g* are points of inflection. Regions *ac* and *gh* are metastable.

Infrequently, cases occur in which more than one double tangent can be drawn to the Gibbs energy curve of a binary mixture. Such a case has been synthesized in Example 7.2. Triple tangents correspond to three phases in equilibrium; an example is synthesized in Problem 7.25.

Plots of Gibbs energies of mixing as represented by several equations for activity coefficients and several values of their parameters are developed in Example 7.3.

Still another way of looking at the phenomenon of phase splitting is in terms of the behavior of the chemical potential or partial molal Gibbs energy, as typified by Figures 7.3 and 7.4. Between the equilibrium compositions (x_1, x_1^*), the chemical potential is constant and equal to the intercept μ_1 on Figure 7.3, so its plot against x_1 is horizontal in that interval. If the range of immiscibility decreases gradually, by reason of temperature change or other factor, eventually it degenerates to a point, as on curve 2 of Figure 7.4. This point of incipient immiscibility or critical solution is mathematically a point of inflection, which is characterized as having

$$\frac{\partial \mu_1}{\partial x_1} = \frac{\partial^2 \mu_1}{\partial x_1^2} = 0. \tag{7.3}$$

Since, by Eq. 2.54

$$\mu_1 = G + (1 - x_1)\frac{\partial G}{\partial x_1}, \tag{7.4}$$

an equivalent condition for incipient miscibility is

$$\frac{\partial^2 G}{\partial x_1^2} = \frac{\partial^3 G}{\partial x_1^3} = 0, \tag{7.5}$$

or

$$\frac{\partial^2 g}{\partial x_1^2} = \frac{\partial^3 g}{\partial x_1^3} = 0. \tag{7.6}$$

These conditions are applied in Example 7.4 to find the range of parameters for which binary mixtures are completely miscible.

These considerations suggest that mutual solubilities of two liquids can be found by plotting either the Gibbs energy or the Gibbs energy of mixing, $G - \Sigma x_i G_i$, against x, and finding the point of double tangency. However, this is feasible only for binary mixtures, whereas application of the principle of equality of partial fugacities is completely general. For a binary mixture still, designating the second phase by an asterisk,

$$\hat{f}_i = f_i^*. \tag{7.7}$$

On applying Eq. 4.57 and canceling out the common variable f_i^0, this becomes

$$\gamma_1 x_1 = \gamma_1^* x_1^*, \tag{7.8}$$

$$\gamma_2(1 - x_1) = \gamma_1^*(1 - x_1^*). \tag{7.9}$$

When the relations between compositions and activity coefficients are known, these equations may be solved for the

Example 7.1. Study of a Gibbs Energy Plot of a Binary Mixture

The binary mixture is described by the van Laar equation,

$$G^{ex}/RT = Ax_1x_2/(Ax_1/B + x_2), \tag{1}$$

with $A = 3$ and $B = 2$ in this particular case. The following plots will be made:

1. The Gibbs energy, $G/RT = G^{ex}/RT + x_1 \ln x_1 + x_2 \ln x_2$ against x_1.
2. The derivative, $d(G/RT)/dx_1$ against x_1.

Also these quantities will be found:

3. The equilibrium compositions of the two liquid phases.
4. The local minima of the Gibbs energy.
5. The points of inflection of the Gibbs energy plot.

The Gibbs energy may be written

$$G/RT = A/D + x_1 \ln x_1 + x_2 \ln x_2, \tag{2}$$

where

$$D = A/Bx_2 + 1/x_1. \tag{3}$$

Letting

$$N = \partial D/\partial x_1 = \frac{A}{Bx_2^2} - \frac{1}{x_1^2}, \tag{4}$$

the derivative becomes

$$\frac{\partial(G/RT)}{\partial x_1} = -\frac{AN}{D^2} + \ln(x_1/x_2), \tag{5}$$

and the second derivative,

$$\frac{\partial^2(G/RT)}{\partial x_1^2} = -A\left[\frac{1}{D^2}\left(\frac{2A}{Bx_2^3} + \frac{2}{x_1^3}\right) - \frac{2N^2}{D^3}\right] + \frac{1}{x_1} + \frac{1}{x_2}. \tag{6}$$

The local minima are found after equating the first derivative to zero, and the points of inflection by equating the second derivative to zero. The compositions at equilibrium are found by solving Eqs. 7.24, 7.25, or by reading Figure 7.7, or as the points of double tangency on the Gibbs energy plot. The points of inflection of the Gibbs energy plot are the extrema of the derivative plot. The summary of the various compositions is:

Miscibility limits, $(x_1, x_1^*) = (0.080, 0.755)$.
Local minima at $x_1 = 0.11185$ and 0.8071.
Inflection points at $x_1 = 0.17178$ and 0.5913.

The same information will be obtained when the symmetrical van Laar equation applies, with $A = B = 2.5$. The Gibbs energy in this case is

$$G/RT = 2.5x_1x_2 + x_1 \ln x_1 + x_2 \ln x_2, \tag{7}$$

and the derivatives are

$$\frac{\partial(G/RT)}{\partial x_1} = 2.5(1 - 2x_1) + \ln(x_1/x_2), \tag{8}$$

$$\frac{\partial^2(G/RT)}{\partial x_1^2} = -2A + \frac{1}{x_1} + \frac{1}{x_2}. \tag{9}$$

The miscibility limits and the local minima are the same, namely $(x_1, x_1^*) = (0.1448, 0.8552)$. The points of inflection are given by

$$x_1 = 0.5[1 \pm \sqrt{(1 - 2/A)}] = 0.2764, 0.7236.$$

These various numbers can be identified on the graphs.

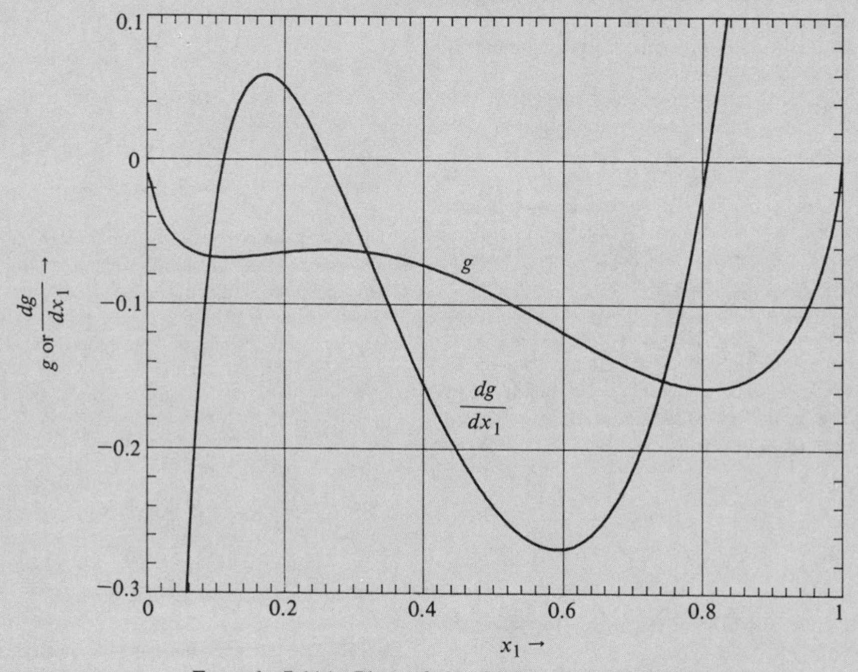

Example 7.1(a). Plots of the Gibbs energy of a mixture represented by the van Laar equation with $A = 3$ and $B = 2$.

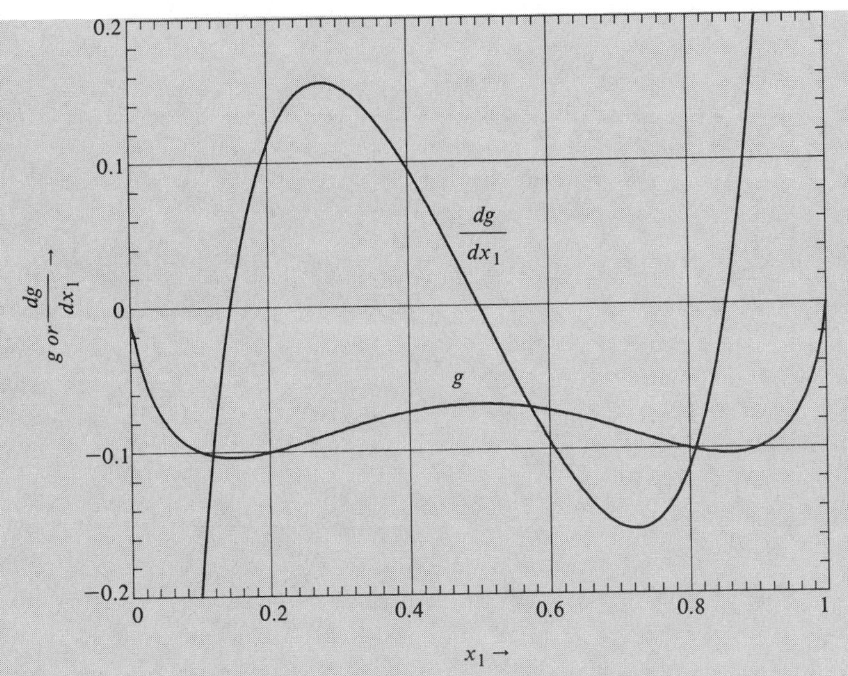

Example 7.1(b). Plots of the Gibbs energy of a mixture represented by the symmetrical van Laar equation with $A = B = 2.5$.

Example 7.2. Multiple Double Tangents

The condition of double tangency is a necessary one for defining the compositions of equilibrium phases, but is not a sufficient one since a given Gibbs energy plot may have several such double tangents. On the first figure, for example, there are six double tangents, but at a particular composition only one of them corresponds to minimum Gibbs energy of the mixture. More than two minima on Gibbs energy plots are not often encountered with the usual equations for excess Gibbs energy, although such behavior does occur with the NRTL model at extreme immiscibilities according to Sorensen (1980). However, no specific example is given there.

A case of this kind is synthesized here. Letting

$$g = (G^{ex} + G^{id})/RT = x_1 x_2 f(x_1) + x_1 \ln x_1 + x_2 \ln x_2, \tag{1}$$

the derivative becomes

$$\frac{dg}{dx_1} = x_1(1 - x_1)\frac{df}{dx_1} + (1 - 2x_1)f(x_1) + \ln\frac{x_1}{1 - x_1}. \tag{2}$$

Curve b is a plot of

$$\frac{dg}{dx_1} = \ln\frac{x_1}{1 - x_1} + 5{,}000(x_1 - 0.1)(x_1 - 0.3)(x_1 - 0.5) \times (x_1 - 0.7)(x_1 - 0.9) = 0. \tag{3}$$

It has five roots corresponding to five positions of minimax on the plot of g against x_1. In this particular case, the function $f(x_1)$ is found by solution of the linear first-order differential equation

$$x_1(1 - x_1)\frac{df}{dx_1} + (1 - 2x_1)f = 5{,}000(x_1 - 0.1)(x_1 - 0.3)(x_1 - 0.5)(x_1 - 0.7)(x_1 - 0.9). \tag{4}$$

The analytical solution of this equation is quoted in Problem 7.22. That solution is employed in making the plot of Eq. 1 in Figure a. Clearly the zeroes of the plot of the derivative, Eq. 3, on Figure b correspond to the extrema on the plot of Eq. 1.

A triple tangent corresponding to three phases in equilibrium is found in Problem 7.25 by replacing the number 5000 in Eq. 4 with 4150.

(Continued next page)

Example 7.2 *(continued)*

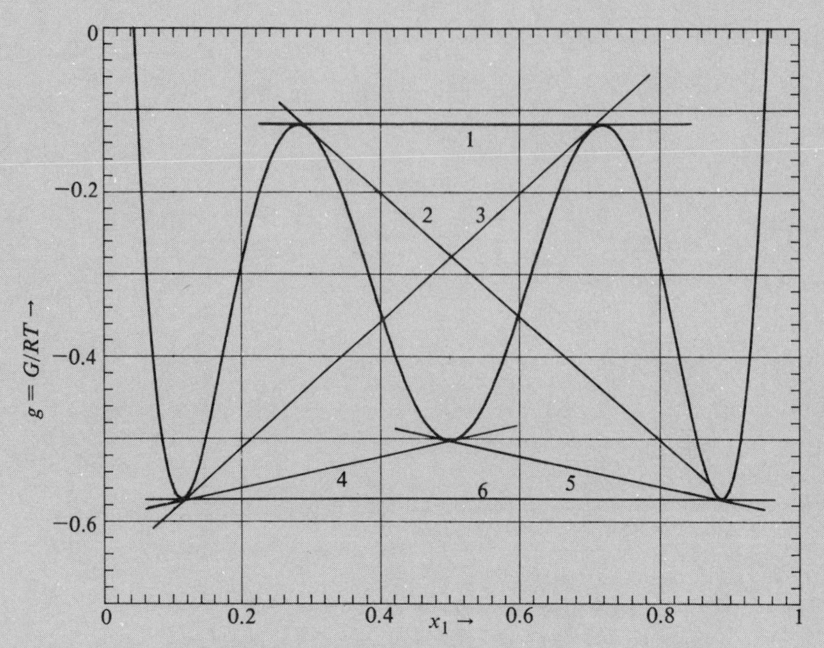

Example 7.2(a) A Gibbs energy plot that has six double tangents. Line 6 represents minimum g in its composition range.

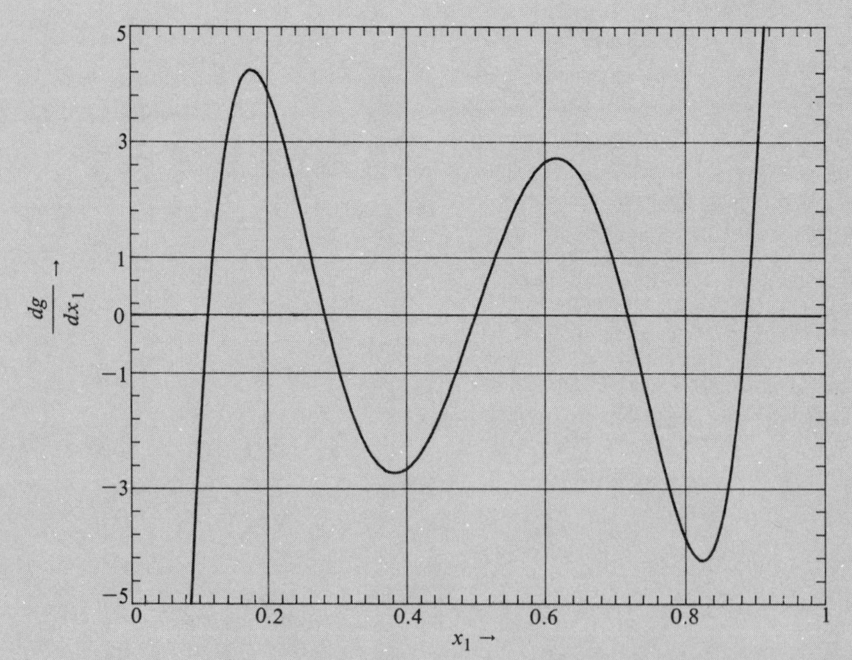

Example 7.2(b). Plot of the derivative of the Gibbs energy showing the location of the zeroes and hence the extrema of $g = G/RT$.

Example 7.3. Gibbs Energies of Mixing, as Represented by Various Equations for the Excess Gibbs Energies

The mixture is defined by the infinite-dilution activity coefficients

$\ln \gamma_1^\infty = 2,$

$\ln \gamma_2^\infty = 3,$

$$g = \frac{G - \Sigma\, x_i G_i^0}{RT} = \frac{G^{ex}}{RT} + \Sigma\, x_i \ln x_i .$$

Formulas for the excess Gibbs energies are in Table 4.2, and the formulas for finding the parameters of the equations in terms of the infinite-dilution activity coefficients are in Table 4.5. The values of the parameters are:

Margules and van Laar: $A_{12} = 2, A_{21} = 3$

Wilson: $A_{12} = 0.3346, A_{21} = 0.0968$

T-K-Wilson: $V_2/V_1 = 3, A_{12} = 0.459, A_{21} = 0.2106$

NRTL: $\alpha_{12} = 0.2, \tau_{12} = 2.5805, \tau_{21} = 0.4598$

NRTL: $\alpha_{12} = -1, \tau_{12} = 0.5881, \tau_{21} = 0.9411$

The equilibrium compositions given by the points of double tangency are approximately:

Margules: (0.20, 0.93)

van Laar (0.23, 0.92)

NRTL 0.2 (0.26, 0.91)

NRTL −1 (0.55, 0.87)

T-K-Wilson (0.38, 0.87)

Wilson (miscible)

These compositions agree closely with those read off Figures 7.7–7.12 but not among themselves. The NRTL equation with $\alpha_{12} = -1$ has three sets of parameters corresponding to the given infinite-dilution activity coefficients. In addition to the set given previously, these are (3.0000, −58.26) and (−11.78, 2.0000). Only the set (0.5881, 0.9411) leads to a prediction of phase split. Plots of the Gibbs energy of mixing are shown for all three cases.

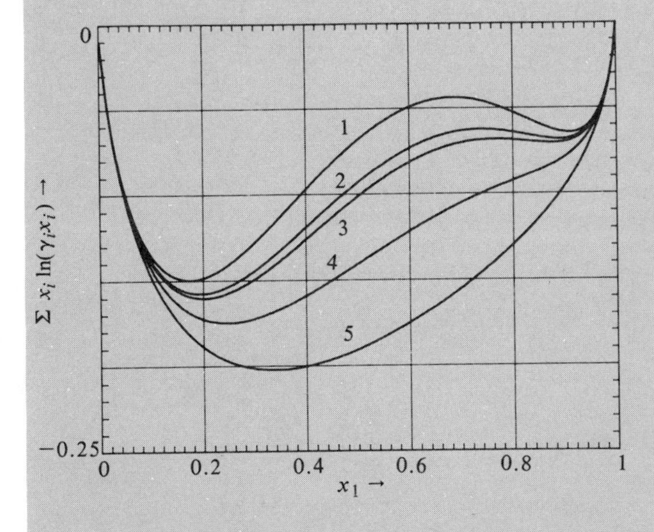

Example 7.3(a). Gibbs energies of mixing represented by several equations for the excess Gibbs energy: (1) Margules. (2) van Laar. (3) NRTL with $\alpha_{12} = 0.2$. (4) TKW with $V_2/V_1 = 3$. (5) Wilson

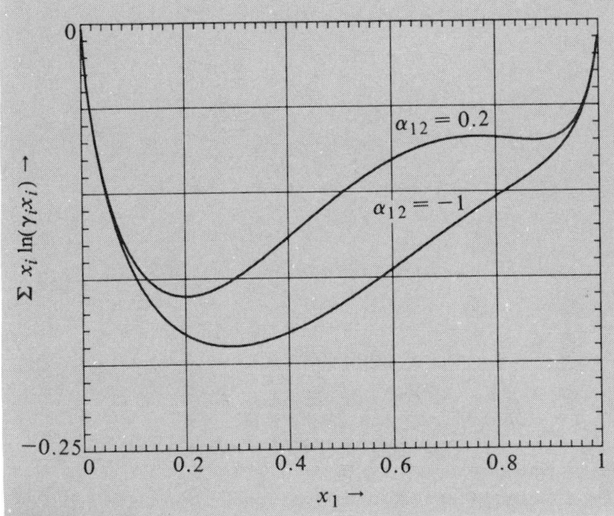

Example 7.3(b). Gibbs energies of mixing represented by the NRTL equation with two different values of the parameter α_{12}.

(Continued next page)

Example 7.3 *(continued)*

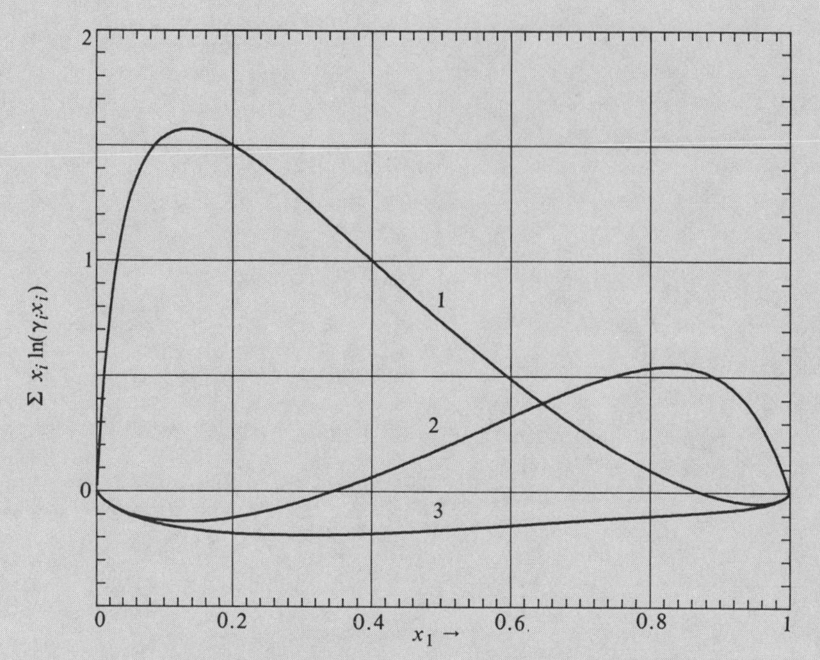

Example 7.3(c). Gibbs energies of mixing represented by the NRTL equation with $\alpha_{12} = -1$. The three sets of parameters satisfying this equation are (1) (3.000, −58.26), (2) (−11.78, 2.000), (3) (0.5881, 0.9411). Only curve 3 shows phase separation.

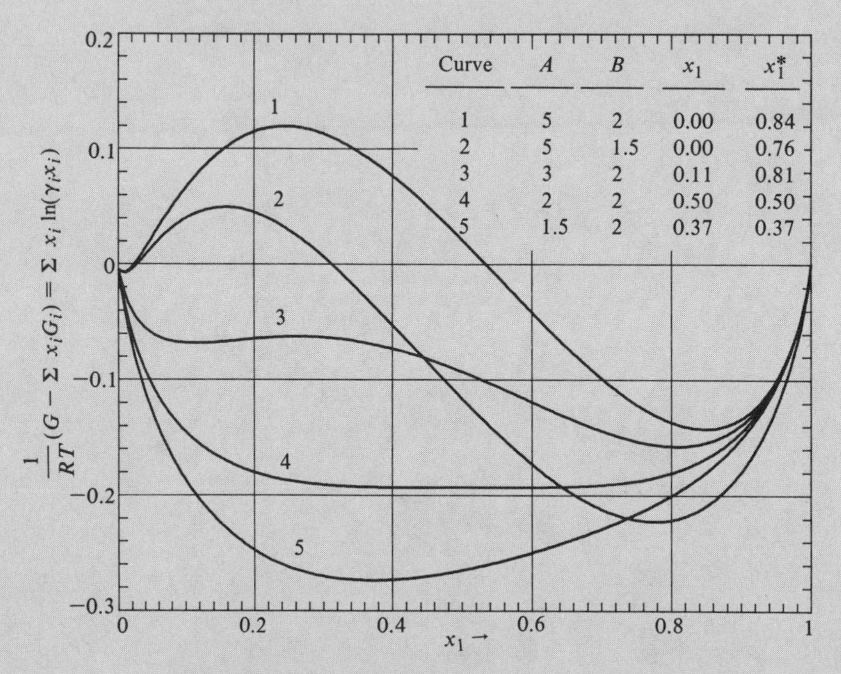

Curve	A	B	x_1	x_1^*
1	5	2	0.00	0.84
2	5	1.5	0.00	0.76
3	3	2	0.11	0.81
4	2	2	0.50	0.50
5	1.5	2	0.37	0.37

Example 7.3(d). Gibbs energies of mixing of binary mixtures whose activity coefficients are represented by van Laar equations with a variety of parameters. The values of the parameters and the miscibility limits are shown in the table; the compositions of the last two entries are those at the minima since there are no phase splits in those cases.

Example 7.3. *(continued)*

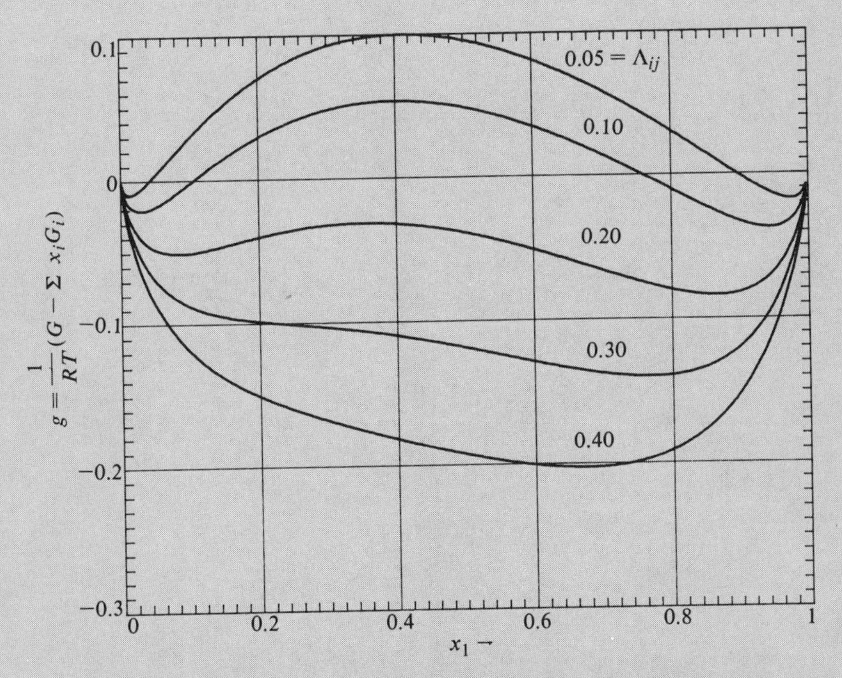

Example 7.3(e). Gibbs mixing energies represented by the T-K-Wilson equation. The chart is for $V_1/V_2 = 3$. Values of the parameters $\Lambda_{12} = \Lambda_{21}$ are indicated on the chart.

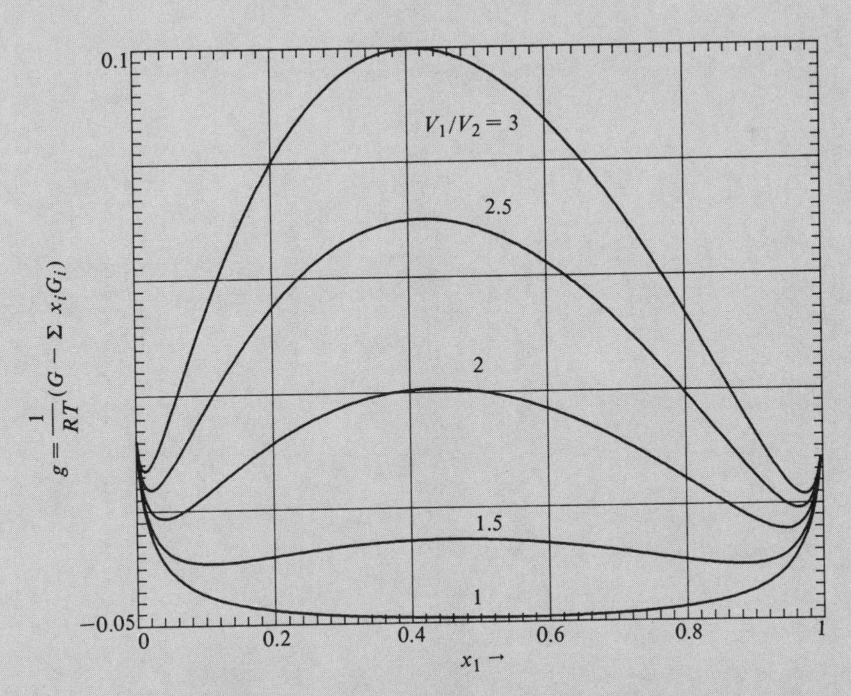

Example 7.3(f). Gibbs mixing energies represented by the T-K-Wilson equation. The graphs are for $\Lambda_{12} = \Lambda_{21} = 0.05$. Values of the volume ratios V_1/V_2 are shown on the graphs.

(Continued next page)

Example 7.3. *(continued)*

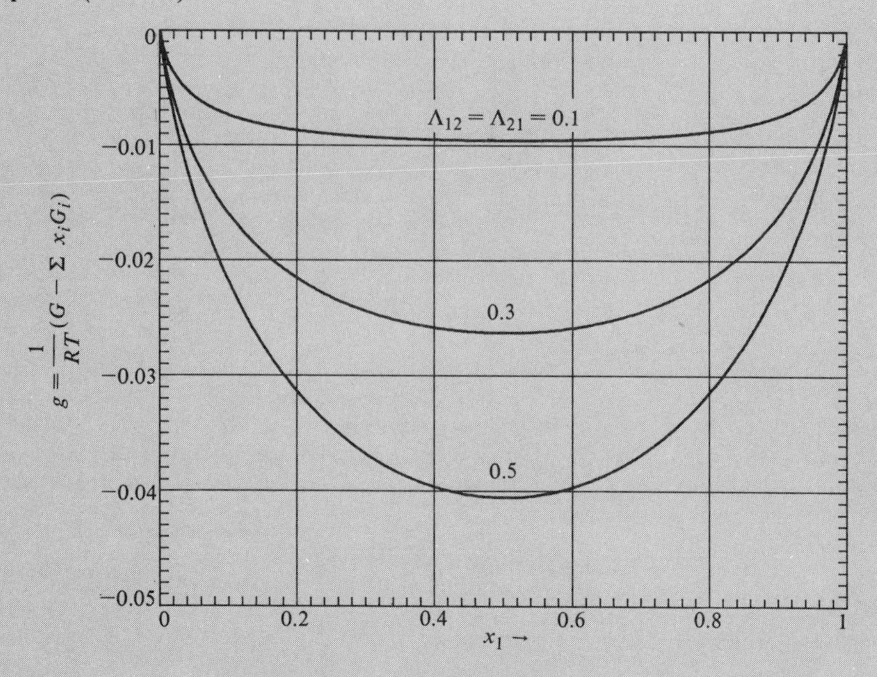

Example 7.3(g). Gibbs mixing energies represented by the T-K-Wilson equation, which, however, reduces to the Wilson equation since $V_1/V_2 = 1$ in this case. Values of the parameters $\Lambda_{12} = \Lambda_{21}$ are shown.

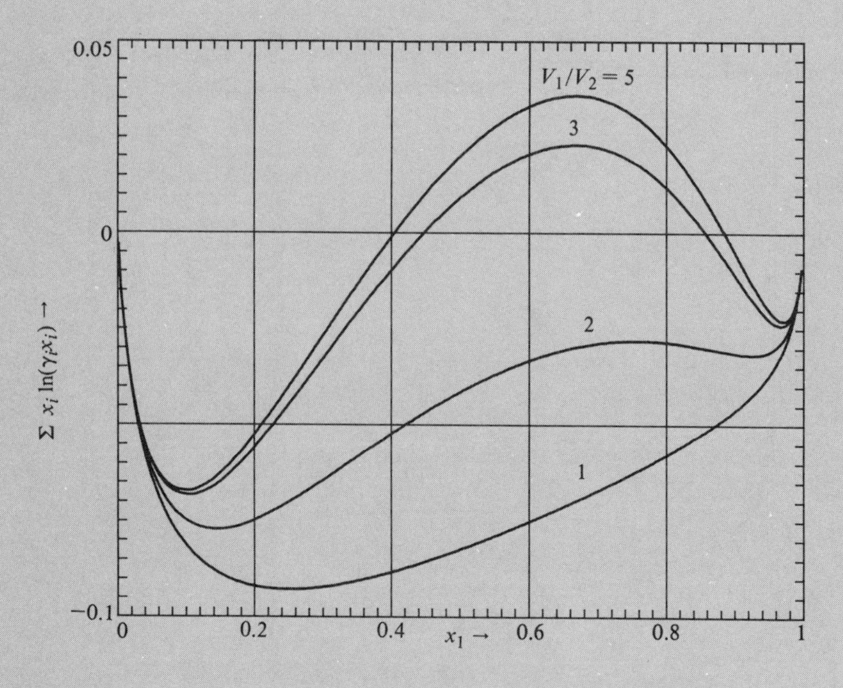

Example 7.3(h). Shapes of Gibbs mixing energy curves as affected by the volume ratio, V_1/V_2, of the T-K-Wilson equation. $\ln \gamma_1^\infty = 3$ and $\ln \gamma_2^\infty = 4$.

Example 7.3. *(continued)*

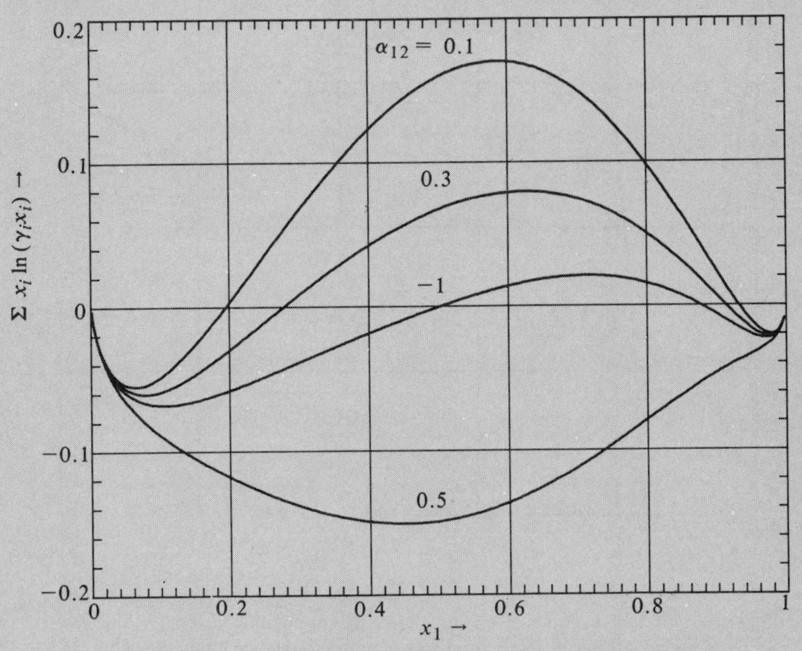

Example 7.3(i). Shapes of Gibbs mixing energy curves as affected by the NRTL parameter α_{12}, with $\ln \gamma_1^\infty = 3$ and $\ln \gamma_1^\infty = 4$.

Example 7.4. Limits of Miscibility Determined with the Margules and van Laar Equations

The limit of miscibility is represented by the inflection point of the plot of μ_1 against x_1. The mathematical condition is

$$\frac{\partial \mu_1}{\partial x_1} = \frac{\partial^2 \mu_1}{\partial x_1^2} = 0, \tag{1}$$

or

$$\frac{\partial^2 g}{\partial x_1^2} = \frac{\partial^3 g}{\partial x_1^3} = 0. \tag{2}$$

For the Margules equation (Table 7.2), these derivatives are:

$$\frac{\partial^2 g}{\partial x_1^2} = 2(3x_1 - 2)A + 2(1 - 3x_1)B + 1/x_1x_2 = 0, \tag{3}$$

$$\frac{\partial^3 g}{\partial x_1^3} = 6A - 6B - 1/x_1^2 + 1/x_2^2 = 0. \tag{4}$$

This system of two equations is solved readily by determinants:

$$A = \frac{-9x_1^2 + 8x_1 - 1}{6x_1^2(1 - x_1)^2}, \tag{5}$$

$$B = \frac{-9x_1^2 + 10x_1 - 2}{6x_1^2(1 - x_1)^2}. \tag{6}$$

For the van Laar equation (Table 7.2) the derivatives are

$$\frac{\partial^2 g}{\partial x_1^2} = \frac{-2A^2B^2}{(Ax_1 + Bx_2)^3} + 1/x_1x_2 = 0, \tag{7}$$

$$\frac{\partial^3 g}{\partial x_1^3} = \frac{6A^2B^2(A - B)}{(Ax_1 + Bx_2)^4} - 1/x_1^2 + 1/x_2^2 = 0. \tag{8}$$

The solution of these equations is given by Kogan (1968, p. 248) as follows:

$$A = \frac{13.5(1 - x_1)}{(2 - x_1)^2(1 + x_1)}, \tag{9}$$

$$B = \frac{13.5x_1}{(2 - x_1)(1 + x_1)^2}. \tag{10}$$

Regions of complete miscibility exist for which the values of A and B are less than given by these equations. The results are plotted in Figures 7.7 and 7.8.

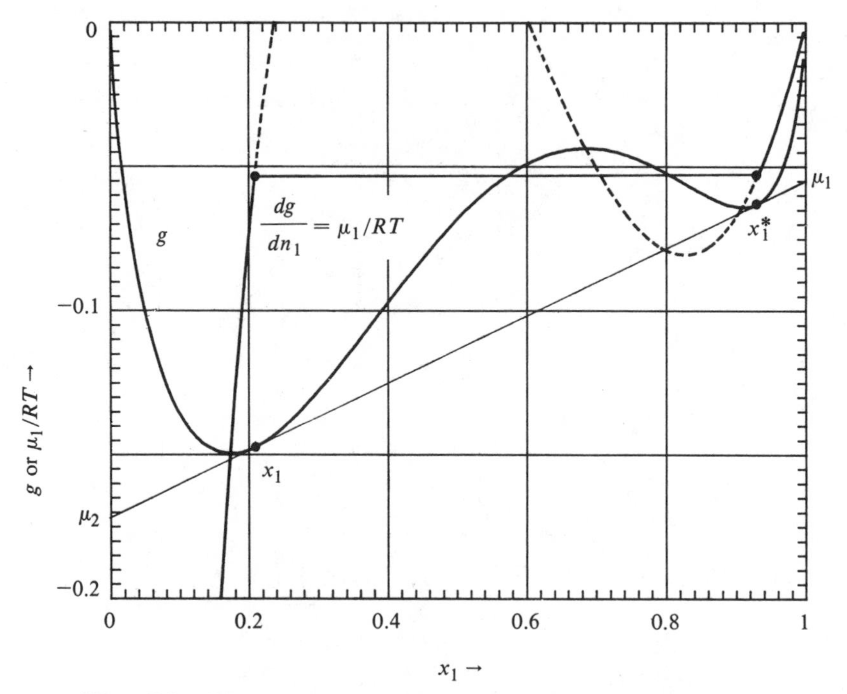

Figure 7.3. Gibbs energies of mixing and chemical potentials. Plots of Gibbs mixing energy and the corresponding chemical potential, showing that the latter is constant between the liquid-liquid equilibrium compositions. Margules equation with $A = 2$ and $B = 3$.

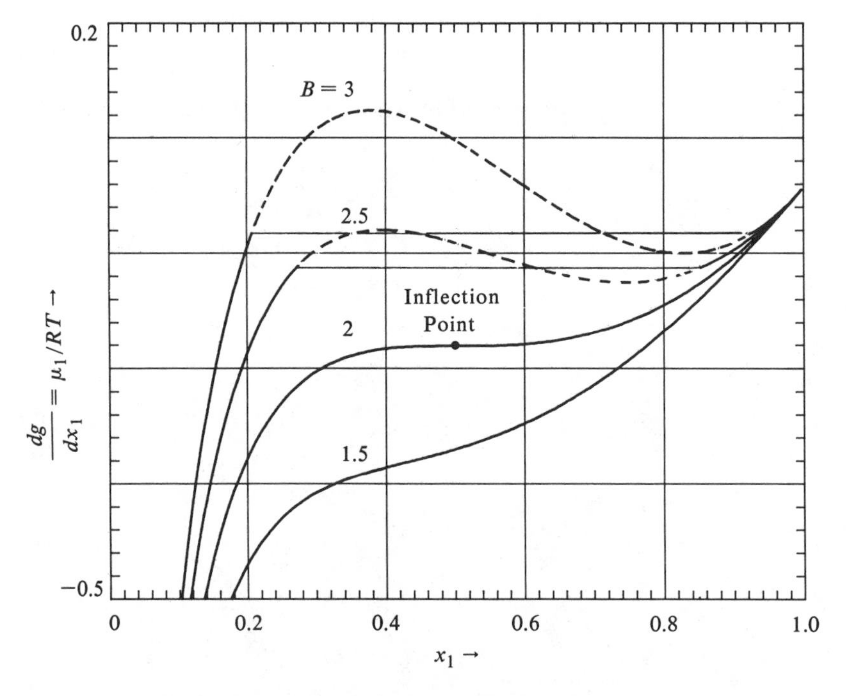

Figure 7.4. Plots of chemical potentials with Margules equations having $A = 2$ and B as marked on the diagram. The range of compositions in which μ_1 is constant narrows as B decreases and becomes simply the inflection point when $B = 2$. At the inflection point, $d\,\mu_1/dx_1 = d^2\mu_1/dx_1^2 = 0$.

equilibrium compositions. As an example, take the symmetrical equation

$$G^{ex}/RT = Ax_1(1-x_1), \tag{7.10}$$

from which are obtained

$$\ln \gamma_1 = A(1-x_1)^2, \tag{7.11}$$

$$\ln \gamma_1^* = A(1-x_1^*)^2, \tag{7.12}$$

$$\ln \gamma_2 = Ax_1^2, \tag{7.13}$$

$$\ln \gamma_2^* = Ax_1^{*2}. \tag{7.14}$$

First explore the condition of convexity. Let

$$g = (G - \Sigma x_i G_i)/RT = G^{ex}/RT + \Sigma x_i \ln x_i. \tag{7.15}$$

The derivatives are:

$$\frac{\partial g}{\partial x_1} = A(1-2x_1) + \ln\left(\frac{x_1}{1-x_1}\right), \tag{7.16}$$

$$\frac{\partial^2 g}{\partial x_1^2} = -2A + \frac{1}{x_1(1-x_1)} \overset{?}{<} 0. \tag{7.17}$$

Clearly the inequality is satisfied for all values of x_1 when

$$A \geq 2. \tag{7.18}$$

Two-phase and one-phase regions in terms of this parameter are identified in Figure 7.5. Corresponding Gibbs energy plots in Figure 7.6 imply that double tangents exist when $A > 2$. Since the curves are symmetrical, the points of double tangency coincide with the local minima and consequently are found by solving

$$A(1-2x_1) + \ln[x_1/(1-x_1)] = 0. \tag{7.19}$$

The same result is found by applying Eqs. 7.8 and 7.9 with Eqs. 7.11–7.14 with the results

$$x_1 \exp[A(1-x_1)^2] = x_1^* \exp[A(1-x_1^*)^2], \tag{7.20}$$

$$(1-x_1)\exp(Ax_1^2) = (1-x_1^*)\exp(Ax_1^{*2}). \tag{7.21}$$

The symmetries of these equations reveal that

$$x_1 = 1 - x_1^*, \tag{7.22}$$

so that Eq. 7.20 may be written

$$x_1 \exp[A(1-x_1)^2] = (1-x_1)\exp(Ax_1^2), \tag{7.23}$$

and may be rearranged to Eq. 7.19 by taking logarithms.

The condition of double tangency also may be applied as a method of numerical solution for the equilibrium compositions. For the symmetrical case, the double tangents are at the local minima, but in other cases it is necessary to solve the pair of equations

$$\frac{g - g^*}{x_1 - x_1^*} = \frac{\partial g}{\partial x_1} \tag{7.24}$$

$$= \frac{\partial g^*}{\partial x_1^*}, \tag{7.25}$$

where g^* is the function g written in terms of x_1^* as the variable. If only approximate values of the equilibrium compositions are needed the double tangent may be found most readily by graphical means.

In most cases the numerical solution is easier in terms of equality of fugacities, Eqs. 7.7ff, which is the method to be emphasized in this chapter. Eqs. 7.24 and 7.25 are applied in Example 7.8 and another application is asked for in Problem 7.24.

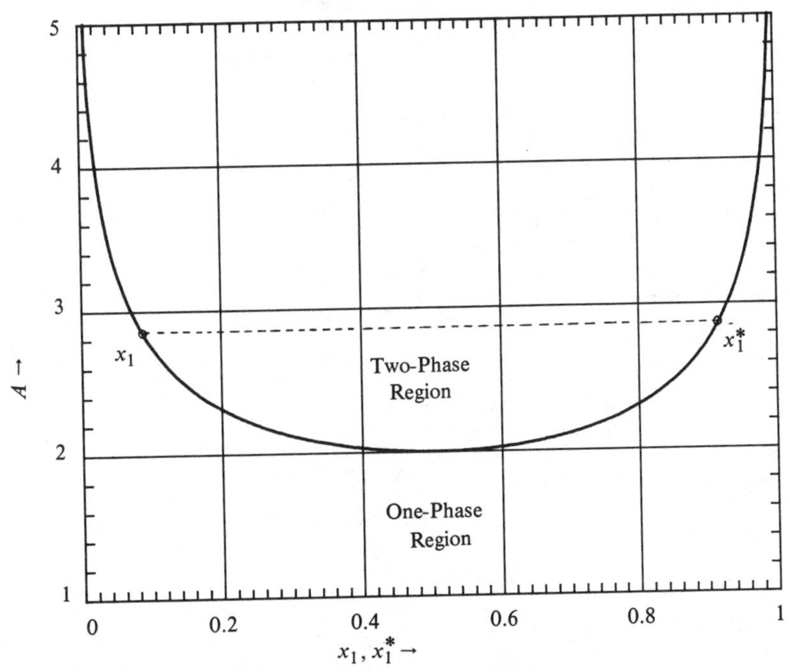

Figure 7.5. Miscibility limits of a binary mixture with a symmetrical equation for the excess Gibbs energy, $G^{ex}/RT = Ax_1x_2$. Miscibility range and the magnitude of A.

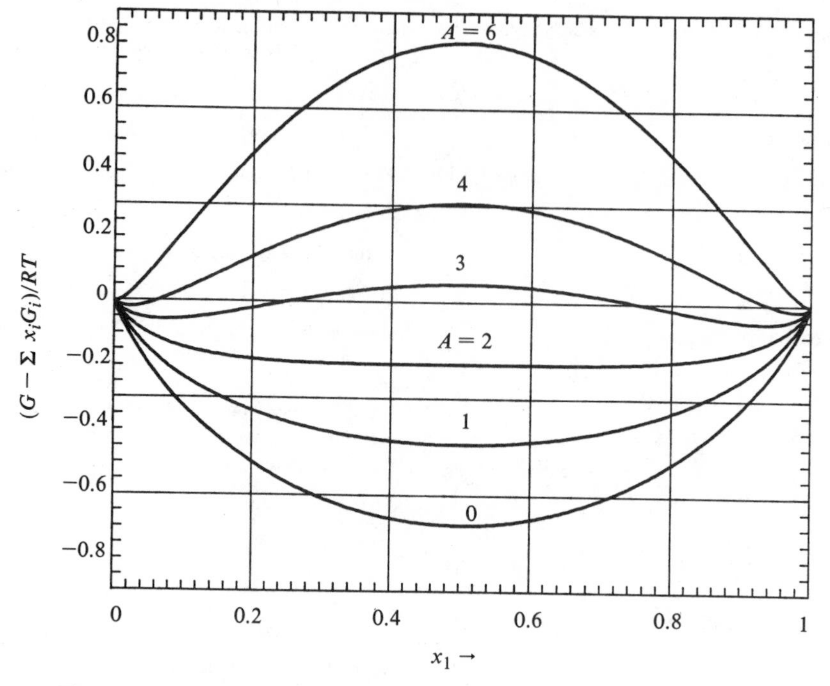

Figure 7.6. Gibbs energies of mixing at several values of A in the symmetrical equation for the excess Gibbs energy, $G^{ex}/RT = Ax_1x_2$. Phase splits occur only when A is greater than 2.

7.4. BINARY MIXTURES

Convexity: Plots of most of the common activity coefficient correlations can have portions that are convex for some values of the compositions and parameters and hence are able to represent liquid-liquid equilibria. The mathematical condition for convexity,

$$\partial^2 g/\partial x_1^2 < 0, \tag{7.26}$$

requires that at least some of the terms of the second derivatives of $g = G^{ex}/RT + \Sigma x_i \ln x_i$ be negative. Equations for these second derivatives of some common correlations are collected in Table 7.1. It is clear that except for the Wilson equation, they all contain positive and negative terms and hence can satisfy Eq. 7.26. The UNIQUAC equation is not included in the table, but it too can represent liquid-liquid equilibria. With the Margules equation, for instance, at $x_1 = 0.5$ the condition becomes

$$-(A + B) + 4 < 0, \tag{7.27}$$

which is clearly satisfied when the sum of the parameters is greater than 4; this conclusion is compatible with Eq. 7.18 for the symmetrical equation.

The fact that the several activity coefficient correlations may differ somewhat in their predictions of liquid-liquid equilibria is brought out in Example 7.7.

7.4.1. Finding Parameters

Since all common activity-coefficient correlations give $\ln \gamma_i$ directly, it is convenient to work with Eqs. 7.8 and 7.9 in the forms

$$p = \ln \gamma_1 - \ln \gamma_1^* - \ln(x_1^*/x_1) \to 0, \tag{7.28}$$

$$q = \ln \gamma_2 - \ln \gamma_2^* - \ln[(1 - x_1^*)/(1 - x_1)] \to 0. \tag{7.29}$$

When the activity coefficients are known as functions of composition, these equations may be solved for the equilibrium compositions, usually by trial because of their complexity; or when the equilibrium compositions are known from experiment, the equations may be solved for the two parameters of a correlating equation.

For the latter problem, analytical solutions of the Margules and van Laar equations are available. The Margules equations become

$$\ln(\gamma_1/\gamma_1^*) = \ln(x_1^*/x_1) = [A + 2(B - A)x_1]x_2^2 - [A + 2(B - A)x_1^*]x_2^{*2} \tag{7.30}$$

$$= A\left[x_2^2\left(1 + 2\left(\frac{B}{A} - 1\right)x_1\right) - x_2^{*2}\left(1 + 2\left(\frac{B}{A} - 1\right)x_1^*\right)\right], \tag{7.31}$$

$$\ln(\gamma_2/\gamma_2^*) = \ln\frac{1 - x_1^*}{1 - x_1} = (B + 2(A - B)x_2)x_1^2 - (B + 2(A - B)x_2^*)x_1^{*2} \tag{7.32}$$

$$= B\left[x_1^2\left(1 + 2\left(\frac{A}{B} - 1\right)x_2\right) - x_1^{*2}\left(1 + 2\left(\frac{A}{B} - 1\right)x_2^*\right)\right]. \tag{7.33}$$

Table 7.1. Second Derivatives of $g = (G - \Sigma\, x_i G_i)/RT = G^{ex}/RT + \Sigma\, x_i \ln x_i$ of Several Equations for G^{ex} of Binary Mixtures

G^{ex}/RT	$\partial^2 g/\partial x_1^2$
Symmetrical	$-2A + 1/x_1x_2$
Margules	$2(-2+3x_1)A + 2(1-3x_1)B + 1/x_1x_2$
van Laar	$\dfrac{-2A^2B^2}{(Ax_1+Bx_2)^3} + 1/x_1x_2$
Wilson	$\dfrac{1}{x_1}\left[\dfrac{1}{\dfrac{x_1}{\Lambda_{12}}+x_2}\right]^2 + \dfrac{1}{x_2}\left[\dfrac{1}{x_1+\dfrac{x_2}{\Lambda_{21}}}\right]^2 + 1/x_1x_2$
T-K-Wilson	$\dfrac{1}{x_1}\left\{\left[\dfrac{1}{\dfrac{x_1}{\Lambda_{12}}+x_2}\right]^2 - \left[\dfrac{1}{\dfrac{V_2x_1}{V_1}+x_2}\right]^2\right\} + \dfrac{1}{x_2}\left\{\left[\dfrac{1}{x_1+\dfrac{x_2}{\Lambda_{21}}}\right]^2 - \left[\dfrac{1}{x_1+\dfrac{V_1x_2}{V_2}}\right]^2\right\} + 1/x_1x_2$
NRTL	$-2\left[\dfrac{\tau_{21}G_{21}}{x_1+G_{21}x_2} + \dfrac{\tau_{12}G_{12}}{G_{12}x_1+x_2}\right] - (1-2x_1)\left[\dfrac{\tau_{21}G_{21}(1-G_{21})}{(x_1+G_{21}x_2)^2} + \dfrac{\tau_{12}G_{12}(1+G_{12})}{(G_{12}x_1+x_2)^2}\right] + 2x_1x_2\left[\dfrac{\tau_{21}G_{21}(1-G_{21})^2}{(x_1+G_{21}x_2)^3} + \dfrac{\tau_{12}G_{12}(1+G_{12})^2}{(G_{12}x_1+x_2)^3}\right) + \dfrac{1}{x_1x_2}$

When these equations are ratioed, A/B remains the only variable, which has been solved for by Carlson & Colburn (1942) as

$$\frac{A}{B} = \frac{2\alpha_2 \ln(x_2^*/x_2) + \beta_1 \ln(x_1^*/x_1)}{2\alpha_1 \ln(x_1^*/x_1) + \beta_2 \ln(x_2^*/x_2)}, \tag{7.34}$$

where

$$\alpha_i = x_i^2 - x_i^{*2} - x_i^3 + x_i^{*3}, \tag{7.35}$$

$$\beta_i = x_i^2 - x_i^{*2} - 2x_i^3 + 2x_i^{*3}. \tag{7.36}$$

After a numerical value of A/B has been found, it is substituted into Eq. 7.31 to find A and then into Eq. 7.33 to find B, or $B = A/(A/B)$.

The solution of the van Laar equations is found in a similar manner. Thus,

$$\ln(\gamma_1/\gamma_1^*) = \ln(x_1^*/x_1) = AB^2\left[\left(\frac{x_2}{Ax_1+Bx_2}\right)^2 - \left(\frac{x_2^*}{Ax_1^*+Bx_2^*}\right)^2\right], \tag{7.37}$$

$$\ln(\gamma_2/\gamma_2^*) = \ln(x_2^*/x_2) = A^2B\left[\left(\frac{x_1}{Ax_1+Bx_2}\right)^2 - \left(\frac{x_1^*}{Ax_1^*+Bx_2^*}\right)^2\right]. \tag{7.38}$$

The ratio of the parameters is

$$\frac{A}{B} = \frac{\left(\dfrac{x_1}{x_2}+\dfrac{x_1^*}{x_2^*}\right)\left(\dfrac{\ln(x_1^*/x_1)}{\ln(x_2/x_2^*)}\right) - 2}{\dfrac{x_1}{x_2}+\dfrac{x_1^*}{x_2^*} - \dfrac{2x_1x_1^*\ln(x_1^*/x_1)}{x_2x_2^*\ln(x_2/x_2^*)}} \tag{7.39}$$

and the value of A is

$$A = \frac{\ln(x_1^*/x_1)}{\dfrac{1}{\left(1+\dfrac{A}{B}(x_1/x_2)\right)^2} - \dfrac{1}{\left(1+\dfrac{A}{B}(x_1^*/x_2^*)\right)^2}}. \tag{7.39a}$$

When more than one set of equilibrium compositions is at hand, some method of nonlinear regression may be used to find the parameters. A suitable objective function is

$$F(A, B) = \Sigma\,(\gamma_1x_1 - \gamma_1^*x_1^*)_i^2 + \Sigma\,(\gamma_2x_2 - \gamma_2^*x_2^*)_i^2 = \text{minimum}. \tag{7.40}$$

Example 7.5. Finding Margules Parameters when Liquid-Liquid Equilibrium Compositions Are Known

Let x and x^* represent the two phase compositions. In terms of the Margules equations the equilibrium conditions,

$$\ln(\gamma_1/\gamma_1^*) = \ln(x^*/x),$$

$$\ln(\gamma_2/\gamma_2^*) = \ln[(1-x^*)/(1-x)],$$

become

$$[A + 2(B-A)x](1-x)^2 - [A + 2(B-A)x^*](1-x^*)^2 - \ln(x^*/x) = 0,$$

$$[B + 2(A-B)(1-x)]x^2 - [B + 2(A-B)(1-x^*)]x^{*2} - \ln[(1-x^*)/(1-x)] = 0.$$

Collect the coefficients of A and B,

$$[(1-2x)(1-x)^2 - (1-2x^*)(1-x^*)^2]A + [2x(1-x)^2 - 2x^*(1-x^*)^2]B - \ln(x^*/x) = 0,$$

$$[2x^2(1-x) - 2x^{*2}(1-x^*)]A + [x^2(2x-1) - x^{*2}(2x^*-1)]B - \ln[(1-x^*)/(1-x)] = 0.$$

These equations are of the mathematical form

$$a_1A + b_1B = c_1,$$

$$a_2A + b_2B = c_2,$$

whose solutions are

$$A = (b_2c_1 - b_1c_2)/D,$$

$$B = (a_1c_2 - a_2c_1)/D,$$

$$D = a_1b_2 - a_2b_1.$$

The BASIC program and solutions for $x_1 = 0.95$ and several values of x_1^* are shown.

```
10 ! LIQUID-LIQUID EQUILIBRIA WITH THE
     MARGULES EQNS
20 X=.95
30 Y=.05
40 A1=(1-2*X)*(1-X)^2-(1-2*Y)*(1-Y)^2
50 B1=2*X*(1-X)^2-2*Y*(1-Y)^2
60 A2=2*X^2*(1-X)-2*Y^2*(1-Y)
70 B2=X^2*(2*X-1)-Y^2*(2*Y-1)
80 C1=LOG(Y/X)
90 C2=LOG((1-Y)/(1-X))
100 D=A1*B2-A2*B1
110 A=(C1*B2-C2*B1)/D
120 B=(A1*C2-A2*C1)/D
130 PRINT USING 140 ; Y,A,B
140 IMAGE .DD,2X,DDD.DDDD,2X,DDD.DDDD
150 Y=Y+.05
160 GOTO 40
170 END
```

x^*	A	B
.05	3.2716	3.2716
.10	2.6668	3.2889
.15	2.2649	3.2795
.20	1.9165	3.2553
.25	1.5714	3.2170
.30	1.2012	3.1618
.35	.7820	3.0847
.40	.2888	2.9781
.45	−.3101	2.8310
.50	−1.0575	2.6268
.55	−2.0155	2.3407
.60	−3.2791	1.9331
.65	−5.004	1.3397
.70	−7.4392	.4479
.75	−11.0735	−.9522
.80	−16.8752	−3.2943
.85	−27.1285	−7.6122
.90	−48.5854	−17.0037
.95		

After differentiation,

$$\frac{\partial F}{\partial A} = \frac{\partial F}{\partial B} = 0, \quad (7.41)$$

the parameters may be solved for by the Newton-Raphson or Simplex methods. The former method was adopted, for example, for the DECHEMA LLE Collection.

Margules parameters are found in Example 7.5 in terms of liquid-liquid equilibrium data.

7.4.2. Finding Compositions

When the activity-coefficient equation is known, any one of several methods may be used to figure equilibrium compositions of binary mixtures. A perfectly general scheme is to solve equations such as Eqs. 7.28 and 7.29 simultaneously by a numerical method, but less accurate methods sometimes may be more convenient and often adequate. Graphical methods are summarized in Example 7.7.

a. Simultaneous equations: Since Eqs. 7.28, and 7.29 are transcendental and fairly complex, direct solution for the composition (x_1, x_1^*) usually is not possible. The numerical procedure based on the Newton-Raphson method will be reviewed in this application. Starting with trial values of the compositions, the values of the functions p and q and the four partial derivatives with respect to the compositions are figured at these trial values, then corrections h and k to x_1 and x_1^* are found by solving the linear equations

$$hp_x + kp_{x^*} + p = 0, \quad (7.42)$$

$$hq_x + kq_{x^*} + q = 0. \quad (7.43)$$

The compositions for the next trial will be $(x_1 + h, x_1^* + k)$. The equations of the derivatives are

Example 7.6. Use of the Newton-Raphson Method to Find Liquid-Liquid Equilibrium Compositions with the Margules and van Laar Equations

For this problem the parameters are $A = 3$ and $B = 2$. The Margules activity-coefficient equations are

$$\ln \gamma_1 = [A + 2(B - A)x_1]x_2^2 = (3 - 2x_1)(1 - x_1)^2, \quad (1)$$

$$\ln \gamma_2 = [B + 2(A - B)x_2]x_1^2 = 2(2 - x_1)x_1^2. \quad (2)$$

Define

$$f = \ln(\gamma_1/\gamma_1^*) - \ln\frac{x_1^*}{x_1} \quad (3)$$

$$= (3 - 2x_1)(1 - x_1)^2 - (3 - 2x_1^*)(1 - x_1^*)^2 - \ln(x_1^*/x_1) = 0, \quad (4)$$

$$g = \ln(\gamma_2/\gamma_2^*) - \ln\frac{1 - x_1^*}{1 - x_1} \quad (5)$$

$$= 2(2 - x_1)x_1^2 - 2(2 - x_1^*)x_1^{*2} - \ln\frac{1 - x_1^*}{1 - x_1} = 0. \quad (6)$$

Partial derivatives of these quantities are

$$f_x = -(1 - x_1)(2 - x_1), \quad (7)$$

$$f_{x*} = -(1 - x_1^*)(2 - x_1^*), \quad (8)$$

$$g_x = 2x_1(4 - 3x_1), \quad (9)$$

$$g_{x*} = 2x_1^*(4 - 3x_1^*). \quad (10)$$

The equations to be solved for the corrections h and k to estimated values of x_1 and x_1^* are

$$hf_x + kf_{x*} + f = 0, \quad (11)$$

$$hg_x + kg_{x*} + g = 0, \quad (12)$$

of which the solutions are

$$h = -(fg_{x*} - gf_{x*})/D, \quad (13)$$

$$k = -(f_x g - g_x f)/D \quad (14)$$

$$D = f_x g_{x*} - f_{x*} g_x. \quad (15)$$

The convergence criterion is step #200 of the program. When starting with the estimates (0.1, 0.8), 15 iterations are required. The results, $x_1 = 0.07195$, $x_1^* = 0.79206$, check values read off Figure 7.7 within the accuracy of reading that graph.

Programs for both Margules and van Laar equations are shown. With the van Laar the miscibility limits are $x_1 = 0.0786$, $x_1^* = 0.7574$. The two equations give somewhat different results, even though their parameters are evaluated from the same γ_i^∞.

```
10 ! NEWTON-RAPHSON SOLUTION OF
     MARGULES EQUATION
20 SHORT F,G,X,X1,H,K,D
30 READ X,X1
40 DATA .1, .9
50 F=(3-2*X)*(1-X)^2-(3-2*X1)*(1-X1)^2-
     LOG(X1/X)
60 G=2*(2-X)*X^2-2*(2-X1)*X1^2-LOG
     ((1-X1)/(1-X))
70 F1=-((1-X)*(2-X))+1/X
80 F2=-((1-X1)*(2-X1))-1/X1
90 G1=2*X*(4-3*X)-1/(1-X)
100 G2=2*X1*(4-3*X1)+1/(1-X1)
110 D=F1*G2-F2*G1
120 H=-((F*G2-G*F2)/D)
130 K=-((F1*G-G1*F)/D)
140 DISP F;G
150 DISP X;X1
160 DISP H;K;D
170 DISP
180 X1=X1+K
190 X=X+H
200 IF ABS(H/X)+ABS(K/X1)<=.0001/2 THEN 220
210 GOTO 50
220 END
```

```
10 ! NEWTON-RAPHSON FOR VAN LAAR, WITH
     A=3, B=2
20 READ A,B
30 DATA 3,2
40 INPUT X,X1
50 H1=A*B^2*((1-X)/(A*X+B*(1-X)))^2
60 H2=A^2*B*(X/(A*X+B*(1-X)))^2
70 H3=A*B^2*((1-X1)/(A*X1+B*(1-X1)))^2
80 H4=A^2*B*(X1/(A*X1+B*(1-X1)))^2
90 F=H1-H3-LOG(X1/X)
100 G=H2-H4-LOG((1-X1)/(1-X))
110 F3=-(2*A^2*B^2*(1-X)/(A*X+B*(1-X))^3)
120 F1=F3+1/X
130 F4=-(2*A^2*B^2*(1-X1)/(A*X1+B*(1-X1))
     ^3)
140 F2=F4-1/X1
150 G1=-(F3*X/(1-X))-1/(1-X)
160 G2=-(F4*X1/(1-X1))+1/(1-X1)
170 D=F1*G2-F2*G1
180 H=(G*F2-F*G2)/D
190 K=(G1*F-F1*G)/D
200 PRINT USING 210 ; X,X1
210 IMAGE .DDDD,3X,.DDDD
220 X=X+H
230 X1=X1+K
240 IF ABS(H/X)+ABS(K/X1)<=.00005 THEN 260
250 GOTO 50
260 END
```

$$p_x = \frac{\partial p}{\partial x_1} = \frac{\partial \ln \gamma_1}{\partial x_1} + \frac{1}{x_1}, \tag{7.44}$$

$$p_{x*} = \frac{\partial p}{\partial x_1^*} = -\frac{\partial \ln \gamma_1^*}{\partial x_1^*} - \frac{1}{x_1^*}, \tag{7.45}$$

$$q_x = \frac{\partial q}{\partial x_1} = \frac{\partial \ln \gamma_2}{\partial x_1} - \frac{1}{1-x_1}, \tag{7.46}$$

$$q_{x*} = \frac{\partial q}{\partial x_1^*} = -\frac{\partial \ln \gamma_2^*}{\partial x_1^*} + \frac{1}{1-x_1^*}. \tag{7.47}$$

The pertinent derivatives of the Margules equation are

$$\frac{\partial \ln \gamma_1}{\partial x_1} = 2x_2[(B-A)(1-3x_1)-A], \tag{7.48}$$

$$\frac{\partial \ln \gamma_2}{\partial x_1} = 2x_1[(B-A)(1-3x_2)+A], \tag{7.49}$$

and those of the van Laar equations are

$$\frac{\partial \ln \gamma_1}{\partial x_1} = \frac{-2A^2B^2x_2}{(Ax_1+Bx_2)^3}, \tag{7.50}$$

$$\frac{\partial \ln \gamma_2}{\partial x_1} = \frac{2A^2B^2x_1}{(Ax_1+Bx_2)^3} = -\frac{x_1}{x_2}\frac{\partial \ln \gamma_1}{\partial x_1}. \tag{7.51}$$

This procedure is employed in Example 7.6, where simple computer programs for the Margules and van Laar equations are developed. A program for two, three, or four components based on the T-K-Wilson equation is in Appendixes C.9 and C.10.

b. Charts: Either equilibrium compositions or equation parameters may be found with the charts of Figures 7.7–7.13. The UNIQUAC equation cannot be represented in this simple fashion. Since the parameters of all the equations are known when the infinite-dilution activity coefficients are known, the latter quantities are plotted along the axes in all cases for the sake of uniformity. A pair of these values ($\ln \gamma_1^\infty$, $\ln \gamma_2^\infty$) determines a pair of compositions (x_1, x_1^*) in the network. Table 4.5 relates the parameters of the equations to the $\ln \gamma_i^\infty$.

Predictions of equilibrium compositions by the several equations disagree considerably in certain ranges. Differences are evident in Table 7.2, which shows $\ln \gamma_i^\infty$ corresponding to several combinations (x_1, x_1^*). As another example, for $\ln \gamma_1^\infty = 2$ and $\ln \gamma_2^\infty = 3$, equilibrium compositions off the charts are:

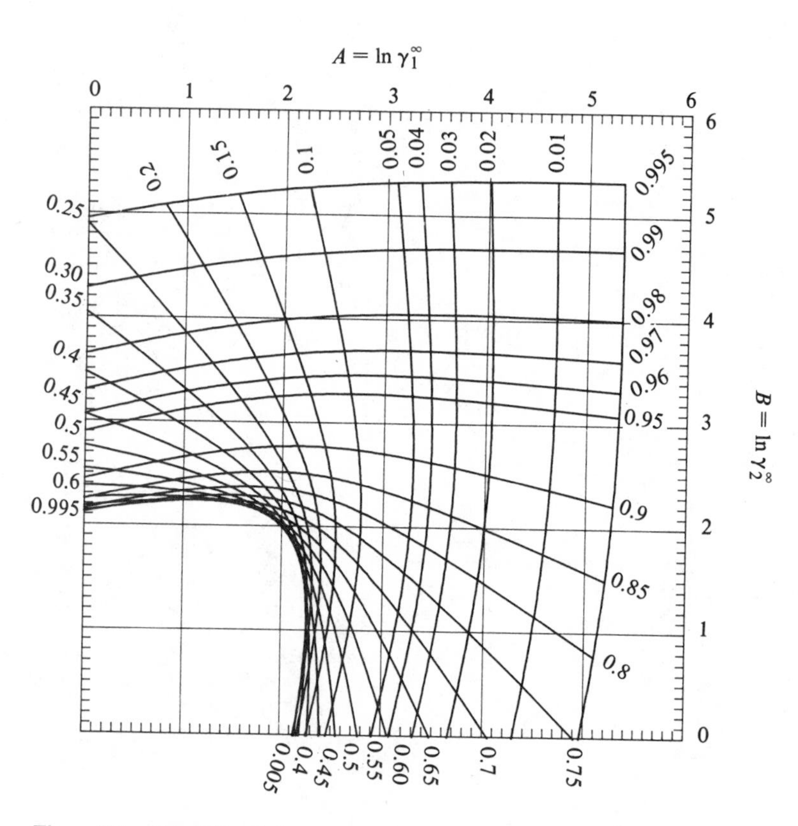

Figure 7.7. Miscibility limits with the Margules equation (After Kogan, 1968). An intersection within the network represents the equilibrium mol fractions (x_1, x_1^*) corresponding to each pair of logarithms of infinite-dilution activity coefficients or equation parameters A, B. For example, $(x_1, x_1^*) = (0.10, 0.98)$ corresponds to $(A, B) = (2.5, 4.03)$.

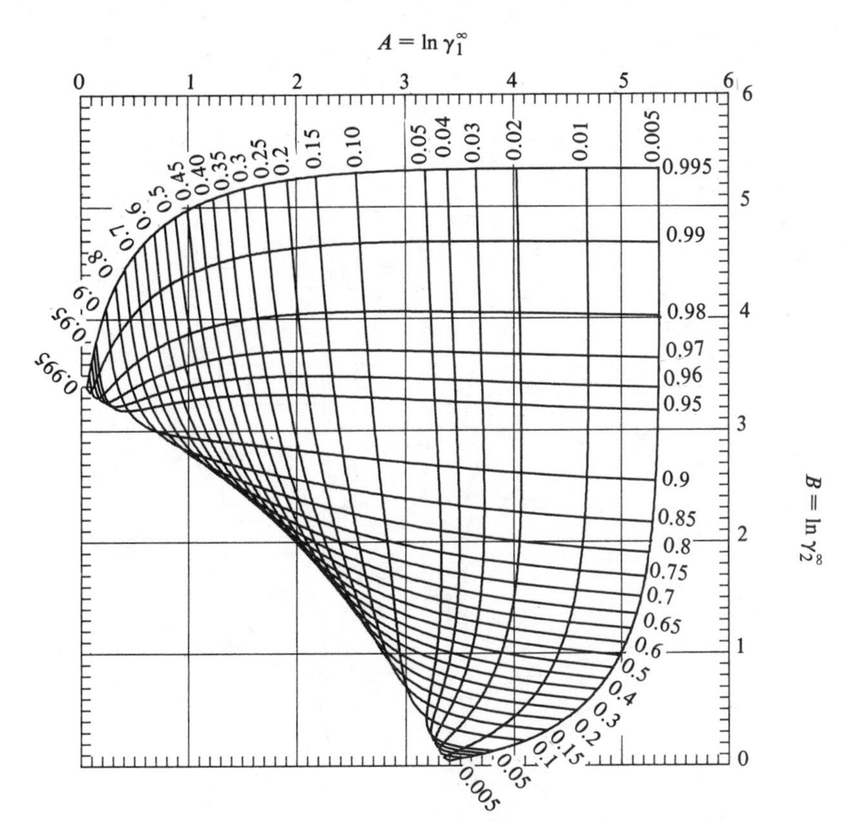

Figure 7.8. Miscibility limits with the van Laar and Scatchard-Hildebrand equations. See the legend of Figure 7.7 for use of the chart.

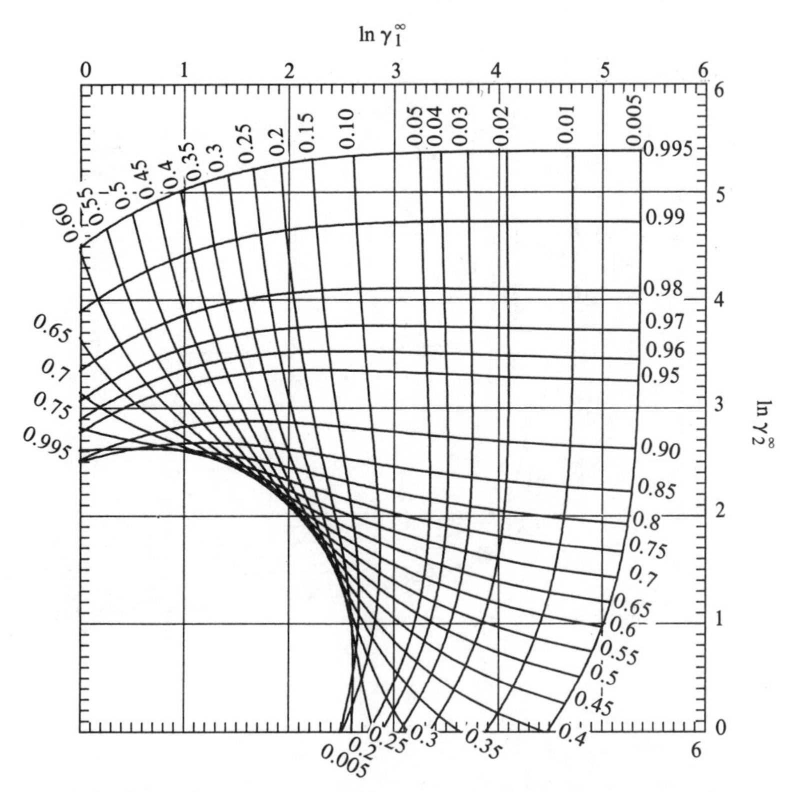

Figure 7.9. Miscibility limits with the NRTL equation with $\alpha_{12} = 0.2$ (Drawn by S. Negahban). $\ln \gamma_1^\infty = \tau_{21} + \tau_{12} \exp(-0.2\tau_{12})$, $\ln \gamma_2^\infty = \tau_{12} + \tau_{21} \exp(-0.2\tau_{21})$. See legend of Figure 7.7 for use of the chart.

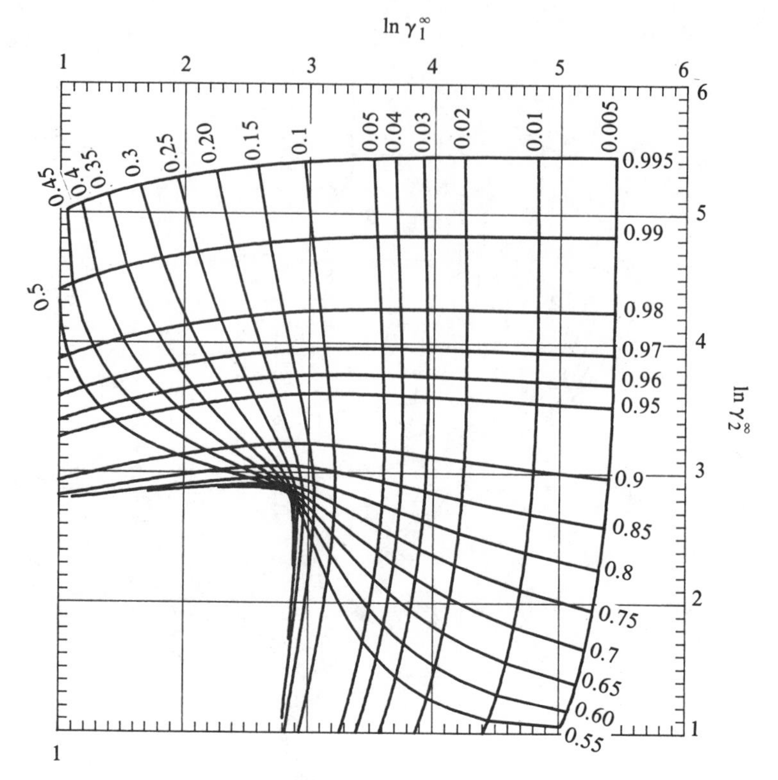

Figure 7.10. Miscibility limits with the NRTL equation with $\alpha_{12}=-1$ (Drawn by S. Negahban). $\ln\ \gamma_1^\infty=\tau_{21}+\tau_{12}\ \exp(\tau_{12})$, $\ln\ \gamma_2^\infty=\tau_{12}+\tau_{21}\ \exp(\tau_{21})$. See legend of Figure 7.7 for use of the chart.

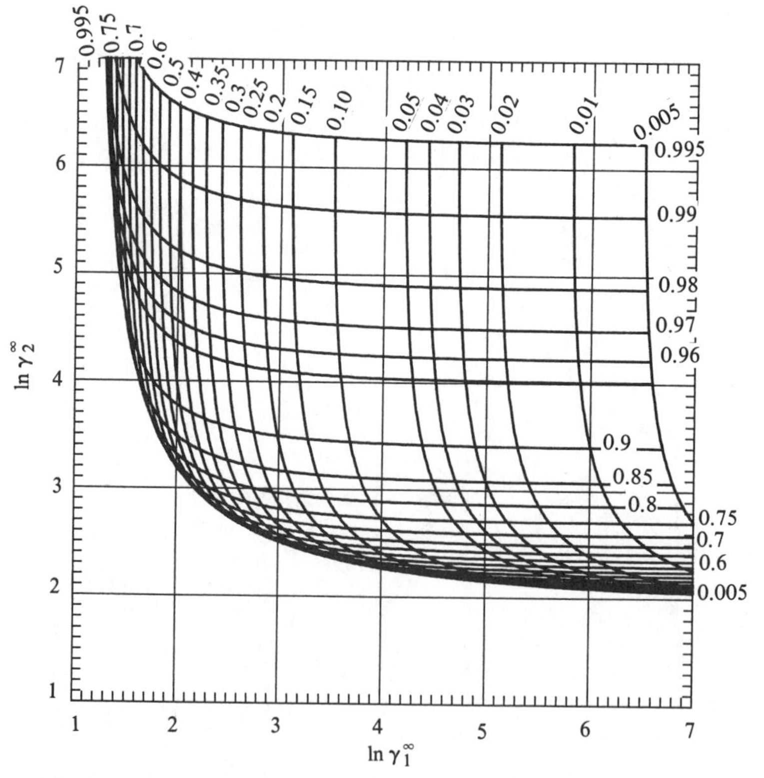

Figure 7.11. Miscibility limits with the T-K-Wilson equation for $V_1/V_2=2$ (Drawn by S. Negahban). Lines are drawn for the same values of x_1 and x_1^* as on Figure 7.7. See the legend of that figure for use of the chart.

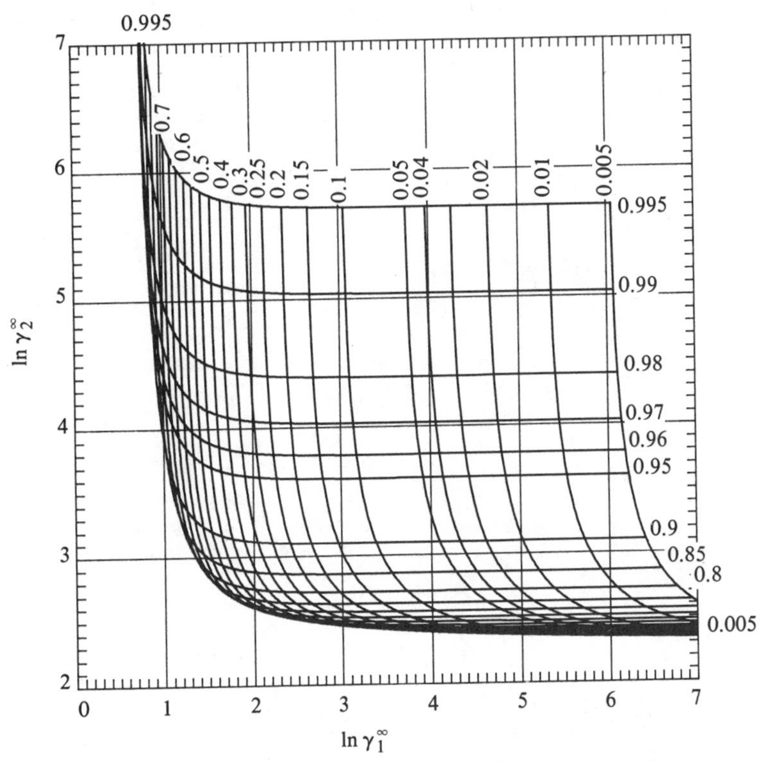

Figure 7.12. Miscibility limits with the T-K-Wilson equation for $V_1/V_2 = 3$ (Drawn by S. Negahban). Lines are drawn for the same values of x_1 and x_1^* as on Figure 7.7. See the legend of that figure for use of the chart.

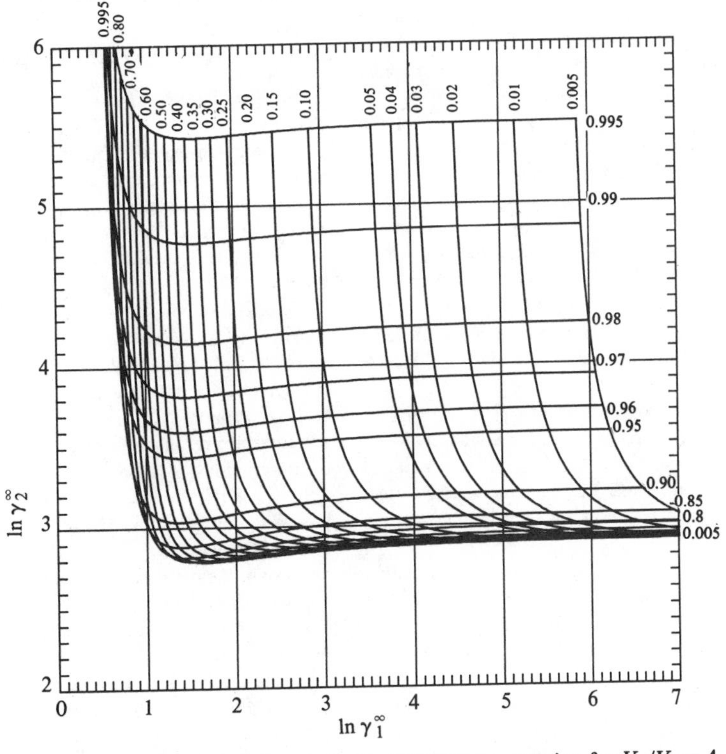

Figure 7.13. Miscibility limits with the T-K-Wilson equation for $V_1/V_2 = 4$ (Drawn by S. Negahban). Lines are drawn for the same values of x_1 and x_1^* as on Figure 7.7. See the legend of that figure for use of the chart.

Table 7.2. Miscibility Limits Computed from Equations Whose Parameters are Derived from Infinite-Dilution Activity Coefficients

(x_1, x_1^*) / Equation	(0.05, 0.90)		(0.05, 0.70)		(0.05, 0.50)		(0.5, 0.5)	
	$\ln \gamma_1^\infty$	$\ln \gamma_2^\infty$	$\ln \gamma_1^\infty$	$\ln \gamma_2^\infty$	$\ln \gamma_1^\infty$	$\ln \gamma_2^\infty$	$\ln \gamma_1^\infty$	$\ln \gamma_2^\infty$
Margules	3.29	2.67	3.16	1.20	2.63	−0.11	2	2
van Laar	3.30	2.69	3.33	1.75	3.31	1.26	2	2
TKW, 2	4.39	3.42	4.81	2.60	5.30	2.31	2.71	2.63
TKW, 3	4.10	3.09	4.78	2.55	5.43	2.42	2.72	2.48
TKW, 4	4.25	3.20	5.52	2.93	6.69	2.91	3.41	2.87
NRTL, 0.2	3.35	2.75	3.56	1.76	3.23	1.09	2.05	2.05
NRTL, −1	3.61	3.17	3.33	2.39	3.39	1.45	2.84	2.85

The numbers alongside TKW are V_2/V_1; those alongside NRTL are α_{12}.

Margules	0.21, 0.92
van Laar	0.24, 0.91
NRTL, $\alpha_{12} = 0.2$	0.27, 0.91
NRTL, $\alpha_{12} = -1$	0.57, 0.84
T-K-W, $V_2/V_1 = 2$	Miscible
T-K-W, $V_2/V_1 = 3$	0.39, 0.87
T-K-W, $V_2/V_1 = 4$	0.40, 0.86

Clearly, predictions based on $\ln \gamma_i^\infty$ may be only roughly qualitative, although it is known that in particular cases any of these equations can be fitted closely in a desired range of concentrations. When there is no additional information, the usual engineering rule of choosing the more conservative result is to be followed. Correlation of binary data in the DECHEMA LLE Data Collection is in terms of only the UNIQUAC equation and the NRTL equation with $\alpha_{12} = 0.2$, since such data are applicable to ternary and more complex mixtures. Values of $\alpha_{12} = 0.1$, 0.3, and −1 are stated to be less satisfactory, and the T-K-Wilson equation apparently was not examined at length in this work.

Figures 7.11–7.13 reveal considerable sensitivity of the TKW equation to the volume ratio. In fact when $V_2/V_1 = 1$, the equation reduces to the Wilson, which of course cannot represent immiscibility; yet there are immiscible binaries whose component molal volumes are within a few percentage points of each other. Also, as pointed out by Maurer & Prausnitz (1978), the UNIQUAC equation reduces to the Wilson when $R_1 = R_2$ and $Q_1 = Q_2$; but in practice this equation has not failed in correlating immiscibility, at least with the 2,000-odd LLE data sets of the DECHEMA Collection.

Nor is the issue of the proper value of α_{12} of the NRTL equation entirely closed. A recent paper by Simonetty et al. (1982) checked 10 ternaries made up of 28 different binaries and found $\alpha_{12} = -1$ somewhat better on the average than NRTL with specific positive values for each system or than the UNIQUAC.

c. Method of double tangency: As explained in Section 7.3, equilibrium compositions (x_1, x_1^*) are points of double tangency on a plot of Gibbs energy of mixing, $\Sigma x_i \ln \gamma_i x_i$, against x_1. With proper computer graphics the plot is made quite easily and affords results that are accurate enough for many purposes, as well as starting values for an accurate numerical solution. Example 7.7(a) uses this method of finding equilibrium compositions. Some effects of variation in the NRTL α_{12} and the TKW ratio V_2/V_1 are shown in Example 7.3.

d. Rectangular construction: This method is applied in Example 7.7(b). Separate plots of $\gamma_1 x_1$ and $\gamma_2 x_2$ against x_1 are made. The rectangular construction shown is found by trial in such a way that $\gamma_1 x_1 = \gamma_1^* x_1^*$ and $\gamma_2(1 - x_1) = \gamma_2^*(1 - x_1^*)$ are satisfied simultaneously. For such a construction to be possible, each curve must have a minimum and a maximum.

e. Loop method: If there is a phase split, a plot of $\gamma_1 x_1$ against $\gamma_2(1 - x_1)$ with x_1 as parameter, will loop on itself at the equilibrium compositions (x_1, x_1^*), as shown in Example 7.7(c) and Figure 7.14. The intersection gives the two values of x_1 at which $\gamma_1 x_1 = \gamma_1^* x_1^*$ and $\gamma_2(1 - x_1) = \gamma_2^*(1 - x_1^*)$. The need for calibrating the looped curve with values of x_1 corresponding to each value of the ordinate and abscissa is obviated by contructing a curve of $\gamma_1 x_1$ against x_1 on the same diagram. This method is used by Null (1970).

7.4.3. The Effect of Temperature

For a given liquid system the temperature is the only property that affects mutual solubilities appreciably, although impurities sometimes play a minor role. Representative binary temperature-composition diagrams are shown in Figure 5.20. The maximum temperature at which two phases can exist is called the *upper critical solution temperature* (UCST). A *lower critical solution temperature* (LCST) is defined analogously.

The shapes of ternary diagrams often are extremely sensitive to temperature, as evidenced by Figures 5.34–5.39. Clearly, such sensitivity to temperature can be represented mathematically only if activity coefficients are known accurately in terms of temperature, which unfortunately is not the situation at present with respect to the usual activity-coefficient equations. At least part of the reason that VLE data-derived activity coefficients often do not represent LLE behavior well is that VLE usually are measured at relatively elevated temperatures, whereas LLE usually are of interest near room temperature. In connection with ternary mixtures, one way of getting around the difficulty is mentioned under *finding parameters* in Section 7.5.

Incorporation of temperature effects when they are known into phase diagrams is relatively straightforward. Several theoretical examples of binary systems with upper and lower critical solution temperatures are examined in Example 7.10. The best simple representation of the effect of temperature on equation parameters is a linear variation with reciprocal temperature.

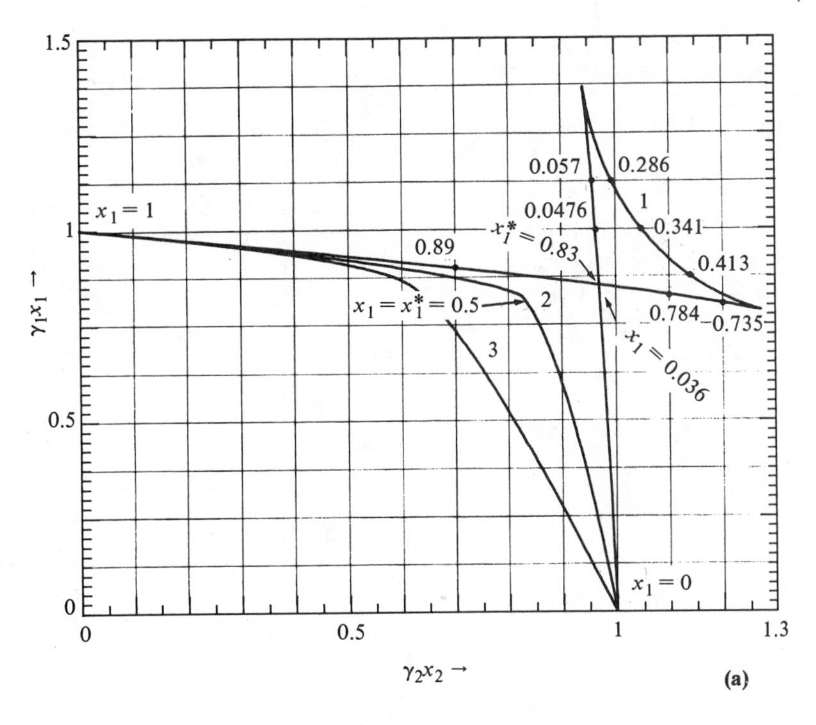

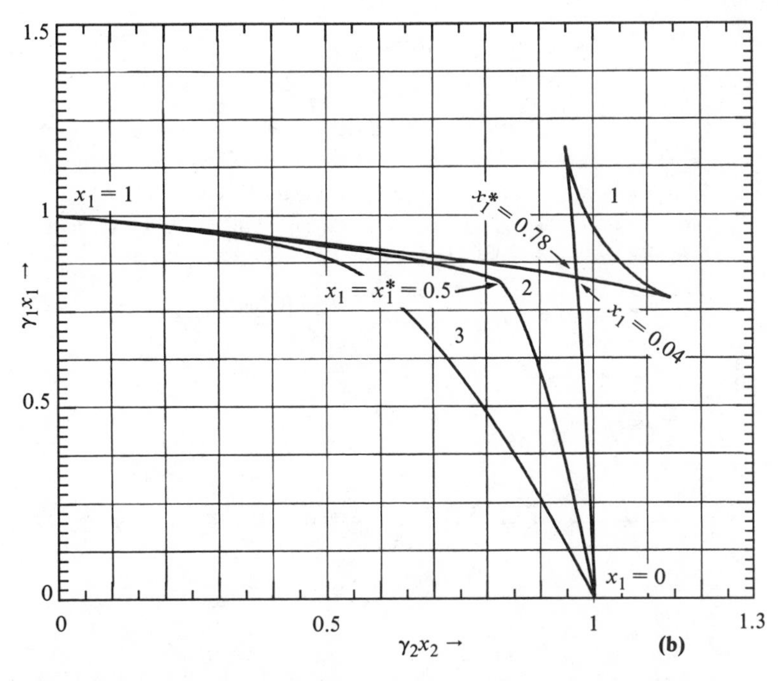

Figure 7.14(a). Looping behavior of plots of $\gamma_1 x_1$ against $\gamma_2 x_2$ with the Margules equation. Values of the parameters are (A, B): 1(3.5, 2), 2(2, 2), 3(1, 2). Both axes are plotted with x_1 as parameter, starting with $x_1 = 0$ at bottom right and ending with $x_1 = 1$ at top left. The two values of x_1 corresponding to the intersection are the compositions of phases in equilibrium. **(b)** Looping behavior of plots of $\gamma_1 x_1$ against $\gamma_2 x_2$ with the van Laar equation. Values of the parameters are (A, B): 1(3.5, 2), 2(2, 2), 3(1, 2). See legend of Figure 7.14(a) for the mode of construction and use, and Example 7.7(c) for avoidance of the need for calibrating these curves.

Example 7.7. Graphical Methods for Finding Binary Liquid-Liquid Equilibrium Compositions

The particular case is with the van Laar equation, with $A_{12} = 3$, $A_{21} = 2$ but the methods are applicable with any equation.

a. The points of double tangency on the plot of Gibbs energy of mixing give the equilibrium compositions as (0.08, 0.75).

b. On the plots of x_1 against $\gamma_1 x_1$ and $\gamma_2 x_2$, the rectangular construction locates the compositions at which $\gamma_i x_i = \gamma_i^* x_i^*$.

c. The horizontal projection from the intersection on the plot of $\gamma_1 x_1$ against $\gamma_2 x_2$ intersects the plot of $\gamma_1 x_1$ against x_1 at the equilibrium compositions.

d. The network chart of the van Laar equation, Figure 7.8, gives the same compositions (0.08, 0.75).

e. The numerical solution with the Newton-Raphson method as in Example 7.6 gives the values (0.07855, 0.75628).

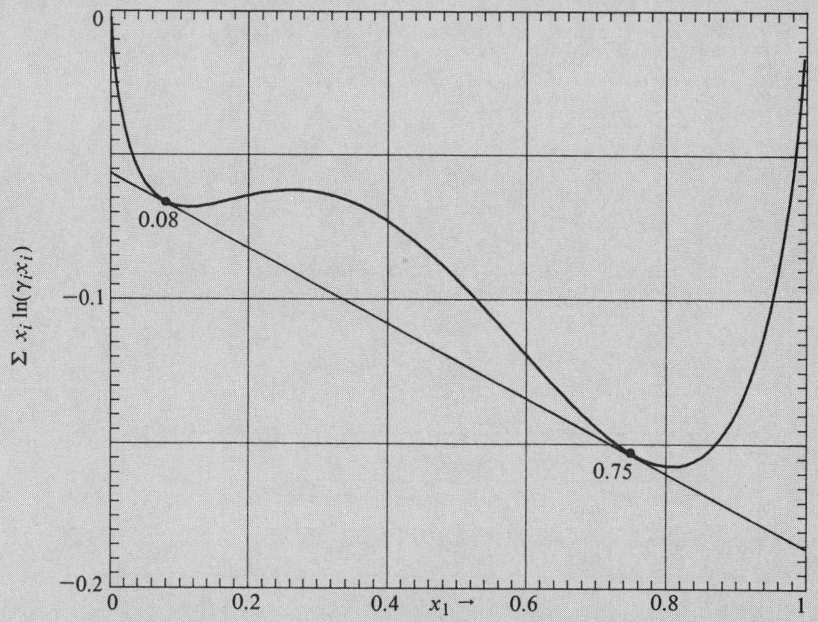

Example 7.7(a). Finding the compositions in liquid-liquid equilibria as the points of double tangency on a Gibbs energy plot.

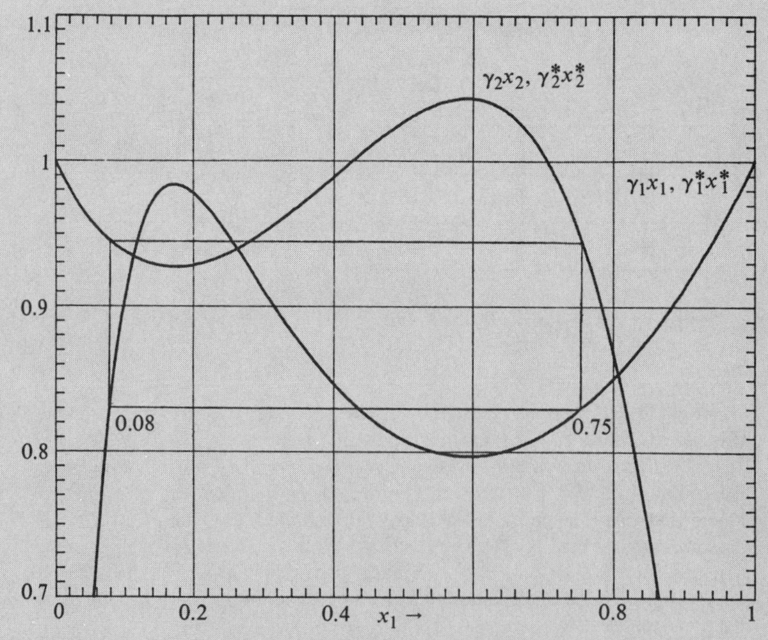

Example 7.7(b). Finding the compositions in liquid-liquid equilibria by rectangular construction on plots of $\gamma_i x_i$ and $\gamma_i^* x_i^*$ against x_1.

Example 7.7. *(continued)*

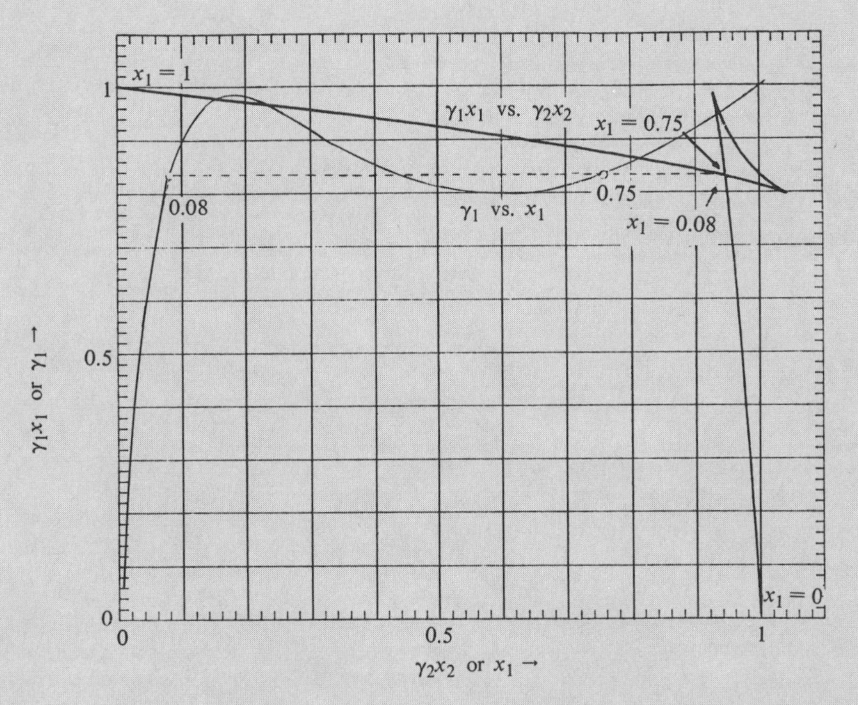

Example 7.7(c). The looped curve is drawn with x_1 as parameter, starting with $x_1 = 0$ at the bottom and ending with $x_1 = 1$ at the left. At just entering the loop the value of $x_1 = 0.08$ and at just leaving the loop $x_1 = 0.75$. Construction of the curve of the γ_1 vs. x_1 obviates the need for calibrating the looped curve.

For an interpretation of Figures 7.7(a) and (b), see Problem 7.27.

Example 7.8. Algebraic Determination of Equilibrium Compositions

The case of Example 7.7 will be solved by direct application of Eqns. 7.24 and 7.25. In terms of the definitions

$$D = 1.5/x_2 + 1/x_1$$

$$N = 1.5/x_2^2 - 1/x_1^2$$

the Gibbs energy function g and its derivative are

$$g = 3/D + x_1 \ln x_1 + x_2 \ln x_2$$

$$dg/dx_1 = -3N/D^2 + \ln(x_1/x_2)$$

Upon substitution into

$$\frac{g - g^*}{x_1 - x_1^*} = \frac{dg}{dx_1} = \frac{dg^*}{dx_1^*}$$

the equations to be solved are

$$\frac{(3/D + x_1 \ln x_1 + x_2 \ln x_2) - (3/D^* + x_1^* \ln x_1^* + x_2^* \ln x_2^*)}{x_1 - x_1^*}$$

$$= -3N/D^2 + \ln(x_1/x_2)$$

$$= -3N^*/D^{*2} + \ln(x_1^*/x_2^*)$$

Since good estimates of x_1 and x_1^* are available by the graphical solutions of the previous Example, these rather complex equations are readily solved by trial. The results are

$$x_1 = 0.0785$$

$$x_1^* = 0.7560$$

$$(g - g^*)/(x_1 - x_1^*) = -0.1296$$

$$dg/dx_1 = -0.1293$$

$$dg^*/dx_1^* = -0.1295$$

which check closely.

Example 7.9. Finding the Condition of Liquid-Liquid Equilibrium by Minimization of the Gibbs Energy

Example 7.7 shows that for a mixture with van Laar parameters $A = 3$ and $B = 2$, the mutual solubilities are $(x_1, x_1^*) = (0.08, 0.755)$. These values will be confirmed by minimization of the Gibbs energy of the mixture. Let

$$D = 1.5/x_2 + 1/x_1, \quad (1)$$

$$N = 1.5/x_2^2 - 1/x_1^2. \quad (2)$$

Then the excess Gibbs energy and its derivative are:

$$G^{ex}/RT = 3/D, \quad (3)$$

$$\frac{\partial G^{ex}/RT}{\partial x_1} = -3N/D^2.$$

The material balance is

$$z_1 = \alpha x_1 + (1-\alpha)x_1^*, \quad (4)$$

and

$$\alpha = (z_1 - x_1^*)/(x_1 - x_1^*) \quad (5)$$

is the fraction of the mixture in phase 1. The Gibbs function, $g = G/RT$, of the mixture is

$$g_t = \alpha g + (1-\alpha)g^* \quad (6)$$

$$= \alpha\left(\frac{3}{D} + \Sigma\, x_i \ln x_i\right) + (1-\alpha)\left(\frac{3}{D^*} + \Sigma x_i^* \ln x_i^*\right). \quad (7)$$

Since the material balance constraint on the minimum is of such simple form, Eq. 5, it is not necessary to employ Lagrange Multipliers. Instead, α is simply eliminated from Eq. 7. The condition for equilibrium is

$$\frac{\partial g_t}{\partial x_1} = \frac{\partial g_t^*}{\partial x_1^*} = 0. \quad (8)$$

These intermediate quantities will be needed:

$$\frac{\partial \alpha}{\partial x_1} = -\frac{(z_1 - x_1^*)}{(x_1 - x_1^*)^2}, \quad (9)$$

$$\frac{\partial \alpha}{\partial x_1^*} = \frac{z_1 - x_1}{(x_1 - x_1^*)^2}, \quad (10)$$

$$\frac{\partial g}{\partial x_1} = -\frac{3N}{D^2} + \ln[x_1/(1-x_1)], \quad (11)$$

$$\frac{\partial g^*}{\partial x_1^*} = \frac{3N^*}{D^{*2}} + \ln[x_1^*/(1-x_1^*)]. \quad (12)$$

The derivatives of the total Gibbs energy function are

$$\frac{\partial g_t}{\partial x_1} = \alpha\frac{\partial g}{\partial x_1} + (g - g^*)\frac{\partial \alpha}{\partial x_1} = 0, \quad (13)$$

$$\frac{\partial g_t}{\partial x_1^*} = (1-\alpha)\frac{\partial g^*}{\partial x_1^*} + (g - g^*)\frac{\partial \alpha}{\partial x_1^*} = 0. \quad (14)$$

The phase split α and the several derivatives on the right can be eliminated with appropriate equations in terms of x_1 and x_1^*. Those unknowns then can be found by simultaneous solution of Eqs. 13 and 14. However, the validity of the equilibrium compositions, $x_1 = 0.07855$ and $x_1^* = 0.75595$, can be verified by substitution into these equations and noting how nearly zero the RHSs are. The results should be independent of the overall composition, z_1. The tabulated values, 10,000 (RHS) show that the zero condition is essentially satisfied.

z_1	α	*Eq. 13*	*Eq. 14*
0.07855	1.000	−0.2	0.0
0.10000	0.968	−0.2	−0.0
0.20000	0.821	−0.2	−0.1
0.30000	0.673	−0.2	−0.1
0.40000	0.525	−0.1	−0.1
0.50000	0.378	−0.1	−0.2
0.60000	0.230	−0.1	−0.2
0.75595	0.000	0.0	−0.3

7.5. TERNARY AND MULTICOMPONENT MIXTURES

At fixed temperature and pressure, the maximum number of phases that can be present at equilibrium equals the number of components; in a practical sense this is the situation for condensed systems even at moderate variable pressures. Thus a ternary mixture can form one, two, or three phases, of which the three-phase types are by far the least common, although some examples are shown in Figures 5.38 and 5.39.

A stable system of six components in six liquid phases—mercury, gallium, phosphorus, aniline, water, and benzene—is cited as a curiosity by Bowden (1938). Shapes of the boundaries and appearances and disappearances of phases are markedly influenced by temperature, as evidenced by Figure 5.38, for instance. Solid and vapor phases also may be present in contact with multiple liquid phases, as in Figure 5.39, a kind of multiplicity that occurs often in cryogenic processing.

7.5.1. Finding Parameters

Since they are able to use binary data directly, the T-K-Wilson, UNIQUAC, and NRTL equations are the only ones of general value in representing multicomponent LLE behavior. Although the NRTL equation has nine parameters for a ternary mixture, it turns out that a universal value $\alpha_{ij} = 0.2$ is satisfactory, so all three of these equations can be regarded as having only six parameters at a given temperature for a ternary mixture. At present the effect of temperature is not taken into account in a systematic fashion, except in the ASOG system and by Renon et al. (1971), who represent NRTL parameters for a number of systems as linear functions of temperature.

In the absence of ternary experimental data, the three sets of binary parameters can be used as an approximate representation of the ternary system; but use of some ternary data in addition usually can improve the overall representation markedly. The marked differences between binary parameters and those determined from ternary data for the same system are indicated by the typical results of Example 7.12. Since there are six parameters to be evaluated, at least two sets of ternary tie-line data, or six equilibrium compositions, are needed for their evaluation; but a greater amount of data can be handled with regression techniques and is, of course, desirable. A combination of ternary LLE and binary VLE data also can be used. The objective function in such cases is made up of VLE and LLE terms:

$$\begin{aligned} OF = &\Sigma[(\Delta x_1)_i^2 + (\Delta x_2)_i^2] && \text{(binary system \#1)},\\ &+ \Sigma[(\Delta x_1)_i^2 + (\Delta x_2)_i^2] && \text{(binary system \#2)},\\ &+ \Sigma[(\Delta x_1)_i^2 + (\Delta x_2)_i^2 + (\Delta x_3)_i^2] && \text{(ternary system)}, \end{aligned} \quad (7.52)$$

where each difference is defined as

$$\Delta x = x_{\text{calculated}} - x_{\text{measured}}. \quad (7.53)$$

When binary parameters based on mutual solubilities are known, they may be assumed valid in the ternary mixture; in that case a single tie-line, together with VLE data on the remaining pairs, suffices to determine all the parameters. Many refinements of regression techniques are given in the literature. The DECHEMA LLE Data Collection evaluates the six parameters directly from ternary data in most cases; when the same pair occurs in several mixtures, they have been able to develop universal parameters for those pairs, which they designate *common.* For extended discussions of procedures for the evaluation of parameters, these papers may be consulted: Anderson & Prausnitz (1978); Fredenslund et al. (1980); Sorensen (thesis, 1980); Sorensen et al. (1979); Sorensen & Arlt (DECHEMA LLE Data Collection, 1980); and Varhegyi & Eon (1977).

7.5.2. Finding Compositions

The problem is the calculation of tie-lines and binodal curves when the parameters of some activity coefficient equation are known. Theoretically, all kinds of equilibria can be determined by finding the minimum Gibbs energy corresponding to a given overall composition, but special procedures have been worked out for particular cases such as two liquid phases, or three liquid phases, or two liquid and one vapor phase. These will be presented in turn.

a. Two phases: The fugacity equalities of Eqs. 7.4 and 7.5 are simply extended to all components. For each component, therefore,

$$\gamma_i x_i = \gamma_i^* x_i^*, \quad (7.54)$$

which is rearranged to

$$x_i^* = (\gamma_i/\gamma_i^*)x_i = K_i x_i, \quad (7.55)$$

where

$$K_i = \gamma_i/\gamma_i^* \quad (7.56)$$

is the distribution ratio. In terms of the overall composition, z_i, and the fraction, β, of the total material that is present in the first phase, the material balance is

$$z_i = \beta x_i + (1-\beta)x_i^* = [\beta + K_i(1-\beta)]x_i. \quad (7.57)$$

On summing these fractions the condition to be solved at equilibrium becomes

$$f(\beta) = \Sigma \frac{z_i}{\beta + K_i(1-\beta)} - 1 \to 0. \quad (7.58)$$

The calculation is performed in much the same way as a vapor-liquid flash with values of K_i dependent on the composition. A convenient set of steps is:

1. Estimate the composition, $x_1, x_2, \ldots,$ of the primary phase.
2. Estimate the phase split, β.
3. With the material balance calculate the composition of the other phase,

$$x_i^* = (z_i - \beta x_i)/(1-\beta). \quad (7.59)$$

4. Figure all the activity coefficients, γ_i and γ_i^*, and the distribution ratios, K_i.
5. By trial from Eq. 7.58, find the phase split, β.
6. Once β has been found, the individual terms of the summation are the x_i. With these new values of the composition, return to step 3 and continue until adequate convergence has been attained.

Even a two-phase, three-component calculation is not to be entered on lightly, since the possibility of convergence depends strongly on the starting estimates of the variables. A computer program using the T-K-Wilson equation for up to four components is given as Tables C.8 and C.9 in the Appendix. A program with the NRTL equation is given by Renon et al. (1971). Up to five components in two phases can be handled by the UNIQUAC program of Fredenslund et al. (1977), and twenty components can be handled with the UNIQUAC program of Prausnitz et al. (1980).

The objective of program C.8 and C.9 is to find compositions of phases in equilibrium. Perhaps the most significant difference between this kind of calculation and a vapor-liquid flash is that the solution cannot be started by assuming ideality, since there is no separation of liquids in that case. The overall composition also is arbitrary when finding tie-line data, although experience indicates that convergence is most rapid if the overall composition results in approximately equal amounts of the two phases, $\beta = 0.5$. These additional rules have been found helpful:

1. Identify the least miscible pair as consisting of components 1 and 3, and call them solvents; component 2 then is the solute.
2. As starting values, take: $x_1 = 0.90, x_2 = x_3 = 0.05$, and $x_1^* = 0.1$, although these values may be varied somewhat to improve convergence.
3. As a starting overall composition, put $z_1 = z_3 = 0.5$, $z_2 = 0$, or take some composition intermediate between the mutual solubilities of the two solvents.
4. Convergence of the Newton-Raphson routine may be improved by taking only a fraction of a calculated increment for the next trial; that is, take $h = \alpha h$, etc., where $0 < \alpha \leq 1$.

Example 7.11 compares several equations on a ternary mixture.

Example 7.10. Temperature Dependence of Mutual Solubilities of Binary Mixtures Having Symmetrical Activity Coefficient Equations

The activity coefficients are represented by

$$\ln \gamma_1 = A x_2^2 \tag{1}$$

$$\ln \gamma_2 = A x_1^2 \tag{2}$$

The plots of temperature against composition are necessarily symmetrical with critical solution temperatures at $x_1 = 0.5$. The temperature dependence of A can be established for parabolic and elliptic shapes of the temperature-composition plots. Schematics of such plots with upper and lower critical solution temperatures are shown as Figures a, b and c.

The relation between the parameter A and the composition is given by Eq. 7.19.

$$A = \frac{1}{1 - 2x_1} \ln(1/x_1 - 1). \tag{3}$$

The equation of a parabola with a lower CST is

$$T = T_L + a(x_1 - 0.5)^2, \tag{4}$$

and of one with an upper CST,

$$T = T_U - a(x_1 - 0.5)^2. \tag{5}$$

The equation of an ellipse with upper and lower critical solution temperatures is

$$\left(\frac{T - T_M}{T_M - T_L}\right)^2 + \left(\frac{x_1 - 0.5}{x_M - 0.5}\right)^2 = 1, \tag{6}$$

where $T_M = 0.5\,(T_L + T_U)$.

In Figure a the three kinds of temperature-composition plots corresponding to Eqs. 4, 5, and 6 are shown for these values of the constants: $T_U = 400$, $T_L = 300$, $a = 400$, and $X_M = 0.35$. Corresponding temperature dependencies of parameter A are shown in Figures b and c. The parameters of activity-coefficient equations often vary linearly with the reciprocal of the temperature.

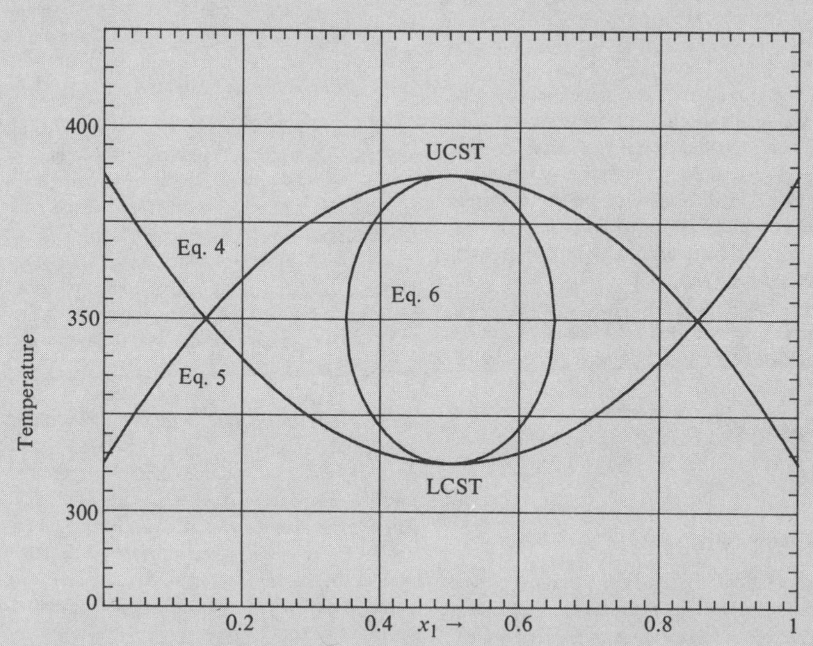

Example 7.10(a). Simulated plots of mutual solubilities of binary mixtures, showing systems with upper and lower critical solution temperatures, and a system with both UCST and LCST.

(Continued next page)

Example 7.10. *(continued)*

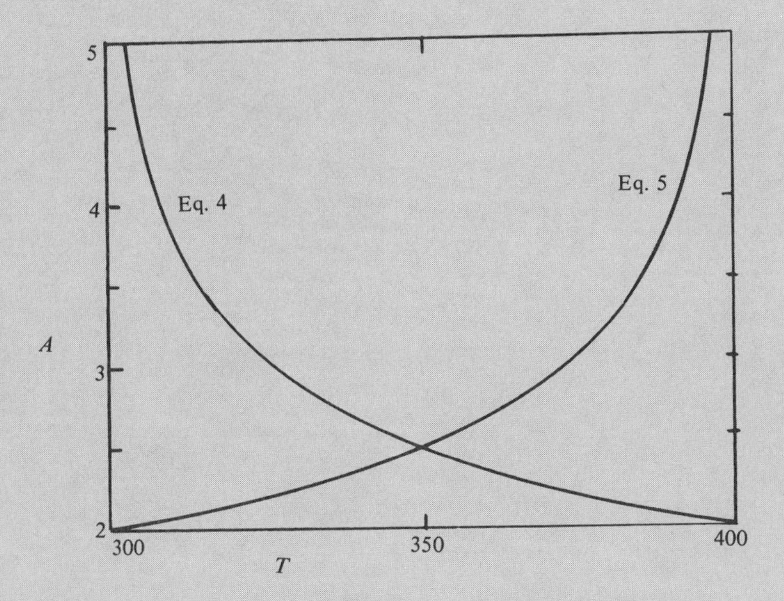

Example 7.10(b). Temperature dependence of A when the temperature-composition plots are parabolic.

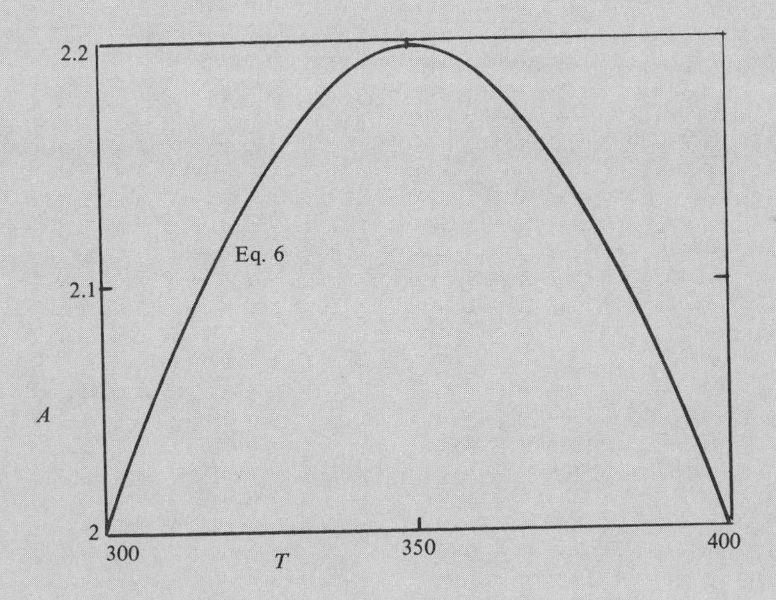

Example 7.10(c). Temperature dependence of A when the temperature-composition plot is elliptic.

Example 7.11. Liquid-Liquid Equilibria in the System of Water + Acetonitrile + Ethyl Acetate with the T-K-Wilson, NRTL, LEMF, and UNIQUAC Equations

Data are given for this system at 60 C by Sugi & Katayama (1978). One of the tie-lines will be checked by evaluatiing the objective function,

$$\Sigma[|\gamma_i x_i - \gamma_i^* x_i^*|/(\gamma_i x_i + \gamma_i^* x_i^*)],$$

with each of the four equations. The tie-line data are:

		x_i	x_i^*	V ml/gmol
1	water	0.9049	0.4733	18.07
2	acetonitrile	0.0800	0.2749	173.0
3	ethyl acetate	0.0151	0.2518	286.0

Equation Parameters:

	TKW	NRTL		LEMF	UNIQUAC			
	Λ_{ij} / Λ_{ji}	τ_{ij} / τ_{ji}	α	τ_{ij} / τ_{ji}	τ_{ij} / τ_{ji}	R	Q	
acetonitrile + water	0.4920	0.6750	0.30	0.5361	0.6146	0.92	1.40	(water)
	0.3532	1.8559		0.8222	0.6490			
ethyl acetate + acetonitrile	0.6326	0.1791	0.30	0.1755	0.5461	1.87	1.72	(ACN)
	1.2060	0.2391		0.1928	1.2834			
water + ethyl acetate	0.2802	4.3581	0.20	0.0621	0.6174	3.48	3.12	(EA)
	0.6039	0.1464		1.2593	0.4052			

Activity coefficients and the objective function are tabulated following. In this case, the NRTL equation with the given α and with $\alpha = -1$ (the LEMF equation) are definitely superior.

		γ_i	γ_i^*	Objective function
TKW	water	1.20034	4.1603	0.2890
	ACN	33.5505	1.5457	0.7253
	EA	488.196	2.2260	0.8587
				1.8730
NRTL	water	1.0225	2.0321	0.0193
	ACN	4.0178	1.3967	0.0887
	EA	31.421	1.9615	0.0201
				0.1281
LEMF	water	1.0424	1.9783	0.0037
	ACN	5.0790	1.3642	0.0401
	EA	20.931	1.4792	0.0819
				0.1257
UNIQUAC	water	7.2666	3.9606	0.5563
	ACN	66.080	3.8030	0.6698
	EA	2393.9	7.4611	0.90119
				2.1273

b. More than two liquid phases: A system of three phases with three components will be taken as an example. A specific case, with a Fortran program, is worked out by Null (1970). Let β_1 and β_2 be the fractions of two of the phases. The overall material balance of each component is

$$z_i = \beta_1 x_i + \beta_2 x_i^* + (1 - \beta_1 - \beta_2) x_i^{**}. \quad (7.60)$$

Rearrange the equilibrium relations, analogous to Eqs. 7.8 and 7.9,

$$x_i = \frac{\gamma_i^{**}}{\gamma_i} x_i^{**} = K_i x_i^{**} \quad (7.61)$$

$$x_i^* = \frac{\gamma_i^{**}}{\gamma_i^*} x_i^{**} = K_i^* x_i^{**} \quad (7.62)$$

The composition of the third phase then may be written,

$$x_i^{**} = \frac{z_i}{1 + (K_i - 1)\beta_1 + (K_i^* - 1)\beta_2}. \quad (7.63)$$

The requirement that the mol fractions sum to unity in each phase may be written in the equivalent fashion,

$$\psi_1 = \Sigma(x_i^{**} - x_i) = \Sigma\, x_i^{**}(1 - K_i) = 0, \quad (7.64)$$

$$\psi_2 = \Sigma(x_i^{**} - x_i^*) = \Sigma\, x_i^{**}(1 - K_i^*) = 0. \quad (7.65)$$

The calculation can proceed as follows:

1. Make trial assumptions β_1 and β_2 of the phase splits and the compositions x_i and x_i^*.
2. Obtain the composition of the third phase from the material balance, Eq. 7.60.
3. Find all the activity coefficients from the appropriate correlations.
4. Evaluate x_i^{**} from Eq. 7.63, then x_i and x_i^* from the equilibrium relations, Eqs. 7.61, 7.62.
5. Find the values of ψ_1 and ψ_2 with Eqs. 7.64, 7.65. If they are not close enough to zero,
6. Find corrections $\Delta\beta_1$ and $\Delta\beta_2$ to the phase splits by the N-R method, using the linear relations,

$$\psi_1 + \frac{\partial\psi_1}{\partial\beta_1}\Delta\beta_1 + \frac{\partial\psi_1}{\partial\beta_2}\Delta\beta_2 = 0, \quad (7.66)$$

$$\psi_2 + \frac{\partial\psi_2}{\partial\beta_1}\Delta\beta_1 + \frac{\partial\psi_2}{\partial\beta_2}\Delta\beta_2 = 0. \quad (7.67)$$

7. Use approximate values of the derivatives made on the assumption that the compositions in Eqs. 7.64, 7.65 depend to a first approximation only on β_1 and β_2. Accordingly, these derivatives are:

$$\frac{\partial\psi_1}{\partial\beta_1} = \Sigma \frac{(K_i - 1)^2 z_i}{D_i^2}, \quad (7.68)$$

$$\frac{\partial\psi_1}{\partial\beta_2} = \frac{\partial\psi_2}{\partial\beta_1} = \Sigma \frac{(K_i - 1)(K_i^* - 1) z_i}{D_i^2}, \quad (7.69)$$

$$\frac{\partial\psi_2}{\partial\beta_2} = \Sigma \frac{(K_i^* - 1)^2 z_i}{D_i^2}, \quad (7.70)$$

where

$$D_i = 1 + (K_i - 1)\beta_1 + (K_i^* - 1)\beta_2. \quad (7.71)$$

This calculation method is not very positive, in view of the six assumptions that must be made in step 1, and convergence often is very slow. Cases of three liquid phases are not overly frequent, although Example 7.12 is one instance.

A quaternary mixture with two phases is analyzed by the UNIQUAC method in Example 7.13.

7.5.3. Metastable and Stable Equilibria

A state of supersaturation sometimes is established and can persist indefinitely in the absence of nuclei on which bubbles or droplets or crystals can grow. It is pointed out by Modell & Reid (Chapter 13, 1983), for example, that small nuclei have higher vapor pressures or chemical potentials than large masses and hence are stable only in contact with supersaturated phases.

States of metastable equilibrium can be described thermodynamically. It has been indicated that true or absolute stability is characterized by equality of chemical potentials of individual components of mixtures in all phases. For binary mixtures the condition of metastable equilibrium is characterized thermodynamically by the statement $(\partial^2 G/\partial x_1^2)_{TP} > 0$. For example, a metastable region of a binary mixture is identified in Figure 7.2. Problem 7.26 also is related to this topic.

For multicomponent mixtures the mathematical description of metastability in terms of derivatives of G with respect to mol fractions is more complex and reference for details must be made to other sources, for instance the books of Lupis (Chapter 11, 1983) and Modell & Reid (Chapter 9, 1983). With the relations found there, Example 7.14 develops the metastable region of a ternary mixture and for comparison the stable region also.

7.5.4. Literature of Calculations of Liquid-Liquid Equilibria

Many examples of calculations and correlations of ternary and some quaternary LLE have been published. The *DECHEMA Collection of LLE Data* (1979, 1980) is the most comprehensive compilation; it gives NRTL and UNIQUAC parameters for more than 1,000 data sets. Table 7.3 is a sample. The background of this work is discussed by Sorensen (thesis, 1979) and by Sorensen et al. (1979). A few examples are given by Prausnitz et al. (1980). Predictions from binary data by several methods are made by Tsuboka & Katayama (1975) and Sugi & Katayama (1978). The books of Renon et al. (1971) and Fredenslund et al. (1977) give a number of examples.

Binary parameters of the NRTL and UNIQUAC equations may be used to predict multicomponent LLE, but the results are seldom of high accuracy. When not even binary data are available, the parameters of the UNIQUAC equation can be obtained from the UNIFAC method of group contributions. A special table of parameters based on liquid-liquid data has been prepared by Magnussen et al. (1981); it is given here as Table E.13. These authors made a study of the accuracy of predictions by UNIQUAC and UNIFAC on seventeen LLE systems. Mean absolute deviations, d mol%, were evaluated. With all six parameters of a ternary system fitted to the data, $d = 0.5$; with binary parameters obtained from VLE, $d = 3.7$;

Example 7.12. The UNIQUAC Parameters of a Ternary System that Forms Three Liquid Phases

The system is 1-hexanol (1) + nitromethane (2) + water (3) for which the data are shown in DECHEMA V/2 (p. 69). UNIQUAC parameters of the three pairs are known. Fifteen experimental tie-lines are available, from which the six ternary parameters are determined directly. A two-stage procedure, recommended by Sorensen et al. (1979, pp. 60ff), is applied. Starting estimates are the binary parameters. In the first stage the objective function is

$$(\mathrm{OJ})_1 = \Sigma\Sigma[(x^{*}_{ik}\gamma^{*}_{ik} - x^{**}_{ik}\gamma^{**}_{ik})^2 + (x^{*}_{ik}\gamma^{*}_{ik} - x^{***}_{ik}\gamma^{***}_{ik})^2 + (x^{**}_{ik}\gamma^{**}_{ik} - x^{***}_{ik}\gamma^{***}_{ik})^2] \rightarrow \text{Minimum.}$$

Meeting this criterion provides improved estimates of the parameters. In the second stage, the objective function minimizes the differences between measured and calculated compositions of all the tie-lines.

$$(\mathrm{OJ})_2 = \Sigma\Sigma\Sigma[(x_{ijk})_{\mathrm{exp}} - (x_{ijk})_{\mathrm{calc}}]^2 \rightarrow \text{Minimum.}$$

i refers to the component.
j refers to the phase.
k refers to the tie-line data.

A simplex procedure for the three-phase case was devised by S. Negahban. The three sets of UNIQUAC parameters, $A_{ij} = (u_{ij} - u_{jj})/R$ K, are:

	Binary sets		*(OJ)₁*		*(OJ)₂*	
ij	A_{ij}	A_{ji}	A_{ij}	A_{ji}	A_{ij}	A_{ji}
12	302.15	59.312	359.80	26.843	331.76	468.898
13	109.38	276.54	81.979	321.77	64.177	360.85
23	421.70	229.72	403.03	198.85	499.65	159.11

The ternary plot is made with the last set of parameters; they satisfy most nearly the measured phase compositions. The points are experimental; the lines are the data fit. The shaded region is three-phase. The Roman numerals identify the number of phases in the particular region.

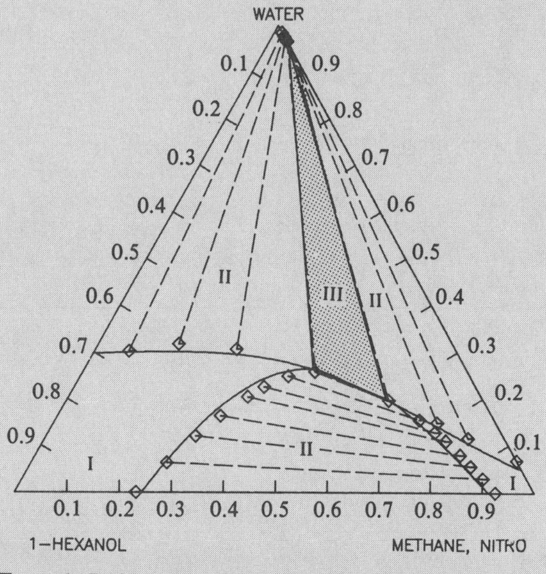

Example 7.12. A ternary diagram with a three-phase region constructed with UNIQUAC parameters (S. Negahban).

Example 7.13. A Quaternary System of Two Phases

Two experimental tie-line data and all binary interaction parameters are known for the quaternary system identified in the following tables. The data are regressed subject to the objective functions used in Example 7.12. The original binary parameters and the regressed set of twelve parameters are used to predict the tie-line data. The four ternary diagrams were drawn on the basis of the regressed parameters. All calculations were made with a UNIQUAC program developed by S. Negahban. A(I,J) is defined in Example 7.12.

Properties of the components:

	R	Q
1. Heptane	$R = 5.1742$	$Q = 4.396$
2. Benzene	3.1828	2.400
3. 2-amino ethanol	2.5736	2.360
4. 1-methyl, 2-pyrrolidone	3.9810	3.200

The original and finally regressed interaction parameters, °K:

		Original		*Final*	
I	J	A(I.J) KELVIN	A(J,I) KELVIN	A(I,J) KELVIN	A(J,I) KELVIN
1	2	−108.50	−120.28	−105.46	−115.85
1	3	344.41	132.92	527.51	179.76
1	4	234.59	17.402	298.39	−41.402
2	3	447.87	29.002	554.16	2.7937
2	4	−191.05	−98.426	−135.53	−371.20
3	4	15.163	−145.01	−18.649	−178.21

(Continued next page)

Example 7.13. *(continued)*

Experimental and calculated tie-line data:

Experimental data

LEFT PHASE				RIGHT PHASE			
(1)	(2)	(3)	(4)	(1)	(2)	(3)	(4)
70.870	24.240	0.290	4.600	2.190	11.560	54.420	31.830
49.590	40.200	1.650	8.560	2.930	19.700	50.520	26.850

Calculated from original binary parameters

LEFT PHASE				RIGHT PHASE			
(1)	(2)	(3)	(4)	(1)	(2)	(3)	(4)
70.870	22.113	2.510	4.507	2.421	4.509	61.239	31.830
49.590	40.121	3.247	7.042	2.021	10.257	60.872	26.850

Calculated from finally regressed set of 12 parameters

LEFT PHASE				RIGHT PHASE			
(1)	(2)	(3)	(4)	(1)	(2)	(3)	(4)
70.870	23.956	0.702	4.473	2.455	11.491	54.224	31.830
49.590	40.288	1.628	8.493	2.547	19.876	50.727	26.850

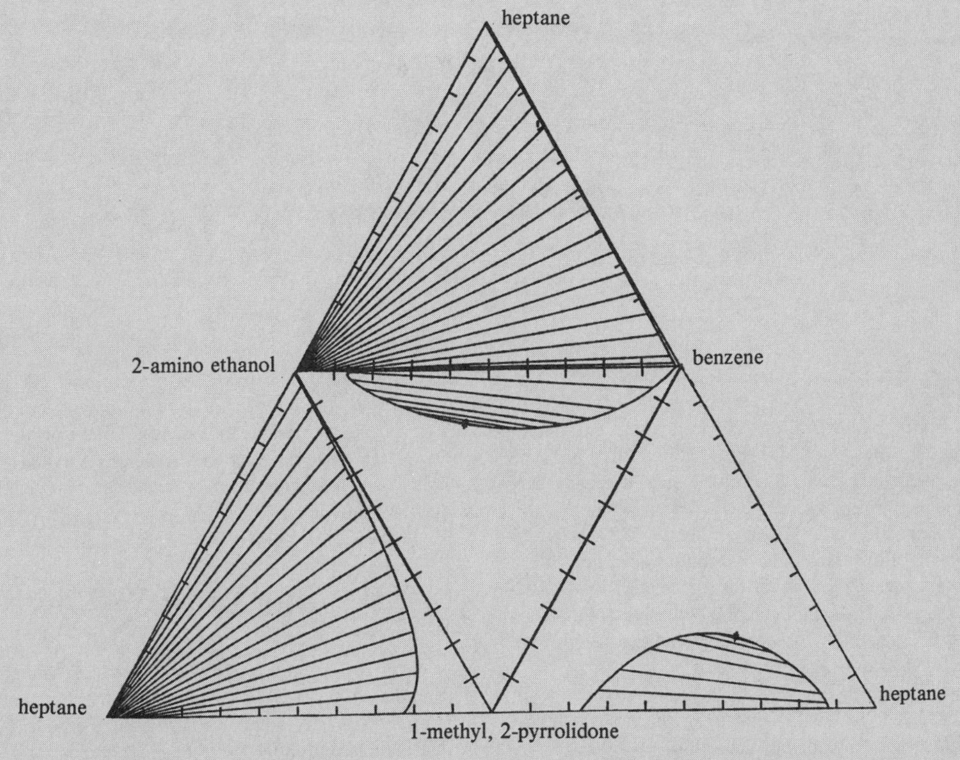

Example 7.13. The four ternary diagrams of the quaternary system. To construct the tetrahedron, fold along the edges of the inner triangle and join the three vertices labelled "heptane." Experimental data of the four ternary systems may be found in the *DECHEMA LLE Data Collection,* V/2, pp. 438, 439, 440; V/3, p. 241.

Example 7.14. Limits of Phase Metastability of a Ternary Mixture

The thermodynamic condition for a ternary mixture to be metastable is shown by Lupis (p. 307, 1983) to be

$$\psi = \left(1 + X_2 \frac{\partial \ln \gamma_2}{\partial X_2}\right)\left(1 + X_3 \frac{\partial \ln \gamma_3}{\partial X_3}\right) - X_2 X_3 \frac{\partial \ln \gamma_2}{\partial X_3}\frac{\partial \ln \gamma_3}{\partial X_2} > 0$$

When $\psi = 0$ the limit of metastability is reached; on the accompanying figure it is denoted by the binodal curve bounding the darkened area. Within that area separation into two phases will be instantaneous. The outer binodal curve is the limit of true stability. Between the two binodal curves phase separation may or may not occur immediately depending on the presence or absence of nuclei.

The binodal curves of this system of methanol + morpholine + heptane were constructed with the aid of UNIQUAC parameters from DECHEMA LLE Data Collection, volume V/2, 96, 1980).

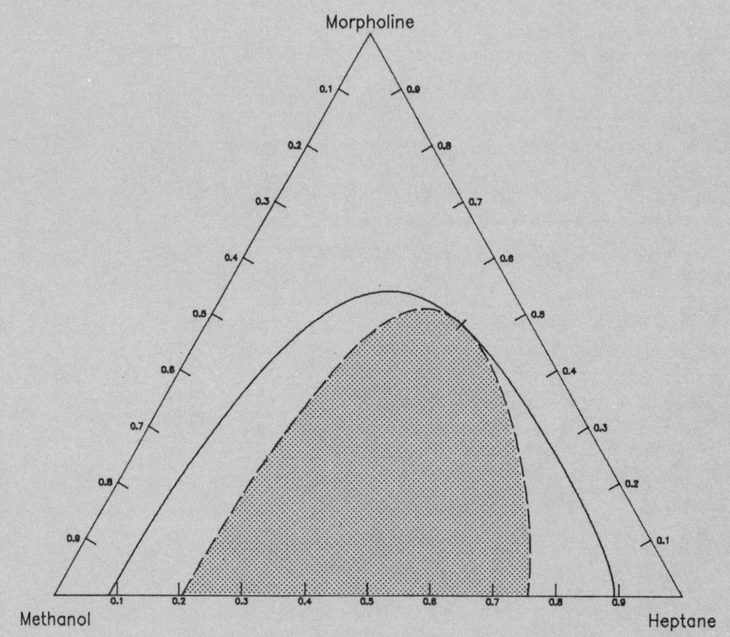

Example 7.14. Metastable and stable conditions of the system methanol + morpholine + heptane (S. Negahban).

with UNIFAC parameters based on VLE data, $d = 9.2\%$; and with the new UNIFAC parameters based on LLE, $d = 1.7\%$. For many applications the predictions with binary parameters and the UNIFAC correlations can only be regarded as semiquantitative. These studies of accuracy may be compared with those of some aqueous ternary vapor-liquid equilibrium data of the DECHEMA collection: 87 sets have a mean deviation of 1.8 mol% and a standard deviation of 1.2 mol%; these are cases for which the parameters were fitted directly to the data. It should be noted, however, that aqueous systems usually are difficult to correlate, and that most of the recorded LLE data are of aqueous mixtures.

Extraction processes are next to distillation processes in their applicability to separating mixtures. Recently the literature of LLE has been growing substantially, although one problem that has not been addressed sufficiently is the effect of temperature; the temperature levels of distillation and extraction processes, both industrially and in laboratory investigations, are much different, those of LLE being low and of VLE being high. Much has been discovered recently, however, and predictability of LLE is in a much better state than with earlier vague classifications based on chemical type and tendency to form hydrogen bonds. It is now possible to decide by calculation whether particular systems are worth investigating experimentally.

7.6. PREDICTION OF LLE WITH UNIFAC OR ASOG

Most of the results found so far show that LLE data are not as well predicted as VLE data with activity coefficients figured from the two available group contribution methods, but work in this area is currently active and improvements are to be expected. The book by Fredenslund et al. (1977) examines fifty ternary systems with UNIFAC. Although a phase split was overlooked in only one of those systems, the accuracy of prediction of tie-lines and binodal curves was somewhat

Table 7.3. Sample Page from Sorensen & Arlt, DECHEMA Liquid-Liquid Equilibrium Data Collection, vol 5, part 2 (1980).

```
(1) C7H8        TOLUENE
---------------------------------------------------------
(2) C3H6O       2-PROPANONE
---------------------------------------------------------
(3) H2O         WATER
---------------------------------------------------------

HACKL A., SOLAR W., ZIEBLAND G.
EUR.FED.CHEM.ENG.,RECOMM.SYST.LIQ.EXTR.STUD.,EDITOR:T.MISEK
(1978)

TEMPERATURE =  10.0 DEG C      TYPE OF SYSTEM = 1
```

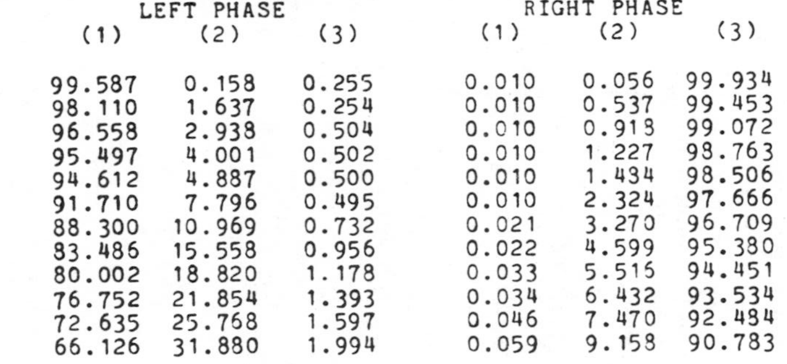

EXPERIMENTAL TIE LINES IN MOLE PCT (GRAPH.INTERPOL.)

LEFT PHASE (1)	(2)	(3)	RIGHT PHASE (1)	(2)	(3)
99.587	0.158	0.255	0.010	0.056	99.934
98.110	1.637	0.254	0.010	0.537	99.453
96.558	2.938	0.504	0.010	0.918	99.072
95.497	4.001	0.502	0.010	1.227	98.763
94.612	4.887	0.500	0.010	1.434	98.506
91.710	7.796	0.495	0.010	2.324	97.666
88.300	10.969	0.732	0.021	3.270	96.709
83.486	15.558	0.956	0.022	4.599	95.380
80.002	18.820	1.178	0.033	5.515	94.451
76.752	21.854	1.393	0.034	6.432	93.534
72.635	25.768	1.597	0.046	7.470	92.484
66.126	31.880	1.994	0.059	9.158	90.783

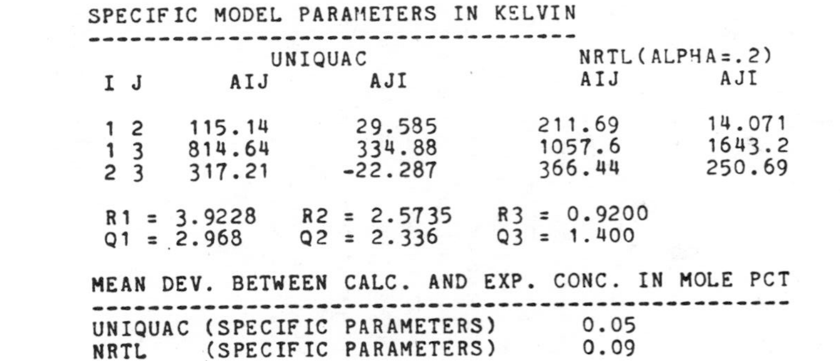

SPECIFIC MODEL PARAMETERS IN KELVIN

I J	UNIQUAC AIJ	AJI	NRTL(ALPHA=.2) AIJ	AJI
1 2	115.14	29.585	211.69	14.071
1 3	814.64	334.88	1057.6	1643.2
2 3	317.21	-22.287	366.44	250.69

```
R1 = 3.9228     R2 = 2.5735     R3 = 0.9200
Q1 = 2.968      Q2 = 2.336      Q3 = 1.400
```

MEAN DEV. BETWEEN CALC. AND EXP. CONC. IN MOLE PCT

UNIQUAC	(SPECIFIC PARAMETERS)	0.05
NRTL	(SPECIFIC PARAMETERS)	0.09

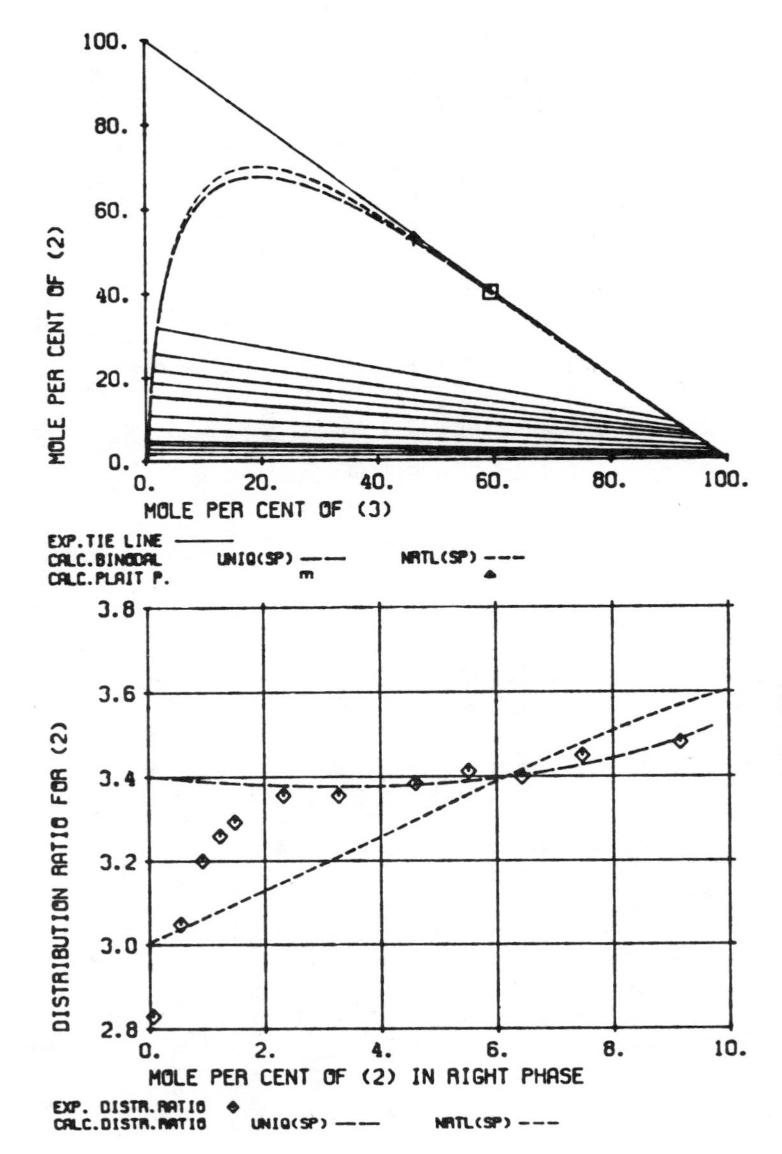

variable. The new set of UNIFAC parameters based on LLE measurements prepared by Magnussen et al. (1981) was cited in the previous section. Further details of this study are given in another paper by Magnussen et al. (1980). These persons all are associated with the *DECHEMA LLE Data Collection.*

Somewhat less attention is being paid to ASOG calculation of LLE. In an examination of thirty-one ternaries, Tochigi et al. (1980) found accurate prediction of tie-lines in only six cases, accurate binodal curves in twenty-four cases, and the occurrence of phase splitting in all cases. The ASOG system has the advantage of including a strong dependence of the parameters on the temperature, a factor that UNIFAC investigators are considering incorporating in their system. Nevertheless, on the basis of published studies, the ASOG predictions can be regarded as only qualitative while those by UNIFAC are at least semiquantitative. Including a proper temperature dependence of the parameters also is regarded an important step forward in developing parameters applicable to both LLE and VLE. From the dates of the papers cited here, it is clear that work in this area is current, much of it in association with the DECHEMA LLE and VLE data collections. The number of systems tested with UNIFAC or ASOG in the literature, however, is a tiny fraction of the 2,000 or so data sets in that collection.

7.7. SOLUBILITY OF LIQUIDS

The distinction between solute and solvent is fairly clear for solid solutes and also for gaseous ones when the solvent is substantially nonvolatile under the conditions of interest, but may be somewhat arbitrary in the case of liquid mixtures where the term *solute* is applied to the component that is present in relatively small concentration, even though mutual solubilities of limited extent may exist. Thus it is convenient to speak of the solubilities of hydrocarbons in water and of water in hydrocarbons as though no mutual solubilities existed.

Mutual solubilities of liquids are, of course, the main topic of this chapter and have been related to activity coefficients and their correlations. For instance, Example 7.10 is a theoretical study of the temperature dependence of solubility with the symmetrical activity-coefficient equation.

In the present section, however, some attention will be devoted to cases in which the mutual solubilities are low and the solute may be regarded as remaining substantially pure. Such one-sided solubility is represented by somewhat simplified mathematics since the fugacity of the solute remains constant and both its mol fraction and activity coefficient in the solid phase remain unity. Therefore, the condition of equality of fugacities may be written

$$\gamma_2 x_2 = \gamma_2^* x_2^* = 1,$$

where the subscript identifies the solute and the asterisk the pure solute phase. In terms of the Scatchard-Hildebrand equation, the solubility may be written

$$x_2 = 1/\gamma_2 = \exp\left(-\frac{V_2}{RT}(\delta_1-\delta_2)^2\phi_1^2\right) \tag{7.72}$$

$$= \left[\exp\left(-\frac{V_2}{RT}(\delta_1-\delta_2)^2\right)\right]^{\phi_1^2}. \tag{7.73}$$

The effect of the volume fraction, ϕ_1, is appreciable when the volume ratio, V_2/V_1, is large, as shown in Figure 7.15.

One important type of one-sided solubility behavior is that of liquid polymers that dissolve ordinary liquids but are not themselves soluble. Interactions of water and hydrocarbons also can be regarded as one-sided solubilities. Much work has been done on this topic, of which an extensive bibliography is included in the *API Technical Data Book* (1970, Chap. 9). Solubilities of water in hydrocarbons usually can be predicted within an order of magnitude (Example 7.15), but the converse estimates may be off by several orders of magnitude, although the numbers in both cases are small. Usually rather special forms of equations are fitted to solubility data. For example, API equations are, for water in hydrocarbons,

$$\log x_w = -(4200H/C + 1050)(1.8/T^0R - 0.0016) \tag{7.74}$$

and for hydrocarbons in water,

$$-\log x_{hc} = a + bC^2 \tag{7.75}$$

where H/C is the weight ratio of hydrogen to carbon and C is carbon number of the hydrocarbon. Constants a and b are specific to different classes of hydrocarbons.

Much effort has been expended on relating solubilities to solubility parameters with limited success. Figs. 7.16(a) and (b) are two such correlations. Sometimes the correlation can be improved by adjusting the solubility parameters to fit solubility data, as in Example 6.8, but no generalizations have been developed.

With regard to the usually limited solubilities of organic substances in water, empirical correlations have been expressed in terms of the ratios of their solubilities in octanol and in water, which is called the *octanol/water partition coefficient.* That coefficient is obtainable from molecular group contributions. Equations and many data from this point of view are presented in the book of Lyman et al. (1982).

Sources of data: In this section a sketchy effort has been made to represent solubility behavior in terms of a simple theory, with results ranging from mediocre to bad. No universally valid theory appears to have been developed, so it is necessary to depend on empirical equations to represent the effects of temperature, pressure, and composition on solubilities. The reasons for deviations of solubility behavior from simple theory are similar to those influencing *PVT* behavior—namely, varying degrees of intermolecular attractions of solutes and solvents. When the important topics of polymer-solvent interactions and surfactant behavior are taken into account, the theory appears even less adequate. A great abundance of solubility data has been published, and several extensive compilations have been made recently or are under way—namely:

1. Landolt-Börnstein, II/2b, c (1975).
2. Freier, *Aqueous Solutions: Data for Inorganic and Organic Compounds* (1975ff).
3. Stephen, Stephen, & Silcock, *Solubilities of Inorganic and Organic Compounds* (7 parts, 1979).
4. *IUPAC Solubility Data Series* (1979ff): The intention is to collect all published data; by 1980 about twenty volumes had appeared.

7.8. VAPOR-LIQUID-LIQUID EQUILIBRIA

At constant values of either T or P, the maximum possible number of phases equals the number of components, but fewer

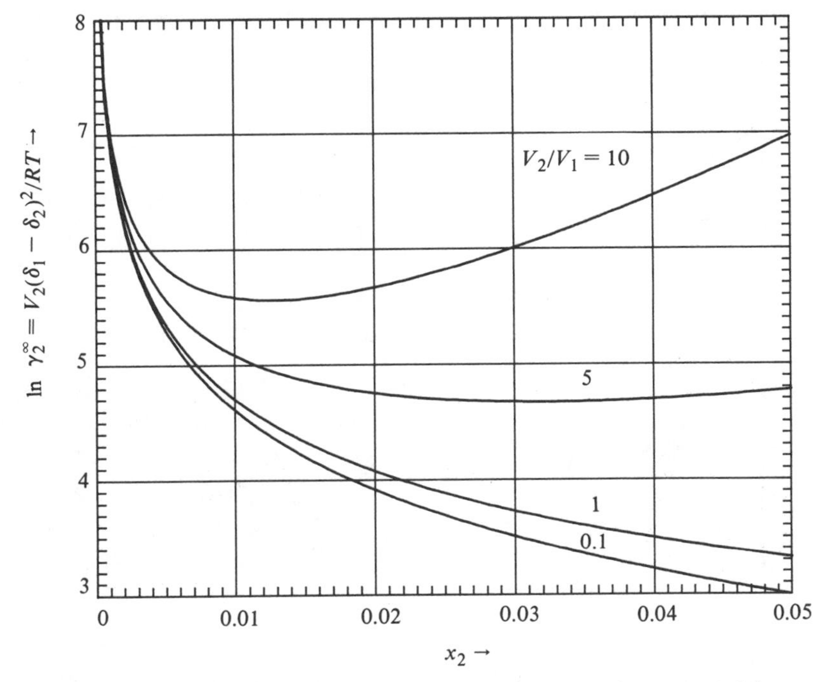

Figure 7.15. Solubility, as mol fraction x_2, in a system represented by the Scatchard-Hildebrand equation. The solute is assumed to remain pure. $\ln \gamma_2^\infty = V_2(\delta_1 - \delta_2)^2/RT$ is the ordinate. See Eq. 7.72.

often exist. The problem is to find the compositions of all phases that may be in equilibrium at a certain overall composition and a specified T or P or H or S. Particular cases may be:

1. Bubblepoint at specified T or P.
2. Dewpoint at specified T or P.
3. Flash at specified T and P.
4. Flash at specified H and either T or P.
5. Flash at specified S and either T or P.

As with problems in LLE or VLE, two distinct methods are usable for solving those of VLLE:

1. Solving directly the system of equations representing material balances and equilibria between phases.
2. Finding the minimum Gibbs energy of the overall mixture. The variables are the amounts and compositions of all the phases and the minimum is a constrained one, subject to the mol fractions summing to unity in each phase.

7.8.1. Binary Mixtures

When only two substances are present, the combination of material balances and equilibria is easy to apply. An iso-

Example 7.15. Mutual Solubilities of Water and Some Hydrocarbons

Solubilities figured from Eq. 7.72 are compared with the the smoothed data of the *API Technical Data Book* (1970, Chap. 9). The adjusted values of the solubility parameter of water required to match the API data are marked δ^*: they compare with the true value 23.53.

	V	δ
hexane	131.6	7.27
benzene	89.4	9.16
water	18.1	23.53

The solubilities at 300 K are:

	Calculated	*API*	δ^*
water in hexane	0.00032	0.00070	13.0
water in benzene	0.0019	0.0033	15.3
hexane in water	4(E-26)	1(E-6)	15.2
benzene in water	4(E-14)	4(E-4)	16.4

There does not seem to be a universal adjusted value of the solubility parameter of water to fit these data. A more extended study might be worthwhile.

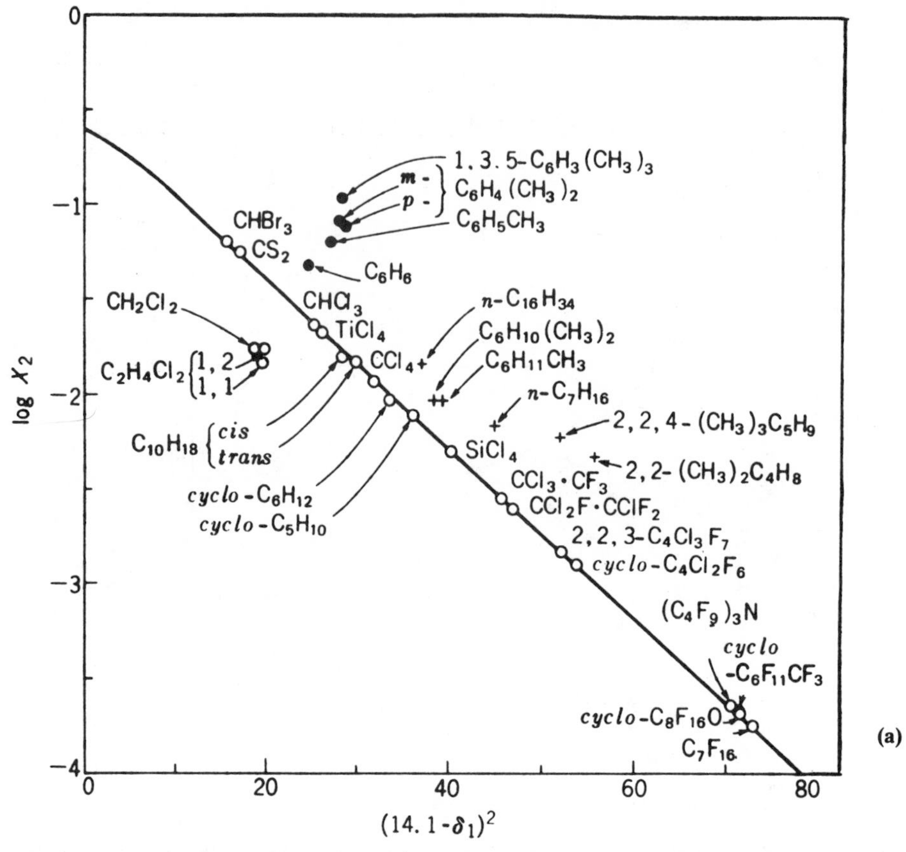

Figure 7.16(a). The solubility of iodine in solvents of various solubility parameters. The solubility parameter of iodine is 14.1, which appears in the abscissa (Shinoda, 1978).

thermal case may be taken for illustration, such as that of Example 7.16. Nomenclature is shown in Figure 7.17. The first step is to figure the mutual miscibilities, as in Section 7.4.2. Then the equilibrium vapor composition is calculable from

$$y_i = \frac{\gamma_i \phi_i^{\text{sat}} P_i^{\text{sat}}}{\hat{\phi}_i P} x_i = K_i x_i \tag{7.76}$$

$$= \frac{\gamma_i^* \phi_i^{\text{sat}} P_i^{\text{sat}}}{\hat{\phi}_i P} x_i^* = K_i^* x_i^*. \tag{7.77}$$

The $P-x_1$ diagram is established by a series of bubblepoint calculations using

$$P = \gamma_1 x_1 P_1^{\text{sat}} + \gamma_2 x_2 P_2^{\text{sat}}. \tag{7.78}$$

The $P-y_1$ relation then is found with the aid of the formula,

$$y_1 = \gamma_1 x_1 P_1^{\text{sat}}/P. \tag{7.79}$$

At constant pressure, the variation of activity coefficients with temperature must be known, or it may be assumed that the built-in temperature compensations of the NRTL or UNIQUAC or TKW equations are adequate. Vapor-pressure dependence on temperature also must be known, for example by the Antoine equation. A suitable procedure for isobaric conditions is:

1. Find the mutual solubilities over a range of temperatures.
2. Find the bubblepoint pressure at each of these temperatures from

$$P = \Sigma \gamma_i x_i P_i^{\text{sat}}, \tag{7.80}$$

and identify the temperature at which the specified pressure is attained. This temperature is the lowest at which a vapor phase can exist at the given pressure.

3. Above minimum temperature and at compositions in the miscibility ranges, the temperature corresponding to a given composition is the bubblepoint, and is found with Eq. 7.78; since the activity coefficients and vapor pressures are functions of temperature, a solution by trial is called for at each liquid composition. The corresponding values of the vapor composition are found with Eq. 7.79. The two branches of the $P-y_1$ curves come together at the minimum temperature.

Another way of finding the equilibrium condition is to take advantage of the fact that the $\overline{G}_i$ are the same for a particular component in all phases. In accordance with the method of tangent intercepts, therefore, equilibrium compositions are

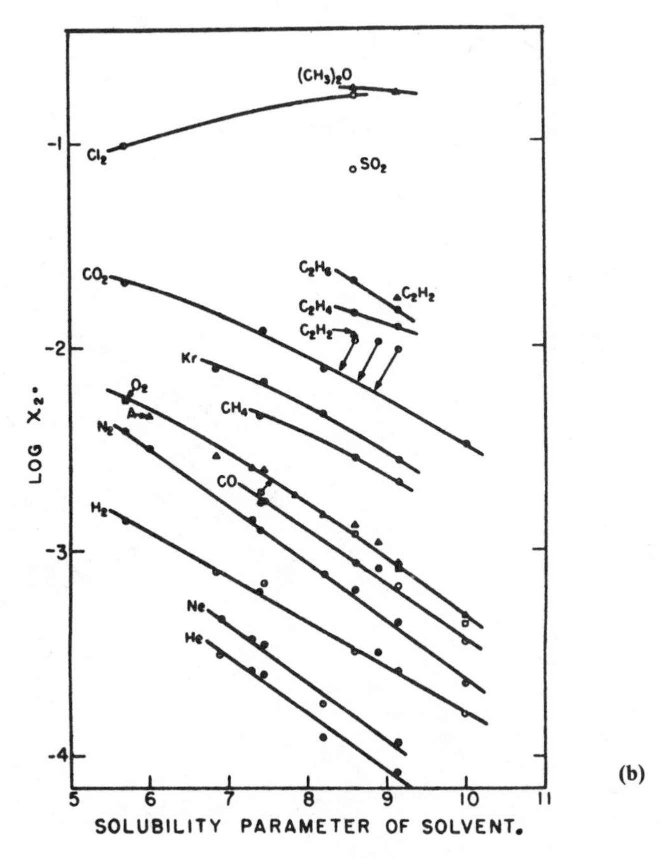

Figure 7.16(b). Solubilities of gases as a function of solubility parameter, δ_1, of the solvent, at 25 C and 1 atm. (Jolley & Hildebrand, JACS 80, 1051 (1958)). Copyright American Chemical Society; used with permission.

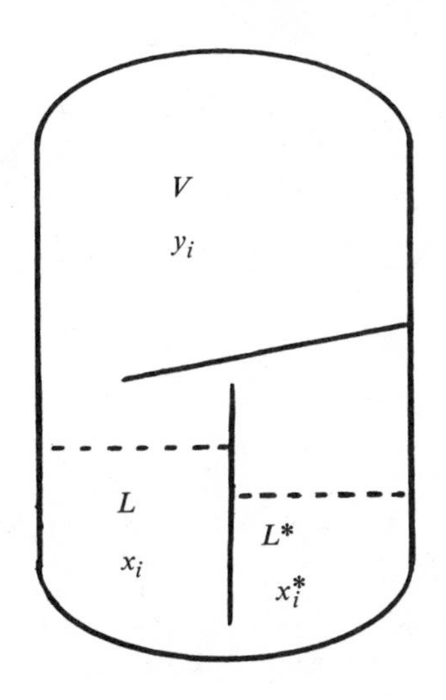

Figure 7.17. Nomenclature for three-phase equilibria:

$$F = V + L + L^*$$
$$\alpha = V/F$$
$$\beta = L/(L + L^*)$$
$$L = (1 - \alpha)\beta F$$
$$L^* = (1 - \alpha)(1 - \beta)F$$
$$V = \alpha F$$

points of common tangency of plots of the Gibbs energies of the vapor and liquid phases. When two liquid phases are in equilibrium with a vapor phase, a triple tangent exists on the Gibbs energy diagram at the proper T and P (see Problem 7.25).

7.8.2. Ternary Systems

The methods to be discussed are readily extendable to multicomponent mixtures. Both calculation methods mentioned at the beginning of this section are applicable to multicomponent systems; they will be described primarily for isobaric conditions and for two liquid phases.

Material balance and equilibria: The LLE equilibria are described by

$$\gamma_i x_i = \gamma_i^* x_i^*, \tag{7.81}$$

and the VLE by Eqs. 7.76, 77. The material balance of a flash is

$$z_i = \alpha y_i + (1 - \alpha)[\beta x_i + (1 - \beta)x_i^*], \tag{7.82}$$

where α is the fraction of the total material in the vapor phase and β is the fraction of the first liquid phase in the total liquid. Let $K_i = y_i/x_i$, $K_i^* = y_i/x_i^*$ and $K_i/K_i^* = \gamma_i/\gamma_i^*$.

Solving for individual phase compositions,

$$x_i = \frac{z_i}{\alpha K_i + (1 - \alpha)[\beta + (1 - \beta)K_i/K_i^*]}, \tag{7.83}$$

$$x_i^* = \frac{z_i}{\alpha K_i^* + (1-\alpha)[\beta K_i^*/K_i + 1 - \beta]} \cdot \tag{7.84}$$

Upon substituting these expressions into the mol fraction balances,

$$\Sigma\, y_i - \Sigma\, x_i = 0, \tag{7.85}$$

$$\Sigma\, y_i - \Sigma\, x_i^* = 0, \tag{7.86}$$

the material balances become,

$$\psi = \Sigma \frac{(K_i - 1)z_i}{\alpha K_i + (1-\alpha)[\beta + (1-\beta)K_i/K_i^*]} = 0, \tag{7.87}$$

$$\psi^* = \Sigma \frac{(K_i^* - 1)z_i}{\alpha K_i^* + (1-\alpha)[\beta K_i^*/K_i + 1 - \beta]} = 0. \tag{7.88}$$

For the bubblepoint,

$$\Sigma\, y_i = \Sigma\, K_i x_i = \Sigma\, K_i^* x_i^* = 1, \tag{7.89}$$

and for the dewpoint

$$\Sigma\, x_i = \Sigma\, x_i^* = \Sigma\, y_i/K_i = \Sigma\, y_i/K_i^* = 1. \tag{7.90}$$

Either the plain or starred equations may be used to find the dew- or bubblepoint conditions. Practical procedures for making these calculations are described in Chapter 6.

At a given temperature, the tie-lines and binodal curve of the liquid phase are established as in Section 7.5. For each tie-line, the bubblepoint pressure is obtained with

$$P = \Sigma \gamma_i x_i P_i^{sat} = \Sigma \gamma_i^* x_i^* P_i^{sat}, \tag{7.91}$$

in which the correct temperature is the one corresponding to the specified system pressure. Since all overall compositions along a tie-line have the same composition and temperature, a two-phase, liquid-liquid region on a bubblepoint temperature-composition (T-x_i) surface corresponds to a single line on the dewpoint T-y_i surface.

Before embarking on a flash calculation for a specified overall composition, it is advisable first to find the bubble- and dewpoints. In the flash Eqs. 7.87, 7.88, the primary unknowns are the phase splits α and β, but since the vaporization-equilibrium-ratios K_i and K_i^* are functions of the composition, they must be estimated at the start of the calculation. Then the values of α and β are found and substituted into Eqs. 7.83, 7.84, 7.76, and 7.77 to find the compositions and hence the improved values of K_i and K_i^*, and so on. Such a procedure is used in Example 7.17. Figure 7.18 is an example from the literature.

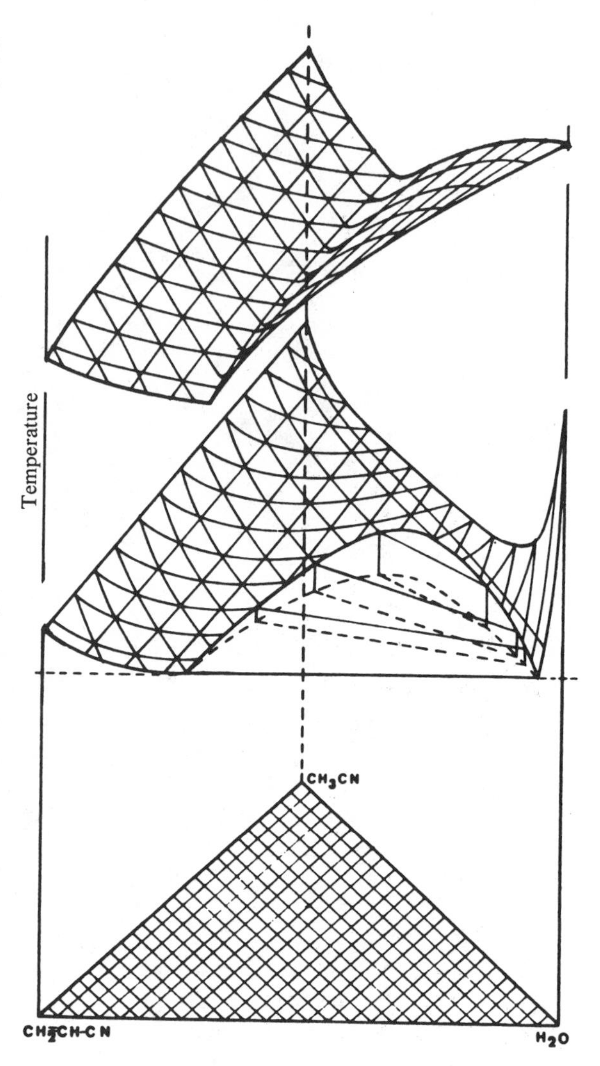

Figure 7.18. Dewpoint and bubblepoint temperature-composition surfaces of a partially miscible ternary mixture: acrylonitrile + acetonitrile + water (Mauri 1980). Copyright American Chemical Society; used with permission.

The two surfaces actually join at the vertices of the triangle but are shown separated for clarity. All compositions along a particular tie-line have the same bubblepoint temperature and equilibrium vapor composition, and thus correspond to a single point on the dewpoint surface. The family of tie-lines corresponds to a single line on the dewpoint surface.

7.8.3. Minimizing the Gibbs Energy

The conditions of equilibrium when various combination of independent properties are held constant are discussed in Section 2.10. In terms of independent properties T and P, the condition of equilibrium is characterized by having a minimum value of the Gibbs energy. Finding this minimum value is a

Example 7.16. Vapor-Liquid-Liquid Equilibria of a Binary Mixture

The correlation of LLE at 30 C in the system methanol (1) + heptane (2) + morpholine (3) is given in DECHEMA V/2, p. 97. From these data, mutual solubilities of the binary of methanol + heptane are $(x_1, x_1^*) = (0.1681, 0.8835)$. The NRTL parameters are

$\Delta g_{12}/R = 565.80$,

$\Delta g_{21}/R = 345.19$,

$\alpha_{12} = 0.2$

These diagrams will be drawn for this binary:

1. The Gibbs energy of mixing against the composition, x_1.
2. The activity coefficients against x_1.
3. The pressure-composition diagram at 30 C.
4. The x-y diagram.

Vapor pressures are $P_1^0 = 164.3$, $P_2^0 = 58.7$ Torr. The given NRTL parameters are converted into

$\tau_{12} = 1.8661$,

$\tau_{21} = 1.1385$,

$G_{12} = 0.6885$,

$G_{21} = 0.7964$.

The equations for the activity coefficients are

$$\ln \gamma_1 = x_2^2 \left(\frac{0.7221}{(x_1 + 0.7964x_2)^2} + \frac{1.2848}{(0.6885x_1 + x_2)^2} \right),$$

$$\ln \gamma_2 = x_1^2 \left(\frac{0.8846}{(x_1 + 0.6885x_2)^2} + \frac{0.9067}{(0.7964x_1 + x_2)^2} \right).$$

1. The points of double tangency on the Gibbs energy plot are $(x_1, x_1^*) = (0.14, 0.88)$, which is a rough check of the quoted experimental data.

2. The activity coefficients in the range $0.14 < x_1 < 0.88$ given by the equations are hypothetical since there are no homogeneous compositions in this range.

3. The pressure is given by $P = \gamma_1 x_1 P_1^0 + \gamma_2 x_2 P_2^0$ and is plotted against x_1. The P-y_1 plot is made with the aid of

$$y_1 = \gamma_1 x_1 P_1^0 / P.$$

The vapor-liquid-liquid equilibrium condition read off this diagram is $P = 200$ Torr, $y_1 = 0.73$. The portions of the P-x_1 and P-y_1 curves above this pressure are nonequilibrium values.

d. The x-y diagram is obtained from

$$y_1 = \gamma_1 x_1 P_1^0 / (\gamma_1 x_1 P_1^0 + \gamma_2 x_2 P_2^0).$$

It has a horizontal portion at $y_1 = 0.73$ corresponding to the region where two liquid phases exist. The dashed portion given by the equation is nonequilibrium.

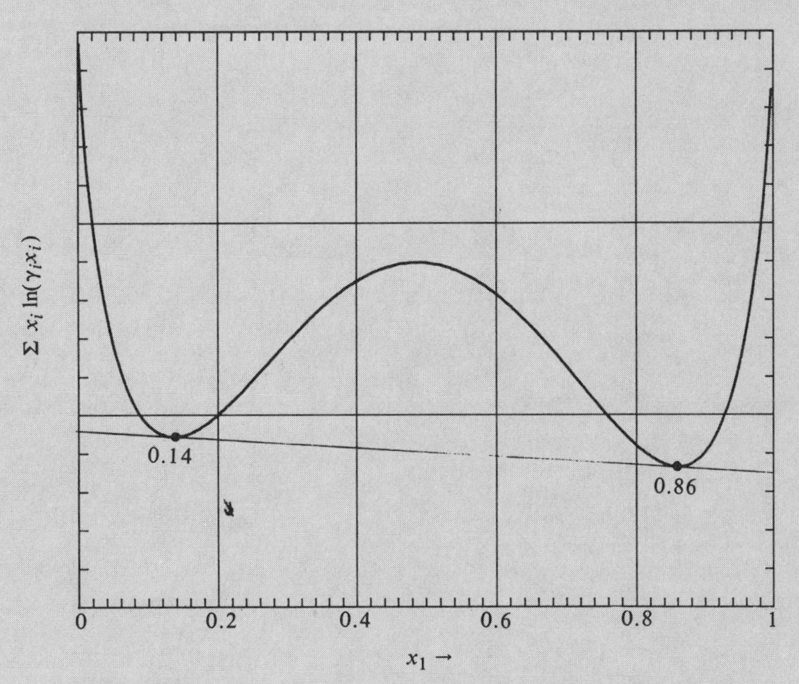

Example 7.16a. Check of experimental mutual solubilities of methanol and heptane.

Example 7.16 (*Continued*)

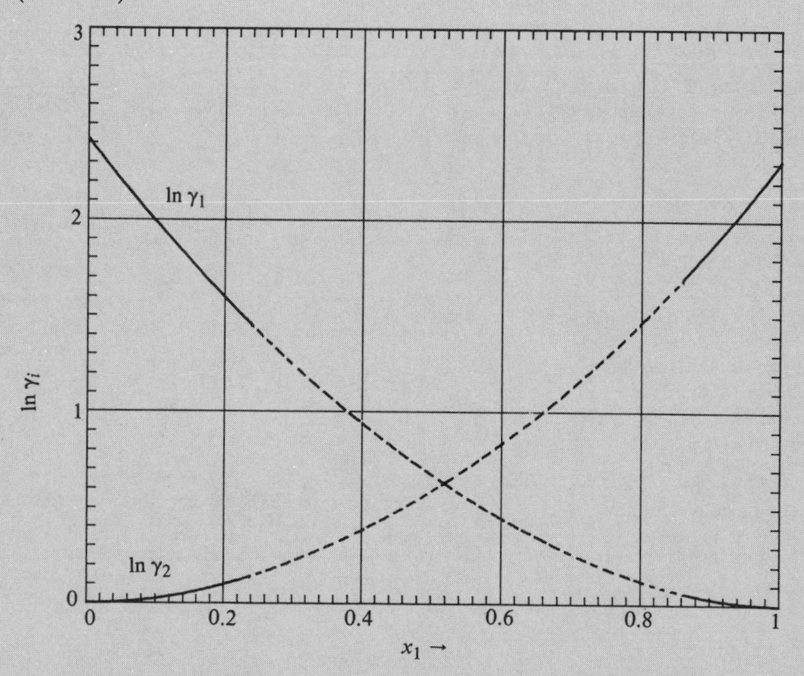

Example 7.16b. Activity coefficients as a function of composition. The region encompassed by the dashed lines is two-phase, so the plotted values have no physical significance there.

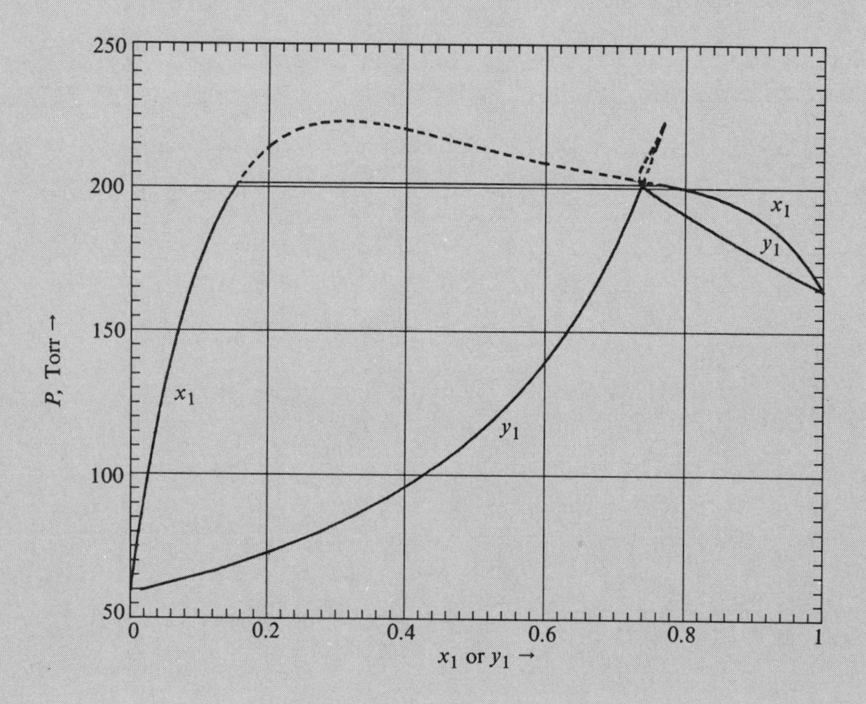

Example 7.16c. Pressure-composition diagram at 30 C. The dashed lines correspond to the two-phase region, so the indicated pressures are not real.

Example 7.16 (*Continued*)

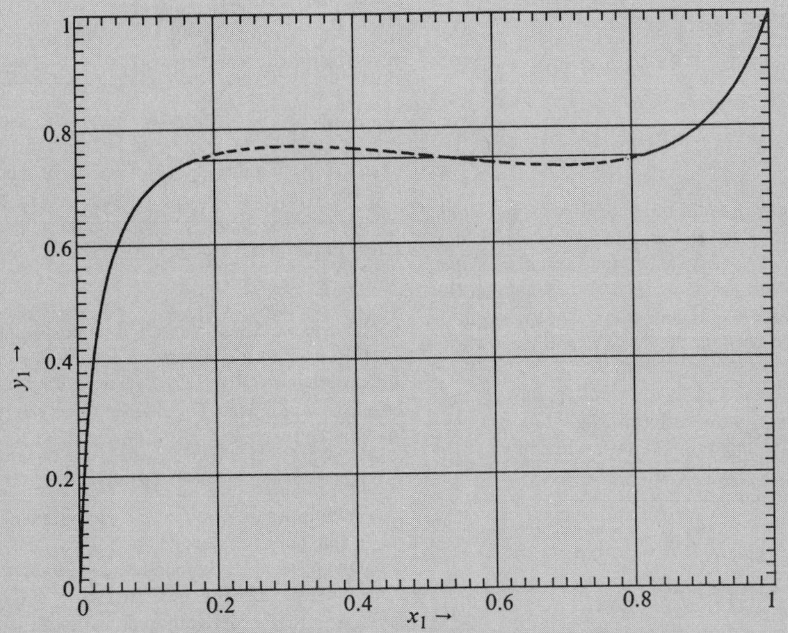

Example 7.16d. Vapor-liquid composition diagram at 30 C. The horizontal line in the range of the dashed curve represents the true vapor composition in the two-phase liquid region.

Example 7.17. A Step in the VLLE Calculation for a Ternary Mixture with Two Liquid Phases

The phase splits and compositions will be determined with trial values of the vaporization equilibrium ratios of the two phases. These are

Component	K_i	K_i^*
1	10	20
2	0.2	2
3	5	0.2

A graphical solution of Eqs. 7.87, 7.88 is shown. The phase splits are $\alpha = 0.475$, $\beta = 0.470$, from which the fractional amounts of the three phases are: $V = 0.475$, $L = 0.247$, and $L^* = 0.278$. The compositions of the phases are:

	x	x^*	y	y^*
1	0.065	0.032	0.649	0.649
2	0.903	0.080	0.181	0.159
3	0.035	0.871	0.174	0.174
	1.003	0.983	1.004	0.982

For the next trial the compositions are to be normalized and used to evaluate the K_i and K_i^* from

$$K_i = \gamma_i \phi_i^{sat} P_i^{sat} / \hat{\phi}_i P$$

and similarly for K_i^*. Clearly, a full-fashioned computer program is advisable for this kind of problem.

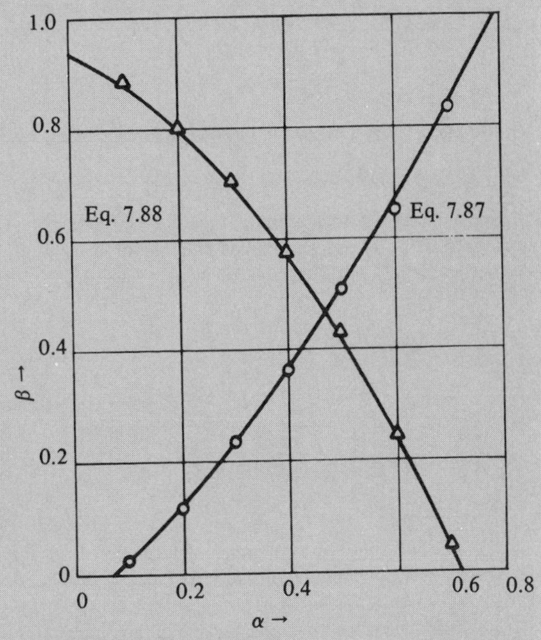

Example 7.17. Graphical solution to find vapor-liquid-liquid proportions of a ternary mixture with specified VERs.

feasible method of establishing the equilibrium condition. Example 7.9 is a simple case.

For a component of the vapor phase,

$$\bar{G}_i = G_i + RT \ln(\hat{f}_i/f_i) = G_i + RT \ln(\hat{\phi}_i y_i/\phi_i), \quad (7.92)$$

and for one of the liquid phases

$$\bar{G}_i = G_i + RT \ln(\hat{f}_i/f_i) = G_i + RT \ln(\gamma_i \phi_i^{sat} P_i^{sat} x_i/\phi_i P). \quad (7.93)$$

Here the Gibbs energies, G_i, of the pure components are those in a gas phase reference state and are the same for both the liquid and vapor phases. Consequently, the fugacities of the pure components also are the same in both phases, that is, $f_i = \phi_i P$. When the same equation of state can be applied to both phases, Eq. 7.92 is used for the liquid phase also with x_i substituted for y_i.

When the mixture contains n_i mols overall, v_i mols in the vapor phase, and l_i and l_i^* mols in the two liquid phases, the total Gibbs energy of the mixture is represented by

$$\frac{n_t G}{RT} = \Sigma \frac{n_i G_i}{RT} + \Sigma\, v_i \ln(\hat{\phi}_i y_i/\phi_i) + \Sigma\, l_i \ln(\gamma_i \phi_i^{sat} P_i^{sat} x_i/\phi_i P) + \Sigma\, l_i^* \ln(\gamma_i^* \phi_i^{sat} P_i^{sat} x_i^*/\phi_i P). \quad (7.94)$$

In the absence of chemical change,

$$n_i = v_i + l_i + l_i^*. \quad (7.95)$$

In terms of the fraction vaporized,

$$\alpha = \Sigma\, v_i/\Sigma\, n_i = \Sigma\, v_i/n_t, \quad (7.96)$$

and the liquid phase split,

$$\beta = \Sigma\, l_i/(\Sigma\, l_i + \Sigma\, l_i^*), \quad (7.97)$$

the Gibbs energy function may be written

$$g = \frac{G - \Sigma\, z_i G_i}{RT} \quad (7.98)$$

$$= \alpha \Sigma\, y_i \ln(\hat{\phi}_i y_i/\phi_i) + (1-\alpha)[\beta \Sigma\, x_i \ln(\gamma_i \phi_i^{sat} P_i^{sat} x_i/\phi_i P) + (1-\beta)\Sigma\, x_i^* \ln(\gamma_i^* \phi_i^{sat} P_i^{sat} x_i^*/\phi_i P)]. \quad (7.99)$$

A simpler form of this equation results when the pure component Gibbs energies are taken as those of gases in the standard state of unit fugacity, in which case Eq. 7.99 becomes:

$$g = \frac{G - \Sigma\, z_i G_i^0}{RT} \quad (7.100)$$

$$= \alpha \Sigma\, y_i \ln(\hat{\phi}_i P y_i) + (1-\alpha)[\beta \Sigma\, x_i \ln(\gamma_i \phi_i^{sat} P_i^{sat} x_i) + (1-\beta)\Sigma\, x_i^* \ln(\gamma_i^* \phi_i^{sat} P_i^{sat} x_i^*)] \quad (7.101)$$

The G_i^0 are the usual Gibbs energies of formation from the elements as listed in compilations such as those of Stull et al. (1969).

For a given overall composition, z_i, one of the phase splits, α or β, can be eliminated with the material balance,

$$z_i = \alpha y_i + (1-\alpha)[\beta x_i + (1-\beta) x_i^*] \quad (7.102)$$

and the vapor mol fractions can be eliminated in terms of the equilibrium relations,

$$y_i = K_i x_i = \frac{\gamma_i \phi_i^{sat} P_i^{sat} x_i}{\hat{\phi}_i P}. \quad (7.103)$$

This can be done directly when the ratio, $\phi_i^{sat}/\hat{\phi}_i$, is substantially unity and must be done by iteration in the alternate event since $\hat{\phi}_i$ is a function of the vapor composition. A final requirement on the system is that the mol fractions of the liquid phases sum to unity, which may be stated in the form

$$\Sigma\, x_i - \Sigma\, x_i^* = 0. \quad (7.104)$$

Problems similar to finding a minimum value of g in Eq. 7.101 subject to a composition constraint such as Eq. 7.104 are covered in Chapter 10. One method of solution uses Lagrange multipliers. That theory states that the minimum of the function,

$$\psi = g + \lambda(\Sigma\, x_i - \Sigma\, x_i^*) \rightarrow \text{minimum}, \quad (7.105)$$

also determines the minimum of the system described by Eqs. 7.101 and 7.104; λ is called the Lagrange multiplier, which is unknown to start but is found by the minimization procedure along with the unknown quantities x_i, x_i^* and one of the phase splits, α or β. The derivatives of ψ with respect to each of the system unknowns, equated to zero, together with Eq. 7.104 constitute a system from which the unknowns and the parameter λ can be found, for example by the Newton-Raphson or comparable method.

Many methods of search for extrema and for solving sets of nonlinear equations have been developed. No single method appears to be capable of coping with all the difficulties that can arise, such as slow or failed convergence, or appearance of negative mol fractions at intermediate stages of calculations, or the intermediate disappearance of phases that were assumed to be present at the start. The solution of such a problem is an exercise of an art.

A simple case of Gibbs energy minimization is studied in Example 7.9. More elaborate cases are covered in Chapter 10.

7.8.4. Literature on Multiphase Equilibria

Methods of solving multiphase equilibria are closely related to methods for solving multireaction equilibria. Among the books that are devoted to the latter topic primarily are those by Holub & Vonka (1976), Rosenbrock & Storey (1966), Van Zeggeren & Storey (1970), and Smith & Missen (1982). Many articles and journal reviews of this topic have been published recently, of which some will be described briefly. References to purely chemical equilibria are made in Chapter 10.

1. Castillo & Grossmann (1981): A nonlinear programming method is applied to finding the identities of phases in simultaneous phase and chemical equilibria.
2. Deam & Maddox (1969) solve a three-phase system with vaporization equilibrium ratios independent of composition.
3. George et al. (1976) find phase compositions by minimizing the Gibbs energy with Powell's method; four phases and fifteen components can be handled by their program.

4. Heidemann (1979) uses a modified Redlich-Kwong equation applicable to both liquid and vapor phases to find three-phase equilibria in water and oil mixtures.
5. In their book, Henley & Rosen (1969, Chap. 8) discuss both phase and chemical equilibria. Algorithms in the forms of flowcharts are given for multiliquid flashes at fixed T and P and at fixed H and P.
6. Lahiri (1979) employs the Newton-Raphson method to find the minimum Gibbs energy and hence the phase compositions in high-temperature metallurgy.
7. Mauri (1980) solves the system of equations representing material balances and equilibria when three phases are present. Computations of dewpoints, bubblepoints, and flashes are described.
8. Peng & Robinson (1976) apply their equation of state to both phases for figuring equilibria in water and hydrocarbon systems. They use the criterion of equality of partial fugacities of individual components in all phases directly.
9. Sanderson & Chien (1973) formulate phase and chemical equilibria relations and solve the resulting nonlinear equations by the method of Marquardt.
10. Gautam & Seider (1979) compare several methods for computation of phase and chemical equilibria, including mixtures of electrolytes. They reach some conclusions as to suitable methods.
11. Seider, Gautam, & White (1980) review the recent literature.
12. Ross & Seider (1980) develop a scheme for simulation of three-phase distillation towers.
13. Tinoco-Garcia & Cano-Dominguez (1979) employ an iterative matching of the fugacities of individual components in all phases to solve the LLE and VLE equations.
14. Niedzwiecki et al. (1980) describe three-phase distillation of hydrocarbons in the presence of water.

7.9. VAPOR-LIQUID-SOLID EQUILIBRIA

Mixtures of substances with a wide range of volatilities can exist as three phases in equilibrium, vapor + liquid + solid. In natural gases at low temperatures, for instance, the solid phase may be CO_2, H_2S, heavy hydrocarbons or hydrates. In such cases it is usually permissible to assume that the solid phase is pure, which simplifies the calculation somewhat. Moreover, the fugacities can be determined from an equation of state such as the BWR, Soave, or Peng-Robinson.

At equilibrium, the partial fugacities of individual components are equal throughout,

$$\hat{f}_i^V = \hat{f}_i^L = f_i^s, \tag{7.106}$$

where the partial fugacity of the solid equals that of the pure component and is given by

$$f_i^s = \phi_i^{\text{sat}} P_i^{\text{sat}} (PF)_i \tag{7.107}$$

in terms of the vapor pressure and the Poynting factor. Fugacities in the fluid phases are obtained directly from an equation of state, for instance from the Soave equation (Eq. 22 of Table 3.4),

$$\ln \hat{\phi}_i = \frac{b_i}{b}(z-1) - \ln z\left(1 - \frac{b}{V}\right)$$

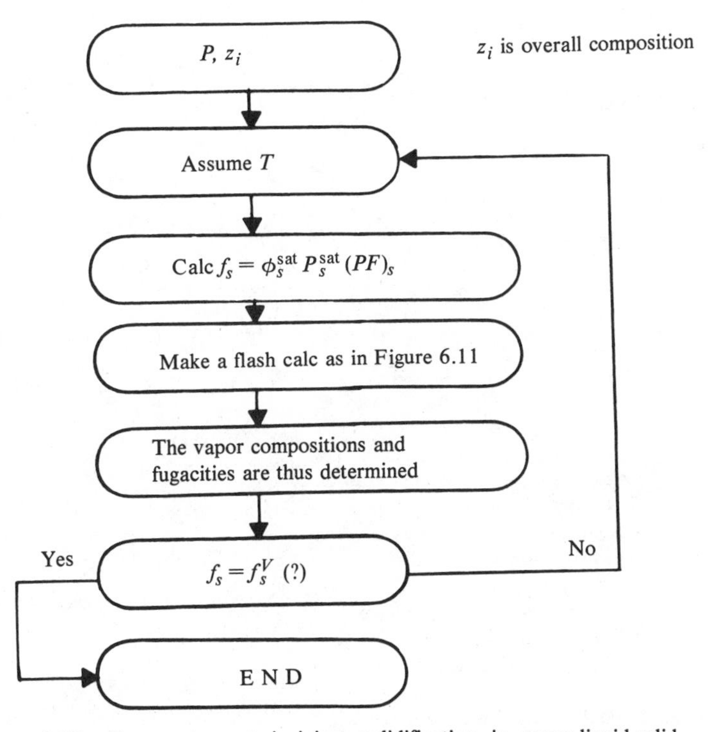

Figure 7.19. Temperature at incipient solidification in vapor-liquid-solid equilibria.

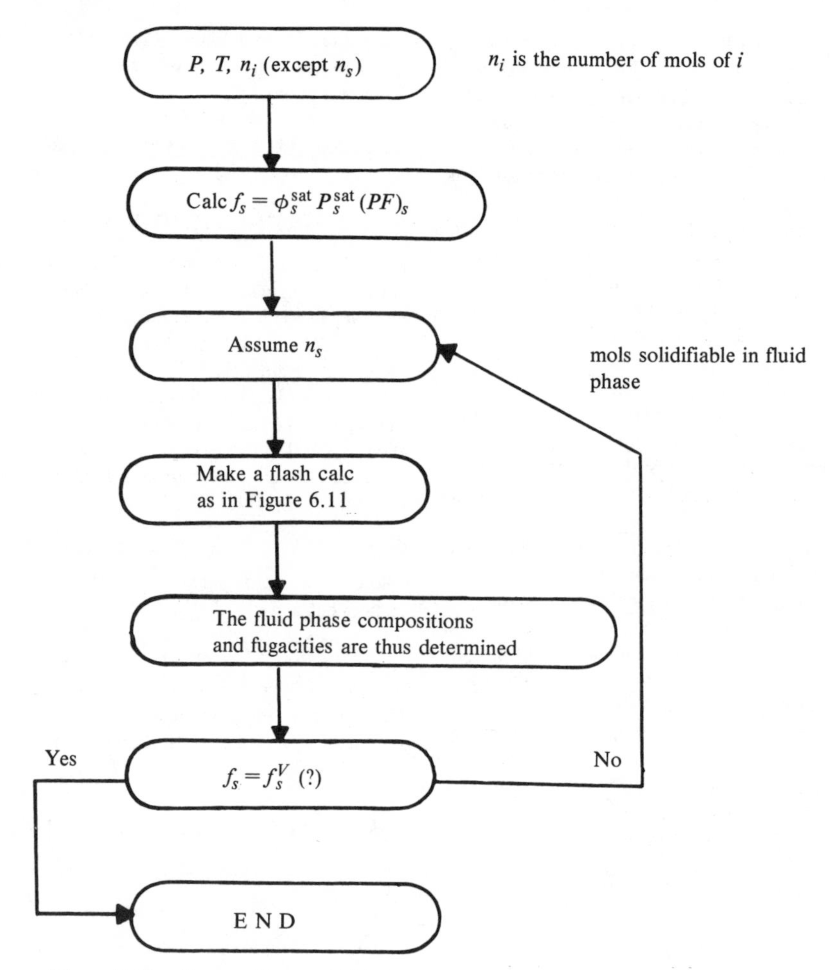

Figure 7.20. Compositions of fluid phases at specified T and P in vapor-liquid-solid equilibria.

$$+\frac{a\alpha}{bRT}\left[\frac{b_i}{b}-\frac{2}{a\alpha}\sum_j(a\alpha)_{ij}\right]\ln\left(1+\frac{b}{V}\right). \tag{7.108}$$

From this equation, the vapor coefficient is obtained after substitution of the largest compressibility, z, and that of the liquid after substitution of the smallest of the three roots of the Soave equation.

The condition at which solid first begins to separate from a given fluid phase composition is found by comparing the solid fugacity figured from Eq. 7.107 with that of the same component in the vapor phase from Eq. 7.108. The pressure or temperature is adjusted until this condition is satisfied. Figure 7.19 outlines the calculation procedure. When the fugacity of the solid is less than the partial fugacity of the vapor phase, some precipitation has occurred. In order to find the amount of this separation and the compositions of the fluid phases, further trials are made after reducing the amount of the solid-forming component in the fluid phases. The procedure is outlined in Figure 7.20. A systematic procedure for performing these reductions is described by Peng & Robinson (1969).

7.10. PROBLEMS

7.1. The excess Gibbs energy of a mixture as a function of temperature, K, is

$$G^{ex}/RT = [2 + 0.009(T - 310)]x_1x_2.$$

Plot the temperature against the mutual solubilities.

7.2. The symmetrical activity coefficient parameter as a function of temperature, K, is

$$A = \begin{cases} 2 + 0.0016(T - 325)^2, & T < 325, \\ 2.4 - 0.0016(T - 325)^2, & T > 325. \end{cases}$$

Construct the plot of temperature against mutual solubilities of the components.

7.3. The van Laar parameters depend on temperature according to

$$A_{12} = 2 + 0.009(T - 310),$$

$$A_{21} = 3 - 0.008(T - 310).$$

Plot the temperature against the mutual solubilities.

7.4. Apply the algebraic method of double tangency, Eqs. 7.24 and 7.25, to find the equilibrium compositions for the Margules equation with $A = 3$, $B = 2.5$, for which

$$g = x_1x_2(Ax_2 + Bx_1) + x_1\ln x_1 + x_2\ln x_2,$$

$$dg/dx_1 = Ax_2(1 - 3x_1) + Bx_1(3x_2 - 1) + \ln(x_1/x_2).$$

Plots of these equations are shown, to assist in starting estimates of the equilibrium compositions.

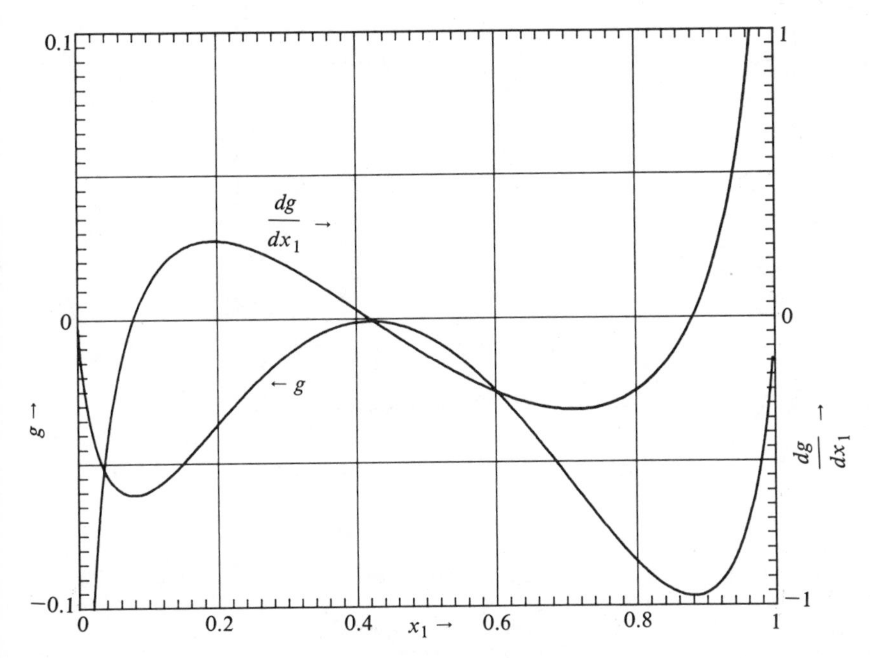

Problem 7.4

7.5. Data and correlations for the system hexane (1) + ethanol (2) + acetonitrile (3) at 40 C were obtained by Sugi & Katayama (1978). Use the given T-K-Wilson parameters to find a number of tie-lines with program C-3, and locate them with reference to the experimental tie-lines and binodal curve.

$\Lambda_{12} = 0.0818$, $\Lambda_{21} = 0.4308$, $\Lambda_{13} = 0.1002$,

$\Lambda_{31} = 0.1641$, $\Lambda_{23} = 0.4159$, $\Lambda_{32} = 0.5677$

Binodal curve coordinates

x_1	x_2	x_1	x_2	x_1	x_2
0.0857	0.0000	0.1966	0.2682	0.5118	0.2142
0.0943	0.0410	0.2525	0.2884	0.5595	0.2005
0.1077	0.0971	0.3040	0.2832	0.6092	0.1723
0.1213	0.1614	0.3525	0.2680	0.6931	0.1255
0.1181	0.2060	0.3845	0.2578	0.7587	0.0848
0.1657	0.2414	0.4315	0.2422	0.9015	0.0000

Tie-line data

Left		*Right*	
x_1	x_2	x_1	x_2
0.8831	0.0166	0.0968	0.0879
0.8674	0.0251	0.1003	0.1299
0.8546	0.0332	0.1054	0.1622
0.8372	0.0433	0.1250	0.1942
0.8101	0.0567	0.1367	0.2169
0.7821	0.0772	0.1522	0.2331
0.7272	0.1086	0.2053	0.2695
0.6912	0.1269	0.2249	0.2748

7.6. Follow the procedure of Example 7.6 to find equilibrium compositions with the T-K-Wilson equation when $\ln \gamma_1^\infty = 4$, $\ln \gamma_2^\infty = 3$, and $V_1/V_2 = 2$. The parameters calculated from these data are $\Lambda_{12} = 0.0579$ and $\Lambda_{21} = 0.1566$. The plot of $G^{ex}/RT + \Sigma\, x_i \ln x_i$ is shown as a guide to trial values of the compositions. (See figure on next page).

7.7. The mutual solubilities of n-butanol and water are $x_1 = 0.9765$ and $x_1^* = 0.6770$. Find the parameters of the Margules, van Laar, and T-K-Wilson equations. (*Ans:* The van Laar parameters are $A_{12} = 0.334$, $A_{21} = 1.60$.)

7.8. For a mixture with volume ratio $V_1/V_2 = 2$, the Wilson parameters are $\Lambda_{12} = 0.2$ and $\Lambda_{21} = 1.5$. Estimate the values of the parameters of the T-K-Wilson equation.

7.9. For the system aniline (1)-methylcyclohexane (2),

$\gamma_1^\infty = 16.1,$

$\gamma_2^\infty = 11.14,$

$V_1/V_2 = 0.7137$

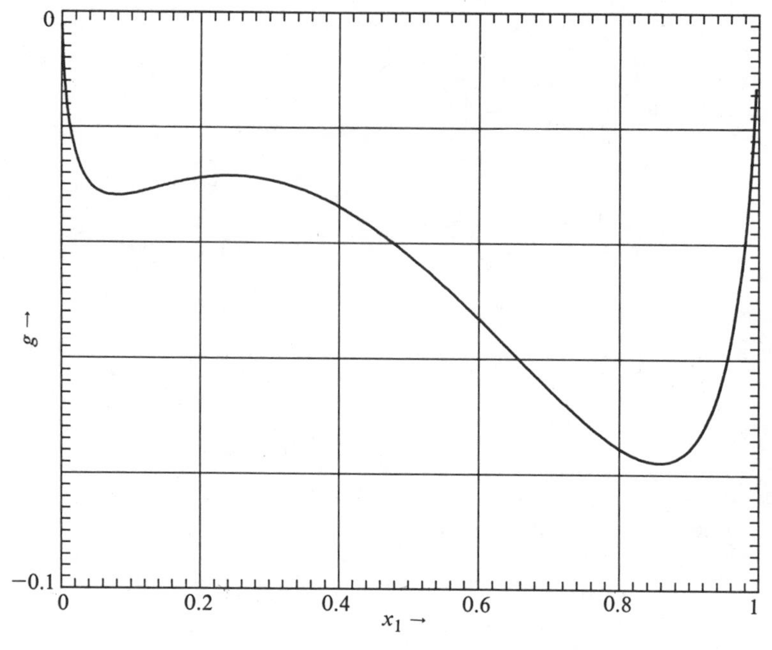

Problem 7.6

Find the mutual solubilities with the van Laar and T-K-Wilson equations by the methods of Examples 7.3 and 7.5.

7.10. For two liquid phases of a binary mixture in equilibrium, show that the composition of one phase is given in terms of the activity coefficients of both components in both phases by

$$x_1 = \frac{\gamma_1^*(\gamma_2 - \gamma_2^*)}{\gamma_1\gamma_2^* - \gamma_1^*\gamma_2}.$$

7.11. The mutual solubilities of $CHCl_3$ and C_7F_{16} at 60 C are $x_1 = 0.1$ and $x_1^* = 0.8$. The solubility parameters are $\delta_1 = 9.26$ $(cal/ml)^{0.5}$ and $\delta_2 = 5.9$; the molal volumes are $V_1 = 80.7$ ml/gmol, $V_2 = 227.3$.

1. Find the infinite dilution activity coefficients
2. Find the mutual solubilities with the Scatchard-Hildebrand equation.
3. With the same $\ln \gamma_i^\infty$, find the mutual solubilities with the equations of van Laar, Margules, NRTL with $\alpha_{12} = 0.2$, NRTL with $\alpha_{12} = -1$, and T-K-Wilson.

7.12. Construct a graphical tieline correlation of the attached data and find the compositions of phases in equilibrium with a phase containing 10, 20, or 30 mol% acetaldehyde. The composition of the plait point is estimated as $x_{water} = 0.33$ and $x_{acetaldehyde} = 0.49$.

(1) $C_4H_{10}O$	ether, diethyl
(2) C_2H_4O	acetaldehyde
(3) H_2O	water

(Suska J. Collect.Czech.Chem.Commun. 44(1979)1999)

Temperature = 15.1 deg C
Type of system = 1
Experimental tie lines in mole pct

Left phase			*Right phase*		
(1)	*(2)*	*(3)*	*(1)*	*(2)*	*(3)*
76.700	15.500	7.800	2.500	8.300	89.200
63.700	27.400	8.900	3.300	17.500	79.200
56.600	32.700	10.700	4.500	23.800	71.700
46.100	40.600	13.300	5.700	29.500	64.800
37.300	46.100	16.600	7.000	34.600	58.400

7.13. The system of Hand (J Phys Chem 34, 1961 (1930) plots $\log(x_{solute}/x_{solvent\ 1})$ of the phase that is rich in solvent 1 against $\log(x_{solute}/x_{solvent\ 2})$ of the phase that is rich in solvent 2. The result is often linear and may be useful in detecting experimental errors. Do this analysis for the data of Problem 7.12, plotting $\log(x_2/x_1)$ of the left phase against $\log(x_2/x_3)$ of the right phase.

7.14 Data of the following system are given by DECHEMA LLE (V/2, p. 326): Octane (1) + 2,4,4-trimethyl pentane (2) + nitroethane (3). Binary miscibilities are estimated as

$x_1 = 0.81, x_3 = 0.19,$

$x_2 = 0.20, x_3 = 0.80.$

One ternary tie-line is known:

Phase 1: $x_1 = 0.48157, x_2 = 0.30806, x_3 = 0.21037,$

Phase 2: $x_1 = 0.08476, x_2 = 0.05863, x_3 = 0.85661.$

For the NRTL ($\alpha_{12} = 0.2$) equation,

1. Find τ_{13}, τ_{31} from the binary data.
2. Find τ_{23}, τ_{32} from the binary data.
3. Hence find τ_{12} and τ_{21} by using the ternary data.

7.15. From the compilation of Renon et al. (1971, p. 168) the NRTL parameters of cyclohexane (1) + furfural (2) are: $\tau_{12} = 2.2531$, $\tau_{21} = 1.3772$, $\alpha_{12} = 0.1973$ at 30 C. Find the mutual solubilities at this temperature. (Note: Experimental values are (0.15, 0.94).)

7.16. In the following table of mutual solubilities, the equilibrium mol fractions in the two phases are those of the first named substance. The page numbers are those of the DECHEMA LLE Data Collection, V/2. Find the Margules parameters and the NRTL τ_{12} and τ_{21} with $\alpha_{12} = 0.2$.

	x_1	x_1^*	°C	page
1,3 dioxane, 4,4-dimethyl + water	0.80369	0.03590	20	59
nitromethane + water	0.93873	0.03039	26.7	66
1-nonanol + nitromethane	0.82952	0.01821	20	72
methanol + hexane	0.88637	0.10361	5	86
methanol + hexane	0.80164	0.23226	25	88
cyclohexane + methanol	0.87090	0.17520	25	115
2-butanone + water	0.63617	0.07311	26.7	218
cyclohexane + ethane, 1,2-diamino	0.90960	0.07050	40	443
cyclohexane + ethane, 1,2-diamino	0.80680	0.133310	60	444
1-butanol, 3-methyl + water	0.65418	0.00517	25	535

7.17. The Antoine coefficients of methanol and heptane are:

	A	B	C
methanol	18.5875	3626.55	−34.29
heptane	15.8737	2911.32	−56.51

Other data are in Example 7.16. Construct the T-x_1 and T-y_1 diagrams of this mixture at 400 Torr.

7.18. Show that inclusion of a multiplier C in the Wilson equation for excess Gibbs energy, thus

$$G^{ex}/RT = -C[x_1 \ln(x_1 + Ax_2) + x_2 \ln(Bx_1 + x_2)]$$

allows the convexity requirement,

$$\frac{\partial^2 G}{\partial x_1^2} < 0,$$

to be satisfied for some values of A, B, and x_1 when $C \neq 1$:

$$G = G^{ex} + RT(x_1 \ln x_1 + x_2 \ln x_2).$$

7.19. Henry's Law coefficient, $H = P/x$, of H_2S in water varies with temperature as shown (*Int. Crit. Tables* 3, 259).

a. Find the constants of Valentiner's equation,

$$\ln H = A + B \ln T + C/T,$$

that best represent these data, over the entire range or over partial ranges.

b. Similarly find the constants of Hildebrand's equation,

$$\ln H = A + B/(T + C).$$

°C	$10^{-4}H$ atm/mol fraction
0	2.24
10	2.97
20	3.76
30	4.49
40	5.20
50	5.77
60	6.26
70	6.66
80	6.82
90	6.92
100	7.01

[*For Problem 7.19*]

7.20. Mutual solubilities and parameters of the UNIQUAC and NRTL equations are given at several temperatures (DECHEMA V/1, p. 205) for 3-butene-2-one (1) and water (2). Show that these data are consistent with the condition, $\gamma_i x_i = \gamma_i^* x_i^*$. The UNIQUAC parameters are $\tau_{ij} = \Delta u_{ij}/R$ and the NRTL are $\tau_{ij} = \Delta g_{ij}/R$, with $\alpha = 0.2$.

			UNIQUAC		NRTL	
°C	x_1	x_1^*	τ_{12}	τ_{21}	τ_{12}	τ_{21}
$R_1 = 3.0178$		$Q_1 = 2.664$				
$R_2 = 0.9200$		$Q_2 = 1.400$				
30.0	18.7	66.8	194.17	26.799	−158.67	967.16
40.0	13.2	59.7	177.67	50.291	−166.13	1038.2
50.0	12.3	58.6	175.50	59.128	−174.13	1084.2
60.0	11.8	60.2	148.64	81.307	−207.84	1157.4
70.0	12.0	64.8	97.844	119.48	−268.92	1256.6
80.0	13.1	74.4	−7.8069	214.27	−393.29	1435.0

7.21. A mixture contains 10% CO_2 and equal amounts of methane and ethane at 27.2 atm. Use the Soave equation to find the temperature at which solid begins to form. The data cited by Peng & Robinson (1979, Adv Chem Series 182, pp. 185–196) indicate this to be about 177 K.

7.22. Find the analytical solution of the differential equation of Example 7.2 for the condition $f(x_1) = 1$ when $x_1 = 0.05$. The plot of the solution is shown.

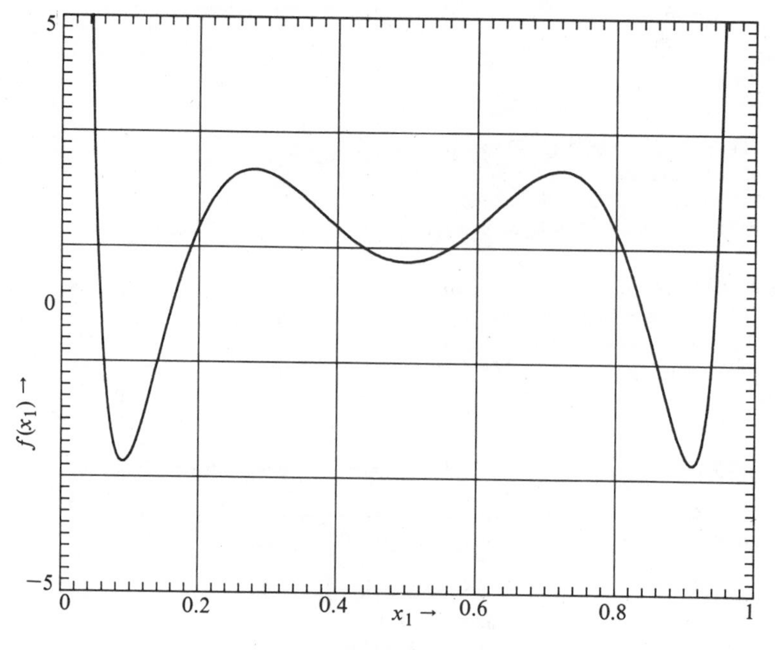

Problem 7.22

Ans. $f(x) = \dfrac{1}{x(1-x)}[1.5876 + 5000(x^6/6 - 2.5x^5/5$

$+ 2.3x^4/4 - 0.95x^3/3$

$+ 0.1689x^2/2 - 0.00945x)]$.

7.23. Construct the isothermal x-y diagrams of mixtures with $P_1^0 = 3$, $P_2^0 = 1$ and Margules parameters,

(a) $A = 2.5$, $B = 0.5$,

(b) $A = 4$, $B = 0.5$.

Identify the miscibility limits on the curves. Note that (a) is an example of a partially miscible system that does not form an azeotrope, and that (b) forms both heterogeneous and homogeneous azeotropes. Mixtures of water + methyl ethyl ketone and water + phenol are actual examples of type (a).

7.24. Show that the magnitudes of the Gibbs energies of the pure components of a binary mixture do not affect the compositions found by the double tangent method. That is, show that

$$g_a = \Sigma\ x_i g_i^0 + \Sigma\ x_i \ln x_i + g^{ex}$$

and

$$g_b = \Sigma\ x_i \ln x_i + g^{ex}$$

have the same abscissas of double tangency.

7.25. In the function $f(x_1)$ of Problem 7.22, find by trial a replacement for the number 5000 that will result in a triple tangent on the plot of g against x_1.

7.26. The composition of a binary mixture at the limit of metastability is given by solution of the equation $\partial^2 g/\partial x_1^2 = 0$, where

$$g = (G - \Sigma\ x_i G_i^0)/RT = G^{ex}/RT + \Sigma\ x_i \ln x_i$$

The locus of these compositions as a function of temperature is called the spinodal curve. Plot such a curve when the excess Gibbs energy is represented by the Margules equation with parameters dependent on temperature:

$$G^{ex}/RT = x_1 x_2 (Bx_1 + Ax_2)$$

with $A = 700/T$°K and $B = 1000/T$°K. Also show the stable limits of miscibility. Carry these results to the temperature of incipient phase separation.

7.27. Examination of the plots of Figures 7.7(a) and (b) suggests the following conclusions:

a. g has a minimax on the upper curve when $\gamma_1 x_1 = \gamma_2 x_2$ on the lower plots.
b. Minimaxes of $\gamma_1 x_1$ and $\gamma_2 x_2$ occur at the same values of x_1.
c. Minimaxes of $\gamma_1 x_1$ and $\gamma_2 x_2$ occur at the same values of x_1 as the points of inflection of the g-plot.

Prove these conclusions analytically with the aid of the Gibbs-Duhem equation.

8 Liquid-Solid Equilibrium

8.1. INTRODUCTION

Two kinds of equilibria between liquids and solids are of particular importance:

1. *Solution equilibria*, which are between liquids and solids of different chemical species.
2. *Melt equilibria,* which are between molten and solid forms of the same chemical species.

The behavior is expressed quantitatively in terms of activity coefficients. In principle, liquid-phase activity coefficients that have been deduced from measurements of VLE or LLE are applicable to the liquid phases of liquid-solid equilibria when they are compensated for temperature. In practice this is fairly accurate for describing solubilities but less so for melt equilibria because they are often complicated by several factors—namely:

1. Solids may possess more than one crystalline form depending on temperature and pressure. Examples are the many crystalline forms of water (Fig. 5.11) and the system of Figure 5.25(h).
2. Intermolecular compounds may be present, for instance between CCl_4 and many hydrocarbons.
3. Solid phases may be either substantially pure or partially miscible.
4. Temperatures of melt equilibria may be out of the safe extrapolation range of VLE or LLE measurements.

For these and possibly other reasons, the behavior of melt equilibria can be far more complex than that of even multiliquid-vapor equilibria. Some of these characteristics serve as a primary basis for the classification of melt equilibria, as in Tables 5.1 and 5.2, where compound formation and solid-phase miscibilities are the major identifying criteria. The areas of technology that have been most active in the study of melt equilibria are metals and alloys, ceramics, and systems of water with inorganic salts; but a substantial body of data on organic systems has been accumulated. The largest single compilation of those data is in Landolt-Börnstein (II/2b, 1962; II/2c, 1964; II/3, 1956).

Various types and several actual examples of binary phase diagrams are illustrated in Figures 5.21–5.26 and of ternary equilibria in Figures 5.40–5.41 and in this chapter. Fascinating stereoscopic drawings of many ternary systems have been made by Tamas & Pal (1970), and very clear drawings in several colors are in the book of Prince (1966).

For the most part, correlations of only the simplest kinds of melt equilibria have been attempted so far in terms of thermodynamic principles and data, except for classification by the phase rule. In the metallurgical field the complications of solid solutions have been tackled with regular solution theory; references to that work are made by Lupis (1983). There are important industrial cases of organic systems to which theory can make valuable contributions.

8.2. SOLUBILITY OF SOLIDS IN LIQUIDS

When the solvent does not enter the solid phase, the fugacity of the solid solute remains that of the pure solid, so the condition of equality of partial fugacities at equilibrium becomes

$$\hat{f}_{2(\text{solid})} = f_{2(\text{solid})} = \gamma_2 x_2 f_{2(\text{subcooled liquid})}. \tag{8.1}$$

Rearranging in simplified notation,

$$x_2 = \frac{f_{2s}}{\gamma_2 f_{2(\text{scl})}}, \tag{8.2}$$

where x_2 is the mol fraction of the solute in the solution and $f_{2(\text{scl})}$ represents the fugacity of the pure solid solute in a subcooled or hypothetical liquid state below its melting point. In Figure 4.1 point b is the location of such a subcooled liquid.

The ratio $f_{2s}/f_{2(\text{scl})}$ of the fugacities of the solid and its subcooled liquid can be evaluated in terms of conditions at the triple point. When the fundamental equation,

$$d\ln f = -\frac{\Delta H}{RT^2}dT + \frac{\Delta V}{RT}dP, \tag{8.3}$$

is applied to each phase and the results subtracted, the conclusion is

$$d\ln\frac{f_{2s}}{f_{2(\text{scl})}} = \frac{H_L - H_s}{RT^2}dT - \frac{V_L - V_s}{RT}dP. \tag{8.4}$$

For practical purposes the difference in specific volumes of condensed phases may be taken independent of pressure, but the enthalpy of fusion $H_L - H_s$ may vary appreciably with temperature. That behavior is described by

$$H_L - H_s = (H_L - H_s)_{tp} + \int_{T_{tp}}^{T} (C_{pL} - C_{ps})dT \tag{8.5}$$

$$= \Delta H_{tp} + \int_{T_{tp}}^{T} \Delta C_p dT \tag{8.6}$$

$$\simeq \Delta H_{tp} + \Delta C_p(T - T_{tp}), \tag{8.7}$$

where the subscript tp designates the triple point. The last form of the equation applies when the heat capacity difference is relatively insensitive to temperature.

After Eq. 8.7 is substituted into Eq. 8.4 and the result is integrated between (T_{tp}, P_{tp}) and the temperature and pressure of the system (T, P), the ratio of the fugacities becomes

$$\ln\frac{f_2}{f_{2(\text{scl})}} = \frac{\Delta H_{tp}}{R}\left(\frac{1}{T_{tp}} - \frac{1}{T}\right) - \frac{\Delta C_p}{R}\left(\ln\frac{T_{tp}}{T} - T_{tp}/T + 1\right) - \frac{\Delta V}{RT}(P - P_{tp}). \tag{8.8}$$

Substitution of this relation into Eq. 8.2 results in the *general solubility equation,*

$$x_2 = \frac{1}{\gamma_2}\exp\left[\frac{\Delta H_{tp}}{R}\left(\frac{1}{T_{tp}} - \frac{1}{T}\right) - \frac{\Delta C_p}{R}\left(\ln\frac{T_{tp}}{T} - T_{tp}/T + 1\right) - \frac{\Delta V}{RT}(P - P_{tp})\right]. \tag{8.9}$$

One or more simplifications of this equation are sometimes adequate.

1. The pressure correction may be negligible, usually the case.

2. Although it is more substantial than the correction for pressure, the contribution of the heat-capacity difference also is often minor, and when it too is dropped the solubility equation becomes

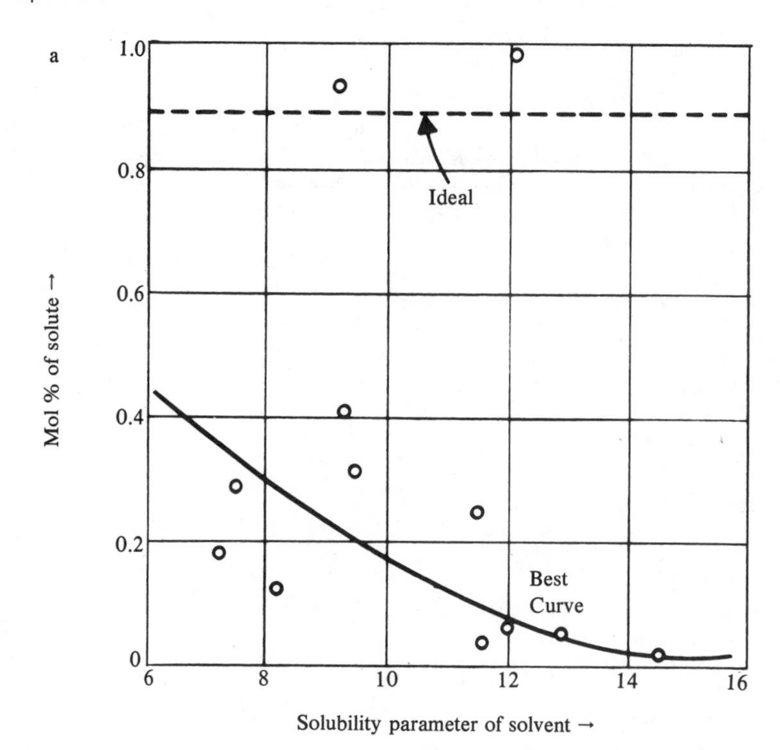

Figure 8.1(a). Solubilities of anthracene in solvents with various solubility parameters. The solvents and their parameters are tabulated following.

Substance	δ	*Substance*	δ
anthracene	9.9	cyclohexane	8.19
naphthalene	9.38	cyclohexanol	11.4
phenanthrene	10.52	diethyl ether	7.54
acetic acid	10.05	ethanol	12.92
acetone	9.51	*n*-hexane	7.27
aniline	11.46	methanol	14.51
1-butanol	11.44	phenol	12.11
carbon disulfide	9.86	1-propanol	12.05
carbon tetrachloride	9.34	2-propanol	11.57
chloroform	9.24		

$$x_2 = \frac{1}{\gamma_2} \exp\left[\frac{\Delta H_{tp}}{R}\left(\frac{1}{T_{tp}} - \frac{1}{T}\right)\right]$$

$$= \frac{1}{\gamma_2} \exp\left[\frac{\Delta S_{tp}}{R}(1 - T_{tp}/T)\right], \quad (8.10)$$

where $\Delta S_{t_p} = \Delta H_{t_p}/T_{t_p}$ is the entropy of fusion at the triple point.

3. Since triple-point temperatures usually are very nearly the same as atmospheric melting points and the latter are more often known (see Table 3.5, for example), the solubility equation becomes, with this substitution,

$$x_2 = \frac{1}{\gamma_2} \exp\left[\frac{\Delta H_m}{R}\left(\frac{1}{T_m} - \frac{1}{T}\right)\right]$$

$$= \frac{1}{\gamma_2} \exp\left[\frac{\Delta S_m}{R}(1 - T_m/T)\right], \quad (8.11)$$

where the subscript m identifies conditions at the atmospheric melting point.

4. A compact solubility equation that requires knowledge only of properties of the pure components is obtained in terms of the Scatchard-Hildebrand equation for the activity coefficient of the solute

$$\gamma_2 = \exp\left[\frac{\Delta S_m}{R}\left(1 - \frac{T_m}{T}\right) - \frac{V_2\phi_1^2(\delta_1 - \delta_2)^2}{RT}\right], \quad (8.12)$$

where ϕ_1 is the volume fraction of the solvent,

$$\phi_1 = V_1x_1/(V_1x_1 + V_2x_2). \quad (8.13)$$

Sometimes the Flory-Huggins correction can give superior results, but neither this extension nor the basic S-H equation does particularly well for the nonhydrocarbon mixtures of Example 8.4. Mixed solvents are represented by the multicomponent form of the S-H equation.

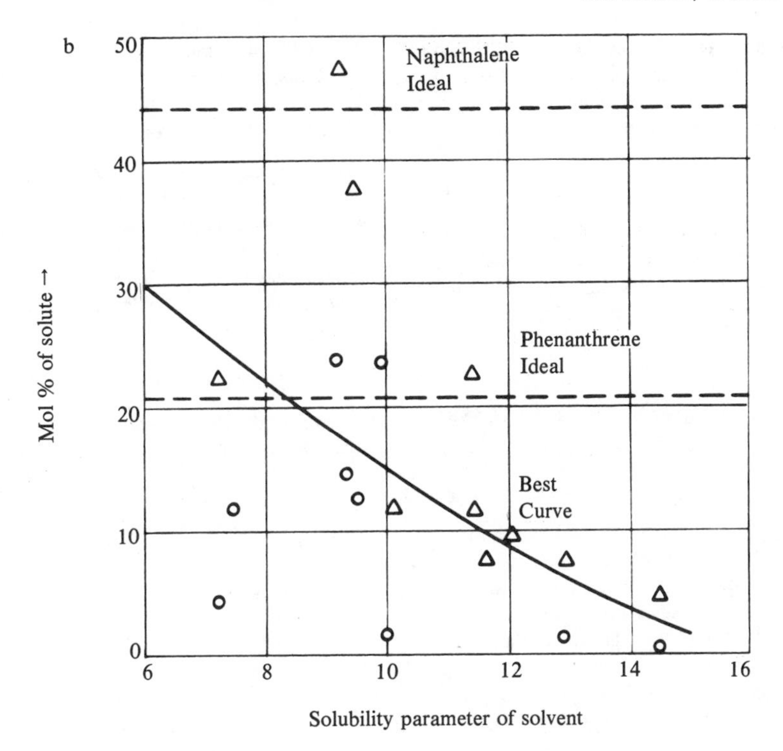

Figure 8.1(b). Solubilities of naphthalene (○) and phenanthrene (△) in solvents with various solubility parameters. Ideal solubilities and best fits to the data are shown.

5. The version of the solubility equation for ideal solutions, with unit activity coefficients,

$$x_2 = \exp\left[\frac{\Delta H_m}{R}\left(\frac{1}{T_m} - \frac{1}{T}\right)\right]$$

$$= \exp\left[\frac{\Delta S_m}{R}\left(1 - \frac{T_m}{T}\right)\right], \qquad (8.14)$$

is associated with the names of Schröder (1893) and van Laar (1908) (Haase, 1956, p. 345), and may be regarded as being based on the Clausius-Clapeyron (1850, 1834) or the van't Hoff equations.

Eq. 8.13 and the other versions of the solubility equation apply also to melt equilibria. When the activity coefficients are ignored, the accuracy frequently is poor. In Figure 8.1(b), for instance, the ideal solubility of naphthalene is 44.1%, but the actual value depends on the nature of the solvent and ranges in this example from 4.4 to 44.1%. The other cases of Figure 8.1 behave similarly.

The nature of the solvent and the enthalpy and temperature of fusion of the solute are the main factors influencing solubility. Even chemically similar compounds may differ substantially in their fusion properties and consequently in their solubilities. For instance, the pertinent properties and the ideal solubilities of the three isomers of chloronitrobenzene at 300 K are:

Isomer	ΔH_m *cal/gmol*	T_m, °*K*	*x*
Ortho	4,546	307.5	0.8303
Meta	4,629	317.6	0.6503
Para	4,965	356.7	0.2661

Since activity coefficients of isomers in the same solvent often are much the same, calculations of ideal solubilities can identify the trend of solubility behaviors, although they may not give absolutely correct values.

The effect of the heat-capacity difference, ΔC_p, is illustrated by the results of Problem 8.1 for the solubility of paraxylene over a range of temperature. Clearly the heat-capacity term contributes little to the calculated solubility at modest displacements from the fusion temperature, 286.39 K.

	x	
T, °*K*	*Complete*	*Neglecting* ΔC_p *term*
200	0.0558	0.0448
250	0.3616	0.3513
286.39	1	1

When vapor pressures are low, they may be substituted for the fugacities in Eq. 8.2. The vapor pressure of the solid must be known, and that of the subcooled liquid may be found by extrapolation of a vapor-pressure curve or equation. Then Eq. 8.2 becomes

$$x_2 = P_s^{sat}/\gamma_2 P_L^{sat}. \tag{8.15}$$

When the amount of extrapolation of the vapor-pressure equation is modest, the accuracy of this simplified method may be fair. It is tried in Examples 8.2 and 8.3, where the solubility of naphthalene in benzene is found to be about 20% high.

8.2.1. Effects of Temperature and Pressure

Eq. 8.14 may be written in the form

$$\ln x_2 = A - B/T, \tag{8.16}$$

where A and B may be sufficiently constant over a moderate temperature range with values characteristic of particular mixtures. This is a widely recognized form of equation for the temperature dependence of other properties; for example, vapor pressures (derived from the Clausius-Clapeyron equation) and viscosities as well as solubilities of gases and liquids. A compact representation of solubilities in nomographic form based on Eq. 8.16 has been constructed by Germann & Germann (1944).

Although the ideal equation, Eq. 8.14, predicts the solubility to increase as the temperature rises, the effect of temperature on the activity coefficient in Eq. 8.11 could be such as to reverse this behavior. Temperature effects like those of Figure 5.20(a)–(e) of liquid-liquid mixtures are conceivable for liquid-solid mixtures, but no specific examples come to mind.

Pressure exerts its effect on solubility primarily through the term $(V_L - V_s)/RT$ of Eq. 8.9, although the activity coefficient also may be affected. When Eqs. 8.2 and 8.4 are combined at constant temperature,

$$d \ln \gamma_2 x_2 = -\frac{V_L - V_s}{RT} dP. \tag{8.17}$$

Introducing

$$d \ln \gamma_2 = \frac{\bar{V}_L - V_L}{RT} dP \tag{8.18}$$

from Table 4.1, the variation of solubility with pressure becomes

$$\frac{\partial \ln x_2}{\partial P} = -\frac{V_L - V_s}{RT} - \frac{\bar{V}_L - V_L}{RT} = -\frac{\bar{V}_L - V_s}{RT} \tag{8.19}$$

$$\simeq -\frac{V_L - V_s}{RT} \tag{8.20}$$

The last expression in this series applies when the partial molal volume of the solute is substantially the same as that of the subcooled liquid, an approximation that may be close for dilute solutions. The smallness of the pressure effect is illustrated by the case of Problem 8.3 in which $dx/dP = -2.7(E-4)$ mol fraction/atm and by the case of hydroxylamine hydrochloride in Figure 5.28(d), which has a pressure coefficient of $-0.4(E-4)$ above 40 C and $+1.34(E-4)$ below 40 C. Effects of really high pressure are illustrated in Figure 8.7.

8.2.2. Mixed Solvents

In ideal cases solubility is independent of the composition of the solvent, but in general it depends on activity coefficients, which are in turn dependent on the nature of the solvent. Use of the multicomponent form of the S-H equation with mixed solvents has been mentioned and is made in Example 8.2. When the data are available, the NRTL or Wilson equation is much preferable.

8.2.3. Immiscible Solvents

In terms of the S-H equation, the ratio of concentrations of a solute in two liquid phases, one of which is identified by an asterisk (*), becomes

$$K = \frac{x_2}{x_2^*} = \frac{\gamma_2^*}{\gamma_2}$$

$$= \exp\left\{\frac{V_2}{RT}[\phi_1^{*2}(\delta_1^* - \delta_2^*)^2 - \phi_1^2(\delta_1 - \delta_2)^2]\right\}, \tag{8.21}$$

where

$$\phi_1 = V_1 x_1/(V_1 x_1 + V_2 x_2), \tag{8.22}$$

$$\phi_1^* = V_1^* x_1^*/(V_1^* x_1^* + V_2^* x_2^*). \tag{8.23}$$

Since activity coefficients depend on composition, the distribution ratio K likewise in general depends on composition, although in dilute solutions or over narrow ranges of composition it is often permissible to assume $K = x_2/x_2^*$ to be constant. In Figure 8.2, for instance, K of methanol is substantially constant over a wide range, but those of ethanol and formic acid are not.

In terms of the symmetric Margules equation, the distribution ratio is

$$K = x_2/x_2^* = \gamma_2^*/\gamma_2$$

$$= (\gamma_2^{*\infty}/\gamma_2^\infty) \exp[(1 - x_2^*)^2 - (1 - x_2)^2] \tag{8.24}$$

$$= K^\infty \exp[(1 - x_2^*)^2 - (1 - Kx_2^*)^2], \tag{8.25}$$

where

$$K^\infty = \gamma^*{}_2^\infty/\gamma_2^\infty. \tag{8.26}$$

Some plots of these results are shown in Example 8.1.

8.2.4. Activity Coefficients

In Chapter 4 the activity coefficients are shown to be correlated by several different kinds of equations, of which the Scatchard-Hildebrand has been used for illustration in the present chapter. Further exploration of the general effects of activity coefficients on solubility can be made, still in an approximate fashion, with the symmetrical Margules equation,

$$\ln \gamma_2 = (1 - x_2)^2 \ln \gamma_2^\infty. \tag{8.27}$$

The ratio of real and ideal solubilities then is given by

$$\ln(x_2/x_{2\,\text{ideal}}) = -\ln \gamma_2 = -(1 - x_2)^2 \ln \gamma_2^\infty. \tag{8.28}$$

Infinite-dilution activity coefficients γ_i^∞ are particularly useful characteristics of an equilibrium system since the parameters of many common correlations are expressible in terms of these

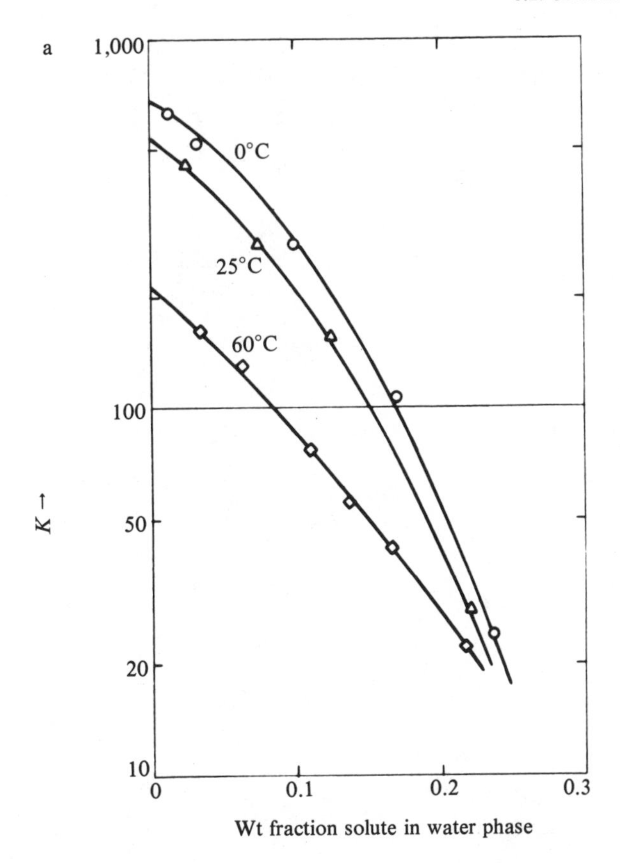

Figure 8.2(a). Distribution coefficient of formic acid between water and benzene at several temperatures. Data from Landolt-Börnstein.

$$K = \frac{\text{gmol solute/liter in water phase}}{\text{gmol solute/liter in benzene phase}}$$

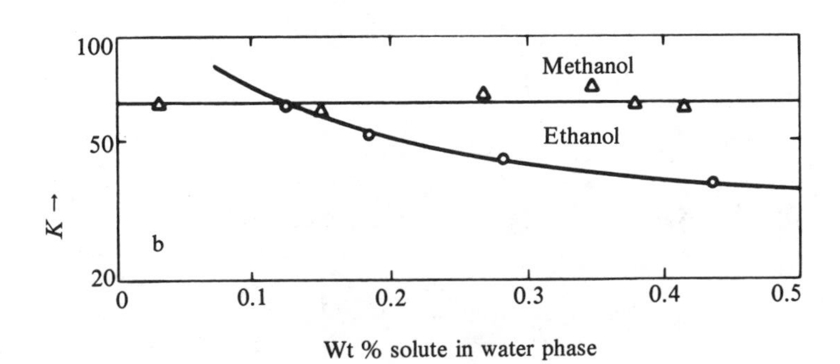

(b). Distribution coefficients of alcohols between water and cyclohexane at 25 C. Data from Landolt-Börnstein.

$$K = \frac{\text{gmol alcohol/liter in water phase}}{\text{gmol alcohol/liter in cyclohexane phase}}$$

coefficients. This form of the solubility equation is plotted in Example 8.1, from which it appears that the solubility relative to the ideal value is enhanced when $\gamma^\infty < 1$ and reduced in the contrary case; the ratio of solubilities also depends on the magnitude of the ideal solubility.

Besides the S-H equation, several others have been used for correlation and prediction of solubility, with varying degrees of success. Some of the literature on this topic will be cited.

1. Activity coefficients obtained with UNIFAC were used to find solubilities and eutectic conditions by Gmehling et al.

Example 8.1. Solubility and Distribution Coefficient under Nonideal Conditions

Since the activity coefficient depends on the composition of the solution, the solubility relative to that under ideal conditions (when the activity coefficient is unity) is involved implicitly in the equation,

$$x_2/x_{id} = 1/\gamma_2 = f(x_2).$$

The specific case to be examined here is when

$$\ln \gamma_2 = A(1-x_2)^2 = \ln \gamma_2^\infty (1-x_2)^2.$$

In these terms the relative solubility becomes

$$x_2/x_{id} = (1/\gamma_2)^{(1-x_2)^2}.$$

The solution for x_2 accomplished with the aid of the N-R method is shown plotted for several values of the ideal solubility, x_{id}. Activity coefficients of solutes often are less than unity, so that range of the plot is of some interest.

When the solute is in contact with two solvents in which the infinite-dilution activity coefficients are γ_2^∞ and $\gamma_2^{*\infty}$, the distribution coefficient is given by Eq. 8.25. That relation is plotted at several values of the solubility in one of the phases, x_2^*, to which the result appears to be relatively insensitive at smaller concentrations.

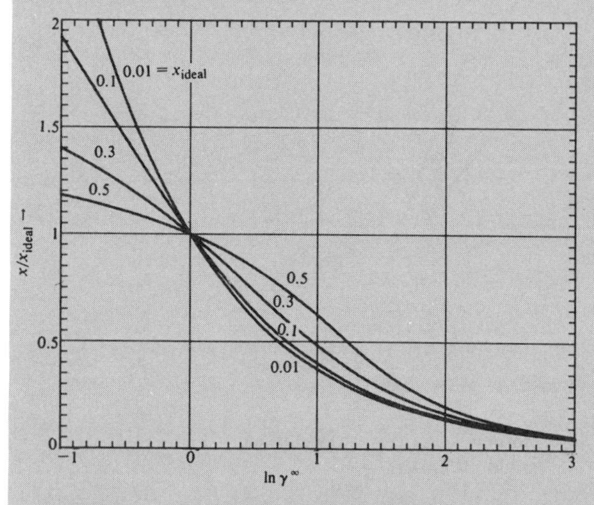

Example 8.1(a). Ratio of real and ideal solubilities when the activity coefficient of the solute is represented by the symmetrical Margules equation.

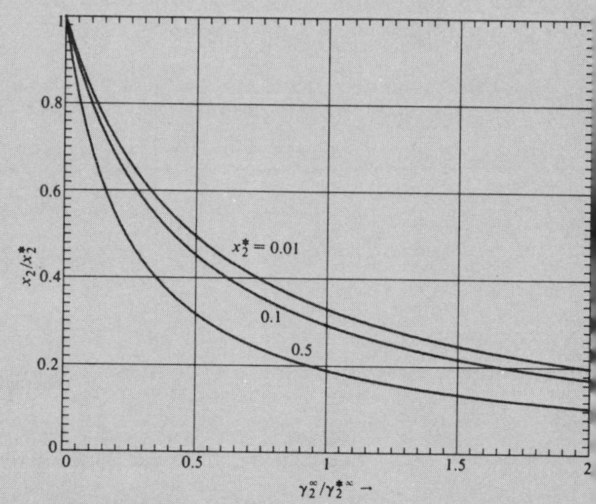

Example 8.1(b). Distribution of a solute between two immiscible solvents in which the infinite-dilution activity coefficients of the solute are known.

Example 8.2. The Solubility of Naphthalene in Benzene, *n*-Hexane, and a Blend Containing 10% Ethanol and 90% Benzene, at 16 C

The properties of the substances are:

Naphthalene: $\delta = 9.9$, $V = 111.5$, $\Delta H_m = 4558$, $T_m = 353.4$.
Benzene: $\delta = 9.2$, $V = 89.4$.
n-Hexane: $\delta = 7.3$, $V = 131.6$.
Ethanol: $\delta = 12.8$, $V = 58.7$.

Units are cal, ml, gmol, °K. One experimental value is known; in benzene, $x_{naph} = 0.217$. $R = 1.987$ cal/(gmol)(K).

Use Eq. 8.13, 8.14:

$$x_2 = \exp\left\{-\left[\frac{\Delta H_m}{R}\left(\frac{1}{T}-\frac{1}{T_m}\right) + \frac{V_2}{RT}\left[\frac{\delta_1-\delta_2}{1+\dfrac{V_2x_2}{V_1(1-x_2)}}\right]^2\right]\right\}. \quad (1)$$

On substitution of the data, the equation becomes, in benzene,

$$x_2 = \exp\left[-\quad 1.4409 + \frac{0.0762}{\left(1+\dfrac{1.2472x_2}{1-x_2}\right)^2}\right] = 0.2273, \quad (2)$$

and in *n*-hexane,

$$x_2 = \exp\left[-\quad 1.4409 + \frac{1.5481}{\left(1+\dfrac{0.8473x_2}{1-x_2}\right)^2}\right] = 0.0585. \quad (3)$$

For the blend the equation is

$$x_2 = \exp\left[-\left(1.4409 + \frac{V_2(\delta_2-\bar{\delta})^2}{RT}\right)\right]. \quad (4)$$

Example 8.2 *(continued)*

The mol fractions of the solvents in terms of that of naphthalene are:

$x_1 = 0.9(1 - x_2)$, benzene

$x_3 = 0.1(1 - x_2)$, ethanol.

The mean solubility parameter is

$$\bar{\delta} = \frac{\Sigma\, x_i V_i \delta_i}{\Sigma\, x_i V_i}$$

$$= \frac{(89.4)(0.9)(1 - x_2)9.2 + 111.5x_2(9.9) + 58.7(0.1)(1 - x_2)12.8}{89.4(0.9)(1 - x_2) + 111.5x_2 + 58.7(0.1)(1 - x_2)}$$

$$= \frac{815.37 + 288.48x_2}{86.33 + 25.7x_2}. \quad (5)$$

The solubility in the blend thus becomes:

$$x_2 = \exp\left[-1.4409 - \frac{111.5}{1.987(289.2)}(9.9 - \bar{\delta})^2\right] \quad (6)$$

$= 0.2315$, by trial solution of Eqs. 5 & 6.

Summary of values of x_2:

		Calculated	
	Experimental	*Correct*	*Approximate*
In benzene	0.217	0.2273	0.2190
In *n*-hexane		0.0585	0.0503
In the blend		0.2315	0.2273

The "approximate" values are obtained by neglecting x_2 on the RHS of Eq. 1.

Example 8.3. Vapor Pressure of Naphthalene and its Solubility in Benzene at 16 C

The vapor pressures, Torr, of solid and liquid naphthalene are

$\ln P_S = 26.708 - 8712/T$,

$\ln P_L = 16.1426 - 3992.01/(T - 71.29)$.

At these low pressures, the ratio of fugacities equals the ratio of the vapor pressures. Accordingly,

$$\gamma_2 x_2 = \frac{f_{2S}}{f_{2(\mathrm{scl})}} = \frac{P_S^{\mathrm{sat}}}{P_L^{\mathrm{sat}}} = \frac{0.0328}{0.1134} = 0.2892.$$

From Example 8.2,

$$\gamma_2 = \exp \frac{0.0762}{\left[1 + \dfrac{1.2472x_2}{1 - x_2}\right]^2}.$$

Therefore,

$$x_2 = 0.2892 \exp \frac{-0.0762}{\left[1 + \dfrac{1.2472x_2}{1 - x_2}\right]^2}$$

$= 0.2793$, by trial.

If x_2 is neglected on the RHS, the direct solution gives $x_2 = 0.2680$, a fair agreement. The result obtained in Example 8.2 is 0.2273.

Example 8.4. Solubility of Naphthalene by the UNIFAC and Solubility Parameter Methods

Experimental data have been collected and UNIFAC calculations made by Gmehling, Anderson, & Prausnitz (1978, pp. 269–273) and are shown in the following table. Data for naphthalene are

$\Delta H_m = 4{,}494$ cal/gmol,

$T_m = 80.2$ C,

$V = 111.5$ ml/gmol,

$\delta = 9.738\ (\text{cal/ml})^{1/2}$.

The temperature is 40 C. The solubility is given by Eq. 8.12:

$$x_2 = \exp\left[\frac{\Delta H_m}{R}\left(\frac{1}{T_m} - \frac{1}{T}\right) - \frac{V_2}{RT}\phi_1^2(\delta_1 - 9.738)^2\right]$$

$$= \exp\left\{-0.8214 - 0.1792\left[\frac{\delta_1 - 9.738}{1 + \dfrac{111.5x_2}{V_1(1-x_2)}}\right]^2\right\}$$

Calculations also are made by adding the Flory-Huggins correction, $\ln\theta_2 + 1 - \theta_2$, to the S-H equation, where $\theta_2 = (1-\phi_1)/x_2$.

As the summary table shows, the UNIFAC calculations are within 10% or so of the experimental data, but neither the S-H nor S-H-F-H come that close in many instances. There does not seem to be any explanation in terms of polarity differences as measured by the dipole moments shown in the last column.

	Solubility, mol %						
Substance	*Exptl*	*UNIFAC*	*S-H*	*S-H-F-H*	*V*	*δ*	*Dipole moment*
methanol	4.4	4.8	0.64	0.64	40.5	14.51	1.7
ethanol	7.3	5.4	4.9	4.9	58.4	12.92	1.7
1-propanol	9.4	9.3	11.3	11.3	74.7	12.05	1.7
2-propanol	7.6	9.3	16.3	16.3	76.5	11.57	1.7
1-butanol	11.6	11.1	18.8	18.7	91.5	11.44	1.8
n-hexane	22.2	25.9	11.5	11.5	131.6	7.27	0
cyclohexanol	22.5	20.5	20.0	20.0	160.0	11.40	1.7
acetic acid	11.7	12.5	40.1	38.3	57.2	10.05	1.3
acetone	37.8	35.8	42.2	41.4	73.5	9.51	2.9
chloroform	47.3	47.0	37.8	37.5	80.2	9.24	1.1
IDEAL	44.1						

(1978), generally in good agreement with experiment. Some of their data are quoted in Example 8.4, Figure 8.1, and Figure 8.5(a).

2. The fugacity ratio $f_{2s}/f_{2(\text{scl})}$ of Eq. 8.2 was found directly from the Lee-Kesler equation by Masuoka et al. (1979) and applied to calculation of solubilities and eutectic diagrams. It was necessary to introduce two interaction parameters for each pair of substances; these were given for 32 pairs and were correlated roughly in terms of the ratios of critical volumes. Figure 8.5(c) is from their paper.

3. Solid-liquid phase equilibria were calculated with the Wilson equation by Morimi & Nakanishi (1977). They found that agreement with experiment was better with this equation than with the S-H equation. Measurements of binary liquid-solid equilibria were applied to the prediction of ternary equilibria of molten salts. Figure 8.9(b) is a result of this work.

4. The T-K-Wilson equation with parameters from VLE data was used by Muir & Howat (1982) to predict a ternary liquid-solid binodal curve, which is given as Figure 8.9(a).

5. Several molten salt systems were analyzed by Null (1967) with the Wilson equation; account was taken of the effect of temperature on the parameters. Activity coefficients were evaluated from liquidus measurements at low concentrations and then applied to predicting both liquidus and solidus behavior over the entire composition range. Figure 8.5(b) is his example.

6. Equilibria of mixtures of carbon dioxide with hydrocarbons in the vapor-liquid-solid range were predicted accurately by Soave (1979) with his version of the Redlich-Kwong equation of state. Binary interaction parameters were derived from measurements of carbon dioxide equilibria with individual hydrocarbons.

7. The group contribution method, ASOG, was employed by Unno et al. (1979) to predict liquid-solid equilibria in several organic systems, of which Figure 8.5(d) is one case.

8.3. MELT EQUILIBRIA

From a thermodynamic point of view, the processes of dissolving a solid directly or of melting it first and then dissolving it are identical, provided that the temperatures, pressures, and compositions are the same in both cases. Accordingly, the solubility equations of Section 8.2 apply also to melt equilibria. The effect of temperature is of perhaps greater importance to melt equilibria, partly because of the high levels and wide range of temperature that are frequently encountered. A variety of minimum and maximum temperatures can occur. Complications can occur because of intermolecular compound formation. Equilibria between several solid phases may be involved; here experimental efforts are particularly demanding because of the slowness of diffusion in solid phases. Figure 12.14 and others in that chapter show

Example 8.5. Activity Coefficients from Eutectic Data

Intermolecular compounds are formed in the systems acetic acid + urea and phenol + acetamide, as shown on the diagrams. Enthalpies of fusion of the compounds are estimated by taking weighted means of the entropies of fusion of the parent compounds.

	ΔH_m	T_m	$\Delta H_m/T_m$
acetic acid (A)	2,755	289.8	9.51
urea (U)	3,472	406.0	8.55
A_2U	2,888	314.2	9.19
phenol (P)	2,695	314.1	8.58
acetamide (Ac)	3,396	355.1	9.56
P_2Ac	2,810	315.5	8.91
PAc	2,786	307.2	9.07

The activity coefficients at the eutectic compositions are determined from

$\ln \gamma_i = \ln(1/x_i) - (1 - T/T_m)\,\Delta H_m/RT.$

The results are tabulated:

Pair	T	x_1	x_2	γ_1	γ_2
$A + A_2U$	284.2	0.85	0.15	1.0706	4.0911
$A_2U + U$	310.2	0.87	0.13	1.0829	2.0361
$P + P_2Ac$	300.7	0.61	0.39	1.3524	2.0565
$PAc + Ac$	305.7	0.82	0.18	1.1925	2.5524

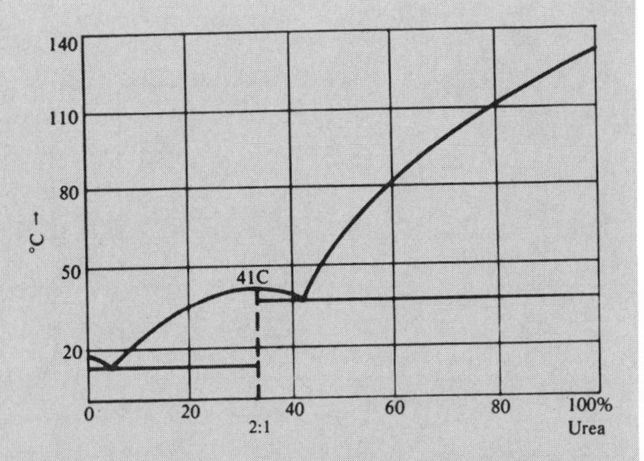

Example 8.5(a). The system acetic acid + urea forms a 2:1 intermolecular compound that melts at 41 C.

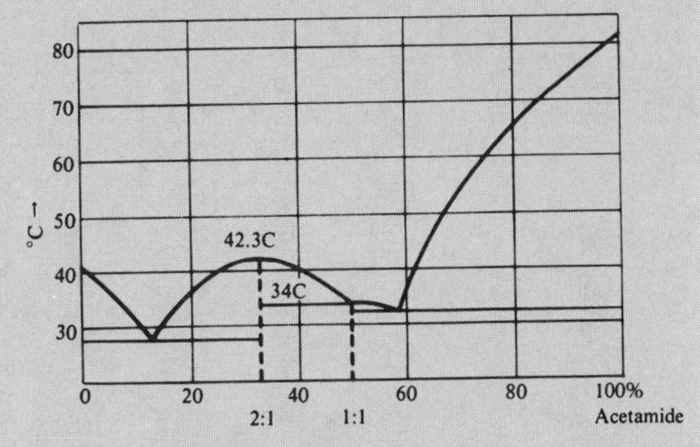

Example 8.5(b). The system phenol + acetamide forms a 2:1 intermolecular compound that melts at 42.3 C and a 1:1 compound that melts near 34 C. Data for both diagrams are taken from Landolt-Börnstein.

experimental heating/cooling curves; on the diagram for establishing the freezing point of benzene, for example, the dip in the curve perhaps could have been avoided at a slower rate of cooling.

In the development of solubility relations, starting with Eq. 8.1, both the mol fraction and activity coefficient of the solid phase were taken as unity; but under melt conditions various degrees of miscibility of solid phases may occur, from complete miscibility to complete immiscibility. In such cases the condition of equality of partial fugacities of the solute becomes

$$\gamma_{2s}x_{2s}f_{2s} = \gamma_{2L}x_{2L}f_{2(\mathrm{scl})}, \tag{8.29}$$

and the equivalent of Eq. 8.9 without the pressure term becomes

$$\ln\frac{\gamma_{2L}x_{2L}}{\gamma_{2s}x_{2s}} = \frac{\Delta H_m}{R}\left(\frac{1}{T_m} - \frac{1}{T}\right) - \frac{\Delta C_p}{R}[\ln(T_m/T) - T_m/T + 1], \tag{8.30}$$

on substitution from Eq. 8.4. This equation could be used to find activity coefficients in the solid phase when all other data have been established in some way. In some cases liquid-phase activity coefficients obtained from VLE data (as in Figure 8.3 for some molten metals) may be extrapolatable down to the temperature range at which solidification can occur. Although comparatively few data of solid phase activity coefficients appear to have been published, the pattern may be similar to that of liquid phase behavior, as the data of Figure 8.4 of CsCl + CsBr suggest. Examples of complete miscibilities of organic solids are shown in Figure 5.22 and of partial miscibilities on Figure 5.25(f) and (g). The most common behaviors that occur and have been investigated in the organic field are complete immiscibility of the solid phases; several such examples are shown in Chapter 5 and in this chapter.

Ideal behavior: Since melt equilibria often deviate substantially from ideal behavior, data of activity coefficients are necessary for accurate work. In the absence of such information, however, calculations based on the Schröder or related equations can be made for purposes of orientation, or for fun. Several examples of this kind of calculation are provided in this chapter or asked for in the problem section.

8.4. BINARY MIXTURES

Classification of types of melt equilibria is based primarily on the number and extent of miscibilities of the liquid and solid

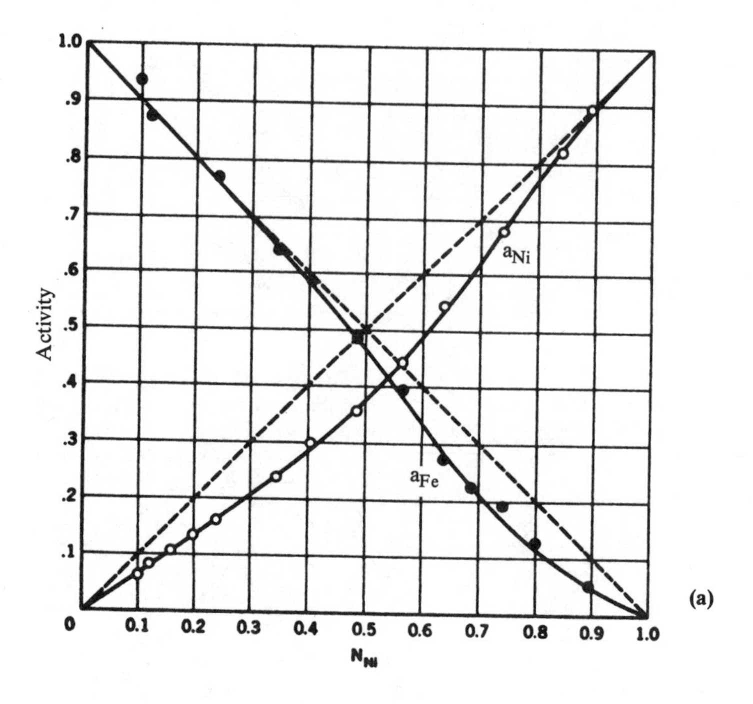

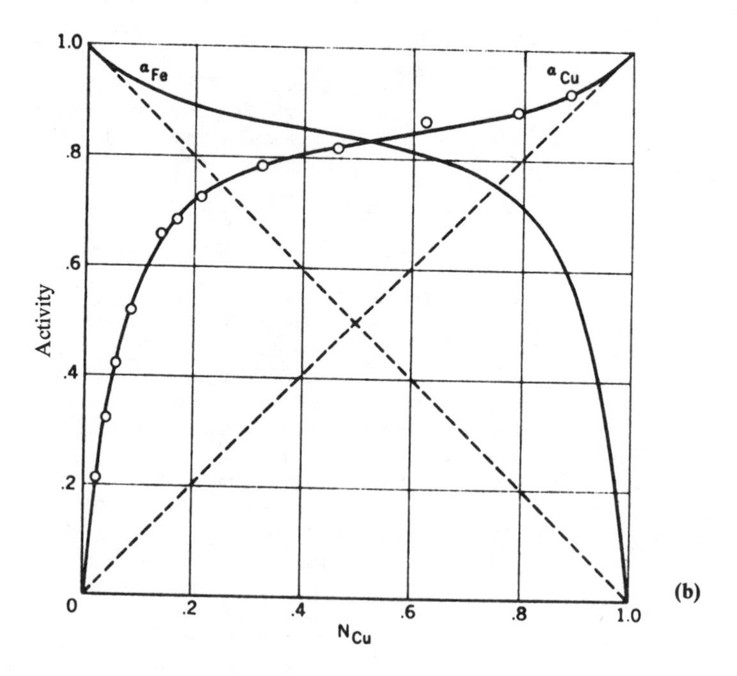

Figure 8.3. Thermodynamic data of some molten metals at elevated temperatures. (a) Activities of iron and nickel at 1,600 C (Zellars et al., 1959). (b) Activities of copper and iron at 1,550 C (Morris & Zellars, 1956). (c) Activity coefficients of mixtures of molten thallium and lead. The unusual behavior is due in part to the presence of the compound $PbTl_2$ (Hildebrand & Sharma, 1929). (d) Partial and total enthalpies of mixing in the iron-nickel system (Belton & Fruehan, 1967). (e) Entropies of mixing of some mixtures of metals (Kubashewski & Alcock, "Metallurgical Thermochemistry, 1979).

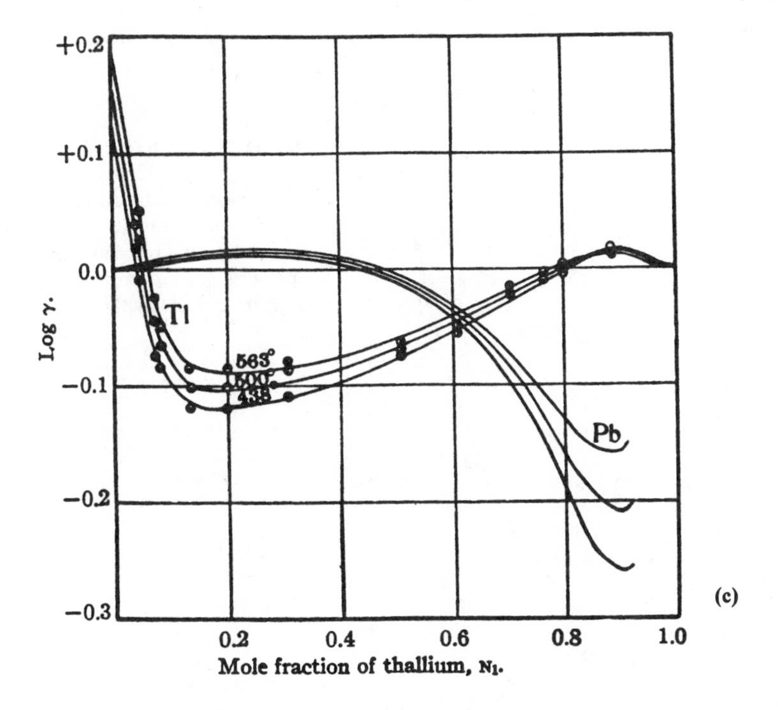

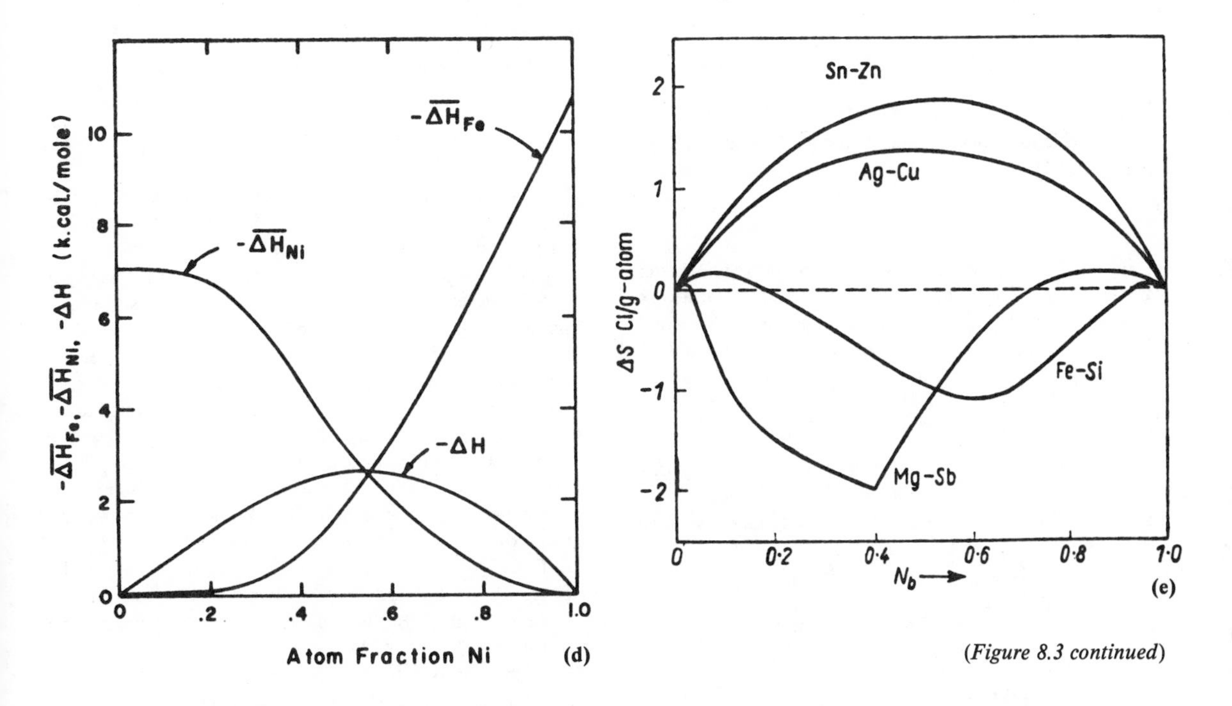

(Figure 8.3 continued)

phases that are present and the nature and melting temperature of intermolecular compounds that may form.

Compound formation: Diagrams of several systems with intermolecular compounds are shown with Problem 8.17. Such phase diagrams may be regarded as composed of several simpler binary diagrams. In Figure 8.6 the system of acetic acid (A) and urea (B) consists of the two binary systems of $A + A_2B$ and $A_2B + B$. Information about thermal and other properties of intermolecular compounds is not often known. Moreover, efforts to estimate entropies of fusion and eutectic points with the Schröder equation often do not appear to give reasonable values, as the results of Problem 8.17 show.

Completely miscible solid and liquid phases: The T-x diagrams of completely miscible systems—for example, Figure 8.5(b)—resemble those of vapor and homogeneous liquid mixtures. Activity coefficients at infinite dilution of both phases can be found by a method analogous to those described in Section 4.15. This was done, for instance, by Null (1967), who then deduced the Wilson parameters and used them to calculate the complete liquidus and solidus.

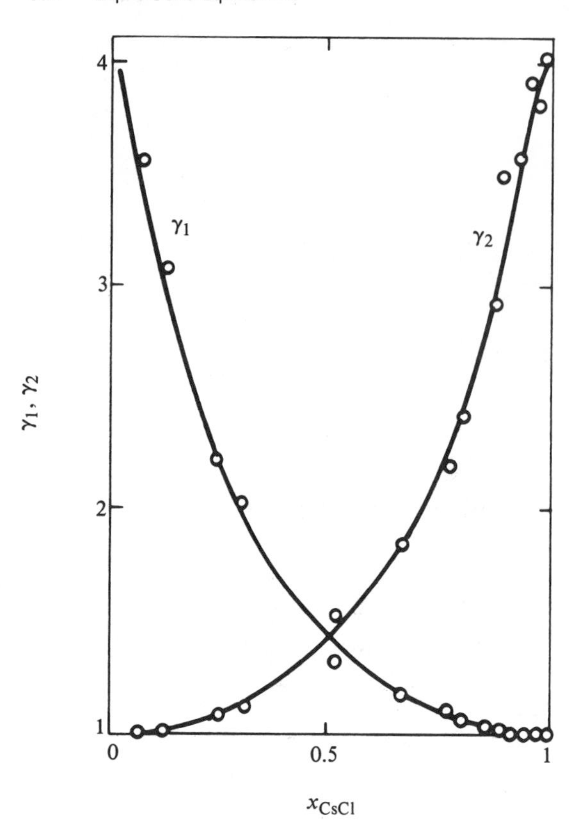

Figure 8.4. Activity coefficients in solid solutions of caesium chloride (1) + caesium bromide (2) at 25 C (Kogan, 1968, p. 274).

Immiscible solid phases: Most common organic systems are in this category since even isomers rarely form mixed crystals, even the optical isomers cited in Problem 8.14. Calculations for immiscible systems can be made readily and with often worthwhile results.

8.4.1. Eutectic Compositions

For ideal mixtures, the eutectic temperature and composition can be found when the enthalpies and temperatures of fusion are known. When the Schröder equations of each component,

$$x_1 = \exp\left[\frac{\Delta H_{m1}}{R}\left(\frac{1}{T_{m1}} - \frac{1}{T}\right)\right], \tag{8.31}$$

$$1 - x_1 = \exp\left[\frac{\Delta H_{m2}}{R}\left(\frac{1}{T_{m2}} - \frac{1}{T}\right)\right], \tag{8.32}$$

are plotted as T against x_1, the curves will intersect at the eutectic condition (T_e, x_e) and will be, moreover, the melt curves. When only the eutectic conditions are wanted, the two equations are combined into one as

$$1 = \exp\left[\frac{\Delta H_{m1}}{R}\left(\frac{1}{T_{m1}} - \frac{1}{T_e}\right)\right]$$

$$+ \exp\left[\frac{\Delta H_{m2}}{R}\left(\frac{1}{T_{m2}} - \frac{1}{T_e}\right)\right], \tag{8.33}$$

from which T_e can be found by trial, then substituted into one of the individual Schröder equations to find the composition, x_e. A similar calculation can be made to find the eutectic conditions for any number of components.

When the activity coefficients are known as functions of temperature and composition, the following equations can be solved simultaneously for the eutectic conditions:

$$\gamma_1 x_1 = \exp\left[\frac{\Delta H_{m1}}{R}\left(\frac{1}{T_{m1}} - \frac{1}{T_e}\right)\right], \tag{8.34}$$

$$\gamma_2(1 - x_1) = \exp\left[\frac{\Delta H_{m2}}{R}\left(\frac{1}{T_{m2}} - \frac{1}{T_e}\right)\right]. \tag{8.35}$$

In one of the simplest cases the activity coefficients are represented by the equations

$$\ln \gamma_1 = A(1 - x_1)^2/T, \tag{8.36}$$

$$\ln \gamma_2 = Ax_1^2/T, \tag{8.37}$$

and the equations to be solved by trial for the eutectic conditions are

$$\frac{A(1 - x_1)^2}{T_e} + \ln x_1 = \frac{\Delta H_{m1}}{R}\left(\frac{1}{T_{m1}} - \frac{1}{T_e}\right), \tag{8.38}$$

$$\frac{Ax_1^2}{T_e} + \ln(1 - x_1) = \frac{\Delta H_{m2}}{R}\left(\frac{1}{T_{m2}} - \frac{1}{T_e}\right). \tag{8.39}$$

Since $1/T_e$ occurs linearly in these equations, it can be eliminated readily, and the problem becomes one of a solution of an equation in one unknown, x_1.

Although calculations for ideal mixtures are easily performed, the results should be used with discretion, since many mixtures are significantly nonideal. For instance, the occurrence of partial miscibility of solid phases is not detected by ideal calculations, and they do form occasionally in organic systems. Figure 8.5 presents comparisons of experimental and calculated melt diagrams. No way has been devised of establishing from molecular structure how nearly ideal a particular mixture will be, since even mixtures of isomers can be nonideal; systems 2 and 3 of Problem 8.16 are such examples. The best predictive methods are the ASOG and UNIFAC, which were applied on Figures 8.5(a) and (d), but neither is yet able to treat mixtures of isomers, and the problem of substantial extrapolation of temperature has not been solved.

Extreme pressures can affect eutectic compositions and temperatures to a substantial degree; some examples are in Figures 5.28 and 8.7. The lowering of the temperature at an assumed fixed composition could be estimated with the Clausius-Clapeyron equation, but in ordinary chemical processing the effect is small and is usually ignored.

8.4.2. Activity Coefficients

Both activity coefficients can be evaluated from a knowledge of the eutectic conditions by solving

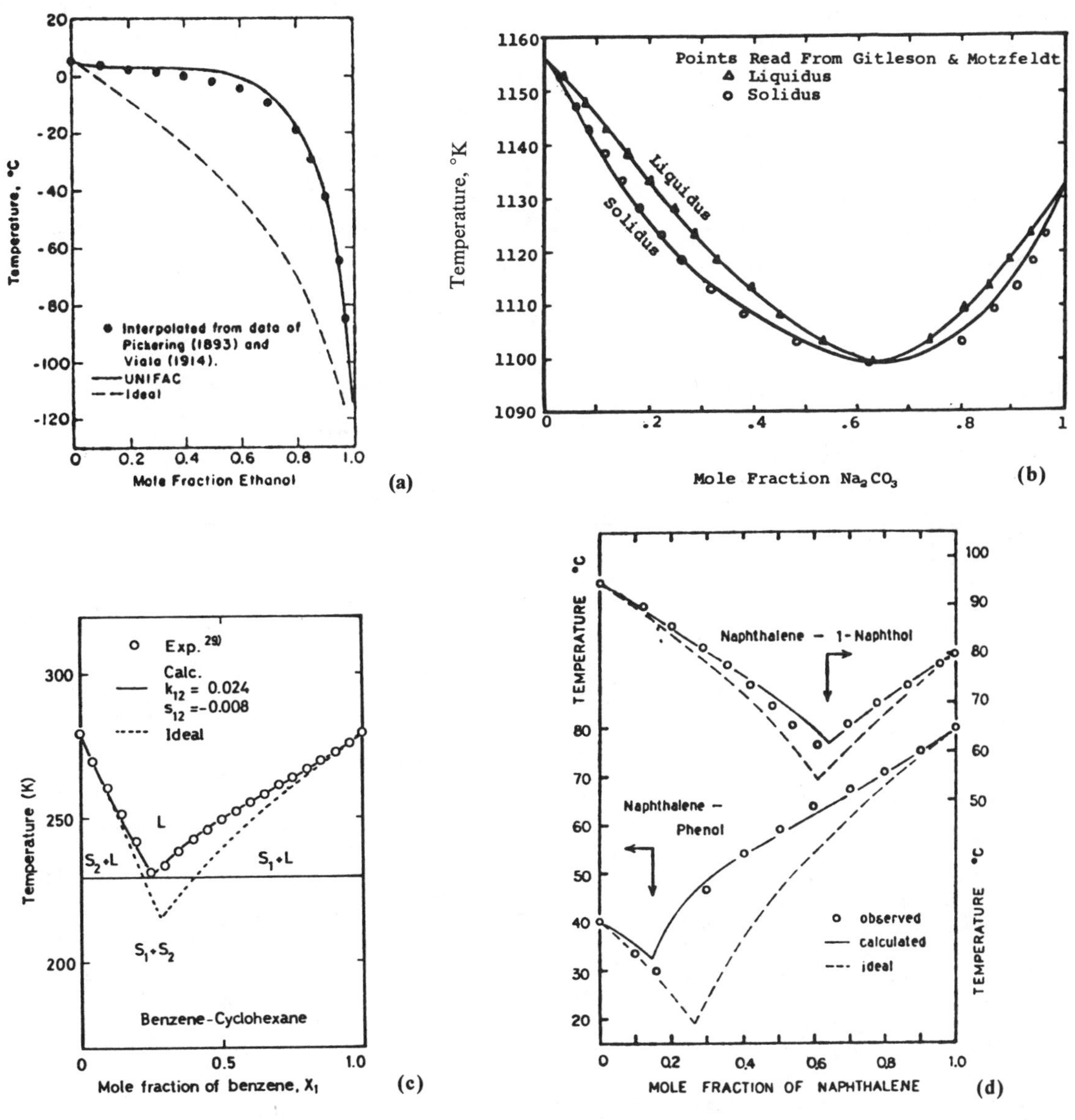

Figure 8.5(a). Prediction of liquid-solid equilibria of ethanol + benzene with UNIFAC activity coefficients. (Gmehling et al., 1978). Copyright American Chemical Society; used with permission. (b). Prediction of the phase diagram of sodium carbonate + sodium sulfate with the aid of the Wilson equation (Null, 1967). (c) Prediction of liquid-solid equilibria for mixtures of benzene and cyclohexane with the aid of the Lee-Kesler equation (Masuoka et al., 1979). (d) Prediction of liquid-solid equilibria of the systems naphthalene + phenol and naphthalene + 1-naphthol with the aid of the ASOG method and employing vapor-liquid data of the benzene + phenol system (Unno et al., 1979).

$$\gamma_i = \frac{1}{x_i} \exp\left[\frac{\Delta H_{mi}}{R}\left(\frac{1}{T_{mi}} - \frac{1}{T_e}\right)\right]. \tag{8.40}$$

Then they can be used to find the parameters of a correlating equation such as the NRTL or Wilson so that the complete melting curves can be determined. With the Wilson equation, practical difficulties can arise when the parameters are found from a limited amount of data such as a eutectic condition, since the process can lead to negative values of the parameters that are unacceptable if the equation is to represent activity coefficients over the entire composition range. For instance, the activity coefficients and the Wilson and NRTL parameters

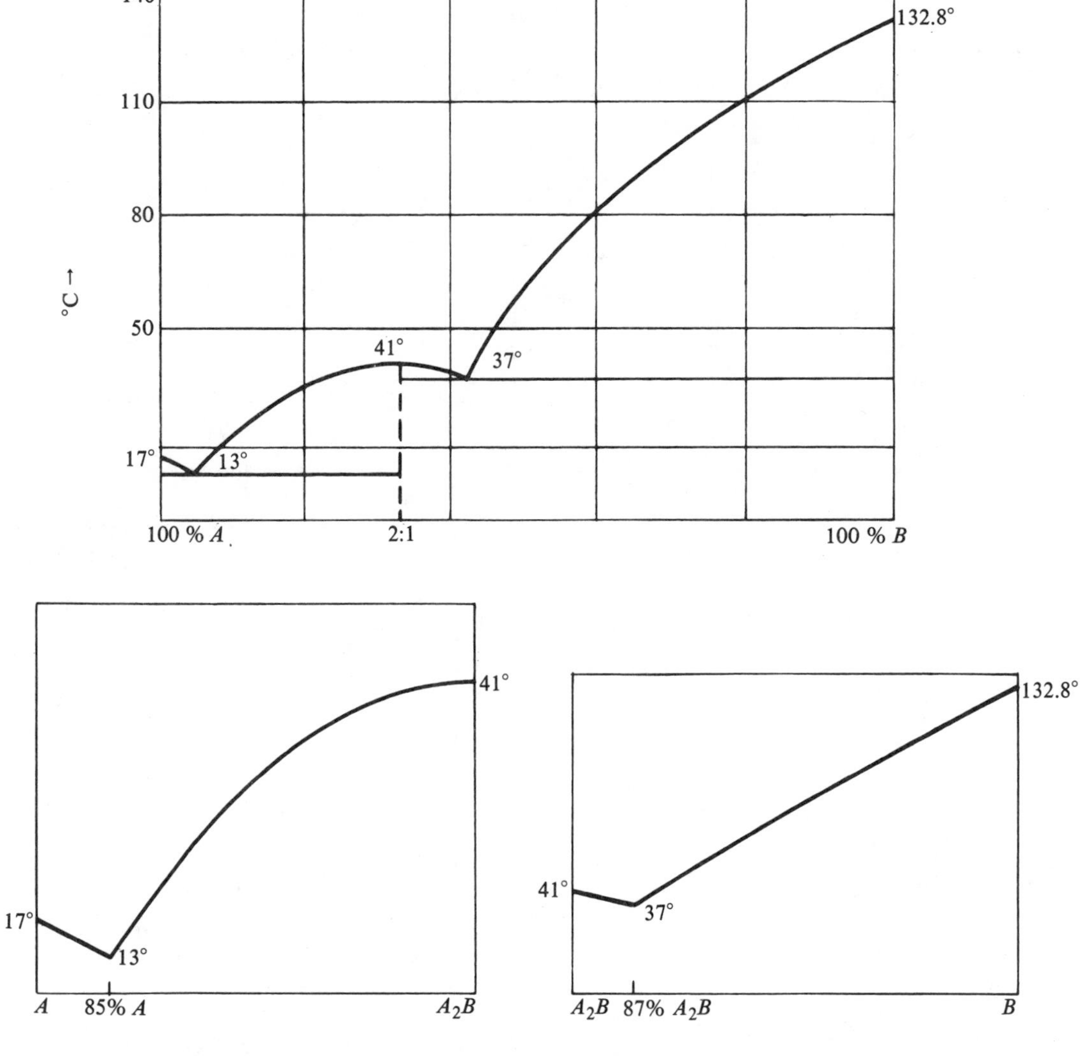

Figure 8.6. The system acetic acid (A) + urea (B), which forms an intermolecular compound A_2B. The overall diagram is made up of two parts, of the compound with each of the pure substances.

based on the eutectic conditions of the nitroaniline isomers of Problem 8.16 are:

			Wilson		NRTL, $\alpha = 0.2$	
γ_1	γ_2	γ_3	Λ_{ij}	Λ_{ji}	τ_{ij}	τ_{ji}
0.9539	0.7039	—	1.6472	1.0582	3.0059	−2.2628
0.9394	—	1.6371	−1.2463	10.2530	2.1591	−1.0630
—	0.6123	1.6523	−0.9108	17.3190	3.4799	−1.2842

Plots of activity coefficients based on these equations are shown in Figure 8.8. Clearly the Wilson equation cannot represent the activity coefficients over the full range of conditions of two of these pairs. Similar plots based on VLE data are shown in Chapter 4.

The application of binary eutectic data to the prediction of ternary melting behavior is described later in this chapter.

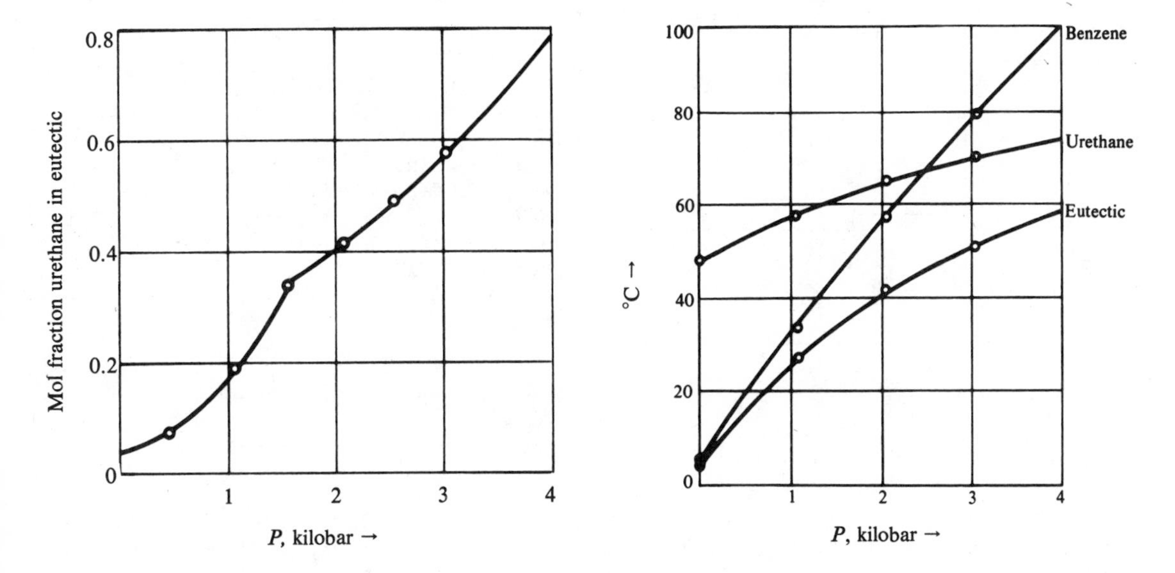

Figure 8.7. Effect of pressure on the melting temperatures of benzene and urethane and on the temperature and composition of their eutectic (data of Pushin & Grebenshchikof, Z physik Chem 118, 276, 447 (1925).

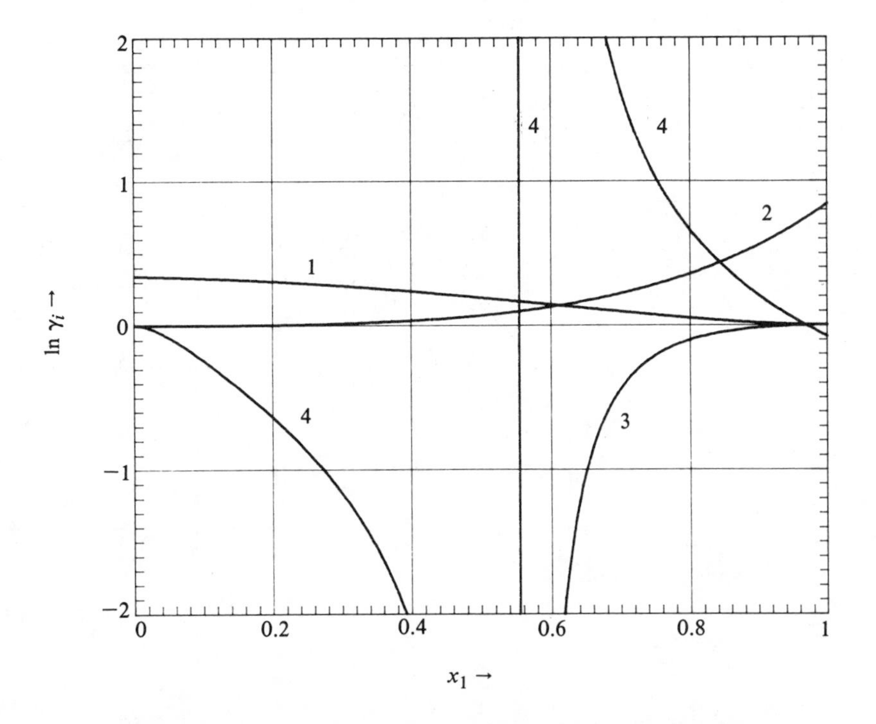

Figure 8.8. Plots of the Wilson and NRTL activity coefficients with parameters deduced from the eutectic composition of mixtures of ortho- and para-nitroaniline. The eutectic composition is $x_1 = 0.83$. The Wilson parameters are $\Lambda_{12} = -1.2463$ and $\Lambda_{21} = 10.2530$. The NRTL parameters are $\alpha_{12} = 0.2$, $\tau_{12} = 2.1591$, and $\tau_{21} = -1.063$. Composition data are cited in Problem 8.16.

Curve 1 is of ln γ_1 with the NRTL equation, curve 2 that of ln γ_2 with the NRTL equation, curve 3 that of ln γ_1 with Wilson, and curve 4 that of ln γ_2 with Wilson. Clearly the Wilson correlation is not valid, at least over most of the concentration range; this result emphasizes the danger of basing Wilson parameters on a single equilibrium measurement.

Example 8.6. Phase Diagram of Ethanol + Water from the Eutectic Conditions

The eutectic is at 150.2 K, 79.61 mol% ethanol. Enthalpies and temperatures of fusion are

	cal/gmol	*°K*
Ethanol	1,461.9	158.7
Water	1,436.0	273.2

The Wilson equation will be used to take advantage of its built-in temperature effect. Activity coefficients are found with Eqs. 8.34, 8.35.

$$\ln 0.7961\,\gamma_1 = -\frac{1{,}461.9}{RT}\left(1-\frac{T}{158.7}\right),$$

$$\ln 0.2039\,\gamma_2 = -\frac{1{,}436}{RT}\left(1-\frac{T}{273.2}\right).$$

At $T = 150.2$, the activity coefficients are $\gamma_1 = 0.9663$, $\gamma_2 = 0.5621$. By the method of Example 4.7 and Table C.5, the Wilson parameters are $\Lambda_{12} = 1.3400$, $\Lambda_{21} = 1.7104$. The exponential Wilson parameters are

$$\lambda_{12}/R = -150.2 \ln \frac{58.4(1.3400)}{18.1} = -219.90,$$

$$\lambda_{21}/R = -150.2 \ln \frac{18.1(1.7104)}{58.4} = 95.33.$$

The equations for the Wilson parameters as functions of temperature become:

$$\Lambda_{12} = 0.3099 \exp(219.90/T), \tag{1}$$

$$\Lambda_{21} = 3.2265 \exp(-95.33/T). \tag{2}$$

The equations for the activity coefficients are

$$\ln \gamma_1 = -\ln(x_1 + \Lambda_{12}x_2) + \beta x_2, \tag{3}$$

$$\ln \gamma_2 = -\ln(\Lambda_{21}x_1 + x_2) - \beta x_1, \tag{4}$$

$$\beta = \Lambda_{12}/(x_1 + \Lambda_{12}x_2) - \Lambda_{21}/(\Lambda_{21}x_1 + x_2). \tag{5}$$

At concentrations above that of the eutectic, the temperature at a specified concentration, x_1, is found by trial with

$$\ln \gamma_1 x_1 = -\frac{1{,}461.9}{RT}\left(1-\frac{T}{158.7}\right), \tag{6}$$

and at concentrations below those of the eutectic from

$$\ln \gamma_2 x_2 = -\frac{1{,}436}{T}\left(1-\frac{T}{273.2}\right), \tag{7}$$

together with the Wilson equations. Trial temperatures are put in from the keyboard until the condition for the left branch, $F1 = 0$, or for the right branch, $F2 = 0$, is satisfied. The results are tabulated and plotted along with data from Lange's *Handbook* (p. *10*.69, 11th ed.). Some improvement possibly could be obtained by incorporating the effect of heat-capacity difference.

```
10 INPUT X1
20 X2=1-X1
30 SHORT F1,F2
40 INPUT T
50 A=18.1/58.4*EXP(219.904/T)
60 B=58.4/18.1*EXP(-(95.33/T))
70 C=A/(X1+A*X2)+B/(B*X1+X2)
80 G1=-LOG(X1+A*X2)+C*X2
90 G2=-LOG(B*X1+X2)-C*X1
100 F2=-G2-LOG(X2)-1436/1.987/T*(1-T/273.2)
110 F1=-G1-LOG(X1)-1461.9/1.987/T*(1-T/158.7)
120 DISP F1;F2
130 GOTO 40
140 END
```

	T	
x_1	*Left*	*Right*
0	273.2	
0.1	261.4	
0.2	247.8	
0.3	233.4	
0.4	218.4	
0.5	203.0	
0.6	187.0	
0.7	169.5	
0.75	159.9	147.8
0.7961	150.2	150.2
0.85	137.1	152.75
0.90	121.8	154.9
0.95		156.9
1		158.7

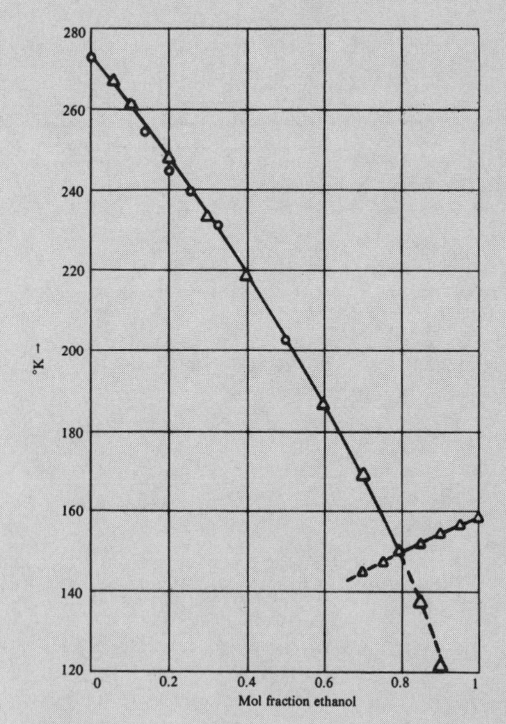

Example 8.6. Freezing curves of mixtures of ethanol + water. The points (○) are from Lange's *Handbook*; the others are calculated.

Example 8.7. Finding the Eutectic Condition with the Ideal, Margules, and Wilson Equations

Two substances have fusion entropies of 13.0, fusion temperatures $T_1 = 350$ K and $T_2 = 425$ K, volume ratio $V_1/V_2 = 0.8$, and infinite-dilution activity coefficients $\ln \gamma_1^\infty = 2$ and $\ln \gamma_2^\infty = 0.5$. The eutectic temperature and composition will be found with (a) the ideal equation; (b) the Margules equation; (c) the Wilson equation.

a. Ideal mixture:

$$x_1 + x_2 = 1 = \exp\left[\frac{\Delta S_1}{R}\left(1 - \frac{T_1}{T}\right)\right] + \exp\left[\frac{\Delta S_2}{R}\left(1 - \frac{T_2}{T}\right)\right]$$

$$= \exp\left[\frac{13}{R}\left(1 - \frac{350}{T}\right)\right] + \exp\left[\frac{13}{R}\left(1 - \frac{425}{T}\right)\right]. \qquad (1)$$

By trial, $T = 339.1$, $x_1 = 0.8103$.

b. Margules equations.

Assume the parameters to be inversely proportional to the temperature and that the given γ_i^∞ are at the melting temperatures.

$$A = \frac{425}{T} \ln \gamma_1^\infty = \frac{850}{T},$$

$$B = \frac{350}{T} \ln \gamma_2^\infty = \frac{175}{T}.$$

The equations of the activity coefficients are

$$\ln \gamma_1 = [A + 2(B - A)x_1]x_2^2,$$

$$\ln \gamma_2 = [B + 2(A - B)x_2]x_1^2.$$

The melt equations are

$$x_1 = \frac{1}{\gamma_1} \exp\left[\frac{13}{R}\left(1 - \frac{350}{T}\right)\right], \qquad (2)$$

$$(1 - x_1) = \frac{1}{\gamma_2} \exp\left[\frac{13}{R}\left(1 - \frac{425}{T}\right)\right]. \qquad (3)$$

This system of equations is solved graphically as shown, with the result: $T = 344.0$, $x_1 = 0.897$.

c. Wilson equations.

The values of the parameters are found from the infinite-dilution activity coefficients as

$$\Lambda_{12} = 0.0804 \text{ at } 425 \text{ K},$$

$$\Lambda_{21} = 1.5204 \text{ at } 350 \text{ K}.$$

Accordingly, the exponential Wilson parameters are

$$\lambda_{12} = -425 \ln 0.8(0.0804) = 1{,}166.15,$$

$$\lambda_{21} = -350 \ln 1.25(1.5204) = -224.74,$$

and the Wilson parameters as functions of temperature are

$$\Lambda_{12} = 1.25 \exp(-1{,}166.15/T),$$

$$\Lambda_{21} = 0.8 \exp(224.74/T).$$

The melt equations are Eqs. 2 and 3. The solution again is obtained graphically, with the results $T = 343.1$, $x_1 = 0.871$.

The Margules and Wilson predict the eutectic conditions to be similar, but the ideal result is much different.

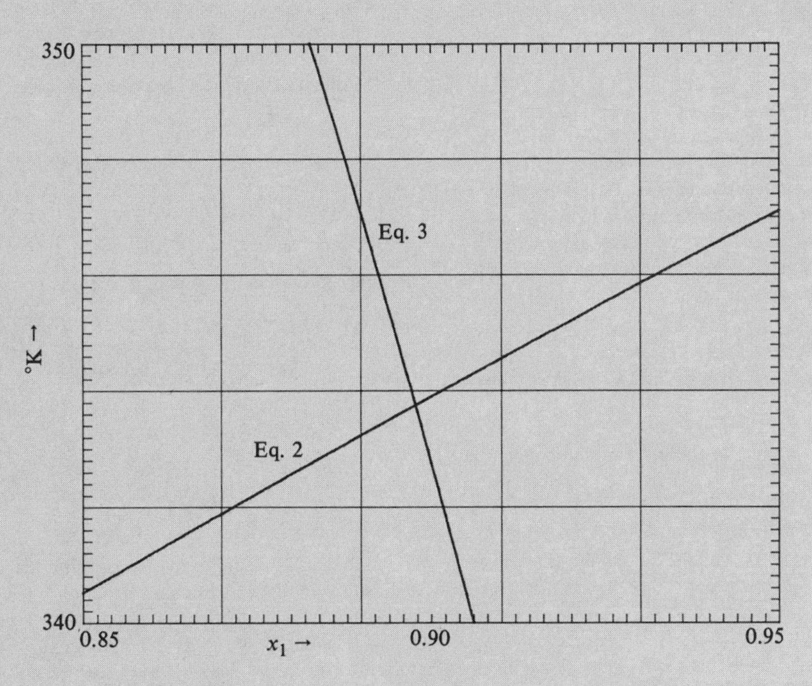

Example 8.7(a). Graphical solution for the eutectic point with the Margules equations. $T = 344$ K, $x_1 = 0.897$.

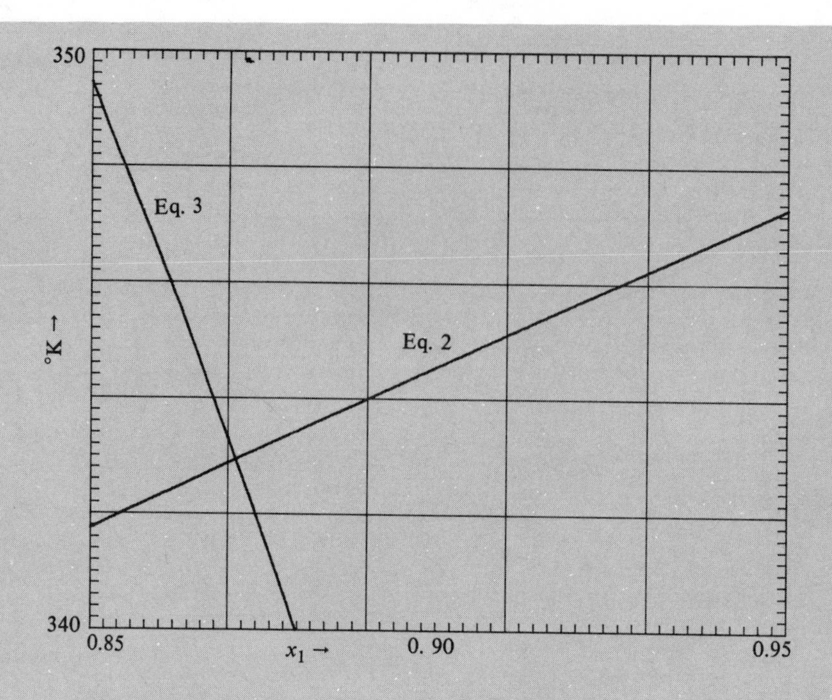

Example 8.7(b). Graphical solution for the eutectic point with the Wilson equation. $T = 343.1, x_1 = 0.871$.

8.4.3. Enthalpies of Fusion of Intermolecular Compounds

If the unlikely assumption of ideal behavior of the intermolecular compound is made, the enthalpy of fusion can be found from a eutectic condition with the Schröder equation,

$$\Delta H_m = \frac{R \ln(1/x_e)}{1/T_e - 1/T_m}. \tag{8.41}$$

This does not appear to be a particularly useful calculation, however. In the cases examined in Problem 8.17, the enthalpies of fusion calculated from the two eutectics in which a particular compound participates disagree by a factor of two or so, and the entropies of fusion range from 20 to 140, compared with a normal range of the parent compounds of 10–15 or so. Walden's Rule is that all substances have the same entropy of fusion, approximately 13.

8.4.4. Partially Miscible Phases and Other Complexities

Instances of two liquid phases in equilibrium with solids are shown in Figures 5.23(d) and (e). Partially miscible solid organic phases are less common, but Figures 5.25(f) and (g) are two cases. A sound theory is especially desirable for correlation and extension of solid-solid equilibrium data; because the processes are very slow, experimental data are difficult to obtain and are usually few in number for specific systems.

In the area of metals some phase diagrams have been calculated from limited measurements by application of regular solution theory (the S-H equation); examples are given by Kaufman & Bernstein (1970), Rudman (1970), and Lupis (1983). The first of these references also considers the effects of intermolecular compound formation on binary phase diagrams.

When solid solutions occur, the fugacity equality of Eq. 8.1 must be modified to include activity coefficients in the solid phase and applied to both components. Identify the solid phase with an asterisk. The simplified equations that obtain when the effects of ΔC_p and pressure are neglected are

$$\gamma_1^* x_1^* = (\gamma_1 x_1)(f_{1(\text{scl})}/f_1)$$
$$= \gamma_1 x_1 \exp[(\Delta S_{m1}/R)(1 - T_{m1}/T)], \tag{8.42}$$

$$\gamma_2^* x_2^* = (\gamma_2 x_2)(f_{2(\text{scl})}/f_2)$$
$$= \gamma_2 x_2 \exp[(\Delta S_{m2}/R)(1 - T_{m2}/T)]. \tag{8.43}$$

Summing up the mol fractions leads to the two equations

$$x_1 + x_2 = (\gamma_1/\gamma_1^*) x_1^* \exp[(\Delta S_{m1}/R)(1 - T_{m1}/T)] + (\gamma_2^*/\gamma_2) x_2^* \exp[(\Delta S_{m2}/R)(1 - T_{m2}/T)] = 1, \tag{8.44}$$

$$x_1^* + x_2^* = (\gamma_1/\gamma_1^*) x_1 \exp[(\Delta S_{m1}/R)(1 - T_{m1}/T)] + (\gamma_2/\gamma_2^*) x_2 \exp[(\Delta S_{m2}/R)(1 - T_{m2}/T)] = 1. \tag{8.45}$$

When the activity coefficients are known as functions of the composition by a two-parameter equation, those parameters can be found by solution of the last two equations, as explained in Chapter 4, for example. In the metallurgical work referred to, both the Scatchard-Hildebrand and symmetrical Margules equations were used.

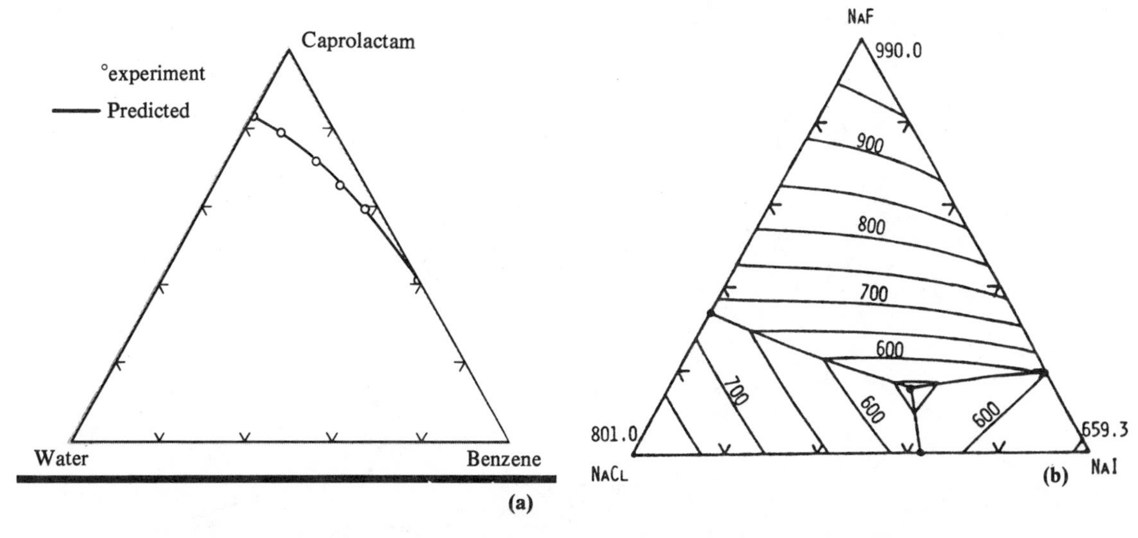

Figure 8.9(a). Prediction of ternary liquid-solid equilibria. The binodal curve of benzene + caprolactam + water at 20 C with the aid of the T-K-Wilson equation (Muir & Howat, 1982). **(b).** Prediction of ternary liquid-solid equilibria. Computer-generated phase diagram of NaF + NaCl + Nal with the aid of the Wilson equation (Morimi & Nakanishi, 1977).

8.5. MULTICOMPONENT SYSTEMS

For the most part the ensuing discussion will be phrased in terms of ternary mixtures, but extension to more components usually will be straightforward. Figures 5.40 and 5.41 are representative ternary T-x diagrams. In the absence of intermolecular compounds, only one ternary eutectic can be formed. Since, in contrast to metallic systems, partial miscibility of organic solid phases is uncommon, only the simplest cases of solid-phase immiscibility and a single ternary eutectic will be treated. The main characteristics of such systems are depicted in Figure 8.10, which shows binary eutectics, a ternary eutectic, and eutectic troughs. In a system of several components, each one is represented by an equation of the form

$$x_i = \frac{1}{\gamma_i} \exp[(\Delta S_{mi}/R)(1 - T_{mi}/T)], \tag{8.46}$$

where $\Delta S_{mi} = \Delta H_{mi}/T_{mi}$ and each activity coefficient depends on the temperature and the overall composition of the liquid phase.

8.5.1. Ternary Eutectics

The ternary eutectic is the single condition at which a liquid is in equilibrium with three solid phases. Its temperature is a minimum on the liquidus surface of the T-x_i diagram. The conditions at the eutectic are found by simultaneous solution of Eq. 8.46 written for each component:

$$x_1 = \frac{1}{\gamma_1} \exp[(\Delta S_{m1}/R)(1 - T_{m1}/T)], \tag{8.47}$$

$$x_2 = \frac{1}{\gamma_2} \exp[(\Delta S_{m2}/R)(1 - T_{m2}/T)], \tag{8.48}$$

$$x_3 = 1 - x_1 - x_2 = \frac{1}{\gamma_3} \exp[(\Delta S_{m3}/R)(1 - T_{m3}/T)], \tag{8.49}$$

for the temperature and the mol fractions x_1 and x_2. Clearly this method is extendable to any number of components. In case the mixture is ideal, the temperature can be found from the single equation

$$\Sigma\, x_i = \Sigma\, \exp[(\Delta S_{mi}/R)(1 - T_{mi}/T)] = 1. \tag{8.50}$$

Then the mol fractions may be found from individual Schröder equations. Example 8.10 employs the Margules and Wilson equations in the solution of this kind of problem.

8.5.2. Eutectic Troughs

Valleys or troughs connect individual binary eutectics to the ternary eutectic point. Along any one of these troughs, two solid phases are in equilibrium with the liquid phase. Along line E_1E_4 of Figure 8.10, for instance, substances A and B are the solid phases. The connecting lines often are nearly straight so they can be approximated directly if the eutectic points are known. However, the conditions along E_1E_4, for instance, can be explored at specified concentrations x_c between 0 and the concentration at the ternary eutectic, as follows. The applicable equations are:

$$x_a = \frac{1}{\gamma_a} \exp[(\Delta S_{ma}/R)(1 - T_{ma}/T)], \tag{8.51}$$

$$x_b = \frac{1}{\gamma_b} \exp[(\Delta S_{mb}/R)(1 - T_{mb}/T)] = 1 - x_a - x_c. \tag{8.52}$$

After x_c is specified, the remaining variables x_a and T are found by simultaneous solution of these equations. For ideal mixtures the problem is reduced to the solution of a single equation for the temperature,

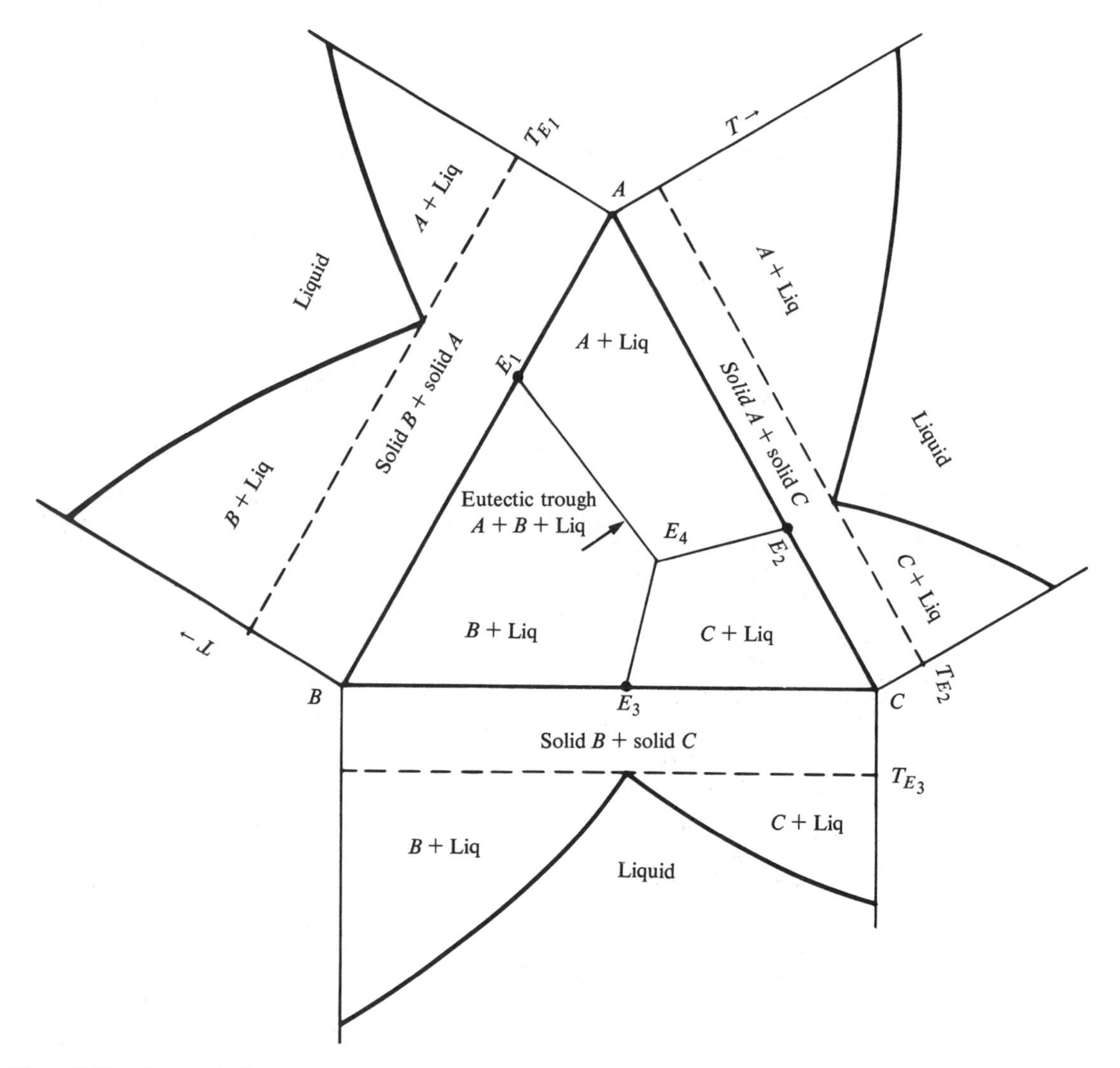

Figure 8.10. Schematic diagram of a ternary and associated binary liquid-solid equilibrium diagrams. The binary eutectics are labeled E_1, E_2, and E_3; the ternary eutectic is E_4. Along the eutectic trough E_1E_4, solid components A and B and a liquid phase are in equilibrium; along E_2E_4, $A + C +$ Liquid; along E_3E_4, $B + C +$ Liquid.

$$1 - x_c = x_a + x_b = \exp[(\Delta S_{ma}/R)(1 - T_{ma}/T)] + \exp[(\Delta S_{mb}/R)(1 - T_{mb}/R)], \qquad (8.53)$$

at each specified value of x_c.

8.5.3. Isotherms

Approximations of isotherms may be obtained by connecting like temperatures on pairs of sides of the triangle with straight lines, as in Example 8.8(b), for instance. However, the real systems of Figures 5.40(b), 5.40(c), 5.41, and 8.11 have substantially curved isotherms. The temperature at which a mixture of a given composition begins to crystallize is the maximum value of those given for individual components by the equation

$$T_i = \frac{T_{mi}}{1 - (R/\Delta S_{mi}) \ln(\gamma_i x_i)}. \qquad (8.54)$$

The component with the maximum temperature is the one that comes out of the melt first. The complete isotherms may be constructed by cross-plotting a series of such calculations. In Example 8.8 the trough curvatures are slight, but they are evident in Example 8.9.

8.5.4. Cooling Path

The progress of solidification of a mixture as the temperature is lowered may be followed in Figure 8.12, where the starting composition is represented by point D. In this case component C is the first to precipitate, starting at temperature T_D, which is found as explained in the preceding section. C alone continues to precipitate along the line CG, which corresponds to the fixed proportion, $\beta = x_{a0}/x_{b0}$, of the other components. They remain in the liquid phase until trough E_3E_4 is reached at point F, at which component B also begins to precipitate along with C.

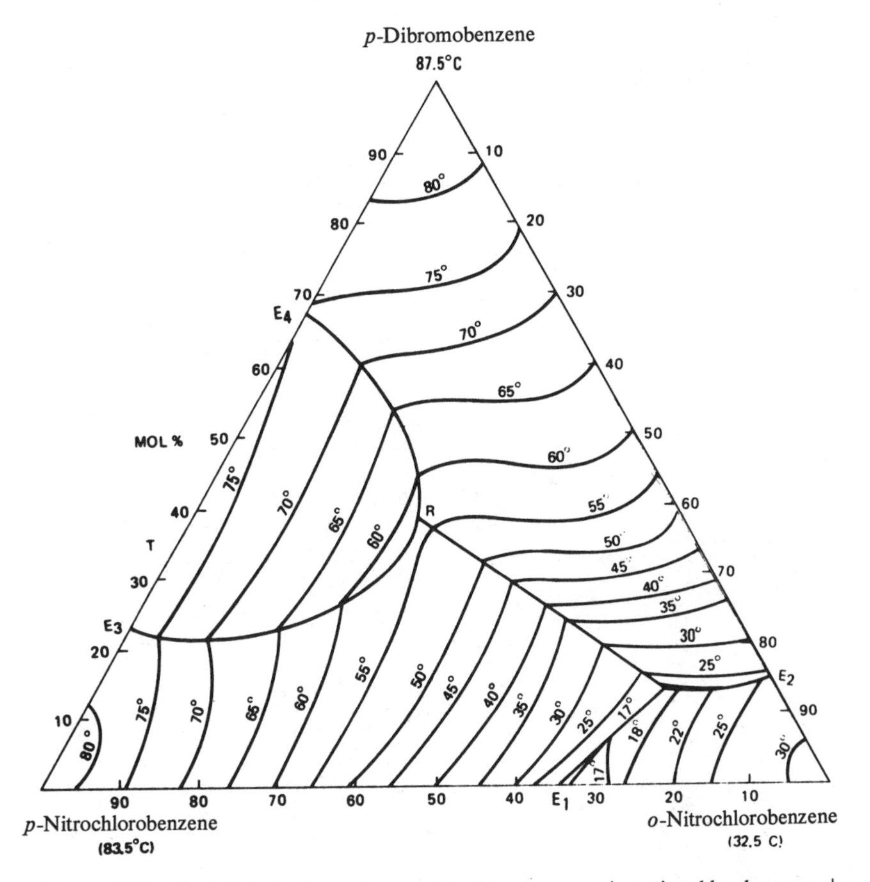

Figure 8.11. Equilibria in the system *p*-nitrochlorobenzene + o-nitrochlorobenzene + *p*-dibromobenzene. The first and last of these form a molecular compound, with eutectics at E_3 and E_4 (Dale, 1981).

The temperature and composition at point F are found by solution of the material balance

$$(x_{a0}/x_{b0} + 1)x_b + x_c - 1 = 0, \qquad (8.55)$$

By introduction of Schröder equations this becomes

$$(x_{a0}/x_{b0} + 1)\exp[(\Delta S_{mb}/R)(1 - T_{mb}/T_F)] + \exp[(\Delta S_{mc}/R)(1 - T_{mc}/T_F)] - 1 = 0. \qquad (8.55)$$

Solution of this equation automatically provides T_F, x_b, and x_c, and subsequently

$$x_a = (x_{a0}/x_{b0})x_b. \qquad (8.55a)$$

Note that the components involved in Eq. 8.53 are the ones that precipitate along trough E_3E_4. If the crystallization path had crossed E_2E_4, the equations would have been

$$(x_{b0}/x_{a0} + 1)x_a + x_c - 1 = 0, \qquad (8.56)$$

and

$$(x_{b0}/x_{a0} + 1)\exp[(\Delta S_{ma}/R)(1 - T_{ma}/T_F)] + \exp[(\Delta S_{mc}/R)(1 - T_{mc}/T_F)] - 1 = 0. \qquad (8.56a)$$

Conditions along path FE_4 are found by solution of Eq. 8.53.

When activity coefficients are known as functions of temperature and compositions, they can be included in these analyses.

8.5.5. Compositions and Amounts of the Two Phases

The problem is to find the compositions and amounts of each component in each phase at every temperature from the onset of crystallization until complete solidification at the ternary eutectic. The following notation will be adopted:

L = fraction of the total mixture that is liquid.
n_i = mols of component i in the liquid/mol of total mixture.
$n_{io} - n_i$ = mols of component i in the solid phases/mol of initial mixture.

The process will be described with reference to Figure 8.12, where key locations are identified as D, E, F, and G. Example 8.9 is a numerical illustration of the process.

1. The onset of crystallization at point D is explained in connection with Eq. 8.54.

2. Pure crystals of component C form until point F is reached. The temperature and compositions, x_{iF}, are found

Example 8.8. Construction of Binary and Ternary Phase Diagrams of Mixtures of the Ortho, Meta, and Para Isomers of Chloronitrobenzene

The fusion temperatures and enthalpies are:

Isomer	T_m, °K	ΔH_m, *cal/gmol*	$\Delta S_m/R$
Ortho	307.5	4546	7.4383
Meta	317.6	4629	7.3327
Para	356.7	4965	7.0055

Binary liquidi are found as in Example 8.6. The binary eutectic conditions are obtained by solving

$$1 = x_o + x_m = \exp\left[7.4383\left(1 - \frac{307.5}{T}\right)\right] + \exp\left[7.3327\left(1 - \frac{317.6}{T}\right)\right],$$

for the ortho-meta mixture, and similarly for the other pairs, with the results

pair	T	x_o	x_m	x_p
$o+m$	285.41	0.5628	0.4372	0
$m+p$	303.34	0	0.7084	0.2916
$o+p$	296.51	0.7590	0	0.2410

The ternary eutectic temperature is found from

$$1 = \exp\left[7.4383\left(1 - \frac{307.5}{T}\right)\right] + \exp\left[7.3327\left(1 - \frac{317.6}{T}\right)\right] + \exp\left[7.0055\left(1 - \frac{356.7}{T}\right)\right].$$

By trial, $T = 279.97$, $x_o = 0.4811$, $x_m = 0.3731$, and $x_p = 0.1458$.

Equations of the three eutectic troughs are:

$$1 - x_p = x_o + x_m = \exp\left[7.4383\left(1 - \frac{307.5}{T}\right)\right] + \exp\left[7.3327\left(1 - \frac{317.6}{T}\right)\right],$$

$$1 - x_o = x_m + x_p = \exp\left[7.3327\left(1 - \frac{317.6}{T}\right)\right] + \exp\left[7.0055\left(1 - \frac{356.7}{T}\right)\right],$$

$$1 - x_m = x_o + x_p = \exp\left[7.4383\left(1 - \frac{307.5}{T}\right)\right] + \exp\left[7.0055\left(1 - \frac{356.7}{T}\right)\right].$$

Calculations are made for values of x_i on the left ranging from zero to the ternary eutectic compositions; pairs of mol fractions are automatically determined when the temperatures have been evaluated. The results are tabulated as follows. The plots of these results show that the ortho-meta and ortho-para troughs are slightly curved. Isotherms are constructed as straight lines connecting corresponding temperatures on the binary phase diagrams and on pairs of troughs.

x_p	T_{om}	x_o	x_m
0	285.4	0.562	0.438
0.04	283.9	0.540	0.420
0.10	281.7	0.546	0.394
0.1458	279.9	0.481	0.3732

x_o	T_{mp}	x_m	x_p
0	303.4	0.708	0.292
0.1	299.3	0.639	0.261
0.2	294.9	0.569	0.231
0.3	290.1	0.499	0.201
0.4	284.8	0.429	0.171
0.4811	279.9	0.372	0.1469

x_m	T_{op}	x_o	x_p
0	296.5	0.759	0.241
0.1	292.6	0.684	0.216
0.2	288.3	0.610	0.190
0.3	283.6	0.535	0.165
0.3731	279.9	0.480	0.1469

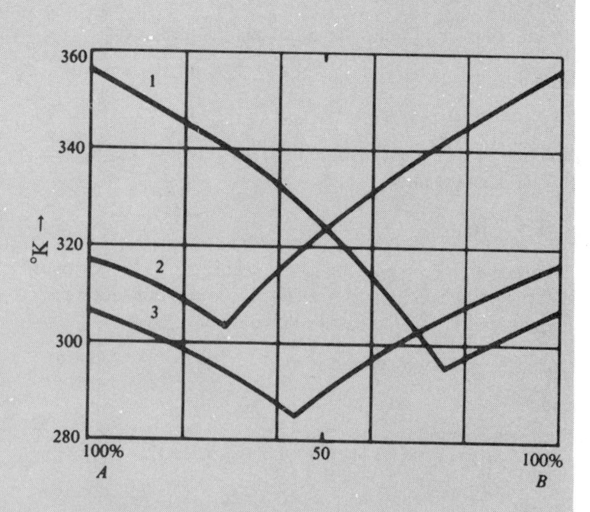

Example 8.8(a). The three binary melt diagrams.
(1). Para (A) + ortho (B) CNB
(2). Meta (A) + para (B) CNB
(3). Ortho (A) + meta (B) CNB

Example 8.8 (*continued*)

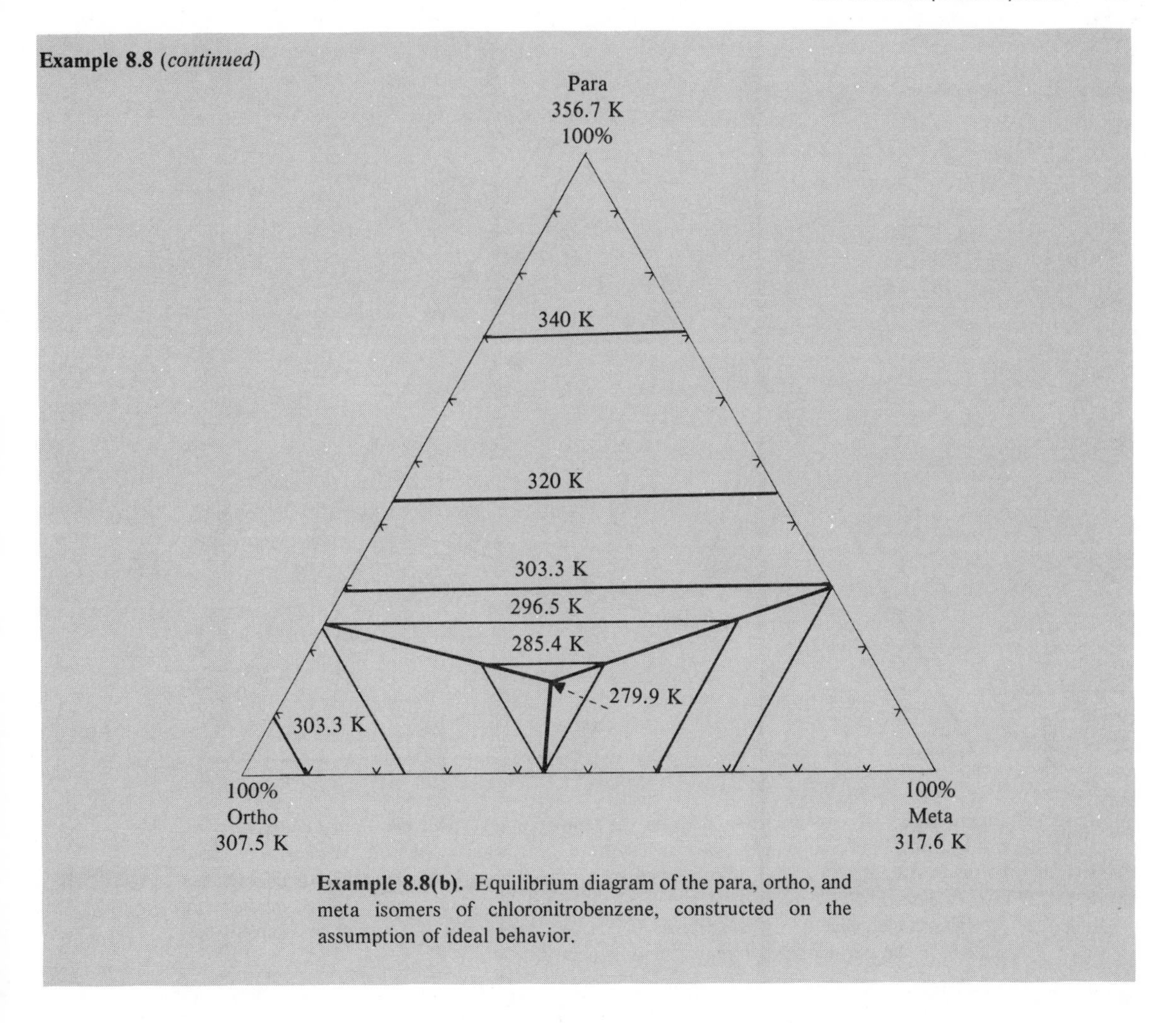

Example 8.8(b). Equilibrium diagram of the para, ortho, and meta isomers of chloronitrobenzene, constructed on the assumption of ideal behavior.

with Eqs. 8.55–8.56(a). The amount of C remaining in the liquid is

$$n_{cF} = (n_{a0} + n_{b0})x_{cF}/(1 - x_{cF}). \quad (8.57)$$

Also

$$n_{aF} = n_{ao},$$

$$n_{bF} = n_{bo}.$$

3. The fraction of pure crystals that is recoverable is

$$\frac{n_{cF}}{n_{c0}} = \frac{n_{a0} + n_{b0}}{n_{c0}} \left(\frac{x_{cF}}{1 - x_{cF}} \right) \quad (8.58)$$

4. At a specified intermediate temperature, T_E, between T_D and T_F, the composition x_{cE} in the liquid is found with Eq. 8.59,

$$x_{cE} = \exp[(\Delta S_{mc}/R)(1 - T_{mc}/T_E)]. \quad (8.59)$$

The number of mols then is found analogously to Eq. 8.57,

$$n_{cE} = (n_{a0} + n_{b0})x_{cE}/(1 - x_{cE}), \quad (8.60)$$

whereas $n_{aE} = n_{a0}$ and $n_{bE} = n_{b0}$.

5. At any point G along the eutectic trough FE_4, the temperature and composition are found with Eq. 8.53 by specifying the composition of the component that does not precipitate, which is component A in Figure 8.12. Thus Eq. 8.53 becomes, in this application,

$$1 - x_a = x_b + x_c = \exp[(\Delta S_{mb}/R)(1 - T_{mb}/T)] + \exp[(\Delta S_{mc}/R)(1 - T_{mc}/T)], \quad (8.61)$$

which is solved for T at specified values of x_a.

The amount of the liquid phase is

$$L = n_{a0}/(1 - x_{aG}), \quad (8.62)$$

and the amounts of the other components are

$$n_{cG} = n_{c0} - Lx_{cG}, \quad (8.63)$$

$$n_{bG} = n_{b0} - Lx_{bG}. \quad (8.64)$$

The amounts in the solid phase are found by overall balances,

$$(n_{cG})_{\text{solid}} = n_{c0} - n_{cG}, \quad (8.65)$$

$$(n_{bG})_{\text{solid}} = n_{b0} - n_{bG}, \quad (8.66)$$

$$(n_{aG})_{\text{solid}} = 0. \quad (8.67)$$

6. The amount of liquid remaining just before solidification of the ternary eutectic is

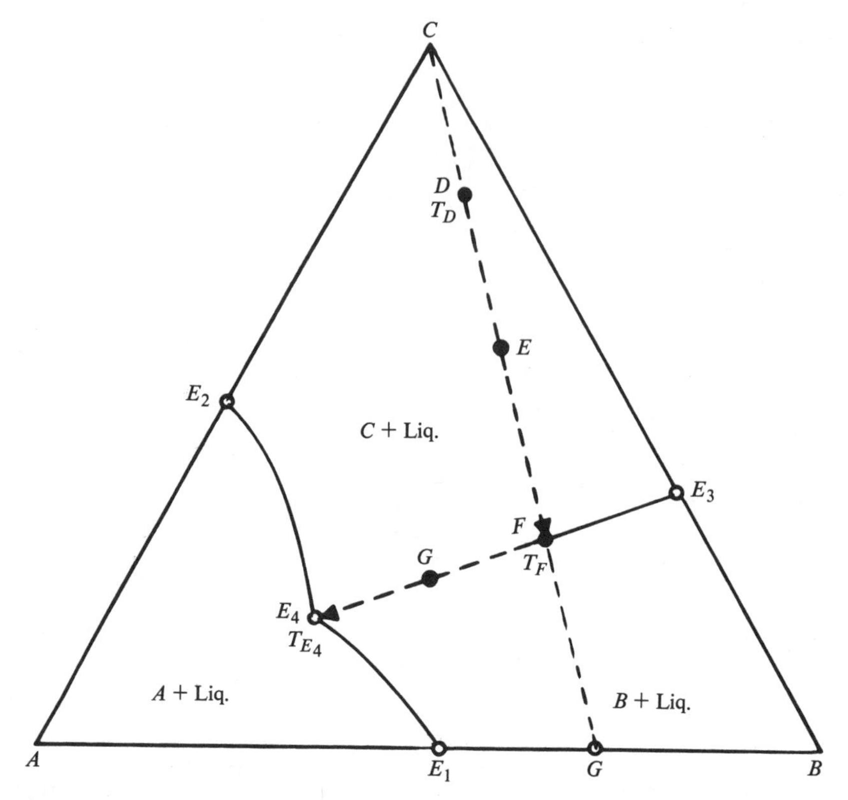

Figure 8.12. A crystallization path as the temperature is lowered. Precipitation of pure component C begins at point D, whose temperature is T_D, and continues until point F is reached on the eutectic trough E_3E_4. Both components C and B precipitate along this trough until the ternary eutectic E_4 is reached, at which complete solidification occurs. Only component C precipitates along line CF, which represents constant proportions of the other components, the same as in the starting mixture represented by point G.

$$L_{E_4} = n_{a0}/x_{aE_4}, \tag{8.68}$$

where the eutectic composition is found with Eq. 8.50.

8.6. SOLUBILITY OF SOLIDS AND LIQUIDS IN SUPERCRITICAL GASES

Except for the fact that highly compressed gases completely fill any confined space into which they are placed, their properties, including solvent power, resemble those of liquids. At moderate pressures the content of a condensable substance in a contacting gas is determined by its vapor or sublimation pressure, and will decrease as the system pressure rises. Beyond pressures in the vicinity of the critical pressure of the gas, however, the solvent power of the gas increases sharply with pressure, just as it ordinarily does with liquid solvents. This enhancement of solubility is interpreted in terms of the sharp decline of the fugacity coefficient of the gaseous solute with increasing pressure. Such behavior is well represented by modern equations of state. Some data are given in Figure 8.13(a)–(d).

Assuming the solute phase to remain pure, the partial fugacity balance of the solute will be

$$\hat{f}_{2V} = \hat{f}_{2L} = f_{2L}. \tag{8.69}$$

On making the usual substitutions, this relationship is transformed to

$$y_2\hat{\phi}_2 P = \phi_2^{\text{sat}} P_2^{\text{sat}} (PF)_2$$
$$= \phi_2^{\text{sat}} P_2^{\text{sat}} \exp[(V_2/RT)(P - P_2^{\text{sat}})], \tag{8.70}$$

which is solved for the content of solute in the vapor phase as

$$y_2 = (\phi_2^{\text{sat}} P_2^{\text{sat}}/\hat{\phi}_2 P)\exp[(V_2/RT)(P - P_2^{\text{sat}})]. \tag{8.71}$$

The pressure term in the denominator usually is dominant over that in the Poynting factor, but in some instances the partial fugacity coefficient will fall so sharply with increasing pressure that the solubility will rise markedly. The solubility of several heavy hydrocarbons in propane is analyzed with the B-truncated virial equation in Example 8.11. Even at pressures below 100 atm in these cases, the ratio of the highest to the lowest solubilities reaches 10–20 or so. Other examples are shown in Figure 5.27 and in Chapter 9.

Enhanced solubility of condensed phases in gases at high pressures was observed more than a hundred years ago, but interest in this phenomenon was revived only recently; the first industrial application was reported by Zhuse (1955, 1960). Current developments are in the recovery or separation of substances that are too unstable for distillation. For such purposes, gases whose critical temperatures are near atmos-

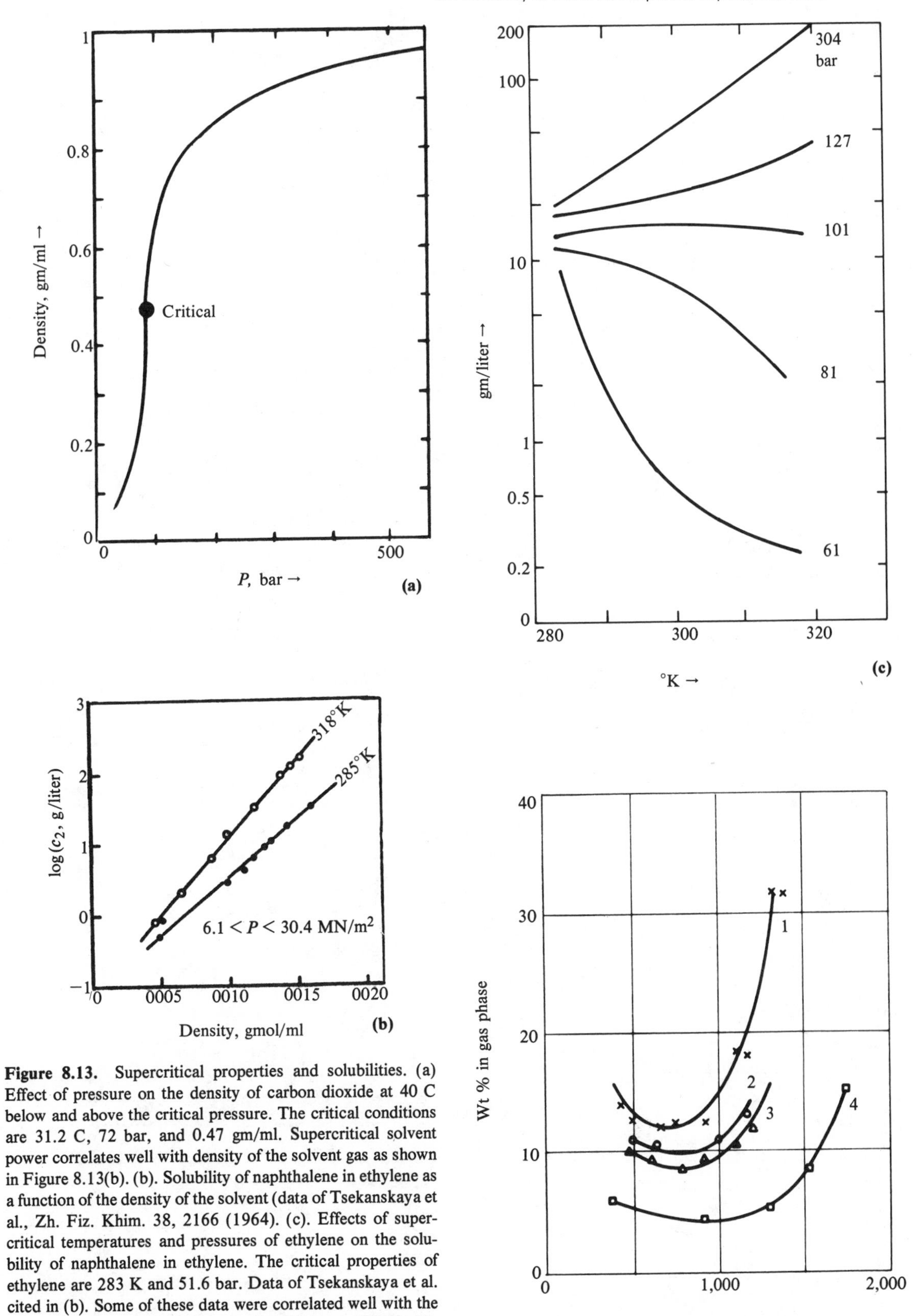

Figure 8.13. Supercritical properties and solubilities. (a) Effect of pressure on the density of carbon dioxide at 40 C below and above the critical pressure. The critical conditions are 31.2 C, 72 bar, and 0.47 gm/ml. Supercritical solvent power correlates well with density of the solvent gas as shown in Figure 8.13(b). (b). Solubility of naphthalene in ethylene as a function of the density of the solvent (data of Tsekanskaya et al., Zh. Fiz. Khim. 38, 2166 (1964). (c). Effects of supercritical temperatures and pressures of ethylene on the solubility of naphthalene in ethylene. The critical properties of ethylene are 283 K and 51.6 bar. Data of Tsekanskaya et al. cited in (b). Some of these data were correlated well with the Lee-Kesler equation by Masuoka and Yorizane (1982). (d). Solubilities in compressed ethylene at 75 C. (1) *n*-hexane, (2) cyclohexane, (3) benzene, (4) ethanol (after Ellis, 1971).

Example 8.9. Solidification Path of a Ternary Mixture

A separation is to be effected between *m*-benzene dinitrile, $m\text{-}C_6H_4(CN)_2$, and *m*-cyanobenzeneamide, C_6H_4 CN.COHN$_2$, from a 90/10 mixture in the presence of *m*-xylene to improve separation of the crystals. The analysis of the process will be made on the assumption of ideal melt behavior since activity coefficients are not available. Physical data are given in the first table.

Equations for the mol fractions are,

$$x_a = \exp[6.4821(1 - 435/T)], \quad (m\text{-}bd)$$

$$x_b = \exp[6.5425(1 - 495T)], \quad (m\text{-}cba)$$

$$x_c = \exp[6.1600(1 - 225.4/T)]. \quad (\text{xylene})$$

The binary eutectic is found with Eq. 8.50 with the results

$$x_a + x_b - 1 = 0,\ T = 414.1,\ x_a = 0.7213,\ x_b = 0.2787.$$

The ternary eutectic also is found with Eq. 8.50, with the results,

$$x_a + x_b + x_c - 1 = 0,\ T = 225.30,\ x_a = 0.0024,$$
$$x_b = 0.0004,\ x_c = 0.9972.$$

The onset of crystallization is found with Eq. 8.54. For all the contents of *m*-xylene investigated here, component *A* is the one that comes out pure. The temperatures of initial crystallization are found from

$$T_{\text{initial}} = 435\Big/\left[1 - \frac{\ln x_{ao}}{6.4821}\right].$$

The results are recorded in the second table. Substance *A* continues to precipitate along the straight line representing the constant proportions of the other substances in the original mixture. The case that will be studied in detail is that of 20% initial *m*-xylene content for which

$$\beta = x_{co}/x_{bo} = 0.20/0.08 = 2.5.$$

Apply Eq. 8.55,

$$(\beta + 1)x_b + x_a - 1 = 0,$$

or

$$(2.5 + 1)\exp[6.5425(1 - 495/T)]$$
$$+ \exp(6.4821(1 - 435/T)] - 1 = 0.$$

The solution is

$$T = 386.45,\ x_a = 0.4429,\ x_b = 0.1592,\ x_c = 0.3980.$$

As the temperature is reduced further, impure crystals precipitate. The crystallization path is along the eutectic trough of substances *A* and *B*, which is represented by Eq. 8.53 and which assumes the form

$$1 - x_c = \exp[6.4821(1 - 435/T)]$$
$$+ \exp[6.5425(1 - 495/T)].$$

Values of x_c in the range from the condition of initial impure crystal formation, 0.3980, to the ternary eutectic, 0.9972, are substituted to find the temperatures and compositions along the trough. The results are shown in the third table and as points on the ternary diagram. Clearly the eutectic trough deviates from linearity, which is represented by the dashed line from the binary eutectic of *A* and *B* to the ternary eutectic, which is virtually pure *m*-xylene.

1. *Thermal properties*

		T_f	$\dfrac{\Delta S_f}{R}$
A	$m\text{-}C_6H_4(CN)_2$	435	6.4821
B	$m\text{-}C_6H_4.CN.CONH_2$	495	6.5425
C	$m\text{-}C_6H_4(CH_3)_2$	225.4	6.1600

2. *Initial and final conditions for formation of pure crystals:*

Feed				*At end of pure crystallization*			
x_{ao}	x_{bo}	x_{co}	$T_{initial}$	T	x_a	x_b	x_c
0.9	0.1	0	428.04	414.1	0.7222	0.2778	0
0.81	0.09	0.1	421.3	399.29	0.5601	0.2084	0.2315
0.72	0.08	0.2	414.0	386.45	0.4429	0.1592	0.3979
0.63	0.07	0.3	406.1	374.73	0.3526	0.1225	0.5249

3. *Eutectic trough of* A *and* B:

x_c	T	x_a	x_b	x_a/x_b
0	414.1	0.7213	0.2787	2.5881
0.1	408.06	0.6518	0.2481	2.6274
0.2	401.49	0.5822	0.2179	2.6718
0.3	394.28	0.5120	0.1880	2.7233
0.3979	386.45	0.4429	0.1592	2.7825
0.5	377.2	0.3704	0.1296	2.8575
0.6	366.65	0.2987	0.1012	2.9503
0.7	353.9	0.2264	0.0736	3.0742
0.8	337.3	0.1529	0.0469	3.2586
0.9	312.3	0.0783	0.0218	3.5991
0.9972	255.30	0.0024	0.0004	6.0410

The composition of the liquid phase as a function of temperature over the entire crystallization path is summarized:

T	x_c	x_b	x_a
414	0.20	0.08	0.72
400	0.3093	0.1237	0.567
386.5	0.3979	0.1592	0.4429
366.7	0.6000	0.1012	0.2988
337.3	0.8000	0.0469	0.1529
225.3	0.9972	0.0004	0.0024

On the basis of one mol of initial mixture, the amounts of the two phases and of the individual components in them are obtained with material balances and the liquid phase compositions tabulated preceding:

			Liquid			*Solid*		
T	L	S	n_c	n_b	n_a	n_c	n_b	n_a
414	1	0	0.2	0.08	0.72	0	0	0
400	0.6467	0.3533	0.2	0.08	0.3667	0	0	0.3533
386.5	0.5027	0.4973	0.2	0.08	0.2227	0	0	0.4973
366.7	0.3333	0.6667	0.2	0.0337	0.0997	0	0.0463	0.6203
337.3	0.2500	0.7500	0.2	0.0018	0.0383	0	0.0683	0.6818
225.3	0.2006	0.7994	0.2	0.0001	0.0005	0	0.0799	0.7195

For example, at 366.7 K and $x_c = 0.6$ in the liquid phase,

$L = n_c/0.6 = 0.2/0.6 = 0.3333,$

$n_c = n_{co} = 0.2,$

$n_b = 0.3333(0.1012) = 0.0337,$

$n_a = 0.3333(0.2987) = 0.0997.$

$S = 1 - L = 0.6667,$

$n_c = 0.2 - 0.2 = 0,$

$n_b = 0.08 - 0.0337 = 0.0463,$

$n_a = 0.72 - 0.0997 = 0.6203.$

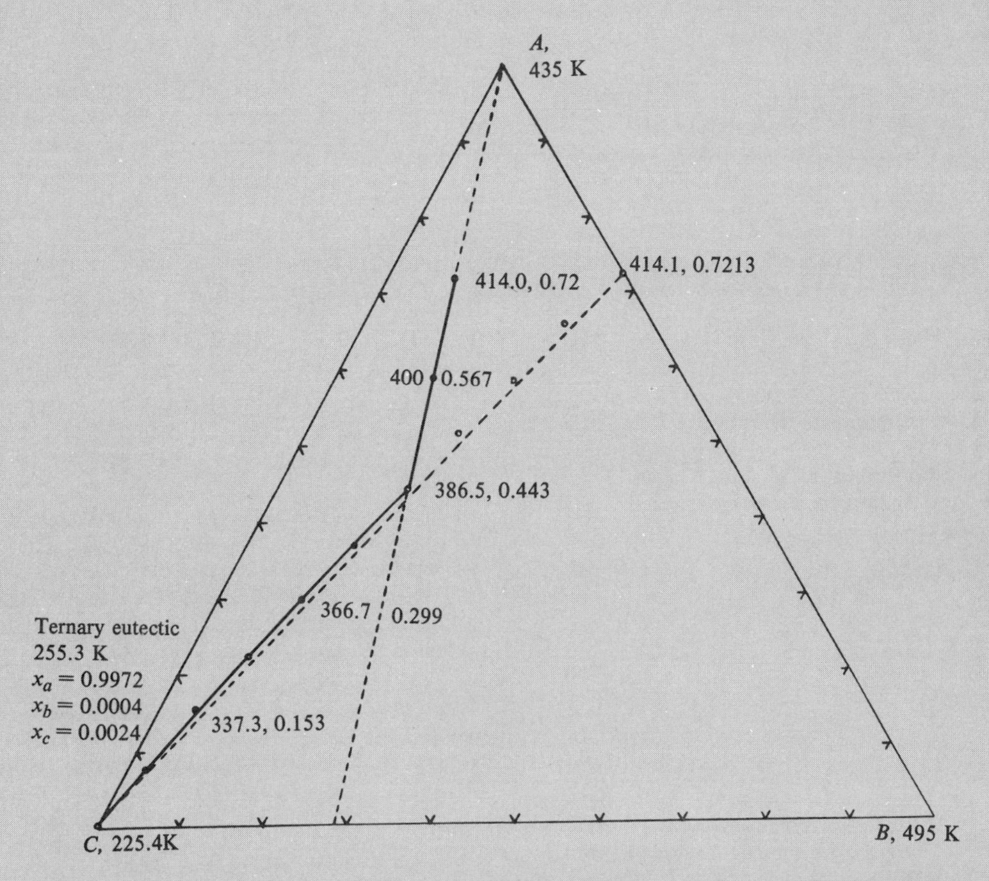

Example 8.9. Precipitation of pure component A begins at 414.0 K and continues until the temperature falls to 386.5 K and the mol fraction $x_a = 0.443$. Both A and B precipitate together along the line connecting the binary eutectic of A and B and the ternary eutectic. The second number of each group is the mol fraction, x_a, of component A.

Example 8.10. Ternary Eutectic with the Wilson Equation

The properties of the components of the mixture are

	V, *ml/gmol*	ΔS_m	T_m
1	150	13	450
2	200	13	375
3	125	13	400

Wilson parameters, $L_{ij} = \lambda_{ij}/R$,

$L_{12} = 1{,}141.2,\quad L_{21} = -364.3,$

$L_{13} = 221.5,\quad L_{31} = 144.5,$

$L_{23} = -281.6,\quad L_{32} = 493.5.$

The ideal eutectic conditions are found by solution of

$$\Sigma\, x_i = 1 = \Sigma \exp\left[-\frac{\Delta S_i}{R}\left(\frac{T_i}{T} - 1\right)\right]$$

$$= \exp\left[-\frac{13}{R}\left(\frac{450}{T} - 1\right)\right] + \exp\left[-\frac{13}{R}\left(\frac{375}{T} - 1\right)\right] + \exp\left[-\frac{13}{R}\left(\frac{400}{T} - 1\right)\right].$$

By trial,

$T = 342.54,$

$x_1 = 0.1284,$

$x_2 = 0.5379,$

$x_3 = 0.3357.$

With these values as estimates, the activity coefficients are found with the equations of Table 4.7. New values of T and x_i then are found by trial from

$$\Sigma\, x_i = 1 = \Sigma \frac{1}{\gamma_i} \exp\left[-\frac{\Delta S_i}{R}\left(\frac{T_i}{T} - 1\right)\right].$$

The results of five successive trials are tabulated and appear to be very nearly converged.

	Trial				
	1	*2*	*3*	*4*	*5*
γ_1	1	1.5832	1.7316	1.7788	1.7932
γ_2	1	1.1136	1.1079	1.1066	1.1062
γ_3	1	1.1548	1.1382	1.1328	1.1311
T	342.54	349.99	350.01	350.014	350.013
x_1	0.1284	0.0974	0.0891	0.0867	0.0860
x_2	0.5379	0.5627	0.5658	0.5665	0.5667
x_3	0.3337	0.3401	0.3451	0.3468	0.3473

Example 8.11. Effect of Pressure on Solvent Power of Propane and on Fugacity Coefficients of Heavy Hydrocarbons

Properties of the three solutes are given in the table. The temperature is 400 K. The Tsonopoulos-virial equation is used to evaluate the partial fugacity coefficients. Propane is designated as component 1, the solutes as component 2 in each case. At first the solubility drops off as the pressure is raised, but later the effects of great decrease in fugacity coefficient and the Poynting Factor counteract this tendency, and the solubility rises.

$$B = y_1^2 B_{11} + y_2^2 B_{22} + 2y_1 y_2 B_{12},$$

$$\ln \hat{\phi}_2 = \frac{P}{RT}(2y_1 B_{12} + 2y_2 B_{22} - B),$$

$$y_2 = \frac{\phi_2^{\text{sat}} P_2^{\text{sat}} (PF)_2}{\hat{\phi}_2 P}$$

$$= \frac{\phi_2^{\text{sat}} P_2^{\text{sat}}}{\hat{\phi}_2 P} \exp\left[\frac{V_2(P - P_2^{\text{sat}})}{RT}\right].$$

The plots are not carried beyond 100 atm because the B-truncated virial equation is not accurate enough. Actually, that equation begins to lose accuracy at $P/P_c > 0.5T/T_c$, which is only about 20 atm for propane. However, the results are qualitatively correct in that they predict an initial fall in solubility, and then a substantial rise with pressure.

	T_c	P_c	V_c	z_c	ω	B_{ij}	P^s	V_L
Propane	369.8	41.9	203	0.281	0.152	−0.2069	—	—
Naphthalene	748.4	40.0	410	0.267	0.302	−2.746	0.0717	132.0
Hexadecane	717.0	14.0	828	0.236	0.708	−10.060	0.00437	294.1
Phenanthrene	878	28.6	594	0.228	0.440	−10.369	0.00153	151.2
$P+N$	526.1	40.16			0.227	−0.7044		
$P+H$	514.9	24.55			0.430	−1.2222		
$P+P$	569.8	32.67			0.246	−1.1467		

Example 8.11 (*continued*)

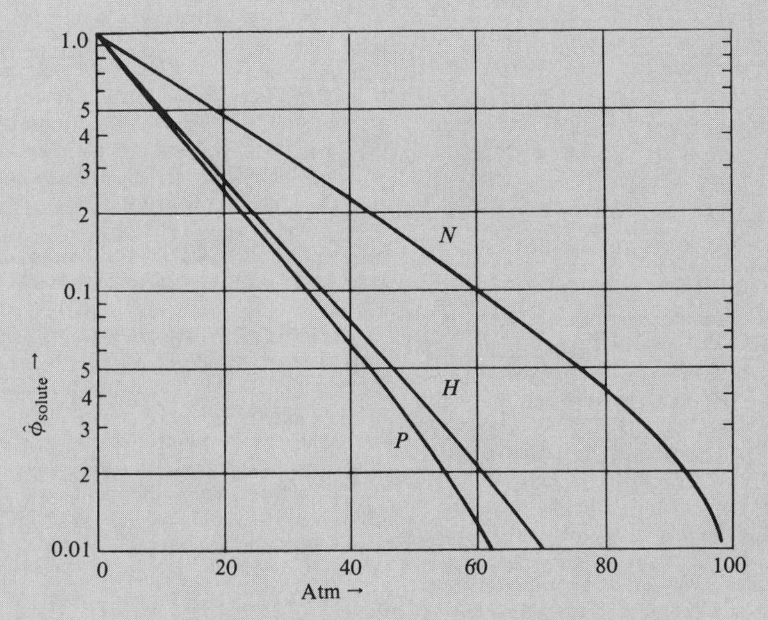

Example 8.11(a). Effect of pressure at 400 K on partial fugacity coefficients of naphthalene (N), hexadecane (H), and phenanthrene (P) in propane.

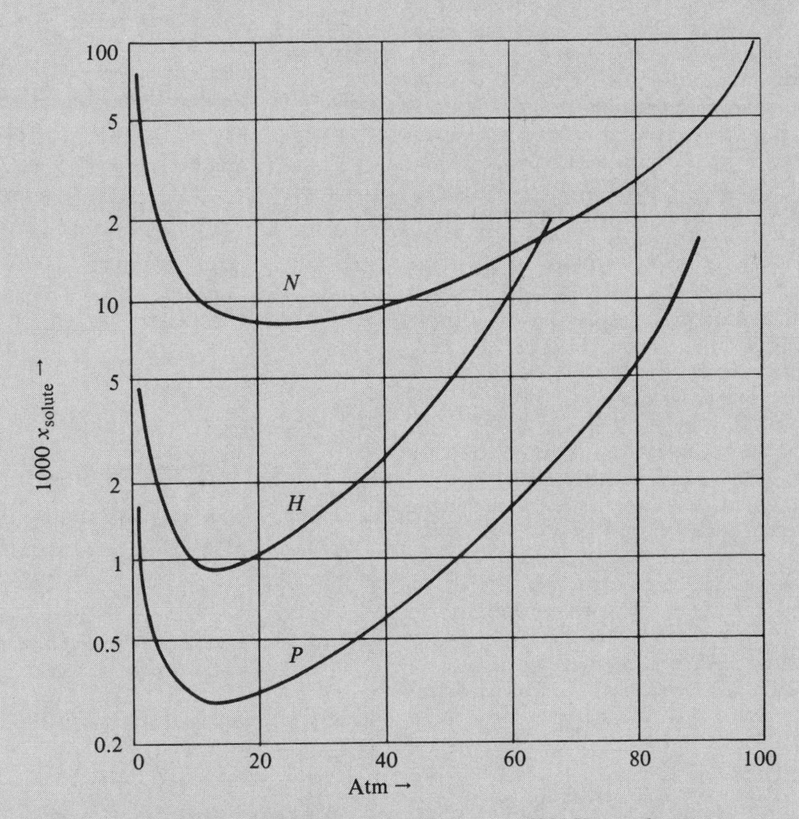

Example 8.11(b). Effect of pressure at 400 K on the solubility of naphthalene (N), hexadecane (H), and phenanthrene (P) in propane.

pheric temperature are most suitable. Some of these are listed in the following table:

	T_c, °C	P_c, bar
carbon dioxide	31.0	73.9
ethylene	9.2	50.6
nitrous oxide	36.5	72.5
ethane	32.3	48.8
$CClF_3$ (R-13 refrigerant)	28.9	38.6
CHF_3 (R-23 refrigerant)	25.9	48.2
propane	96.8	42.0

A list of about a dozen industrial supercritical extractions is given by Stahl & Schilz (1976), who also describe the use of supercritical CO_2 and N_2O with thin-film chromatography of about 100 plant drugs. Several industrial and analytical applications are described by Peter et al. (1974). A review of the literature with particular emphasis on extraction of coal is by Gangoli & Thodos (1977). Several symposia that have been published recently are cited in Chapter 9, where phase behavior at supercritical conditions is discussed in more detail.

Besides its advantages for low-temperature separation of low-volatility substances, supercritical extraction often is more efficient thermally than ordinary separation processes because the solubility can be controlled by relatively small adjustments of pressure and temperature without requiring change of phase. One example of such a thermally efficient process is the separation of ethanol and water with supercritical CO_2, although this process still is in the pilot plant stage. Two industrial applications about which some details are available will be described briefly.

1. Supercritical deasphalting of crude oil vacuum residua is accomplished with light hydrocarbons such as pentane. In one particular case, at 39.5 atm and 468 K and a solvent density of 19 lb/cuft, only the asphalt remains undissolved and is separated from the solution by settling. When the conditions are changed to 37 atm and 489 K, the density of the solvent falls to 7.7 lb/cuft and the oil and solvent separate. These small variations of pressure and temperature in the vicinity of the critical conditions of the solvent (33.3 atm, 469.6 K) correspond to appreciably lower energy requirements in comparison with conventional extraction, which requires somewhat higher solvent/oil ratios.

2. Deashing of coal can be accomplished with a solvent consisting of about 80% pyridine and 20% "anthracene oil," which is made in process. Critical properties of the pyridine are 620 K, 55.6 atm. Solution of more than 90% of the ash-free coal is effected at 617 K and 62.2 atm. After the undissolved coal and ash are removed mechanically from the solution, the solvent and dissolved coal are separated by changing the conditions to 620 K and 60 atm.

Although the B-truncated virial equation was adopted for illustrative purposes in Example 8.9, it is not accurate at high pressures in the critical region. Other equations are better. Soave (1979) applied his modification of the R-K equation to equilibria between hydrocarbons and solid carbon dioxide, but binary interaction parameters were needed for accuracy. A more extensive study with the Redlich-Kwong equation in supercritical applications was made by Stephan & Shaber (1979); they also found a need for binary interaction parameters—for example, $k_{12} = 0.11$ for CO_2 with alkanes.

In general, solubility behavior can be estimated with Eq. 8.71 and a suitable relation for the partial fugacity coefficient. However, a few qualitative rules also can be stated.

1. In many cases the enhancement of the solubility is especially significant near the critical temperature of the solvent gas. In Figure 16(a), for instance, the enhancement is greatest with CO_2 whose critical temperature, 31.3 C, is the nearest of the three solvents to the test temperature of 0 C.

2. Solubility tends to a minimum roughly in the vicinity of the critical pressure. All the solutes in Figure 8.13(d) have minimum solubilities near the critical pressure of ethylene, 744 psia. Incidentally, those curves would be much closer together if the solubilities were expressed in mol % rather than wt %.

3. Density of the solvent is perhaps the prime factor in determining solubility. Figure 8.13(b) shows approximately linear variation of log(solubility) with density in the high pressure range. Figures 13(b) and 13(c) together, which are the same data in different forms, also show that density correlates the effects of both temperature and pressure quite neatly. As illustrated for CO_2 in Figure 8.13(a), the variation of density itself with pressure is highly nonlinear in the intermediate and high ranges.

8.7. PROBLEMS

8.1. Find the ideal solubility of p-xylene in the temperature range 200–275 K, given that $\Delta H_m = 4{,}090$ cal/gmol. $T_m = 286.39$ K, and $\Delta C_p = 5.96$ cal/(gmol)(K).

8.2. Solubility of naphthalene in benzene is 0.14 mol fraction at 0 C and 0.478 at 45 C. Its melting point is 80.1 C. Calculate the heat of fusion for each of these data. For comparison, the experimental value is 4,550 cal/gmol.

8.3. Find the pressure coefficient of solubility, $(\partial x/\partial P)_T$, using the Scatchard-Hildebrand equation for activity coefficient, given the data

$T = 300$ K,

$V_1 = 50$ ml/gmol,

$V_2 = 18$ ml/gmol

$x_1 = x_2 = 0.5$,

$\delta_1 - \delta_2 = 3\ (\text{cal/ml})^{0.5}$,

$\Delta V_f = 5$ ml/gmol.

8.4. For the mixture of $SnCl_4(1)$ and $SnI_4(2)$ an experimental value is $V_2(\delta_2 - \bar{\delta})^2/RT = 3.8$ at 298 K. The melting point and heat of fusion of the iodide are 417.7 K and 4,480 cal/gmol. Find the solubility of iodine in this mixture at 298 K.

8.5. The phase diagram of ethylene glycol (G) + water (W) shows the existence of molecular compounds GW_2 and $GW_{2/3}$. Data on the freezing temperatures and compositions are given in the table. The heats of fusion are 2,685 cal/gmol for ethylene glycol and 1,436 for water. Estimate the heats of fusion of the two hydrates.

Substance or eutectic	*Freezing T, °C*	*Wt % Glycol*
Ice	0	0
Ice + GW_2	−51.2	57.3

Substance or eutectic	Freezing T, °C	Wt % Glycol
GW_2	−49.6	61.6
$GW_2 + GW_{2/3}$	−63.3	75.6
$GW_{2/3}$	−39.7	84.4
$GW_{2/3} + G$	−49.4	87.0
G	−12.8	100

8.6. Use the data and results of Example 8.8 for mixtures of the isomers of chloronitrobenzene for the following cases:

a. Find the temperatures of incipient solidification and identify the substance which crystallizes first when $x_{ortho} = 0.25$ or 0.5 or 0.75 for the three sets of ratios $x_{meta}/x_{para} = 0.25$, 0.5, and 2.0.

b. For the mixture of composition $x_o = 0.3$, $x_m = 0.1$, and $x_p = 0.6$, trace the cooling path from initial crystallization to the point at which two substances begin to precipitate and continuing to the ternary eutectic.

c. Find the composition of the liquid and the amount of crystallization that has taken place when the temperature falls to 310 K.

d. Find the compositions of the solid and liquid phases and the temperature when $x_{meta} = 0.3$ in the liquid phase.

8.7. a. Benzene melts at 5.5° C with a latent heat of fusion of 2,350 cal. mole^{-1}; chlorobenzene melts at −45° C with a latent heat of 1,800 cal. mole^{-1}. They form a simple eutectic system. Assuming that they form ideal mixtures, construct the phase diagram for the condensed system, labelling each area, and determine the eutectic temperature and composition.

(Ans. Eutectic is at $x_{benzene} = 0.263$ and $T = -61$ C.)

b. The melting points and latent heats of fusion of *p*-nitrobenzene (*A*) and *p*-nitroaniline (*B*) are:

	A	*B*
t/°C	172·8	147·5
Δh/cal. mole^{-1}	6250	5135

These two substances form a simple eutectic; construct the ideal phase diagram and compare it with the following experimental results:

t° C	x_A	t° C	x_A
172.8	1.00	122.2	0.374
165.2	0.865	(119.3)	Eutectic
154.0	0.694	125.2	0.294
142.0	0.551	135.4	0.176
139.0	0.525	142.3	0.084
131.3	0.444	147.5	0.000

(Ans. Eutectic is at $x_B = 0.635$ and $T = 118$ C.)

8.8. Heptane (1) and 1,2-ethanediol (3) are substantially immiscible in the presence of less than 10% or so of hexanol (2). Experimental values of the distribution ratio, K = concn in heptane phase/concn in ethanediol phase, are known as functions of the concentration in the ethanediol phase:

Mol % in ethanediol	K
0.0432	1.20
0.0667	2.12
0.0932	2.43

Check these values of K with the NRTL equation for which $\alpha = 0.20$ and

$\tau_{12} = 2.9188$ $\tau_{21} = -0.6747$

$\tau_{23} = 0.8866$ $\tau_{32} = 1.3954$

The agreement will not be close, possibly because of the effect of some mutual ternary solubilities (DECHEMA, V/2, p. 431).

8.9. Find Margules and Wilson parameters of the system naphthalene + benzene from the eutectic conditions, 270.2 K and $x_1 = 0.125$. Enthalpies and temperatures of fusion are: naphthalene, 4,550 cal/gmol, 353.3 K; benzene, 2,379 cal/gmol, 278.7 K.

8.10. Construct the ideal solidification curve for mixtures of meta- and ortho- dibromobenzenes. The enthalpies and temperatures of fusion are: meta, 13.38 cal/gm, −6.9 C; ortho, 12.78 cal/gm, 1.8 C.

8.11. Anthracene and phenanthrene are completely miscible in the liquid and solid phases. Heats of fusion are 6,898 and 4,456 cal/gmol. Equilibrium compositions of anthracene in the solid and liquid phases are tabulated.

a. Find the ratios, γ_S/γ_L, of each component at each temperature.

b. At 160 C, find the Margules parameters and hence the individual activity coefficients.

°C	x_S	x_L
218	0	0
200	0.10	0.26
180	0.21	0.47
160	0.32	0.62
140	0.45	0.76
120	0.71	0.88
101	1	1

8.12. Freezing points of methanol + water mixtures are given. The enthalpies and temperatures of fusion are: methanol, 759.4 cal/gmol, 175.4 K; water, 1,436 cal/gmol, 273.2 K. Find the activity coefficients of both substances at each temperature.

x	T
.0273	271.00
.0472	268.20
.0725	264.90
.0994	261.50
.1273	257.60
.1572	253.20
.1883	248.20
.2215	243.20
.2563	237.60

8.13. Freezing points of mixtures of glycerol + water are given. The enthalpies of fusion are: glycerol, 4,421 cal/gmol; water, 1,436 cal/gmol.

a. Find the activity coefficients at each temperature.
b. Compare the ideal freezing curves with the data.
c. Find the Wilson parameters from the eutectic conditions, $x = 0.28$ and $T = 228.7$ K, and compute the freezing curve with their aid. Molal volumes are: glycerol, 116.13; water, 18.02.

x	T
.0000	273.2
.0213	271.6
.0466	268.4
.0774	263.7
.1154	257.7
.1646	251.2
.2269	239.6
.2800	228.7
.3134	235.4
.4390	254.0
.6378	271.6
1.0000	291.2

8.14. Find the Margules and van Laar parameters corresponding to the eutectic compositions shown on the two diagrams of

a. the levo- and dextro- optical isomers of β-benzylhydratropic acid (Bickel & Peaslee, JACS 70, 1790 (1948).

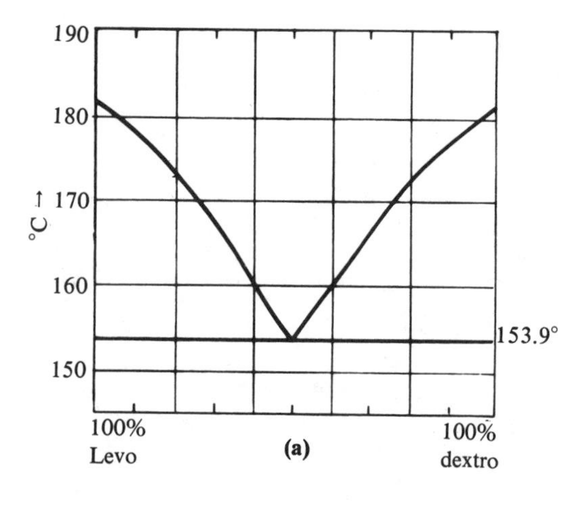

(a)

b. the system of ortho- and para-chloronotrobenzene. Enthalpies and temperatures of fusion are: para-CNB, 31.51 cal/gm, 83.5 C; ortho-CNB, 28.85 cal/gm, 32.5 C. For the optical isomers, the entropies of fusion may be taken as 13 in each case unless better data can be found.

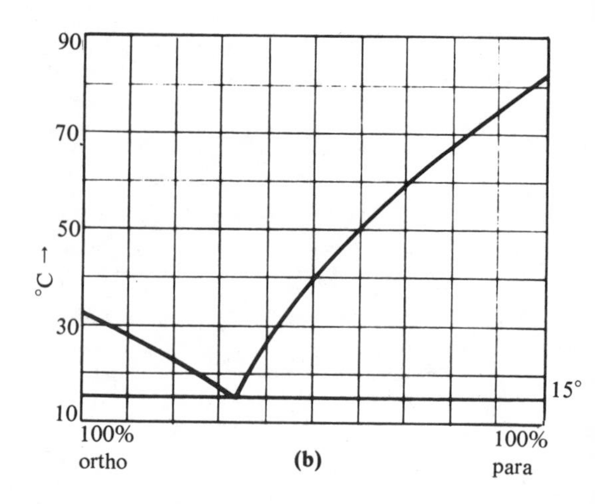

(b)

8.15. Enthalpies and temperatures of fusion and heat capacities of biphenyl and 1,2-bibenzyl are given. Find the ideal eutectic temperatures by (a) including the heat capacity terms; (b) neglecting the heat capacity terms. Units are calories, gmols and °C.

	T_m	ΔH_m	$C_p(l)$	$C_p(s)$
Biphenyl	69.0	4235	63.5	54.8
Bibenzyl (1,2)	51.25	5390	74.75	67.0

8.16. The four ternary mixtures listed form simple binary and ternary eutectics. In each case evaluate the Wilson and NRTL parameters from the binary eutectic data. Use those results to find the temperatures of the ternary eutectics and compare them with the experimental data.

Enthalpies of fusion of the first group of substances are estimated by Walden's Rule, with $\Delta H/T = 14$. The eutectic data are from Landolt-Börnstein (1956, II/3). Compositions are in wt%.

		Menthol $C_{10}H_{20}O$	*Phenacetin* $C_{10}H_{13}O_2N$	*Antipyrine* $C_{11}H_{a12}ON_2$
Tern. E	31 °C	82.5	1.5	16
Bin. e_1	75 °C	—	40	60
Bin. e_2	40 °C	96.8	3.2	—
Bin. e_3	33 °C	84	—	16

		o-Nitroaniline $C_6H_6O_2N_2$	*m-Nitroaniline* $C_6H_6O_2N_2$	*p-Nitroaniline* $C_6H_6O_2N_2$
Tern. E	43.3 °C	67	20	13
Bin. e_1	52.2 °C	75.5	24.5	—
Bin. e_2	56.6 °C	83	—	17
Bin. e_3	89.7 °C	—	63	37

		o-Dinitro-benzene $C_6H_4O_4N_2$	*m-Dinitro-benzene* $C_6H_4O_4N_2$	*p-Dinitro-benzene* $C_6H_4O_4N_2$
Tern. E	60 °C	30	60	10
Bin. e_1	64 °C	66	34	—
Bin. e_2	82 °C	—	84	16
Bin. e_3	101 °C	73.5	—	26.5

		1,3-Dinitro-benzene $C_6H_4O_4N_2$	2,4-Dinitro-toluene $C_7H_4O_4N_3$	2,4,6-Trinitro-toluene $C_7H_5O_4N_3$
Tern. E	29 °C	35.5	31	33.5
Bin. e_1	43.2 °C	43.5	56.5	—
Bin. e_2	51 °C	45.5	—	54.5
Bin. e_3	45.8 °C	—	52	48

Substance	°C	*cal/gmol*	$\Delta H/T$
menthol	42.5	4420	(14)
phenacetin	134.5	5708	(14)
antipyrine	113.0	5407	(14)
o-nitroaniline	71.2	3847	11.17
m-nitroaniline	147.0	5037	11.99
p-nitroaniline	114.0	5654	14.60
o-dinitrobenzene	116.93	5418	13.89
m-dinitrobenzene	89.7	4150	11.44
p-dinitrobenzene	173.5	6718	15.04
1,3-dinitrobenzene	89.7	4150	11.44
2,4-dinitrotoluene	70.14	4752	13.84
2,4,6-trinitrotoluene	80.83	5071	14.32

8.17. The four systems shown form the molecular compounds indicated in the figures.

a. Find the enthalpies of fusion of the molecular compounds from the eutectic conditions, assuming ideal behavior.

b. On the assumption that the entropy of fusion of a molecular compound is the weighted mean of those of the monomers, find the activity coefficients from the eutectic conditions.

Substance	°C	ΔH_m, *cal/gmol*	$\Delta H_m/T_m$
acetamide	81.9	3396	9.56
acetic acid	16.6	2755	9.51
tetrachlormethane	−23.0	784	3.13
cinnamic acid	133.0	5402	13.30
dioxane	11.8	3067	10.76
phenol	40.1	2695	8.58
picric acid	122.0	4663	11.80
urea	132.6	3472	8.56

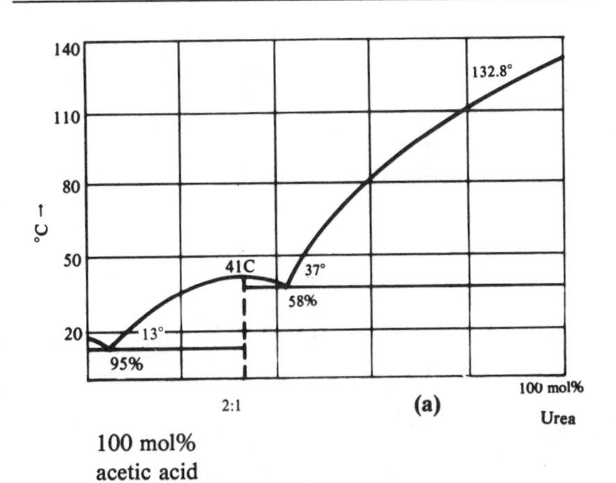

(a)

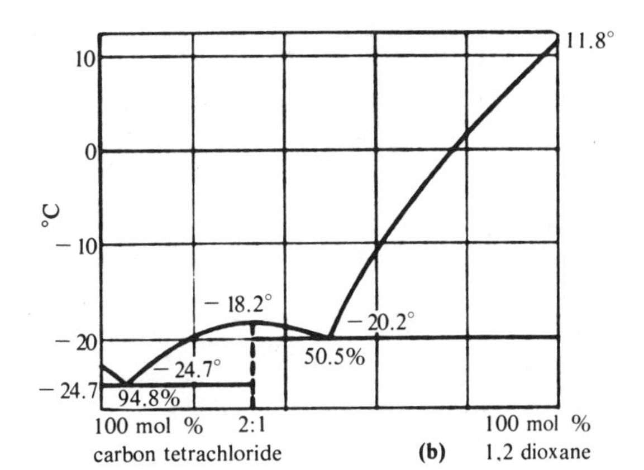

(b)

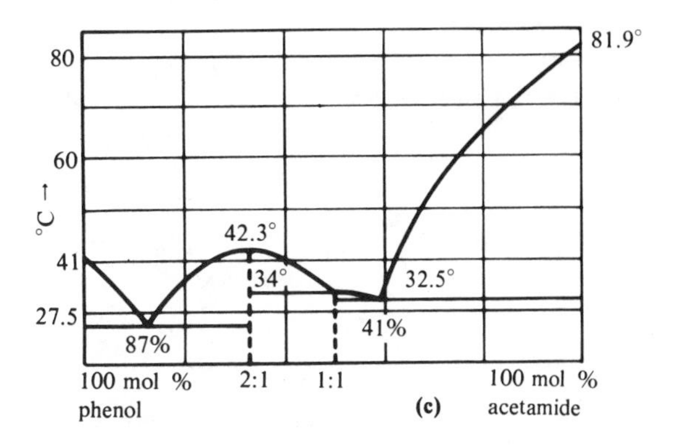

(c)

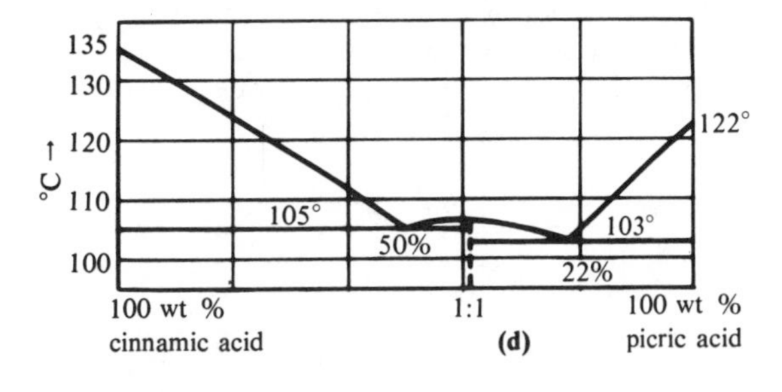

(d)

8.18. These are data of simple eutectic systems. The given eutectic mol fraction is that of the first named substance.

a. Where the enthalpy of fusion is missing, find it on the assumption of ideal behavior.

b. Where enthalpies of fusion are known, find the activity coefficients.

	T_m	ΔH_m	ΔS_m	x_1	γ
pyridine	−41.5				1
formamid	2.2				1
eutectic	−56.7			0.677	
acenaphthene	94.0				1
3,5-dinitrotoluene	63.5				1
eutectic	46.0			0.3042	

	T_m	ΔH_m	ΔS_m	x_1	γ
p-dichlorbenzene	52.9	4,273	13.10		
p-chlorphenol	42.9				1
eutectic	27.2			0.266	
camphor	175				1
o-nitrophenol	44.5	3,720	11.71		
eutectic	14.5			0.518	
camphor	175				1
picric acid	121.5	4,663	11.81		
eutectic	70			0.693	

8.19. A binary mixture containing 90% of A has the properties $T_a = 400$, $T_b = 475$, $\Delta S_a = \Delta S_b = 13$. Find the fraction liquid and its composition at a temperature halfway between that of the eutectic and the melting point of the component that precipitates.

8.20. These data are given for a ternary mixture whose solid phases are immiscible. Find the temperature at which precipitation just begins and identify the substance that precipitates first.

	x	ΔS_m	T_m	γ
1	0.3	13	400	2.5
2	0.5	12	425	1.2
3	0.2	13	450	4

8.21. Solubilities of naphthalene in several solvents are given at 40 C in Example 8.4. The temperature and enthalpy of fusion of naphthalene are 353.4 K, 4,558 cal/gmol. Find the solubilities

a. Assuming ideality.

b. Using the Scatchard-Hildebrand equation.

c. Using the Flory-Huggins modification of the S-H equation.

	δ	ml/gmol
Chlorobenzene	9.62	101.8
toluene	8.91	106.8
hexane	7.27	131.6
acetone	9.51	73.5
methanol	14.51	40.5
naphthalene	9.90	—

8.22. Use the Soave equation to find the solubility of phenanthrene in propane at 400 K and pressures from 1 to 1,000 atm. Physical properties are cited in Example 8.11.

8.23. Para-xylene is to be recovered in a continuous crystallizer from a 50-50 mixture with the meta isomer at 80 F. Recovery is to be 95% of theoretical with purity 99.5%. In this kind of unit, the effluent crystals are washed free of occlusions by countercurrent flow of molten material of essentially the same composition; the reflux ratio in this instance is 1. Data are (in the units, Btu, lb, °F):

	T_m	ΔH_m	C_p
para-xylene	55.8	68.09	0.41
meta-xylene	−54.0	46.82	0.41
Eutectic	−70.6		

The eutectic contains 86% meta isomer. Make the material and heat balances of the process, basis 100 lb mixture. See figure at the left, below.

Note: A process for complete separation of the isomers is described by Dale, *Encyclopedia of Chemical Processing and Design 13,* 456–506 (1981). Pentane is added to form eutectics with each of the isomers, and a combination of crystallization and fractionation is used.

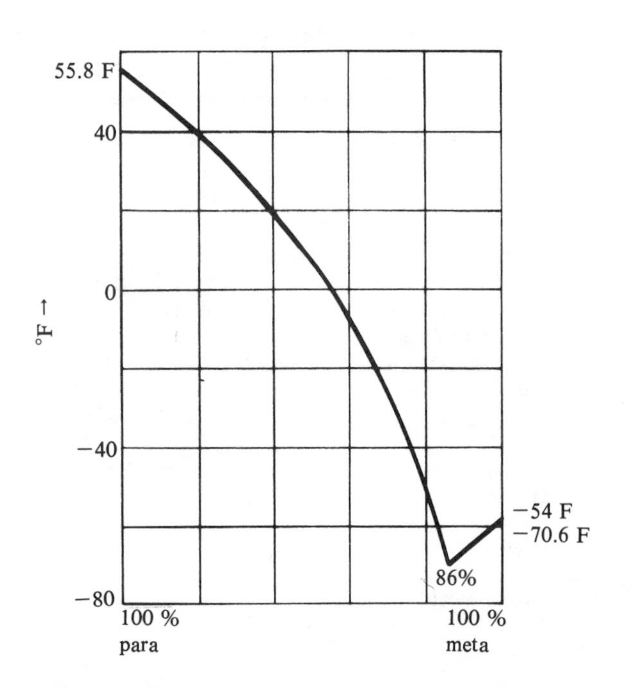

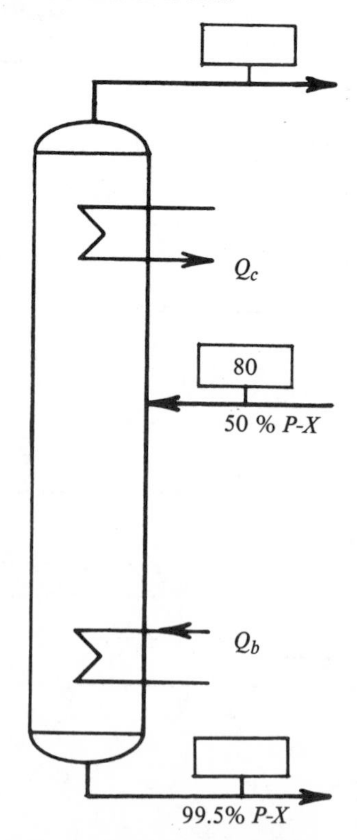

8.24. Equilibrium data for anthracene + phenanthrene were obtained by Bradley & Marsh [J Chem Soc (London) 650 (1933)]. An equimolal mixture at 218 C is to be separated in a column into products of 99% purities. The reflux ratio is to be 1.5 times the minimum reflux ratio. Find the value of the reflux ratio and the number of theoretical stages needed for this separation.

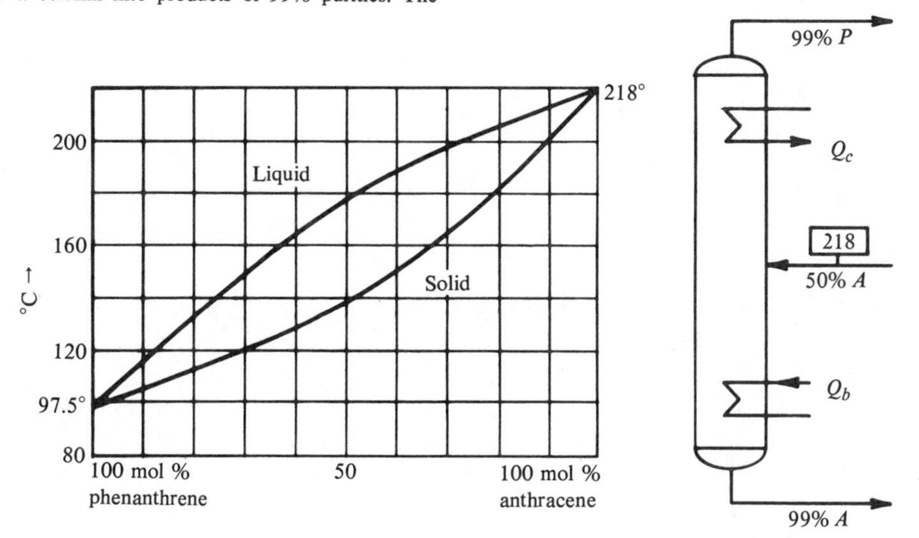

8.25. Thermal data of solid and liquid para and meta xylenes are given.

a. Construct the T-x diagram.

b. A mixture containing 30% para at 80 F is charged to a separating column where the products are essentially pure para and eutectic. Bottoms reflux ratio is 1. Make the heat balance. (Use table at right.)

	T_m, °F	ΔH_m, Btu/lb mol	C_p, Btu/lb°F
Para, solid	55.9	7362	$0.322 + 0.00053\,T$°F
Para, liquid			$0.377 + 0.00049\,T$
Meta, solid	−54.2	4977	$0.313 + 0.00055\,T$
Meta, liquid			$0.389 + 0.00038\,T$

8.26. A para isomer is to be recovered from a mixture with the meta containing 40% para. As shown on the phase diagram, the eutectic contains 15% para. After recovery of crystals, the eutectic is recycled to a soaker in which equilibrium of 35% para is reached; this mixture is recycled to the crystallizer. For a production rate of 100 mols/hr of para, find the composition, x_p, of the feed to the crystallizer and the amount of eutectic recycle, E. Some data are shown on the flow sketch.

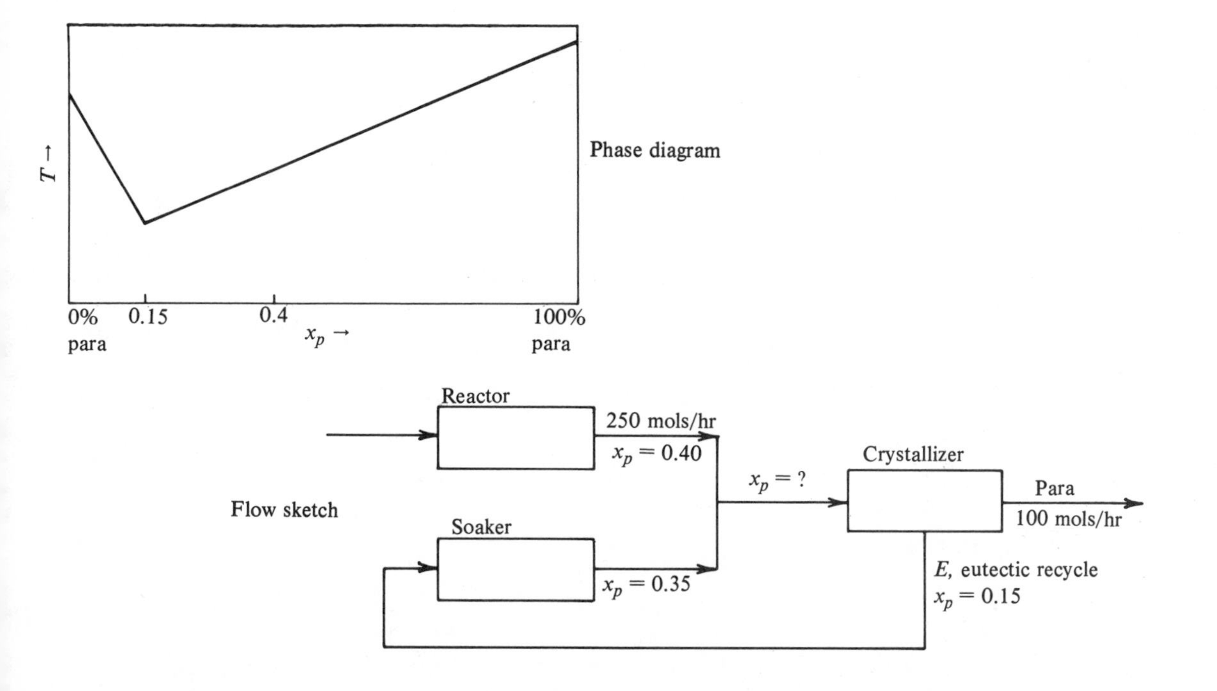

8.27. Melting temperatures and heats of fusion of two substances are $T_1 = 320$ K, $\Delta H_1 = 3500$ cal/gmol, $T_2 = 380$ K, and $\Delta H_2 = 2100$ cal/gmol. The parameters of the Margules equation are $A_{12} = 1.2$ and $A_{21} = 1.8$. Find the eutectic conditions when the behavior is ideal and also with the given activity relations.

9 Other Topics

Several other topics of phase equilibria will be discussed briefly in this chapter, with references to the recent literature where they may be studied in depth. Quantitative aspects of such equilibria have not been developed fully in all cases because of their intractability or their lesser importance to technology.

9.1. ADSORPTION

Solids have a selective affinity for fluids that becomes quantitatively significant when their specific surfaces are large, of the order of several hundred m^2/gm. Such materials are masses of numerous fine pores that have been generated by expulsion of volatile substances. The most important adsorbents are activated carbon prepared by partial volatilization or combustion of a carbonaceous body; and activated alumina, silica gel, and molecular sieves, which are all formed by expulsion of water vapor from a solid. The starting material for silica gel is a coagulated silicic acid, and that for molecular sieves is hydrated aluminosilicate crystals that end up as porous crystal structures.

The extent and selectivity of adsorption from mixtures depends on the chemical natures of the fluids and solids as well as the physical condition defined by T and P. Equilibrium will be established between the bulk of the fluid phase and an adsorbed liquid phase that may be regarded as existing in two dimensions and whose properties are influenced by the nature of the solid surface. When the phase behavior is not ideal, these phase equilibria, like any others, are expressed quantitatively in terms of fugacity and activity coefficients. The effect of the nature of the solid is typified by the case of silica gel, for example, which has the greatest affinity for polycyclic aromatics and less in decreasing order for monocyclic aromatics, diolefins, olefins, and paraffins; the order is different on activated carbons. Figure 9.1 shows the effects of T, P, and nature of adsorbent on adsorption of ethane. The amount of adsorption of vapors can be quite substantial, as much as 0.5–1.0 g/g adsorbent of substances such as benzene and halogenated and oxygenated organics (Mantell, 1951, p. 162). Generally the amount of adsorption increases with pressure, decreases as the temperature rises, and decreases as the concentration in the ambient fluid becomes smaller. These variables are controlled in regulating the adsorption process and regeneration of adsorbents.

Analytical separation of substances by adsorption, the process of *chromatography*, is a tremendously important operation. Some of the many adsorbents that are suitable with different classes of substances to be analyzed for or separated are shown in Table 9.1. Commercially, adsorption is used mostly for recovering or removing small amounts of contaminants, such as heavier hydrocarbons or water from natural gas, noxious gases from air, solvents from air in printing and painting establishments, and biologically harmful organics such as phenol from wastewaters. Some applications are made to separating substances of comparable concentrations, of which an outstanding example is the separation of aromatics and paraffins and some isomers with molecular sieves. At one time ethylene and ethane were separated commercially in a moving bed adsorber (the *hypersorption* process; Mantell, 1951, Chap. 16), but the process has been abandoned. With the advent of molecular sieves and their greater selectivity in some cases, interest has been revived in using adsorbents for such separations as propylene from propane and butenes from saturates. The energy costs of such processes are less than those of distillation or extractive distillation, but the unusual equipment needed for continuous operation has retarded most installations of this kind.

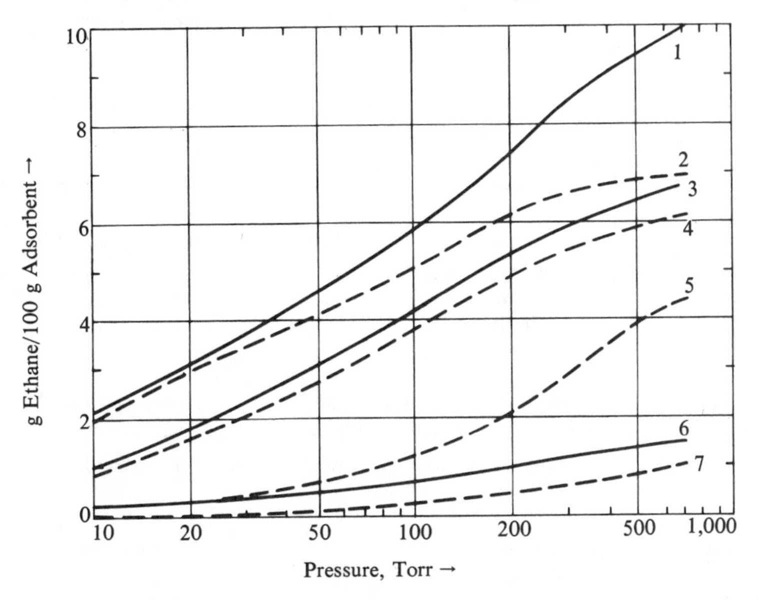

Figure 9.1. Effects of temperature, pressure, and kind of adsorbent on the amount of ethane adsorbed (data of Union Carbide Corporation).

1. Activated carbon at 25 C.
2. Type 4A molecular sieve (MS) at 0 C.
3. Type 5A MS at 25 C.
4. Type 4A MS at 25 C.
5. Type 4A MS at 75 C.
6. Silica gel at 25 C.
7. Type 4A MS at 150 C.

Table 9.1. Typical Chromatographic Systems
(Selected from Meites (editor), "Handbook of Analytical Chemistry," McGraw-Hill (1963)).

Adsorbed Substances	*Adsorbent*	*Solvent*
Alcohols:		
Aliphatic:	(see also Esters)	
Monoalcohols	Alumina	Petroleum ether
Xanthates	Paper	*n*-BuOH + KOH
Polyalcohols	Alumina	Petroleum ether
Saturated:		
Mono- or polyhydroxy compounds	Paper	$Me_2CO + H_2O$
Unsaturated:		
Phenols	Paper	$Me_2CO + H_2O$
Aldehydes and ketones:		
Aliphatic:		
Homologous series	Silica gel	C_6H_6
	Alumina	Petroleum ether + C_6H_6
Aromatic	Magnesol	Petroleum ether + C_6H_6
	Alumina	Petroleum ether + C_6H_6
Amines:		
Acylated	Alumina	C_6H_6
Aromatic	Silica gel + Celite	Petroleum ether contg. 20% Et_2O
Heterocyclic	Alumina	Petroleum ether
Nitroaromatic	Lime	Petroleum ether
	Alumina	Petroleum ether
Quaternary	Paper	EtOH + aq. NH_3
Esters:		
Monoesters:		
Aliphatic	Alumina	Petroleum ether
Aromatic	Alumina	Petroleum ether + C_6H_6
Polyesters:		
Fats:	Alumina	Petroleum ether + C_6H_6
	Silica gel	Petroleum ether + ET_2O + trichloroethylene
Ethers		
Aliphatic	Alumina or magnesia	Petroleum ether
Aromatic	Silica gel	Petroleum ether
Hydrocarbons:		
Aliphatic hydrocarbons:		
Carotenes	Alumina	Petroleum ether
	Lime	Petroleum ether
		Petroleum ether + anisole
	Magnesia	Petroleum ether
		Petroleum ether + Me_2CO
		CCl_4
cis-trans isomers	Lime or magnesia	Petroleum ether
Paraffins	Floridin	Petroleum ether
n- and *i*-	Charcoal	Petroleum ether
Trimethyl-butane and -pentane	Silica gel	Petroleum ether
Natural bioparaffins (non-saponifiable)	Alumina or magnesia	Petroleum ether
Petroleum fractions	Alumina, charcoal, or silica gel	Light petroleum
Terpenes	Alumina, charcoal, magnesia, or silica gel	Petroleum ether
Aromatic hydrocarbons:		
Low-molecular-weight	Alumina	Petroleum ether
	Silica gel	Pentane
Condensed rings	Alumina or charcoal	Petroleum ether
Spatial isomers	Alumina	Petroleum ether
	Silica gel	Petroleum ether + C_6H_6
Substituted ethylenes (*cis-trans* isomers)	Alumina	Petroleum ether

9.1.1. Measurements

Although knowledge of rates of adsorption is important in practice, it is not of concern here. The kind of information that is of concern is the equilibrium amounts and compositions as functions of T, P, and concentration of the fluid phase. Finding equilibrium data for pure vapors or vapors mixed with a relatively unadsorbable medium such as air is a relatively straightforward operation: known amounts of gas and solid are placed in contact in a closed space, the amount of adsorption is found by weighing the solid and the compositions of the gas, and adsorbates are found by an appropriate analytical technique. Many such data are compiled, for example, in Landolt-Börnstein (1972, IV/4b, pp. 121–187).

Because of occlusion of unadsorbed material in the pores of the adsorbent, indirect methods are required for measuring liquid adsorptive equilibria. One way is to maintain separate open containers of the liquid and adsorbent in a closed space; the vapor that is gradually formed is adsorbed until eventually equilibrium is attained between the liquid, vapor, and adsorbate phases. Periods of four to six weeks have been mentioned as required for attainment of equilibrium under such conditions. Removal of occluded material by centrifuging also has been used. In a chromatographic method used for example by Eagle & Scott (1950), a liquid of known composition and amount is allowed to reach equilibrium in a column containing a known weight of solid. Then a solvent is pumped through the column, and some property of the effluent is monitored for analytical purposes. Plateaus—for example, of refractive index—are obtained on a composition-time plot that correspond to individual pure components of the mixture. From the lengths of the plateaus and proper calibration of the equipment, the composition and amount of the adsorbate can be determined.

Since the external surface of a porous adsorbent granule is a tiny fraction of its total surface, a natural limit to the amount of adsorption that can occur is the volume of the pore space, which is commonly in the range of 30–60%. Usually such limits are not approached closely except at high pressures since the attraction of the solid for the liquid diminishes as successive layers of molecules accumulate on the surface. In one often-cited case, however, that of chlorine on activated carbon at atmospheric conditions, the volume of adsorbate measured at ambient conditions is greater than the pore space, indicating a degree of compression of liquid chlorine.

9.1.2. Equilibrium in Gas Adsorption

Data are usually recorded as a mass ratio, r, of adsorbate to adsorbent as a function of T and P. No generally valid relation,

$$r = f(T, P), \tag{9.1}$$

is known, but many kinds of equations with parameters characteristic of individual systems have been used. The simplest, known by the name of Freundlich, who popularized it, is

$$r = k_1 P^{1/k_2}, \tag{9.2}$$

where k_2 usually is greater than unity. Clearly only the straight portions of Figure 9.1 conform to this rule. An equation with some theoretical basis is that of Langmuir,

$$r = k_1 P/(1 + k_2 P). \tag{9.3}$$

Strictly it was developed for unimolecular layer adsorption, but it has been applied more widely. A particular advantage is its capability for extension to mixtures, which is, for component i,

$$r_i = k_i \theta_v P_i, \tag{9.4}$$

where θ_v is the fraction of the surface that is *not* covered by any adsorbate,

$$\theta_v = 1/(1 + \Sigma\, k_i P_i). \tag{9.5}$$

The summation extends over all substances present. For evaluation of the constants of Eq. 9.4 from experimental data, the equation is handled most conveniently in linearized form,

$$P_i/r_i = (1 + \Sigma\, k_i P_i)/k_i. \tag{9.6}$$

A survey of studies with the Langmuir equation is made by Young & Crowell (1962). At high pressures, some improvement in fit of the Freundlich or Langmuir equations could be expected if fugacities were used instead of pressures, but there are not many published high-pressure adsorption data that can be checked. Lewis et al. (1950) obtained data up to about 20 atm, but since the results are presented only in graphical form and the smallest fugacity coefficient is about 0.95, nothing can be proved. The nitrogen data of Ray & Box (1950) go to 15 atm, but the fugacity coefficients range only from 0.999 to 1.009.

A sort of combination of the Freundlich and Langmuir relations was made by Yon & Turock (1971), who use the expression

$$r = (kP)^{1/n}/[1 + (kP)^{1/n}], \tag{9.7}$$

or

$$\ln P = -\ln K + n \ln \frac{r}{1 - r}, \tag{9.8}$$

for pure substances, and

$$r_i = \frac{(k_i P_i)^{1/n_i}}{1 + \Sigma (k_i P_i)^{1/n_i}} \tag{9.9}$$

for mixtures. In some cases the effective parameters in a mixture may be obtained from measurements on pure substances, but usually interaction parameters also are needed.

Binary mixture equilibria are conveniently recorded on an x-y diagram which brings out the analogies between adsorptive and other phase equilibria (Fig. 9.2). When the curves are reasonably symmetrical, they can be represented by a constant *relative volatility*,

$$\alpha = \frac{(1 - y_1)}{y_1} \bigg/ \frac{(1 - x_1)}{x_1}, \tag{9.10}$$

or

$$y_1 = \frac{\alpha x_1}{1 + (\alpha - 1) x_1}. \tag{9.11}$$

Usually, however, α is not constant but depends on composition. Such dependence sometimes is represented by an equation of the form,

$$\alpha = \frac{\alpha_0 (1 + k_1 x_1)}{1 + k_2 x_1 + k_3 x_1^2}, \tag{9.12}$$

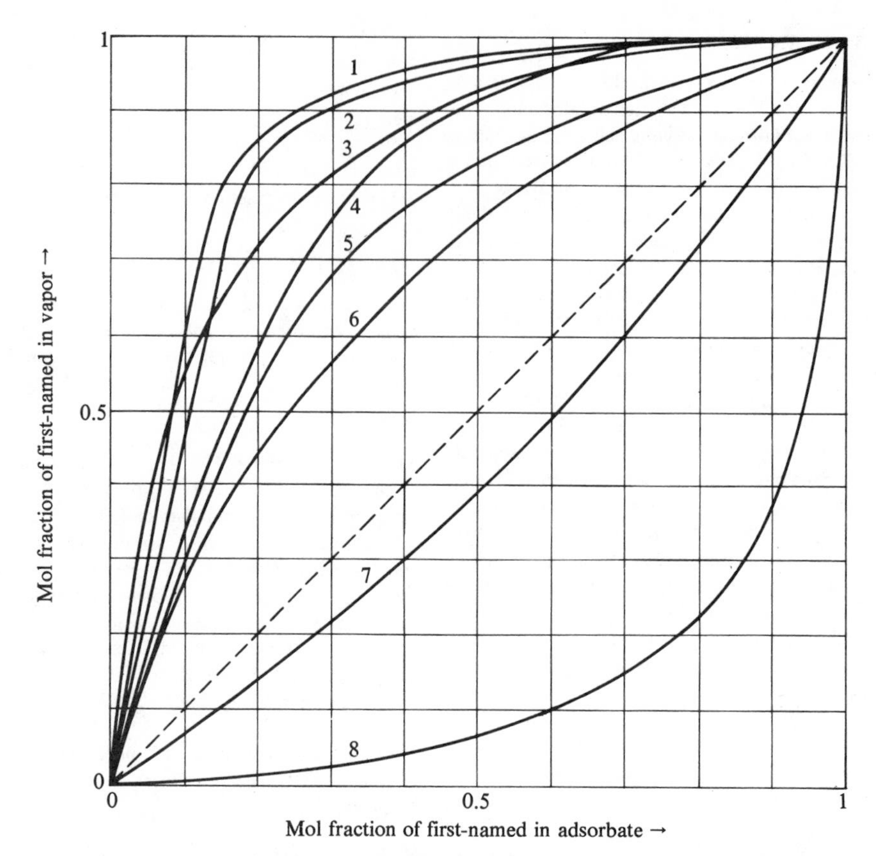

Figure 9.2. Adsorption of Binary Mixtures (Data of Union Carbide Corporation).

1. Ethane + ethylene, type 4A MS, 25 C, 250 Torr.
2. Ethane + ethylene, type 4A MS, 25 C, 730 Torr.
3. Ethane + ethylene, type 4A MS, 75 C, 730 Torr.
4. Carbon dioxide + hydrogen sulfide, type 5A MS, 27 C, 760 Torr.
5. *n*-Pentane + *n*-hexane, type 5A MS, 100 C, 760 Torr.
6. Ethane + ethylene, silica gel, 25 C, 760 Torr.
7. Ethane + ethylene, Columbia G carbon, 25 C, 760 Torr.
8. Acetylene + ethylene, type 4A MS, 31 C, 740 Torr.

which is discussed, for example, by Hala et al. (1967); but no general correlations have been developed for the constants. A relation used by Eagle & Scott (1950),

$$\frac{y}{1-y} = k_1 \left[\frac{x}{1-x}\right]^{k_2} \tag{9.13}$$

has been found applicable to some gas mixtures and could be tried for liquid ones also.

9.1.3. Thermodynamics of Adsorption

A fundamental equation of the fluid phase is

$$d\underline{U} = Td\underline{S} - Pd\underline{V} + \Sigma\ \mu_{fi} dn_{fi}, \tag{9.14}$$

where the chemical potential and composition are identified with a subscript, f, for the fluid phase. If the adsorbate phase is visualized to be essentially two-dimensional, the term $Pd\underline{V}$ is replaced by a surface energy term, πdA, in the energy balance,

$$d\underline{U} = Td\underline{S} - \pi dA + \Sigma\ \mu_{ai} dn_{ai}, \tag{9.15}$$

where A is the total surface of the adsorbent and π is a surface potential, energy per unit area, commonly called the *spreading pressure* by analogy with the corresponding term in the energy balance of the fluid phase. Subscript a identifies the adsorbate phase. From this equation, the Gibbs-Duhem equation of the adsorbate phase is obtained as

$$\underline{S}dT - Ad\pi + \Sigma\ n_{ai} d\,\mu_{ai} = 0, \tag{9.16}$$

and the isothermal form,

$$Ad\pi = \Sigma\ n_{ai} d\,\mu_{ai}. \tag{9.17}$$

At equilibrium the chemical potentials of the fluid and adsorbate phases are the same,

$$\mu_{ai} = \mu_{fi} = RT \ln \hat{f}_i. \tag{9.18}$$

So the equation for the spreading pressure becomes,

$$d\pi = \frac{RT}{A}\ \Sigma\ n_{ai}\, d \ln \hat{f}_i. \tag{9.19}$$

Since the partial fugacities can be related to the pressure by some equation of state, integration of this equation can be performed and the spreading pressure found when the experimental relation between system pressure and the amount and composition of adsorbate is known. The work is simplified further when the gas phase is an ideal solution and the Lewis & Randall Rule is invoked,

$$\hat{f}_i = y_i f_i. \tag{9.20}$$

Moreover, at moderate pressures the fugacity equals the pressure, so this simplified equation is obtained:

$$(A/RT)\, d\pi = \Sigma\, n_{ai}\, d \ln(y_i P) \tag{9.21}$$

$$= n_t\, d \ln P + \Sigma\, n_{ai} d \ln y_i. \tag{9.22}$$

For a pure substance the last term drops out, and the spreading pressure is given by

$$A\,\pi/RT = \int_0^P (n/P) dP. \tag{9.23}$$

With the Freundlich equation,

$$n = k_1 P^{1/k_2}, \tag{9.24}$$

the integral is

$$A\,\pi/RT = k_1 k_2 P^{1/k_2}, \tag{9.25}$$

and with the Langmuir equation,

$$n = k_1 P/(1 + k_2 P), \tag{9.26}$$

the integral is,

$$A\,\pi/RT = (k_1/k_2) \ln(1 + k_2 P). \tag{9.27}$$

More complex empirical adsorption equations may be needed in practice. For example Snyder & Chao (1970) use equations with five constants in their analysis of enthalpies of adsorption and spreading pressures of light hydrocarbons.

Letting

$$x_i = n_{ai}/n_t, \tag{9.28}$$

Eq. 9.22 becomes, for a binary mixture,

$$\frac{A}{n_t RT} d\pi + d \ln P + \frac{x_1}{y_1} dy_1 + \frac{x_2}{y_2} dy_2 = 0, \tag{9.29}$$

or, in terms of the first component alone,

$$\frac{A}{n_t RT} d\pi + d \ln P + \frac{x_1 - y_1}{y_1(1 - y_1)} dy_1 = 0. \tag{9.30}$$

At constant pressure, the composition of the liquid is given by

$$x_1 = y_1 + \frac{y_1(1 - y_1)}{n_t} \frac{\partial(\pi A/RT)}{\partial y_1}. \tag{9.31}$$

Clearly this result obviates the need for finding adsorbate compositions experimentally when appropriate other data have been measured. A procedure for doing this on the basis of measurements of pressure versus total adsorption at constant T and P and a series of constant vapor compositions is described by van Ness (1969), but no suitable data were available to that author for testing purposes.

Adsorption equilibria of mixtures are represented by Myers (1968) with Raoult's law in the form

$$y_i = P_i^* x_i/P \tag{9.32}$$

Here P_i^* is an effective vapor pressure of the pure component adsorbed at the same temperature and spreading pressure as that of the adsorbed mixture. The composition of a binary mixture, for example, is given by

$$x_1 = (P - P_2^*)/(P_1^* - P_2^*). \tag{9.33}$$

For instance if the spreading pressure relationship is derived from Freundlich isotherms, the effective vapor pressures are

$$P_1^* = \frac{1}{a_1} (\pi A/RT)^{b_1}, \tag{9.34}$$

$$P_2^* = \frac{1}{a_2} (\pi A/RT)^{b_2} \tag{9.35}$$

and the mol fraction of adsorbate at a specified pressure and temperature becomes

$$x_1 = \frac{\left[P - \frac{1}{a_2}(\pi A/RT)^{b_2}\right]}{\left[\frac{1}{a_1}(\pi A/RT)^{b_1} - \frac{1}{a_2}(\pi A/RT)^{b_2}\right]}. \tag{9.36}$$

When the molar area A/n, the surface of adsorbent per unit amount of adsorbate, can be approximated as the weighted sum of the molar areas of the pure components,

$$1/n = \Sigma\, x_i/n_i. \tag{9.37}$$

Thus the total adsorption of a mixture can be found when the adsorption amounts of the pure components are known at the same pressure and the mol fractions have been found with Eq. 9.36. Example 9.1 illustrates the correlation of spreading pressures of pure substances and their use in predicting binary adsorption behavior.

Nonideality of the adsorbate phase is taken into account by introducing activity coefficients, so that Raoult's law is modified to

$$y_i = \gamma_i P_i^* x_i/P. \tag{9.38}$$

This is done by Costa et al. (1981) who use measurements on pure substances and binary mixtures to predict ternary equilibria. Spreading pressures of binary mixtures are obtained by integration of

$$\frac{A}{RT}(\pi - \pi_1) = \int_1^{y_1} \frac{n_t(x_1 - y_1)}{y_1(1 - y_1)} dy_1. \tag{9.39}$$

Then corresponding values of effective vapor pressures P_1^* and P_2^* are found from pure component data. Activity coefficients are calculated from Eq. 9.38 and correlated by either the Wilson or the UNIQUAC equation. Data for mixtures of ethylene and propylene are shown in Figure 9.3. Equilibria in two ternary hydrocarbon systems were predicted with the two correlations; mol fractions generally were within 0.01. At higher pressures it would be necessary to recognize vapor-phase nonidealities, using instead of Raoult's Law,

$$\hat{\phi}_i P y_i = \gamma_i P_i^* x_i, \tag{9.40}$$

or the more nearly complete relation

$$\hat{\phi}_i P y_i = \gamma_i \phi_i^s P_i^* (PF)_i x_i \tag{9.41}$$

but no high pressure adsorption appear to have been analyzed this way.

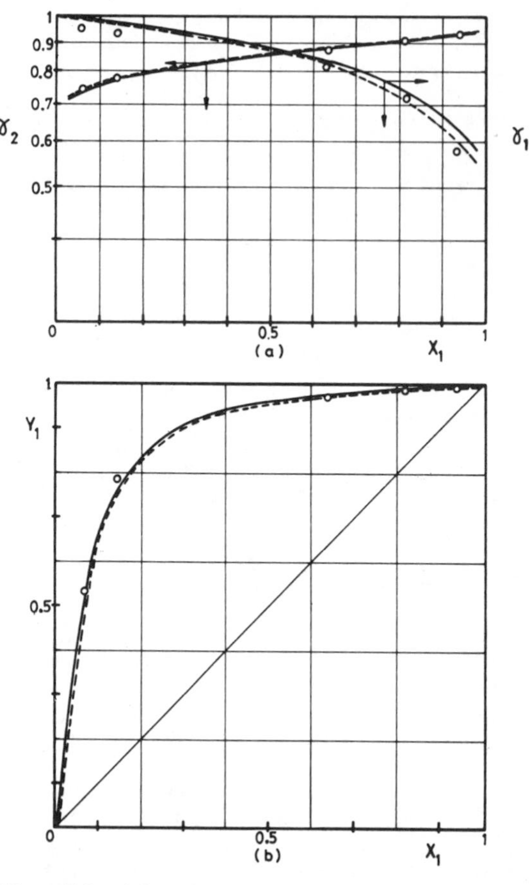

Figure 9.3. Adsorptive equilibrium data and activity coefficients of mixtures of ethylene and propylene (Costa et al. 1981).

The conditions are 20 C and 75 Torr. (a) Activity coefficients (o exp., —— WILSON,--- UNIQUAC). (b) y-x diagram (molar fractions) (o exp., —— WILSON, ---UNIQUAC).

9.1.4. Liquids

Nearly all adsorption studies of liquids have been with pure substances or binary mixtures. Those cases in which a solute is present in small concentration can be regarded as one component systems. Either the Freundlich or the Langmuir equation holds about as well for dilute liquid solutions as for dilute gases. The Freundlich constants are compiled for about 100 liquid systems in Landolt-Börnstein (II/3, 525, 1956); for mixtures only references to the literature are made. Adsorption of a pure liquid is essentially independent of pressure; temperature and the nature of the liquid are the variables with a particular adsorbent.

Equilibria of miscible binary liquid mixtures are represented conveniently on x-x^* diagrams, where the asterisk designates the adsorbate. The shapes of such diagrams depend on the nature of the adsorbent as well as those of the liquids. The comparisons of vapor-liquid and vapor-adsorbate equilibria in Figure 9.4 bring out this effect. No particularly simple method of predicting x-x^* diagrams or giving them a general mathematical form has been discovered. Some diagrams can be represented by a single *relative volatility*, Eq. 9.10. Eagle & Scott (1950) used Eq. 9.13 for liquid mixtures of hydrocarbons; for toluene + iso-octane the exponent k_2 ranged from 0.628 to 0.881, depending on the adsorbent. The equation can be rearranged to express the relative volatility as a function of the liquid composition,

$$\alpha = k_1 \left[\frac{x_1}{1-x_1}\right]^{k_2-1}, \tag{9.42}$$

which may be superior in some instances to Eq. 9.12, but the equations have not been compared in specific instances. Figures 9.5 and 9.6 present some data on adsorption of liquid mixtures.

A thermodynamic analysis of liquid-adsorbate equilibria starts with equality of chemical potentials or partial fugacities of the two phases, which may be written

$$\gamma_i x_i f_i = \gamma_i^* x_i^* f_i^*. \tag{9.43}$$

On neglecting the Poynting Factor, the pure component fugacity on the left becomes equal to that of the saturated vapor,

$$f_i = \phi_i^s P_i^s, \tag{9.44}$$

whereas for f_i^* the same replacement is made as in Eq. 9.38,

$$f_i^* = P_i^*, \tag{9.45}$$

where P_i^* is the effective vapor pressure corresponding to the spreading pressure of the mixture. On making both substitutions, the equilibrium condition becomes

$$\gamma_n \phi_i^s P_i^s x_i = \gamma_i^* P_i^* x_i^*. \tag{9.46}$$

Two kinds of problems can be analyzed by these means.

1. From a single measurement of liquid-adsorbate equilibrium composition, or only a few such measurements, the parameters of an activity-coefficient equation can be found, which can be used to calculate the adsorption equilibrium over the full composition range.
2. The parameters of an activity-coefficient equation can be found by regression of a full range of equilibrium measurements as part of a program for prediction of multicomponent adsorption equilibria.

Besides the liquid-adsorbate equilibrium compositions, certain other data are needed to perform these calculations—namely, activity coefficients of the liquid phase, spreading pressures of the pure components, and the spreading pressure of the mixture as a function of the composition. Activity coefficients of the liquid can be obtained by independent vapor-liquid measurements or by UNIQUAC or ASOG.

Upon substitution of

$$d \ln f = (V/RT)\, dP \tag{9.47}$$

into Eq. 9.19 for a pure component, the spreading pressure becomes

$$\frac{A\pi}{RT} = \frac{nV}{RT} P, \tag{9.48}$$

since both the specific volume and the amount of liquid adsorption are essentially independent of pressure. For ideal solutions, $\hat{f}_i = x_i f_i$ and Eq. 9.19 becomes

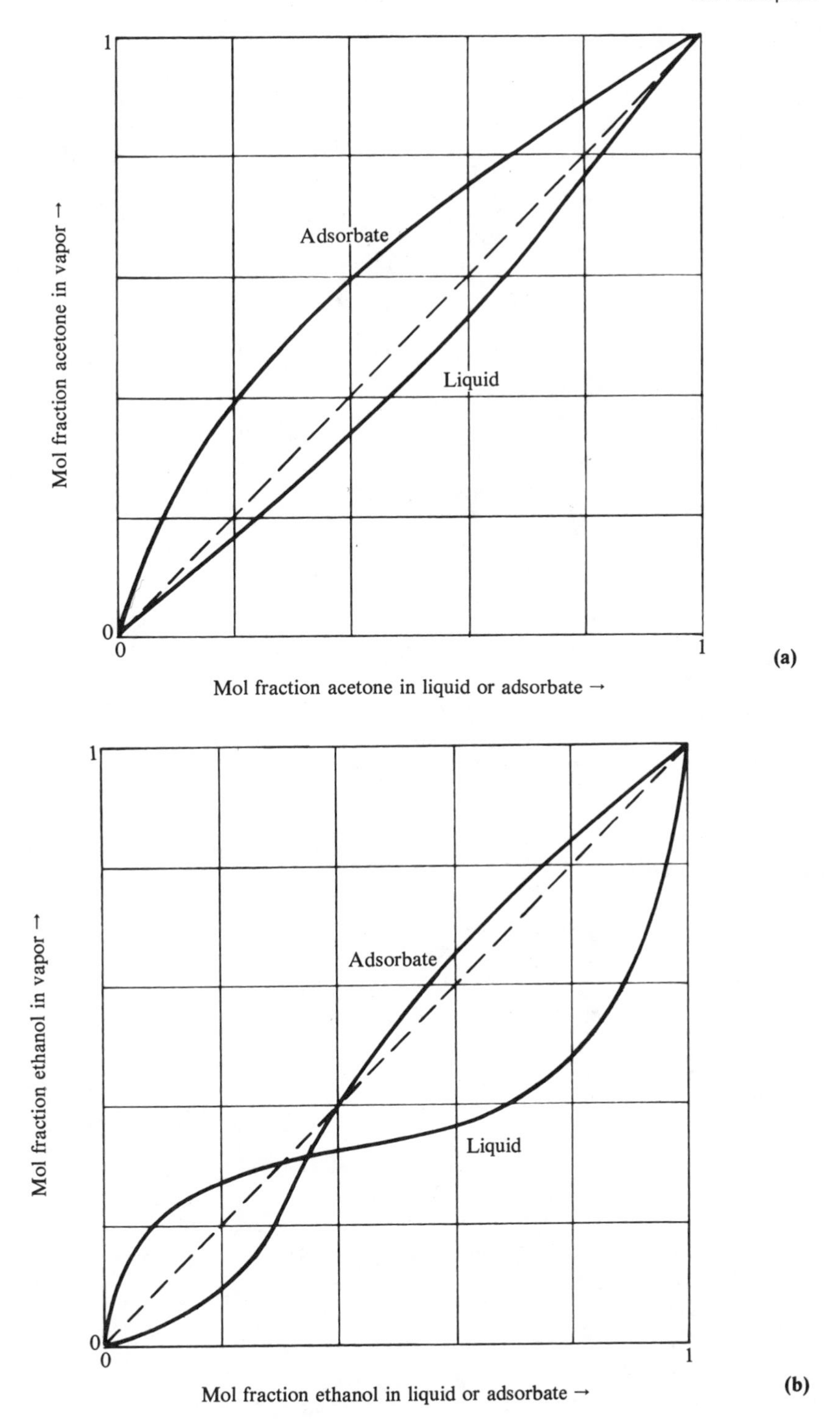

Figure 9.4. Equilibria of vapor with liquid and of vapor with adsorbate on charcoal (Data of Tryhorn & Wyatt, 1926, 1928): (a) acetone + benzene; (b) ethanol + benzene.

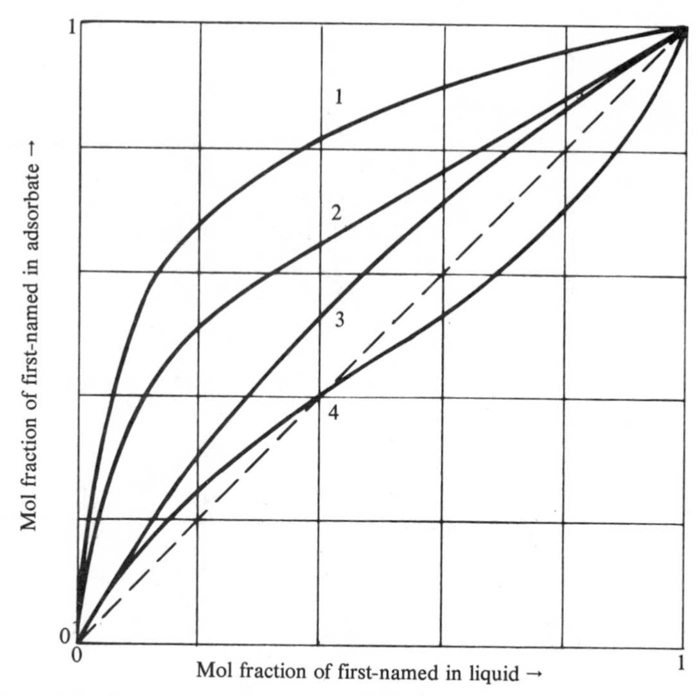

Figure 9.5. Binary liquid adsorption equilibria on X-Y diagrams

1. Toluene + iso-octane on silica gel (Eagle & Scott, 1950).
2. Toluene + iso-octane on charcoal (Eagle & Scott, 1950).
3. Ethylene dichloride + benzene on boehmite (Kipling).
4. Ethylene dichloride + benzene on charcoal (Kipling).

Kipling, in "Proceedings of the Second International Congress of Surface Activity," III 462 (1957).

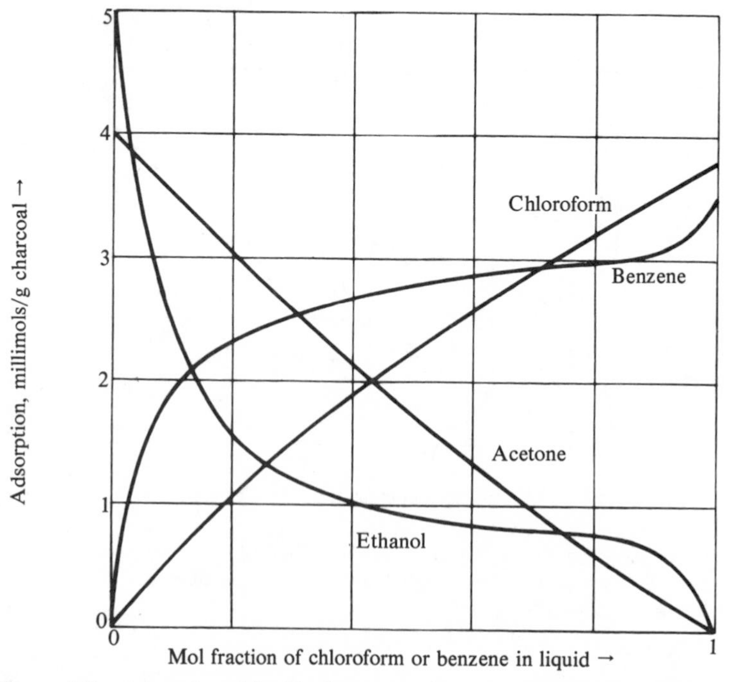

Figure 9.6. Adsorption of liquid mixtures on charcoal. Chloroform + acetone and benzene + ethanol. The ordinate gives the amount of each individual substance that is adsorbed, the abscissa the mol fraction of chloroform (mixed with acetone) or the mol fraction of benzene (mixed with ethanol). (Data gathered by Kipling, "Adsorption from Solutions of Non-Electrolytes," 1965.)

$$\frac{A}{RT}d\pi = \Sigma\ n_i d\ \ln f_i + \Sigma\ n_i d\ \ln x_i. \quad (9.49)$$

Since the differentials of the pure component fugacities drop out when P and T are constant,

$$\frac{A}{RT}d\pi = \Sigma\ n_i d\ \ln x_i. \quad (9.50)$$

Thus for a binary mixture, the adaptation of Eq. 9.39 to liquid-adsorbate equilibrium is

$$\frac{A}{RT}(\pi - \pi_1) = \int_1^{x_1} \frac{n_t(x_1^* - x_1)}{x_1(1-x_1)}dx_1. \quad (9.51)$$

The procedure for finding the adsorbate activity coefficients is analogous to that in connection with Eq. 9.38ff. The activity coefficient of the adsorbate is:

$$\gamma_i^* = \frac{\gamma_i \phi_i^{sat} P_i^{sat} x_i}{P_i^* x_i^*}, \quad (9.52)$$

with P_i^* evaluated at the spreading pressure of a mixture of composition x_i found from Eq. 9.51. All the data needed for this kind of calculation apparently have not been found for any specific case. However, Example 9.2 carries through with a highly artificial set of data.

9.2. AQUEOUS ELECTROLYTE SYSTEMS

Besides its ubiquity, water has two main properties that distinguish it from other common substances: a high degree of molecular association and a high dielectric strength. By reason of the former property, it is not well represented by reduced equations of state; by reason of the latter, dissolved inorganic substances dissociate into ions to varying degrees. Accordingly, accurate description of phase equilibria of aqueous systems should take into account the simultaneous occurrence of the chemical equilibria of molecular association and ionic dissociation.

A resurgence of research effort in this area has been caused by pollution problems generally and by development of coal conversion processes where contaminated waste waters are formed in abundance. Such processes as the removal of SO_2, H_2S, and CO_2 from gases by scrubbing with aqueous solutions are technologically well established, but their design still is amenable to improvement, particularly for complex

Example 9.1. Equilibrium Adsorption of a Binary Mixture from Data of Adsorption of the Pure Components

Adsorption data of carbon dioxide (1) and ethylene (2) over a range of pressures at 20 C are quoted by Myers (1968) and used for predicting adsorption of mixtures of these substances. For each substance the spreading pressures, π, are calculated with Eq. 9.32,

$$\frac{A}{RT}\pi = \int_0^{P^*} \frac{n}{P}dP.$$

For convenience of calculation here, Myers' results in the form of small graphs are fitted by the purely empirical equations,

$$P_1^* = (6.375 + 0.0656\ \pi A/RT)\pi A/RT,$$

$$P_2^* = (0.500 + 0.0438\ \pi A/RT)\pi A/RT.$$

The relation between the equilibrium compositions of vapor and adsorbate is found as follows. The pressure is 760 Torr and the temperature is 20 C.

1. Assume a value of $\pi A/RT$ that will ultimately give a value for x_1 in the range from 0 to 1.
2. Find the corresponding effective vapor pressures from Eqs. 1 and 2.
3. Calculate the mol fraction of component 1 in the adsorbate phase from

   $$x_1 = (760 - P_2^*)/(P_1^* - P_2^*)$$

   and that of the vapor phase from

   $$y_1 = P_1^* x_1/760.$$
4. Make a sufficient number of trials of $\pi A/RT$ to cover the desired composition range.

The results for x_1 in the range of 0 to 1 are summarized in this table:

$\frac{A}{RT}\pi$	P_1^*	P_2^*	x_1	y_1
69.51	760.2	246.1	1.000	1.000
70	767.8	249.4	0.985	0.995
80.00	930.0	320.0	0.721	0.883
90.00	1105.4	399.4	0.511	0.743
100.00	1293.8	487.5	0.338	0.575
110.00	1495.4	584.4	0.193	0.379
120.00	1710.1	690.0	0.069	0.154
125.00	1822.3	746.1	0.013	0.031
126.20	1849.8	759.9	0.000	0.000

The amount of total adsorption at a particular adsorbate composition is found with Eq. 9.37:

$$n = 1/(x_1/n_1 + x_2/n_2).$$

The values of n_1 and n_2 at corresponding values of P_1^* and P_2^* are experimental values and are read off a graph from Myers' paper. The second table in this example summarizes the results over a range of compositions. The units of n are ml at STP/g of carbon.

y_1	x_1	$\frac{A}{RT}\pi$	P_1^*	P_2^*	n_1	n_2	n
1	1	69.5	760		48	0	48
0.883	0.721	80	930	320	55	44	51.4
0.575	0.338	100	1294	488	64	57	59.2
0	0	126.2	0	760	0	64	64

Example 9.2. Activity Coefficients of the Adsorbate from Binary Liquid Adsorption Data

Activity coefficients of the adsorbate will be determined for a particular mol fraction in the bulk liquid phase, $x_1 = 0.7$. Simplified equilibrium relations are assumed.

1. The operating pressure is 20 atm. The quantity $A/RT = 1$.

2. Vapor pressures are $P_1^s = 12$ atm, $P_2^s = 20$ atm; specific volumes are given by $V_1/RT = 0.002$, $V_2/RT = 0.0012$; fugacity coefficients of saturated vapors are $\phi_1^s = \phi_2^s = 1$.

3. Equilibrium adsorptions of the pure components are $n_1 = 0.5$ gmol/kg, $n_2 = 0.3$.

4. Equilibrium adsorption of the mixture is represented by a Freundlich equation,

$$n = n_1 x_1^{b_1} + n_2(1 - x_1)^{b_2} = 0.5x_1^{0.4} + 0.3(1 - x_1)^{0.5}.$$

5. van Laar parameters of the liquid-phase activity coefficients are $A = 1.8939, B = 3.8055$. When $x_1 = 0.7, \gamma_1 = 1.5$, $\gamma_2 = 3.0$.

6. Equilibrium between compositions of liquid and adsorbate is given by

$$x_1^* = \frac{x_1}{\alpha - (\alpha - 1)x_1}.$$

With $\alpha = 1.5$, $x_1^* = 0.6087$ when $x_1 = 0.7$.

7. Since $A/RT = 1$ in this case, the pure component spreading pressures from Eq. 9.48 are

$$\pi_1 = (V_1/RT)P = 0.002\,P = 0.04 \text{ at 20 atm},$$

$$\pi_2 = 0.0012\,P = 0.024.$$

8. Applying Eq. 9.51 to the mixture,

$$\pi = \pi_1 + \int_1^{x_1} \frac{n(x_1^* - x_1)}{x_1(1 - x_1)}\,dx_1$$

$$= \pi_1 + (\alpha - 1)\int_{x_1}^{1} \frac{n}{\alpha - (\alpha - 1)x_1}\,dx_1$$

$$= 0.4 + \int_{0.7}^{1} \frac{0.5x_1^{0.4} + 0.3(1 - x_1)^{0.5}}{3 - 2x_1}\,dx_1 = 0.1205.$$

9. At this spreading pressure, the effective vapor pressures are

$$P_1^* = \frac{\pi}{V_1/RT} = \frac{0.1205}{0.002} = 60.3,$$

$$P_2^* = \frac{\pi}{V_2/RT} = \frac{0.1205}{0.0012} = 100.4.$$

10. Substituting into Eq. 9.52,

$$\gamma_1^* = \frac{\gamma_1 \phi_1^{sat} P_1^{sat} x_1}{P_1^* x_1^*} = \frac{1.5(12)(0.7)}{60.3(0.6087)} = 0.3433$$

$$\gamma_2^* = \frac{3(20)(0.3)}{100.4(0.3913)} = 0.4582$$

11. The van Laar parameters of the adsorbate phase are found with Eqs. 4.93, 4.94.

$$A^* = \ln \gamma_1^* \left[1 + \frac{x_2^* \ln \gamma_2^*}{x_1^* \ln \gamma_1^*}\right]^2 = -2.3080,$$

$$B^* = \ln \gamma_2^* \left[1 + \frac{x_1^* \ln \gamma_1^*}{x_2^* \ln \gamma_2^*}\right]^2 = -7.6510.$$

12. Adsorbate compositions, x_1^*, over the full range of bulk liquid compositions, x_1, are found from rearranged Eq. 9.52,

$$x_1^* = \frac{\gamma_1}{\gamma_1^*}\frac{P_1^{sat}}{P_1^*}x_1 = \frac{\gamma_1}{\gamma_1^*}\frac{12}{\pi/0.002}x_1.$$

Each value of x_1 is specified, the corresponding spreading pressures found by step 8, the activity coefficient, γ_1, obtained in terms of A and B. Since γ_1^* depends on x_1^*, the final step is a trial solution for x_1^*, using the values of A^* and B^* found in step 11.

The (x_1, x_1^*) results found this way could be compared with the relation of step 6.

mixtures. The influence of dissolved substances on the solubilities of gases and liquids and on the relative volatilities of liquid mixtures is also of technological interest and importance.

A valuable entry into the existing literature is provided by a recently published symposium, ACS Symposium Series No. 133 (Newman, ed., 1980). A detailed review of correlation methods of electrolyte equilibria also is made by Renon (1980). Most of the references in this section are to the cited ACS Symposium.

Varying extents of empiricism and fundamental basis have been adopted by different investigators. Aqueous systems with light hydrocarbons and inorganic gases have been brought into the general pattern of phase equilibria by Peng & Robinson (Newman, 1980, pp. 393–414), who employed their equation of state with special interaction parameters of binaries with water. These are highly temperature- and component-dependent, ranging from about −0.7 to 0.4, compared with 0–0.2 of nonaqueous binaries. The results for important industrial cases seem to be quite acceptable.

For evaluating systems of low mutual solubilities, several molecular association models of water were tested by Baumgaertner et al. (Newman, 1980, pp. 415–434), of which the most successful was a combination of 1, 2, 4, 8, and 12 water oligomers. These oligomerization equilibria, together with a cubic equation of state, were used to represent VLE up to 500 bar in some cases. Association in the vapor phase was quite small, less than 2%; some data for the liquid phase are:

T, °K	Mol fraction of oligomer				
	1	2	4	8	12
273	0.362	0.031	0.127	0.349	0.132
$647(T_c)$	0.866	0.082	0.052	0.0003	—

In their analysis of residual enthalpies of mixtures of steam with hydrocarbons, Wormald & Colling (Newman, 1980, pp. 435–447) reviewed the relations between virial coefficients and association equilibrium constants. For dimerization this relation is developed in Section 1.4. For the first few coefficients the relations are,

$$B = B_0 - k_2RT, \tag{9.53}$$

$$C = C_0 - (2k_3 - 4k_2^2)(RT)^2, \tag{9.54}$$

$$D = D_0 - (3k_4 - 18k_2k_3 + 20k_2^3)(RT)^3, \tag{9.55}$$

and so on. B_0, C_0, D_0 are virial coefficients in the absence of association; k_2 is the equilibrium constant for dimerization, k_3 that for trimerization, and so on. Because of the complexity of these relations, the temperature dependence of the parameters, and the difficulty of formulating cross-coefficients, the direct use of these equations was abandoned by these investigators. Instead, they assumed that the residual enthalpy is made up of contributions from a nonassociating reference vapor and a correction due to association. Thus,

$$\Delta H' = \Delta H'(B_0, C_0, D_0, \ldots) + \Delta H'(k_2, k_3, k_4, \ldots). \tag{9.56}$$

They found this method superior to the Peng-Robinson for mixtures of steam with heptane, with the best binary interaction parameter, $k_{ij} = -1.3$, that they could develop.

The pattern of accounting for the effect of ionic equilibria on thermodynamic properties of mixtures was developed by Pitzer et al. and is reviewed by Pitzer (Newman, 1980, pp. 451–456). The basic premise is that the Gibbs excess energy is representable by a virial-type expansion involving interactions between pairs and triplets of components together with a term for infinite dilution, and involving ionic strength, I, as a major property. Thus,

$$G^{ex}/RT = f(I) + \Sigma\Sigma\ x_ix_j\ \lambda_{ij} + \Sigma\Sigma\Sigma\ x_ix_jx_k\mu_{ijk}. \tag{9.57}$$

The infinite-dilution term, $f(I)$, is a function of the ionic strength only and is represented by an empirical modification of the Debye-Hückel equation. λ_{ij} also depends on the ionic strength. Both λ_{ij} and μ_{ijk} are correlated in terms of measurable parameters for cations and anions of different charges.

An extension of this work to include interactions of molecules with ions and with other molecules was made by Chen et al. (Newman 1980, pp. 61–89); good representations of systems like $KHCO_3 + K_2CO_3 + CO_2 + H_2O$ were made with six adjustable parameters.

As stated by Renon (1980), some of the limitations of the Pitzer method are: (1) Water should remain the main component, (2) the temperature dependence of the parameters is substantial and not simple, (3) the covariance between parameters is high, and (4) the parameters are highly specific to each system. In an effort to get around these difficulties, ionic equilibrium effects on the excess Gibbs energy were combined with "local composition" effects as in the development of the NRTL model by Cruz & Renon (1978) and Chen et al. (Newman 1980, pp. 61–89). Good fits to experimental data of binary and ternary systems were obtained with only binary parameters. Three methods, but not the Pitzer-NRTL, for phase equilibria of volatile electrolytes such as NH_3, CO_2, SO_2, and H_2S were compared in detail by Maurer (Newman 1980, pp. 139–172).

Experimental data on electrolyte systems are of course essential for developing correlation methods. A bibliography of 145 items was prepared by Wilhoit (Newman 1980, pp. 467–493). Original data are also reported in the symposium on ammonia volatility (Newman 1980, pp. 187–226), on SO_2 scrubbing with magnesia-lime slurries (Newman 1980, pp. 247–268), and with sodium citrate solutions (Newman 1980, pp. 269–294).

Taking into account simultaneous phase equilibrium and equilibria of several chemical reactions makes the numerical effort quite involved. Simple cases of this type are treated in Chapter 10. The outline of a computer program for solving VLE with volatile electrolytes (CO_2, SO_2, H_2S) is given by Renon (1980).

9.3. UNDEFINED COMPOUNDS AND MIXTURES

Although the principle of corresponding states is a cornerstone of applied thermodynamics, its application does require knowledge of critical properties, which is not always available. Some substances decompose before the critical temperature can be reached experimentally, and mixtures may consist of many close boiling substances whose individual compositions may not be known, as in the case of petroleum fractions, coal tar liquids, and other natural mixtures.

As a result of many years' effort, it has become possible to regard petroleum mixtures as made up of pseudocomponents that are characterized by the average of boiling points extending over a range of 5–10 C and the density of such a fraction. From these two basic properties, correlations have been developed for the determinations of molecular weights, acentric factors, critical temperature, and critical pressure; and an indication of the proportions of aromatic, naphthenic, and paraffinic constituents; some of these are shown in Table 9.2. Since these correlations are based on data below 650 C, they may not be safely applicable to heavy residua nor perhaps to coal liquids, which are made up largely of ring compounds. For making flash calculations of petroleum fractions, the method based on the Soave equation was found to be the most accurate of several methods analyzed by Sims & Daubert (1980), although all methods were poor when the vaporization was less than 20% or so.

The validity of commonly used correlations for molecular weights, critical temperature and pressure, and vapor pressure when applied to some ring compounds of coal tar liquids was tested with varying degrees of success by Newman (1981). If a simple conclusion can be drawn, it is that the breakdown of each fraction into proportions of aromatics, naphthenics, and paraffinics should be measured or estimated; otherwise, errors of 10–15% in the vapor pressures of liquids are to be expected from methods such as Soave, Lee-Kesler, or Pitzer-Curl. Predictions for solids are much poorer. Other examples of correlations for vapor pressure, liquid density, surface tension, and viscosity of coal liquids are given by Wilson (1980). A

Table 9.2. Properties of Narrow-Boiling Range Petroleum Fractions

1. T_b = atmospheric boiling point, °K.
2. S = specific gravity, 15/15 C.
3. M = molecular weight
   $= 219.06\ T_b^{0.118}\ S^{1.88}\ \exp(0.00392\ T_b - 3.07\ S)$.
4. P_c = critical pressure, atm
   $= 802.1\ (10^6)\ T_b^{-2.3125}\ S^{2.3201}$.
5. T_c = critical temperature, °K
   $= 34.3122\ T_b^{0.58848}\ S^{0.3596}$.
6. ω = acentric factor
   $= -1 + 0.1861 \ln P_c/(T_c/T_b - 1)$
   (Edmister, *Petroleum Refiner* 37(4):173–179 (1958). A more accurate equation for hydrocarbons due to Lee & Kesler is quoted in Section 1.3.1.
7. For narrow-boiling-range fractions, the distinction between the different kinds of average boiling points vanishes, and the pseudocritical and true critical properties become the same.

Items 3, 4, and 5 are from the *API Data Book.*

three-parameter (not T_c, P_c, and ω) BWR equation for prediction of phase equilibria and thermodynamic properties in general has been developed by Brule et al. (1982); unfortunately, only one experimental data point was available for checking purposes. The semifinal word on the properties of coal-derived liquids may be the data book that has been in preparation for several years under the sponsorship of the Institute of Gas Technology (Project 8979, several annual reports).

Work is still active on petroleum fractions. A comprehensive study of phase equilibria using the API pseudocomponent method was made by Yarborough (1979). The basis is the Redlich-Kwong equation with parameters established as functions of the acentric factor and reduced temperature:

$$a = f_a(\omega, T_r)\, R^2 T_c^{2.5}/P_c, \tag{9.58}$$

$$b = f_b(\omega, T_r) R T_c/P_c, \tag{9.59}$$

with emphasis on reservoir fluids and conditions.

A similar development apparently was made by Robinson et al. (1978) for use with the Peng-Robinson equation. The required properties are obtained from ASTM distillations or carbon numbers from chromatographic analyses. Actual details are available in a computer program from the Gas Processors Association, Tulsa. The Soave equation is the basis of the method adopted for the *API Technical Data Book*, and is based on pseudocomponents with properties figured from formulas like those of Table 9.2. Binary interaction parameters with inorganic gases are known only through n-decane and trimethylbenzene (Table 1.12), but extrapolations perhaps can be made with some assurance. Chromatographic analysis can detect substances with as many as 50 carbon atoms.

Certain properties can be evaluated with some accuracy from structural contributions. Methods for normal boiling points and critical properties are given in the books of Lyman et al. (1982); Reid et al. (1966, 1977); and Sterbacek et al. (1979). The acentric factor follows from those properties, with Edmister's formula, for example. The least certain of these determinations from structure is that of the normal boiling point; but usually, if anything is known about a substance of definite structure, it is the NBP, or at least a vapor pressure at some temperature.

An equation of state whose parameters are found directly from group contributions is in the process of development, the PFGC equation (parameters from group contributions), which promises to be of special value in dealing with undefined components or mixtures. Critical properties and acentric factors are not needed. The groups can be determined by advanced analytical techniques such as C^{13} NMR spectroscopy. Several papers devoted to this project have been published—for example, one by Moshfeghian et al. (1980)—but the values of the group parameters will not be divulged until they have been firmed up. Some of the types of molecules to which the method is being extended are aromatic and other molecules containing N, O, and S that occur in coal-liquefaction processes, halogenated refrigerants, and organic compounds in general. The referenced paper is concerned with the system methanol + water + hydrocarbons + acid gases; however, the PFGC and Soave equations do equally well in this application.

9.4. POLYMERS

Phase equilibrium behavior involving polymers is of concern in several areas.

1. Solution processes of polymers with plasticizers, monomers, and other liquids.
2. Evaporation behavior of monomers from solution.
3. Mutual miscibilities of different polymers.
4. Fusion behavior of polymers.

At present none of these topics has been well quantified in thermodynamic terms, but efforts are being made. The fact that polymers ordinarily are not subjected to multistage separation processes has reduced the economic incentive for research on their phase behavior. Also, the problem is somewhat complicated by the variety of possible polymer states: They are mixtures of a range of molecular weights, they may be amorphous glassy or rubbery, or they may have more than one crystalline form depending on the temperature and the history. To the point is a comment by Bondi (1977) that his literature review "examines the current state of ignorance" of the thermodynamics of polymer phase behavior.

The large difference in molecular sizes of monomer and polymer was recognized as a significant property in determining the mixing properties of a solution by Flory (1941) and Huggins (1941). The results of this work will be described briefly. Accessible treatments may be found in the books of Huggins (1958) where earlier literature is discussed, and Chao & Greenkorn (1975). When there is a great disparity in molecular sizes of the components of a mixture, the entropy of mixing is more accurately expressed in terms of volume fractions rather than mol fractions. Thus,

$$\Delta S^{\text{mix}}/R = -n_1 \ln \phi_1 - n_2 \ln \phi_2, \tag{9.60}$$

instead of

$$\Delta S^{\text{mix}}/R(\text{ideal}) = -x_1 \ln x_1 - x_2 \ln x_2, \tag{9.61}$$

for a mixture containing n_1 mols of monomer and n_2 mols of polymer with a molecular weight m times that of the monomer. The volume fractions are defined as

$$\phi_1 = n_1/(n_1 + mn_2), \tag{9.62}$$

$$\phi_2 = mn_2/(n_1 + mn_2). \tag{9.63}$$

The enthalpy change on mixing is, first of all, proportional to the mass of the system, which is in turn proportional to $(n_1 + mn_2)$. Moreover, the function representing this enthalpy must vanish when $\phi_1 = 0$ or $\phi_2 = 0$. The simplest equation with these properties is

$$\Delta H^{\text{mix}}/RT = \chi(n_1 + mn_2)\phi_1\phi_2, \tag{9.64}$$

where χ is an empirical parameter named after Flory, the value of which depends on the nature and composition of the system. Some numerical values are given in the cited books of Huggins and of Chao & Greenkorn. At 25 C, in the Huggins table they range from 0.14 to 0.91 out of about 40 data. The Flory parameter is related to the solubility parameters; thus,

$$\chi = \chi_0 + \frac{V_1}{RT}(\delta_1 - \delta_2)^2. \tag{9.65}$$

On combining the entropy and enthalpy equations, the equation for the Gibbs energy of mixing results:

$$\Delta G^{\text{mix}}/RT = n_1 \ln \phi_1 + n_2 \ln \phi_2 + \chi(n_1 + mn_2)\phi_1\phi_2. \tag{9.66}$$

This is readily extended to multicomponent mixtures. Details of the derivation of equations for the activities and activity coefficient are given in Example 9.3, with the results,

$$\ln a_1 = \ln \phi_1 + (1 - 1/m)\phi_2 + \chi\phi_2^2, \tag{9.67}$$

$$\ln \gamma_1 = \ln(\phi_1/x_1) + (1 - 1/m)\phi_2 + \chi\phi_2^2, \tag{9.68}$$

$$\ln a_2 = \ln \phi_2 - (m - 1)\phi_1 + \chi\phi_1^2. \tag{9.69}$$

Mol fractions are not a satisfactory unit with polymer mixtures since their molecular weights are large and usually not known accurately, so activity coefficients do not have their usual utility. Activities expressed in volume fractions as given here are the useful terms. The condition of equality of partial fugacities between phases at equilibrium implies equality of activities when the same reference state is used for all phases, which is an argument in favor of the use of activities.

Graphical representation of the effects of the molecular weight ratio and the Flory parameter is shown in Figure 9.7. Changes in m beyond 100 or so have little effect on the activity of the monomer. In practical cases, where m usually is greater than 1,000, the activity of the polymer is very small. This is consistent with the fact that high-molecular-weight polymers are essentially insoluble, but do imbibe monomers and other liquids; that is, they assume the role of solvent for low-molecular-weight substances. A further result of the theory is that complete miscibility occurs below a critical value of the Flory parameter,

$$\chi_c \leq 0.5(1 + 1/m)^2 \simeq 0.5, \tag{9.70}$$

and only partial miscibility above this value. From the relation of χ to solubility parameters, this result implies that for maximum miscibility the solubility parameters of the components of the mixture should be nearly the same.

Many polymers do not mix, and when some pairs are mixed with water they form two aqueous phases. For example, a mixture with 8% each of polyethylene glycol and dextran in water forms two phases with compositions 2% PEG + 20% dextran and 12% PEG with less than 0.01% dextran. Such behavior is useful for extraction of enzymes for which organic solvents cannot be used. By control of polymer concentrations, pH and salt concentrations, partition coefficients of five or more can be attained (Kula, 1979).

Experimental methods for the determination of activities in polymer mixtures include measurements of vapor pressures, osmotic pressures, freezing-point depressions, and mutual solubilities; but usually not vapor-liquid equilibria because of the essentially nonvolatile quality of the polymers. These topics are discussed at length, for example, by Hildebrand & Scott (1950, 1964).

In principle, the derived expressions for activities may be used to describe solubilities, melt equilibria, and vapor-liquid equilibria of polymers and low-molecular-weight liquids. With respect to solubilities, the theory seems to be applicable only to the dilute range. This aspect of the subject is treated by Shinoda (1978, Chap. 8). Efforts have been made to extend the theory to concentrated solutions—for example, by Beret et al. (1980), who also mention an extension of the UNIFAC group method to such cases. Applications to melt equilibria, to VLE and other topics are referred to in the review paper of Bondi (1977); generally the results are poor.

Unfortunately, mostly negative conclusions have been drawn in this section with regard to the thermodynamics of polymer phase equilibria. The field seems to be open for much more theoretical and experimental work.

9.5. LIQUID CRYSTALS AND MICELLES

Liquid crystals: Some pure liquids (more than 6,000 when counted in 1978) and more mixtures have certain properties that are more often associated with solid crystals, such as optical anisotropy, X-ray diffraction, and distinctive electrical and thermal properties. Moreover, they display abrupt changes in their properties at particular temperatures, indicating that some changes occur in internal structure, although the general characteristics of liquids are preserved. For instance, solid ethyl-*p*-azoxycinnamate changes sharply at 140 C into a turbid liquid, which in turn changes sharply at 249 C into a clear liquid; such changes are reproducible and reversible.

Structured phases of these kinds are called *mesomorphs* (intermediate forms) or *liquid crystals,* and have been termed a fourth state of matter by enthusiasts. Organic compounds that behave in this way have long, narrow molecules with a polar group such as -OR or -COOR at one or both ends and often a mildly active group such as -C=C-, -C=N-, or -N-O-N- in the middle. Only para positions are conducive to mesomorphism, not meta or ortho. The molecules tend to arrange themselves parallel to each other into swarms or "crystals" of the order of 100,000 molecules each. The crystals are not composed of entirely fixed numbers of molecules having a definite geometrical shape; they are statistical in nature, in equilibrium with surrounding groups of molecules and fluctuating in size and shape. As the temperature rises, the swarms decrease in size and eventually become too small to scatter light and become amorphous liquids in the common sense. Because they are strongly temperature-sensitive, liquid crystals are termed *thermotropic* (heat-turning) aggregates, in contrast with another type of aggregate, the micelle, whose structural makeup depends also on its concentration, and consequently is called *lyotropic* (liquid-turning).

Specialists in this area have found it convenient to make a fairly elaborate classification of the different kinds of clus-

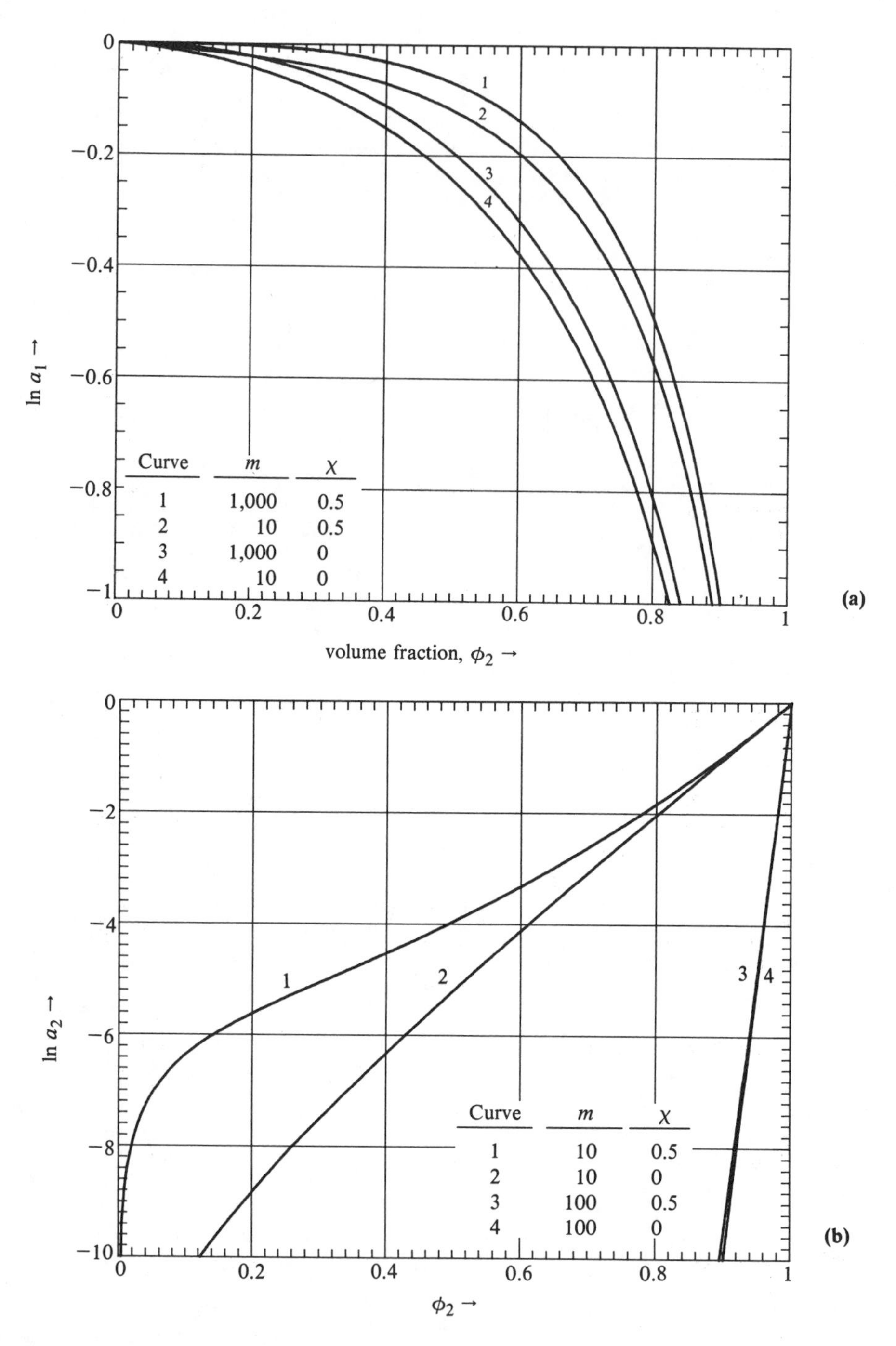

Figure 9.7(a). Activities of polymer (2) and monomer (1) as functions of the ratio of their molecular weights, m, and the flory parameter; beyond $m = 100$ on this scale, the curves of $\ln a_1$ are virtually indistinguishable. **(b)** $\ln a_2$ against ϕ_2.

terings that can occur. *Smectic* (soaplike) structures are made up of parallel lamellae one molecule thick the long way; there are at least seven smectic subclasses. *Nematic* (threadlike) structures are made up of long, straight, parallel rods of cylindrical or other cross-sections. *Cholesteric* structures (after cholesterol, which has this form) are made up of twisted rods, sometimes approaching helical shapes. A book by Demus & Richter (1978) has 212 micrographs, many in color, of a variety of these substances.

Transition from one type of structure to another is effected primarily by temperature change and is definite and reproducible. Table 9.4 shows a few such transitions, with temperatures and enthalpy and entropy changes which were found with differential thermal analysis (DTA, Chap. 12)

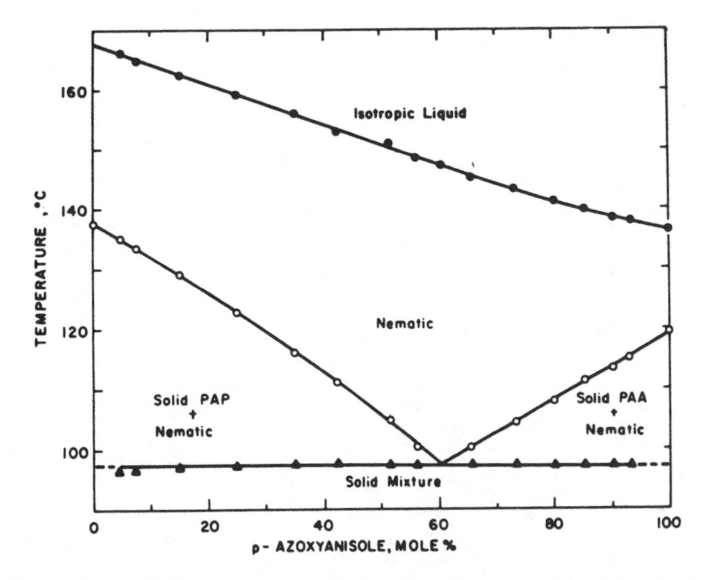

Figure 9.8. Phase diagram of *p*-azoxyanisole + *p*-azoxyphenetole, showing eutectic formation with the nematic phase (Hsu & Johnson, 1974).

equipment. These temperatures also can be observed visually, preferably with polarized light, on the heated stage of a microscope. A variety of other observation techniques is used for phase studies, as described for example by Gray & Winsor (1974).

Mixtures of liquid crystals of different kinds exhibit solidification behaviors like those of ordinary liquids. Figure 9.8 is a melt diagram with a simple eutectic, whereas Figure 5.23(f) shows the equivalent of a minimum azeotrope between the amorphous and crystalline liquids. The Schröder equation (Section 8.4.1) has been found by a number of investigators to represent eutectic temperatures within a degree or two in many cases, of which Table 9.3 is a sample. The largest energy changes are associated with the transition from the solid to the first liquid crystalline phase; the other phase transitions have minor energy changes. Other thermal data and correlations will be mentioned at the end of this section.

Micelles: Although they do not have liquid crystal character, certain long molecules with a polar end and a nonpolar end do have the property of forming aggregates in the presence of water and some organic liquids. These aggregates, called micelles (Latin, crumbs), may be spherical, cylindrical, or

Table 9.3. Enthalpies and Entropies of Phase Changes of Several di-*n*-alkyl *p*-azoxy-α-methyl cinnamates. Barrall & Johnson, in Gray and Winsor, eds., *Liquid Crystals and Plastic Crystals,* Vol. 2, pp. 254–306 (1974).

$$C_nH_{2n+1}\text{—O—}\overset{\overset{\displaystyle O}{\|}}{C}\text{—}\underset{\underset{\displaystyle CH_3}{|}}{C}\text{=CH—}\langle\bigcirc\rangle\text{—}\underset{\underset{\displaystyle O}{\downarrow}}{N}\text{=N—}\langle\bigcirc\rangle\text{—CH=}\underset{\underset{\displaystyle CH_3}{|}}{C}\text{=}\overset{\overset{\displaystyle O}{\|}}{C}\text{—O—}C_nH_{2n+1}$$

Note that Walden's Rule for fusion, $\Delta S \simeq 13$, does not hold in these cases.

Compound	*Transition*	T (°C)	ΔH (*kcal/mole*)	ΔS (*cal/mole/°K*)
Ethyl	solid → smectic *A*	110	7.89	20.6
	smectic *A* → nematic	125	0.26	0.65
	nematic → isotropic	141	0.29	0.70
Propyl	solid → smectic *A*	72	6.69	19.4
	smectic *A* → nematic	119	0.26	0.66
	nematic → isotropic	131	0.29	0.72
Hexyl	solid → smectic *A*	55	5.98	18.2
	smectic *A* → isotropic liquid	91	0.96	2.6
Dodecyl	solid → smectic *C*	79	15.7	44.6
	smectic *C* → smectic *A*	83	≈0.24	≈0.71
	smectic *A* → isotropic liquid	87	2.01	5.58

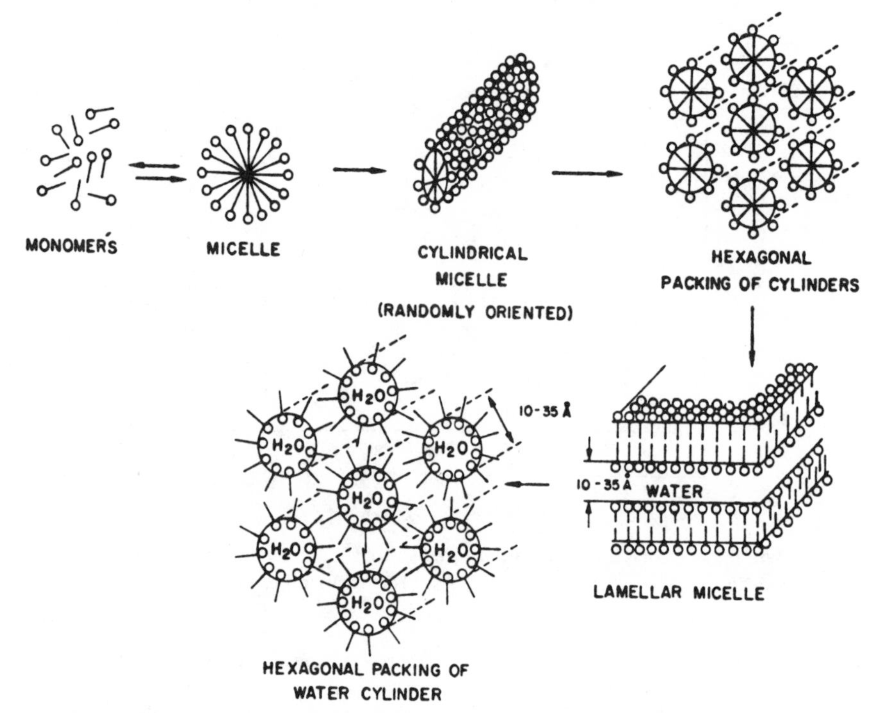

Figure 9.9. Successive stages in the formation of micellar structures as the concentration of surfactant in water increases in the direction of the arrows. (Bansal & Shah, 1977).

lamellar in shape, with the polar ends pointing toward the water phase if that is the surrounding medium. Figure 9.9 illustrates these structures. If the surrounding medium is predominantly organic, the polar ends will point inward and the micelles are called inverted. A soap or a nonionic surfactant and water, with or without another organic substance, is a common example of a micellar system. Micelles have the interesting and useful property of "solubilizing" organic substances that are not ordinarily miscible with water; they do this by incorporating the solute within the micelle and insulating it from the water with a barrier of the polar ends of the soap.

Like liquid crystals, micelles are merely aggregates of a statistical nature, but only of a few hundred molecules on average—with sizes and shapes that are variable and fluctuating and strongly dependent on the concentration and temperature. A basic difference between micellar aggregates and liquid crystals is that the former exist in contact with an amorphous liquid phase, whereas the latter fill the whole space by themselves.

In small concentrations, surfactants form true solutions in water; but at a certain value, known as the *critical micellar concentration* (cmc), micelles begin to form and continue to do so until some appreciable concentration obtains. The cmc is quite small in most cases, of the order of 0.01 molar for ionic surfactants such as soaps, and 0.0001 molar for nonionic ones such as ethylene oxide ethers of alkyl benzenes. It is not detectable optically since the micelles are smaller than the wavelength of light and remain in colloidal solution. Other physical properties of the system, however, may change abruptly at the cmc. Thus Figure 9.10 shows measurement of the surface tension as a function of soap concentration. At 0.07 wt% the plot becomes horizontal at least for some concentration increase beyond that, indicating that two phases exist beyond the cmc—the micelle and the true solution of 0.07 wt%.

A particular chemical system can have several kinds of micellar structures depending on the concentration and the temperature. A representative binary micellar system is shown in Figure 9.11. Depending on the temperature and concentration, one of five different structures may prevail: solid, amorphous liquid, optically isotropic but structured liquid, a neat (lamellar) phase, and a middle (rodlike) phase. The peaks of some of the regions are reminiscent of the peaks corresponding to compound formation on the freezing diagrams of amorphous liquids (Chapters 5 and 8). The ternary system of Figure 9.12 also displays a variety of pure aggregates and mixtures of aggregates. More than 100 such ternary and binary diagrams are given by Ekwall (1975). So far, apparently, no phase diagram of a micellar system, even a simple eutectic like Figure 9.8, has been constructed on the basis of thermal data.

Thermodynamic properties: Liquid crystals are simpler entities than micellar aggregates, so more success has been achieved in correlating the changes in properties accompanying phase transitions. Some such data are quoted in Table 9.3. The source of that table also has many data of transition entropies of groups of chemically similar substances with alkyl chains of various lengths, with which the entropy changes increase quite regularly in most cases. The success in predicting freezing behavior with the Schröder equation also has been mentioned earlier.

Aqueous micellar systems involve a number of complicating factors, even at the cmc or in the dilute range where the micelles can be assumed spherical. The degree of aggregation depends primarily on the concentration, temperature, and

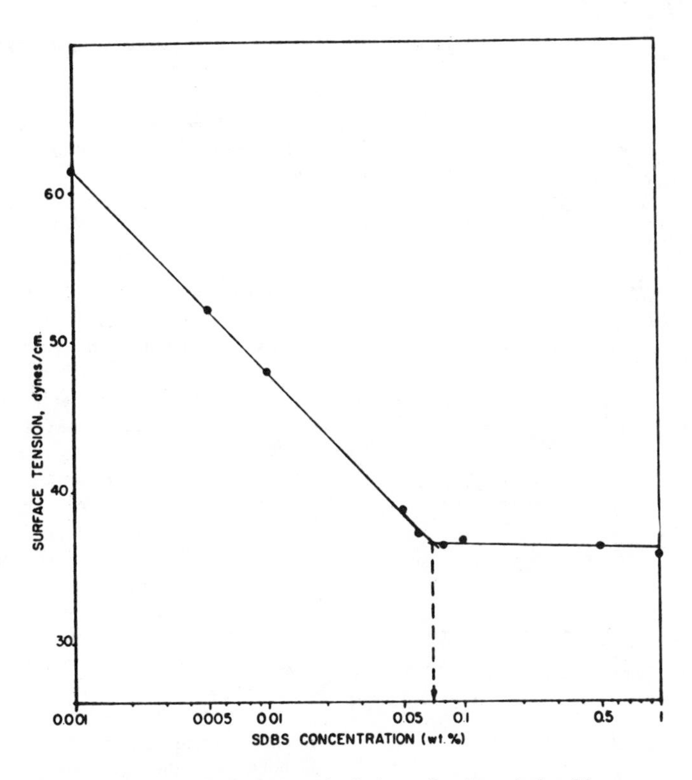

Figure 9.10. Surface tension of solutions of sodium dodecyl benzene sulfonate in water at 25 C (Shah et al., 1979). The break in the curve at 0.07 wt% is at the critical micellar concentration (cmc).

ionic strength. Figure 9.13 and Table 9.4 show these effects at the cmc, but they have not been correlated in any form useful for general prediction.

Several investigators have addressed themselves to the problem of the equilibrium of micelle formation, which is written as a stoichiometric equation,

Table 9.4. The Critical Micelle Concentration and Degree of Aggregation of Several Surfactants with and without Added Salt (Phillips 1955).

Surfactant	*Solution*	*Critical micelle concentration (moles liter^{-1})*	*Aggregation number, n*
Sodium dodecyl sulfate	Water	0.0081	80
	0.02 *M* NaCl	0.00382	94
	0.03 *M* NaCl	0.00309	100
	0.10 *M* NaCl	0.00139	112
	0.20 *M* NaCl	0.00083	118
	0.40 *M* NaCl	0.00052	126
Dodecylamine hydrochloride	Water	0.0131	56
	0.0157 *M* NaCl	0.0104	93
	0.0237 *M* NaCl	0.00925	101
	0.0460 *M* NaCl	0.00723	142
Decyl trimethyl ammonium bromide	Water	0.0680	36
	0.013 *M* NaCl	0.0634	38
Dodecyl trimethyl ammonium bromide	Water	0.0153	50
	0.013 *M* NaCl	0.0107	56
Tetradecyl trimethyl ammonium bromide	Water	0.00302	75
	0.013 *M* NaCl	0.00180	96

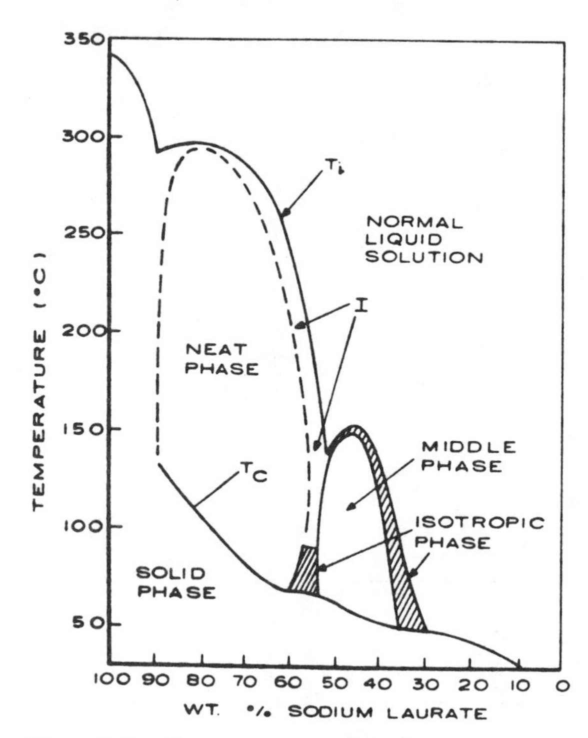

Figure 9.11. Temperature-composition phase diagram of sodium laurate + water (Wojtowicz, 1975).

Three kinds of two-phase regions are present:

1. *Isotropic solution,* made up of spherical micelles, which form the basic unit of a liquid crystal structure and are deployed on a face- or body-centered cubic arrangement within the aqueous solution.
2. *Middle phase,* made up of rodlike micelles of cylindrical (sometimes square or rectangular) cross-section of considerable length, deployed in a hexagonal arrangement within the aqueous medium. Similar in character to the nematic phase of liquid crystals.
3. *Neat phase,* made of lamellar micelles c, similar to the smectic phase of liquid crystals except that the layers are separated by aqueous solution.

$$Nm \text{ (monomer)} \rightleftharpoons M \text{ (micelle)}, \tag{9.71}$$

where N is the degree of aggregation. The standard Gibbs energy change of the process is

$$\Delta G^0 = \frac{RT}{N} \ln C_M + RT \ln C_m, \tag{9.72}$$

which may be written

$$\Delta G^0 = RT \ln(\text{cmc}) + \frac{RT}{N} \ln N - \frac{RT}{N} \ln(NC_M). \tag{9.73}$$

Of the quantities on the right, the cmc often is obtained from measurements of surface tension, the molecular weight of the micelle by membrane osmometry, and the degree of aggregation consequently by material balance.

The Gibbs energy change of the process was analyzed by Birdi (1977) as made up of the changes of three fundamental processes, but he was not able to evaluate all the individual contributions.

With the cmc correlated as a function of concentration, temperature, and ionic strength, and with measurements of the degree of aggregation, it is possible to correlate the Gibbs energy change of the micellization process in terms of the same variables. Once this has been achieved, the enthalpy change with temperature is obtainable with the van't Hoff isochore,

$$\Delta H^\circ = -RT^2 \frac{\partial(\Delta G^\circ/T)}{\partial T}. \tag{9.74}$$

Such correlations in terms of aggregation number are analogous to those for the Gibbs energy change of reaction in terms of group contributions from the participants.

Another problem to which a solution is desirable is the construction of complete phase diagrams like those of Figure 9.11 and 9.12, but such a hope is perhaps Utopian since it has not been realized even with amorphous and unassociated liquids in the areas of organic crystallization, ceramics, or metallurgy, as discussed in Chapter 8. Solubilization by means of surfactants is another topic for correlations. Its counterpart in the field of amorphous liquids has achieved some measure of correlation in terms of activity coefficients, which are obtainable from structural contributions, but the effect of temperature is not well accounted for.

Perhaps the main reason that more progress has not been achieved on the thermodynamic aspects of micelle behavior is the richness of the field, with its diversity of constituents and influences, which still provides ample opportunities for the collection of useful and interesting facts. Correlations may come later.

9.6. MIXTURES AT SUPERCRITICAL CONDITIONS

The critical properties of many binary mixtures vary quite regularly with composition, and form a continuous *P-x* locus from pure component to pure component, with homogeneous conditions above the locus and liquid + vapor regions below. Such behavior is shown, for instance, in Figures 1.31 and 1.32. However, substances that differ markedly in molecular size, shape, volatility, or polarity may have discontinuities in the critical loci to the extent that critical points of the pure substances are not connected by the locus, and phase splitting may occur at temperatures and pressures well above the critical values of the components. Such behavior usually is called an instance of gas-gas equilibria, although the densities of the supercritical phases are typical of liquid phases. As will be indicated later, however, the distinction between liquid and vapor phases at very high pressures is not always clear or meaningful. One effect of high pressure, that of solubility in compressed gases, was discussed in Section 8.6.

A binary critical locus emanating from the critical point of the less volatile substance may have a characteristic shape depending on the chemical nature of the pair. In Figure 9.14, some of the critical loci have maximum temperatures or pressures, sometimes approaching the vicinity of the left critical point, but continually diverging in the case of $He + CH_4$. In some instances there is a possibility of convergence if the pressure is raised sufficiently, but often the curves terminate at the formation of a solid phase. Supercritical phase splitting is shown for ammonia + nitrogen in Figure 9.15 and for carbon tetrafluoride–n-heptane in Figure 9.16.

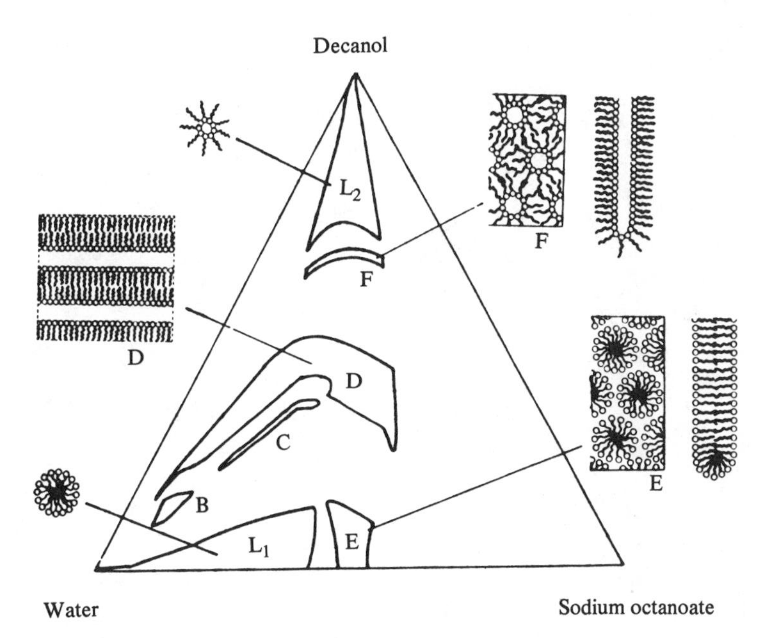

Figure 9.12. Phase diagram of the ternary system sodium octanoate + decanol + water at 20 C. Structures of the aggregates are shown schematically: L_1 and L_2 denote isotropic solution phases; B, C, and D are lamellar phases; E is a hexagonal phase; and F is an inverted hexagonal phase. This drawing is by Lindman (1977), based on original work by Ekwall (1975). Since then B and C have been shown to be single-phase regions. A revised diagram has been prepared by Friman & Daniellson (1982).

Phase behavior in the vicinity of the lower of the pure component critical points is illustrated by the example of Figure 9.17 of methane + hexane, of which the full critical locus is shown in Figure 9.14. In this case the critical loci emanating from the critical points of the pure components are interrupted, but they do connect to the termini of a three-phase locus. The associated P-x diagrams also show that the three-phase line still is present at 193.2 K but has terminated by the time the temperature has risen to 198 K. Several other examples of this type are given in the book of McGlashan (1979).

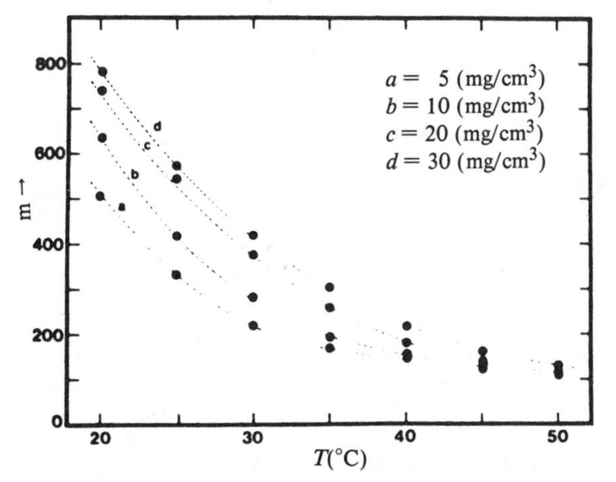

Figure 9.13. The number of molecules, m, per micelle (aggregation number) of sodium dodecyl sulfate in 0.6 molar NaCl as a function of temperature and concentration (Corti & Digiorgio, 1979).

Supercritical phase separation often is associated with mixtures that have only limited mutual solubilities. Both phenomena are common when the molecules differ greatly in various respects. The shapes of regions of immiscibility are examined in Section 7.4, primarily with respect to the effect of temperature, which is illustrated in Figure 5.21 and Example 7.11. Some effects of pressure are shown in Figures 5.28 and 9.18. From the latter figure it appears that the effects of increasing pressure and of decreasing salt concentration on solubility are similar, but in general the effect of added salts is not as simple as in this particular case. Details of such studies may be found in a paper by Schneider (1978).

As already mentioned, the distinction between liquid and gas phases is not always easy to make. Thus in Figure 9.19(a) of ethylene + nitromethane, below the convex portion of the critical locus, a liquid + vapor mixture usually is understood to be present; but to the left of the nearly vertical lines of both Figures 9.19(a) and 9.19(b), the phases usually are called gas + gas. Where along the curve does the liquid phase become a gas phase?

Although supercritical behavior does seem unusual, it is amenable to theoretical treatment. In fact, gas-gas equilibria had been predicted by van der Waals, and some experimental evidence had been found by about 1900, but modern experimental discoveries began with the work of Krichevskii (1940). Modified Redlich-Kwong equations have been applied to

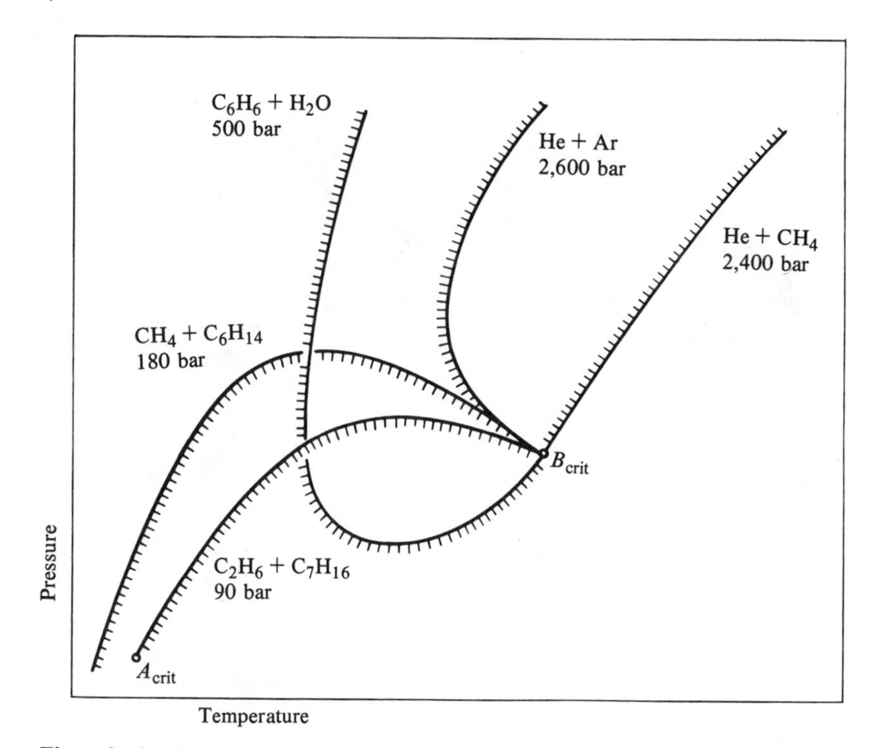

Figure 9.14. Representative critical loci in the vicinity of the critical point of the less volatile of the components of the mixture. The dashes on the lines point to the two-phase regions. Curves are not to scale. The indicated pressures are the greatest that were measured in individual cases.

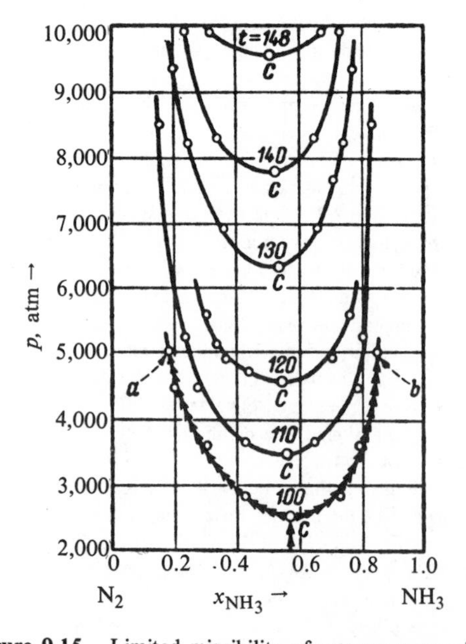

Figure 9.15. Limited miscibility of gaseous ammonia and nitrogen mixtures. At 100 C, for example, a mixture containing 57% NH_3 begins to stratify at 2,600 atm; whereas at 5,000 atm mixtures between 18 and 84.5% NH_3 are two-phase. The critical properties of the components are NH_3 (111.3 atm, 132.6 C); N_2(33.5 atm, −151 C) (Karapetyants, 1978).

high-pressure phase equilibria by Deiters & Schneider (1976) and Stephan & Schaber (1979). In another study, Alwani & Schneider examined the effects of polarity on critical loci (1976). This field of research is quite active. The literature is reviewed at length by Schneider (1978) and by Rowlinson & Swinton (1982).

In addition to the purely scientific interest, phase behavior at supercritical conditions has technological relevance in such areas as migration of petroleum and natural gas in reservoirs, reactor performance, chromatographic separations of substances with low migration rates or low temperature stability, desalination of water, and separation processes in general. Some of these topics were discussed in Section 8.6. The background and some applications of this technology are described in recent symposia edited by Schneider et al. (1980) and Paulaitis et al. (1983).

An essential feature of such processes is that the vapor pressures and hence the solubilities of condensed phases may be enhanced greatly by raising the pressure above the critical pressure of the gas and near its critical temperature. After a desired dissolution has been performed, recovery of solvent is accomplished with minimum energy requirement by lowering the pressure, or sometimes by lowering the temperature also to a small extent. Because of its acceptability in food handling, much work has been done with CO_2 for extracting thermally sensitive natural products. The range of thermodynamic conditions for such processes and typical temperature and pressure profiles in the extraction vessel are shown in Figure 9.20. Operating data of several commercial supercritical extractions

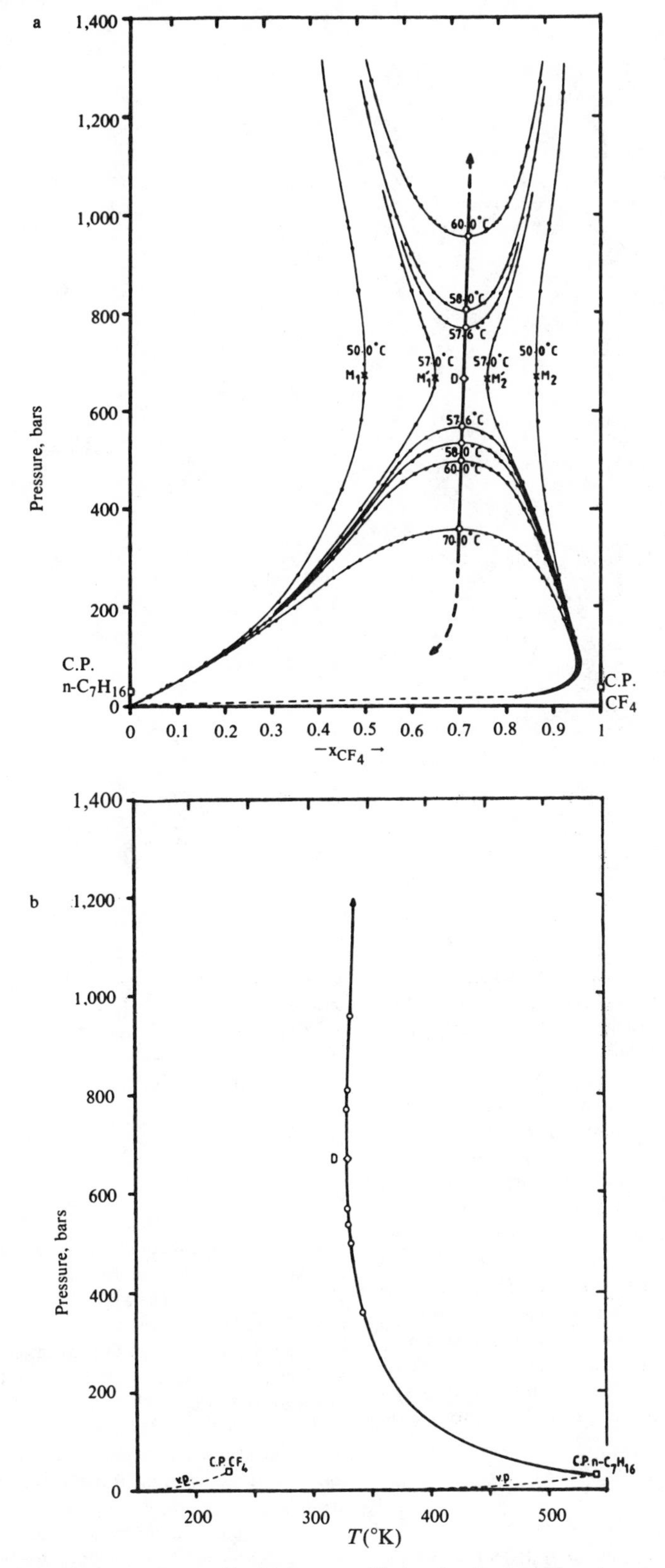

Figure 9.16(a). Supercritical phase equilibria of mixtures of carbon tetrafluoride and *n*-heptane. The regions bounded by the parabolic-shaped curves and the pairs of lines marked with the same temperatures are two-phase (Mendonca & Bett, in Newman, ed., 1983). **(b).** The critical locus, P_c against T_c, of the mixture of part (a). It starts at the critical point of *n*-heptane but does not return to that of the carbon tetrafluoride (Mendonca & Bett, in Newman, ed., 1983). Critical properties are CF_4 (37.5 bar, 227.6 K); C_7H_{16} (27.4 bar, 540.2 K).

Example 9.3. The Flory-Huggins Activity Equations

Some details are given of the derivation of the equations of the activities of the monomer (1) and polymer (2) from the basic equation for the Gibbs energy of mixing. m is the number of monomer molecules of type 2 in the polymer. See Section 9.4.

1. $\phi_1 = \dfrac{n_1}{n_1 + m n_2}.$

2. $\phi_2 = \dfrac{m n_2}{n_1 + m n_2}.$

3. $\dfrac{\partial \phi_1}{\partial n_1} = -\dfrac{\partial \phi_2}{\partial n_1} = \dfrac{\phi_1 \phi_2}{n_1}.$

4. $\dfrac{\partial \phi_1}{\partial n_2} = -\dfrac{\partial \phi_2}{\partial n_2} = -\dfrac{\phi_1 \phi_2}{n_2}.$

5. $\underline{G}^{\text{mix}}/RT = n_1 \ln \phi_1 + n_2 \ln \phi_2 + \chi(n_1 + m n_2)\phi_1 \phi_2$

6. $\qquad = n_1 \ln \phi_1 + n_2 \ln \phi_2 + \dfrac{\chi m n_1 n_2}{n_1 + m n_2}.$

7. $\ln a_1 = \dfrac{\partial}{\partial n_1} (\underline{G}^{\text{mix}}/RT)$

8. $\qquad = \ln \phi_1 + \left[\dfrac{n_1}{\phi_1} - \dfrac{n_2}{\phi_2}\right] \dfrac{d\phi_1}{dn_1} + \dfrac{\chi(m n_2)^2}{(n_1 + m n_2)^2}$

9. $\qquad = \ln \phi_1 + \left[\dfrac{n_1}{\phi_1} - \dfrac{n_2}{\phi_2}\right] \dfrac{\phi_1 \phi_2}{n_1} + \chi \phi_2^2$

10. $\qquad = \ln \phi_1 + \left(1 - \dfrac{1}{m}\right) \phi_2 + \chi \phi_2^2.$

11. $\ln \gamma_1 = \ln a_1 - \ln x_1$

12. $\qquad = \ln(\phi_1/x_1) + \left(1 - \dfrac{1}{m}\right) \phi_2 + \chi \phi_2^2.$

13. $\ln a_2 = \ln \phi_2 + \left[\dfrac{n_1}{\phi_1} - \dfrac{n_2}{\phi_2}\right] \left[-\dfrac{\phi_1 \phi_2}{n_2}\right] + \dfrac{\chi m n_1^2}{(n_1 + m n_2)^2}$

14. $\qquad = \ln \phi_2 - (m-1)\phi_1 + \chi m \phi_1^2.$

are shown in Table 9.6. In the source of this table, there are given some rules for selection of proper solvents and the kinds of substances that can be extracted effectively. For instance, weakly polar substances such as hydrocarbons, esters, ethers, and lactones are extractable below 100 bar; those containing single strongly polar groups such as -OH or -COOH are difficult to extract; and more highly polar substances such as sugars and aminoacids are not extractable below 500 bar.

Several real and potential applications are cited by Paul & Wise (1971) and by Ellis (1971). Much attention has been devoted to the recovery of coal liquids, some of which is reviewed by Gagnoli & Thodos (1977). Pilot plant results with toluene and *o*-cresol are reported by Maddocks et al. (1979), although no clear-cut economic superiority was established. An important advantage of supercritical solvents in comparison with liquid solvents in such cases is their ease of recovery from residues because of their relatively low densities and high volatilities. A list of commercial extraction processes that employ supercritical gases as solvents is given in Table 9.5.

9.7. HYDRATES

Certain gases form solid phases with water that are stable well above the freezing point of pure water. Industrially, such effects are deleterious in wet natural gas transport lines because of plugging with solid hydrates at certain conditions of temperature and pressure. A beneficial effect of hydrate formation is used in the desalination of seawater with propane, as analyzed, for example, by Sherwood (1963); a phase diagram of this system is shown in Figure 9.21. Dissolved salts or other relatively nonvolatile substances lower the melting points of hydrates, as shown in Figure 9.21 and 9.22(b). Other data of dissociation pressures as a function of temperature are shown in Figure 9.22(a). The relations are well represented by the equation

$$\ln P = k_1 + k_2/T, \tag{9.75}$$

but when the temperature range is small, as on the figures shown, the form $\ln P = k_1 + k_2 T$ is practically equivalent.

The history and theory of hydrate formation is summarized by Davidson (1973) (in Franks, *Water: A Comprehensive Treatise,* Vol. 2, p. 115) who lists about 70 known hydrates. It is recognized that hydrates are not stoichiometric compounds but are solid solutions of the gas in ice. In the presence of a stabilizing gas, the ice forms a crystalline structure with uniform cavities large enough to accommodate one molecule each of the gas. The chemical nature of the gas as well as the molecular size are important; for instance, butane does not form a hydrate, whereas isobutane does. Two main types of crystalline lattices have been identified, one corresponding to a maximum gas content represented by the molecular formula M.17 H_2O when all the cavities are filled, and the other to either M.7-2/3 H_2O or M.5-3/4 H_2O, where M signifies one mole of the gas. The type of lattice formed depends on the nature of the gas. For instance, ethane forms type I structure and propane type II. The conditions under which either type of structure prevails in the presence of mixtures depends on the conditions as shown in Figure 9.23.

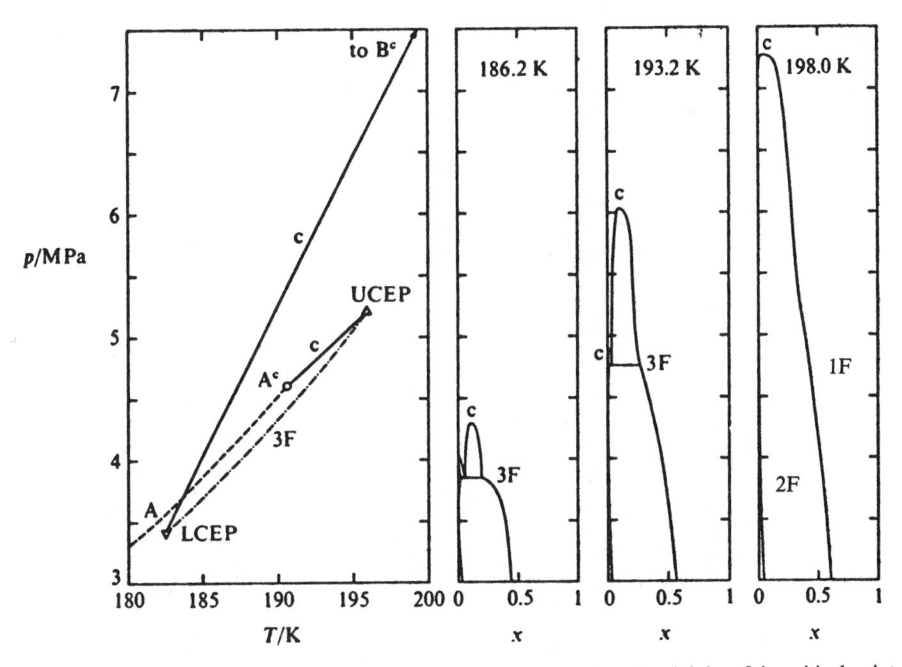

Figure 9.17. $P-T$ and $P-x$ diagrams of hexane + methane in the vicinity of the critical point of methane.

A = vapor pressure curve of methane,

A^c = critical point of methane, 190.6 K, 4.60 MPa,

c = critical loci,

UCEP = upper critical endpoint, terminus of the three-phase line,

LCEP = lower critical endpoint.

In some instances the critical locus from B^c can terminate at UCEP instead of at the LCEP as in this example (McGlashan, Chap. 17 1979).

Table 9.5. Commercial Processes Employing Extraction with Supercritical Gases (Stahl & Schilz 1976)

			Extraction conditions		
Charge	*Extract*	*Solvent*	*P, atm*	*T, °C*	*Recovery conditions*
Petroleum	light HC	propane, propylene	100–150	100	50 bar 100 C
Wool fat	lanolin	propane, propylene	60–110	100–105	1 bar
Petroleum	HC fractions	ethylene	40–120	20	40 bar
Coffee beans	caffeine	CO_2, N_2O	120–180	40–80	adsorption on act. carbon
Tobacco	nicotine	CO_2, N_2O, SF_6, aromatics, halog. HC	65–1000	35–100	subcritical adsorption
Hops	resins, α- and β-acids	CO_2	70–400	45–50	subcritical
Black pepper	piperine, ethereal and fatty oils	CO_2	<400	40–60	65 bar 25–60 C stripping with CO_2
Cloves	eugenol	CO_2	<400	40–60	"
Cinnamon	cinnamic aldehyde	CO_2	<400	40–60	"
Vanilla beans	vanillan	CO_2	<400	40–60	"
Roast coffee	aromatic substances, glycerides, fatty acids	CO_2	<320	-50	65 bar
Fermented black tea	aromatic substances	CO_2	<300	-50	50–70 bar

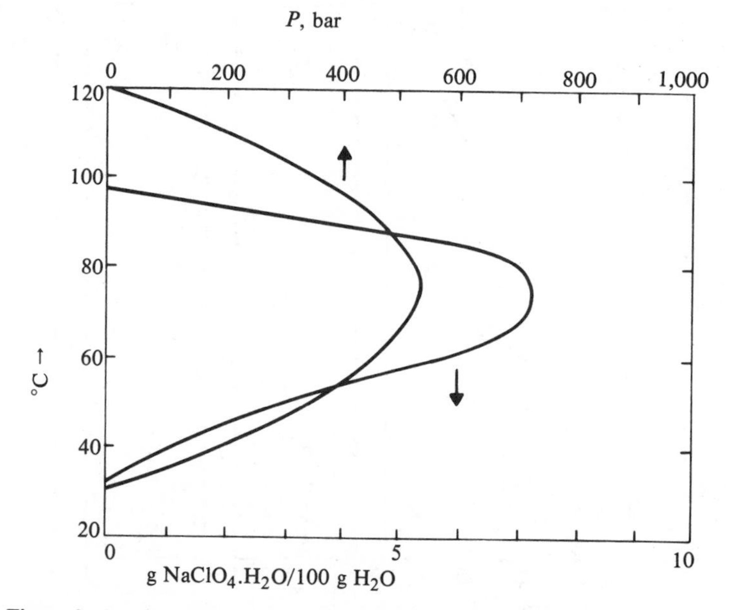

Figure 9.18. The effects of pressure and dissolved salt on the miscibilities of isobutanol and water (Schneider & Russo, 1966).

The effect of concentration of $NaClO_4.H_2O$ is shown at a pressure of 450 bar. The effect of pressure is shown on a solution containing 5.81 g $NaClO_4.H_2O$/100 g H_2O. The salt-free solution contains 40% isobutanol and 60% water. Two phase regions are within the parabolic shapes.

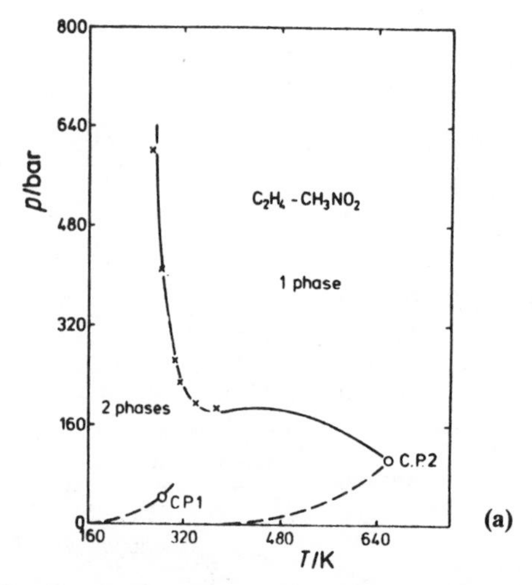

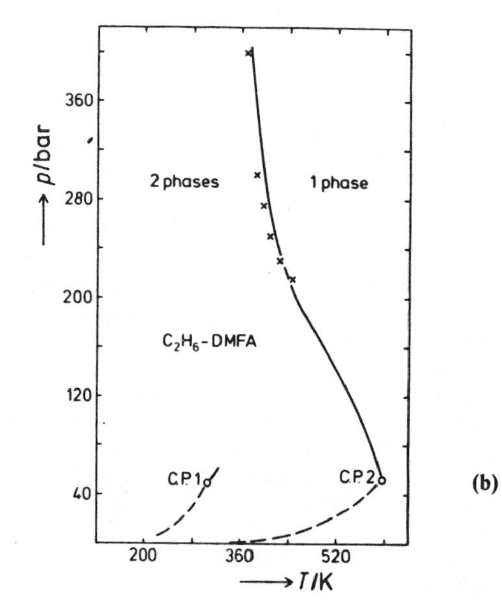

Figure 9.19. Two binary mixtures that have phase separation above the critical points. Ethylene + nitromethane and ethane + dimethyl-formamide (Deiters & Schneider 1976). CP1 and CP2 are the critical points of the pure substances. (a) Ethylene + nitromethane; (b) Ethane + dimethyl-formamide.

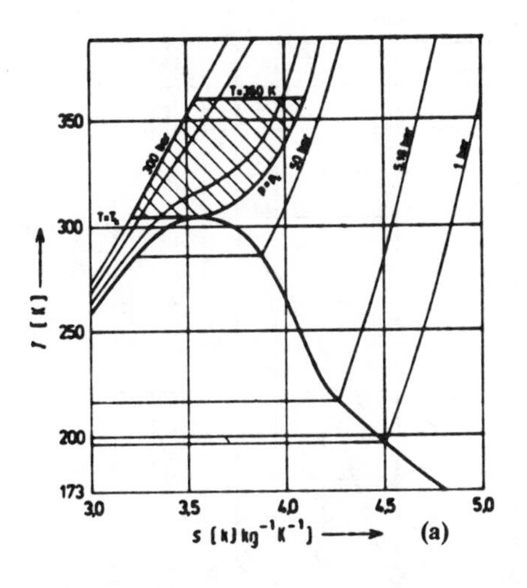

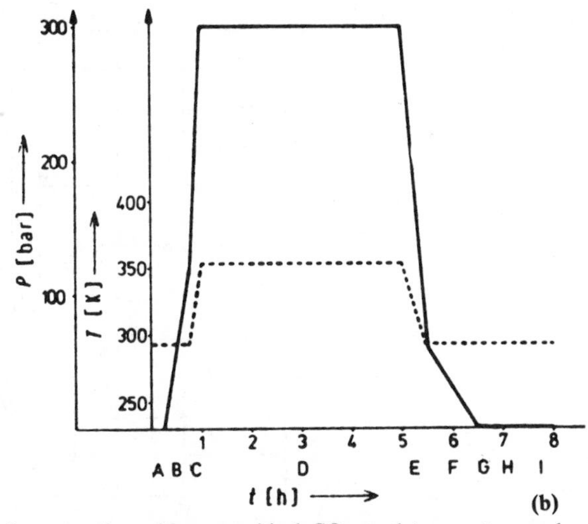

Figure 9.20. Range of thermodynamic operating conditions for extraction with supercritical CO_2, and temperature and pressure profiles of the extraction vessel (Eggers & Tschiersch, 1980). (a) Range of operating conditions for extraction of natural products with supercritical CO_2; (b) Pressure and temperature profiles in an extraction vessel. *A*: feeding solids into extraction vessel. *B*: filling plant with solvent. *C*: heating to extraction conditions. *D*: withdrawal and separation of extract. *E*: venting to tank pressure. *F*: suction down to residual pressure. *G*: venting of residual gas. *H*: removal of spent solids and extract. *I*: regeneration of the absorbent.

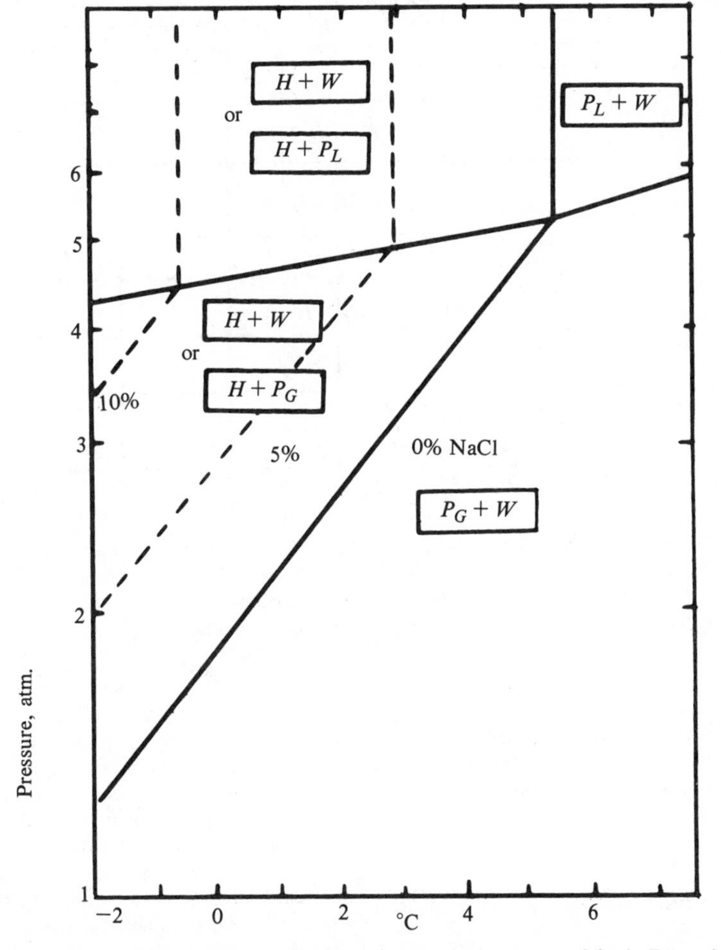

Figure 9.21. Phase diagram of propane + water in the presence of salt (Data of Knox, Hess, Jones, & Smith 1971). P_G = propane gas, P_L = propane liquid, H = hydrate, W = water. Which alternate exists depends on the overall amounts of propane and water present.

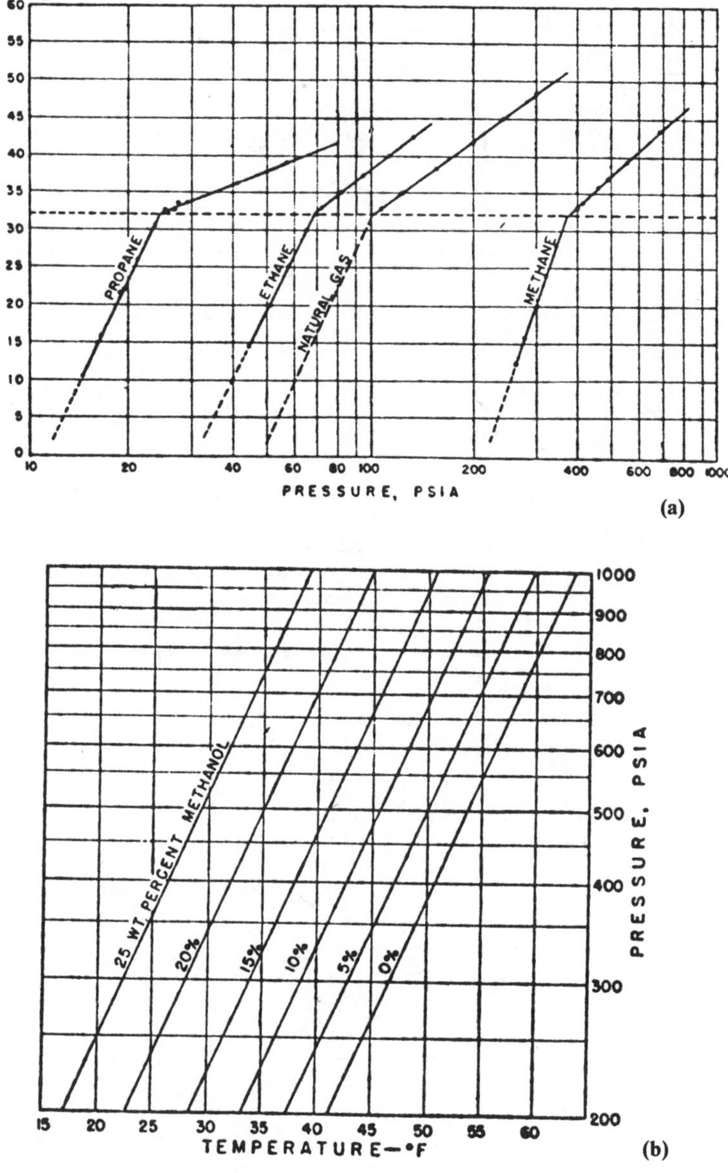

Figure 9.22. Dissociation pressures of hydrates of hydrocarbons and the effect of methanol concentration (Deaton & Frost, 1946). (a) Dissociation pressures below and above the freezing point of water; (b) Dissociation pressures of natural gas hydrates in the presence of several concentrations of methanol.

A theory due to van der Waals & Platteeuw (1959) expresses the fraction of the cavities occupied by gas molecules by means of a Langmuir adsorption-type equation,

$$\theta_i = \frac{C_i \hat{f}_i}{1 + \Sigma\ C_j \hat{f}_j} = \frac{C_i y_i \hat{\phi}_i P}{1 + \Sigma\ C_j y_j \hat{\phi}_j P}, \qquad (9.76)$$

where $\hat{f}_i$ and $\hat{\phi}_i$ are the partial fugacity and fugacity coefficient.

For a pure gas,

$$\theta = \frac{C\phi P}{1 + C\phi P}.$$

A method of evaluating the fugacity coefficients of the solid phase was developed by Parrish & Prausnitz (1972); this enables the prediction of dissociation pressures of mixtures of hydrate-forming and non-hydrate-forming gases. A pressure-composition diagram in the hydrate-forming range of propane + methane + water is shown as Figure 9.24. In order to prevent hydrate formation in natural gas lines, it is common practice to inject methanol, glycol, or ammonia. Recent work has made this kind of design more nearly quantitative. Menten, Parrish, & Sloan (unpublished, 1982) analyzed the effects of inhibitors through their influence on the activity

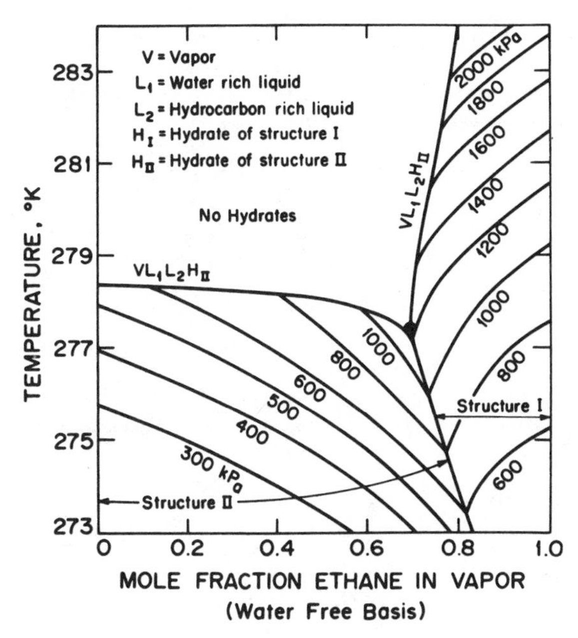

Figure 9.23. Hydrate equilibria in the ethane + propane + water system. (Holder & Hand, 1982).

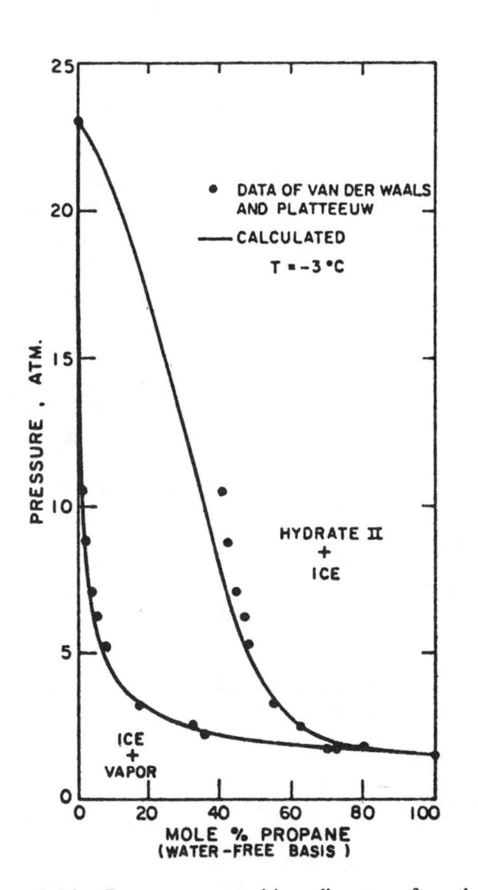

Figure 9.24. Pressure composition diagram of methane + propane + water (Parrish & Prausnitz 1972).

coefficient of water. A book by Makogon (1981) surveys current practice in dealing with gas hydrate problems.

9.8. PROBLEMS

9.1. Adsorptive equilibrium data of ethane + ethylene on silica gel and type 4A molecular sieves at 25C and 730–760 Torr are read off Figure 9.2 as given.

	y	
x	SiO_2	*4A*
1	0.275	0.470
2	0.422	0.830
3	0.563	0.902
4	0.670	0.940
5	0.752	0.962
6	0.822	0.979
7	0.880	0.985
8	0.928	0.990
9	0.968	0.995

a. Find average relative volatilities of the two sets of data, plot the corresponding *x-y* diagrams, and locate the original data for comparison on these graphs.

b. Find the best values of the parameters for each set of data from the equation

$$\frac{y}{1-y} = k_1 \left(\frac{x}{1-x}\right)^{k_2}.$$

Which correlation is best in these cases?

9.2. For pure ethylene and propylene, adsorption data and calculated spreading pressures are given by Costa et al (1981). The pressure is 75 Torr.

N_t (*gmol*)	*p* (*mm Hg*)	$\frac{\pi^\circ A}{RT}$
	Ethylene	
$6.500\ 10^{-3}$	4.10	$6.010\ 10^{-3}$
$1.592\ 10^{-2}$	17.99	$2.471\ 10^{-2}$
$3.029\ 10^{-2}$	52.09	$4.847\ 10^{-2}$
$4.399\ 10^{-2}$	125.00	$8.217\ 10^{-2}$
$5.252\ 10^{-2}$	181.44	$1.005\ 10^{-1}$
$6.056\ 10^{-2}$	242.32	$1.169\ 10^{-1}$
$6.796\ 10^{-2}$	309.34	$1.326\ 10^{-1}$
$7.459\ 10^{-2}$	384.10	$1.481\ 10^{-1}$
$8.060\ 10^{-2}$	465.00	$1.630\ 10^{-1}$
$8.659\ 10^{-2}$	546.12	$1.764\ 10^{-1}$
$9.209\ 10^{-2}$	632.00	$1.895\ 10^{-1}$
$9.658\ 10^{-2}$	728.00	$2.002\ 10^{-1}$
	Propylene	
$1.390\ 10^{-2}$	0.24	$1.166\ 10^{-2}$
$3.105\ 10^{-2}$	2.56	$3.366\ 10^{-2}$

(*continued on next page*)

N_t (gmol)	p (mm Hg)	$\frac{\pi^\circ A}{RT}$
5.090 10^{-2}	12.80	1.161 10^{-1}
7.005 10^{-2}	30.03	1.705 10^{-1}
9.095 10^{-2}	64.90	2.356 10^{-1}
1.095 10^{-1}	122.19	3.014 10^{-1}
1.252 10^{-1}	209.14	3.666 10^{-1}
1.377 10^{-1}	327.70	4.268 10^{-1}
1.468 10^{-1}	463.78	4.773 10^{-1}
1.575 10^{-1}	618.47	5.217 10^{-1}
1.620 10^{-1}	713.63	5.447 10^{-1}

a. Verify the given spreading pressures.
b. Find a range of equilibrium compositions by Myers method and compare with the experimental data that follow. These data also are plotted on Figure 9.3.

x_1	y_1
0.800	0.993
0.614	0.980
0.253	0.858
0.141	0.696
0.055	0.436
0.028	0.248

9.3. From the data of Costa & others (1981) of adsorbed mixtures of ethylene and propylene, activity coefficients of the adsorbed phase are calculated as given in the table with values at infinite dilution obtained by graphical extrapolation. From these infinite-dilution coefficients, find the parameters of the Margules and Wilson equations, and also the activity coefficients from those equations at the adsorbate compositions listed.

x_1	γ_1	γ_2
0	(0.757)	1
0.021	0.7580	1.0119
0.064	0.7637	0.9835
0.126	0.7922	0.9621
0.277	0.8407	0.8917
0.619	0.9398	0.8624
0.807	0.9991	0.7772
1	1	(0.69)

9.4. For the case of Example 9.2, find the van Laar and Margules parameters of the adsorbate phase at a liquid composition of $x_1 = 0.3$. That is, take $x_1 = 0.3$ in step 5.

9.5. Adsorption data of acetic acid from aqueous solution by charcoal are given as equilibrium molarity against amount adsorbed per gm of charcoal. Check these data against the Freundlich and Langmuir equations.

Molarity	*gm/gm C*
0.434	0.209
0.202	0.152
0.0899	0.1083
0.0347	0.0818
0.0113	0.060
0.00333	0.037

9.6. Check the data of adsorption of carbon monoxide on carbon at 0°C against the Freundlich and Langmuir equations.

Torr	*ml @ STP per gram*
100	3.38
200	6.15
300	8.44
400	10.39
500	12.21
600	13.77
700	15.25

9.7. Adsorptions of propylene, ethylene and their 2:1 mixture on carbon at 20 C are read off a small graph by Costa et al. (1981) as follows, with pressures in Torr and amounts in centimols/gm.

	n		
P	*propylene*	*ethylene*	*mixture*
100	0.210	0.080	0.120
200	0.245	0.110	0.165
300	0.270	0.135	0.200
400	0.285	0.155	0.220
500	0.300	0.165	0.240
600	0.310	0.180	0.255
700	0.320	0.185	0.265

a. For each of the three sets of data, find the constants of the Freundlich equation, $n_i = a_i P^{b_i}$.
b. Compare the mixture data with predictions by these three rules:

1. $n = \Sigma\, x_i n_i$.
2. $1/n = \Sigma\, x_i/n_i$.
3. $\ln n = \Sigma\, x_i \ln n_i$.

Note that none of these rules may be applicable over the whole range of compositions because the composition of the vapor phase is not taken into account explicitly.

10 Chemical Equilibria

10.1. INTRODUCTION

Equilibria between phases may involve substances that interact chemically in some or all of the phases in contact. In such cases the equilibrium state depends on the chemical natures of the participants and their coefficients of distribution between phases, as well as the temperature, pressure, and overall composition. The rate of attainment of chemical equilibrium is of practical importance. Equilibrium is attained more readily at higher temperatures, sometimes by adjustment of pressure and commonly in the presence of catalysts. Temperature often is a double-edged parameter: its rise will accelerate reaching equilibrium but the composition then may be unfavorable. Practically it is essential first to find the conditions under which an equilibrium condition is favorable and then to look for ways to make the reaction proceed at a satisfactory rate. Here only the compositions and distributions of phases at equilibrium as functions of the temperature and pressure will be of concern. The discussion will begin at the beginning, with the theory of single and multiple chemical reactions in a single phase; the study of multiphase processes will follow.

For reacting systems as well as inert ones, the equilibrium state is characterized as having a minimum Gibbs energy at specified T and P. Thus,

$$n_t G = \Sigma\ n_i(G_i^0 + RT\ln(\hat{f}_i/f_i^0) = \Sigma\ n_i(G_i^0 + RT\ln a_i). \rightarrow \text{minimum} \tag{10.1}$$

Similarly, at specified T and V, the condition in terms of Helmholtz energy is

$$n_t A = \Sigma\ n_i(A_i^0 + RT\ln a_i) \rightarrow \text{minimum}. \tag{10.2}$$

Since T and P are practically a more important group of variables than T and V, the ensuing discussion will be entirely in terms of Gibbs rather than Helmholtz energy. The composition enters these equations through the partial fugacity. Thus for gases, Eq. 10.1 becomes

$$n_t G = \Sigma\ n_i[G_i^0 + RT\ln(y_i\hat{\phi}_i P)] \rightarrow \text{minimum}. \tag{10.3}$$

When the mixture is represented by a single stoichiometric relation, all the compositions are expressible in terms of the fractional conversion, ε, of a key component so the minimum in that event can be found after setting

$$\frac{d(n_t G)}{d\varepsilon} = 0. \tag{10.4}$$

Similarly, when two stoichiometric relations are sufficient, equilibrium is found by solving for two variables, ε_1 and ε_2, from

$$\left[\frac{\partial(n_t G)}{\partial\varepsilon_1}\right]_{\varepsilon_2} = \left[\frac{\partial(n_t G)}{\partial\varepsilon_2}\right]_{\varepsilon_1} = 0 \tag{10.5}$$

and so on for any number of stoichiometric relations. When it is not convenient to find a proper set of independent stoichiometric relations, equilibrium can be found directly from Eq. 10.1, together with material balances on the chemical elements present as constraints on the minimization process taking advantage of the fact that the content of each element in the equilibrium mixture is the same as the corresponding content in the starting material. Minimization of Gibbs energy subject to balances on the chemical elements is handled most readily by the method of Lagrange multipliers to be described later, although other methods for complex equilibria also are available and may be simpler in particular cases than the general method. For single reactions, plots of the Gibbs energy as a function of the conversion and the location of the minima are shown in Examples 2.10 and 10.1.

10.2. THE DEGREE OF ADVANCEMENT OF A CHEMICAL REACTION

The stoichiometric equation of a chemical reaction involving the participants A_i with stoichiometric coefficients ν_i may be written

$$\sum_R \nu_i A_i \rightleftarrows \sum_P \nu_i A_i \tag{10.6}$$

where the summations are over the reactants, Σ_R, and over the products, Σ_P. The difference between the sums of the stoichiometric coefficients on the right and on the left is

$$\Delta\nu_i = \sum_P \nu_i - \sum_R \nu_i. \tag{10.7}$$

At a given conversion of one of the participants, all of the amounts present can be expressed in terms of a single quantity called the degree of advancement,

$$\varepsilon = \frac{n_i - n_{io}}{\pm\nu_i} \quad \begin{cases} + \text{ for products (RHS)} \\ - \text{ for reactants (LHS)} \end{cases} \tag{10.8}$$

so that for each participant,

$$n_i = n_{io} \pm \nu_i\varepsilon \quad \begin{cases} + \text{ for products} \\ - \text{ for reactants} \end{cases} \tag{10.9}$$

In differential form,

$$dn_i = \pm\nu_i d\varepsilon \tag{10.10}$$

As an example, for the reaction

$$A_1 + 2A_2 \rightarrow 3A_3 + 4A_4, \tag{10.11a}$$

these quantities are

$$\Delta\nu_i = 3 + 4 - 1 - 2 = 4, \tag{10.11b}$$

$$\varepsilon = \frac{A_3 - A_{30}}{3} = \frac{A_4 - A_{40}}{4} = \frac{A_1 - A_{10}}{-1} = \frac{A_2 - A_{20}}{-2}, \tag{10.11c}$$

$$A_1 = A_{10} - \varepsilon, \tag{10.11d}$$

$$A_2 = A_{20} - 2\varepsilon, \tag{10.11e}$$

$$A_3 = A_{30} + 3\varepsilon, \tag{10.11f}$$

$$A_4 = A_{40} + 4\varepsilon. \tag{10.11g}$$

10.3. THE EQUILIBRIUM EQUATION

In differential form, the Gibbs energy of a chemically reacting mixture is

$$d(nG) = -nSdT + nVdP + \Sigma\ \bar{G}_i dn_i = -nSdT + nVdP + \Sigma\ \bar{G}_i\nu_i d\varepsilon, \tag{10.12}$$

$$\nu = \begin{cases} + \text{ for products} \\ - \text{ for reactants} \end{cases} \tag{10.13}$$

Similarly, when T and V are independent variables,

$$d(nA) = -\ nSdT + Pd(nV) + \Sigma\ \bar{A}_i\nu_i d\varepsilon. \tag{10.14}$$

At constant temperature and pressure and equilibrium, Eq. 10.13 becomes

$$\left(\frac{\partial(nG)}{\partial\varepsilon}\right)_{TP} = \Sigma\ \bar{G}_i\nu_i = 0,$$

$$\nu = \begin{cases} + \text{ for products} \\ - \text{ for reactants} \end{cases} \qquad (10.15)$$

which may be written in the form

$$\sum_P \bar{G}_i\nu_i - \sum_R \bar{G}_i\nu_i = 0. \qquad (10.16)$$

For each component, the partial molal Gibbs energy depends on the Gibbs energy in the pure state and the fugacity in the mixture,

$$\bar{G}_i = G_i^0 + RT\ln(\hat{f}_i/f_i^0) = G_i^0 + RT\ln a_i, \qquad (10.17)$$

where G_i^0 is the Gibbs energy of formation of substance i from the elements. Substitution of Eq. 10.17 allows Eq. 10.16 to be written in several forms:

$$\Delta G^0 = \sum_P \nu_i G_i^0 - \sum_R \nu_i G_i^0 \qquad (10.18)$$

$$= RT\left[\sum_P \ln(\hat{f}_i/f_i^0)^{\nu_i} - \sum_R \ln(f/f_i^0)^{\nu_i}\right] \qquad (10.19)$$

$$= RT\left[\sum_P \ln(a_i)^{\nu_i} - \sum_R \ln(a_i)^{\nu_i}\right] \qquad (10.20)$$

$$= RT\left[\ln\left(\prod_P a_i^{\nu_i}\right) - \ln\left(\prod_R a_i^{\nu_i}\right)\right]. \qquad (10.21)$$

ΔG^0, as defined by Eq. 10.18, is the Gibbs energy change accompanying the reaction with participants in their standard states.

The chemical equilibrium constant is defined as

$$K_{ce} = \exp(-\Delta G^0/RT) \qquad (10.22)$$

$$= \frac{\prod_P (f_i/f_i^0)^{\nu_i}}{\prod_R (f_i/f^0)^{\nu_i}} = \prod_P a_i^{\nu_i} \Big/ \prod_R a_i^{\nu_i}. \qquad (10.23)$$

For the reaction of Eq. 10.11a, for example,

$$\underline{\Delta G} = 3G_3^0 + 4G_4^0 - G_1^0 - 2G_2^0 \qquad (10.24)$$

$$K_{ce} = \frac{a_3^3 a_4^4}{a_1 a_2^2}. \qquad (10.25)$$

Data of equilibrium constants of formation of many substances are tabulated by Stull et al. (1969) over a range of temperatures. Some of this information has been regressed by the equation

$$\log_{10} K_p = A + \frac{B}{T}. \qquad (10.26)$$

Values of the constants A and B are given in Table D.1. This form of the temperature dependence of K_p derives from the assumption of a suitable average enthalpy of formation (see the derivation of Eq. 10.71). The closeness of the relation is indicated by the agreement of the lines in Figure 10.6 with the points representing the best data for several systems, the most intensively investigated one of which is the formation of ethanol. A convenient nomogram of equilibrium data of 51 reactions has been based on Eq. 10.26 by Luft (1957).

Example 10.1. Chemical Equilibrium as the Condition of Minimum Gibbs Energy

For the participants in the reaction, $CH_4 + H_2O \rightleftarrows CO + 3\,H_2$, Gibbs energies and equilibrium constants of formation at 800*K* are,

	ΔG_f	$\log_{10} K$
1 CH_4	−0.56	0.154
2 H_2O	−48.65	13.290
3 CO	−43.68	11.933

The given equilibrium constants combine to $K_{ce} = 0.03083$ and will be used to verify the condition of equilibrium found by minimization of the Gibbs energy. Ideal behavior of vapor will be assumed. For each component, $f_i^0 = 1$ and

$$G_i = G_i^0 + RT\ln\hat{f}_i/f_i^0 = G_i^0 + RT\ln\hat{f}_i = G_i^0 + RT\ln y_i\hat{\phi}_i P$$

$$\rightarrow G_i^0 + RT\ln y_i P.$$

The partial fugacity coefficients have been assumed unity and the G_i^0 are evaluated at $f_i^0 = 1$. The mol fractions are given in terms of R and ε,

R = ratio of H_2O to CH_4 at the start,

ε = fractional conversion of CH_4,

$n_t = 1 + R + 2\varepsilon$,

$y_1 = (1-\varepsilon)/n_t$,

$y_2 = (R-\varepsilon)/n_t$,

$y_3 = \varepsilon/n_t$,

$y_4 = 3\varepsilon/n_t$.

	ε	
R	*Graph*	*Eq 2*
1	0.21	0.2076
1.5	0.26	0.2555
2	0.29	0.2976
2.5	0.33	0.3357

The Gibbs energy function of the mixture is

$$g = \frac{n_t G}{RT} = n_t\,\Sigma\, y_i\left(\frac{G_i^0}{RT} + \ln y_i P\right)$$

$$= n_t\left[\ln P + \Sigma\, y_i\left(\frac{G_i^0}{RT} + \ln y_i\right)\right]. \qquad (1)$$

g is plotted against ε for $P = 1.5$ at several values of R. The set of larger curves is drawn with expanded and different scales of ordinates to identity the minima more readily. The equilibrium

Example 10.1 *(continued)*

conversions read off the graphs are compared with those obtained from the equilibrium equation,

$$K = 0.0308319 = \frac{y_3 y_4^3 P^2}{\bar{y}_1 y_2}$$

$$= \frac{\varepsilon(3\varepsilon)^3 P^2}{(1-\varepsilon)(R-\varepsilon)(1+R+2\varepsilon)^2}. \qquad (2)$$

Algebraic minimization of Eq. 1 will result in Eq. 2 with $RT \ln K = G_3^0 + 3G_4^0 - G_1^0 - G_2^0$. Comparison of results off the graph and from Eq. 2 is in the table.

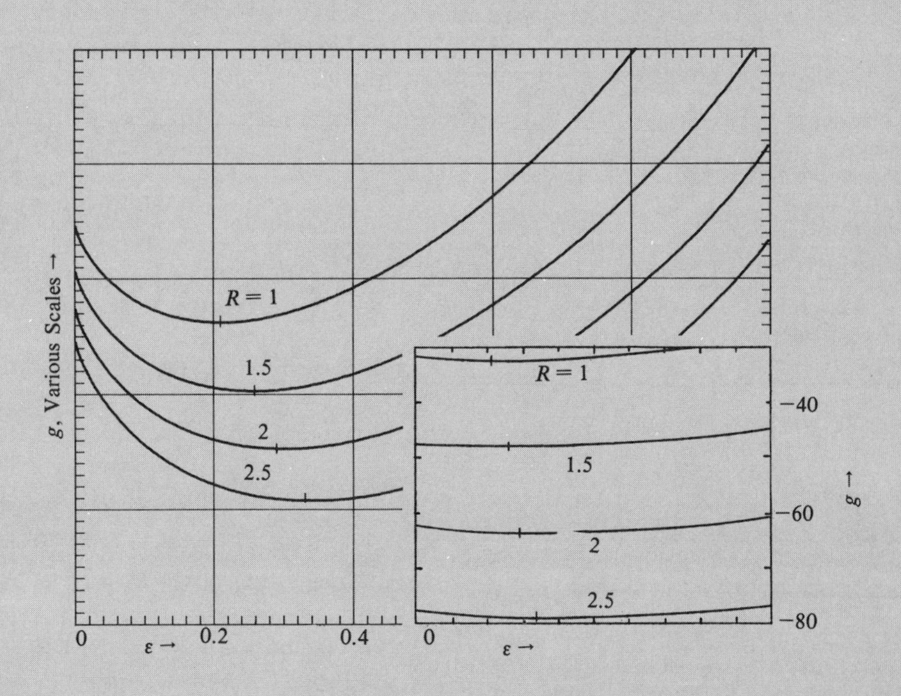

Example 10.1. Gibbs energy of a reacting mixture as a function of fractional conversion, ε. Minima are indicated.

The equilibrium constant of a particular reaction is obtained by appropriate combination of the individual constants of formation. Thus, corresponding to Eq. 10.24,

$$\ln K_{ce} = 3 \ln K_3 + 4 \ln K_4 - \ln K_1 - 2 \ln K_2. \qquad (10.27)$$

10.4. THE EQUILIBRIUM CONSTANT AND REFERENCE STATES

Basically numerical values of the chemical equilibrium constant, K_{ce}, are obtained experimentally, but in the present state of the art they may be usually calculated from readily available data of Gibbs energies of formation of the participants of the reaction,

$$-RT \ln K_{ce} - \Delta G_f^0 = \Delta H_f^0 - T\Delta S_f^0. \qquad (10.28)$$

Enthalpies of formation are measured calorimetrically, whereas the absolute entropies in this equation are figured with measurements of heat capacity down to absolute zero or obtained from spectroscopic data. Several extensive compilations of such imformation are available—for example:

1. Stull, Westrum, & Sinke, "Chemical Thermodynamics of Organic Compounds" (1969). Tabulations for 917 substances, 298–1000 K.
2. Landolt-Börnstein, II/4 (1961), IV/4a (1967), IV/4b (1972).
3. Glushko, *Termodinamicheskie Svoistva* (8 vols., 1978 ff).
4. Texas A&M University TRC, "Selected Values of Hydrocarbon Properties" (current, looseleaf); "Selected Values of Chemical Compound Properties" (current, looseleaf).
5. Chemetron Corporation, "Physical and Thermodynamic Properties" (1969). Tabulation for 68 gases, 32-2200 F.

A sample data sheet from Stull is given with Example 10.15.

The reference states at which ΔG_f^0 and f_i^0 are evaluated are arbitrary, but certain standards have been adopted for convenience and will be described. The several common forms of activities and corresponding chemical equilibrium constants are summarized in Table 10.1.

Table 10.1. Summary of Expressions for Activities and Chemical Equilibrium Constants, $K_{ce} = K_a$ for the Reaction $\Sigma_R \, \nu_i A_i \rightleftarrows \Sigma_P \, \nu_i A_i$.

$$K_J = \prod_P (J_i)^{\nu_i} / \sum_R (J_i)^{\nu_i},$$

$$(PF)_i = \exp \int_1^P \frac{V_i}{RT} dP \simeq \exp \left[\frac{V_i}{RT}(P-1) \right].$$

System	Activity	Equilibrium Constant
1. General	a	K_a
1. General, in terms of fugacity	$\hat{f}_i / f_i^0$	K_f / K_{f^0}
2. Reference state is ideal gas at 1 atm	$\hat{f}_i = \hat{\phi}_i y_i P$	$K_{\hat{\phi}} K_y P^{\Delta\nu}$
3. Pure condensed phase	$(PF)_i$	$K_{(PF)}$
4. Condensed-phase solution, reference state pure substance at 1 atm	$\gamma_i x_i f_i / f_i^0 = \gamma_i x_i (PF)_i$	$K_\gamma K_x K_{(PF)}$

10.4.1. GASES

The usual reference state is the gas at unit fugacity, which is in many instances the same as the real gas at one atmosphere. In such cases the equilibrium constant becomes

$$K_{ce} = K_{\hat{f}} = \prod_P \hat{f}_i^{\nu_i} / \prod_R \hat{f}_i^{\nu_i}. \tag{10.29}$$

The listings of ΔG_f^0 of the gas phase in Stull (1969) and in most other collections are for this reference state. Introducing fugacity coefficients,

$$K_{ce} = K_{\hat{f}} = \prod_P (\hat{\phi}_i y_i P)^{\nu_i} / \prod_R (\hat{\phi}_i y_i P)^{\nu_i} \tag{10.30}$$

$$= K_{\hat{\phi}} K_y \, P^{\Delta\nu}. \tag{10.31}$$

When all the participants are ideal gases, the fugacity coefficients are unity and the equilibrium constant becomes

$$K_{ce} = K_p = K_y \, P^{\Delta\nu}. \tag{10.32}$$

When the Lewis-Randall Rule for ideal solutions is invoked, Eq. 10.31 becomes

$$K_{ce} = K_\phi K_y P^{\Delta\nu}. \tag{10.33}$$

Some examples of numerical values of K_ϕ are shown in Figures 10.2 and 10.3. Equilibrium compositions of several gas-phase reacting systems are represented as functions of temperature and pressure in Figures 10.1–10.5.

10.4.2. Pure Condensed Phases

The reference state is the pure substance at one atm. The activity is corrected to the system pressure with the Poynting Factor:

$$a_i = f_i / f_i^0 = \exp \int_1^P \frac{V_i}{RT} dP. \tag{10.34}$$

In some applications it is acceptable to ignore this effect of pressure and to take unity as the activity of all condensed phases. In the case of Example 10.3, however, the effect of pressure at the higher values is quite substantial.

Standard Gibbs energies of formation most often are tabulated for the gas state. The relation between these values and those for a condensed phase of a substance A may be found by summing up the changes for the following sequence of steps, starting with the formation of the gas from the elements and finishing with the formation of the liquid (or solid) at the same temperature:

$$\sum \text{Elements} \rightarrow A(\text{gas}, P=1); \quad \Delta G_f^0(\text{gas}), \tag{10.35a}$$

$$A(\text{gas}, P=1) \rightarrow A(\text{gas}, P^{\text{sat}}); \quad \Delta G_1 = \exp \int_1^{ps} V_G \, dP, \tag{10.35b}$$

$$A(\text{gas}, P^{\text{sat}}) \rightarrow A(\text{liq}, P^{\text{sat}}); \quad \Delta G_2 = 0, \tag{10.35c}$$

$$A(\text{liq}, P^{\text{sat}}) \rightarrow A(\text{liq}, P=1); \quad \Delta G_3 = \exp \int_{ps}^1 V_L \, dP, \tag{10.35d}$$

$$\sum \text{Elements} \rightarrow A(\text{liq}, P=1); \quad \Delta G_f^0(\text{liq}). \tag{10.35e}$$

The net value is

$$\Delta G_f^0(\text{liq}) = \Delta G_f^0(\text{gas}) + \int_1^{p\text{sat}} (V_G - V_L) \, dP. \tag{10.36}$$

When the gas is ideal and the liquid volume is constant,

$$\Delta G_f^0(\text{liq}) = \Delta G_f^0(\text{gas}) + RT \ln P^{\text{sat}} - V_L(P^{sat} - 1). \tag{10.37}$$

When the last term is small the result is

$$\Delta G_f^0(\text{liq}) = \Delta G_f^0(\text{gas}) + RT \ln P^{\text{sat}}. \tag{10.38}$$

In Example 10.4, Eq. 10.38 is checked against some data. A similar derivation for the enthalpy yields

$$\Delta H_f^0(\text{liq}) = \Delta H_f^0(\text{gas}) - \Delta H_v^{\text{sat}} + \int_1^{p\text{sat}} \left[\left(\frac{\partial H_G}{\partial P} \right)_T - \left(\frac{\partial H_L}{\partial P} \right) \right] dP, \tag{10.39}$$

and approximately

$$\Delta H_f^0(\text{liq}) = \Delta H_f^0(\text{gas}) - \Delta H_v^{\text{sat}}, \tag{10.40}$$

where ΔH_v^{sat} is the enthalpy of vaporization.

Combining Eqs. 10.38 and 10.22 for a reaction gives a relation between the equilibrium constants $K(\text{gas})$, based on unit fugacity as the reference state of each participant, and $K(\text{liq})$, based on condensed phases of pure participants at one atmosphere as reference states:

$$K(\text{gas}) = K(\text{liq}) K_p^{\text{sat}} \tag{10.41}$$

where

$$K_p^{\text{sat}} = \prod_P (P_i^{\text{sat}})^{\nu_i} \Big/ \prod_R (P_i^{\text{sat}})^{\nu_i}. \tag{10.42}$$

Example 10.5 examines equilibrium compositions of the same reaction occurring in either the liquid phase or the vapor phase and finds them to be different. Such differences arise because

Example 10.2. The Various Forms of the Equilibrium Constant for a Gas-Phase Reaction, $A + 2B \rightleftharpoons C$

The pressure is 3 atm and other data are:

Substance	y_i	$\hat{\phi}_i$	$\hat{f}_i$	P_i
A	0.3	0.9	0.81	0.9
B	0.5	0.8	1.20	1.5
C	0.2	0.7	0.42	0.6

The various terms occurring in the definition of K_{ce} are:

$$K_{ce} = K_{\hat{f}} = 0.42/(0.81)(1.2)^2 = 0.3601,$$

$$K_y = 0.2/(0.3)(0.5)^2 = 2.6667,$$

$$K_{\hat{\phi}} = 0.7/(0.9)(0.8)^2 = 1.2153,$$

$$K_p = 0.6/(0.9)(1.5)^2 = 0.2963,$$

$$P^{\Delta\nu} = 3^{(1-1-2)} = 0.1111,$$

$$K_{ce} = K_y K_{\hat{\phi}} P^{\Delta\nu} = 2.6667(1.2153)(0.1111) = 3601.$$

Example 10.3. Effect of Pressure on Equilibrium of a Reaction Involving Solid Phases

For the reaction, $CaCO_3 \rightleftharpoons CaO + CO_2$, the decomposition temperatures will be found at several pressures:

a. Assuming the activities of the solids to be unity.

b. Applying the Poynting corrections to the activities.

Because of the high temperatures, the fugacity coefficient of CO_2 is taken as unity, but see Problem 10.9. The activities of the solids are ($R = 82.05$),

$$a_{CaCO_3} = \exp[(271.2/RT)(P-1)],$$

$$a_{CaO} = \exp[(190.7/RT)(P-1)].$$

Accordingly, the equilibrium constant is

$$K = a_{CO_2} a_{CaO}/a_{CaCO_3} = P \exp[(190.7 - 271.2)(P-1)/RT]. \quad (1)$$

The equilibrium constant as a function of temperature is given by Hougen, Watson & Ragatz (1959, p. 1025),

$$\ln K = -22{,}265/T + 29.26 - 1.424 \ln T + 0.000373\,T - 0.2076(10^{-6})T^2 + 50{,}266/T^2.$$

These equations are combined and solved for the temperature by trial at several pressures, with the following results:

	T, °K		*a*	
P, atm	*correct*	*unit activities*	*CaO*	*CaCO₃*
1	1,146.4	1,146.4	1	1
10	1,317.9	1,318.5	1.016	1.023
100	1,550.9	1,558.9	1.160	1.235
500	1,749.6	1,797.1	1.940	2.567

Clearly, at the higher pressures the Poynting Factor does affect the prediction of the decomposition temperature. The activities are strongly affected by the pressure, but those for the two solids tend to cancel each other out, so the effect on temperature is much less than it would be if only one condensed phase were present.

Example 10.4. Gibbs Energy of Formation of the Liquid from that of the Vapor

Data for several substances are taken from Stull et al. (1969), and values of the liquid (or solid) are calculated for comparison from the gas-phase value with

$$\Delta G_f^0(l) = \Delta G_f^0(g) + RT \ln P^s.$$

In all these cases, the term $\int_1^{P^s} V^L dP < 0.003$. Vapor pressures are from Table D2. Small differences in vapor pressure can account for the incomplete correspondence between calculated and tabulated ΔG_f^0 (1); the values of vapor pressure corresponding to the differences between Stull's tabulated vapor and liquid Gibbs energies of formation are given in the last column. In some cases there is fair agreement.

	ΔG_f^0 *(Stull)*		$\Delta G_f^0(l)$		
	g	*l(s)*	*calc*	P^s	P^s *(Stull)*
propanol	−38.95	−40.79	−41.09	0.027	0.0448
ethyl acetate	−78.25	−79.52	−79.47	0.128	0.117
valeric acid	−85.37	−89.25	−90.31	2.4 (E-4)	0.14(E-4)
benzene	30.99	29.72	29.76	0.126	0.117
naphthalene	53.44	(48.05)	48.65	3.1(E-4)	1.1(E-4)

Example 10.5. Comparison of Equilibria in Liquid and Vapor Phases

For the participants of the reaction,

$$CH_3COOH + C_2H_5OH \rightleftharpoons CH_3COOC_2H_5 + H_2O.$$

Gibbs energies of formation at 298 K as liquids and as gases are:

	ΔG_f			
	liquid	*gas*	P^s *(data)*	P^s *(Antoine)*
acetic acid	−93.06	−90.03	0.0060	0.0201
ethanol	−41.62	−40.22	0.0940	0.0780
ethyl acetate	−79.52	−78.25	0.1171	0.1277
water	−56.69	−54.64	0.0314	0.03111
ΔG_f^0	−1.5300	−2.6400		
K_{ce}	13.2488	86.3587		

Vapor pressures corresponding to the liquid and vapor Gibbs energies, calculated from Eq. 10.38, are given in column 4, and those from the Antoine equation in the last column. The two vapor-pressure values of acetic acid disagree somewhat; this may be accounted for by molecular association (data are given in Problem 1.11, for instance). The two K_ps are: 6.5183, from the Gibbs energies; 2.5301, from the Antoine equation.

For equimolal reactants, the conversion is given by

$$K = \varepsilon^2/(1-\varepsilon)^2,$$

so that

$$\varepsilon = \sqrt{K}/(1+\sqrt{K}) = \begin{cases} 0.7845, \text{ liquid phase, } K = 13.2488, \\ 0.9028, \text{ vapor phase, } K = 86.3587. \end{cases}$$

For this reaction to occur in the vapor phase at 298 K, the pressure would need to be reduced substantially below atmospheric.

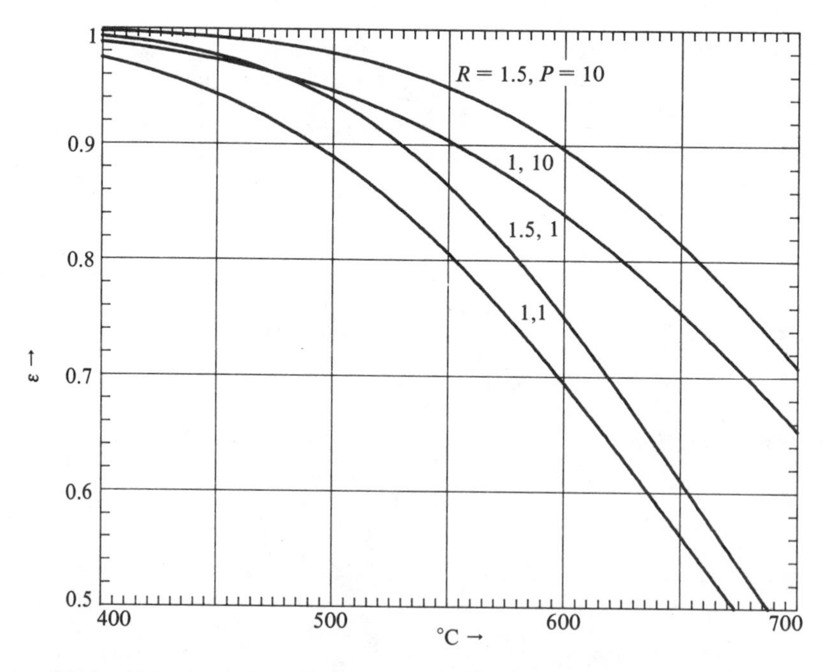

Figure 10.1. Equilibrium in the oxidation of SO_2 with air. For the stoichiometric equation, $SO_2 + 0.5\ O_2 \rightleftharpoons SO_3$, the equilibrium equation is

$$\frac{\varepsilon}{1-\varepsilon}\sqrt{\frac{2(1+2.38R-0.5\varepsilon)}{P(R-\varepsilon)}} = \exp\left[\frac{11{,}846.5}{T+273.2} - 11.256\right],$$

where ε = fractional conversion of SO_2, $R/2$ = initial ratio of O_2/SO_2, P = pressure, atm.

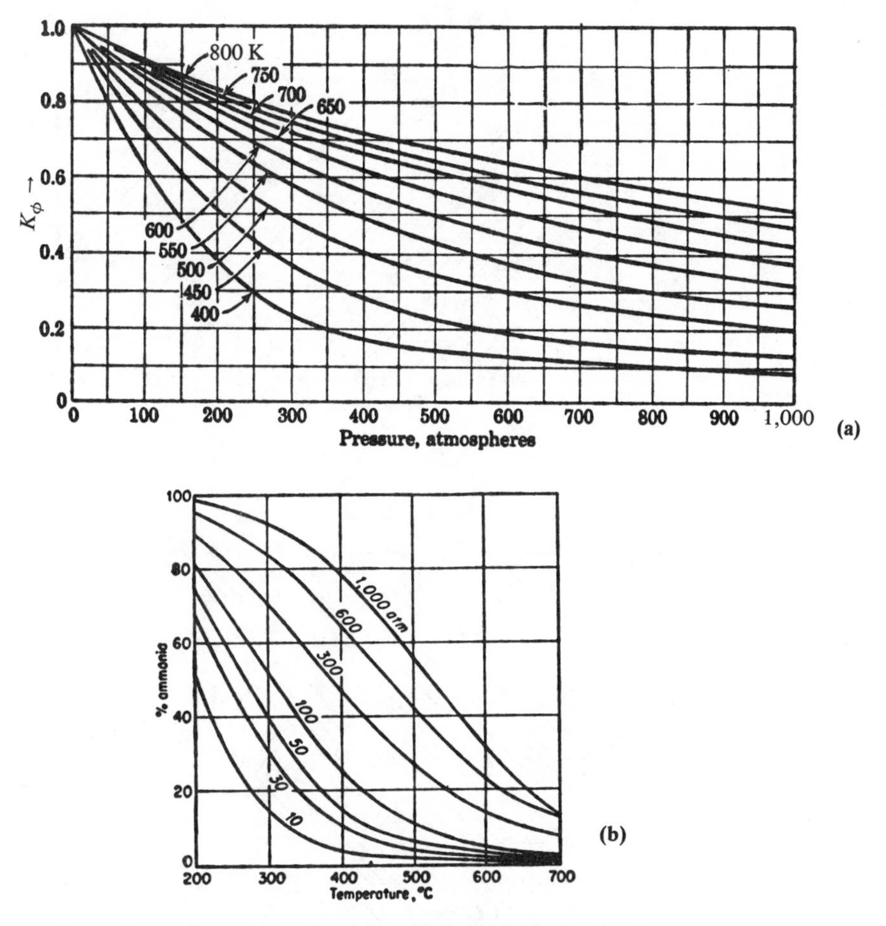

Figure 10.2 Equilibrium in the synthesis of ammonia. (a) Fugacity coefficient function. $K_\phi = \phi_{NH_3}/(\phi_{N_2})^{0.5}(\phi_{H_2})^{1.5}$ (Hougen et al., 1959). (b) Percentage of ammonia at equilibrium from stoichiometric proportions of nitrogen and ammonia (Comings, 1956).

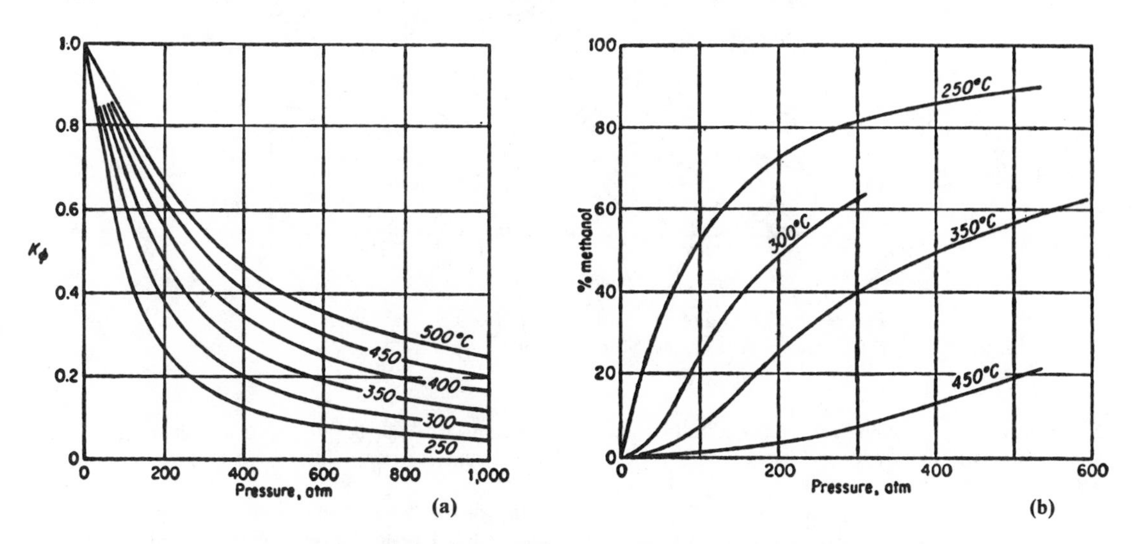

Figure 10.3. Chemical equilibrium in the synthesis of methanol from hydrogen and carbon monoxide. $2\,H_2 + CO \rightleftharpoons CH_3OH$, $K_{ce} = K_\phi K_p / P^2$, $\log_{10} K_p = -12.5173 + 5{,}090.5/T$, data of Table D.1. (a) Values of the fugacity coefficient function, $K_\phi = \phi_{CH_3OH}/(\phi_{H_2})^2\phi_{CO}$ (Newton & Dodge, 1935); (b) Mol% of methanol at equilibrium from stoichiometric proportions of hydrogen and carbon monoxide (Comings, 1956).

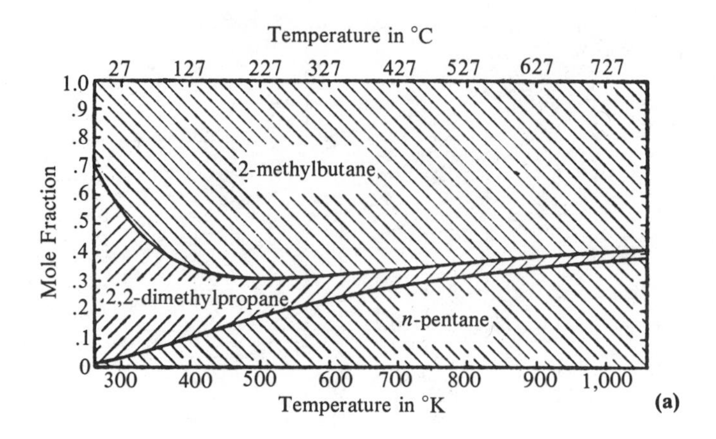

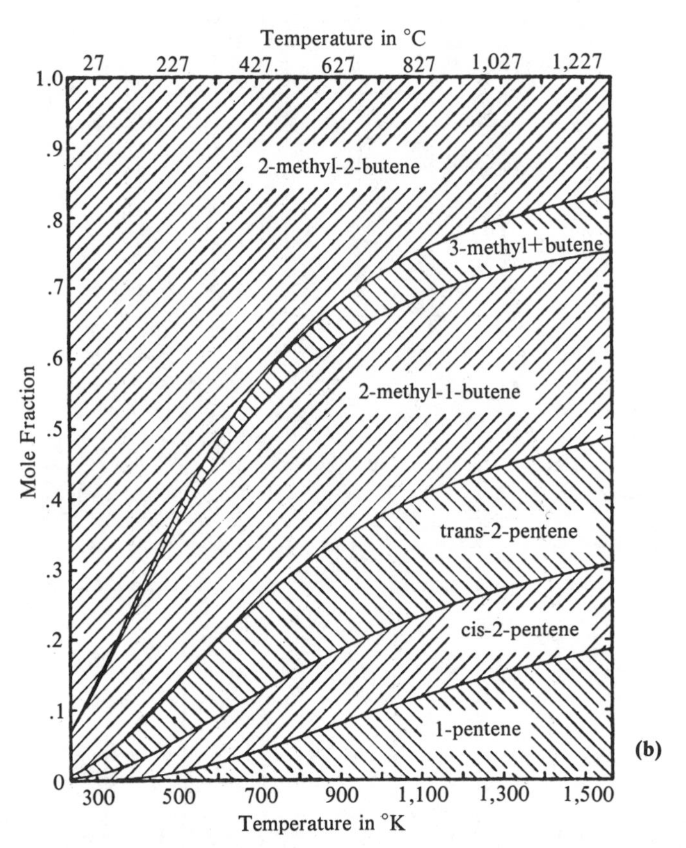

Figure 10.4. Equilibria between C_5 isomers. (a) pentanes; (b) pentenes; (c) pentynes. (Rossini, 1950).

of varying extents of deviation from ideality in the two phases that are represented by activity coefficients and the effect of pressure on the gas phase.

10.4.3. Condensed Phase Solutions

The discussion will be phrased in terms of liquid solutions, but extension to solid solutions is straightforward. When an equation of state is applicable to both liquid and vapor phases, the activity coefficients can be figured as in Section 4.5. The standard state will be the vapor at unit fugacity, and the equilibrium constant will be given by Eq. 10.29, as in the vapor case.

In general, however, the fugacity behavior of condensed phases must be expressed in terms of activity coefficients, with standard states as pure condensed phases at 1 atm, and determined by direct experiment or calculated from molecular structure. Thus,

$$a_i = \hat{f}_i/f_i^0 = \gamma_i x_i f_i/f_i^0 = \gamma_i x_i \exp(V_{iL}(P-1)/RT). \tag{10.43}$$

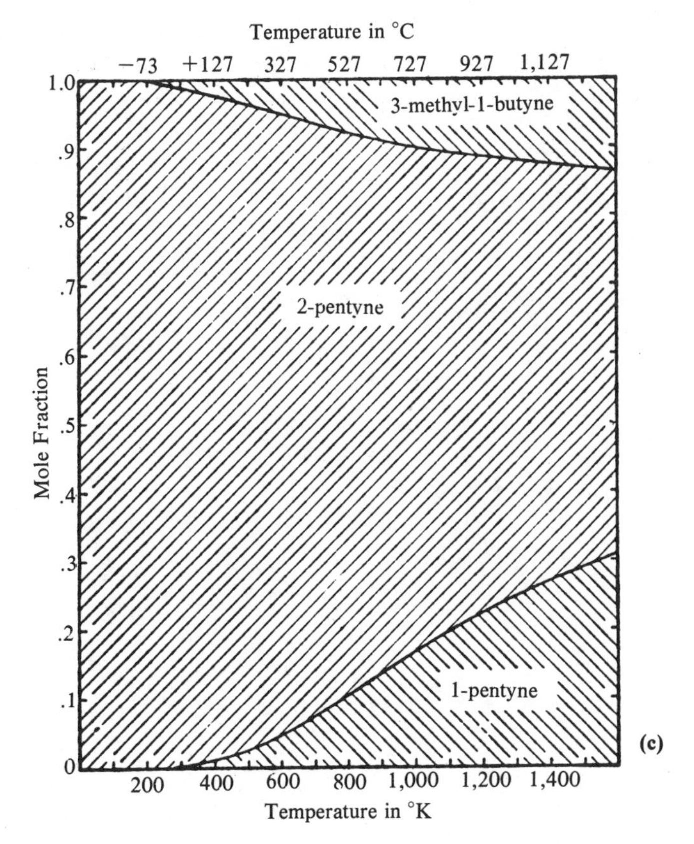

Figure 10.4 *(continued)*

When the Poynting Factor can be neglected, the equilibrium constant becomes simply

$$K_{ce} = K_\gamma K_x. \tag{10.44}$$

For ideal solutions, K_γ is unity, and

$$K_{ce} = K_x = \prod_P (x_i)^{\nu_i} / \sum_R (x_i)^{\nu_i}. \tag{10.45}$$

Liquid-phase reactions may involve a solvent and several solutes. Because of the complex relations existing between activity coefficients and concentrations in multicomponent mixtures, data are scarce and the application of solution thermodynamics to nonideal reacting systems has been minimal. When accurate Gibbs formation energies are available, experimental determinations of equilibrium conversions will provide values of K_γ corresponding to each equilibrium mixture; different equilibrium compositions may be obtained with different starting proportions of reactants. From a sufficient number of measurements, the parameters of a suitable correlating equation for the activity coefficients of the participants can be determined. Thus Example 10.6 illustrates the method with a two-parameter equation for a two-component mixture. The minimum number of measurements of K_γ needed equals the number of parameters to be determined. For instance, for three components with the Wilson equation, six parameters are involved, and at least six equilibrium compositions are needed.

10.4.4. Mixed Phases

When liquid and vapor phases are present together, the problem is to find the amounts and compositions of the phases under conditions of flash and chemical equilibrium. For any particular substance it is desirable to adopt the same reference state in each phase, since the chemical equilibrium constant then will be numerically the same in each phase. Because ΔG_f^0 data are most commonly available for gases at unit fugacity, this reference state is favored for all phases. In view of the fact that the partial fugacities are the same throughout for individual components at equilibrium, and noting the assumption, $f_i^0 = 1$, several ways of writing the activities are:

$$a_i = \hat{f}_{iV} = \hat{\phi}_i y_i P \tag{10.46}$$

$$= \hat{f}_{iL} = \gamma_i x_i f_i = \gamma_i x_i \phi_i^s P_i^s \exp(V_{iL}(P - P^s)/RT)$$

$$= \gamma_i x_i \phi_i^s P_i^s (PF)_i. \tag{10.47}$$

In these various terms the chemical equilibrium constants become,

$$K_{ce} = K_{\hat{f}} = K_{\hat{\phi}} K_y P^{\Delta\nu}, \text{ for gases} \tag{10.48}$$

$$= K_\gamma K_x K_{\phi^s} K_{P^s} K_{\text{Poynting}}, \text{ for liquids} \tag{10.49}$$

If there were any advantage to it, the activities expressions could be mixed. For instance, for the reaction, $A \rightleftarrows 2B$, there are four equivalent forms:

$$K_{ce} = \frac{(\hat{\phi}y)_b^2 P}{(\hat{\phi}y)_a} \tag{10.50}$$

$$= \frac{[\gamma x \phi^{\text{sat}} P^{\text{sat}}(PF)]_b^2}{[\gamma x \phi^{\text{sat}} P^{\text{sat}}(PF)]_a} \tag{10.51}$$

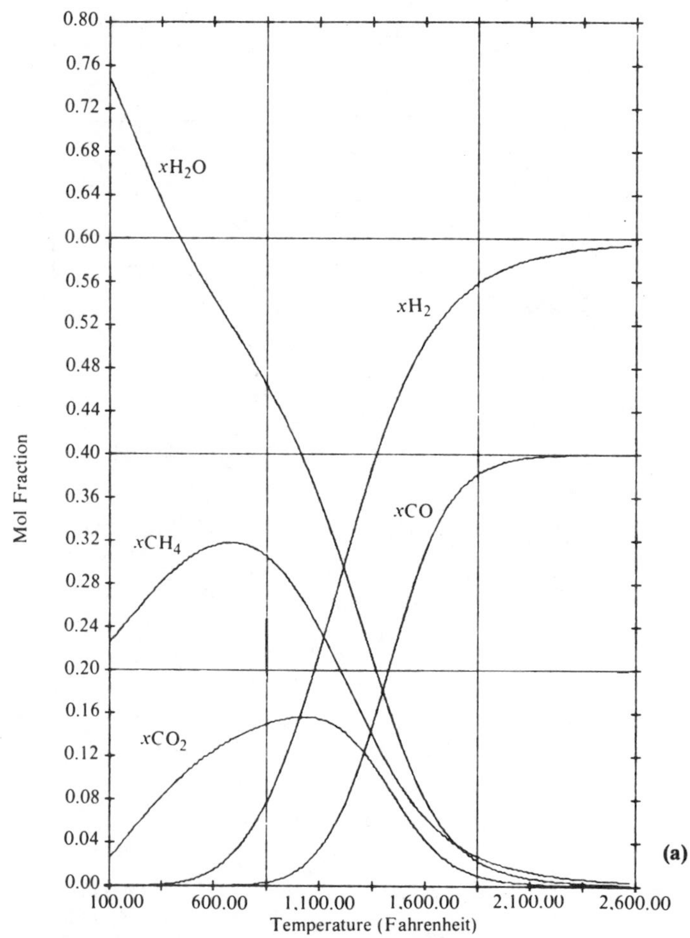

Figure 10.5. Equilibrium compositions in the reaction of methane and water (Baron, Porter, & Hammond, 1975). (a) Effect of temperature at 10 atm when H/O = 3. (b) Effect of pressure at 1,200 F with H/O = 3.

Example 10.6. Activity Coefficients in a Liquid Phase Reaction Mixture

For the reaction, $A \rightleftharpoons 2B$, the equilibrium constant from Gibbs energies of formation is

$$K_{ce} = K_\gamma K_x = 0.7.$$

Measurements of equilibrium compositions with two different starting proportions of the participants and the corresponding calculated K_γs are:

x_1	K_γ	γ_1	γ_2
0.5	1.4	(1.55)	(1.72)
0.6	2.625	(1.43)	(1.94)

The parameters of the Margules equation will be found from these data.

$$\ln \gamma_1 = [A + 2(B - A)x_1]x_2^2,$$

$$\ln \gamma_2 = [B + 2(A - B)x_2]x_1^2.$$

For $x_1 = 0.5$ the activity coefficients are given by:

$$\ln \gamma_1 = 0.25B,$$

$$\ln \gamma_2 = 0.25A,$$

and for $x_1 = 0.6$,

$$\ln \gamma_1 = -0.032A + 0.192B,$$

$$\ln \gamma_2 = 0.288A + 0.072B.$$

When these pairs of activity coefficients are substituted into the definition,

$$K_\gamma = \gamma_2^2/\gamma_1,$$

two equations are obtained for the parameters:

$$1.4 = \frac{(\exp 0.25A)^2}{\exp 0.25B},$$

$$2.625 = \frac{[\exp(0.288A + 0.072B)]^2}{\exp(-0.032A + 0.192B)}.$$

The solution is $A = 1.7588$, $B = 2.1717$. The corresponding activity coefficients are shown in parentheses in the table.

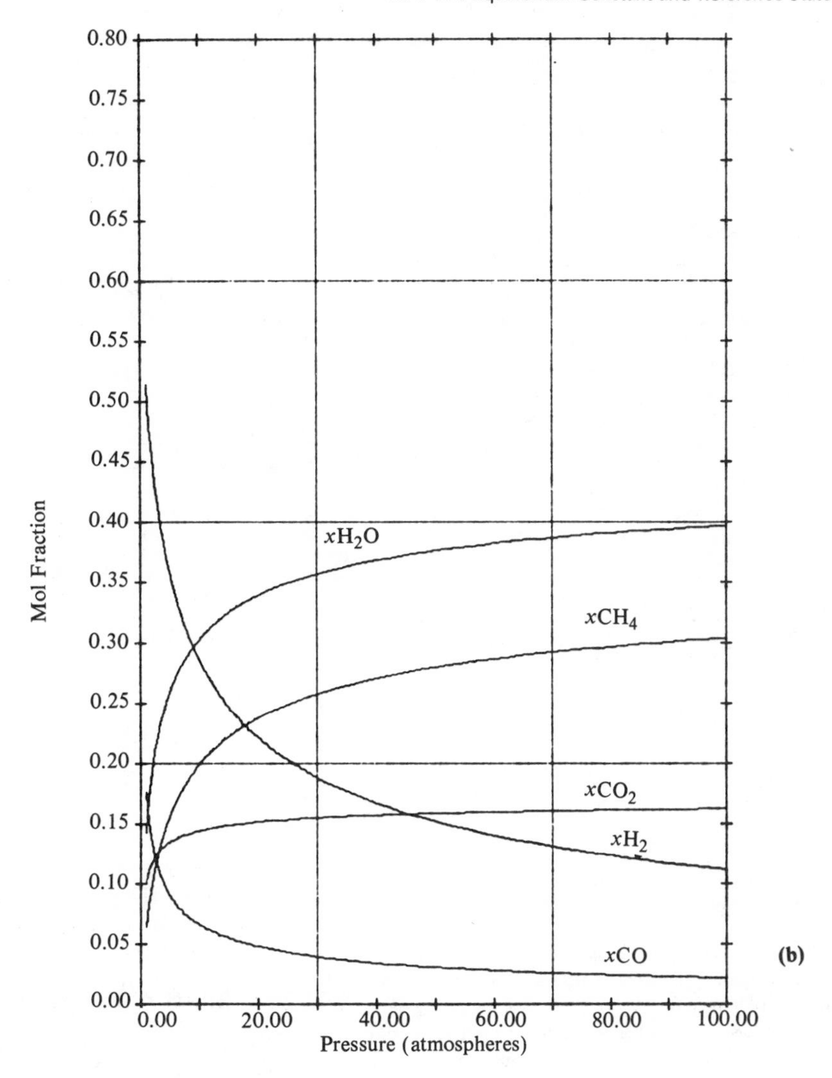

Figure 10.5 (*continued*)

Example 10.7. Compositions at Equilibrium of a Single Reaction

Data for the reaction, $A \rightleftarrows B + 2C$, are:

$K_{ce} = 4.2,$

$K_{\hat{\phi}} = 0.8$ (estimated in advance),

$P = 5,$

$n_{a0} = 2,$

$n_{b0} = 1,$

$n_{c0} = 0.$

Accordingly,

$$K_y = \frac{K_{ce}}{K_{\hat{\phi}} P^2} = \frac{4.2}{0.8(25)} = 0.2100$$

$$= \frac{(n_{b0} + \varepsilon)(n_{c0} + 2\varepsilon)^2}{(n_{a0} - \varepsilon)(n_{t0} + 2\varepsilon)^2} = \frac{(1 + \varepsilon)(2\varepsilon)^2}{(2 - \varepsilon)(3 + 2\varepsilon)^2}.$$

By trial, $\varepsilon = 0.8492$. Hence the numbers of mols and the mol fractions are

$n_a = 1.1508, \quad y_a = 0.2449,$

$n_b = 1.8492, \quad y_b = 0.3936,$

$n_c = 1.6984, \quad y_c = 0.3615.$

In general, the estimated $K_{\hat{\phi}}$ should be verified after the trial equilibrium composition has been found, and the calculations repeated if there is an appreciable difference. In the present case, however, the basis for calculating $K_{\hat{\phi}}$ is not given.

Example 10.8. A Nonideal Liquid-Phase Reaction

For the reaction, $A + 2B \rightleftarrows C$, the equilibrium constant is 0.5 and the Wilson parameters are:

$A_{12} = 0.6, A_{21} = 1.4,$

$A_{13} = 0.7, A_{31} = 0.5,$

$A_{23} = 0.8, A_{32} = 0.9.$

The subscripts 1, 2, and 3 refer to substances *A*, *B*, and *C*, in that order. The equilibrium composition with stoichiometric proportions of *A* and *B* will be found. Equations for ternary activity coefficients are in Table 4.7. The solution will be started by assuming $K_\gamma = 1$, and the compositions found in this way will be used to find an improved value.

$n_t = 3 - 2\varepsilon,$

$x_a = (1 - \varepsilon)/n_t,$

$x_b = 2x_a,$

$x_c = \varepsilon/n_t,$

$$K_x = \frac{K}{K_\gamma} = \frac{0.5}{K_\gamma} = \frac{\varepsilon(3-2\varepsilon)^2}{4(1-\varepsilon)^3},$$

$K_\gamma = \gamma_c/\gamma_a\gamma_b^2.$

For a first trial, assume $K_\gamma = 1$. The value of $\varepsilon = 0.16375$. Mol fractions and activity coefficients are tabulated. With $K_\gamma = 0.9085$, $\varepsilon = 0.17575$ and the next trial become $K_\gamma = 0.9080$. Linear interpolation between the trials is adequate and is shown in the last column of the table.

	$\varepsilon = 0.16375$		$\varepsilon = 0.17575$		$\varepsilon = 0.17580$
	x	γ	x	γ	x
A	0.3129	1.6367	0.3112	1.6329	0.3112
B	0.6257	1.0113	0.6224	1.0102	0.6224
C	0.0614	1.5208	0.0664	1.5131	0.0664
K_γ		0.9085		0.9080	

$$= \frac{(\hat{\phi}y)_b^2 P^2}{[\gamma x \phi^{sat} P^{sat}(PF)]_a} \quad (10.52)$$

$$= \frac{[\gamma x \phi^{sat} P^{sat}(PF)]_b^2}{(\hat{\phi}y)_a P}. \quad (10.53)$$

In all these expressions the reference states are at unit fugacity. When figuring multiphase equilibria (Section 10.7), it is usual to employ K_{ce} in terms of gas-phase fugacities only. However, fugacities of both phases are needed in formulating the flash equation, which must be solved simultaneously with chemical equilibrium to find the overall equilibrium conditions.

10.4.5. Components Not Present in Both Phases

Those components that are essentially nonvolatile under reaction conditions, of course, do not participate in the gas-phase reaction and so are not involved in the gas-phase K_{ce}. Moreover, in the formulation of the flash equation, such substances have VER = 0, whereas gases that are insoluble in the liquid have infinite VERs. As an example, take the multiphase reaction,

$A(\text{gas}) + B\,(\text{liquid}) \rightleftarrows 2\,C\,(\text{gas and liquid}),$

where only component *C* is present in both phases. With *B* absent from the vapor phase, the chemical equilibrium constant is

$$K_{ce} = (\hat{\phi}_c y_c)^2 \mathrm{P}/\hat{\phi}_a y_a, \quad (10.54)$$

and with *A* absent from the liquid phase ($K_a \to \infty$, $K_b = 0$), the flash equation is

$$\Sigma\, x_i = \frac{z_b}{1-\beta} + \frac{z_c}{1+\beta(K_c - 1)} = 1, \quad (10.55)$$

$$y_i = K_i x_i, \quad (10.56)$$

$$\beta = V/F. \quad (10.57)$$

The mechanism of the reaction of this example must involve an interface where both *A* and *B* are present and can achieve contact. Section 10.7 details some methods for solving combined phase and chemical equilibria.

10.4.6. Effect of Pressure

Since the energies of formation, ΔG_f^0, are defined at fixed pressure (generally 1 atm) or unit fugacity which is substantially equivalent to 1 atm, the derivatives with respect to pressure of both ΔG^0 and K_{ce} are zero,

$$\left(\frac{\partial G^0}{\partial P}\right)_T = \left(\frac{\partial K_e}{\partial P}\right)_T = 0. \quad (10.58)$$

However, the equilibrium compositions of gas-phase reactions can be changed by pressure, which is evident from the denominator of

$$K_y = K_{ce}/K_\phi P^{\Delta\nu} \quad (10.59)$$

when $\Delta\nu \neq 0$ or because of the variation of K_ϕ with pressure. The latter effect can be quite substantial, for instance in the synthesis of methanol at 300 C, where some results based on Figure 10.3a are:

		ε	
P, atm	$K_{\hat{\phi}}$	*Correct*	$@K_{\hat{\phi}} = 1$
100	0.62	0.488	0.406
200	0.39	0.737	0.628
400	0.20	0.878	0.778
600	0.13	0.921	0.836

These data are for stoichiometric proportions of the reactants.

In liquid-phase reactions, the effect of pressure is modest and can arise from either K_γ or K_{Poynting}. The derivative of activity coefficient with pressure,

$$\left(\frac{\partial \ln \gamma_i}{\partial P}\right)_T = \frac{\bar{V}_i - V_i}{RT}, \tag{10.60}$$

usually is small; moreover, data rarely are available, particularly over a range of pressure, so that effect usually is taken as zero. Since data for evaluation of the Poynting Factors are readily available, this effect can be taken account of when it is significant (see Example 10.3).

10.4.7. The Effect of Temperature

Derivation of the equation for the effect of temperature on the chemical equilibrium constant begins with the definition

$$G = H - TS, \tag{10.61}$$

or

$$G/T = H/T - S = H/T + (\partial G/\partial T)_P. \tag{10.62}$$

Eq. 10.62 is a first-order linear differential equation with the solution

$$\frac{G}{T} = -\int \frac{H}{T^2}\,dT + \text{Constant}. \tag{10.63}$$

Differentiation of this solution gives the useful result,

$$\left[\frac{\partial(G/T)}{\partial T}\right]_P = -\frac{H}{T^2}. \tag{10.64}$$

When the Gibbs formation energies of all participants are combined stoichiometrically and substituted into Eq. 10.64, the result is

$$-\left[\frac{\partial(\Delta G^0/RT)}{\partial T}\right]_P = \left(\frac{\partial \ln K_{ce}}{\partial T}\right)_P = \frac{\Delta H_r^0}{T^2}. \tag{10.65}$$

This is called the van't Hoff (1884) isochore.

The standard enthalpy change of reaction, ΔH_r^0, depends on the temperature. When there is no phase change within the temperature interval of interest, this dependence is represented by

$$\Delta H_r^0 = \Delta H_{298}^0 + \int_{298}^{T} \Delta C_p\,dT, \tag{10.66}$$

where

$$\Delta C_p = \sum_P \nu_i C_{pi} - \sum_R \nu_i C_{pi}. \tag{10.67}$$

When there is a phase change, appropriate changes in Eq. 10.66 are readily made. After substitution of Eq. 10.66 into Eq. 10.65, integration gives,

$$\ln\frac{K}{K_{298}} = \frac{1}{R}\int_{298}^{T}\frac{1}{T^2}\left[\Delta H_{298}^0 + \int_{298}^{T}\Delta C_p\,dT\right]dT. \tag{10.68}$$

If the heat of reaction, ΔH_r^0 is known at T_1, and K is known at T_2 (instead of both at 298 K, which is the usual case), the integral becomes,

$$\ln\frac{K}{K_{T_2}} = \frac{1}{R}\int_{T_2}^{T}\frac{1}{T^2}\left[\Delta H_{T_1}^0 + \int_{T_1}^{T}\Delta C_p\,dT\right]dT. \tag{10.69}$$

The effect of temperature on methanol synthesis is explored in Example 10.9. The marked effect in this case and also that for sulfur oxidation, as shown in Figure 10.1, are representative. Equilibrium yields of many desired reactions are favored by low temperatures, so it is unfortunate that the kinetics often is unfavorable at those conditions. In ammonia synthesis, for instance, acceptable rates of reaction are achieved under conditions of only about 25% conversion per pass, necessitating extensive recycling of unconverted material. In other cases, staged reactors at decreasing temperature levels are effective.

When the heat of reaction is essentially constant over a practical temperature interval, direct integration of Eq. 10.65 gives the useful result,

$$\ln\frac{K}{K_1} = \Delta H_r^0\left(\frac{1}{T_1} - \frac{1}{T}\right), \tag{10.70}$$

or

$$\ln K = A + \frac{B}{T}. \tag{10.71}$$

This linear variation of ln K with reciprocal temperature is mentioned in Section 10.3, where some data are cited.

10.5. EQUILIBRIUM COMPOSITION

Consider a reaction that is represented by a single stoichiometric equation,

$$\sum_R \nu_i A_i \rightleftarrows \sum_P \nu_i A_i. \tag{10.72}$$

The amounts of all substances present are expressible in terms of a single quantity, ε, the degree of advancement that was introduced in Section 10.2.

$$\varepsilon = \frac{n_i - n_{io}}{\pm \nu_i} \quad \begin{cases} + \text{ for products,} \\ - \text{ for reactants.} \end{cases} \tag{10.73}$$

The total number of mols present at any time is

$$n_t = n_{to} + \varepsilon\,\Delta\nu = n_{to} + \varepsilon\left[\sum_P \nu_i - \sum_R \nu_i\right], \tag{10.74}$$

and the individual mol fractions are

$$y_i = \frac{n_{io} \pm \nu_i\varepsilon}{n_t} \quad \begin{cases} + \text{ for products,} \\ - \text{ for reactants.} \end{cases} \tag{10.75}$$

Thus the mol fraction form of the equilibrium constant is:

$$K_y = \frac{\prod_P (n_{io} + \nu_i\varepsilon)^{\nu_i}}{\prod_R (n_{io} - \nu_i\varepsilon)^{\nu_i}\,(n_{to} + \varepsilon\Delta\nu)^{\Delta\nu}}. \tag{10.76}$$

The equation may be complex enough to require a solution for ε by trial, as in Example 10.7.

For the bimolecular reaction,

$$A + B \rightleftarrows C + D, \tag{10.77}$$

the equilibrium equation assumes the standard quadratic form,

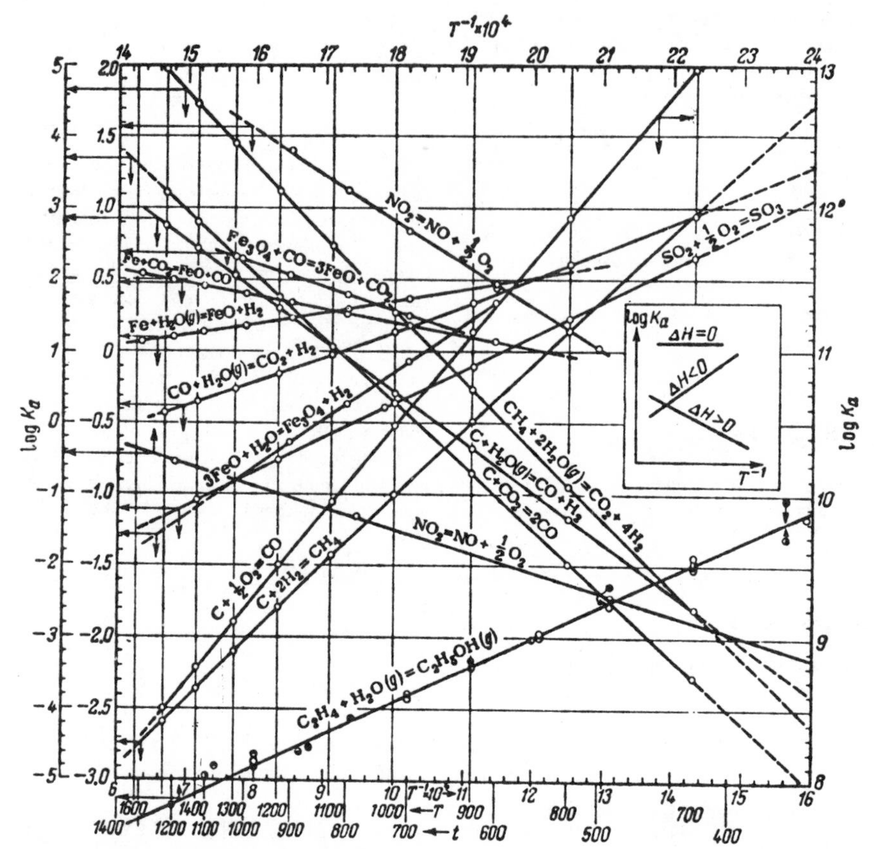

Figure 10.6. Equilibrium constants of some reactions as a function of temperature (Karapetyants, 1978). See also Table D.1 (in Appendix D).

Example 10.9. The Equilibrium Constant of Methanol Synthesis as a Function of Temperature

The reaction is $CO + 2H_2 = CH_3OH$. Heat capacities and enthalpies and Gibbs energies of formation at 298 K are taken from Table D.2.

$$\Delta H^0_{298} = -48{,}080 - (-26{,}420) - 2(0) = -21{,}660 \text{ cal/gmol},$$

$$\Delta G^0_{298} = -38{,}840 - (-32{,}180) - 2(0) = -6{,}030 \text{ cal/gmol},$$

$$K_{298} = \exp\frac{6{,}030}{298\,R} = 26{,}466.7,$$

$$\Delta C_p = C_{p(CH_3OH)} - C_{p(CO)} - 2\,C_{p(H_2)}$$
$$= -15.287 + 0.01558T + 6.113(E-6)T^2 + 6.195(E-9)T^3.$$

The heat of reaction as a function of temperature is

$$\Delta H^0_T = -21{,}660 + \int_{298}^{T} \Delta C_p\, dT$$
$$= -17{,}862 - 15.287T + 0.00779T^2 + 2.038(E-6)T^3 + 1.549(E-9)T^4,$$

and the equilibrium constant as a function of temperature is given by

$$\ln\frac{K_T}{26{,}466.7} = \int_{298}^{T} \frac{\Delta H^0_T}{RT^2}\, dT$$
$$= 8{,}989.4\left(\frac{1}{T} - \frac{1}{298}\right) - 7.965\ln\frac{T}{298} + 0.00392(T-298)$$
$$+ 0.5128(E-6)(T^2 - 298^2) + 0.2599(E-9)(T^3 - 298^3). \tag{1}$$

As the following tabulation shows, calculated values from this equation compare only moderately well with the results of Ewell (1940), and with the equation,

$$\log_{10} K = -12.5173 + 5090.5/T, \tag{2}$$

with coefficients from Table D.1.

	10^6K		
T, °C	*Eq. 1*	*Ewell*	*Eq. 2*
100	14.65(*E*6)	10.84(*E*6)	13.27(*E*6)
200	22,110	16,950	17,390
300	288	232	231
400	13.0	10.9	11.1
500	1.302	1.134	1.165

$$(K_y - 1)\varepsilon^2 - [K_y(n_{ao} + n_{bo}) + n_{co} + n_{do}]\varepsilon$$
$$+ (K_y n_{ao} n_{bo} - n_{co} n_{do}) = 0. \quad (10.78)$$

A reacting system that is represented by several simultaneous stoichiometric equations can be handled in terms of one degree of advancement, ε_i, for each equation, and an equilibrium constant for each reaction. In order to establish equations relating equilibrium to the compositions, the reactions are arranged in any convenient order and are assumed to occur consecutively, with the starting amounts of the participants in any particular reaction being the amounts present after the degrees of advancement of all the preceding reactions. Elimination of intermediate quantities will result in a set of equations for the mols of all substances present in terms of the starting quantities and all the degrees of advancement.

The procedure is illustrated with the example of the simultaneous reactions,

$$2A + B \rightleftarrows 2C, \quad (10.79)$$

$$A + 2C \rightleftarrows 3D. \quad (10.80)$$

The two degrees of advancement are

$$\varepsilon_1 = \frac{n_{a0} - n_{a1}}{2} = n_{b0} - n_b = \frac{n_{c1} - n_{c0}}{2}, \quad (10.81)$$

$$\varepsilon_2 = n_{a1} - n_a = \frac{n_{c1} - n_c}{2} = \frac{n_d - n_{d0}}{3}. \quad (10.82)$$

On elimination of n_{a1} and n_{c1} from these equations, the numbers of mols of all of substances in terms of initial amounts and the two degrees of advancement are

$$n_a = n_{a0} - 2\varepsilon_1 - \varepsilon_2, \quad (10.83a)$$

$$n_b = n_{b0} - \varepsilon_1, \quad (10.83b)$$

$$n_c = n_{c0} + 2\varepsilon_1 - 2\varepsilon_2, \quad (10.83c)$$

$$n_d = n_{d0} + 3\varepsilon_2, \quad (10.83d)$$

$$n_t = n_{t0} - \varepsilon_1. \quad (10.83e)$$

Accordingly, the two equilibrium equations are:

$$K_1/K_{\phi 1} = \frac{n_c^2 n_t}{n_a^2 n_b P} = f_1(\varepsilon_1, \varepsilon_2), \quad (10.84)$$

$$K_2/K_{\phi 2} = \frac{n_d^3}{n_a n_c^2} = f_2(\varepsilon_1, \varepsilon_2). \quad (10.85)$$

Any favorite method of solving simultaneous nonlinear equations may be used to find the degrees of advancement and hence the equilibrium compositions. Example 10.10 employs a graphical method of solution of such a problem. More details about computation methods are covered in the next section, where less cumbersome methods of formulating solutions for multireaction equilibrium also are treated.

10.6. MULTIPLE REACTIONS

Even in apparently simple chemical systems a large number of substances may be present at equilibrium in addition to the principal ones, made from the latter by various series or parallel reactions. For instance, in the isomerization of isopropanol to *n*-propanol, side reactions may produce some acetaldehyde, acetone, hydrogen, carbon dioxide, water, and possibly other substances in significant or trace amounts. The pyrolysis of butane, which might be expected to take the path, $C_4H_{10} \rightleftarrows C_2H_4 + C_2H_6$, actually results in some propylene, several butenes, butadiene, methane, hydrogen, and even benzene.

Of the stoichiometric equations that can be written to relate the chief and by-product participants, some may turn out to be combinations of others, but a certain minimum number always can be found to represent the reacting system. Such independent stoichiometric equations lead to corresponding equilibrium equations, which, together with material balances on the chemical elements present, constitute a mathematical system from which the equilibrium composition can be found. An alternative method of locating the equilibrium condition is to find the minimum Gibbs energy of the reacting system. Both methods will be described in detail. Surveys of the literature of these topics and many details are covered in the books of van Zeggeren & Storey (1970), Holub & Vonka (1976), and Smith & Missen (1982). Other references are in Section 7.8.3.

Independent and derived components and reactions: A systematic method of finding a set of independent reactions from which all participants can be formed, starting with a postulated mixture, was devised by Brinkley (1946) and elaborated by Kandiner & Brinkley (1950). The first step is to form a composition matrix with substances as rows and chemical elements as columns, as in Example 10.13. The number of independent components equals the rank of the matrix, which is the largest nonzero determinant that can be formed from the matrix. Mathematically, the rank is the maximum number of linearly independent rows and columns of the matrix, as explained, for instance, by Gere & Weaver (1965). There are systematic computerized procedures for finding the rank, but when only three or four chemical elements are present, it usually can be found readily by inspection.

10.6.1. The Stoichiometric Method

The method of Brinkley for setting up and solving the simultaneous equations of multiple reaction equilibria comprises seven steps, as follows. Some comments are made after the listing of the steps, and the method is applied in Example 10.13.

1. Select the chemical species that may be present at equilibrium.
2. Determine the *number* of independent components.
3. Choose a set of *independent* components from among the species present, and identify the remaining species as *derived* components.
4. By inspection, write a set of chemical equations by which each derived constituent is made individually from only the independent ones.
5. Write the chemical equilibrium equation of each of these reactions.
6. Write the material balances on the individual chemical elements: weighted sum of the element in all of the substances present at equilibrium = amount present in the initial mixture.
7. The equilibrium equations of step 5 and the element balances of step 6 make up a system that is solved for the composition of the equilibrium mixture.

Example 10.10. Equilibria in the Synthesis of Methanol from CO_2 and H_2. Graphical Solution. $T = 350\,K, P = 500$ atm.

The simultaneous reactions in this process are:

$$\underset{1-\varepsilon_1}{CO_2} + \underset{3-\varepsilon_1}{H_2} \rightleftharpoons \underset{\varepsilon_1}{CO} + \underset{\varepsilon_1}{H_2O}; \; K_1 = 0.05,$$

$$\underset{\varepsilon_1-\varepsilon_2}{CO} + \underset{3-\varepsilon_1-2\varepsilon_2}{2\,H_2} \rightleftharpoons \underset{\varepsilon_2}{CH_3OH}; \; K_2 = 4.458(E-5).$$

The degrees of advancement, ε_1 and ε_2, are introduced below the chemical symbols. The amounts of the participants in terms of these factors are, with $H_2/CO_2 = 3$ at the start,

CO_2	$1 - \varepsilon_1$
H_2	$3 - \varepsilon_1 - 2\varepsilon_2$
CO	$\varepsilon_1 - \varepsilon_2$
H_2O	ε_1
CH_3OH	ε_2
n_t	$4 - 2\varepsilon_2$

Consequently, the equilibrium constant equations are

$$K_1/K_{\hat{\phi}} = \frac{(\varepsilon_1 - \varepsilon_2)\varepsilon_1}{(1 - \varepsilon_1)(3 - \varepsilon_1 - 2\varepsilon_2)} = 0.05,$$

$$K_2/K_{\hat{\phi}} = \frac{\varepsilon_2(4 - 2\varepsilon_2)^2}{(\varepsilon_1 - \varepsilon_2)(3 - \varepsilon_1 - 2\varepsilon_2)^2 P^2} = 4.458(E-5).$$

The equations are solved graphically. Curves of ε_1 against ε_2 are made for each equation by assuming values of ε_1 and finding corresponding values of ε_2 by the N-R method. The intersection of the curves is identified on the graph.

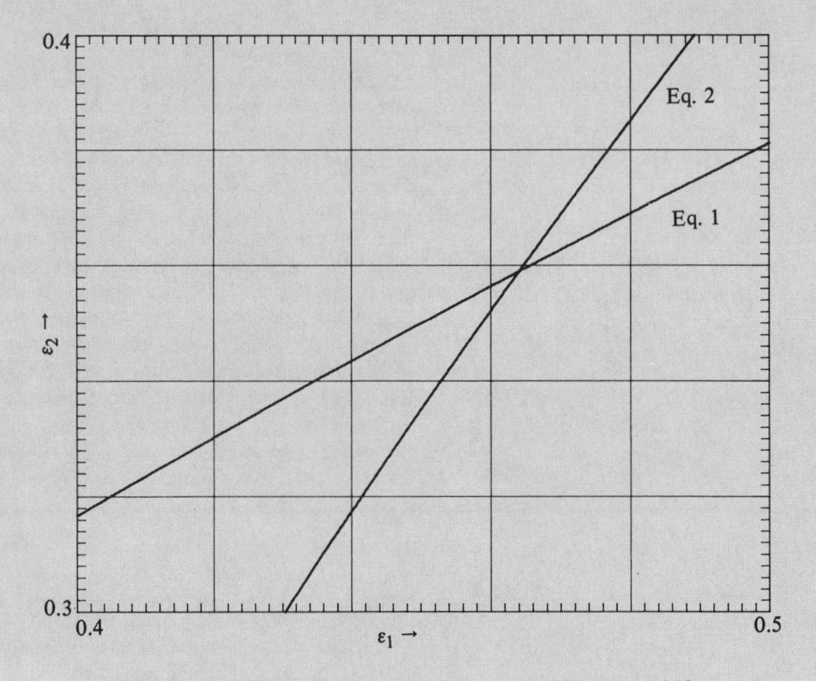

Example 10.10. The intersection is at $\varepsilon_1 = 0.463$, $\varepsilon_2 = 0.359$.

Comments are made on several of these steps, as follows:

1. Choice of the species that may be present at equilibrium is a matter of judgment and may be based on analogy. It is better to err on the side of too many components than on too few, since on completion of the calculation any redundant ones will be found to have negligible concentrations.

2. The number of independent components usually equals the number of chemical elements present, but may be smaller when the reactions include isomerizations, polymerizations, or additions. Such cases are revealed when finding the rank of the composition matrix. A discussion of this point is made by Aris & Mah (1963). One of the systems of Example 10.11 has independent components fewer than the number of chemical elements present.

3. Although several combinations of independent components usually can be formed, numerical calculations are speeded up by choosing those species that are expected to have greatest concentrations at equilibrium, subject to the following limitations:

a. The components must be stoichiometrically independent of each other—that is, with no isomers, polymers, or additions.

b. All the chemical elements present at the start must be included in the independent set.

c. Any component whose concentration is not expressible by a chemical equilibrium constant—for example, a pure condensed phase of unit activity—must be included as an independent one.

4. There is only one possible way in which each derived

component can be formed from the independent ones.

5. Equilibrium constants must be known numerically and may be derived from the Gibbs energies of formation of the components.

7. Since the system of equations usually includes some nonlinear ones, some approximation method of solution needs to be employed. Much attention has been devoted to this problem in connection with multireaction equilibria. For example, there may be difficulties with convergence and with the appearance of negative mol fractions at intermediate stages of successive approximations of the roots. Reduction of the equations to linear systems before or during application of the Newton-Raphson method, the use of linear programming to handle the chemical element constraints, and the method of steepest descent and other optimization techniques have been used.

When a solution by successive approximation is employed and requires a starting estimate of the composition, that estimate must be consistent with the element balances. Example 10.13 uses this principle in applying the direct Newton-Raphson method with the Brinkley procedure.

10.6.2. The Relaxation Method

After a set of independent stoichiometric equations has been found and the material balances formulated, the major problem of finding the individual degrees of advancement remains. In some cases this effort can be eased by a relaxation method, called the *series reactor method* by Meissner, Kusik, & Dalzell (1969), which solves the equilibrium equations one at a time, serially and repeatedly until convergence is attained.

They adopt a hypothetical system of reactors equal in number to the independent stoichiometric equations, hooked up as in Figure 10.7, which is of a system for three reactions. Each reactor operates batchwise and forces equilibrium of the reaction with the same number as the reactor. The process is started by charging the original mixture into the first reactor, where reaction #1 reaches equilibrium. That product is then charged to the second reactor, where reaction #2 reaches equilibrium, followed by transfer to the third reactor, then back to the first reactor, and so on. The feed to each reactor is the equilibrium product from the preceding one and only one reaction occurs in each vessel.

Numerically, the advantage of the series reactor method is that each time an equation in only one unknown, a degree of advancement, needs to be solved—a simpler process than solving a number of nonlinear equations simultaneously. In the cases available, convergence seems to be attained in less than a dozen passes through the reactor system. By comparison, the N-R method for simultaneous reactions needs starting estimates of all the component amounts, and convergence often is quite sensitive to those estimates. One of the authors' examples, a seven-reaction process, stabilized adequately after eight cycles with results comparable to those obtained by Gibbs energy minimization.

This method of solution of a two-reaction process is used in Example 10.14 and of a three-reaction one in Example 10.15. Also, comparisons are made in the first case with direct iteration and a graphical method, which are feasible when the number of reactions is small.

10.6.3. Minimization of Gibbs Energy

The condition that the Gibbs energy of a system at a given T and P be a minimum at equilibrium can be applied directly to finding the equilibrium composition of a reacting mixture without prior specification of any stoichiometric relations. The only information needed is the Gibbs energies of formation of all the chemical species expected to be present at equilibrium, as well as their fugacity or activity coefficient behaviors when they cannot be assumed unity. The method is flexible with respect to possible species, since redundant ones will be seen to have negligible concentrations when the calculations are completed.

Mathematically, what is involved is called a conditional or constrained minimum, since the Gibbs energy of the reacting mixture is a minimum at equilibrium, subject to conservation

Example 10.11. Independent Components and Independent Reactions

The reaction between H_2S *and* SO_2: The components expected to be present at equilibrium are identified in the composition matrix. The first three rows form a nonzero determinant, so there are three independent components. Choose H_2S, SO_2, and H_2O as independent.

		H	O	S
1	H_2S	2	0	1
2	SO_2	0	2	1
3	H_2O	2	1	0
4	H_2	2	0	0
5	S_2	0	0	2

For H_2, $2\,H_2O + H_2S \rightarrow 3\,H_2 + SO_2$

For S_2, $2\,H_2S + SO_2 \rightarrow 3/2\,S_2 + 2H_2O$

Production of ethanol from ethylene and water: Only these three substances are expected to be present at equilibrium. The rank of the composition matrix is 2. Choose C_2H_4 and H_2O as independent.

		C	H	O
1	C_2H_4	2	4	0
2	H_2O	0	2	1
3	C_2H_5OH	2	6	1

For C_2H_5OH: $C_2H_4 + H_2O \rightarrow C_2H_5\,OH$.

The rank of the matrix is less than the number of elements. Only two of the three element balances are independent; the hydrogen balance is obtained by adding multiples of the carbon and oxygen balances. The carbon, oxygen, and total balances are used together with the equilibrium equation to find the four unknowns n_1, n_2, n_3, and n.

Example 10.12. Equilibrium in the Reaction between H_2S and SO_2, Graphical and Direct Iterative Solutions

The initial mixture contains 35% H_2S, 15% SO_2, and 50% inert N_2 at 1,500 K and 0.8 atm. Substances expected to be present at equilibrium are H_2S, H_2, S_2, H_2O, and N_2. Two independent reactions and their equilibrium constants are:

$H_2S \rightleftharpoons H_2 + 0.5S_2; K_1 = 0.334,$

$2\,H_2S + SO_2 \rightleftharpoons 2\,H_2O + 1.5\,S_2; K_2 = 30.2.$

In terms of the individual degrees of advancement, the material balances are as shown at the right.

1. H_2S	$0.35 - \varepsilon_1 - 2\varepsilon_2$
2. H_2	ε_1
3. S_2	$0.5\,\varepsilon_1 + 1.5\,\varepsilon_2$
4. SO_2	$0.15 - \varepsilon_2$
5. H_2O	$2\,\varepsilon_2$
6. N_2	0.5
n_t	$1 + 0.5\varepsilon_1 + 0.5\varepsilon_2$

The equilibrium equations are:

$$f_1 = -0.334 + \frac{\varepsilon_1[0.5(\varepsilon_1 + 3\varepsilon_2)]^{0.5}(0.8)^{0.5}}{(0.35 - \varepsilon_1 - 2\varepsilon_2)(1 + 0.5\varepsilon_1 + 0.5\varepsilon_2)^{0.5}} = 0, \quad (1)$$

$$f_2 = -30.2 + \frac{(2\varepsilon_2)^2[0.5(\varepsilon_1 + 3\varepsilon_2)]^{1.5}(0.8)^{0.5}}{(0.35 - \varepsilon_1 - 2\varepsilon_2)^2(0.15 - \varepsilon_2)(1 + 0.5\varepsilon_1 + 0.5\varepsilon_2)^{0.5}} = 0. \quad (2)$$

1. The two equations are plotted by first specifying ε_1 and then finding ε_2 by single variable N-R. The intersection is at $\varepsilon_1 = 0.0549$ and $\varepsilon_2 = 0.1158$.

2. The equilibrium equations are rearranged for solution by iteration into the forms

$\varepsilon_1 = g_1(\varepsilon_1, \varepsilon_2),$

$\varepsilon_2 = g_2(\varepsilon_1, \varepsilon_2),$

which in the present case are

$$\varepsilon_1 = \frac{0.5281(0.35 - \varepsilon_1 - 2\varepsilon_2)(1 + 0.5\varepsilon_1 + 0.5\varepsilon_2)^{0.5}}{(\varepsilon_1 + 3\varepsilon_2)^{0.5}}, \quad (3)$$

$$\varepsilon_2 = \sqrt{\frac{23.8752(0.35 - \varepsilon_1 - 2\varepsilon_2)^2(0.15 - \varepsilon_2)(1 + 0.5\varepsilon_1 + 0.5\varepsilon_2)^{0.5}}{(\varepsilon_1 + 3\varepsilon_2)^{1.5}}}. \quad (4)$$

Direct iteration proceeds by substituting a pair $(\varepsilon_1^{(0)}, \varepsilon_2^{(0)})$ on the RHS of the first equation to find $\varepsilon_1^{(1)}$ on the left; then substituting $(\varepsilon_1^{(1)}, \varepsilon_2^{(0)})$ on the right of Eq. 2 to find $\varepsilon_2^{(1)}$, and so on until convergence. In the present example, however, direct iteration does not converge, so the Wegstein method is used to force convergence. The calculations start with $\varepsilon_2 = 0$, find ε_1 from Eq. 3 by Wegstein iteration, substitute this value of ε_1 into Eq. 2, find $\varepsilon_2 = 0.0063$ from that equation by Wegstein iteration, back to Eq. 4, and so on, as given in the tabulation.

Wegstein Trials	
ε_1	ε_2
	0
0.1947	0.0063
0.1083	0.0987
0.0725	0.1105
0.0603	0.1142
0.0566	0.1153
0.0555	0.1157
0.0551	0.1158
0.0550	0.1158

	Mols @ Equilib.		
	Graph	*Wegstein*	*Relaxation (Example 10.13)*
H_2S	0.064	0.0634	0.065
H_2	0.058	0.0550	0.055
S_2	0.200	0.2012	0.201
SO_2	0.036	0.0342	0.034
H_2O	0.228	0.2316	0.231
N_2	0.500	0.5000	0.500
Total	1.086	1.0854	1.084

3. The Newton-Raphson method can be applied to solving the simultaneous equations. In cases like the present, it is usually preferable to find the necessary derivatives numerically, for instance,

$$\frac{\partial f}{\partial \varepsilon_1} = \frac{f(1.0001\varepsilon_1, \varepsilon_2) - f(\varepsilon_1, \varepsilon_2)}{0.0001\varepsilon_1}, \quad (5)$$

which is readily evaluated by calculator or computer. Since there are other examples of the use of the N-R method for simultaneous equation solution elsewhere in this book it will not be used here.

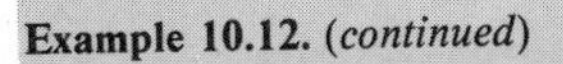

Example 10.12. (*continued*)

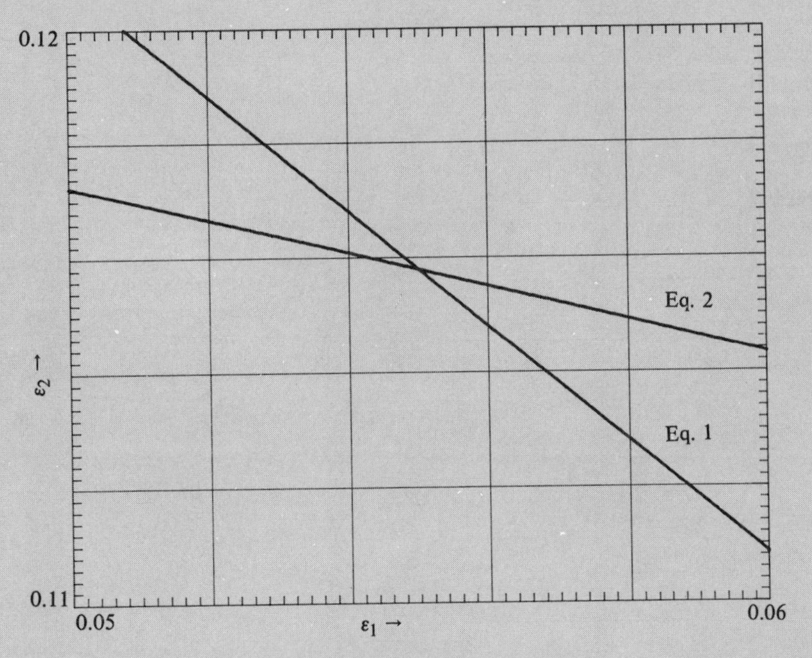

Example 10.12. The intersection is at $\varepsilon_1 = 0.0549$, $\varepsilon_2 = 0.1158$.

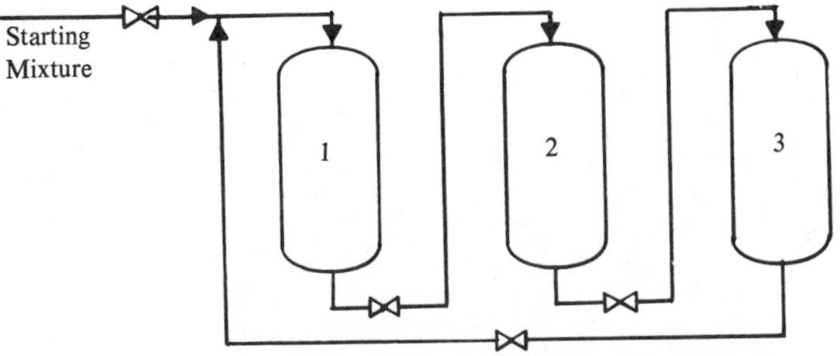

Figure 10.7. Diagram of the relaxation or series-reactor method for equilibrium of multiple reactions. Each reactor operates under batch conditions and attains equilibrium of only its numbered reaction. The process is started by charging the starting mixture to reactor 1 where reaction 1 attains equilibrium. That product is transferred to reactor 2 where only reaction 2 comes to equilibrium. Transfer of that product to reactor 3 is made and equilibrium of reaction 3 is forced. Then the material is returned to reactor 1 and the process is continued until the effluent composition from any reactor becomes constant.

Example 10.13. Reaction between C_3H_8 and SO_2 by the Brinkley Method

When propane and sulfer dioxide react at 500 C and atmospheric pressure, reaction products are expected to include C_3H_8, SO_2, C_3H_6, C_2H_4, CH_4, H_2O, H_2S, and H_2. Derive the following:

1. the composition matrix
2. a suitable set of independent components
3. chemical equations for the formation of derived components
4. the algebraic equations from whose solution the composition of the system at equilibrium can be determined.

	C	*H*	*O*	*S*	*log K_f*
1 C_3H_6	3	6	0	0	−9.519
2 H_2S	0	2	0	1	3.516
3 H_2O	0	2	1	0	13.9660
4 H_2	0	2	0	0	0
5 C_3H_8	3	8	0	0	−8.030
6 C_2H_4	2	4	0	0	6.788
7 CH_4	1	4	0	0	0.3527
8 SO_2	0	0	2	1	20.674

The determinant based on the first four rows $= -6 \neq 0$; therefore four independent components are present. Take C_3H_6, H_2S, H_2O, and H_2. Equations of formation of the derived constituents are:

$$5\quad C_3H_6 + H_2 = C_3H_8;\ K_5 = (\pi/n)^{-1} n_5/n_1 n_4,$$

$$6\quad 2\,C_3H_6 = 3C_2H_4;\ K_6 = (\pi/n) n_6^3/n_1^2,$$

$$7\quad C_3H_6 + 3H_2 = 3CH_4;\ K_7 = (\pi/n)^{-1} n_7^3/n_1 n_4^3,$$

$$8\quad H_2S + 2H_2O = 3H_2 + SO_2;\ K_8 = (\pi/n) n_4^3 n_8/n_2 n_3^2,$$

$$9\quad n = n_1 + n_2 + n_3 + n_4 + n_5 + n_6 + n_7 + n_8.$$

Assuming 1 lbmol each of C_3H_8 and SO_2 to be present initially, the balances on C, H, S, and O are:

$$1\quad 3 = 3n_1 + 3n_5 + 2n_6 + n_7,$$

$$2\quad 8 = 6n_1 + 2n_2 + 2n_3 + 2n_4 + 8n_5 + 4n_6 + 4n_7,$$

$$3\quad 1 = n_2 + n_8,$$

$$4\quad 2 = n_3 + 2n_8.$$

For the species present, log K_f of formation at 500 C are tabulated in the foregoing matrix. The nine equations are arranged into standard forms as preliminary to solution by the N-R or equivalent method.

$$f_1 = 3n_1 + 3n_5 + 2n_6 + n_7 - 3 = 0,$$

$$f_2 = 6n_1 + 2n_2 + 2n_3 + 2n_4 + 8n_5 + 4n_6 + 4n_7 - 8 = 0,$$

$$f_3 = n_2 + n_8 - 1 = 0,$$

$$f_4 = n_3 + 2n_8 - 2 = 0,$$

$$f_5 = n\, n_5 - K_5 P n_1 n_4 = 0,$$

$$f_6 = P n_6^3 - K_6 n\, n_1^4 = 0,$$

$$f_7 = n\, n_7^3 - K_7 P n_1 n_4^3 = 0,$$

$$f_8 = P n_4^3 n_8 - K_8\, n\, n_2 n_3^2 = 0,$$

$$f_9 = n - n_1 - n_2 - n_3 - n_4 - n_5 - n_6 - n_7 - n_8 = 0.$$

Let f_{i0} represent the value of f_i at estimated initial values $n_{10}, n_{20}, \ldots$. According to the N-R method, corrected values of the estimates are $n_j = n_{j0} + \delta_j$, where the δ_j are obtained from the system of nine linear equations,

$$f_{i0} + \sum_{j=1}^{9} \delta_j \left(\frac{\partial f_i}{\partial n_j}\right)_0 = 0; \qquad i = 1, 2, \ldots, 9.$$

The estimated initial values must satisfy the material balances. One such set consists of $n = 3$ and $n_6 = n_7 = n_8 = 1$. The speed of convergence does depend on such starting values. This problem is solved by Kandiner & Brinkley (1950).

Example 10.14. Equilibrium in the Reaction between H_2S and SO_2, Relaxation Method of Solution

The reaction of Example 10.12 is solved by the relaxation method by Meissner, Kusik, & Dalzell (1969). Their tabulation of the steps in the solution is given below. On the assumption that the two reactions occur independently, the equilibrium equations in terms of the individual degrees of advancement are:

$$K_1 = \frac{n_2 n_3^{0.5}}{n_1}\left[\frac{P}{n_{t1}}\right]^{0.5}$$

$$= \frac{(n_{20} + \varepsilon_1)(n_{30} + 0.5\varepsilon_1)^{0.5}}{n_{10} - \varepsilon_1}\left[\frac{P}{n_{t0} + 0.5\varepsilon_1}\right]^{0.5}$$

$$= 0.334, \tag{1}$$

$$K_2 = \frac{n_5^2 n_3^{1.5}}{n_1^2 n_4}\left[\frac{P}{n_{t2}}\right]^{0.5}$$

$$= \frac{(n_{50} + 2\varepsilon_2)^2 (n_{30} + 1.5\varepsilon_2)^{1.5}}{(n_{10} - 2\varepsilon_2)^2 (n_{40} - \varepsilon_2)}\left[\frac{P}{n_{t0} + 0.5\varepsilon_2}\right]^{0.5}$$

$$= 30.2. \tag{2}$$

The quantity $n_{t0} = \Sigma n_{i0}$ changes as the successive approximations are made. The initial quantities, n_{i0}, for any reaction are established by allowing the preceding reaction to reach equilibrium, in the cyclic arrangement shown in Figure 10.7. Solutions for ε_1 and ε_2 are found by Wegstein's method. Equations for the first two cycles are:

(*continued on next page*)

Example 10.14. *(continued)*

$$0.334 = \frac{\varepsilon_1(0.5\varepsilon_2)^{0.5}(0.8)^{0.5}}{(0.35-\varepsilon_1)(1+0.5\varepsilon_1)^{0.5}}; \qquad \varepsilon_1 = 0.194$$

$$30.2 = \frac{4\varepsilon_2^2(0.097+1.5\varepsilon_2)^{1.5}(0.8)^{0.5}}{(0.156-2\varepsilon_2)^2(0.15-\varepsilon_2)(1.097+0.5\varepsilon_2)^{0.5}}; \qquad \varepsilon_2 = 0.066$$

$$0.334 = \frac{(0.194+\varepsilon_1)(0.196+0.5\varepsilon_1)^{0.5}(0.8)^{0.5}}{(0.023-\varepsilon_1)(1.129+0.5\varepsilon_1)^{0.5}}; \qquad \varepsilon_1 = -0.086$$

$$30.2 = \frac{(0.132+2\varepsilon_2)^2(0.153+1.5\varepsilon_2)^{1.5}(0.8)^{0.5}}{(0.108-2\varepsilon_2)^2(0.084-\varepsilon_2)(1.085+0.5\varepsilon_2)^{0.5}}; \qquad \varepsilon_2 = 0.033$$

The following table is the solution of Meissner et al.

Cycle		*1*		*2*		*3*		*4*		*5*		*6*		
Reactor		*1*	*2*	*3*	*4*	*5*	*6*	*7*	*8*	*9*	*10*	*11*	*12*	*12*
Equation		1	2	1	2	1	2	1	2	1	2	1	2	2
ε_1		0.194	...	−0.086	...	−0.035	...	−0.009	...	−0.007	...	−0.002	...	
ε_2		...	0.066	...	0.033	...	0.012	...	0.00	...	0.001	...	0.001	
	Feed, Moles	*Moles*	*Moles*	*Moles*	*Moles*	*Moles*	*Moles*	*Moles*	*Moles*	*Moles*	*Moles*	*Moles*	*Moles*	*Partial Pressure, Atm.*
H_2S	0.35	0.156	0.023	0.108	0.044	0.080	0.056	0.069	0.061	0.065	0.063	0.064	0.063	0.0465
H_2	0	0.194	0.194	0.108	0.108	0.073	0.073	0.064	0.064	0.057	0.057	0.055	0.055	0.0406
S_2	0	0.097	0.196	0.153	0.202	0.184	0.202	0.195	0.201	0.199	0.201	0.200	0.201	0.1483
SO_2	0.15	0.15	0.084	0.084	0.051	0.051	0.039	0.039	0.036	0.036	0.035	0.035	0.034	0.0251
H_2O	0	0	0.132	0.132	0.197	0.197	0.221	0.221	0.228	0.228	0.230	0.230	0.231	0.1706
N_2	0.50	0.50	0.50	0.50	0.50	0.50	0.50	0.50	0.50	0.50	0.50	0.500	0.50	0.3690
Total		1.097	1.129	1.085	1.102	1.085	1.091	1.088	1.090	1.085	1.086	1.084	1.084	0.8001

Example 10.15. Equilibrium in the Dissociation of Isopropanol by the Relaxation Method

The reaction will be checked at 1 atm and 400 K and 800 K. Species assumed to be present are isopropanol, n-propanol, acetone, propionaldehyde, and hydrogen, and are numbered in this order. There are three independent reactions:

$$\underset{(1)}{i\ C_3H_7\ OH} \rightleftarrows \underset{(3)}{n\ C_3H_7OH}, \tag{1}$$

$$\underset{(1)}{i\ C_3H_7OH} \rightleftarrows \underset{(4)}{(CH_3)_2CO} + \underset{(2)}{H_2}, \tag{2}$$

$$\underset{(1)}{i\ C_3H_7OH} \rightleftarrows \underset{(5)}{C_2H_5CHO} + \underset{(2)}{H_2}. \tag{3}$$

Equilibrium constants from the accompanying table are,

T, °K	K_1	K_2	K_3
400	0.0640	0.0760	0.00012
800	1.9055	0.0025	0.0301

The equilibrium equations are

$$\frac{n_{30}+\varepsilon_1}{n_{10}-\varepsilon_1} = K_1, \tag{4}$$

$$\frac{(n_{40}+\varepsilon_2)(n_{20}+\varepsilon_2)\,P}{(n_{10}-\varepsilon_2)\,(\Sigma\, n_{i0}+\varepsilon_2)} = K_2, \tag{5}$$

(continued on next page)

Example 10.15. *(continued)*

$$\frac{(n_{50} + \varepsilon_3)(n_{20} + \varepsilon_3) P}{(n_{10} - \varepsilon_3)(\Sigma n_{i0} + \varepsilon_3)} = K_3. \qquad (6)$$

The case worked out is for 1 mol of pure isopropanol initially. At 400 K three cycles are sufficient to make the degree of advancement of a stage less than 10^{-4}. At 800 K six cycles are needed for the same performance, but that work is not shown here. The formulation of the isopropanol dissociation is not entirely realistic because some possibilities have been neglected—for instance, the formation of propylene and subsequent hydrogenation to propane. However, the intent of the problem has been to illustrate the method in a simple way. Thermodynamic data are taken from Stull (1969).

	Cycle	*1*			*2*			*3*		
	Reaction	*1*	*2*	*3*	*1*	*2*	*3*	*1*	*2*	*3*
1	*i* propanol	1	0.9398	0.6843	0.6837	0.6992	0.6968	0.6968	0.6969	0.6969
2	hydrogen	0	0	0.2555	0.2561	0.2561	0.2585	0.2585	0.2585	0.2585
3	*n* propanol	0	0.0602	0.0602	0.0602	0.0447	0.0447	0.0446	0.0446	0.0446
4	acetone	0	0	0.2555	0.2555	0.2555	0.2579	0.2579	0.2579	0.2579
5	propionald	0	0	0	0.0006	0.0006	0.0006	0.0006	0.0006	0.0006
	n_t	1.00	1.00	1.2555	1.2561	1.2561	1.2585	1.2585	1.2585	1.2585
	ε_1	0.0602			−0.0155			−0.0001		
	ε_2		0.2555			0.0024			$2(10^{-6})$	
	ε_3			0.0006			$3(10^{-5})$			$3(10^{-5})$

No. 519 Propyl Alcohol C_3H_3O (Ideal Gas State)

	cal/(mole°K)			kcal/mole			
T, °K	*Cp°*	*S°*	*$(G°-H°_{298})/T$*	*$H°-H°_{298}$*	*ΔHf°*	*ΔGf°*	*Log Kp*
298	20.82	77.63	77.63	0.00	−61.55	−38.95	28.550
300	20.91	77.76	77.64	0.04	−61.58	−38.81	28.273
400	25.86	84.46	78.52	2.38	−63.11	−30.98	16.926
500	30.51	90.75	80.34	5.21	−64.40	−22.79	9.963
600	34.56	96.68	82.57	8.47	−65.46	−14.37	9.235
700	38.03	102.27	84.99	12.10	−66.29	−5.79	1.807
800	41.04	107.55	87.48	16.06	−66.94	2.90	−0.791
900	43.65	112.54	89.99	20.29	−67.41	11.66	−2.831
1000	45.93	117.26	92.49	24.78	−67.73	20.47	−4.474

No. 520 Isopropyl Alcohol C_3H_8O (Ideal Gas State)

	cal/(mole°K)			kcal/mole			
T, °K	*Cp°*	*S°*	*$-(G°-H°_{298})/T$*	*$H°-H°_{298}$*	*ΔHf°*	*ΔGf°*	*Log Kp*
298	21.21	74.07	74.07	0.00	−65.15	−41.49	30.411
300	21.31	74.21	74.08	0.04	−65.18	−41.34	30.118
400	26.78	81.09	74.98	2.45	−66.65	−33.17	18.120
500	31.89	87.64	76.86	5.40	−67.82	−24.66	10.776
600	35.76	93.81	79.18	8.78	−68.74	−15.94	5.804
700	39.21	99.58	81.68	12.53	−69.46	−7.07	2.208
800	42.13	105.01	84.26	16.61	−69.99	1.87	−0.511
900	44.63	110.12	86.86	20.95	−70.36	10.88	−2.642
1000	46.82	114.94	89.43	25.52	−70.58	19.93	−4.355

Example 10.15. (*continued*)

No. 546 Propionaldehyde C_3H_6O (Ideal Gas State)

	cal/(mole°K)			kcal/mole			
T, °A	Cp°	S°	$-(G°-H°_{298})/T$	$H°-H°_{298}$	ΔHf°	ΔGf°	Log Kp
298	18.80	72.83	72.83	0.00	−45.90	−31.18	22.851
300	18.87	72.95	72.84	0.04	−45.92	−31.09	22.644
400	23.09	78.97	73.63	2.14	−47.00	−25.97	14.188
500	26.89	84.54	75.26	4.64	−47.91	−20.60	9.005
600	30.22	89.74	77.24	7.50	−48.66	−15.07	5.489
700	33.03	94.62	79.38	10.67	−49.26	−9.42	2.941
800	35.45	99.19	81.57	14.10	−49.73	−3.70	1.011
900	37.55	103.49	83.77	17.75	−50.08	2.08	−0.505
1000	39.27	107.54	85.95	21.59	−50.32	7.89	−1.725

No. 554 Acetone C_3H_6O (Ideal Gas State)

	cal/(mole°K)			kcal/mole			
T, °A	Cp°	S°	$-(G°-H°_{298})/T$	$H°-H°_{298}$	ΔHf°	ΔGf°	Log Kp
298	17.90	70.49	70.49	0.00	−52.00	−36.58	26.811
300	17.97	70.61	70.50	0.04	−52.02	−36.48	26.577
400	22.00	76.33	71.25	2.04	−53.20	−31.12	17.001
500	25.89	81.67	72.80	4.44	−54.22	−25.48	11.135
600	29.34	86.70	74.70	7.20	−55.07	−19.65	7.156
700	32.34	91.45	76.76	10.29	−55.74	−13.69	4.273
800	34.93	95.94	78.88	13.65	−56.28	−7.64	2.088
900	37.19	100.19	81.01	17.26	−56.67	−1.54	0.373
1000	39.15	104.21	83.13	21.08	−56.93	4.61	−1.007

of the total amounts of individual chemical elements making up the chemical species present. Several ways of incorporating the constraints into the body of the problem are possible, but one of the more convenient ones is the method of Lagrange Multipliers, which will be used here.

Lagrange multipliers: These are factors in a general technique for finding the minimax of a function of n variables,

$$f(x_1, x_2, \ldots, x_n) \rightarrow \text{minimax}, \tag{10.86}$$

when there are m additional relations, with $m < n$, between the variables,

$$\phi_k(x_1, x_2, \ldots, x_n) = 0 \qquad k = 1, 2, \ldots, m, \tag{10.87}$$

which are called *constraints* or *conditions.* According to this principle, minimaxing the composite function,

$$\psi = f + \Sigma\ \lambda_k\phi_k \rightarrow \text{minimax}, \tag{10.88}$$

is equivalent to doing the same with the original function, f. The Lagrange multipliers, λ_k, are constants determined by the process. The derivatives equated to zero,

$$\left(\frac{\partial\psi}{\partial x_i}\right) = 0 \qquad i = 1, 2, \ldots, n, \tag{10.89}$$

together with the constraints

$$\phi_k = 0 \qquad k = 1, 2, \ldots, m, \tag{10.90}$$

constitute a system of $n + m$ equations from which the x_i and λ_k at minimax are evaluated.

Alternatively, when the process is algebraically feasible, m of the variables can be substituted out to make f a function of $n-m$ variables, which can be minimaxed in the usual fashion. Often, however, this is a more laborious procedure and is course not possible when the ϕ_k are at all mathematically complex. Formal proof of the method of Lagrange multipliers may be found in standard mathematical books—for example, Smirnov (1964) (*Course in Higher Mathematics,* Vol. 1). Example 10.16 deals with several cases of minimaxing with Lagrange multipliers and shows the results to be the same as those obtained by perhaps more familiar methods when those are feasible.

The basic equations: The total Gibbs energy of a mixture is made up of contributions of the pure components and of mixing effects,

$$nG = \Sigma\ n_i\bar{G}_i = \Sigma\ n_i(G_i^0 + \bar{G}_i - G_i^0)$$
$$= \Sigma\ n_i(G_i^0 + RT\ln a_i), \tag{10.91}$$

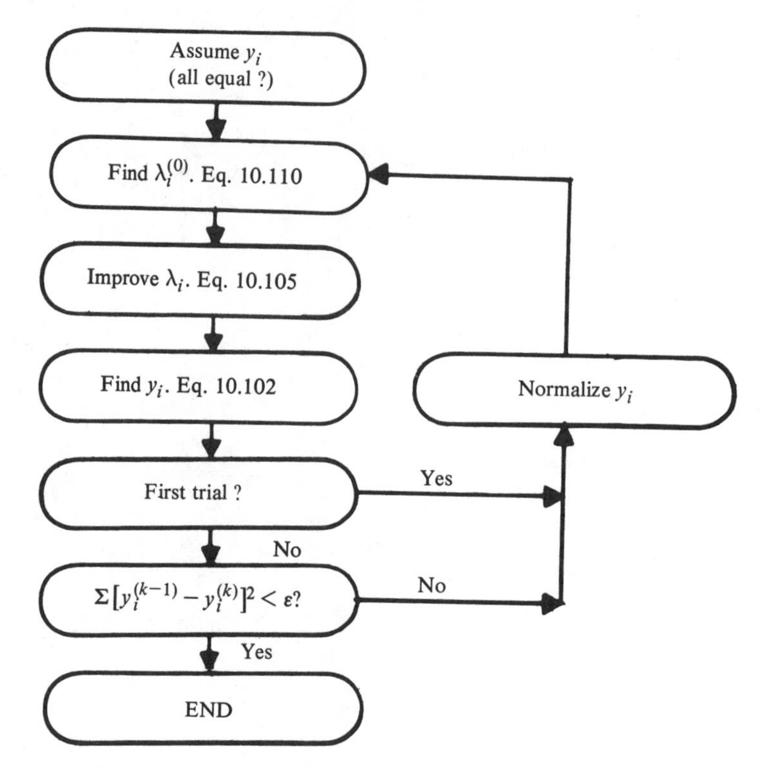

Figure 10.8. Equilibrium by minimization of Gibbs energy.

or, more compactly,

$$g = nG/RT = \Sigma\ n_i(g_i^0 + \ln a_i). \tag{10.92}$$

The activities of liquids are expressed in terms of activity coefficients,

$$a_i = \hat{f}_i/f_i^0 = \gamma_i x_i = \gamma_i n_i/n, \tag{10.93}$$

with pure liquid at 1 atm as the standard state, and those of gases in terms of fugacity coefficients,

$$a_i = \hat{f}_i/f_i^0 = \hat{\phi}_i y_i P = \hat{\phi}_i P n_i/n, \tag{10.94}$$

with unit fugacity as the standard state.

Since activity and fugacity coefficients depend on the compositions of the phases, which are unknown in advance, solutions of problems involving nonideal mixtures are started by assuming ideality and thus finding as estimate of the composition which is in turn used to find estimates of the activity and fugacity coefficients. The process is repeated as many times as necessary to achieve a desired accuracy. Since the main patterns of solving ideal and nonideal individual cases are the same, for simplicity the ensuing discussion will be phrased only in terms of ideal gases.

Along with reading the following description of the method, Example 10.17 should be studied. In the ideal case the Gibbs energy function is

$$g = \Sigma\ n_i[g_i^0 + \ln(n_i P/n)] = \Sigma\ n_i[c_i + \ln(n_i/n)], \tag{10.95}$$

where

$$c_i = g_i^0 + \ln P. \tag{10.96}$$

The derivative will be needed,

$$\frac{\partial g}{\partial n_i} = c_i + \ln(n_i/n) + \Sigma\, n_i\left(\frac{1}{n_i} - \frac{1}{n}\right) = c_i + \ln(n_i/n). \tag{10.97}$$

The material balances on the m chemical elements are

$$b_j - \Sigma\ a_{ij}n_i = 0 \qquad j = 1, 2, \ldots, m, \tag{10.98}$$

for m elements and c components, where a_{ij} is the number of atoms of element j in molecule i, and b_j is the total amount of element j in the mixture. In terms of Lagrange multipliers, the constrained function to be minimized is

$$f = g + \sum_{j=1}^{m} \lambda_j(b_j - \sum_{i=1}^{c} a_{ij}n_i) \tag{10.99}$$

$$= \sum_{i=1}^{c} n_i[c_i + \ln(n_i/n)] + \sum_{j=1}^{m} \lambda_j(b_j - \sum_{i=1}^{c} a_{ij}n_i). \tag{10.100}$$

The derivatives with respect to compositions are,

$$\partial f/\partial n_i = c_i + \ln(n_i/n) - \sum_{k=1}^{m} a_{ik}\lambda_k = 0 \qquad i = 1, 2, \ldots, c. \tag{10.101}$$

These derivatives together with the material balances, Eq. 10.98, and $\Sigma\ n_i = n$, constitute a system of $c + m + 1$ equations from which the n_i, λ_k and n can be evaluated, in principle.

Solving the equations: Any procedure for solving a system of nonlinear equations numerically requires starting estimates of the (in this case) $c + m + 1$ unknowns. However, special characteristics of these equations allow considerable simplification of general procedures so that fewer than $c + m + 1$ estimates need be made in advance, as will be explained.

The derivatives, Eq. 10.101, are solved for the mol fractions as,

$$y_i = n_i/n = \exp\left[\sum_{k=1}^{m} a_{ik}\lambda_k - c_i\right]. \qquad (10.102)$$

On dividing the element balances, Eq. 10.98, by the total number of mols and then substituting the preceding values of y_i into them, the result is

$$\sum_{i=1}^{c} a_{ij} \exp\left[\sum_{k=1}^{m} a_{ik}\lambda_k - c_i\right] - \frac{b_j}{n} = 0. \qquad (10.103)$$

The variable, n, can be eliminated from this equation at each value of i by first setting $j = 1$ in Eq. 10.103; thus,

$$\frac{1}{n} = \frac{1}{b_1} \sum_{i=1}^{c} a_{i1} \exp[a_{ik}\lambda_k - c_i], \qquad (10.104)$$

and substituting these results into Eq. 10.103, so that

$$\sum_{i=1}^{c} q_{ij} \exp\left[\sum_{k=1}^{m} a_{ik}\lambda_k - c_i\right] = d_j \qquad j = 1, 2, \ldots, m, \qquad (10.105)$$

where

$$q_{i1} = 1, d_1 = 1, \qquad (10.106)$$

$$q_{ij} = a_{ij} - a_{i1}b_j/b_1, d_j = 0 \qquad j = 2, 3, \ldots, m. \qquad (10.107)$$

Eq. 10.105 is a set of m nonlinear equations that is to be solved numerically for the m unknown λ_k.

Finding the multipliers: The Newton-Raphson method will be applied to finding the λ_k. Starting estimates of these quantities are obtained by the following procedure:

1. Estimate starting values of the mol fractions, y_i^0; if no better choice is justifiable, all mol fractions may be taken equal.
2. Expand $\ln y_i$ in a truncated Taylor series,

$$\ln y_i = \ln y_i^0 + (y_i - y_i^0)/y_i^0. \qquad (10.108)$$

3. Isolate the y_i and replace $\ln y_i$ from Eq. 10.102,

$$y_i = y_i^0(1 - \ln y_i^0 + \ln y_i) \qquad (10.109)$$

$$= y_i^0(1 - \ln y_i^0 + \sum_{k=1}^{m} a_{ik}\lambda_k^0 - c_i) \quad i = 1, 2, \ldots, c.$$

4. Substitute these mol fractions into Eq. 10.105, with the result,

$$\sum_{i=1}^{c} q_{ij} y_i^0[1 - \ln y_i^0 + \sum_{k=1}^{m} a_{ik}\lambda_k^0 - c_i] = d_j$$

$$j = 1, 2, \ldots, m, \qquad (10.110)$$

where $d_1 = 1$ and $d_j = 0$ when $j \neq 1$.

5. Solve the set of linear relations, Eq. 10.110, for the m unknown starting values, λ_k^0.

Improved values of the λ_k then are found from Eq. 10.105 with the N-R procedure and subsequently improved values of the mol fractions are found from Eq. 10.102. The alternating sequence of determinations of the λ_k and the y_i is continued until satisfactory convergence is achieved. For evaluating convergence, the mass balance relations Eq. 10.98 may be used in the form

$$D = \sum_{j=1}^{m} \left|1 - \sum_{i=1}^{c} a_{ij}n_i/b_j\right| < \varepsilon. \qquad (10.111)$$

Alternatively, the difference in successive evaluations of the total number of mols

$$|n^{(t+1)} - n^{(t)}| < \varepsilon. \qquad (10.112)$$

The overall number of mols may be found at each stage of the calculation from

$$n = \sum_{j=1}^{m} b_j / \sum_{j=1}^{m} \sum_{i=1}^{c} a_{ij} y_i. \qquad (10.113)$$

An outline of this method of finding equilibrium compositions is given as Figure 10.8.

10.6.4. Comparison of the Three Methods

The need for solving simultaneous nonlinear equations is encountered frequently, and a considerable variety of techniques has been applied to these problems, with the objectives of assuring or accelerating convergence of successive trials. Beyond the basics touched on in the present treatment, various degrees of sophistication are covered in the references cited at the beginning of Section 10.6 and in their bibliographies. It is clear that for most problems with more than three reactions or more than six to eight substances, computer solution is desirable.

1. Relaxation (series reactor) method: In this method a complete stoichiometry must be specified, but no estimate of the equilibrium composition is required. It is a relatively straightforward procedure, that evaluates one unknown at a time. A calculator or small computer is quite adequate when the number of stoichiometries is small, say six or less. Results are comparable to those with other methods, but no information is available on the relative amounts of computer time needed by the various procedures.

2. Simultaneous stoichiometric equations (Brinkley) method: The several variants of this method require the solution of a number of simultaneous equations equal to the number of chemical species plus 1. Direct iteration, Newton-Raphson, and various optimization methods have been applied. The speed and even the possibility of convergence often are strongly dependent on the initial estimates, which must be consistent with the material balances on the chemical elements. The obvious method of making all components zero except three or four, which can be made to fit the material balance by inspection, is not always satisfactory. A method utilizing random number generation of degrees of advancement of all the reactions is described qualitatively by Holub & Vonka (1976, p. 123).

3. Gibbs energy minimization: The particular method with Lagrange multipliers described here reduces the number of equations to be solved simultaneously to the rank of the composition matrix, which is at most equal to the number of chemical elements present, which in turn is rarely more than five or six. The method does not require stoichiometric analysis, simply a list of the chemical species that are expected to be present at equilibrium and their thermodynamic data, and components may be added very simply

Example 10.16. Cases of Minimax with the Aid of Lagrange Multipliers

a. Maximize $f = xyz$ with the constraint $\phi = x + y + z - 1 = 0$.

$$\psi = f + \lambda\phi,$$

$$\frac{\partial\psi}{\partial x} = yz + \lambda = 0,$$

$$\frac{\partial\psi}{\partial y} = xz + \lambda = 0,$$

$$\frac{\partial\psi}{\partial z} = xy + \lambda = 0.$$

Therefore,

$$-\lambda = yz = xz = xy,$$

and consequently,

$$x = y = z = \frac{1}{3} \text{ and } \lambda = -\frac{1}{9}.$$

Alternatively, eliminate $z = 1 - x - y$,

$$f = xy(1 - x - y) \rightarrow \text{maximum.}$$

The two derivatives of f are zero at minimax; thus,

$$\frac{\partial f}{\partial x} = y(1 - 2x + y) = 0,$$

$$\frac{\partial f}{xy} = x(1 - 2y - x) = 0,$$

whence $x = y = z = 1/3$, as before.

b. Maximize $f = x^2 + y^2 + z^2$ with two constraints,

$$\phi_1 = \frac{x^2}{4} + \frac{y^2}{6} + \frac{z^2}{8} - 1 = 0,$$

$$\phi_2 = x + y + z - 1 = 0.$$

The Lagrange function is

$$\psi = f + \lambda_1\phi_1 + \lambda_2\phi_2 \rightarrow \text{Maximum.}$$

The derivatives are

$$\frac{\partial\psi}{\partial x} = 2x + \lambda_1 x/2 + \lambda_2 = 0,$$

$$\frac{\partial\psi}{\partial y} = 2y + \lambda_1 y/3 + \lambda_2 = 0,$$

$$\frac{\partial\psi}{\partial z} = 2z + \lambda_1 z/4 + \lambda_2 = 0.$$

The last three equations can be solved for y and z in terms of λ_1 and x:

$$y = x(2 + \lambda_1/2)/(2 + \lambda_1/3),$$

$$z = x(2 + \lambda_1/2)/(2 + \lambda_1/4).$$

On substituting into the equation for ϕ_2,

$$x\left[1 + \frac{2 + \lambda_1/2}{2 + \lambda_1/3} + \frac{2 + \lambda_1/2}{2 + \lambda_1/4}\right] = 1,$$

and by using the equation for ϕ_1,

$$x^2\left[\frac{1}{4} + \frac{1}{6}\left(\frac{2 + \lambda_1/2}{2 + \lambda_1/3}\right)^2 + \frac{1}{8}\left(\frac{2 + \lambda_1/2}{2 + \lambda_1/4}\right)^2\right] = 1.$$

The solution by trial gives the results

$$\lambda_1 = -4.333,$$

$$\lambda_2 = -0.321,$$

$$x = 1.9277,$$

$$y = -0.5766,$$

$$z = -0.3501.$$

Elimination of y and z, for example, to obtain a function of x alone, and subsequent differentiation of f to find the extremum is a much more cumbersome process, but should give the same result ultimately.

c. Minimize the Gibbs energy of the reaction $C_2H_6 \rightleftarrows C_2H_4 + H_2$.

The carbon and hydrogen balances are

$$\phi_1 = n_1 + n_2 - 1 = 0,$$

$$\phi_2 = 3n_1 + 2n_2 + n_3 - 3 = 0.$$

At atmospheric pressure, assuming ideal gases, the Gibbs energy function is

$$g = \frac{nG}{RT} = \Sigma\ n_i[g_i^0 + \ln(n_i/n)].$$

Accordingly, the function to be minimized is

$$\psi = \Sigma\ n_i[g_i^0 + \ln(n_i/n)] + \lambda_1\phi_1 + \lambda_2\phi_2.$$

The derivatives are:

$$\frac{\partial g}{\partial n_i} = g_i^0 + \ln(n_i/n) + \Sigma\, n_i\left(\frac{1}{n_i} - \frac{1}{n}\right) = g_i^0 + \ln(n_i/n),$$

$$\frac{\partial\psi}{\partial n_1} = g_1^0 + \ln(n_1/n) + \lambda_1 + 3\lambda_2 = 0,$$

$$\frac{\partial\psi}{\partial n_2} = g_2^0 + \ln(n_2/n) + \lambda_1 + 2\lambda_2 = 0,$$

$$\frac{\partial\psi}{\partial n_3} = g_3^0 + \ln(n_3/n) + \lambda_2 = 0.$$

Upon eliminating λ_1 and λ_2 from the last three equations, the result is

$$g_2^0 + g_3^0 - g_1^0 = \frac{\Delta G^0}{RT} = -\ln\frac{n_2 n_3}{n_1 n},$$

which is clearly correct. Alternatively, the mol numbers all can be expressed in terms of a degree of advancement, leaving an equation for the Gibbs energy in only one variable, whose minimum can be found simply. Thus,

$$g = (1 - \varepsilon)\left(g_1^0 + \ln\frac{1-\varepsilon}{1+\varepsilon}\right) + \varepsilon\left(g_2^0 + g_3^0 + 2\ln\frac{\varepsilon}{1+\varepsilon}\right).$$

Minimization of this function gives the same result as before.

without overburdening the computation unduly. The method can be readily formulated to take any two of the variables T, P, V, H, or S as fixed, although most of the literature is concerned with fixed T and P. Gibbs energy minimization is amenable to solution by various optimization methods such as maximum descent or linear or convex programming, and much literature is devoted to these topics. The strong arguments in favor of this method over the Brinkley types is the smaller number of simultaneous nonlinear equations to be solved, equal to the number of elements as compared with the number of chemical species, and the fact that the solution can be started with quite arbitrary choice of mol fractions.

Examples of numerical solutions in the literature:

1. Oliver, Stephanou, & Baier (1962) (*Chem Eng* 121–128, Feb 19). $CH_4 + H_2O$.
2. Kandiner & Brinkley (1950) (*IEC 42* 850–855, 1526). $C_3H_8 + 5\ O_2 + 20\ N_2$.
3. White, Johnson, & Dantzig (1958) (*J. Chem Phys 28* 751–755). Hydrazine (N_2H_4) + O_2.
4. Meissner, Kusik, & Dalzell (1969) (*IEC Fund 8* 659–665). $SO_2 + H_2S$, $CH_4 + H_2O$, $N_2H_4 + O_2$.
5. Balzhizer, Samuels, & Eliassen (1972) (*Chemical Engineering Thermodynamics,* 513–527). $C_2H_6 + H_2O$.
6. Zeleznik & Gordon (1968) (*IEC 60* 27–57). $C_2H_8N_2 + 2\ ClO_3F$, $H_2 + 0.5\ O_2$.
7. Holub & Vonka (1976) (*Chemical Equilibrium of Gaseous Systems*). $C_3H_8 + 5\ O_2 + 20\ N_2$, $CH_4 + H_2O$, $CH_3OH + NH_3$, $CH_4 + O_2$.
8. Smith & Missen (1982) (*Chemical Reaction Equilibrium Analysis*). A variety of systems with BASIC, FORTRAN, and HP-41C programs.

10.7. COMBINED CHEMICAL AND PHASE EQUILIBRIA

Reactions of substances distributed between several phases are not uncommon. Some industrial cases are: aqueous acids plus organic liquids, hydrogenation of organic liquids, saponification with aqueous alkalis, hydration of ethylene and propylene, biodegradation of organics in water, and combustion of liquid and solid fuels. When the condensed phases are pure, their activities may be taken constant or subject to a small correction for pressure; equilibrium computation then is quite straightforward. In the usual situation, however, the several phases are mixtures. The cases to be considered here are of two liquid phases and of a vapor phase with one liquid phase. Extension to systems with more than two phases is direct in principle but more complex computationally.

The general pattern of solving multiphase chemical equilibria is to formulate the chemical equilibrium in one phase and then to relate the compositions of the phases by phase-equilibrium relations. In the case of vapor-liquid reactions, it is preferable to formulate the chemical equilibrium in the vapor phase and to relate the compositions of the phases with a flash equation. For two liquid phases, it is somewhat better to formulate the chemical equilibrium in the phase that has the greater concentrations of reactants, if that can be estimated in advance. For instance, in cases like the nitration of aromatic liquids with aqueous acid, the reaction is known to occur primarily in the organic phase. In all cases of multiphase reaction equilibria, the relative amounts of the two phases is an unknown to be evaluated.

10.7.1. Two Liquid Phases

Chemical equilibrium is represented by

$$K_{ce} = \prod_P (a_i)^{\nu_i} / \prod_R (a_i)^{\nu_i} = K_\gamma K_x, \tag{10.114}$$

and phase equilibrium by

$$x_i^* = K_i x_i, \tag{10.115}$$

where

$$K_i = \gamma_i / \gamma_i^* \tag{10.116}$$

is the distribution ratio and the asterisk designates the second phase. Chemical equilibrium is formulated for the primary phase. Since the activity coefficients and distribution ratios in general depend on the compositions of the phases, several trials may be needed in which successive improvements in the evaluation of the phase compositions are made. The phase split is written in terms of the total number of mols in each phase,

$$\beta = \frac{n_t}{n_t + n_t^*}. \tag{10.117}$$

The degree of advancement is

$$\varepsilon = \frac{n_i + n_i^* - n_{io}}{\pm \nu_i} \quad \begin{cases} + \text{ for products,} \\ - \text{ for reactants.} \end{cases} \tag{10.118}$$

The amount of any component in both phases is

$$n_i + n_i^* = n_{io} \pm \varepsilon \nu_i \quad \begin{cases} + \text{ for products,} \\ - \text{ for reactants,} \end{cases} \tag{10.119}$$

and the total of both phases is

$$n_t + n_t^* = n_{to} + \varepsilon \Delta \nu. \tag{10.120}$$

The overall mol fractions are

$$z_i = \frac{n_i + n_i^*}{n_t + n_t^*} = \frac{n_{io} \pm \varepsilon \nu_i}{n_{to} + \varepsilon \Delta \nu}. \tag{10.121}$$

The material balance between phases is

$$z_i = \beta x_i + (1 - \beta) x_i^* = [\beta + K_i(1 - \beta)] x_i, \tag{10.122}$$

where β is the fraction of the total material that is present in the primary phase. Solving for the x_i and summing gives the phase split equation,

$$\Sigma\, x_i = \Sigma \frac{z_i}{\beta + K_i(1 - \beta)} = 1, \tag{10.123}$$

or the standard form,

$$\psi(\beta, \varepsilon) = -1 + \Sigma \frac{n_{io} \pm \varepsilon \nu_i}{(n_{to} + \varepsilon \Delta \nu)[\beta + K_i(1 - \beta)]} = 0. \tag{10.124}$$

When the chemical equilibrium constant and the equations relating the activity coefficients to the composition are known, the remaining problem is to find the equilibrium conversion and the amounts and compositions of the two phases. A suitable procedure with this objective is block-diagramed in Figure 10.9 and is described in the following steps.

Example 10.17. Minimization of the Gibbs Energy of the Reaction between 1 mol CH_4 and 5 mols H_2O at 900 K and 2 atm.

The five substances assumed to be present at equilibrium and their properties are:

No.	Substance	G_f^0	C	H	O	$c = \frac{G^0}{RT} + \ln P$
1	CH_4	1.99	1	4	0	1.806
2	H_2O	−47.36	0	2	1	−25.790
3	CO_2	−94.58	1	0	2	−52.195
4	CO	−50.61	1	0	1	−27.607
5	H_2	0	0	2	0	0.693

Material balances on the elements are:

C: $\phi_1 = 1 - n_1 - n_3 - n_4 = 0,$

H: $\phi_2 = 14 - 4n_1 - 2n_2 - 2n_5 = 0,$

O: $\phi_3 = 5 - n_2 - 2n_3 - n_4 = 0,$

$$n = n_1 + n_2 + n_3 + n_4 + n_5.$$

The function to be minimized is

$$f = \sum_1^5 n_i(c_i + \ln(n_i/n)) + \sum_1^3 \lambda_k \phi_k \rightarrow \text{Minimum.}$$

The derivatives are:

$$\frac{\partial f}{\partial n_1} = c_1 + \ln(n_1/n) + \lambda_1 + 4\lambda_2 = 0,$$

$$\frac{\partial f}{\partial n_2} = c_2 + \ln(n_2/n) + 2\lambda_2 + \lambda_3 = 0,$$

$$\frac{\partial f}{\partial n_3} = c_3 + \ln(n_3/n) + \lambda_1 + 2\lambda_3 = 0,$$

$$\frac{\partial f}{\partial n_4} = c_4 + \ln(n_4/n) + \lambda_1 + \lambda_3 = 0,$$

$$\frac{\partial f}{n_5} = c_5 + \ln(n_5/n) + 2\lambda_2 = 0.$$

These nine equations may be solved simultaneously for the six n_i and three λ_k by the N-R or equivalent method. This procedure requires that initial estimates be made of all nine unknowns. However, the method described in the text requires estimates only of the five mol fractions which can be taken equal without killing the calculation. Eq. 10.102 to 10.110 will be set up for this case.

The various coefficients are:

i \ j	a_{ij} 1	2	3	q_{ij} 1	2	3	c_i
1	1	4	0	1	−10	−5	1.806
2	0	2	1	1	2	1	−25.770
3	1	0	2	1	−14	−3	−52.195
4	1	0	1	1	−14	−4	−27.607
5	0	2	0	1	2	0	0.693
b_j	1	14	5				
d_j	1	0	0				

Substitute Eq. 10.103,

$$\sum_1^3 a_{ik}\lambda_k = a_{i1}\lambda_1 + a_{i2}\lambda_2 + a_{i3}\lambda_3,$$

into Eq. 10.110 and write out the latter for each value of j.

$j = 1$:

$$q_{11}y_1(1 - \ln y_1 + a_{11}\lambda_1 + a_{12}\lambda_2 + a_{13}\lambda_3 - c_1)$$
$$+ q_{21}y_2(1 - \ln y_2 + a_{21}\lambda_1 + a_{22}\lambda_2 + a_{23}\lambda_3 - c_2)$$
$$+ q_{31}y_3(1 - \ln y_3 + a_{31}\lambda_1 + a_{32}\lambda_2 + a_{33}\lambda_3 - c_3)$$
$$+ q_{41}y_4(1 - \ln y_4 + a_{41}\lambda_1 + a_{42}\lambda_2 + a_{43}\lambda_3 - c_4)$$
$$+ q_{51}y_5(1 - \ln y_5 + a_{51}\lambda_1 + a_{52}\lambda_2 + a_{53}\lambda_3 - c_5) = 1.$$

$j = 2$:

$$q_{12}y_1(1 - \ln y_1 + a_{11}\lambda_1 + a_{12}\lambda_2 + a_{13}\lambda_3 - c_1)$$
$$+ q_{22}y_2(1 - \ln y_2 + a_{21}\lambda_1 + a_{22}\lambda_2 + a_{23}\lambda_3 - c_2)$$
$$+ q_{32}y_3(1 - \ln y_3 + a_{31}\lambda_1 + a_{32}\lambda_2 + a_{33}\lambda_3 - c_3)$$
$$+ q_{42}y_4(1 - \ln y_4 + a_{41}\lambda_1 + a_{42}\lambda_2 + a_{43}\lambda_3 - c_4)$$
$$+ q_{52}y_5(1 - \ln y_5 + a_{51}\lambda_1 + a_{52}\lambda_2 + a_{53}\lambda_3 - c_5) = 0.$$

$j = 3$:

$$q_{13}y_1(1 - \ln y_1 + a_{11}\lambda_1 + a_{12}\lambda_2 + a_{13}\lambda_3 - c_1)$$
$$+ q_{23}y_2(1 - \ln y_2 + a_{21}\lambda_1 + a_{22}\lambda_2 + a_{23}\lambda_3 - c_2)$$
$$+ q_{33}y_3(1 - \ln y_3 + a_{31}\lambda_1 + a_{32}\lambda_2 + a_{33}\lambda_3 - c_3)$$
$$+ q_{43}y_4(1 - \ln y_4 + a_{41}\lambda_1 + a_{42}\lambda_2 + a_{43}\lambda_3 - c_4)$$
$$+ q_{53}y_5(1 - \ln y_5 + a_{51}\lambda_1 + a_{52}\lambda_2 + a_{53}\lambda_3 - c_5) = 0.$$

Assume the starting mol fractions to be equal, $y_i = 0.2$. Substitute these values and the values of a_{ij}, q_{ij}, and c_i from the table of coefficients into the preceding equations. The results simplify to

$$0.6\lambda_1 + 1.6\lambda_2 + 0.8\lambda_3 = -22.2204,$$
$$-7.6\lambda_1 - 6.4\lambda_2 - 8\lambda_3 = 227.5390,$$
$$-2.4\lambda_1 - 3.6\lambda_2 - 1.8\lambda_3 = 52.1794.$$

The solution of these three linear equations gives the starting values of the Lagrange multipliers, which are

$$\lambda_1^{(0)} = -2.0631 \qquad \lambda_2^{(0)} = 0.2039 \qquad \lambda_3^{(0)} = -26.6455$$

Example 10.17. *(continued)*

With these as starting values, the N-R method is used to find improved values with Eq. 10.105,

$$\sum_{i=1}^{c} q_{ij}\exp\left(\sum_{k=1}^{m} a_{ik}\lambda_k - c_i\right) = d_j; \qquad j = 1, 2, 3.$$

These equations are written out for each value of j.

$j = 1$:

$$\exp(\lambda_1 + 4\lambda_2 - c_1) + \exp(2\lambda_2 + \lambda_3 - c_2)$$
$$+ \exp(\lambda_1 + 2\lambda_3 - c_3) + \exp(\lambda_1 + \lambda_3 - c_4)$$
$$+ \exp(2\lambda_2 - c_5) = 1.$$

$j = 2$:

$$-10\exp(\lambda_1 + 4\lambda_2 - c_1) + 2\exp(2\lambda_2 + \lambda_3 - c_2)$$
$$- 14\exp(\lambda_1 + 2\lambda_3 - c_3) - 14\exp(\lambda_1 + \lambda_3 - c_4)$$
$$+ 2\exp(2\lambda_2 - c_5) = 0.$$

$j = 3$:

$$-5\exp(\lambda_1 + 4\lambda_2 - c_1) + \exp(2\lambda_2 + \lambda_3 - c_2)$$
$$- 3\exp(\lambda_1 + 2\lambda_3 - c_3) - 4\exp(\lambda_1 + \lambda_3 - c_4) = 0$$

After the improved values, λ_k, have been found by solution of the three preceding equations, improved values of the mol fractions are found with Eq. 10.102, which are written out as follows:

$$y_i = \exp\left(\sum_{k=1}^{m} a_{ik}\lambda_k - c_i\right),$$

$$y_1 = \exp(a_{11}\lambda_1 + a_{12}\lambda_2 + a_{13}\lambda_3 - c_1),$$
$$y_2 = \exp(a_{21}\lambda_1 + a_{22}\lambda_2 + a_{23}\lambda_3 - c_2),$$
$$y_3 = \exp(a_{31}\lambda_1 + a_{32}\lambda_2 + a_{33}\lambda_3 - c_3),$$
$$y_4 = \exp(a_{41}\lambda_1 + a_{42}\lambda_2 + a_{43}\lambda_3 - c_4),$$
$$y_5 = \exp(a_{51}\lambda_1 + a_{52}\lambda_2 + a_{53}\lambda_3 - c_5).$$

With these improved values of mol fractions, further improved values of the Lagrange multipliers may be found, and so on.

Example 10.18. A Reaction Involving Two Liquid Phases

The reaction is $2A + B \rightleftarrows C$, starting with two mols of A and one of B. The equilibrium constant is $K_{ce} = 1.2$. Preliminary calculations show that approximate values of the activity and distribution coefficients are

	γ	γ^*	$K = \gamma/\gamma^*$
A (1)	1.5	2	0.75
B (2)	2.1	1.5	1.40
C (3)	1.25	1	1.25

With these values, $K_\gamma = 1.25/(1.5)^2(2.1) = 0.2646$. The composition of the primary phase is given by

$$x_1 = \frac{2(1-\varepsilon)}{(3-2\varepsilon)[\beta + 0.75(1-\beta)]}, \tag{1}$$

$$x_2 = \frac{1-\varepsilon}{(3-2\varepsilon)[\beta + 1.4(1-\beta)]}, \tag{2}$$

$$x_3 = \frac{\varepsilon}{(3-2\varepsilon)[\beta + 1.25(1-\beta)]}. \tag{3}$$

Sets of the variables ε and β, the degree of advancement and the phase split, are found that satisfy the phase equilibrium relation

$$\psi = f(\varepsilon,\beta) = -1 + x_1 + x_2 + x_3 = 0. \tag{4}$$

A value of ε is assumed, β is found from Eq. 4, and the results checked against the expression for the equilibrium constant,

$$K_x = x_3/x_1^2 x_2 = K_{ce}/K_\gamma = 1.2/0.2646 = 4.5351. \tag{5}$$

From the tabulation it appears that this value of K_x is obtained when $\varepsilon = 0.5637$ and $\beta = 0.3551$. The corresponding phase compositions are:

$$x_1 = 0.5556, \qquad x_1^* = 0.4167,$$
$$x_2 = 0.1852, \qquad x_2^* = 0.2593,$$
$$x_3 = 0.2592, \qquad x_3^* = 0.3240.$$

Activity coefficients corresponding to these phase compositions should be recalculated from appropriate equations, such as the NRTL, to verify the estimate made at the beginning of this calculation. If there are significant differences between the new and the old, the calculations should be repeated until convergence is reached.

ε	β	K_x	K_{ce}
0.2900	0.9933	1.1835	0.3132
0.3000	0.9771	1.2493	0.3306
0.4000	0.7913	2.0728	0.5485
0.5000	0.5490	3.3403	0.8838
0.5637	0.3551	4.5348	1.1999
0.6000	0.2269	5.4224	1.4348
0.6100	0.1891	5.7012	1.5085
0.6200	0.1501	5.9971	1.5868
0.6300	0.1097	6.3114	1.6700
0.6400	0.0682	6.6457	1.7585

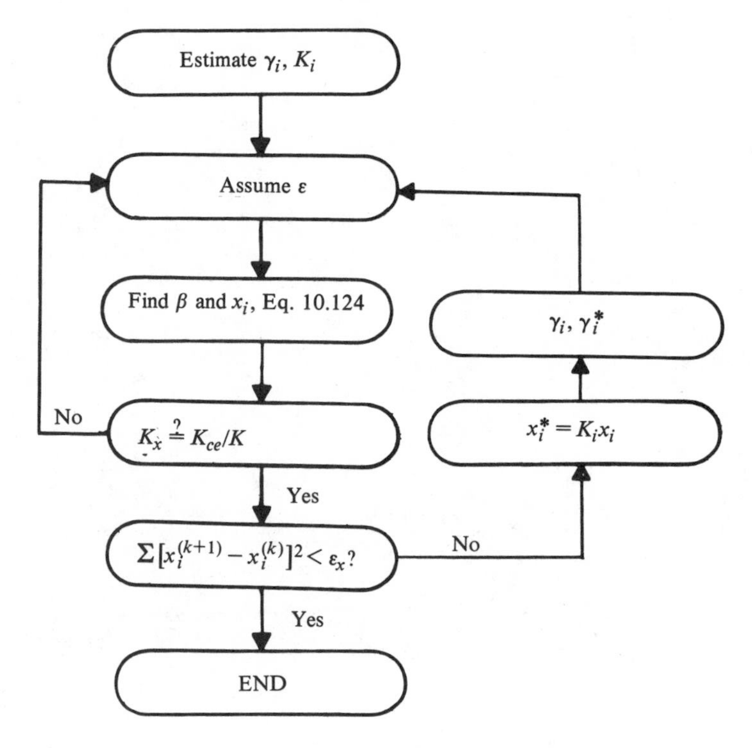

Figure 10.9. Chemical equilibrium in two liquid phases.

1. Assume values of the activity coefficients in both phases. These will be checked after the compositions have been determined on this basis, and the calculations will be repeated if close agreement has not been found.
2. Assume ε (or β).
3. Find β (or ε) from the phase split equation, Eq. 10.124 and hence the compositions x_i of the primary phase.
4. Evaluate K_x with these data and compare with $K_x = K_{ce}/K_\gamma$.
5. Return to step (2) with other trial values until agreement is reached in step (4).
6. Evaluate $x_i^* = K_i x_i$ and find the activity coefficients γ_i and γ_i^* of all substances. Compare the new $K_i = \gamma_i/\gamma_i^*$ and the K_γ with those of step (1) and repeat the calculations as many times as required to convergence.

Example 10.18 follows this procedure for one cycle of the calculation of a three-component, two liquid-phase reaction.

10.7.2. Liquid and Vapor Phases

Since Gibbs energies of formation most often are known for the vapor phase, it is preferable to formulate the chemical equilibrium equation for that phase. The interrelation between Gibbs energies of formation in the two phases is examined in Example 10.4. The vapor phase chemical equilibrium constant is

$$K_{ce} = K_{\hat{f}} = K_{\hat{\phi}} K_y P^{\Delta\nu}. \tag{10.125}$$

Phase equilibrium is expressed in terms of the vaporization equilibrium ratios,

$$K_i = \frac{y_i}{x_i} = \frac{\gamma_i \phi_i^{\text{sat}} P_i^{\text{sat}} (PF)_i}{\hat{\phi}_i P}. \tag{10.126}$$

Formulation of the flash equation must take into account the fact that the relative amounts of the two phases depend on the extent of reaction. The amounts of individual substances in both phases are given in terms of the degree of advancement,

$$n_i = n_{io} \pm \varepsilon \nu_i \qquad \begin{cases} + \text{ for products,} \\ - \text{ for reactants,} \end{cases} \tag{10.127}$$

and the total amount in both phases,

$$n_t = n_{to} + \varepsilon \Delta\nu = V + L. \tag{10.128}$$

Individual material balances are

$$n_i = Lx_i + Vy_i = [L + K_i(n_t - L)]x_i, \tag{10.129}$$

or

$$x_i = \frac{n_i}{L + K_i(n_t - L)}. \tag{10.130}$$

The mol fractions sum to unity,

$$\Sigma\, x_i = \Sigma \frac{n_i}{L + K_i(n_t - L)} = 1. \tag{10.131}$$

The final form of the flash equation is

$$\psi(\varepsilon, L) = -1 + \Sigma \frac{n_{io} \pm \varepsilon \nu_i}{L + K_i(n_{to} + \varepsilon \Delta\nu - L)}$$

$$= 0 \qquad \begin{cases} + \text{ for products,} \\ - \text{ for reactants.} \end{cases} \tag{10.132}$$

At the dewpoint, $L = 0$, and,

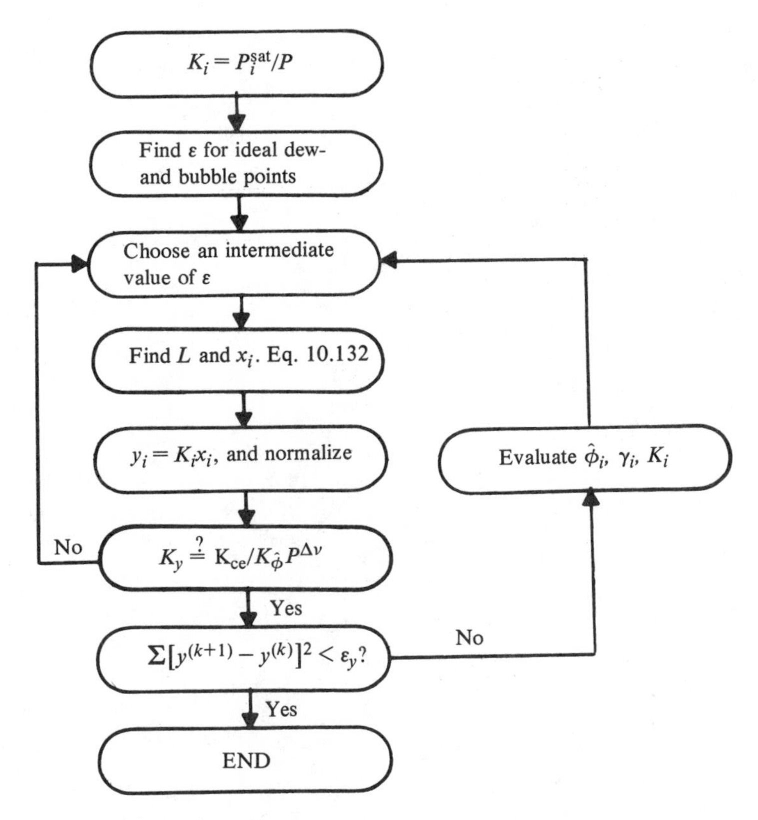

Figure 10.10. Chemical equilibrium in liquid and vapor phases.

$$\psi(\varepsilon, 0) = -1 + \Sigma \frac{n_{io} \pm \varepsilon\nu_i}{K_i(n_{to} + \varepsilon\Delta\nu)} = 0, \tag{10.133}$$

while at the bubblepoint, $L = n_{to}$, so that

$$\psi(\varepsilon, n_{to}) = -1 + \Sigma \frac{n_{io} \pm \varepsilon\nu_i}{n_{to} + K_i\varepsilon\Delta\nu} = 0. \tag{10.134}$$

A two-phase region exists between the degrees of advancement calculated by these two equations. It is advisable to check the degrees of advancement of the single-phase reactions before proceeding on to analyze the two-phase process. The problem is to find the degree of advancement and the phase split, represented by the amount of the liquid phase, L, and the corresponding phase compositions that satisfy the chemical equilibrium equation, Eq. 10.125. A procedure for the case of specified temperature, pressure, and overall initial composition is outlined in Figure 10.10 and described following.

1. Initially, assume that both phases are ideal, so that $\hat{\phi}_i = \gamma_i = 1$ and $K_i = P_i^s/P$.
2. Find the degrees of advancement corresponding to the dewpoint and the bubblepoint. These values will be only approximate in view of the assumption of step (1).
3. Choose a value of ε between those found in step (2).
4. Find L by trial from the flash equation, Eq. 10.132, and hence the liquid phase compositions, x_i, and the vapor phase compositions, $y_i = K_i x_i$, using ideal K_i's.
5. Evaluate K_y and compare with $K_y = K_{ce}/K_{\hat{\phi}}P^{\Delta\nu}$. If there is insufficient agreement, take another value of ε, return to step (4) and repeat this process as necessary until agreement is reached.
6. With the converged values of the compositions, refigure the γ_i, $\hat{\phi}_i$, K_i and $K_{\hat{\phi}}$, then return to step (2) or step (3).

Continue until successive determinations of the phase compositions meet a suitable criterion, for instance,

$$\Sigma[x^{(k+1)} - x^{(k)}]^2 \le \text{tolerance}, \tag{10.135}$$

or until the degree of advancement stabilizes.

One-phase components: Nonvolatile substances do not appear in the formulation of the vapor-phase chemical-equilibrium equation and their vaporization equilibrium ratios are zero. Insoluble gases have infinite vaporization equilibrium ratios.

Ideal systems: When the phases are so nearly ideal that the fugacity and activity coefficients are substantially unity at all compositions, the vaporization equilibrium ratios likewise will be independent of composition and steps (5) and (6) are not needed.

Example 10.19 is of a system with ideal vapor-phase and constant vaporization equilibrium ratios; the calculation represents one cycle of those for a system in which the extent of deviation from ideality does depend on the composition.

10.8. PROBLEMS

10.1. The reaction $A \rightleftharpoons 3B$ takes place at constant temperature, starting with pure A. At equilibrium the dewpoint pressure is 12 atm. Pure component vapor pressures are

Example 10.19. Reaction Involving Liquid and Vapor Phases

The reaction $A + 2B \rightleftarrows C$ occurs at a pressure of 4 atm. The equilibrium constant is $K_{ce} = K_p = K_y P^{\Delta v} = 3.5$ and the vaporization equilibrium ratios are $K_a = 5$, $K_b = 1$, and $K_c = 0.8$. Stoichiometric proportions of A and B are used. Find the amounts and compositions of the phases at equilibrium.

In terms of the degree of advancement,

$$n_a = 1 - \varepsilon,$$

$$n_b = 2(1 - \varepsilon),$$

$$n_c = \varepsilon,$$

$$n_t = 3 - 2\varepsilon = L + V,$$

$$x_i = \frac{n_i}{L + K_i V} = \frac{n_i}{L + K_i(n_t - L)} = \frac{n_i}{L + K_i(3 - 2\varepsilon - L)},$$

$$y_i = K_i x_i,$$

$$K_{ce} = K_y/P^2 = \frac{y_c}{y_a y_b^2 P^2}.$$

The flash equation is

$$\psi(L, \varepsilon) = -1 + \sum x_i = -1 + \sum \frac{n_i}{L + K_i(n_t - L)}$$

$$= -1 + \sum \frac{n_i}{L + K_i(3 - 2\varepsilon - L)}.$$

Rearrange for solution by iteration.

$$L = 0.25\left[5(3 - 2\varepsilon) - \frac{1 - \varepsilon}{1 - \frac{2(1 - \varepsilon)}{3 - 2\varepsilon} - \frac{\varepsilon}{0.2L + 0.8(3 - 2\varepsilon)}}\right].$$

L is solved for at specified values of ε. The results are tabulated. The degree of advancement when $K_{ce} = 3.5$ is $\varepsilon = 0.8429$.

The bubblepoint is at $\varepsilon = 0.7619$ and the dewpoint at about $\varepsilon = 0.9524$. If K_{ce} had been less than 1.9215, the reaction would have been single-phase vapor; if more than about 166.7, single-phase liquid.

ε	L	V	y_a	y_b	y_c	K_{ce}
0.7619	0	1.4762	0.1613	0.3226	0.5161	1.9215
0.80	0.2800	1.1200	0.1701	0.2857	0.5442	2.4497
0.8429	0.5588	0.7554	0.1813	0.2391	0.5796	3.4968
0.85	0.6012	0.6988	0.1831	0.2308	0.5861	3.7566
0.90	0.8700	0.3300	0.1984	0.1667	0.6349	7.1979
0.95	1.0859	0.0141	0.2162	0.0909	0.6927	24.23
0.9524	1.0933	0.0019	0.0432	0.0869	0.8699	166.66

$P_a^0 = 20$, $P_b^0 = 9$ atm. What is the equilibrium constant, K_p?

10.2. The reaction $2A \rightleftharpoons B$ takes place at 5 atm, starting with 1 mol of pure A in the reactor. At equilibrium the amount of the liquid phase was 0.14 mol. The vaporization equilibrium ratios are $K_a = 5$, $K_b = 0.5$. Ideal gas and solution behavior may be assumed. Find

a. The chemical equilibrium constant, K_{ce}.

b. The degrees of advancement corresponding to the bubblepoint and dewpoint conditions.

10.3. When starting out with 1 mol of pure A, the equilibrium composition of the ideal gas-phase reactions

$$A \rightleftharpoons 2B + C,$$

$$2B \rightleftharpoons D,$$

is $y_a = 0.25$, $y_d = 0.40$. Find the equilibrium constants, K_y, of the two reactions.

10.4. When starting with pure substance A, the fractional conversion at equilibrium of the reaction, $A \rightleftarrows 3B$, is 0.30. The corresponding value of K_γ is 2.7. Find the parameter of the symmetrical Margules equation.

10.5. A reaction, $A + 2B \rightleftarrows C$, is conducted with various starting ratios, $r = n_{bo}/n_{ao}$. The equilibrium constant is $K_x = 5$. At what ratio is the conversion of substance A a maximum?

10.6. For the ideal gas reaction, $A \rightleftarrows 2B$, data are:

$$\Delta H_{298}^0 = 3{,}000 \text{ cal/gmol } A,$$

$$\Delta G_{298}^0 = -400 \text{ cal/gmol } A,$$

$$\Delta C_p = 3.5 \text{ cal/(gmol } A)(K).$$

The pressure is 1.2 atm. Find the temperatures at which the conversions of substance A are 25%, 50%, 75% and 95%.

10.7. For the case of Example 10.7, the fugacity relation is given by $K_{\hat{\phi}} = 0.6 + 0.35\varepsilon^{0.3}$. Find the conversion and the numerical value of $K_{\hat{\phi}}$.

10.8. Find the degree of advancement and equilibrium composition in the decomposition of methanol according to $CH_3OH \rightleftarrows CO + 2\,H_2$ at 200 C and 5 atm. The equilibrium constant is $K_p = 42$ and $K_{\hat{\phi}}$ may be assumed unity at this low pressure.

10.9. At 1,800 K and 450 atm the fugacity coefficient of CO_2 is about 1.10 by the Lee-Kesler table. Other data are given in Example 10.3. What pressure is needed to suppress dissociation of $CaCO_3$ at 1,800 K?

10.10. Find the equilibrium composition of the dimerization, $2A \rightleftarrows B$, at 20 atm and 800 K. The equilibrium constant is 1.25. Second virial coefficients are $B_1 = -1.802$, $B_2 = -5.871$, and $B_{12} - 3.361$ cuft/lbmol. Starting amounts are $n_{ao} = 2$, $n_{bo} = 0.25$. A is component 1, B component 2. Equations for the fugacity coefficients are in Table 3.4.

10.11. For the dissociation of *n*-propanol to propionaldehyde and hydrogen, find the equilibrium constant as a function of temperature and compare the results with those quoted from Stull at several temperatures.
Data: $\Delta H^0_{298} = 15{,}650$ cal/gmol,
$K_{298} = 2.00\ (E{-}6)$,
$C_p = a + bT + cT^2 + dT^3$.

	a	*b(E2)*	*c(E6)*	*d(E8)*
H_2	6.483	0.2215	−3.298	0.1826
C_2H_6CHO	2.800	6.244	−31.050	0.5078
C_3H_7OH	0.590	7.942	−44.310	−0.3356

	K	
T	*(Stull)*	*(Calc.)*
298	2(*E*−6)	2(*E*−6)
500	0.1102	
800	63.36	
1,000	561.1	

10.12. For the reactions

$$CH_3Cl(g) + H_2O(g) \rightarrow CH_3OH(g) + HCl(g),$$

$$2\ CH_3OH(g) \rightarrow (CH_3)_2O(g) + H_2O(g),$$

the equilibrium constants at 600 K are $K_1 = 0.00154$, and $K_2 = 10.6$. For an initial mixture of 1 gmol each of methyl chloride and water at 600 K and 1 atm, show that the equilibrium partial conversion of methyl chloride is 0.0482 and that the amount of dimethyl ether formed is 0.0094 gmols.

10.13. The simultaneous gas-phase reactions at 1 atm,

$$2A + B \rightleftarrows 2C;\ K_1 = 2,$$

$$A + C \rightleftarrows D;\ K_2 = 0.5,$$

take place at 1 atm, beginning with stoichiometric proportions of *A* and *B*. Find the equilibrium composition of the mixture, either graphically or by N-R. (Answer: $\varepsilon_1 = 0.4215$, $\varepsilon_2 = 0.1452$.)

10.14. Find the composition of the equilibrium mixture resulting from the decomposition of propane according to the following reactions at 1,400 F and one atm:

$$C_3H_8 \rightleftarrows C_3H_6 + H_2 \qquad K_p = 7.88,$$

$$C_3H_8 \rightleftarrows C_2H_4 + CH_4 \qquad K_p = 755.$$

10.15. At 600 C, the principal reactions between steam and methane are

$$CH_4 + H_2O \rightleftarrows CO + 3\ H_2 \qquad K_p = 0.574,$$

$$CO + H_2O \rightleftarrows CO_2 + H_2 \qquad K_p = 2.21.$$

Find the composition of the equilibrium mixture when 1 mol of methane is mixed with 5 mols of steam at 1 atm. Use the Newton-Raphson method.

10.16. In Example 10.15 for the dissociation of isopropanol, take into account the additional substances C_3H_6, C_3H_8, and H_2O. Find a complete set of independent stoichiometric relations and their equilibrium constants, given these data for equilibrium constants of formation, $\log_{10}K$:

	400 K	*800 K*
1. isopropanol	18.120	−0.511
2. hydrogen		
3. *n*-propanol	16.926	−0.791
4. acetone	17.001	2.088
5. propionaldehyde	14.188	1.011
6. propylene	−10.175	−9.511
7. propane	−0.651	−8.18
8. water	29.241	13.290

10.17. The partial oxidation of methane with oxygen is stated to involve the following reactions (Wellman & Katell, 1964):

$$CH_4 + 2O_2 \rightarrow CO_2 + 2H_2O,$$

$$CH_4 + CO_2 \rightleftharpoons 2CO + 2H_2,$$

$$CH_4 + H_2O \rightleftharpoons CO + 3H_2,$$

$$H_2O + CO \rightleftharpoons CO_2 + H_2,$$

$$2CO \rightleftharpoons CO_2 + C.$$

Assuming that only the substances appearing in the foregoing equations will be present at equilibrium:

a. Find a set of independent components.

b. Write stoichiometric equations for the formation of the derived components and compare with the given equations.

10.18. Hydrogen can be made by steam reforming of butane at about 1,400 F and 5 atm. The substances listed possibly may be present at equilibrium. Take the water/butane ratio to be *R*. Formulate the algebraic equations from which the equilibrium composition could be calculated.

Substance	$\log_{10}K$
CH_4	−1.1581
CO	10.2895
CO_2	20.015
C_2H_4	−6.110
C_4H_{10}	−4.40
H_2	0
H_2O	9.6453

10.19. The product from the pyrolysis of methane with a 3:1 ratio of steam to methane is charged to a catalytic reactor maintained at about 800 F and 8 atm, with the objective of making higher-molecular-weight hydrocarbons eventually. As an exercise, assume that the following substances may be present at equilibrium: CO, H_2, CH_4, C_4H_{10}, C_8H_{18}, H_2O, CO_2, CH_3OH, and CH_3CHO. Equilibrium constants of formation are assumed known. Write the algebraic equations from which the equilibrium composition could be determined.

10.20. Two mol of ethylene are reacted with one mol of oxygen at moderate pressure and elevated temperature in the presence of an appropriate catalyst. Some of the stoichiometric equations that can be written for this system are given following. Identify suitable independent components and reactions and write the algebraic equations from which the composition of the equilibrium mixture could be found.

$$C_2H_4 + 3O_2 = 2CO_2 + 2H_2O,$$

$$2C_2H_4 + O_2 = 2C_2H_4O,$$

$$2C_2H_4O = HCHO + C_2H_4,$$

$$6C_2H_4O = 5C_2H_4 + 2CO_2 + 2H_2O,$$
$$2C_2H_4O + O_2 = 2CO_2 + 2CH_4,$$
$$C_2H_4 + H_2O = HCHO + CH_4,$$
$$CH_4 + O_2 = 2CO_2 + 2H_2O,$$
$$HCHO + O_2 = HCOOH,$$
$$2HCOOH + O_2 = CO_2 + H_2O.$$

10.21. The system consisting originally of one mol each of ethylene and water is expected to contain the following substances listed with their ln K_f at 400 K: C_2H_4, −22.257; H_2O, 67.330; C_2H_5OH, 43.528; CH_3COOH, 107.112; H_2, 0; C_2H_6, 4.343. Write the equations from which the equilibrium composition could be determined,

a. By the method of Brinkley.

b. By minimization of Gibbs energy, employing Lagrange multipliers.

10.22. For the reaction, $A + B \rightleftarrows 2C$, the Gibbs energies of formation are known. Designate the fractional conversion of A by ε.

a. Following Example 10.1, write the total Gibbs energy of the reacting mixture,

$$g = n_t G/RT = n_t \Sigma\, y_i[G_i^0/RT + \ln(y_i P)],$$

in terms of ε.

b. By algebraic minimization of g show that equilibrium is represented by

$$K = [2\varepsilon/(1-\varepsilon)]^2,$$

where

$$-RT \ln K = 2G_c^0 - G_a^0 - G_b^0.$$

10.23. Use Lagrange multipliers to solve these minimax problems.

a. $f = x^2 + y^2 \rightarrow$ minimax, with $y + 10x \ln x + 5 = 0$.

b. $f = xyz \rightarrow$ minimax, with $(x/a)^2 + (y/b)^2 + (z/c)^2 - 1 = 0$.

c. $f = \cos^2 x + \cos^2 y \rightarrow$ minimax, with $y - x - \pi/4 = 0$.

d. Equilibrium of the reaction, $CO_2 + 3\ H_2 \rightleftharpoons CH_3OH + H_2O$, with stoichiometric proportions on the left and at atmospheric pressure.

10.24. In the reaction of one mol of C_3H_8 with 5 mols of O_2 and 20 mols of N_2, these substances are expected to be present at equilibrium: CO_2, H_2O, N_2, CO, H_2, H, OH, O, NO, and O_2.

a. Find the rank of the composition matrix and hence the number of independent components.

b. Taking CO_2, H_2O, N_2, and CO as the independent components, write stoichiometric equations for the formations of the remaining substances.

c. Write the equilibrium equations and chemical element balances from which the composition of the equilibrium mixture could be found.

d. Find another set of independent components and the corresponding equations for the formation of the remaining substances. (Kandiner & Brinkley, 1950).

10.25. The substances that are expected to be present at equilibrium from the oxidation of ammonia by air are: NH_3, NO_2, NO, H_2O, O_2, and N_2. The numbers of mols of N, H, and O present initially are designated by q_N, q_H, and q_O.

a. Verify that NH_3, H_2O, and O_2 are satisfactory as independent components.

b. Formulate the equilibrium equations and the element balances required for the solution of this problem by the method of Brinkley.

10.26. Use the relaxation method to find equilibrium compositions resulting from the reactions, $2A \rightleftharpoons 3B$ and $A + B \rightleftharpoons C$. Data are $n_{a0} = 2$, $n_{bo} = n_{co} = 0$, $K_1 = 3$, $K_2 = 2$, and $P = 0.5$.

10.27. A mixture containing one part nitrogen, three parts hydrogen, and five parts water is maintained at 25 C and 100 Torr until chemical and phase equilibrium is attained. The chemical equilibrium condition is expressed by

$$K_{ce} = y_{NH_3}/y_{N_2}^{0.5}\, y_{H_2}^{1.5}\, P = 725.7,$$

where P is in atmospheres. Phase equilibria are given in terms of Henry's Law, $P_i = H_i x_i$, with

$$H_{N_2} = 91{,}053 \text{ atm}, H_{H_2} = 70{,}658 \text{ atm}, H_{NH_3} = 0.960 \text{ atm}.$$

The vapor pressure of water is 25.76 Torr. Find the proportions and compositions of the two phases at equilibrium (Sandler, "Chemical and Engineering Thermodynamics," 1977).

10.28. A gas-phase dimerization $2A \rightleftharpoons B$ occurs in the presence of a nonvolatile solvent. Starting amounts are one gmol each of A and the solvent. Vaporization equilibrium ratios are $K_a = 5$ and $K_b = 0.5$. The chemical equilibrium constant is $K_y = 0.080$. Find:

a. The fractional conversion of substance A.

b. The amounts and compositions of the two phases.

Note: The fractional conversion is expected to be in the range from 0.3 to 0.5.

10.29. A reaction $2A \rightleftharpoons B$ proceeds in both liquid and vapor phases. At atmospheric pressure and 350 K the overall fractional conversion of A was measured as 0.3. Van Laar constants are $A = 2.3026$, $B = 1.6094$. Other data are

	A	B
P^{sat}, Torr	952	86.98
ϕ^{sat}	0.98	0.98
$\hat{\phi}$	0.95	1.05
(PF)	1.002	0.998

Find:

a. The chemical equilibrium constant.

b. The amounts and compositions of the two phases.

10.30. For the reaction $0.5\ N_2 + 1.5\ H_2 \rightleftarrows NH_3$ the equilibrium constant at 873 K is $K_f = 0.00156$. The reaction is conducted at 873 K and 200 atm with stoichiometric proportions of the reactants.

Verify that the fractional conversion at equilibrium is about 0.161.

The virial equation may be assumed to apply, with the following values of the second virial coefficients, B_{ij} liters/gmol, figured from the Tsonopoulos equations:

ij		B_{ij}
11	N_2	0.044647
22	H_2	0.025626
33	NH_3	0.047731
12		0.033751
13		0.046201
23		0.035809

11 Evaluation of Changes in Enthalpy and Entropy

11.1. BASIC RELATIONS

In practice, the analysis of phase equilibria also usually requires evaluation of changes in enthalpy or entropy or other properties that may result from changes in primary variables such as temperature, pressure, and composition. Thus the design of separation equipment requires enthalpy balances as well as equilibrium relations and the analysis of the performance of compressors or expanders requires knowledge of the entropy behavior.

For a few pure substances and a very few mixtures, charts or tables of energy functions are available that are based on more or less complete experimental data. This class of substances includes the lighter hydrocarbons, common gases, and refrigerants. Less precise correlations have been devised for heavier petroleum fractions. Because data are so limited, particularly for mixtures, the only feasible method for evaluating changes in energy functions is based on equations of state to find the effect of pressure and on ideal-gas heat capacities to find the effect of temperature. Many such heat-capacity data have been measured, and satisfactory estimates usually can be made from molecular group contributions.

The analysis of deviations of energy functions from ideal behavior is similar in some respects to the development of equations for fugacity coefficients and vaporization equilibrium ratios in earlier chapters. The basic thermodynamic equations relating various thermodynamic properties to (P, V, T, C_p^{id}) data are collected in Tables 2.2 and A.3 and in the tables of this chapter.

Evaluation of residual property differences requires integration of equations of state. Since those can be explicit in either pressure or volume, interchange of the independent variables P and V sometimes is necessary. The following is a list of some pertinent formulas:

$$\int_{P_1}^{P_2} V dP = P_2 V_2 - P_1 V_1 - \int_{V_1}^{V_2} P dV, \tag{11.1}$$

$$\int_0^P V dP = PV - RT - \int_\infty^V P dV, \tag{11.2}$$

$$\int \left(\frac{\partial P}{\partial T}\right)_V dV = -\int \left(\frac{\partial V}{\partial T}\right)_P dP, \text{ when integration is at constant } T, \tag{11.3}$$

$$\left(\frac{\partial P}{\partial V}\right)_T \left(\frac{\partial T}{\partial P}\right)_V \left(\frac{\partial V}{\partial T}\right)_P = -1, \text{ chain rule}, \tag{11.4}$$

$$\left(\frac{\partial M}{\partial V}\right)_T = \left(\frac{\partial M}{\partial P}\right)_T \left(\frac{\partial P}{\partial V}\right)_T. \tag{11.5}$$

Differential formulas relating energy functions to T and P or V are derivable from the original definitions of the functions, or often more conveniently and perhaps more safely with the aid of the Bridgman Table A.7. The pertinent formulas are:

$$dH = \left(\frac{\partial H}{\partial T}\right)_P dT + \left(\frac{\partial H}{\partial P}\right)_T dP \tag{11.6}$$

$$= C_p dT + \left[V - T\left(\frac{\partial V}{\partial T}\right)_P\right] dP \tag{11.7}$$

$$= C_p dT + d(PV) + \left[T\left(\frac{\partial P}{\partial T}\right)_V - P\right] dV \tag{11.8}$$

$$= C_p dT + \left[T\left(\frac{\partial P}{\partial T}\right)_V + V\left(\frac{\partial P}{\partial V}\right)_T\right] dV \tag{11.9}$$

$$= C_p dT - \frac{RT^2}{P}\left(\frac{\partial z}{\partial T}\right)_P dP, \tag{11.10}$$

$$dS = \left(\frac{\partial S}{\partial T}\right)_P dT + \left(\frac{\partial S}{\partial P}\right)_T dP$$

$$= \left(\frac{\partial S}{\partial T}\right)_V dT + \left(\frac{\partial S}{\partial V}\right)_T dV \tag{11.11}$$

$$= \frac{C_p}{T} dT - \left(\frac{\partial V}{\partial T}\right)_P dP \tag{11.12}$$

$$= \frac{C_v}{T} dT + \left(\frac{\partial P}{\partial T}\right)_V dV \tag{11.13}$$

$$= \frac{C_p}{T} dT - \frac{R}{P}\left[z + T\left(\frac{\partial z}{\partial T}\right)_P\right] dP, \tag{11.14}$$

$$C_p - C_v = -T\left(\frac{\partial P}{\partial T}\right)_V^2 \Big/ \left(\frac{\partial P}{\partial V}\right)_T \tag{11.15}$$

$$= -T\left(\frac{\partial V}{\partial T}\right)_P^2 \Big/ \left(\frac{\partial V}{\partial P}\right)_T, \tag{11.16}$$

$$\left(\frac{\partial C_p}{\partial P}\right)_T = -T\left(\frac{\partial^2 V}{\partial T^2}\right)_P, \tag{11.17}$$

$$\left(\frac{\partial C_v}{\partial V}\right)_T = T\left(\frac{\partial^2 P}{\partial T^2}\right)_V, \tag{11.18}$$

$$dU = dH - d(PV), \tag{11.19}$$

$$dG = dH - d(TS), \tag{11.20}$$

$$dA = dU - d(TS). \tag{11.21}$$

Since heat capacities depend on both P and T, care must be exercised in evaluating the effects of changes in P and T on the energy properties. In most instances, heat capacities are

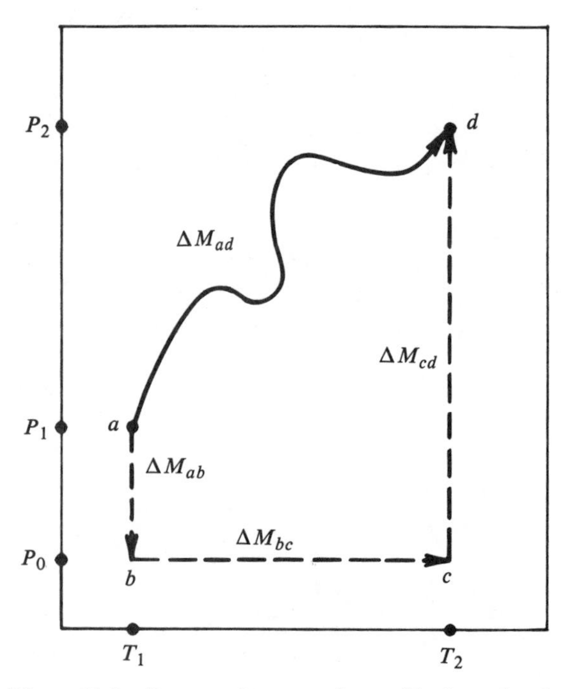

Figure 11.1. Property changes made up of isothermal and isobaric changes.

$$\Delta M_{ad} = \Delta M_{ab} + \Delta M_{bc} + \Delta M_{cd}.$$

known only at atmospheric or near zero pressures when they are termed *ideal-gas heat capacities,* C_p^{id}. Advantage can be taken of such knowledge in proceeding from condition (P_1, T_1) to condition (P_2, T_2) by the sequence of isothermal and isobaric steps represented by *abcd* in Figure 11.1. The equivalent processes are:

1. Reduction of the initial pressure P_1 to the low value, P_0, at which the substance behaves like an ideal gas, at constant temperature T_1.
2. Integration from T_1 to T_2 at constant pressure P_0.
3. Raising the pressure from P_0 to P_2 at constant temperature T_2.

The changes of steps 1 and 3 are called residual or departure properties. For a general property, M, it is designated by

$$\Delta M' = M^{\text{id}} - M. \tag{11.22}$$

It is the difference in values of the property between the ideal-gas and real states at the same pressure and temperature.

11.2. EVALUATION OF RESIDUAL PROPERTIES

Values of these properties can be obtained from direct experimental data. Thus calorimetric measurements of enthalpies and heat capacities lead directly to residual enthalpies, $\Delta H'$, and residual entropies, $\Delta S'$. However, such data are not abundant, particularly for mixtures. Figures 1.3 and 1.4 show some extreme effects of pressure on heat capacities. Some measured values of residual properties of ethylene are graphed in Figure 11.2. Experimental *PVT* data of nitrogen are employed in Example 11.4 to evaluate residual enthalpies and entropies, and the results are compared with those calculated from some equations of state.

In practice, residual properties are found with the aid of equations of state which are, of course, based on experimental *PVT* and other data. The starting points are the differential expressions of the preceding section with $dT = 0$.

For the most part, EOS are of volume explicit form,

$$V = V(P, T), \tag{11.23}$$

or

$$V = \frac{RT}{P} z(P, T), \tag{11.24}$$

or of pressure explicit form,

$$P = P(V, T), \tag{11.25}$$

or

$$P = \frac{RT}{V} z(V, T). \tag{11.26}$$

The virial equation is the only common example of a volume-explicit EOS, and the Pitzer-Curl and Lee-Kesler of the z-forms.

Equations will be developed for residual enthalpies and entropies in terms of both types of EOS. For residual enthalpy, the following series of relations is obtained:

$$\Delta H' = H^{\text{id}} - H = \int_0^P \left[\left(\frac{\partial H^{\text{id}}}{\partial P} \right)_T - \left(\frac{\partial H}{\partial P} \right)_T \right] dP \tag{11.27}$$

$$= \int_0^P \left[T \left(\frac{\partial V}{\partial T} \right)_P - V \right] dP \tag{11.28}$$

$$= -\int_\infty^V \left[T \left(\frac{\partial P}{\partial T} \right)_V + V \left(\frac{\partial P}{\partial V} \right)_T \right] dV \tag{11.29}$$

$$= (U + PV)^{\text{id}} - (U + PV) \tag{11.30}$$

$$= RT - PV + \int_\infty^V \left[P - T \left(\frac{\partial P}{\partial T} \right)_V \right] dV \tag{11.31}$$

$$= -RT(z-1) - RT^2 \int_\infty^V \frac{1}{V} \left(\frac{\partial z}{\partial T} \right)_V dV. \tag{11.32}$$

Since pressure-explicit EOS are most common, Eq. 11.31 is the most useful one.

The starting point for the development of expressions for the residual entropy is the Maxwell relation,

$$\left(\frac{\partial S}{\partial V} \right)_T = \left(\frac{\partial P}{\partial T} \right)_V, \tag{11.33}$$

or

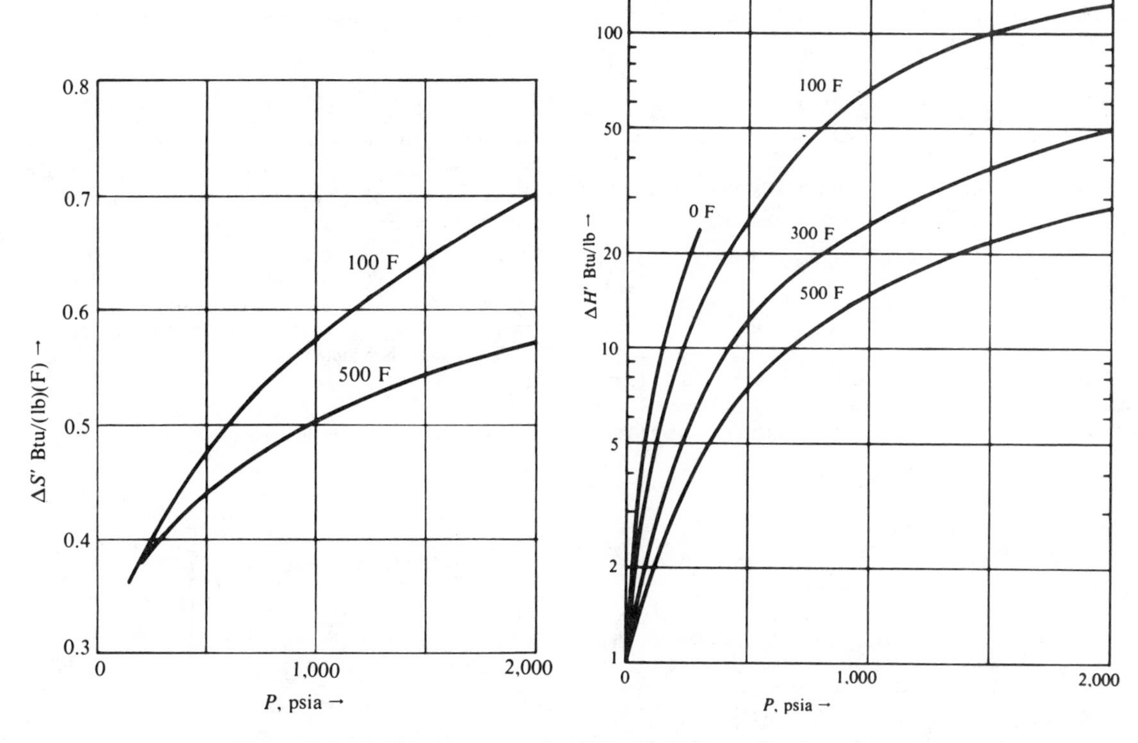

Figure 11.2. Residual entropy and enthalpy of ethylene as functions of pressure and temperature. Data from the book of Starling (1973).

$$dS = \left(\frac{\partial P}{\partial T}\right)_V dV. \tag{11.34}$$

Since the entropy of even ideal gases depends on pressure and becomes infinite as $P \to 0$ or $V \to \infty$, the residual property cannot be obtained by direct integration of Eq. 11.34. In order to circumvent this problem, introduce the quantity,

$$V^{\text{id}} = \frac{RT}{P}, \tag{11.35}$$

which is the volume of an ideal gas at the system T and P. At values of V greater than V^{id}, the gas also behaves ideally. Accordingly integration of Eq. 11.34 over this range results in

$$S(V^{\text{id}}) - S(\infty) = \int_\infty^{V^{\text{id}}} \left(\frac{\partial P}{\partial T}\right)_V dV = \int_\infty^{V^{\text{id}}} \frac{R}{V} dV. \tag{11.36}$$

Also, over the whole range of volumes, the integral is symbolized by

$$S(\infty) - S(V) = -\int_\infty^V \left(\frac{\partial P}{\partial T}\right)_V dV. \tag{11.37}$$

In order to obtain a finite integrand for the evaluation of the residual entropy, combine Eq. 11.36 and 11.37 with the following,

$$0 = \int_\infty^V \frac{R}{V} dV - \int_\infty^V \frac{R}{V} dV. \tag{11.38}$$

The result is

$$\Delta S' = S^{\text{id}} - S = S(V^{\text{id}}) - S(V)$$
$$= \int_\infty^V \left[\frac{R}{V} - \left(\frac{\partial P}{\partial T}\right)_V\right] dV + \int_V^{V^{\text{id}}} \frac{R}{V} dV. \tag{11.39}$$

The second integral on the right is

$$\int_V^{V^{\text{id}}} \frac{R}{V} dV = R \ln \frac{V^{\text{id}}}{V} = R \ln \frac{RT}{PV} = -R \ln z, \tag{11.40}$$

so that finally,

$$\Delta S' = S^{\text{id}} - S = \int_\infty^V \left[\frac{R}{V} - \left(\frac{\partial P}{\partial T}\right)_V\right] dV - R \ln z. \tag{11.41}$$

With volume explicit EOS, the starting point is another Maxwell relation,

$$\left(\frac{\partial S}{\partial P}\right)_T = -\left(\frac{\partial V}{\partial T}\right)_P. \tag{11.42}$$

When this is combined with the corresponding expression for an ideal gas, the differential of the residual entropy is obtained,

$$d(S^{id} - S) = \left[\frac{R}{P} - \left(\frac{\partial V}{\partial T}\right)_P\right] dP. \tag{11.43}$$

Since the values of $S^{id} - S$ and of the integrand both are zero at $P = 0$, integration gives,

$$\Delta S' = S^{id} - S = \int_0^P \left[\frac{R}{P} - \left(\frac{\partial V}{\partial T}\right)_P\right] dP. \tag{11.44}$$

Equivalents of Eqs. 11.41 and 11.44 in terms of the z-forms of the EOS are given in summary Table 11.1. There also are gathered formulas for other residual properties that were found by appropriate combination of formulas for residual enthalpy and entropy or by analogous derivations.

11.3. SPECIFIC EQUATIONS OF STATE

Formulas for residuals from the virial EOS are summarized in Table 11.2 and those from some others in Table 11.3. Details of the development of the virial forms are given later in this section. For some EOS, numerical tables or graphs have been prepared. A graphical comparison of residuals with three different correlations of the second virial coefficient is shown in Figures 11.3 and 11.4, and indicates substantial differences in prediction. Since the Tsonopoulos correlation represents PVT data best, it is adopted for the construction of Figures 11.5 and 11.6.

Tabulations of the residuals were prepared by Pitzer et al. and are shown, for example, in the book of Lewis & Randall et

Example 11.1. Equations for the Isothermal Effect of Pressure on the Thermodynamic Properties of a Gas Represented by the Clausius EOS

The equations of Table 11.1 are applied to the Clausius equation, which is

$$P = \frac{RT}{V - b} - \frac{a}{TV^2}. \tag{1}$$

Pertinent derivatives are:

$$\left(\frac{\partial P}{\partial T}\right)_V = \frac{R}{V - b} + \frac{a}{T^2V^2}, \tag{2}$$

$$\left(\frac{\partial P}{\partial V}\right)_T = -\frac{RT}{(V - b)^2} + \frac{2a}{TV^3}, \tag{3}$$

$$\left(\frac{\partial V}{\partial T}\right)_P = -\left(\frac{\partial P}{\partial T}\right)_V \left(\frac{\partial V}{\partial P}\right)_T. \tag{4}$$

Accordingly,

$$H^{id} - H = RT - PV + \int_\infty^V \left[P - T\left(\frac{\partial P}{\partial T}\right)_V\right] dV$$

$$= RT - PV - \int_\infty^V \frac{2a}{TV^2} dV$$

$$= RT - PV + \frac{2a}{TV} = \frac{3a}{TV} - \frac{bRT}{V - b}, \tag{5}$$

$$S^{id} - S = \int_\infty^V \left[\frac{R}{V} - \left(\frac{\partial P}{\partial T}\right)_V\right] dV - R \ln z$$

$$= \int_\infty^V \left[\frac{R}{V} - \frac{R}{V - b} - \frac{a}{T^2V^2}\right] dV - R \ln z \tag{6}$$

$$= -R\left[\ln\left[z\left(1 - \frac{b}{V}\right)\right] - \frac{a}{RT^2V}\right], \tag{7}$$

$$U^{id} - U = H^{id} - H - (RT - PV) = \frac{2a}{TV}, \tag{8}$$

$$A^{id} - A = U^{id} - U - T(S^{id} - S) = \frac{a}{TV} + RT \ln\left[z\left(1 - \frac{b}{V}\right)\right], \tag{9}$$

$$G^{id} - G = H^{id} - H - T(S^{id} - S) = \frac{2a}{RT} + RT\left[\ln\left[z\left(1 - \frac{b}{V}\right)\right] - \frac{b}{V - b}\right], \tag{10}$$

$$C_v^{id} - C_v = T\int_\infty^V \frac{2a}{T^3V^2} dV = -\frac{2a}{T^2V}, \tag{11}$$

$$C_p^{id} - C_p = R - \frac{2a}{T^2V} + T\left[\frac{R}{V - b} + \frac{a}{T^2V^2}\right]^2 \Big/ \left(-\frac{RT}{V - b} + \frac{2a}{T^2V^3}\right). \tag{12}$$

Table 11.1. Residual Properties. General Equations for Equations of State that are Explicit in either P or V.

The residual property, $\Delta M' = M^{\text{id}} - M$, is the difference between its values in the ideal-gas state and the real state at the same temperature and pressure.

Equations of state explicit in volume are of the forms: $V = V(P, T)$ or $V = (RT/P)z(P, T)$. Those explicit in pressure are of the forms: $P = P(V, T)$ or $P = (RT/V)z(V, T)$.

A. For EOS explicit in volume,

$$\Delta H' = \int_0^P \left[T\left(\frac{\partial V}{\partial T}\right)_P - V\right] dP \tag{1}$$

$$= RT^2 \int_0^P \left(\frac{\partial z}{\partial T}\right)_P \frac{dP}{P}, \tag{2}$$

$$\Delta U' = \Delta H' + PV - RT, = \Delta H' - RT(1 - z), \tag{3}$$

$$\Delta S' = \int_0^P \left[\left(\frac{\partial V}{\partial T}\right)_P - \frac{R}{P}\right] dP \tag{4}$$

$$= R \int_0^P \left[z - 1 + T\left(\frac{\partial z}{\partial T}\right)_P\right] \frac{dP}{P}, \tag{5}$$

$$\Delta A' = \Delta U' - T\Delta S', \tag{6}$$

$$\Delta G' = \Delta H' - T\Delta S', \tag{7}$$

$$\ln \phi = \frac{\Delta S'}{R} - \frac{\Delta H'}{RT} = \int_0^P (z - 1)\frac{dP}{P}, \tag{8}$$

$$\Delta C_p' = T \int_0^P \left(\frac{\partial^2 V}{\partial T^2}\right)_P dP \tag{9}$$

$$= RT \int_0^P \left[2\left(\frac{\partial z}{\partial T}\right)_P + T\left(\frac{\partial^2 z}{\partial T^2}\right)_P\right] \frac{dP}{P}, \tag{10}$$

$$C_p - C_v = -T\left(\frac{\partial P}{\partial T}\right)_V^2 \Big/ \left(\frac{\partial P}{\partial V}\right)_T \tag{11}$$

$$= -T\left[\frac{RT}{V}\left(\frac{\partial z}{\partial T}\right)_V + \frac{zR}{V}\right]^2 \Big/ \left[\frac{RT}{V}\left(\frac{\partial z}{\partial V}\right)_T - \frac{zRT}{V^2}\right], \tag{12}$$

$$\Delta C_v' = \Delta C_p' - R + (C_p - C_v). \tag{13}$$

B. For EOS explicit in pressure,

$$\Delta H' = -\int_\infty^V \left[V\left(\frac{\partial P}{\partial V}\right)_T + T\left(\frac{\partial P}{\partial T}\right)_V\right] dV \tag{14}$$

$$= \Delta U' + RT - PV = \Delta U' + RT(1 - z) \tag{15}$$

$$= -RT^2 \int_\infty^V \left(\frac{\partial z}{\partial T}\right)_V \frac{dV}{V} + RT - PV, \tag{16}$$

$$\Delta U' = \int_\infty^V \left[P - T\left(\frac{\partial P}{\partial T}\right)_V\right] dV \tag{17}$$

$$= -RT^2 \int_\infty^V \left(\frac{\partial z}{\partial T}\right)_V \frac{dV}{V}, \tag{18}$$

$$\Delta S' = \int_\infty^V \left[\frac{R}{V} - \left(\frac{\partial P}{\partial T}\right)_V\right] dV - R\ln z \tag{19}$$

$$= -\int_\infty^V \left[z - 1 + T\left(\frac{\partial z}{\partial T}\right)_V\right] \frac{dV}{V} - R\ln z, \tag{20}$$

$$\Delta A' = \Delta U' - T\Delta S', \tag{21}$$

$$\Delta G' = \Delta H' - T\Delta S', \tag{22}$$

$$\ln \phi = \frac{\Delta S'}{R} - \frac{\Delta H'}{RT} = z - 1 - \ln z - \int_\infty^V (z - 1)\frac{dV}{V}, \tag{23}$$

$$\Delta C_v' = -T \int_\infty^V \left(\frac{\partial^2 P}{\partial T^2}\right)_V dV \tag{24}$$

$$= -RT \int_\infty^V \left[2\left(\frac{\partial z}{\partial T}\right)_V + T\left(\frac{\partial^2 z}{\partial T^2}\right)_V\right] \frac{dV}{V}, \tag{25}$$

$$C_p - C_v = -T\left(\frac{\partial V}{\partial T}\right)_P^2 \Big/ \left(\frac{\partial V}{\partial P}\right)_T \tag{26}$$

$$= \frac{-T\left[\frac{zR}{P} + \frac{RT}{P}\left(\frac{\partial z}{\partial T}\right)_P\right]^2}{\frac{RT}{P}\left[-\frac{1}{P} + \left(\frac{\partial z}{\partial P}\right)_T\right]}, \tag{27}$$

$$\Delta C_p' = \Delta C_v' + R - (C_p - C_v). \tag{28}$$

$$= T \int_0^P \left(\frac{\partial^2 V}{\partial T^2}\right)_P dP \tag{29}$$

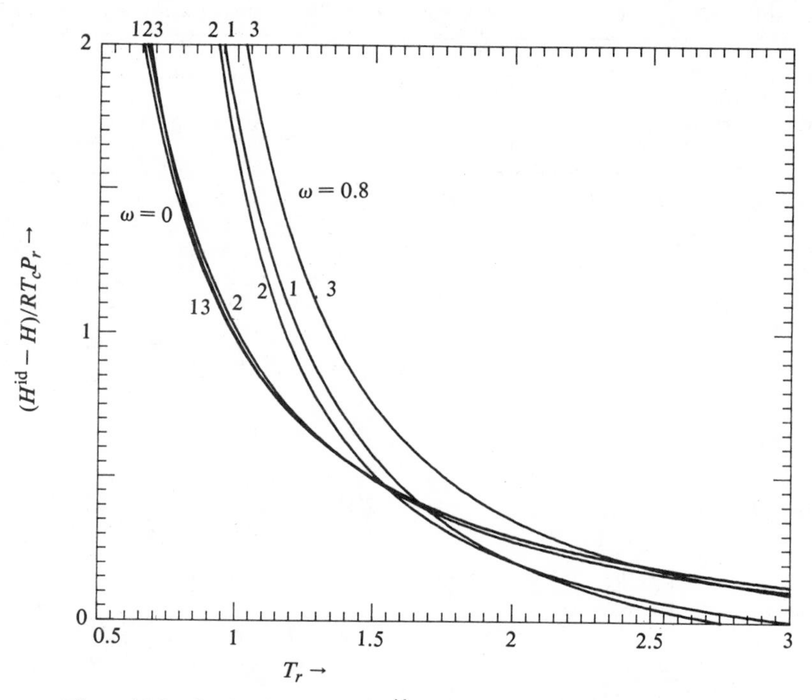

Figure 11.3. Residual enthalpies, $H^{id} - H$, based on several correlations of the second virial coefficient. 1. Pitzer-Curl; 2. Abbott; 3. Tsonopoulos.

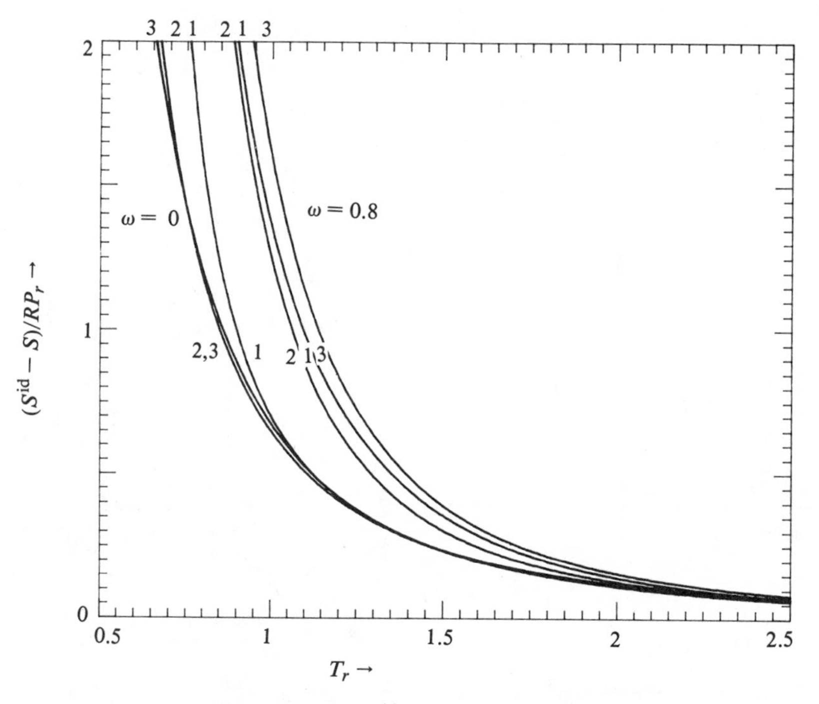

Figure 11.4. Residual entropies, $(S^{id} - S)_P$, based on several correlations of the second virial coefficient. 1. Pitzer-Curl; 2. Abbott; 3. Tsonopoulos.

al. (1961). That information was graphed by Edmister (1967, pp. 165–170. More accurate data based on the correlation of Lee & Kesler are given as Tables E.15–E.18; these data are also given in graphical form in the *API Data Book.* Graphs based on the Redlich-Kwong equation also were constructed by Edmister (1968, pp. 145–149. Such charts and graphs are useful for spot checking, but any values are readily obtainable with a calculator or minicomputer even from EOS as complex as the Lee-Kesler or BWRS.

For the most part, the derivations of the equations of Table 11.3 are quite straightforward. As an illustration, formulas for the residuals from the Soave equation are developed in Example 11.3.

The virial equation: Formulas for residuals in terms of several correlations of the second virial coefficient are shown in Table 11.2. The correlation of Orbey & Vera (1983) of the third virial coefficient has been cited in Table 1.8. It is of the Pitzer form,

$$C = (RT_c/P_c)^2(C^{(0)} + \omega C^{(1)}), \tag{11.45}$$

where the superscripted items are functions of the reduced temperature.

The z-forms, Eqs. 2 and 5 of Table 11.1, are the starting points. The derivative is

$$\left(\frac{\partial z}{\partial T}\right)_P = \frac{dB'}{dT}P + \frac{dC'}{dT}P^2 \tag{11.46}$$

$$= \frac{P}{RT}\left(\frac{dB}{dT} - \frac{B}{T}\right) + \left(\frac{P}{RT}\right)^2\left[\frac{dC}{dT} - 2B\frac{dB}{dT} - \frac{2}{T}(C - B^2)\right]. \tag{11.47}$$

The formula for $\Delta H'$ is Eq. 7 of Table 11.2. That for $\Delta S'$ is derived now:

$$\frac{\Delta S'}{R} = \int_0^P \left[z - 1 + T\left(\frac{dz}{dT}\right)\right]\frac{dP}{P} \tag{11.48}$$

$$= \frac{dB}{dT}\frac{P}{R} + \frac{1}{2}\left(T\frac{dC}{dT} - 2BT\frac{dB}{dT} + B^2 - C\right)\left(\frac{P}{RT}\right)^2, \tag{11.49}$$

where

$$\frac{dB}{dT} = (RT_c/P_c)\left(\frac{dB^{(0)}}{dT} + \omega\frac{dB^{(1)}}{dT}\right) = \frac{R}{P_c}\left(\frac{dB^{(0)}}{dT_r} + \omega\frac{dB^{(1)}}{dT_r}\right), \tag{11.50}$$

$$\frac{dC}{dT} = (RT_c/P_c)^2\left(\frac{dC^{(0)}}{dT} + \omega\frac{dC^{(1)}}{dT}\right) = (R^2T_c/P_c^2)\left(\frac{dC^{(0)}}{dT_r} + \omega\frac{dC^{(1)}}{dT_r}\right). \tag{11.51}$$

Apparently no simple plot like that of Figure 11.5 or 11.6 can be made if the third virial coefficient is to be included.

Liquids: Normally pressure has a smaller effect on residual properties of liquids than on those of gases, but in the vicinity of the critical point and at quite high pressures the distinction between liquid and vapor becomes blurred and the residuals become comparable. The following excerpts from the data on which Figure 11.7a is based make this point clearly:

P_r	0.4	1.0	2.0	10.0
$\Delta H'/T_c$ @ $T_r = 0.9$	1.48	9.94	10.29	8.74
$\Delta H'/T_c$ @ $T_r = 1.1$	0.74	2.42	5.22	6.94

The effects are determined by the isothermal compressibility coefficient and the thermal expansion coefficient, both of which usually are small for liquids. Particularly simple expressions for the effect of pressure on the enthalpy and entropy of liquids are obtained when these coefficients are constant, namely,

$$dH = \left[V - T\left(\frac{\partial V}{\partial T}\right)_P\right]dP = V(1 - \beta T)\,dP \tag{11.52}$$

$$= V_0(1 - \beta T)\exp(-kP)\,dP, \tag{11.53}$$

$$dS = -\left(\frac{\partial V}{\partial T}\right)_P dP = -\beta dP = V_0\beta\exp(-kP)\,dP, \tag{11.54}$$

where the compressibility is

$$k = -\frac{1}{V}\left(\frac{\partial V}{\partial P}\right)_T, \tag{11.55}$$

and the thermal expansion coefficient is

$$\beta = \frac{1}{V}\left(\frac{\partial V}{\partial T}\right)_P. \tag{11.56}$$

Examination of liquid isotherms on pressure-enthalpy diagrams like those in the book of Starling (1973) also will prove informative. At lower temperatures, $(\partial H/\partial P)_T$ is small and positive; but as the critical temperature is approached, the values become negative and much larger in absolute value. Example 11.6 examines the effect of pressure on several liquids.

Both liquids and gases are represented in the charts of Figure 11.7. At present the best ways of finding residual properties of nonpolar liquids are from equations of state that are applicable to liquids as well as gases. A check of six of these methods, covering both pure hydrocarbons and some binary mixtures, was made against about 1,800 experimental liquid enthalpies by Tarakad & Danner (1976). The Lee-Kesler equation proved to be the best, slightly better than the BWRS equation, but the simpler Soave EOS also proved to be quite reliable. Its use is illustrated in Example 11.11.

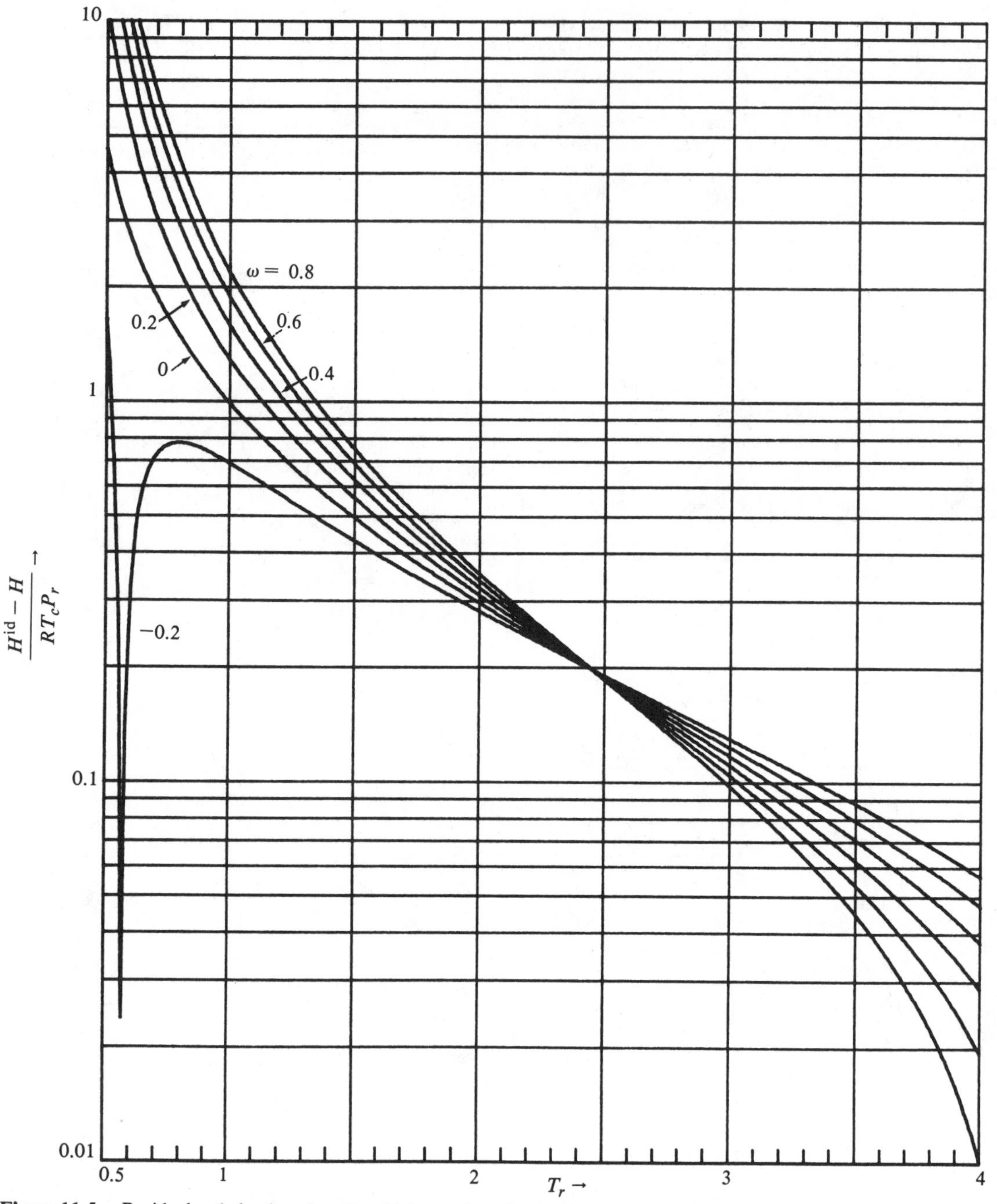

Figure 11.5. Residual enthalpy based on the virial equation with the Tsonopoulos correlation. The parameters on the curves are the acentric factors, ω.

Table 11.2. Residual Properties from the Virial Equation

$$z = 1 + B/V + C/V^2 + \ldots = 1 + B'P + C'P^2 + \ldots, \tag{1}$$

$$B' = B/RT, \tag{2}$$

$$C' = (C - B^2)/(RT)^2, \tag{3}$$

$$\frac{dB'}{dT} = \frac{1}{RT}\left(\frac{dB}{dT} - \frac{B}{T}\right), \tag{4}$$

$$\frac{dC'}{dT} = \left(\frac{1}{RT}\right)^2 \left(\frac{B^2 - C}{T} + \frac{dC}{dT} - 2B\frac{dB}{dT}\right), \tag{5}$$

$$H^{id} - H = RT^2\left[P\frac{dB'}{dT} + \frac{P^2}{2}\frac{dC'}{dT} + \ldots\right] \tag{6}$$

(continued on next page)

Table 11.2 *(continued)*

$$= PT\left(\frac{dB}{dT} - \frac{B}{T}\right)$$

$$+ \frac{P^2}{2R}\left[\frac{B^2 - C}{T} + \frac{dC}{dT} - 2B\frac{dB}{dT} + \ldots\right], \quad (7)$$

$$S^{id} - S = R\left[B'P + \frac{C'P^2}{2} + \ldots + T\left(P\frac{dB'}{dT} + \frac{P^2}{2}\frac{dC'}{dT} + \ldots\right)\right]. \quad (8)$$

With the Tsonopoulos correlation for B:

$$H^{id} - H = RT_cP_r[-0.1445 + 0.660/T_r + 0.416/T_r^2 + 0.0484/T_r^3 + 0.0055/T_r^8 + \omega(-0.0637 - 0.993/T_r^2 + 1.692/T_r^3 + 0.072/T_r^8)], \quad (9)$$

$$S^{id} - S = \frac{RP_r}{T_r^2}[0.330 + 0.277/T_r + 0.036/T_r^2 + 0.0049/T_r^7 + \omega(-0.662 + 1.269/T_r^2 + 0.064/T_r^7)]. \quad (10)$$

With the Abbott correlation for B:

$$H^{id} - H = RT_cP_r\left[\frac{1.097}{T_r^{1.6}} - 0.083 + \omega\left(\frac{0.894}{T_r^{4.2}} - 0.139\right)\right], \quad (11)$$

$$S^{id} - S = \frac{RP_r}{T_r^{2.6}}(0.675 + 0.722\,\omega/T_r^{2.6}). \quad (12)$$

With the Pitzer-Curl correlation for B:

$$H^{id} - H = RT_cP_r[-(0.1445 + 0.073\omega) + (0.660 - 0.92\omega)/T_r + (0.4155 + 1.50\omega)/T_r^2 + (0.0484 + 0.388\omega)/T_r^3 + 0.0657\omega/T_r^8], \quad (13)$$

$$S^{id} - S = RP_r[(0.330 - 0.46\omega)/T_r^2 + (0.2770 + 1.00\omega)/T_r^3 + (0.0363 + 0.29\omega)/T_r^4 + 0.0584/T_r^9]. \quad (14)$$

Table 11.3. Residual Properties from Some Equations of State

van der Waals (Table 1.3)

$$\Delta H' = RT\left[\frac{2a}{RTV} - \frac{b}{(V-b)}\right], \quad (1)$$

$$\Delta S' = -R\ln[z(1 - b/V)]. \quad (2)$$

Redlich-Kwong (Table 1.9)

$$\Delta H' = RT\left[1 - z + \frac{1.5a}{bRT^{1.5}}\ln(1 + b/V)\right], \quad (3)$$

$$\Delta S' = -R\left[\ln[z(1 - b/V)] - \frac{a}{2bRT^{1.5}}\ln(1 + b/V)\right]. \quad (4)$$

Soave (Table 1.11)

$$D_i = -T\frac{d(a\alpha)_i}{dT} = [m(a\alpha)\sqrt{T_r/\alpha}\,]_i, \quad (5)$$

$$D = \sum_i\sum_j y_iy_jm_j(1 - k_{ij})\sqrt{a_i\alpha_i}\sqrt{a_jT_{rj}}, \quad (6)$$

$$\Delta\frac{H'}{RT} = 1 - z + \frac{1}{bRT}(a\alpha + D)\ln(1 + b/V), \quad (7)$$

$$= 1 - z + \frac{A}{B}\left(1 + \frac{D}{a\alpha}\right)\ln(1 + B/z), \quad (8)$$

$$\Delta\frac{S'}{R} = -\ln[z(1 - b/V)] + \frac{D}{bRT}\ln(1 + b/V), \quad (9)$$

$$= -\ln(z - B) + \frac{BD}{Aa\alpha}\ln(1 + B/z). \quad (10)$$

For pure substances, use D_i and E_i instead of D and E in Eqs. 9 and 10.

Peng-Robinson (Table 1.13)

Note that the coefficients in the definitions of k_i, A_i, and B_i differ slightly between the Soave and P-R equations, but the D and D_i equations have the same forms.

$$\Delta\frac{H'}{RT} = 1 - z + \frac{A}{2.828B}\left(1 + \frac{D}{a\alpha}\right)\ln\frac{z + 2.414B}{z - 0.414B} \quad (11)$$

$$\Delta\frac{S'}{R} = -\ln(z - B) + \frac{BD}{2.828Aa\alpha}\ln\frac{z + 2.414B}{z - 0.414B} \quad (12)$$

For pure substances, use D_i instead of D in Eqs. 7 to 10.

BWR (Table 1.16)

BWRS (Table 1.17)

Lee-Kesler: See Table 1.18. A detailed procedure is given in the *API Data Book*, pp. 6B1.8 and 7B3.7. One of the alternatives described there as acceptable is the use of Kay's rules for pseudocritical properties instead of the more complex Lee-Kesler rules.

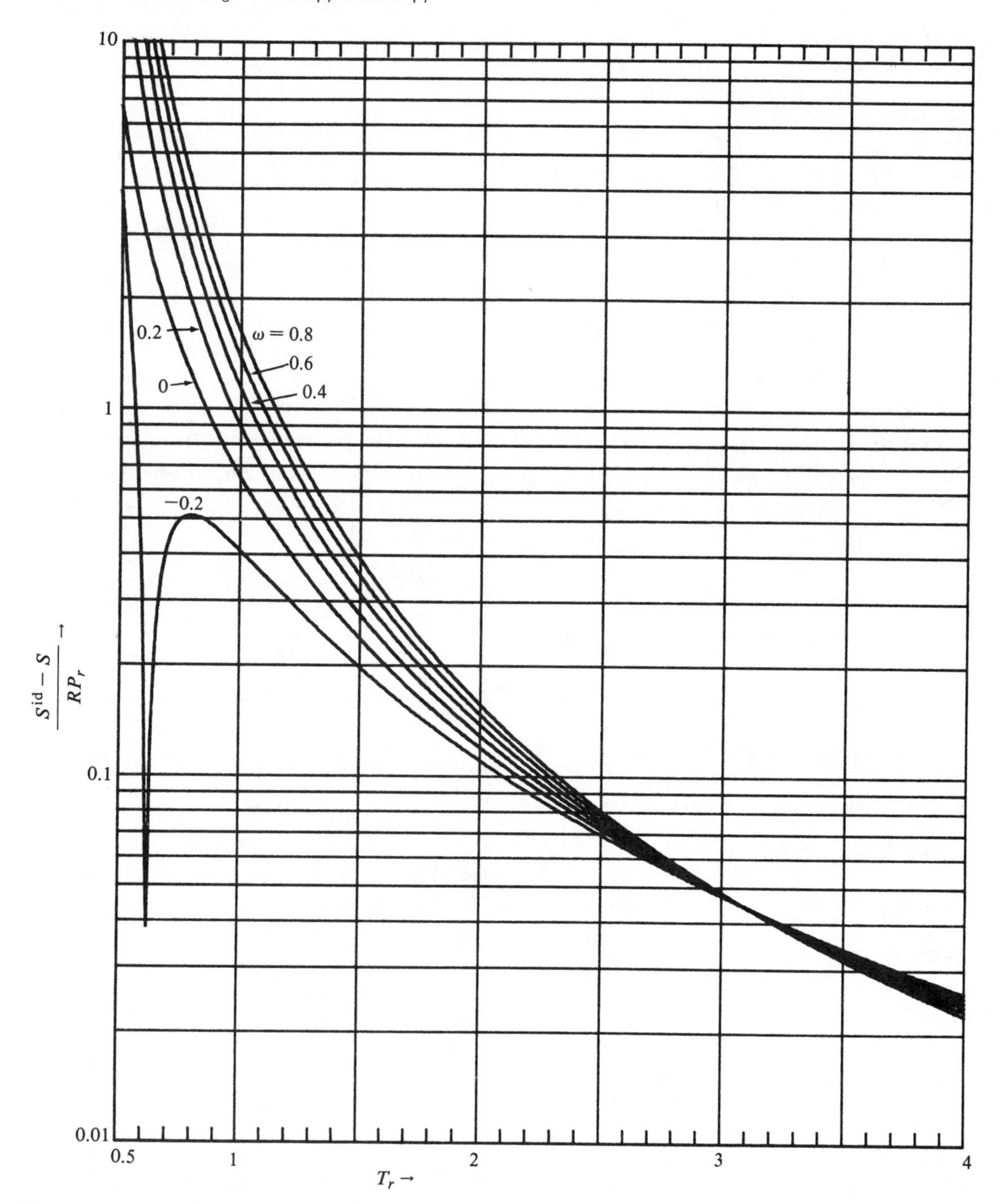

Figure 11.6. Residual entropy based on the virial equation with the Tsonopoulos correlation. The parameters on the curves are the acentric factors, ω.

Example 11.2. Residual Properties with the *B*-truncated Virial Equation by Several Methods

The equation

$$z = \frac{PV}{RT} = 1 + \frac{BP}{RT} \tag{1}$$

can be written in either pressure- or volume-explicit form as well as in the z-form, so it can be used as a check on the consistency of the various equations for $\Delta H'$ and $\Delta S'$ of Table 11.1. The required derivatives are:

$$\left(\frac{\partial V}{\partial T}\right)_P = \frac{R}{P} + \frac{dB}{dT}, \tag{2}$$

$$\left(\frac{\partial P}{\partial T}\right)_V = \frac{R}{V-B} + \frac{RT}{(V-B)^2}\frac{dB}{dT}, \tag{3}$$

$$\left(\frac{\partial z}{\partial T}\right)_P = \frac{P}{RT}\left(\frac{dB}{dT} - \frac{B}{T}\right). \tag{4}$$

Equation numbers cited are those of Table 11.1. With Eq. 1,

$$\Delta H' = \int_0^P \left[T\left(\frac{R}{P} + \frac{dB}{dT}\right) - \left(\frac{RT}{P} + B\right)\right] dP$$

$$= P\left(T\frac{dB}{dT} - B\right). \tag{5}$$

With Eq. 2,

$$\Delta H' = RT^2 \int_0^P \frac{P}{RT}\left(\frac{dB}{dT} - \frac{B}{T}\right)\frac{dP}{P} = P\left(T\frac{dB}{dT} - B\right). \tag{6}$$

With Eqs. 15 and 17,

$$\Delta H' = RT - PV + \int_\infty^V \left[\frac{RT}{V-B} - T\left(\frac{R}{V-B} + \frac{RT}{(V-B)^2}\frac{dB}{dT}\right)\right] dV \tag{7}$$

$$= -BP + \frac{RT^2}{V-B}\frac{dB}{dT} = P\left(T\frac{dB}{dT} - B\right). \tag{8}$$

With Eq. 4,

$$\Delta S' = \int_0^P \left(\frac{R}{P} + \frac{dB}{dT} - \frac{R}{P}\right) dP = P\frac{dB}{dT}. \tag{9}$$

With Eq. 5,

$$\Delta S' = \int_0^P \left[\frac{BP}{RT} + \frac{P}{R}\left(\frac{dB}{dT} - \frac{B}{T}\right)\right]\frac{dP}{P} = P\frac{dB}{dT}. \tag{10}$$

With Eq. 19,

$$\Delta S' = \int_\infty^V \left[\frac{R}{V} - \frac{R}{V-B} - \frac{RT}{(V-B)^2}\frac{dB}{dT}\right] dV - R\ln z$$

$$= R\ln\frac{V}{V-B} + \frac{RT}{V-B}\frac{dB}{dT} - R\ln z$$

$$= R\ln z + P\frac{dB}{dT} - R\ln z = P\frac{dB}{dT}. \tag{11}$$

In all cases, there is complete agreement.

Example 11.3. Residual Properties from the Soave EOS

The equation and the necessary derivatives are:

$$P = \frac{RT}{V-b} - \frac{a\alpha}{V(V+b)}, \tag{1}$$

$$\alpha = \left[1 + m\left(1 - \sqrt{\frac{T}{T_c}}\right)\right]^2, \tag{2}$$

$$\left(\frac{\partial P}{\partial T}\right)_V = \frac{R}{V-b} - a\frac{d\alpha}{dT}\frac{1}{V(V+b)}, \tag{3}$$

$$\frac{d\alpha}{dT} = -m\sqrt{\frac{\alpha}{TT_c}} = -\frac{m}{T}\sqrt{\alpha T_r} \tag{4}$$

Using Eq. 17 of Table 11.1,

$$\Delta U' = \int_\infty^V \left[\frac{RT}{V-b} - \frac{a\alpha}{V(V+b)} - \frac{RT}{V-b} + aT\frac{d\alpha}{dT}\frac{1}{V(V+b)}\right] dV \tag{5}$$

$$= \int_\infty^V a\left(T\frac{d\alpha}{dT} - \alpha\right)\frac{1}{V(V+b)} dV$$

(*continued on next page*)

Example 11.3. (*continued*)

$$= \frac{a}{b}\left(\alpha - T\frac{d\alpha}{dT}\right)\ln(1 + b/V). \tag{6}$$

Substituting into Eq. 15 of Table 11.1,

$$\Delta H' = RT - PV + \Delta U'$$

$$= RT\left[1 - z + \frac{a}{bRT}\left(\alpha - T\frac{d\alpha}{dT}\right)\ln(1 + b/V)\right]. \tag{7}$$

With Eq. 19 of Table 11.1,

$$\Delta S' = \int_\infty^V \left[\frac{R}{V} - \frac{R}{V-b} + a\frac{d\alpha}{dT}\frac{1}{V(V+b)}\right]dV$$

$$- R\ln z \tag{8}$$

$$= R\ln\frac{V}{V-b} + \frac{a}{b}\frac{d\alpha}{dT}\ln\frac{V}{V+b} - R\ln z \tag{9}$$

$$= -R\left[\ln[z(1 - b/V)] + \frac{a}{bR}\frac{d\alpha}{dT}\ln(1 + b/V)\right]. \tag{10}$$

For mixtures, derivation of the formula for $d(a\alpha)/dT$ is given in Section 11.5.

Example 11.4. Residual Properties Directly from Volumetric Data and from the Virial and Soave Equations

The specific volumes of nitrogen were measured over a range of pressures and temperatures by Deming & Shupe (1931). These data are converted to compressibilities and differentiated with the Lagrange unequal interval formula to find $(dz/dT)_P/P$ at 0 C. Then equations 2 and 5 of Table 11.1 are applied to find the residuals $\Delta H'$ and $\Delta S'$.

$$\Delta H' = RT^2 \int_0^P \frac{1}{P}\left(\frac{\partial z}{\partial T}\right)_P dP,$$

$$\Delta S' = R\int_0^P \frac{1}{P}\left(z - 1 + T\left(\frac{\partial z}{\partial T}\right)_P\right]dP.$$

These are evaluated at 0°C.

P	*V, ml/gm*		
atm	*−25*	*0*	*20 °C*
20	35.75	39.67	42.74
40	17.65	19.72	21.33
60	11.66	13.11	14.23
80	8.694	9.828	10.70
100	6.950	7.886	8.604
200	3.645	4.139	4.524
300	2.704	3.020	3.271
400	2.287	2.510	2.693
500	2.046	2.217	2.360
600	1.890	2.032	2.146
800	1.695	1.798	1.880
1,000	1.567	1.650	1.717
1,100	1.516	1.591	1.654
1,200	1.473	1.543	1.601

P	*z*						
	−25	*0*	*20 °C*	(*)	(**)	$\Delta H'$	$\Delta S'$
0						0	0
20	.9831	.9910	.9949	12.4389	5.8683	37.8	0.12
40	.9707	.9853	.9930	11.8361	5.6950	—	0.23
60	.9619	.9825	.9937	11.2889	5.5486	108.3	0.35
80	.9563	.9821	.9963	10.6639	5.3443	—	0.46
100	.9556	.9850	1.0014	9.7822	5.0122	171.4	0.56
200	1.0023	1.0340	1.0531	5.4706	3.3075	—	0.97
300	1.1153	1.1317	1.1421	1.9348	1.9226	337.2	1.24
400	1.2578	1.2541	1.2538	−.1853	1.1617	—	1.29
500	1.4065	1.3846	1.3734	−1.4009	.7679	336.2	1.39
600	1.5592	1.5229	1.4986	−2.7006	.2657	305.8	1.44
800	1.8644	1.7967	1.7505	−3.1086	.2913	219.7	1.50
1000	2.1545	2.0610	1.9984	−3.4011	.2619	123.1	1.55
1100	2.2928	2.1861	2.1176	−3.4542	.2674	72.3	1.58
1200	2.4303	2.3128	2.2361	−3.5162	.2650	20.6	1.61

Example 11.4. *(continued)*

$$* \frac{10^6}{P}\left(\frac{\partial z}{\partial T}\right)_P \quad @\ 0°\text{C}.$$

$$** \ 10^3 \frac{R}{P}\left[z - 1 + 273.2\left(\frac{\partial z}{\partial T}\right)\right] \quad @\ 0°\text{C}.$$

Nitrogen data:

$T_c = 126.2,\ P_c = 33.5,\ \omega = 0.040,\ T = 273.2.$

With the *Tsonopoulos-Virial,* substitute into Eqs. 9 and 10 of Table 11.2

$$\Delta H' = \frac{RT_cP}{P_c} f_H(T_r) = 7.4854\ f_H(T_r)P,$$

$$\Delta S' = \frac{RP}{P_c} f_S(T_r) = 0.0593\ f_S(T_r)P.$$

Both $\Delta H'$ and $\Delta S'$ vary linearly with pressure,

$$\Delta H' = \begin{cases} 0 & @\ P = 0, \\ 560 & @\ P = 300, \end{cases}$$

$$\Delta S' = \begin{cases} 0 & @\ P = 0, \\ 1.79 & @\ P = 300. \end{cases}$$

Soave method: (See Table 1.11 for formulas)

$R = 1.987,$

$a = 0.42747\ R^2T_c^2/P_c = 802.37,$

$b = 0.08664\ R\ T_c/P_c = 0.6485,$

$m = 0.48508 + 1.55171\omega - 0.15613\omega^2 = 0.5469,$

$\alpha = [1 + k(1 - \sqrt{T/T_c}]^2 = 0.5509,$

$$\frac{d\alpha}{dT} = -m\sqrt{\frac{\alpha}{TT_c}} = -\ 2.186(10^{-3}),$$

$$A = 0.42747\ \alpha \frac{P}{P_c}\left(\frac{T_c}{T}\right)^2 = 1.4998(10^{-3})P,$$

$$B = 0.08664\left(\frac{P}{P_c}\right)\left(\frac{T_c}{T}\right) = 1.1946(10^{-3})P,$$

$AB = 1.7917(10^{-6})P^2,$

$A - B - B^2 = 0.3052(10^{-3})P - 1.4271(10^{-6})P^2.$

$z^3 - z^2 + (A - B - B^2)z - AB = 0,$

$$b/V = \frac{bP}{zRT} = 1.1946(10^{-3})\frac{P}{z}.$$

First solve for z at each pressure. Then substitute into Eqs. 8 and 10 of Table 11.3.

$$\Delta H' = 542.85(1 - z) + 1420.52 \ln\left[1 + \frac{1.1946(10^{-3})P}{z}\right],$$

$$\Delta S' = -\ 1.987 \ln(z - 1.1946(10^{-3})P) + 2.7047 \ln\left(1 + 1.1946(10^{-3})\frac{P}{z}\right).$$

The results are tabulated. The plot compares the three methods. The full line is from the experimental data, the broken straight line is the Tsonopoulos virial, and the circles are the Soave.

P	z	$\Delta H'$	$\Delta S'$
20	0.9952	36.31	0.1220
60	0.9933	102.61	0.3506
100	1.0016	159.19	0.5539
300	1.1532	301.25	1.1882
500	1.3822	302.73	1.4527
800	1.7541	208.44	1.6236
1,000	2.0051	118.28	1.6816

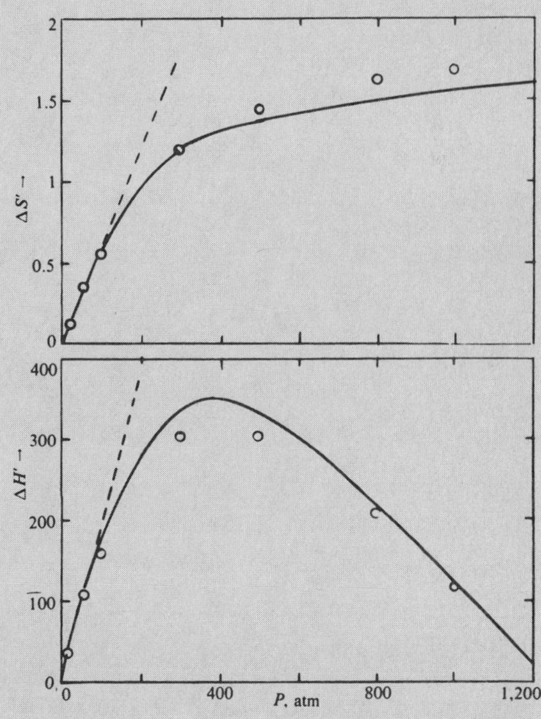

Example 11.4. Residual enthalpy and entropy of nitrogen.

Example 11.4A. Residual Enthalpies with the Soave and Peng-Robinson Eqs. and the Lee-Kesler Tables.

Eqs. 9 and 11 of Table 11.3 and Tables E.15 and E.16 will be used. For the pure substances, values are read off API charts for comparison. The specific case for which data are entered in the sample program is that of mixtures of ethane, propane, and n-butane at 30 atm and 478 K (400 F). For the pure components and for an equimolal mixture, the results are:

			z		$\Delta H'$, *cal/gm*			
y_1	y_2	y_3	*Soave*	*P-R*	*API*	*L-K*	*Soave*	*P-R*
1	0	0	0.960	0.944	5.6	5.7	6.1	6.8
0	1	0	0.899	0.872	8.9	8.5	8.8	9.6
0	0	1	0.798	0.748	12.8	12.8	12.4	13.6
⅓	⅓	⅓	0.899	0.872	—	9.1	8.8	9.6

	ω	T_c	P_c	T_r	P_r	$\Delta H^{(0)}/RT_c$	$\Delta H^{(1)}/RT_c$	$\Delta H'$, *cal/gm*
C_2	0.098	305.4	48.2	1.57	0.62	0.2811	0.0217	5.73
C_3	0.152	369.8	41.9	1.29	0.72	0.4793	0.1738	8.45
n-C_4	0.198	425.2	37.5	1.12	0.80	0.789	0.458	12.78
mix	0.148	366.8	42.5	1.30	0.71	0.479	0.161	9.06

Kay's rules are used for the pseudocritical properties with the L-K equation.

The Soave predictions compare moderately well with the API and L-K values, the Peng-Robinson less well in this example. For a total of about 5,000 data points, Tarakad & Danner (1976) find the Soave and Lee-Kesler methods to agree fairly well with each other and on an average within about 2 cal/gm with experimental data.

```
 10 ! RESIDUAL ENTHALPY, ΔH', WITH SOAVE
     AND P-R EQNS. EQNS. 8 & 11 OF TABLE 11.3.
     "HRESD"
 20 ! WITH SOAVE, GOSUB 420 IN LINE 130, PRINT
     H IN LINE 360
 30 ! WITH P-R, GOSUB 520 IN LINE 130, PRINT H1
     IN LINE 360
 40 ! T1=Tc, P1=Pc, W=omega, A1=alpha
 50 OPTION BASE 1
 60 DIM A(3), B(3), M(3), W(3), T1(3), P1(3), Y(3),
     A1(3)
 70 N=3
 80 READ T,P
 90 DATA 478, 30
100 MAT INPUT Y
110 MAT READ T1, P1, W
120 DATA 305.4, 369.8, 425.2, 48.2, 41.9, 37.5, .098,
     .152, .193
130 GOSUB 420
140 A=0
150 FOR I=1 TO N
160 FOR J=1 TO N
170 A=A+Y(I)*Y(J)*(A(I)*A(J))^.5
180 NEXT J
190 NEXT I
200 INPUT Z ! Trial value
210 GOSUB 380
220 H=F/F1
230 Z=Z-H
240 IF ABS(H/Z)<=.0001 THEN 260
250 GOTO 210
260 D=0
270 FOR I=1 TO N
280 FOR J=1 TO N
290 D=D+Y(I)*Y(J)*(A(I)*A(J))^.5*(1+.5*M(I)*(T/A
     1(I)/T1(I))^.5+.5*M(J)*(T/A1(J)/T1(J))^.5)
300 NEXT J
310 NEXT I
320 H=1.987*T*(1-Z+D/B*LOG(1+B/Z))
330 H1=1.987*T*(1-Z+D/B/8^.5*LOG((Z+2.414*B)
     /(Z-.414*B)))
340 PRINT"T=";T "P="; P; "Y1="; Y(1); "Y2=";
     Y(2); "Y3="; Y(3)
350 PRINT "Z=";Z
360 PRINT "ΔH'=";H
370 END
380 ! SR FOR Z
390 F=Z^3-Z^2+(A-B-B^2)*Z-A*B
400 F1=3*Z^2-2*Z+A-B-B^2
410 RETURN
420 ! SR FOR SOAVE
430 B=0
440 FOR I=1 TO N
450 M(I)=.48508+1.55171*W(I)-.15613*W(I) ^2
460 A1(I)=(1+M(I)*(1-(T/T1(I))^5))^2
470 A(I)=.42747*A1(I)*P/P1(I)*(T1(I)/T)^2
480 B(I)=0.08664*P/P1(I)*T1(I)/T
490 B=B+Y(I)*B(I)
500 NEXT I
510 RETURN
520 ! SR FOR PENG-ROBINSON
530 B=0
540 FR I=1 TO N
550 M(I)=.3746+1.54226*W(I)-.26992*W(I)^2
560 A1(I)=(1+M(I)*(1-(T/T1(I))^5))^2
570 A(I)=.45724*A1(I)*P/P1(I)*(T1(I)/T)^2
580 B(I)=.0778*P/P1(I)*T1(I)/T
590 B=B+Y(I)*B(I)
600 NEXT I
610 RETURN
```

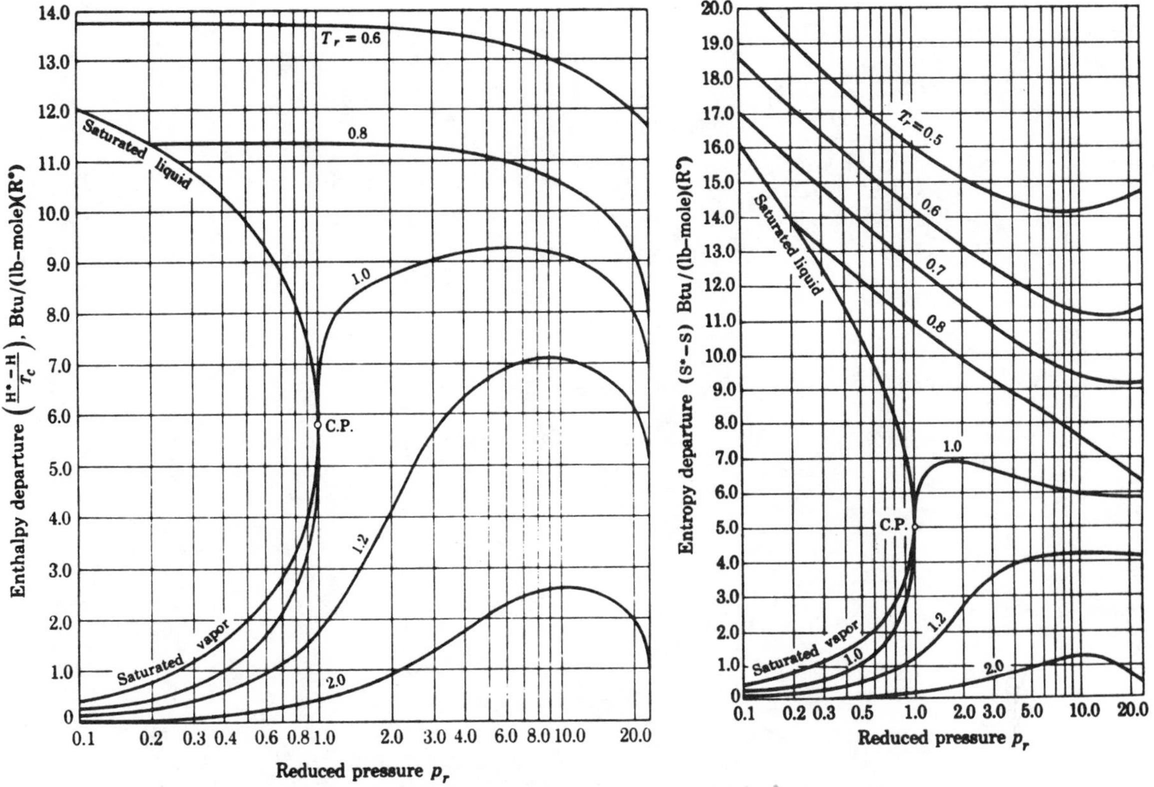

Figure 11.7. Enthalpy and entropy residuals of liquids and gases. The data are at values of the critical compressibility $z_c = 0.27$; corrections are given for other compressibilities in the range 0.23–0.29 in the original (Hougen et al. 1959).

Example 11.5. Numerical Evaluation of Isothermal Changes of Properties with the van der Waals EOS

The case is that of ethane which is to be compressed at 500 K from 10 atm to 50 atm. The constants of the van der Waals equation are $a = 5.49$ and $b = 0.064$, in the units (liters/gmol, atm, °K).

Method a will use the formulas of Table 11.3 for the residuals of H and S. Method b will proceed from the basic formulas of Section 11.1.

The terminal volumes and compressibilities are found by trial from the EOS: $V_1 = 4.0325$, $V_2 = 0.7507$, $z_1 = 0.9829$, and $z_2 = 0.9149$.

(a)

$$\Delta H' = \frac{2a}{V} - \frac{0.08205\,bT}{V-b} = \begin{cases} 2.0613, \text{ initially,} \\ 10.8028, \text{ finally,} \end{cases}$$

$$H_2 - H_1 = \Delta H'_1 - \Delta H'_2 = -8.4175 \text{ liter atm/gmol,}$$

$$\Delta S' = -0.08205 \ln\left[z\left(1 - \frac{b}{V}\right)\right] = \begin{cases} 0.0027, \text{ initially,} \\ 0.0146, \text{ finally,} \end{cases}$$

$$S_2 - S_1 = -0.08205 \ln(P_2/P_1) + \Delta S'_1 - \Delta S'_2$$
$$= -0.1321 + 0.0027 - 0.0146 = -0.1440 \text{ liter atm/(gmol)(K).}$$

(b) The enthalpy as a function of the volume is

$$\left(\frac{\partial H}{\partial V}\right)_T = V\left(\frac{\partial P}{\partial V}\right)_T + T\left(\frac{\partial P}{\partial T}\right)_V$$
$$= -\frac{RTV}{(V-b)^2} + \frac{2a}{V^2} + \frac{RT}{V-b}.$$

The integral is

$$H_2 - H_1 = -RT\left[\frac{-b}{V-b} + \ln(V-b) - \frac{2a}{V}\right] + RT\ln(V-b)\,\Bigg|_{V_1 = 4.0325}^{V_2 = 0.7507}$$
$$= -8.4175 \text{ liter atm/gmol.}$$

The direct operation for the entropy change is

$$S_2 - S_1 = \int_{V_1}^{V_2} \left(\frac{\partial P}{\partial T}\right)_V dV = R \ln\frac{V_2 - b}{V_1 - b}$$
$$= -0.1440 \text{ liter atm/(gmol)(K)}$$

Example 11.5. *(continued)*

Thus the results for both enthalpy and entropy change check by the two methods. The changes of other properties are found by combining the ΔH and ΔS. Thus,

$$U_2 - U_1 = H_2 - H_1 - (P_2V_2 - P_1V_1)$$
$$= -5.6275 \text{ liter atm/gmol},$$

$$A_2 - A_1 = U_2 - U_1 - T(S_2 - S_1)$$
$$= 66.3725 \text{ liter atm/gmol},$$
$$G_2 - G_1 = H_2 - H_1 - T(S_2 - S_1)$$
$$= 63.5825 \text{ liter atm/gmol}.$$

Example 11.6. Effect of Pressure on Enthalpy and Entropy of Some Liquids

The effects of a change of pressure from 0 to 100 atm at 300 K will be found, in the units cal, gm, K.

$$k = -\frac{1}{V}\left(\frac{\partial V}{\partial P}\right)_T,$$

$$V = V_0 e^{-kP},$$

$$\Delta H = \int_0^P V(1-\beta T)dP = V_0(1-\beta T)\int_0^P e^{-kp}\,dP$$
$$= \frac{V_0(1-\beta T)}{k}(1-e^{-kP}),$$

$$\Delta S = -\int_0^P \left(\frac{\partial V}{\partial T}\right)_P dP = -\int_0^P \beta V dP$$
$$= \beta V_0 \int_0^P e^{-kP}\,dP = -\frac{\beta V_0}{k}(1-e^{-kP}).$$

The values of V_0, k, β and the residuals are

	V_0 ml/gm	10^3k	$10^3\beta$	ΔH	$-10^3\Delta S$
acetic acid	.953	.090	1.10	1.54	2.53
acetone	1.266	.125	1.50	1.67	4.57
benzene	1.130	.095	1.30	1.66	3.54
ethanol	1.267	.114	1.10	2.04	3.35
glycerol	.793	.025	.54	1.61	1.03
methanol	1.264	.120	1.40	1.76	4.26
octane	1.422	.120	.72	2.68	2.46
phenol	.944	.050	.90	1.66	2.05
water	1.000	.046	2.00	.97	4.83

On a unit mass basis, the effect of pressure is fairly nearly the same for all these substances. All the changes are comparatively small. The effect of changing the pressure 100 atm is roughly equivalent to changing the temperature 1 to 3 K.

11.4. RESULT OF CHANGES IN BOTH *P* AND *T*

In terms of ideal-gas heat capacities and the residual properties, changes in the energy functions become simply

$$H_2 - H_1 = \int_{T_1}^{T_2} C_p^{\text{id}}\,dT + \Delta H_1' - \Delta H_2', \tag{11.57}$$

$$S_2 - S_1 = \int_{T_1}^{T_2} \frac{C_p^{\text{id}}}{T}\,dT - R\ln\frac{P_2}{P_1} + \Delta S_1' - \Delta S_2'. \tag{11.58}$$

In the last equation the first two terms on the right are the entropy change of ideal gas for the process from (P_1, T_1) to (P_2, T_2). Slight complications arise when it is required to find changes in Gibbs and Helmholtz energies when both T and P change; these are discussed in Example 11.7.

When only isothermal changes are needed, it may be simpler to ignore the formulas for the residual properties and to proceed directly with integration between the desired pressure or volume levels. In terms of pressure-explicit EOS, the appropriate equations are

$$H_2 - H_1 = \int_{V_1}^{V_2}\left(\frac{\partial H}{\partial V}\right)_T dV$$
$$= \int_{V_1}^{V_2}\left[T\left(\frac{\partial P}{\partial T}\right)_V + V\left(\frac{\partial P}{\partial V}\right)_T\right]dV, \tag{11.59}$$

and

$$S_2 - S_1 = \int_{V_1}^{V_2}\left(\frac{\partial P}{\partial T}\right)_V dV. \tag{11.60}$$

The evaluation of entropy changes, particularly, is simpler by this procedure than with residuals. The results, of course, should be the same. Example 11.5 compares the two methods with the van der Waals equation. The use of residuals in analyzing a typical problem is demonstrated in Example 11.8, which deals with turbine performance.

Example 11.7. Changes in Properties Resulting from Changes in both P and T

The ideal gas heat capacity, C_p^{id}, is assumed to be known as a function of temperature. Changes will be found by following the path *abcd* of Figure 11.1. $\Delta M'_1 = M_1^{id} - M_1$ is evaluated at (P_1, T_1) and $\Delta M'_2$ at (P_2, T_2)

For enthalpy,

$$H_2 - H_1 = \int_{T_1}^{T_2} C_p^{id}\,dT + \Delta H'_1 - \Delta H'_2, \quad (1)$$

and for internal energy,

$$U_2 - U_1 = H_2 - H_1 - (P_2V_2 - P_1V_1)$$

$$= \int_{T_1}^{T_2} (C_p^{id} - R)\,dT + \Delta U'_1 - \Delta U'_2. \quad (2)$$

Since ideal-gas entropy depends on pressure, that factor is included in the equation for entropy change, with the result,

$$S_2 - S_1 = \int_{T_1}^{T_2} \frac{C_p^{id}}{T}\,dT - R\ln\frac{P_2}{P_1} + \Delta S'_1 - \Delta S'_2. \quad (3)$$

For Gibbs energy,

$$G_2 - G_1 = H_2 - H_1 - (T_2S_2 - T_1S_1). \quad (4)$$

The individual entropies in the last equation must be on the absolute basis; for example,

$$S_1 = S_0^{abs} + \int_{T_0}^{T_1} \frac{C_p^{id}}{T}\,dT - R\ln\frac{P_1}{P_0} - \Delta S'_1. \quad (5)$$

The reference condition usually is taken as $T_0 = 298$ K and $P_0 = 1$ atm. Values of absolute entropies at these conditions are known for several thousand substances; some data, for example, are in the book by Stull et al. (1969). Estimates also can be made from group contributions, for which the method of Anderson, Beyer, & Watson is described in the books of Hougen et al. (*Thermodynamics,* 1959, p. 1004) and Reid et al. (1977, p. 266).

Substitution of Eqs. 1 and 5 into 4 will give the desired result. Similarly, the Helmholtz energy change is evaluated from

$$A_2 - A_1 = U_2 - U_1 - (T_2S_2 - T_1S_1). \quad (6)$$

Evaluation of changes in the Planck function, G/T, and of the Massieu function, A/T, requires knowledge of absolute enthalpies. Thus,

$$\left(\frac{G}{T}\right)_2 - \left(\frac{G}{T}\right)_1 = \left(\frac{H}{T}\right)_2 - \left(\frac{H}{T}\right)_1 - (S_2 - S_1), \quad (7)$$

where $S_2 - S_1$ is given by Eq. 4 and the enthalpy of state 1, for example, by

$$H_1 = H_0^{abs} + \int_{T_0}^{T_1} C_p^{id}\,dT - \Delta H'_1. \quad (8)$$

Numerical values of absolute enthalpies are given for many substances in the book of Stull et al. (1969) and in other compilations.

Example 11.8. Expansion of a Nonideal Gas in an Adiabatic Turbine

The effluent of a cumene reactor is at 588.9 K and 600 psia and has the composition: 9.64% cumene, 3.79% propylene, and 86.57% benzene. It is to be expanded to atmospheric pressure in an adiabatic turbine with an isentropic efficiency of 70%. The effluent temperature will be found.

The Tsonopoulos-virial equation will be the basis for finding residual properties from Table 11.2 or Figures 11.5 and 11.6. The weighted ideal-gas heat capacities and the pseudocritical properties with Kay's rules are

$$C_p^{id} = -7.890 + 0.1183T - 7.515(E{-}5)T^2 + 1.792(E{-}8)T^3,$$

$T_c = 562$ K, $P_c = 46.6$ atm, $\omega = 0.221$.

At the starting conditions,

$T_r = 1.048$, $P_r = 0.876$,

$\Delta S'_1 = 0.68(1.987)(0.876) = 1.1888$ cal/(gmol)(K),

$\Delta H'_1 = 1.032(1.987)(562)(0.876) = 1009.9$ cal/gmol.

At isentropic conditions,

$$S_2 - S_1 = 0 = \int_{588.9}^{T_{2s}} \frac{C_p^{id}}{T}\,dT - R\ln(14.7/600) - \Delta S'_2 + 1.1888.$$

The value of the isentropic temperature, T_{2s}, will be found by trial.

With $T_{2s} = 465.2$, $T_r = 0.828$, $P_r = 0.0215$, $\Delta S'_2 = 0.0656$, $\Delta H'_2 = 42.5$, and the value of the integral is -8.4923. Checking the balance,

$$-8.4923 + 7.3699 - 0.0656 + 1.1888 = 0.0008 \simeq 0.$$

The isentropic enthalpy change is

$$(H_2 - H_1)_s = \int_{588.9}^{465.2} C_p^{id}\,dT - \Delta H'_2 + \Delta H'_1$$

$$= -4{,}471.5 - 42.5 + 1{,}009.9$$

$$= -3{,}504.1 \text{ cal/gmol}.$$

At 70% efficiency,

$$0.7(-3{,}504.1) = \int_{588.9}^{T_2} C_p^{id}\,dT - \Delta H'_2 + \Delta H'_1.$$

Try $T_2 = 496.3$, $T_r = 0.883$:

$\Delta H'_2 = 1.51(1.987)(562)(0.0215) = 36.3.$

Checking the balance,

$$-0.7(3{,}504.1) = -2{,}452.9 \stackrel{?}{=} -3{,}426.5 - 36.3 + 1{,}009.9$$

$$= -2{,}452.9,$$

which checks exactly. Note that the virial coefficient properly should have been evaluated with use of the cross-coefficient instead of with Kay's rule, but in this case the change would have made only little difference.

11.5. MIXTURES

Finding the enthalpy or some property M at some temperature and an elevated pressure may be accomplished in terms of excess properties M^{ex} or residual properties $\Delta M'$. The case of enthalpy may be examined first. For a mixture of ideal gases,

$$H^{id} = \sum y_i H_i^{id} = \sum y_i \left(H_{i0} + \int_{T_0}^{T} C_{pi}^{id}\, dT\right). \tag{11.61}$$

From the definitions of H^{ex} and $\Delta H'$ the enthalpy at P and T may be written

$$H = H^{ex} + \sum y_i H_i^{id} \tag{11.62}$$

or

$$H = H^{id} - \Delta H' = \sum y_i H_i^{id} - \Delta H' \tag{11.63}$$

Clearly these are identical since $H^{ex} = -\Delta H'$, but one or the other expression may be more convenient in particular cases. Both quantities are evaluated from equations of state, the former by Eq. 11.64, for instance, and the latter by the equations of Table 11.2 and 11.3.

Both processes are laborious when the correct combining rules for mixtures are employed. Problem 11.20, for example, applies the B-truncated virial equation and correlations. Since the corrections of enthalpy for nonideality usually are not large, however, some simplifications are permissible. For instance, the method in the API data book for evaluation of both $\Delta H'$ and $\Delta S'$ employs the Lee-Kesler equation but with Kay's rules for the pseudocritical properties and the acentric factor. Presumably this procedure is equally valid with other equations of state.

Excess properties likewise are found by means of equations of state. By application of the rule, $M = \Sigma\, y_i \bar{M}_i$, Eqs. 3b and 7b of Table 3.2 can be converted into:

$$H^{ex} = -RT^2 \left(\frac{\partial \ln \hat{\phi}}{\partial T}\right)_P, \tag{11.64}$$

$$G^{ex} = RT(\ln \phi - \Sigma\, y_i \ln \hat{\phi}_i), \tag{11.65}$$

$$S^{ex} = \frac{H^{ex} - G^{ex}}{T}$$

$$= -R\, \Sigma\, y_i \left[T\left(\frac{\partial \ln \hat{\phi}_i}{\partial T}\right)_{Py_j} + \ln \hat{\phi}_i - \ln \phi_i\right]. \tag{11.66}$$

When no EOS is applicable to the liquid state, excess properties are expressed in terms of activity coefficients by Eqs. 11.67–11.69.

$$H^{ex} = -RT^2\, \Sigma\, x_i \left(\frac{\partial \ln \gamma_i}{\partial T}\right)_{Px}, \tag{11.67}$$

$$G^{ex} = RT\, \Sigma\, x_i \ln \gamma_i, \tag{11.68}$$

$$S^{ex} = -R\, \Sigma\, x_i \left[T\left(\frac{\partial \ln \gamma_i}{\partial T}\right)_{Px} + \ln \gamma_i\right]. \tag{11.69}$$

The last three equations will be developed in the next section.

Forming the derivatives of the equations for partial fugacity coefficients from most equations of state that are needed in Eqs. 11.64–11.69 may be an involved process. However, a numerical method is always usable with a computer on the principle that

$$\frac{df}{dT} \simeq \frac{f(1.0001\,T) - f(T)}{0.0001\,T} \tag{11.70}$$

by going through the evaluation of $\ln \hat{\phi}_i$ twice with slightly different temperatures. The factor 1.0001 can be varied as desired.

For example, with the Soave equation, whose fugacity coefficient equations are in Table 3.4,

$$\ln \hat{\phi}_i = \frac{B_i}{B}(z-1) - \ln(z-B) + \frac{A}{B}\left[\frac{B_i}{B} - \frac{2}{a\alpha}\sum_j y_j (a\alpha)_{ij}\right] \ln\left(1 + \frac{B}{z}\right), \tag{11.71}$$

the derivative becomes

$$\left(\frac{\partial \ln \hat{\phi}_i}{\partial T}\right) \simeq \frac{(\ln \hat{\phi}_i)_2 - (\ln \hat{\phi}_i)_1}{T_2 - T_1} \tag{11.72}$$

$$= \frac{\Delta \ln \hat{\phi}_i}{\Delta T}, \tag{11.73}$$

The finite difference equivalent, Eq. 11.73, of the derivative is found upon substitution of the two values of the temperature into Eq. 11.71; there the quantities $A_i, B_i, A, B, z, \alpha_i$, and α are temperature dependent. Example 11.10 employs this method.

In practical work the enthalpy change of mixing at the system pressure often is neglected so that the enthalpy of a mixture becomes the weighted sum of those of the components, thus

$$H = \sum y_i \bar{H}_i \simeq \sum y_i H_i \tag{11.74}$$

and

$$\Delta H' \simeq \sum y_i \Delta H_i' \tag{11.75}$$

The risk of this approximation is hard to appraise. That the result is not consistent with combining rules of even the simplest equations of state can be demonstrated with the example of the van der Waals equation for which the residual enthalpy of a mixture is given by,

$$\Delta H' = -RT\left[\frac{b}{V-b} - \frac{2a}{RTV}\right], \tag{11.76}$$

but the parameters are derived from those of the components with the rules,

$$a = (y_i \sqrt{a_i})^2, \tag{11.77}$$

$$b = \Sigma\, y_i b_i. \tag{11.78}$$

If the enthalpies are assumed additive, the result is

$$\Delta H' = RT \sum y_i \left(\frac{b_i}{V - b_i} \right) - \frac{2}{V} \sum y_i a_i. \qquad (11.79)$$

Clearly, the parameters a and b of the mixture are not formed in the same way from those of the pure components in Eqs. 11.76–11.79. Similar conclusions will apply to more complicated equations of state and their combining rules.

When the parameters of an equation of state are independent of temperature, the same equations apply for $\Delta H'$ and $\Delta S'$ of pure substances as of mixtures with the correct combining rules. Of the EOS covered in this chapter, this conclusion is valid for the van der Waals, Redlich-Kwong, BWR, BWRS, and Lee-Kesler equations; but not for the virial, Soave, or Peng-Robinson equations.

The virial equation: The virial coefficients are strong functions of the temperature, and this fact has been taken into account in the derivations of Eqs. 11.46–11.51. The combining rules are stated in Table 1.8. For the second virial coefficient,

$$B = \sum\sum y_i y_j B_{ij} = \sum y_i^2 B_i + 2 \sum\sum_{i \neq j} y_i y_j B_{ij}, \qquad (11.80)$$

where, in terms of the Pitzer-Curl type of correlation,

$$B_{ij} = \frac{RT_{cij}}{P_{cij}} (B_{ij}^{(0)} + \omega_{ij} B_{ij}^{(1)}). \qquad (11.81)$$

When $i \neq j$, the cross-critical properties usually are evaluated with the Lorentz-Berthelot rules. The derivative is

$$\frac{dB}{dT} = R \sum \frac{y_i^2 T_{ci}}{P_{ci}} \left(\frac{dB_i^{(0)}}{dT} + \omega_i \frac{dB_i^{(1)}}{dT} \right) + 2R \sum\sum_{i \neq j} y_i y_j \frac{T_{cij}}{P_{cij}} \left(\frac{dB_{ij}^{(1)}}{dT} + \omega_{ij} \frac{dB_{ij}^{(1)}}{dT} \right). \qquad (11.82)$$

Substitution of Eqs. 11.80–11.82 into Eqs. 11.46–11.51 will yield the equation for the departure properties. Equations in terms of the Tsonopoulos, Abbott, and Pitzer-Curl correlations are in Table 11.2.

Soave and Peng-Robinson equations: Both these equations have the same form of dependence of the parameter α on the temperature

$$\alpha_i = [1 + m_i(-\sqrt{T/T_{ci}})]^2, \qquad (11.83)$$

and the same combining rules for the parameter combination, $a\alpha$, and for b:

$$a\alpha = \sum_i \sum_j y_i y_j (1 - k_{ij}) \sqrt{(a\alpha)_i (a\alpha)_j}, \qquad (11.84)$$

$$b = \sum y_i b_i. \qquad (11.85)$$

The m_i are not exactly the same in the two equations. The derivative of the Soave equation is

$$\left(\frac{\partial P}{\partial T} \right)_V = \frac{R}{V - b} - \frac{1}{V(V + b)} \frac{d(a\alpha)}{dT}, \qquad (11.86)$$

and that of the Peng-Robinson equation,

$$\left(\frac{\partial P}{\partial T} \right)_V = \frac{R}{V - b} - \frac{1}{V^2 + 2bV - b^2} \frac{d(a\alpha)}{dT}. \qquad (11.87)$$

From Eq. 11.83,

$$\frac{d(a\alpha)_i}{dT} = -\frac{m_i}{T} (a\alpha)_i \sqrt{T_{ri}/\alpha_i}. \qquad (11.86)$$

For the individual terms,

$$\frac{d}{dT} \sqrt{(a_i\alpha_i)(a_j\alpha_j)} = \sqrt{a_i a_j} \frac{d}{dT} \sqrt{\alpha_i \alpha_j} \qquad (11.89)$$

$$= 0.5 \sqrt{\frac{a_i a_j}{\alpha_i \alpha_j}} \left[\alpha_i \frac{d\alpha_j}{dT} + \alpha_j \frac{d\alpha_i}{dT} \right] \qquad (11.90)$$

$$= -\frac{1}{2\sqrt{T}} [k_j \sqrt{a_j} \sqrt{a_i \alpha_i / T_{cj}} + k_i \sqrt{a_i} \sqrt{a_j \alpha_j / T_{ci}}] \qquad (11.91)$$

$$= -\frac{1}{2T} \sqrt{(a\alpha)_i (a\alpha)_j} [k_j \sqrt{T_{rj}/\alpha_j} + k_i \sqrt{T_{ri}/\alpha_i}]. \qquad (11.92)$$

Since $A_i = (a\alpha)_i P/R^2T^2$ and $B = bP/RT$, the following replacement can be made in Eq. 11.92 and results in simpler final expressions for $\Delta H'$ and $\Delta S'$:

$$\sqrt{(a\alpha)_i (a\alpha)_j} = bRT \sqrt{A_i A_j / B}. \qquad (11.92a)$$

For the mixture, the derivative is

$$\frac{d(a\alpha)}{dT} = -\frac{1}{2\sqrt{T}} \sum_i \sum_j y_i y_j (1 - k_{ij}) \times \left[m_j \sqrt{a_j} \sqrt{\frac{a_i \alpha_i}{T_{cj}}} + m_i \sqrt{a_i} \sqrt{\frac{a_j \alpha_j}{T_{ci}}} \right]. \qquad (11.93)$$

$$= -\frac{1}{T} \sum_i \sum_j y_i y_j m_j (1 - k_{ij}) \sqrt{a_i \alpha_i} \sqrt{a_j T_{rj}}. \qquad (11.93a)$$

Substitution into Eqs. 11.31 and 11.41 will result in the equations for $\Delta H'$ and $\Delta S'$ that are summarized in Table 11.3. Example 11.4A develops a computer program for $\Delta H'$ with the Soave equation for a mixture.

Studies in the literature: A comprehensive check of several equations of state for the prediction of enthalpies of pure and mixed substances was made in the API study by Danner et al. (1978). The Soave and Peng-Robinson were less satisfactory than the Lee-Kesler but were as good as the BWR. An earlier study by Sehgal et al. (1968) found the BWR equation satisfactory. A satisfactory comparison of binary excess enthalpy data with predictions from the Wilson equation was obtained by Nagata & Yamada (1974); they used a quadratic dependence of the exponential parameters on temperature. Nagata & Ohta (1978) also found the UNIFAC method reliable for prediction of excess enthalpies of several binary

Example 11.9. Excess Properties with the Wilson and Scatchard-Hildebrand Equations

The system is ethanol (1) + benzene (2). Solubility parameters and molal volumes are: ethanol, 12.915 H, 58.4 ml/gmol; benzene, 9.158 H, 89.4 ml/gmol. The Wilson parameters are $\lambda_{12} = 1{,}297.9$ cal/gmol, $\lambda_{21} = 131.47$.

By the S-H equation,

$$H^{ex} = RT\Sigma\ x_i \ln \gamma_i = (\delta_1 - \delta_2)^2[x_1V_1(1-\phi_1)^2 + x_2V_2\phi_1^2],$$

$$\phi_1 = 1/(1 + x_2V_2/x_1V_1),$$

With the S-H method, H^{ex} is independent of temperature and $S^{ex} = 0$.

By the Wilson equation, use Eqs. 9 and 10 of Example 11.11. The plots reveal poor agreement which is not surprising since the activity coefficients predicted by the two methods also are quite different, as shown in Figure 4.19(a), for instance.

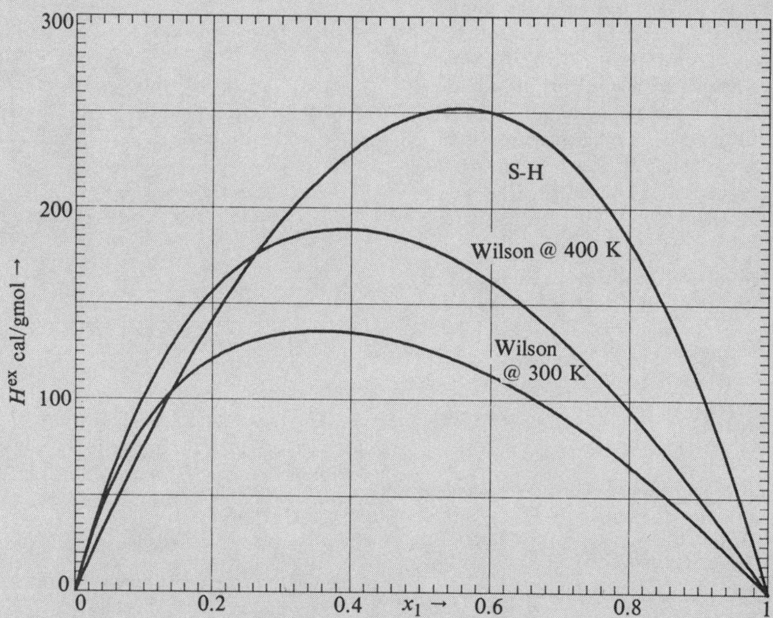

Example 11.9(a). Excess enthalpy with the Wilson and Scatchard-Hildebrand equations. The S-H prediction is independent of temperature.

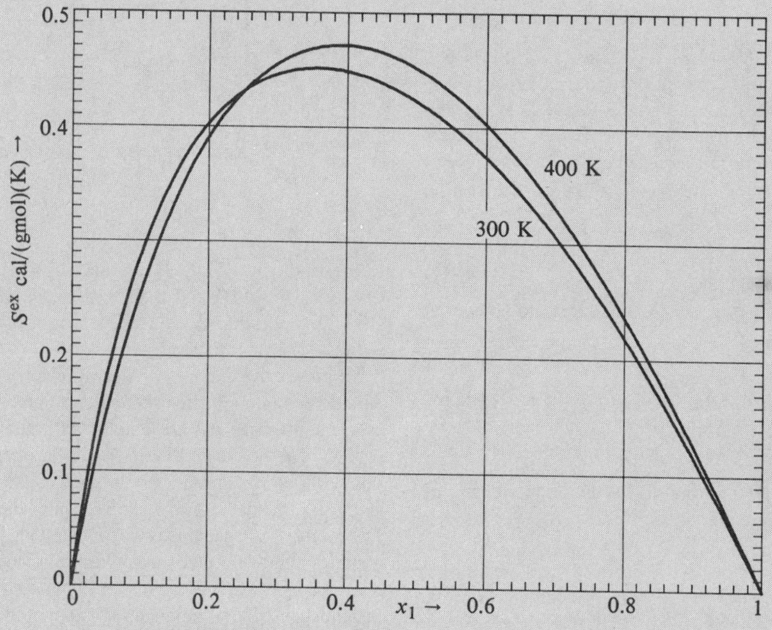

Example 11.9(b). Excess entropy with the Wilson equation. The S-H prediction is $S^{ex} = 0$

Example 11.10. Excess Properties of Liquid Mixtures and Vapor Mixtures with the Soave Equation

An equimolal mixture of H_2S (1) and *n*-heptane (2) will be examined in the vapor phase at 422 K and 5 atm, and in the liquid phase at 310.9 K and 30 atm.

The Soave equation is in Table 1.11 and fugacity coefficients are in Tables 3.3 & 3.4. The excess properties H^{ex} and S^{ex} will be found with Eqs. 11.64 and 11.66. The derivatives are found with Eq. 11.72, using a relatively large temperature difference, 1%, in order to exaggerate the effect on the parameters for illustrative purposes.

From Table 1.12, the binary interaction parameter is $k_{12} = 0.0737$. The constants a and b are obtained with $R = 0.08205$.

	T_c	P_c	ω	a	b
H_2S	373.2	88.2	0.100	4.544	0.0301
n-C_7	540.2	27	0.351	31.103	0.1422
Mix				14.418	0.0862

The calculations of compressibilities and fugacity coefficients are summarized in the table. The excess properties are obtained as follows:

At 422 K for the vapor phase,

$$H^{ex} = -1.987(422)^2(0.5)(-66.7 - 309.5)(E-6)$$
$$= 66.6 \text{ cal/gmol},$$

$$S^{ex} = \frac{H^{ex}}{T} + R\ \Sigma\ y_i \ln(\phi_i/\hat{\phi}_i)$$
$$= 66.6/422 + 1.987\,(0.5)(-0.0223 - 0.0312)$$
$$= 0.105 \text{ cal/(gmol)(K)}.$$

At 310.9 K for the liquid phase,

$$H^{ex} = -1.987(310.9)^2(0.5)(0.0158 + 0.0139)$$
$$= -2{,}775 \text{ cal/gmol},$$
$$S^{ex} = -2{,}775/310.9 + 1.987(0.5)(-0.2548 - 0.0752)$$
$$= -9.3 \text{ cal/(gmol)(K)}.$$

	α	$a\alpha$	10^3A	10^3B	z	$\ln\phi$	$\ln\hat{\phi}$	$\ln\frac{\phi}{\hat{\phi}}$	$10^6\frac{d\ln\hat{\phi}}{dT}$
At 422 K,									
H_2S	0.9207	4.184	17.45	4.347	0.9868	−0.0132	0.00907	−0.0223	−66.7
n-C_7	1.2485	38.832	161.9	20.53	0.8363	−0.1514	−0.1212	−0.0312	−309.5
mix		16.658	69.47	12.45	0.9405				
At 426.2 K,									
H_2S	0.9142	4.154	16.99	4.304	0.9872	−0.0128	0.00879	−0.0216	
n-C_7	1.2386	38.524	157.5	20.33	0.8421	−0.1461	−0.1215	−0.0246	
mix		16.529	67.58	12.325	0.9425				
At 310.9 K,									
H_2S	1.1139	5.062	233.35	35.40	0.0582	−0.2929	−0.0381	−0.2548	0.0158
n-C_7	1.5473	48.126	2220.1	167.23	0.1989	−5.4491	−5.328	−0.1211	0.0139
mix		20.525	946.3	101.37	0.1335				
At 314.0 K,									
H_2S	1.1079	5.034	227.53	35.05	0.0586	−0.2358	0.0108	−0.2466	
n-C_7	1.5473	48.126	2175.1	165.58	0.1975	−5.3602	−5.285	−0.0752	
mix		20.499	926.5	100.37	0.1327				

and ternary mixtures. An early review of enthalpy prediction methods was made by Nathan (1967).

11.6. MIXING OF LIQUIDS

When no equation of state is sufficiently accurate for the liquid phase, other methods must be used to evaluate enthalpies and other properties of such mixtures. The useful quantities for this purpose are the excess properties that were discussed in Chapter 2. They are differences between the properties of real substances and those of ideal mixtures. For instance, the excess enthalpy is

$$H^{ex} = H - \Sigma\ x_i H_i = \Sigma\ x_i(\bar{H}_i - H_i) = \Sigma\ x_i\,\bar{H}_i^{ex}. \tag{11.94}$$

It is related to other properties for which data may be readily obtainable,

$$H^{ex} = -T^2 \int_0^P \left[\frac{\partial(V^{ex}/T)}{\partial T}\right]_P dP \tag{11.95}$$

$$= -T^2 \left[\frac{\partial(G^{ex}/T)}{\partial T}\right]_P . \tag{11.96}$$

Dilatometric methods for measurement of excess volumes are well developed and have been used for the evaluation of excess enthalpies, but the Gibbs energy relation is much the more useful one. The excess Gibbs energy is determined in terms of activity coefficients by the relation,

$$G^{ex} = RT \Sigma\, x_i \ln \gamma_i . \tag{11.97}$$

The activity coefficients are obtained experimentally by measurements of phase equilibria or by other methods explained in Chapter 4. The excess enthalpy may be written,

$$H^{ex} = -RT^2 \Sigma\, x_i \left[\frac{\partial \ln \gamma_i}{\partial T}\right]_P , \tag{11.98}$$

and the excess entropy,

$$S^{ex} = \frac{H^{ex} - G^{ex}}{T} = -R \Sigma\, x_i \left[T\left(\frac{\partial \ln \gamma_i}{\partial T}\right)_P + \ln \gamma_i\right]. \tag{11.99}$$

The properties of the mixture are represented by,

$$H = H^{ex} + \Sigma\, x_i H_i , \tag{11.100}$$

$$S = S^{ex} + \Sigma\, x_i[S_i + \ln(1/x_i)], \tag{11.101}$$

where the pure component properties are found as functions of temperature and pressure as explained earlier in this chapter.

The reverse procedure of finding activity coefficients from calorimetric data of partial molal enthalpies with the relation

$$\bar{H}_i^{ex} = -RT^2 \left(\frac{\partial \ln \gamma_i}{\partial T}\right)_P \tag{11.102}$$

has been used occasionally, for instance by Hanks et al. (1979) and by Tai et al. (1972).

Any activity coefficient correlation with built-in temperature compensation can be used to derive equations for the excess properties. With the Scatchard-Hildebrand equation, for instance,

$$\ln \gamma_i = \frac{V_i}{RT}(\delta_i - \bar{\delta})^2 , \tag{11.103}$$

and

$$\frac{\partial \ln \gamma_i}{\partial T} = -\frac{\ln \gamma_i}{T} , \tag{11.104}$$

so that

$$H^{ex} = RT \Sigma\, x_i \ln \gamma_i = \Sigma\, x_i V_i(\delta_i - \bar{\delta})^2 , \tag{11.105}$$

and

$$S^{ex} = 0. \tag{11.106}$$

The last result is not novel since zero excess entropy is the assumption under which the theory of "regular" solutions and the S-H equation were developed.

The Wilson, NRTL, and UNIQUAC equations also have built-in, approximate temperature dependence. For the Wilson parameters,

$$\Lambda_{kj} = \frac{V_j}{V_i} \exp\left(-\frac{\lambda_{ij}}{RT}\right), \tag{11.107}$$

the derivative is

$$\frac{d\Lambda_{ij}}{dT} = \frac{\lambda_{ij}\Lambda_{ij}}{RT^2} . \tag{11.108}$$

A similar result is obtained if λ_{ij} is assumed to vary linearly with temperature. The derivative of the activity coefficient is

$$\frac{d \ln \gamma_i}{dT} = -\frac{\sum_j x_i \frac{d\Lambda_{ij}}{dT}}{\sum_j x_i \Lambda_{ij}} - \sum_k x_k \left[\frac{\frac{d\Lambda_{ki}}{dT}\sum_j x_j \Lambda_{kj} - \Lambda_{ki}\sum_j x_j \frac{d\Lambda_{kj}}{dT}}{\left(\sum_j x_j \Lambda_{kj}\right)^2}\right]. \tag{11.109}$$

A complete derivation of equations for excess enthalpy and entropy from the binary Wilson equation is made in Example 11.11. Both the S-H and the Wilson equations are employed in the numerical solution of Example 11.9.

Derivations with the NRTL equation were made by Renon et al. (1971) whose results are given as Table 11.4. They assumed linear temperature variation of the two sets of parameters, α_{ij} and τ_{ij}.

Liquid-phase enthalpies were obtained with the Redlich-Kwong equation by Joffe & Zudkevitch (1970); they employed a modified form of this equation with temperature-dependent parameters, although they were not able to generalize this temperature-dependence. The report by Tarakad & Danner (1976) of the API-sponsored work covers six methods, including the Soave and BWRS equations, and finds the Lee-Kesler method to be the most satisfactory for hydrocarbon systems.

11.7. VAPORIZATION ENTHALPY

The enthalpy change accompanying material transfer between phases is obtained by differencing the enthalpies of the terminal states found by the methods of this chapter. For pure substances, the enthalpy of vaporization can be found by empirical methods that are simpler than application of equations of state to both phases even when that is valid.

A starting basis of many of such methods is the Clausius-Clapeyron equation, which relates the vapor pressure and the enthalpy of vaporization,

$$\Delta H_v = RT^2 \Delta z (d \ln P^0/dT), \tag{11.110}$$

where Δz is the difference between the compressibilities of liquid and vapor phases. This equation often is combined with one of the many available vapor pressure correlations. For instance, in terms of the Antoine equation,

$$\ln P^0 = A - B/(T + C), \tag{11.111}$$

the enthalpy of vaporization becomes

Table 11.4. Excess Enthalpy and Entropy with the NRTL Equation (Renon et al., *Calcul sur Ordinateur des Equilibres*, 13–14, 1971).

The parameters α_{ji} and C_{ji} are taken to vary linearly with temperature. The excess entropy is found by combining Eqs. 6 and 7 according to

$$S^{ex} = \frac{1}{T}(H^{ex} - \Sigma x_i \ln \gamma_i), \quad (1)$$

$$C_{ji} = g_{ji} - g_{ii} = C_{ji}^C + C_{ji}^T(T - 273.15), \quad (2)$$

$$\alpha_{ji} = \alpha_{ji}^C + \alpha_{ji}^T(T - 273.15), \quad (3)$$

$$G_{ji} = \exp\left(-\alpha_{ji}\frac{C_{ji}}{RT}\right), \quad (4)$$

$$\tau_{ji} = \frac{C_{ji}}{RT} \quad (5)$$

$$\ln \gamma_i = \frac{\Sigma \tau_{ji}G_{ji}x_j}{\sum_k G_{ki}x_k} + \sum_j \frac{x_jG_{ij}}{\sum_k G_{kj}x_k}\left(\tau_{ij} = \frac{\sum_l \tau_{lj}G_{lj}x_l}{\sum_k G_{kj}x_k}\right), \quad (6)$$

$$H^{ex} = \sum_i \frac{x_i}{\sum_k x_kG_{ki}} \sum_j x_jG_{ji}\left[\left(C_{ji}^C - 273.15\, C_{ji}^T\right)\left(1 - \alpha_{ij}\left(\tau_{ji} - \frac{\sum_k \tau_{ki}G_{ki}x_k}{\sum_k G_{ki}x_k}\right)\right) + RT^2\alpha_{ji}^T\tau_{ji}\left(\tau_{ji} - \frac{\sum_k \tau_{ki}G_{ki}x_k}{\sum_k G_{ki}x_k}\right)\right]. \quad (7)$$

Example 11.11. Derivation of the Equations for Excess Enthalpy and Entropy from the Wilson Equation for Binary Activity Coefficients

$$\ln \gamma_1 = -\ln(x_1 + \Lambda_{12}x_2) + \beta x_2, \quad (1)$$

$$\ln \gamma_2 = -\ln(\Lambda_{21}x_1 + x_2) - \beta x_1, \quad (2)$$

$$\beta = \frac{\Lambda_{12}}{x_1 + \Lambda_{12}x_2} - \frac{\Lambda_{21}}{\Lambda_{21}x_1 + x_2}, \quad (3)$$

$$\Lambda_{ij} = \frac{V_j}{V_i}\exp(-\lambda_{ij}/RT), \quad (4)$$

$$\frac{d\Lambda_{ij}}{dT} = \frac{\lambda_{ij}\Lambda_{ij}}{RT^2}, \quad (5)$$

$$\frac{d\beta}{dT} = \frac{x_1}{(x_1 + \Lambda_{12}x_2)^2}\frac{d\Lambda_{12}}{dT} - \frac{x_2}{(\Lambda_{21}x_1 + x_2)^2}\frac{d\Lambda_{21}}{dT} \quad (6)$$

$$\frac{d\ln \gamma_1}{dT} = -\frac{x_2}{x_1 + \Lambda_{12}x_2}\frac{d\Lambda_{12}}{dT} + x_2\frac{d\beta}{dT}, \quad (7)$$

$$\frac{d\ln \gamma_2}{dT} = -\frac{x_1}{\Lambda_{21}x_1 + x_2}\frac{d\Lambda_{21}}{dT} - x_1\frac{d\beta}{dT}, \quad (8)$$

$$H^{ex} = -RT^2\left[x_1\frac{d\ln \gamma_1}{dT} + x_2\frac{d\ln \gamma_2}{dT}\right], \quad (9)$$

$$S^{ex} = H^{ex}/T + R[x_1 \ln(x_1 + \Lambda_{12}x_2) + x_2 \ln(\Lambda_{21}x_1 + x_2)] \quad (10)$$

$$= \frac{H^{ex}}{T} - R(x_1 \ln \gamma_1 + x_2 \ln \gamma_2). \quad (11)$$

See also Eq. 4.268.

$$\Delta H_v = BRT^2\Delta z/(T + C)^2. \quad (11.111a)$$

Only the Antoine constants and an equation of state for evaluation of the compressibilities of the vapor and liquid are required, but Δz is often taken as unity.

An often satisfactory equation requiring knowledge of only the atmospheric boiling point, T_b, and the critical temperature and pressure is a combination of results by Riedel and by

$$\Delta H_v = \frac{1.1093RT_b(-1 + \ln P_c)}{0.930 - \frac{T_b}{T_c}}\left[\frac{T_c - T}{T_c - T_b}\right]^{0.38}. \quad (11.112)$$

An equation based on the Pitzer-Curl concept of corresponding states was developed by Carruth & Kobayashi (1972). They state:

$$\frac{\Delta H_v}{RT_c} = 7.08(1 - T_r)^{0.354} + 10.95\,\omega(1 - T_r)^{0.456}. \tag{11.113}$$

These three equations are used in Example 11.13 and compared with the result found with the Soave equation in Example 11.12 and with real data. Except with the Clausius-Clapeyron-Antoine equation, the results are within about 2%, which is about the best that is claimed for any of the empirical methods. Some of the many other empirical expressions that have been proposed are described in the books of Reid et al. (1977) and Sterbacek et al. (1979). Somewhat different is a group contribution method worked out by Sastri et al. (1969). An extensive literature survey of heats of vaporization is the book of Tamir & Stephan (1983).

11.8. HYDROCARBON MIXTURES

The standard method of treating mixtures of large numbers of hydrocarbons is to take them as an equivalent of a smaller number of pseudocomponents characterized by average boiling points and densities in appropriate proportions deduced from a true boiling-point distillation. From those properties, the critical properties and molecular weights can be estimated, and finally the enthalpies from the Lee-Kesler or similar correlations can be evaluated.

Less laborious methods often are adequate—for instance, curve fits of graphical correlations such as those presented in the *API Technical Data Book.* A simple formula of this type is due to Zhvanetskii & Platonov (1972) for liquids,

$$H = \frac{0.3897t + 0.0004638t^2}{\rho^{2/3}} \times \left(0.3265 + \frac{0.4515}{\rho}\,(T_b/273.2 + 1)^{1/3}\right), \tag{11.114}$$

in cal/gmol, where

t = temperature, °C,

T_b = average boiling point of the pseudocomponent, °K,

ρ = density, gm/ml.

Example 11.12. Enthalpy Residual of Liquid Benzene and its Enthalpy of Vaporization with the Soave Equation

The conditions are 505.9 K and $P^{sat} = 23.5$ atm. The properties are $T_c = 562.1$ K, $P_c = 48.3$ atm, $\omega = 0.212$. Use the formulas of Table 1.11 in the units (ml/gmol, atm, °K). Various constants associated with the Soave equation are

$R = 82.05,$

$a = 18.825(E6),$

$b = 82.73,$

$k = 0.8070,$

$\alpha = 1.0845,$

$$\frac{d\alpha}{dT} = \frac{-k[1 + k(1 - \sqrt{T/T_c})]}{\sqrt{T_c T}} = -0.00157,$$

$$A = \frac{18.825(E6)(1.0845)(23.5)}{(82.05)^2(505.9)^2} = 0.2785,$$

$$B = \frac{82.73(23.5)}{82.05(505.9)} = 0.0468,$$

$$z^3 - z^2 + 0.2295 - 0.0130 = 0.$$

The three values of the compressibility are $z = 0.08629$ (liquid), 0.21590, and 0.69781 (vapor). For the liquid phase,

$V = zRT/P = 152.4$ ml/gmol

$$H^{id} - H = RT(1 - z) + \frac{a}{b}\left(\alpha - T\frac{d\alpha}{dT}\right)\ln\left(1 + \frac{b}{V}\right)$$

$$= 82.05(505.9)(1 - 0.08629) + \frac{18.825(E6)}{82.73}[1.0845 + 505.9(0.00157)] \times \ln\left(1 + \frac{82.73}{152.4}\right)$$

$$= 223{,}307 \text{ ml atm/gmol, } 5{,}407.8 \text{ cal/gmol.}$$

For the gas phase,

$z = 0.69781,$

$V = 1{,}232.4$ ml/gmol,

$$H^{id} - H = 82.05(505.9)(0.30219) + \frac{18.825(E6)}{82.73} \times [1.0845 + 505.9(0.00157)]\ln\left(1 + \frac{82.73}{1{,}232.4}\right)$$

$$= 40{,}319.6 \text{ ml atm/gmol, } 976.4 \text{ cal/gmol.}$$

Since H^{id} is the same for both phases, the latent heat of vaporization is the difference of the departure functions,

$$\Delta H_v = H_v - H_L = (H^{id} - H)_L - (H^{id} - H)_V$$

$$= 5{,}407.8 - 976.4$$

$$\doteq 4{,}431.4 \text{ cal/gmol.}$$

The API data book gives 4,512 cal/gmol, which is a fairly good check. The *API Data Book* $(H^{id} - H)_V = 998$ cal/gmol also is a fair check.

Example 11.13. Enthalpy of Vaporization of Benzene at $T_r = 0.9$ with Empirical Equations

Data are:

$\ln P^{sat} = 15.9008 - 2{,}788.51/(T - 52.36)$, Torr, °K,

$T_b = 353.3$ K,

$T_c = 562.1$ K,

$P_c = 48.3$ atm,

$\omega = 0.212$.

From Example 11.12, the compressibility difference is

$z = 0.69781 - 0.08629 = 0.61152.$

The Clausius-Clapeyron equation gives

$$\Delta H_v = \frac{2{,}788.51(1.987)(505.9)^2(0.61152)}{(505.9 - 52.36)^2}$$

$= 4{,}215$ cal/gmol.

Note that without the Δz correction the result would have been greatly in error.

The Riedel-Watson equation:

$$\Delta H_v = \frac{2.17(353.3)(-1 + \ln 48.3)}{0.930 - 353.3/562.1}\left[\frac{562.1 - 505.9}{562.1 - 353.3}\right]^{0.38}$$

$= 4{,}444$ cal/gmol.

Carruth & Kobayashi:

$$\Delta H_v = 1.987(562.1)[7.08(1 - 0.9)^{0.354} + 10.95(0.212) \times (1 - 0.9)^{0.456}]$$

$= 4{,}407$ cal/gmol.

The value found with the Soave equation in Example 11.12 was 4431. The *API Data Book* shows 4,512 cal/gmol, with which all but the Clausius-Clapeyron result agree fairly closely.

A much more complex curve fit was developed by Carli (1974) for both liquid and vapor fractions as a function of T, P, and characterization factor; thirty-six coefficients are involved.

For computer work it is common to input values at three or four temperatures from the API charts and to utilize an interpolation routine to find intermediate values. Either a simple polynomial in temperature or a relation such as

$$\sqrt{H} = k_1 + k_2 T + k_3 T^2 \qquad (11.115)$$

is frequently employed.

11.9. PROBLEMS

11.1. Develop expressions for the departure properties, $\Delta H'$ and $\Delta S'$, from these equations of state:

a. Clausius, $(P + a/T(V + c)^2)(V - b) = RT$.

b. Lorentz, $(P + a/V^2)V = RT(1 + b/V)$.

c. Dieterici, $P(V - b) = RT \exp(-a/RTV)$.

d. Berthelot, $z = 1 + \dfrac{9}{128}\dfrac{P}{Pc}\dfrac{Tc}{T}\left[1 - 6\left(\dfrac{Tc}{T}\right)^2\right]$.

11.2. Compare departure properties $\Delta H'$ and $\Delta S'$ figured from the Redlich-Kwong, Soave, and Peng-Robinson equations for saturated toluene vapor at 502 F and 290 psia. For comparison, the *API Data Book* gives $\Delta H' = 22.5$ Btu/lb.

11.3. Margules parameters of water (1) + propionic acid (2) are given at these temperatures (DECHEMA VL Equilib Data, Part 1a, 221, 223):

$T = 60$ C, $A_{12} = 1.3410$, $A_{21} = 1.9777$,

$T = 90$ C, $A_{12} = 0.8593$, $A_{21} = 1.9916$.

On the assumption that $A_{12} = a_0 + a_1 T$ and $A_{21} = b_0 + b_1 T$, find equations for the excess enthalpy and the excess entropy. What are the numerical values of H^{ex} and S^{ex} of an equimolal mixture at 75 C?

11.4. An equation for the reduced vapor pressure, $P°/P_c$, in terms of the reduced temperature for hydrocarbons is

$$\ln P_r = C(3 - 4/T_r + 1/T_r^2) + \frac{0.63278 - C}{18.744}(T_r^8 - 64/T_r + 63).$$

Find the equation for the heat of vaporization (Zia & Thodos, Can J Chem Eng 52, 530 (1974); Henrich et al. Int Chem Eng 20, 77 (1980)).

11.5. The van der Waals equation for n-heptane is

$(P + 31.51/V^2)(V - 0.2065) = 0.08205\ T$

atm, liters/gmol, °K.

At 500 K and 14.3 atm the roots of the equation are $V = 2.1483$, 0.5191, and 0.4080. The Antoine equation holds for the vapor pressure. Find the heat of vaporization at the given temperature. The accepted value is 4,398 cal/gmol.

11.6. The specific volume, cuft/lb, of ammonia is given as a function of the temperature, T °F, and the pressure, P psia, in the table. Find the pressure correction of enthalpy, $\Delta H' = H' - H$, from the residual volumes, and compare with calculations from the Virial-Tsonopoulos and Soave equations. The true values of $H_{10} - 2_{200\ \text{psia}}$ are also given in the table.

	V, cuft/lb				
T \ P	*10*	*50*	*100*	*200*	*$H_{10} - H_{200}$, Btu/lb*
100	35.07	6.843	3.304	1.520	34.4
150	38.26	7.521	3.672	1.740	24.0
200	41.45	8.185	4.021	1.935	18.3

11.7. Show that the virial equation,

$$z = 1 + B\rho + C\rho^2 + \ldots,$$

leads to these equations for the heat capacity and the Joule-Thomson coefficient:

$$c_p - c_p^0 = -(RT^2B'')\rho + R\left[(B - TB')^2 - C + TC' - \frac{1}{2}T^2C''\right]\rho^2 + \ldots,$$

$$\mu c_p^0 = -(B - TB') + [2B^2 - 2TBB' - 2C + TC' - (RT^2/c_p^0)(B - TB')B'']\rho + \ldots,$$

when starting with the thermodynamic relations

$$C_p = C_p^0 - T\int_0^P (\partial^2 V/\partial T^2)_p\, dp, \quad \mu C_p = T(\partial V/\partial T)_p - V,$$

and where $B' = dB/dT$, $B'' = d^2B/dT^2$ and so on.

11.8. Find the excess enthalpy and entropy of an equimolal mixture of ethanol (1) + trichlorethylene (2) at 1 atmosphere. Data are Hirata et al. (no. 487, 1976):

$T = 70.8$ C,

$\Lambda_{12} = 0.12141$ (Wilson parameter)

$\Lambda_{21} = 0.53404$ (Wilson parameter)

$V_1 = 58.39$ ml/gmol

$V_2 = 89.62$ ml/gmol

11.9. Gibbs excess energies and chemical potentials of water + methylethyldiamine, $C_5H_{13}N$, have been correlated by the equations shown (*Int. Data Series,* Series B, 120, 121 (July 31, 1980). From each of the three correlations, find H^{ex} and S^{ex} at 308.15 K at $x_1 = 0.25$, 0.5, and 0.75.

Smoothing Equation

$$G^E/RT = Ax_1(1 - x_1) + x_1(1 - x_1)(2x_1 - 1) + [C_1 + C_2(2x_1 - 1) + C_3(2x_1 - 1)x_1^6]$$

$$\mu_1^E/RT = A(1 - x_1)^2 - C_1(1 - 6x_1 + 9x_1^2 - 4x_1^3) + C_2(1 - 10x_1 + 29x_1^2 - 32x_1^3 + 12x_1^4) + C_3(7x_1^6 - 46x_1^7 + 107x_1^8 - 104x_1^9 + 36x_1^{10})$$

$$\mu_2^E/RT = Ax_1^2 - C_1(3x_1^2 - 4x_1^3) + C_2(5x_1^2 - 16x_1^3 + 12x_1^4) - C_3(6x_1^7 - 35x_1^8 + 64x_1^9 - 36x_1^{10})$$

	T, °K					
	283.15	293.15	303.15	308.15	313.15	320.15
A	1.7983	1.9409	2.0656	2.1477	2.2067	2.2647
C_1	0.01811	0.08909	0.15728	0.20630	0.24316	0.27895
C_2	0.7496	0.6312	0.5597	0.5351	0.4808	0.4527
C_3	0.1105	0.5906	0.8575	0.8690	0.8946	0.9173

Wilson equation: $G^E/RT = x_1 \ln (x_1 + A_{12}x_2) + x_2 \ln (x_2 + A_{21}x_1)$

T, °K	A_{12}	A_{21}	$\sigma(G^E)$/J mol^{-1}
283.15	0.2611	0.2170	35.7
293.15	0.2795	0.1292	31.9
303.15	0.2750	0.0777	25.6
308.15	0.2477	0.0615	18.7
313.15	0.2507	0.0448	11.3
320.15	0.2364	0.0340	5.5

NRTL equation: $G^E/RT = x_1x_2[A_{21}G_{21}/(x_1 + x_2G_{21}) + A_{12}G_{12}/(x_2 + x_1G_{12})]$ where $G_{12} = \exp(-\alpha A_{21}/RT)$, $G_{21} = \exp(-\alpha A_{21}/RT)$ and A_{12}, A_{21} and α are adjustable parameters.

T, °K	A_{12}	A_{21}	α	$\sigma(G^E)$/J mol^{-1}
283.15	5166.79	4922.48	0.615	3.5
293.15	6211.77	4826.83	0.568	2.8
303.15	7067.01	4812.98	0.528	4.6
308.15	7363.92	4734.57	0.503	4.1
313.15	7528.72	4651.36	0.486	5.3
320.15	7791.81	4596.28	0.467	5.3

11.10. Acetone at the rate of 2 gmol/sec is subjected to a reversible cycle consisting of an isothermal expansion (*ab*) from 30 atm to 10 atm at 1,000 K, followed by isobaric (*bc*) and isochoric (*ca*) processes. The Redlich-Kwong equation applies, of which the parameters are $a = 360.5$ and $b = 0.0778$, in the units atm, °K, and liters/gmol. The heat capacity of the ideal gas state is

$$C_p' = 1.505 + 0.06224\,T - 2.992(E-5)T^2 + 4.867(E-9)T^3, \text{ cal/gmol°K.}$$

For the processes *ab, bc, ca,* and *abca,* find the changes in enthalpy and the work and heat effects.

11.11. Carbon monoxide is compressed isentropically at the rate of 1 kgmol/sec from 300 K and 10 atm to 50 atm. The van der Waals parameters are $a = 1.485$, $b = 0.03985$, in the units atm, °K, liters/gmol. The ideal gas heat capacity is

$$C_p' = 6.79 + 0.0021T - 2.06(E5)/T^2, \text{ cal/(gmol)(K).}$$

Find the exit temperature and the enthalpy change.

11.12. Carbon monoxide is expanded isenthalpically from 300 atm and 500 K to 10 atm. The Redlich-Kwong parameters are: $a = 16.985$, $b = 0.0274$, in the units, atm, °K, liters/gmol. Ideal gas heat capacity is given in Problem 11.11. Find the exit temperature.

11.13. An equimolal mixture of water + ethanol is to be prepared at 80 C from the pure components at 0 C.

a. What is the enthalpy input to the system, cal/gmol, if the final pressure is essentially one atm?

b. What further change in enthalpy is required if the pressure is to be raised to 1,000 atm and the temperature is to remain at 80 C?

Excess enthalpy data are known as functions of temperature as follows (*Int. Data Series,* Series B, #56, #57 (September 26, 1978):

T, °C	25	50	58	70	90	110
H^{ex}, J/gmol	−407	−121.7	−18.5	129.2	378.5	598.7

Heat capacity of ethanol is 0.68 cal/gmol K. The volume average compressibility is

$$k = -\frac{1}{V}\frac{dV}{dP} = 0.0001/\text{atm},$$

and the thermal expansion coefficient is

$$\beta = \frac{1}{V}\frac{dV}{dT} = 0.00085/\text{K}.$$

Specific gravity of the mixture is 0.86.

11.14. An equimolal mixture of water and 1,4-dioxane is prepared at 50 C. Find the enthalpy change, cal/gmol, by several methods.

a. Use the Scatchard-Hildebrand equation with $\delta_1 = 23.53$, $\delta_2 = 10.13$, $V_1 = 18$, and $V_2 = 85.3$.

b. Use the Margules equation with $A_{12} = 1.7692$, $A_{21} = 1.8814$, and assume that $T \ln \gamma_i =$ constant.

c. Use the Wilson equation, given that $\lambda_{12} = 1{,}811.85$ and $\lambda_{21} = -137.85$ cal/gmol.

d. Use the NRTL equation with $\alpha_{12} = 0.44$, $\Delta g_{12} = 825.83$ and $\Delta g_{21} = 765.13$ cal/gmol.

11.15. A substance has the equation of state

$$P = 0.082T/(V - 0.035) - 7.89/(V^2 + 0.08V - 0.0002),$$

where the units are atm, liters/gmol, °K. Find

a. the enthalpy at 400 K and 30 atm relative to that at 400 K and 1 atm.

b. $C_p - C_v$ at 400 K and 30 atm.

Some solutions of the equation of state are

P	V	$(\partial V/\partial T)_P$
1	32.60	0.08265
15	1.898	0.006046
30		0.003758

11.16. These data are known for a gas at 500 K: When $P = 10$ atm, $V = 1.0516$ liters/gmol; when $P = 50$, $V = 0.4132$.

a. Find the parameters of the Clausius equation, $P = RT/(V - b) - a/TV^2$.

b. Find the changes in entropy, internal energy, and enthalpy when the pressure is raised from 10 to 50 atm at 500 K.

11.17. The enthalpy change of the reaction, $CO + H_2O \rightleftarrows CH_3OH$, is −50,880 cal/gmol at 700 K and zero pressure. What is the enthalpy change at 200 atm? Assume the Redlich-Kwong equation to apply and neglect enthalpy change of mixing.

11.18. Show that Eq. 12 of Table 11.3 can be written,

$$\Delta S'/R = -\ln[z(1 - B/z)] + \frac{1}{2B}\ln(1 + B/z) \times \sum_i \sum_j [y_i y_j (1 - k_{ij})\sqrt{A_i A_j}(m_i\sqrt{T_{ri}/\alpha_i} + m_j\sqrt{T_{rj}/\alpha_j})].$$

Extend the program of Example 11.4A to cover the evaluation of $\Delta S'$.

11.19. In Example 11.4A, explore the effect of temperature in the range of 300 to 600 K on $\Delta H'$ of the liquid and vapor phases of that mixture.

12 Principles of Experimental Methods for Phase Equilibria

12

Principles of Experimental Methods for Phase Equilibria

In concept the measurement of phase equilibria involves simply the measurements of (1) pressure, (2) temperature, (3) phase compositions, and (4) phase amounts. In practice, however, it is not a simple matter to obtain data of sufficient accuracy. Care must be taken to assure that equilibrium really exists, that the temperature and pressure are measured at the position where equilibrium exists, and that the taking of samples for analysis does not disturb the equilibrium appreciably. These problems are serious enough to have given rise to a considerable variety of equipment and techniques for the study of the several kinds of phase equilibria. Some typical ones will be described in this chapter.

A change of phase is accompanied by changes in physical properties, some of which may be measured conveniently and adequately. Depending on the particular system, even properties such as the speed of sound or the electrical conductivity may be utilized to detect phase changes, but the most generally useful techniques are the following:

1. Visual or with optical instruments: The sample is heated or cooled in a transparent vessel, and abrupt changes in refractivity, turbidity, color, or bubble or droplet formation are noted.

2. Dilatometric: An abrupt change in volume of a sample often accompanies phase transformations. Urethane, of which the phase diagram is shown as Figure 5.10(b), is an example; at 1 atm the specific volume is 1.014 ml/g; on melting at 47.9 C, its volume increases by 0.0599 ml/g. Transformations at the three triple points also are accompanied by sizable changes in specific volume, as shown in Table 12.1. A review of the abundant literature on dilatometric methods may be found in the book of Daniels (1973, Chap. 5).

3. Thermal methods that study the temperature or heat transfer behavior as a function of time: There are three distinct methods:

a. Thermal analysis (TA), or the method of heating or cooling curves, which observes the variation of the temperature of a sample with time under conditions of approximately constant heat input or output.

b. Differential thermal analysis (DTA), in which the difference in temperature between the sample and a reference material that is not subject to phase change is noted as a function of time as heat is transferred in or out.

c. Differential scanning calorimetry (DSC), in which the temperatures of the sample and the reference material are maintained the same and the variation of the enthalpy transfer with time required to maintain this condition is measured. This method detects not only the temperatures at which phase transformations occur but also the accompanying enthalpy changes.

DTA and DSC are highly refined techniques for which several commercial instruments are available. They are used primarily for the study of condensed phases.

It is not within the scope of this chapter to describe the measurement and control of temperature and pressure, for which a very broad range of equipment and techniques has been developed, from simple to sophisticated. These topics are covered in detail in standard works on laboratory technique—for example, Weissberger & Rossiter (1971) and Hala et al. (1967). Both these texts also cover the measurement of vapor pressure.

Table 12.1. Volume Changes Accompanying Phase Transformations of Urethane. See Figure 5.10b and LB II/2a, 254.

Pressure, atm	*Temp, C*	*Phase Change*	*ΔV, ml/gm*
2,350	66.2	I to Liquid	0.0253
		II to Liquid	0.0355
		I to II	−0.0102
3,400	25.5	I to II	−0.00922
		II to III	0.04820
		I to III	0.05742
4,230	76.8	II to Liquid	0.01840
		III to Liquid	0.06396
		II to III	0.04556

12.1. SOLUBILITY IN LIQUIDS

In simple terms, solubilities are determined by keeping solute and solvent in contact until equilibrium is attained, then measuring the amounts and compositions of the phases by appropriate means. Attainment of equilibrium is accelerated by mechanical agitation, shaking, or vibration of the vessel or by circulating one phase through the other by pumping. The proper contact time may be only a few minutes or up to several weeks with viscous liquids and materials that are soluble only with difficulty. Whether or not an equilibrium condition has been reached can be checked by: (1) Approaching it from opposite directions, that is, from an undersaturated state or a supersaturated state like that obtained at a temperature much different from the desired value; (2) periodic sampling of the solution until the conditions are found to be steady.

Care must be taken to avoid metastable equilibrium. For example, supersaturation is more likely to occur if solute is being expelled from solution than if it is being dissolved. If the solubility is being found by lowering the temperature of the solution several experiments at different cooling rates should be performed. At a sufficiently low cooling rate and with agitation, supersaturation should not occur.

Gases: Since the early work on the solubility of gases by Henry (1803), Bunsen (1855), and Ostwald (1890), a large variety of equipment has been developed. The operating pressure and temperature and of course the natures of the solute and solvent as well as the accuracy required are the factors influencing the kind of equipment and procedures to use. Two functions are required of the equipment: (1) saturation of the liquid with the gas, and (2) analysis of the amount of dissolution that occurs.

Figure 12.1 is a schematic representation of the main features of practicable equipment, although it does not represent any specific design that actually has been used. A weighed amount of solvent is placed in the dissolver, which then is charged with solute to the desired pressure. The shrouded, high-speed impeller recirculates the gas through the solution. Any foreign gases that may have been dissolved in the solvent are removed by evacuating the solvent initially. Then the process of saturation with the solute is performed. The pressure is kept constant by adjustment of input of the solute gas. The amount of gas dissolved is found by noting the change of volume in the graduated tube and applying an appropriate *PVT* equation. If the solvent is essentially involatile, a small sample of the saturated solution may be withdrawn from the dissolver into an evacuated burette, as in Figure 12.2. After the gas has flashed from the solution, the liquid is drawn off while the remaining gas is compressed to

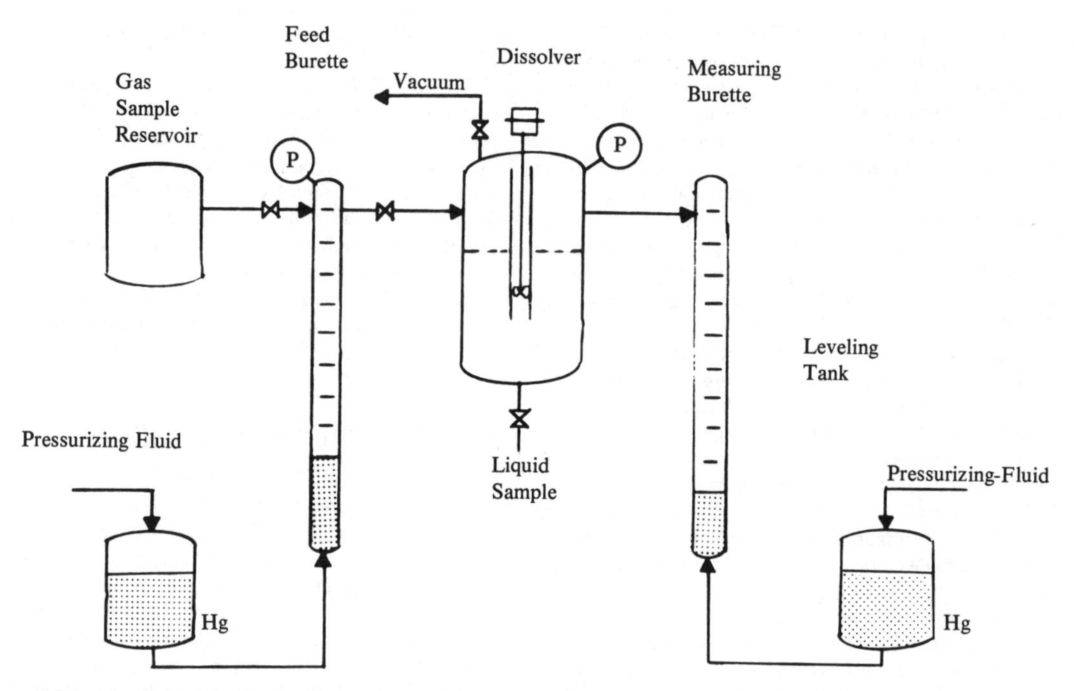

Figure 12.1. Apparatus for dissolving a gas in a liquid and for measuring the amount dissolved. The amount of gas introduced into the dissolver becomes known when the initial and final pressures and mercury levels in the feed cylinder are measured. The amount of gas remaining undissolved in the vessel of known volume is found by noting the amount of gas space in the measuring cylinder after the mercury has been leveled there.

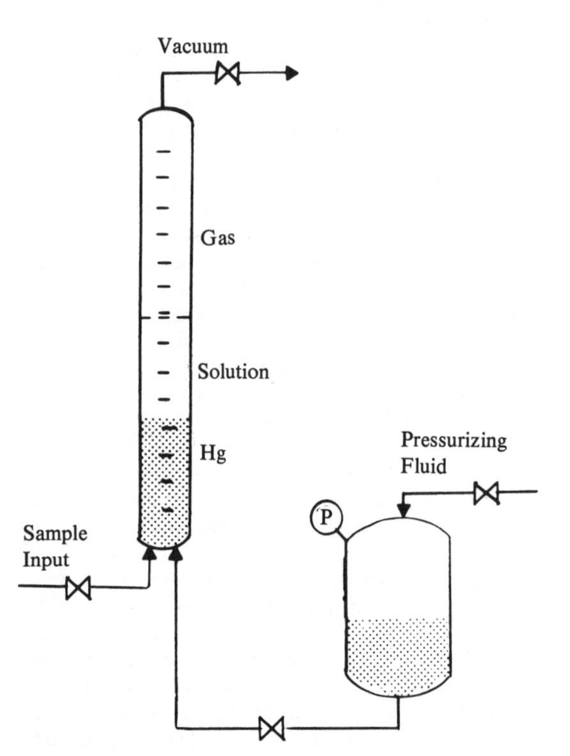

Figure 12.2. Apparatus for finding the amount of gas that has been dissolved. The solution is introduced into an evacuated burette, the liquid is withdrawn, and the gas is recompressed to some desired amount so that its mass can be found by reading the burette.

atmospheric pressure with a column of mercury through a leveling bulb and its volume noted.

Appropriate thermostating and strength for high pressures can be designed into the equipment, and the agitation can be made intense enough to handle immiscible liquids if such are being tested. Abundant references to the extensive literature are in the books edited by Dack (1976) and Weissberger & Rossiter (1971).

Liquids: Mutual solubilities of two liquids are determined by mixing weighed amounts until equilibrium is attained, separating the phases, and measuring their amounts and compositions. Many analytical schemes are feasible: chemical analysis, chromatographic analysis, or determination of refractive index or density when these are known as functions of composition. When there is a tendency to emulsification, stirring must be gentle, so it may take several hours for attainment of equilibrium.

Solubilities over a range of temperatures are determined by preparing solutions of known compositions at elevated temperatures, then cooling slowly and noting the temperature at which the isotropic system becomes cloudy. In some cases when the cloud point is hard to detect, even with photoelectric means, minute traces of dyes that are preferentially soluble in the precipitating phase have been incorporated in the solution.

Hill's method (1923) avoids the need for analysis by performing two experiments simultaneously with different proportions of the two liquids and in contact with the same vapor phase. In such cases the compositions of the two liquids are independent of the proportions of the phases, since they are in equilibrium with the same gas phase. In terms of the following notation:

m_1 = mass of component 1 in experiment 1.
m_2 = mass of component 1 in experiment 2.
x = concentration of component 1 in the lower phase.
y = concentration of component 1 in the upper phase.
L_1 = volume of the lower phase in experiment 1.
V_1 = volume of the upper phase in experiment 1.
L_2 = volume of the lower phase in experiment 2.
V_2 = volume of the upper phase in experiment 2.

the material balance on component 1 becomes

$$m_1 = L_1 x + V_1 y,$$

$$m_2 = L_2 x + V_2 y,$$

which is readily solved for the compositions of the phases, x and y. A similar set of equations is written and solved for component 2, or the amounts of component 2 may be calculated from x and y and the densities of the two phases.

Solids: If great accuracy is not required and the pressure and temperature are near atmospheric, simple methods and equipment suffice for measuring the solubility of solids in liquids: The two phases are shaken together until equilibrium is attained; the liquid phase is sampled after decantation or filtration and then analyzed chemically or by evaporation. Cooling a hot concentrated solution gradually and noting the temperature at which precipitation occurs is another widely used method.

The problem of sampling at extreme temperatures and pressures has been solved in a number of ways, some of which are described by Weissberger & Rossiter (1971). Usually they involve withdrawing a sample from the supernatant liquid through a valved line and bringing it to a temperature suitable for analysis in a thermostat. When the separation of the phases does not take place readily and filtration is necessary, a double-chambered apparatus like that of Figure 12.3 may be used. After the solution becomes saturated, the vessel is inverted and the solution filters through the tube while gas escapes from chamber B to A through the capillary.

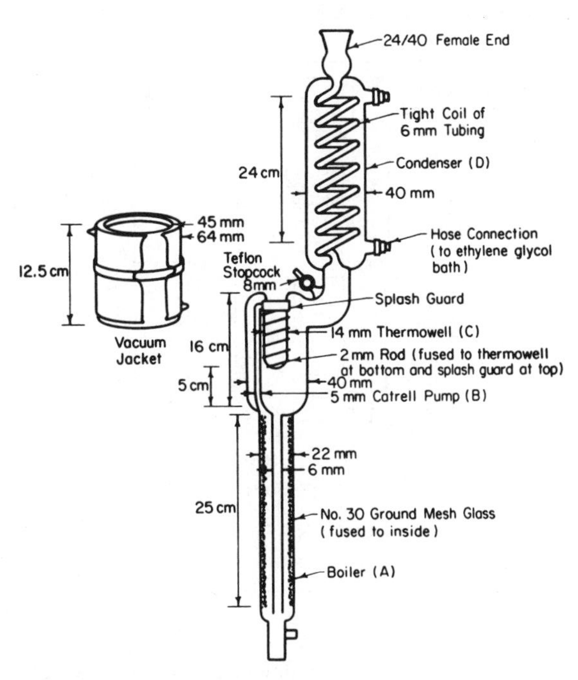

Figure 12.4. Ebulliometer design. It consists of a boiler (A), a Cottrell tube (B), a thermowell (C) that is splashed with the vapor-liquid mixture from the Cottrell, and a condenser (D) that is supplied with refrigerant cooling at −20C (Thomas et al. 1982). Copyright American Chemical Society; used with permission.

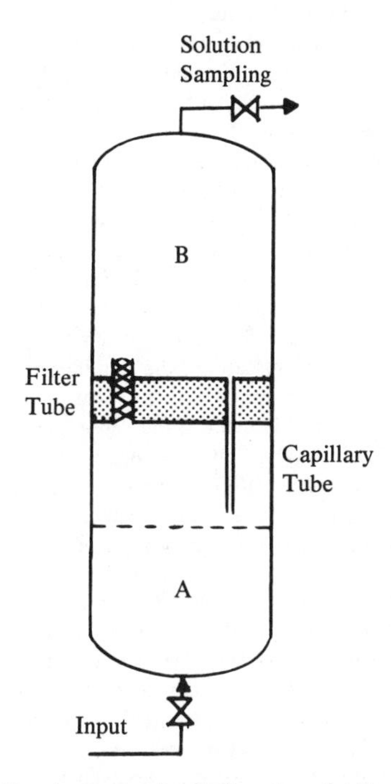

Figure 12.3. Apparatus for finding the solubility of a solid. Dissolution is accomplished in chamber A. After inversion of the equipment, the filtered solution is sampled from chamber B.

Materials that are soluble only with difficulty sometimes require days or weeks for attainment of equilibrium, and analysis of the resulting dilute solutions may require methods peculiar to the individual systems—for example, electrical conductivity. Infterferometry has been proposed as a general technique for dilute systems (Mitchell 1926).

12.2. BOILING POINT AND VAPOR PRESSURE

The obvious way to measure a boiling temperature by noting the reading of a thermometer located in either the vapor or the liquid phase may be subject to appreciable error. Superheating is caused by surface tension and hydrostatic head, whereas small amounts of volatile inpurities such as water, which is often present in trace amounts in organic liquids, make the vapor-phase thermometer reading too low because the condensation temperature of a mixture is being measured rather than that of the pure vapor.

At present, most satisfactory ebulliometers employ Cottrell's principle, which is to circulate a mixture of bubbles and entrained liquid against a thermometer in the vapor phase, as shown in Figures 12.4 and 12.8. Refinements to this basic instrument include means to slow down the runoff of liquid from the thermometer and control of the vaporization rate,

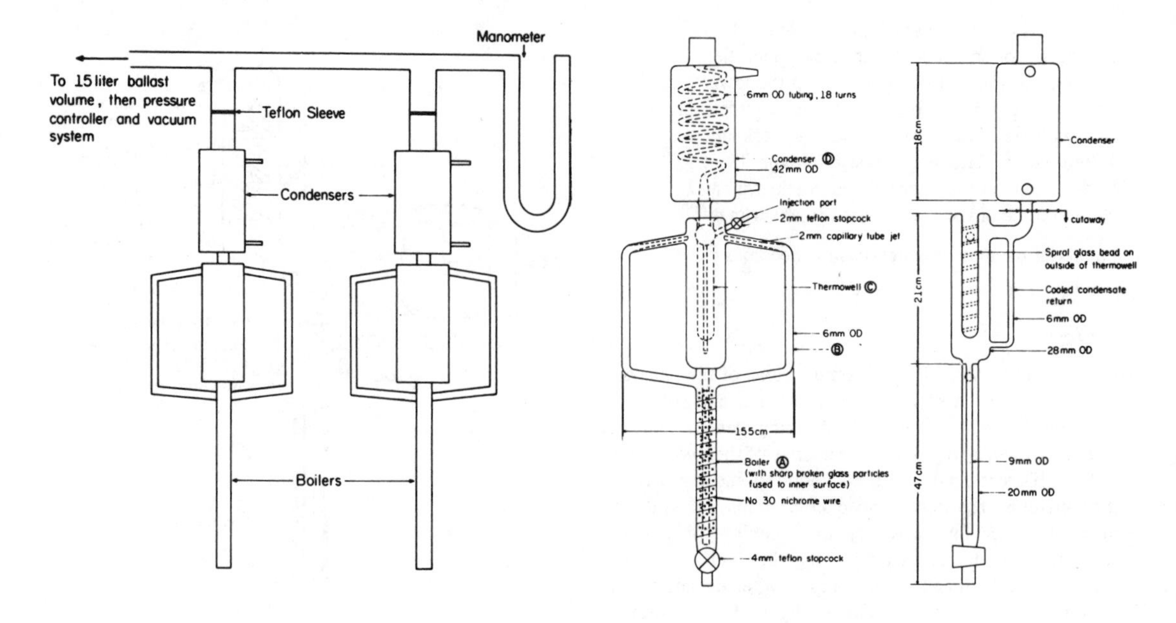

Figure 12.5. A differential ebulliometer for measuring boiling points of dilute solutions. One of the cells is loaded with pure solvent and the other with the dilute solution. Temperature differences are accurate to 0.001 K. Details of an improved ebulliometer are shown in Figure 12.4 (Eckert et al. 1981).

which must be kept in a certain range—depending to some extent on the nature of the liquid—for accurate determination of the boiling point. Special arrangements required for heterogeneous liquids are discussed by Hala et al. (1967).

Ebulliometers constructed to operate over a range of pressures are used to find vapor pressures. The pressure is set and the boiling temperature is found. A device called an isoteniscope (Smith & Menzies, 1910) is used to find the vapor pressure at a specified temperature. In Figure 12.6, the bulb and the attached U-tube are filled partially with the liquid being tested. After boiling to remove any dissolved gases from the liquid, the device is thermostatted. Then the pressure is adjusted with a mercury column until bubbles begin to form. Instrumental instead of visual means to detect the bubble formation may be desirable.

12.3. VAPOR-LIQUID EQUILIBRIUM

Recirculating stills: An early, reasonably satisfactory device for finding the equilibrium compositions of liquid and vapor is that of Othmer, shown in Figure 12.7. Vapor is continuously generated in vessel A, then condensed and collected in reservoir B. Overflow of condensate returns to the boiler. When the thermometer in the vapor phase indicates steady conditions, the boiling liquid and the condensate are sampled and analyzed.

Inaccurate temperature readings may be obtained, as mentioned for ebulliometer operation, so modern devices usually incorporate Cottrell pumping in their design. The first instrument of this kind was the Gillespie (1946) Still, shown in Figure 12.8. Operation is identical to that of the Othmer Still except that the thermometer is splashed with a mixture of bubbles and entrained liquid.

Over the years many improvements have been made or attempted in the design of vapor-liquid equilibrium stills in order to improve accuracy, or for greater ease of operation, or to handle particular mixtures or two-phase liquids, or for extreme temperatures and pressures. Hala et al. (1967) have diagrams of about thirty different designs of such equipment.

Static methods: A simple method, in principle, of determining vapor-liquid equilibrium is to maintain the mixture under conditions of temperature and pressure at which liquid and vapor phases coexist, to mix them thoroughly until equilibrium is established, and then to sample and analyze the two phases without disturbing the equilibrium appreciably. One of the problems that may arise at low or moderate pressures is the sizable mass of the vapor sample needed for analysis, which may be large enough to upset equilibrium when it is withdrawn. With equipment of this kind, accordingly, it is desirable to employ chromatography or mass spectrometry for analysis since samples of a milliliter or so then suffice.

One working scheme for withdrawing small samples employs plug-type sampling valves that have cutouts of 1 ml or so in volume on the face of the plug. The cutout faces the cell contents until equilibrium is established, then it is turned 180 degrees so that the entrained sample can proceed to the chromatograph. At elevated pressures, samples of such volumes also may be sufficient for analysis by other means.

A differential ebulliometer capable of measuring temperatures of very dilute solutions can be employed to find activity coefficients at infinite dilution. In the apparatus shown in Figure 12.5, one of the ebulliometers contains the pure solvent

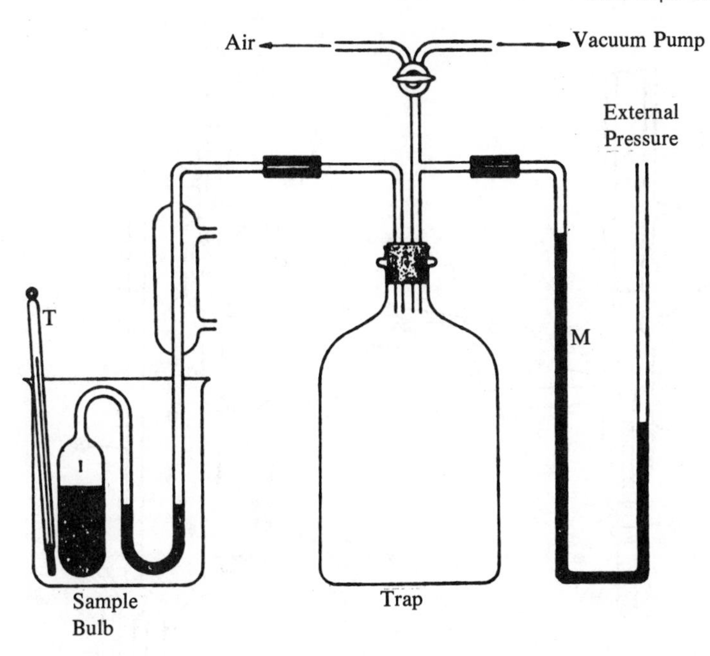

Figure 12.6. Measurement of vapor pressure by isoteniscope of Smith & Menzies (1910). The bulb and attached U-tube are partially filled with the sample. The sample is boiled until all dissolved gas has been expelled, then the temperature is reduced to the desired value. Finally, the external pressure is adjusted until the liquid in the U-tube is leveled. That is the vapor pressure.

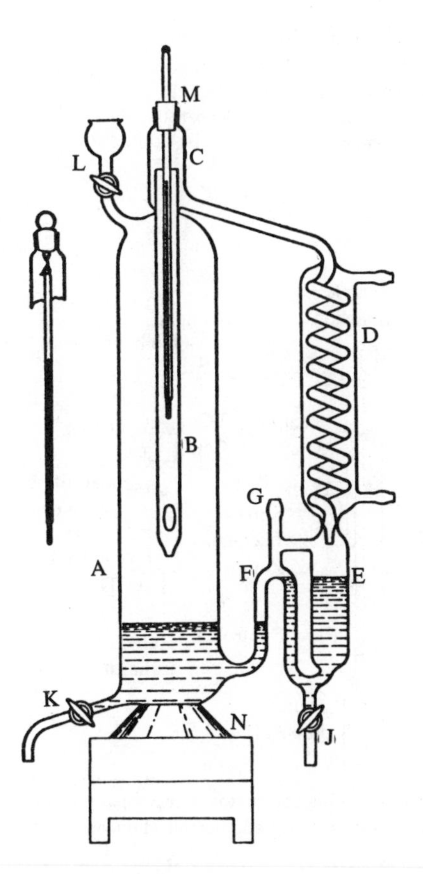

Figure 12.7. Vapor-liquid equilibrium still employing recirculation of condensate until equilibrium is established (Othmer 1928). *B* is a vapor shield to prevent condensation and rectification on the thermometer. Condensate sample is withdrawn at *J* and liquid sample at *K*. The equipment can be pressurized at *L*, but originally it was designed for operation at atmospheric pressure only.

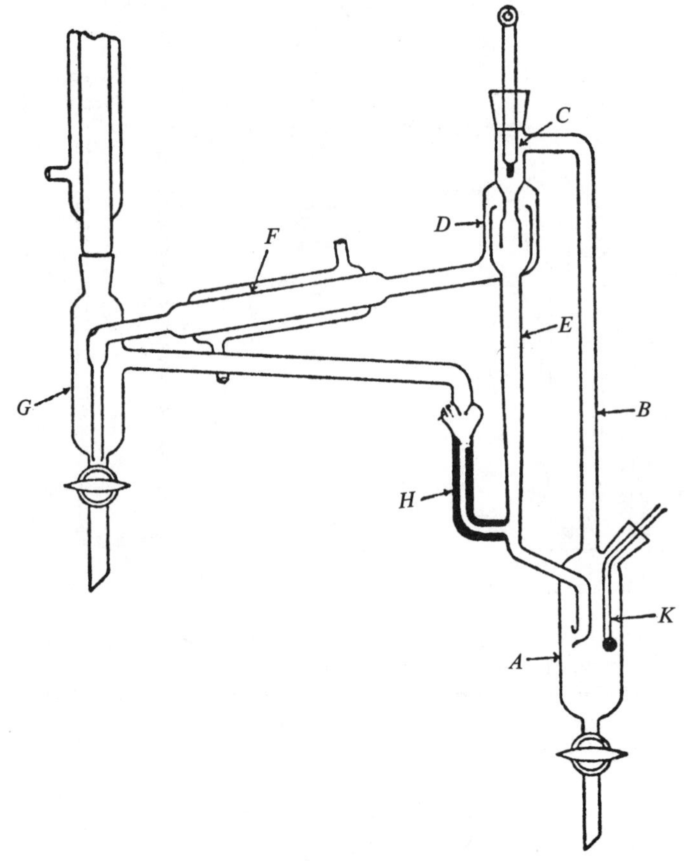

Figure 12.8. Vapor-liquid equilibrium still with recirculation of both phases (D.T.C. Gillespie, *Ind. Eng. Chem. Anal. Ed. 18,* 1946, 575). A = boiler; B = Cottrell pump; C = thermometer; D = vapor-liquid separator; E = entrained liquid return; F = condenser; G = condensate receiver; H = condensate overflow return; K = heater.

and the other the dilute solution. Temperature differences accurate to within 0.001 K can be measured, and the effect of concentrations of the order of 0.005 mol fraction can be detected. It also may be mentioned here that gas chromatographic techniques are being used to measure activity coefficients of solvents or pairs of solvents in heavier liquids. An apparatus of this kind is illustrated in Figure 12.9. Such methods are reviewed at length by Letcher (1978), and some results have been obtained recently by Eckert et al. (1981).

Somewhat elaborate assemblies of equipment are required for accurate work at elevated pressures or low temperatures. The equilibrium cells usually are built with a window so that bubble- or dewpoints can be observed visually. Two typical assemblies are represented in Figures 12.10 and 12.11. The experimental substances are first charged under their vapor pressures or by displacement with mercury into loading cells made of precision bore tubing so that a transfer of mass can be figured from the change in level and the liquid density. Levels are read with a cathetometer for accuracy. Mixing of the charge in the equilibrium cell is accomplished with a falling and rising ball actuated by an external magnet or with a magnetic stirrer. Any dissolved foreign gases are vented from the liquids in the loading cells. The temperature is adjusted

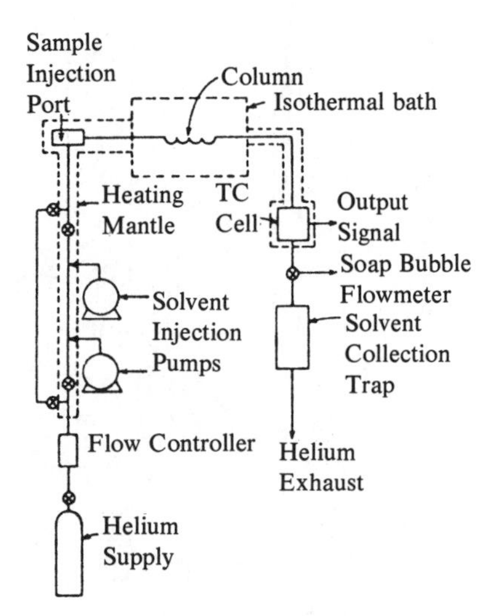

Figure 12.9. Gas chromatographic apparatus for measuring infinite-dilution activity coefficients (Dincer et al. 1981).

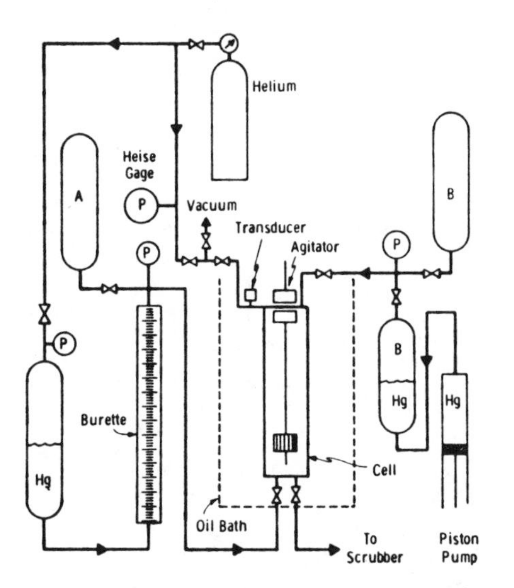

Figure 12.10. Equipment for the measurement of vapor-liquid equilibria at elevated pressures and moderate temperatures (Schotte 1980).

with the thermostatic bath and the pressure through mercury supplied with a piston pump. Samples of the liquid phase alone or of both phases can be taken for analysis after equilibrium has been achieved.

Schotte's equipment of Figure 12.10 is designed for pressures to 55–70 atm. The cell is 10 cm dia by 15 cm. The metering accuracy is 0.4 ml/component, the temperature is within 0.2 C, pressure accuracy is within 0.8%, and vapor composition analysis within 1–2%. The overall error for a typical set of measurements and activity coefficient derived with the best kinds of correlations is at most 7%, often as good as 5% according to this author. Some detail of a cell used by other investigators at pressures up to 100 bar is given in Figure 12.12.

Pressure-composition measurements: Static equilibrium cells like those of Figures 12.10–12.12 are especially useful for finding bubble- and dewpoint pressure at controlled temperatures and known overall compositions. Such data can be transformed into vapor-liquid equilibrium data by the methods described in Section 4.17.4. Although assumptions are needed about the forms of suitable correlating equations for fugacity and activity coefficients—such as the virial and Wilson equations, for instance—very accurate results are obtainable without the bother and uncertainty of sampling and analyzing either or both phases.

Glass cells are usable at pressures of tens of atmospheres. With them, the formation of bubbles or droplets can be observed visually, but photoelectric means of detection are readily implemented when preferable or necessary. Many variations of static equipment design and operation have been used. The equilibrium cells usually are charged by displacement with mercury or with pumps. Liquids also may be charged as weighed closed ampoules that are broken in place by a magnetically controlled drop. This technique has been used by McGlashan & Williamson (1961) and Pemberton & Mash (1978). Among the many instrumental methods that have been employed to detect bubble- and dewpoints may be mentioned differential thermal analysis, a technique of very wide application in the study of phase transitions and well supplied with commercial equipment of debugged design.

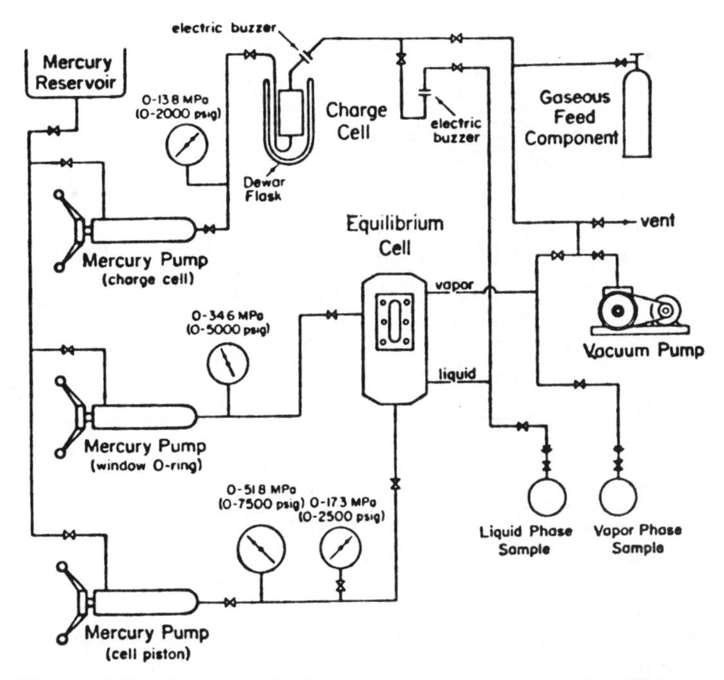

Figure 12.11. Equipment for the measurement of vapor-liquid equilibria at elevated pressures and moderate temperatures (Gomez-Nieto & Thodos 1978).

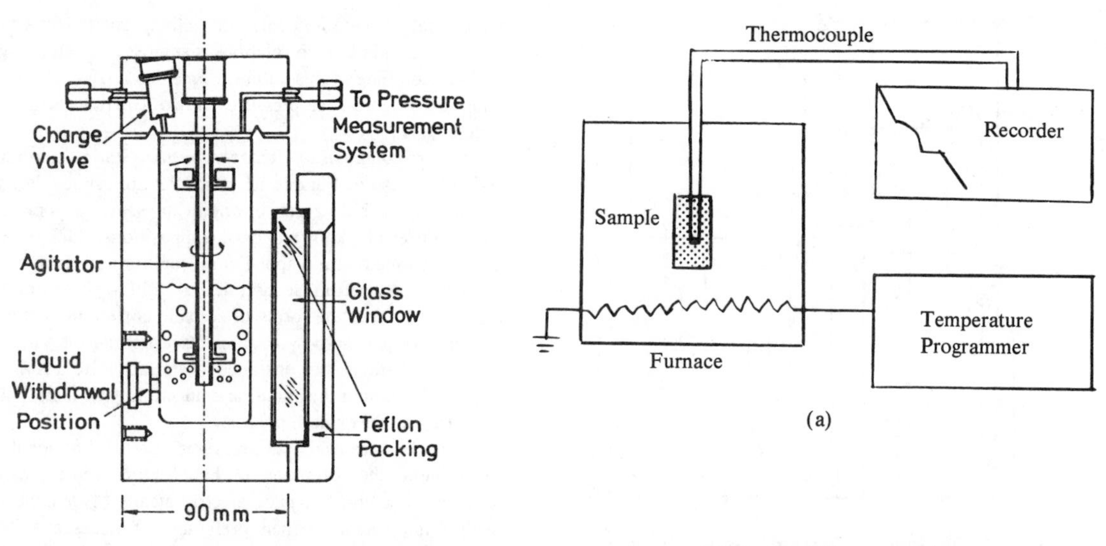

Figure 12.12. A visual equilibrium cell for measurement of vapor-liquid equilibria at up top 100 bar, with a magnetic stirrer (Bae et al. 1981).

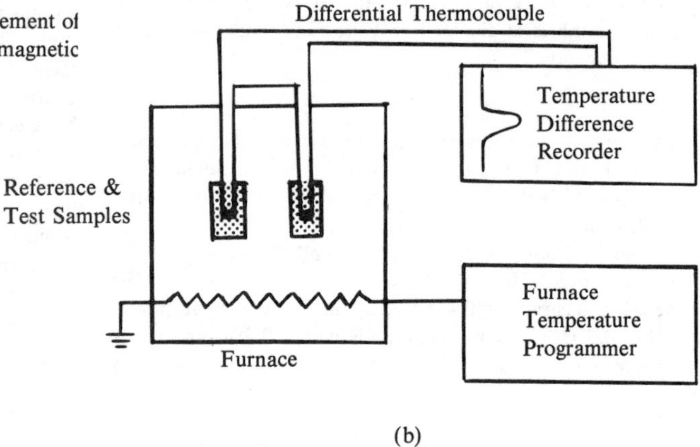

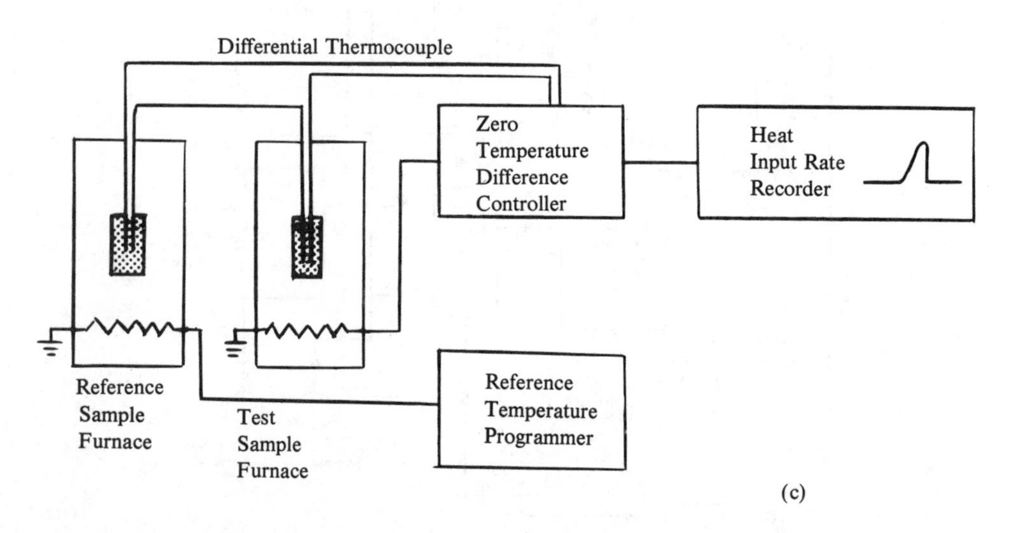

Figure 12.13. Schematic diagrams of equipment for the three basic types of thermal analysis. (a) Conventional thermal analysis. The temperature of the test sample is recorded as the furnace temperature is changed at a constant rate. (b) Differential thermal analysis. The difference between the temperatures of the test and reference samples is recorded as the furnace temperature is changed at a constant rate. (c) Differential scanning calorimetry. Measures the rate of heat transfer to the test sample needed to maintain a zero temperature difference between the test and reference samples as the temperature of the reference sample is changed at a constant rate.

12.4. LIQUID-SOLID (MELT) EQUILIBRIUM

Melting points of pure substances are determined routinely in laboratories as a means of identifying them, and many commercial instruments are available for this purpose. Mostly they employ a temperature-controlled surface on which the solid sample is placed. The temperature of fusion is noted visually under magnification, or photoelectrically in completely automatic instruments.

The same kind of equipment ordinarily cannot be used for mixtures since they liquefy over a range of temperatures (excepting eutectics) and the beginning and end of the liquefaction cannot be observed with any accuracy. In the study of liquid-solid equilibria, moreover, it may be necessary to identify the occurrence of other kinds of transitions, such as those entirely in the solid state, and the formation of intermolecular compounds.

Since phase transitions and solution phenomena usually are accompanied by appreciable thermal effects, measurement of these effects is an obvious technique for the study of phase behavior. When the thermal effects are small or the thermal measurements are difficult to interpret because of complexity of the system, many other techniques may be used to supplement or confirm them. Changes in density, electrical conductivity, magnetic properties, or sonic velocity may be observed, as well as X-ray diffraction patterns, depending on the peculiarities of the individual system. Thermal methods have the broadest utility, however, so the discussion will be confined to them.

Method of heating or cooling curves: Such curves are plots of the time variation of the temperature of a sample located in a furnace whose temperature is varied gradually, preferably linearly. A diagram of heating apparatus is in Figure 12.13(a); for cooling, a stream of gas is employed. In modern thermal-analysis equipment, the sample sizes are small, 5–100 mg, and the rate of change of temperature may be as much as 2–10 C/min.

In the absence of a phase change with appreciable thermal effect, the temperature of the sample will follow that of the furnace, but otherwise the temperature trace will show a break. Figure 12.14(a) is a hypothetical cooling curve of a pure substance with constant heat capacities; the temperature of the horizontal portion is the freezing point. The curve for benzene, Figure 12.14(b), shows a momentary subcooling of the liquid phase and a not-quite-horizontal temperature line during the period of freezing. Even with great care and excellent equipment, there is a tendency for such overshooting to occur. On the plot for a eutectic system of Figure 12.14(e), benzoic acid freezes at point A, a mixture containing 20% cinnamic acid begins to freeze at point B, and the lowest freezing point of any mixture is the eutectic at C. The lengths of the horizontal displacements of temperature traces are proportional to the enthalpy changes of the phase transitions.

Both heating and cooling curves are shown for the systems of Figures 12.14(e) and 12.15(b). For a given rate of change of furnace temperature, the two types of curves should be identical, as they are shown; but cooling curves are more likely to be in error because of a greater tendency of all systems to subcooling than to superheating. Accordingly, when cooling, smaller sizes of sample and lower temperature rates of change are preferable. Transitions in the solid state are particularly sluggish and the thermal effects are small, so greater care must be taken to identify them.

With complex systems like those of Figure 12.15, a large number of temperature traces may be needed to establish a phase diagram. Even so, interpretation of the data may be difficult and other supplementary techniques may need to be used, as mentioned in the opening of this section.

Differential thermal analysis (DTA): Ordinarily this is a more sensitive technique than straight thermal analysis, and commercial equipment is available. It was developed at about the same time as the method of cooling and heating curves (Le Chatelier 1887; Roberts-Austen 1899) but has been refined in recent years as the remaining problems have become harder and interest in thermal methods has expanded. A diagram of such apparatus is shown in Figure 12.13(b). DTA equipment measures the difference between the temperature of a test sample and a reference sample that has no phase transitions in the temperature range of interest. The temperature of the furnace is varied uniformly at from 2–10°C/min for a sample size of a few milligrams. The data are plotted as the temperature difference against the temperature of the reference sample, which is essentially the temperature of the furnace.

When there is a phase transition, a pulse develops in such a plot, as shown, for instance, in Figure 12.16 for anisaldine, which has two phase transitions. When the pulse is narrow, which will be the case at a low rate of change of temperature, the easily identified peak will essentially identify the transition temperature. The characteristics of a particular instrument may affect these interpretations, so it is usual practice to make a calibration with substances of accurately known transition temperatures.

Although the shape of a DTA pulse will depend on the heating rate, the size of the sample and temperature gradients in it, its area is always proportional to the thermal effect of the transition. Quantitative data are obtainable after calibration of the instrument with standard substances.

A simple type and a moderately complex system are shown with their DTA traces in Figures 12.17 and 12.18. As many as three pulses occur on some of the curves.

DTA and DSC are used widely for the identification of substances and their purities. The diagrams are called thermograms, and many are catalogued (for example, by Sadtler Research Laboratories).

Differential scanning calorimetry (DSC): Magnitudes of the thermal effects of phase transitions are evaluated by integrating an area of a pulse and also taking into account the heating rate. Some commercial DTA equipment has built-in integrators, useful in routine testing. A related testing method is differential scanning calorimetry, for which equipment was developed by Perkin-Elmer Corporation (1964). As appears in Figure 12.13(c), the test and reference samples have separate heat inputs. The temperature of the reference is programmed to a uniform rate, whereas heat input to the test sample is varied to maintain a zero temperature difference between the samples. The rate of energy input required to do this is recorded as a function of the sample temperature and indirectly of the time, since the heating rate also is known. The heat effect of a transition is obtained by integration, for which commercial instruments have built-in facilities.

Both DTA and DSC may accomplish the same end results, although some claims of greater sensitivity are made for the latter. When obtained under standardized conditions, the thermograms are highly characteristic; catalogs of both kinds are available. These methods are particularly useful for the detection of impurities in otherwise pure substances and for

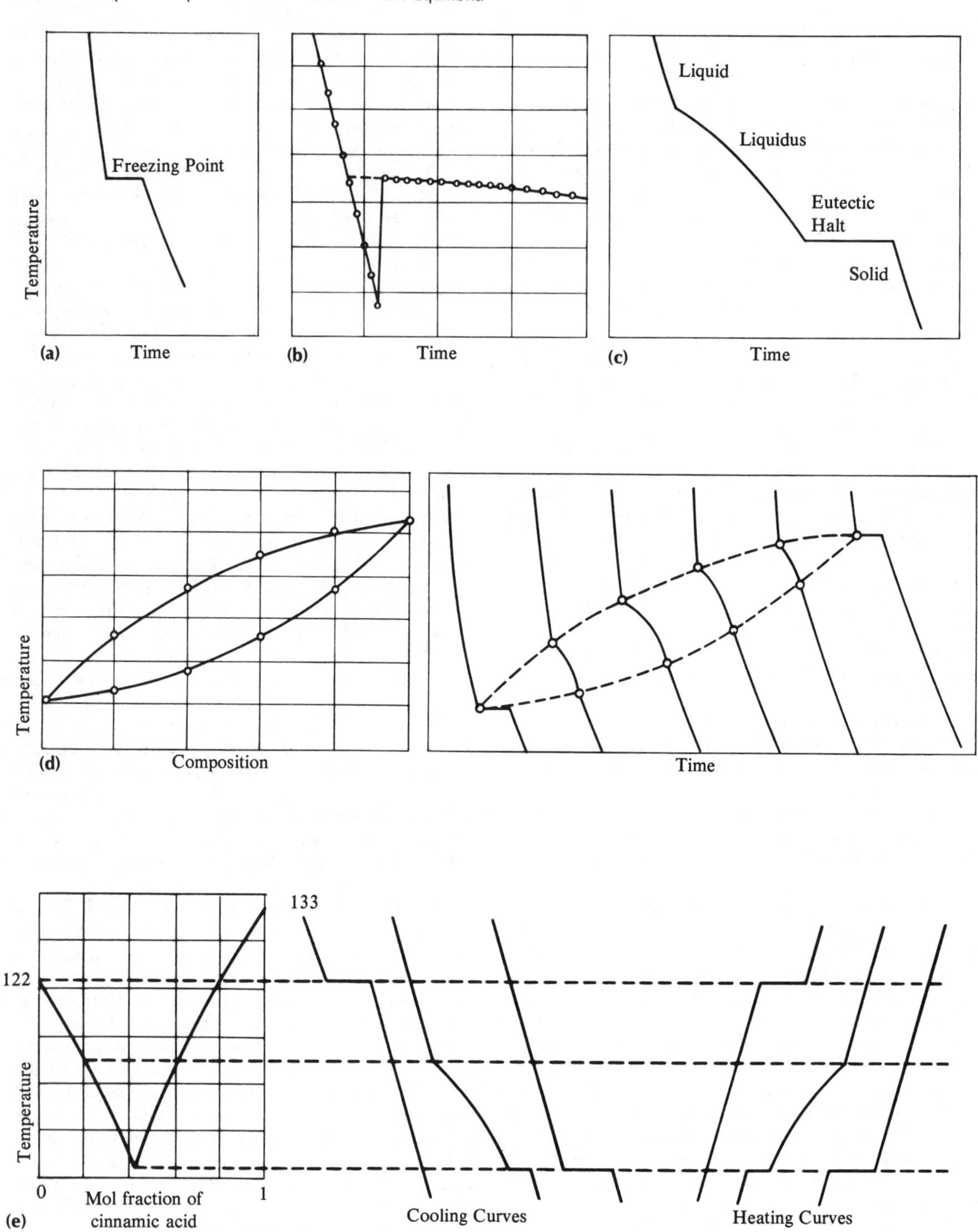

Figure 12.14. Cooling/heating curves of some simple systems: (a) curve of a pure substance with constant heat capacity; (b) cooling curve for determining the freezing point of benzene; divisions of the temperature scale are 0.1 C, so the overshoot is about 0.3 C; (c) temperature trace with eutectic formation; (d) phase diagram and temperature traces of a system with completely miscible liquid and solid phases; (e) the simple eutectic system benzoic acid + cinnamic acid.

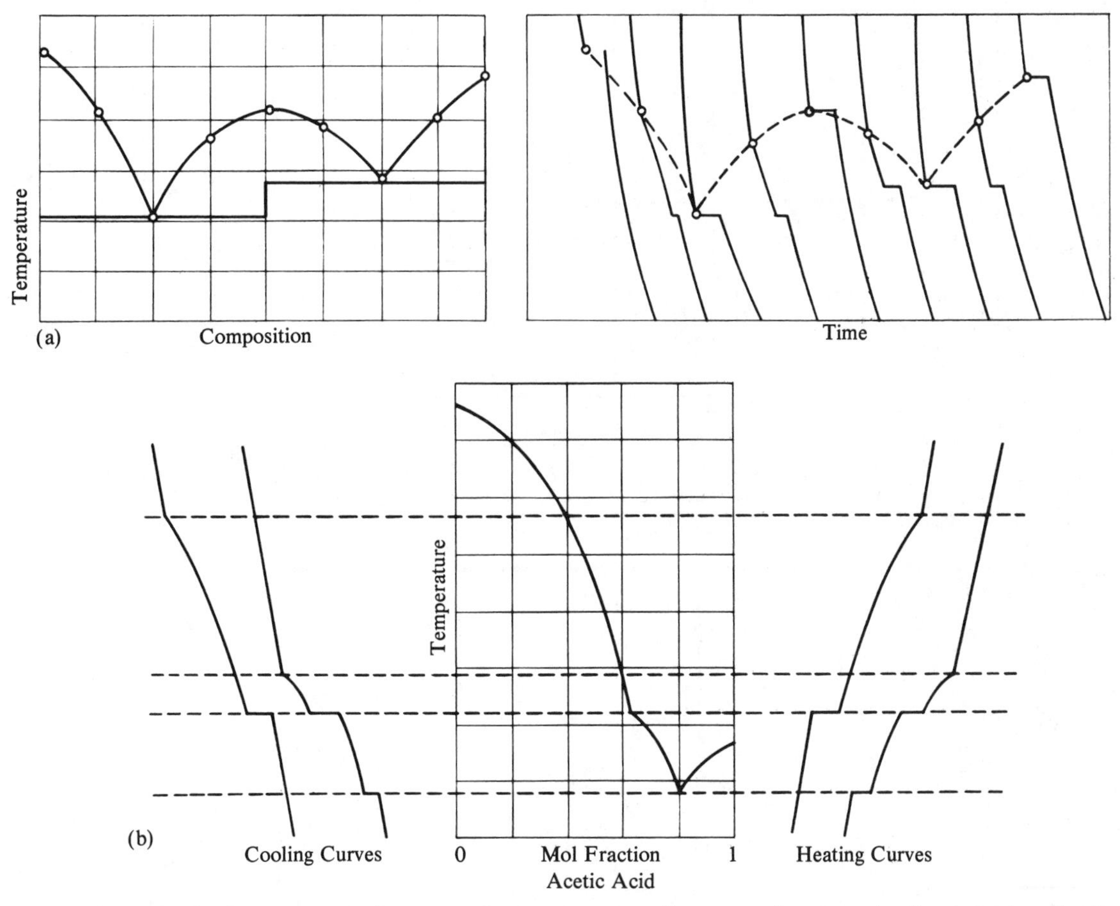

Figure 12.15. Cooling/heating curves of more complex systems: (a) cooling curves of a system that forms an intermolecular compound that is stable at its melting point; (b) the peritectic system, acetic acid, + dimethylpyrrone.

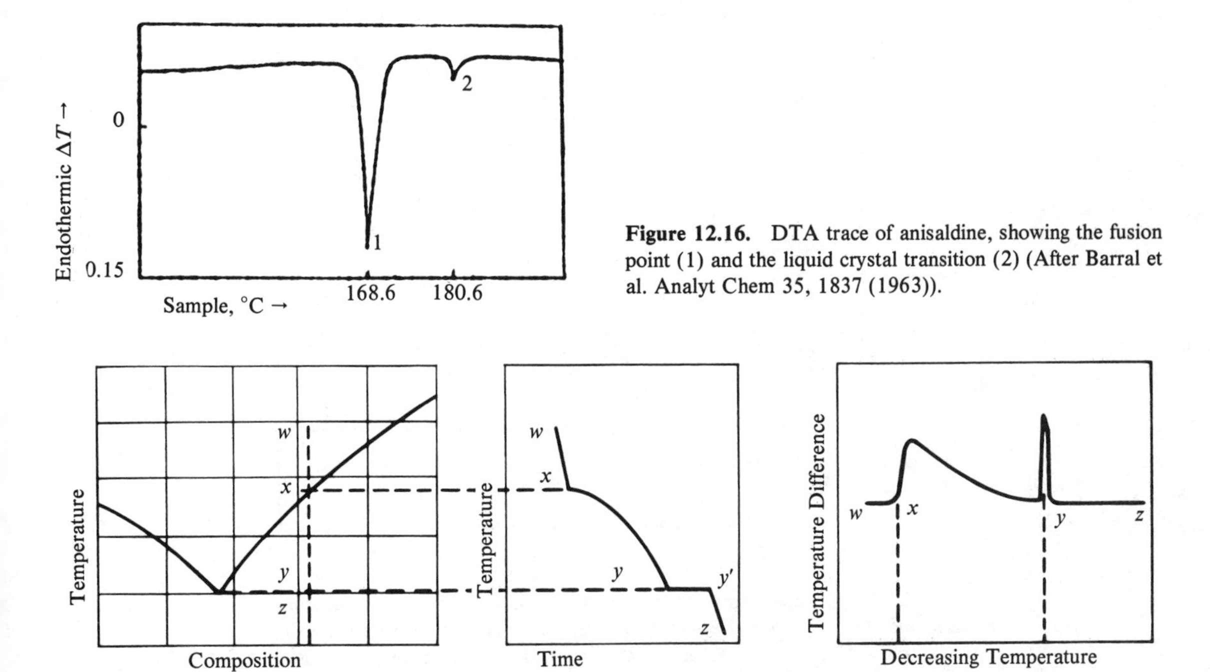

Figure 12.16. DTA trace of anisaldine, showing the fusion point (1) and the liquid crystal transition (2) (After Barral et al. Analyt Chem 35, 1837 (1963)).

Figure 12.17. A binary eutectic system with a cooling curve and a DTA temperature trace at a composition marked *w*.

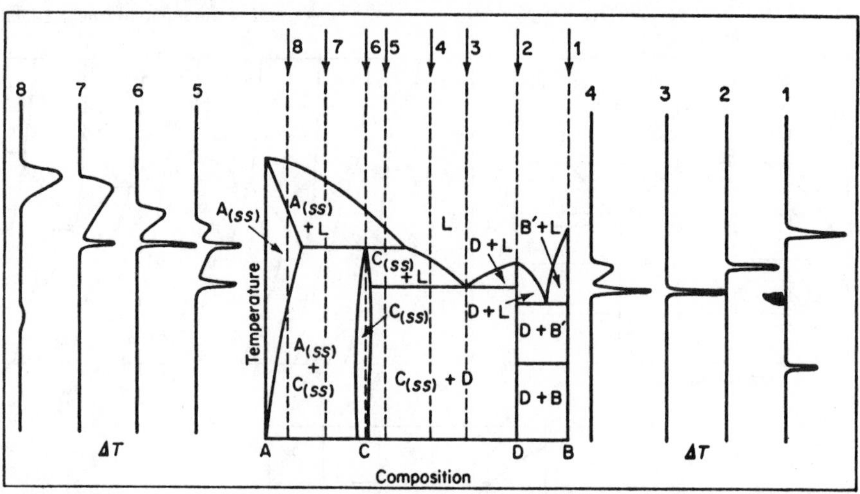

Figure 12.18. A hypothetical binary system $A + B$, with DTA traces at eight compositions. (Gutt & Majumdar, in Mackenzie (editor), *Differential Thermal Analysis,* Vol. 2, 1972. Subscript (SS) designates solid solution.

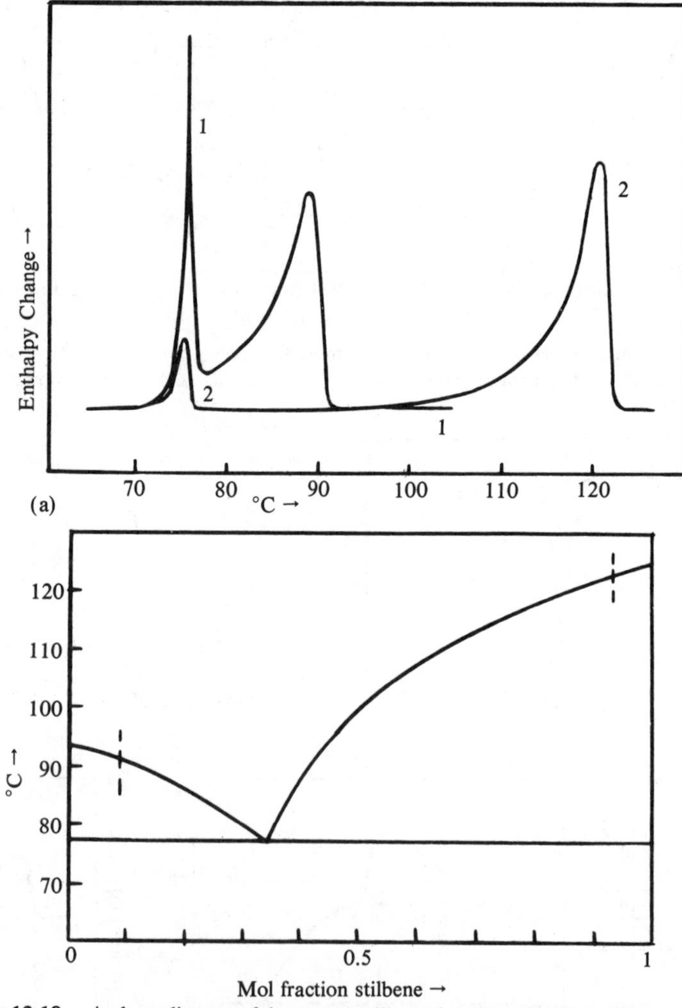

Figure 12.19. A phase diagram of the system stilbene + triphenylmethane from measurements with a Perkin-Elmer differential scanning calorimeter: (a) Two thermal traces; No. 1 is at 0.083 mol fraction stilbene, 8.4 mg sample; No. 2 is at 0.931 mol fraction stilbene, 5.5 mg sample. The two peaks on each curve correspond to start of fusion and to the eutectic temperature: heating rate 5°C/min; (b) the complete phase diagram established with a few hours operation.

the characterization of mixtures and polymers. DSC has been applied to the determination of melt diagrams. Figure 12.19 is an example. The amount of sample needed is measured in milligrams, and a few hours of operating time are sufficient for establishing completely a diagram as simple as the one shown, as well as determining the transitional enthalpy changes.

12.5. GENERAL REFERENCES ON EXPERIMENTAL METHODS FOR PHASE EQUILIBRIA

A few books will be listed that will provide direct information on this topic and also entries into the extensive journal literature.

1. Dack, M.R.J. (editor), *Solutions and solubilities,* Part 1, Chap. 7: The solubility of gases in liquids (1976).
2. Hala, E.; J. Pick; V. Fried; & O. Vilim, *Vapor-liquid equilibrum* (1967).
3. Le Neindre, B., & B. Vodar (editors). *Experimental thermodynamics,* Vol. 2: *Experimental thermodynamics of non-reacting fluids,* Chap. 6: Measurement of PVT properties of gases and gas mixtures at low pressures; Chap. 8: PVT relationships in gases at high pressures and high temperatures; Chap. 16: Phase equilibria of liquid-gas mixtures; Chap. 17: Liquid-solid phase equilibria (1975).
4. McGlashan, M. L. *Chemical thermodynamics,* Vol. 2: *A specialist periodical report,* Chap. 3: Experimental methods for studying phase behavior of mixtures at high temperatures and pressures (1978).
5. Pope, M. I., & M. D. Judd. *Differential thermal analysis* (1977).
6. Weissberger, A., & B. W. Rossiter (editors). *Physical methods of chemistry,* Part V: *Determination of thermodynamic and surface properties* (1971).
7. Wendlandt, W. W. *Thermal methods of analysis* (1976).

Appendixes

Appendixes

Appendix A. General Thermodynamic Formulas

Systematic collections of thermodynamic formulas dissociated from their derivations and possible restrictions must be used with caution, but they are nevertheless of value to the initiate. Although no comprehensive summary of such formulas appears to have been published, many tables of limited scope are available, for instance, in:

1. Bridgman, P.W. *A Condensed Collection of Thermodynamic Formulas.* Harvard University Press (1925). Covers first and second derivatives of other thermodynamic quantities in terms of P, V, T, S, C_p, and their derivatives. (See Table A.7.)
2. Goranson, R.W. *Thermodynamic Relations in Multicomponent Systems.* Carnegie Institution of Washington (1930).
3. Shaw, A. N. The derivation of thermodynamical relations for a simple system. *Phil. Trans. Roy. Soc.* (London) *A234*, 299–328 (1935). Covers first and second derivatives of one-component systems. The method is much more flexible than that of Bridgman in that any derivative also may be expressed in terms other than those involving only P, V, T, S, and C_p. The treatment relating to first derivatives also is presented by T.K. Sherwood and C.E. Reed, *Mathematics in Chemical Engineering,* McGraw-Hill (1939), and by Carroll (1965) (*J. Chem. Education 42*(4), 218–221).
4. Tunnell, G. *Relations between Intensive Thermodynamic Quantities and their First Derivatives in a Binary System of One Phase.* W.H. Freeman and Company (1961).
5. Sage, B.H. *Thermodynamics of Multicomponent Systems.* Reinhold Publishing Corporation (1965). An appendix lists formulas for systems of variable mass and composition in Bridgman style.
6. Van Ness, H.C., M.M. Abbott. *Classical Thermodynamics of Non-Electrolyte Solutions.* McGraw-Hill (1982). Tables of formulas relating to mixtures are scattered throughout the book and appendixes.

Specialized tables of thermodynamic formulas are located throughout the book, particularly in Chapters 2, 3, 4, and 11. The basic tables of this appendix are:

Table A.1. Definitions and symbols of thermodynamic quantities,
Table A.2. Some named relations in thermodynamics,
Table A.3. Thermodynamic formulas,
Table A.4. Directly measurable properties,
Table A.5. Properties of the ideal gas and values of the gas constant R,
Table A.6. Ideal gas processes,
Table A.7. A Bridgman method for first order derivatives.

Table A1. Definitions and Symbols of Thermodynamic Quantities

Thermal Quantity	*Symbol*	*Definition*
Heat	Q	
Internal energy	U	
Entropy	S	
Enthalpy (also called heat content, heat function, total heat)	H	$U + PV$
Helmholtz function (also called free energy and work function)		$U - TS$
Gibbs function (also called free energy, free enthalpy, thermodynamic potential).	G	$H - TS$
Volume expansivity (coefficient of volume expansion).	β	$\frac{1}{V}\left(\frac{\partial V}{\partial T}\right)_P$
Isothermal bulk modulus	B	$-V\left(\frac{\partial P}{\partial V}\right)_T$
Adiabatic bulk modulus	B_s	$-V\left(\frac{\partial P}{\partial V}\right)_S$
Isothermal compressibility	k_T	$-\frac{1}{V}\left(\frac{\partial V}{\partial P}\right)_T$
Adiabatic compressibility	k_s	$-\frac{1}{V}\left(\frac{\partial V}{\partial P}\right)_S$
Heat capacity at constant volume	C_v	$\left(\frac{dU}{dT}\right)_V = T\left(\frac{\partial S}{\partial T}\right)_V$
Heat capacity at constant pressure	C_P	$\left(\frac{dH}{dT}\right)_P = T\left(\frac{\partial S}{\partial T}\right)_P$
Ratio of heat capacities	γ	$\frac{C_P}{C_V}$
Joule coefficient	η	$\left(\frac{\partial T}{\partial V}\right)_U$
Joule-Thomson (Kelvin) coefficient	μ	$\left(\frac{\partial T}{\partial P}\right)_H$
Massieu function	ψ	$\frac{A}{T} = \frac{U}{T} - S$
Planck function	Φ	$\frac{G}{T} = \frac{H}{T} - S$

Table A.2. Some Named Relations in Thermodynamics

1. First law: $dQ = dU + dW$.
2. Second law: $dQ = TdS$.
3. Third law: $\lim_{T \to 0} S = 0$
4. Ideal-gas law: $PV = RT$.
5. Raoult's law: $P_i = y_i P_i^{sat}$.
6. Henry's law: $P_i = k_H x_i$ or $\hat{f}_i = k_H x_i$.
7. Lewis & Randall rule: $\hat{f}_i = y_i f_i$ or $\hat{\phi}_i = \phi_i$.
8. Clapeyron equation: $\dfrac{dP}{dT} = \dfrac{\Delta H}{T \Delta V}$.
9. Clapeyron-Clausius equation: $\dfrac{d \ln P}{dT} = \dfrac{\Delta H_{vap}}{RT^2}$ (ideal vapor).
10. Gibbs-Duhem equation: $\Sigma\ x_i d\bar{M}_i = \left(\dfrac{\partial M}{\partial T}\right)_P dT + \left(\dfrac{\partial M}{\partial P}\right)_T dP$.
11. Gibbs-Helmholtz equation: $T\left(\dfrac{\partial \Delta G}{\partial T}\right)_P = \Delta G - \Delta H$.
12. Hess's law: Enthalpy change of a reaction is independent of the path.
13. Kirchhoff equation: $\left(\dfrac{\partial \Delta H}{\partial T}\right)_P = \Delta C_p$.
14. Lorentz-Berthelot rules: For the parameters of cubic EOS such as the van der Waals, $\sqrt{a} = \Sigma\ y_i \sqrt{a_i}$ and $b = \Sigma\ y_i b_i$ (see Rowlinson & Swinton, *Liquids and Liquid Mixtures,* 1982).
15. van't Hoff isochore: $(\partial \ln K_{ce}/\partial T)_P = \Delta H^0/RT^2$. (For chemical reaction.)

Table A.3. Basic Equations and Relations

Fundamental Equations:

$nU = f(S, V, N_1, n_2, \ldots, n_k)$ internal energy,

$nH = f(S, P, n_1, n_2, \ldots, n_k)$

$= n(U + PV)$ enthalpy,

$nA = f(T, V, n_1, n_2, \ldots, n_k)$

$= n(U - TS)$ Helmholtz energy,

$nG = f(T, P, n_1, n_2, \ldots, n_k)$

$= n(U - TS + PV)$ Gibbs energy.

$$P = -\left(\frac{\partial U}{\partial V}\right)_{Sn} = -\left(\frac{\partial A}{\partial V}\right)_{Tn},$$

$$S = -\left(\frac{\partial A}{\partial T}\right)_{Vn} = -\left(\frac{\partial G}{\partial T}\right)_{Pn} = -\left(\frac{\partial \mu}{\partial T}\right)_P,$$

$$V = \left(\frac{\partial H}{\partial P}\right)_{Sn} = \left(\frac{\partial G}{\partial P}\right)_{Tn} = \left(\frac{\partial \mu}{\partial P}\right)_T,$$

Differential Relations:

$d(nU) = Td(nS) - Pd(nV) + \Sigma\ \mu_i dn_i,$

$d(nH) = Td(nS) + nVdP + \Sigma\ \mu_i dn_i,$

$d(nA) = -nSdT - Pd(nV) + \Sigma\ \mu_i dn_i,$

$d(nG) = -nSdT + nVdP + \Sigma\ \mu_i dn_i.$

$$U = \left(\frac{\partial (A/T)}{\partial (1/T)}\right)_V,$$

$$H = \left(\frac{\partial (G/T)}{\partial (1/T)}\right)_P,$$

Intensive Properties:

$$T = \left(\frac{\partial U}{\partial S}\right)_{Vn} = \left(\frac{\partial H}{\partial S}\right)_{Pn},$$

$$\mu_i = \left(\frac{\partial nU}{\partial n_i}\right)_{SVn_j} = \left(\frac{\partial nH}{\partial n_i}\right)_{SPn_j} = \left(\frac{\partial nA}{\partial n_i}\right)_{TVn_j}$$

$$= \left(\frac{\partial nG}{\partial n_i}\right)_{TPn_j}, \; j \neq i.$$

Maxwell Relations:

$$\left(\frac{\partial T}{\partial V}\right)_S = -\left(\frac{\partial P}{\partial S}\right)_V,$$

$$\left(\frac{\partial T}{\partial P}\right)_S = \left(\frac{\partial V}{\partial S}\right)_P$$

$$\left(\frac{\partial S}{\partial V}\right)_T = \left(\frac{\partial P}{\partial T}\right)_V,$$

$$\left(\frac{\partial S}{\partial P}\right)_T = -\left(\frac{\partial V}{\partial T}\right)_P.$$

Chemical Reaction Equilibrium:

$$K_{eq} = \exp(-\Delta G^0/RT),$$

$$-R\left(\frac{\partial \ln K_{eq}}{\partial T}\right) = \left(\frac{\partial \Delta G^0/T}{\partial T}\right)_P = -\Delta H^0/T^2.$$

Table A.4. Directly Measurable Properties

1. Isothermal compressibility: $k_T = -\dfrac{1}{V}\left(\dfrac{\partial V}{\partial P}\right)_T.$

2. Adiabatic compressibility: $k_S = -\dfrac{1}{V}\left(\dfrac{\partial V}{\partial P}\right)_S.$

3. Coefficient of thermal expansion. $\beta = \dfrac{1}{V}\left(\dfrac{\partial V}{\partial T}\right)_P.$

Heat capacity or specific heat:

4. at constant pressure, $C_p = \left(\dfrac{\partial H}{\partial T}\right)_P = T\left(\dfrac{\partial S}{\partial T}\right)_P = T\left(\dfrac{\partial V}{\partial T}\right)_P\left(\dfrac{\partial P}{\partial T}\right)_S.$

5. at constant volume, $C_V = \left(\dfrac{\partial U}{\partial T}\right)_V = T\left(\dfrac{\partial S}{\partial T}\right)_V = -T\left(\dfrac{\partial P}{\partial T}\right)_V\left(\dfrac{\partial V}{\partial T}\right)_S.$

6. $C_p - C_V = T\left(\dfrac{\partial P}{\partial T}\right)_V\left(\dfrac{\partial V}{\partial T}\right)_P = \dfrac{TV\beta^2}{K_T} = -T\left(\dfrac{\partial V}{\partial T}\right)_P^2\left(\dfrac{\partial P}{\partial V}\right)_T = -T\left(\dfrac{\partial P}{\partial T}\right)_V^2 \Big/ \left(\dfrac{\partial P}{\partial V}\right)_T.$

7. $\left(\dfrac{\partial C_V}{\partial V}\right)_T = T\left(\dfrac{\partial^2 P}{\partial T^2}\right)_V.$

8. $\left(\dfrac{\partial C_P}{\partial P}\right)_T = -T\left(\dfrac{\partial^2 V}{\partial T^2}\right)_P.$

9. Joule-Thomson coefficient $= \left(\dfrac{\partial T}{\partial P}\right)_H = \dfrac{1}{C_p}\left[T\left(\dfrac{\partial V}{\partial T}\right)_P - V\right].$

Table A.5. Properties of the Ideal Gas and Values of the Gas Constant in Various Units

$$pV = nRT$$

$$\left(\frac{\partial V}{\partial T}\right)_P = \frac{V}{T}, \quad \left(\frac{\partial P}{\partial T}\right)_V = \frac{P}{T}$$

$$dG = nC_p dT - nTC_p \frac{dT}{T} + nRT\frac{dp}{P}$$

$$dU = nC_V\, dT$$

$$\left(\frac{\partial U}{\partial P}\right)_T = 0, \quad \left(\frac{\partial T}{\partial V}\right)_U = 0, \quad \left(\frac{\partial C_V}{\partial V}\right)_T = 0$$

$$k_T = 1/p$$

$$k_S = C_V/pC_p$$

$$\beta = 1/T$$

$$dH = nC_p dT$$

$$\left(\frac{\partial H}{\partial P}\right)_T = 0, \quad \left(\frac{\partial C_p}{\partial P}\right)_T = 0, \quad \left(\frac{\partial T}{\partial P}\right)_H = 0$$

$$C_p - C_V = R, \quad C_p = \frac{Rk}{k-1}, \quad k = C_p/C_V$$

$$dS = \frac{nC_p}{T}\, dT - nR\frac{dp}{P}$$

$$dA = nC_V dT - nTC_V\frac{dT}{T} - nRT\frac{dV}{V}$$

Energy	*Temperature*	*Mole*	*R*
lb-ft^2/sec^2	°Rankine	lb	4.969×10^4
ft lbf	°Rankine	lb	1544
cu ft atm	°Rankine	lb	0.7302
cu ft (lbf/sq in.)	°Rankine	lb	10.73
Btu	°Rankine	lb	1.987
hp-hr	°Rankine	lb	7.805×10^{-4}
kwhr	°Rankine	lb	5.819×10^{-4}
joule(abs)	°Kelvin	gm	8.314
kg-m^2/sec^2	°Kelvin	kg	8.314×10^3
kgf m	°Kelvin	kg	8.478×10^2
cu cm atm	°Kelvin	gm	82.0562
calorie	°Kelvin	gm	1.987

Table A.6. Ideal Gas Processes (Weber & Meissner, *Thermodynamics for Chemical Engineers,* Wiley, 1957). $PV = RT$, constant heat capacities C_p and C_v, $k = C_p/C_v$.

Condition of restraint	$V = C$	$P = C$	$T = C$	$Q = 0$ *Reversible adiabatic*
PV relations	$V_1 = V_2$	$P_1 = P_2$	$P_1 V_1 = P_2 V_2$ $\frac{P_2}{P_1} = \frac{V_1}{V_2}$	$P_1 V_1^k = P_2 V_2^k$ $\frac{P_2}{P_1} = \left(\frac{V_1}{V_2}\right)^k$
PT relations	$\frac{P_2}{P_1} = \frac{T_2}{T_1}$	....	$T_1 = T_2$	$\frac{T_2}{T_1} = \left(\frac{P_2}{P_1}\right)^{(k-1)/k}$
TV relations	....	$\frac{V_2}{V_1} = \frac{T_2}{T_1}$	....	$\frac{T_2}{T_1} = \left(\frac{V_1}{V_2}\right)^{k-1}$
$\Delta(PV)$	$= V(P_2 - P_1)$	$= P(V_2 - V_1)$ $= PV_1\left(\frac{T_2}{T_1} - 1\right)$	$= 0$	$= P_1 V_1\left[\left(\frac{P_2}{P_1}\right)^{(k-1)/k} - 1\right]$ $= RT_1\left[\left(\frac{P_2}{P_1}\right)^{(k-1)/k} - 1\right]$
ΔT	$= T_1\left(\frac{P_2}{P_1} - 1\right)$	$= T_1\left(\frac{V_2}{V_1} - 1\right)$	$= 0$	$= T_1\left[\left(\frac{P_2}{P_1}\right)^{(k-1)/k} - 1\right]$ $= T_1\left[\left(\frac{V_1}{V_2}\right)^{k-1} - 1\right]$
Q	$= \Delta U = c_p \Delta T$ $= \frac{\Delta(PV)}{k-1}$	$= \Delta H = c_p \Delta T$ $= \frac{k-1}{k}\Delta(PV)$	$= -RT \ln\frac{P_2}{P_1}$ $= -RT \ln\frac{V_1}{V_2}$	$= 0$
$W_{nonflow} = \int P\,dV$ Reversible	$= 0$	$= \Delta(PV)$	$= -RT \ln\frac{P_2}{P_1}$	$= -\Delta U = -c_v \Delta T$ $= \frac{-RT_1}{k-1}\left[\left(\frac{P_2}{P_1}\right)^{(k-1)/k} - 1\right]$

(Continued next page)

Table A.6. *(continued)*

$W_{\text{non-flow}}$ Reversible		$= P(V_2 - V_1)$	$= -RT \ln \dfrac{V_1}{V_2}$	$= \dfrac{-\Delta(PV)}{k-1}$ $= \dfrac{-P_1V_1}{k-1}\left[\left(\dfrac{P_2}{P_1}\right)^{(k-1)/k} - 1\right]$
W_{flow} $= -\int V dP$ Reversible	$= V(\Delta P) = V(P_2 - P_1)$	$= 0$	$= -RT \ln \dfrac{P_2}{P_1}$ $= -RT \ln \dfrac{V_1}{V_2}$	$= -\Delta H = -c_p \Delta T$ $= -\dfrac{kRT_1}{k-1}\left[\left(\dfrac{P_2}{P_1}\right)^{(k-1)/k} - 1\right]$ $= -\dfrac{k}{k-1}\Delta(PV)$ $= -\dfrac{kP_1V_1}{k-1}\left[\left(\dfrac{P_2}{P_1}\right)^{(k-1)/k} - 1\right]$
ΔU	$= c_v \Delta T$ $= \dfrac{\Delta(PV)}{k-1}$	$= c_v \Delta T$ $= \dfrac{\Delta(PV)}{k-1}$	$= 0$	$= -W_{\text{nonflow}} = c_v \Delta T$ $= \dfrac{RT_1}{k-1}\left[\left(\dfrac{P_2}{P_1}\right)^{(k-1)/k} - 1\right]$ $= \dfrac{\Delta(PV)}{k-1}$ $= \dfrac{P_1V_1}{k-1}\left[\left(\dfrac{P_2}{P_1}\right)^{(k-1)/k} - 1\right]$
ΔH	$= c_p \Delta T$ $= \dfrac{k}{k-1}\Delta(PV)$	$= c_p \Delta T$ $= \dfrac{k}{k-1}\Delta(PV)$	$= 0$	$= -W_n = c_p \Delta T$ $= \dfrac{kRT_1}{k-1}\left[\left(\dfrac{P_2}{P_1}\right)^{(k-1)/k} - 1\right]$ $= \dfrac{k}{k-1}\Delta(PV)$ $= \dfrac{kP_1V_1}{k-1}\left[\left(\dfrac{P_2}{P_1}\right)^{(k-1)/k} - 1\right]$

Table A.7. A Bridgman-Type Table for First Derivatives of Thermodynamic Quantities. After P.W. Bridgman, *A Condensed Collection of Thermodynamic Formulas,* Harvard University Press (1926).

$$(\partial T)_P = -(\partial P)_T = 1$$

$$(\partial V)_P = -(\partial P)_V = \left(\frac{\partial V}{\partial T}\right)_P$$

$$(\partial S)_P = -(\partial P)_S = \frac{C_p}{T}$$

$$(\partial U)_P = -(\partial P)_U = C_p - P\left(\frac{\partial V}{\partial T}\right)_P$$

$$(\partial H)_P = -(\partial P)_H = C_p$$

$$(\partial G)_P = -(\partial P)_G = -S$$

$$(\partial A)_P = -(\partial P)_A = -\left[S + P\left(\frac{\partial V}{\partial T}\right)_P\right]$$

$$(\partial V)_T = -(\partial T)_V = -\left(\frac{\partial V}{\partial P}\right)_T$$

$$(\partial S)_T = -(\partial T)_S = \left(\frac{\partial V}{\partial T}\right)_P$$

$$(\partial U)_T = -(\partial T)_U = T\left(\frac{\partial V}{\partial T}\right)_P + P\left(\frac{\partial V}{\partial P}\right)_T$$

$$(\partial H)_T = -(\partial T)_H = -V + T\left(\frac{\partial V}{\partial T}\right)_P$$

$$(\partial G)_T = -(\partial T)_G = -V$$

$$(\partial A)_T = -(\partial T)_A = P\left(\frac{\partial V}{\partial P}\right)_T$$

$$(\partial S)_V = -(\partial V)_S = \frac{1}{T}\left[C_p\left(\frac{\partial V}{\partial P}\right)_T + T\left(\frac{\partial V}{\partial T}\right)_P^2\right]$$

$$(\partial U)_V = -(\partial V)_U = C_P\left(\frac{\partial V}{\partial P}\right)_T + T\left(\frac{\partial V}{\partial T}\right)_P^2$$

$$(\partial H)_V = -(\partial V)_H = C_p\left(\frac{\partial V}{\partial P}\right)_T + T\left(\frac{\partial V}{\partial T}\right)_P^2 - V\left(\frac{\partial V}{\partial T}\right)_P.$$

$$(\partial G)_V = -(\partial V)_G = -\left[V\left(\frac{\partial V}{\partial T}\right)_P + S\left(\frac{\partial V}{\partial P}\right)_T\right]$$

$$(\partial A)_V = -(\partial V)_A = -S\left(\frac{\partial V}{\partial P}\right)_T$$

$$(\partial U)_S = -(\partial S)_U = \frac{P}{T}\left[C_P\left(\frac{\partial V}{\partial P}\right)_T + T\left(\frac{\partial V}{\partial T}\right)_P^2\right]$$

$$(\partial H)_S = -(\partial S)_H = -\frac{VC_p}{T}$$

$$(\partial G)_S = -(\partial S)_G = -\frac{1}{T}\left[VC_p - ST\left(\frac{\partial V}{\partial T}\right)_P\right]$$

$$(\partial A)_S = -(\partial S)_A = \frac{1}{T}\left\{P\left[C_P\left(\frac{\partial V}{\partial P}\right)_T + T\left(\frac{\partial V}{\partial T}\right)_P^2\right] + ST\left(\frac{\partial V}{\partial T}\right)_P\right\}$$

$$(\partial H)_U = -(\partial U)_H = -V\left[C_p - P\left(\frac{\partial V}{\partial T}\right)_P\right] - P\left[C_p\left(\frac{\partial V}{\partial P}\right)_T + T\left(\frac{\partial V}{\partial T}\right)_P^2\right]$$

$$(\partial G)_U = -(\partial U)_G = -V\left[C_p - P\left(\frac{\partial V}{\partial T}\right)_P\right] + S\left[T\left(\frac{\partial V}{\partial T}\right)_P + P\left(\frac{\partial V}{\partial P}\right)_T\right]$$

$$(\partial A)_U = -(\partial U)_A = P\left[(C_p + S)\left(\frac{\partial V}{\partial P}\right)_T + T\left(\frac{\partial V}{\partial T}\right)_P^2\right] + ST\left(\frac{\partial V}{\partial T}\right)_P$$

$$(\partial G)_H = -(\partial H)_G = -V(C_p + S) + TS\left(\frac{\partial V}{\partial T}\right)_P$$

$$(\partial A)_H = -(\partial H)_A = -\left[S + P\left(\frac{\partial V}{\partial T}\right)_P\right] \times \left[V - T\left(\frac{\partial V}{\partial T}\right)_P\right] + PC_p\left(\frac{\partial V}{\partial P}\right)_T$$

$$(\partial A)_G = -(\partial G)_A = -S\left[V + P\left(\frac{\partial V}{\partial P}\right)_T\right] - PV\left(\frac{\partial V}{\partial T}\right)_P$$

Appendix B. Numerical Methods

Few of the relations useful in the study and representation of phase equilibria are tractable enough to be solvable by analytical means, so numerical methods must be resorted to if the job is to proceed into "iron and steel." Although those methods can be extremely sophisticated and demanding of computer time, the simpler methods of this appendix are economical of time and are adequate for illustrative purposes in this text. Ample details of methods of professional caliber may be found in the literature—for example:

1. Gerald, *Applied Numerical Analysis* (1978).
2. Himmelblau, *Process Analysis by Statistical Methods* (1970).
3. Lapidus, *Digital Computation for Chemical Engineers* (1962).
4. Rosenbrock & Storey, *Computational Techniques for Chemical Engineers* (1966).
5. Scarborough, *Numerical Mathematical Analysis,* 6th ed. (1966).
6. Wolberg, *Prediction Analysis* (1967).

The sections of this appendix are:

B.1. Rules for partial differentiation,
B.2. Root finding:
 a. Analytical Newton-Raphson method,
 b. Numerical Newton-Raphson method,
 c. Wegstein's method,
 d. Example,
B.3. Simultaneous nonlinear equations:
 a. Newton-Raphson,
 b. Hill climbing,
B.4. Interpolation:
 a. Linear interpolation,
 b. Double linear interpolation,
 c. Interpolation with uneven intervals,
B.5. Curve fitting:
 a. Linearization,
 b. Method of least squares,
 c. Nonlinear least squares,
B.6. Numerical differentiation:
 a. Unequal and equal intervals,
B.7. Numerical integration:
 a. Simpson's Rule,
 b. Trapezoidal Rule,
 c. Uneven intervals.

B.1. RULES FOR PARTIAL DIFFERENTIATION

a. The differential of $u = u(x, y, z, \dots)$ is

$$du = \left(\frac{\partial u}{\partial x}\right)_{y,z} dx + \left(\frac{\partial u}{\partial y}\right)_{x,z} dy + \left(\frac{\partial u}{\partial z}\right)_{x,y} dz.$$

b. For the function $u = u(x, y)$,

1. The chain rule is

$$\left(\frac{\partial u}{\partial x}\right)_y \left(\frac{\partial x}{\partial y}\right)_u \left(\frac{\partial y}{\partial u}\right)_x = -1.$$

2. The rule for cross-derivatives is

$$\frac{\partial}{\partial x}\left(\frac{\partial u}{\partial y}\right)_x = \frac{\partial}{\partial y}\left(\frac{\partial u}{\partial x}\right)_y.$$

3. When the variables that is held constant is changed,

$$\left(\frac{\partial u}{\partial x}\right)_z = \left(\frac{\partial u}{\partial x}\right)_y + \left(\frac{\partial u}{\partial y}\right)_x \left(\frac{\partial y}{\partial x}\right)_z.$$

4. When u is constant,

$$dx = \left(\frac{\partial x}{\partial y}\right)_u dy.$$

B.2. ROOT FINDING

The problem is to find the value of x that satisfies the equation $y = f(x) = 0$. All available numerical methods require a starting estimate of the root, as close as possible to the correct value when there is more than one root.

a. Analytical Newton-Raphson method: If x_0 is an estimate of the root, an improved value is

$$x = x_0 - k\, f(x_0)/f'(x_0),$$

where $f'(x_0)$ is the derivative and k is a positive fraction that is selected arbitrarily to improve convergence; $k = 1$ is often acceptable. The process is repeated with newly determined values until the desired accuracy is obtained.

b. Numerical Newton-Raphson method: When the derivative is awkward to find analytically, a ratio of finite differences can be used. For example,

$$f'(x) = \frac{f(1.0001x) - f(x)}{0.0001x},$$

and the improved estimate of the root is

$$x = x_0 - k\frac{0.0001x_0 f(x_0)}{f(1.0001x_0) - f(x_0)}.$$

The factor 1.0001 is arbitrary and can be made larger or smaller to speed up convergence or to improve accuracy.

c. Wegstein's method: The equation is rearranged into the form,

$$x = f(x).$$

An improved estimate of the root is

$$x = \frac{x_0 f(x_1) - f(x_0)^2}{x_0 + f(x_1) - 2f(x_0)},$$

where x_0 is an initial estimate and

$$x_1 = f(x_0).$$

The process is repeated with the new estimate for a start.

d. Example: The equation to be solved is

$$y = -1 + 0.7x \exp(0.5(1 - x)^2) + 1.2(1 - x)\exp(0.5x^2) = 0,$$

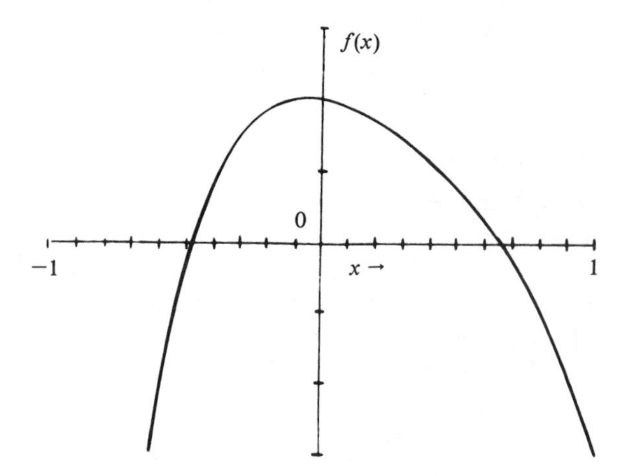

Figure B.1. Graphical Solution.

or

$$x = \frac{1 - 1.2(1 - x)\exp(0.5x^2)}{0.7\exp(0.5(1 - x)^2)}.$$

The derivative is

$$\frac{dy}{dx} = (x^2 - x + 1)[0.7\exp(0.5(1 - x)^2 - 1.2\exp(0.5x^2)].$$

The plot of the equation, Figure B.1, shows it to have two roots. Successive approximations of the roots with three different starting values are shown by the three methods. With a starting value of 0.1, the Wegstein method converges to the negative root, while the others go to the positive root. The effect of the starting value on the number of iterations and the root found by the Wegstein method is shown in the second table.

Successive Iterations by the Three Methods

N-R Analytical	*N-R Numerical*	*Wegstein*
0.9	0.9	0.9
0.7152	0.71252	0.76691
0.65784	0.65686	0.68119
0.65443	0.65458	0.65610
0.65442	0.65444	0.65443
	0.65443	
0.5	0.68002	0.5
0.68001	0.65580	0.73760
0.65509	0.65452	0.66818
0.65442	0.65443	0.65489
0.65442	0.65443	0.65443
0.1	0.1	0.1
1.4367	1.4368	−1.3707
1.114	0.97308	−0.49429
0.8412	0.75575	−0.47079
0.68865	0.66976	
0.65562	0.65532	
0.65442	0.65448	
	0.65443	

Number of Iterations with the Wegstein Method and the Root Found

$x^{(0)}$	*−0.4708*	*0.6544*
0	3	
0.1	3	
0.15	4	
0.2	5	
0.25		12
0.3		38
0.35		8
0.4		6
0.5		4

```
 10 ! N-R SOLUTION OF Y=F(X)=0 WITH
      ANALYTICAL DERIVATIVE
 20 SHORT X
 30 INPUT X
 40 PRINT X
 50 GOSUB 120
 60 H=Y/Y1
 70 X=X-H
 80 PRINT X
 90 IF ABS(H/X)<=.0001 THEN 110
100 GOTO 50
110 END
120 ! SR FOR Y AND Y1
130 Y=-1+.7*X*EXP(.5*(1-X)^2)+1.2*(1-X)*EXP
      (.5*X^2)
140 Y1=(X^2-X+1)*(.7*EXP(.5*(1-X)^2)-1.2*
      EXP(.5*X^2))
150 RETURN
```

```
 10 ! N-R SOLUTION OF Y=F(X)=0 WITH
      NUMERICAL DERIVATIVE
 20 SHORT X
 30 INPUT X
 40 PRINT X
 50 GOSUB 160
 60 Y1=Y
 70 X=1.0001*X
 80 GOSUB 160
 90 Y2=Y
100 H=.0001*X*Y1/(Y2-Y1)
110 X=X/1.0001-H
120 PRINT X
130 IF ABS(H/X)<=.0001 THEN 150
140 GOTO 50
150 END
160 ! SR FOR Y=F(X)
170 Y=-1+.7*X*EXP(.5*(1-X)^2)+1.2*(1-X)*EXP
      (.5*X^2)
180 RETURN
```

```
10.! WEGSTEIN SOL'N OF X=F(X)
20 SHORT X
30 INPUT X1
40 PRINT X1
50 X=X1
60 X2=X
70 GOSUB 160
80 X=Y1
```

```
90 GOSUB 160
100 X=(X2*Y1-X^2)/(X1+Y1-2*X)
110 IF ABS((X-Y1)/(X+Y1))<=.0001 THEN 150
120 PRINT X
130 X1=X
140 GOTO 60
150 END
160 ! SR FOR Y1=X=F(X)
170 Y1=(1-1.2*(1-X)*EXP(.5*X^2))/.7/EXP(.5*
    (1-X)^2)
180 RETURN
```

B.3. SIMULTANEOUS NONLINEAR EQUATIONS

Only the Newton-Raphson method for the solution of sets of nonlinear equations will be described. It is adequate in many instances, it is simple in concept, and it suffices for the purposes of this text. In practical work it sometimes may be necessary to use more sophisticated techniques, for which computer programs are available in some libraries. Some of these methods are based on converting the problem of solving a set of equations,

$$\phi_i(x_1, x_2, \ldots, x_n) = 0 \qquad i = 1, 2, \ldots, n,$$

into finding the minimum of a function,

$$f(x_1, x_2, \ldots, x_n) = \Sigma\,(\phi_i)^2 \qquad \text{minimum},$$

which can be done by so-called hill-climbing techniques that are described, for instance, in the books of Rosenbrock & Storey (1966) and Himmelblau (1970). A relatively complex minimization problem is discussed in Section 10.6.3.

When only two variables are involved, a graphical method may be convenient. Given the equations,

$$f(x, y) = 0,$$

$$g(x, y) = 0,$$

points on the plots are found by specifying values of one of the variables, say y, and finding corresponding values of the xs from each equation, by one variable N-R for instance. This is done for enough values of y to give an intersection of the curves. The graphical solution can serve as the starting values of a numerical solution of greater accuracy. Examples 4.18, 10.10, and 10.12 are done this way.

It is assumed that a method of solving systems of *linear* equations is known, with determinants if there are only a few variables, or by matrix methods in general. The N-R method reduces the set of nonlinear equations in the primary variables into a set of linear ones in corrections to trial values of the primary variables. Taking, for simplicity, a system with three unknowns,

$$f(x, y, z) = 0,\ g(x, y, z) = 0,\ h(x, y, z) = 0,$$

estimates of the roots are (x_0, y_0, z_0) and corrections are (a, b, c), so that

$$x = x_0 + a,\ y = y_0 + b,\ z = z_0 + c$$

are improved values. The linear equations to be solved for a, b, and c are

$$f_0 + a\,f_x + b\,f_y + c\,f_z = 0,$$

$$g_0 + a\,g_x + b\,g_y + c\,g_z = 0,$$

$$h_0 + a\,h_x + b\,h_y + c\,h_z = 0.$$

where, for example,

$$f_0 = f(x_0, y_0, z_0),$$

$$f_x = \partial f/\partial x, \text{ evaluated at } (x_0, y_0, z_0),$$

and so on.

The solution in terms of determinants is

$$a = \frac{-\begin{vmatrix} f_0 & f_y & f_z \\ g_0 & g_y & g_z \\ h_0 & h_y & h_z \end{vmatrix}}{D}$$

$$b = \frac{-\begin{vmatrix} f_x & f_0 & f_z \\ g_x & g_0 & g_z \\ h_x & h_0 & h_z \end{vmatrix}}{D}$$

$$c = \frac{-\begin{vmatrix} f_x & f_y & f_0 \\ g_x & g_y & g_0 \\ h_x & h_y & h_0 \end{vmatrix}}{D}$$

$$D = \begin{vmatrix} f_x & f_y & f_z \\ g_x & g_y & g_z \\ h_x & h_y & h_z \end{vmatrix}$$

In the next trial the starting values are $x = x_0 + a$, $y = y_0 + b$, and $z = z_0 + c$. This method is used in Example 4.7, for which the computer program is in Table C.5.

B.4. INTERPOLATION

a. Linear interpolation for the solution of $x = f(x) = y$ based on two trials.

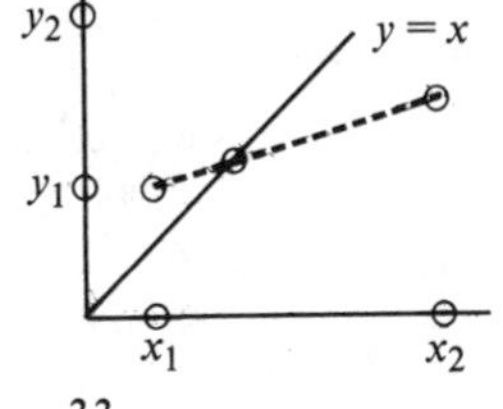

$$m = \frac{(y_2 - y_1)}{(x_2 - x_1)},$$

$$x = \frac{(y_1 - m x_1)}{(1 - m)}$$

Example: $V = 0.2 + \dfrac{33}{20 + 15/V^2} = y.$

Two trials are $(V, y) = (1, 1.1429), (1.5, 1.4375)$,

$m = (1.4375 - 1.1429)/(1.5 - 1) = 0.5892,$

$V = (1.1429 - 0.5892(1))/0.4108 = 1.3479.$

The correct answer is $V = 1.3873$, so further trials should be made for improved accuracy by this method.

b. Double linear interpolation: Given four tabular entries, z_1, z_2, z_3, z_4, to find the value $z(x, y)$ at intermediate values of the principal variables x and y.

	x		
y	x_1	x	x_2
y_1	z_1		z_3
y		z	
y_2	z_2		z_4

$$z = \frac{(x_2 - x)[z_1(y_2 - y) + z_2(y - y_1)] + (x - x_1)[(z_3(y_2 - y) + z_4(y - y_1)]}{(x_2 - x_1)(y_2 - y_1)}.$$

Example: The specific volume of superheated steam, given the values in the table. The linear interpolations are shown in the printout and compared with the correct values in parentheses.

T	P 290	x	310
500	1.8324		1.7042
y		(z)	
600	2.0770		1.9360

```
10 READ X1,X2, Y1, Y2, Z1, Z2, Z3, Z4
20 DATA 290,310,500,600,1.8324,2.077,1.7042,1.936
30 INPUT X,Y
40 Z=(X2-X)*(Z1*(Y2-Y)+Z2*(Y-Y1))+(X-X1)*
   (Z3*(Y2-Y)+Z4*(Y-Y1))
50 Z=Z/(X2-X1)/(Y2-Y1)
60 PRINT "X=";X;"Y=";Y;"Z=";Z
70 END
```

X = 305 Y = 570 Z = 1.90075 (1.9012)

X = 295 Y = 570 Z = 1.96933 (1.9696)

c. Interpolation with uneven intervals: Lagrange's interpolation polynomial for three points is:

$$y = \frac{(x - x_1)(x - x_2)}{(x_0 - x_1)(x_0 - x_2)} y_0 + \frac{(x - x_0)(x - x_2)}{(x_1 - x_0)(x_1 - x_2)} y_1 + \frac{(x - x_0)(x - x_1)}{(x_2 - x_0)(x_1 - x_2)} y_2.$$

Example: Given the values of entropy of steam at 875 psia at several temperatures, to find the value at 700 F.

T	540	700	800	920
S	1.4181	(1.5292)	1.5849	1.6377

The interpolated value is

$$S = 0.2227(1.4181) + 1.1282(1.5849) - 0.3509(1.6377)$$
$$= 1.5292.$$

The correct value from the steam tables is 1.5339.

B.5. CURVE FITTING

a. Linearization: When the equation, $f(x, y) = 0$, has two constants, they can be found most readily if the equation can be rearranged into a linear form,

$$g(x, y) = a + b\, h(x, y).$$

Examples: 1. The virial equation, $z = 1 + B/V + C/V^2$, rearranges to $V(z - 1) = B + C/V$, a linear relation between $V(z - 1)$ and $1/V$.

2. The van Laar equations of Section 4.8 are linearized to

$$\sqrt{\ln \gamma_1} = \sqrt{A} + \sqrt{A/B}\, \sqrt{\ln \gamma_2},$$

from which the parameters A and B can be found when at least two sets of values of $\ln \gamma_1$ and $\ln \gamma_2$ are known.

3. The equation $y = a \exp(b/T)$ is widely used in the linear form, $\ln y = A + B/T$, to represent the effect of temperature on vapor pressure, vaporization equilibrium ratios, chemical equilibrium constants, and other properties.

b. Method of least squares: The best values of the parameters $a, b, \ldots$ relating a group of variables,

$$y = f(x_1, x_2, \ldots, x_n, a, b, \ldots),$$

are obtained when the sum of the squares of the residuals is a minimum,

$$\Sigma (y_i - f_i)^2 = \text{minimum},$$

which requires that the derivatives with respect to the unknown parameters all be zero,

$$\frac{\partial f}{\partial a} = \frac{\partial f}{\partial b} = \frac{\partial f}{\partial c} = \ldots = 0.$$

Except for polynomial relations, the differentiations lead to nonlinear equations for the unknowns $a, b, c, \ldots$, which can be solved for by the N-R method. For a polynomial,

$$y = a_0 + a_1 x + a_2 x^2 + a_3 x^3 + \ldots + a_k x^k,$$

the least-squares principle is

$$\Sigma (-y_i + a_0 + a_1 x_i + a_2 x_i^2 + a_3 x_i^3 + \ldots)^2$$
$$= \text{minimum}.$$

Differentiation with respect to the a_i results in the set of equations,

$$n a_0 + a_1 \Sigma\, x_i + a_2 \Sigma\, x_i^2 + \ldots = \Sigma\, y_i,$$
$$a_0 \Sigma\, x_i + a_1 \Sigma\, x_i^2 + a_2 \Sigma\, x_i^3 + \ldots = \Sigma\, x_i y_i,$$
$$a_0\, \Sigma\, x_i^2 + a_1\, \Sigma\, x_i^3 + a_2\, \Sigma\, x_i^4 + \ldots = \Sigma\, x_i^2 y_i,$$

and so on. n is the number of sets (x_i, y_i), of data. These equations are solvable for the unknown a_i by any standard method for linear systems.

Example: The excess Gibbs energy function, $y = G^{ex}/RTx(1 - x)$, of water + triethylamine is known as a function of composition (*Int. Data Series, Prop. Aq. Solutions,* sheet No. 110 (June 3, 1980). A quadratic polynomial will be fitted to the data.

x	y	y_{calc}
0.05	10.2847	10.45
0.25	9.5148	9.30
0.45	9.3282	9.14
0.65	9.7665	9.96
0.85	11.3528	11.77
0.95	13.4083	13.04

For the equation, $y = a + bx + cx^2$, the derivative equations become

$$6a + 3.2b + 2.315d = 63.6553,$$

$3.2a + 2.315b + 1.853c = 35.8266,$

$2.315a + 1.853b + 1.5599 = 26.9391.$

The solution of these linear equations gives the result

$y = 10.8896 - 9.4422\,x + 12.3252\,x^2.$

The entries in the last column are back calculated from this equation. A more complex correlating equation of greater accuracy is used in the reference cited.

c. Nonlinear least squares: A computer program for determining the parameters of the Wilson and T-K-Wilson equations that are involved nonlinearly in the derived equations is given in Table C.6. Wolberg (1967, Chap. 3) gives a straightforward account of this method.

B.6. NUMERICAL DIFFERENTIATION

Equations for this purpose are based on interpolating formulas in terms of finite differences. The Lagrange formula for several numbers of intervals is given by Milne (*Numerical Calculus,* 1949). Formulas for higher derivatives are summarized, for example, in *CRC Handbook of Mathematical Sciences,* 5th ed. (1978). Here the three-point unequal interval Lagrange equation will be used for illustration.

The derivative of the Lagrange interpolation formula is

$$\frac{dy}{dx} = \frac{2x - x_1 - x_2}{(x_0 - x_1)(x_0 - x_2)} y_0 + \frac{2x - x_0 - x_2}{(x_1 - x_0)(x_1 - x_2)} y_1 + \frac{2x - x_0 - x_1}{(x_2 - x_0)(x_2 - x_1)} y_2.$$

When the spacings are uniform, $h = x_2 - x_1 = x_1 - x_0$,

$$y'_0 = (-3y_0 + 4y_1 - y_2)/2h,$$

$$y'_1 = (-y_0 + y_2)/2h,$$

$$y'_2 = (y_0 - 4y_1 + 3y_2)/2h.$$

Example: Excess Gibbs energies of water + 1,4-dioxane are (*Int. Data Series, Thermo. Prop Aq. Org. Systems,* No. 97 (April 23, 1979):

x_1	0	0.05	0.15
G^{ex}	0	242.5	637.8

The chemical potential is

$$\mu_1 = G^{ex} + (1 - x_1)\frac{\partial G^{ex}}{\partial x_1}.$$

The differentiations are performed with the Lagrange formula at the reference points and two others, with the results shown. The last column is of precise values given in the reference paper.

```
10 READ X0,X1,X2,Y0,Y1,Y2
20 DATA 0,.05,.15,0,242.5,637.8
30 INPUT X,Y
40 Z=(2*X−X1−X2)*Y0/(X0−X1)/(X0−X2)
50 Z=Z+(2*X−X0−X2)*Y1/(X1−X0)/(X1−X2)
60 Z=Z+(2*X−X0−X1)*Y2/(X2−X0)/(X2−X1)
70 M=Y+(1−X)*Z
80 PRINT "X=";X;"dG/dX=";Z;"μ=";M
90 GOTO 30
100 END
```

x	dG/dx	μ Calc.	μ Correct
0	5,149	5,149	5,166
0.05	4,551	4,566	4,554
0.10	3,953	4,012	4,008
0.15	3,355	3,490	3,520
0.20	2,757	3,000	3,084

Cubic splines: Data may be fitted by passing cubic curves through pairs of adjacent points and smoothing them at the individual points (x_i, y_i) by making the slopes and curvatures the same whether the point is approached from the x_{i-1} or x_{i+1} direction. For the ith interval, which extends from (x_i, y_i) to (x_{i+1}, y_{i+1}), the equation of the cubic is

$$y = a_i(x - x_i)^3 + b_i(x - x_i)^2 + c_i(x - x_i) + d_i.$$

The intervals between abscissae are designated by

$$h_i = x_{i+1} - x_i.$$

The coefficients for each interval are given in terms of the second derivatives, S_i, by the equations,

$$a_i = (S_{i+1} - S_i)/6h_i,$$

$$b_i = S_i/2,$$

$$c_i = (y_{i+1} - y_i)/h_i - h_i(2S_i + S_{i+1})/6,$$

$$d_i = y_i.$$

Equations for the derivatives are

$$y' = 3a_i(x - x_i)^2 + 2b_i(x - x_i) + c_i,$$

$$y'' = 6a_i(x - x_i) + 2b_i.$$

The S_i are found by solving the set of linear equations,

$$h_{i-1}S_{i-1} + 2(h_{i-1} + h_i)S_i + h_iS_{i+1} = 6[(y_{i+1} - y_i)/h_i - (y_i - y_{i-1})/h_{i-1}],$$

with $i = 2$ to $n - 1$, and $S_1 = S_n = 0$. Other specifications on the end values are sometimes used and are described, for example, by Gerald (1978, p. 479). When the intervals are even, $h = h_i = h_{i+1}$, the equations for S_i become

$$hS_{i-1} + 4hS_i + hS_{i+1} = 6(y_{i+1} + y_{i-1})/h;$$

$$i = 2 \text{ to } n - 1.$$

The solution of the set of linear equations is relatively simple because the matrix of the coefficients on the left is tridiagonal.

Example 2: Application of cubic splines to finding the chemical potentials, given excess Gibbs energies.

The data are taken from the reference cited in Example 1 of this section. The equations for the determination of the S_i are:

$$0.5S_2 + 0.1S_3 = -8{,}470,$$

$$0.1S_2 + 0.5S_3 + 0.1S_4 = -7{,}544,$$

$$0.1S_3 + 0.3S_4 = -4{,}212.$$

The S_i and the corresponding values of the coefficients of the cubic splines are tabulated, together with the original data. The second table summarizes the calculation of the derivatives at the given and intermediate points, and also the chemical potentials figured from

$\mu = G^{ex} + (1-x)dG^{ex}/dx.$

The comparison with the values of the μs given in the reference is not particularly favorable, nor is the meshing of the derivatives at common points of adjacent intervals. Perhaps a better fit would be obtained if some condition other than $S_1 = S_5 = 0$ were imposed on the calculation. In most other cases the method of cubic splines is significantly more accurate than simple interpolating formulas.

i	x	G^{ex}	h	S
1	0	0	0.1	0
2	0.10	454.5	0.15	−14,949
3	0.25	924.5	0.10	−9,954
4	0.35	1,112.1	0.05	−10,722
5	0.40	1,170.8		0

i	a	b	c	d
1	—	—	—	—
2	−24,915	0	4,794	0
3	5,551	7,475	4,005	454.5
4	−1,281	4,977	2,399	924.5
5	35,740	5,361	1,063	1,112.1

			μ	
x	G^{ex}	dg^{ex}/dx	*Calc.*	*Data*
0	0	4,794	4,794	5,166
0.05	242.5	4,607	4,619	4,554
0.10	454.5	4,047	4,097	4,008
0.10	—	4,005	4,059	
0.15	637.8	3,299	3,442	3,520
0.20	794.1	2,676	2,935	3,084
0.25	924.5	2,137	2,527	2,693
0.25	—	2,400	2,724	
0.30	1,030.2	1,892	2,355	2,342
0.35	1,112.1	1,365	2,000	2,025
0.35	—	1,063	1,803	
0.40	1,170.8	795	1,648	1,738

B.7. NUMERICAL INTEGRATION

This operation is built into even inexpensive calculators and is always available as a canned computer program so it is rarely necessary to program it. The most common method is Simpson's rule which obtains any desired accuracy with mathematical functions by using small enough intervals. When the intervals are fixed, more accurate formulas in terms of higher differences may be desirable (see for example, *CRC Handbook of Mathematical Science,* 653, 1978).

a. Simpsons Rule

$$\int_{x_0}^{x_n} y\,dx = \frac{\Delta x}{3}\left(y_0 + y_n + 2\sum_{\text{even}} y_i + 4\sum_{\text{odd}} y_i\right),$$

where n is an even number of equal intervals and the summations over the y_i with even subscripts and over those with odd subscripts are indicated.

b. Trapezoidal Rule

$$\int_{x_0}^{x_n} y\,dx = \frac{\Delta x}{2}(y_0 + y_n + 2\Sigma\, y_i).$$

c. The Lagrange formula with three unevenly spaced points is

$$\int_a^b y\,dx = \frac{(b-a)}{6}(A_0 y_0 + A_1 y_1 + A_2 y_2),$$

where

$$A_0 = \frac{2(b^2+ab+a^2) - 3(x_1+x_2)(b+a) + 6x_1x_2}{(x_0-x_1)(x_0-x_2)},$$

$$A_1 = \frac{2(b^2+ab+a^2) - 3(x_0+x_2)(b+a) + 6x_0x_2}{(x_1-x_0)(x_1-x_2)},$$

$$A_2 = \frac{2(b^2+ab+a^2) - 3(x_0+x_1)(b+a) + 6x_0x_1}{(x_2-x_0)(x_2-x_1)}.$$

When $a = x_0$ and $b = x_2$, the coefficients are:

$$A_0 = \frac{2x_0^2 - 3x_0x_1 - x_0x_2 + 3x_1x_2 - x_2^2}{(x_1-x_0)(x_2-x_0)},$$

$$A_1 = \frac{(x_2-x_0)^2}{(x_1-x_0)(x_2-x_1)},$$

$$A_2 = \frac{-x_0^2 + 3x_0x_1 - x_0x_2 - 3x_1x_2 + 2x_2^2}{(x_2-x_0)(x_2-x_1)}.$$

Example: Compressibilities of isobutane at 361 K at several pressures are:

P, psia	0	100	225
z	1	0.8815	0.7099
$(z-1)/P$	0	−0.001185	−0.001289

The fugacity coefficient is evaluated from

$$\ln\phi = \int_0^P \frac{z-1}{P}\,dP.$$

```
10 READ X0,X1,X2,Y0,Y1,Y2,A
20 DATA 0,100,225,0,−.001185,−.001289,0
30 INPUT B
40 C=B ^ 2+A*B+A ^ 2
50 A0=(2*C−3*(X1+X2)*(B+A)+6*X1*X2)/
   (X0−X1)/(X0−X2)
```

```
 60 A1=(2*C-3*(X0+X2)*(B+A)+6*X0*X2)/
    (X1-X0)/(X1-X2)
 70 A2=(2*C-3*(X0+X1)*(B+A)+6*X0*X1)/
    (X2-X0)/(X2-X1)
 80 I=(B-A)/6*(A0*Y0+A1*Y1+A2*Y2)
 90 F=EXP(I)
100 PRINT USING 110 ; B,F
110 IMAGE DDD,5X. DDD
120 GOTO 30
130 END
```

P	ϕ
50	0.981
100	0.935
150	0.875
200	0.815
225	0.788
250	0.765

Appendix C. Computer Programs

Use of a computer is essential for repeated application of equations and algorithms pertaining to phase equilibria. Many short programs appear throughout the book in connection with examples and problem solutions. Some longer programs of general utility are collected here as Tables C.1 to C.11. They are written in BASIC for the HP-85 minicomputer and may be regarded as indicative of even more complex programs suitable for practical work.

Many algorithms related to phase equilibria are given in the literature. Reference to computer programs that are available for rent or purchase is made in the book of Leesley (1982) and the articles by Peterson et al. (1978–1979) that are cited in Appendix D. There are fewer actual programs in the open literature. The principal collections in the area of interest, all written in FORTRAN of varying degrees of intelligibility, may be listed:

1. Fredenslund et al., *Vapor-Liquid Equilibria Using UNIFAC* (1977). Estimation of UNIFAC parameters from VLE data. Fugacity coefficients with the virial equation. Activity coefficients with UNIFAC-UNIQUAC. Prediction of VLE and LLE.
2. Kojima & Tochigi, *Prediction of Vapor-Liquid Equilibria by the ASOG Method* (1979). Evaluation of activity coefficients from molecular structure. Prediction of VLE when the vapor phase is ideal. A program for determining ASOG parameters from experimental data may be obtainable from the authors.
3. Shahin Negahban, Department of Chemical and Petroleum Engineering, University of Kansas (1982). Regression for parameters from binary and ternary LLE data. Calculation of tie-lines of binary and ternary two liquid-phase systems. Ternary three-liquid-phase systems. Quaternary two-phase systems. Utilizes both NRTL and UNIQUAC equations.
4. Null, *Phase Equilibria in Process Design* (1970). LLE in ternary two-phase and three-phase systems, utilizing binary van Laar activity coefficient equations.
5. Prausnitz et al., *Computer Calculations for Multicomponent Vapor-Liquid and Liquid-Liquid Equilibria* (1980). Regression for the parameters of the UNIQUAC equation from experimental data. Calculation of second virial coefficients, fugacities, activity coefficients, and enthalpy and entropy residuals. Routines for VLE and LLE.
6. Renon et al. *Calcul sur Ordinateur des Equilibres Liquide-Vapeur et Liquide-Liquide* (1971). Regression for the parameters of the NRTL equation from VLE experimental data. Routines for VLE and LLE.
7. Smith & Missen, *Chemical Reaction Equilibrium Analysis: Theory and Algorithms.* (Wiley, 1982). Computer programs for the HP-41C calculator, and in BASIC and FORTRAN.
8. Sorensen, *Correlation of Liquid-Liquid Equilibrium Data,* doctoral thesis, Instituttet for Kemiteknik, Lyngby, Denmark (1980). Description of a program for regression of NRTL and UNIQUAC parameters from experimental ternary LLE data and the calculation of the binodal curve with two liquid phases. The actual program, prepared in its final form by M.L. Michelsen, is obtainable from the Instituttet.
9. Sorensen & Arlt, *Liquid-Liquid Equilibrium Data Collection: Ternary Systems,* DECHEMA Chemistry Data Series, V-2 (1980). Binodal curves of ternary two-phase systems with the NRTL and UNIQUAC equations.

The list of computer programs in this appendix is:

Table C.1.	Plot of the Harmens-Knapp equation of state,
Table C.2.	Soave equation of state: Compressibility and specific volume of a binary mixture,
Table C.3.	Soave equation of state: Fugacity coefficients in a binary mixture,
Table C.4.	Soave equation of state: Enthalpy and entropy departures of a binary mixture,
Table C.5.	Wilson or T-K-Wilson parameters from one set of activity coefficients by the Newton-Raphson method,
Table C.6.	Wilson and T-K-Wilson parameters by non-linear regression,
Table C.7.	Determination of activity coefficients from molecular structure by the ASOG method,
Table C.8.	Solubility limits with the van Laar or Scatchard-Hildebrand equations,
Table C.9.	Equilibrium compositions of two liquid phases comprised of two or three components, with the T-K-Wilson equation,
Table C.10.	Equilibrium compositions of two liquid phases comprised of four components, with the T-K-Wilson equation,
Table C.11.	Activity coefficients of multicomponent mixtures in terms of the Wilson equation.

Table C.1. Plot of the Harmens-Knapp Equation of State

```
   10 ! PLOTS OF THE HARMENS-KNAPP EOS
   20 ! DATA FOR ACETONE: PC=46.4 atm,
      TC=508.1 K, W=0.309
   30 T=525
   40 SCALE 0,1,5,-30.70
   50 XAXIS 0,.1
   60 YAXIS 1.5,10
   70 YAXIS 0,10
   80 READ T1,P1,W
   90 DATA 508.1,46.4,.309
 ζ100 X1=.3211-.08*W+.0384*W^2
 β110 B1=.1077+.76405*X1-1.2428*X1^2+.9621*X1
      ^3
Ωa120 Y1=1-3*X1+3*X1^2+B1*X1*(3-6*X1+B1*
      X1)
Ωb130 Z=B1*X1
 b140 B2=.08205*T1*Z/P1
 c150 C2=1+(1-3*X1)/Z
  160 IF T<=T1 THEN 170 ELSE 260
  170 IF A1<=2 THEN 180 ELSE 220
 A180 A=.5+.27767*W+2.17225*W^2
  190 B=-.022+.338*W-.854*W^2
 α200 A3=(1+A*(1-(T/T1) .5)-B*(1-T1/T))^2
  210 GOSUB 300
  220 A=.4131+1.1465*W
  230 B=.0118
  240 A3=(1+A*(1-(T/T1) .5)-B*(1-T1/T))^2
  250 GOSUB 300
```

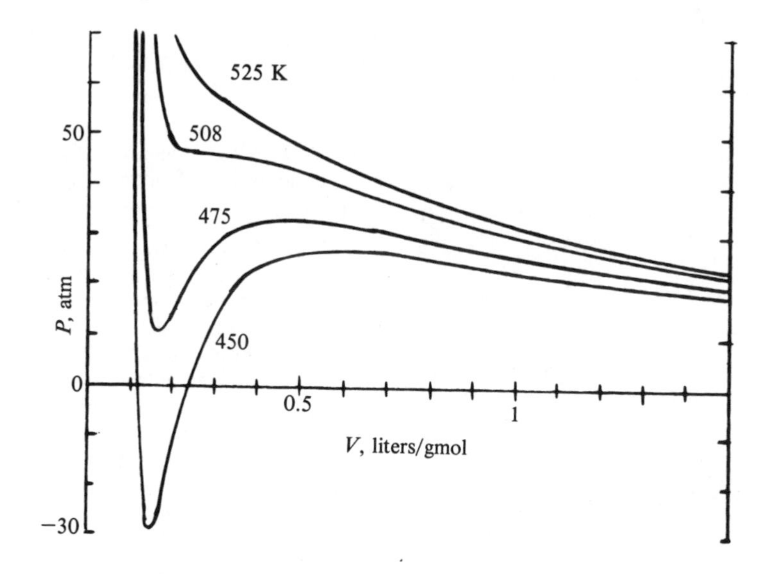


```
260 A3=1−(.6258+1.5527*W)*LOG(T/T1)+(.1533+
    .41 *W)*LOG(T/T1)^2
270 GOSUB 300
280 END
290 ! SR FOR X,Y
a300 A2=A3*Y1*.08205^2*T1^2/P1
310 FOR X=B2+EPS TO 1.5 STEP .01
P320 Y=.08205*T/(X−B2)−A2/(X^2+B2*C2*X−
    (C2−1)*B2^2)
330 DRAW X,Y
V 340 NEXT X
350 MOVE 0,0
360 BEEP
370 RETURN
```

Table C.2. Soave Equation of State: Compressibility and Specific Volume of a Binary Mixture (Table 1.11)

```
10 ! PROGRAM "S−1"
20 ! SPECIFIC VOLUME OF A BINARY MIX WITH
    THE SOAVE EQN. P=atm, V=liters/gmol, T=deg K
30 ! EXAMPLE DATA IN LINES 70 & 90 ARE FOR
    CO2+C3H6
40 SHORT Z,V
50 OPTION BASE 1
60 DIM P(2), T(2), W(2)
70 DATA 304.2,364.9,72.9,45.45,.225,.148
80 MAT READ T,P,W
90 K=.0914
100 R=.08205
110 INPUT T,P,Y
120 FOR J=1 TO 2
130 A(J)=.42747*R^2*T(J)^2/P(J)*(1+(.48508+
    1.55171*W(J)−.15613*W(J)^2)*(1−(T/T(J))^5))
    ^2
140 B(J)=.08664*R*T(J)/P(J)
150 NEXT J
160 A=Y^2*A(1)+(1−Y)^2*A(2)+2*Y*(1−Y)*(1−
    K)*(A(1)*A(2))^.5
170 B=Y*B(1)+(−Y)*B(2)
180 A=A*P/R^2/T^2
190 B=B*P/R/T
200 Z=1
210 GOSUB 310
220 H=F/F1
230 Z=Z−H
240 IF ABS(H/Z)<=.0001 THEN 260
250 GOTO 210
260 V=Z*R*T/P
270 PRINT "P=";P;"T=";T;"Y=";Y
280 PRINT "Z=";Z
290 PRINT "V=";V;"liters/gmol"
300 END
310 ! SR FOR COMPRESSIBILITY,Z
320 F=Z^3−Z^2+(A−B−B^2)*Z−A*B
330 F1=3*Z^2−2*Z+A−B−B^2
340 RETURN

P= 10   T=300   Y=.5
Z= .91282
V= 22.2469 liters/gmol

P= 50   T=300   Y=.5
Z= .16708
V= .082253 liters/gmol

P= 25.5   T=303   Y=.5
Z= .75335
V= .73448 liters/gmol
```

Table C.3. Soave Equation of State: Fugacity Coefficients in a Binary Mixture (Table 3.4)

```
10 ! PROGRAM "S−2"
20 ! FUGACITY COEFFICIENTS WITH THE
    SOAVE EQN. TABLE 3.4.
```

(Continued next page)

Table C.3. *(continued)*

```
30 ! EXAMPLE DATA IN LINES 50 & 70 ARE FOR
   CO2+C3H6
40 OPTION BASE 1
50 DIM P(2),T(2),W(2),A(2),B(2),D(2),K(2)
60 DATA 304.2,364.9,72.9,45.45,.225,.148
70 MAT READ T,P.W
80 K=.0914
90 R=.08205
100 INPUT P,T,Y1
105 PRINT "P=";P;"T=";T;"Y=";Y1
110 Y2=1-Y1
120 FOR J=1 TO 2
130 K(J)=.48508+1.55171*W(J)-.15163*W(J)^2
140 A(J)=(1+K(J)*(1-(T/T(J))^.5))^2
150 B(J)=.08664*R*T(J)/P(J)
160 NEXT J
170 FOR J=1 TO 2
180 D(J)=T(J)*(A(J)/P(J))^.5/(Y1*T(1)*(A(1)/P(1))^
    .5+Y2*T(2)*(A(2)/P(2))^.5)
190 A(J)=A(J)*.42747*R^2*T(J)^2/P(J)
200 NEXT J
210 A=Y1^2*A(1)+Y2^2*A(2)+2*Y1*Y2*(1-K)*(A
    (1)*A(2)) .5
220 A=A*P/R^2/T^2
230 C=Y1*B(1)+Y2*B(2)
240 B=C*P/R/T
250 Z=1
260 GOSUB 380
270 H=F/F1
280 Z=Z-H
290 IF ABS(H/Z)<=.0001 THEN 310
300 GOTO 260
310 FOR J=1 TO 2
320 F(J)=B(J)/C*(Z-1)-LOG(Z-B)-A/B*(2*D(J)-B
    (J)/C)*LOG(1+B/Z)
330 F(J)=EXP(F(J))
340 PRINT "FUG COEFF=";F(J)
350 NEXT J
360 PRINT "Z=";Z
370 END
380 ! SR FOR COMPRESSIBILITY, Z
390 F=Z^3-Z^2+(A-B-B^2)*Z-A*B
400 F1=3*Z^2-2*Z+A-B-B^2
410 RETURN

P= 50  T=300  Y=.5
FUG COEFF=.943593853229
FUG COEFF=1.24057606753
Z=.167077288477

P= 30  T=300  Y=.5
FUG COEFF=.897040939721
FUG COEFF=.84238281801
Z= .675247249859

P= 10  T=300  Y=.5
FUG COEFF=.96038164773
FUG COEFF=.942323379853
Z= .912818464905
```

Table C.4. Soave Equation of State: Enthalpy and Entropy Departures of a Binary Mixture (Table 11.3)

```
10 ! PROGRAM "S-3"
20 ! ENTHALPY & ENTROPY DEPARTURES WITH
   THE SOAVE EQN. units are cal, gmol, deg K
30 ! EXAMPLE DATA IN LINES 60 & 80 ARE FOR
   CO2+C3H6
40 OPTION BASE 1
50 DIM P(2), T(2), W(2)
60 DATA 304.2,364.9,72.9,45.45,.225,.148
70 MAT READ T,P,W
80 K=.0914
90 R=.08205
100 INPUT P,T,Y
110 FOR J=1 TO 2
120 K(J)=.48508+1.55171*W(J)-.15163*W(J)^2
130 A1(J)=.42747*R^2*T(J)^2/P(J)
140 A(J)=(1+K(J)*(1-(T/T(J))^.5))^2
150 A(J)=A(J)*A1(J)
160 B(J)=.08664*R*T(J)/P(J)
170 NEXT J
180 A1=Y^2*A1(1)+(1-Y)^2*A1(2)+2*Y*(1-Y)*
    (A1(1)*A1(2))^.5
190 C=Y^2*A(1)+(1-Y)^2*A(2)+2*Y*(1-Y)*(1-K)
    *(A(1)*A(2))^.5
200 A=C*P/R^2/T^2
210 B1=Y*B(1)+(1-Y)*B(2)
220 B=B1*P/R/T
230 Z=1
240 GOSUB 450
250 H=F/F1
260 Z=Z-H
270 IF ABS(H/Z)<=.0001 THEN 290
280 GOTO 240
290 V=Z*R*T/P
300 FOR J=1 TO 2
310 D(J)=-(K(J)*(A1(J)*(A(J)/T/T(J)) ^.5))
320 NEXT J
330 D1=-(1/2/T ^.5*(K(2)*(A1(2)*A(1)/T(2))^.5+K
    (1)*(A1(1)*A(2)/T(1)) ^.5))
340 D=Y^2*D(1)+(1-Y)^2*D(2)+2*Y*(1-Y)*(1-
    K)*D1
350 H1=1.987*T*(1-Z)+1/B1*(C-T*D)*LOG(1+B1/
    V)
360 S1=1.987*LOG(V/(V-B1))-1/B1*D*LOG(1+B1/
    V)
370 PRINT "P=";P;"atm"
380 PRINT "T=";T;"K"
390 PRINT "Y=";Y;"mol fraction"
400 PRINT "Z=";Z
410 PRINT "V=";V;"liters/gmol"
420 PRINT "ΔH=";H1;"cal/gmol"
430 PRINT "ΔS=";S1;"cal/gmol/K"
440 END
450 ! SR FOR COMPRESSIBILITY, Z
460 F=Z^3-Z^2+(A-B-B^2)*Z-A*B
470 F1=3*Z^2-2*Z+A-B-B^2
480 RETURN

P= 30 atm
T= 300 K
Y= .5 mol fraction
Z= .675247249859
```

Table C.4. *(continued)*

```
V = .55404036851 liters/gmol
ΔH = 217.212185573 cal/gmol
ΔS=.20582271096 cal/gmol/K

P = 10 atm
T = 300 K
Y = .5 mol fraction
Z = .912818464904
V = 2.24690265136 liters/gmol
ΔH = 57.9623661825 cal/gmol
ΔS = 4.98532833145E−2 cal/gmol/K
```

Table C.5. The Wilson or T-K-Wilson Parameters from One Set of Activity Coefficients by the Newton-Raphson Method (see Example 4.7)

```
10 ! FINDING WILSON OR T-K-WILSON
     PARAMETERS FROM ONE SET OF ACTIVITY
     COEFFS, USING THE N-R METHOD
20 ! FOR WILSON EQN, MAKE R=V2/V1=1
30 READ X,Z1,Z2,R
40 DATA .736,1.0224,1.3983,.5107
45 SHORT A,B
50 INPUT A,B
60 X2=1−X
70 C=1/(X2+X/A)−1/(X+X2/B)
80 C1=1/(X2+X/R)−1/(X+X2*R)
90 F=LOG(Z1)+LOG((X+X2*A)/(X+X2*R))−X2*
     (C−C1)
100 G=LOG(Z2)+LOG((X2+X*B)/(X2+X/R))+X*
     (C−C1)
110 F1=A*(X2/(X+X2*A))^2
120 F2=(X2/(X2+X*B))^2
130 G1=(X/(X+A*X2))^2
140 G2=B*(X/(X2+B*X))^2
150 D=G1*F2−F1*G2
160 H=(F*G2−G*F2)/D
170 K=(F1*G−G1*F)/D
180 PRINT USING 190 ; A,B
190 IMAGE D.DDDD,2X,D.DDDD
200 A=A+H
210 B=B+K
220 IF ABS(H/A)+ABS(K/B)>=.0001 THEN 70
230 PRINT
240 PRINT "X1=";X
250 PRINT "A12=";A
260 PRINT "A21=";B
270 END

WILSON PARAMETERS, R=1

 .5000     .5000
 .5715     .8399
 .3140    1.1717
 .2953    1.2602
 .2924    1.2671

X1=.736
A12=.29235
A21=1.2672
```

Table C.5. *(continued)*

```
T-K-W PARAMETERS, R=.5107

 .5000     .5000
 .7220     .8575
 .3668    1.2910
 .3314    1.4531
 .3211    1.4782
 .3210    1.4785

X1=.736
A12=.32104
A21=1.4785
```

Table C.6. Wilson and T-K-Wilson Parameters by Nonlinear Regression

Example is water + ethanol @ 60 C, $V_2/V_1 = 3.244$. Successive approximations are shown.

```
10 ! "W&TKW" LEAST SQUARES NEWTON-
     RAPHSON METHOD FOR WILSON AND T-K
     WILSON PARAMETERS.
20 ! IN STEP #60, PLACE V=V2/V1 FOR T-K-W,
     AND V=1 FOR WILSON.!
30 ! N=# OF DATA POINTS. ENTER X1 (OR X2)
     AND G1 and G2 IN STEPS 80-89. CHANGE #110
     IF NEEDED.
40 OPTION BASE 1
50 SHORT A,B
60 INPUT N,A,B,V
70 DIM X1(11),X2(11),G1(11),G2(11),Q(11),Q1(11),
     R(11),R1(11),R2(11)
80 DATA .051,.086,.197,.375,.509,.527,.545,.808,.851,
     .86,.972
81 DATA 1.028,1.079,1.161,1.366,1.591,1.608,1.621,
     2.135,2.19,2.26,2.35
82 DATA 3.845,3.135,2.155,1.421,1.198,1.181,1.171,
     1.022,1.015,1.01,.998
90 MAT READ X2,G1,G2
100 FOR I=1 TO N
110 X1(I)=1−X2(I)
120 NEXT I
130 GOSUB 520
140 FOR I=1 TO N
150 Q(I)=X1(I)*LOG(G1(I))+X2(I)*LOG(G2(I))
160 R(I)=Q(I)+Q1(I)
170 NEXT I
180 F=0
190 FOR I=1 TO N
200 F=F+R(I)^2
210 R1(I)=X1(I)*X2(I)/(X1(I)+A*X2(I))
220 R2(I)=X1(I)*X2(I)/(B*X1(I)+X2(I))
230 NEXT I
240 S=0
250 S1=0
260 S2=0
270 T=0
280 T1=0
290 T2=0
300 FOR I=1 TO N
310 S=S+R(I)*R1(I)
320 T=T +R(I)*R2(I)
```

Table C.6. *(continued)*

```
330 S1=S1+R1(I)^2
340 S2=S2+R1(I)*R2(I)
350 T1=S2
360 T2=T2+R2(I)^2
370 NEXT I
380 D=S1*T2-S2*T1
390 H=(S*T2-T*T1)/D
400 K=(S1*T-T1*S)/D
403 PRINT USING 406 ; A,B
406 IMAGE D.DDDD,2X,D.DDDD
410 IF ABS(H/A)+ABS(K/B)<=.0001 THEN 460 ELSE
    430
430 A=A-H
440 B=B-K
450 GOTO 130
460 PRINT "A=";A;"B=";B
470 END
520 FOR I=1 TO N
530 Q1(I)=X1(I)*LOG((X1(I)+A*X2(I))/(X1(I)+V*X2
    (I)))+X2(I)*LOG((X1(I)*B+X2(I))/(X1(I)/V+X2
    (I)))
540 NEXT I
550 RETURN
560 END

T-K-WILSON PARAMETERS, N=11,V2/V1=3.244

 .2000      .2000
 .6913      .5029
1.2177      .2351
1.4352      .1949
1.4608      .1890
1.4617      .1887
A=1.4617    B=.1887

 .5000      .5000
 .9964      .3797
1.4144      .1703
1.4630      .1877
1.4618      .1886
1.4617      .1887
A=1.4617    B=.18868

WILSON PARAMETERS, N=11, V2/V1=1

 .1000      .1000
 .4931      .2511
 .8751      .0753
 .9492      .0845
 .9515      .0842
 .9514      .0842
A=.95143    B=.084168

 .4000      .4000
 .8502      .0317
 .9604      .0758
 .9514      .0841
 .9514      .0842
A=.95143    B=.084168
```

(Wilson params do not converge when starting values are 0.5 or more in this example)

Table C.7. Introduction: ASOG Method for Activity Coefficients from Molecular Structure

The code numbers of the groups of which individual molecules are made up are given in the triangular chart. The number of atoms other than H in the total of any one kind of group is ν_{ki}; exceptions are $\nu_{H_2O} = 1.6$, $\nu_{CH} = 0.8$, and $\nu_C = 0.5$ in alkanes. Lines 190 and 260 of the program are the data for this particular example, which is a mixture of ethanol (1) and benzene (2). The interaction parameters a_{kL} have been entered as steps 270–3080 in the program.

Group k / Molecule i	ν_{ki} N(K,J)			$\sum_k \nu_{ki}$	ν_k
	CH_2 (#1)	ArCH (#3)	OH (#6)	S(J)	N1(J)
1	2	0	1	3	3
2	0	6	0	6	6

(Identification of the groups is made in the diagram on the following page.)

```
T=345.0
X=0.00  GAM=  7.4671
X=1.00  GAM=  1.0000

T=345.0
X= .20  GAM=  2.6628
X= .80  GAM=  1.1059

T=345.0
X= .40  GAM=  1.6235
X= .60  GAM=  1.7962

T=345.0
X= .60  GAM=  1.2258
X= .40  GAM=  1.7962

T=345.0
X= .80  GAM=  1.0518
X= .20  GAM=  2.5645

T=345.0
X=1.00  GAM=  1.0000
X=0.00  GAM=  4.0507
```

Table C.7A. Determination of Activity Coefficients from Molecular Structure by the ASOG Method

```
10 REM *ASOG METHOD FOR ACTIVITY
   COEFFICIENTS
20 REM *ENTER DATA ON 190: T, P, Q, N(K,J),
   S(J), N1(J)
30 REM *ENTER GROUP NUMBERS K AND L ON
   260
40 REM *INPUT MOL FR X(J) ON RUN
50 REM *T=TEMP, P=#OF GROUPS, Q=# OF
   COMPOUNDS
60 SHORT A(31, 31)
70 READ T,P,Q
```

(Continued next page)

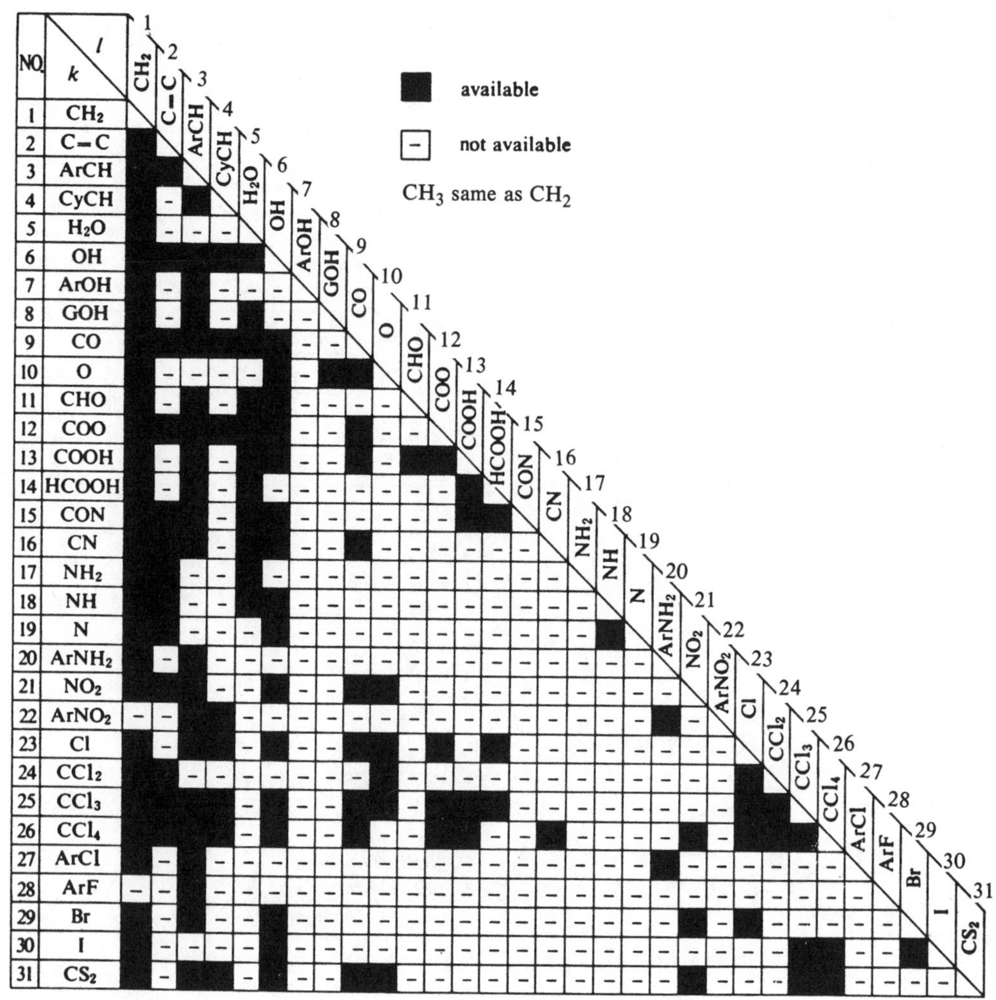

Table C.7A. *(continued)*

```
 80 FOR K=1 TO P
 90 FOR J=1 TO Q
100 READ N(K,J)
110 NEXT J
120 NEXT K
130 FOR J=1 TO Q
140 READ S(J)
150 NEXT J
160 FOR J=1 TO Q
170 READ N1(J)
180 NEXT J
190 DATA 345,3,2,2,0,0,6,1,0,3,6,3,6
200 FOR K=1 TO P
210 READ K1(K)
220 NEXT K
230 FOR L=1 TO P
240 READ L1(L)
250 NEXT L
260 DATA 1,3,6,1,3,6
270 A(2,1)=EXP(-1.524+713.8/T)
280 A(3,1)=EXP(.7297-176.8/T)
290 A(4,1)=EXP(-.1842+.3/T)
300 A(5,1)=EXP(.5045-2382.3/T)
310 A(6,1)=EXP(4.7125-3060/T)
320 A(7,1)=EXP(3.5403-2282.8/T)
```

Table C.7A. *(continued)*

```
330 A(8,1)=EXP(-12.9277+4063.3/T)
340 A(9,1)=EXP(-1.7588+169.6/T)
350 A(10,1)=EXP(-.5097+165.7/T)
360 A(11,1)=EXP(-.4274-560/T)
370 A(12,1)=EXP(-.3699+162.6/T)
380 A(13,1)=EXP(-10.9719+4022/T)
390 A(14,1)=EXP(-.0721-264.8/T)
400 A(15,1)=EXP(-3.7607+849.8/T)
410 A(16,1)=EXP(-12.2577+3268.8/T)
420 A(17,1)=EXP(-1.1005-346.7/T)
430 A(18,1)=EXP(.2778-274.9T)
440 A(19,1)=EXP(.1993+2.8/T)
450 A(20,1)=EXP(-1.9291-615.1/T)
460 A(21,1)=EXP(-1.4089+228.5/T)
470 A(23,1)=EXP(-1.3346+116/T)
480 A(24,1)=EXP(.2849-151.1/T)
490 A(25,1)=EXP(-.1134+41.1/T)
500 A(26,1)=EXP(.6926-358.5/T)
510 A(27,1)=EXP(-1.8796+482.5/T)
520 A(29,1)=EXP(-7.0135+1842.7/T)
530 A(30,1)=EXP(-2.0131+2.8/T)
540 A(31,1)=EXP(-.0033+6/T)
550 A(1,2)=EXP(1.3296-995/T)
560 A(3,2)=EXP(-1.5842+247.4/T)
570 A(6,2)=EXP(10.5763-4545.3/T)
```

Table C.7A. *(continued)*

```
580 A(9,2)=EXP(-1.6404+109.8/T)
590 A(12,2)=EXP(-3.4011+1159.1/T)
600 A(15,2)=EXP(-.7601+.4/T)
610 A(16,2)=EXP(-2.1511+2.7/T)
620 A(17,2)=EXP(-1.6397+4.3/T)
630 A(18,2)=EXP(-.9317+1.9/T)
640 A(19,2)=EXP(-2.7491+2.6/T)
650 A(21,2)=EXP(-1.1622+80.2/T)
660 A(24,2)=EXP(-.8108+1.7/T)
670 A(25,2)=EXP(-.8444+1.1/T)
680 A(26,2)=EXP(-1.0915+1.9/T)
690 A(1,3)=EXP(-.7457+146/T)
700 A(2,3)=EXP(-.5754+338.7/T)
710 A(4,3)=EXP(.5301-251/T)
720 A(6,3)=EXP(-.5859-939.1/T)
730 A(7,3)=EXP(-2.0123-478/T)
740 A(8,3)=EXP(.9602+5.6/T)
750 A(9,3)=EXP(-.4021-216.8/T)
760 A(11,3)=EXP(-.1261+312/T)
770 A(12,3)=EXP(-.1541+97.5/T)
780 A(13,3)=EXP(-.2256-213.7/T)
790 A(14,3)=EXP(-.3-232.2/T)
800 A(15,3)=EXP(.8736-98/T)
810 A(16,3)=EXP(.864+2.3/T)
820 A(20,3)=EXP(-2.6417+62.5/T)
830 A(21,3)=EXP(-.1225-161.1/T)
840 A(22,3)=EXP(6.783-2325.8/T)
850 A(23,3)=EXP(-.324+7.1/T)
860 A(25,3)=EXP(.2511+1/T)
870 A(26,3)=EXP(.8304-374.5/T)
880 A(27,3)=EXP(-1.5456+191.7/T)
890 A(28,3)=EXP(145.3566-51745.1/T)
900 A(29,3)=EXP(-1.5507+253.8/T)
910 A(31,3)=EXP(-1.3705+279.3/T)
920 A(1,4)=EXP(.153+2.1/T)
930 A(3,4)=EXP(-.3288+156.3/T)
940 A(6,4)=EXP(5.6308-3221.4/T)
950 A(9,4)=EXP(-2.7194+428/T)
960 A(12,4)=EXP(-.0991+2.4/T)
970 A(22,4)=EXP(-4.4362+1354.8/T)
980 A(23,4)=EXP(1.1119-640/T)
990 A(25,4)=EXP(-.2819+3.9/T)
1000 A(26,4)=EXP(.0193-78.1/T)
1010 A(31,4)=EXP(.3599-178.5/T)
1020 A(1,5)=EXP(-.2727-277.3/T)
1030 A(6,5)=EXP(-5.8341+1582.5/T)
1040 A(8,5)=EXP(2.2157-450.2/T)
1050 A(9,5)=EXP(.3198-91.2/T)
1060 A(11,5)=EXP(2.0292-556.8/T)
1070 A(12,5)=EXP(-2.5548+659.9/T)
1080 A(13,5)=EXP(-2.1113+779.7/T)
1090 A(14,5)=EXP(1.5229-921.5/T)
1100 A(15,5)=EXP(.2189+226.8/T)
1110 A(16,5)=EXP(-4.5933+1450.9/T)
1120 A(17,5)=EXP(4.4468-1847.2/T)
1130 A(18,5)=EXP(-2.6892+919.5/T)
1140 A(1,6)=EXP(-41.2503+7686.4/T)
1150 A(2,6)=EXP(-.1595-248.2/T)
1160 A(3,6)=EXP(2.2682-1111.5/T)
1170 A(4,6)=EXP(-11.9939-2231.6/T)
1180 A(5,6)=EXP(1.4318-280.2/T)
1190 A(9,6)=EXP(-.3283+1.3/T)
```

Table C.7A. *(continued)*

```
1200 A(10,6)=EXP(.9348-152.2/T)
1210 A(11,6)=EXP(-.8049+4.4/T)
1220 A(12,6)=EXP(-.0296+2.6/T)
1230 A(13,6)=EXP(1.7-664.5/T)
1240 A(15,6)=EXP(-.5965+2.2/T)
1250 A(16,6)=EXP(-.1045-91.9/T)
1260 A(18,6)=EXP(5.9417-1834.8/T)
1270 A(19,6)=EXP(2.6807-115.6/T)
1280 A(21,6)=EXP(2.8755-916.7/T)
1290 A(23,6)=EXP(8.9682-2889.4/T)
1300 A(25,6)=EXP(-2.2978+605.7/T)
1310 A(26,6)=EXP(-9.7985+2539/T)
1320 A(29,6)=EXP(.7523-589.5/T)
1330 A(30,6)=EXP(-.8586-421.2/T)
1340 A(31,6)=EXP(.9985-4056.7/T)
1350 A(1,7)=EXP(-6.5482+2719.5/T)
1360 A(3,7)=EXP(.6483-252/T)
1370 A(1,8)=EXP(13.6669-5554/T)
1380 A(3,8)=EXP(-7.6896-7.5/T)
1390 A(5,8)=EXP(-7.9975+2397/T)
1400 A(10,8)=EXP(-4.2085+2478.6/T)
1410 A(1,9)=EXP(2.6172-865.1/T)
1420 A(2,9)=EXP(.1444+40.5/T)
1430 A(3,9)=EXP(.9273-185.8/T)
1440 A(4,9)=EXP(3.2821-1042.6/T)
1450 A(5,9)=EXP(.0585-278.8/T)
1460 A(6,9)=EXP(-.7262+2.9/T)
1470 A(10,9)=EXP(1.0836+1.1/T)
1480 A(12,9)=EXP(-.1212+180/T)
1490 A(13,9)=EXP(1.8864-543/T)
1500 A(16,9)=EXP(.013-152.5/T)
1510 A(21,9)=EXP(.0617+1.9/T)
1520 A(23,9)=EXP(.9482-408/T)
1530 A(25,9)=EXP(.3823+3.1/T)
1540 A(26,9)=EXP(.8583-252.5/T)
1550 A(31,9)=EXP(-.1141-88.2/T)
1560 A(1,10)=EXP(-.09+32.4/T)
1570 A(6,10)=EXP(-.671-150.8/T)
1580 A(8,10)=EXP(-5.3133+1673/T)
1590 A(9,10)=EXP(-.8071+1.8224/T)
1600 A(21,10)=EXP(.2035+1.4/T)
1610 A(23,10)=EXP(16.7667-5200.2/T)
1620 A(24,10)=EXP(.5476-122.6/T)
1630 A(25,10)=EXP(.8597+.6/T)
1640 A(31,10)=EXP(.0428+.3/T)
1650 A(1,11)=EXP(2.7287-563.7/T)
1660 A(3,11)=EXP(2.2781-14146.9/T)
1670 A(5,11)=EXP(-4.2492+825/T)
1680 A(6,11)=EXP(-1.1238+2.7/T)
1690 A(13,11)=EXP(3.32-775.3/T)
1700 A(1,12)=EXP(-15.2623+515/T)
1710 A(2,12)=EXP(-5.8807-9/T)
1720 A(3,12)=EXP(-.5812-249.3/T)
1730 A(4,12)=EXP(-2.0465+10.5/T)
1740 A(5,12)=EXP(-2.4686+565.7/T)
1750 A(6,12)=EXP(.0583-455.3/T)
1760 A(9,12)=EXP(-2.5152+489.5/T)
1770 A(13,12)=EXP(-2.132+228.5/T)
1780 A(23,12)=EXP(-23.0752+7174.8/T)
1790 A(25,12)=EXP(-16.8658-3637.1/T)
1800 A(26,12)=EXP(-2.8851-27.1/T)
1810 A(1,13)=EXP(9.7236-3797.5/T)
```

Table C.7A. *(continued)*

```
1820 A(3,13)=EXP(1.4405-492.9/T)
1830 A(5,13)=EXP(-.4492+7.4/T)
1840 A(6,13)=EXP(3.8786-1712/T)
1850 A(9,13)=EXP(1.0434-626/T)
1860 A(11,13)=EXP(-5.329+579.7/T)
1870 A(12,13)=EXP(6.4321-2243.2/T)
1880 A(14,13)=EXP(0+0/T)
1890 A(15,13)=EXP(4.6766-1737.4/T)
1900 A(25,13)=EXP(-.3039+3.7/T)
1910 A(26,13)=EXP(5.2636-2014.2/T)
1920 A(1,14)=EXP(.2365-95.3/T)
1930 A(3,14)=EXP(.1022+12/T)
1940 A(5,14)=EXP(-.64+423.3/T)
1950 A(15,14)=EXP(4.2244-1417.6/T)
1960 A(23,14)=EXP(.0373-274/T)
1970 A(25,14)=EXP(-.1506+3.3/T)
1980 A(13,14)=EXP(0+0/T)
1990 A(1,15)=EXP(6.46-2141.4/T)
2000 A(2,15)=EXP(-.1753+2.8/T)
2010 A(3,15)=EXP(.7817-38/T)
2020 A(5,15)=EXP(-1.2646+18/T)
2030 A(6,15)=EXP(-.26+1.4/T)
2040 A(13,15)=EXP(-6.6366-1946.4/T)
2050 A(14,15)=EXP(4.387-1281.9/T)
2060 A(1,16)=EXP(-4.7726+1515.4/T)
2070 A(2,16)=EXP(.2814+1.5/T)
2080 A(3,16)=EXP(.0646+.6/T)
2090 A(5,16)=EXP(4.1422-1771.9/T)
2100 A(6,16)=EXP(-.694-222.4/T)
2110 A(9,16)=EXP(.7995-115.1/T)
2120 A(26,16)=EXP(-.6321+251/T)
2130 A(1,17)=EXP(.6435-159.8/T)
2140 A(2,17)=EXP(.1075+1.9/T)
2150 A(5,17)=EXP(4.2174-1082.2/T)
2160 A(1,18)=EXP(-12.3803-785.1/T)
2170 A(2,18)=EXP(.0688+2/T)
2180 A(5,18)=EXP(-.5958+489.1/T)
2190 A(6,18)=EXP(-3.5886+1484/T)
2200 A(19,18)=EXP(-12.4922+231.6/T)
2210 A(1,19)=EXP(-14.6389+5.9/T)
2220 A(2,19)=EXP(.7286+1.7/T)
2230 A(6,19)=EXP(-2.1231+231.6/T)
2240 A(18,19)=EXP(2.6204-769/T)
2250 A(1,20)=EXP(-5.8144-209/T)
2260 A(3,20)=EXP(-16.4849+922.8/T)
2270 A(22,20)=EXP(1.0286+1.1/T)
2280 A(27,20)=EXP(-66.005+202.8/T)
2290 A(1,21)=EXP(-.3228-70.8/T)
2300 A(2,21)=EXP(2.0923-678.1/T)
2310 A(3,21)=EXP(-.407+131.4/T)
2320 A(6,21)=EXP(.4399-821.2/T)
2330 A(9,21)=EXP(.0734+.8/T)
2340 A(10,21)=EXP(.3906+.1/T)
2350 A(26,21)=EXP(-1.0435+229.9/T)
2360 A(29,21)=EXP(.1008+2.3/T)
2365 A(31,21)=EXP(-2.0453+344.4/T)
2370 A(3,22)=EXP(-42.7459-291.9/T)
2380 A(4,22)=EXP(-70.4174+619/T)
2390 A(20,22)=EXP(-1.5563+2.1/T)
2400 A(1,23)=EXP(-16.9612+108.9/T)
2410 A(3,23)=EXP(-2.5514+421/T)
2420 A(4,23)=EXP(-1.89+29.8/T)
```

Table C.7A. *(continued)*

```
2425 A(6,23)=EXP(-3.5535+3861/T)
2430 A(6,23)=EXP(-3.638+1211.9/T)
2440 A(10,23)=EXP(-.9031-10128.3/T)
2450 A(12,23)=EXP(-6.8581+2592.5/T)
2460 A(14,23)=EXP(-4.0112+911.5/T)
2470 A(24,23)=EXP(-7.5005-694.2/T)
2480 A(25,23)=EXP(-11.2727+3813.5/T)
2490 A(26,23)=EXP(-2.2388+3/T)
2500 A(29,23)=EXP(.3796+146.6/T)
2510 A(1,24)=EXP(.2805-96.9/T)
2520 A(2,24)=EXP(.4032+2.1/T)
2530 A(10,24)=EXP(1.2256-114.1/T)
2540 A(23,24)=EXP(.325-358.8/T)
2550 A(26,24)=EXP(.073+2.1/T)
2560 A(1,25)=EXP(.2352-119.8/T)
2570 A(2,25)=EXP(.4347+2/T)
2580 A(3,25)=EXP(-.2599+1.9/T)
2590 A(4,25)=EXP(.139+1.9/T)
2600 A(6,25)=EXP(2.804-1898.7/T)
2610 A(9,25)=EXP(-.118-101.3/T)
2620 A(10,25)=EXP(.6112+1.6/T)
2630 A(12,25)=EXP(-.4238+380.9/T)
2640 A(13,25)=EXP(-.607+168.3/T)
2650 A(14,25)=EXP(-.565+2.1/T)
2660 A(23,25)=EXP(3.532-1537/T)
2670 A(26,25)=EXP(.0195+24.1/T)
2680 A(30,25)=EXP(-2.7033+694.7/T)
2690 A(31,25)=EXP(-1.3831+269.1/T)
2700 A(1,26)=EXP(-.3917+227.9/T)
2710 A(2,26)=EXP(.4222+1.9/T)
2720 A(3,26)=EXP(-.5769+270.4/T)
2730 A(4,26)=EXP(.0103+55.7/T)
2740 A(6,26)=EXP(5.9993-3241/T)
2750 A(9,26)=EXP(.6643-536.3/T)
2760 A(12,26)=EXP(-.5166+253.3/T)
2770 A(13,26)=EXP(-5.0606+1679.3/T)
2780 A(16,26)=EXP(-.8584-241.2/T)
2790 A(21,26)=EXP(-.436-79.1/T)
2800 A(23,26)=EXP(-.5089+.8/T)
2810 A(24,26)=EXP(-.1924+1.2/T)
2820 A(25,26)=EXP(-.0063-43.9/T)
2830 A(30,26)=EXP(-1.3672+3.4/T)
2840 A(31,26)=EXP(-.7212+218.8/T)
2850 A(1,27)=EXP(3.6114-1148.9/T)
2860 A(3,27)=EXP(2.74-670.3/T)
2870 A(20,27)=EXP(-240.8051+165.1/T)
2880 A(3,28)=EXP(-322.4732+114126.3/T)
2890 A(1,29)=EXP(-10.7655-1628.9/T)
2900 A(3,29)=EXP(-4.6626+1154.8/T)
2910 A(6,29)=EXP(1.67-1217.7/T)
2920 A(21,29)=EXP(-.792+2.1/T)
2930 A(23,29)=EXP(-2.0545+258.6/T)
2940 A(30,29)=EXP((.9094-441.6/T)
2950 A(1,30)=EXP(.3598-1.1/T)
2960 A(6,30)=EXP(.5339-836.5/T)
2970 A(25,30)=EXP(-.4744+130.2/T)
2980 A(26,30)=EXP(.4361-1.2/T)
2990 A(29,30)=EXP(.3107+74.6/T)
3000 A(1,31)=EXP(-.3104+11.8/T)
3010 A(3,31)=EXP(-.2253+128.7/T)
3020 A(4,31)=EXP(-.2922+114/T)
3030 A(6,31)=EXP(5.3401-2977/T)
```

Table C.7A. *(continued)*

```
3040 A(9,31)=EXP(-.0999-247.9/T)
3050 A(10,31)=EXP(.137+2.5/T)
3060 A(21,31)=EXP(-1.5109+244.8/T)
3070 A(25,31)=EXP(-.2393+138.5/T)
3080 A(26,31)=EXP(.7392-255.8/T)
3090 FOR K=1 TO 31
3100 A(K,K)=1
3110 NEXT K
3120 J=1
3130 FOR K=1 TO P
3140 FOR L=1 TO P
3150 B(K,L)=A(K1(K),L1(L))
3160 NEXT L
3170 NEXT K
3180 FOR J=1 TO Q
3190 INPUT X(J)
3200 NEXT J
3210 REM *CALC OF X1(K)
3220 K=1
3230 GOSUB 4140
3240 K=K+1
3250 IF K>=P+1 THEN 3280 ELSE 3260
3260 GOSUB 4140
3270 GOTO 3240
3280 S2=0
3290 FOR J=1 TO Q
3300 S2=S2+X(J)*S(J)
3310 NEXT J
3320 FOR K=1 TO P
3330 X1(K)=S1(K)/S2
3340 NEXT K
3350 REM *CALC OF D(K)
3360 K=1
3370 GOSUB 4200
3380 K=K+1
3390 IF K>=P+1 THEN 3420 ELSE 3400
3400 GOSUB 4200
3410 GOTO 3380
3420 REM *CALC OF C(K),G(K)
3430 L=1
3440 GOSUB 4260
3450 L=L+1
3460 IF L>=P+1 THEN 3490 ELSE 3470
3470 GOSUB 4260
3480 GOTO 3450
3490 FOR K=1 TO P
3500 G(K)=1-C(K)-LOG(D(K))
3510 NEXT K
3520 REM *CALC OF X2(K,J)
3530 J=1
3540 GOSUB 4320
3550 J=J+1
3560 IF J>=Q+1 THEN 3590 ELSE 3570
3570 GOSUB 4320
3580 GOTO 3550
3590 REM *CALC OF D1(K,J)
3600 J=1
3610 GOSUB 4370
3620 J=J+1
3630 IF J>=Q+1 THEN 3660 ELSE 3640
3640 GOSUB 4370
3650 GOTO 3620
```

Table C.7A. *(continued)*

```
3660 REM *CALC OF C1(L,J)
3670 J=1
3680 L=1
3690 GOSUB 4460
3700 L=L+1
3710 GOSUB 4460
3720 L=L+1
3730 IF L>=P+1 THEN 3760 ELSE 3740
3740 GOSUB 4460
3750 GOTO 3720
3760 J=J+1
3770 IF J>=Q+1 THEN 3780 ELSE 3680
3780 REM *CALC OF G1(K,J)
3790 FOR K=1 TO P
3800 FOR J=1 TO Q
3810 G1(K,J)=1-C1(K,J)-LOG(D1(K,J))
3820 NEXT J
3830 NEXT K
3840 REM *CALC OF G2(J)
3850 J=1
3860 G2(J)=0
3870 FOR K=1 TO P
3880 G2(J)=G2(J)+N(K,J)*(G(K)-G1(K,J))
3890 NEXT K
3900 J=J+1
3910 IF J>=Q+1 THEN 3920 ELSE 3860
3920 REM *CALC OF G3(J),G4(J)
3930 S3=0
3940 FOR J=1 TO Q
3950 S3=S3+N1(J)*X(J)
3960 NEXT J
3970 FOR J=1 TO Q
3980 R(J)=N1(J)/S3
3990 NEXT J
4000 FOR J=1 TO Q
4010 G3(J)=1-R(J)+LOG(R(J))
4020 NEXT J
4030 FOR J=1 to Q
4040 G4(J)=EXP(G3(J)+G2(J))
4050 NEXT J
4060 REM *PRINT RESULTS
4070 PRINT USING 4110 ; T
4080 FOR J=1 TO Q
4090 PRINT USING 4120 ; X(J),G4(J)
4100 NEXT J
4110 IMAGE "T="DDD.D
4120 IMAGE "X="D.DD,2X,"GAM="DDD.DDDD
4130 END
4140 REM *SR FOR S1(K)
4150 S1(K)=0
4160 FOR J=1 TO Q
4170 S1(K)=S1(K)+X(J)*N(K,J)
4180 NEXT J
4190 RETURN
4200 REM *SR FOR D(K)
4210 D(K)=0
4220 FOR L=1 TO P
4230 D(K)=D(K)+X1(L)*B(K,L)
4240 NEXT L
4250 RETURN
4260 REM *SR FOR C(K)
4270 C(L)=0
```

(Continued next page)

Table C.7A. *(continued)*

```
4280 FOR K=1 TO P
4290 C(L)=C(L)+X1(K)*B(K,L)/D(K)
4300 NEXT K
4310 RETURN
4320 REM *SR FOR X2(K,J)
4330 FOR K=1 TO P
4340 X2(K,J)=N(K,J)/S(J)
4350 NEXT K
4360 RETURN
4370 REM *SR FOR D1(K,J)
4380 K=1
4390 D1(K,J)=0
4400 FOR L=1 TO P
4410 D1(K,J)=D1(K,J)+X2(L,J)*B(K,L)
4420 NEXT L
4430 K=K+1
4440 IF K>=P+1 THEN 4450 ELSE 4400
4450 RETURN
4460 REM *SR FOR C1(L,J)
4470 C1(L,J)=0
4480 FOR K=1 TO P
4490 C1(L,J)=C1(L,J)+X2(K,J)*B(K,L)/D1(K,J)
4500 NEXT K
4510 RETURN
4520 END
```

Table C.8. Solubility Limits with the van Laar or Scatchard-Hildebrand Equations. See Example 7.5 for application of the Margules equation.

```
10 ! SOLUBILITY LIMITS WITH THE VAN LAAR
   EQNS "VLAAR"
20 ! SPECIFY PARAMETERS A & B, AND FIND
   CORRESPONDING X & X1 BY TRIAL WITH
   THE N-R METHOD
30 SHORT X,X1
40 INPUT A,B
50 REM *INPUT TRIAL VALUES OF X&X1
60 INPUT X,X1
70 G1=EXP(A/(1+A*X/B/(1-X))^2)
80 G2=EXP(B/(1+B*(1-X)/(A*x))^2)
90 G3=EXP(A/1+A*X1/B/(1-X1))^2)
100 G4=EXP(B/(1+B*(1-X1)/(A*X1))^2)
110 D1=-(2*A^2/(B*(1-X)^2*(1+A*X/(B*(1-X)))
   3)*G1)
120 D2=2*B^2/(A*X^√*(1+B*(1-X)/(A*X))^3)*G2
130 D3=-(2*A^2/(B*(1-X1)^2*(1+A*X1/(B*(1-X1)))
   3)*G3)
140 D4=2*B^2/(A*X1^2*(1+B*(1-X1)/(A*X1))^3)*G4
150 M1=G1+X*D1
160 M2=-G3-X1*D3
170 M3=-G2+(1-X)*D2
180 M4=G4-(1-X1)*D4
190 F=G1*X-G3*X1
200 G=G2*(1-X)-G4*(1-X1)
210 D=M1*M4-M2*M3
220 H=(F*M4-G*M2)/D
230 K=(G*M1-F*M3)/D
```

Table C.8. *(continued)*

```
240 IF ABS(H/X)+ABS(K/X1)<=.0001 THEN 280
    ELSE 250
250 X=X-H
260 X1=X1-K
270 GOTO 70
280 PRINT "A=";A
290 PRINT "B="; B
300 PRINT "X=";X
310 PRINT "X*=";X1
320 END
```

```
A= 4              A= 2
B= 2.5            B= 2.5
X= .021689        X= .28881
X*= .88404        X* =.83242
```

Table C.9. Discussion of Equilibrium between two Liquid Phases Involving Two, Three, or Four Components and Using T-K-Wilson Parameters

The parameters and the volume ratios are entered in the sequences: $A_{11}, A_{12}, A_{13}, \ldots, A_{21}, A_{22}, A_{23}, \ldots, \ldots, V_{11}, V_{12}, V_{13}, \ldots, V_{21}, V_{22}, V_{23}, \ldots, \ldots$. Separate programs are given for three and for four components. Two components are handled by the three-component program. $V_{ij} = V_j/V_i$.

Three components:

1. Identify the least miscible pair as consisting of components 1 and 3.

2. As starting values, take: $x_1 = 0.9$, $x_2 = x_3 = 0.05$, $x_1^* = 0.1$, although these values may need to be varied to improve convergence.

3. Input the starting overall composition: $z_1 = z_3 = 0.5$, $z_2 = 0$, or some composition intermediate between the mutual solubilities of the two solvents.

4. Convergence seems to be most rapid if the overall composition results in an approximately equal split of the phases ($F = 0.5$).

5. Convergence may be optimized by adjusting α to some value between 0 and 1, but the programs use $\alpha = 1$ to start (step #260, $F = F + \alpha H$).

6. The program displays successive approximations of ϕ and the final differences ($\gamma_i x_i - \gamma_i' x_i'$), which should be nearly zero.

Two components: Input full 9-element A_{ij} and V_{ij} matrices, but write unity for the parameters of the third (pseudo-) component. Thus:

$$[A] = \begin{bmatrix} 1 & A_{12} & 1 \\ A_{21} & 1 & 1 \\ 1 & 1 & 1 \end{bmatrix}, \quad [V] = \begin{bmatrix} 1 & V_{12} & 1 \\ V_{21} & 1 & 1 \\ 1 & 1 & 1 \end{bmatrix}.$$

On line #100, input $z_3 = 0$. Make x_i data on line #120: 0.9, 0.1, 0; or leave as is.

Four components:

1. Identify component 2 as the solute.

2. As starting values, take: $x_1 = 0.91$, $x_2 = x_3 = x_4 = 0.03$; $x_1^* = 0.1$, although these values may need to be adjusted to improve convergence.

3. The overall composition to start may be taken as: $z_2 = 0$, $z_1 = z_3 = z_4 = 1/3$, although this may need to be adjusted if it proves be in the one-phase region.

Table C.9. *(continued)*

Example of hexane (1) + ethanol (2) + acetonitrile (3). The input data are,

$$[A] = \begin{bmatrix} 1 & 0.4308 & 0.1641 \\ 0.0818 & 1 & 0.5677 \\ 0.1002 & 0.4159 & 1 \end{bmatrix}$$

$$[V] = \begin{bmatrix} 1 & 0.4465 & 0.4159 \\ 2.2400 & 1 & 0.8991 \\ 2.4909 & 1.1122 & 1 \end{bmatrix}$$

The phase splits, F, overall compositions, z_i, and corresponding compositions for several cases are:

	1	*2*	*3*
	F = .49228		
z	.5000	0.0000	.5000
x	.9149	0.0000	.0851
x^*	.0978	0.0000	.9022
	F = .46986		
z	.4500	.1000	.4500
x	.8156	.0689	.1155
x^*	.1260	.1275	.7465
	F = .42381		
z	.4000	.2000	.4000
x	.7194	.1364	.1442
x^*	.1651	.2468	.5882
	F = .33479		
z	.3500	.3000	.3500
x	.7028	.1759	.1213
x^*	.1724	.3625	.4651
	F = .25318		
z	.3000	.4000	.3000
x	.8012	.1397	.0591
x^*	.1301	.4882	.3817

Table C.9A. Equilibrium Compositions of Two Liquid Phases Comprising Two or Three Components, with the T-K-Wilson Equation

```
10 ! "W-14" LIQ-LIQ EQUILIB WITH T-K-WILSON
   FOR 3 CPDS.
30 OPTION BASE 1
40 N=3
50 DIM A(3,3),V(3,3),X(3),X1(3),Z(3)
60 X1(1)=.1
70 MAT READ A,V
80 DATA 1,.0818,.1002,.4308,1,.4159,.1641,.5677,1
90 DATA 1,2.24,2.491,.4465,1,1.1122,.4015,.8991,1
100 MAT INPUT Z
110 MAT READ X
120 DATA .9,.05,.05
130 F=(Z(1)-X1(1))/(X(1)-X1(1))
140 FOR I=2 TO N
150 X1(I)=(Z(I)-X(I)*F)/(1-F)
160 NEXT I
170 GOSUB 550
180 S=0
190 T=0
200 FOR I=1 TO N
```

Table C.9A. *(continued)*

```
210 K(I)=G1(I)/G(I)
220 S=S+Z(I)*(K(I)-1)/(1+(K(I)-1)*F)
230 T=T+Z(I)*((K(I)-1)/(1+(K(I)-1)*F)) 2
240 NEXT I
250 H=S/T
260 F=F+H
270 DISP F
280 S1=0
290 T1=0
300 FOR I=1 TO N
310 X1(I)=Z(I)/(1+(K(I)-1)*F)
320 IF X1(I)<0 THEN 330 ELSE 340
330 X1(I)=0
340 X(I)=K(I)*X1(I)
350 S1=S1+X(I)
360 T1=T1+X1(I)
370 NEXT I
380 FOR I=1 TO 3
390 X(I)=X(I)/S1
400 X1(I)=X1(I)/T1
410 NEXT I
420 IF ABS(H/F)<=.001 THEN 430 ELSE 170
430 PRINT "F=";F
435 PRINT
440 MAT PRINT USING 450 ; COL Z
450 IMAGE D.DDDD,2X
460 MAT PRINT USING 470 ; COL X
470 IMAGE D.DDDD,2X
480 MAT PRINT USING 490 ; COL X1
490 IMAGE D.DDDD,2X
500 FOR J=1 TO 3
510 Y(J)=G(J)*X(J)-G1(J)*X1(J)
520 DISP Y(J)
530 NEXT J
540 END
550 ! SR FOR ACT COEFFS
560 A=0
570 B=0
580 C=0
590 A1=0
600 B1=0
610 C1=0
620 A2=0
630 B2=0
640 C2=0
650 A3=0
660 B3=0
670 C3=0
680 FOR K=1 TO N
690 A=A+X(K)*A(1,K)
700 A1=A1+X(K)*V(1,K)
710 A2=A2+X1(K)*A(1,K)
720 A3=A3+X1(K)*V(1,K)
730 B=B+X(K)*A(2,K)
740 B1=B1+X(K)*V(2,K)
750 B2=B2+X1(K)*A(2,K)
760 B3=B3+X1(K)*V(2,K)
770 C=C+X(K)*A(3,K)
780 C1=C1+X(K)*V(3,K)
790 C2=C2+X1(K)*A(3,K)
800 C3=C3+X1(K)*V(3,K)
810 NEXT K
```

Table C.9A *(continued)*

```
820 G=LOG(A1/A)+X(1)*(1/A1-1/A)+X(2)*(V(2,1)/
    B1-A(2,1)/B)
830 G(1)=EXP(G+X(3)*(V(3,1)/C1-A(3,1)/C))
840 G=LOG(B1/B)+X(1)*(V(1,2)/A1-A(1,2)/A)+X(2)
    *(1/B1-1/B)
850 G(2)=EXP(G+X(3)*(V(3,2)/C1-A(3,2)/C))
860 G=LOG(C1/C)+X(1)*(V(1,3)/A1-A(1,3)/A)+X(2)
    *(V(2,3)/B1-A(2,3)/B)
870 G(3)=EXP(G+X(3)*(1/C1-1/C))
880 G=LOG(A3/A2)+X1(1)*(1/A3-1/A2)+X1(2)*(V
    (2,1)/B3-A(2,1)/B2)
890 G1(1)=EXP(G+X1(3)*(V(3,1)/C3-A(3,1)/C2))
900 G=LOG(B3/B2)+X1(1)*(V(1,2)/A3-A(1,2)/A2)+
    X1(2)*(1/B3-1/B2)
910 G1(2)=EXP(G+X1(3)*(V(3,2)/C3-A(3,2)/C2))
920 G=LOG(C3/C2)+X1(1)*(V(1,3)/A3-A(1,3)/A2)+
    X1(2)*V(2,3)/B3-A(2,3)/B2)
930 G1(3)=EXP(G+X1(3)*(1/C3-1/C2))
940 RETURN
950 END
```

Table C.10. Equilibrium Compositions of Two Liquid Phases Comprising Four Components, with the T-K-Wilson Equation

```
10 ! "W-13" LIQ-LIQ EQUILIB WITH T-K-WILSON
   EQN FOR 4 CPDS
20 SHORT F
30 OPTION BASE 1
40 N=4
50 DIM A(4,4),V(4,4),X(4),X1(4),Z(4)
60 X1(1)=.1
70 MAT READ A,V
80 ! DATA FOR A, 16 ITEMS
90 ! DATA FOR V, 16 ITEMS
100 MAT INPUT Z
110 MAT READ X
120 DATA .9, .05, .03, .02
130 F=(Z(1)-X1(1))/(X(1)-X1(1))
140 FOR I=2 TO N
150 X1(I)=(Z(I)-X(I)*F)/(1-F)
160 NEXT I
170 GOSUB 550
180 S=0
190 T=0
200 FOR I=1 TO N
210 K(I)=G1(I)/G(I)
220 S=S+Z(I)*(K(I)-1)/(1+(K(I)-1)*F)
230 T=T+Z(I)*((K(I)-1)/(1+(K(I)-1)*F))^2
240 NEXT I
250 H=S/T
260 F=F+H
270 DISP F
280 S1=0
290 T1=0
300 FOR I=1 TO N
310 X1(I)=Z(I)/(1+(K(I)-1)*F)
320 IF X1(I)<0 THEN 330 ELSE 340
330 X1(I)=0
340 X(I)=K(I)*X1(I)
350 S1=S1+X(I)
```

Table C.10. *(continued)*

```
360 T1=T1+X1(I)
370 NEXT I
380 FOR I=1 TO 3
390 X(I)=X(I)/S1
400 X1(I)=X1(I)/T1
410 NEXT I
420 IF ABS(H/F)<=.001 THEN 430 ELSE 170
430 PRINT "F=";F
440 MAT PRINT USING 450 ; COL Z
450 IMAGE D.DDD,3X
460 MAT PRINT USING 470 ; COL X
470 IMAGE D.DDDD,2X
480 MAT PRINT USING 490 ; COL X1,
490 IMAGE D.DDDD,2X
500 FOR J=1 TO 3
510 Y(J)=G(J)*X(J)-G1(J)*X1(J)
520 DISP Y(J)
530 NEXT J
540 END
550 ! SR FOR ACT COEFFS
560 A=0
570 B=0
580 C=0
590 D=0
600 A1=0
610 B1=0
620 C1=0
630 D1=0
640 A2=0
650 B2=0
660 C2=0
670 D2=0
680 A3=0
690 B3=0
700 C3=0
710 D3=0
720 FOR K=1 TO N
730 A=A+X(K)*A(1,K)
740 A1=A1+X(K)*V(1,K)
750 A2=A2+X1(K)*A(1,K)
760 A3=A3+X1(K)*V(1,K)
770 B=B+X(K)*A(2,K)
780 B1=B1+X(K)*V(2,K)
790 B2=B2+X1(K)*A(2,K)
800 B3=B3+X1(K)*V(2,K)
810 C=C+X(K)*A(3,K)
820 C1=C1+X(K)*V(3,K)
830 C2=C2+X1(K)*A(3,K)
840 C3=C3+X1(K)*V(3,K)
850 D=D+X(K)*A(4,K)
860 D1=D1+X(K)*V(4,K)
870 D2=D2+X(K)*A(4,K)
880 D3=D3+X(K)*V(4,K)
890 NEXT K
900 G=LOG(A1/A)+X(1)*(1/A1-1/A)+X(2)*(V(2,1)/
    B1-A(2,1)/B)
910 G(1)=EXP(G+X(3)*(V(3,1)/C1-A(3,1)/C)+X(4)*
    (V(4,1)/D1-A(4,1)/D))
920 G=LOG(B1/B)+X(1)*(V(1,2)/A1-A(1,2)/A)+X
    (2)*(1/B1-1/B)
930 G(2)=EXP(G+X(3)*(V(3,2)/C1-A(3,2)/C)+X(4)*
    (V(4,2)/D1-A(4,2)/D))
```

Table C.10. *(continued)*

```
940 G=LOG(C1/C)+X(1)*(V(1,3)/A1-A(1,3)/A)+X
    (2)*(V(2,3)/B1-A(2,3)/B)
950 G(3)=EXP(G+X(3)*(1/C1-1/C)+X(4)*(V(4,3)/
    D1-A(4,3)/D))
960 G=LOG(D1/D)+X(1)*(V(1,4)/A1-A(1,4)/A)+X
    (2)*(V(2,4)/B1-A(2,4)/B)
970 G(4)=EXP(G+X(3)*(V(3,4)/C1-A(3,4)/C)+X(4)*
    (1/D1-1/D))
980 G=LOG(A3/A2)+X1(1)*(1/A3-1/A2)+X1(2)*(V
    (2,1)/B3-A(2,1)/B2)
990 G1(1)=EXP(G+X1(3)*(V(3,1)/C3-A(3,1)/C2)+
    X1(4)*(V(4,1)/D3-A(4,1)/D2))
1000 G=LOG(B3/B2)+X1(1)*(V(1,2)/A3-A(1,2)/A2)+
    X1(2)*(1/B3-1/B2)
1010 G1(2)=EXP(G+X1(3)*(V(3,2)/C3-A(3,2)/C2)+
    X1(4)*(V(4,2)/D3-A(4,2)/D2))
1020 G=LOG(C3/C2)+X1(1)*(V(1,3)/A3-A(1,3)/A2)+
    X1(2)*(V(2,3)/B3-A(2,3)/B2)
1030 G1(3)=EXP(G+X1(3)*(1/C3-1/C2)+X1(4)*(V
    (4,3)/D3-A(4,3)/D2))
1040 G=LOG(D3/D2)+X1(1)*(V(1,4)/A3-A(1,4)
    /A2)+X1(2)*(V(2,4)/B3-A(2,4)/B2)
1050 G1(4)=EXP(G+X1(3)*(V(3,4)/C3-A(3,4)/C2)+
    X1(4)*(1/D3-1/D2))
1060 RETURN
1070 END
```

Table C.11. Activity Coefficients of Multicomponent Mixtures with the Wilson Equation

```
10 ! ACT COEFF WITH WILSON EQN, "C.11"
20 ! ENTER X(J) ON LINE 90
30 ! ENTER MATRIX OF PARAMS, A(I,J), AS
   SHOWN AT END. I IS THE ROW NO., J THE
   COLUMN NO.
40 ! EXAMPLE IS OF A FOUR COMPONENT
   SYSTEM. CHANGE LINES 60 AND 70 FOR
   OTHER CASES
50 OPTION BASE 1
```

Table C.11. *(continued)*

```
60 N=4
70 DIM X(4),A(4,4),G(4)
80 MAT READ X
90 DATA .1,.2,.2,.5
100 MAT READ A
110 DATA 1,2.2,3.3,.44,.22,1,1.1,.11,.22,.55,1,.33,.66,
    .77,.88,1
120 FOR K=1 TO N
130 G1=0
140 FOR J=1 TO N
150 G1=G1+X(J)*A(K,J)
160 NEXT J
170 G2=0
180 FOR I=1 TO N
190 G3=0
200 FOR J=1 TO N
210 G3=G3+X(J)*A(I,J)
220 NEXT J
230 G2=G2+X(I)*A(I,K)/G3
240 NEXT I
250 G(K)=EXP(1-LOG(G1)-G2)
260 PRINT USING 270 ; X(K),G(K)
270 IMAGE D.DDDD,4X,DD.DDDD
280 NEXT K
290 END
```

x_i	γ_i
.1000	1.0341
.2000	1.6336
.2000	1.1395
.5000	1.4102

I \ *J*	*1*	*2*	*3*	*4*
1	1	2.2	3.3	.44
2	.22	1	1.1	.11
3	.22	.55	1	.33
4	.66	.77	.88	1

Appendix D. Physical Properties

Physical property data may be found directly in the literature or in standard data compilations, or they may be estimated from correlations or molecular structure. The last method is described for example in the books of Bretsznajder (1962), Lyman et al. (1982), Reid et al. (1977), and Sterbacek et al. (1979). According to one estimate, about 2 million data are published annually in some 35,000–50,000 articles, so keeping up with the literature requires computer retrieval, which is implemented in several systems. Several helpful lists of compilations of data are available—for example:

1. Fratzcher, W., H.P. Picht, & H.J. Bittrich. The acquisition, collection and tabulation of substance data on fluid systems for calculations in chemical engineering. *Int. Chem. Eng. 20* (1), 19–28 (1980).
2. Maizell, R.E. *How to Find Chemical Information.* Wiley (1978).
3. Mellon, M.G., *Chemical Publications: Their Nature and Use,* McGraw-Hill, (1982).
4. Rasmussen, P., & A. Fredenslund. *Data Banks for Chemical Engineers.* Kemiingeniorgruppen (Lyngby, Denmark) (1980).
5. Torres-Marchal, C. Getting chemical engineering information from non–English-speaking countries. *Chem. Eng. 88* (14), 117–125 (July 13, 1981).
6. *Physico-Chemical Properties for Chemical Engineering.* Prepared by the Society of Chemical Engineers of Japan. Approximately annual volumes since 1977. Literature references with abstracts of theoretical articles and indications as to the nature of the data in the source. Categories are *PVT* and liquid density, vapor pressure and enthalpy of vaporization, heat capacity and enthalpy, vapor-liquid equilibria, solubility and heat of solution, viscosity, thermal conductivity, and molecular diffusivity. Maruzen, Tokyo (1977–date).

Several proprietary data banks provide information on demand for a fee. Some of these sources are identified in the book of Rasmussen (loc. cit.) and also in these references:

7. Peterson, J.N., C.-C. Chen, & L. B. Evans, Computer programs for chemical engineers. *Chem. Eng.* 145–154 (June 5, 1978); 69–82 (July 3, 1978); 79–86 (July 31, 1978); 107–115 (August 28, 1978); 167–173 (May 21, 1979). Prepared in connection with the ASPEN project.
8. Hilsenrath, J. Summary of on-line or interactive physico-chemical numerical data systems. National Bureau of Standards Technical Note 1122 (1980).
9. Leesley, M.E., *Computer-Aided Process Plant Design.* Gulf Publishing Company (1982).

A service of this kind is being provided now to subscribers by the American Institute of Chemical Engineers and will be made accessible to the public eventually.

The data on 72 substances in this appendix are excerpted largely from a list of 468 substances, identified following. Another useful compilation on 176 substances is taken from the FLOWTRAN simulation program and is reproduced in:

10. Henley, E.J., & J.D. Seader. *Equilibrium-Stage Separation Processes in Chemical Engineering.* Wiley (1981).

Phase .equilibrium data have been collected recently in a comprehensive manner in the books of Gmehling & Onken (1977 ff)., Kogan et al. (1966), Landolt-Börnstein (1966–date), Maczynski et al. (1976ff.), Hirata et al. (1976), and Sorensen & Arlt (1979–1980). A bibliography of equilibrium properties of fluids has been prepared by Hiza, Kidnay, & Miller (1975, 1982).

In Table D.1, the constants of the equation of the equilibrium constants of formation,

$$\log_{10} K_p = A + B/T,$$

were obtained by regression of the tabulated data of Stull et al. (1969) over the temperature range 298–1,000 K. The general data of Table D.2 are excerpted from a revision of a table that appears in Reid et al. (1977), except for the solubility parameters, which are gathered from other sources.

Table D.1. Equilibrium Constants of Formation, $\log_{10} K_p = A + 1{,}000\ B/T$ over the temperature range 298–1,000 K

Formula	*Name*	*A*	*B*
HCl	hydrogen chloride	0.4285	4.852
H_2O	water	−2.6322	12.7401
H_2S	hydrogen sulfide	1.1627	1.4390
H_3N	ammonia	−5.7596	2.5853
H_4N_2	hydrazine	−11.8129	−4.7331
SO_2	sulfur dioxide	−0.2973	15.8165
SO_3	sulfur trioxide	−5.1933	20.9378
$COCl_2$	phosgene	−2.4317	11.5237
CCl_4	carbon tetrachloride	−7.0782	5.1369
CO	carbon monoxide	4.7272	5.7622
CO_2	carbon dioxide	0.1165	20.5658
CS_2	carbon disulfide	7.3293	−5.5939
$CHCl_3$	chloroform	5.8676	5.3323
CH_2O_2	formic acid	−7.0563	20.5541
CH_4	methane	−5.1030	4.2138
CH_4O	methanol	−7.7901	10.8527
C_2H_2	acetylene	2.9533	−11.8005
C_2H_3N	acetonitrile	−3.7480	−4.3729
C_2H_4	ethylene	−3.6279	−2.4422
C_2H_4O	acetaldehyde	−6.7327	9.0089
$C_2H_4O_2$	acetic acid	−11.1003	23.0204
C_2H_6	ethane	−10.4383	4.8874
C_2H_6O	ethanol	−13.0030	12.7215
C_3H_6	propylene	−8.7127	−0.6258
C_3H_6O	acetone	−12.6811	11.8304
C_3H_8	propane	−15.8040	6.0076
C_3H_8O	1-propanol	−18.8086	14.5270
C_3H_8O	isopropanol	−18.9711	14.7867
C_4H_6	1,2-butadiene	−7.4546	−8.0979
C_4H_6	1,3-butadiene	−8.1031	−5.4136
C_4H_8	1-butene	−14.1659	0.5660
C_4H_8	cis-2-butene	−14.6256	0.9963
C_4H_8	trans-2-butene	−14.6446	1.1448
C_4H_8O	methyl ethyl ketone	−17.7890	12.9993
$C_4H_8O_2$	1,4-dioxane	−25.4654	17.1068
$C_4H_8O_2$	ethyl acetate	−21.8841	23.6913
C_4H_{10}	*n*-butane	−21.1552	7.2844
C_4H_{10}	isobutane	−21.9428	7.7140
$C_4H_{10}O$	*n*-butanol	−23.7886	15.0466
$C_4H_{10}O$	isobutanol	−23.8537	15.9293
$C_4H_{10}O$	tert-butyl alcohol	−25.5398	17.6692
$C_4H_{10}O$	ethyl ether	−24.8400	13.8636

Table D.1 *(continued)*

C_5H_5N	pyridine	−9.9614	−6.9212
C_5H_8	isoprene	−13.5832	−3.5175
C_5H_{12}	*n*-pentane	−26.5977	8.4631
C_5H_{12}	2-methyl butane	−26.8757	8.8811
C_5H_{12}	2,2-dimethyl propane	−28.6731	9.4316
$C_5H_{12}O$	1-pentanol	−29.2003	16.6242
C_6H_6	benzene	−9.5963	−3.8557
C_6H_6O	phenol	−12.2574	5.4163
C_6H_7N	aniline	−15.3245	−4.0887
C_6H_{12}	cyclohexane	−29.8950	7.3581
C_6H_{14}	*n*-hexane	−32.0048	9.6646
$C_7H_6O_2$	benzoic acid	−15.8402	15.7865
C_7H_8	toluene	−14.3874	−2.0150
C_7H_{16}	*n*-heptane	−37.4102	10.8575
C_8H_8	styrene	−13.1807	−7.1768
C_8H_{10}	ethyl benzene	−19.7509	−0.8499
C_8H_{18}	*n*-octane	−42.0549	12.0549

Table D.2. Property Data Bank

Selected from Appendix A of *The Properties of Gases and Liquids*, by Reid, Prausnitz, & Sherwood, published by McGraw-Hill Book Company (1977), in the version prepared for CAST, Massachusetts Institute of Technology, by J.F. Boston. Conversion factors to other common units are shown in parentheses following.

Codes and Units of the Properties

MW	MOLECULAR WEIGHT
TFP	NORMAL FREEZING POINT, K
TB	NORMAL BOILING POINT, K
TC	CRITICAL TEMPERATURE, K
PC	CRITICAL PRESSURE, N/M**2 (× 1/101,325 = atm)
VC	CRITICAL VOLUME, M**3/KG-MOLE (= liters/gmol)
ZC	CRITICAL COMPRESSIBILITY
OMEGA	PITZER ACENTRIC FACTOR
VL	LIQUID DENSITY AT TREF, KG/M**3 (= grams/liter)
TREF	REFERENCE TEMPERATURE FOR VL. K
MUP	DIPOLE MOMENT, (N M**4)**.5 (× 0.3162D25 = Debye)
CPIG1 TO CPIG4	CONSTANTS FOR IDEAL GAS HEAT CAPACITY EQUATION. CP = CPIG1 + (CPIG2*T) + (CPIG3*T**2) + (CPIG4*T**3) IN (N M/KG-MOLE K) AND T IN K (× 2.388D-4 = cal(gmol)(K)
VISC 1	LIQUID VISCOSITY EQUATION CONSTANT, Ns/m^2 (× 1,000 = centipoise)
VISC 2	LIQUID VISCOSITY EQUATION CONSTANT, K LOG(VISC) = (VISC1)* ((1/T)*(1/VISC2))
DHFORM	STANDARD ENTHALPY OF FORMATION AT 298 K, N M/KG-MOLE (× 2.388D−4=cal/gmol)
DGFORM	STANDARD GIBBS ENERGY OF FORMATION AT 298 K. IDEAL GAS, 1 ATM, N M/KG-MOLE (× 2.388D-4 = cal/gmol)

Table D.2 *(continued)*

ANT 1 TO ANT7	ANTOINE VAPOUR PRESSURE EQUATION COEFFICIENTS, LN(VP) = ANT1 + ANT2/(T + ANT3) + (ANT4*T) + (ANT5*LN(T)) + (ANT6*T** ANT7) WITH VAPOUR PRESSURE IN PA, AND TEMPERATURE IN K (× 1/101325 = atm)
TMIN	MINIMUM TEMPERATURE FOR ANTOINE EQUATION, K
TMAX	MAXIMUM TEMPERATURE FOR ANTOINE EQUATION, K
DHVB	HEAT OF VAPOURIZATION AT NORMAL BOILING POINT, N M/KG-MOLE (× 2.388D-4 = cal/gmol)
δ	SOLUBILITY PARAMETER AT 25 C, $(CAL/ML)^{0.5}$
δ_1	SOLUBILITY PARAMETER AT 25 C, $(MPa)^{0.5}$

	ARGON	*CHLORINE*	*HYDROGEN CHLORIDE*	*HYDROGEN (NORMAL)*	*WATER*	*HYDROGEN SULFIDE*
FORMULA	AR	CL2	HCL	H2	H2O	H2S
MW	0.39948000D+02	0.70906000D+02	0.36461000D+02	0.20160000D+01	0.18015000D+02	0.34080000D+02
TFP	0.83800000D+02	0.17220000D+03	0.15900000D+03	0.14000000D+02	0.27320000D+03	0.18760000D+03
TB	0.87300000D+02	0.23870000D+03	0.18810000D+03	0.20400000D+02	0.37320000D+03	0.21280000D+03
TC	0.15080000D+03	0.41700000D+03	0.32460000D+03	0.33200000D+02	0.64730000D+03	0.37320000D+03
PC	0.48737325D+07	0.77007000D+07	0.83086500D+07	0.12969600D+07	0.22048320D+08	0.89368650D+07
VC	0.74900000D−01	0.12400000D+00	0.81000000D−01	0.65000000D−01	0.56000000D−01	0.98500000D−01
ZC	0.29100000D+00	0.27500000D+00	0.24900000D+00	0.30500000D+00	0.22900000D+00	0.28400000D+00
OMEGA	−0.40000000D−02	0.73000000D−01	0.12000000D+00	−22000000D+00	0.34400000D+00	0.10000000D+00
VL	0.13730000D+04	0.15630000D+04	0.11930000D+04	0.71000000D+02	0.99800000D+03	0.99300000D+03
TREF	0.90000000D+02	0.23910000D+03	0.18810000D+03	0.20000000D+02	0.29300000D+03	0.21360000D+03
MUP	0.0	0.63245552D−25	0.34785054D−24	0.0	0.56920997D−24	0.28460498D−24
CPIG1	0.20804209D+05	0.26929498D+05	0.30291498D+05	0.27143024D+05	0.32242547D+05	0.31941097D+05
CPIG2	−0.32112756D−01	0.33837718D+02	−0.72012960D+01	0.92737620D+01	0.19238346D+01	0.14364911D+01
CPIG3	0.51665112D−04	−0.38690219D−01	0.12459917D−01	−0.13808066D−01	0.10554923D−01	0.24321121D−01
CPIG4	0.0	0.15470226D−04	−0.38979108D−05	0.76450968D−05	−0.35964612D−05	−0.11764908D−04
VISC1	0.10757000D+03	0.19196000D+03	0.37278000D+03	0.13820000D+02	0.65825000D+03	0.34279000D+03
VISC2	0.22268146D+02	0.46662679D+02	0.85850678D+02	0.24838213D+01	0.12362298D+03	0.67601597D+02
DHFORM	0.0	0.0	−0.92360808D+08	0.0	−0.24199704D+09	−0.20180376D+08
DGFORM	0.0	0.0	−0.95333436D+08	0.0	−0.22876675D+09	−0.33075720D+08
ANT1	−0.73078490D+04	−0.36143310D+04	−0.64603120D+04	−0.51752110D+04	−0.31397400D+02	−0.93201210D+02
ANT2	0.17633770D+06	0.19292300D+06	0.28430820D+06	0.38036580D+05	−0.20463660D+04	−0.82474460D+00
ANT3	0.39162290D+02	0.10600010D+03	0.82323240D+02	0.82040620D+01	−0.75402240D+02	−0.18128770D+03
ANT4	−0.63849410D+01	−0.10226150D+01	−0.23612360D+01	−0.30307020D+02	−0.12054280D−01	−0.55504840D−01
ANT5	0.14404790D+04	0.60082960D+03	0.11128860D+04	0.14664860D+04	0.91657510D+01	0.21720190D+02
ANT6	0.58241780D−02	0.36163370D−03	0.10312470D−02	0.12734970D+00	0.48791950D−17	0.66817760D−10
ANT7	0.20000000D+01	0.20000000D+01	0.20000000D+01	0.20000000D+01	0.60000000D+01	0.40000000D+01
TMIN	0.83800000D+02	0.17220000D+03	0.15900000D+03	0.15660000D+02	0.27320000D+03	0.18760000D+03
TMAX	0.15080000D+03	0.41700000D+03	0.32460000D+03	0.33200000D+02	0.64730000D+03	0.37320000D+03
DHVB	0.65314080D+07	0.20431584D+08	0.16161048D+08	0.90434880D+06	0.40683136D+08	0.18673128D+08
δ	5.33	8.708	7.0110		23.53	8.8
δ_1	10.91	17.818	14.346		48.15	18.0

	AMMONIA	*HYDRAZINE*	*HELIUM-4*	*NITROGEN*	*OXYGEN*	*SULFUR DIOXIDE*
FORMULA	H3N	H4N2	HE(4)	N2	O2	O2S
MW	0.17031000D+02	0.32045000D+02	0.40030000D+01	0.28013000D+02	0.31999000D+02	0.64063000D+02
TFP	0.19540000D+03	0.27470000D+03	0.0	0.63300000D+02	0.54400000D+02	0.19770000D+03
TB	0.23970000D+03	0.38670000D+03	0.42100000D+01	0.77400000D+02	0.90200000D+02	0.26300000D+03
TC	0.40560000D+03	0.65300000D+03	0.51900000D+01	0.12620000D+03	0.15460000D+03	0.43080000D+03
PC	0.11277473D+08	0.14692125D+08	0.22696800D+06	0.33943875D+07	0.50459850D+07	0.78830850D+07
VC	0.72500000D+01	0.96100000D−01	0.57300000D−01	0.89500000D−01	0.73400000D−01	0.12200000D+00
ZC	0.24200000D+00	0.26000000D+00	0.30100000D+00	0.29000000D+00	0.28800000D+00	0.26800000D+00
OMEGA	0.25000000D+00	0.32800000D+00	−0.38700000D+00	0.40000000D−01	0.21000000D−01	0.25100000D+00
VL	0.63900000D+03	0.10080000D+04	0.12300000D+03	0.80400000D+03	0.11490000D+04	0.14550000D+04
TREF	0.27320000D+03	0.29300000D+03	0.43000000D+01	0.78100000D+02	0.90000000D+02	0.26300000D+03
MUP	0.47434164D−24	0.94868328D−24	0.0	0.0	0.0	0.50596442D−24
CPIG1	0.27314683D+05	0.97678044D+04	0.0	0.31149792D+05	0.28105988D+05	0.23852200D+05
CPIG2	0.23831266D+02	0.18945270D+03	0.0	−0.13565232D+02	−0.36801972D−02	0.66988800D+02
CPIG3	0.17073770D−01	−0.16571354D+00	0.0	0.26795520D−01	0.17458956D−01	−0.49613580D−01
CPIG4	−0.11848644D−04	0.60248052D−04	0.0	−0.11681172D−04	−0.10651219D−04	0.13280530D−04
VISC1	0.34904000D+03	0.52498000D+03	0.0	0.90300000D+02	0.85680000D+02	0.39785000D+03
VISC2	0.69012222D+02	0.10926159D+03	0.0	0.18216343D+02	0.18371721D+02	0.81046903D+02
DHFORM	−0.45719856D+08	0.95249700D+08	0.0	0.0	0.0	−0.29705346D+09
DGFORM	−0.16161048D+08	0.15863785D+09	0.0	0.0	0.0	−0.30036103D+09
ANT1	−0.36891000D+02	−0.81937220D+04	0.17144170D+02	−0.19496430D+04	−0.14079980D+04	−0.18537110D+02
ANT2	−0.80343580D+03	0.67364210D+06	−0.33732900D+02	0.47772050D+05	0.39737770D+05	−0.15622620D+04
ANT3	−0.66161870D+02	0.17465370D+03	0.17900000D+01	0.39737620D+02	0.48151280D+02	−0.45780700D+02
ANT4	−0.20785260D−01	−0.12760300D+01	0.0	−0.13499100D+01	−0.79895110D+00	−0.17722480D−01
ANT5	0.10587760D+02	0.12518600D+04	0.0	0.38108330D+03	0.26744380D+03	0.75143910D+01
ANT6	0.77121770D−16	0.27092930D−03	0.0	0.13866070D−07	0.49293240D−08	0.35731280D−13
ANT7	0.60000000D+01	0.20000000D+01	0.0	0.40000000D+01	0.40000000D+01	0.50000000D+01
TMIN	0.19540000D+03	0.35915000D+03	0.30000000D+01	0.63300000D+02	0.61840000D+02	0.19770000D+03
TMAX	0.40560000D+03	0.65300000D+03	0.50000000D+01	0.12620000D+03	0.15460000D+03	0.43080000D+03
DHVB	0.23362344D+08	0.44798760D+08	0.92109600D+05	0.55810044D+07	0.68244840D+07	0.24932394D+08
δ	12.408	18.11		4.40	4.0	6.0
δ_1	25.389	37.06		9.00	8.18	12.28

	SULFUR TRIOXIDE	PHOSGENE	CARBON TETRACHLORIDE	CARBON MONOXIDE	CARBON DIOXIDE	CARBON DISULFIDE
FORMULA	O3S	CCL20	CCL4	CO	CO2	CS2
MW	0.80058000D+02	0.98916000D+02	0.15382300D+03	0.28010000D+02	0.44010000D+02	0.76131000D+02
TFP	0.29000000D+03	0.14500000D+03	0.25000000D+03	0.68100000D+02	0.21660000D+03	0.16130000D+03
TB	0.31800000D+03	0.28080000D+03	0.34970000D+03	0.81700000D+02	0.19470000D+03	0.31940000D+03
TC	0.49100000D+03	0.45500000D+03	0.55640000D+03	0.13290000D+03	0.30420000D+03	0.55200000D+03
PC	0.82073250D+07	0.56742000D+07	0.45596250D+07	0.34957125D+07	0.73764600D+07	0.79033500D+07
VC	0.13000000D+00	0.19000000D+00	0.27600000D+00	0.93100000D−01	0.94000000D−01	0.17000000D+00
ZC	0.26000000D+00	0.28000000D+00	0.27200000D+00	0.29500000D+00	0.27400000D+00	0.29300000D+00
OMEGA	0.41000000D+00	0.20400000D+00	0.19400000D+00	0.49000000D−01	0.22500000D+00	0.11500000D+00
VL	0.17800000D+04	0.13810000D+04	0.15840000D+04	0.80300000D+03	0.77700000D+03	0.12930000D+04
TREF	0.31800000D+03	0.29300000D+03	0.29800000D+03	0.81000000D+02	0.29300000D+03	0.27300000D+03
MUP	0.0	0.34785054D−24	0.0	0.31622776D−25	0.0	0.0
CPIG1	0.45904D+5	0.28089241D+05	0.40716630D+05	0.30869276D+05	0.19795190D+05	0.27444474D+05
CPIG2	0.5240D+2	0.13607100D+03	0.20486012D+03	−0.12853476D+02	0.73436472D+02	0.81265788D+02
CPIG3	−0.2731D−1	−0.13736891D+00	−0.22696643D+00	0.27892462D−01	−0.56019384D−01	−0.76660308D−01
CPIG4	0.5560D−5	0.50702148D−04	0.88425216D−04	−0.12715312D−04	0.17153320D−04	0.26728531D−04
VISC1	0.13728000D+04	0.0	0.54015000D+03	0.94060000D+02	0.57808000D+03	0.27408000D+03
VISC2	0.18691687D+03	0.0	0.11120589D+03	0.19104228D+02	0.94446586D+02	0.62734410D+02
DHFORM	−0.39552700D+09	−0.22106304D+09	−0.10048320D+09	−0.11061526D+09	−0.39376854D+09	0.11714666D+09
DGFORM	−0.37061554D+09	−0.20691166D+09	−0.58280256D+08	−0.13736891D+09	−0.39464777D+09	0.66946932D+08
ANT1	0.79498460D+05	0.20649270D+02	−0.34210080D+02	−0.30121770D+04	−0.52840260D+04	−0.18102110D+04
ANT2	−0.51465900D+07	−0.21673100D+04	−0.11587170D+04	0.62507530D+05	0.30915310D+06	0.14750340D+06
ANT3	0.12721910D+03	−0.43150000D+02	−0.87403690D+02	0.33139810D+02	0.11114140D+03	0.17363370D+03
ANT4	0.16615650D+02	0.0	−0.13500030D−01	−0.33345090D+01	−0.10099290D+01	−0.23123160D+00
ANT5	−0.12628050D+05	0.0	0.93655890D+01	0.61935790D+03	0.85020940D+03	0.27678890D+03
ANT6	−0.44255980D−02	0.0	0.10799190D−16	0.37100650D−02	0.18866410D−14	0.31644950D−10
ANT7	0.20000000D+01	0.0	0.60000000D+01	0.20000000D+01	0.60000000D+01	0.40000000D+01
TMIN	0.29000000D+03	0.21300000D+03	0.25000000D+03	0.68100000D+02	0.16731000D+03	0.22080000D+03
TMAX	0.49100000D+03	0.34100000D+03	0.55640000D+03	0.13290000D+03	0.30420000D+03	0.55200000D+03
DHVB	0.40678949D+08	0.24409044D+08	0.30019356D+08	0.60457392D+07	0.17165880D+08	0.26753652D+08
δ	15.329	7.1200	9.338	3.1300	7.1200	9.8640
δ_1	31.366	14.5691	19.108	6.4047	14.5691	20.1839

	CHLOROFORM	FORMIC ACID	METHANE	METHANOL	ACETYLENE	VINYL CHLORIDE
FORMULA	CHCL3	CH202	CH4	CH40	C2H2	C2H3CL
MW	0.11937800D+03	0.46025000D+02	0.16043000D+02	0.32042000D+02	0.26038000D+02	0.62499000D+02
TFP	0.20960000D+03	0.28150000D+03	0.90700000D+02	0.17550000D+03	0.19240000D+03	0.11940000D+03
TB	0.33430000D+03	0.37380000D+03	0.11170000D+03	0.33780000D+03	0.18920000D+03	0.25980000D+03
TC	0.53640000D+03	0.58000000D+03	0.19060000D+03	0.51260000D+03	0.30830000D+03	0.42970000D+03
PC	0.54715500D+07	0.0	0.46001550D+07	0.80958675D+07	0.61402950D+07	0.56032725D+07
VC	0.23900000D+00	0.0	0.99000000D−01	0.11800000D+00	0.11300000D+00	0.16900000D+00
ZC	0.29300000D+00	0.0	0.28800000D+00	0.22400000D+00	0.27100000D+00	0.26500000D+00
OMEGA	0.21600000D+00	0.0	0.80000000D−02	0.55900000D+00	0.18400000D+00	0.12200000D+00
VL	0.14890000D+04	0.12260000D+04	0.42500000D+03	0.79100000D+03	0.61500000D+03	0.96900000D+03
TREF	0.29300000D+03	0.28800000D+03	0.11170000D+03	0.29300000D+03	0.18900000D+03	0.25900000D+03
MUP	0.34785054D−24	0.47434164D−24	0.0	0.53758719D−24	0.0	0.47434164D−24
CPIG1	0.24002924D+05	0.11714666D+05	0.19250906D+05	0.21151714D+05	0.26820641D+05	0.59494428D+04
CPIG2	0.18932710D+03	0.13577792D+03	0.52125660D+02	0.70924392D+02	0.75781080D+02	0.20192936D+03
CPIG3	−0.18409360D+00	0.84112812D−01	0.11974248D−01	0.25870237D−01	−0.50074128D−01	−0.15361369D+00
CPIG4	0.66570120D−04	0.20167816D−04	−0.11316920D−04	−0.28516295D−04	0.14122076D−04	0.47729520D−04
VISC1	0.39481000D+03	0.72935000D+03	0.11414000D+03	0.55530000D+03	0.0	0.27690000D+03
VISC2	0.85797238D+02	0.13921037D+03	0.22912330D+02	0.10823454D+03	0.0	0.59450112D+02
DHFORM	−0.10132056D+09	0.37886353D+09	−0.74901852D+08	−0.20130134D+09	0.22688269D+09	0.35169120D+08
DGFORM	−0.68579784D+08	0.35123065D+09	−0.50869620D+08	−0.16261531D+09	0.20934000D+09	0.51539508D+08
ANT1	−0.26501960D+02	0.21880970D+02	−0.18436630D+04	0.43744470D+02	−0.14634450D+03	0.38131010D+04
ANT2	−0.14830630D+04	0.35995800D+04	0.62033300D+05	−0.46749420D+04	0.12143980D−01	−0.25799530D+06
ANT3	−0.67757450D+02	0.26090000D+02	0.58912410D+02	−0.27120530D+02	−0.19209530D+03	0.12676080D+03
ANT4	−0.12913810D−01	0.0	−0.80123750D+00	−0.68086370D−03	−0.18888080D+00	0.61228460D+00
ANT5	0.82398520D+01	0.0	0.33519300D+03	−0.29141350D+01	0.35883170D+02	−0.59201660D+03
ANT6	0.13523580D−16	0.0	0.24853960D−08	0.17842010D−16	0.15316000D−03	−0.95496840D−07
ANT7	0.60000000D+01	0.0	0.40000000D+01	0.60000000D+01	0.20000000D+01	0.30000000D+01
TMIN	0.21456000D+03	0.27100000D+03	0.90700000D+02	0.20504000D+03	0.19240000D+03	0.23633500D+03
TMAX	0.53640000D+03	0.40900000D+03	0.19060000D+03	0.51260000D+03	0.30830000D+03	0.42970000D+03
DHVB	0.29726280D+08	0.21938832D+08	0.81851940D+07	0.35277977D+08	0.16956540D+08	0.20640924D+08
δ	9.236	12.1	5.680	14.510	5.329	
δ_1	18.899	24.8	11.623	29.691	10.904	

	ACETONITRILE	*ETHYLENE*	*ACETALDEHYDE*	*ACETIC ACID*	*ETHANE*	*ETHANOL*
FORMULA	C2H3N	C2H4	C2H4O	C2H4O2	C2H6	C2H6O
MW	0.41053000D+02	0.28054000D+02	0.44054000D+02	0.60052000D+02	0.30070000D+02	0.46069000D+02
TFP	0.22930000D+03	0.10400000D+03	0.15020000D+03	0.28980000D+03	0.89900000D+02	0.15910000D+03
TB	0.35480000D+03	0.16940000D+03	0.29360000D+03	0.39110000D+03	0.18450000D+03	0.35150000D+03
TC	0.54800000D+03	0.28240000D+03	0.46100000D+03	0.59440000D+03	0.30540000D+03	0.51620000D+03
PC	0.48332025D+07	0.50358525D+07	0.55728750D+07	0.57856575D+07	0.48838650D+07	0.63834750D+07
VC	0.17300000D+00	0.12900000D+00	0.15400000D+00	0.17100000D+00	0.14800000D+00	0.16700000D+00
ZC	0.18400000D+00	0.27600000D+00	0.22000000D+00	0.20000000D+00	0.28500000D+00	0.24800000D+00
OMEGA	0.32100000D+00	0.85000000D−01	0.30300000D+00	0.45400000D+00	0.98000000D−01	0.63500000D+00
VL	0.78200000D+03	0.57700000D+03	0.77800000D+03	0.10490000D+04	0.54800000D+03	0.78900000D+03
TREF	0.29300000D+03	0.16300000D+03	0.29300000D+03	0.29300000D+03	0.18300000D+03	0.29300000D+03
MUP	0.11067972D−23	0.0	0.79056940D−24	0.41109609D−24	0.0	0.53758719D−24
CPIG1	0.20481826D+05	0.38058012D+04	0.77162724D+04	0.48399408D+04	0.54093456D+04	0.90141804D+04
CPIG2	0.11961688D+03	0.15658632D+03	0.18225140D+03	0.25485052D+03	0.17810647D+03	0.21407108D+03
CPIG3	−0.44924364D−01	−0.83484792D−01	−0.10065067D+00	−0.17530132D+00	−0.69375276D−01	−0.83903472D−01
CPIG4	0.32029020D−05	0.17551066D−04	0.23801958D−04	0.49487976D−04	0.87127308D−05	0.13732704D−05
VISC1	0.33491000D+03	0.16898000D+03	0.36870000D+03	0.60094000D+03	0.15660000D+03	0.68664000D+03
VISC2	0.72894789D+02	0.35212913D+02	0.75058843D+02	0.12109599D+03	0.33760263D+02	0.12999361D+03
DHFORM	0.87922800D+08	0.52335000D+08	−0.16646717D+09	−0.43513412D+09	−0.84740832D+08	−0.23496322D+09
DGFORM	0.10567483D+09	0.68161104D+08	−0.13339145D+09	−0.37693760D+09	−0.32950116D+08	−0.16839310D+09
ANT1	−0.30057280D+04	−0.17884590D+04	0.21140870D+02	−0.14271780D+03	−0.20410290D+04	−0.75760900D+02
ANT2	0.25039050D+06	0.77818890D+05	−0.24651500D+04	0.22656960D+00	0.94395410D+05	−0.31006470D+04
ANT3	0.17508420D+03	0.83119340D+02	−0.37150000D+02	−0.33091450D+03	0.88739680D+02	−0.40500640D+02
ANT4	−0.36956270D+00	−0.57861500D+00	0.0	−0.46303860D−01	−0.60049080D+00	−0.88140770D−01
ANT5	0.45560340D+03	0.30942770D+03	0.0	0.28820290D+02	0.34802740D+03	0.20812080D+02
ANT6	0.46555470D−10	0.36372090D−06	0.0	0.13776960D−10	0.31743870D−06	0.50453330D−04
ANT7	0.40000000D+01	0.30000000D+01	0.0	0.40000000D+01	0.30000000D+01	0.20000000D+01
TMIN	0.22930000D+03	0.11296000D+03	0.21000000D+03	0.32692000D+03	0.12216000D+03	0.20648000D+03
TMAX	0.54800000D+03	0.28240000D+03	0.32000000D+03	0.59440000D+03	0.30540000D+03	0.51620000D+03
DHVB	0.31401000D+08	0.13552672D+08	0.25748820D+08	0.23697288D+08	0.14716602D+08	0.38769768D+08
δ	12.049	5.801	9.844	10.051	6.050	12.915
δ_1	24.655	11.870	20.143	20.567	12.380	26.427

	PROPYLENE	ACETONE	PROPIONIC ACID	METHYL ACETATE	PROPANE	1-PROPANOL
FORMULA	C3H6	C3H6O	C3H6O2	C3H6O2	C3H8	C3H8O
MW	0.42081000D+02	0.58080000D+02	0.74080000D+02	0.74080000D+02	0.44097000D+02	0.60096000D+02
TFP	0.87900000D+02	0.17820000D+03	0.25250000D+03	0.17500000D+03	0.85500000D+02	0.14690000D+03
TB	0.22540000D+03	0.32940000D+03	0.41400000D+03	0.33010000D+03	0.23110000D+03	0.37040000D+03
TC	0.36500000D+03	0.50810000D+03	0.61200000D+03	0.50680000D+03	0.36980000D+03	0.53670000D+03
PC	0.46204200D+07	0.47014800D+07	0.53702250D+07	0.46913475D+07	0.42455175D+07	0.51675750D+07
VC	0.18100000D+00	0.20900000D+00	0.23000000D+00	0.22800000D+00	0.20300000D+00	0.21850000D+00
ZC	0.27500000D+00	0.23200000D+00	0.24200000D+00	0.25400000D+00	0.28100000D+00	0.25300000D+00
OMEGA	0.14800000D+00	0.30900000D+00	0.53600000D+00	0.32400000D+00	0.15200000D+00	0.62400000D+00
VL	0.61200000D+03	0.79000000D+03	0.99300000D+03	0.93400000D+03	0.58200000D+03	0.80400000D+03
TREF	0.22300000D+03	0.29300000D+03	0.29300000D+03	0.29300000D+03	0.23100000D+03	0.29300000D+03
MUP	0.12649110D−24	0.91706050D−24	0.47434164D−24	0.53758719D−24	0.0	0.53758719D−24
CPIG1	0.37095048D+04	0.63011340D+04	0.56689272D+04	0.16550420D+05	−0.42244812D+04	0.24702120D+04
CPIG2	0.23454454D+03	0.26058643D+03	0.36889895D+03	0.22453808D+03	0.30626442D+03	0.33251566D+03
CPIG3	−0.11601623D+00	−0.12526906D+00	−0.28646086D+00	−0.43417116D−01	−0.15863785D+00	−0.18551711D+00
CPIG4	0.22047689D−04	0.20377156D−04	0.98766612D−04	0.29144315D−04	0.32146250D−04	0.42956568D−04
VISC1	0.27384000D+03	0.36725000D+03	0.53504000D+03	0.40862000D+03	0.22267000D+03	0.95104000D+03
VISC2	0.53901514D+02	0.77291732D+02	0.11175727D+03	0.84706479D+02	0.47690488D+02	0.16116547D+03
DHFORM	0.20431584D+08	−0.21771360D+09	−0.45544010D+09	−0.40972025D+09	0.10391638D+09	−0.25656710D+09
DGFORM	0.62760132D+08	−0.15315314D+09	−0.36956884D+09	0.0	−0.23487948D+08	−0.16190356D+09
ANT1	−0.24957960D+04	0.21544070D+02	0.15382540D+04	0.11504590D+03	−0.21284520D+04	0.39826350D+03
ANT2	0.11799810D+06	−0.29404600D+04	−0.12824810D+06	−0.77691770D+04	0.11710080D+06	−0.22098850D+05
ANT3	0.94078850D+02	−0.35930000D+02	0.14256410D+03	0.10521710D+02	0.10987330D+03	0.28134120D+02
ANT4	−0.85284220D+00	0.0	0.16885370D+00	0.11846230D−01	−0.49471660D+00	0.55274220D−01
ANT5	0.42674050D+03	0.0	−0.22662320D+03	−0.14594400D+02	0.35067670D+03	−0.59550090D+02
ANT6	0.35782860D−03	0.0	−0.85203420D−08	0.13296100D−16	0.17788010D−06	0.33038320D−05
ANT7	0.20000000D+01	0.0	0.30000000D+01	0.60000000D+01	0.30000000D+01	0.20000000D+01
TMIN	0.14600000D+03	0.24100000D+03	0.33660000D+03	0.20272000D+03	0.14792000D+03	0.21468000D+03
TMAX	0.36500000D+03	0.35000000D+03	0.61200000D+03	0.50680000D+03	0.36980000D+03	0.53670000D+03
DHVB	0.18421920D+08	0.29140128D+08	0.32238360D+08	0.30144960D+08	0.18786172D+08	0.41784264D+08
δ	6.208	9.566	12.385	9.014	6.400	12.050
δ_1	12.703	19.574	25.342	18.445	13.096	24.657

	ISOPROPYL ALCOHOL	*GLYCEROL*	*1,2-BUTADIENE*	*1,3-BUTADIENE*	*1-BUTENE*	*CIS-2-BUTENE*
FORMULA	C3H8O	C3H8O3	C4H6	C4H6	C4H8	C4H8
MW	0.60096000D+02	0.92095000D+02	0.54092000D+02	0.54092000D+02	0.56108000D+02	0.56108000D+02
TFP	0.18470000D+03	0.29100000D+03	0.13700000D+03	0.16430000D+03	0.87800000D+02	0.13430000D+03
TB	0.35540000D+03	0.56300000D+03	0.28400000D+03	0.26870000D+03	0.26690000D+03	0.27690000D+03
TC	0.50830000D+03	0.72600000D+03	0.44370000D+03	0.42500000D+03	0.41960000D+03	0.43560000D+03
PC	0.47622750D+07	0.66874500D+07	0.44988300D+07	0.43265775D+07	0.40226025D+07	0.42049875D+07
VC	0.22000000D+00	0.25500000D+00	0.21900000D+00	0.22100000D+00	0.24000000D+00	0.23400000D+00
ZC	0.24800000D+00	0.28000000D+00	0.26700000D+00	0.27000000D+00	0.27700000D+00	0.27200000D+00
OMEGA	0.72400000D+00	0.0	0.25500000D+00	0.19500000D+00	0.18700000D+00	0.20200000D+00
VL	0.78600000D+03	0.12610000D+04	0.65200000D+03	0.62100000D+03	0.59500000D+03	0.62100000D+03
TREF	0.29300000D+03	0.29300000D+03	0.29300000D+03	0.29300000D+03	0.29300000D+03	0.29300000D+03
MUP	0.53758719D−24	0.94868328D−24	0.12649110D−24	0.0	0.94868328D−25	0.94868328D−25
CPIG1	0.32426766D+05	0.84238416D+04	0.11199690D+05	−0.16872804D+04	−0.29935620D+04	0.43961400D+03
CPIG2	0.18848974D+03	0.44421948D+03	0.27235134D+03	0.34185222D+03	0.35319845D+03	0.29533687D+03
CPIG3	0.64058040D−01	−0.31589406D+00	−0.14683108D+00	−0.23400025D+00	−0.19904047D+00	−0.10178111D+00
CPIG4	−0.92612016D−04	0.93784320D−04	0.30890210D−04	0.63346284D−04	0.44631288D−04	−0.61545960D−06
VISC1	0.11397000D+04	0.33371000D+04	0.0	0.30059000D+03	0.25630000D+03	0.26894000D+03
VISC2	0.17470193D+03	0.29743861D+03	0.0	0.62070056D+02	0.54674549D+02	0.56842739D+02
DHFORM	−0.27260255D+09	−0.58531464D+09	0.16232224D+09	0.11023844D+09	−0.12560400D+06	−0.69919560D+07
DGFORM	−0.17350099D+09	0.0	0.19857992D+09	0.15076667D+09	0.71343072D+08	0.65900232D+08
ANT1	0.23585670D+02	0.22131970D+02	0.20996670D+02	0.20665470D+02	−0.26389800D+04	−0.23513930D+04
ANT2	−0.36402000D+04	−0.44870400D+04	−0.23972600D+04	−0.21426600D+04	0.14105410D+06	0.12905430D+06
ANT3	−0.53540000D+02	−0.14020000D+03	−0.30880000D+02	−0.34300000D+02	0.10929260D+03	0.11409570D+03
ANT4	0.0	0.0	0.0	0.0	−0.76149810D+00	−0.65874770D+00
ANT5	0.0	0.0	0.0	0.0	0.44014500D+03	0.39071810D+03
ANT6	0.0	0.0	0.0	0.0	0.27743420D−03	0.23470370D−03
ANT7	0.0	0.0	0.0	0.0	0.20000000D+01	0.20000000D+01
TMIN	0.27300000D+03	0.44000000D+03	0.24500000D+03	0.21500000D+03	0.16784000D+03	0.17424000D+03
TMAX	0.37400000D+03	0.60000000D+03	0.30500000D+03	0.29000000D+03	0.41960000D+03	0.43560000D+03
DHVB	0.39858336D+08	0.61127280D+08	0.24283440D+08	0.22483116D+08	0.21930458D+08	0.23362344D+08
δ	11.572	17.69	7.950	6.940	6.766	6.760
δ_1	23.679	36.20	16.267	14.201	13.845	13.832

	METHYL ETHYL KETONE	*TETRA-HYDROFURAN*	*1,4-DIOXANE*	*ETHYL ACETATE*	*N-BUTANE*	*ISOBUTANE*
FORMULA	C4H8O	C4H8O	C4H8O2	C4H8O2	C4H1O	C4H1O
MW	0.72107000D+02	0.72107000D+02	0.88107000D+02	0.88107000D+02	0.58124000D+02	0.58124000D+02
TFP	0.18650000D+03	0.16470000D+03	0.28500000D+03	0.18960000D+03	0.13480000D+03	0.11360000D+03
TB	0.35280000D+03	0.33910000D+03	0.37450000D+03	0.35030000D+03	0.27270000D+03	0.26130000D+03
TC	0.53560000D+03	0.54020000D+03	0.58700000D+03	0.52320000D+03	0.42520000D+03	0.40810000D+03
PC	0.41543250D+07	0.51878400D+07	0.52081050D+07	0.38300850D+07	0.37996875D+07	0.36477000D+07
VC	0.26700000D+00	0.22400000D+00	0.23800000D+00	0.28600000D+00	0.25500000D+00	0.26300000D+00
ZC	0.24900000D+00	0.25900000D+00	0.25400000D+00	0.25200000D+00	0.27400000D+00	0.28300000D+00
OMEGA	0.32900000D+00	0.0	0.28800000D+00	0.36300000D+00	0.19300000D+00	0.17600000D+00
VL	0.80500000D+03	0.88900000D+03	0.10330000D+04	0.90100000D+03	0.57900000D+03	0.55700000D+03
TREF	0.29300000D+03	0.29300000D+03	0.29300000D+03	0.29300000D+03	0.29300000D+03	0.29300000D+03
MUP	0.10435516D−23	0.53758719D−24	0.12649110D−24	0.60083274D−24	0.0	0.31622776D−25
CPIG1	0.10944295D+05	−0.19104368D+05	−0.53574293D+05	0.72347904D+04	0.94872888D+04	−0.13900176D+04
CPIG2	0.35591987D+03	0.51623244D+03	0.59871240D+03	0.40716630D+03	0.33130148D+03	0.38472505D+03
CPIG3	−0.18999698D+00	−0.41315342D+00	−0.40850608D+00	−0.20917253D+00	−0.11082460D+00	−0.18459601D+00
CPIG4	0.39196822D−04	0.14540756D−03	0.10621912D−03	0.28545602D−04	−0.28219032D−05	0.28951722D−04
VISC1	0.42384000D+03	0.41979000D+03	0.66036000D+03	0.42738000D+03	0.26584000D+03	0.30251000D+03
VISC2	0.87760659D+02	0.88991097D+02	0.12850773D+03	0.88826671D+02	0.57054241D+02	0.63321324D+02
DHFORM	−0.23852200D+09	−0.18434480D+09	−0.31526604D+09	−0.44321465D+09	−0.12623202D+09	−0.13460562D+09
DGFORM	−0.14616119D+09	0.0	−0.18091163D+09	−0.32761710D+09	−0.17165880D+08	−0.20892132D+08
ANT1	−0.78357580D+04	0.20999670D+02	0.21025470D+02	0.48610870D+02	−0.20775660D+04	−0.22805130D+04
ANT2	0.53123170D+06	−0.27683800D+04	−0.29668800D+04	−0.82999240D+04	0.13037520D+06	0.13705180D+06
ANT3	0.13869530D+03	−0.46900000D+02	−0.62150000D+02	0.28708980D+02	0.12967170D+03	0.12196710D+03
ANT4	−0.15758780D+01	0.0	0.0	−0.47696360D−01	−0.40673270D+00	−0.46894860D+00
ANT5	0.12395040D+04	0.0	0.0	−0.46184520D+00	0.33409570D+03	0.36914510D+03
ANT6	0.41455620D−03	0.0	0.0	0.34456160D−04	0.11135780D−06	0.13719500D−06
ANT7	0.20000000D+01	0.0	0.0	0.20000000D+01	0.30000000D+01	0.30000000D+01
TMIN	0.26780000D+03	0.27000000D+03	0.27500000D+03	0.20928000D+03	0.17008000D+03	0.16324000D+03
TMAX	0.53560000D+03	0.37000000D+03	0.41000000D+03	0.52320000D+03	0.42520000D+03	0.40810000D+03
DHVB	0.31233528D+08	0.29600676D+08	0.36383292D+08	0.32238360D+08	0.22407754D+08	0.21310812D+08
δ	9.199	9.10	10.13	8.974	6.634	6.730
δ_1	18.823	18.62	20.73	18.363	13.575	13.771

	N-BUTANOL	2-BUTANOL	ISOBUTANOL	TERT-BUTYL ALCOHOL	ETHYL ETHER	DIETHYLENE GLYCOL
FORMULA	C4H10O	C4H10O	C4H10O	C4H10O	C4H10O	C4H10O3
MW	0.74123000D+02	0.74123000D+02	0.74123000D+02	0.74123000D+02	0.74123000D+02	0.10612200D+03
TFP	0.18390000D+03	0.15850000D+03	0.16520000D+03	0.29880000D+03	0.15690000D+03	0.26500000D+03
TB	0.39090000D+03	0.37270000D+03	0.38100000D+03	0.35560000D+03	0.30770000D+03	0.51900000D+03
TC	0.56290000D+03	0.53600000D+03	0.54770000D+03	0.50620000D+03	0.46670000D+03	0.68100000D+03
PC	0.44177700D+07	0.41948550D+07	0.42961800D+07	0.39719400D+07	0.36375675D+07	0.46609500D+07
VC	0.27400000D+00	0.26800000D+00	0.27300000D+00	0.27500000D+00	0.28000000D+00	0.31600000D+00
ZC	0.25900000D+00	0.25200000D+00	0.25700000D+00	0.25900000D+00	0.26200000D+00	0.26000000D+00
OMEGA	0.59000000D+00	0.57600000D+00	0.58800000D+00	0.61800000D+00	0.28100000D+00	0.0
VL	0.81000000D+03	0.80700000D+03	0.80200000D+03	0.78700000D+03	0.71300000D+03	0.11160000D+04
TREF	0.29300000D+03	0.29300000D+03	0.29300000D+03	0.29300000D+03	0.29300000D+03	0.29300000D+03
MUP	0.56920997D−24	0.53758719D−24	0.53758719D−24	0.53758719D−24	0.41109609D−24	0.0
CPIG1	0.32657040D+04	0.57526632D+04	−0.77078988D+04	−0.48612935D+05	0.21423856D+05	0.73059660D+05
CPIG2	0.41801011D+03	0.42454152D+03	0.46892160D+03	0.71719884D+03	0.33586510D+03	0.34608089D+03
CPIG3	−0.22416127D+00	−0.23282795D+00	−0.28838678D+00	−0.70840656D+00	−0.10353956D+00	−0.14678921D+00
CPIG4	0.46850292D−04	0.47729520D−04	0.72306036D−04	0.29198743D−03	−0.93574980D−05	0.18463788D−04
VISC1	0.98454000D+03	0.14417000D+04	0.11991000D+04	0.97210000D+03	0.35314000D+03	0.19430000D+04
VISC2	0.16726246D+03	0.19617583D+03	0.18483874D+03	0.17129030D+03	0.72767733D+02	0.24155823D+03
DHFORM	−0.27486342D+09	−0.29282479D+09	−0.28340449D+09	−0.31262836D+09	−0.25238030D+09	−0.57149820D+09
DGFORM	−0.15089227D+09	−0.16772321D+09	−0.16743013D+09	−0.17777153D+09	−0.12242203D+09	0.0
ANT1	0.22108770D+02	0.22102970D+02	0.21763970D+02	0.21747570D+02	0.98234300D+01	0.21925370D+02
ANT2	−0.31370200D+04	−0.30260300D+04	−0.28747344D+04	−0.26582900D+04	−0.25462230D+04	−0.41225200D+04
ANT3	−0.94430000D+02	−0.86650000D+02	−0.10030000D+03	−0.95500000D+02	−0.35442790D+02	−0.12250000D+03
ANT4	0.0	0.0	0.0	0.0	−0.68873780D−02	0.0
ANT5	0.0	0.0	0.0	0.0	0.22952660D+01	0.0
ANT6	0.0	0.0	0.0	0.0	0.28640650D−16	0.0
ANT7	0.0	0.0	0.0	0.0	0.60000000D+01	0.0
TMIN	0.28800000D+03	0.29800000D+03	0.29300000D+03	0.29300000D+03	0.18668000D+03	0.40200000D+03
TMAX	0.40400000D+03	0.39300000D+03	0.38800000D+03	0.37600000D+03	0.46670000D+03	0.56000000D+03
DHVB	0.43124040D+08	0.40821300D+08	0.42077340D+08	0.39062844D+08	0.26711784D+08	0.57233556D+08
δ	11.440	11.08	10.949	10.316	7.544	13.551
δ_1	23.409	22.67	22.404	21.109	15.437	27.728

	PYRIDINE	*N-PENTANE*	*2-METHYL BUTANE*	*2,2-DIMETHYL PROPANE*	*1-PENTANOL*	*BENZENE*
FORMULA	C5H5N	C5H12	C5H12	C5H12	C5H12O	C6H6
MW	0.79102000D+02	0.72151000D+02	0.72151000D+02	0.72151000D+02	0.88150000D+02	0.78114000D+02
TFP	0.23150000D+03	0.14340000D+03	0.11330000D+03	0.25660000D+03	0.19500000D+03	0.27870000D+03
TB	0.38850000D+03	0.30920000D+03	0.30100000D+03	0.28260000D+03	0.41100000D+03	0.35330000D+03
TC	0.62000000D+03	0.46960000D+03	0.46040000D+03	0.43380000D+03	0.58600000D+03	0.56210000D+03
PC	0.56336700D+07	0.33741225D+07	0.33842550D+07	0.32018700D+07	0.38503500D+07	0.48939975D+07
VC	0.25400000D+00	0.30400000D+00	0.30600000D+00	0.30300000D+00	0.32600000D+00	0.25900000D+00
ZC	0.27700000D+00	0.26200000D+00	0.27100000D+00	0.26900000D+00	0.26000000D+00	0.27100000D+00
OMEGA	0.24000000D+00	0.25100000D+00	0.22700000D+00	0.19700000D+00	0.58000000D+00	0.21200000D+00
VL	0.98300000D+03	0.62600000D+03	0.62000000D+03	0.59100000D+03	0.81500000D+03	0.88500000D+03
TREF	0.29300000D+03	0.29300000D+03	0.29300000D+03	0.29300000D+03	0.29300000D+03	0.28900000D+03
MUP	0.72732385D−24	0.0	0.31622776D−25	0.0	0.53758719D−24	0.0
CPIG1	0.39791347D+05	−0.36257688D+04	−0.95249700D+04	−0.16592288D+05	0.38686032D+04	−0.33917267D+05
CPIG2	0.49278636D+03	0.48734352D+03	0.50660280D+03	0.55516968D+03	0.50450940D+03	0.47436444D+03
CPIG3	−0.35579426D+00	−0.25803248D+00	−0.27293749D+00	−0.33063160D+00	−0.26393587D+00	−0.30170081D+00
CPIG4	0.10044133D−03	0.53046756D−04	0.57233556D−04	0.76325364D−04	0.51204564D−04	0.71301204D−04
VISC1	0.61850000D+03	0.31366000D+03	0.36732000D+03	0.35554000D+03	0.11511000D+04	0.54564000D+03
VISC2	0.12077243D+03	0.66469257D+02	0.74699240D+02	0.73905376D+02	0.18293405D+03	0.10791118D+03
DHFORM	0.14025780D+09	−0.14653800D+09	−0.15457666D+09	−0.16609036D+09	−0.29893752D+09	0.82982376D+08
DGFORM	0.19033193D+09	−0.83736000D+07	−0.14821272D+08	−0.15239952D+08	−0.14611932D+09	0.12974893D+09
ANT1	0.20983770D+02	−0.31144360D+02	−0.25965910D+04	−0.53893750D+04	0.21419770D+02	−0.37675460D+02
ANT2	−0.30951300D+04	−0.12065390D+04	0.15024030D+06	0.31478200D+06	−0.30268900D+04	−0.10975230D+04
ANT3	−0.61150000D+02	−0.64139850D+02	0.12090470D+03	0.11741390D+03	−0.10500000D+03	−0.93532510D+02
ANT4	0.0	−0.16082770D−01	−0.67404480D+00	−0.13554940D+01	0.0	−0.14047730D−01
ANT5	0.0	0.91627870D+01	0.42657630D+03	0.87916190D+03	0.0	0.99482220D+01
ANT6	0.0	0.31313440D−16	0.22501220D−03	0.43847210D−03	0.0	0.10416500D−16
ANT7	0.0	0.60000000D+01	0.20000000D+01	0.20000000D+01	0.0	0.60000000D+01
TMIN	0.28500000D+03	0.18784000D+03	0.18416000D+03	0.25660000D+03	0.31000000D+03	0.27870000D+03
TMAX	0.42500000D+03	0.46960000D+03	0.46040000D+03	0.43380000D+03	0.41100000D+03	0.56210000D+03
DHVB	0.35169120D+08	0.25790688D+08	0.24702120D+08	0.22767818D+08	0.44380080D+08	0.30781354D+08
δ	10.62	7.020	7.020	7.020	11.12	9.158
δ_1	21.73	14.364	14.364	14.364	22.75	18.739

	PHENOL	ANILINE	CYCLOHEXANE	METHYL ISOBUTYL-KETONE	BENZOIC ACID	TOLUENE
FORMULA	C6H6O	C6H7N	C6H12	C6H12O	C7H6O2	C7H8
MW	0.94113000D+02	0.93129000D+02	0.84162000D+02	0.10016100D+03	0.12212400D+03	0.92141000D+02
TFP	0.31400000D+03	0.26700000D+03	0.27970000D+03	0.18900000D+03	0.39560000D+03	0.17800000D+03
TB	0.45500000D+03	0.45750000D+03	0.35390000D+03	0.38960000D+03	0.52300000D+03	0.38380000D+03
TC	0.69420000D+03	0.69900000D+03	0.55340000D+03	0.57100000D+03	0.75200000D+03	0.59170000D+03
PC	0.61301625D+07	0.53094300D+07	0.40732650D+07	0.32727975D+07	0.45596250D+07	0.41137950D+07
VC	0.22900000D+00	0.27000000D+00	0.30800000D+00	0.37100000D+00	0.34100000D+00	0.31600000D+00
ZC	0.24000000D+00	0.24700000D+00	0.27300000D+00	0.26000000D+00	0.25000000D+00	0.26400000D+00
OMEGA	0.44000000D+00	0.38200000D+00	0.21300000D+00	0.40000000D+00	0.62000000D+00	0.25700000D+00
VL	0.10590000D+04	0.10220000D+04	0.77900000D+03	0.80100000D+03	0.10750000D+04	0.86700000D+03
TREF	0.31300000D+03	0.29300000D+03	0.29300000D+03	0.29300000D+03	0.40300000D+03	0.29300000D+03
MUP	0.50596442D−24	0.50596442D−24	0.94868328D−25	0.88543773D−24	0.53758719D−24	0.12649110D−24
CPIG1	−0.35843195D+05	−0.40515664D+05	−0.54541444D+05	0.38937240D+04	−0.51292487D+05	−0.24354616D+05
CPIG2	0.59829372D+03	0.63848700D+03	0.61127280D+03	0.56563668D+03	0.62927604D+03	0.51246432D+03
CPIG3	−0.48273804D+00	−0.51330168D+00	−0.25233844D+00	−0.33180390D+00	−0.42370416D+00	−0.27653814D+00
CPIG4	0.15269260D−03	0.16332707D−03	0.13213541D−04	0.82312488D−04	0.10621912D−03	0.49111164D−04
VISC1	0.14055000D+04	0.10746000D+04	0.65362000D+03	0.47365000D+03	0.26176000D+04	0.46733000D+03
VISC2	0.20675411D+03	0.17885216D+03	0.12456186D+03	0.98093576D+02	0.27794845D+03	0.96736798D+02
DHFORM	−0.96422004D+08	0.86917968D+08	−0.12321752D+09	−0.28403251D+09	−0.29039645D+09	0.50032260D+08
DGFORM	−0.32908248D+08	0.16680211D+09	0.31777812D+08	0.0	−0.21055417D+09	0.12208709D+09
ANT1	0.21054660D+03	−0.19266030D+03	−0.16959900D+03	0.20609270D+02	0.22056170D+02	−0.22847860D+02
ANT2	−0.22555420D+05	−0.75564260D+02	−0.13416300D+00	−0.28936600D+04	−0.41907000D+04	−0.18741860D+04
ANT3	0.74616710D+02	−0.24438150D+03	−0.27659790D+03	−0.70750000D+02	−0.12520000D+03	−0.72921440D+02
ANT4	−0.10845870D−01	−0.90804580D−01	−0.10627640D+00	0.0	0.0	−0.10836630D−01
ANT5	−0.25260010D+02	0.39052960D+02	0.36245390D+02	0.0	0.0	0.74851470D+01
ANT6	0.14980730D−04	0.32747950D−04	0.48013040D−04	0.0	0.0	0.75672290D−17
ANT7	0.20000000D+01	0.20000000D+01	0.20000000D+01	0.0	0.0	0.60000000D+01
TMIN	0.31400000D+03	0.38445000D+03	0.27970000D+03	0.28500000D+03	0.40500000D+03	0.23668000D+03
TMAX	0.69420000D+03	0.69900000D+03	0.55340000D+03	0.42500000D+03	0.56000000D+03	0.59170000D+03
DHVB	0.45636120D+08	0.41868000D+08	0.29977488D+08	0.35587800D+08	0.50660280D+08	0.33201324D+08
δ	12.106	11.461	8.193	8.58		8.914
δ_1	24.772	23.452	16.765	17.557		18.240

	STYRENE	*O-XYLENE*	*M-XYLENE*	*P-XYLENE*	*ETHYLBENZENE*	*N-OCTANE*
FORMULA	C8H8	C8H10	C8H10	C8H10	C8H10	C8H18
MW	0.10415200D+03	0.10616800D+03	0.10616800D+03	0.10616800D+03	0.10616800D+03	0.11423200D+03
TFP	0.24250000D+03	0.24800000D+03	0.22530000D+03	0.28640000D+03	0.17820000D+03	0.21640000D+03
TB	0.41830000D+03	0.41760000D+03	0.41230000D+03	0.41150000D+03	0.40930000D+03	0.39880000D+03
TC	0.64700000D+03	0.63020000D+03	0.61700000D+03	0.61620000D+03	0.61710000D+03	0.56880000D+03
PC	0.39922050D+07	0.37287600D+07	0.35463750D+07	0.35159775D+07	0.36071700D+07	0.24824625D+07
VC	0.0	0.36900000D+00	0.37600000D+00	0.37900000D+00	0.37400000D+00	0.49200000D+00
ZC	0.0	0.26300000D+00	0.26000000D+00	0.26000000D+00	0.26300000D+00	0.25900000D+00
OMEGA	0.25700000D+00	0.31400000D+00	0.33100000D+00	0.32400000D+00	0.30100000D+00	0.39400000D+00
VL	0.90600000D+03	0.88000000D+03	0.86400000D+03	0.86100000D+03	0.86700000D+03	0.70300000D+03
TREF	0.29300000D+03	0.29300000D+03	0.29300000D+03	0.29300000D+03	0.29300000D+03	0.29300000D+03
MUP	0.31622776D−25	0.15811388D−24	0.94868328D−25	0.31622776D−25	0.12649110D−24	0.0
CPIG1	−0.28248340D+05	−0.15851225D+05	−0.29165249D+05	−0.25091492D+05	−0.43098919D+05	−0.60959808D+04
CPIG2	0.61587828D+03	0.59620032D+03	0.62969472D+03	0.60415524D+03	0.70715052D+03	0.77120856D+03
CPIG3	−0.40230961D+00	−0.34432243D+00	−0.37471860D+00	−0.33737234D+00	−0.48106332D+00	−0.41951736D+00
CPIG4	0.99352764D−04	0.75278664D−04	0.84782700D−04	0.68202972D−04	0.13008388D−03	0.88550820D−04
VISC1	0.52864000D+03	0.51354000D+03	0.45342000D+03	0.47516000D+03	0.47282000D+03	0.47370000D+03
VISC2	0.10765617D+03	0.10594135D+03	0.95195399D+02	0.98626941D+02	0.98720249D+02	0.97031344D+02
DHFORM	0.14745910D+09	0.19008072D+08	0.17249616D+08	0.17961372D+08	0.29810016D+08	−0.20858638D+09
DGFORM	0.21394548D+09	0.12217082D+09	0.11894699D+09	0.12120786D+09	0.13067003D+09	0.16412256D+08
ANT1	0.20912070D+02	0.80811040D+01	−0.32447180D+04	−0.43637890D+04	−0.12931300D+02	−0.11737850D+03
ANT2	−0.33285700D+04	−0.34240540D+04	0.24302390D+06	0.33397640D+06	−0.33902620D+04	−0.15641240D+04
ANT3	−0.63720000D+02	−0.50815350D+02	0.16336200D+03	0.16450780D+03	−0.36508840D+02	−0.60739230D+02
ANT4	0.0	−0.52463410D−02	−0.58847630D+00	−0.76178230D+00	−0.15539180D−01	−0.86680810D−01
ANT5	0.0	0.24766890D+01	0.50685500D+03	0.67750020D+03	0.66026580D+01	0.26971140D+02
ANT6	0.0	0.48165370D−17	0.14444320D−03	0.18044380D−03	0.71440690D−11	0.41398210D−04
ANT7	0.0	0.60000000D+01	0.20000000D+01	0.20000000D+01	0.40000000D+01	0.20000000D+01
TMIN	0.30500000D+03	0.25208000D+03	0.24680000D+03	0.28640000D+03	0.24684000D+03	0.22752000D+03
TMAX	0.46000000D+03	0.63020000D+03	0.61700000D+03	0.61620000D+03	0.61710000D+03	0.56880000D+03
DHVB	0.36843840D+08	0.36843840D+08	0.36383292D+08	0.36006480D+08	0.35587800D+08	0.34436430D+08
δ	9.211	8.987	8.818	8.769	8.783	7.551
δ_1	18.848	18.389	18.044	17.943	17.972	15.451

Appendix E. Parameters and Correlations

The numerical tables of this appendix are supplementary to the material of Chapters 1, 3, 4, and 11. The tables included in this appendix are:

Table E.1. Values of $Z^{(0)}$ (Lee-Kesler)

	PR														
TR	*0.010*	*0.050*	*0.100*	*0.200*	*0.400*	*0.600*	*0.800*	*1.000*	*1.200*	*1.500*	*2.000*	*3.000*	*5.000*	*7.000*	*10.000*
0.30	0.0029	0.0145	0.0290	0.0579	0.1158	0.1737	0.2315	0.2892	0.3470	0.4335	0.5775	0.8648	1.4366	2.0048	2.8507
0.35	0.0026	0.0130	0.0261	0.0522	0.1043	0.1564	0.2084	0.2604	0.3123	0.3901	0.5195	0.7775	1.2902	1.7987	2.5539
0.40	0.0024	0.0119	0.0239	0.0477	0.0953	0.1429	0.1904	0.2379	0.2853	0.3563	0.4744	0.7095	1.1758	1.6373	2.3211
0.45	0.0022	0.0110	0.0221	0.0442	0.0882	0.1322	0.1762	0.2200	0.2638	0.3294	0.4384	0.6551	1.0841	1.5077	2.1338
0.50	0.0021	0.0103	0.0207	0.0413	0.0825	0.1236	0.1647	0.2056	0.2465	0.3077	0.4092	0.6110	1.0094	1.4017	1.9801
0.55	0.9804	0.0098	0.0195	0.0390	0.0778	0.1166	0.1553	0.1939	0.2323	0.2899	0.3853	0.5747	0.9475	1.3137	1.8520
0.60	0.9849	0.0093	0.0186	0.0371	0.0741	0.1109	0.1476	0.1842	0.2207	0.2753	0.3657	0.5446	0.8959	1.2398	1.7440
0.65	0.9881	0.9377	0.0178	0.0356	0.0710	0.1063	0.1415	0.1765	0.2113	0.2634	0.3495	0.5197	0.8526	1.1773	1.6519
0.70	0.9904	0.9504	0.8958	0.0344	0.0687	0.1027	0.1366	0.1703	0.2038	0.2538	0.3364	0.4991	0.8161	1.1241	1.5729
0.75	0.9922	0.9598	0.9165	0.0336	0.0670	0.1001	0.1330	0.1656	0.1981	0.2464	0.3260	0.4823	0.7854	1.0787	1.5047
0.80	0.9935	0.9669	0.9319	0.8539	0.0661	0.0985	0.1307	0.1626	0.1942	0.2411	0.3182	0.4690	0.7598	1.0400	1.4456
0.85	0.9946	0.9725	0.9436	0.8810	0.0661	0.0983	0.1301	0.1614	0.1924	0.2382	0.3132	0.4591	0.7388	1.0071	1.3943
0.90	0.9954	0.9768	0.9528	0.9015	0.7800	0.1006	0.1321	0.1630	0.1935	0.2383	0.3114	0.4527	0.7220	0.9793	1.3496
0.93	0.9959	0.9790	0.9573	0.9115	0.8059	0.6635	0.1359	0.1664	0.1963	0.2405	0.3122	0.4507	0.7138	0.9648	1.3257
0.95	0.9961	0.9803	0.9600	0.9174	0.8206	0.6967	0.1410	0.1705	0.1998	0.2432	0.3138	0.4501	0.7092	0.9561	1.3108
0.97	0.9963	0.9815	0.9625	0.9227	0.8338	0.7240	0.5580	0.1779	0.2055	0.2474	0.3164	0.4504	0.7052	0.9480	1.2968
0.98	0.9965	0.9821	0.9637	0.9253	0.8398	0.7360	0.5887	0.1844	0.2097	0.2503	0.3182	0.4508	0.7035	0.9442	1.2901
0.99	0.9966	0.9826	0.9648	0.9277	0.8455	0.7471	0.6138	0.1959	0.2154	0.2538	0.3204	0.4514	0.7018	0.9406	1.2835
1.00	0.9967	0.9832	0.9659	0.9300	0.8509	0.7574	0.6353	0.2901	0.2237	0.2583	0.3229	0.4522	0.7004	0.9372	1.2772
1.01	0.9968	0.9837	0.9669	0.9322	0.8561	0.7671	0.6542	0.4648	0.2370	0.2640	0.3260	0.4533	0.6991	0.9339	1.2710

1.02	0.9969	0.9842	0.9679	0.9343	0.8610	0.7761	0.6710	0.5146	0.2629	0.2715	0.3297	0.4547	0.6980	0.9307	1.2650
1.05	0.9971	0.9855	0.9707	0.9401	0.8743	0.8002	0.7130	0.6026	0.4437	0.3131	0.3452	0.4604	0.6956	0.9222	1.2481
1.10	0.9975	0.9874	0.9747	0.9485	0.8930	0.8323	0.7649	0.6880	0.5984	0.4580	0.3953	0.4770	0.6950	0.9110	1.2232
1.15	0.9978	0.9891	0.9780	0.9554	0.9081	0.8576	0.8032	0.7443	0.6803	0.5798	0.4760	0.5042	0.6987	0.9033	1.2021
1.20	0.9981	0.9904	0.9808	0.9611	0.9205	0.8779	0.8330	0.7858	0.7363	0.6605	0.5605	0.5425	0.7069	0.8990	1.1844
1.30	0.9985	0.9926	0.9852	0.9702	0.9396	0.9083	0.8764	0.8438	0.8111	0.7624	0.6908	0.6344	0.7358	0.8998	1.1580
1.40	0.9988	0.9942	0.9884	0.9768	0.9534	0.9298	0.9062	0.8827	0.8595	0.8256	0.7753	0.7202	0.7761	0.9112	1.1419
1.50	0.9991	0.9954	0.9909	0.9818	0.9636	0.9456	0.9278	0.9103	0.8933	0.8689	0.8328	0.7887	0.8200	0.9297	1.1339
1.60	0.9993	0.9964	0.9928	0.9856	0.9714	0.9575	0.9439	0.9308	0.9180	0.9000	0.8738	0.8410	0.8617	0.9518	1.1320
1.70	0.9994	0.9971	0.9943	0.9886	0.9775	0.9667	0.9563	0.9463	0.9367	0.9234	0.9043	0.8809	0.8984	0.9745	1.1343
1.80	0.9995	0.9977	0.9955	0.9910	0.9823	0.9739	0.9659	0.9583	0.9511	0.9413	0.9275	0.9118	0.9297	0.9961	1.1391
1.90	0.9996	0.9982	0.9964	0.9929	0.9861	0.9796	0.9735	0.9678	0.9624	0.9552	0.9456	0.9359	0.9557	1.0157	1.1452
2.00	0.9997	0.9986	0.9972	0.9944	0.9892	0.9842	0.9796	0.9754	0.9715	0.9664	0.9599	0.9550	0.9772	1.0328	1.1516
2.20	0.9998	0.9992	0.9983	0.9967	0.9937	0.9910	0.9886	0.9865	0.9847	0.9826	0.9806	0.9827	1.0094	1.0600	1.1635
2.40	0.9999	0.9996	0.9991	0.9983	0.9969	0.9957	0.9948	0.9941	0.9936	0.9935	0.9945	1.0011	1.0313	1.0793	1.1728
2.60	1.0000	0.9998	0.9997	0.9994	0.9991	0.9990	0.9990	0.9993	0.9998	1.0010	1.0040	1.0137	1.0463	1.0926	1.1792
2.80	1.0000	1.0000	1.0001	1.0002	1.0007	1.0013	1.0021	1.0031	1.0042	1.0063	1.0106	1.0223	1.0565	1.1016	1.1830
3.00	1.0000	1.0002	1.0004	1.0008	1.0018	1.0030	1.0043	1.0057	1.0074	1.0101	1.0153	1.0284	1.0635	1.1075	1.1848
3.50	1.0001	1.0004	1.0008	1.0017	1.0035	1.0055	1.0075	1.0097	1.0120	1.0156	1.0221	1.0368	1.0723	1.1138	1.1834
4.00	1.0001	1.0005	1.0010	1.0021	1.0043	1.0066	1.0090	1.0115	1.0140	1.0179	1.0249	1.0401	1.0747	1.1136	1.1773

Note: This information is given in graphical form in the *API Data Book*, Figures 6B1.4–6B1.7 and as Figure 1.22 in this book.

Table E.2. Values of $Z^{(1)}$ (Lee-Kesler)

	PR														
Tr	*0.010*	*0.050*	*0.100*	*0.200*	*0.400*	*0.600*	*0.800*	*1.000*	*1.200*	*1.500*	*2.000*	*3.000*	*5.000*	*7.000*	*10.000*
0.30	−0.0008	−0.0040	−0.0081	−0.0161	−0.0323	−0.0484	−0.0645	−0.0806	−0.0966	−0.1207	−0.1608	−0.2407	−0.3996	−0.5572	−0.7915
0.35	−0.0009	−0.0046	−0.0093	−0.0185	−0.0370	−0.0554	−0.0738	−0.0921	−0.1105	−0.1379	−0.1834	−0.2738	−0.4523	−0.6279	−0.8863
0.40	−0.0010	−0.0048	−0.0095	−0.0190	−0.0380	−0.0570	−0.0758	−0.0946	−0.1134	−0.1414	−0.1879	−0.2799	−0.4603	−0.6365	−0.8936
0.45	−0.0009	−0.0047	−0.0094	−0.0187	−0.0374	−0.0560	−0.0745	−0.0929	−0.1113	−0.1387	−0.1840	−0.2734	−0.4475	−0.6162	−0.8606
0.50	−0.0009	−0.0045	−0.0090	−0.0181	−0.0360	−0.0539	−0.0716	−0.0893	−0.1069	−0.1330	−0.1762	−0.2611	−0.4253	−0.5831	−0.8099
0.55	−0.0314	−0.0043	−0.0086	−0.0172	−0.0343	−0.0513	−0.0682	−0.0849	−0.1015	−0.1263	−0.1669	−0.2465	−0.3991	−0.5446	−0.7521
0.60	−0.0205	−0.0041	−0.0082	−0.0164	−0.0326	−0.0487	−0.0646	−0.0803	−0.0960	−0.1192	−0.1572	−0.2312	−0.3718	−0.5047	−0.6928
0.65	−0.0137	−0.0772	−0.0078	−0.0156	−0.0309	−0.0461	−0.0611	−0.0759	−0.0906	−0.1122	−0.1476	−0.2160	−0.3447	−0.4653	−0.6346
0.70	−0.0093	−0.0507	−0.1161	−0.0148	−0.0294	−0.0438	−0.0579	−0.0718	−0.0855	−0.1057	−0.1385	−0.2013	−0.3184	−0.4270	−0.5785
0.75	−0.0064	−0.0339	−0.0744	−0.0143	−0.0282	−0.0417	−0.0550	−0.0681	−0.0808	−0.0996	−0.1298	−0.1872	−0.2929	−0.3901	−0.5250
0.80	−0.0044	−0.0228	−0.0487	−0.1160	−0.0272	−0.0401	−0.0526	−0.0648	−0.0767	−0.0940	−0.1217	−0.1736	−0.2682	−0.3545	−0.4740
0.85	−0.0029	−0.0152	−0.0319	−0.0715	−0.0268	−0.0391	−0.0509	−0.0622	−0.0731	−0.0888	−0.1138	−0.1602	−0.2439	−0.3201	−0.4254
0.90	−0.0019	−0.0099	−0.0205	−0.0442	−0.1118	−0.0396	−0.0503	−0.0604	−0.0701	−0.0840	−0.1059	−0.1463	−0.2195	−0.2862	−0.3788
0.93	−0.0015	−0.0075	−0.0154	−0.0326	−0.0763	−0.1662	−0.0514	−0.0602	−0.0687	−0.0810	−0.1007	−0.1374	−0.2045	−0.2661	−0.3516
0.95	−0.0012	−0.0062	−0.0126	−0.0262	−0.0589	−0.1110	−0.0540	−0.0607	−0.0678	−0.0788	−0.0967	−0.1310	−0.1943	−0.2526	−0.3339
0.97	−0.0010	−0.0050	−0.0101	−0.0208	−0.0450	−0.0770	−0.1647	−0.0623	−0.0669	−0.0759	−0.0921	−0.1240	−0.1837	−0.2391	−0.3163
0.98	−0.0009	−0.0044	−0.0090	−0.0184	−0.0390	−0.0641	−0.1100	−0.0641	−0.0661	−0.0740	−0.0893	−0.1202	−0.1783	−0.2322	−0.3075
0.99	−0.0008	−0.0039	−0.0079	−0.0161	−0.0335	−0.0531	−0.0796	−0.0680	−0.0646	−0.0715	−0.0861	−0.1162	−0.1728	−0.2254	−0.2989
1.00	−0.0007	−0.0034	−0.0069	−0.0140	−0.0285	−0.0435	−0.0588	−0.0879	−0.0609	−0.0678	−0.0824	−0.1118	−0.1672	−0.2185	−0.2902
1.01	−0.0006	−0.0030	−0.0060	−0.0120	−0.0240	−0.0351	−0.0429	−0.0223	−0.0473	−0.0621	−0.0778	−0.1072	−0.1615	−0.2116	−0.2816

1.02	−0.0005	−0.0026	−0.0051	−0.0102	−0.0198	−0.0277	−0.0303	−0.0062	−0.0227	−0.0524	−0.0722	−0.1021	−0.1556	−0.2047	−0.2731
1.05	−0.0003	−0.0015	−0.0029	−0.0054	−0.0092	−0.0097	−0.0032	−0.0220	−0.1059	−0.0451	−0.0432	−0.0838	−0.1370	−0.1835	−0.2476
1.10	−0.0000	0.0000	0.0001	0.0007	0.0038	0.0106	0.0236	0.0476	0.0897	0.1630	0.0698	−0.0373	−0.1021	−0.1469	−0.2056
1.15	0.0002	0.0011	0.0023	0.0052	0.0127	0.0237	0.0396	0.0625	0.0943	0.1548	0.1667	0.0332	−0.0611	−0.1084	−0.1642
1.20	0.0004	0.0019	0.0039	0.0084	0.0190	0.0326	0.0499	0.0719	0.0991	0.1477	0.1990	0.1095	−0.0141	−0.0678	−0.1231
1.30	0.0006	0.0030	0.0061	0.0125	0.0267	0.0429	0.0612	0.0819	0.1048	0.1420	0.1991	0.2079	0.0875	0.0176	−0.0423
1.40	0.0007	0.0036	0.0072	0.0147	0.0306	0.0477	0.0661	0.0857	0.1063	0.1383	0.1894	0.2397	0.1737	0.1008	0.0350
1.50	0.0008	0.0039	0.0078	0.0158	0.0323	0.0497	0.0677	0.0864	0.1055	0.1345	0.1806	0.2433	0.2309	0.1717	0.1058
1.60	0.0008	0.0040	0.0080	0.0162	0.0330	0.0501	0.0677	0.0855	0.1035	0.1303	0.1729	0.2381	0.2631	0.2255	0.1673
1.70	0.0008	0.0040	0.0081	0.0163	0.0329	0.0497	0.0667	0.0838	0.1008	0.1259	0.1658	0.2305	0.2788	0.2628	0.2179
1.80	0.0008	0.0040	0.0081	0.0162	0.0325	0.0488	0.0652	0.0816	0.0978	0.1216	0.1593	0.2224	0.2846	0.2871	0.2576
1.90	0.0008	0.0040	0.0079	0.0159	0.0318	0.0477	0.0635	0.0792	0.0947	0.1173	0.1532	0.2144	0.2848	0.3017	0.2876
2.00	0.0008	0.0039	0.0078	0.0155	0.0310	0.0464	0.0617	0.0767	0.0916	0.1133	0.1476	0.2069	0.2819	0.3097	0.3096
2.20	0.0007	0.0037	0.0074	0.0147	0.0293	0.0437	0.0579	0.0719	0.0857	0.1057	0.1374	0.1932	0.2720	0.3135	0.3355
2.40	0.0007	0.0035	0.0070	0.0139	0.0276	0.0411	0.0544	0.0675	0.0803	0.0989	0.1285	0.1812	0.2602	0.3089	0.3459
2.60	0.0007	0.0033	0.0066	0.0131	0.0260	0.0387	0.0512	0.0634	0.0754	0.0929	0.1207	0.1706	0.2484	0.3009	0.3475
2.80	0.0006	0.0031	0.0062	0.0124	0.0245	0.0365	0.0483	0.0598	0.0711	0.0876	0.1138	0.1613	0.2372	0.2915	0.3443
3.00	0.0006	0.0029	0.0059	0.0117	0.0232	0.0345	0.0456	0.0565	0.0672	0.0828	0.1076	0.1529	0.2268	0.2817	0.3385
3.50	0.0005	0.0026	0.0052	0.0103	0.0204	0.0303	0.0401	0.0497	0.0591	0.0728	0.0949	0.1356	0.2042	0.2584	0.3194
4.00	0.0005	0.0023	0.0046	0.0091	0.0182	0.0270	0.0357	0.0443	0.0527	0.0651	0.0849	0.1219	0.1857	0.2378	0.2994

Table E.3. Binary Interaction Parameters, k_{12}, for Evaluating the Cross-Pseudocritical Temperatures,

$T_{c12} = (1 - k_{12})(T_{c1} T_{c2})^{0.5}$,

for Evaluating Cross-Coefficients of the RK, PR, and Soave Equations or the B_{ij} of the Virial Equation. The complete formulas are in Table 1.5. Values in parentheses are interpolated or estimated (Prausnitz & Chueh 1968).

System			*System*			*System*		
(1)	(2)	$k_{12} \times 10^2$	(1)	(2)	$k_{12} \times 10^2$	(1)	(2)	$k_{12} \times 10^2$
Methane	Ethylene	1	*n*-Pentane	iso-Pentane	0	Nitrogen	Methane	3
	Ethane	1	(or	*n*-Hexane	0		Ethylene	4
	Propylene	2	iso-Pentane)	Cyclohexane	0		Ethane	5
	Propane	2		*n*-Heptane	0		Propylene	(7)
	n-Butane	4		*n*-Octane	0		Propane	(9)
	iso-Butane	4		Benzene	(1)		*n*-Butane	12
	n-Pentane	6		Toluene	(1)			
	iso-Pentane	6				Argon	Methane	2
	n-Hexane	8	*n*-Hexane	*n*-Heptane	0		Ethylene	3
	Cyclohexane	8	(or	*n*-Octane	0		Ethane	3
	n-Heptane	10	Cyclohexane)	Benzene	(1)		Oxygen	1
	n-Octane	(12)		Toluene	1		Nitrogen	0
	Benzene	(8)						
	Toluene	(8)	*n*-Heptane	*n*-Octane	0	Tetrafluoro-	Methane	7
	Naphthalene	14		Benzene	(1)	methane	Nitrogen	2
				Toluene	(1)		Helium	(16±2)
Ethylene	Ethane	0						
(or	Propylene	0	*n*-Octane	Benzene	(1)	Hydrogen	Nitrogen	0
Ethane)	Propane	0		Toluene	(1)		Argon	0
	n-Butane	1					Methane	3
	iso-Butane	1	Benzene	Toluene	(0)		Ethane	(5)
	n-Pentane	2					Propane	(7)

	iso-Pentane	2	Carbon	Methane	(5±2)		*n*-Butane	(8)
	n-Hexane	3	dioxide	Ethylene	6		iso-Butane	(8)
	Cyclohexane	3		Ethane	8		*n*-Pentane	(9)
	n-Heptane	4		Propylene	10		iso-Pentane	(9)
	n-Octane	(5)		Propane	11±1		*n*-Hexane	10
	Benzene	3		*n*-Butane	16±2			
	Toluene	(3)		iso-Butane	(16±2)	Helium	Nitrogen	16
	Naphthalene	8		*n*-Pentane	(18±2)		Argon	5±1
				iso-Pentane	(18±2)		Methane	(46)
Propylene	Propane	0		Naphthalene	24			
(or	*n*-Butane	0				Neon	Methane	28
Propane)	iso-Butane	0	Hydrogen	Methane	5±1		Krypton	20±2
	n-Pentane	1	sulfide	Ethylene	(5±1)			
	iso-Pentane	0		Ethane	6	Krypton	Methane	1
	n-Hexane	(1)		Propylene	(7)			
	Cyclohexane	(1)		Propane	8			
	n-Heptane	(2)		*n*-Butane	(9)			
	n-Octane	(3)		iso-Butane	(9)			
	Benzene	2		*n*-Pentane	10			
	Toluene	(2)		iso-Pentane	(10)			
				Carbon dioxide	8			
n-Butane	iso-Butane	0						
(or iso-	*n*-Pentane	0	Acetylene	Methane	(5)			
Butane)	iso-Pentane	0		Ethylene	6			
	n-Hexane	0		Ethane	8			
	Cyclohexane	0		Propylene	7			
	n-Heptane	0		Propane	9			
	n-Octane	(1)		*n*-Butane	(10)			
	Benzene	(1)		iso-Butane	(10)			
	Toluene	(1)		*n*-Pentane	(11)			
				iso-Pentane	(11)			

Table E.4. Constants of Some Equations of State of Some Substances

van der Waals: $(P + a/V^2)(V - b) = RT,\ a = 27R^2T_c^2/64P_c,\ b = RT_c/8P_c.$

Berthelot: $(P + a/TV^2)(V - b) = RT,\ a = 27R^2T_c^3/64P_c,\ b = 9RT_c/128P_c.$

Dieterici: $P(V - b) = RT\exp(-a/RTV),\ a = 4R^2T_c^2/e^2,\ b = RT_c/e^2P_c,\ e = 2.718.$

Beattie-Bridgman: $P = \frac{RT}{V} + \frac{\beta}{V^2} + \frac{\gamma}{V^3} + \frac{\delta}{V^4}$

$$\beta = RTB_0 - A_0 - \frac{Rc}{T^2}$$

$$\gamma = -RTB_0b + A_0a - \frac{RcB_0}{T^2}$$

$$\delta = \frac{RB_0bc}{T^2}$$

Units: P atm, V liters/gmol, T, °K, $R = 0.08206$.

			van der Waals		*Berthelot*		*Dieterici*	
Gas	T_c	P_c	*a*	*b*	*a*	*b*	*a*	*b*
H_2	33.2	12.8	0.24463	0.02661	8.1217	0.01497	0.3139	0.02881
He	5.19	2.24	.034161	0.023766	0.1773	0.013369	0.04383	0.025731
Ar	150.8	48.1	1.3431	0.032159	202.54	0.018089	1.7235	0.034818
N_2	126.2	33.5	1.3506	0.03864	170.45	0.02174	1.7331	0.04184
O_2	154.6	49.8	1.3634	0.03184	210.78	0.01791	1.7496	0.03448
CO_2	304.2	72.8	3.60111	0.04286	1098.5	0.02411	4.6337	0.04641
CH_4	190.6	45.4	2.2732	0.04306	433.27	0.02422	2.9169	0.04662
C_2H_5OH	516.2	63.0	12.016	0.08405	6202.7	0.04728	15.418	0.09100
C_6H_6	562.1	48.3	18.583	0.11937	10446.0	0.06715	23.845	0.12924

Beattie-Bridgeman

Gas	A_0	*a*	B_0	*b*	$10^{-4}C$
He	0.0216	0.05984	0.01400	0	0.004
Ne	0.2125	0.2196	0.02060	0	0.101
Ar	1.2907	0.02328	0.03931	0	5.99
H_2	0.1975	−0.00506	0.02096	−0.04359	0.050
N_2	1.3445	0.02617	0.05046	−0.00691	4.20
O_2	1.4911	0.02562	0.04624	0.004208	4.80
Air	1.3012	0.01931	0.04611	−0.01101	4.34
CO_2	5.0065	0.07132	0.10476	0.07235	66.00
CH_4	2.2769	0.01855	0.05587	−0.01587	12.83
$(C_2H_5)_2O$	31.278	0.12426	0.45446	0.11954	33.33

Table E.5. Values of the Constants of the BWR Equation of State. *P* atm, *V* liters/gmol, *T*, °*K*, $R = 0.08206$. Reproduced from Holub & Vonka, *The Chemical Equilibrium of Gaseous Systems*, Reidel, 1976. Originally from Novak, Malijevsky, Matous, & Sobr, *Gases and Gas Mixtures, Behavior of State and Thermodynamic Properties*, Czechoslovak Academy of Sciences, 1972.

Substance	A_0	B_0	C_0	*a*	*b*
Hydrogen[a]	9.7319×10^{-2}	1.8041×10^{-2}	3.8914×10^{2}	-9.2211×10^{-3}	1.7976×10^{-4}
Nitrogen	1.1925	0.0458	5.8891×10^{3}	0.0149	1.98154×10^{-3}
Nitrogen	0.872086	2.81066×10^{-2}	7.81375×10^{3}	3.12319×10^{-2}	3.2351×10^{-3}
Oxygen	1.4988	4.6524×10^{-2}	3.8617×10^{3}	-4.0507×10^{-2}	-2.7963×10^{-4}
CO	1.34122	5.45425×10^{-2}	8.562×10^{3}	3.665×10^{-2}	2.6316×10^{-3}
CO	1.03115	0.040	1.124×10^{4}	3.665×10^{-2}	2.6316×10^{-3}
CO_2	2.7374	4.9909×10^{-2}	1.38564×10^{5}	1.3681×10^{-1}	7.2105×10^{-3}
CO_2	2.51604	4.48842×10^{-2}	1.474405×10^{5}	1.3681×10^{-1}	4.12381×10^{-3}
CO_2[a]	2.7634	4.5628×10^{-2}	1.1333×10^{5}	5.1689×10^{-2}	3.0819×10^{-3}
SO_2[a]	7.08538	0.10896	4.43966×10^{5}	6.87046×10^{-2}	1.93727×10^{-3}
SO_2	2.12042	2.61817×10^{-2}	7.93840×10^{5}	0.844680	1.46531×10^{-2}
N_2O[a]	3.0868	5.1953×10^{-2}	1.2725×10^{5}	0.10946	3.7755×10^{-3}
H_2S	3.10377	3.48471×10^{-2}	1.9721×10^{5}	0.144984	4.42477×10^{-3}
NH_3	3.78928	5.16461×10^{-2}	1.78567×10^{5}	0.10354	7.19561×10^{-4}
CH_3Cl	4.56359	5.07705×10^{-2}	5.83918×10^{5}	0.180052	5.19665×10^{-3}
Methane	1.8550	4.2600×10^{-2}	2.257×10^{4}	0.0494	3.38004×10^{-3}
Methane	1.79894	4.54625×10^{-2}	3.18382×10^{4}	0.04352	2.52033×10^{-3}
Ethane	4.15556	6.27724×10^{-2}	1.79592×10^{5}	0.34516	1.1122×10^{-2}
Ethylene	3.33958	5.56833×10^{-2}	1.31140×10^{5}	0.259	0.00860
Acetylene	1.5307	5.5851×10^{-3}	2.1586×10^{5}	−0.10001	-3.7810×10^{-5}
Propane	6.87225	9.7313×10^{-2}	5.08256×10^{5}	0.9477	0.0225
Propene	6.11220	8.50647×10^{-2}	4.39182×10^{5}	7.74056×10^{-1}	1.87059×10^{-2}
Propyne[a]	5.10806	6.9779×10^{-2}	6.40624×10^{5}	0.69714	1.4832×10^{-2}
Isobutane	10.23264	1.37544×10^{-1}	8.49943×10^{5}	1.93763	4.24352×10^{-2}
Butane	10.0847	1.24361×10^{-1}	0.99283×10^{6}	1.88231	3.99983×10^{-2}
1-Butene[a]	9.05497	0.116019	9.27248×10^{5}	1.68197	3.4815×10^{-2}
2-Butene, cis	9.82266	0.121971	1.0719×10^{6}	1.91732	3.8444×10^{-2}
1,3-Butadiene	7.41998	9.5452×10^{-2}	1.03999×10^{6}	1.39146	2.8002×10^{-2}
Isobutylene	8.95325	1.16025×10^{-1}	9.2728×10^{5}	1.6927	3.48156×10^{-2}
Pentane	12.1794	0.156751	2.12121×10^{6}	4.0748	6.6812×10^{-2}
Isopentane	12.7959	0.160053	1.74632×10^{6}	3.7562	6.6812×10^{-2}
Neopentane[a]	14.9413	0.19534	1.07186×10^{6}	2.72334	5.71607×10^{-2}
Neopentane[a]	7.06955	5.17798×10^{-2}	1.62085×10^{6}	2.06202	4.62003×10^{-2}
1-Pentene	11.05352	1.27921×10^{-1}	1.38870×10^{6}	2.262816	4.2286×10^{-2}
Hexane	14.4373	1.77813×10^{-1}	3.31935×10^{6}	7.11671	1.09131×10^{-1}
Heptane	17.5206	1.99005×10^{-1}	4.75574×10^{6}	10.36475	1.51954×10^{-1}
Nonane	−41.456199	-9.64946×10^{-1}	2.75136×10^{6}	37.17914	6.04989×10^{-1}
Decane	−19.38795	-9.46923×10^{-1}	3.43152×10^{6}	59.87797	9.86288×10^{-1}

[a]Constants calculated with the aid of the standard program at the Department of Physical Chemistry, Institute of Chemical Technology, Prague, using literature data.

Table E.5 *(continued)*

Substance	*c*	α	γ	*Range of validity* Temp., °C	to d_r	P_{max} (atm)
Hydrogen[a]	$-2{,}4613 \times 10^2$	-0.34215×10^{-5}	1.89×10^{-3}	(0)–(150)	2.5	2500
Nitrogen	5.48064×10^2	2.91545×10^{-4}	7.5×10^{-3}	(−163)–(200)	1.25	600
Nitrogen	5.47364×10^2	7.093×10^{-5}	4.5×10^{-3}	(−170)–(100)	2.0	
Oxygen	-2.0376×10^2	8.641×10^{-6}	3.59×10^{-3}	(−110)–(125)	0.8	
CO	1.04×10^3	1.350×10^{-4}	0.006	(−140)–(−25)		1000
CO	1.04×10^3	1.350×10^{-4}	0.006	(−25)–(200)		1000
CO_2	1.49183×10^4	8.4658×10^{-5}	5.393×10^{-3}	(10)–(150)		700
CO_2	1.49183×10^4	8.4658×10^{-5}	5.253×10^{-3}	(150)–(250)		700
CO_2[a]	7.0672×10^3	1.1271×10^{-4}	4.94×10^{-3}	(0)–(275)	2.1	700
SO_2[a]	5.85038×10^4	5.86479×10^{-4}	8.687×10^{-3}	(10)–(250)	2.0	200
SO_2	1.13356×10^5	7.1951×10^{-5}	5.923×10^{-3}	(10)–(250)	2.0	200
N_2O[a]	1.3794×10^4	9.377×10^{-5}	5.301×10^{-3}	(−30)–(150)	2.0	200
H_2S	1.87032×10^4	7.0316×10^{-5}	4.555×10^{-3}	(5)–(170)	2.2	700
NH_3	1.57536×10^2	4.651890×10^{-6}	1.980×10^{-2}	(0)–(300)	1.5	1100
CH_3Cl	6.87309×10^4	4.13840×10^{-4}	1.131×10^{-2}	(40)–(220)	2.1	300
Methane	2.545×10^3	1.24359×10^{-4}	0.006	(−70)–(200)	1.8	400
Methane	3.5878×10^3	3.30×10^{-4}	1.05×10^{-2}	(0)–(350)	1.8	400
Ethane	3.2767×10^4	2.43389×10^{-4}	1.18×10^{-2}	(0)–(275)	1.6	300
Ethylene	2.112×10^4	1.78×10^{-4}	9.23×10^{-3}	(0)–(200)	1.6	300
Acetylene	6.0162×10^3	-5.549×10^{-5}	7.14×10^{-3}	(20)–(250)	1.6	150
Propane	1.29×10^5	6.07175×10^{-4}	0.022	(100)–(275)	1.75	
Propene	1.02611×10^5	4.55696×10^{-4}	1.829×10^{-2}	(25)–(300)	1.45	
Propyne[a]	1.09855×10^5	2.7363×10^{-4}	1.245×10^{-2}	(50)–(200)		300
Isobutane	2.8601×10^5	1.07408×10^{-3}	0.034	(100)–(240)	1.8	
Butane	3.1640×10^5	1.10132×10^{-3}	3.4×10^{-2}	(150)–(300)	1.8	
1-Butene[a]	2.7493×10^5	9.1084×10^{-4}	2.96×10^{-2}	(150)–(250)		250
2-Butene, cis	3.33972×10^5	1.05693×10^{-3}	3.27×10^{-2}			
1,3-Butadiene	2.45052×10^5	7.09881×10^{-4}	2.35×10^{-2}			
Isobutylene	2.7492×10^5	9.10889×10^{-4}	2.96×10^{-2}	(150)–(275)	1.8	
Pentane	8.2417×10^5	1.810×10^{-3}	4.75×10^{-2}	(140)–(280)	1.5	200
Isopentane	0.695×10^6	1.70×10^{-3}	4.63×10^{-2}	(130)–(280)	1.5	200
Neopentane[a]	4.73969×10^5	2.24898×10^{-3}	5.352×10^{-2}	(160)–(275)	2.1	250
Neopentane[a]	4.31017×10^5	2.51254×10^{-3}	5.342×10^{-2}	(30)–(200)	1.7	70
1-Pentene	4.53779×10^5	1.219208×10^{-3}	3.595×10^{-2}			
Hexane	1.51276×10^6	2.81086×10^{-3}	6.668×10^{-2}	(275)–(350)	1.8	
Heptane	2.47×10^6	4.35611×10^{-3}	9×10^{-2}	(275)–(350)	1.8	
Nonane	2.516085×10^6	3.230516×10^{-3}	12.23×10^{-2}	(40)–(250)		700
Decane	7.822297×10^6	4.35394×10^{-3}	15.3×10^{-2}	(40)–(250)		700

[a]Constants calculated with the aid of the standard program at the Department of Physical Chemistry, Institute of Chemical Technology, Prague, using literature data.

Table E.6. Values of $[\log(f/p)]^{(0)}$ (Lee-Kesler)

	PR														
TR	*0.010*	*0.050*	*0.100*	*0.200*	*0.400*	*0.600*	*0.800*	*1.000*	*1.200*	*1.500*	*2.000*	*3.000*	*5.000*	*7.000*	*10.000*
0.30	−3.708	−4.402	−4.696	−4.985	−5.261	−5.412	−5.512	−5.584	−5.638	−5.697	−5.759	−5.810	−5.782	−5.679	−5.461
0.35	−2.471	−3.166	−3.461	−3.751	−4.029	−4.183	−4.285	−4.359	−4.416	−4.479	−4.547	−4.611	−4.608	−4.530	−4.352
0.40	−1.566	−2.261	−2.557	−2.848	−3.128	−3.283	−3.387	−3.463	−3.522	−3.588	−3.661	−3.735	−3.752	−3.694	−3.545
0.45	−0.879	−1.575	−1.871	−2.162	−2.444	−2.601	−2.707	−2.785	−2.845	−2.913	−2.990	−3.071	−3.104	−3.063	−2.938
0.50	−0.344	−1.040	−1.336	−1.628	−1.912	−2.070	−2.177	−2.256	−2.317	−2.387	−2.468	−2.555	−2.601	−2.572	−2.468
0.55	−0.008	−0.614	−0.911	−1.204	−1.488	−1.647	−1.755	−1.835	−1.897	−1.969	−2.052	−2.145	−2.201	−2.183	−2.096
0.60	−0.007	−0.269	−0.566	−0.859	−1.144	−1.304	−1.413	−1.494	−1.557	−1.630	−1.715	−1.812	−1.878	−1.869	−1.795
0.65	−0.005	−0.026	−0.283	−0.576	−0.862	−1.023	−1.132	−1.214	−1.278	−1.352	−1.439	−1.539	−1.612	−1.611	−1.549
0.70	−0.004	−0.021	−0.043	−0.341	−0.627	−0.789	−0.899	−0.981	−1.045	−1.120	−1.208	−1.312	−1.391	−1.396	−1.344
0.75	−0.003	−0.017	−0.035	−0.144	−0.430	−0.592	−0.703	−0.785	−0.850	−0.925	−1.015	−1.121	−1.204	−1.215	−1.172
0.80	−0.003	−0.014	−0.029	−0.059	−0.264	−0.426	−0.537	−0.619	−0.685	−0.760	−0.851	−0.958	−1.046	−1.062	−1.026
0.85	−0.002	−0.012	−0.024	−0.049	−0.123	−0.285	−0.396	−0.479	−0.544	−0.620	−0.711	−0.819	−0.911	−0.930	−0.901
0.90	−0.002	−0.010	−0.020	−0.041	−0.086	−0.166	−0.276	−0.359	−0.424	−0.500	−0.591	−0.700	−0.794	−0.817	−0.793
0.93	−0.002	−0.009	−0.018	−0.037	−0.077	−0.122	−0.214	−0.296	−0.361	−0.437	−0.527	−0.637	−0.732	−0.756	−0.735
0.95	−0.002	−0.008	−0.017	−0.035	−0.072	−0.113	−0.176	−0.258	−0.322	−0.398	−0.488	−0.598	−0.693	−0.719	−0.699
0.97	−0.002	−0.008	−0.016	−0.033	−0.067	−0.105	−0.148	−0.223	−0.287	−0.362	−0.452	−0.561	−0.657	−0.683	−0.665
0.98	−0.002	−0.008	−0.016	−0.032	−0.065	−0.101	−0.142	−0.206	−0.270	−0.344	−0.434	−0.543	−0.639	−0.666	−0.649
0.99	−0.001	−0.007	−0.015	−0.031	−0.063	−0.098	−0.137	−0.191	−0.254	−0.328	−0.417	−0.526	−0.622	−0.649	−0.633
1.00	−0.001	−0.007	−0.015	−0.030	−0.061	−0.095	−0.132	−0.176	−0.238	−0.312	−0.401	−0.509	−0.605	−0.633	−0.617
1.01	−0.001	−0.007	−0.014	−0.029	−0.059	−0.091	−0.127	−0.168	−0.224	−0.297	−0.385	−0.493	−0.589	−0.617	−0.602
1.02	−0.001	−0.007	−0.014	−0.028	−0.057	−0.088	−0.122	−0.161	−0.210	−0.282	−0.370	−0.477	−0.573	−0.601	−0.588
1.05	−0.001	−0.006	−0.013	−0.025	−0.052	−0.080	−0.110	−0.143	−0.180	−0.242	−0.327	−0.433	−0.529	−0.557	−0.546
1.10	−0.001	−0.005	−0.011	−0.022	−0.045	−0.069	−0.093	−0.120	−0.148	−0.193	−0.267	−0.368	−0.462	−0.491	−0.482
1.15	−0.001	−0.005	−0.009	−0.019	−0.039	−0.059	−0.080	−0.102	−0.125	−0.160	−0.220	−0.312	−0.403	−0.433	−0.426
1.20	−0.001	−0.004	−0.008	−0.017	−0.034	−0.051	−0.069	−0.088	−0.106	−0.135	−0.184	−0.266	−0.352	−0.382	−0.377
1.30	−0.001	−0.003	−0.006	−0.013	−0.026	−0.039	−0.052	−0.066	−0.080	−0.100	−0.134	−0.195	−0.269	−0.296	−0.293
1.40	−0.001	−0.003	−0.005	−0.010	−0.020	−0.030	−0.040	−0.051	−0.061	−0.076	−0.101	−0.146	−0.205	−0.229	−0.226
1.50	−0.000	−0.002	−0.004	−0.008	−0.016	−0.024	−0.032	−0.039	−0.047	−0.059	−0.077	−0.111	−0.157	−0.176	−0.173
1.60	−0.000	−0.002	−0.003	−0.006	−0.012	−0.019	−0.025	−0.031	−0.037	−0.046	−0.060	−0.085	−0.120	−0.135	−0.129
1.70	−0.000	−0.001	−0.002	−0.005	−0.010	−0.015	−0.020	−0.024	−0.029	−0.036	−0.046	−0.065	−0.092	−0.102	−0.094

(Continued next page)

Table E.6 *(continued)*

	PR														
TR	*0.010*	*0.050*	*0.100*	*0.200*	*0.400*	*0.600*	*0.800*	*1.000*	*1.200*	*1.500*	*2.000*	*3.000*	*5.000*	*7.000*	*10.000*
1.80	−0.000	−0.001	−0.002	−0.004	−0.008	−0.012	−0.015	−0.019	−0.023	−0.028	−0.036	−0.050	−0.069	−0.075	−0.066
1.90	−0.000	−0.001	−0.002	−0.003	−0.006	−0.009	−0.012	−0.015	−0.018	−0.022	−0.028	−0.038	−0.052	−0.054	−0.043
2.00	−0.000	−0.001	−0.001	−0.002	−0.005	−0.007	−0.009	−0.012	−0.014	−0.017	−0.021	−0.029	−0.037	−0.037	−0.024
2.20	−0.000	−0.000	−0.001	−0.001	−0.003	−0.004	−0.005	−0.007	−0.008	−0.009	−0.012	−0.015	−0.017	−0.012	0.004
2.40	−0.000	−0.000	−0.000	−0.001	−0.001	−0.002	−0.003	−0.003	−0.004	−0.004	−0.005	−0.006	−0.003	0.005	0.024
2.60	−0.000	−0.000	−0.000	−0.000	−0.000	−0.001	−0.001	−0.001	−0.001	−0.001	−0.001	0.001	0.007	0.017	0.037
2.80	0.000	0.000	0.000	0.000	0.000	0.000	0.001	0.001	0.001	0.002	0.003	0.005	0.014	0.025	0.046
3.00	0.000	0.000	0.000	0.000	0.001	0.001	0.002	0.002	0.003	0.003	0.005	0.009	0.018	0.031	0.053
3.50	0.000	0.000	0.000	0.001	0.001	0.002	0.003	0.004	0.005	0.006	0.008	0.013	0.025	0.038	0.061
4.00	0.000	0.000	0.000	0.001	0.002	0.003	0.004	0.005	0.006	0.007	0.010	0.016	0.028	0.041	0.064

Table E.7 Values of $[\log(f/p)]^{(1)}$ (Lee-Kesler)

	PR														
TR	*0.010*	*0.050*	*0.100*	*0.200*	*0.400*	*0.600*	*0.800*	*1.000*	*1.200*	*1.500*	*2.000*	*3.000*	*5.000*	*7.000*	*10.000*
0.30	−8.778	−8.779	−8.781	−8.785	−8.790	−8.797	−8.804	−8.811	−8.818	−8.828	−8.845	−8.880	−8.953	−9.022	−9.126
0.35	−6.528	−6.530	−6.532	−6.536	−6.544	−6.551	−6.559	−6.567	−6.575	−6.587	−6.606	−6.645	−6.723	−6.800	−6.919
0.40	−4.912	−4.914	−4.916	−4.919	−4.929	−4.937	−4.945	−4.954	−4.962	−4.974	−4.995	−5.035	−5.115	−5.195	−5.312
0.45	−3.726	−3.728	−3.730	−3.734	−3.742	−3.750	−3.758	−3.766	−3.774	−3.786	−3.806	−3.845	−3.923	−4.001	−4.114
0.50	−2.838	−2.839	−2.841	−2.845	−2.853	−2.861	−2.869	−2.877	−2.884	−2.896	−2.915	−2.953	−3.027	−3.101	−3.208
0.55	−0.013	−2.163	−2.165	−2.169	−2.177	−2.184	−2.192	−2.199	−2.207	−2.218	−2.236	−2.273	−2.342	−2.410	−2.510
0.60	−0.009	−1.644	−1.646	−1.650	−1.657	−1.664	−1.671	−1.677	−1.684	−1.695	−1.712	−1.747	−1.812	−1.875	−1.967
0.65	−0.006	−0.031	−1.242	−1.245	−1.252	−1.258	−1.265	−1.271	−1.278	−1.287	−1.304	−1.336	−1.397	−1.456	−1.539
0.70	−0.004	−0.021	−0.044	−0.927	−0.934	−0.940	−0.946	−0.952	−0.958	−0.967	−0.983	−1.013	−1.070	−1.124	−1.201
0.75	−0.003	−0.014	−0.030	−0.675	−0.682	−0.688	−0.694	−0.700	−0.705	−0.714	−0.728	−0.756	−0.809	−0.858	−0.929

0.80	−0.002	−0.010	−0.020	−0.043	−0.481	−0.487	−0.493	−0.499	−0.504	−0.512	−0.526	−0.551	−0.600	−0.645	−0.709
0.85	−0.001	−0.006	−0.013	−0.028	−0.321	−0.327	−0.332	−0.338	−0.343	−0.351	−0.364	−0.388	−0.432	−0.473	−0.530
0.90	−0.001	−0.004	−0.009	−0.018	−0.039	−0.199	−0.204	−0.210	−0.215	−0.222	−0.234	−0.256	−0.296	−0.333	−0.384
0.93	−0.001	−0.003	−0.007	−0.013	−0.029	−0.048	−0.141	−0.146	−0.151	−0.158	−0.170	−0.190	−0.228	−0.262	−0.310
0.95	−0.001	−0.003	−0.005	−0.011	−0.023	−0.037	−0.103	−0.108	−0.114	−0.121	−0.132	−0.151	−0.187	−0.220	−0.265
0.97	−0.000	−0.002	−0.004	−0.009	−0.018	−0.029	−0.042	−0.075	−0.080	−0.087	−0.097	−0.116	−0.149	−0.180	−0.223
0.98	−0.000	−0.002	−0.004	−0.008	−0.016	−0.025	−0.035	−0.059	−0.064	−0.071	−0.081	−0.099	−0.132	−0.162	−0.203
0.99	−0.000	−0.002	−0.003	−0.007	−0.014	−0.021	−0.030	−0.044	−0.050	−0.056	−0.066	−0.084	−0.115	−0.144	−0.184
1.00	−0.000	−0.001	−0.003	−0.006	−0.012	−0.018	−0.025	−0.031	−0.036	−0.042	−0.052	−0.069	−0.099	−0.127	−0.166
1.01	−0.000	−0.001	−0.003	−0.005	−0.010	−0.016	−0.021	−0.024	−0.024	−0.030	−0.038	−0.054	−0.084	−0.111	−0.149
1.02	−0.000	−0.001	−0.002	−0.004	−0.009	−0.013	−0.017	−0.019	−0.015	−0.018	−0.026	−0.041	−0.069	−0.095	−0.132
1.05	−0.000	−0.001	−0.001	−0.002	−0.005	−0.006	−0.007	−0.007	−0.002	0.008	0.007	−0.005	−0.029	−0.052	−0.085
1.10	−0.000	−0.000	0.000	0.000	0.001	0.002	0.004	0.007	0.012	0.025	0.041	0.042	0.026	0.008	−0.019
1.15	0.000	0.000	0.001	0.002	0.005	0.008	0.011	0.016	0.022	0.034	0.056	0.074	0.069	0.057	0.036
1.20	0.000	0.001	0.002	0.003	0.007	0.012	0.017	0.023	0.029	0.041	0.064	0.093	0.102	0.096	0.081
1.30	0.000	0.001	0.003	0.005	0.011	0.017	0.023	0.030	0.038	0.049	0.071	0.109	0.142	0.150	0.148
1.40	0.000	0.002	0.003	0.006	0.013	0.020	0.027	0.034	0.041	0.053	0.074	0.112	0.161	0.181	0.191
1.50	0.000	0.002	0.003	0.007	0.014	0.021	0.028	0.036	0.043	0.055	0.074	0.112	0.167	0.197	0.218
1.60	0.000	0.002	0.003	0.007	0.014	0.021	0.029	0.036	0.043	0.055	0.074	0.110	0.167	0.204	0.234
1.70	0.000	0.002	0.004	0.007	0.014	0.021	0.029	0.036	0.043	0.054	0.072	0.107	0.165	0.205	0.242
1.80	0.000	0.002	0.003	0.007	0.014	0.021	0.028	0.035	0.042	0.053	0.070	0.104	0.161	0.203	0.246
1.90	0.000	0.002	0.003	0.007	0.014	0.021	0.028	0.034	0.041	0.052	0.068	0.101	0.157	0.200	0.246
2.00	0.000	0.002	0.003	0.007	0.013	0.020	0.027	0.034	0.040	0.050	0.066	0.097	0.152	0.196	0.244
2.20	0.000	0.002	0.003	0.006	0.013	0.019	0.025	0.032	0.038	0.047	0.062	0.091	0.143	0.186	0.236
2.40	0.000	0.002	0.003	0.006	0.012	0.018	0.024	0.030	0.036	0.044	0.058	0.086	0.134	0.176	0.227
2.60	0.000	0.001	0.003	0.006	0.011	0.017	0.023	0.028	0.034	0.042	0.055	0.080	0.127	0.167	0.217
2.80	0.000	0.001	0.003	0.005	0.011	0.016	0.021	0.027	0.032	0.039	0.052	0.076	0.120	0.158	0.208
3.00	0.000	0.001	0.003	0.005	0.010	0.015	0.020	0.025	0.030	0.037	0.049	0.072	0.114	0.151	0.199
3.50	0.000	0.001	0.002	0.004	0.009	0.013	0.018	0.022	0.026	0.033	0.043	0.063	0.101	0.134	0.179
4.00	0.000	0.001	0.002	0.004	0.008	0.012	0.016	0.020	0.023	0.029	0.038	0.057	0.090	0.121	0.163

Note: This information is given in graphical form in the *API Data Book*, Figures 7G1.4, 7G1.5, 7G1.6, and 7G1.7.

Table E.8. Selected Wilson Parameters (Holmes & Van Winkle 1970) (Copyright American Chemical Society; used with permission.)

Components 1	2	$\lambda_{12}-\lambda_{11}$, *cal/g mol*	$\lambda_{12}-\lambda_{22}$, *cal/g mol*	*Press., mm Hg*	*Temp., °C*
Acetone	Benzene	494.92	−167.91	760	..
	Carbon tetrachloride	651.76	−12.67	760	..
	Chloroform	−72.20	−332.23	760	..
	2,3-Dimethylbutane	948.29	234.96	760	..
	Ethanol	38.17	418.96	760	..
	Methanol	−214.95	664.08	760	..
		−203.03	666.99	...	55
	n-Pentane	996.75	262.74	760	..
	2-Propanol	127.43	284.99	760	..
		429.17	53.40	...	55
	Water	439.64	1405.49	760	..
Acetonitrile	Water	694.08	1610.07	760	..
Benzene	Acetone	−167.91	494.93	760	..
	1-Butanol	160.12	817.67	760	..
	Carbon tetrachloride	−103.41	204.82	760	..
	Chloroform	141.62	−204.22	760	..
	Cyclohexane	187.23	80.02	760	..
	Cyclopentane	266.56	−24.18	760	..
	Ethanol	131.47	1297.90	760	..
	n-Heptane	99.35	292.94	760	..
		73.63	364.63	...	75
	n-Hexane	173.93	169.92	760	..
	Methanol	153.86	1620.36	760	..
	Methyl acetate	229.25	−23.84	760	..
	Methylcyclohexane	−4.15	360.92	760	..
	Methylcyclopentane	161.44	97.33	760	..
	1-Propanol	−73.91	1370.32	760	..
		67.14	1222.07	...	75
	2-Propanol	160.53	1007.94	760	..
		272.35	1066.93	500	..
1-Butanol	Benzene	817.67	160.12	760	..
	Toluene	887.80	104.68	760	..
Butyl cellosolve	Ethylcyclohexane	643.51	636.11	400	..
	n-Octane	1070.54	298.62	400	..
Carbon tetrachloride	Acetone	−12.67	651.76	760	..
	Benzene	204.82	−103.41	760	..
	2-Propanol	111.11	1232.94	760	..
Cellosolve	Ethylbenzene	755.77	121.89	760	..
	n-Hexane	834.86	656.23	760	..
	Hexene-1	370.05	705.47	760	..
	n-Octane	989.04	622.77	760	..
Chloroform	Acetone	−332.23	−72.20	760	..
	Benzene	−204.22	141.62	760	..
	2,3-Dimethylbutane	213.88	223.69	760	..
	Ethyl acetate	−367.50	−92.50	760	..
	Methanol	−373.30	1703.68	760	..
	Methyl acetate	−451.09	113.24	760	..
	Methyl-ethyl-ketone	−231.61	−235.12	760	..
Cyclohexane	Benzene	80.02	187.23	760	..
	Ethanol	303.42	2151.01	760	..
	Methyl acetate	345.11	691.65	760	..
	2-Propanol	69.02	1734.12	760	..
		223.13	1590.51	500	..
	Toluene	−414.68	909.36	760	..
Cyclopentane	Benzene	−24.18	266.56	760	..
2,3-Dimethylbutane	Acetone	234.96	948.29	760	..
	Chloroform	223.69	213.88	760	..
	Methanol	449.08	2771.85	760	..

Table E.8 (*continued*)

Components 1	2	$\lambda_{12}-\lambda_{11}$, *cal/g mol*	$\lambda_{12}-\lambda_{22}$, *cal/g mol*	*Press., mm Hg*	*Temp., °C*
1,4-Dioxane	*n*-Hexane	806.80	164.58	760	..
	Hexene-1	495.19	176.39	760	..
Ethanol	Acetone	418.96	38.17	760	..
	Benzene	1297.90	131.47	760	..
	Cyclohexane	2151.01	303.42	760	..
	Ethyl acetate	844.69	−178.81	760	..
		822.03	−62.43	...	40
		744.81	−52.14	...	60
	n-Heptane	2096.50	617.57	760	..
	n-Hexane	2281.99	283.63	760	..
	Methanol	−511.39	598.44	760	..
	Methylcyclopentane	2221.47	161.53	760	..
	Toluene	1238.70	251.93	756	..
	Water	382.30	955.45	760	..
Ethyl acetate	Chloroform	−92.50	−367.50	760	..
	Ethanol	−178.81	844.69	760	..
		−62.43	822.03	...	40
		−52.14	744.81	...	60
	Methanol	−200.36	985.69	760	60
		−316.92	1203.57	...	40
		20.32	866.15	...	50
		−173.45	1030.15	...	60
	1-Propanol	−198.72	661.24	760	..
		42.39	558.40	...	40
		−25.10	519.67	...	60
	2-Propanol	60.99	289.68	760	..
		39.77	664.42	...	40
		45.72	488.11	...	60
Ethylbenzene	Cellosolve	121.89	755.77	760	..
	Ethylcyclohexane	396.01	−240.92	400	..
	Hexylene glycol	52.43	1601.04	400	..
	n-Octane	304.31	−134.87	760	..
Ethylcyclohexane	Butyl cellosolve	636.11	643.51	400	..
	Ethylbenzene	−240.92	396.01	400	..
	Hexylene glycol	76.95	3592.40	400	..
n-Heptane	Benzene	292.94	99.35	760	..
		364.63	73.63	...	75
	Ethanol	617.57	2096.50	760	..
	1-Propanol	316.22	1353.98	...	75
n-Hexane	Benzene	169.92	173.93	760	..
	Cellosolve	656.23	834.86	760	..
	1,4-Dioxane	164.58	806.80	760	..
	Ethanol	283.63	2281.99	760	..
	Hexane-1	415.18	−279.86	760	..
	Methylcyclopentane	272.09	−175.70	760	..
	1-Propanol	834.85	812.66	760	..
	1,2,3-Trichloropropane	116.39	1106.54	760	..
Hexane-1	Cellosolve	705.47	370.05	760	..
	1,4-Dioxane	176.39	495.19	760	..
	n-Hexane	−279.86	415.18	760	..
	1,2,3-Trichloropropane	156.93	570.31	760	..
Hexylene glycol	Ethylbenzene	1601.04	52.43	400	..
	Ethylcyclohexane	3592.40	76.95	400	..
Methanol	Acetone	664.08	−214.95	760	..
		666.99	−203.03	...	55
	Benzene	1620.36	153.86	760	..
	Chloroform	1703.68	−373.30	760	..
	2,3-Dimethylbutane	2771.85	449.08	760	..

Table E.8. *(continued)*

Components 1	Components 2	$\lambda_{12}-\lambda_{11}$, cal/g mol	$\lambda_{12}-\lambda_{22}$, cal/g mol	Press., mm Hg	Temp., °C
Methanol *(continued)*	Ethanol	598.44	−511.39	760	..
	Ethyl acetate	985.69	−200.36	760	..
		1203.57	−316.92	...	40
		866.15	20.32	...	50
		1030.15	−173.45	...	60
	Methyl acetate	834.06	−78.81	760	..
	2-Propanol	88.02	−30.19	760	..
	Water	205.30	482.16	760	..
Methyl acetate	Benzene	−23.84	229.25	760	..
	Chloroform	113.24	−451.09	760	..
	Cyclohexane	691.65	345.11	760	..
	Methanol	−78.81	834.06	760	..
Methylcyclohexane	Benzene	360.92	−4.15	760	..
	2-Propanol	209.75	1831.76	500	..
Methylcyclopentane	Benzene	97.33	161.44	760	..
	Ethanol	161.53	2221.47	760	..
	n-Hexane	−175.70	272.09	760	..
	Toluene	−451.92	957.61	760	..
Methyl-ethyl-ketone	Chloroform	−235.12	−231.61	760	..
n-Octane	Butyl cellosolve	298.62	1070.54	400	..
	Cellosolve	622.77	989.04	760	..
	Ethylbenzene	−134.87	304.31	760	..
	2-Propanol	422.41	1391.09	400	..
n-Pentane	Acetone	262.74	996.75	760	..
1-Propanol	Benzene	1370.32	−73.91	760	..
		1222.07	67.14	...	75
	Ethyl acetate	661.24	−198.72	760	..
		558.40	42.39	...	40
		519.67	−25.10	...	60
	n-Heptane	1353.98	316.22	...	75
	n-Hexane	812.66	834.85	760	..
	Water	1015.80	1284.61	760	..
		1942.36	1144.00	...	40
		1051.44	1188.52	...	60
2-Propanol	Acetone	284.99	127.43	760	..
		53.40	429.17	...	55
	Benzene	1007.94	160.53	760	..
		1066.93	272.35	500	..
	Carbon tetrachloride	1232.94	111.11	760	..
	Cyclohexane	1734.12	69.02	760	..
		1590.51	223.13	500	..
	Ethyl acetate	289.68	60.99	760	..
		664.42	39.77	...	40
		418.11	45.72	...	60
	Methanol	−30.19	88.02	760	..
	Methylcyclohexane	1831.76	209.75	500	..
	n-Octane	1391.09	422.41	400	..
	2,2,4-Trimethylpentane	1231.69	183.12	760	..
Toluene	1-Butanol	104.68	887.80	760	..
	Cyclohexane	909.36	−414.68	760	..
	Ethanol	251.93	1238.70	756	..
	Methylcyclopentane	957.61	−452.92	760	..
1,2,3-Trichloropropane	*n*-Hexane	1106.54	116.93	760	..
	Hexene-1	570.31	156.39	760	..
2,2,4-Trimethylpentane	2-Propanol	183.12	1231.69	760	..
water	Acetone	1405.49	439.64	760	..
	Acetonitrile	1610.07	694.08	760	..

Table E.8. *(continued)*

Components		$\lambda_{12}-\lambda_{11}$, *cal/g mol*	$\lambda_{12}-\lambda_{22}$, *cal/g mol*	*Press., mm Hg*	*Temp., °C*
1	*2*				
2,2,4-Trimethylpentane water *(continued)*	Ethanol	955.45	382.30	760	..
	Methanol	482.61	205.30	760	..
	1-Propanol	1284.61	1015.80	760	..
		1144.00	1942.36	...	40
		1188.52	1051.44	...	60

Table E.9. Selected Margules Parameters (Holmes & Van Winkle, 1970) (Copyright American Chemical Society; used with permission.)

Components				*Press.,*	*Temp.,*
1	*2*	A_{12}	A_{21}	*mm Hg*	*°C*
Acetone	Benzene	0.2012	0.1533	760	..
	Carbon tetrachloride	0.3874	0.3282	760	..
	Chloroform	−0.3051	−0.2676	760	..
	2,3-Dimethylbutane	0.6345	0.6358	760	..
	Ethanol	0.2569	0.2870	760	..
	Methanol	0.2634	0.2798	760	..
		0.2762	0.2877	...	55
	n-Pentane	0.7386	0.6329	760	..
	2-Propanol	0.2152	0.2688	760	..
		0.3154	0.2428	...	55
	Water	0.9709	0.5576	760	..
Acetonitrile	Water	1.0489	0.8231	760	..
Benzene	Acetone	0.1533	0.2012	760	..
	1-Butanol	0.3449	0.5651	760	..
	Carbon tetrachloride	0.0359	0.0488	760	..
	Chloroform	−0.0824	−0.0532	760	..
	Cyclohexane	0.1462	0.1640	760	..
	Cyclopentane	0.1634	0.1290	760	..
	Ethanol	0.5718	0.7883	760	..
	n-Heptane	0.0842	0.1899	760	..
		0.0953	0.2088	...	75
	n-Hexane	0.1430	0.2010	760	..
	Methanol	0.7494	0.8923	760	..
	Methyl acetate	0.1219	0.0939	760	..
	Methylcyclohexane	0.0760	0.1760	760	..
	Methylcyclopentane	0.1342	0.1606	760	..
	1-Propanol	0.3251	0.7332	760	..
		0.4303	0.7286	...	75
	2-Propanol	0.4523	0.6551	760	..
		0.5392	0.7527	500	..
1-Butanol	Benzene	0.5651	0.3449	760	..
	Toluene	0.5340	0.3699	760	..
Butyl cellosolve	Ethylcyclohexane	0.5814	0.5784	400	..
	n-Octane	0.6903	0.5227	400	..
Carbon tetrachloride	Acetone	0.3282	0.3874	760	..
	Benzene	0.0488	0.0359	760	..
	2-Propanol	0.4763	0.7656	760	..
Cellosolve	Ethylbenzene	0.4379	0.3750	760	..
	n-Hexane	0.6633	0.7183	760	..
	Hexene-1	0.4228	0.5818	760	..
	n-Octane	0.6117	0.7467	760	..

Table E.9. *(continued)*

Components					
1	*2*	A_{12}	A_{21}	*Press., mm Hg*	*Temp., °C*
Chloroform	Acetone	−0.2676	−0.3051	760	..
	Benzene	−0.0532	−0.0824	760	..
	2,3-Dimethylbutane	0.1637	0.2677	760	..
	Ethyl acetate	−0.2726	−0.4275	760	..
	Methanol	0.3702	0.7767	760	..
	Methyl acetate	−0.2112	−0.3270	760	..
	Methyl-ethyl-ketone	−0.2938	−0.3507	760	..
Cyclohexane	Benzene	0.1640	0.1462	760	..
	Ethanol	0.7743	1.0699	760	..
	Methyl acetate	0.5789	0.5313	760	..
	2-Propanol	0.5006	0.9539	760	..
		0.5883	0.9795	500	..
	Toluene	0.0689	0.1563	760	..
Cyclopentane	Benzene	0.1290	0.1634	760	..
2,3-Dimethylbutane	Acetone	0.6358	0.6345	760	..
	Chloroform	0.2677	0.1637	760	..
	Methanol	1.1265	1.5255	760	..
1,4-Dioxane	*n*-Hexane	0.5230	0.4857	760	..
	Hexene-1	0.3577	0.3755	760	..
Ethanol	Acetone	0.2870	0.2569	760	..
	Benzene	0.7883	0.5718	760	..
	Cyclohexane	1.0699	0.7743	760	..
	Ethyl acetate	0.3925	0.3313	760	..
		0.4816	0.4130	...	40
		0.4080	0.3849	...	60
	n-Heptane	1.0806	1.0226	760	..
	n-Hexane	1.1738	0.8337	760	..
	Methanol	0.0081	0.0189	760	..
	Methylcyclopentane	1.1965	0.7065	760	..
	Toluene	0.7066	0.6933	756	..
	Water	0.6848	0.3781	760	..
Ethyl acetate	Chloroform	−0.4275	−0.2726	760	..
	Ethanol	0.3313	0.3925	760	..
		0.4130	0.4816	...	40
		0.3849	0.4080	...	60
	Methanol	0.4463	0.4229	760	..
		0.4213	0.5626	...	40
		0.5324	0.4482	...	50
		0.4737	0.4767	...	60
	1-Propanol	0.1982	0.2849	760	..
		0.3412	0.3923	...	40
		0.2749	0.3094	...	60
	2-Propanol	0.2112	0.1961	760	..
		0.3717	0.4573	...	40
		0.3037	0.3256	...	60
Ethylbenzene	Cellosolve	0.3750	0.4379	760	..
	Ethylcyclohexane	0.0800	0.0626	400	..
	Hexylene glycol	0.3105	0.7358	400	..
	n-Octane	0.0889	0.0903	760	..
Ethylcyclohexane	Butyl cellosolve	0.5784	0.5814	400	..
	Ethylbenzene	0.0626	0.0800	400	..
	Hexylene glycol	0.4293	0.9448	400	..
n-Heptane	Benzene	0.1899	0.0842	760	..
		0.2088	0.0953	...	75
	Ethanol	1.0226	1.0806	760	..
	1-Propanol	0.7719	0.7548	...	75
n-Hexane	Benzene	0.2010	0.1430	760	..
	Cellosolve	0.7183	0.6633	760	..

Table E.9. *(continued)*

Components					
1	*2*	A_{12}	A_{21}	*Press., mm Hg*	*Temp., °C*
n-Hexane *(continued)*	1,4-Dioxane	0.4857	0.5238	760	..
	Ethanol	0.8337	1.1738	760	..
	Hexene-1	0.0283	0.0078	760	..
	Methylcyclopentane	0.0188	0.0014	760	..
	1-Propanol	0.8511	0.5763	760	..
	1,2,3-Trichloropropane	0.4298	0.6916	760	..
Hexene-1	Cellosolve	0.5818	0.4228	760	..
	1,4-Dioxane	0.3755	0.3577	760	..
	n-Hexane	0.0078	0.0283	760	..
	1,2,3-Trichloropropane	0.3382	0.4307	760	..
Hexylene glycol	Ethylbenzene	0.7358	0.3105	400	..
	Ethylcyclohexane	0.9448	0.4293	400	..
Methanol	Acetone	0.2798	0.2634	760	..
		0.2877	0.2762	...	55
	Benzene	0.8923	0.7494	760	..
	Chloroform	0.7767	0.3702	760	..
	2,3-Dimethylbutane	1.5255	1.1265	760	..
	Ethanol	0.0189	0.0081	760	..
	Ethyl acetate	0.4229	0.4463	760	..
		0.5626	0.4213	...	40
		0.4482	0.5324	...	50
		0.4767	0.4737	...	60
	Methyl acetate	0.4393	0.4261	760	..
	2-Propanol	−0.0326	−0.0329	760	..
	Water	0.3794	0.2211	760	..
Methyl acetate	Benzene	0.0939	0.1219	760	..
	Chloroform	−0.3270	−0.2112	760	..
	Cyclohexane	0.5313	0.5789	760	..
	Methanol	0.4261	0.4393	760	..
Methylcyclohexane	Benzene	0.1760	0.0760	760	..
	2-Propanol	0.6785	1.0343	500	..
Methylcyclopentane	Benzene	0.1606	0.1342	760	..
	Ethanol	0.7065	1.1965	760	..
	n-Hexane	0.0014	0.0188	760	..
	Toluene	0.0694	0.1627	760	..
Methyl-ethyl-ketone	Chloroform	−0.3507	−0.2938	760	..
n-Octane	Butyl cellosolve	0.5227	0.6903	400	..
	Cellosolve	0.7467	0.6117	760	..
	Ethylbenzene	0.0903	0.0889	760	..
	2-Propanol	0.8524	0.8044	400	..
n-Pentane	Acetone	0.6329	0.7386	760	..
1-Propanol	Benzene	0.7332	0.3251	760	..
		0.7286	0.4303	...	75
	Ethyl acetate	0.2849	0.1982	760	..
		0.3923	0.3412	...	40
		0.3094	0.2749	...	60
	n-Heptane	0.7548	0.7719	...	75
	n-Hexane	0.5763	0.8511	760	..
	Water	1.0536	0.4393	760	..
		1.0748	0.4507	...	40
		1.0825	0.4653	...	60
2-Propanol	Acetone	0.2688	0.2152	760	..
		0.2428	0.3154	...	55
	Benzene	0.6551	0.4523	760	..
		0.7527	0.5392	500	..
	Carbon tetrachloride	0.7656	0.4763	760	..
	Cyclohexane	0.9539	0.5006	760	..
		0.9795	0.5883	500	..

Table E.9. *(continued)*

Components					
1	*2*	A_{12}	A_{21}	*Press., mm Hg*	*Temp., °C*
2-Propanol *(continued)*	Ethyl acetate	0.1961	0.2112	760	..
		0.4573	0.3717	...	40
		0.3256	0.3037	...	60
	Methanol	−0.0329	−0.0326	760	..
	Methylcyclohexane	1.0343	0.6785	500	..
	n-Octane	0.8044	0.8524	400	..
	2,2,4-Trimethylpentane	0.6601	0.6924	760	..
Toluene	1-Butanol	0.3699	0.5340	760	..
	Cyclohexane	0.1563	0.0689	760	..
	Ethanol	0.6933	0.7066	756	..
	Methylcyclopentane	0.1627	0.0694	760	..
1,2,3-Trichloropropane	*n*-Hexane	0.6916	0.4298	760	..
	Hexene-1	0.4307	0.3382	760	..
2,2,4-Trimethylpentane	2-Propanol	0.6924	0.6601	760	..
Water	Acetone	0.5576	0.9709	760	..
	Acetonitrile	0.8231	1.0489	760	..
	Ethanol	0.3781	0.6848	760	..
	Methanol	0.2211	0.3794	760	..
	1-Propanol	0.4393	1.0536	760	..
		0.4507	1.0748	...	40
		0.4653	1.0825	...	60

Table E.10. Selected van Laar Parameters (Holmes & Van Winkle, 1970) (Copyright American Chemical Society; used with permission.)

Components					
1	*2*	$\bar{A}_{12}$	$\bar{A}_{21}$	*Press., mm Hg*	*Temp., °C*
Acetone	Benzene	0.2039	0.1563	760	..
	Carbon tetrachloride	0.3889	0.3301	760	..
	Chloroform	−0.3045	−0.2709	760	..
	2,3-Dimethylbutane	0.6345	0.6358	760	..
	Ethanol	0.2574	0.2879	760	..
	Methanol	0.2635	0.2801	760	..
		0.2763	0.2878	...	55
	n-Pentane	0.7403	0.6364	760	..
	2-Propanol	0.2186	0.2690	760	..
		0.3158	0.2495	...	55
	Water	0.9972	0.6105	760	..
Acetonitrile	Water	1.0680	0.8207	760	..
Benzene	Acetone	0.1563	0.2039	760	..
	1-Butanol	0.3594	0.5865	760	..
	Carbon tetrachloride	0.0360	0.0509	760	..
	Chloroform	−0.0858	−0.0556	760	..
	Cyclohexane	0.1466	0.1646	760	..
	Cyclopentane	0.1655	0.1302	760	..
	Ethanol	0.5804	0.7969	760	..
	n-Heptane	0.0985	0.2135	760	..
		0.1072	0.2361	...	75
	n-Hexane	0.1457	0.2063	760	..
	Methanol	0.7518	0.8975	760	..
	Methyl acetate	0.1292	0.0919	760	..
	Methylcyclohexane	0.0910	0.1901	760	..
	Methylcyclopentane	0.1360	0.1605	760	..
	1-Propanol	0.3772	0.7703	760	..

Table E.10. *(continued)*

Components					
1	*2*	$\bar{A}_{12}$	$\bar{A}_{21}$	*Press., mm Hg*	*Temp., °C*
Benzene *(continued)*	1-Propanol *(continued)*	0.4508	0.7564	...	75
	2-Propanol	0.4638	0.6723	760	..
		0.5455	0.7716	500	..
1-Butanol	Benzene	0.5865	0.3594	760	..
	Toluene	0.5430	0.3841	760	..
Butyl cellosolve	Ethylcyclohexane	0.5814	0.5784	400	..
	n-Octane	0.6967	0.5318	400	..
Carbon tetrachloride	Acetone	0.3301	0.3889	760	..
	Benzene	0.0509	0.0360	760	..
	2-Propanol	0.4918	0.7868	760	..
Cellosolve	Ethylbenzene	0.4402	0.3762	760	..
	n-Hexane	0.6629	0.7206	760	..
	Hexene-1	0.4367	0.5860	760	..
	n-Octane	0.6158	0.7507	760	..
Chloroform	Acetone	−0.2709	−0.3045	760	..
	Benzene	−0.0556	−0.0858	760	..
	2,3-Dimethylbutane	0.1736	0.2790	760	..
	Ethyl acetate	−0.2868	−0.4478	760	..
	Methanol	0.4104	0.8263	760	..
	Methyl acetate	−0.2249	−0.3343	760	..
	Methyl-ethyl-ketone	−0.2990	−0.3486	760	..
Cyclohexane	Benzene	0.1646	0.1466	760	..
	Ethanol	0.7811	1.1031	760	..
	Methyl acetate	0.5799	0.5317	760	..
	2-Propanol	0.5322	1.0162	760	..
		0.6156	1.0158	500	..
	Toluene	0.0702	0.2578	760	..
Cyclopentane	Benzene	0.1302	0.1655	760	..
2,3-Dimethylbutane	Acetone	0.6358	0.6345	760	..
	Chloroform	0.2790	0.1736	760	..
	Methanol	1.1276	1.5408	760	..
1,4-Dioxane	*n*-Hexane	0.5260	0.4850	760	..
	Hexene-1	0.3578	0.3757	760	..
Ethanol	Acetone	0.2879	0.2574	760	..
	Benzene	0.7969	0.5804	760	..
	Cyclohexane	1.1031	0.7811	760	..
	Ethyl acetate	0.3972	0.3311	760	..
		0.4833	0.4151	...	40
		0.4093	0.3842	...	60
	n-Heptane	1.0832	1.0208	760	..
	n-Hexane	1.2005	0.8422	760	..
	Methanol	0.0088	0.0254	760	..
	Methylcyclopentane	1.2330	0.7332	760	..
	Toluene	0.7067	0.6932	756	..
	Water	0.7292	0.4104	760	..
Ethyl acetate	Chloroform	−0.4478	−0.2868	760	..
	Ethanol	0.3311	0.3972	760	..
		0.4151	0.4833	...	40
		0.3842	0.4093	...	60
	Methanol	0.4470	0.4227	760	..
		0.4278	0.5741	...	40
		0.5399	0.4476	...	50
		0.4736	0.4768	...	60
	1-Propanol	0.2051	0.2893	760	..
		0.3444	0.3913	...	40
		0.2762	0.3092	...	60
	2-Propanol	0.2113	0.1964	760	..
		0.3747	0.4604	...	40

Table E.10. *(continued)*

Components					
1	*2*	$\bar{A}_{12}$	$\bar{A}_{21}$	*Press., mm Hg*	*Temp., °C*
Ethyl acetate *(continued)*	2-Propanol *(continued)*	0.3039	0.3261	...	60
Ethylbenzene	Cellosolve	0.3762	0.4402	760	..
	Ethylcyclohexane	0.0821	0.0628	400	..
	Hexylene glycol	0.3719	0.8383	400	..
	n-Octane	0.0890	0.0902	760	..
Ethylcyclohexane	Butyl cellosolve	0.5784	0.5814	400	..
	Ethylbenzene	0.0628	0.0821	400	..
	Hexylene glycol	0.4770	1.1219	400	..
n-Heptane	Benzene	0.2135	0.0985	760	..
		0.2361	0.1072	...	75
	Ethanol	1.0208	1.0832	760	..
	1-Propanol	0.7719	0.7550	...	75
n-Hexane	Benzene	0.2063	0.1457	760	..
	Cellosolve	0.7206	0.6629	760	..
	1,4-Dioxane	0.4850	0.5260	760	..
	Ethanol	0.8422	1.2005	760	..
	Hexene-1	0.0393	0.0114	760	..
	Methylcyclopentane	0.0023	0.0226	760	..
	1-Propanol	0.8734	0.5952	760	..
	1,2,3-Trichloropropane	0.4520	0.7257	760	..
Hexene-1	Cellosolve	0.5860	0.4367	760	..
	1,4-Dioxane	0.3757	0.3578	760	..
	n-Hexane	0.0114	0.0393	760	..
	1,2,3-Trichloropropane	0.3419	0.4372	760	..
Hexylene glycol	Ethylbenzene	0.8383	0.3719	400	..
	Ethylcyclohexane	1.1219	0.4770	400	..
Methanol	Acetone	0.2801	0.2635	760	..
		0.2878	0.2763	...	55
	Benzene	0.8975	0.7518	760	..
	Chloroform	0.8263	0.4104	760	..
	2,3-Dimethylbutane	1.5408	1.1276	760	..
	Ethanol	0.0254	0.0088	760	..
	Ethyl acetate	0.4227	0.4470	760	..
		0.5741	0.4278	...	40
		0.4476	0.5399	...	50
		0.4768	0.4736	...	60
	Methyl acetate	0.4394	0.4262	760	..
	2-Propanol	−0.0325	−0.0329	760	..
	Water	0.3861	0.2439	760	..
Methyl acetate	Benzene	0.0919	0.1292	760	..
	Chloroform	−0.3343	−0.2249	760	..
	Cyclohexane	0.5317	0.5799	760	..
	Methanol	0.4262	0.4394	760	..
Methylcyclohexane	Benzene	0.1901	0.0910	760	..
	2-Propanol	0.6886	1.0659	500	..
Methylcyclopentane	Benzene	0.1605	0.1360	760	..
	Ethanol	0.7332	1.2330	760	..
	n-Hexane	0.0226	0.0023	760	..
	Toluene	0.0717	0.2475	760	..
Methyl-ethyl-ketone	Chloroform	−0.3486	−0.2990	760	..
n-Octane	Butyl cellosolve	0.5318	0.6967	400	..
	Cellosolve	0.7507	0.6158	760	..
	Ethylbenzene	0.0902	0.0890	760	..
	2-Propanol	0.8535	0.8043	400	..
n-Pentane	Acetone	0.6364	0.7403	760	..
1-Propanol	Benzene	0.7703	0.3772	760	..
		0.7564	0.4508	...	75

Table E.10. *(continued)*

Components					
1	*2*	$\bar{A}_{12}$	$\bar{A}_{21}$	*Press., mm Hg*	*Temp., °C*
1-Propanol *(continued)*	Ethyl acetate	0.2893	0.2051	760	..
		0.3913	0.3444	...	40
		0.3092	0.2762	...	60
	n-Heptane	0.7550	0.7719	...	75
	n-Hexane	0.5952	0.8734	760	..
	Water	1.1433	0.5037	760	..
		1.2315	0.5305	...	40
		1.1879	0.5224	...	60
2-Propanol	Acetone	0.2690	0.2186	760	..
		0.2495	0.3158	...	55
	Benzene	0.6723	0.4638	760	..
		0.7716	0.5455	500	..
	Carbon tetrachloride	0.7868	0.4918	760	..
	Cyclohexane	1.0162	0.5322	760	..
		1.0158	0.6156	500	..
	Ethyl acetate	0.1964	0.2113	760	..
		0.4604	0.3747	...	40
		0.3261	0.3039	...	60
	Methanol	−0.0329	−0.0325	760	..
	Methylcyclohexane	1.0659	0.6886	500	..
	n-Octane	0.8043	0.8535	400	..
	2,2,4-Trimethylpentane	0.6603	0.6927	760	..
Toluene	1-Butanol	0.3841	0.5430	760	..
	Cyclohexane	0.2578	0.0702	760	..
	Ethanol	0.6932	0.7067	756	..
	Methylcyclopentane	0.2475	0.0717	760	..
1,2,3-Trichloropropane	*n*-Hexane	0.7257	0.4520	760	..
	Hexene-1	0.4372	0.3419	760	..
2,2,4-Trimethylpentane	2-Propanol	0.6927	0.6603	760	..
Water	Acetone	0.6105	0.9972	760	..
	Acetonitrile	0.8207	1.0680	760	..
	Ethanol	0.4104	0.7292	760	..
	Methanol	0.2439	0.3861	760	..
	1-Propanol	0.5037	1.1433	760	..
		0.5305	1.2315	...	40
		0.5224	1.1879	...	60

Table E.11. UNIFAC Group Volume and Surface-Area Parameters for Prediction of Vapor-Liquid Equilibria (Gmehling, Rasmussen, & Fredenslund 1982). Copyright American Chemical Society; used with permission.

Main Group	*Subgroup*	*No.*	R_k	Q_k	*Sample Group Assignment*	
1	CH_3	1	0.9011	0.848	hexane:	2 CH_3, 4 CH_2
"CH_2"	CH_2	2	0.6744	0.540	2-methylpropane:	3 CH_3, 1 CH
	CH	3	0.4469	0.228	2,2-dimethylpropane:	4 CH_3, 1 C
	C	4	0.2195	0.000		
2	CH_2=CH	5	1.3454	1.176	1-hexene:	1 CH_3, 3 CH_2, 1 CH_2=CH
"C=C"	CH=CH	6	1.1167	0.867	2-hexene:	2 CH_3, 2 CH_2, 1 CH=CH
	CH_2=C	7	1.1173	0.988	2-methyl-1-butene:	2 CH_3, 1 CH_2, 1 CH_2=C
	CH=C	8	0.8886	0.676	2-methyl-2-butene:	3 CH_3, 1 CH=C
	C=C	9	0.6605	0.485	2,3-dimethylbutene-2:	4 CH_3, 1 C=C
3	ACH	10	0.5313	0.400	benzene:	6 ACH
"ACH"	AC	11	0.3652	0.120	styrene:	1 CH_2=CH, 5 ACH, 1 AC
4	$ACCH_3$	12	1.2663	0.968	toluene:	5 ACH, 1 $ACCH_3$
"$ACCH_2$"	$ACCH_2$	13	1.0396	0.660	ethylbenzene:	1 CH_3, 5 ACH, 1 $ACCH_2$
	ACCH	14	0.8121	0.348	cumene:	2 CH_3, 5 ACH, 1 ACCH
5	OH	15	1.000	1.200	2-propanol:	2 CH_3, 1 CH, 1 OH

Table E.11. *(continued)*

Main Group	*Subgroup*	*No.*	R_k	Q_k	*Sample Group Assignment*	
"OH"						
6 "CH_3OH"	CH_3OH	16	1.4311	1.432	methanol:	1 CH_3OH
7 "H_2O"	H_2O	17	0.92	1.40	water:	1 H_2O
8 "ACOH"	ACOH	18	0.8952	0.680	phenol:	5 ACH, 1 ACOH
9 "CH_2CO"	CH_3CO	19	1.6724	1.488	ketone group is 2nd carbon; 2-butanone:	1 CH_3, 1 CH_2, 1 CH_3CO
	CH_2CO	20	1.4457	1.180	ketone group is any other carbon; 3-pentanone:	2 CH_3, 1 CH_2, 1 CH_2CO
10 "CHO"	CHO	21	0.9980	0.948	acetaldehyde:	1 CH_3, 1 CHO
11 "CCOO"	CH_3COO	22	1.9031	1.728	butyl acetate:	1 CH_3, 3 CH_2, 1 CH_3COO
	CH_2COO	23	1.6764	1.420	butyl propanoate:	2 CH_3, 3 CH_2, 1 CH_2COO
12 "HCOO"	HCOO	24	1.2420	1.188	ethyl formate:	1 CH_3, 1 CH_2, 1 HCOO
13 "CH_2O"	CH_3O	25	1.1450	1.088	dimethyl ether:	1 CH_3, 1 CH_3O
	CH_2O	26	0.9183	0.780	diethyl ether	2 CH_3, 1 CH_2, 1 CH_2O
	CH—O	27	0.6908	0.468	diisopropyl ether:	4 CH_3, 1 CH, 1 CH—O
	FCH_2O	28	0.9183	1.1	tetrahydrofuran:	3 CH_2, 1 FCH_2O
14 "CNH_2"	CH_3NH_2	29	1.5959	1.544	methylamine:	1 CH_3NH_2
	CH_2NH_2	30	1.3692	1.236	propylamine:	1 CH_3, 1 CH_2, 1 CH_2NH_2
	$CHNH_2$	31	1.1417	0.924	isopropylamine:	2 CH_3, 1 $CHNH_2$
15 "CNH"	CH_3NH	32	1.4337	1.244	dimethylamine:	1 CH_3, 1 CH_3NH
	CH_2NH	33	1.2070	0.936	diethylamine:	2 CH_3, 1 CH_2, 1 CH_2NH
	CHNH	34	0.9795	0.624	diisopropylamine:	4 CH_3, 1 CH, 1 CHNH
16 "$(C)_3N$"	CH_3N	35	1.1865	0.940	trimethylamine:	2 CH_3, 1 CH_3N
	CH_2N	36	0.9597	0.632	triethylamine:	3 CH_3, 2 CH_2, 1 CH_2N
17 "$ACNH_2$"	$ACNH_2$	37	1.0600	0.816	aniline:	5 ACH, 1 $ACNH_2$
18 "pyridine"	C_5H_5N	38	2.9993	2.113	pyridine:	1 C_5H_5N
	C_5H_4N	39	2.8332	1.833	3-methylpyridine:	1 CH_3, 1 C_5H_4N
	C_5H_3N	40	2.667	1.553	2,3-dimethylpyridine:	2 CH_3, 1 C_5H_3N
19 "CCN"	CH_3CN	41	1.8701	1.724	acetonitrile:	1 CH_3CN
	CH_2CN	42	1.6434	1.416	propionitrile:	1 CH_3, 1 CH_2CN
20 "COOH"	COOH	43	1.3013	1.224	acetic acid:	1 CH_3, 1 COOH
	HCOOH	44	1.5280	1.532	formic acid:	1 HCOOH
21 "CCl"	CH_2Cl	45	1.4654	1.264	1-chlorobutane:	1 CH_3, 2 CH_2, 1 CH_2Cl
	CHCl	46	1.2380	0.952	2-chloropropane:	2 CH_3, 1 CHCl
	CCl	47	1.0060	0.724	2-chloro-2-methylpropane:	3 CH_3, 1 CCl
22 "CCl_2"	CH_2Cl_2	48	2.2564	1.988	dichloromethane:	1 CH_2Cl_2
	$CHCl_2$	49	2.0606	1.684	1,1-dichloroethane:	1 CH_3, 1 $CHCl_2$
	CCl_2	50	1.8016	1.448	2,2-dichloropropane:	2 CH_3, 1 CCl_2
23 "CCl_3"	$CHCl_3$	51	2.8700	2.410	chloroform:	1 $CHCl_3$
	CCl_3	52	2.6401	2.184	1,1,1-trichloroethane:	1 CH_3, 1 CCl_3
24 "CCl_4"	CCl_4	53	3.3900	2.910	tetrachloromethane:	1 CCl_4
25 "ACCl"	ACCl	54	1.1562	0.844	chlorobenzene:	5 ACH, 1 ACCl
26 "CNO_2"	CH_3NO_2	55	2.0086	1.868	nitromethane:	1 CH_3NO_2
	CH_2NO_2	56	1.7818	1.560	1-nitropropane:	1 CH_3, 1 CH_2, 1 CH_2NO_2
	$CHNO_2$	57	1.5544	1.248	2-nitropropane:	2 CH_3, 1 $CHNO_2$
27 "$ACNO_2$"	$ACNO_2$	58	1.4199	1.104	nitrobenzene:	5 ACH, 1 $ACNO_2$
28 "CS_2"	CS_2	59	2.057	1.65	carbon disulfide:	1 CS_2
29 "CH_3SH"	CH_3SH	60	1.8770	1.676	methanethiol:	1 CH_3SH
	CH_2SH	61	1.6510	1.368	ethanethiol:	1 CH_3, 1 CH_2SH

Table E.11. *(continued)*

Main Group	*Subgroup*	*No.*	R_k	Q_k	*Sample Group Assignment*	
30 "furfural"	furfural	62	3.1680	2.481	furfural:	1 furfural
31 "DOH"	$(CH_2OH)_2$	63	2.4088	2.248	1,2-ethanediol:	$1(CH_2OH)_2$
32 "I"	I	64	1.2640	0.992	1-iodoethane:	1 CH_3, 1 CH_2, 1 I
33 "Br"	Br	65	0.9492	0.832	1-bromoethane:	1 CH_3, 1 CH_2, 1 Br
34 "C≡C"	CH≡C	66	1.2920	1.088	1-hexyne:	1 CH_3, 3 CH_2, 1 CH≡C
	C≡C	67	1.0613	0.784	2-hexyne:	2 CH_3, 2 CH_2, 1 C≡C
35 "Me_2SO"	Me_2SO	68	2.8266	2.472	dimethyl sulfoxide:	1 Me_2SO
36 "ACRY"	ACRY	69	2.3144	2.052	acrylonitrile:	1 ACRY
37 "ClCC"	Cl(C=C)	70	0.7910	0.724	trichloroethylene:	1 CH=C, 3 Cl(C=C)
38 "ACF"	ACF	71	0.6948	0.524	hexafluorobenzene:	6 ACF
39 "DMF"	DMF-1	72	3.0856	2.736	dimethylformamide:	1 DMF-1
	DMF-2	73	2.6322	2.120	diethylformamide:	2 CH_3, 1 DMF-2
40 "CF_2"	CF_3	74	1.4060	1.380	perfluorohexane:	2 CF_3, 4 CF_2
	CF_2	75	1.0105	0.920		
	CF	76	0.6150	0.460	perfluoromethylcyclohexane:	1 CH_3, 5 CH_2, 1 CF

Table E.12. UNIFAC Interaction Parameters for Prediction of Vapor-Liquid Equilibria (Gmehling, Rasmussen, & Fredenslund 1982). (Copyright American Chemical Society; used with permission.)

	1	*2*	*3*	*4*	*5*	*6*	*7*	*8*
1 CH_2	0.0	−200.0	61.13	76.50	986.5	697.2	1318.0	1333.0
2 C=C	2520.0	0.0	340.7	4102.0	693.9	1509.0	634.2	547.4
3 ACH	−11.12	−94.78	0.0	167.0	636.1	637.3	903.8	1329.0
4 $ACCH_2$	−69.70	−269.7	−146.8	0.0	803.2	603.2	5695.0	884.9
5 OH	156.4	8694.0	89.60	25.82	0.0	−137.1	353.5	−259.7
6 CH_3OH	16.51	−52.39	−50.00	−44.50	249.1	0.0	−181.0	−101.7
7 H_2O	300.0	692.7	362.3	377.6	−229.1	289.6	0.0	324.5
8 ACOH	275.8	1665.0	25.34	244.2	−451.6	−265.2	−601.8	0.0
9 CH_2CO	26.76	−82.92	140.1	365.8	164.5	108.7	472.5	−133.1
10 CHO	505.7	n.a.	n.a.	n.a.	−404.8	−340.2	232.7	n.a.
11 CCOO	114.8	269.3	85.84	−170.0	245.4	249.6	10000.0	−36.72
12 HCOO	90.49	91.65	n.a.	n.a.	191.2	155.7	n.a.	n.a.
13 CH_2O	83.36	76.44	52.13	65.69	237.7	339.7	−314.7	n.a.
14 CNH_2	−30.48	79.40	−44.85	n.a.	−164.0	−481.7	−330.4	n.a.
15 CNH	65.33	−41.32	−22.31	223.0	−150.0	−500.4	−448.2	n.a.
16 $(C)_3N$	−83.98	−188.0	−223.9	109.9	28.60	−406.8	−598.8	n.a.
17 $ACNH_2$	5339.0	n.a.	650.4	979.8	529.0	5.182	−339.5	n.a.
18 pyridine	−101.6	n.a.	31.87	49.80	−132.3	−378.2	−332.9	−341.6
19 CCN	24.82	34.78	−22.97	−138.4	185.4	157.8	242.8	n.a.
20 COOH	315.3	349.2	62.32	268.2	−151.0	1020.0	−66.17	n.a.
21 CCl	91.46	−24.36	4.680	122.9	562.2	529.0	698.2	n.a.
22 CCl_2	34.01	−52.71	121.3	n.a.	747.7	669.9	708.7	n.a.
23 CCl_3	36.70	−185.1	288.5	33.61	742.1	649.1	826.7	n.a.
24 CCl_4	−78.45	−293.7	−4.700	134.7	856.3	860.1	1201.0	10000.
25 ACCl	−141.3	−203.2	−237.7	375.5	246.9	661.6	920.4	n.a.
26 CNO_2	−32.69	−49.92	10.38	−97.05	341.7	252.6	417.9	n.a.

Table E.12. *(continued)*

	1	*2*	*3*	*4*	*5*	*6*	*7*	*8*
27 $ACNO_2$	5541.0	n.a.	1824.0	−127.8	561.6	n.a.	360.7	n.a.
28 CS_2	−52.65	16.62	21.50	40.68	823.5	914.2	1081.0	n.a.
29 CH_3SH	−7.481	n.a.	28.41	n.a.	461.6	382.8	n.a.	n.a.
30 furfural	−25.31	n.a.	157.3	404.3	521.6	n.a.	23.48	n.a.
31 DOH	140.0	n.a.	221.4	150.6	267.6	n.a.	0.0	838.4
32 I	128.0	n.a.	58.68	n.a.	501.3	n.a.	n.a.	n.a.
33 Br	−31.52	n.a.	155.6	291.1	721.9	n.a.	n.a.	n.a.
34 C≡C	−72.88	−184.4	n.a.	n.a.	n.a.	n.a.	n.a.	n.a.
35 Me_2SO	50.49	n.a.	−2.504	−143.2	−25.87	695.0	−240.0	n.a.
36 ACRY	−165.9	n.a.	n.a.	n.a.	n.a.	n.a.	386.6	n.a.
37 ClCC	41.90	−3.167	−75.67	n.a.	640.9	726.7	n.a.	n.a.
38 ACF	−5.132	n.a.	−237.2	−157.3	649.7	645.9	n.a.	n.a.
39 DMF	−31.95	37.70	−133.9	−240.2	64.16	172.2	−287.1	n.a.
40 CF_2	147.3	n.a.	n.a.	n.a.	n.a.	n.a.	n.a.	n.a.

	9	*10*	*11*	*12*	*13*	*14*	*15*	*16*
1 CH_2	476.4	677.0	232.1	741.4	251.5	391.5	255.7	206.6
2 C=C	524.5	n.a.	71.23	468.7	289.3	396.0	273.6	658.8
3 ACH	25.77	n.a.	5.994	n.a.	32.14	161.7	122.8	90.49
4 $ACCH_2$	−52.10	n.a.	5688.0	n.a.	213.1	n.a.	−49.29	23.50
5 OH	84.00	441.8	101.1	193.1	28.06	83.02	42.70	−323.0
6 CH_3OH	23.39	306.4	−10.72	193.4	−180.6	359.3	266.0	53.90
7 H_2O	−195.4	−257.3	14.42	n.a.	540.5	48.89	168.0	304.0
8 ACOH	−356.1	n.a.	−449.4	n.a.	n.a.	n.a.	n.a.	n.a.
9 CH_2CO	0.0	−37.36	−213.7	n.a.	5.202	n.a.	n.a.	n.a.
10 CHO	128.0	0.0	n.a.	n.a.	304.1	n.a.	n.a.	n.a.
11 CCOO	372.2	n.a.	0.0	372.9	−235.7	n.a.	−73.50	n.a.
12 HCOO	n.a.	n.a.	−261.1	0.0	n.a.	n.a.	n.a.	n.a.
13 CH_2O	52.38	−7.838	461.3	n.a.	0.0	n.a.	141.7	n.a.
14 CNH_2	n.a.	n.a.	n.a.	n.a.	n.a.	0.0	63.72	−41.11
15 CNH	n.a.	n.a.	136.0	n.a.	−49.30	108.8	0.0	−189.2
16 $(C)_3N$	n.a.	n.a.	n.a.	n.a.	n.a.	38.89	865.9	0.0
17 $ACNH_2$	−399.1	n.a.	n.a.	n.a.	n.a.	n.a.	n.a.	n.a.
18 pyridine	−51.54	n.a.	n.a.	n.a.	n.a.	n.a.	n.a.	n.a.
19 CCN	−287.5	n.a.	−266.6	n.a.	n.a.	n.a.	n.a.	n.a.
20 COOH	−297.8	n.a.	−256.3	312.5	−338.5	n.a.	n.a.	n.a.
21 CCl	286.3	−47.51	n.a.	n.a.	225.4	n.a.	n.a.	n.a.
22 CCl_2	423.2	n.a.	−132.9	n.a.	−197.7	n.a.	n.a.	−141.4
23 CCl_3	552.1	n.a.	176.5	488.9	−20.93	n.a.	n.a.	−293.7
24 CCl_4	372.0	n.a.	129.5	n.a.	113.9	261.1	91.13	−126.0
25 ACCl	128.1	n.a.	−246.3	n.a.	n.a.	203.5	−108.4	1088.0
26 CNO_2	−142.6	n.a.	n.a.	n.a.	−94.49	n.a.	n.a.	n.a.
27 $ACNO_2$	n.a.	n.a.	n.a.	n.a.	n.a.	n.a.	n.a.	n.a.
28 CS_2	303.7	n.a.	243.8	n.a.	112.4	n.a.	n.a.	n.a.
29 CH_3SH	160.6	n.a.	n.a.	239.8	63.71	106.7	n.a.	n.a.
30 furfural	317.5	n.a.	−146.3	n.a.	n.a.	n.a.	n.a.	n.a.
31 DOH	n.a.	n.a.	152.0	n.a.	9.207	n.a.	n.a.	n.a.
32 I	138.0	n.a.	21.92	n.a.	476.6	n.a.	n.a.	n.a.
33 Br	−142.6	n.a.	n.a.	n.a.	736.4	n.a.	n.a.	n.a.
34 C≡C	443.6	n.a.	n.a.	n.a.	n.a.	n.a.	n.a.	n.a.
35 Me_2SO	110.4	n.a.	41.57	n.a.	−122.1	n.a.	n.a.	n.a.
36 ACRY	n.a.	n.a.	n.a.	n.a.	n.a.	n.a.	n.a.	n.a.
37 ClCC	−8.671	n.a.	−18.87	n.a.	−209.3	n.a.	n.a.	n.a.
38 ACF	n.a.	n.a.	n.a.	n.a.	n.a.	n.a.	n.a.	n.a.
39 DMF	97.04	n.a.	n.a.	n.a.	−158.2	n.a.	n.a.	n.a.
40.CF_2	n.a.	n.a.	n.a.	n.a.	n.a.	n.a.	n.a.	n.a.

Table E.12. *(continued)*

	17	*18*	*19*	*20*	*21*	*22*	*23*	*24*
1 CH_2	1245.0	287.7	597.0	663.5	35.93	53.76	24.90	104.3
2 C=C	n.a.	n.a.	405.9	730.4	99.61	337.1	4584.0	5831.0
3 ACH	668.2	−4.449	212.5	537.4	−18.81	−144.4	−231.9	3.000
4 $ACCH_2$	764.7	52.80	6096.0	603.8	−114.1	n.a.	−12.14	−141.3
5 OH	−348.2	170.0	6.712	199.0	75.62	−112.1	−98.12	143.1
6 CH_3OH	335.5	580.5	36.23	−289.5	−38.32	−102.5	−139.4	−67.80
7 H_2O	213.0	459.0	112.6	−14.09	325.4	370.4	353.7	497.5
8 ACOH	n.a.	−305.5	n.a.	n.a.	n.a.	n.a.	n.a.	1827.0
9 CH_2CO	937.9	165.1	481.7	669.4	−191.7	−284.0	−354.6	−39.20
10 CHO	n.a.	n.a.	n.a.	n.a.	751.9	n.a.	n.a.	n.a.
11 CCOO	n.a.	n.a.	494.6	660.2	n.a.	108.9	−209.7	54.47
12 HCOO	n.a.	n.a.	n.a.	−356.3	n.a.	n.a.	−287.2	n.a.
13 CH_2O	n.a.	n.a.	n.a.	664.6	301.1	137.8	−154.3	47.67
14 CNH_2	n.a.	n.a.	n.a.	n.a.	n.a.	n.a.	n.a.	−99.81
15 CNH	n.a.	n.a.	n.a.	n.a.	n.a.	n.a.	n.a.	71.23
16 $(C)_3N$	n.a.	n.a.	n.a.	n.a.	n.a.	−73.85	−352.9	−8.283
17 $ACNH_2$	0.0	n.a.	−216.8	n.a.	n.a.	n.a.	n.a.	8455.0
18 pyridine	n.a.	0.0	−169.7	−153.7	n.a.	−351.6	−114.7	−165.1
19 CCN	617.1	134.3	0.0	n.a.	n.a.	n.a.	−15.62	−54.86
20 COOH	n.a.	−313.5	n.a.	0.0	44.42	−183.4	76.75	212.7
21 CCl	n.a.	n.a.	n.a.	326.4	0.0	108.3	249.2	62.42
22 CCl_2	n.a.	587.3	n.a.	1821.0	−84.53	0.0	0.0	56.33
23 CCl_3	n.a.	18.98	74.04	1346.0	−157.1	0.0	0.0	−30.10
24 CCl_4	1301.0	309.2	492.0	689.0	11.80	17.97	51.90	0.0
25 ACCl	323.3	n.a.	356.9	n.a.	−314.9	n.a.	n.a.	−255.4
26 CNO_2	n.a.	n.a.	n.a.	n.a.	n.a.	n.a.	n.a.	−34.68
27 $ACNO_2$	5250.0	n.a.	n.a.	n.a.	n.a.	n.a.	n.a.	514.6
28 CS_2	n.a.	n.a.	335.7	n.a.	−73.09	n.a.	−26.06	−60.71
29 CH_3SH	n.a.	n.a.	125.7	n.a.	−27.94	n.a.	n.a.	n.a.
30 furfural	n.a.	n.a.	n.a.	n.a.	n.a.	n.a.	48.48	−133.1
31 DOH	164.4	n.a.	n.a.	n.a.	n.a.	n.a.	n.a.	n.a.
32 I	n.a.	n.a.	n.a.	n.a.	n.a.	−40.82	21.76	48.49
33 Br	n.a.	n.a.	n.a.	n.a.	1169.0	n.a.	n.a.	225.8
34 C≡C	n.a.	n.a.	329.1	n.a.	n.a.	n.a.	n.a.	n.a.
35 Me_2SO	n.a.	n.a.	n.a.	n.a.	n.a.	−215.0	−343.6	−58.43
36 ACRY	n.a.	n.a.	−42.31	n.a.	n.a.	n.a.	n.a.	n.a.
37 ClCC	n.a.	n.a.	298.4	2344.0	201.7	n.a.	85.32	143.2
38 ACF	n.a.	n.a.	n.a.	n.a.	n.a.	n.a.	n.a.	−124.6
39 DMF	335.6	n.a.	n.a.	n.a.	n.a.	n.a.	n.a.	−186.7
40 CF_2	n.a.	n.a.	n.a.	n.a.	n.a.	n.a.	n.a.	n.a.

	25	*26*	*27*	*28*	*29*	*30*	*31*	*32*
1 CH_2	321.5	661.5	543.0	153.6	184.4	354.5	3025.0	335.8
2 C=C	959.7	542.1	n.a.	76.30	n.a.	n.a.	n.a.	n.a.
3 ACH	538.2	168.0	194.9	52.07	−10.43	−64.69	210.4	113.3
4 $ACCH_2$	−126.9	3629.0	4448.0	−9.451	n.a.	−20.36	4975.0	n.a.
5 OH	287.8	61.11	157.1	477.0	147.5	−120.5	−318.9	313.5
6 CH_3OH	17.12	75.14	n.a.	−31.09	37.84	n.a.	n.a.	n.a.
7 H_2O	678.2	220.6	399.5	887.1	n.a.	188.0	0.0	n.a.
8 ACOH	n.a.	n.a.	n.a.	n.a.	n.a.	n.a.	−687.1	n.a.
9 CH_2CO	174.5	137.5	n.a.	216.1	−46.28	−163.7	n.a.	53.59
10 CHO	n.a.	n.a.	n.a.	n.a.	n.a.	n.a.	n.a.	n.a.
11 CCOO	629.0	n.a.	n.a.	183.0	n.a.	202.3	−101.7	148.3
12 HCOO	n.a.	n.a.	n.a.	n.a.	4.339	n.a.	n.a.	n.a.
13 CH_2O	n.a.	95.18	n.a.	140.9	−8.538	n.a.	−20.11	−149.5
14 CNH_2	68.81	n.a.	n.a.	n.a.	−70.14	n.a.	n.a.	n.a.

Table E.12. *(continued)*

	25	*26*	*27*	*28*	*29*	*30*	*31*	*32*
15 CNH	4350.0	n.a.	n.a.	n.a.	n.a.	n.a.	n.a.	n.a.
16 $(C)_3N$	−86.36	n.a.	n.a.	n.a.	n.a.	n.a.	n.a.	n.a.
17 $ACNH_2$	699.1	n.a.	−62.73	n.a.	n.a.	n.a.	125.3	n.a.
18 pyridine	n.a.	n.a.	n.a.	n.a.	n.a.	n.a.	n.a.	n.a.
19 CCN	52.31	n.a.	n.a.	230.9	21.37	n.a.	n.a.	n.a.
20 COOH	n.a.	n.a.	n.a.	n.a.	n.a.	n.a.	n.a.	n.a.
21 CCl	464.4	n.a.	n.a.	450.1	59.02	n.a.	n.a.	n.a.
22 CCl_2	n.a.	n.a.	n.a.	n.a.	n.a.	n.a.	n.a.	177.6
23 CCl_3	n.a.	n.a.	n.a.	116.6	n.a.	−64.38	n.a.	86.40
24 CCl_4	475.8	490.9	534.7	132.2	n.a.	546.7	n.a.	247.8
25 ACCl	0.0	−154.5	n.a.	n.a.	n.a.	n.a.	n.a.	n.a.
26 CNO_2	794.4	0.0	533.2	n.a.	n.a.	n.a.	139.8	304.3
27 $ACNO_2$	n.a.	−85.12	0.0	n.a.	n.a.	n.a.	n.a.	n.a.
28 CS_2	n.a.	n.a.	n.a.	0.0	n.a.	n.a.	n.a.	n.a.
29 CH_3SH	n.a.	n.a.	n.a.	n.a.	0.0	n.a.	n.a.	n.a.
30 furfural	n.a.	n.a.	n.a.	n.a.	n.a.	0.0	n.a.	n.a.
31 DOH	n.a.	481.3	n.a.	n.a.	n.a.	n.a.	0.0	n.a.
32 I	n.a.	64.28	n.a.	n.a.	n.a.	n.a.	n.a.	0.0
33 Br	224.0	125.3	n.a.	n.a.	n.a.	n.a.	n.a.	n.a.
34 C≡C	n.a.	174.4	n.a.	n.a.	n.a.	n.a.	n.a.	n.a.
35 Me_2SO	n.a.	n.a.	n.a.	n.a.	85.70	n.a.	535.8	n.a.
36 ACRY	n.a.	n.a.	n.a.	n.a.	n.a.	n.a.	n.a.	n.a.
37 ClCC	n.a.	313.8	n.a.	167.9	n.a.	n.a.	n.a.	n.a.
38 ACF	n.a.	n.a.	n.a.	n.a.	n.a.	n.a.	n.a.	n.a.
39 DMF	n.a.	n.a.	n.a.	n.a.	−71.00	n.a.	−191.7	n.a.
40 CF_2	n.a.	n.a.	n.a.	n.a.	n.a.	n.a.	n.a.	n.a.

	33	*34*	*35*	*36*	*37*	*38*	*39*	*40*
1 CH_2	479.5	298.9	526.5	689.0	−0.505	125.8	485.3	−2.859
2 C=C	n.a.	523.6	n.a.	n.a.	237.3	n.a.	320.4	n.a.
3 ACH	−13.59	n.a.	169.9	n.a.	69.11	389.3	245.6	n.a.
4 $ACCH_2$	−171.3	n.a.	4284.0	n.a.	n.a.	101.4	5629.0	n.a.
5 OH	133.4	n.a.	−202.1	n.a.	253.9	44.78	−143.9	n.a.
6 CH_3OH	n.a.	n.a.	−399.3	n.a.	−21.22	−48.25	−172.4	n.a.
7 H_2O	n.a.	n.a.	−139.0	160.8	n.a.	n.a.	319.0	n.a.
8 ACOH	n.a.	n.a.	n.a.	n.a.	n.a.	n.a.	n.a.	n.a.
9 CH_2CO	245.2	−246.6	−44.58	n.a.	−44.42	n.a.	−61.70	n.a.
10 CHO	n.a.	n.a.	n.a.	n.a.	n.a.	n.a.	n.a.	n.a.
11 CCOO	n.a.	n.a.	52.08	n.a.	−23.30	n.a.	n.a.	n.a.
12 HCOO	n.a.	n.a.	n.a.	n.a.	n.a.	n.a.	n.a.	n.a.
13 CH_2O	−202.3	n.a.	172.1	n.a.	145.6	n.a.	254.8	n.a.
14 CNH_2	n.a.	n.a.	n.a.	n.a.	n.a.	n.a.	n.a.	n.a.
15 CNH	n.a.	n.a.	n.a.	n.a.	n.a.	n.a.	n.a.	n.a.
16 $(C)_3N$	n.a.	n.a.	n.a.	n.a.	n.a.	n.a.	n.a.	n.a.
17 $ACNH_2$	n.a.	n.a.	n.a.	n.a.	n.a.	n.a.	−293.1	n.a.
18 pyridine	n.a.	n.a.	n.a.	n.a.	n.a.	n.a.	n.a.	n.a.
19 CCN	n.a.	−203.0	n.a.	81.57	−19.14	n.a.	n.a.	n.a.
20 COOH	n.a.	n.a.	n.a.	n.a.	−90.87	n.a.	n.a.	n.a.
21 CCl	−125.9	n.a.	n.a.	n.a.	−58.77	n.a.	n.a.	n.a.
22 CCl_2	n.a.	n.a.	215.0	n.a.	n.a.	n.a.	n.a.	n.a.
23 CCl_3	n.a.	n.a.	363.7	n.a.	−79.54	n.a.	n.a.	n.a.
24 CCl_4	41.94	n.a.	337.7	n.a.	−86.85	215.2	498.6	n.a.
25 ACCl	−60.70	n.a.	n.a.	n.a.	n.a.	n.a.	n.a.	n.a.
26 CNO_2	10.17	−27.70	n.a.	n.a.	48.40	n.a.	n.a.	n.a.
27 $ACNO_2$	n.a.	n.a.	n.a.	n.a.	n.a.	n.a.	n.a.	n.a.
28 CS_2	n.a.	n.a.	n.a.	n.a.	−47.37	n.a.	n.a.	n.a.
29 CH_3SH	n.a.	n.a.	31.66	n.a.	n.a.	n.a.	78.92	n.a.
30 furfural	n.a.	n.a.	n.a.	n.a.	n.a.	n.a.	n.a.	n.a.
31 DOH	n.a.	n.a.	−417.2	n.a.	n.a.	n.a.	302.2	n.a.
32 I	n.a.	n.a.	n.a.	n.a.	n.a.	n.a.	n.a.	n.a.
33 Br	0.0	n.a.	n.a.	n.a.	n.a.	n.a.	n.a.	n.a.
34 C≡C	n.a.	0.0	n.a.	n.a.	n.a.	n.a.	−119.8	n.a.

Table E.12. *(continued)*

	33	34	35	36	37	38	39	40
35 Me_2SO	n.a.	n.a.	0.0	n.a.	n.a.	n.a.	−97.71	n.a.
36 ACRY	n.a.	n.a.	n.a.	0.0	n.a.	n.a.	n.a.	n.a.
37 ClCC	n.a.	n.a.	n.a.	n.a.	0.0	n.a.	n.a.	n.a.
38 ACF	n.a.	n.a.	n.a.	n.a.	n.a.	0.0	n.a.	n.a.
39 DMF	n.a.	6.699	136.6	n.a.	n.a.	n.a.	0.0	n.a.
40 CF_2	n.a.	n.a.	n.a.	n.a.	n.a.	n.a.	n.a.	0.0

n.a. = not available.

UNIFAC Parameters for Interactions with the ACOH Group

main group, *m*	$a_{m,\,ACOH}$, K	$a_{ACOH,m}$, K
CH_2	1333.0	275.8
C=C	547.4	1665.0
ACH	1329.0	25.34
$ACCH_2$	884.9	244.2
OH	−259.7	−451.6
CH_3OH	−101.7	−265.2
H_2O	324.5	−601.8
CH_2CO	−133.1	−356.1
CCOO	−36.72	−449.4
pyridine	−341.6	−305.5
CCl_4	10000.0	1827.0
DOH	838.4	−687.1

UNIFAC Group-Interaction Parameters for Already Existing Groups, a_{mn}, K (Previously Marked "n.a., Non-available").

main group *m*	main group *n*	a_{mn}, K	a_{nm}, K
C=C	CCOO	71.23	269.3
C=C	HCOO	468.7	91.65
C=C	ACCl	959.7	−203.2
ACH	CCl_2	−144.4	121.3
$ACCH_2$	$(C)_3N$	23.50	109.9
$ACCH_2$	$ACNO_2$	4448.0	−127.8
OH	$ACNO_2$	157.1	561.6
CH_3OH	$(C)_3N$	53.90	−406.8
CH_3OH	$ACNH_2$	335.5	5.182
CH_3OH	ACCl	17.12	661.6
CH_2CO	$ACNH_2$	937.9	−399.1
CH_2CO	pyridine	165.1	−51.54
CH_2CO	ACCl	174.5	128.1
CHO	CH_2O	304.1	−7.838
CCOO	DOH	−101.7	152.0
HCOO	CCl_3	−287.2	488.9
CH_2O	DOH	−20.11	9.207
CH_2O	Br	−202.3	736.4
CNH_2	$(C)_3N$	−41.11	38.99
CNH_2	CCl_4	−99.81	261.1
CNH	$(C)_3N$	−189.2	865.9
$ACNH_2$	CCN	−216.8	617.1
$ACNH_2$	ACCl	699.1	323.3
pyridine	COOH	−153.7	−313.5
pyridine	CCl_2	−351.6	587.3
pyridine	CCl_4	−165.1	309.2
CCN	ACCl	52.31	356.9
COOH	CCl_3	76.75	1346.0
CCl	ACCl	464.4	−314.9
CNO_2	$ACNO_2$	533.2	−85.12
CNO_2	I	304.3	64.28
H_2O	DOH	0.0	0.0
OH	CH_3SH	147.5[a]	461.6

[a] This parameter was misprinted in Skjold-Jørgensen et al. (1979).

UNIFAC Group-Interaction Parameters for New Groups, a_{mn}, K

main group *m*	main group *n*	a_{mn}, K	a_{nm}, K
CH_2	C≡C	298.9	−72.88
C=C	C≡C	523.6	−184.4
CH_2CO	C≡C	−246.6	443.6
CCN	C≡C	−203.0	329.1
CNO_2	C≡C	−27.70	174.4
CH_2	Me_2SO	526.5	50.49
ACH	Me_2SO	169.9	−2.504
$ACCH_2$	Me_2SO	4284.	−143.2
OH	Me_2SO	−202.1	−25.87
CH_3OH	Me_2SO	−399.3	695.0
H_2O	Me_2SO	−139.0	−240.0
CH_2CO	Me_2SO	−44.58	110.4
CCOO	Me_2SO	52.08	41.57
CH_2O	Me_2SO	172.1	−122.1
CCl_2	Me_2SO	215.0	−215.0
CCl_3	Me_2SO	363.7	−343.6
CCl_4	Me_2SO	337.7	−58.43
CH_3SH	Me_2SO	31.66	85.70
DOH	Me_2SO	−417.2	535.8
CH_2	ACRY	689.0	−165.9
H_2O	ACRY	160.8	386.6
CCN	ACRY	81.57	−42.31
CH_2	ClCC	−0.505	41.90
C=C	ClCC	237.3	−3.167
ACH	ClCC	69.11	−75.67
OH	ClCC	253.9	640.9
CH_3OH	ClCC	−21.22	726.7
CH_2CO	ClCC	−44.42	−8.671
CCOO	ClCC	−23.30	−18.87
CH_2O	ClCC	145.6	−209.3
CCN	ClCC	−19.14	298.4
COOH	ClCC	−90.87	2344.

UNIFAC Group-Interaction Parameters for New Groups, a_{mn}, K (continued)

main group m	main group n	a_{mn}, K	a_{nm}, K
CCl	ClCC	−58.77	201.7
CCl_3	ClCC		85.32
CCl_4	ClCC	−86.85	143.2
CNO_2	ClCC	48.40	313.8
CS_2	ClCC	−47.37	167.9
CH_2	ACF	125.8	−5.132
ACH	ACF	389.3	−237.2
$ACCH_2$	ACF	101.4	−157.3
OH	ACF	44.78	649.7
CH_3OH	ACF	−48.25	645.9
CCl_4	ACF	215.2	−124.6
CH_2	DMF	485.3	−31.95
C=C	DMF	320.4	37.70
ACH	DMF	245.6	−133.9
$ACCH_2$	DMF	5629.0	−240.2
OH	DMF	−143.9	64.16
CH_3OH	DMF	−172.4	172.2
H_2O	DMF	319.0	−287.1
CH_2CO	DMF	−61.70	97.04
CH_2O	DMF	254.8	−158.2
$ACNH_2$	DMF	−293.1	335.6
CCl_4	DMF	498.6	−186.7
CH_3SH	DMF	78.92	−71.00
DOH	DMF	302.2	−191.7
C≡C	DMF	−119.8	6.699
Me_2SO	DMF	−97.71	136.6
CH_2	CF_2	−2.859	147.3

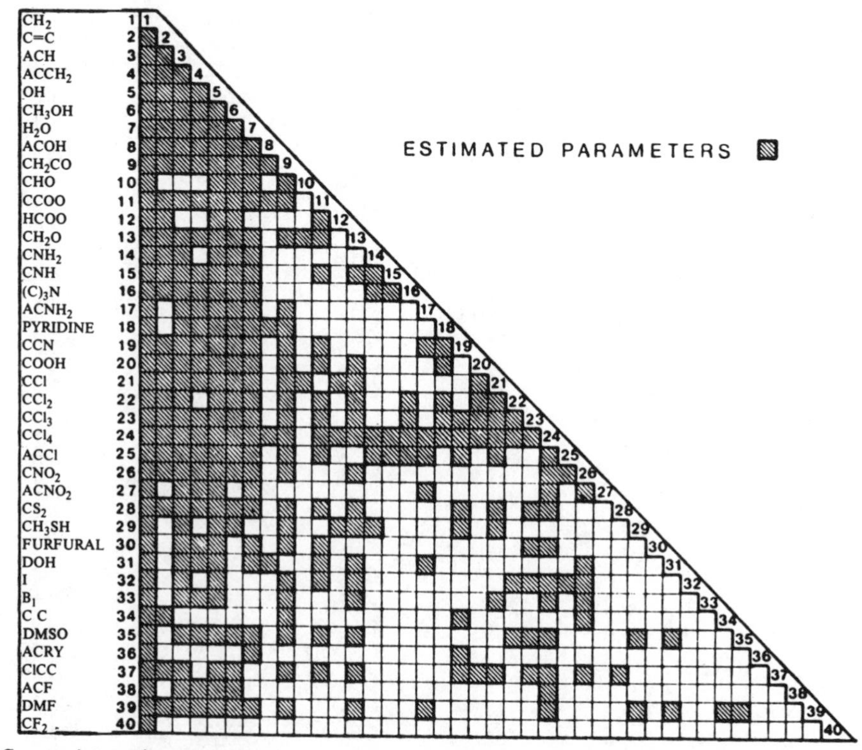

UNIFAC group-interaction parameters.

Table E.13. UNIFAC Group Volume and Surface-Area Parameters for Prediction of Liquid-Liquid Equilibria (Magnussen, Rasmussen, & Fredenslund 1981). (Copyright American Chemical Society; used with permission.)

Main Group	*Subgroup*	*No.*	R_k	Q_k	*Sample Group Assignment*
1 "CH_2"	CH_3	1	0.9011	0.848	butane: 2 CH_3, 2 CH_2
	CH_2	2	0.6744	0.540	
	CH	3	0.4469	0.228	2-methylpropane: 3 CH_3, 1 CH
	C	4	0.2195	0.000	2,2-dimethylpropane: 4 CH_3, 1 C
2 "C=C"	CH_2=CH	5	1.3454	1.176	1-hexene: 1 CH_3, 3 CH_2, 1 CH_2=CH
	CH=CH	6	1.1167	0.867	2-hexene: 2 CH_3, 2 CH_2, 1 CH=CH
	CH=C	7	0.8886	0.676	2-methyl-2-butene: 3 CH_3, 1 CH=C
	CH_2=C	8	1.1173	0.988	2-methyl-1-butene: 2 CH_3, 1 CH_2, 1 CH_2=C
3 "ACH"	ACH	9	0.5313	0.400	benzene: 6 ACH
	AC	10	0.3652	0.120	styrene: 1 CH_2=CH, 5 ACH, 1 AC
4 "$ACCH_2$"	$ACCH_3$	11	1.2663	0.968	toluene: 5 ACH, 1 $ACCH_3$
	$ACCH_2$	12	1.0396	0.660	ethylbenzene: 1 CH_3, 5 ACH, 1 $ACCH_2$
	ACCH	13	0.8121	0.348	cumene: 2 CH_3, 5 ACH, 1 ACCH
5 "OH"	OH	14	1.0000	1.200	2-butanol: 2 CH_3, 1 CH_2, 1 CH, 1 OH
6	P1	15	3.2499	3.128	1-propanol: 1 P1
7	P2	16	3.2491	3.124	2-propanol: 1 P2
8	H_2O	17	0.92	1.40	water: 1 H_2O
9	ACOH	18	0.8952	0.680	phenol: 5 ACH, 1 ACOH
10 "CH_2CO"	CH_3CO	19	1.6724	1.488	ketone group is 2nd carbon; 2-butanone: 1 CH_3, 1 CH_2, 1 CH_3CO
	CH_2CO	20	1.4457	1.180	ketone group is any other carbon: 3-pentanone: 2 CH_3, 1 CH_2, 1 CH_2CO
11	CHO	21	0.9980	0.948	acetaldehyde: 1 CH_3, 1 CHO
12	furfural	22	3.168	2.484	furfural: 1 furfural
13 "COOH"	COOH	23	1.3013	1.224	acetic acid: 1 CH_3, 1 COOH
	HCOOH	24	1.5280	1.532	formic acid: 1 HCOOH
14 "COOC"	CH_3COO	25	1.9031	1.728	butyl acetate: 1 CH_3, 3 CH_2, 1 CH_3COO
	CH_2COO	26	1.6764	1.420	butyl propanoate: 2 CH_3, 3 CH_2, 1 CH_2COO
15 "CH_2O"	CH_3O	27	1.1450	1.088	dimethyl ether: 1 CH_3, 1 CH_3O
	CH_2O	28	0.9183	0.780	diethyl ether: 2 CH_3, 1 CH_2, 1 CH_2O
	CH—O	29	0.6908	0.468	diisopropyl ether: 4 CH_3, 1 CH, 1 CH—O
	FCH_2O	30	0.9183	1.1	tetrahydrofuran: 3 CH_2, 1 FCH_2O
16 "CCl"	CH_2Cl	31	1.4654	1.264	1-chlorobutane: 1 CH_3, 2 CH_2, 1 CH_2Cl
	CHCl	32	1.2380	0.952	2-chloropropane: 2 CH_3, 1 CHCl
	CCl	33	1.0060	0.724	2-chloro-2-methylpropane: 3 CH_3, 1 CCl
17 "CCl_2"	CH_2Cl_2	34	2.2564	1.988	dichloromethane: 1 CH_2Cl_2
	$CHCl_2$	35	2.0606	1.684	1,1-dichloroethane: 1 CH_3, 1 $CHCl_2$
	CCl_2	36	1.8016	1.448	2,2-dichloropropane: 2 CH_3, 1 CCl_2
18 "CCl_3"	$CHCl_3$	37	2.8700	2.410	chloroform: 1 $CHCl_3$
	CCl_3	38	2.6401	2.184	1,1,1-trichloroethane: 1 CH_3, 1 CCl_3
19	CCl_4	39	3.3900	2.910	tetrachloromethane: 1 CCl_4
20	ACCl	40	1.1562	0.844	chlorobenzene: 5 ACH, 1 ACCl
21 "CCN"	CH_3CN	41	1.8701	1.724	acetonitrile: 1 CH_3CN
	CH_2CN	42	1.6434	1.416	propionitrile: 1 CH_3, 1 CH_2CN
22	$ACNH_2$	43	1.0600	0.816	aniline: 5 ACH, 1 $ACNH_2$
23 "CNO_2"	CH_3NO_2	44	2.0086	1.868	nitromethane: 1 CH_3NO_2
	CH_2NO_2	45	1.7818	1.560	1-nitropropane: 1 CH_3, 1 CH_2, 1 CH_2NO_2
	$CHNO_2$	46	1.5544	1.248	2-nitropropane: 2 CH_3, 1 $CHNO_2$
24	$ACNO_2$	47	1.4199	1.104	nitrobenzene: 5 ACH, 1 $ACNO_2$
25 "DOH"	$(CH_2OH)_2$	48	2.4088	2.248	1,2-ethanediol: 1 $(CH_2OH)_2$
26 "DEOH"	$(HOCH_2CH_2)_2O$	49	4.0013	3.568	diethylene glycol: 1 $(HOCH_2CH_2)_2O$
27 "pyridine"	C_5H_5N	50	2.9993	2.113	pyridine: 1 C_5H_5N
	C_5H_4N	51	2.8332	1.833	3-methylpyridine: 1 CH_3, 1 C_5H_4N
	C_5H_3N	52	2.667	1.553	2,3-dimethylpyridine: 2 CH_3, 1 C_5H_3N
28 "TCE"	CCl_2=CHCl	53	3.3092	2.860	trichloroethylene: 1 CCl_2=CHCl
29 "MFA"	$HCONHCH_3$	54	2.4317	2.192	methylformamide: 1 $HCONHCH_3$
30 "DMFA"	$HCON(CH_3)_2$	55	3.0856	2.736	dimethylformamide: 1 $HCON(CH_3)_2$
31 "TMS"	$(CH_2)_4SO_2$	56	4.0358	3.20	tetramethylenesulfone: 1 $(CH_2)_4SO_2$
32 "DMSO"	$(CH_2)_2SO$	57	2.8266	2.472	dimethyl sulfoxide: 1 $(CH_2)_2SO$

Table E.14. UNIFAC Group-Interaction Parameters for Prediction of Liquid-Liquid Equilibria (Superscript 1 = *P1* or *P2* estimated from "generated" data; superscript 2 = parameter based on VLE data (Magnussen, Rasmussen, & Fredenslund 1981). (Copyright American Chemical Society; used with permission.)

	1 CH_2	2 $C{=}C$	3 ACH	4 $ACCH_2$	5 OH	6 $P1$	7 $P2$	8 H_2O
1 CH_2	0	74.54	−114.8	−115.7	644.6	329.6	310.7	1300.0
2 C=C	292.3	0	340.7[2]	4102.0[2]	724.4	1731.0	1731.0	896.0
3 ACH	156.5	−94.78[2]	0	167.0[2]	703.9	511.5	577.3	859.4
4 $ACCH_2$	104.4	−269.7[2]	−146.8[2]	0	4000.0	136.6	906.8	5695.0
5 OH	328.2	470.7	−9.210	1.270	0	937.3	991.3	28.73
6 P1	−136.7	−135.7	−223.0	−162.6	−281.1	0	0	−61.29
7 P2	−131.9	−135.7	−252.0	−273.6	−268.8	0	0	5.890
8 H_2O	342.4	220.6	372.8	203.7	−122.4	247.0	104.9	0
9 ACOH	−159.8		−473.2	−470.4	−63.15	−547.2	−547.2	−595.9
10 CH_2CO	66.56	306.1	−78.31	−73.87	216.0	401.7[1]	−127.6[1]	634.8
11 CHO	146.1	517.0	−75.30	223.2	−431.3	643.4[1]	231.4[1]	623.7
12 FURF	14.78		−10.44	−184.9	444.7	−94.64	732.3	211.6
13 COOH	1744.0	−48.52	75.49	147.3	118.4	728.7[1]	349.1[1]	652.3
14 COOC	−320.1	485.6	114.8[2]	−170.0[2]	180.6	−76.64	−152.8	385.9
15 CH_2O	1571.0	76.44[2]	52.13[2]	65.69[2]	137.1	−218.1	−218.1	212.8
16 CCl	73.80	−24.36[2]	4.680[2]	122.9[2]	455.1	351.5	351.5	770.0
17 CCl_2	27.90	−52.71[2]			669.2	−186.1[1]	−401.6[1]	740.4
18 CCl_3	21.23	−185.1[2]	288.5[2]	33.61[2]	418.4	−465.7	−465.7	793.2
19 CCl_4	89.97	−293.70[2]	−4.70[2]	134.7[2]	713.5	−260.3	512.2	1205.0
20 ACCl	−59.06		777.8	−47.13	1989.0			390.7
21 CCN	29.08	34.78[2]	56.41	−53.29	2011.0			63.48
22 $ACNH_2$	175.8		−218.9	−15.41	529.0[2]			−239.8
23 CNO_2	94.34	375.4	113.6	−97.05	483.8	264.7	264.7	13.32
24 $ACNO_2$	193.6		7.180	−127.1	332.6			439.9
25 DOH	108.5		247.3	453.4	−289.3			−424.3
26 DEOH	81.49		−50.71	−30.28	−99.56			
27 PYR	−128.8		−225.3	−124.6	−319.2			203.0
28 TCE	147.3				837.9			1153.0
29 MFA	−11.91	176.7	−80.48					−311.0
30 DMFA	14.91	132.1	−17.78					−262.6
31 TMS	67.84	42.73	59.16	26.59				1.110
32 DMSO	36.42	60.82	29.77	55.97				

Table E.14. *(continued)*

	9 *ACOH*	10 *CH_2CO*	11 *CHO*	12 *FURF*	13 *COOH*	14 *COOC*	15 *CH_2O*	16 *CCl*
1 CH_2	2255.0	472.6	158.1	383.0	139.4	972.4	662.1	42.14
2 C=C		343.7	−214.7		1647.0	−577.5	289.3[2]	99.61[2]
3 ACH	1649.0	593.7	362.3	31.14	461.8	6.0[2]	32.14[2]	−18.81[2]
4 $ACCH_2$	292.6	916.7	1218.0	715.6	339.1	5688.0[2]	213.1[2]	−114.1[2]
5 OH	−195.5	67.07	1409.0	−140.3	−104.0	195.6	262.5	62.05
6 P1	−153.2	−47.41[1]	−344.1[1]	299.3	244.4[1]	19.57	1970.0	−166.4
7 P2	−153.2	353.8[1]	−338.6[1]	−241.8	−57.98[1]	487.1	1970.0	−166.4
8 H_2O	344.5	−171.8	−349.9	66.95	−465.7	−6.320	64.42	315.9
9 ACOH	0	−825.7				−898.3		
10 CH_2CO	−568.0	0	−37.36[2]	120.3	1247.0	258.7	5.202[2]	1000.0
11 CHO		128.0[2]	0	1724.0	0.750	−245.8		751.8[2]
12 FURF		48.93	−311.6	0	1919.0	57.70		
13 COOH		−101.3	1051.0	−115.7	0	−117.6	−96.62	19.77
14 COOC	−337.3	58.84	1090.0	−46.13	1417.0	0	−235.7[2]	
15 CH_2O		52.38[2]			1402.0	461.3[2]	0	301.1[2]
16 CCl		483.9	−47.51[2]		337.1		225.4[2]	0
17 CCl_2		550.6		808.8	437.7	−132.9[2]	−197.7[2]	−21.35
18 CCl_3		342.2		203.1	370.4	176.5[2]	−20.93[2]	−157.1[2]
19 CCl_4	1616.0[2]	550.0		70.14	438.1	129.5[2]	113.9[2]	11.80[2]
20 AACl		190.5			1349.0	−246.3[2]		
21 CCN		−349.2				2.410		
22 $ACNH_2$	−860.3	857.7			681.4			
23 CNO_2		377.0			152.4		−94.49[2]	
24 $ACNO_2$	−230.4	211.6						
25 DOH	523.0[2]	82.77		−75.23	−1707.0	29.86		
26 DEOH								
27 PYR	−222.7[2]			−201.9				
28 TCE		417.4		123.2	639.7			
29 MFA								
30 DMFA				−281.9				
31 TMS								
32 DMSO								

	17 *CCl_2*	18 *CCl_3*	19 *CCl_4*	20 *ACCl*	21 *CCN*	22 *$ACNH_2$*	23 *CNO_2*	24 *$ACNO_2$*
1 CH_2	−243.9	7.5	−5.550	924.8	696.8	902.2	556.7	575.7
2 C=C	337.1[2]	4583.0[2]	5831.0[2]		405.9[2]		425.7	
3 ACH		−231.9[2]	3000.0[2]	−878.1	29.13	1.640	−1.770	−11.19
4 $ACCH_2$		−12.14[2]	−141.3[2]	−107.3	1208.0	689.6	3629.0[2]	−175.6
5 OH	272.2	−61.57	−41.75	−597.1	−189.3	−348.2[2]	−30.70	−159.0
6 P1	128.6[1]	1544.0	224.6				150.8	
7 P2	507.8[1]	1544.0	−207.0				150.8	
8 H_2O	370.7	356.8	502.9	−97.27	198.3	−109.8	1539.0	32.92
9 ACOH			4894.0[2]			−851.6		−16.13
10 CH_2CO	−301.0	12.01	−10.88	902.6	430.6	1010.0	400.0	−328.6
11 CHO								
12 FURF	−347.9	−249.3	61.59					
13 COOH	1670.0	48.15	43.83	874.3		942.2	446.3	
14 COOC	108.9[2]	−209.7[2]	54.57[2]	629.0[2]	−149.2			
15 CH_2O	137.8[2]	−154.3[2]	47.67[2]				95.18[2]	
16 CCl	110.5	249.2[2]	62.42[2]					
17 CCl_2	0		56.33[2]					
18 CCl_3		0	−30.10[2]		70.04[2]	−75.50		
19 CCl_4	17.97[2]	51.90[2]	0	475.8[2]	492.0[2]	1302.0[2]	490.9[2]	534.7[2]
20 ACCl			−255.4[2]	0	346.2		−154.5[2]	
21 CCN		−15.62[2]	−54.86[2]	−465.2	0			
22 $ACNH_2$		−216.3	8455.0[2]			0		179.9
23 CNO_2			−34.68[2]	794.4[2]			0	

Table E.14. *(continued)*

	17 CCl_2	18 CCl_3	19 CCl_4	20 ACCl	21 CCN	22 $ACNH_2$	23 CNO_2	24 $ACNO_2$
24 $ACNO_2$			514.6[2]			175.8		0
25 DOH				−241.7		164.4[2]	481.3[2]	−246.0
26 DEOH								
27 PYR		−114.7[2]		−906.5	−169.7[2]	−944.9		
28 TCE								
29 MFA								
30 DMFA								
31 TMS								
32 DMSO								

	25 DOH	26 DEOH	27 PYR	28 TCE	29 MFA	30 DMFA	31 TMS	32 DMSO
1 CH_2	527.5	269.2	−300.0	−63.6	928.3	331.0	561.4	956.5
2 C=C					500.7	115.4	784.4	265.4
3 ACH	358.9	363.5	−578.2		364.2	−58.10	21.97	84.16
4 $ACCH_2$	337.7	1023.0	−390.7				238.0	132.2
5 OH	536.6	53.37	183.3	−44.44				
6 P1								
7 P2								
8 H_2O	−269.2		−873.6	1429.0	−364.2	−117.4	18.41	
9 ACOH	−538.6		−637.3[2]					
10 CH_2CO	211.6			148.0				
11 CHO								
12 FURF	−278.2		−208.4	−13.91		173.8		
13 COOH	572.7			−2.160				
14 COOC	343.1							
15 CH_2O								
16 CCl								
17 CCl_2								
18 CCl_3			18.98[2]					
19 CCl_4								
20 ACCl	124.8		−387.7					
21 CCN			134.3					
22 $ACNH_2$	125.3[2]		924.5					
23 CNO_2	139.8[2]							
24 $ACNO_2$	963.0							
25 DOH	0							
26 DEOH		0						
27 PYR			0					
28 TCE				0				
29 MFA					0			
30 DMFA						0		
31 TMS							0	
32 DMSO								0

Table E.15. Values of $\left[\dfrac{H^0 - H}{RT_c}\right]^{(0)}$ (Lee-Kesler)

	PR														
TR	*0.010*	*0.050*	*0.100*	*0.200*	*0.400*	*0.600*	*0.800*	*1.000*	*1.200*	*1.500*	*2.000*	*3.000*	*5.000*	*7.000*	*10.000*
0.30	6.045	6.043	6.040	6.034	6.022	6.011	5.999	5.987	5.975	5.957	5.927	5.868	5.748	5.628	5.446
0.35	5.906	5.904	5.901	5.895	5.882	5.870	5.858	5.845	5.833	5.814	5.783	5.721	5.595	5.469	5.278
0.40	5.763	5.761	5.757	5.751	5.738	5.726	5.713	5.700	5.687	5.668	5.636	5.572	5.442	5.311	5.113
0.45	5.615	5.612	5.609	5.603	5.590	5.577	5.564	5.551	5.538	5.519	5.486	5.421	5.288	5.154	4.950
0.50	5.465	5.463	5.459	5.453	5.440	5.427	5.414	5.401	5.388	5.369	5.336	5.270	5.135	4.999	4.791
0.55	0.032	5.312	5.309	5.303	5.290	5.278	5.265	5.252	5.239	5.220	5.187	5.121	4.986	4.849	4.638
0.60	0.027	5.162	5.159	5.153	5.141	5.129	5.116	5.104	5.091	5.073	5.041	4.976	4.842	4.704	4.492
0.65	0.023	0.118	5.008	5.002	4.991	4.980	4.968	4.956	4.945	4.927	4.896	4.833	4.702	4.565	4.353
0.70	0.020	0.101	0.213	4.848	4.838	4.828	4.818	4.808	4.797	4.781	4.752	4.693	4.566	4.432	4.221
0.75	0.017	0.088	0.183	4.687	4.679	4.672	4.664	4.655	4.646	4.632	4.607	4.554	4.434	4.303	4.095
0.80	0.015	0.078	0.160	0.345	4.507	4.504	4.499	4.494	4.488	4.478	4.459	4.413	4.303	4.178	3.974
0.85	0.014	0.069	0.141	0.300	4.309	4.313	4.316	4.316	4.316	4.312	4.302	4.269	4.173	4.056	3.857
0.90	0.012	0.062	0.126	0.264	0.596	4.074	4.094	4.108	4.118	4.127	4.132	4.119	4.043	3.935	3.744
0.93	0.011	0.058	0.118	0.246	0.545	0.960	3.920	3.953	3.976	4.000	4.020	4.024	3.963	3.863	3.678
0.95	0.011	0.056	0.113	0.235	0.516	0.885	3.763	3.825	3.865	3.904	3.940	3.958	3.910	3.815	3.634
0.97	0.011	0.054	0.109	0.225	0.490	0.824	1.356	3.658	3.732	3.796	3.853	3.890	3.856	3.767	3.591
0.98	0.010	0.053	0.107	0.221	0.478	0.797	1.273	3.544	3.652	3.736	3.806	3.854	3.829	3.743	3.569
0.99	0.010	0.052	0.105	0.216	0.466	0.773	1.206	3.376	3.558	3.670	3.758	3.818	3.801	3.719	3.548
1.00	0.010	0.051	0.103	0.212	0.455	0.750	1.151	2.584	3.441	3.598	3.706	3.782	3.774	3.695	3.526
1.01	0.010	0.050	0.101	0.208	0.445	0.728	1.102	1.796	3.283	3.516	3.652	3.744	3.746	3.671	3.505

Table E.15 *(continued)*

TR	*PR*														
	0.010	*0.050*	*0.100*	*0.200*	*0.400*	*0.600*	*0.800*	*1.000*	*1.200*	*1.500*	*2.000*	*3.000*	*5.000*	*7.000*	*10.000*
1.02	0.010	0.049	0.099	0.203	0.434	0.708	1.060	1.627	3.039	3.422	3.595	3.705	3.718	3.647	3.484
1.05	0.009	0.046	0.094	0.192	0.407	0.654	0.955	1.359	2.034	3.030	3.398	3.583	3.632	3.575	3.420
1.10	0.008	0.042	0.086	0.175	0.367	0.581	0.827	1.120	1.487	2.203	2.965	3.353	3.484	3.453	3.315
1.15	0.008	0.039	0.079	0.160	0.334	0.523	0.732	0.968	1.239	1.719	2.479	3.091	3.329	3.329	3.211
1.20	0.007	0.036	0.073	0.148	0.305	0.474	0.657	0.857	1.076	1.443	2.079	2.807	3.166	3.202	3.107
1.30	0.006	0.031	0.063	0.127	0.259	0.399	0.545	0.698	0.860	1.116	1.560	2.274	2.825	2.942	2.899
1.40	0.005	0.027	0.055	0.110	0.224	0.341	0.463	0.588	0.716	0.915	1.253	1.857	2.486	2.679	2.692
1.50	0.005	0.024	0.048	0.097	0.196	0.297	0.400	0.505	0.611	0.774	1.046	1.549	2.175	2.421	2.486
1.60	0.004	0.021	0.043	0.086	0.173	0.261	0.350	0.440	0.531	0.667	0.894	1.318	1.904	2.177	2.285
1.70	0.004	0.019	0.038	0.076	0.153	0.231	0.309	0.387	0.466	0.583	0.777	1.139	1.672	1.953	2.091
1.80	0.003	0.017	0.034	0.068	0.137	0.206	0.275	0.344	0.413	0.515	0.683	0.996	1.476	1.751	1.908
1.90	0.003	0.015	0.031	0.062	0.123	0.185	0.246	0.307	0.368	0.458	0.606	0.880	1.309	1.571	1.736
2.00	0.003	0.014	0.028	0.056	0.111	0.167	0.222	0.276	0.330	0.411	0.541	0.782	1.167	1.411	1.577
2.20	0.002	0.012	0.023	0.046	0.092	0.137	0.182	0.226	0.269	0.334	0.437	0.629	0.937	1.143	1.295
2.40	0.002	0.010	0.019	0.038	0.076	0.114	0.150	0.187	0.222	0.275	0.359	0.513	0.761	0.929	1.058
2.60	0.002	0.008	0.016	0.032	0.064	0.095	0.125	0.155	0.185	0.228	0.297	0.422	0.621	0.756	0.858
2.80	0.001	0.007	0.014	0.027	0.054	0.080	0.105	0.130	0.154	0.190	0.246	0.348	0.508	0.614	0.689
3.00	0.001	0.006	0.011	0.023	0.045	0.067	0.088	0.109	0.129	0.159	0.205	0.288	0.415	0.495	0.545
3.50	0.001	0.004	0.007	0.015	0.029	0.043	0.056	0.069	0.081	0.099	0.127	0.174	0.239	0.270	0.264
4.00	0.000	0.002	0.005	0.009	0.017	0.026	0.033	0.041	0.048	0.058	0.072	0.095	0.116	0.110	0.061

Table E.16. Values of $\left[\frac{H^0 - H}{RT_c}\right]^{(1)}$ (Lee-Kesler)

	PR														
TR	*0.010*	*0.050*	*0.100*	*0.200*	*0.400*	*0.600*	*0.800*	*1.000*	*1.200*	*1.500*	*2.000*	*3.000*	*5.000*	*7.000*	*10.000*
0.30	11.098	11.096	11.095	11.091	11.083	11.076	11.069	11.062	11.055	11.044	11.027	10.992	10.935	10.872	10.781
0.35	10.656	10.655	10.654	10.653	10.650	10.646	10.643	10.640	10.637	10.632	10.624	10.609	10.581	10.554	10.529
0.40	10.121	10.121	10.121	10.120	10.121	10.121	10.121	10.121	10.121	10.121	10.122	10.123	10.128	10.135	10.150
0.45	9.515	9.515	9.516	9.517	9.519	9.521	9.523	9.525	9.527	9.531	9.537	9.549	9.576	9.611	9.663
0.50	8.868	8.869	8.870	8.872	8.876	8.880	8.884	8.888	8.892	8.899	8.909	8.932	8.978	9.030	9.111
0.55	0.080	8.211	8.212	8.215	8.221	8.226	8.232	8.238	8.243	8.252	8.267	8.298	8.360	8.425	8.531
0.60	0.059	7.568	7.570	7.573	7.579	7.585	7.591	7.596	7.603	7.614	7.632	7.669	7.745	7.824	7.950
0.65	0.045	0.247	6.949	6.952	6.959	6.966	6.973	6.980	6.987	6.997	7.017	7.059	7.147	7.239	7.381
0.70	0.034	0.185	0.415	6.360	6.367	6.373	6.381	6.388	6.395	6.407	6.429	6.475	6.574	6.677	6.837
0.75	0.027	0.142	0.306	5.796	5.802	5.809	5.816	5.824	5.832	5.845	5.868	5.918	6.027	6.142	6.318
0.80	0.021	0.110	0.234	0.542	5.266	5.271	5.278	5.285	5.293	5.306	5.330	5.385	5.506	5.632	5.824
0.85	0.017	0.087	0.182	0.401	4.753	4.754	4.758	4.763	4.771	4.784	4.810	4.872	5.008	5.149	5.358
0.90	0.014	0.070	0.144	0.308	0.751	4.254	4.248	4.249	4.255	4.268	4.298	4.371	4.530	4.688	4.916
0.93	0.012	0.061	0.126	0.265	0.612	1.236	3.942	3.934	3.937	3.951	3.987	4.073	4.251	4.422	4.662
0.95	0.011	0.056	0.115	0.241	0.542	0.994	3.737	3.712	3.713	3.730	3.773	3.873	4.068	4.248	4.497
0.97	0.010	0.052	0.105	0.219	0.483	0.837	1.616	3.470	3.467	3.492	3.551	3.670	3.885	4.077	4.336
0.98	0.010	0.050	0.101	0.209	0.457	0.776	1.324	3.332	3.327	3.363	3.434	3.568	3.795	3.992	4.257
0.99	0.009	0.048	0.097	0.200	0.433	0.722	1.154	3.164	3.164	3.223	3.313	3.464	3.705	3.909	4.178
1.00	0.009	0.046	0.093	0.191	0.410	0.675	1.034	2.471	2.952	3.065	3.186	3.358	3.615	3.825	4.100
1.01	0.009	0.044	0.089	0.183	0.389	0.632	0.940	1.375	2.595	2.880	3.051	3.251	3.525	3.742	4.023

Table E.16 *(continued)*

TR	*PR*														
	0.010	*0.050*	*0.100*	*0.200*	*0.400*	*0.600*	*0.800*	*1.000*	*1.200*	*1.500*	*2.000*	*3.000*	*5.000*	*7.000*	*10.000*
1.02	0.008	0.042	0.085	0.175	0.370	0.594	0.863	1.180	1.723	2.650	2.906	3.142	3.435	3.661	3.947
1.05	0.007	0.037	0.075	0.153	0.318	0.498	0.691	0.877	0.878	1.496	2.381	2.800	3.167	3.418	3.722
1.10	0.006	0.030	0.061	0.123	0.251	0.381	0.507	0.617	0.673	0.617	1.261	2.167	2.720	3.023	3.362
1.15	0.005	0.025	0.050	0.099	0.199	0.296	0.385	0.459	0.503	0.487	0.604	1.497	2.275	2.641	3.019
1.20	0.004	0.020	0.040	0.080	0.158	0.232	0.297	0.349	0.381	0.381	0.361	0.934	1.840	2.273	2.692
1.30	0.003	0.013	0.026	0.052	0.100	0.142	0.177	0.203	0.218	0.218	0.178	0.300	1.066	1.592	2.086
1.40	0.002	0.008	0.016	0.032	0.060	0.083	0.100	0.111	0.115	0.108	0.070	0.044	0.504	1.012	1.547
1.50	0.001	0.005	0.009	0.018	0.032	0.042	0.048	0.049	0.046	0.032	−0.008	−0.078	0.142	0.556	1.080
1.60	0.000	0.002	0.004	0.007	0.012	0.013	0.011	0.005	−0.004	−0.023	−0.065	−0.151	−0.082	0.217	0.689
1.70	0.000	0.000	0.000	−0.000	−0.003	−0.009	−0.017	−0.027	−0.040	−0.063	−0.109	−0.202	−0.223	−0.028	0.369
1.80	−0.000	−0.001	−0.003	−0.006	−0.015	−0.025	−0.037	−0.051	−0.067	−0.094	−0.143	−0.241	−0.317	−0.203	0.112
1.90	−0.001	−0.003	−0.005	−0.011	−0.023	−0.037	−0.053	−0.070	−0.088	−0.117	−0.169	−0.271	−0.381	−0.330	−0.092
2.00	−0.001	−0.003	−0.007	−0.015	−0.030	−0.047	−0.065	−0.085	−0.105	−0.136	−0.190	−0.295	−0.428	−0.424	−0.255
2.20	−0.001	−0.005	−0.010	−0.020	−0.040	−0.062	−0.083	−0.106	−0.128	−0.163	−0.221	−0.331	−0.493	−0.551	−0.489
2.40	−0.001	−0.006	−0.012	−0.023	−0.047	−0.071	−0.095	−0.120	−0.144	−0.181	−0.242	−0.356	−0.535	−0.631	−0.645
2.60	−0.001	−0.006	−0.013	−0.026	−0.052	−0.078	−0.104	−0.130	−0.156	−0.194	−0.257	−0.376	−0.567	−0.687	−0.754
2.80	−0.001	−0.007	−0.014	−0.028	−0.055	−0.082	−0.110	−0.137	−0.164	−0.204	−0.269	−0.391	−0.591	−0.729	−0.836
3.00	−0.001	−0.007	−0.014	−0.029	−0.058	−0.086	−0.114	−0.142	−0.170	−0.211	−0.278	−0.403	−0.611	−0.763	−0.899
3.50	−0.002	−0.008	−0.016	−0.031	−0.062	−0.092	−0.122	−0.152	−0.181	−0.224	−0.294	−0.425	−0.650	−0.827	−1.015
4.00	−0.002	−0.008	−0.016	−0.032	−0.064	−0.096	−0.127	−0.158	−0.188	−0.233	−0.306	−0.442	−0.680	−0.874	−1.097

Note: This information is given in graphical form in the *API Data Book*, Figures 7B3.4 and 7B3.5.

Table E.17. Values of $\left[\frac{S^0 - S}{R}\right]^{(0)}$ (Lee-Kesler)

	PR														
TR	*0.010*	*0.050*	*0.100*	*0.200*	*0.400*	*0.600*	*0.800*	*1.000*	*1.200*	*1.500*	*2.000*	*3.000*	*5.000*	*7.000*	*10.000*
0.30	11.614	10.008	9.319	8.635	7.961	7.574	7.304	7.099	6.935	6.740	6.497	6.182	5.847	5.683	5.578
0.35	11.185	9.579	8.890	8.205	7.529	7.140	6.869	6.663	6.497	6.299	6.052	5.728	5.376	5.194	5.060
0.40	10.802	9.196	8.506	7.821	7.144	6.755	6.483	6.275	6.109	5.909	5.660	5.330	4.967	4.772	4.619
0.45	10.453	8.847	8.157	7.472	6.794	6.404	6.132	5.924	5.757	5.557	5.306	4.974	4.603	4.401	4.234
0.50	10.137	8.531	7.841	7.156	6.479	6.089	5.816	5.608	5.441	5.240	4.989	4.656	4.282	4.074	3.899
0.55	0.038	8.245	7.555	6.870	6.193	5.803	5.531	5.324	5.157	4.956	4.706	4.373	3.998	3.788	3.607
0.60	0.029	7.983	7.294	6.610	5.933	5.544	5.273	5.066	4.900	4.700	4.451	4.120	3.747	3.537	3.353
0.65	0.023	0.122	7.052	6.368	5.694	5.306	5.036	4.830	4.665	4.467	4.220	3.892	3.523	3.315	3.131
0.70	0.018	0.096	0.206	6.140	5.467	5.082	4.814	4.610	4.446	4.250	4.007	3.684	3.322	3.117	2.935
0.75	0.015	0.078	0.164	5.917	5.248	4.866	4.600	4.399	4.238	4.045	3.807	3.491	3.138	2.939	2.761
0.80	0.013	0.064	0.134	0.294	5.026	4.649	4.388	4.191	4.034	3.846	3.615	3.310	2.970	2.777	2.605
0.85	0.011	0.054	0.111	0.239	4.785	4.418	4.166	3.976	3.825	3.646	3.425	3.135	2.812	2.629	2.463
0.90	0.009	0.046	0.094	0.199	0.463	4.145	3.912	3.738	3.599	3.434	3.231	2.964	2.663	2.491	2.334
0.93	0.008	0.042	0.085	0.179	0.408	0.750	3.723	3.569	3.444	3.295	3.108	2.860	2.577	2.412	2.262
0.95	0.008	0.039	0.080	0.168	0.377	0.671	3.556	3.433	3.326	3.193	3.023	2.790	2.520	2.362	2.215
0.97	0.007	0.037	0.075	0.157	0.350	0.607	1.056	3.259	3.188	3.081	2.932	2.719	2.463	2.312	2.170
0.98	0.007	0.036	0.073	0.153	0.337	0.580	0.971	3.142	3.106	3.019	2.884	2.682	2.436	2.287	2.148
0.99	0.007	0.035	0.071	0.148	0.326	0.555	0.903	2.972	3.010	2.953	2.835	2.646	2.408	2.263	2.126
1.00	0.007	0.034	0.069	0.144	0.315	0.532	0.847	2.178	2.893	2.879	2.784	2.609	2.380	2.239	2.105
1.01	0.007	0.033	0.067	0.139	0.304	0.510	0.799	1.391	2.736	2.798	2.730	2.571	2.352	2.215	2.083

Table E.17 *(continued)*

TR	*PR*														
	0.010	*0.050*	*0.100*	*0.200*	*0.400*	*0.600*	*0.800*	*1.000*	*1.200*	*1.500*	*2.000*	*3.000*	*5.000*	*7.000*	*10.000*
1.02	0.006	0.032	0.065	0.135	0.294	0.491	0.757	1.225	2.495	2.706	2.673	2.533	2.325	2.191	2.062
1.05	0.006	0.030	0.060	0.124	0.267	0.439	0.656	0.965	1.523	2.328	2.483	2.415	2.242	2.121	2.001
1.10	0.005	0.026	0.053	0.108	0.230	0.371	0.537	0.742	1.012	1.557	2.081	2.202	2.104	2.007	1.903
1.15	0.005	0.023	0.047	0.096	0.201	0.319	0.452	0.607	0.790	1.126	1.649	1.968	1.966	1.897	1.810
1.20	0.004	0.021	0.042	0.085	0.177	0.277	0.389	0.512	0.651	0.890	1.308	1.727	1.827	1.789	1.722
1.30	0.003	0.017	0.033	0.068	0.140	0.217	0.298	0.385	0.478	0.628	0.891	1.299	1.554	1.581	1.556
1.40	0.003	0.014	0.027	0.056	0.114	0.174	0.237	0.303	0.372	0.478	0.663	0.990	1.303	1.386	1.402
1.50	0.002	0.011	0.023	0.046	0.094	0.143	0.194	0.246	0.299	0.381	0.520	0.777	1.088	1.208	1.260
1.60	0.002	0.010	0.019	0.039	0.079	0.120	0.162	0.204	0.247	0.312	0.421	0.628	0.913	1.050	1.130
1.70	0.002	0.008	0.017	0.033	0.067	0.102	0.137	0.172	0.208	0.261	0.350	0.519	0.773	0.915	1.013
1.80	0.001	0.007	0.014	0.029	0.058	0.088	0.117	0.147	0.177	0.222	0.296	0.438	0.661	0.799	0.908
1.90	0.001	0.006	0.013	0.025	0.051	0.076	0.102	0.127	0.153	0.191	0.255	0.375	0.570	0.702	0.815
2.00	0.001	0.006	0.011	0.022	0.044	0.067	0.089	0.111	0.134	0.167	0.221	0.325	0.497	0.620	0.733
2.20	0.001	0.004	0.009	0.018	0.035	0.053	0.070	0.087	0.105	0.130	0.172	0.251	0.388	0.492	0.599
2.40	0.001	0.004	0.007	0.014	0.028	0.042	0.056	0.070	0.084	0.104	0.138	0.201	0.311	0.399	0.496
2.60	0.001	0.003	0.006	0.012	0.023	0.035	0.046	0.058	0.069	0.086	0.113	0.164	0.255	0.329	0.416
2.80	0.000	0.002	0.005	0.010	0.020	0.029	0.039	0.048	0.058	0.072	0.094	0.137	0.213	0.277	0.353
3.00	0.000	0.002	0.004	0.008	0.017	0.025	0.033	0.041	0.049	0.061	0.080	0.116	0.181	0.236	0.303
3.50	0.000	0.001	0.003	0.006	0.012	0.017	0.023	0.029	0.034	0.042	0.056	0.081	0.126	0.166	0.216
4.00	0.000	0.001	0.002	0.004	0.009	0.013	0.017	0.021	0.025	0.031	0.041	0.059	0.093	0.123	0.162

Table E.18. Values of $\left[\frac{S^0 - S}{R}\right]^{(1)}$ (Lee-Kesler)

TR	*PR* 0.010	0.050	0.100	0.200	0.400	0.600	0.800	1.000	1.200	1.500	2.000	3.000	5.000	7.000	10.000
0.30	16.782	16.774	16.764	16.744	16.705	16.665	16.626	16.586	16.547	16.488	16.390	16.195	15.837	15.468	14.925
0.35	15.413	15.408	15.401	15.387	15.359	15.333	15.305	15.278	15.251	15.211	15.144	15.011	14.751	14.496	14.153
0.40	13.990	13.986	13.981	13.972	13.953	13.934	13.915	13.896	13.877	13.849	13.803	13.714	13.541	13.376	13.144
0.45	12.564	12.561	12.558	12.551	12.537	12.523	12.509	12.496	12.482	12.462	12.430	12.367	12.248	12.145	11.999
0.50	11.202	11.200	11.197	11.192	11.182	11.172	11.162	11.153	11.143	11.129	11.107	11.063	10.985	10.920	10.836
0.55	0.115	9.948	9.946	9.942	9.935	9.928	9.921	9.914	9.907	9.897	9.882	9.853	9.806	9.769	9.732
0.60	0.078	8.828	8.826	8.823	8.817	8.811	8.806	8.799	8.794	8.787	8.777	8.760	8.736	8.723	8.720
0.65	0.055	0.309	7.832	7.829	7.824	7.819	7.815	7.810	7.807	7.801	7.794	7.784	7.779	7.785	7.811
0.70	0.040	0.216	0.491	6.951	6.945	6.941	6.937	6.933	6.930	6.926	6.922	6.919	6.929	6.952	7.002
0.75	0.029	0.156	0.340	6.173	6.167	6.162	6.158	6.155	6.152	6.149	6.147	6.149	6.174	6.213	6.285
0.80	0.022	0.116	0.246	0.578	5.475	5.468	5.462	5.458	5.455	5.453	5.452	5.461	5.501	5.555	5.648
0.85	0.017	0.088	0.183	0.408	4.853	4.841	4.832	4.826	4.822	4.820	4.822	4.839	4.898	4.969	5.082
0.90	0.013	0.068	0.140	0.301	0.744	4.269	4.249	4.238	4.232	4.230	4.236	4.267	4.351	4.442	4.578
0.93	0.011	0.058	0.120	0.254	0.593	1.219	3.914	3.894	3.885	3.884	3.896	3.941	4.046	4.151	4.300
0.95	0.010	0.053	0.109	0.228	0.517	0.961	3.697	3.658	3.647	3.648	3.669	3.728	3.851	3.966	4.125
0.97	0.010	0.048	0.099	0.206	0.456	0.797	1.570	3.406	3.391	3.401	3.437	3.517	3.661	3.788	3.957
0.98	0.009	0.046	0.094	0.196	0.429	0.734	1.270	3.264	3.247	3.268	3.318	3.412	3.569	3.701	3.875
0.99	0.009	0.044	0.090	0.186	0.405	0.680	1.098	3.093	3.082	3.126	3.195	3.306	3.477	3.616	3.796
1.00	0.008	0.042	0.086	0.177	0.382	0.632	0.977	2.399	2.868	2.967	3.067	3.200	3.387	3.532	3.717
1.01	0.008	0.040	0.082	0.169	0.361	0.590	0.883	1.306	2.513	2.784	2.933	3.094	3.297	3.450	3.640

Table E.18 *(continued)*

TR	*PR*														
	0.010	*0.050*	*0.100*	*0.200*	*0.400*	*0.600*	*0.800*	*1.000*	*1.200*	*1.500*	*2.000*	*3.000*	*5.000*	*7.000*	*10.000*
1.02	0.008	0.039	0.078	0.161	0.342	0.552	0.807	1.113	1.655	2.557	2.790	2.986	3.209	3.369	3.565
1.05	0.007	0.034	0.069	0.140	0.292	0.460	0.642	0.820	0.831	1.443	2.283	2.655	2.949	3.134	3.348
1.10	0.005	0.028	0.055	0.112	0.229	0.350	0.470	0.577	0.640	0.618	1.241	2.067	2.534	2.767	3.013
0.15	0.005	0.023	0.045	0.091	0.183	0.275	0.361	0.437	0.489	0.502	0.654	1.471	2.138	2.428	2.708
1.20	0.004	0.019	0.037	0.075	0.149	0.220	0.286	0.343	0.385	0.412	0.447	0.991	1.767	2.115	2.430
1.30	0.003	0.013	0.026	0.052	0.102	0.148	0.190	0.226	0.254	0.282	0.300	0.481	1.147	1.569	1.944
1.40	0.002	0.010	0.019	0.037	0.072	0.104	0.133	0.158	0.178	0.200	0.220	0.290	0.730	1.138	1.544
1.50	0.001	0.007	0.014	0.027	0.053	0.076	0.097	0.115	0.130	0.147	0.166	0.206	0.479	0.823	1.222
1.60	0.001	0.005	0.011	0.021	0.040	0.057	0.073	0.086	0.098	0.112	0.129	0.159	0.334	0.604	0.969
1.70	0.001	0.004	0.008	0.016	0.031	0.044	0.056	0.067	0.076	0.087	0.102	0.127	0.248	0.456	0.775
1.80	0.001	0.003	0.006	0.013	0.024	0.035	0.044	0.053	0.060	0.070	0.083	0.105	0.195	0.355	0.628
1.90	0.001	0.003	0.005	0.010	0.019	0.028	0.036	0.043	0.049	0.057	0.069	0.089	0.160	0.286	0.518
2.00	0.000	0.002	0.004	0.008	0.016	0.023	0.029	0.035	0.040	0.048	0.058	0.077	0.136	0.238	0.434
2.20	0.000	0.001	0.003	0.006	0.011	0.016	0.021	0.025	0.029	0.035	0.043	0.060	0.105	0.178	0.322
2.40	0.000	0.001	0.002	0.004	0.008	0.012	0.015	0.019	0.022	0.027	0.034	0.048	0.086	0.143	0.254
2.60	0.000	0.001	0.002	0.003	0.006	0.009	0.012	0.015	0.018	0.021	0.028	0.041	0.074	0.120	0.210
2.80	0.000	0.001	0.001	0.003	0.005	0.008	0.010	0.012	0.014	0.018	0.023	0.035	0.065	0.104	0.180
3.00	0.000	0.001	0.001	0.002	0.004	0.006	0.008	0.010	0.012	0.015	0.020	0.031	0.058	0.093	0.158
3.50	0.000	0.000	0.001	0.001	0.003	0.004	0.006	0.007	0.009	0.011	0.015	0.024	0.046	0.073	0.122
4.00	0.000	0.000	0.001	0.001	0.002	0.003	0.005	0.006	0.007	0.009	0.012	0.020	0.038	0.060	0.100

Note: This information is given in graphical form in the *API Data Book*, Figures 7F1.4 and 7F1.5.

Appendix F. Notation Index

This list covers the notation used more or less throughout the text. Other notation is defined and used locally. Individual letters, $a, b, c, \ldots, k, \ldots, A, B, C, \ldots \alpha, \beta, \gamma, \ldots$, often are used to represent empirical constants. Physical dimensions are $[E]$ for thermal energy, $[F]$ for force, $[L]$ for length, $[M]$ for mass, $[T]$ for temperature, and $[t]$ for time. Typical units also are given in some instances, such as: V, $[L^3]/[M]$, liters/gmol. Sometimes references are made to places in the text where the symbol is introduced first or where it is defined more completely and in context. The symbols are grouped in these categories:

F.1. Acronyms,
F.2. Letters,
F.3. Greek letters,
F.4. Subscripts,
F.5. Superscripts,
F.6. Special symbols.

F.1. ACRONYMS

ASOG	*A*nalytical *S*olution of *G*roups
BWR	*B*enedict-*W*ebb-*R*ubin equation of state
BWRS	*B*enedict-*W*ebb-*R*ubin-*S*tarling equation of state
EOS	*E*quation *o*f *S*tate
LEMF	*L*ocal *E*quivalent *M*ol *F*raction, a special case of the NRTL equation, with $\alpha_{12} = -1$
L-K	*L*ee-*K*esler equation of state
L-K-P	*L*ee-*K*esler-*P*löcker equation of state
LLE	*L*iquid-*L*iquid *E*quilibrium
N-R	*N*ewton-*R*aphson method
NRTL	*N*on-*R*andom-*T*wo-*L*iquid, a correlation of activity coefficients.
PCS	*P*rinciple of *C*orresponding *S*tates
P-R	*P*eng-*R*obinson equation of state
R-K	*R*edlich-*K*wong equation of state
SH	*S*catchard & *H*ildebrand, solubility parameter activity coefficient
SHFH	*S*catchard-*H*ildebrand-*F*lory-*H*uggins equation
T-K-W	*T*suboka-*K*atayama-*W*ilson equation
UNIFAC	*UNI*QUAC *F*unctional Group *A*ctivity *C*oefficients, a method of calculating activity coefficients from molecular structure
UNIQUAC	*UN*iversal *QUA*si *C*hemical, a correlation of activity coefficients
VER	*V*aporization *E*quilibrium *R*atio $= K_i = y_i/x_i$
VLE	*V*apor-*L*iquid *E*quilibrium

F.2. LETTERS

a	attraction parameter of the van der Waals, Soave or Peng-Robinson EOS, $(L^6)/M^2$, (liters/gmol)^2
a	attraction parameter of the Redlich-Kwong EOS, $(L^6)(T^{0.5})/(M^2)$, (liter)^2(K)$^{0.5}$/(gmol)^2
a, b, c	parameters of some cubic equations of state
A	$= aP/R^2T^{2.5}$, derived parameter of the R-K EOS
A	$= a\alpha P/R^2T^2$, derived parameter of the Soave or Peng-Robinson EOS
A	specific Helmholtz energy = U-TS, $(E)/(M)$, joules/gmol
a_i	$= \hat{f}_i/f^0$, activity of a species in a mixture
A_{ij}, A_{ji}	parameters of the Margules or van Laar equations
b	residual volume parameter of the van der Waals, Redlich-Kwong, Soave, Peng-Robinson, and some other cubic EOS, $(L^3)/(M)$, liters/gmol
B	$= bP/RT$, derived parameter of the Redlich-Kwong, Soave, and P-R EOS
C_i	concentration of the species, $(M)/(L^3)$, gmol/liter
c_{ij}	binary interaction parameter for the cross-attraction parameter of the Redlich-Kwong, Soave, and Peng-Robinson EOS, Section 1.3.6.
C_p	heat capacity at constant pressure, $(E)/(M)(T)$, joules/(gmol)(K)
C_v	heat capacity at constant volume, $(E)/(M)(T)$, joules/(gmol)(K)
f	fugacity, $(F)/(L^2)$, bar, Torr
$\hat{f}_i$	partial fugacity of species i, $(F)/(L^2)$, bar, Torr
F	total mass or mol number of a system, in flash calculations, Chap. 6
g	$= n\,G/R\,T = \underline{G}/RT$, Gibbs energy function
G	specific Gibbs energy, H-TS, $(E)/(M)$, joules/gmol
$\Delta G'$	see $\Delta M'$
g_c	conversion factor of mass and force units, $(M)(L)/(F)(t^2)$. Numerical values in Table G.1
ΔG_{f298}	Gibbs energy of formation in the standard state from the elements at 298 K, $(E)/(M)$, joules/gmol
g_{ij}, g_{ji}	parameters of the NRTL equation, Table 4.4
G_{ij}, G_{ji}	parameters of the NRTL equation, Table 4.4
ΔG_{rxn}	Gibbs energy change accompanying reaction, $(E)/(M)$, joules/(gmol of reference substance or for the reaction as written)
H	specific enthalpy, $U + PV$, $(E)/(M)$, joules/gmol
$\Delta H'$	see $\Delta M'$
ΔH_{f298}	enthalpy of formation in the standard state from the elements at 298 K, $(E)/(M)$, joules/gmol
ΔH_m	enthalpy of melting, (E)/(M), joules/gmol
ΔH_{rxn}	enthalpy change accompanying a reaction, $(E)/(M)$, joules/(gmol of reference substance or for the reaction as stoichiometrically written)
ΔH_{vap}	enthalpy of vaporization, $(E)/(M)$, joules/gmol
k	$= C_p/C_v$, ratio of heat capacities
K_a	$= K_{ce} = K_{rxn}$, equilibrium constant of a chemical reaction expressed in terms of activities. For K_f, K_x, K_y, K, see Table 10.2.
K_{ce}	$= K_{rxn}$. See K_a.
k_{Hi}	Henry's law coefficient, $k_{Hi} = \hat{f}_i/x_i$, or $k_{Hi} = P_i/x_i$, $(F)/(L^2)$, bar
k_{ij}	binary interaction parameter for cross-critical temperature, Section 1.3.6
K_i	$= y_i/x_i$, vaporization equilibrium ratio, VER, Chapter 6
L	amount of a liquid phase, mass, or mol number, (M), gmols, Chapter 6
l_i	term in the UNIQUAC and UNIFAC methods, Table 4.4
M	any intensive thermodynamic property, used for making generalized statements, such as

	$M = \Sigma \bar{M}_i x_i$, which could mean, for example, $G = \Sigma \bar{G}_i x_i$
M	molecular weight
M'	see $\Delta M'$
$\Delta M'$	$= M^{id} - M$, departure or residual of a property of a substance from the value it would have if it behaved as an ideal gas
M_{mix}	$= M - \Sigma M_i x_i$, property change on mixing, the difference between the values of the property of the mixture and the weighted value of the properties of the components of the mixture
n_i	number of mols of the species in the system
P	pressure of a system, $(F)/(L^2)$, bar, Torr
P_c	critical pressure, $(F)/(L^2)$, bar, Torr
P_i	partial pressure of a species i, $(F)/(L^2)$, bar, Torr
P_{pc}	pseudocritical pressure of a mixture, Section 1.3.4
P_r	$= P/P_c$, reduced pressure
P^0	$= P^{sat}$, vapor pressure, (F/L^2), bar, Torr
P^{osm}	osmotic pressure, $(F)/(L^2)$, bar, Torr
P^{sat}	$= P^0$, vapor pressure, $(F)/L^2)$, bar, Torr
Q	amount of heat transfer, (E), joules or Btu
q_i	area parameter of the UNIQUAC and UNIFAC methods, Table 4.4
Q_k	area parameter contribution of molecular group k, Tables E.11, E.13
R	gas constant, $(E)/(M)(T)$. Numerical values are given in Table A.5.
r_i	volume parameter of the UNIQUAC and UNIFAC methods, Table 4.4
R_k	volume parameter contribution of molecular group k, Table E.11. E.13
S	specific entropy, $(E)/(M)(T)$, joules/(gmol)(K)
$\Delta S'$	see $\Delta M'$
ΔS_m	entropy of melting $= \Delta H_m/T_m$
T	temperature, (t), °K, °R, °C, °F
T_{bp}	temperature of boiling at atmospheric pressure, (T)
T_c	critical temperature
T_m	melting temperature, (T)
T_{pc}	pseudocritical temperature of a mixture, Section 1.3.4
T_r	$= T/T_c$, reduced temperature
T_{tp}	temperature of the triple point, (T)
U	specific internal energy, $(E)/(M)$, joules/gmol
$\Delta U'$	see $\Delta M'$
u_{ij}, u_{ji}	parameters of the UNIQUAC equation, Table 4.4
V	amount of a vapor phase, (M), gmols or grams, Chapter 6
V	specific volume, $(L^3)/(M)$, the reciprocal of the density, ρ, liters/gmol, liters/kg
V_c	critical volume
V_{pc}	pseudocritical volume of a mixture, Section 1.3.4
V_r	$= V/V_c$, reduced volume
W	work transfer, (E), joules, Btu, ft-lbf
x_i	mol fraction of a species in a liquid-phase mixture
y_i	mol fraction of a species in a vapor-phase mixture
z	$= PV/RT$, compressibility factor
z_c	$= P_c V_c/RT_c$, critical compressibility
z_i	mol fraction of a species in all phases of a system

F.3. GREEK LETTERS

α	parameter of Soave and Peng-Robinson equations, Tables 1.11 and 1.13
α_{ij}	volatility of species i relative to that of species j
α_{ij}, α_{ji}	parameters in the NRTL equation
β	group of terms in the Wilson equation, Table 4.4
γ_i	$= a_i/x_i$, activity coefficient of species i
γ_i^∞	limiting value of activity coefficient at zero concentration or infinite dilution.
δ_i	solubility parameter of species i, $([E]/[L^3])^{0.5}$, $(cal/ml)^{0.5}$
$\bar{\delta}$	solubility parameter of the mixture, same units as δ_i.
Δ	a difference, as in $\Delta U = U_{final} - U_{initial}$
ε	$= \pm(n_i - n_{i0})/\nu_i$, + for products and − for reactants, degree of advancement of a reaction, gmol/unit of stoichiometric coefficient
θ_i	area fraction of species i in UNIQUAC equation
λ_{ij}, λ_{ji}	exponential parameters in Wilson or T-K-Wilson equations, $[E]/[M]$, cal/gmol, Table 4.4
Λ_{ij}, Λ_{ji}	Wilson or T-K-Wilson parameters, Table 4.4
μ	$= (\partial T/\partial P)_H$, Joule-Thomson coefficient
μ_i	$= \bar{G}_i$, chemical potential of species i
ν_i	stoichiometric coefficient of species i
ρ	density, $[M]/[L^3]$, gmol/liter
τ_{ij}, τ_{ji}	parameters of the NRTL equation, Table 4.4
ϕ_i	$= f_i/P$, fugacity coefficient
$\hat{\phi}_i$	$= \hat{f}_i/y_i P$, partial fugacity coefficient of species i in a mixture
Φ_i	$= x_i V_i/\Sigma x_i V_i$, volume fraction of species i in a mixture
ω	acentric factor

F.4. SUBSCRIPTS

$i, j, k, \ldots$	identifying a species in a mixture, as in n_i
b	at the normal boiling point, as in T_b
c	critical property, as in T_c
ce	chemical equilibrium, as in K_{ce}
m	at the melting point, as in T_m
mix	of a mixture, as in $T_{c,mix}$; or of mixing, as in ΔH_{mix}
r	reduced property, as in $T_r = T/T_c$
rxn	of reaction as in ΔH_{rxn}, enthalpy change of reaction
surr	surroundings
syst	system
bp	at the atmospheric boiling point
pc	pseudocritical, Section 1.3.4
tp	at the triple point, as in T_{tp}

F.5. SUPERSCRIPTS

0	property of a standard state, as in f^0
*	limiting value of a property, as $P \to 0$ or $V \to \infty$
id	property of an ideal gas, as in H^{id}
liq	property of a liquid, as in f_i^{liq}
sat	property at saturation conditions, as in P^{sat} = vapor pressure
sol	property of a solid phase, as in f_i^{sol}
vap	property of a vapor phase, as in f_i^{vap}

F.6. SPECIAL SYMBOLS

$-$ overscore, partial molal property, as in $\bar{H}_i = (\partial H/\partial n_i)_{T,P,n_{j\neq i}}$

$\_$ underscore, identifying the property of the total mass of a system, as in $\underline{H} = nH$

$\hat{}$ carat, used only in $\hat{f}_i$ and $\hat{\phi}_i$ to identify these properties of a species in a mixture

$'$ prime, designating a residual property, as in $\Delta H' = H^{id} - H$

$\sum_{i=1}^{n}$ mathematical summation, as in $\sum_1^n x_i = x_1 + x_2 + \ldots + x_n$

$\prod_{i=1}^{n}$ mathematical product, as in $\prod_1^n x_i = x_1 x_2 \ldots x_n$

Appendix G. Units and Conversions

Although there is a strong movement toward the adoption of a consistent, international set of units for physical quantities, the SI (*système internationale*), most data still are recorded in older units, so it is necessary to be familiar with formerly common terms. It may be that long-established terms such as atm, Torr, and mm Hg will persist indefinitely. The SI system is readily extended to new concepts and is adaptable to a range of magnitudes of units. Thus Homer's use of the unit of beauty, the Helen, as the amount of beauty that will launch 1,000 ships, is inconveniently large; but the micro-Helen, the amount that will launch a rowboat, and the micro-Nelly, the amount that will launch a rowboat in the opposite direction, are logically defined and quite practical units.

Comprehensive tables of equivalents of various units are widely available. Some particularly useful literature is:

1. Eshbach, O.W., & M. Souders (editors). *Handbook of Engineering Fundamentals.* Wiley (1975)
2. Mullin, J.W. Recent developments in the changeover to the international system of units. *The Chemical Engineer,* 353 (October 1971).
3. *SI for AIChE.* Publication #X-95 of the American Institute of Chemical Engineers (1979).

Table G.1 suffices for most of the needs of this book.

Table G.1. Conversion of Units

Multiples
micro = 10(E-6)
milli = 10(E-3)
centi = 10(E-2)
deci = 10(E-1)
kilo = 10(E3)
mega = 10(E6)

Length
1 ft = 0.3048 m = 30.48 cm = 304.8 mm

Volume
1 cu ft = 0.0283 cu m = 7.481 U.S. Gals
1 cu m = 35.34 cu ft = 1000 liters

Standard Gas Volume
22.414 liters/gmol @ 0°C and 1 atm
359.05 cuft/lbmol @ 32°F and 1 atm

Gas constant R
see Table A.5

Table G.1. *(continued)*

Mass
1 lb = 0.4536 kg
1 kg = 2.2046 lb

Density
1 lb/cu ft = 16.018 kg/cu m
1 gm/cu cm = 62.43 lb/cu ft

Force
1 lb force = 0.4536 kg force = 4.448 Newtons

Pressure
1 atm = 760 Torr = 760 mm Hg = 101,325 Newtons/sq m = 1.01325 bar = 10,330 kg/sq m = 14.696 lbf/sq in = 2,116.2 lbf/sq ft
1 bar = 100,000 Newtons/sq m
1 Pascal = 1 Newton/sq m

Energy, Work, and Heat
1 Btu = 252.16 cal = 1055.06 joules = 0.2930 watt-hrs = 10.41 liter-atm
1 HP-hr = 0.7457 kwatt-hr = 778 ft lbf = 2545 cal
1 cal = 4.1868 joules
1 joule = 1 Newton-meter = 1 watt-sec = 0.2388 cal = 0.000948 Btu

Power
1 ft lbf/sec = 0.0018182 HP = 1.356 watts = 0.0012856 Btu/sec = 0.3238 cal/sec
1 watt = 1 joule/sec = 1 Newton-meter/sec

Temperature
°K (Kelvin) = °C (centigrade) + 273.16 = [°F (Fahrenheit) + 459.6]/1.8 = °R (Rankine)/1.8
°R = 1.8 °K = °F + 459.6
°C = (°F − 32)/1.8

Temperature Difference
1 °C = 1 °K = 1.8 °R = 1.8 °F

Heat Capacity and Entropy
1 cal/(gram)(°K) = 4.1868 joules/(gram)(°K) = Btu/(lb)(°R)

Specific Energy
1 cal/gram = 4.1868 joules/gram = 1.8 Btu/lb

Volumetric Flow
1 cu ft/sec = 0.028316 cu m/sec = 28.316 liters/sec

Gravitational Constant
g_c = 1 kg mass/Newton-sec^2
= 1 gram cm/dyne-sec^2
= 9.806 kg mass/kg force-sec^2
= 32.174 lb mass/lb force-sec^2

Appendix H. References to the Literature

The references are grouped as: (1) collections of papers, primarily as symposia; (2) books; and (3) individual papers in journals or in the collections. Citations in the text are by the name of the first author and the date; they are not identified as being in one of the three given categories. The references are identified with the chapter number in brackets to which their subject matter is most pertinent. Special bibliographies appear on pages 101, 294, 336, 388, 543, 563, 577, and 638.

It may be of interest to note the journals that are cited most frequently for phase equilibria: (1) *AIChE Journal;* (2) *Chemical Engineering Science;* (3) *Fluid Phase Equilibria;* (4) *Industrial & Engineering Chemistry, Fundamentals;* and (5) *Industrial & Engineering Chemistry, Process Design & Development.*

The journal, *Theoretical Chemical Engineering Abstracts,* includes phase equilibria and may be more convenient to use than *Chemical Abstracts.* The approximately annual publication, *Physico-Chemical Properties for Chemical Engineering* (Maruzen Company, Tokyo, since 1977) also has abstracts; its content is described in Appendix D.

COLLECTIONS OF PAPERS AND BIBLIOGRAPHIES

Anon. "Phase Equilibrium and Fluid Properties in the Chemical Industry." *Proceedings*, Second International Conference, Berlin, March 17, 1980, European Federation of Chemical Engineers (EFCE), DECHEMA (1980).

Chao, K.C. (chairman). *Applied Thermodynamics.* Based on a symposium sponsored by Industrial & Engineering Chemistry, June 12–14, 1967. American Chemical Society Publications (1968).

Chao, K.C. & R.L. Robinson (editors). "Equations of State in Engineering and Research." Advances in Chemistry Series *182*. American Chemical Society (1979).

Hiza, M.J., A.J. Kidnay, & R.C. Miller. *Equilibrium Properties of Fluid Mixtures.* 1. *A Bibliography of Data on Fluids of Cryogenic Interest.* 2. *A Bibliography of Experimental Data on Selected Fluids.* Plenum Press (1975, 1982).

Kehiaian, H.V., & B.J. Zwolinski (editors). *International Data Series.* Series A: *Selected Data on Mixtures.* Series B: *Thermodynamic Properties of Aqueous Organic Systems.* Thermodynamics Research Center, Texas A & M University (1971–date).

Leesley, M.E. (editor). *Computer-Aided Process Plant Design*. Gulf Publishing Company (1982).

Le Neindre, B., & B. Vodar (editors). *Experimental Thermodynamics.* II. *Experimental Thermodynamics of Non-reacting Fluids.* International Union of Pure and Applied Chemistry. Butterworths (1975)

McGlashan, M.L. (editor). *Chemical Thermodynamics: A Specialist Periodical Report.* Vol. I (1973). Vol. II (1978). The Chemical Society, London.

Mah, R.S.H., & W.D. Seider (editors). *Foundations of Computer-Aided Process Design.* Vols. I and II. American Institute of Chemical Engineers (1980)

Mittal, K.L. (editor). *Micellization, Solubilization and Microemulsions*, Vols. 1 & 2. Plenum Press (1977).

Mittal, K.L. (editor). *Solution Chemistry of Surfactants*, Vols. 1 & 2. Plenum Press (1977).

Mittal, K.L., & E.J. Fendler (editors). *Solution Behavior of Surfactants.* Vols. 1 & 2. Plenum Press (1982).

Newman, S.A. (editor). *Chemical Engineering Thermodynamics*. From the Second World Congress of Chemical Engineering, Montreal, Oct 4–9, 1981. Ann Arbor Science (1982).

Newman, S.A., H.E. Barner, M. Klein, & S.I. Sandler (editors). "Thermodynamics of Aqueous Systems with Industrial Applications." ACS Symposium Series *133.* American Chemical Society (1980).

Paulaitis, M.E., J.M.L. Penninger, R.D. Gray Jr., & P. Davidson (editors). *Chemical Engineering at Supercritical Fluid Conditions*. Ann Arbor Science (1983).

Skinner, H.A. (editor). *International Review of Science: Physical Chemistry*. Series 2, Vol. 10. *Thermochemistry and Thermodynamics*. Butterworths (1975).

Storvick, T.S. & S. I. Sandler (editors). "Phase Equilibria and Fluid Properties in the Chemical Industry." ACS Symposium Series *60*. American Chemical Society (1977).

Tamir, A., & K. Stephan. *Heats of Phase Change of Pure Components and Mixtures: A Literature Guide.* Elsevier, (1983)

Wichterle, I., J. Linek, & E. Hala. *Vapor-Liquid Equilibrium Bibliography & Supplements 1, 2, and 3.* Elsevier (1973, 1976, 1979, 1982).

Wisniak, J. & A. Tamir. *Liquid-Liquid Equilibria and Extraction* (1980). *Mixing and Excess Thermodynamic Properties: A Literature Source Book and Supplement* (1978, 1982). *Phase Diagrams: A Literature Source Book*, Parts A & B (1981). Elsevier.

BOOKS

Bevington, P.R. *Data Reduction and Error Analysis for the Physical Sciences.* McGraw-Hill (1969).

Blasdale, W.C., *Equilibria in Saturated Salt Solutions.* Reinhold (1927).

Blinder, S.M. *Advanced Physical Chemistry.* Macmillan (1969).

Bowden, S.T. *The Phase Rule and Phase Reactions.* Macmillan (1938).

Bretsznajder, S. *Prediction of Transport and other Physical Properties of Fluids*. Pergamon (1971); *Wlasnosci Gazow i Cieczy.* Wydawnictwa Naukowo-Techniczne (Warsaw) (1962).

Bridgman, P.W. *Condensed Collection of Thermodynamic Formulas*. Harvard University Press (1926).

Chao, K.C., & R.A. Greenkorn. *Thermodynamics of*

Fluids: An Introduction to Equilibrium Theory. Marcel Dekker (1975).

Christensen, J.J., R.W. Hanks, & R.M. Izatt. *Handbook of Heats of Mixing*. Wiley (1982).

Dack, M.R.J. (editor). *Solutions and Solubilities*, Parts I & II. Technique of Organic Chemistry, Vol. 8, A. Weissberger (editor). Wiley-Interscience (1976).

Daniels, T.C., *Thermal Analysis*, Wiley (1973).

Daubert, T.E., M.S. Graboski, & R.P. Danner. *Documentation of the Basis for Selection of the Contents of Chapter 8—Vapor-Liquid Equilibrium K-Values in Technical Data Book—Petroleum Refining*. No. 8-78, American Petroleum Institute (1978).

Deaton, W.M., & E.M. Frost, Jr. *Gas Hydrates and Their Relation to the Operation of Natural-Gas Pipe Lines*. U.S. Bureau of Mines Monograph 8 (1946).

Dickerson, R.E. *Molecular Thermodynamics*. Benjamin (1969).

Doring, R., H. Knapp, L.R. Oellrich, U.J. Plöcker, & J.M. Prausnitz. *Vapor-Liquid Equilibria for Mixtures of Low-Boiling Substances*. DECHEMA (1982).

Durell, C.V. *Homogeneous Coordinates*. G. Bell & Sons (1961).

Dymond, J.H., & E.B. Smith. *The Virial Coefficients of Pure Gases and Mixtures: A Critical Compilation*. Clarendon Press, Oxford (1980).

Edmister, W.C. *Applied Hydrocarbon Thermodynamics*, Vols. I & II. Gulf Publishing Company (1984, 1974).

Ferguson, F.D., & T.K. Jones. *The Phase Rule*. Butterworths (1966).

Francis, A.W. *Liquid-Liquid Equilibriums*. Wiley-Interscience (1963).

Fredenslund, A., J. Gmehling, & P. Rasmussen. *Vapor-Liquid Equilibria Using UNIFAC*. Elsevier (1977).

Fried, V., H.F. Hameka, & U. Blukis. *Physical Chemistry*. Macmillan (1977).

Gaskell, D.R. *Introduction to Metallurgical Thermodynamics*. McGraw-Hill (1973).

Gerald, C.F. *Applied Numerical Analysis*. Addison-Wesley (1978).

Gerasimov, Ya. (editor). *Physical Chemistry*, Vol. 1. Mir Publishers, Moscow (1974).

Glasstone, S. *Textbook of Physical Chemistry*. Van Nostrand (1946).

Gmehling, J., & U. Onken. *Vapor-Liquid Equilibrium Data Collection*. DECHEMA Chemistry Data Series, Vol. 1, DECHEMA (1977ff).

Guggenheim, E.A. *Thermodynamics*, 7th ed. North-Holland (1977).

Haase, R., *Thermodynamik der Mischphasen*. Springer (1956).

Haase, R., & H. Schonert. *Solid-Liquid Equilibrium*. Pergamon (1969).

Hala, E., J. Pick, V. Fried, & O. Vilim, *Vapor-Liquid Equilibrium*. Pergamon (1967).

Henley, E.J., & E.M. Rosen. *Material and Energy Balance Computations*. Wiley (1969).

Henley, E.J., & J.D. Seader. *Equilibrium-Stage Separation Operations in Chemical Engineering*. Wiley (1981).

Hildebrand, J.H., J.M. Prausnitz, & R.L. Scott. *Regular and Related Solutions*. Van Nostrand (1970).

Hildebrand, J.H., & R.L. Scott. *The Solubility of Nonelectrolytes*. Dover (1964).

Himmelblau, D.M. *Process Analysis by Statistical Methods*. Wiley (1970).

Himmelblau, D.M., B.L. Brady, & J.J. McKetta, Jr. *Survey of Solubility Diagrams for Ternary and Quaternary Liquid Systems*. Special Publication No. 30, Bureau of Engineering Research, University of Texas, Austin (1959).

Hirata, M., S. Ohe, & K. Nagahama. *Computer-Aided Data Book of Vapor-Liquid Equilibria*. Elsevier (1976).

Holub, R., & P. Vonka. *Chemical Equilibrium of Gaseous Systems*. Reidel (1976).

Hougen, O.A., K.M. Watson, & R.A. Ragatz. *Chemical Process Principles*. Part II. *Thermodynamics*. Wiley (1959).

Huggins, M.L., *Physical Chemistry of High Polymers*. Wiley (1958).

Karapetyants, M. Kh. *Chemical Thermodynamics*. Mir Publishers, Moscow (1978).

Kaufman, L., & H. Bernstein, *Computer Calculation of Phase Diagrams*. Academic Press (1970).

Keenan, J.H., F.G. Keyes, P.G. Hill, & J.G. Moore. *Steam Tables*, English Units (1969), SI Units (1978). Wiley.

King, M.B. *Phase Equilibrium in Mixtures*. Pergamon (1969).

Kipling, J.J. *Adsorption from Solutions of Nonelectrolytes*. Academic Press (1965).

Klotz, I.M., & R.M. Rosenberg. *Chemical Thermodynamics. Basic Theory and Methods*. Benjamin-Cummings (1972).

Kofler, L. *Thermomicromethoden zur Kennzeichnung organischer Stoffe und Stoffgemische*. Verlag Chemie (1954).

Kogan, V.B. *Heterogeneous Equilibria* (in Russian). Izdatelstvo "Khimiya," Leningrad (1968).

Kogan, V.B., V.M. Fridman, & V.V. Kafarov. *Equilibrium between Liquid and Vapor*, 2 vols. (in Russian). Izdatelstvo "Nauka," Moscow (1966).

Kojima, K., & K. Tochigi. *Prediction of Vapor-Liquid Equilibria by the ASOG Method*. Elsevier (1979).

Kubashewski, O., & C.B. Alcock. *Metallurgical Thermochemistry*, 5th ed. Pergamon Press (1979).

Kubo, R. *Thermodynamics, an Advanced Course with Problems and Solutions*. North-Holland (1968).

Landolt-Börnstein, *Numerical Data and Functional Relationships in Science and Technology*. II/2a: *Vapor-Liquid Equilibria and Osmotic Phenomena* (1960); II/2b: *Solution Equilibria* (1962); II/2c: *Solution Equilibria* (1964); II/3: *Melt Equilibria and Surface Phenomena* (1956); IV/4c1: *Absorption of Gases, Low Vapor Pressure Liquids* (76); IV/4c2: *Absorption of Gases in High Vapor Pressure Liquids* (in preparation); NS IV/2: *Heats of Mixing and Solution* (1976); NS IV/3: *Thermodynamic Equilibria of Boiling Mixtures* (1975). Springer Verlag (1950ff).

Lapidus, L. *Digital Computation for Chemical Engineers*. McGraw-Hill (1962).

Larkin, J.V. (editor). *Selected Data on Mixtures*. International Data Series B, Thermodynamic Properties of Aqueous Organic Systems. Engineering Sciences Data Unit, Ltd. (1978ff).

Lipka, J., *Graphical and Mechanical Computation*. Wiley (1918).

Loney, S.L. *The Elements of Coordinate Geometry*. Part II. *Trilinear Coordinates, etc*. Macmillan (1923).

Lupis, C.H.P., *Chemical Thermodynamics of Materials*. North Holland (1983).

Lydersen, A.L., R.A. Greenkorn, & O.A. Hougen. *Generalized Thermodynamic Properties of Pure Fluids*. University of Wisconsin Eng. Exp. Station, Report 4 (October 1955).

Lyman, W.J., W.F. Reehl, & D.H. Rosenblatt. *Handbook of Chemical Property Estimation Methods: Environmental Behavior of Organic Compounds*. McGraw-Hill (1982).

McGlashan, M.L. *Chemical Thermodynamics*. Academic Press (1979).

Mackenzie, R.C. (editor). *Differential Thermal Analysis*. Vo. 2: *Applications*. Academic Press (1972).

Maczynski, A. (editor). *Verified Vapor-Liquid Equilibrium Data: Thermodynamical Data for Technology*, several parts. PWN-Polish Scientific Publishers (1976ff.).

Makogon, Yu F. *Hydrates of Natural Gas*, PennWell (1981).

Mason, E.A., & T.H. Spurling. *The Virial Equation of State*. International Encyclopedia of Physical Chemistry and Chemical Physics, Topic 10, Vol. 2. Pergamon Press (1969).

Modell, M., & R.C. Reid. *Thermodynamics and its Applications*. Prentice-Hall (1983).

Nielsen, L.E. *Predicting the Properties of Mixtures*. Marcel Dekker (1978).

Null, H.R. *Phase Equilibrium in Process Design*. Wiley-Interscience (1970).

Nyvlt, J. *Solid-Liquid Phase Equilibria*. Elsevier (1977).

Ohe, S. *Computer-Aided Data Book of Vapor Pressure*. Data Publishing Company, Tokyo (1976).

Opfell, J.B., C.J. Pings, & B.H. Sage. *Equation of State for Hydrocarbons*. Monograph on API Research Project 37, American Petroleum Institute (1959).

Partington, J.R. *An Advanced Treatise on Physical Chemistry*. Vol. 1: *Fundamental Principles and Properties of Gases*; Vol. II: *The Properties of Liquids*. Longmans (1950, 1951).

Partington, J.R., & W.G. Shilling. *The Specific Heat of Gases*. E. Benn (1924).

Paul, R.F.M., & W.S. Wise. *Principles of Gas Extraction*. Mills & Boon Ltd. (1971).

Pickering, S.F. *Relations between the Temperatures, Pressures and Densities of Gases*. Circular No. 279, U.S. Bureau of Standards (1925).

Pope, M.I., & M.D. Judd. *Differential Thermal Analysis: A Guide to the Technique and its Applications*. Heyden (1977).

Prausnitz, J.M. *Molecular Thermodynamics of Fluid Phase Equilibria*. Prentice-Hall (1969).

Prausnitz, J.M., T. Anderson, E. Grens, C. Eckert, R. Hsieh, & J. O'Connell. *Computer Calculations for Multicomponent Vapor-Liquid and Liquid-Liquid Equilibria*. Prentice-Hall (1980).

Prausnitz, J.M., & P.L. Chueh. *Computer Calculations for High Pressure Vapor Liquid Equilibria*. Prentice-Hall (1968).

Prausnitz, J.M., C.A. Eckert, R.V. Orye, & J.P. O'Connell. *Computer Calculations for Multicomponent Vapor Liquid Equilibria*. Prentice-Hall (1967).

Priestley, E.B., P.J. Wojtowicz, & P. Sheng. *Introduction to Liquid Crystals*. Plenum Press (1975).

Prigogine, I., & R. Defay. *Chemical Thermodynamics*. Longmans (1954).

Prince, A. *Alloy Phase Equilibria*. Elsevier (1966).

Rasmussen, P., & A. Fredenslund. *Data Banks for Chemical Engineers*. Kemiigeniorgruppen, Lyngby, Denmark (1980).

Redlich, O. *Thermodynamics Fundamentals and Applications*. Elsevier (1978).

Reed, T.M., & K.E. Gubbins. *Applied Statistical Mechanics: Thermodynamic and Transport Properties of Fluids*. McGraw-Hill (1973).

Reid, R.C., J.M. Prausnitz, & T.K. Sherwood. *The Properties of Gases and Liquids*. McGraw-Hill (1958, 1966, 1977).

Reisman, A. *Phase Equilibria, Basic Principles, Applications, Experimental Techniques*. Academic Press (1970).

Renon, H., L. Asselineau, G. Cohen, & C. Raimbault. *Calcul sur Ordinateur des Equilibres Liquide-Vapeur et Liquide-Liquide*. Editions Technip (1971).

Ricci, J.E. *The Phase Rule and Heterogeneous Equilibrium*. Van Nostrand (1951).

Rosenbrock, H.H., & C. Storey. *Computational Techniques for Chemical Engineers*. Pergamon Press (1966).

Rowlinson, J.S. *The Perfect Gas*. International Encyclopedia of Physical Chemistry and Chemical Physics, Topic 10, Vol. 5, Pergamon Press (1963).

Rowlinson, J.S., & F.L. Swinton. *Liquids and Liquid Mixtures*, 3rd ed. Butterworths (1982).

Sage, B.H. *Thermodynamics of Multicomponent Systems*. Reinhold Publishing Company (1965).

Sage, B.H., & W.N. Lacey. *Volumetric and Phase Behavior of Hydrocarbons*. Stanford University Press (1939).

Salzer, H.E., C.H. Richards, & I. Arsham. *Tables for the Solution of Cubic Equations*. McGraw-Hill (1958).

Scarborough, J.B. *Numerical Mathematical Analysis*. Johns Hopkins University Press (1966).

Schneider, G.M., E. Stahl, & G. Wilke. *Extraction with Supercritical Gases*. Verlag Chemie (1980).

Sherwood, T.K. *A Course in Process Design*. MIT Press (1963).

Sherwood, T.K., & C.E. Reed. *Applied Mathematics in Chemical Engineering*. McGraw-Hill (1939).

Shinoda, K. *Principles of Solution and Solubility*. Marcel Dekker (1978).

Smith, J.M., & H.C. Van Ness. *Introduction to Chemical Engineering Thermodynamics*. McGraw-Hill (1975).

Smith, W.R., & D.W. Missen. *Chemical Reaction Analysis: Theory and Algorithms*. Wiley (1982).

Sorensen, J.M. *Correlation of Liquid-Liquid Equilibrium Data*. Instituttet for Kemiteknik, Danmarks Tekniske Hojskole (1980).

Sorensen, J.M., & W. Arlt. *Liquid-Liquid Equilibrium Data Collection*, 3 parts. DECHEMA (1979, 1980).

Stanley, H.E. *Introduction to Phase Transitions and Critical Phenomena*. Clarendon Press, Oxford (1971).

Starling, K.E. *Fluid Thermodynamic Properties for Light Petroleum Systems*. Gulf Publishing Company (1973).

Stephen, H., T. Stephen, & H. Silcock. *Solubilities of Inorganic and Organic Compounds*, 7 parts. Pergamon (1979).

Sterbacek, A., B. Biskup, & P. Tausk. *Calculation of Properties Using Corresponding States Methods*. Elsevier (1979).

Stull, D.R., E.F. Westrum, & G.C. Sinke. *The Chemical Thermodynamics of Organic Compounds*. Wiley (1969).

Tabor, D. *Gases, Liquids and Solids*. Cambridge University Press (1979).

Tamas, F., & I. Pal. *Phase Equilibrium Spatial Diagrams*. Illiffe Books (1970).

Teeple, J.E. *Industrial Development of Searles Lake Brines*. Reinhold (1929).

Temperley, H.N.V., & D.H. Trevena. *Liquids and their Properties*. Ellis Horwood Ltd. (1978).

Treybal, R. *Liquid Extraction*. McGraw-Hill (1963).

Van Ness, H.C., & M.M. Abbott. *Classical Thermodynamics of Non-Electrolyte Solutions with Applications to Phase Equilibria*. McGraw-Hill (1982).

Van Zeggeren, F., & S.H. Storey. *The Computation of Chemical Equilibria*. Cambridge University Press (1970).

Vogel, R. *Heterogeneous Equilibria* (in German). Akademische Verlagsgesellschaft (1959).

Vukalovich, M.P., & I.I. Novikov. *Equations of State of Real Gases* (in Russian). Gosenergoizdat, Moscow (1948). The Library of Congress has a copy.

Wendlandt, W.W. *Thermal Methods of Analysis*. Wiley (1974).

West, D.R.F. *Ternary Equilibrium Diagrams*, 2nd ed. Chapman & Hall (1982).

Wolberg, J.R. *Prediction Analysis*. Van Nostrand (1967).

Yaws, C.L. *Physical Properties: A Guide to the Physical, Thermodynamic and Transport Properties of Industrially Important Chemical Compounds*. McGraw-Hill (1977).

Yeremin, E.N. *Fundamentals of Chemical Thermodynamics*. Mir Publishers, Moscow (1981).

Young, D.M., A.D. Crowell. *Physical Adsorption of Gases*. Butterworths (1962).

Zernike, J. *Chemical Phase Theory*. Kluwer, Deventer, Holland (1955).

INDIVIDUAL PAPERS

The numbers in brackets following each reference are those of the chapters on the same topic.

Abbott, M.M. Cubic Equations of State (review). *AIChE Journal 19* 596–601 (1973). [1]

Abbott, M.M. Cubic Equations of State: An Interpretive Review, K.C. Chao & R.L. Robinson (editors). *Equations of State in Engineering and Research*. 47–70, *Advances in Chemistry Series 182*, American Chemical Society (1979). [1]

Abbott, M.M., J.K. Floess, G.E. Walsh, Jr., & H.C. Van Ness. Vapor-liquid equilibrium. Part IV: Reduction of *P-x* data for ternary systems. *AIChE Journal 21*, 72–76 (1975). [4]

Abbott, M.M., & H.C. Van Ness. Vapor-liquid equilibrium. Part III: Data reduction with precise expressions for G^E. *AIChE Journal 21*, 62–71 (1975). [4]

Abbott, M.M., & H.C. Van Ness. An extension of Barker's method for reduction of VlE data. *Fluid Phase Equilibria 1*, 3–11 (1977). [4]

Abrams, D.S., & J.M. Prausnitz. Statistical Thermodynamics of liquid mixtures: A new expression for the excess Gibbs energy of partly or completely miscible systems. *AIChE Journal 21*, 116–128 (1975). [7]

Abrams, D.S., F. Seneci, P.L. Chueh, & J.M. Prausnitz. Thermodynamics of multicomponent liquid mixtures containing subcritical and supercritical components. *Ind. Eng. Chem. Fundamen. 14*, 52–54 (1975). [6]

Adler, S.B., C.F. Spencer, H. Ozkardesh, & C.M. Kuo. Industrial uses of equations of state: A state of the art review, T.S. Storvick & S.I. Sandler (editors). *Phase equilibria and fluid properties in the chemical industry*, 150–199, *ACS Symposium Series 60*, American Chemical Society (1977). [1]

Alwani, Z., & G.M. Schneider. Fluid mixtures at high pressure: Phase separation and critical phenomena in binary mixtures of a polar compound with supercritical carbon dioxide, ethane and ethene up to 1000 bar. *Ber Bunsenges Physik Chemie 80*, 1310–1315 (1976). [9]

Ambrose, D. The correlation and estimation of vapor pressures. *Proceedings of the NPL Conference: Chemical Thermodynamic Data on Fluids and Fluid Mixtures*. IPC Science and Technology Press (1979). [1]

Ambrose, D., Counsell & Davenport, *J. Chem Thermodynamics 2* 283 (1970). [1]

Anderson, T.F., & J.M. Prausnitz. Computational methods for high-pressure phase equilibria and other fluid-phase properties using a partition function. 1: Pure fluids; 2: Mixtures; *Ind. Eng. Chem. Process Des. Dev. 19*, 1–4 (1980). [6]

Anderson, T.F., & J.M. Prausnitz. Application of the UNIQUAC equation to calculation of multicomponent phase equilibria. 1: Vapor-liquid equilibria; 2: Liquid-liquid equilibria. *Ind. Eng. Chem. Process Des. Dev. 17*, 552–567 (1978). [7]

Antezane, F.J., J.Y. Cheh. Component fugacities in hydrogen-ammonia-propane mixtures. 1: The fugacity of hydrogen; 2: The fugacity of ammonia. *Ind. Eng. Chem. Fundamen. 14*, 224–232 (1975); *15*, 95–99 (1976). [3]

Arich, G., I. Kikic, & P. Allessi. The liquid-liquid equilibrium for activity coefficient determination. *Chem. Eng. Science 30*, 187–191 (1975). [7]

Aristovich, V. Yu, & E.I. Stepanova. Determination of the existence and composition of multicomponent azeotropes by calculation from data for binary systems. *J. Applied Chem. (USSR) 43*, 2217–2223 (1970). [4]

Ashraf, F.A., & J.H. Vera. A simplified group method analysis. *Fluid Phase Equilibria 4*, 211–228 (1980). [4]

Asselineau, L., G. Bogdanic, & J. Vidal. Calculation of thermodynamic properties and vapor-liquid equilibria of refrigerants. *Chem. Eng. Science 33*, 1269–1276 (1978). [11]

Asselineau, L., G. Bogdanic, & J. Vidal. A versatile algorithm for calculating vapor-liquid equilibria. *Fluid Phase Equilibria 3*, 273–290 (1979). [6]

Avet'yan, V.S., L.E. Karp, F.B. Petlyuk, & L.A. Serafimov. A mathematical description of the phase equilibrium in polyazeotropic mixtures. *Russian J. Phys. Chem. 52*, 1425–1427 (1978). [4]

Bae, H.K., K. Nagahama, & M. Hirata. Measurement and correlation of high pressure vapor-liquid equilibria for the systems ethylene+1-butene and ethylene+propylene. *J. Chem. Eng. Japan 14*, 1–6 (1981). [12]

Baker, L.E., & K.D. Luks. Critical Point and saturation pressure calculations for multicomponent systems. Paper SPE 7478, presented at 53rd Annual Fall Technical Conference of Society of Petroleum Engineers of AIME, held in Houston, Texas, October 1–3, 1978. [1]

Bansal, V.K., & D.O. Shah. Micellar solutions for improved oil recovery, K.L. Mittal (editor). *Solution Chemistry of Surfactants*, Vol. 1, 87–113 (1977). [9]

Barieau, R.E. The calculation of fugacities from volumetric and phase equilibrium data by means of the Gibbs-Duhem equation. *Ind. Eng. Chem. Fundamen. 10*, 428–433 (1971). [3]

Barker, J.A. Determination of activity coefficients from total pressure measurements. *Australian J. Chem. 6*, 207–210 (1953). [4]

Barner, H.E., & S.B. Adler. Low temperature B-W-R applications. *Hydrocarbon Processing 47* (10), 150–156 (1968). [1]

Barner, H.E., & S.B. Adler. Three-parameter formulation of the Joffe equation of state. *Ind. Eng. Chem. Fundamen. 9*, 521–530 (1970). [1]

Barner, H.E., & S.B. Adler. Calculation of solution nonideality from binary *T-x* data. *Ind. Eng. Chem. Process Des. Dev. 12*, 71–75 (1973). [4]

Baron, R.E., J.H. Porter, & O.H. Hammond. Chemical Equilibria in Carbon-Hydrogen-Oxygen Systems. MIT Press (1975). [10]

Barton, A.F. Solubility parameters. *Chem. Reviews 75*, 73–753 (1975). [4]

Barton, P., R.E. Holland, & R.H. McCormick. Correlation of vapor-liquid equilibria of C_3-C_5 hydrocarbons using solubility parameters. *Ind. Eng. Chem. Process Des. Dev. 13*, 378–383 (1974). [4]

Beattie, J.A., & W.H. Stockmayer. The thermodynamics and statistical mechanics of real gases. *Treatise on Physical Chemistry* Vol. 2, H.S. Taylor & S. Glasstone (editors), 187–352. Van Nostrand (1951). [2]

Belton, G.R., & R.J. Fruehan. The determination of activities by mass spectrometry. I: The metallic systems iron-nickel and iron-cobalt. *J. Phys. Chem. 71*, 1403–1409 (1967).

Bender, E. Equations of state for ethylene and propylene. *Cryogenics 15* (11), 667–673 (1975).[1]

Benedict, M., G.B. Webb, & L.C. Rubin. An empirical equation for thermodynamic properties of light hydrocarbons and their mixtures. *J. Chem. Physics 8*, 334–345 (1940); *10* 747–758 (1942). [1]

Benedict, M., G.B. Webb, & L.C. Rubin. An empirical equation for thermodynamic properties of light hydrocarbons and their mixtures: Fugacities and liquid-vapor equilibria. *Chem. Eng. Progress 47* (8), 419 (1951); *47* (9), 449–454 (1951). [1]

Beret, S., & J.M. Prausnitz. Perturbed hard-chain theory: An equation of state for fluids containing small or large molecules. *AIChE Journal 21*, 1123–1132 (1975). [1]

Beutier, D., & H. Renon. Representation of NH_3-H_2S-H_2O, NH_3-CO_2-H_2O and NH_3-SO_2-H_20 vapor-liquid equilibria. *Ind. Eng. Chem. Process Des. Dev. 17*, 220–230 (1978). [9]

Birdi, K.S. Thermodynamics of micelle formation, K.L. Mittal (editor). *Solution Chemistry of Surfactants*, Vol. 1, 151–169 (1977). [9]

Bishnoi, P.R., R.D. Miranda, & D.B. Robinson. BWR Applied to NG/SNG needs. *Hydrocarbon Processing 54* (11), 197–201 (1974). [6]

Bishnoi, P.R., & D.B. Robinson. New mixing rules for the BWR parameters to predict mixture properties. *Canad. J. Chem. Eng. 50*, 101–107 (1972). [1]

Bishnoi, P.R., & D.B. Robinson. Mixing rules improve BWR use. *Hydrocarbon Processing 51* (11), 152–156 (1972). [1]

Black, C. Phase equilibria in binary and multicomponent systems: Modified van Laar-type equation. *Ind. Eng. Chem. 50*, 403–412 (1958). [7]

Bondi, A. Polymer equilibria, Storvick, T.S. & S.I. Sandler (editors). *Phase Equilibria and Fluid Properties in the Chemical Industry*. ACS Symposium Series *60*, 118–140 (1977). [9]

Boston, J.F., & P.M. Mathias. Phase equilibria in a third generation process simulator. *Phase Equilibria and Fluid Properties in the Chemical Industry*, 2nd Int. Conf., Berlin, March 17–21, 1980, EFCE DECHEMA (1980). [6]

Boublik, T. Progress in statistical thermodynamics applied to fluid phase. *Fluid Phase Equilibria I*, 37–87 (1977). [1]

Bradley, G., & J.K. Marsh. The system anthracene-phenanthrene. *J. Chem. Soc.* (London), 136–138, 650–652 (1933). [5]

Brancker, A.V., T.G. Hunter, & A.W. Nash. The quaternary system acetic acid–chloroform–acetone–water at 25 C. *J. Phys. Chem.* 683–698 (1940). [5]

Brandani, V. Use of infinite-dilution activity coefficients for predicting azeotrope formation at constant temperature and partial miscibility in binary liquid mixtures. *Ind. Eng. Chem. Fundamen. 13*, 154–156 (1974); *14*, 73 (1975). [4]

Brandani, V., G. Di Giacomo, & P.U. Foscolo. Isothermal vapor-liquid equilibria for the water-formaldehyde system: A predictive thermodynamic model. *Ind. Eng. Chem. Process Des. Dev. 19*, 179–185 (1980). [4]

Brandani, V., & J.M. Prausnitz. Thermodynamics of gas solubility in liquid solvents and solvent mixtures. *Fluid Phase Equilibria 7*, 259–274 (1981). [1]

Breedveld, G.J.F., & J.M. Prausnitz. Thermodynamic properties of supercritical fluids and their mixtures at very high pressures. *AIChE Journal 19*, 783–796 (1973). [6]

Brelvi, S.W. Fugacities of supercritical hydrogen, helium, nitrogen and methane in binary liquid mixtures. *Ind. Eng. Chem. Process Des. Dev. 19*, 80–84 (1980). [3]

Brielles, J., A. Dedit, M. Lallemand, B. Le Neindre, Y. Lerou, J. Vermese, & D. Vidal, Equation of state of gases at high pressures and low or moderate temperatures, B. Le Neindre & B. Vodar (editors). *Experimental Thermodynamics*, Vol. 2, 347–383, Butterworths (1975). [1]

Bril', Zh A., A.S. Mmozzhukhin, T.V. Petrova, F.B. Petlyuk, & L.A. Serafimov. Computer modelling of liquid-liquid-vapor equilibria in multicomponent mixtures. I: Liquid-vapor equilibria in binary mixtures; II: Liquid-liquid-vapor equilibrium in binary mixtures; III: Heterogeneous liquid equilibrium diagrams for ternary mixtures. *Russ. J. Phys. Chem. 47*, 1466–1472, 1556–1559 (1973). [7]

Brinkley, S.R. Note on the conditions of equilibrium for systems of many constituents. *J. Chem. Physics 14*, 563–564 (1946). [10]

Brinkman, N.D., L.C. Tao, & J.H. Weber, The calculation of the parameters for the Wilson equation for a ternary system. *Ind. Eng. Chem. Fundamen. 13*, 156–157 (1974). [4]

Brown, G.G., M. Souders, Jr., & R.L. Smith. Pressure-volume-temperature relations of paraffin hydrocarbons. *Ind. Eng. Chem. 24*, 513–515 (1931). [1]

Bruin, S. Activity coefficient relations in miscible and partially miscible multicomponent systems. *Ind. Eng. Chem. Fundamen. 9*, 305–314 (1970). [7]

Brule, M.R., C.T. Lin, L.L. Lee, & K.E. Starling. Multiparameter corresponding-states correlation of coal-fluid thermodynamic properties. *AIChE Journal 28*, 616–625 (1982). [9]

Byer, S.M., R.E. Gibbs, & H.C. Van Ness. Vapor-liquid equilibrium. II: Correlations from *P-x* data for 15 systems. *AIChE Journal 19*, 245–252 (1973). [6]

Budantseva, L.S., & T.M. Lesteva. A method for the calculation of the composition of the vapor phase from the concentration variation of the boiling points in binary systems. *Russ. J. Phys. Chem. 50*, 1019–1021 (1976). [6]

Cajander, B.C., H.G. Hipkin, & J.M. Lenoir. Prediction of equilibrium ratios from nomograms of improved accuracy. *J. Chem. Eng. Data 5*, 251–259 (1960). [6]

Carli, A. A correlation for enthalpy of petroleum fractions. *Chemical Processing*, 87–88 (April 1974). [11]

Carlson, H.C., & A.P. Colburn. Vapor-liquid equilibria of nonideal solutions. *Ind. Eng. Chem. 34*, 581–589 (1942). [4]

Carnahan, N.F., & K.E. Starling. Equation of state for nonattracting spheres. *J. Chem. Physics 51*, 635–636 (1969). [1]

Carnahan, N.F., & K.E. Starling. Intermolecular repulsions and the equation of state for fluids. *AIChE Journal 18*, 1184–1189 (1972). [1]

Carniti, P., L. Cori, & V. Ragaini. A critical analysis of the Hand and Othmer-Tobias correlations. *Fluid Phase Equilibria 2*, 39–47 (1978).

Carroll, B. On the use of Jacobians in thermodynamics. *J. Chem. Education 42* (4), 218–221 (1965). [2]

Castillo, J., & I.E. Grossmann. Computation of phase and chemical equilibria. *Computers and Chemical Engineering 5*, 99–108 (1981). [10]

Chang, S-D.,& B. C-Y. Yu. A generalized virial equation of state and its application to vapor-liquid equilibria at low temperatures. *Advances in Cryogenic Engineering 17*, 255–268 (1972). [1]

Chao, K.C., & J.D. Seader. A general correlation of vapor-liquid equilibria in hydrocarbon mixtures. *AIChE Journal 7*, 598–605 (1961). [6]

Chen, C.C., H.I. Britt, J.F. Boston, & L.B. Evans. Two new activity coefficient models for the vapor-liquid equilibrium of electrolyte systems. In Newman, S.A. (editor), *Thermodynamics of Aqueous Systems with Industrial Applications.* ACS Symposium Series *133*, American Chemical Society (1980). [9]

Cheng, S-I., & T.L. Chan. Estimation of Wilson parameters by non-linear regression. *The Chemical Engineering Journal 4*, 282–286 (1972). [4]

Chiu, J. Visual observation in differential thermal analysis. *Analytical Chemistry 35*, 933–934 (1963). [12]

Christiansen, L.J., M.L. Michelsen, & A. Fredenslund. Successive approximation in distillation calculations using the Soave-Redlich-Kwong equation of state. *12th Symposium on Computer Applications in Chemical Engineering* (1979) (CACE). [6]

Chueh, P.L., & J.M. Prausnitz. Vapor-liquid equilibria at high pressures: Vapor-phase fugacity coefficients in non-polar and quantum-gas mixtures. *Ind. Eng. Chem. Fundamen. 6*, 492–498 (1967). [1, 3]

Chueh, P.L., & J.M. Prausnitz. Calculation of high-pressure vapor-liquid equilibria. *Ind. Eng. Chem. 60*, 34–52 (1968). [6, 1]

Chung, W.K., S.E.M. Hamam, & B. C-Y. Lu. A modified Redlich-Kwong equation of state capable of representing the liquid state. *Ind. Eng. Chem. Fundamen. 16*, 494–495 (1977). [7]

Clark, F.G., & C.R. Koppany. Method tests V/L correlations. *Hydrocarbon Processing 57* (11), 282–286 (1978). [6]

Clark, P.E., A. Clow, E.K. Easterbrook, H.M. Haendler, & H.A. Iddles. Ternary systems with ammonium nitrate and guanidine nitrate. *J. Phys. & Colloid Chemistry 53*, 1009–1015 (1949). [5, 8]

Colburn, A.P. in *Chemical Engineers' Handbook*, 528 (1950).

Collatz, L. in *Handbuch der Physik 2*, 353 (1955).

Comings, E.W. *High Pressure Technology*, McGraw-Hill (1956).

Cooper, H.W., & J.C. Goldfrank. B-W-R constants and new correlations. *Hydrocarbon Processing 46*, (12), 141–146 (1967). [1]

Cope, J.Q., W.K. Lewis, & H.C. Weber. Generalized thermodynamic properties of higher hydrocarbon vapors. *Ind. Eng. Chem. 28*, 887–892 (1931). [1]

Corti, M., & V. Degiorgio. Investigation of aggregation phenomena in aqueous sodium dodecyl sulfate solutions at high NaCl concentration by quasielastic light scattering. In K.L. Mittal (editor), *Solution Chemistry of Surfactants*, Vol. 1, 377–390. Plenum Press (1979). [9]

Costa, E., J.L. Sotelo, G. Calleja, & C. Marron. Adsorption of binary and ternary hydrocarbon gas mixtures on activated carbon: Experimental determination and theoretical prediction of the ternary equilibrium data. *AIChE Journal 27*, 5–12 (1981). [9]

Coward, I., S.E. Gale & D.R. Webb. Process engineering calculations with equations of state. *Trans. I. Chem. E. 56*, 19–27 (1978). [6]

Coward, I., & D.R. Webb. Computational difficulties with the Chao and Seader correlation. *Computers and Chemical Engineering 2*, 177–187. (1978). [6]

Cox, J.D., & I.J. Lawrenson. The PVT behavior of single gases, M.L. McGlashan (editor). *Chemical Thermodynamics*, Vol. 1, A Specialist Periodical Report, 162–203. The Chemical Society, London (1973). [1]

Cragoe, C.S. in *International Critical Tables III* 228 (1928).

Cysewski, G.R., & J.M. Prausnitz. Estimation of gas solubilities in polar and nonpolar solvents. *Ind. Eng. Chem. Fundamen. 15*, 304–309 (1976). [6]

Dale, G.H. in *Encyclopedia of Chemical Processing and Design 13*, 464, Dekker (1981).

Danner, R.P., M.P. Nicoletti, & R.S. Al-Ameer. Determination of gas mixture adsorption equilibria by the tracer-pulse technique. *Chem. Eng. Science 35*, 2129–2133 (1980). [9]

Davidson, D.W. Clathrate hydrates. *Water, a Comprehensive Treatise*, edited by F. Franks, Vol. 2, Chap. 3, Plenum Press (1973). [9]

Deam, J.R., & R.N. Maddox. How to figure three-phase flash. *Hydrocarbon Processing 48* (7), 163–164 (1969). [7]

Deaton, W.M., & E.M. Frost. *Gas Hydrates*, U.S.B.M. Monograph 8 (1946).

Deiters, U., & G.M. Schneider. Fluid mixtures at high pressures: Computer calculations of the phase equilibria and the critical phenomena in fluid binary mixtures from the Redlich-Kwong equation of state. *Ber Bunsenges Physik Chemie 80*, 1316–1321 (1976). [9]

de Ligny, C.L., N.G. Van der Veen, & J.C. Van Houwellingen. Correlation and prediction of solubility and entropy of solution of gases in liquids by means of factor analysis. *Ind. Eng. Chem. Fundamen. 15*, 336–340 (1976). [6]

Deming, W.E., & L.E. Shupe. Some physical properties of compressed gases. *I. Nitrogen, Physical Review*, Second Series *37*, 638–654 (1931).

De Santis, R., F. Gironi, & L. Marrelli. Vapor-liquid equilibrium from a hard-sphere equation of state. *Ind. Eng. Chem. Fundamen. 15*, 183–189 (1976). [1]

De Santis, R., & B. Grande. An equation for predicting third virial coefficients of nonpolar gases. *AIChE Journal 25*, 931–938 (1979). [1]

De Santis, R., L. Marrelli, & P.N. Muscetta. Liquid-liquid equilibria in water-aliphatic alcohol system in the presence of sodium chloride. *The Chem. Eng. Journal 11*, 207–214 (1976). [7]

Dincer, S., D.C. Bonner, & R.A. Elefritz. Vapor-liquid equilibria in the benzene-polybutadiene-cyclohexane system. *Ind. Eng. Chem. Fundamen. 16*, 54–59 (1979). [6]

Djordjevic, B.D., A.N. Mihajlov, D.K. Grozdanic, A.Z. Tasic, & A.L. Horvath. Applicability of the Redlich-Kwong equation of state and its modifications to polar gases. *Chem. Eng. Science 32*, 1103–1107 (1977). [1]

Djordjevic, B.D., A.N. Mihajlov-Dudukovic, & A.Z. Tasic. Correlation of second virial coefficients of polar gases by Redlich-Kwong equation of state. *AIChE Journal 26*, 858–862 (1980). [1]

Domina, E.V., & N.D. Zakharov. A thermodynamic method for prediction of the critical parameters of ternary systems. *Russ. J. Phys. Chem. 52*, 1427–1429 (1978). [1]

Donohue, M.D., & J.M. Prausnitz. Statistical thermodynamics of solutions in natural gas and petroleum refining. *Research Report RR-26* (1977), Gas Processors Association, 1812 First Place, Tulsa, Oklahoma 74103. *AIChE Journal 24*, 849–860 (1978). [1]

Douslin, D.R. The pressure, volume and temperature properties of fluids. *International Review of Science, Physical Chemistry*. Series 2, Vol. 10: *Thermochemistry and Thermodynamics*, H.A. Skinner (editor), 191–246 (1975). [1]

Dowling, G.R., & W.G. Todd, Comparing vapor-liquid equilibrium correlations. *Chemical Engineering 80* (6), 115–120 (March 19, 1973). [6]

Drahos, J., I. Wichterle, & E. Hala. A generalized method for calculation and prediction of vapor-liquid equilibria at high pressures. *Fluid Phase Equilibria 1*, 173–184 (1978). [6]

Eagle, S., & J.W. Scott. Liquid phase adsorption equilibria and kinetics. *Ind. Eng. Chem. 42*, 1287–1294 (1950). [9]

Eckert, C.A., B.A. Newman, G.L. Nicolaides, & T.C. Long. Measurement and application of limiting activity coefficients. *AIChE Journal 27*, 33–40 (1981). [4]

Eckert, C.A., E.R. Thomas, & K.P. Johnston. Nonelectrolyte solutions: State of the art review. Proc. 2nd International Conf. on Phase Equilibria and Fluid Properties in the Chemical Industry. West Berlin, March 1980. DECHEMA. [4]

Eckert, C.A., E.R. Thomas, B.A. Newman, & G.L. Nicolaides. Limiting activity coefficients from differential ebulliometry. *J. Chem. Eng. Data 27*, 233–240 (1982). [4]

Edmister, W.C. Pseudo-critical for mixtures. *Petroleum Refiner 27* (4), 134–142 (1948). [1]

Edmister, W.C. Convergence correction to vapor-liquid equilibrium ratios. *Pet. Refiner 28* (9), 95–102 (September 1949). [6]

Edmister, W.C. Compressibility factors and equations of state. *Petroleum Refiner 37* (4), 173–179 (1958). [1]

Edmister, W.C. Isothermal pressure corrections to the enthalpy and entropy. *Hydrocarbon Processing 46* (4), 165–170 (1967). [11]

Edmister, W.C. Compressibility factors and fugacity coefficients from the Redlich-Kwong equation of state. *Hydrocarbon Processing 47* (9), 239–244 (1968). [1, 3]

Egan, C.J., & R.V. Luthy. Separation of xylenes: Selective solid compound formation with carbon tetrachloride. *Ind. Eng. Chem. 47*, 250–253 (1955). [8]

Eggers, R., & R. Tschiersch. Development and design of plant for high-pressure extraction of natural products. In G.M. Schneider, E. Stahl, & G. Wilke (editors), *Extraction with Supercritical Gases*, 176–188. Verlag Chemie (1980). [9]

Ekwall, P. Composition, properties and structures of liquid crystalline phases in systems of amphilic compounds, G.H. Brown (editor). *Advances in Liquid Crystals*, Vol. 1, 1–142 (1975). [9]

Ellis, S.R.M. Vapour phase extraction processes. *British Chem. Eng. 16*, 358–361 (1971). [9]

Ellis, S.R.M., & D.A. Jonah. Prediction of activity coefficients at infinite dilution. *Chem. Eng. Science 17*, 971–976 (1962). [4]

Elwell, R.H., & L.M. Welch. Rectification in ternary systems containing azeotropes. *Ind. Eng. Chem. 37*, 1224–1231 (1945). [4]

Elshayal, I.M. & B.C-Y. Lu. Prediction of vapor-liquid equilibria by means of a modified Clausius equation of state. *Canad. J. Chem. Eng. 51*, 76–81 (1973). [6]

Elshayal, I.M., & B.C-Y. Lu. Measurement of total pressures for ethylene-propane mixtures. *Canad. J. Chem. Eng. 53*, 83–87 (1975). [6]

Epstein, L.F. Redlich-Kwong equation of state: Exact critical constant relations. *Chem. Eng. Science 31*, 87 (1976). [1]

Evelein, K.A., & R.G. Moore. Prediction of phase equilibria in sour natural gas systems using the Soave-Redlich-Kwong equation of state. *Ind. Eng. Chem. Process Des. Dev. 18*, 618–624 (1979). [6]

Ewell, R.H., J.M. Harrison, & L. Berg. Azeotropic distillation. *Ind. Eng. Chem. 36*, 871–875 (1944). [1, 6]

Ezekwe, J. Ph. D. Thesis, University of Kansas (1982). [1]

Fair, J. Sorption processes. *Chem. Eng. 76* (14), 90–110 (July 14, 1969). [9]

Ferrell, R.A. in W.S. Goree & F. Chilton. *Conference on Fluctuations in Superconductors*, Stanford Research Institute (1968). [1]

Findlay, R.A., & J.A. Weedman. Separation and purification by crystallization. *Advances in Petroleum Chemistry and Refining*, Vol. 1, 119–208 (1958). [9]

Fishtine, S.H. Estimates of saturated fluid densities and critical constants. *Ind. Eng. Chem. Fundamen. 2*, 148–155 (1963). [11]

Francis, A.W. Ternary systems of liquid carbon dioxide. *J. Phys. Chem. 50*, 1099–1114 (1954). [5,7]

Francis, A.W. Ternary systems with three separate binodal curves. *J. Phys. Chem. 60*, 20–27 (1956). [5]

Fredenslund, A., J. Gmehling, M.L. Michelsen, P.

Rasmussen, & J.M. Prausnitz. Computerized design of multicomponent distillation columns using the UNIFAC group contribution method for calculation of activity coefficients. *Ind. Eng. Chem. Process Des. Dev. 16*, 450–462 (1977). [6]

Fredenslund, A., T. Jensen, T. Magnussen, M.L. Michelsen, & P. Rasmussen. Phase equilibria, flash and distillation calculations. *Computer-Aided Process Plant Design*, edited by M.E. Leesley. Gulf Publishing Company (1982). [6]

Fredenslund, A., R.L. Jones, & J.M. Prausnitz. Group-contribution estimation of activity coefficients in nonideal liquid mixtures. *AIChE Journal 21*, 1086–1098 (1975). [4]

Fredenslund, A., M.L. Michelsen, & J.M. Sorensen. Liquid-liquid equilibrium calculations using activity coefficient models. *Proceedings,* 2nd International Conference, Berlin, March 21, 1980, EFCE, DECHEMA, 433–444. [7]

Fredenslund, A., P. Rasmussen, & M.L. Michelsen. Recent progress in the computation of equilibrium ratios. *Symposium on Distillation, AIChE,* 87th National Meeting, Boston, August 19–22, 1979. [6]

Fredenslund, A., P. Rasmussen, & M.L. Michelsen. Recent progress in the computation of equilibrium ratios. *Chem. Eng. Commun. 4*, 485–500 (1980). [6]

Fredenslund, A., P. Rasmussen, & J. Mollerup. Thermophysical and transport properties for chemical process design. *Foundations of Computer-Aided Chemical Process Design*, edited by R.S.H. Mah & W.D. Seider, Vol. 2, 1–29 (1980) *AIChE*. [6]

Friman, R., & I. Daniellson, Lamellar mesophase with high contents of water—X-ray investigations of sodium octanoate-decanol-water system. *J. Colloid Interface Sci. 86*, 501–514 (1982). [9]

Fuller, G.G. A modified Redlich-Kwong equation of state capable of representing the liquid state. *Ind. Eng. Chem. Fundamen. 15*, 254–257 (1976). [1]

Gangoli, N., & G. Thodos. Liquid fuels and chemical feedstocks from coal by supercritical gas extraction: A review. *Ind. Eng. Chem. Prod. Res. 16*, 208–216 (1977). [9]

Gas Processors Association. GPA Peng-Robinson Programs, including new C_7+ Subroutine. Gas Processors Association, 1812 First Place, 15 East Fifth Street, Tulsa, Oklahoma 74103. [6]

Gautam, R., & W.D. Seider. Computation of phase and chemical equilibrium. I: Local and constrained minima in Gibbs free energy; II: Phase splitting; III: Electrolytic solutions. *AIChE Journal 25*, 991–1015 (1979). [10]

George, B., L.P. Brown, C.H. Farmer, P. Buthod, & F.S. Manning. Computation of multicomponent, multiphase equilibrium. *Ind. Eng. Chem. Process Des. Dev. 15*, 372–377 (1976). [7]

Gere, J.M., & W. Weaver, Matrix Algebra for Engineers. Van Nostrand (1965). [10]

Germann, F.E.E., & R.P. Germann. Line coordinate representation of solubility curves. *Ind. Eng. Chem. 36*, 93–96 (1944). [6]

Ghosh, S.K., & S.J. Chopra. Activity coefficients from the Wilson equation. *Ind. Eng. Chem. Process Des. Dev. 14*, 304–308 (1975). [4]

Gibbs, R.E., & H.C. Van Ness. Vapor-liquid equilibria from total-pressure measurements: A new apparatus. *Ind. Eng. Chem. Fundamen. 11*, 410–413 (1972). [12]

Gibbons, R.M. The equation of state of neon between 27 and 70 K. *Cryogenics 9* (8), 251–260 (1969). [1]

Gibbons, R.M. Equations for the second virial coefficient of polar molecules. *Cryogenics 14* (7), 399–403 (1974). [1]

Gibbons, R.M., J.R. Coulthurst, D. Farrell, D. Gough, & J. Gillett. Industrial uses of thermodynamic data. *Proceedings*, NPL Conference Chemical Thermodynamic Data on Fluids and Fluids Mixtures. IPC Science and Technology Press (1979). [11]

Gillespie, D.T. Vapor-liquid equilibrium still for miscible liquids. *Ind. Eng. Chem. Analytical Ed. 18*, 575–577 (1946). [12]

Gmehling, J.G., T.F. Anderson, & J.M. Prausnitz. Solid-liquid equilibria using UNIFAC. *Ind. Eng. Chem. Fundamen. 17*, 269–273 (1978). [8]

Gmehling, J., P. Rasmussen, & F. Fredenslund. Vapor liquid equilibria by UNIFAC group contribution revision and extension. 2. *Ind. Eng. Chem. Process Des. Dev. 21*, 118–127 (1982). [4]

Gmehling, J., & U. Onken. Calculation of activity coefficients from structural group contributions. *Int. Chem. Eng. 19*, 566–570 (1979). [4]

Goldwasser, S.R. Basis for calculating equilibrium gas composition on a digital computer. *Ind. Eng. Chem. 51*, 595–596 (1959). [10]

Gomez-Nieto, M., & G. Thodos. Benedict-Webb-Rubin parameters for ethanol. *Canad. J. Chem. Eng. 54*, 438–440 (1976). [1]

Gomez-Nieto, M., & G. Thodos. Generalized vapor pressure equation for nonpolar substances. *Ind. Eng. Fundamen. 17*, 45–52 (1978). [1]

Gomez-Nieto, M., & G. Thodos. Vapor-liquid equilibrium behavior for the propane-acetone system at elevated pressures. *Chem. Eng. Science 33*, 1589–1595 (1978). [6]

Gothard, F.A., M.F. Codrea Ciobanu, D.G. Breban, C.I. Bucur, & G.V. Sorescu. Predicting the parameters in the Wilson equations for activity coefficients in binary hydrocarbon systems. *Ind. Eng. Chem. Process Des. Dev. 15*, 333–337 (1976). [4]

Graboski, M.S., & T.E. Daubert. A modified Soave equation of state for phase equilibrium calculations. 1: Hydrocarbon systems, *Ind. Eng. Chem. Process Des. Dev. 17*, 443–448 (1978). 2: Systems containing CO_2, H_2S, N_2 and CO. Ibid, 448–454 (1978). 3: Systems containing hydrogen. Ibid, *18*, 300–306, (1979). [1]

Gray, R.D. Industrial experience in applying the Redlich-Kwong equation to vapor-liquid equilibria. K.C. Chao & R.L. Robinson (editors); Equations of state in engineering and research, 253–270. *Advances in Chemistry Series 182*. American Chemical Society (1979). [1]

Gray, R.D. Case studies of difficult industrial problems: Computation of phase equilibria in the critical region. *Proc., 2nd Int. Conf.*, Berlin, March 21, 1980. EFCE, DECHEMA, 919–928. [6]

Grayson, H.G., & C.W. Streed. Vapor-liquid equilibrium for high temperature, high pressure hydrogen-hydrocarbon mixtures. *Sixth World Petroleum Congress Proceedings*, Sect. III, 233–245 (1963). [6]

Griffiths, R.B., & J.C. Wheeler. Critical points in multicomponent systems. *Physical Review* A *2* 1047–1063 (1970). [1]

Grossmann, W.E., & J. Davidson. Computation of restricted chemical equilibria. *Computers and Chem. Eng. 6*, 181–184 (1982). [10]

Guerreri, G. Vapour-liquid equilibria in non-polar multicomponent systems, 1. *British Chemical Engineering 15*, 927–931 (1970). [6]

Gunn, R.D. Corresponding states: Theoretical development for mixtures. *AIChE Journal 18*, 183–188 (1972). [1]

Hadden, S.T., & H.G. Grayson. New charts for hydrocarbon vapor-liquid equilibria. *Hydrocarbon Processing 40* (9), 207–218 (1961). [6]

Hala, E. Vapor-liquid equilibrium in the critical region: Concentration limits of binary systems. *Ind. Eng. Chem. Fundamen. 14*, 136–137 (1975). [6]

Halm, R.L., & L.I. Stiel. A fourth parameter for the vapor pressure and entropy of vaporization of polar fluids. *AIChE Journal 13*, 351– 355 (1967). [1]

Hammick, D.L., G.M. Hills, & J. Howard. The composition of the compounds of picryl chloride and of *s*-trinitrobenzene and benzene. *J. Chem. Soc. London*, 1530–1532 (1932). [5]

Hamam, S.E., W.K. Chung, I.M. Elshayal, & B.C-Y Lu. Generalized temperature-dependent parameters of the Redlich-Kwong equation of state for vapor-liquid equilibrium calculations. *Ind. Eng. Chem. Process Des. Dev. 16*, 51–59 (1977). [6]

Handa, Y.P., & G.R. Benson. Volume changes on mixing two liquids: A review of the experimental techniques and the literature data. *Fluid Phase Equilibria 3*, 185–249 (1979). [11]

Hankinson, R.W., T.A. Coker, & H.G. Thomson. Get accurate LNG densities with COSTALD. *Hydrocarbon Processing 61* (4), 207–208 (1982). [1]

Hankinson, R.W., B.D. Langfitt, & D.P. Tassios. A single parameter equation for the prediction of multicomponent vapor-liquid data from binary isobaric data. *Canad. J. Chem. Eng. 50*, 511–514 (1972). [6]

Hankinson, R.W., & G.H. Thomson. A new correlation for saturated densities of liquids and their mixtures. *AIChE Journal 25*, 653–663 (1979). [1]

Hanks, R.W., T.K. O'Neill, & J.J. Christensen. The prediction of vapor-liquid equilibrium from heat of mixing for binary hydrocarbon-alcohol mixtures. *Ind. Eng. Process Des. Dev. 18*, 408–414 (1979). [11]

Hansen, C., & A. Beerbower. Solubility parameters. *Encyclopedia of Chemical Technology*, Supplement (1971). [4]

Harmens, A., & H. Knapp. Three-parameter cubic equation of state for normal substances. *Ind. Eng. Fundamen. 19*, 291–294 (1980). [1]

Hayden, J.G., & J.P. O'Connell. A generalized method for predicting second virial coefficients. *Ind. Eng. Chem. Process Des. Dev. 14*, 209–216 (1975). [1]

Hecht, G., & C. Holste. Limitations of density calculations with reduced equations of state (in German). *Chem. Technik 17*, 518–524 (1965). [11]

Helpinstill, J.G., & M. Van Winkle. Prediction of infinite dilution activity coefficients for polar-polar binary systems. *Ind. Eng. Chem. Process Des. Dev. 7*, 213–220 (1968). [4]

Herington, E.F.G. Tests for consistency of experimental isobaric vapor liquid equilibrium data. *Institute of Petroleum Journal 37*, 457–470 (1951); symmetrical-area tests for the examination of the reliability of vapour-liquid equilibrium data. *Inst. Chem. Engineers* (London), International Symposium on Distillation, September 8–10, 1969, *3*, 17–24. [4]

Herskovitz, M., & M. Gottlieb. UNIFAC group contribution method for silicone compounds. *Ind. Eng. Chem. Process Des. Dev. 20*, 407–409 (1981). [4]

Heidemann, R.A. Predict three-phase equilibria. *Hydrocarbon Processing 53* (11), 167–170 (1974). [7]

Heidemann, R.A. Three-phase equilibria using equations of state. *AIChE Journal 20*, 847–855 (1974). [7]

Heidemann, R.A. Use of infinite-dilution activity coefficients for predicting azeotrope formation at constant temperature and partial miscibility in binary liquid mixtures. *Ind. Eng. Chem. Fundamen. 14*, 72–73 (1975). [4]

Heidemann, R.A., & A.M. Khalil. The calculation of critical points. *AIChE Journal 26*, 769–779 (1980). [1]

Hildebrand, J.H., & R.H. Lamoreaux. Solubility of gases in liquids: Fact and theory. *Ind. Eng. Chem. Fundamen. 13*, 110–115 (1974). [6]

Hildebrand, J.H., & J.N. Sharma. The activities of molten alloys of thallium with tin and lead. *JACS*, 462–471 (1929). [8]

Hill, A.E., The mutual solubility of liquids. I. Mutual solubility of ethyl ether and water. II. The solubility of water in benzene. *JACS 45* 1143–1155 (1923) [9]

Hiranuma, M. A new expression similar to the three-parameter Wilson equation. *Ind. Eng. Chem. Fundamen. 13*, 219–222 (1974). [4]

Hiranuma, M. The contribution of non-neighbor pairs to free energy of mixing. *J. Chem. Eng. Japan 8*, 69–77 (1975). [1]

Hiza, M.J., & A.G. Duncan. A correlation for the prediction of interaction energy parameters for mixtures of small molecules. *AIChE Journal 16*, 733–737 (1970). [1]

Hlavaty, K. Correlation of the binodal curve in a ternary liquid mixture with one pair of immiscible liquids. *Coll. Czech. Chem. Commun. 37*, 4005–4007 (1972). [7]

Hlavaty, K. Semirational method for the construction of equations of state for liquids and gases. *Coll. Czech. Chem. Commun. 39*, 2927–2934 (1974). [1]

Holder, G.H., & J.H. Hand. Multiple-phase equilibria in hydrates from methane, ethane, propane and water. *AIChE Journal 28*, 440–447 (1982). [9]

Holland, R.E. Correlation of vapor-liquid equilibria using solubility parameters. Ph. D. thesis, Pennsylvania State University (1971). [4]

Holmes, M.J., & M. Van Winkle. Prediction of ternary vapor-liquid equilibria from binary data. *Ind. Eng. Chem. 62*, 21–31 (1970). [6]

Holste, J.C., M.Q. Watson, M.T. Bellomy, P.T. Eubank, & K.R. Hall. Determination of interaction second virial coefficients: He-CO_2 system. *AIChE Journal 26*, 954–964 (1980). [1]

Horry, S.E., & J.M. Prausnitz. Molecular thermodynamics of monolayer gas adsorption on homogeneous and heterogeneous solid surfaces. *AIChE Symposium Series 74*, 3–9 (1967). [9]

Hopke, S.W. Application of equations of state in Exxon's production operations. *ACS Symposium Series 60*, 221–223 (1977). [1]

Horvath, A.L. Redlich-Kwong equation of state: review for chemical engineering calculations. *Chem. Eng. Science 29*, 1334–1340 (1974). [1]

Horvath, C., & H.-J. Lin. A simple three-parameter equation of state with critical compressibility-factor correlation. *Canad. J. Chem. Eng. 55*, 450–456 (1977). [1]

Hoy, K.L. New values of the solubility parameters from vapor pressure data. *J. Paint Technology 42*, No. 541, 76–118 (1970). [4]

Hrynakowski, K., & M. Szmyt. Solid-liquid equilibria in systems in which incongruent melting binary compounds occur. III (in German). *Z. Physik Chem. 175A*, 83–98 (1935). [5]

Hsu, E.C.-H., & J.F. Johnson. Prediction of eutectic temperatures, compositions and phase diagrams for binary mesophase systems. *Mol. Cryst. Liq. Cryst. 27*, 95–104 (1974). [9]

Hulme, D.S., & E.P. Raynes. Eutectic mixtures of nematic 4′-substituted 4-cyanobiphenyls. *J.C.S. Chem. Comm.*, 98–99 (1974). [9]

Huron, M.J. Use of the Soave equation and of the stability conditions for calculating the critical points of mixtures. *Chem. Eng. Science 31*, 837–839 (1976). [1]

Huron, M.J., G.N. Dufour, & J. Vidal. Vapour-liquid equilibrium and critical locus curve calculations with the Soave equation for hydrocarbon systems with carbon dioxide and hydrogen sulfide. *Fluid Phase Equilibria 1*, 247–265 (1978). [6]

Huron, M.J., & J. Vidal. New mixing rules in simple equations of state for representing vapour-liquid equilibria of strongly non-ideal mixtures. *Fluid Phase Equilibria 3*, 255–271 (1979). [1]

Hutchins, R.A. Liquid phase adsorption—Maximizing performance. *Chem. Eng. 87* (4), 101–110 (February 25, 1980). [9]

Inoue, M., K. Azumi, & N. Suzuki. A new vapor pressure assembly for static vapor-liquid equilibrium. *Ind. Eng. Chem. Fundamen. 14*, 312–314 (1975). [12]

Ishikawa, T., W.K. Chung, & B.C.-Y. Lu. A cubic perturbed, hard sphere equation of state for thermodynamic properties and vapor-liquid equilibrium calculations. *AIChE Journal 26*, 372–378 (1980). [1]

Jänecke, E. The system $H_2O + CO_2 + NH_3$. *Z für Elektrochem 35*, 716–727 (1929). [5], [7]

Jelinek, J., & V. Hlavacek. Compute boiling points faster. *Hydrocarbon Processing 50* (8), 135–136 (1971). [6]

Jenkins, J.D., & M. Gibson-Robinson. Vapour-liquid equilibrium in systems with association in both phases: The simultaneous evaluation of liquid phase models and thermodynamic consistency testing for the system acetic acid-toluene. *Chem. Eng. Science 32*, 931–938 (1977). [6]

Joffe, J. Fugacities in gas mixtures. *Ind. Eng. Chem. 40*, 1738–1741, 2439–2442 (1948). [3]

Joffe, J. Combining rules for the third parameter in the pseudocritical method for mixtures. *Ind. Eng. Chem. Fundamen. 10*, 532–533 (1971). [1]

Joffe, J. Vapor-liquid equilibria by the pseudocritical method. *Ind. Eng. Chem. Fundamen. 15*, 298–303 (1976). [6]

Joffe, J. Vapor-liquid equilibria and densities with the Martin equation of state. *Ind. Eng. Chem. Process Des. Dev. 20*, 168–172 (1981). [1]

Joffe, J., H. Joseph, & D. Tassios. Vapor-liquid equilibria with a modified Martin equation, S.A. Newman (editor). *Chemical Engineering Thermodynamics*, 211–220 (1982), Ann Arbor Science. [1,6]

Joffe, J., & D. Zudkevitch. Prediction of critical properties of mixtures: Rigorous procedure for binary mixtures. *AIChE Symposium Series 81*, 43–51 (1967). [1]

Joffe, J., & D. Zudkevitch. Prediction of liquid-phase enthalpies with the Redlich-Kwong equation of state. *Ind. Eng. Chem. Fundamen. 9*, 545–548 (1970). [11]

Johnson, D.W., & C.P. Colver. Mixture properties by computer. *Hydrocarbon Processing 47* (12), 79–83 (1968); *48* (1), 127–133 (1969). [1,11]

Jolls, K.R., & G.P. Willers. Computer generated phase diagrams for ethylene and propylene. *Cryogenics 18* (6), 329–336 (1978). [2]

Jolls, K.R., G.P. Willers, & L.D. Jenson. The computer generation of thermodynamic phase diagrams. *Trans. of the Computers in Education Division of ASEE 8* (10) (1976). [1,2]

Kalra, H., & D.B. Robinson. Vapor-liquid equilibrium in a six-component simulated sour natural gas system at sub-ambient temperatures. *Fluid Phase Equilibria 3*, 133–144 (1979). [6]

Kandiner, H.J., & S.R. Brinkley, Jr. Calculation of complex equilibrium relations. *Ind. Eng. Chem. 42*, 850–855, 1526 (1950). [10]

Kappallo, W., N. Lund & K. Schaffer. Intermolecular forces between equal and different molecules from the virial coefficient. *Z. phys. Chem.* (Frankfurt) *37* (3, 4) 196–209 (1963).

Kata, K., K. Nagahama, & M. Hirata. Generalized interaction parameters for the Peng-Robinson equation of state: carbon dioxide-*n*-paraffin binary systems. *Fluid Phase Equilibria 7*, 219–231 (1981). [1]

Kato, M., W.K. Chung, & B.C.-Y. Lu. Modified parameters for the Redlich-Kwong equation of state. *Canad. J. Chem. Eng. 54*, 441–445 (1976). [1]

Kato, M., W.K. Chung, & B. C.-Y. Lu. Binary interaction coefficients of the Redlich-Kwong equation of state. *Chem. Eng. Science 31*, 773–776 (1976). [1]

Katz, D.L., & A. Firoozabadi. Predicting phase behavior of condensate/crude oil systems using methane interaction coefficients. *J. Pet. Tech.* 1649–1655 (November 1978). [1]

Katz, D.L., & F. Kurata. Retrograde condensation. *Ind. Eng. Chem. 32*, 817–827 (1940). [6]

Kaul, B.K., & J.M. Prausnitz. Second virial coefficients of gas mixtures containing simple fluids and heavy hydrocarbons. *Ind. Eng. Chem. Fundamen. 16*, 335–339 (1977). [1]

Kay, W.B. Density of hydrocarbon gases and vapors at high temperature and pressure. *Ind. Eng. Chem. 28*, 1014–1019 (1936). [1]

Kay, W.B. Liquid-vapor phase equilibrium relations in the ethane-*n*-heptane system. *Ind. Eng. Chem. 30*, 459–465 (1938). [6]

Kehiaian, H.V., J.P.E. Grolier, M.R. Kechavarz, G.C. Benson, O. Kiyohara, & Y.P. Handa. Thermodynamic properties of binary mixtures containing ketones. VII: Analysis of the properties of *n*-alkanone + *n*-alkane, and *n*-alkanone+*n*-alkanone mixtures in terms of a quasi-chemical group contribution model. *Fluid Phase Equilibria 7*, 95–120 (1981). [4]

Kesler, M.G., & B.I. Lee. On the development of an equation of state for vapor-liquid equilibrium calculations. *ACS Symposium Series 60*, 236–240 (1977). [1]

Kesler, M.G., B.I. Lee, D.W. Benzing, & A. Cruz. Method improves convergence pressure predictions. *Hydrocarbon Processing 48* (6), 177–179 (1969). [6]

Kesler, M.G., B.I. Lee, & S.I. Sandler. A third parameter for use in generalized thermodynamic correlations. *Ind. Eng. Chem. Fundamen. 18*, 49–54 (1979). [1]

Kikic, I., & P. Alessi. Liquid-liquid equilibrium for the activity coefficient determination: Effect of mutual solubility between binary systems and auxiliary solvents. *Canad. J.*

Chem. Eng. 55, 78–81 (1977). [7]

Kikic, I., P. Allessi, P. Rasmussen, & A. Fredenslund. On the combinatorial part of the UNIQUAC and UNIFAC models. *Canad. J. Chem. Eng. 58*, 253–258 (1980). [4]

King, M.B., J. Al-Najjar, & K. Kassim. The solubilities of carbon dioxide, hydrogen sulfide and propane in some normal alkane solvents. I: Experimental determinations in the range 15–70 C and comparison with ideal solution values; II: Correlation of data at 25 C in terms of solubility parameters and regular solution theory. *Chem. Eng. Science 32*, 1241–1252 (1977). [6]

King, M.B., H.S.H. Al-Najjar, & J.K. Ali. Integration of the Jonah-King equation to predict gas solubilities over a range of temperatures when the solubility at one temperature is known. *Chem. Eng. Science 34*, 1080–1082 (1979). [6]

Knobler, C.M. Volumetric properties of gaseous mixtures, M.L. McGlashan (editor). *Chemical Thermodynamics*, Vol. 2, a Specialist Periodical Report, 199–237. The Chemical Society, London (1978). [1]

Knox, W.G., M. Hess, G.E. Jones, & H.B. Smith. The hydrate process. *Chem. Eng. Progress 57* (2), 66–71 (1961). [9]

Kobayashi, R., & D.L. Katz. Vapor-liquid equilibria for binary hydrocarbon-water systems. *Ind. Eng. Chem. 45*, 440–451 (1953). [9]

Kogan, V.B., & I.N. Tsiparis. Verification of liquid-vapor equilibrium data for ternary systems with chemical interaction between the components; systems in which the liquid phases are saturated solutions of nonvolatile substances. *J. Applied Chem. USSR 41*, 2517–2521 (1968). [6]

Kohman, G.T., & D.H. Andrews. Solubility relations in isomeric organic compounds. V: The construction of the ideal ternary solubility diagram and its use in analysis. *J. Phys. Chem. 29*, 1317–1324 (1925).

Koningsveld, R. Thermodynamics of polymer solutions. *Int. Chem. Eng. 19*, 420–429 (1979). [4]

Konstam, A.H., & W.R. Feairheller, Jr. Calculation of solubility parameters of polar compounds. *AIChE Journal 16*, 837–840 (1970). [4]

Krolikowski, T.S. Industrial view of the state of the art in phase equilibria, T.S. Storvick & S.I. Sandler (editors); Phase equilibria and fluid properties in the chemical industry, 62–86 *ACS Symposium Series 60*. American Chemical Society (1977). [1]

Kruis, A., in *Landolt-Börnstein* IV/4c/5 (1976). [9]

Krumins, A.E., A.K. Rastogi, M.E. Rusak, & D. Tassios. Prediction of binary vapor-liquid equilibrium from one-parameter equations. *Canad. J. Chem. Eng. 58*, 663–669 (1980). [6]

Ku, P.S., & B.F. Dodge. Compressibility of the binary systems, helium + nitrogen and carbon dioxide + ethylene. *J. Chem. Eng. Data 12*, 158–164 (1967). [1]

Kudchaker, A.P. Thermodynamic properties of Refrigerant 500, K.C. Chao & R.L. Robinson (editors); Equations of state in engineering and research. *Advances in Chemistry Series 182*, 305–322. American Chemical Society (1979). [11]

Kula, M-R. Extraction and purification of enzymes using aqueous two-phase systems. In L.B. Wingard et al. (editors), *Applied Biochemistry and Bioengineering*, Vol. 2, 71–195, Academic Press (1979). [9]

Kummel, R., & G. Wilde. Osmotic and activity coefficients in the ternary system water-calcium choloride-urea at 298.15 K. *Fluid Phase Equilibria 2*, 215–223 (1978). [4]

Kurnik, R.T., S.J. Holla, & R.C. Reid. Solubility of solids in supercritical carbon dioxide and ethylene. *J. Chem. Eng. Data 26*, 47–51 (1981). [9]

Labinov, C.D., H.V. Boiko, & H.K. Bolomin. A new method of finding the constants of the B-W-R equation for gaseous mixtures (in Russian). *Zhurnal fizicheskii Khimii 41*, 618–621 (1967).

Ladurelli, A.J., C.H. Eon, & G. Guiochon. Fallibilities inherent in the Wilson equation applied to systems having a negative excess Gibbs energy. *Ind. Eng. Chem. Fundamen. 14*, 191–195 (1975). [4]

Lahiri, A.K. Multicomponent multiphase equilibria. *Fluid Phase Equilibria 3*, 113–121 (1979). [7]

Lambert, J.D., G.A.H. Roberts, J.S. Rowlinson, & V.J. Wilkinson. Second virial coefficients of organic vapors. *Proc. Roy. Soc. London 196A*, 113–125 (1949). [1]

Langmuir, I. Third Colloid Symposium Monograph, Reinhold (1925). [9]

Lapina, R.P. Predict water thermo properties with BWRS. *Hydrocarbon Processing 54* (2), 115–118 (1975). [11]

Larrinaga, L. Graphically determining the Wilson parameters. *Chem. Eng. 88* (7), 87–91 (April 7, 1981). [4]

Leach, M.J. An approach to multiphase vapor-liquid equilibria. *Chemical Engineering 84* (10), 137–140 (May 23, 1977). [7]

Lee, B.I., & M.G. Kesler. A generalized thermodynamic correlation based on three-parameter corresponding states. *AIChE Journal 21*, 510–527 (1975). [1]

Lee, H.H., & J.C. Warner. The ternary system diphenyl-diphenylamine-benzophenone. *J. Am. Chem. Soc. 55*, 4474–4477 (1933). [8]

Lee, H.H., & J.C. Warner. The system biphenyl-bibenzyl-naphthalene. Nearly ideal binary and ternary systems. *J. Am. Chem. Soc. 57*, 318–321 (1935). [8]

Lee, S.-M., P.T. Eubank, & K.R. Hall. Truncation errors associated with the virial equation. *Fluid Phase Equilibria 1*, 219–224 (1978). [1]

Leesley, M.E., & G. Heyen. The dynamic approximation method of handling vapor-liquid equilibrium data in computer calculations for chemical processes. In *Computer-aided Process Plant Design*, edited by M.E. Leesley. Gulf Publishing Company (1982). [6]

Legret, D., D. Richon, & H. Renon. Vapor liquid equilibria up to 100 MPa: A new apparatus. *AIChE Journal 27*, 203–207 (1981). [11]

Leland, T.W. Equations of state for phase equilibrium computations: Present capabilities and future trends. *Proceedings*, 2nd International Conference, Berlin, March 21, 1980. EFCE, DECHEMA, 283–334. [1]

Leland, T.W., & P.S. Chappelear. The corresponding states principle: A review of current theory and practice. *Ind. Eng. Chem. 60*, 15–43 (1968). [1]

Leland, T.W., & W.H. Mueller. Applying the theory of corresponding states to multicomponent systems. *Ind. Eng. Chem. 51*, 597–600 (1959). [1]

Lenoir, J.M. Predict K values at low temperatures. *Hydrocarbon Processing 48* (9), 167–172; 48 (11), 121–124 (1969). [6]

Lenoir, J.M., & C.R. Koppany. Need equilibrium ratios? Do it right. *Hydrocarbon Processing 46* (11), 249–252 (1967). [6]

Lenoir, J.M., & G.A. White. Predicting convergence pressure. *Petroleum Refiner 37* (3), 173–181 (1958). [6]

Letcher, T.M. Activity coefficients at infinite dilution from gas-liquid chromatography, M.L. McGlashan (editor). In

Chemical Thermodynamics, Vol. 2 A Specialist Periodical Report, 46–70. The Chemical Society, London (1978). [4]

Levelt-Sengers, J.M.H. Critical behavior in fluids. In *High Pressure Technology*, Vol. II, edited by I.L. Spain and J. Paauwe. Dekker (1977). [1,9]

Levelt-Sengers, J.M.H. Critical exponents at the turn of the century. *Physica 82A*, 319–351 (1976). [1]

Lewis, W.K., E.R. Gilliland, B. Chertow, & W.P. Cadogan. Adsorption equilibria: Hydrocarbon mixtures. *Ind. Eng. Chem. 42*, 1319–1332 (1950). [9]

Leyendecker, W.R., & R.D. Gunn. Prediction of component fugacities and related properties of mixtures. *AIChE Journal 18*, 188–193 (1972). [3]

Lin, C.-T., & T.E. Daubert. Prediction of the fugacity coefficients of nonpolar hydrocarbon systems from equations of state. *Ind. Eng. Chem. Process Des. Dev. 17*, 544–549 (1978). [3]

Lin, C.T., & T.E. Daubert. Prediction of partial molar volume from the Lee-Kesler equation of state. *AIChE Journal 25*, 365–367 (1979). [11]

Lin, C.-T., & T.E. Daubert. Estimation of partial molar volume and fugacity coefficients in mixtures from the Soave and Peng-Robinson equations of state. *Ind. Eng. Chem. Process Des. Dev. 19*, 51–59 (1980). [11]

Lin, C.-J., & S.W. Hopke. Application of the BWRS equation to methane, ethane, propane and nitrogen systems. *AIChE Symposium Series 140*, 37–47 (1974). [6]

Lindman, B., G. Lindblom, H. Wennerstrom, & H. Gustavvson. Ionic interactions in amphilic systems studied by NMR. K.L. Mittal (editor), *Solution Chemistry of Surfactants*, Vol. 1, 195–227 (1977). [9]

Lu, B. C.-Y, P. Yu, & A.H. Sugie. Prediction of vapor-liquid-liquid equilibria by means of a modified regula falsi. *Chem. Eng. Science 29*, 321–326 (1974). [7]

Lucas, K. Calculation of properties of gases and liquids by the molecular theory—state of knowledge and application in chemical engineering. *Ind. Chem. Eng. 18*, 408–416 (1978). [11]

Luft, N.W. Heats of reaction. *Chem. Eng. 64* (11), 235–237 (1957). [10]

Luna, J.L.F., & F.B. de Castro. Evaluations of various modifications of the Redlich-Kwong equation. *Ind. Chem. Eng. 18*, 611–626 (1978). [1]

Lydersen, A.L., R.A. Greenkorn, & O.A. Hougen. Generalized thermodynamic properties of pure fluids. *Univ. Wisconsin Eng. Exp. Sta. Rept.* (October 4, 1955). (The data are reproduced by Hougen et al., *Thermodynamics*, 1959.)

McCarty, R.D. A modified Benedict-Webb-Rubin equation of state for methane using recent experimental data. *Cryogenics 14* (5), 276–280 (1974). [1]

McGlashan, M.L., & A.G. Williamson. Thermodynamics of mixtures of *n*-hexane and *n*-hexadecane, Part 2: Vapor pressures and activity coefficients. *Trans. Faraday Soc. 57*, 588–600 (1961). [4]

Maddocks, R.R., J. Gibson, & D.F. Williams. Supercritical extraction of coal. *Chem. Eng. Progress 75* (6), 49–55 (1979). [9]

Maffiolo, G., J. Vidal, & L. Asselineau. Vapor-liquid equilibrium constants of hydrocarbon mixtures at moderate pressures (in French). *Chem. Eng. Science 30*, 625–630. [6]

Magnussen, T., P. Rasmussen, & A. Fredenslund. UNIFAC parameter table for prediction of liquid-liquid equilibria. *Ind. Eng. Chem. Process Des. Dev. 20*, 331–339 (1981). [4]

Magnussen, T., J.M. Sorensen, P. Rasmussen, & A. Fredenslund. Liquid-liquid equilibrium data: Prediction. *Fluid Phase Equilibria 4*, 151–163 (1980). [7]

Maher, P.J., & B.D. Smith. Infinite dilution activity coefficient values from total pressure VLE data: Effect of equation of state used. *Ind. Eng. Chem. Fundamen. 18*, 354–357 (1979). [4]

Mair, B.J., J.W. Westhaver, & F.D. Rossini. Theoretical analysis of fractionating processes of adsorption. *Ind. Eng. Chem. 42*, 1279–1286 (1950). [9]

Malbrunot, P. *PVT* relationships in gases at high pressures and high temperatures. B. Le Neindre & B. Vodar (editors), *Experimental Thermodynamics*, Vol. II, 383–420. Butterworths (1975). [1]

Mandell, L., & P. Ekwal. The three-component system sodium caprylate-decanol–water. The phase equilibria at 20 C. *Acta Polytechnica Scandinavica*, Series 74, I-III (1968). [9]

Mantell, C.L. *Adsorption* 162, McGraw-Hill (1951). [9]

Mapstone, G.E. Tie-line interpolation and checking. *British Chem. Eng. 15*, 778–779 (1970). [7]

Marina, J.M., & D.P. Tassios. Effective local compositions in phase equilibrium correlations. *Ind. Eng. Chem. Process Des. Dev. 12*, 67–71 (1973). [4]

Martin, J.J. Equations of state. *Ind. Eng. Chem. 59*, 34–52 (1967). [1]

Martin, J.J. Cubic equations of state—which? *Ind. Eng. Chem. Fundamen. 18*, 81–97 (1979); *19*, 128–129 (1980). [1]

Martinez-Ortiz, J.A., & D.B. Manley. Direct solution of the isothermal Gibbs-Duhem equation by an iterative method for binary systems. *AIChE Journal 23*, 393–395 (1977). [2]

Martinez-Ortiz, J.A., & D.B. Manley. Direct solution of the isothermal Gibbs-Duhem equation for multicomponent systems. *Ind. Chem. Process Des. Dev. 17*, 246–351 (1978). [2]

Masuoka, H., R. Tawaraya, & S. Saito. Calculation of solid-liquid equilibria using the modified BWR equation of state of Lee and Kesler. *J. Chem. Eng. Japan 12*, 257–262 (1979). [8]

Maurer, G., & J.M. Prausnitz. On the derivation and extension of the UNIQUAC equation. *Fluid Phase Equilibria 2*, 9–99 (1978). [4,7]

Mauri, C. Unified procedure for solving multiphase-multicomponent vapor-liquid equilibrium calculation. *Ind. Eng. Chem. Process Des. Dev. 19*, 482–489 (1980). [7]

Mauser, H. Phase diagrams: Their determination and use (in German). Ullmann, *Enzyklopadie der technischen Chemie*, II/1. Verlag Chemie (1972). [5]

Maxwell, J.C. On the dynamical evidence of the molecular constitution of matter, *Nature 11*, 357–359, 374–377 (1975). [1]

Mecke, R. Forces in liquids (in German). *Z. Elektrochem. 52*, 269 (1948). [1]

Medani, M.S., & M.A. Hasan. Phase equilibria calculations with a modified Redlich-Kwong equation of state. *Canad. J. Chem. Eng. 56*, 251–256 (1978). [6]

Medir, M., & F. Giralt. Correlation of activity coefficients of hydrocarbons in water at infinite dilution with molecular parameters. *AIChE Journal 28*, 341–343 (1982). [4]

Meissner, H.P., C.L. Kusik, & W.H. Dalzell. Equilibrium composition with multiple reactions. *Ind. Eng. Chem. Fundamen. 8*, 659–665 (1969). [10]

Mendonca, J.M.M. & K.E. Bett. Phase equilibria in binary mixtures of carbon tetrafluoride and *n*-alkanes at high pressure. S.A. Newman (editor), *Chemical Engineering Thermodynamics*, 117–130. Ann Arbor Science (1982). [9]

Mentzer, R.A., R.A. Greenkorn, & K.-C. Chao. Principle of corresponding states and vapor-liquid equilibria of molecular fluids, and their mixtures with light gases. *Ind. Eng. Chem. Process Des. Dev. 20*, 240–252 (1981). [1]

Michels, A., B. Blaisse, & C. Michels. The isotherms of CO_2 in the neighborhood of the critical point and round the coexistence lines. *Proc. Roy. Soc. A160*, 358–375 (1937). [1]

Michelsen, M.L. Calculation of phase envelopes and critical points for multicomponent mixtures. *Fluid Phase Equilibria 4*, 1–10 (1980). [6]

Miller, D.G. Estimating vapor pressures—a comparison of equations. *Ind. Eng. Chem. 56*, 46–57 (1964). [1]

Miller, D.G. Joule-Thomson Inversion Curve, Corresponding States and Simpler Equations of State. *Ind. Eng. Chem. Fundamen. 9*, 585–589 (1970). [1]

Mills, M.B., M.J. Wills, & V.L. Bhirud. The calculation of density by the BWRS equation of state in process simulation contexts. *AIChE Journal 26*, 902–910 (1980). [11]

Miller, R.C., & M.J. Hiza. Experimental molar volumes for some LNG-related saturated liquid mixtures. *Fluid Phase Equilibria 2*, 49–57 (1978). [11]

Mindovich, E. Ya. Thermal analysis and X-ray investigation of molecular compounds of picric acid with cyclic hydrocarbons. *Zh. Fiz. Khim. 30*, 1082–1087 (1956). [5]

Mitchell, S. A method for determining the solubility of sparingly soluble substances. *J. Chem. Soc.* (London) 1333–1336 (1926). [12]

Miyahara, K., H. Sadotomo, & K. Kitamura. Evaluation of the Wilson parameters by nomographs. *J. Chem. Eng. Japan 3*, 157–160 (1970). [4]

Modell, M. Criteria of criticality. T.S. Storvick & S.I. Sandler (editors), Phase equilibria and fluid properties in chemical engineering, 369–389. *ACS Symposium Series 60*. American Chemical Society (1977).

Moldover, M.R., & J.S. Gallagher. Critical points of mixtures: An analogy with pure fluids. *AIChE Journal 24*, 267–278 (1978). [1]

Mollerup, J. Thermodynamic properties from corresponding states theory. *Fluid Phase Equilibria 4*, 11–34 (1980). [11]

Mollerup, J. A note on excess Gibbs energy models, equations of state and the local composition concept. *Fluid Phase Equilibria 7*, 121–138 (1981). [4]

Morgan, M.S., & R.C. Reid. Generalized partial quantities from Pitzer's expansion and pseudocritical rules. *AIChE Journal 16*, 889–890 (1970). [2]

Morimi, J., & K. Nakanishi. Use of the Wilson equation to calculate solid-liquid phase equilibria in binary and ternary systems. *Fluid Phase Equilibria 1*, 153–160 (1977). [8]

Morris, J.P., & G.R. Zellars. Vapor pressure of liquid copper and activities in liquid Fe-Cu alloys. *Trans. AIME 206*, 1086–1090 (1956). [8]

Moshfegian, M., A. Shariat, & J.H. Erbar. Application of the PEGC-MES equation of state to synthetic and natural gas systems. S.A. Newman (editor), Thermodynamics of aqueous systems with industrial applications, *ACS Symposium Series 133*, 333–360 (1980). [1,6,9]

Muir, R.F., & C.S. Howat III. Predicting solid-liquid equilibrium data from vapor-liquid data. *Chemical Engineering 89* (4), 89–92 (February 22, 1982). [8]

Mundis, C.J., L. Yarborough, & R.L. Robinson, Jr. Vaporization equilibrium ratios for CO_2 and H_2S in paraffinic, naphthenic and aromatic solvents. *Ind. Eng. Chem. Process Des. Dev. 16*, 254–259 (1977). [6]

Myers, A.L. Adsorption of gas mixtures: Thermodynamic approach. *Ind. Eng. Chem. 60*(5), 45–49 (1968). [9]

Myers, A.L., & J.M. Prausnitz. Thermodynamics of mixed-gas adsorption. *AIChE Journal 11*, 121–127 (1965). [9]

Nagata, I., & K. Katoh. Effective UNIQUAC equation in phase equilibrium calculation. *Fluid Phase Equilibria 5*, 225–244 (1980). [6]

Nagata, I., M. Nagashima, & M. Ogura. A comment on an extended form of the Wilson equation to correlation of partially miscible systems. *J. Chem. Eng. Japan 8*, 406–408 (1975). [7]

Nagata, I., M. Nagashima. Correlation and prediction of vapor-liquid and liquid-liquid equilibria. *J. Chem. Eng. Japan 9*, 6–11 (1976). [7]

Nagata, I., & T. Ohta. Calculation of high-pressure phase equilibrium from total pressure-liquid composition data. *Ind. Eng. Chem. Process Des. Dev. 15*, 211–215 (1976). [6]

Nagata, I., & T. Ohta. Prediction of the excess enthalpies of mixing for mixtures using the UNIFAC method. *Chem. Eng. Science 33*, 177–182 (1978). [11]

Nagata, I., & T. Yamada. Correlation and prediction of excess thermodynamic functions of strongly nonideal liquid mixtures. *Ind. Eng. Chem. Process Des. Dev. 13*, 47–53 (1974). [11]

Nagata, I., & S. Yasuda. Fugacity coefficients in binary mixtures containing acetic acid. *J. Chem. Eng. Japan 8*, 398–400 (1975). [3]

Nagata, I., & S. Yasuda. On the Carnahan-Starling equation of state. *J. Chem. Eng. Japan 10*, 64–65 (1977). [1]

Nakamura, R., G.J.F. Breedveld, & J.M. Prausnitz. Thermodynamic properties of gas mixtures containing common polar and nonpolar components. *Ind. Eng. Chem. Process Des. Dev. 15*, 557–564 (1976). [6]

Nakanishi, K., H. Wada, & H. Touhara. Thermodynamic excess functions of methanol + piperidine solutions at 298.15. *Fourth Int. Conf. on Chem. Thermodynamics* (August 26, 1975), Montpellier, France, 169–175. [2]

Nath, J. Acentric factor and the heats of vaporization for unassociated polar and nonpolar organic liquids. *Ind. Eng. Chem. Fundamen. 18*, 297–298 (1979). [1]

Nath, J., S.S. Das, & M.L. Yadava. On the choice of acentric factor. *Ind. Eng. Chem. Fundamen. 15*, 223–225 (1976). [1]

Nathan, D.I. Prediction of mixture enthalpies. *British Chem. Eng. 12* (2), 223–226 (1967). [11]

Nelder, J.A., & R. Mead. A simplex method for function minimization. *Computer Journal 7*, 308–313 (1964). [4]

Newman, S.A. Correlations evaluated for coal-tar liquids. *Hydrocarbon Processing 60* (12), 133–142 (1981). [9]

Newton, R.H., & B.F. Dodge. *Ind. Eng. Chem. 27*, 577 (1935). [10]

Nghiem, L.X., & K. Aziz. A robust iterative method for flash calculations using the Soave-Redlich-Kwong or the

Peng-Robinson equation of state. Paper No. SPE 8285, presented at 54th annual Fall Technical Conference of the Society of Petroleum Engineers, Las Vegas (September 23–26, 1979). [6]

Nghiem, L.X., D.K. Fong, & K. Aziz. Compositional modelling with an equation of state. Paper No. 9306, presented at 55th Annual Fall Technical Conference of the Society of Petroleum Engineers, Dallas (September 21–24, 1980). [6]

Niedzwiecki, J.L., R.D. Springer, & R.G. Wolfe. Multicomponent distillation in the presence of free water. *Chem. Eng. Progress 76* (4), 57–58 (1980). [6]

Nishiumi, H. Thermodynamic property prediction of C_{10} to C_{20} paraffins and their mixtures by the generalized BWR equation of state. *J. Chem. Eng. Japan 13*, 74–76 (1980). [1]

Nishiumi, H. An improved generalized BWR equation of state with three polar parameters applicable to polar substances. *J. Chem. Eng. Japan 13*, 178–183 (1980). [1]

Nishiumi, H., & D.B. Robinson. Compressibility factor of polar substances based on a four-parameter corresponding states principle. *J. Chem. Eng. Japan 14*, 259–266 (1981). [1]

Nishiumi, H., & S. Saito. Correlation of the binary interaction parameter of the modified generalized BWR equation of state. *J. Chem. Eng. Japan 10*, 176–180 (1977). [1]

Nitta, T., & T. Katayama. A new interpretation of the Wilson equation as a short-cut form of the associated solution theory. *J. Chem. Eng. Japan 7*, 381–382 (1974). [4]

Null, H.R. Application of the Wilson equation to solid-liquid equilibria. *AIChE Symposium Series 63*, 52–56 (1967). [8]

Ochi, K., S. Hiraba, & K. Kojima. Prediction of solid-equilibrium using ASOG. *J. Chem. Eng. Japan 15*, 59–61 (1982). [8]

Ochi, K., & B.C.-Y. Lu. Determination and correlation of binary vapor-liquid equilibrium data. *Fluid Phase Equilibria 1*, 185–200 (1978). [12]

O'Connell, J.P. Thermodynamics of gas solubility. *Proceedings*, 2nd International Conference, Berlin (March 21, 1980). EFCE, DECHEMA, 445–456. [6]

Oellrich, L., U. Plöcker, J.M. Prausnitz, & H. Knapp. Methods for calculation of phase equilibria and enthalpies with the aid of equations of state (in German). *Chem. Ing. Tech. 49*, 955–965 (1977). [11]

Oishi, T., & J.M. Prausnitz. Estimation of solvent activities in polymer solutions using a group-contribution method. *Ind. Eng. Chem. Process Des. Dev. 17*, 333–339 (1978). [7]

Oliver, R.C., S.E. Stephanou, & R.W. Baier. Calculating free energy minimization. *Chem. Eng. 69* (4), 121–128 (February 19, 1962). [11]

Orye, R.V. Prediction and correlation of phase equilibria and thermal properties with the BWR equation of state. *Ind. Eng. Chem. Process Des. Dev. 8*, 579–588 (1969). [11]

Orye, R.V., & J.M. Prausnitz. Multicomponent equilibria with the Wilson equation. *Ind. Eng. Chem. 57*, 18–26 (1965). [4]

Osborn, A. How to calculate three-phase flash vaporization. *Chemical Engineering 71* (26), 97–100 (December 21, 1964). [7]

Othmer, D.F. Composition of vapors from boiling binary solutions. *Ind. Eng. Chem. 20*, 743–746 (1928). [12]

Othmer, D.F. Correlating physical and chemical data for chemical engineering use. *Proceedings*, 11th International Congress of Pure and Applied Chemistry, London (1947). [1]

Ott, J.B., J.R. Coates, & H.T. Hall, Jr. Comparison of equations of state in effectively describing *PVT* relations. *J. Chem. Education 48*, 515–517 (1971). [1]

Otto, J. Equations of state (in German). *Handbuch der Experimental Physik*, Bd 8/2, W. Wien & F. Harms (editors), 207–246. Akademische Verlagsgesellschaft (1929). [1]

Ozkardesh, H., R.R. Tarakad, & S.B. Adler. Single-set treatment of all available binary system VLE data leads to simplified correlation procedures. *Ind. Eng. Chem. Fundamen. 17*, 206–209 (1978). [6]

Palmer, D.A. Predicting equilibrium relationships for maverick mixtures. *Chem. Eng.* (1975), 80–85 (June 9, 1975). [6]

Parrish, W.R., & J.M. Prausnitz. Dissociation pressures of gas hydrates formed by gas mixtures. *Ind. Eng. Chem. Process Des. Dev. 11*, 26–35 (1972). [9]

Passut, C.A., & R.P. Danner. Development of a four-parameter corresponding states method: Vapor pressure prediction. *AIChE Symposium Series 140*, 30–36 (1974). [1]

Pavlov, S. Yu, L.L. Karpacheva, V.P. Bubenkov, & A.B. Kirnos. Prediction of activity coefficients in terms of the interaction of structural groups. *Russian J. Phys. Chem. 52*, 508–510 (1978). [4]

Pemberton, R.C., & C.J. Mash. Thermodynamic properties of aqueous non-electrolyte mixtures. II: Vapour pressures and excess Gibbs energies for water + ethanol at 303.15 to 363.15 determined by an accurate static method. *J. Chem. Thermodynamics 10*, 867–888 (1978). [12]

Peng, D.-Y., & D.B. Robinson. A new two-constant equation of state. *Ind. Eng. Chem. Fundamen. 15*, 59–64 (1976). [1]

Peng, D.-Y., & D.B. Robinson. Two and three phase equilibrium calculations for systems containing water. *Canad. J. Chem. Eng. 54*, 595–599 (1976). [7]

Peng, D.-Y., & D.B. Robinson. A rigorous method for predicting the critical properties of multicomponent systems from an equation of state. *AIChE Journal 23*, 137–144 (1977). [1]

Peng, D.-Y., & D.B. Robinson. Calculation of three-phase solid-liquid-vapor- equilibrium. K.C. Chao & R.L. Robinson (editors). *Equations of state in engineering and research*, 185–196, *Advances in Chemistry Series 182*, American Chemical Society (1979). [1]

Peng, D.-Y., & D.B. Robinson. Two- and three-phase equilibrium calculations for coal gasification and related processes. S.A. Newman (editor), *Thermodynamics of aqueous systems with industrial applications, ACS Symposium Series 133*, 393–414 (1980). [7]

Pesult, D.R. Binary interaction constants for mixtures with a wide range in component properties. *Ind. Eng. Chem. Fundamen. 17*, 235–242 (1978). [1]

Peter, S. Thermodynamics of multicomponent systems as a basis for physico-chemical separation processes. *Int. Chem. Eng. 19*, 410–419 (1979). [6]

Peter, S., G. Brunner, & R. Riha. High pressure phase equilibria and their technical application (in German). *Chem. Ing. Technik 46*, 623–668 (1974). [6]

Phillips, J.N. The energetics of micelle formation. *Trans. Faraday Soc. 51*, 561–569 (1955). [9]

Pistorius, C.W.F.T. Melting points and volume changes upon melting. B. Le Neindre & B. Vodar (editors), *Experimental Thermodynamics*, Vol. II, 803–834, Butterworths (1975). [12]

Pitzer, K.S., R.F. Curl. et al. Volumetric and thermodynamic properties of fluids—enthalpy, free energy and entropy. *Ind. Eng. Chem. 50*, 265–274 (1958). (See also *JACS 77*, 3427, 3433; *79*, 2369.) [1]

Pitzer, K.S. Origin of the acentric factor. T.S. Storvick & S.I. Sandler (editors), *Phase equilibria and fluid properties in the chemical industry*, 1–10, *ACS Symposium Series 60*, American Chemical Society (1977). [1]

Pitzer, K.S. Thermodynamics of aqueous electrolytes at various temperatures, pressures and compositions. S.A. Newman, (editor), *Thermodynamics of Aqueous Systems with Industrial Applications*, *ACS Symposium Series 133*, American Chemical Society (1980). [9]

Pitzer, K.S. Thermodynamics of electrolyte solutions over the entire miscibility range. S.A. Newman (editor), *Chemical Engineering Thermodynamics*, 309–322 (1982). Ann Arbor Science. [9]

Pitzer, K.S. Theory: Ion interaction approach. R. Pytkowicz (editor), *Activity Coefficients in Electrolyte Solutions*, Vol. 1, 157–208. CRC Press (1979). [9]

Plöcker, U., H. Knapp, & J.M. Prausnitz. Calculation of high-pressure vapor-liquid equilibria from a corresponding states correlation with emphasis on asymmetric mixtures. *Ind. Eng. Chem. Process Des. Dev. 17*, 324–332 (1978). [6]

Poling, B.E., E.A. Grens II, & J.M. Prausnitz. Thermodynamic properties from a cubic equation of state: Avoiding trivial roots and spurious derivatives. *Ind. Eng. Chem. Process Des. Dev. 20*, 127–130 (1981). [11]

Pollin, A.G., V. Fried, & M. Yorizane. The prediction of excess enthalpy and excess Gibbs energy from volumetric data. *J. Chem. Eng. Japan 11*, 326–327 (1978). [11]

Potts, A.D., & D.W. Davidson. Ethanol hydrate. *J. Phys. Chem. 69* (3), 996–1000 (1965). [9]

Prausnitz, J.M. Solubility thermodynamics in chemical engineering. *J. Phys. Chem. 66*, 640–645 (1962). [6,7]

Prausnitz, J.M. State of the art review of phase equilibria. T.S. Storvick & S.I. Sandler (editors), *Phase equilibria and fluid properties in the chemical industry*, 11–62, *ACS Symposium Series 60*, American Chemical Society (1977). [1]

Prausnitz, J.M. Practical applications of molecular thermodynamics for calculating phase equilibria. *International Chemical Engineering 19*, 401–409 (1979). [6]

Prausnitz, J.M. State of the art review of phase equilibria. *Proceedings*, 2nd International Conference, Berlin (March 21, 1980). EFCE, DECHEMA, 231–282. [1]

Prausnitz, J.M. Calculation of phase equilibria for separation operations. *Trans. I. Chem. E. 59*, 3–16 (1981). [6]

Prausnitz, J.M., & R.D. Gunn. Volumetric properties of nonpolar gaseous mixtures. *AIChE Journal 4*, 430–435 (1958). [1]

Prausnitz, J.M., & R.D. Gunn. Pseudocritical constants from volumetric data for gas mixtures. *AIChE Journal 4*, 494 (1958). [1]

Prausnitz, J.M., & F.H. Shair. A thermodynamic correlation of gas solubilities. *AIChE Journal 7*, 682–687 (1961). [6]

Prausnitz, J.M., & J.H. Targovnik. Salt effects in aqueous vapor-liquid equilibria. *Ind. Eng. Chem. Chem. and Eng. Data Series 3*, 234–239 (1958). [6]

Prins, A. Mixtures of liquid crystals in binary systems. *Z. Physik. Chem. 67*, 689–723 (1909). [9]

Rachford, H.H., & J.D. Rice. Procedure for use of electronic digital computers in calculating flash vaporization hydrocarbon equilibrium. *Petrol. Technol.*, Sect. 1, p. 19, Sect. 2, p. 3 (October 1952). [6]

Rackett, H.G. Equation of state for saturated liquids. *J. Chem. Eng. Data 15*, 514–517 (1970). [1]

Raimondi, L. A modified Redlich-Kwong equation of state for vapour-liquid equilibrium calculations. *Chem. Eng. Science 35*, 1269–1275 (1980). [6]

Rasmussen, P., & A. Fredenslund. Prediction of separation factors using group contribution methods: A review. *Separation and Purification Methods 7* (2), 147–182 (1978). [4]

Ray, G.C., & E.O. Box. Adsorption of gases on activated charcoal. *Ind. Eng. Chem. 42*, 1315–1318 (1950). [9]

Rea, H.E. Effect of pressure and temperature on the liquid densities of pure hydrocarbons. *J. Chem. Eng. Data 18*, 227–230 (1973). [1]

Redlich, O. On the three-parameter representation of the equation of state. *Ind. Eng. Chem. Fundamen. 14*, 257–260 (1975). [1]

Redlich, O., & J.N.S. Kwong. On the thermodynamics of solutions: V: An equation of state. Fugacities of gaseous solutions. *Chem. Review 44*, 233–244 (1949). [1]

Redlich, O., & V.B.T. Ngo. An improved equation of state. *Ind. Eng. Chem. Fundamen. 9*, 287–290 (1970). [1]

Ree, F.H., & W.G. Hoover. Seventh virial coefficients of hard spheres and hard disks. *J. Chem. Physics 46*, 4181–4197 (1967). [1]

Reich, R., W.T. Ziegler, & K.A. Rogers. Adsorption of methane, ethane and ethylene gases and their binary and ternary mixtures and carbon dioxide on activated carbon at 212-301 K and pressures to 35 atmospheres. *Ind. Eng. Chem. Process Des. Dev. 19*, 336–344 (1980). [9]

Reid, R.C., & B.L. Beegle. Critical point criteria in Legendre transform notation. *AIChE Journal 23*, 726–732 (1977). [1]

Renon, H. Deviations from ideality in electrolyte solutions. R.S.H. Mah & W.D. Seider (editors), *Foundations of Computer-Aided Chemical Process Design*, Vol. 2, 53–83 (1980) *AIChE*. [9]

Renon, H., & J.M. Prausnitz. Local compositions in thermodynamic excess functions for liquid mixtures. *AIChE Journal 14*, 135–144 (1968). [4]

Renon, H., & J.M. Prausnitz. Estimation of parameters for the NRTL equation for excess Gibbs energies of strongly nonideal liquid mixtures. *Ind. Eng. Chem. Process Des. Dev. 8*, 413–419 (1969). [4]

Rheinboldt, H., & M. Kircheisen. Melting diagrams of systems with miscibility gaps: Picric acid + triphenylmethane, dinitrotoluene + urea, phenylenediamine + triphenylmethane. *J. für Praktische Chem. 112*, 187–195 (1926). [7,5]

Richon, D., P. Antoine, & H. Renon. Infinite dilution activity coefficients of linear and branched alkanes from C_1 to C_9 in *n*-hexadecane by inert gas stripping. *Ind. Eng. Chem. Process Des. Dev. 19*, 144–147 (1980). [4]

Richter, H.R.D., & B.M. Burnside. A general programme for producing pressure-enthalpy diagrams. *J. Mech. Science 17* (1), 31–39 (1975). [2]

Righter, W.M., & K.R. Hall. Optimal truncation of the virial equation. *AIChE Journal 21*, 406–407 (1975). [1]

Rizzi, A., & J.F.K. Huber. Comparative calculations of activity coefficients in binary liquid mixtures at infinite dilution using the "solution of groups" model. *Ind. Eng. Chem. Process Des. Dev. 20*, 204–210 (1981). [4]

Robinson, D.B., D.-Y. Peng, & H.-J. Ng. Applications of the Peng-Robinson equation of state. T.S. Storvick & S.I. Sandler (editors), *Phase Equilibria and Fluid Properties in the Chemical Industry, ACS Symposium Series 60* (1977). [6]

Robinson, D.B., & D-Y. Peng. The use of equations of state in multiphase equilibrium calculations. *Proceedings*, 2nd International Conference, Berlin (March 21, 1980). EFCE, DECHEMA, 335–354. [7]

Robinson, D.B., D.-Y. Peng, & H.-J. Ng. Applications of the Peng-Robinson equation of state. *ACS Symposium Series 60*, 200–220 (1977). [6]

Robinson, D.B., D.-Y. Peng, & H.-J. Ng. Capability of the Peng-Robinson programs. Part 1: VIE and critical property calculations. *Hydrocarbon Processing 57* (4), 95–98 (1978). [6]

Robinson, R.L., & K.C. Chao. A correlation of vaporization equilibrium ratios for gas processing streams. *Ind. Eng. Chem. Process Des. Dev. 10*, 221–229 (1971). [6]

Robinson, R.L., & K.C. Chao. A correlation of vaporization equilibrium ratios for gas processing systems. *Ind. Eng. Chem. Process Des. Dev. 10*, 221–229 (1979). [6]

Ross, B.A., & W.D. Seider. Simulation of three-phase distillation towers. *Computers and Chemical Engineering 5*, 7–20 (1980). [7]

Rossini, F.D. in *Science of Petroleum* V/1, 153, Oxford (1950). [10]

Rowlinson, J.S. The properties of real gases. *Encyclopedia of Physics*, Vol. XII, S. Flugge (editor). Springer-Verlag (1958). [1]

Rowlinson, J.S. Legacy of van der Waals. *Nature*, 244–417 (August 17, 1973). [1]

Rudman, P.S., in H. Herman (editor), *Advances in Materials Research 4*, 147–194, Interscience (1970). [8]

Russell, R.A. Non-ideal liquid activity coefficients. *Erdöl and Kohle 29*, 407–409 (1976). [4]

Sabylin, I.I., A.G. Polozov, & A.B. Klionskii. Method for the calculation of the liquid-liquid-vapor equilibrium in three-component systems. *J. Applied Chem. USSR 52*, 280–284 (1979). [7]

Sanderson, R.V., & H.H.Y. Chien. Simultaneous chemical and phase equilibrium calculation. *Ind. Eng. Chem. Process Des. Dev. 12*, 81–85 (1973). [10]

Santacesaris, E. Measurement of activity coefficients at infinite dilution by stripping and detention time methods. *Fluid Phase Equilibria 3*, 167–176 (1979). [12]

Sarashina, E., J. Nohka, Y. Arai, & S. Saito. Correlation of critical locus for binary mixtures by the BWR equation. *J. Chem. Eng. Japan 7*, 219–222 (1974). [1]

Sastri, S.R.S., M.V.R. Rao, K.A. Reddy, & L.K. Doraiswamy. A generalized method for estimating the latent heat of organic compounds. *Brit. Chem. Eng. 14* (7), 959–963 (1969). [11]

Saville, G. Measurement of *PVT* properties of gases and gas mixtures at low pressures. B. Le Neindre & B. Vodar (editors), *Experimental Thermodynamics*, Vol. II, 321–346. Butterworths (1975). [1,12]

Sayegh, S.G., & J.H. Vera. Model-free methods for vapor-liquid equilibria calculations. *Chem. Eng. Science 35*, 2247–2256 (1980). [6]

Sayegh, S.G., J.H. Vera, & G.A. Ratcliff. Vapor-liquid equilibria for the ternary system *n*-heptane/*n*-propanol/*l*-chlorobutane and its constituent binaries at 298.15 K. *Canad. J. Chem. Eng. 57*, 513–519 (1979). [6]

Scatchard, G., & W.J. Hamer. The application of equations for chemical potentials of partially miscible solutions. *JACS 57*, 1805–1809 (1935). The application of equations for chemical potentials to equilibrium between solid solution and liquid solution. *JACS 75*, 1809–1811 (1935). [4]

Scheller, W.A., & S.V. Narashmha Rao. Isothermal vapor-liquid equilibrium data for system heptane-2-pentanone at 90 C. *J. Chem. Eng. Data 18*, 223–225 (1973). [6]

Schmidt, G., & H. Wenzel. A modified van der Waals type of equation of state. *Chem. Eng. Science 35*, 1503–1512 (1980). [1]

Schneider, G.M. Phase equilibria of liquid and gaseous mixtures at high pressures. B. Le Neindre & B. Vodar (editors), *Experimental Thermodynamics*, Vol. II, 787–802. Butterworths (1975). [9,12]

Schneider, G.M. High pressure phase diagrams and critical properties of fluid mixtures. M.L. McGlashan (editor), *Chemical Thermodynamics*, Vol. 2: *A Specialist Periodical Report*, 105–146. The Chemical Society, London (1978). [8,9]

Schneider, G.M., & C. Russo. Effect of pressure on miscibilities of liquids. V. Effect of salts on miscibilities of propanol + water, 2-butanol + N_2O and pyridine + water up to 6000 bar, *Ber Bunsengesellschaft 70*, 1008–1014 (1966). [9]

Schotte, W. Collection of phase equilibrium data for separation technology. *Ind. Eng. Chem. Process Des. Dev. 19*, 432–439 (1980). [12]

Schreiber, L.B., & C.A. Eckert. Use of infinite dilution activity coefficients with Wilson's equation. *Ind. Eng. Chem. Process Des. Dev. 10*, 572–576 (1971). [4]

Schreinemakers, F.A.H. Equilibria in the system water + phenol + acetone. *Z. Physik Chem. 33*, 78–98 (1900). [7]

Schulte, H.W., P. Grenzheuser, & J. Gmehling. Application of a modified Wilson equation to liquid mixtures showing phase splitting. *Fluid Phase Equilibria 4*, 185–196 (1980). [7]

Schweitzer, O.R., & C.E. Wales. Phase Equilibria, Part 1: Phase rule and equilibria relations; Part 2: Equilibria in one-component systems. *Chemical Engineering*, 117–120 (May 27, 1963); 111–114 (June 24, 1963). [5]

Scott, R.L. Liquid State. Physical Chemistry. VIIIA, edited by Eyring, Henderson, & Yost, 1–83. Academic Press (1971). [1]

Sehgal, I.J.S., V.F. Yesavage, A.E. Mather, & J.E. Powers. Enthalpy predictions tested by data. *Hydrocarbon Processing 47* (8), 137–143 (1968). [11]

Seider, W.D., R. Gautam, & C.W. White. Computation of phase and chemical equilibrium: A review. R.G. Squires & G.V. Reklaitis (editors), *Computer Applications to Chemical Engineering*, 115–134, *ACS Symposium Series 124,* American Chemical Society (1980). [6,7,10]

Seider, W.D., R. Gautam, & C.W. White, III. Computation of phase and chemical equilibrium: A review. *Foundations of Computer-Aided Chemical Process Design*, 115–134, *AIChE* (1980). [6,7,10]

Sengers, J.V., & A. Levelt-Sengers. The critical state. *Chem. & Eng. News*, 104–118 (June 10, 1968). [1]

Shah, D.O., K.S. Chan, & R.M. Giordano. The effect of dissolved oils and alcohols on the CMC of synthetic and petroleum sulfonates. K.L. Mittal (editor), *Solution Chemistry of Surfactants*, Vol. 1, 391–406 (1979). Plenum. [9]

Shah, K.K., & G. Thodos. A comparison of equations of state. *Ind. Eng. Chem. 57*, 30–37 (1965). [1]

Shah, M.K., & P.R. Bishnoi. Multistage multicomponent separation calculations using thermodynamic properties evaluated by the SRK/PR equation of state. *Canad. J. Chem. Eng. 56*, 478–486 (1978). [6]

Shah, P.N., & C.L. Yaws. Densities of liquids. *Chem. Eng. 83* (21), 131–133 (October 25, 1976). [1]

Shaw, A.N. The derivation of thermodynamical relations for a simple system. *Phil. Trans. Roy. Soc.* (London) A234, 299–328 (1935). [2, A.1]

Signer, R., H. Arm, & H. Daeniker. Vapor pressures, densities, thermodynamic mixing functions and refractive indexes of the binary systems water-tetrahydrofuran and water-diethylether at 25 C (in German). *Helv. Chim. Acta. 52*, 2347–2351 (1969). [2]

Silverman, N., & D. Tassios. The number of roots in the Wilson equation and its effects on vapor-liquid equilibrium calculations. *Ind. Eng. Chem. Process Des. Dev. 16*, 13–20 (1977). [4]

Sim, W.J., & T.E. Daubert. Prediction of vapor-liquid equilibria of undefined mixtures. *Ind. Eng. Chem. Process Des. Dev. 19*, 386–393 (1980). [9]

Simnick, J.J., H.M. Lin, & K.C. Chao. The BACK equation of state and phase equilibria in pure fluids and mixtures. K.C. Chao & R.L. Robinson (editors), *Equations of state in engineering and research*, 209–234, *Advances in Chemistry Series 182*. American Chemical Society (1979). [1]

Simonet, R., & E. Behar. A modified Redlich-Kwong equation of state for accurately representing pure component data. *Chem. Eng. Science 31*, 37–43 (1976). [1]

Simonetty, J., D. Yee, & D. Tassios. Prediction and correlation of liquid-liquid equilibria. *Ind. Eng. Chem. Process Des. Dev. 21*, 174–180 (1982). [7]

Skjold-Jorgensen, S., B. Kolbe, J. Gmehling, & P. Rasmussen. Vapor-liquid equilibria by UNIFAC group contribution: Revision and extension. *Ind. Eng. Chem. Process Des. Dev. 18*, 714–722 (1979). [4]

Skjold-Jorgensen, S., P. Rasmussen, & A. Fredenslund. On the temperature dependence of the UNIQUAC/UNIFAC models. *Chem. Eng. Science 35*, 2389–2403 (1980). [4]

Skjold-Jorgensen, S., P. Rasmussen, & A. Fredenslund. On the concentration dependence of the UNIQUAC/UNIFAC models. *Chem. Eng. Science 37*, 99–111 (1982). [4]

Skolnik, H. Effect of pressure in azeotropy. *Ind. Eng. Chem. 43*, 172–276 (1951). [4]

Sloan, E.D., Jr., & J.C. Mullins. Nonideality of binary adsorbed mixtures of benzene and Freon-11 on highly graphitized carbon at 298.15 K. *Ind. Eng. Chem. Fundamen. 14*, 347–355 (1975). [9]

Slocum, E.W. Multipurpose high-pressure phase-equilibrium apparatus. *Ind. Eng. Chem. Fundamen. 14*, 126–128 (1975). [12]

Smith, A., & A.W.C. Menzies. Studies in vapor pressure. *JACS* 897–914 (1910).[12]

Smith, B.D. Simplified calculation of chemical equilibria in hydrocarbon systems containing isomers. *AIChE Journal 5*, 26–28 (1959). [10]

Smith, W.R. The computation of chemical equilibrium in complex systems: Review. *Ind. Eng. Chem. Fundamen. 19*, 1–10 (1980). [10]

Snyder, C.F., & K.C. Chao. Heat of adsorption of light hydrocarbons and their mixtures on activated carbon. *Ind. Eng. Chem. Fundamen. 9*, 437–443 (1970). [9]

Soave, G. Equilibrium constants from a modified Redlich-Kwong equation of state. *Chem. Eng. Science 27*, 1197–1203 (1972). [6]

Soave, G.S. Application of the Redlich-Kwong equation of state to solid-liquid equilibria calculations. *Chem. Eng. Science 33*, 225–229 (1979). [8]

Soave, G.S. Application of a cubic equation of state to vapour-liquid equilibria of systems containing polar compounds. *I. Chem. E. Symposium Series*, No. 56, 1.2/1–1.2/16 (1979). [1,6]

Soave G. Rigorous and simplified procedures for determining the pure-component parameters in the Redlich-Kwong-Soave equation of state. *Chem. Eng. Science 35*, 1725–1729 (1980). [1]

Sokolov, B.I. Equation of state of a saturated liquid. *Russian J. Phys. Chem. 50*, 668–669 (1976). [1]

Sood, S.K., & G.G. Haselden. Corrections to C_0 for the Benedict-Webb-Rubin equation. *AIChE Journal 16*, 891–892 (1970). [1]

Sorensen, J., T. Magnussen, P. Rasmussen, & A. Fredenslund. Liquid-liquid equilibrium data: Their retrieval, correlation and prediction. *Fluid Phase Equilibria 2*, 297–308 (1979); *3*, 47–82 (1979); *4*, 151–163 (1980). [7]

Spear, R.R., R.L. Robinson, & K.C. Chao. Critical states of mixtures and equations of state. *Ind. Eng. Chem. Fundamen. 8*, 2–8 (1969). [1]

Spear, R.R., R.L. Robinson, & K.C. Chao. Critical states of ternary mixtures and equations of state. *Ind. Eng. Chem. Fundamen. 10*, 588–592 (1971). [1]

Spencer, C.F., & R.P. Danner. Prediction of bubble-point density of mixtures. *J. Chem. Eng. Data 18*, 230–234 (1973). [11]

Stahl, E., & W. Schilz. Extraction with supercritical gases in thin-layer chromatography: Applicability to natural substances (in German). *Chem. Ing. Technik 48*, 773–778 (1976). [9]

Stein, F.P., & E.J. Miller. Extension of the Hayden-O'Connell correlation to the second virial coefficients of some hydrogen-bonding mixtures. *Ind. Eng. Chem. Process Des. Dev. 19*, 123–138 (1980). [1]

Stephan, K., & K. Schaber. Phase equilibria for vapour phase extraction processes. *Ger. Chem. Eng. 2*, 38–45 (1979). [9]

Stiel, L.I. A generalized theorem of corresponding states for the thermodynamic properties of non-polar and polar fluids. *Chem. Eng. Science 27*, 2109–2115 (1972). [1]

Stookey, D.J., & B.D. Smith. Prediction of excess free energy from excess enthalpy and excess volume data for hydrocarbon mixtures. *Ind. Eng. Chem. Process Des. Dev. 12*, 372–376 (1973). [11]

Su, G.-J., & C.-H. Chang. Generalized equation of state for real gases. *Ind. Eng. Chem. 38*, 802–806 (1946). [1]

Su, G.-J., & D.S. Viswanath. Generalized Benedict-Webb-Rubin equation of state for real gases. *AIChE Journal 11*, 205–207 (1965). [1]

Sugi, H., & T. Katayama. Ternary liquid-liquid and miscible binary vapor-liquid equilibrium data for the two systems *n*-hexane + ethanol + acetonitrile and water + acetonitrile + ethyl acetate. *J. Chem. Eng. Japan 11*, 167–172 (1978). [7]

Sutton, T.L., & J.F. MacGregor. The analysis and design of binary vapour-liquid equilibrium experiments. I: Parameter estimation and consistency tests; II: The design of experiments. *Canad. J. Chem. Eng.* 55, 602–613 (1977). [12]

Tai, T.B., R.S. Ramalho, & S. Kaliaguine. Application of Wilson's equation to the determination of vapor-liquid equilibrium data and heats of mixing for nonideal solutions. *Canad. J. Chem. Eng. 50*, 771–776 (1972). [11]

Tajbl, D.G., J.S. Kanofsky, & J.M. Braband. UOP's OLEX process: New applications. *Energy Processing/Canada*, 61–63 (May–June 1980). [9]

Takeo, M., K. Nishii, T. Nitta, & T. Katayama. Isothermal vapor-liquid equilibria for two binary mixtures of heptane with 2-butanone and 4-methyl-2-pentanone measured by a dynamic still with a pressure regulation. *Fluid Phase Equilibria 3*, 123–131 (1979). [12]

Tamir, A., & J. Wisniak. Vapor equilibrium in associating systems (water-formic acid-propionic acid). *Ind. Eng. Chem. Fundamen. 15*, 274–280 (1976). [6]

Tang, Y.P. On the solution of BWR compressibility. *Canad. J. Chem. Eng. 48*, 726–727 (1970). [1]

Tarakad, R.R., & R.P. Danner. A comparison of enthalpy prediction methods. *AIChE Journal 22*, 409–411 (1976). [11]

Tarakad, R.R., & D.P. Danner. An improved corresponding states method for polar fluids: Correlation of second virial coefficients. *AIChE Journal 23,* 685–695 (1977). [1]

Tarakad, R.R., C.F. Spencer, & S.B. Adler. A comparison of eight equations of state to predict gas-phase density and fugacity. *Ind. Eng. Chem. Process Des. Dev. 18*, 726–739 (1979). [1]

Tassios, D.P. Infinite dilution relative volatilities through gas-liquid chromatography. *Ind. Eng. Chem. Process Des. Dev. 11*, 43–46 (1972). [4]

Tassios, D. Limitations in correlating strongly nonideal binary systems with the NRTL and LEMF equations. *Ind. Eng. Chem. Process Des. Dev. 15*, 574–578 (1976). [4]

Tassios, D. The number of roots in the NRTL and LEMF equations and the effect on their performance. *Ind. Eng. Chem. Process Des. Dev. 18*, 182–186 (1979). [4]

Taylor, S.L., & T.M. Reed, III. Virial coefficients and critical properties of perfluorohexanes. *AIChE Journal 16*, 738–741 (1970). [1]

Tee, L.S., S. Gotoh & W.E. Stewart. Molecular parameters for normal fluids, The Lennard-Jones and Kihara potentials, *IEC Fundamen.* 5, 356–357 (1966). [1]

Teja, A., & S.I. Sandler (on Part II). A corresponding states equation for saturated liquid densities. I: Application to LNG; II: Application to calculation of swelling factors of CO_2-crude oil systems. *AIChE Journal 26*, 337–345 (1980). [1]

Teja, A.S., S.I. Sandler, & N.C. Patel. A generalization of the corresponding states principle using two nonspherical reference fluids. *The Chemical Engineering Journal 21*, 21–28 (1981). [1]

Thompson, P.A. An equation for liquid-vapor saturation densities as a function of pressure. K.C. Chao & R.L. Robinson (editors), *Equation of state in engineering and research*, 365–384, *Advances in Chemistry Series 182*. American Chemical Society (1979). [1,11]

Thomson, G.H., K.R. Brobst, & R.W. Hankinson. An improved correlation for densities of compressed liquids and liquid mixtures. *AIChE Journal 28*, 671–676 (1982). [1,11]

Tinoco-Garcia, L., & J. Cano-Dominguez. A new technique for solving multi-phase equilibria. Unpublished, Instituto Mexicano del Petroleo (1979). [7,10]

Tochigi, K., & K. Kojima. The determination of group Wilson parameters to activity coefficients by ebulliometer. *J. Chem. Eng. Japan 9*, 267–273 (1976). [4]

Tochigi, K., M. Hiraga, & K. Kojima. Prediction of liquid-liquid equilibria for ternary systems by the ASOG method. *J. Chem. Eng. Japan 13*, 159–162 (1980). [7]

Tochigi, K., B.C.-Y. Lu, K. Ochi, & K. Kojima. On the temperature dependence of ASOG parameters for VLE calculations. *AIChE Journal 27*, 1022–1024 (1981). [4]

Torres-Marchal, C. Graphical design for ternary distillation systems. *Chemical Engineering 88* (21), 134–155 (October 19, 1981). [6]

Tryhorn, F.G., & W.F. Wyatt. Adsorption. I: Adsorption by coconut charcoal from alcohol-benzene and acetone-benzene mixtures. *Trans. Faraday Soc. 21*, 399–405 (1925). [9]

Tryhorn, F.G., & W.F. Wyatt. Adsorption. II: The adsorption by a coconut charcoal of saturated vapours of some pure liquids. *Trans. Faraday Soc. 22*, 134–145 (1926). [9]

Tryhorn, F.G., & W.F. Wyatt. Adsorption. III: Stages in the adsorption by a coconut charcoal from vapour mixtures of alcohol and benzene, and of acetone and benzene. *Trans. Faraday Soc. 23*, 139–145 (1927). [9]

Tryhorn, F.G., & W.F. Wyatt. Adsorption. IV: Adsorption by coconut charcoal from binary mixtures of saturated vapours: The systems methyl alcohol-benzene, ethyl alcohol-benzene, *n*-propyl alcohol-benzene and *n*-butyl alcohol-benzene. *Trans. Faraday Soc. 24*, 36–47 (1928). [9]

Tsonopoulos, C. An empirical correlation of second virial coefficients. *AIChE Journal 20*, 263–272 (1974). [1]

Tsonopoulos, C. Second virial cross-coefficients: Correlation and prediction of k_{ij}. K.C. Chao & R.L. Robinson (editors), *Equations of state in engineering and research*, 143–162, *Advances in Chemistry Series 182*. American Chemical Society (1979). [1]

Tsonopoulos, C., & J.M. Prausnitz. Equations of state: A review for engineering applications. *Cryogenics 9* (10), 315–327 (1969). [1]

Tsuboka, T., & T. Katayama. Modified Wilson equation for vapor-liquid and liquid-liquid equilibria. *J. Chem. Eng. Japan 8*, 181–187 (1975). [4]

Tsuboka, T., & T. Katayama. Correlations based on local fraction model between new excess Gibbs energy equations. *J. Chem. Eng. Japan 8*, 404–406 (1975). [4]

Unno, Y., D. Hoshino, K. Nagahama, & M. Hirata. Prediction of solid-liquid equilibria from vapor-liquid equilibrium data using the solution of groups model. *J. Chem. Eng. Japan 12*, 81–85 (1979). [8]

van der Waals, J.H. On the continuity of the gaseous and liquid state. Dissertation, Leiden (1973). Physical Memoirs, English translation by Threlfall & Adair, *Physical Society 1*, iii, 333 (1890). [1]

van der Waals, J.H., & J.C. Plateeuw. Clathrate solutions. *Advances Chem. Physics* II, 1–57. Interscience Publishers (1959). [9]

Van Laar, J.J. The vapor pressure of binary mixtures. *Z. Physik Chem. 72*, 723–751 (1910). On the theory of vapor pressures of binary mixtures. *Z. Physik Chem. 83*, 599–608 (1913). [4]

Van Ness, H.C. Adsorption of gases on solids: Review of the role of thermodynamics. *Ind. Eng. Chem. Fundamen. 8*, 464–473 (1969). [9]

Van Ness, H.C. On integration of the coexistence equation for binary vapor-liquid equilibrium. *AIChE Journal 16*, 18–

22 (1970). [2]

Van Ness, H.C. On use of constant-pressure activity coefficients. *Ind. Eng. Chem. Fundamen. 18*, 431–433 (1979). [4]

Van Ness, H.C., & M.M. Abbott. Vapor-liquid equilibrium. Part VI: Standard state fugacities for supercritical components. *AIChE Journal 25*, 645–653 (1979). [3]

Van Ness, H.C., S.M. Byer, & R.E. Gibbs. Vapor-liquid equilibrium. I: An appraisal of data reduction methods. *AIChE Journal 19*, 238–244 (1973). [6]

Van Ness, H.C., F. Pedersen, & P. Rasmussen. Vapor-liquid equilibrium. V: Data reduction by maximum likelihood. *AIChE Journal 24*, 1055–1062 (1978). [6]

Varhegyi, G., & C.H. Eon. Calculation of the free energy parameters from ternary liquid-liquid equilibrium data. *Ind. Eng. Chem. Fundamen. 16*, 182–185 (1977). [7]

Vera, J.H., & J.M. Prausnitz. Generalized van der Waals theory for dense fluids. *The Chemical Engineering Journal 3*, 1–13 (1972). [1]

Verhille, R.P. Effects of equilibrium data correlating equations on the design of continuous rectification columns for binary separations. *I. Chem. E. Symposium Series*, No. 56, 1.3/1–1.3/23 (1979). [6]

Vidal, J. Mixing rules and excess properties in cubic equations of state. *Chem. Eng. Science 33*, 789–791 (1978). [1]

Viiroya, A.K., Yu I. Kallas, & E.E. Siirde. Use of the UNIFAC method for calculation of phase equilibria in multicomponent systems. *J. Applied Chem. USSR 51*, 2356–2359 (1978). [4]

Viswanath, D.S., & G.-J. Su. Generalized *PVT* behavior of gases. *AIChE Journal 11*, 202–204 (1965). [1]

Wagner, W. New vapor pressure measurements for argon and nitrogen and a new method for establishing rational vapor pressure equations. *Cryogenics*, 470–482 (August 1973). [1]

Wales, C.E. Phase equilibria. III: Behavior of one-component systems; IV: Equilibria in two-component systems; V: Phase equilibria in binary systems. *Chemical Engineering*, 141–144 (July 22, 1963); 167–174 (August 19, 1963); 187–192 (September 16, 1963). [5]

Weller, H., H. Schuberth, & E. Leibniz. Vapor-liquid equilibrium of the system phenol/*n*-butylacetate/water at 44.4 C (in German). *J. Prakl. Chemie*, Series 4, *21*, 234–249 (1963). [6]

Wellman, P., & S. Katell. How pressure and temperature affect steam-methane reforming. *Hydrocarbon Processing 42* (6), 135–137 (1963). [10]

Wenzel, H., & W. Rupp. Calculation of phase equilibria in systems containing water and supercritical components. *Chem. Eng. Science 33*, 683–687 (1978). [6]

White, M.G., & K.C. Chao. Principle of corresponding states of liquid solutions. *Ind. Eng. Chem. Fundamen. 14*, 166–171 (1975). [1]

White, W.B., S.M. Johnson, & G.B. Dantzig. Chemical equilibrium in complex mixtures. *J. Chem. Physics 28*, 751–755 (1958). [10]

Wichterle, I. High pressure vapour-liquid equilibrium: A review. *Fluid Phase Equilibria 1*, 161–172 (1977); *1*, 225–245 (1978); *1*, 305–316 (1978); *2*, 59–78 (1978); *2*, 143–159 (1978). [1]

Williamson, A.G. Phase equilibria of two-component systems and multicomponent systems. B. Le Neindre & B. Vodar (editors), *Experimental Thermodynamics*, II, 749–786. Butterworths (1975). [12]

Wilson, G.M. Vapor-liquid equilibrium. XI: A new expression for the excess free energy of mixing. *J. Am. Chem. Soc. 86*, 127–130 (1964). [4]

Wilson, G.M. Areas of research on activity coefficients from group contributions at the Thermochemical Institute. T.S. Storvick & S.I. Sandler (editors), *Phase equilibria and fluid properties in chemical engineering*, 429–444, *ACS Symposium Series 60*. American Chemical Society (1977).[4]

Wilson, G.M. Thermophysical and transport properties of synthetic fuel systems at extreme temperatures and pressures. R.S.H. Mah & W.D. Seider (editors), *Foundations of Computer-Aided Chemical Process Design*, Vol. II, 31–51 (1980). *AIChE.* [9]

Wilson, G.M., & C.H. Deal. Activity coefficients and molecular structure. *Ind. Eng. Chem. Fundamen. 1*, 20–23 (1962). [4]

Winnick, J., & J. Kong. Excess volumes of mixtures containing polar liquids. *Ind. Eng. Chem. Fundamen. 13*, 292–293 (1974). [11]

Witonsky, R.J., & J.G. Miller. The second virial coefficient of the helium-nitrogen system from 175 to 475 C, *JACS 85*, 282–286 (1963). [1]

Wohl, A. Investigations on equations of state. *Z physik Chem 87*, 1–39 (1914); *99*, 207–241 (1921). [1]

Wojtowicz, P.J. in Priestley et al. (editors), *Introduction to Liquid Crystals* 333–350, Plenum Press (1975). [9]

Yamada, T. An improved generalized equation of state. *AIChE Journal 19*, 286–291 (1973). [1]

Yamada, T., & R.D. Gunn. Saturated liquid molar volumes: The Rackett equation. *J. Chem. Eng. Data 18*, 234–235 (1973). [1]

Yarborough, L. Application of a generalized equation of state to petroleum reservoir fluids. K.C. Chao & R.L. Robinson (editors), *Equations of state in engineering and research*, 385–440, *Advances in Chemistry Series 182*. American Chemical Society (1979). [1]

Yen, L.C., & R.E. Alexander. Estimation of vapor and liquid enthalpies. *AIChE Journal 11*, 334–339 (1965). [11]

Yeo, K.O., & S.D. Christian. A thermodynamic method for predicting solubilities of solutes in nonpolar solvents. *Ind. Eng. Chem. Fundamen. 13*, 196–198 (1974). [6]

Yokoyama, C., K. Arai, & S. Saito. Semiempirical equation of state for polar substances on the basis of perturbation theory, in Newman (editor), *Chemical Engineering Thermodynamics*, Ann Arbor Science (1983). [1]

Yon, C.M., & P.H. Turnock. Multicomponent adsorption equilibria on molecular sieves. *AIChE Symposium Series 117*, 75–83 (1971). [9]

Yorizane, M., & Y. Miyano. A generalized correlation for Henry's constants in nonpolar binary systems. *AIChE Journal 24*, 181–186 (1978). [6]

Young, C.L. Experimental methods for studying phase behavior of mixtures at high temperatures and pressures. M.L. McGlashan (editor), *Chemical Thermodynamics*. II: *A Specialist Periodical Report*, 71–104, The Chemical Society, London (1977). [12]

Yu, W.C., H.M. Lee, & R.M. Ligon. Predict high pressure properties. *Hydrocarbon Processing 61* (1), 171–178 (1982). [11]

Zaks, I.A., & I.L. Krupatkin. Use of elements of diagrams of liquid-phase equilibria of three-component systems for the correlation of tie lines. *J. Applied Chem. USSR 51*, 2125–2128 (1978). [7]

Zarkarian, J.A., F.E. Anderson, J.A. Boyd, & J.M. Prausnitz. UNIFAC parameters from gas-liquid chromatographic data. *Ind. Eng. Chem. Process Des. Dev. 18*, 657–661 (1979). [4]

Zeleznik, F.J., & S. Gordon. Calculation of complex chemical equilibria. *Ind. Eng. Chem. 60*, 27–57 (1968). [10]

Zellars, G.R., S.L. Payne, J.P. Morris, & R.L. Kipp. The activities of iron and nickel in liquid Fe-Ni alloys. *Trans. AIME 215*, 181–192 (1959). [8]

Zellner, M.G., L.C. Claitor, & J.M. Prausnitz. Prediction of vapor-liquid equilibria and enthalpies of mixtures at low temperatures. *Ind. Eng. Chem. Fundamen. 9*, 549–564 (1970). [6]

Zhvanetskii, I.B., & V.M. Platonov. Analytical equations for the calculation of enthalpy and heat capacity of liquid petroleum products. Neftepererab Neftekhim (Moscow), No. 5, 51 (1972). [11]

Zhvanetskii, I.B., & V.M. Platonov. Calculation of phase equilibrium constants for petroleum cuts. *Int. Chem. Eng. 18*, 84–85 (1978). [9]

Zief, M., & W.R. Wilcox (editors), Chap. 2: Phase diagrams, G.M. Wolten, W.R. Wilcox; Chap. 11: Column crystallization, R. Albertins, W.C. Gates, J.E. Powers; Chap. 16: Phillips fractional-solidification process, D.L. McKay; *Fractional Solidification*, Vol. 1. Marcel Dekker (1967). [5]

Zhuse, T.P. Compressed hydrocarbon gases as a solvent. *Petroleum* (London), 298–300 (August 1960). [9]

Zudkevitch, D., & J. Joffe. Correlation and prediction of vapor-liquid equilibria with the Redlich-Kwong equation of state. *AIChE Journal 16*, 112–119 (1970). [6]

Substance Index

Data of pure substances and mixtures given in the Appendixes are not cited here. The prefix *P* designates a problem of the text.

Subject Index

The prefix *P* designates reference to a problem in the text.